Hyperbolic Problems: Theory, Numerics, Applications

Proceedings of the Seventeenth International Conference on Hyperbolic Problems (HYP2018)

AIMS Series on Applied Mathematics Volume 10

University Park, Pennsylvania, USA
25 - 29 June 2018

Editors:

Alberto Bressan
Marta Lewicka
Dehua Wang
Yuxi Zheng

ISBN: 978-1-7138-0834-3

Printed from e-media with permission by:

Curran Associates, Inc.
57 Morehouse Lane
Red Hook, NY 12571

Some format issues inherent in the e-media version may also appear in this print version.

Printed with permission by Curran Associates, Inc. (2020)

American Institute of Mathematical Sciences
P.O. Box 2604
Springfield, Missouri
65801-2604

Phone: (417) 351-3204

www.aimsciences.org

Additional copies of this publication are available from:

Curran Associates, Inc.
57 Morehouse Lane
Red Hook, NY 12571 USA
Phone: 845-758-0400
Fax: 845-758-2633
Email: curran@proceedings.com
Web: www.proceedings.com

AIMS Series on Applied Mathematics Volume 10

Hyperbolic Problems: Theory, Numerics, Applications

Proceedings
of the Seventeenth International Conference
on Hyperbolic Problems
held at the Pennsylvania State University, University Park,
June 25-29, 2018

Alberto Bressan,
Marta Lewicka,
Dehua Wang,
Yuxi Zheng
Editors

American Institute of Mathematical Sciences

Preface

This volume contains the Proceedings of the XVII International Conference (HYP2018) on "Hyperbolic Problems: Theory, Numerics, Applications", which was held at the Pennsylvania State University, University Park, on June 25–29, 2018. This conference was the 17-th in a series started in 1986 in St. Etienne (France). Since then, these meetings have become established as the foremost international forums in their discipline, bringing together researchers, students and practitioners, with interest in the theoretical, computational and applied aspects of hyperbolic problems.

The HYP2018 conference was attended by over 230 participants. It featured 7 plenary and 16 invited lectures, while 170 contributed talks were given within special sessions. In addition, a poster session was organized, with prizes given to the best posters among junior participants.

The meeting highlighted a number of topics which have seen vigorous activity and significant progress in recent years. In particular: the theory of complex fluids and multi-phase flow, with applications to biological and engineering problems, the Einstein equations of general relativity, transport equations with rough coefficients, models of traffic flow on a network of roads, and collective dynamics of many-body systems.

Hyperbolic conservation laws are a classical subject that goes back to L. Euler (1755), and has seen contributions by some of the greatest mathematicians. Yet, the mathematical theory is far from complete and many fundamental questions remain unsolved. The depth and complexity of the problems make this a very challenging field, requiring a constant stream of new ideas and methods. Moreover, the ever increasing number of applications provides a strong stimulus to search for more efficient computational methods. For all these reasons, hyperbolic problems remain an extremely active research area.

The contributions collected in this volume cover a wide range of topics. Some of these represent the latest developments on classical multi-dimensional problems, dealing with shock reflections and with the stability of vortices and boundary layers. Other contributions provide sharp results on the structure and regularity of solutions to conservation laws, or discuss the fine line between well-posedness and ill-posedness for transport equations with rough coefficients, and for the equations of inviscid fluid flow. Further progress is reported at the interface between hyperbolic and kinetic models, including the hydrodynamic limit of the Boltzmann equation. Kinetic and macroscopic models for collective dynamics of many-body systems have attracted much interest in the past few years, and are also covered in this volume. Finally, a large number of papers are devoted to advances in computational methods, with diverse applications such as: submarine avalanches, tsunami waves, chemically reacting flows, solitary waves, gas flow on a network of pipelines, traffic flow with multiple types of vehicles, etc…

We believe that the present volume will provide a timely survey of the state of the art, paving the way for further progress in this exciting field.

We take this opportunity to thank all the members of the HYP2018 Scientific Committee (listed as https://www.hyp2018.psu.edu/scientific-committee/) for their expertise in selecting an outstanding group of plenary and invited speakers. We are also extremely thankful to all other members of the Organizing Committee and to the support staff (listed as https://www.hyp2018.psu.edu/organizers/), who contributed in an essential way to the success of the event.

Finally, we gratefully acknowledge the support from the following Institutions:

- National Science Foundation,
- Office of Naval Research,
- Eberly College of Science,
- Penn State University,
- Institute for Mathematics and its Applications, Minneapolis,
- Kenneth P. Dietrich School of Arts & Sciences, University of Pittsburgh,
- Department of Mathematics at Penn State University,
- Department of Mathematics, University of Pittsburgh,
- Institute for CyberScience at Penn State,
- Fluid Dynamics Research Consortium at Penn State,
- Center for Interdisciplinary Mathematics at Penn State.

State College, July 2019

Alberto Bressan
Marta Lewicka
Dehua Wang
Yuxi Zheng

Contents

Part 1

Plenary Lectures

UNIQUENESS AND STABILITY FOR THE SHOCK REFLECTION-DIFFRACTION PROBLEM FOR POTENTIAL FLOW

GUI-QIANG G. CHEN

Mathematical Institute, University of Oxford, Oxford, OX2 6GG, UK

MIKHAIL FELDMAN

Department of Mathematics, University of Wisconsin, Madison, WI 53706-1388, USA

WEI XIANG

City University of Hong Kong, Kowloon Tong, Hong Kong, China

ABSTRACT. When a plane shock hits a two-dimensional wedge head on, it experiences a reflection-diffraction process, and then a self-similar reflected shock moves outward as the original shock moves forward in time. The experimental, computational, and asymptotic analysis has indicated that various patterns occur, including regular reflection and Mach reflection. The von Neumann conjectures on the transition from regular to Mach reflection involve the existence, uniqueness, and stability of regular shock reflection-diffraction configurations, generated by concave cornered wedges for compressible flow. In this paper, we discuss some recent developments in the study of the von Neumann conjectures. More specifically, we present our recent results of the uniqueness and stability of regular shock reflection-diffraction configurations governed by the potential flow equation in an appropriate class of solutions. We first show that the transonic shocks in the global solutions obtained in Chen-Feldman [19] are convex. Then we establish the uniqueness of global shock reflection-diffraction configurations with convex transonic shocks for any wedge angle larger than the detachment angle or the critical angle. Moreover, the stability of the solutions with respect to the wedge angle is also shown. Our approach also provides an alternative way of proving the existence of the admissible solutions established first in [19].

2000 *Mathematics Subject Classification.* Primary: 35M12, 35C06, 35R35, 35L65, 35L70, 35L67, 35J70, 76H05, 35B45, 35B35, 35B40, 35B36, 35B38; Secondary: 35L20, 35J67, 76N10, 76L05, 76J20, 76N20, 76G25.

Key words and phrases. Compressible flow, conservation laws, potential flow equation, transonic shock, nonlinear elliptic equations, mixed-type equation, regular reflection, Mach reflection, shock reflection-diffraction, admissible solutions, free boundary, convexity, uniqueness, stability.

The research of Gui-Qiang G. Chen was supported in part by the UK Engineering and Physical Sciences Research Council Award EP/E035027/1 and EP/L015811/1, and the Royal Society–Wolfson Research Merit Award (UK). The research of Mikhail Feldman was supported in part by the National Science Foundation under Grants DMS-1764278 and DMS-1401490, and the Van Vleck Professorship Research Award by the University of Wisconsin-Madison. The research of Wei Xiang was supported in part by the Research Grants Council of the HKSAR, China (Project No. CityU 21305215, Project No. CityU 11332916, Project No. CityU 11304817, and Project No. CityU 11303518).

1. **Introduction.** We survey some recent developments in the mathematical analysis of the shock reflection-diffraction problem for potential flow and the corresponding von Neumann conjectures on the existence, uniqueness, and stability of regular shock reflection-diffraction configurations for the transition from regular to Mach reflection. The shock reflection-diffraction problem is a lateral Riemann problem and has been not only longstanding open in fluid mechanics but also fundamental in the mathematical theory of multidimensional conservation laws.

When a planar shock hits a concave cornered wedge, the incident shock interacts with the wedge, leading to the occurrence of shock reflection-diffraction (*cf.* [12, 53]). Beginning from the work of E. Mach [45] in 1878, various patterns of shock reflection-diffraction configurations have been observed experimentally and later numerically, including regular reflection and Mach reflection. The existence of the regular reflection solutions for potential flow has been now fully understood mathematically (see [17, 19]), by reducing the shock reflection-diffraction problem to a *free boundary problem*, where the unknown reflected shock is regarded as a free boundary. Then a natural followup fundamental question is to study the uniqueness and stability of the solutions we have obtained.

For the uniqueness problem, it is necessary to restrict to a class of solutions. Recent results [24,25,33,46] show the non-uniqueness of solutions with planar shocks in the class of entropy solutions with shocks of the Cauchy problem (initial value problem) for the multidimensional compressible Euler equations (isentropic and full). Our setup is different – the problem for solutions with shocks for potential flow is on the domain with boundaries, so these non-uniqueness results do not apply directly. However, these results indicate that it is natural to study the uniqueness and stability problems in a more restricted class of solutions. In this paper, we show the uniqueness in the class of self-similar solutions of regular shock reflection-diffraction configurations with convex transonic shocks, which are called admissible solutions; see the detailed definition in §3. Technically, restricting to the class of admissible solutions allows us to reduce the uniqueness problem for shock reflection-diffraction to a corresponding uniqueness problem for solutions of a free boundary problem for a nonlinear elliptic equation, which is degenerate for the supersonic case (see Fig. 2.1 below).

A key property of admissible solutions which we employ in the uniqueness proof is that the admissible solutions converge to the unique normal reflection solution as the wedge angle tends to $\frac{\pi}{2}$. Then the outline of the uniqueness argument is the following: If there are two different admissible solutions, defined by the potential functions φ and φ^*, for some wedge angle $\theta_{\rm w}^* < \frac{\pi}{2}$, then it suffices to:

(i) construct continuous families of solutions parametrized by the wedge angle $\theta_{\rm w} \in [\theta_{\rm w}^*, \frac{\pi}{2}]$, starting from φ and φ^*, respectively, in an appropriate norm;

(ii) prove *local uniqueness*: If two admissible solutions for the same wedge angle are close in the norm mentioned above, then they must be equal.

Combining this with the fact that both families converge to the unique normal reflection as $\theta_{\rm w} \to \frac{\pi}{2}-$, we conclude a contradiction.

Therefore, it remains to perform the two steps described above. Both steps can be achieved if we linearize the free boundary problem around an admissible solution, and then show that the linearization is sufficiently regular so that the solutions for close wedge angles can be constructed by the implicit function theorem. Indeed, this approach works for one regular shock reflection-diffraction case – the subsonic-away-from-sonic case (see §5 for more details).

However, it turns out that the linearization does not have such properties for the other case – the supersonic case, owing to the elliptic degeneracy near the sonic arc and relatively lower regularity of admissible solutions near the corner point between the shock and the sonic arc. For this case, instead, we develop a nonlinear approach: We prove directly the local uniqueness property and employ it to perturb any given admissible solution φ for the wedge angle $\theta_{\rm w}$, that is, to construct an admissible solution close to φ for all wedge angles that are sufficiently close to $\theta_{\rm w}$ by using the Leray-Schauder degree argument in a *small iteration set*. We note that, in [19], the solutions have also been constructed by the Leray-Schauder degree argument, but in a *large* iteration set, *i.e.* a subset in a space determined by some weighted and scaled $C^{k,\alpha}$ norms, with bounds by the constants sufficiently large so that the *apriori* estimates of the admissible solutions assure that a fixed point of the iteration map does not occur at the boundary of the iteration set. In the present case of *small* iteration set, the similar property is shown by using the local uniqueness.

Our proof of the local uniqueness is based on the convexity of the reflected-diffracted transonic shock, established in Chen-Feldman-Xiang [21]. We note that the convexity of the shocks is consistent with physical experiments and numerical simulations; see e.g. [4, 12, 26, 28, 34–39, 42, 47, 50, 52, 54], and the references therein. Also see [10, 11, 41, 43, 44, 48, 50] for the convexity of transonic shocks in numerical Riemann solutions of the Euler equations for compressible fluids. Mathematically, the Rankine-Hugoniot conditions on the shock whose location is unknown, together with the nonlinear equation in the elliptic and hyperbolic regions, enforce a restriction to possible geometric shapes of the transonic shock. Moreover, one of our observations is that the convexity of transonic shocks is not a local property. In fact, the uniform convexity is a result of the interaction of the cornered wedge and the incident shock, since the reflected shock remains flat when the wedge is a flat wall for the normal shock reflection. In addition, for this problem, it seems to be difficult to apply the methods directly as in [5–7, 29, 49], owing to the difference and the more complicated structure of the boundary conditions.

In [21], we have developed an approach in which the global properties of solutions are incorporated in the proof of the convexity of transonic shocks. In particular, we have introduced a general set of conditions and employed the approach to prove the convexity of transonic shocks under these conditions. As a direct corollary, we have proved the uniform convexity of transonic shocks in the two longstanding fundamental shock problems – the shock reflection-diffraction problem by wedges and the reflection problem for supersonic flows past solid ramps.

Moreover, as a byproduct of our uniqueness proof, we have developed a new way of establishing the existence of global solutions of the shock reflection-diffraction problem up to the detachment angle or the critical angle, based on the fine convexity structure. Our approach is also helpful for other related mathematical problems including free boundary problems with degeneracy.

The previous works on unsteady flows with shocks in self-similar coordinates include the following: The problem of shock reflection-diffraction by a concave cornered wedges for potential flow has been systematically analyzed in Chen-Feldman [17, 19] and Bae-Chen-Feldman [1], where the existence of regular shock reflection-diffraction configurations has been established up to the detachment wedge angle or the critical angle for potential flow. For the Mach reflection, S. Chen [23] proved the local stability of flat Mach configuration in self-similar coordinates. Also see [8, 9, 16, 27, 40] for the unsteady transonic small disturbance equation and the nonlinear wave system, [51] for the Chaplygin gas, and [56] for the pressure-gradient system. Meanwhile, other problems have been tackled. For the shock diffraction problem, Chen-Feldman-Hu-Xiang [20] showed that regular shock configurations cannot exist for potential flow. For supersonic flow past a solid ramp, Elling-Liu [30] obtained a first rigorous unsteady result under certain assumptions for potential flow. Then Bae-Chen-Feldman [2,3] succeeded to remove the assumptions in [30] and established the existence theorem for global shock reflection configurations so that the steady supersonic weak shock solution as the long-time behavior of an unsteady flow for all physical parameters, via new mathematical techniques developed first in Chen-Feldman [19]. See also [13–15, 31, 32] and the references therein for the steady transonic shocks over two-dimensional wedges.

The organization of this paper is the following: In §2, we introduce the free boundary problem for the shock reflection-diffraction problem. Then the existence and regularity results obtained in [19] are given in §3. In §4, we describe the result and present the main steps in the proof of the convexity of the regular reflected-diffracted transonic shock based on [21]. In §5, we discuss our recent result and outline the proof on the uniqueness and stability of regular shock reflection-diffraction configurations.

2. The Potential Flow Equation and the Shock Reflection-Diffraction Problem.

In this section we formulate the shock reflection-diffraction problem as a free boundary problem for the potential flow equation in the self-similar coordinates.

2.1. The potential flow equation.

The Euler equations for potential flow consist of the conservation law of mass and Bernoulli's law:

$$\partial_t \rho + \nabla_{\mathbf{x}} \cdot (\rho \mathbf{v}) = 0, \tag{1}$$

$$\partial_t \Phi + \frac{1}{2}|\nabla_{\mathbf{x}}\Phi|^2 + i(\rho) = B_0, \tag{2}$$

where ρ is the density, Φ is the velocity potential so that $\mathbf{v} = \nabla_{\mathbf{x}}\Phi$, B_0 is the Bernoulli constant determined by the incoming flow and/or boundary conditions, $\mathbf{x} = (x_1, x_2) \in \mathbb{R}^2$, and $i(\rho) = \int_1^\rho \frac{p'(s)}{s} ds$ for the pressure function $p = p(\rho)$. For polytropic gas, by scaling,

$$p(\rho) = \frac{\rho^\gamma}{\gamma}, \quad c^2(\rho) = \rho^{\gamma-1}, \quad i(\rho) = \frac{\rho^{\gamma-1} - 1}{\gamma - 1}, \quad \gamma > 1,$$

where $c(\rho)$ is the sound speed.

The system is invariant under the self-similar scaling:

$$(\mathbf{x}, t) \to (\alpha \mathbf{x}, \alpha t), \quad (\rho, u, v, \Phi) \to (\rho, u, v, \frac{\Phi}{\alpha}) \qquad \text{for } \alpha \neq 0.$$

Thus, we can seek self-similar solutions of the form:

$$(\rho, u, v)(\mathbf{x}, t) = (\rho, u, v)(\boldsymbol{\xi}), \quad \Phi(\mathbf{x}, t) = t\big(\varphi(\boldsymbol{\xi}) + \frac{1}{2}|\boldsymbol{\xi}|^2\big) \qquad \text{for } \boldsymbol{\xi} = (\xi_1, \xi_2) = \tfrac{\mathbf{x}}{t},$$

where φ is called a pseudo–velocity potential that satisfies $\nabla_{\boldsymbol{\xi}}\varphi = (u - \xi_1, v - \xi_2) = (U, V)$ which is called a pseudo-velocity. Then the pseudo–potential function φ satisfies the following equation for self–similar solutions:

$$\operatorname{div}(\rho D\varphi) + 2\rho = 0, \tag{3}$$

where the density function $\rho = \rho(|D\varphi|^2, \varphi)$ is determined by

$$\rho(|D\varphi|^2, \varphi) = \big(\rho_0^{\gamma-1} - (\gamma - 1)(\varphi + \frac{1}{2}|D\varphi|^2)\big)^{\frac{1}{\gamma-1}}, \tag{4}$$

and the divergence *div* and gradient D are with respect to the self–similar variables $\boldsymbol{\xi}$, and ρ_0 is a positive constant (to be given in Problem 2.1 below) so that $\rho_0^{\gamma-1} = (\gamma - 1)B_0 + 1$. Therefore, the potential function φ is governed by the following second-order potential flow equation:

$$div\big(\rho(|D\varphi|^2, \varphi)D\varphi\big) + 2\rho(|D\varphi|^2, \varphi) = 0. \tag{5}$$

Equation (5) is a second-order equation of mixed elliptic-hyperbolic type: It is elliptic if and only if $|D\varphi| < c(|D\varphi|^2, \varphi)$, which is equivalent to

$$|D\varphi| < c_\star(\varphi, \gamma) := \sqrt{\frac{2}{\gamma+1}\big(\rho_0^{\gamma-1} - (\gamma - 1)\varphi\big)}. \tag{6}$$

If ρ is a constant, then (3)–(4) imply that the corresponding pseudo-velocity potential φ is of the form:

$$\varphi(\boldsymbol{\xi}) = -\frac{1}{2}|\boldsymbol{\xi}|^2 + (u, v) \cdot \boldsymbol{\xi} + k$$

for constants u, v, and k. Such a solution is called a uniform or constant state.

2.2. **Weak solutions and the Rankine-Hugoniot conditions.** Since shocks are involved in the problem under consideration, we define the notion of weak solutions of equation (5), which admits the shocks.

Definition 2.1. A function $\varphi \in W^{1,1}_{\text{loc}}(\Omega)$ is called a weak solution of (5) if

(i) $\rho_0^{\gamma-1} - \varphi - \frac{1}{2}|D\varphi|^2 \geq 0 \quad a.e.$ in Ω,

(ii) $(\rho(|D\varphi|^2, \varphi), \rho(|D\varphi|^2, \varphi)|D\varphi|) \in (L^1_{\text{loc}}(\Omega))^2$,

(iii) For every $\zeta \in C_c^\infty(\Omega)$,

$$\int_\Omega \big(\rho(|D\varphi|^2, \varphi)D\varphi \cdot D\zeta - 2\rho(|D\varphi|^2, \varphi)\zeta\big)\mathrm{d}\boldsymbol{\xi} = 0.$$

For a piecewise smooth solution φ divided by a shock, it is easy to verify that φ satisfies the conditions in Definition 2.1 if and only if φ is a classic solution of (5) in each smooth subregion and satisfies the following Rankine-Hugoniot conditions across the shock:

$$[\rho(|D\varphi|^2, \varphi)D\varphi \cdot \boldsymbol{\nu}]_S = 0, \tag{7}$$

$$[\varphi]_S = 0, \tag{8}$$

where $\boldsymbol{\nu}$ is a unit normal to S. Condition (7) is due to the conservation of mass, while condition (8) is due to the irrotationality.

There are fairly many weak solutions to the given conservation laws. The physically relevant solutions must satisfy the entropy condition. For potential flow, the discontinuity of $D\varphi$ satisfying the Rankine-Hugoniot conditions (7)–(8) is called a shock if it satisfies the following *entropy condition: The density ρ increases across a shock in the pseudo–flow direction.* From (7), the entropy condition indicates that the normal derivative function $\varphi_{\boldsymbol{\nu}} = D\varphi \cdot \boldsymbol{\nu}$ on a shock always decreases across the shock in the pseudo–flow direction.

2.3. **Shock reflection-diffraction problem.** The incident shock separates two constant states: state (0) with density ρ_0 and velocity $\mathbf{v}_0 = (0,0)$ ahead of the shock, and state (1) with density ρ_1 and velocity $\mathbf{v}_1 = (u_1, 0)$ behind the shock, where the entropy condition holds: $\rho_1 > \rho_0$ on the shock. The incident shock moves from the left to the right and hits the vertex of wedge:

$$W := \{\mathbf{x} : |x_2| < x_1 \tan\theta_{\rm w}, x_1 > 0\}$$

at the initial time. The slip boundary condition $\mathbf{v}\cdot\boldsymbol{\nu} = 0$ is prescribed on the solid wedge boundary.

Then the shock reflection-diffraction problem can be formulated as follows:

Problem 2.1 (Initial-boundary value problem). *Seek a solution of system* (1)–(2) *for* $B_0 = \frac{\rho_0^{\gamma-1}-1}{\gamma-1}$ *with the initial condition at* $t = 0$:

$$(\rho, \Phi)|_{t=0} = \begin{cases} (\rho_0, 0) & \text{for } |x_2| > x_1\tan\theta_{\rm w},\ x_1 > 0, \\ (\rho_1, u_1 x_1) & \text{for } x_1 < 0, \end{cases} \tag{9}$$

and the slip boundary condition along the wedge boundary ∂W:

$$\nabla_{\mathbf{x}}\Phi\cdot\boldsymbol{\nu}|_{\partial W\times\mathbb{R}_+} = 0, \tag{10}$$

where $\boldsymbol{\nu}$ *is the exterior unit normal to* ∂W.

The initial-boundary value problem, Problem 2.1, is a lateral Riemann problem with boundary $\partial W \times \mathbb{R}_+$ in the $(\mathbf{x}, t)$–coordinates. Since state (1) does not satisfy the slip boundary condition, the solution must differ from state (1) behind the shock so that the shock reflection-diffraction configurations occur. These configurations are self-similar, so the problem can be reformulated in the self-similar coordinates $\boldsymbol{\xi} = (\xi_1, \xi_2)$. Depending on the data, there may be various patterns of shock reflection-diffraction configurations, including regular reflection and Mach reflection.

By the symmetry of the problem with respect to the ξ_1–axis, we consider only the upper half-plane $\{\xi_2 > 0\}$ and prescribe the condition $\varphi_{\boldsymbol{\nu}} = 0$ on the symmetry line $\{\xi_2 > 0\}$. Note that state (1) satisfies this condition.

We study self-similar solutions of Problem 2.1. Thus we give a formulation for the solution of Problem 2.1 in the self-similar coordinates $\boldsymbol{\xi} = (\xi_1, \xi_2)$. Let

$$\Lambda = \mathbb{R}^2_+ \setminus \{\boldsymbol{\xi} : \xi_1 > 0,\ 0 < \xi_2 < \xi_1\tan\theta_{\rm w}\},$$

where $\mathbb{R}^2_+ = \mathbb{R}^2 \cap \{\xi_1 > 0\}$. Then, following Definition 2.1, we have

Definition 2.2. $\varphi \in C^{0,1}(\overline{\Lambda})$ is a weak solution of the shock reflection-diffraction problem if φ satisfies equation (5) in Λ in the weak sense, the boundary condition:

$$\partial_{\boldsymbol{\nu}}\varphi = 0 \qquad \text{on } \partial\Lambda, \tag{11}$$

and the asymptotic condition:

$$\lim_{R\to\infty} \|\varphi - \overline{\varphi}\|_{0,\Lambda\setminus B_R(0)} = 0, \tag{12}$$

where

$$\bar{\varphi} = \begin{cases} \varphi_0 & \text{for } \xi_1 > \xi_1^0,\ \xi_2 > \xi_1\tan\theta_{\rm w}, \\ \varphi_1 & \text{for } \xi_1 < \xi_1^0,\ \xi_2 > 0, \end{cases}$$

and $\xi_1^0 > 0$ is the location of the incident shock.

2.4. **Solutions of regular reflection structure.** We will show that, for certain values of parameters, there exist self-similar solutions of the regular reflection structure for the shock reflection-diffraction problem and, moreover, these solutions are unique in the class of self-similar solutions of such a structure.

Figs. 2.1–2.2 show two different regular shock reflection-diffraction configurations in the self-similar coordinates. The regular reflection solutions are piecewise smooth; more precisely, they are smooth away from the incident and reflected-diffracted shocks, as well as the sonic circle (which is a weak discontinuity) for the supersonic case as shown in Fig. 2.1.

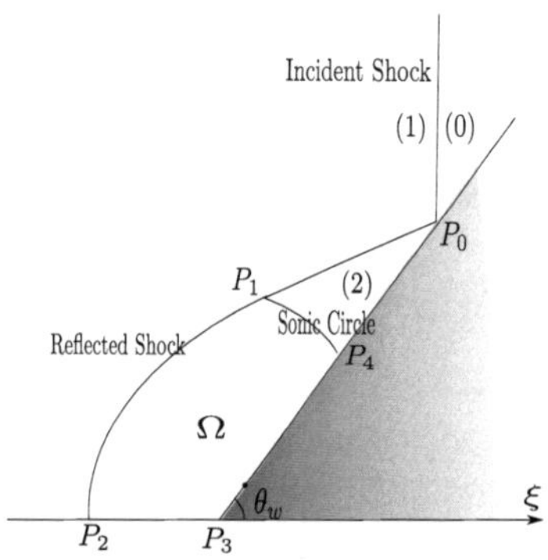

FIGURE 2.1. Supersonic regular shock reflection-diffraction configuration

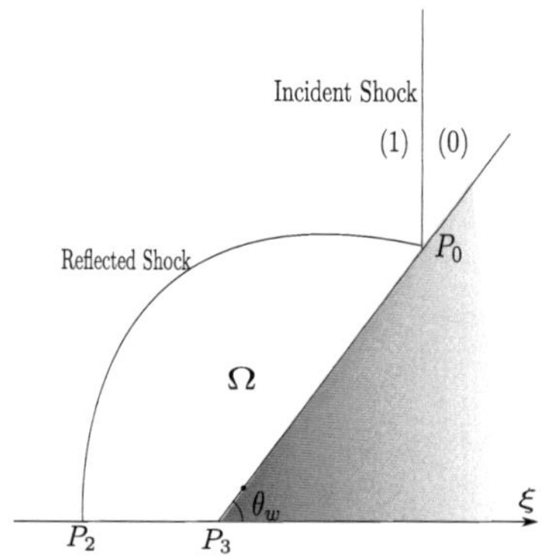

FIGURE 2.2. Subsonic regular shock reflection-diffraction configuration

A necessary condition for the existence of piecewise-smooth regular shock reflection-diffraction configurations is the existence of the constant state (2) with the pseudo-potential φ_2 that satisfies both the slip boundary condition on the wedge and the Rankine-Hugoniot conditions with state (1) across the flat shock $S_1 = \{\varphi_1 = \varphi_2\}$, which passes through point P_0 where the incident shock meets the wedge boundary. Therefore, it requires the following three conditions at P_0:

$$\begin{aligned}
&D\varphi_2 \cdot \boldsymbol{\nu}_{\mathrm{w}} = 0,\\
&\varphi_2 = \varphi_1, \qquad\qquad (13)\\
&\rho(|D\varphi_2|^2, \varphi_2) D\varphi_2 \cdot \boldsymbol{\nu}_{S_1} = \rho_1 D\varphi_1 \cdot \boldsymbol{\nu}_{S_1},
\end{aligned}$$

where $\boldsymbol{\nu}_{\mathrm{w}}$ is the outward normal to the wedge boundary, and $\boldsymbol{\nu}_{S_1} = \frac{D(\varphi_1 - \varphi_2)}{|D(\varphi_1 - \varphi_2)|}$.

It is well-known (see *e.g.* [19]) that, for given parameters (ρ_0, ρ_1) of states (0) and (1), there exists a detachment angle $\theta_{\mathrm{w}}^{\mathrm{d}} \in (0, \frac{\pi}{2})$ such that the algebraic equations (13) have two solutions for each wedge angle $\theta_{\mathrm{w}} \in (\theta_{\mathrm{w}}^{\mathrm{d}}, \frac{\pi}{2})$, which become equal when $\theta_{\mathrm{w}} = \theta_{\mathrm{w}}^{\mathrm{d}}$. Then two two-shock configurations occur at P_0 when $\theta_{\mathrm{w}} \in (\theta_{\mathrm{w}}^{\mathrm{d}}, \frac{\pi}{2})$. For each θ_{w}, state (2) with the smaller density is called a weak state (2). In this paper, state (2) always refers to the weak one, since the strong state (2) is ruled out by the stability/continuity criterion as introduced first by Chen-Feldman in [17]; see also [19]. Depending on the wedge angle, state (2) can be either supersonic or subsonic at P_0. Moreover, for θ_{w} near $\frac{\pi}{2}$ (resp. for θ_{w} near $\theta_{\mathrm{w}}^{\mathrm{d}}$), state (2) is supersonic (resp. subsonic) at P_0. The type of state (2) at P_0 determines the type of reflection, *i.e.* supersonic or subsonic, as shown in Figs. 2.1–2.2.

We consider solutions of the structure shown in Figs. 2.1–2.2. Outside of region Ω, the flow consists of the uniform states (0), (1), and (2) as indicated on the pictures, separated by the straight shocks. In particular, the incident shock separating states (0) and (1) within Λ is the half-line $S_0^+ = \{(\xi_1, \xi_2) : \xi_1 = \xi_{1,P_0}, \xi_2 > \xi_{2,P_0}\}$. The flow is non-uniform and pseudo-subsonic in Ω. Here Ω is an open bounded connected domain, and $\partial\Omega = \overline{\Gamma_{\text{shock}}} \cup \overline{\Gamma_{\text{sonic}}} \cup \overline{\Gamma_{\text{wedge}}} \cup \overline{\Gamma_{\text{sym}}}$, where curve Γ_{shock} with endpoints P_1 and $P_2 \in \{\xi_2 = 0\}$ in the supersonic case (resp. P_0 and $P_2 \in \{\xi_2 = 0\}$ in the subsonic case) is a transonic shock which separates a constant state (1) outside Ω from a pseudo-subsonic (non-constant) state inside Ω, and $\overline{\Gamma_{\text{sonic}}} \cup \overline{\Gamma_{\text{wedge}}} \cup \overline{\Gamma_{\text{sym}}}$ is the fixed boundary with arc Γ_{sonic} between points P_1 and P_4 of the pseudo-sonic circle of state (2) (we also use notation $\overline{\Gamma_{\text{sonic}}} = \{P_0\}$ for the subsonic reflection case as shown in Fig. 2.2), the line segment Γ_{wedge} is the part of $\partial\Omega$ on the wedge boundary, *i.e.* $\Gamma_{\text{wedge}} = P_3P_4$ in the supersonic case and $\Gamma_{\text{wedge}} = P_0P_3$ in the subsonic case, and $\Gamma_{\text{sym}} = P_2P_3$ is the part of $\partial\Omega$ on the symmetry line $\{\xi_2 = 0, \xi_1 < 0\}$.

3. **Existence and Regularity of Regular Shock Reflection-Diffraction Configurations.** We first notice that a key obstacle to the existence of regular shock reflection-diffraction configurations is an additional possibility that, at the critical wedge angle $\theta_{\text{w}}^{\text{c}} \in (\theta_{\text{w}}^{\text{d}}, \frac{\pi}{2})$, the reflected shock P_0P_2 may attach to the wedge vertex P_3, *i.e.* $P_2 = P_3$. We can rule out such a solution if $u_1 \le c_1$. In the opposite case $u_1 > c_1$, there would be a possibility that the reflected shock is attached to the wedge vertex, as the experiments show (*e.g.* [53, Fig. 238]). We note that the condition on (u_1, c_1) can be explicitly expressed through parameters (ρ_0, ρ_1) of states (0) and (1), besides $\gamma \ge 1$, by using (4) and the Rankine-Hugoniot conditions on the incident shock. Recall that $\rho_1 > \rho_0$. It can be shown that there exists $\rho^c > \rho_0$ such that

$$u_1 \le c_1 \quad \text{iff } \rho_1 \in [\rho_0, \rho^c], \qquad u_1 > c_1 \quad \text{iff } \rho_1 \in [\rho^c, \infty).$$

Now we state the existence and regularity results for the solutions of shock reflection-diffraction problem which have the regular reflection structure as on Fig. 2.1–2.2, established in Chen-Feldman [19]. We prove these results in the class of *admissible solutions* of the regular reflection problem, defined as follows:

Definition 3.1. Let $\theta_{\text{w}} \in (\theta_{\text{w}}^{\text{d}}, \frac{\pi}{2})$. A function $\varphi \in C^{0,1}(\overline{\Lambda})$ is an admissible solution of the regular reflection problem (5) and (11)–(12) if φ is a solution in the sense of Definition 2.2 and satisfies the following properties:

(i) The structure of solutions is as follows:

- If $|D\varphi_2(P_0)| > c_2$, then φ is of the *supersonic* regular shock reflection-diffraction configuration described in §2.4 and shown on Fig. 2.1 and satisfies:
  The reflected-diffracted shock Γ_{shock} is C^2 in its relative interior. Curves Γ_{shock}, Γ_{sonic}, Γ_{wedge}, and Γ_{symm} do not have common points except their endpoints.
  φ satisfies the following properties:

$$\varphi \in C^{0,1}(\Lambda) \cap C^1(\Lambda \setminus (\overline{S_0^+} \cup \overline{P_0P_1P_2})),$$
$$\varphi \in C^1(\overline{\Omega}) \cap C^3(\overline{\Omega} \setminus (\overline{\Gamma_{\text{sonic}}} \cup \{P_2, P_3\})),$$

$$\varphi = \begin{cases} \varphi_0 & \text{for } \xi_1 > \xi_1^0 \text{ and } \xi_2 > \xi_1 \tan\theta_{\rm w}, \\ \varphi_1 & \text{for } \xi_1 < \xi_1^0 \text{ and above curve } P_0P_1P_2, \\ \varphi_2 & \text{in region } P_0P_1P_4. \end{cases} \tag{14}$$

- If $|D\varphi_2(P_0)| \leq c_2$, then φ is of the *subsonic* regular shock reflection-diffraction configuration described in §2.4 and shown on Fig. 2.2 and satisfies:
  The reflected-diffracted shock $\Gamma_{\rm shock}$ is C^2 in its relative interior. Curves $\Gamma_{\rm shock}$, $\Gamma_{\rm wedge}$, and $\Gamma_{\rm symm}$ do not have common points except their end-points.
  φ satisfies the following properties:

$$\varphi \in C^{0,1}(\Lambda) \cap C^1(\Lambda \setminus (\overline{S_0^+} \cup \overline{\Gamma_{\rm shock}})),$$
$$\varphi \in C^1(\overline{\Omega}) \cap C^3(\overline{\Omega} \setminus \{P_0, P_3\}),$$

$$\varphi = \begin{cases} \varphi_0 & \text{for } \xi_1 > \xi_1^0 \text{ and } \xi_2 > \xi_1 \tan\theta_{\rm w}, \\ \varphi_1 & \text{for } \xi_1 < \xi_1^0 \text{ and above curve } P_0P_2, \\ \varphi_2(P_0) & \text{at } P_0, \end{cases} \tag{15}$$

$$D\varphi_{|\Omega}(P_0) = D\varphi_2(P_0).$$

Moreover, in both supersonic and subsonic cases, denote $\Gamma_{\rm shock}^{\rm ext} = \Gamma_{\rm shock} \cup \{P_0\} \cup \Gamma_{\rm shock}^-$, where $\Gamma_{\rm shock}^-$ is the reflection of $\Gamma_{\rm shock}$ with respect to the ξ_1-axis. Then curve $\Gamma_{\rm shock}^{\rm ext}$ is C^1 in its relative interior.

(ii) Equation (5) is strictly elliptic in $\overline{\Omega} \setminus \overline{\Gamma_{\rm sonic}}$, *i.e.*

$$|D\varphi| < c(|D\varphi|^2, \varphi) \qquad \text{in } \overline{\Omega} \setminus \overline{\Gamma_{\rm sonic}},$$

where, for the subsonic and sonic cases, we use notation $\overline{\Gamma_{\rm sonic}} = \{P_0\}$.

(iii) $\partial_{\boldsymbol\nu}\varphi_1 > \partial_{\boldsymbol\nu}\varphi > 0$ on $\Gamma_{\rm shock}$, where $\boldsymbol\nu$ is the normal to $\Gamma_{\rm shock}$, pointing to the interior of Ω.

(iv) $\varphi_2 \leq \varphi \leq \varphi_1$ in Ω.

(v) Let $\mathbf{e}_{S_1}$ be the unit vector parallel to $S_1 := \{\varphi_1 = \varphi_2\}$, oriented so that $\mathbf{e}_{S_1} \cdot D\varphi_2(P_0) > 0$:

$$\mathbf{e}_{S_1} = -\frac{(v_2,\, u_1 - u_2)}{\sqrt{(u_1-u_2)^2 + v_2^2}}. \tag{16}$$

Let $\mathbf{e}_{\xi_2} = (0,1)$. Then

$$\partial_{\mathbf{e}_{S_1}}(\varphi_1 - \varphi) \leq 0, \quad \partial_{\xi_2}(\varphi_1 - \varphi) \leq 0 \qquad \text{on } \Gamma_{\rm shock}. \tag{17}$$

Remark 3.1. It can be shown that Definition 3.1 is equivalent to the definition of admissible solutions in [19]; see Definitions 15.1.1–15.1.2 there. Thus, all the estimates and properties of admissible solutions shown in [19] hold for the admissible solutions defined above. In particular, the admissible solutions converge (in an appropriate sense) to the normal reflection solution as $\theta_{\rm w} \to \frac{\pi}{2}-$.

Remark 3.2. For the supersonic case, $\mathbf{e}_{S_1}$ defined by (16) has the expression:

$$\mathbf{e}_{S_1} = \frac{P_1 - P_0}{|P_1 - P_0|}.$$

Moreover, in the supersonic (resp. subsonic/sonic) case, $\mathbf{e}_{S_1}$ is tangential to Γ_{shock} in its upper endpoint P_1 (resp. P_0), because $(\varphi, D\varphi)|_\Omega = (\varphi_2, D\varphi_2)$ at that point, and its orientation at that endpoint of Γ_{shock} is towards the relative interior of Γ_{shock}.

Remark 3.3. Since the admissible solution φ is a weak solution in the sense of Definition 2.2 and is of regularity as in (i) of Definition 3.1, it satisfies (5) classically in Ω with the Rankine-Hugoniot conditions:

$$\varphi = \varphi_1, \quad \rho(|D\varphi|^2, \varphi)D\varphi \cdot \boldsymbol{\nu} = \rho_1 D\varphi_1 \cdot \boldsymbol{\nu} \qquad \text{on } \Gamma_{\text{shock}}, \tag{18}$$

and the boundary condition:

$$\partial_{\boldsymbol{\nu}}\varphi = 0 \qquad \text{on } \Gamma_{\text{wedge}} \cup \Gamma_{\text{sym}}. \tag{19}$$

Remark 3.4. The admissible solution φ is not a constant state in Ω. Indeed, if φ is a constant state in Ω, then $\varphi = \varphi_2$ in Ω: This follows from both (14) for the supersonic case (since φ is C^1 across Γ_{sonic}) and property $(\varphi, D\varphi) = (\varphi_2, D\varphi_2)$ at P_0 for the subsonic case. However, φ_2 does not satisfy (19) on Γ_{sym} since $\mathbf{v}_2 = (u_2, v_2) = (u_2, u_2 \tan\theta_{\text{w}})$ with $u_2 > 0$ and $\theta_{\text{w}} \in (0, \frac{\pi}{2})$.

The following theorem shows that the admissible solution has additional regularity and monotonicity properties.

Theorem 3.1 (Properties of admissible solutions). *There exits a constant $\alpha = \alpha(\rho_0, \rho_1, \gamma) \in (0, \frac{1}{2})$ such that any admissible solution in the sense of Definition 3.1 with wedge angle $\theta_{\text{w}} \in (\theta_{\text{w}}^{\text{d}}, \frac{\pi}{2})$ has the following properties:*

(i) *Additional regularity:*
- *If $|D\varphi_2(P_0)| > c_2$, i.e. when φ is of the* supersonic *regular shock reflection-diffraction configuration as in Fig. 2.1, it satisfies $\varphi \in C^{1,\alpha}(\overline{\Omega}) \cap C^\infty(\overline{\Omega} \setminus (\overline{\Gamma_{\text{sonic}}} \cup \{P_3\}))$, and φ is $C^{1,1}$ across Γ_{sonic}, including endpoints P_1 and P_4. The reflected-diffracted shock $P_0P_1P_2$ is $C^{2,\beta}$ up to its endpoints for any $\beta \in [0, \frac{1}{2})$, and C^∞ except P_1.*
- *If $|D\varphi_2(P_0)| \le c_2$, i.e. when φ is of the* subsonic *regular shock reflection-diffraction configuration as in Fig. 2.2, it satisfies*

$$\varphi \in C^{1,\beta}(\overline{\Omega}) \cap C^{1,\alpha}(\overline{\Omega} \setminus \{P_0\}) \cap C^\infty(\overline{\Omega} \setminus \{P_0, P_3\})$$

for some $\beta = \beta(\rho_0, \rho_1, \gamma, \theta_w) \in (0, \alpha]$ where β is non-decreasing with respect to θ_w, and the reflected-diffracted shock Γ_{shock} is $C^{1,\beta}$ up to its endpoints and C^∞ except P_0.

(ii) *For each $\mathbf{e} \in Con(\mathbf{e}_{S_1}, \mathbf{e}_{\xi_2})$,*

$$\partial_{\mathbf{e}}(\varphi_1 - \varphi) < 0 \qquad \text{in } \overline{\Omega}, \tag{20}$$

where the vectors $\mathbf{e}_{S_1}$ and $\mathbf{e}_{\xi_2}$ are defined in Definition 3.1(v), and

$$Con(\mathbf{e}_{S_1}, \mathbf{e}_{\xi_2}) := \{a\mathbf{e}_{S_1} + b\mathbf{e}_{\xi_2} \; : \; a, b > 0\}. \tag{21}$$

(iii) *Denote by $\boldsymbol{\nu}_{\text{w}}$ the unit interior normal on Γ_{wedge} (with respect to Ω), i.e. $\boldsymbol{\nu}_{\text{w}} = (-\sin\theta_{\text{w}}, \cos\theta_{\text{w}})$. Then $\partial_{\boldsymbol{\nu}_{\text{w}}}(\varphi - \varphi_2) \le 0$ in $\overline{\Omega}$.*

Remark 3.5. $Con(\mathbf{e}_{S_1}, \mathbf{e}_\eta) = \{a\mathbf{e}_{S_1} + b\mathbf{e}_\eta : a, b > 0\}$ is an open set; that is, it does not include the directions of $\mathbf{e}_{S_1}$ and $\mathbf{e}_{\xi_2}$.

Now we state the results on the existence of admissible solutions.

Theorem 3.2 (Global solutions up to the detachment angle for the case: $u_1 \leq c_1$). *Let the initial data (ρ_0, ρ_1, γ) satisfy that $u_1 \leq c_1$. Then, for each $\theta_{\rm w} \in (\theta_{\rm w}^{\rm d}, \frac{\pi}{2})$, there exists an admissible solution of the regular reflection problem in the sense of Definition 3.1. Note that these solutions satisfy the properties stated in Theorem 3.1.*

Theorem 3.3 (Global solutions up to the detachment angle for the case: $u_1 > c_1$). *Let the initial data (ρ_0, ρ_1, γ) satisfy that $u_1 > c_1$. Then there is $\theta_{\rm w}^{\rm c} \in [\theta_{\rm w}^{\rm d}, \frac{\pi}{2})$ such that, for each $\theta_{\rm w} \in (\theta_{\rm w}^{\rm c}, \frac{\pi}{2})$, there exists an admissible solution of the regular reflection problem in the sense of Definition 3.1. Note that these solutions satisfy the properties stated in Theorem 3.1.*

If $\theta_{\rm w}^{\rm c} > \theta_{\rm w}^{\rm d}$, then, for the wedge angle $\theta_{\rm w} = \theta_{\rm w}^{\rm c}$, there exists an attached shock solution φ with all the properties listed in Definition 3.1 and Theorem 3.1(ii)–(iii) except that $P_2 = P_3$. In addition, for the regularity of solution φ, we have

- *For the supersonic case with $\theta_{\rm w} = \theta_{\rm w}^{\rm c}$,*

$$\varphi \in C^\infty(\overline{\Omega}\backslash(\overline{\Gamma_{\rm sonic}} \cup \{P_3\})) \cap C^{1,1}(\overline{\Omega}\backslash\{P_3\}) \cap C^{0,1}(\overline{\Omega}),$$

  and the reflected shock $P_1P_2P_3$ is Lipschitz up to the endpoints, $C^{2,\beta}$ with any $\beta \in [0, \frac{1}{2})$ except point P_3, and C^∞ except points P_1 and P_3.
- *For the subsonic case with $\theta_{\rm w} = \theta_{\rm w}^{\rm c}$,*

$$\varphi \in C^\infty(\overline{\Omega}\backslash\{P_1, P_3\}) \cap C^{1,\beta}(\overline{\Omega}\backslash\{P_3\}) \cap C^{0,1}(\overline{\Omega})$$

  for β as in Theorem 3.1, and the reflected shock $P_1P_2P_3$ is Lipschitz up to the endpoints, $C^{1,\beta}$ except point P_3, and C^∞ except points P_1 and P_3.

In the next two sections, §4–§5, we show how the convexity of the transonic shocks and the uniqueness of the admissible solutions can be achieved.

4. **Convexity of Transonic Shocks in the Shock Reflection-Diffraction Configurations.** We first note that, for an admissible solution, $\Gamma_{\rm shock}$ is a graph in any direction $\mathbf{e} \in Con := Con(\mathbf{e}_{S_1}, \mathbf{e}_\eta)$, where $Con(\mathbf{e}_{S_1}, \mathbf{e}_\eta)$ is defined in (21). For the subsonic/sonic reflections case, we denote $P_1 := P_0$ so that $\Gamma_{\rm shock}$ has endpoints P_1 and P_2 in all cases. More precisely, the following was shown in [19], as a consequence of Theorem 3.1(ii):

Lemma 4.1. *Let φ be an admissible solution. Denote $\phi := \varphi - \varphi_1$. Let $\boldsymbol{\tau}_{P_1}$ be a unit tangent vector to $\Gamma_{\rm shock}$ at P_1, directed into the interior of $\Gamma_{\rm shock}$. Let $\mathbf{e} \in Con$, and let $\mathbf{e}^\perp$ be the orthogonal unit vector to $\mathbf{e}$ with $\mathbf{e}^\perp \cdot \boldsymbol{\tau}_{P_1} > 0$. Let (S, T) be the coordinates with respect to basis $\{\mathbf{e}, \mathbf{e}^\perp\}$ so that $T_{P_2} > T_{P_1}$. Then there exists $f_{\mathbf{e}} \in C^1(\mathbb{R})$ such that*

(a) $\Gamma_{\rm shock} = \{S = f_{\mathbf{e}}(T) : T_{P_1} < T < T_{P_2}\}$, $\Omega \subset \{S < f_{\mathbf{e}}(T) : T \in \mathbb{R}\}$, $P_1 = (f_{\mathbf{e}}(T_{P_1}), T_{P_1})$, $P_2 = (f_{\mathbf{e}}(T_{P_2}), T_{P_2})$, *and* $f_{\mathbf{e}} \in C^\infty(T_{P_1}, T_{P_2})$;

(b) *The directions of the tangent lines to $\Gamma_{\rm shock}$ lie between $\boldsymbol{\tau}_{P_1}$ and $\boldsymbol{\tau}_{P_2}$; that is, in the (S, T)-coordinates,*

$$-\infty < \frac{\boldsymbol{\tau}_{P_2} \cdot \mathbf{e}}{\boldsymbol{\tau}_{P_2} \cdot \mathbf{e}^\perp} = f_{\mathbf{e}}'(T_{P_2}) \leq f_{\mathbf{e}}'(T) \leq f_{\mathbf{e}}'(T_{P_1}) = \frac{\boldsymbol{\tau}_{P_1} \cdot \mathbf{e}}{\boldsymbol{\tau}_{P_1} \cdot \mathbf{e}^\perp} < \infty$$

for any $T \in (T_{P_1}, T_{P_2})$;

(c) $\boldsymbol{\nu}(P) \cdot \mathbf{e} < 0$ *for any* $P \in \Gamma_{\rm shock}$;

(d) $\phi_{\mathbf{e}} > 0$ *on* $\Gamma_{\rm shock}$;

(e) *For any* $T \in (T_{P_1}, T_{P_2})$,

$$\phi_{\boldsymbol{\tau\tau}}(f_{\mathbf{e}}(T), T) < 0 \iff f''_{\mathbf{e}}(T) > 0,$$

and

$$\phi_{\boldsymbol{\tau\tau}}(f_{\mathbf{e}}(T), T) > 0 \iff f''_{\mathbf{e}}(T) < 0.$$

In [21], we provide a framework for the convexity of transonic shocks in the self-similar coordinates. Specifically, for the transonic shocks in the shock reflection-diffraction configurations, we have the following theorem.

Theorem 4.1 (Convexity of transonic shocks). *If a solution of the shock reflection-diffraction problem is admissible in the sense of Definition* 3.1, *then its shock curve* Γ_{shock} *is strictly convex in the following sense: For any* $\mathbf{e} \in Con$, $f_{\mathbf{e}}$ *from Lemma* 4.1 *is concave on* (T_{P_1}, T_{P_2}), *and* $f''_{\mathbf{e}}(T) < 0$ *for all* $T \in (T_{P_1}, T_{P_2})$. *That is,* Γ_{shock} *is uniformly convex on closed subsets of its relative interior. Moreover, for a regular reflection solution in the sense of Definition* 2.2 *with pseudo-potential* $\varphi \in C^{0,1}(\Lambda)$ *satisfying Definition* 3.1(i)–(iv), *the shock is strictly convex if and only if Definition* 3.1(v) *holds.*

Now we discuss the techniques developed in [21] by giving the main steps in the proof of Theorem 4.1. While the argument in [21] is for a general domain Ω, we focus here on the regular shock reflection-diffraction configurations, in which both the solution domain Ω and the solution structure are somewhat simpler.

Outline of the Proof of Theorem 4.1: The proof consists of eight steps, while the first three steps are general properties of shock reflection-diffraction solutions; see [19]. Below we use notation $\phi := \varphi - \varphi_1$.

1. We establish a relation between the extrema of the solution and the geometric shape of the transonic shock. For a fixed unit vector $\mathbf{e} \in \mathbb{R}^2$, denote $w := \partial_{\mathbf{e}}\phi$ in Ω. We show that, if a local minimum (resp. maximum) of w is attained at $P \in \Gamma^0_{\text{shock}}$ and $\boldsymbol{\nu}(P) \cdot \mathbf{e} < 0$, then $\phi_{\boldsymbol{\tau\tau}} > 0$ (resp. $\phi_{\boldsymbol{\tau\tau}} < 0$) at P, where $\boldsymbol{\nu}$ denotes the interior unit normal on Γ_{shock} towards Ω.

2. We establish a nonlocal relation between the values of $\phi_{\mathbf{e}}$ and the positions where these values are taken. Let ϕ be a solution as in Theorem 4.1, and let $\mathbf{e} \in Con$. We use the coordinates from Lemma 4.1. Assume that, for two different points $P = (T, f_{\mathbf{e}}(T))$ and $P_1 = (T_1, f_{\mathbf{e}}(T_1))$ on Γ_{shock},

$$f_{\mathbf{e}}(T) > f_{\mathbf{e}}(T_1) + f'_{\mathbf{e}}(T)(T - T_1), \qquad f'_{\mathbf{e}}(T) = f'_{\mathbf{e}}(T_1).$$

Then

(i) $d(P) := \text{dist}(O_0, L_P) > \text{dist}(O_0, L_{P_1}) =: d(P_1)$, where O_0 is the center of sonic circle of state (0), and L_P and L_{P_1} are the tangent lines of Γ_{shock} at P and P_1, respectively.

(ii) If the unit vector $\mathbf{e} \in Con$, then

$$\phi_{\mathbf{e}}(P) > \phi_{\mathbf{e}}(P_1).$$

3. We show that the shock graph is real analytic.

4. We now develop a minimal/maximal chain argument. Let ϕ be an admissible solution, and let $\mathbf{e} \in \mathbb{R}^2$. Note that $\phi_{\mathbf{e}}$ satisfies the strong maximum principle in Ω. Then we can introduce the minimal (or maximal) chain as follows:

Let $E_1, E_2 \in \partial\Omega$. We say that points E_1 and E_2 are connected by a minimal (resp. maximal) chain with radius r if and only if there exist $r > 0$, integer $k_1 \geq 1$, and a chain of balls $\{B_r(C^i)\}_{i=0}^{k_1}$ such that

(i) $C^0 = E_1$, $C^{k_1} = E_2$, and $C^i \in \overline{\Omega}$ for $i = 0, \dots, k_1$;
(ii) $C^{i+1} \in \overline{B_r(C^i) \cap \Omega}$ for $i = 0, \dots, k_1 - 1$;
(iii) $\phi_{\mathbf{e}}(C^{i+1}) = \min_{\overline{B_r(C^i)\cap\Omega}} \phi_{\mathbf{e}} < \phi_{\mathbf{e}}(C^i)$ (resp. $\phi_{\mathbf{e}}(C^{i+1}) = \max_{\overline{B_r(C^i)\cap\Omega}} \phi_{\mathbf{e}} > \phi_{\mathbf{e}}(C^i)$) for $i = 0, \dots, k_1 - 1$;
(iv) $\phi_{\mathbf{e}}(C^{k_1}) = \min_{\overline{B_r(C^{k_1})\cap\Omega}} \phi_{\mathbf{e}}$ (resp. $\phi_{\mathbf{e}}(C^{k_1}) = \max_{\overline{B_r(C^{k_1})\cap\Omega}} \phi_{\mathbf{e}}$).

For such a chain $\{C^i\}_{i=0}^{k_1}$, we also use the following terminology: The chain starts at E_1 and ends at E_2, or the chain is from E_1 to E_2.

This definition does not rule out the possibility that $B_r(C^i) \cap \partial\Omega \neq \emptyset$, or even $C^i \in \partial\Omega$, for some or all $i = 0, \dots, k_1 - 1$. The radius r is a parameter in the definition of minimal or maximal chains. We do not fix r at this point. In fact, the radii are determined for various chains, respectively.

Then we prove the following results:

(a) *The chains with sufficiently small radius are connected sets.* More precisely, there exists r^* depending only on (ρ_0, ρ_1, γ) such that, for any minimal or maximal chain $\{C^i\}_{i=0}^{k_1}$ with $r \in (0, r^*]$, $\cup_{i=0}^{k_1} B_r(C^i) \cap \Omega$ is connected.

(b) *The existence of the minimal or maximal chain of radius $r < r^*$.* More precisely, if $E_1 \in \partial\Omega$, and E_1 is not a local minimum point (resp. maximum point) of $\phi_{\mathbf{e}}$ with respect to $\overline{\Omega}$, then, for any $r \in (0, r^*)$, there exists a minimal (resp. maximal) chain $\{G^i\}_{i=0}^{k_1}$ for $\phi_{\mathbf{e}}$ of radius r, starting at E_1, *i.e.* $G^0 = E_1$. Moreover, $G^{k_1} \in \partial\Omega$ is a local minimum (resp. maximum) point of $\phi_{\mathbf{e}}$ with respect to $\overline{\Omega}$, and $\phi_{\mathbf{e}}(G^{k_1}) < \phi_{\mathbf{e}}(E_1)$ (resp. $\phi_{\mathbf{e}}(G^{k_1}) > \phi_{\mathbf{e}}(E_1)$).

(c) *The minimal and maximal chains do not intersect.* Specifically, for any $\delta > 0$, there exists $r_1^* \in (0, r^*]$ such that the following holds: Let $\mathcal{C} \subset \partial\Omega$ be connected, let E_1 and E_2 be the endpoints of $\mathcal{C}$, and let there be a minimal chain $\{E^i\}_{i=0}^{k_1}$ of radius $r_1 \in (0, r_1^*]$, which starts at E_1 and ends at E_2. If there exists $H_1 \in \mathcal{C}^0 = \mathcal{C} \setminus \{E_1, E_2\}$ such that

$$\phi_{\mathbf{e}}(H_1) \geq \phi_{\mathbf{e}}(E_1) + \delta,$$

then, for any $r_2 \in (0, r_1]$, any maximal chain $\{H^j\}_{j=0}^{k_2}$ of radius r_2 starting from H_1 satisfies $H^{k_2} \in \mathcal{C}^0$, where $\mathcal{C}^0$ denotes the relative interior of curve $\mathcal{C}$ as before.

Note that, if H_1 is not a local maximum point of $\phi_{\mathbf{e}}$ with respect to $\overline{\Omega}$, then the existence of the maximal chain $\{H^j\}_{j=0}^{k_2}$ of radius r_2 starting from H_1 follows from result (b).

(d) *Result* (c) *also holds if the roles of minimal and maximal chains are interchanged.* For any $\delta > 0$, there exists $r_1^* \in (0, r^*]$ such that the following holds: Let $\mathcal{C} \subset \partial\Omega$ be connected, and let E_1 and E_2 be the endpoints of $\mathcal{C}$. Assume that there exists a maximal chain $\{E^i\}_{i=0}^{k_1}$ of radius $r_1 \in (0, r_1^*]$, which starts at E_1 and ends at E_2. If there exists $H_1 \in \mathcal{C}^0$ such that

$$\phi_{\mathbf{e}}(H_1) \leq \phi_{\mathbf{e}}(E_1) - \delta,$$

then, for any $r_2 \in (0, r_1]$, any minimal chain $\{H^j\}_{j=0}^{k_2}$ of radius r_2, starting from H_1, satisfies that $H^{k_2} \in \mathcal{C}^0$.

(e) *Two minimal chains do not intersect*: For any $r_1 \in (0, r^*]$, there exists $r_2^* = r_2^*(r_1) \in (0, r^*]$ such that the following holds: Let $\mathcal{C} \subset \partial\Omega$ be connected, and let E_1 and E_2 be the endpoints of $\mathcal{C}$. Assume that there exists a minimal chain $\{E^i\}_{i=0}^{k_1}$ of radius $r_1 \in (0, r^*]$, which starts at E_1 and ends at E_2. If there exists $H_1 \in \mathcal{C}^0$ such that

$$\phi_{\mathbf{e}}(H_1) < \phi_{\mathbf{e}}(E_2),$$

then, for any $r_2 \in (0, r_2^*]$, any minimal chain $\{H^j\}_{j=0}^{k_2}$ of radius r_2, starting from H_1, satisfies that $H^{k_2} \in \mathcal{C}^0$.

5. Denote by $\boldsymbol{\nu}_{\rm w}$ the unit normal on $\Gamma_{\rm wedge}$ pointing into Ω. By [19, Lemma 8.2.11], $\boldsymbol{\nu}_{\rm w} \in Con(\mathbf{e}_{S_1}, \mathbf{e}_{\xi_2})$ for any wedge angle $\theta_{\rm w} \in (\theta_{\rm w}^{\rm d}, \frac{\pi}{2})$. We use $\mathbf{e} = \boldsymbol{\nu}_{\rm w}$ for the following four steps below. We work in the corresponding (S,T)–coordinates defined in Lemma 4.1, so it suffices to prove that the graph is concave:

$$f_{\mathbf{e}}''(T) \le 0 \qquad \text{for all } T \in (T_{P_1}, T_{P_2}).$$

If there exists $\hat{P} \in \Gamma_{\rm shock}^0$ with $f_{\mathbf{e}}''(T_{\hat{P}}) > 0$, we prove the existence of a point $C \in \Gamma_{\rm shock}^0$ such that $f_{\mathbf{e}}''(T_C) \ge 0$, and C is a point of strict local minimum of $\phi_{\mathbf{e}}$ along $\Gamma_{\rm shock}$ but is *not* a local minimum point of $\phi_{\mathbf{e}}$ relative to $\overline{\Omega}$.

6. Then we prove that there exists $C_1 \in \Gamma_{\rm shock}^0$ such that there is a minimal chain with radius r_1 from C to C_1.

7. We show that the existence of points C and C_1 described above yields a contradiction (which implies that there is no $\hat{P} \in \Gamma_{\rm shock}^0$ with $f_{\mathbf{e}}''(T_{\hat{P}}) > 0$). This is proved by showing the following facts:

- Let A_2 be a maximum point of $\phi_{\mathbf{e}}$ along $\Gamma_{\rm shock}$ lying between points C and C_1. Then A_2 is a local maximum point of $\phi_{\mathbf{e}}$ relative to Ω, and there is no point between C and C_1 such that the tangent line at this point is parallel to the one at A_2.
- Between C and A_2, or between C_1 and A_2, there exists a local minimum point C_2 of $\phi_{\mathbf{e}}$ along $\Gamma_{\rm shock}$ such that $C_2 \ne C_1$, or $C_2 \ne C$, and C_2 is not a local minimum point of $\phi_{\mathbf{e}}$ relative to domain $\overline{\Omega}$.
- Then, following an argument similar to the one used above and going through several steps, we arrive at the situation that the endpoint of the minimal chain cannot lie anywhere on $\partial\Omega$, which is a contradiction.

This indicates that $f_{\mathbf{e}}'' \le 0$ on $\Gamma_{\rm shock}$; that is, $\Gamma_{\rm shock}$ is convex. In the rest of the argument, we prove that $f_{\mathbf{e}}'' < 0$ on $\Gamma_{\rm shock}^0$.

8. Using the fact that the shock graph is real analytic, we show that, for every $P \in \Gamma_{\rm shock}^0$, either $f_{\mathbf{e}}''(T_P) < 0$ or there exists an even integer $k > 2$ such that $f_{\mathbf{e}}^{(i)}(T_P) = 0$ for all $i = 2, \dots, k-1$, and $f_{\mathbf{e}}^{(k)}(T_P) < 0$. This shows the strict convexity of the shock, which implies that the shock does not contain any straight segment. The above property is equivalent to the facts that $\partial_{\boldsymbol{\tau}}^i \phi(P) = 0$ for all $i = 2, \dots, k-1$, and $\partial_{\boldsymbol{\tau}}^k \phi(P) > 0$.

9. We show the uniform convexity of $\Gamma_{\rm shock}^0$ in the sense that $f_{\mathbf{e}}''(T_P) < 0$ for every $P \in \Gamma_{\rm shock}^0$, or equivalently, $f_{\mathbf{e}}''(T) < 0$ on (T_{P_1}, T_{P_2}), for some (and thus any) $\mathbf{e} \in Con$. In fact, if it is not true, *i.e.* if $\phi_{\boldsymbol{\tau\tau}} = 0$ at some $P_{\rm d}$, then we can obtain a contradiction by proving that there exists a unit vector $\mathbf{e} \in \mathbb{R}^2$ such that $P_{\rm d}$ is a local minimum point of $\phi_{\mathbf{e}}$ along $\Gamma_{\rm shock}^0$, but $P_{\rm d}$ is not a local minimum point of $\phi_{\mathbf{e}}$

in Ω. Then we can construct a minimal chain for $\phi_{\mathbf{e}}$ connecting $P_{\rm d}$ to $C^{k_1} \in \partial\Omega$. We show that

- $C^{k_1} \notin \Gamma_{\rm sonic}$,
- $C^{k_1} \notin \Gamma_{\rm wedge} \cup \Gamma_{\rm sym}$,
- $C^{k_1} \notin \Gamma_{\rm shock}$.

This implies that $\phi_{\boldsymbol{\tau\tau}} > 0$ on $\Gamma^0_{\rm shock}$ so that $f''_{\mathbf{e}}(T) < 0$ on (T_{P_1}, T_{P_2}); see Lemma 4.1.

5. **Uniqueness and Stability of Regular Shock Reflection-Diffraction Configurations.** In this section, we discuss the uniqueness and stability of global regular shock reflection-diffraction configurations. More specifically, we describe the results in Chen-Feldman-Xiang [22].

As indicated earlier, recent results [24,25,33,46] have shown the non-uniqueness of solutions with planar shocks in the class of entropy solutions with shocks of the Cauchy problem for the multidimensional compressible Euler system. Moreover, the uniqueness problem for general self-similar solutions of the Euler system is still open (*cf.* [24]). While these results do not apply directly to our case, they indicate that it be natural to study the uniqueness of solutions in some more restrictive class, instead of general time-dependent solutions (*i.e.* solutions of Problem 2.1), or even general self-similar solutions as in Definition 2.2.

In [22], we have established the uniqueness of regular reflection solutions for each wedge angle in the class of *admissible solutions* introduced in Definition 3.1.

Theorem 5.1 (Uniqueness). *For any wedge angle $\theta_{\rm w} \in (\theta^{\rm d}_{\rm w}, \frac{\pi}{2})$ when $u_1 \le c_1$ and $\theta_{\rm w} \in (\theta^{\rm c}_{\rm w}, \frac{\pi}{2})$ when $u_1 > c_1$, any solution, satisfying all properties* (i)–(iv) *in Definition* 3.1 *and one of the following properties*:

(a) *the transonic shock $\Gamma_{\rm shock}$ is convex, i.e. domain Ω is a convex set,*

(b) *condition* (17) *holds,*

is unique in the class of admissible solutions. Moreover, such solutions are continuous with respect to the wedge angle $\theta_{\rm w}$ in the C^1–norm (more precisely, the continuity with respect to the norm described in Remark 5.1 *below).*

Remark 5.1. For an admissible solution φ with a wedge angle $\theta_{\rm w}$, we define its norm based on its restriction to Ω. Since region Ω depends on the solution, we map a unit square $Q^{\rm iter} = (0,1)^2$ to Ω and use this mapping to define a function u on $Q^{\rm iter}$, which corresponds to $\varphi_{|\Omega}$. Furthermore, the sides of square $Q^{\rm iter}$ are mapped to the boundary parts $\Gamma_{\rm sonic}$, $\Gamma_{\rm wedge}$, $\Gamma_{\rm sym}$, and $\Gamma_{\rm shock}$. The mapping depends on $(\varphi, \theta_{\rm w})$ and is invertible; that is, given a function u on $Q^{\rm iter}$ and $\theta_{\rm w}$, we can recover φ and Ω. Moreover, this mapping and its inverse have appropriate continuity properties. See [19, §12.2 and §17.2] for the details. Then we define function spaces for admissible solutions and "approximate admissible solutions" in terms of the function spaces for the corresponding functions u on $Q^{\rm iter}$. The convergence of admissible solutions $\varphi^{(i)} \to \varphi^{(\infty)}$ in the C^1–norm as the corresponding wedge angles $\theta^{(i)}_{\rm w} \to \theta^{(\infty)}_{\rm w}$, defined in terms of convergence in an appropriate norm for the functions on $Q^{\rm iter}$, implies

$$
\begin{aligned}
&\|\varphi^{(i)}\|_{C^1(\Omega^{(i)})} \le C \qquad \text{for all } i,\\
&\|\varphi^{(i)} - \varphi^{(\infty)}\|_{C^1(\overline{\Omega^{(i)}\cap\Omega^{(\infty)}})} + d_{\rm H}(\overline{\Omega^{(i)}}, \overline{\Omega^{(\infty)}}) \to 0 \qquad \text{as } \theta^{(i)}_{\rm w} \to \theta^{(\infty)}_{\rm w},
\end{aligned}
\tag{22}
$$

where $d_{\rm H}$ denotes the Hausdorff distance between the sets.

Remark 5.2. By Theorem 4.1, conditions (a) and (b) in Theorem 5.1 for the solutions satisfying properties (i)–(iv) in Definition 3.1 are equivalent.

Remark 5.3. We note that, under either one of conditions (a) and (b) in Theorem 5.1, the solution is an admissible solution. Indeed, in both cases, the solution satisfies properties (i)–(iv) in Definition 3.1. If, in addition, condition (b) holds, then the solution is admissible, directly from Definition 3.1. Remark 5.2 shows the same for the case when condition (a) holds.

The proof of Theorem 5.1 is obtained by showing the following proposition on the existence and uniqueness of a family of admissible solutions that are continuous with respect to $\theta_{\rm w}$, containing a given admissible solution.

Proposition 5.1. *Fix (ρ_0, ρ_1, γ). Define interval $I := (\theta_{\rm w}^{\rm d}, \frac{\pi}{2}]$ when $u_1 \le c_1$ and $I := (\theta_{\rm w}^{\rm c}, \frac{\pi}{2}]$ when $u_1 > c_1$. For every admissible solution φ^* with a wedge angle $\theta_{\rm w}^* \in I$, there exists a family*

$$\mathfrak{S} = \{(\varphi, \theta_{\rm w}) \ : \ \theta_{\rm w} \in I, \ \varphi \in C^{0,1}(\Lambda(\theta_{\rm w}))\}$$

such that

$$(\varphi^*, \theta_{\rm w}^*) \in \mathfrak{S}, \tag{23}$$

and $\mathfrak{S}$ satisfies the following properties:

(a) *For each $\theta_{\rm w} \in I$, there exists one and only one pair $(\varphi, \theta_{\rm w}) \in \mathfrak{S}$. Then we can define $\varphi^{(\theta_{\rm w})} := \varphi$ if $(\varphi, \theta_{\rm w}) \in \mathfrak{S}$.*
(b) *Each $\varphi^{(\theta_{\rm w})}$ is an admissible solution corresponding to the wedge angle $\theta_{\rm w}$.*
(c) *$\varphi^{(\frac{\pi}{2})}$ is the normal shock reflection solution* (see §3.1 in [17] for the definition).
(d) *$\varphi^{(\theta_{\rm w})}$ is continuous with respect to the wedge angle $\theta_{\rm w} \in I$ in the C^1–norm as in Remark 5.1.*

Moreover, a family $\mathfrak{S}$ satisfying properties (a)–(d) *listed above* (*but without requiring* (23)) *is unique. That is, if there are two families $\mathfrak{S}_1$ and $\mathfrak{S}_2$ satisfying properties* (a)–(d), *then $\mathfrak{S}_1 = \mathfrak{S}_2$. Thus, the family $\mathfrak{S}$ contains all the admissible solutions for all $\theta_{\rm w} \in I$.*

Proposition 5.1 directly implies Theorem 5.1.

Proposition 5.1 is proved by showing the local uniqueness and existence of admissible solutions.

As we have discussed in the introduction, the outline of the uniqueness proof (*i.e.* Proposition 5.1) is the following: If there are two different admissible solutions, defined by the potential functions φ and $\hat{\varphi}$, for some wedge angle $\theta_{\rm w}^* \in I \setminus \{\frac{\pi}{2}\}$, it suffices to:

(i) construct continuous families of solutions parametrized by the wedge angle $\theta_{\rm w} \in [\theta_{\rm w}^*, \frac{\pi}{2}]$, starting from φ and $\hat{\varphi}$, respectively, in the norm discussed in Remark 5.1;
(ii) prove *local uniqueness*: If two admissible solutions with the same wedge angle are close in the norm given in the second line of (22), then they are equal.

Combining this with the fact that, by Remark 3.1, both families converge to the normal reflection as $\theta_{\rm w} \to \frac{\pi}{2}-$, we obtain a contradiction; see more details in §5.3 below. Furthermore, the continuous family defined above can be extended to all $\theta_{\rm w} \in I$, hence determining the family $\mathfrak{S}$ in Proposition 5.1.

In order to construct the continuous family of solutions $\mathfrak{S}$ described in Proposition 5.1, starting from the given solution $\varphi = \varphi^{(\theta_{\rm w}^*)}$, it suffices to show that any given admissible solution can be perturbed, that is, an admissible solution can be constructed to be close to φ for all wedge angles that are sufficiently close to $\theta_{\rm w}^*$. More precisely, using the mapping of admissible solutions to the functions on the unit square discussed in Remark 5.1, we work in an appropriately weighted and scaled $C^{2,\alpha}$ space on $Q^{\rm iter}$. We choose this function space according to the norms and the other quantities in the *a priori* estimates for admissible solutions in [19], mapped to $Q^{\rm iter}$. Denote the norm in this space by $\|\cdot\|^*$. Thus, we consider space $C_*(Q^{\rm iter})$, which is completion of $C^\infty(\overline{Q^{\rm iter}})$ with respect to norm $\|\cdot\|^*$. This space satisfies

$$C_*(Q^{\rm iter}) \subset C^{1,\alpha}(\overline{Q^{\rm iter}}) \cap C^{2,\alpha}(Q^{\rm iter}).$$

For any admissible solution φ, the corresponding function u on $Q^{\rm iter}$ satisfies $u \in C_*(Q^{\rm iter})$. Now we state the local existence assertion.

Proposition 5.2 (Local existence). *Fix any admissible solution $(\hat{\varphi}, \hat{\theta}_{\rm w})$ with $\hat{\theta}_{\rm w} \in I$. Then, for every sufficiently small $\varepsilon > 0$, there is $\delta > 0$ with the following property: For each $\theta_{\rm w} \in [\hat{\theta}_{\rm w} - \delta, \hat{\theta}_{\rm w} + \delta)] \cap I$, there exists an admissible solution φ such that u and $\hat{u}$ on $Q^{\rm iter}$ corresponding to φ and $\hat{\varphi}$, respectively, satisfy*

$$\|u - \hat{u}\|^* < \varepsilon.$$

Note that, if ε is sufficiently small, the solutions obtained in Proposition 5.2 are unique for each wedge angle, by the local uniqueness.

Thus, to prove Proposition 5.1, it suffices to prove the local uniqueness, as well as the local existence in the sense of Proposition 5.2. See §5.3 below for more details in the proof of Proposition 5.2 from these properties. In fact, from Remark 3.3, we study these questions for the free boundary problem (5) and (18)–(19), where the unknowns are φ in Ω and $\Gamma_{\rm shock}$. Moreover, the admissible solutions satisfy property (ii) in Definition 3.1, from which equation (5) is strictly elliptic in $\overline{\Omega} \setminus \overline{\Gamma_{\rm sonic}}$ in the supersonic and sonic cases $|D\varphi_2(P_0)| \geq c_2$ and uniformly elliptic in $\overline{\Omega}$ in the subsonic case $|D\varphi_2(P_0)| < c_2$.

The proofs of the local existence and uniqueness are different for the following two cases:

(a) Supersonic and subsonic-near-sonic case: $|D\varphi_2(P_0)| > (1-\sigma)c_2$,

(b) Subsonic-away-from-sonic case: $|D\varphi_2(P_0)| \leq (1-\sigma)c_2$,

where $\sigma > 0$ depends on (ρ_0, ρ_1, γ) and is such that, for the wedge angles satisfying $(1-\sigma)c_2 \leq |D\varphi_2(P_0)| \leq 1$ (which are the subsonic-near-sonic and sonic cases), the admissible solutions are $C^{2,\alpha}$ up to P_0 according to [19].

The reason for the different proofs for cases (a) and (b) is that, in the supersonic and sonic case, the degenerate ellipticity of equation (5) near $\Gamma_{\rm sonic}$ or P_0 makes it difficult to use the linearization of problem (5) and (18)–(19) for the application of the implicit function theorem which would imply both the local existence and uniqueness. On the other hand, under the conditions stated in case (a), φ is $C^{1,1}$ up to $\Gamma_{\rm sonic}$ in the supersonic case (by Theorem 3.2(i)), and C^2 up to P_0 in the subsonic-near-sonic and sonic cases; this higher regularity allows us to use the different methods described below. In the subsonic-away-from-sonic case (b), the known regularity up to P_0 is $C^{1,\alpha}$, *i.e.* lower than that in case (a), but equation (5) is uniformly elliptic in $\overline{\Omega}$; this allows to analyze the linearization of problem (5) and (18)–(19) at φ, and thus obtain the local uniqueness and existence by the implicit function theorem.

It remains to discuss the proof of the local uniqueness and existence in the supersonic and subsonic-near-sonic case (a). The outline of this proof is in §5.1–§5.2 below.

5.1. **Local uniqueness in the supersonic and subsonic-near-sonic case** (a). Assume that φ and φ^* are regular shock reflection solutions for the same wedge angle $\theta_{\rm w}$, which are $C^{1,1}$ up to $\overline{\Gamma_{\rm sonic}}$ (where we denote $\overline{\Gamma_{\rm sonic}} = \{P_0\}$ in the subsonic and sonic cases) and satisfy the properties listed in Theorem 5.1. Let Ω and Ω^* be respectively their elliptic regions, and let $\Gamma_{\rm shock}$ and $\Gamma^*_{\rm shock}$ be respectively their reflected shocks. We recall that φ and φ^* satisfy (5) and (18)–(19) in Ω and Ω^*, respectively.

Let $\hat{\Omega} := \Omega \cap \Omega^*$, and let $\hat{\Gamma}_{\rm shock} := \partial\hat{\Omega} \cap (\Gamma^*_{\rm shock} \cup \Gamma_{\rm shock})$. We now show that, under the following assumption:

$$\|\varphi - \varphi^*\|_{C^1(\hat{\Omega})} + \|\varphi - \varphi_1\|_{C^0((\Omega\cup\Omega^*)\setminus\hat{\Omega})} + \|\varphi^* - \varphi_1\|_{C^0((\Omega\cup\Omega^*)\setminus\hat{\Omega})} \leq \delta_2, \tag{24}$$

the function, $\delta\varphi := \varphi - \varphi^*$, satisfies the boundary condition:

$$\mathcal{M}(\delta\varphi) = \beta_{\boldsymbol{\nu}}(\delta\varphi)_{\boldsymbol{\nu}} + \beta_{\boldsymbol{\tau}}(\delta\varphi)_{\boldsymbol{\tau}} + \vartheta\delta\varphi = 0 \qquad \text{on the inner shock } \hat{\Gamma}_{\rm shock}, \tag{25}$$

with

$$\beta_{\boldsymbol{\nu}} > 0, \qquad \vartheta < 0, \tag{26}$$

where $\boldsymbol{\nu}$ is the unit inner normal and $\boldsymbol{\tau}$ is the unit tangent on $\hat{\Gamma}_{\rm shock}$. We note that the property that $\vartheta < 0$ in (26) is obtained by using the convexity of $\Gamma^*_{\rm shock}$ and $\Gamma_{\rm shock}$.

Also, it follows from [18] that $\delta\varphi$ satisfies a homogeneous linear elliptic equation in $\hat{\Omega}$ for which the comparison principles hold. Properties (26), combined with methods of [18], show that Hopf's lemma holds for $\delta\varphi$ on $\hat{\Gamma}_{\rm shock}$. Finally, $\delta\varphi$ satisfies the homogeneous Neumann condition on $(\partial\hat{\Omega} \cap \partial\Lambda) \setminus \{P_3\}$, and $\delta\varphi = 0$ on $\Gamma_{\rm sonic}$.

These facts ensure that $\delta\varphi \equiv 0$ in $\hat{\Omega}$. From this, we can show

$$\Omega^* = \Omega, \qquad \varphi = \varphi^* \quad \text{in } \Omega. \tag{27}$$

This completes the proof of the local uniqueness.

Remark 5.4. We remark that, due to the issue that the regularity of φ at the reflection point P_0 is only $C^{1,\alpha}$ for the subsonic-away-sonic reflection case $|D\varphi_2(P_0)| \leq (1-\sigma)c_2$, we cannot apply this argument. However, as we discussed earlier, the implicit function theorem can be applied in that case.

5.2. **Local existence in the supersonic and subsonic-near-sonic case** (a). Now we discuss the proof of the local existence, Proposition 5.2. The existence of a solution is obtained by the application of the Leray-Schauder degree theory [55, §13.6($A4^*$)]; see also [19, §3.4].

In order to apply the degree theory, the iteration set should be bounded and open in an appropriate function space (in fact, in its product with the parameter space, *i.e.* interval $[\hat{\theta}_{\rm w} - \delta, \hat{\theta}_{\rm w} + \delta] \cap I$ of the wedge angles), the iteration map should be defined and continuous on the closure of the iteration set, and any fixed point of the iteration map should not occur on the boundary of the iteration set. We choose this function space according to the norms and the other quantities in the *a priori* estimates. Moreover, since we have to use the same function space for all values of the parameters, and the functions require to have the same domain, we define the iteration set in terms of the functions on the unit square $Q^{\rm iter}$, which are related to the admissible solutions by the mapping described in Remark 5.1. The function space is $C_*(Q^{\rm iter})$, introduced above. Let $\hat{u}$ be the function on $Q^{\rm iter}$ corresponding to the admissible solution $\hat{\varphi}$ for the wedge angle $\hat{\theta}_{\rm w}$ in Proposition 5.2. In order to prove the existence result in Proposition 5.2 for given ε and δ, we define the iteration set by

$$\mathcal{K}^{(\hat{u},\hat{\theta}_{\rm w})}_{\varepsilon,\delta} := \{(u,\theta_{\rm w}) \in C_*(Q^{\rm iter}) \times ([\hat{\theta}_{\rm w}-\delta,\ \hat{\theta}_{\rm w}+\delta]\cap I) \ : \ \|u-\hat{u}\|^* < \varepsilon\}. \quad (28)$$

From its definition, the iteration set is non-empty, open (in the subspace topology) and bounded in $C_*(Q^{\rm iter}) \times ([\hat{\theta}_w-\delta,\ \hat{\theta}_w+\delta]\cap I)$.

We also define the iteration set for each wedge angle $\theta_{\rm w} \in [\hat{\theta}_{\rm w}-\delta,\ \hat{\theta}_{\rm w}+\delta]\cap I$ by

$$\mathcal{K}^{(\hat{u},\hat{\theta}_{\rm w})}_{\varepsilon}(\theta_{\rm w}) := \{u \in C_*(Q^{\rm iter}) \ : \ (u,\theta_{\rm w}) \in \mathcal{K}^{(\hat{u},\hat{\theta}_{\rm w})}_{\varepsilon,\delta}\}. \quad (29)$$

To prove Proposition 5.2, we need to show the existence of an admissible solution in $\overline{\mathcal{K}^{(\hat{u},\hat{\theta}_{\rm w})}_{\varepsilon,\theta_{\rm w}}}$ for each $\theta_{\rm w} \in [\hat{\theta}_{\rm w}-\delta,\ \hat{\theta}_{\rm w}+\delta]\cap I$ if ε is small, depending on $(\rho_0,\rho_1,\gamma,\hat{\theta}_{\rm w})$, and δ is small, depending on ε and $(\rho_0,\rho_1,\gamma,\hat{\theta}_{\rm w})$.

The iteration map $\mathcal{F}$ is defined as follows:

Given $(u,\theta_{\rm w}) \in \overline{\mathcal{K}^{(\hat{u},\hat{\theta}_{\rm w})}_{\varepsilon,\delta}}$, define the corresponding *elliptic domain* $\Omega = \Omega(u,\theta_{\rm w})$ by mapping from the unit square $Q^{\rm iter}$ to the *physical plane*, as discussed in Remark 5.1. This determines iteration $\Gamma_{\rm shock}$ and function φ in Ω, depending on $(u,\theta_{\rm w})$. We set up a boundary value problem in Ω for a *new iteration potential* $\tilde{\varphi}$ by modifying problem (5) and (18)–(19), by partially substituting φ into the coefficients of (5), and making other modifications including the ellipticity cutoff in the equation.

In the supersonic and sonic cases, the modified equation is elliptic in $\overline{\Omega}\setminus\Gamma_{\rm sonic}$, degenerate near $\Gamma_{\rm sonic}$ (or P_0 in the sonic case), and nonlinear near $\Gamma_{\rm sonic}$. In the subsonic case, the modified equation is linear and uniformly elliptic in $\overline{\Omega}$.

In all the supersonic, sonic, and subsonic cases, we prescribe one condition on $\Gamma_{\rm shock}$, which is an oblique derivative condition, by combining the two conditions in (18) and partially substituting φ into the coefficients of the main terms.

Let $\tilde{\varphi}$ be the solution of the boundary value problem in Ω. We show that $\tilde{\varphi}$ gains the regularity in comparison with φ. Then we define $\tilde{u}$ on $Q^{\rm iter}$ by mapping $\tilde{\varphi}$ back in such a way that the gain-in-regularity of the solution is preserved, which is needed in order to have the compactness of the iteration map. This requires some care, since the original mapping between $Q^{\rm iter}$ and the *physical domain* is defined by u and hence has a lower regularity. Then the iteration map is defined by

$$\mathcal{F}(u,\theta_{\rm w}) = \tilde{u}.$$

The boundary value problem in the definition of $\mathcal{F}$ is defined so that, at the fixed point $u=\tilde{u}$, its solution satisfies the potential flow equation (5) with the ellipticity cutoff in a small neighborhood of $\Gamma_{\rm sonic}$ in the supersonic case, both the Rankine-Hugoniot conditions (18) on $\Gamma_{\rm shock}$, and the boundary condition (19) on $\Gamma_{\rm wedge}\cup\Gamma_{\rm sym}$. On the sonic arc $\Gamma_{\rm sonic}$ in the supersonic case and at P_0 in the subsonic and sonic cases, we need two conditions: $\tilde{\varphi}=\varphi_2$ and $D\tilde{\varphi}=D\varphi_2$. However, we can prescribe only one condition on the fixed boundary. We choose the Dirichlet condition $\tilde{\varphi}=\varphi_2$ on $\Gamma_{\rm sonic}$ in the supersonic case and at P_0 in the subsonic and sonic cases, and prove that $D\tilde{\varphi}=D\varphi_2$ on $\Gamma_{\rm sonic}$ or at P_0 holds for the solution of the iteration problem for the fixed point.

Then we prove the following facts:

(i) Any fixed point $u = \mathcal{F}(u, \theta_{\rm w})$, mapped to the *physical plane*, is an admissible solution φ. For that, we remove the ellipticity cutoff and prove the inequalities and monotonicity properties in the definition of the admissible solutions for the regions and the wedge angles where they are not readily known from the definition of the iteration set.

(ii) The iteration map is continuous on $\overline{\mathcal{K}^{(\hat{u},\hat{\theta}_{\rm w})}_{\varepsilon,\delta}}$ and compact. We prove this by using the gain-in-regularity of the solution of the iteration boundary value problem.

(iii) Any fixed point of the iteration map cannot occur on the boundary of the iteration set if δ is small depending on ε and (ρ_0, ρ_1, γ). Now we discuss this step in more details:

The *small* iteration set (29) is the first key difference between this proof of the local existence and the proof of the existence of admissible solutions in [19], which is also obtained by the Leray-Schauder degree argument. In [19], the continuity of admissible solutions with respect to $\theta_{\rm w}$ was not studied; for this reason, the iteration set is chosen to be *large* for the wedge angles away from $\frac{\pi}{2}$. That is, the iteration set for such a wedge angle is defined by the bounds in the appropriate norms related to the *a priori* estimates and by the lower bounds of certain directional derivatives, corresponding to the strict monotonicity properties so that the actual solution cannot be on the boundary of the iteration set according to the *a priori* estimates. In the present case of *small* iteration set (29), a different approach is developed, based on the local uniqueness and compactness of admissible solutions shown in [19]. That is, fixing small $\varepsilon > 0$, and assuming that, for any $\delta > 0$, there exists an admissible solution $\tilde{\varphi}$ for the wedge angle $\tilde{\theta}_{\rm w}$ such that $|\tilde{\theta}_{\rm w} - \hat{\theta}_{\rm w}| \le \delta$ and $\|\tilde{u} - \hat{u}\|^* = \varepsilon$, we obtain a sequence of admissible solutions and their wedge angles $(\varphi^{(i)}, \theta_{\rm w}^{(i)})$ with $\theta_{\rm w}^{(i)} \to \hat{\theta}_{\rm w}$ and $\|u^{(i)} - \hat{u}\|^* = \varepsilon$. Then, using the compactness of admissible solutions, we can send to a limit for a subsequence so that an admissible solution $\bar{\varphi}$ is obtained for the wedge angle $\hat{\theta}_{\rm w}$ such that $\|\bar{u} - \hat{u}\|^* = \varepsilon$. This contradicts the local uniqueness if ε is small.

Now the Leray-Schauder degree theory guarantees that the fixed point index:

$$\operatorname{Ind}(\mathcal{F}(\cdot, \theta_{\rm w}),\ \overline{\mathcal{K}^{(\hat{u},\hat{\theta}_{\rm w})}_{\varepsilon}(\theta_{\rm w})}) \tag{30}$$

of the iteration map on the iteration set (for given $\theta_{\rm w}$) is independent of the wedge angle $\theta_{\rm w} \in [\hat{\theta}_{\rm w} - \delta,\ \hat{\theta}_{\rm w} + \delta] \cap I$.

It remains to show that, at some wedge angle, index (30) is non-zero. We show that, for the wedge angle $\hat{\theta}_{\rm w}$,

$$\operatorname{Ind}(\mathcal{F}(\cdot, \hat{\theta}_{\rm w}),\ \overline{\mathcal{K}^{(\hat{u},\hat{\theta}_{\rm w})}_{\varepsilon}(\hat{\theta}_{\rm w})}) = 1.$$

We prove this by showing that

$$\mathcal{F}(v, \hat{\theta}_{\rm w}) = \hat{u} \qquad \text{for each } v \in \mathcal{K}^{(\hat{u},\hat{\theta}_{\rm w})}_{\varepsilon}(\hat{\theta}_{\rm w}). \tag{31}$$

This means that the iteration boundary value problem in domain $\Omega(v, \hat{\theta}_{\rm w})$ defined by every $v \in \mathcal{K}^{(\hat{u},\hat{\theta}_{\rm w})}_{\varepsilon}(\hat{\theta}_{\rm w})$ has the unique solution $\hat{\varphi}$ (in fact, its carefully defined extension from $\Omega(\hat{u}, \hat{\theta}_{\rm w})$). This step is another key difference from the existence proof of admissible solutions in [19]. In [19], the iteration set includes the normal reflection $\varphi^{\rm normal}$ for $\theta_{\rm w} = \frac{\pi}{2}$, and property (31) is shown for $\theta_{\rm w} = \frac{\pi}{2}$ and $u^{\rm normal}$ on the right-hand side. Since $\varphi^{\rm normal}$ is an explicitly known uniform state, globally defined, showing (31) is straightforward for the normal reflection, and does not require defining its extension, or any special properties of the coefficients of the iteration problem. In the present case, when $\hat{\varphi}$ is an arbitrary admissible solution, this step is much more involved, and requires an extension of $\hat{\varphi}$ from Ω to a larger region (so that the extension satisfies certain properties) and some careful definition of the coefficients of the iteration equation and the boundary condition on Γ_{shock}, for which we need at least the $C^{1,1}$–regularity of φ near Γ_{sonic}. Thus, our method works for the supersonic and subsonic-near-sonic case; however, it does not readily work for the subsonic-away-from-sonic case (for this reason, in this case, we use a different approach as we discussed above).

This completes the proof of the local existence of supersonic and subsonic-near-sonic reflection solutions.

5.3. **Proof of Proposition 5.1.** Based on the local uniqueness and existence, we employ the compactness of admissible solutions proved in [19] to conclude that, for every admissible solution φ^* with the wedge angle $\theta_{\text{w}}^* \in I$, a family $\mathfrak{S}$ with the properties listed in Proposition 5.1 exists.

It remains to prove the uniqueness of admissible solutions for each wedge angle.

For a given wedge angle θ_{w} as in Theorem 5.1, assume that there are two admissible solutions φ and $\tilde{\varphi}$ corresponding to the wedge angle θ_{w}^*. Let $\mathfrak{S}$ and $\tilde{\mathfrak{S}}$ be the continuous families with $(\varphi, \theta_w^*) \in \mathfrak{S}$ and $(\tilde{\varphi}, \theta_w^*) \in \tilde{\mathfrak{S}}$ in Proposition 5.1. Let $\mathfrak{A}$ be the set of all $\theta_{\text{w}} \in [\theta_{\text{w}}^*, \frac{\pi}{2}]$ such that $\varphi^{\theta_{\text{w}}} = \tilde{\varphi}^{\theta_{\text{w}}}$. Since $\frac{\pi}{2} \in \mathfrak{A}$ by (c) of Proposition 5.1, it follows that $\mathfrak{A} \neq \emptyset$. The continuity of both families $\mathfrak{S}$ and $\tilde{\mathfrak{S}}$ with respect to θ_w implies that $\mathfrak{A}$ is closed. Also, by the assumption above, $\theta_{\text{w}}^* \notin \mathfrak{A}$. Denote $\theta_{\text{w}}^{\text{inf}} := \inf \mathfrak{A}$, then $\theta_{\text{w}}^{\text{inf}} \in (\theta_{\text{w}}^*, \frac{\pi}{2}]$. Now, using the continuity of families $\mathfrak{S}$ and $\tilde{\mathfrak{S}}$, we can show that, choosing $\theta_{\text{w}} \in (\theta_{\text{w}}^*, \theta_{\text{w}}^{\text{inf}})$ to be sufficiently close to $\theta_{\text{w}}^{\text{inf}}$, we obtain that $\varphi^{(\theta_{\text{w}})} = \tilde{\varphi}^{(\theta_{\text{w}})}$ by the local uniqueness property. This contradicts the definition of $\theta_{\text{w}}^{\text{inf}}$.

REFERENCES

[1] M. Bae, G.-Q. Chen, and M. Feldman. Regularity of solutions to regular shock reflection for potential flow. *Invent. Math.* **175** (2009), 505–543.

[2] M. Bae, G.-Q. Chen, and M. Feldman. Prandtl-Meyer reflection for supersonic flow past a solid ramp. *Quart. Appl. Math.* **71** (2013), 583–600.

[3] M. Bae, G.-Q. Chen, and M. Feldman. *Prandtl-Meyer Reflection Configurations, Transonic Shocks, and Free Boundary Problems.* Research Monograph, Preprint `arXiv:1901.05916`.

[4] G. Ben-Dor. *Shock Wave Reflection Phenomena.* Springer-Verlag, New York, 1991.

[5] L. Caffarelli, D. Jerison, and C. E. Kenig. Some new monotonicity theorems with applications to free boundary problems. *Ann. of Math.* (2), **155** (2002), 369–404.

[6] L. Caffarelli and J. Salazar. Solutions of fully non-linear elliptic equations with patches of zero gradient: existence, regularity and convexity of level curves. *Trans. Amer. Math. Soc.* **354** (2002), 3095–3115.

[7] L. Caffarelli and J. Spruck. Convexity properties of solutions to some classical variational problems. *Comm. P.D.E.* **7** (1982), 1337–1379.

[8] S. Čanić, B. L. Keyfitz, and E. H. Kim. Free boundary problems for the unsteady transonic small disturbance equation: transonic regular re ection. *Methods Appl. Anal.* **7** (2000), 313–336.

[9] S. Čanić, B. L. Keyfitz, and E. H. Kim. Free boundary problems for nonlinear wave system: Mach stems for interacting shocks. *SIAM J. Math. Anal.* **37** (2006), 1947–1977.

[10] T. Chang, G.-Q. Chen, and S. Yang. On the Riemann problem for two-dimensional Euler equations I: Interaction of shocks and rarefaction waves. *Discrete Contin. Dynam. Systems,* **1** (1995), 555–584.

[11] T. Chang, G.-Q. Chen, and S. Yang. On the Riemann problem for two-dimensional Euler equations II: Interaction of contact discontinuities. *Discrete Contin. Dynam. Systems*, **6** (2000), 419–430.

[12] C. J. Chapman. *High Speed Flow.* Cambridge University Press: Cambridge, 2000.

[13] G.-Q. Chen. Supersonic flow onto solid wedges, multidimensional shock waves and free boundary problems. *Sci. China Math.* **60** (2017), 1353–1370.

[14] G.-Q. Chen, J. Chen, and M. Feldman. Stability and asymptotic behavior of transonic flows past wedges for the full Euler equations. *Interfaces Free Bound.* **19** (2018), 591–626.

[15] G.-Q. Chen, J. Chen, and M. Feldman. Transonic flows with shocks past curved wedges for the full Euler equations. *Discrete Contin. Dyn. Syst.* **36** (2016), 4179–4211.

[16] G.-Q. Chen, X. Deng, and W. Xiang. Shock diffraction by convex cornered wedges for the nonlinear wave system. *Arch. Rational Mech. Anal.* **211** (2014), 61–112.

[17] G.-Q. Chen and M. Feldman. Global solutions of shock reflection by large-angle wedges for potential flow. *Ann. of Math.* (2), **171** (2010), 1067–1182.

[18] G.-Q. Chen and M. Feldman. Comparison principles for self-similar potential flow. *Proc. Amer. Math. Soc.* **140** (2012), 651–663.

[19] G.-Q. Chen and M. Feldman. *Mathematics of Shock Reflection-Diffraction and Von Neumann's Conjecture.* Research Monograph, Annals of Mathematics Studies, 197, Princeton University Press, Princetion, 2018.

[20] G.-Q. Chen, M. Feldman, J. Hu, and W. Xiang. Loss of regularity of solutions of the Lighthill problem for shock diffraction for potential flow, Preprint `arXiv:1705.06837`.

[21] G.-Q. Chen, M. Feldman and W. Xiang. Convexity of transonic shocks in self-similar coordinates, Preprint `arXiv:1803.02431`.

[22] G.-Q. Chen, M. Feldman, and W. Xiang. Uniqueness and stability of regular shock reflection-diffraction configurations by wedges for potential flow. Preprint 2019.

[23] S. Chen. Mach configuration in pseudo-stationary compressible flow. *J. Amer. Math. Soc.* **21** (2008), 63–100.

[24] E. Chiodaroli, C. De Lellis, and O. Kreml. Global ill-posedness of the isentropic system of gas dynamics. *Comm. Pure Appl. Math.* **68** (2015), 1157–1190.

[25] E. Chiodaroli and O. Kreml. On the energy dissipation rate of solutions to the compressible isentropic Euler system. *Arch. Ration. Mech. Anal.* **214** (2014), 1019–1049.

[26] A. Y. Dem'yanov and A. V. Panasenko. Numerical solution to the problem of the diffraction of a plane shock wave by a convex corner. *Fluid Dynamics*, **16** (1981), 720–725 (Translated from the original Russian).

[27] X. Deng and W. Xiang. Global solutions of shock reflection by wedges for the nonlinear wave equation. *Chinese Ann. Math. B*, **32** (2011), 643–668.

[28] R L. Deschambault and I. I. Glass. An update on non-stationary oblique shock-wave reflections: actual isopycnics and numerical experiments. *J. Fluid Mech.* **131** (1983), 27–57.

[29] J. Dolbeault and R. Monneau. Convexity estimates for nonlinear elliptic equations and application to free boundary problems. *Ann. I.H. Poincaré–A.N.* **19** (2002), 903–626.

[30] V. Elling and T.-P. Liu. Supersonic flow onto a solid wedge. *Comm. Pure. Appl. Math.* **61** (2008), 1347–1448.

[31] B. Fang. Stability of transonic shocks for full Euler system in supersonic flow past a wedge. *Math. Meth. Appl. Sci.* **29** (2006), 1–26.

[32] B. Fang and W. Xiang. The uniqueness of transonic shocks in supersonic flow past a 2-D wedge. *J. Math. Anal. Appl.* **437** (2016): 194–213.

[33] E. Feireisl, C. Klingenberg, O. Kreml, and S. Markfelder. On oscillatory solutions to the complete Euler system, Preprint `arXiv:1710.10918`.

[34] H. M. Glaz, P. Colella, I. I. Glass, and R. L. Deschambault. A numerical study of oblique shock-wave reflection with experimental comparisons. *Proc. Roy. Soc. Lond.* **A398** (1985), 117–140.

[35] H. M. Glaz, P. Colella, I. I. Glass, and R. L. Deschambault. A detailed numerical, graphical, and experimental study of oblique shock wave reflection. Lawrence Berkeley Laboratory Report, LBL-20033, 1985.

[36] H. M. Glaz, P. A. Walter, I. I. Glass, and T. C. J. Hu. Oblique shock wave reflections in SF_6: A comparison of calculation and experiment. *AIAA J. Prog. Astr. Aero.* **106** (1986), 359–387.

[37] J. Glimm and A. Majda. Multidimensional Hyperbolic Problems and Computations. Springer-Verlag: New York, 1991.

[38] R. G. Hindman, P. Kutler, and D. Anderson. A two-dimensional unsteady Euler-equation solver for flow regions with arbitrary boundaries. AIAA Report 79-1465, 1979.

[39] M. S. Ivanov, D. Vandromme, V. M. Formin, A. N. Kudryavtsev, A. Hadjadj, and D. V. Khotyanovsky. Transition between regular and Mach reflection of shock waves: new numerical and experimental results. *Shock Waves*, **11** (2001), 199–207.

[40] E. H. Kim. Global sub-sonic solution to an interacting transonic shock of the self-similar nonlinear wave equation. *J. Diff. Eqs.* **248** (2010), 2906–2930.

[41] A. Kurganov and E. Tadmor. Solution of two-dimensional Riemann problems for gas dynamics without Riemann problem solvers. *Numer. Methods Partial Diff. Eqs.* **18** (2002), 584–608.

[42] P. Kutler and V. Shankar. Diffraction of a shock wave by a compression corner, Part I: Regular reflection. *AIAA J.* **15** (1977), 197–203.

[43] P. D. Lax and X.-D. Liu. Solution of two-dimensional Riemann problems of gas dynamics by positive schemes. *SIAM J. Sci. Comput.* **19** (1998), 319–340.

[44] X. D. Liu and P. D. Lax. Positive schemes for solving multi-dimensional hyperbolic systems of con-servation laws. *J. Comp. Fluid Dynamics*, **5** (1996), 133–156.

[45] E. Mach. Über den verlauf von funkenwellenin der ebene und im raume, *Sitzungsber. Akad. Wiss. Wien*, **78** (1878), 819–838.

[46] S. Markfelder and C. Klingenberg. The 2-d isentropic compressible Euler equations may have infinitely many solutions which conserve energy, Preprint `arXiv:1709.04982`.

[47] G. P. Schneyer. Numerical simulation of regular and Mach reflection. *Phy. Fluids*, **18** (1975), 1119–1124.

[48] C. W. Schulz-Rinne, J. P. Collins, and H. M. Glaz. Numerical solution of the Riemann problem for two-dimensional gas dynamics. *SIAM J. Sci. Comput.* **14** (1993), 1394–1414.

[49] P. I. Plotnikov and J. F. Toland. Convexity of Stokes waves of extreme form. *Arch. Ration. Mech. Anal.* **171** (2014), 349–416.

[50] D. Serre. Shock reflection in gas dynamics. In: *Handbook of Mathematical Fluid Dynamics*, Vol. 4, pp. 39–122, Elsevier: North-Holland, 2007.

[51] D. Serre. Multidimensional shock interaction for a Chaplygin gas. *Arch. Ration. Mech. Anal.* **191** (2009), 539–577.

[52] V. Shankar, P. Kutler, and D. Anderson. Diffraction of a shock wave by a compression corner, Part II: Single Mach reflection. *AIAA J.* **16** (1978), 4–5.

[53] M. Van Dyke. *An Album of Fluid Motion.* The Parabolic Press, Stanford, 1982.

[54] P. Woodward and P. Colella. The numerical simulation of two-dimensional fluid flow with strong shocks. *J. Comp. Phys.* **54** (1984), 115–173.

[55] E. Zeidler, *Nonlinear Functional Analysis and its Applications, I: Fixed-Point Theorems.* Springer-Verlag: Berlin, 1986.

[56] Y. Zheng. Two-dimensional regular shock reflection for the pressure gradient system of conservation laws. *Acta Math. Appl. Sinica* (English Ser), **22** (2006), 177–210.

[57] Y. Zheng. *Systems of Conservation Laws: Two-Dimensional Riemann Problems.* Birkhäuser Boston, Inc.: Boston, MA, 2001.

E-mail address: `chengq@maths.ox.ac.uk`
E-mail address: `feldman@math.wisc.edu`
E-mail address: `weixiang@cityu.edu.hk`

CENTRAL-UPWIND SCHEME FOR A NON-HYDROSTATIC SAINT-VENANT SYSTEM

ALINA CHERTOCK*

Department of Mathematics, North Carolina State University,
Raleigh, NC 27695, USA

ALEXANDER KURGANOV

Department of Mathematics and SUSTech International Center for Mathematics,
Southern University of Science and Technology, Shenzhen, 518055, China

JASON MILLER

Applied Physics Laboratory, Johns Hopkins University,
Laurel, MD 20723, USA

JUN YAN

Department of Mathematics, North Carolina State University,
Raleigh, NC 27695, USA

ABSTRACT. We develop a second-order central-upwind scheme for the non-hydrostatic version of the Saint-Venant system recently proposed in [M.-O. BRISTEAU AND J. SAINTE-MARIE, *Discrete Contin. Dyn. Syst. Ser. B*, 10 (2008), pp. 733–759]. The designed scheme is both well-balanced (capable of exactly preserving the "lake-at-rest" steady state) and positivity preserving. We then use the central-upwind scheme to study ability of the non-hydrostatic Saint-Venant system to model long-time propagation and on-shore arrival of the tsunami-type waves. We discover that for a certain range of the dispersive coefficients, both the shape and amplitude of the waves are preserved even when the computational grid is relatively coarse. We also demonstrate the importance of the dispersive terms in the description of on-shore arrival.

1. **Introduction.** Tsunami waves are characterized by having a relatively low amplitude, large wavelength, and large characteristic wave speed, see, e.g., [7, 27, 31]. In fact, the amplitude of a tsunami wave can be so small that it may not even be noticed by a ship traveling through it in deep water. Because of their speed and wavelength, however, these waves contain a tremendous amount of energy. When the depth of the water decreases (in the beginning of the on-shore arrival stage of tsunami wave propagation), tsunamis undergo a process called wave shoaling, in which the wave slows down and the wavelength decreases. In order to conserve energy, it is transformed from kinetic to potential energy and the wave amplitude increases. This potential energy can then be released in disastrous fashion when

2000 *Mathematics Subject Classification.* 65M08, 76M12, 86-08, 35L55, 35L65, 35L75.

Key words and phrases. hyperbolic systems of balance laws, dispersive shallow water systems, Godunov-type central-upwind schemes.

The first author is supported by NSF grant DMS-1521051 and DMS-1818684.

The second author is supported by NSFC grant 11771201.

* Corresponding author: Alina Chertock.

the wave comes to shore. It is therefore very important to have accurate models and corresponding numerical methods for tsunami waves in order to mitigate any catastrophe that may result.

One model used for shallow water waves is the classical Saint-Venant system [12], which is a depth-averaged system that can be derived from the Navier-Stokes equations (see, e.g., [14]). The Saint-Venant system is a very good simplification for lakes, rivers, and coastal areas in which the typical time and space scales of interest are relatively short. Tsunami waves form in deep water and travel very long distances (thousands of kilometers) before coming to the shore. Over long time, solutions of the Saint-Venant system break down, dissipate in an unphysical manner, shock waves develop, and the system fails to capture small, trailing waves that are seen in nature and laboratory experiments [29]. Thus, it is necessary to use a more sophisticated model in order to preserve the wave characteristics over long time simulations.

Non-hydrostatic models (the celebrated Green-Naghdi equation [17] and several others, see, e.g., [1,3,4] and references therein) work well for long-time propagation of tsunami-like waves because they allow the wave to travel for long distances without decaying in amplitude. In addition, since these systems are dispersive, they give rise to trailing waves that are observed to follow tsunamis in nature. However, it is necessary to achieve some balance between dispersion observed with a non-hydrostatic model and the dissipation seen in the classical Saint-Venant system.

The non-hydrostatic Saint-Venant system presented in [5,6] is given by

$$\begin{cases} h_t + (hu)_x = 0, \\ (hu)_t + M_t + \left(hu^2 + \dfrac{g}{2}h^2\right)_x + N \\ \qquad = -ghB_x + p^a w_x - 4(\nu u_x)_x - \kappa(h,hu)u, \end{cases} \tag{1}$$

where $h(x,t)$ is the water depth measured vertically from the bottom topography, described by function $B(x,t)$, $u(x,t)$ is the vertically averaged velocity, hu is the horizontal momentum or discharge, $p^a = p^a(x,t)$ is the atmospheric pressure function, $w := h + B$ is the free surface, ν is the viscosity coefficient, κ is the friction function, and M and N are defined as

$$M(h,hu,B) = \left(-\frac{1}{3}h^3u_x + \frac{1}{2}h^2B_xu\right)_x + B_x\left(-\frac{1}{2}h^2u_x + B_xhu\right), \tag{2}$$

and

$$\begin{aligned} N(h,hu,B) &= \left((h^2)_t(hu_x - B_xu)\right)_x \\ &\quad + 2B_xh_t(hu_x - B_xu) - B_{xt}\left(-\frac{1}{2}h^2u_x + B_xhu\right). \end{aligned} \tag{3}$$

Here, M and N are terms that arise when the system is derived from the Euler equations and include non-hydrostatic pressure terms [6].

One of the goals of the current work is to numerically study the effects of the dispersion terms present in the non-hydrostatic model (1)–(3). To this end, we introduce the *new scaling parameters* α_M and α_N as coefficients to M and N in (1). For the purpose of this work we will neglect fluid viscosity and friction by setting ν and $\kappa(h,hu)$ to be identically zero and also assume that the bottom topography function is independent of time, i.e., $B = B(x)$. In addition, we follow the approach in [20,24] and rewrite our system in terms of the equilibrium variables $w = h + B$

and $q := hu$:

$$\begin{cases} w_t + q_x = 0, \\ q_t + \alpha_M M_t + \left(\dfrac{q^2}{w-B} + \dfrac{g}{2}(w-B)^2\right)_x + \alpha_N N = -g(w-B)B_x + p^a w_x. \end{cases} \tag{4}$$

When $\alpha_M = \alpha_N = p^a \equiv 0$, (4) reduces to the classical Saint-Venant system, and as we increase these parameters, the amount of dispersion in our model increases and the effects of the lack of the hydrostatic pressure assumption should be apparent.

To study the non-hydrostatic effects, we design a highly accurate and robust numerical method for (4). A good scheme for this model should be well-balanced (it should exactly preserve "lake-at-rest" steady-state solutions at the discrete level), it should preserve positivity of h, and it should be able to properly handle discontinuous/nonsmooth solutions. The system (4) presents challenges in the approximation and treatment of the higher-order mixed derivatives in the non-hydrostatic terms whose semi-discretization leads to stiff terms that require an efficient numerical solver for the resulting system of ODEs. In this paper, we develop a central-upwind scheme for (4) which possesses all of the aforementioned features and use it to examine the effects of the non-hydrostatic pressure terms on the propagation of waves over long times and on their on-shore arrival.

Central-upwind schemes (first introduced in [26] and further developed in [21,23]) are Godunov-type finite volume methods. They belong to the class of Riemann-problem-solver-free central schemes and thus can be applied to a variety of hyperbolic systems of conservation laws as a "black-box" solver. When central-upwind schemes are applied to systems of balance laws, a special treatment of the source terms appearing in the system at hand must be developed. This was done for single- and two-layer shallow water models in [2,9–11,19,20,24,25]. In order to apply the central-upwind scheme to (4), one needs to specify the way the terms on the right-hand side (RHS) of (4) are discretized. As it was mentioned above, this should be done in such a way that physically relevant steady-state solutions are exactly preserved and h is *guaranteed* to be nonnegative.

The physically relevant steady-state solution for (4) is the "lake-at-rest" solution, corresponding to the water surface being perfectly flat and stationary:

$$w = h + B \equiv \text{Const}, \qquad hu \equiv 0. \tag{5}$$

Preserving this particular steady state would guarantee that no artificial surface waves are generated, and also ensure that small perturbations of the water surface will not lead to a "numerical storm". This is achieved by using a special discretization of the geometric source term on the RHS of (4) which is presented in Section 2.1.3.

Preserving positivity of h is essential since solutions containing negative h would not only be unphysical, but will cause the numerical computations to fail. To ensure positivity of h, we follow the idea from [24]. We first replace the bottom topography with its continuous piecewise linear approximation and then adjust the piecewise linear reconstruction of the water heights, ensuring that through each computational cell the depth of each layer is nonnegative. This is presented in Section 2.1.1.

With the numerical method in place, we examine the effect of the non-hydrostatic pressure terms in Section 3, where we try to strike a balance between dissipation and dispersion inherent in the system.

2. Numerical Method.

2.1. **Central-Upwind Scheme.** We develop a new well-balanced positivity preserving scheme for (4), which is based on the semi-discrete central-upwind scheme from [23] (see also [24,25]). For simplicity, we introduce a uniform grid $x_j = j\Delta x$ where Δx is a small spatial scale, and denote the computational cells centered at x_j by $I_j := [x_{j-\frac{1}{2}}, x_{j+\frac{1}{2}}]$.

We rewrite the system (4) in the following form:

$$\mathbf{U}_t + \mathcal{M}(\mathbf{U}, B)_t + \mathbf{F}(\mathbf{U}, B)_x + \mathcal{N}(\mathbf{U}, B) = \mathbf{S}(\mathbf{U}, B), \quad \mathbf{U} := (w, q)^\top \tag{6}$$

where

$$\mathbf{F}(\mathbf{U}, B) = \left(q, \frac{q^2}{w-B} + \frac{g}{2}(w-B)^2\right)^T, \quad \mathbf{S}(\mathbf{U}, B) = (0, -g(w-B)B_x + p^a w_x)^\top,$$

$$\mathcal{M}(\mathbf{U}, B) = (0, \alpha_M M(\mathbf{U}, B))^T, \quad \mathcal{N}(\mathbf{U}, B) = (0, \alpha_N N(\mathbf{U}, B))^T.$$

Using the above notations, a semi-discrete central-upwind scheme for (6) takes the form of the following system of time-dependent ODEs:

$$\frac{d}{dt}\left(\overline{\mathbf{U}}_j(t) + \overline{\mathcal{M}}_j(t)\right) = -\frac{\mathbf{H}_{j+\frac{1}{2}}(t) - \mathbf{H}_{j-\frac{1}{2}}(t)}{\Delta x} + \overline{\mathbf{S}}_j(t) - \overline{\mathcal{N}}_j(t), \tag{7}$$

where $\overline{(\cdot)}_j(t)$ is used to denote the approximated cell averages over the corresponding cells:

$$\overline{\mathbf{U}}_j(t) \approx \frac{1}{\Delta x}\int\limits_{I_j} \mathbf{U}(x,t)\,dx, \qquad \overline{\mathbf{S}}_j(t) \approx \frac{1}{\Delta x}\int\limits_{I_j} \mathbf{S}(\mathbf{U}(x,t), B(x))\,dx,$$

$$\overline{M}_j(t) \approx \frac{1}{\Delta x}\int\limits_{I_j} M(\mathbf{U}(x,t), B(x))\,dx, \qquad \overline{N}_j(t) \approx \frac{1}{\Delta x}\int\limits_{I_j} N(\mathbf{U}(x,t), B(x))\,dx,$$

and $\mathbf{H}_{j+\frac{1}{2}}(t)$ are the central-upwind numerical fluxes $\mathbf{H}_{j+\frac{1}{2}}$ proposed in [24] (see also [21,23]):

$$\begin{aligned}\mathbf{H}_{j+\frac{1}{2}}(t) = {}& \frac{a^+_{j+\frac{1}{2}}\mathbf{F}\left(\mathbf{U}^-_{j+\frac{1}{2}}, B_{j+\frac{1}{2}}\right) - a^-_{j+\frac{1}{2}}\mathbf{F}\left(\mathbf{U}^+_{j+\frac{1}{2}}, B_{j+\frac{1}{2}}\right)}{a^+_{j+\frac{1}{2}} - a^-_{j+\frac{1}{2}}} \\ & + \frac{a^+_{j+\frac{1}{2}}a^-_{j+\frac{1}{2}}}{a^+_{j+\frac{1}{2}} - a^-_{j+\frac{1}{2}}}\left[\mathbf{U}^+_{j+\frac{1}{2}} - \mathbf{U}^-_{j+\frac{1}{2}}\right].\end{aligned} \tag{8}$$

Here, the values $\mathbf{U}^\pm_{j+\frac{1}{2}}$ are the right/left point values at $x = x_{j+\frac{1}{2}}$ of the conservative piecewise linear reconstruction $\widetilde{\mathbf{U}}$,

$$\widetilde{\mathbf{U}}(x) := \overline{\mathbf{U}}_j + (\mathbf{U}_x)_j\,(x - x_j), \quad x_{j-\frac{1}{2}} < x < x_{j+\frac{1}{2}}, \tag{9}$$

which is used to approximate $\mathbf{U}$ at time t, that is,

$$\mathbf{U}^\pm_{j+\frac{1}{2}} := \widetilde{\mathbf{U}}\left(x_{j+\frac{1}{2}} \pm 0\right) = \overline{\mathbf{U}}_{j+\frac{1}{2}\pm\frac{1}{2}} \mp \frac{\Delta x}{2}(\mathbf{U}_x)_{j+\frac{1}{2}\pm\frac{1}{2}}. \tag{10}$$

The numerical derivatives $(\mathbf{U}_x)_j$ are at least first-order accurate component-wise approximations of $\mathbf{U}_x(x_j, t)$, computed using a nonlinear limiter needed to ensure the non-oscillatory nature of the reconstruction (9). The right- and left-sided local speeds $a^\pm_{j+\frac{1}{2}}$ in (8) are obtained from the smallest and largest eigenvalues of the

Jacobian $\frac{\partial \mathbf{F}}{\partial \mathbf{U}}$ (see Section 2.1.1 for details). Notice that the terms $\mathbf{U}^{\pm}_{j+\frac{1}{2}}$, $\overline{\mathbf{U}}_j$, $a^{\pm}_{j+\frac{1}{2}}$, $\widetilde{\mathbf{U}}(x)$ and $(\mathbf{U}_x)_j$ all depend on t, but we suppress this dependence for simplicity.

We also follow the work of [24, 25] and replace $B(x)$ in (8) with its continuous piecewise linear approximation by defining

$$B_{j+\frac{1}{2}} := B(x_{j+\frac{1}{2}}) \quad \text{and} \quad B_j := \frac{1}{2}(B_{j+\frac{1}{2}} + B_{j-\frac{1}{2}}). \tag{11}$$

This will help to ensure the positivity preserving nature of the proposed scheme, as we show below.

2.1.1. *Positivity-Preserving Reconstruction.* The use of a piecewise linear reconstruction (9) requires the computation of slopes $(\mathbf{U}_x)_j$ to obtain the right/left point values defined in (10). It is well-known that in order to ensure the non-oscillatory nature of the reconstruction, the use of a nonlinear limiter is required. We choose to use the generalized minmod limiter:

$$(\mathbf{U}_x)_j = \text{minmod}\left(\theta\frac{\overline{\mathbf{U}}_j - \overline{\mathbf{U}}_{j-1}}{\Delta x}, \frac{\overline{\mathbf{U}}_{j+1} - \overline{\mathbf{U}}_{j-1}}{2\Delta x}, \theta\frac{\overline{\mathbf{U}}_{j+1} - \overline{\mathbf{U}}_j}{\Delta x}\right), \tag{12}$$

where $\theta \in [1, 2]$ and the minmod function defined as

$$\text{minmod}(z_1, z_2, \dots) := \begin{cases} \min\limits_j\{z_j\}, & \text{if } z_j > 0 \ \forall j, \\ \max\limits_j\{z_j\}, & \text{if } z_j < 0 \ \forall j, \\ 0, & \text{otherwise,} \end{cases}$$

is applied in a componentwise manner. The parameter θ can be used the control the amount of numerical viscosity present in the resulting scheme (see, e.g., [28, 30, 33] for more details concerning the generalized minmod and other nonlinear limiters).

Even when all of the cell averages $\overline{h}_j$ are nonnegative, the reconstructed right/left point values at the cell interface $h^{\pm}_{j+\frac{1}{2}}$ may be negative. To guarantee positivity of h throughout the entire computational domain, we follow the procedure from [24] and amend the reconstruction (9), (10), (12) in the following conservative way:

$$\begin{aligned} &\text{if } w^-_{j+\frac{1}{2}} < B_{j+\frac{1}{2}}, \quad \text{then take} \quad (w_x)_j := -\frac{\overline{w}_j}{\Delta x/2} \\ &\qquad\qquad \Longrightarrow w^-_{j+\frac{1}{2}} = B_{j+\frac{1}{2}},\ w^+_{j-\frac{1}{2}} = 2\overline{w}_j, \\ &\text{if } w^+_{j-\frac{1}{2}} < B_{j-\frac{1}{2}}, \quad \text{then take} \quad (w_x)_j := \frac{\overline{w}_j}{\Delta x/2} \\ &\qquad\qquad \Longrightarrow w^-_{j+\frac{1}{2}} = 2\overline{w}_j,\ w^+_{j-\frac{1}{2}} = B_{j-\frac{1}{2}}. \end{aligned} \tag{13}$$

It is necessary to compute the nonconservative quantity $u = q/h$ for the computation of numerical fluxes and local propagation speeds. We follow the desingularization procedure outlined in [24, 25] to avoid possible division by small values of h:

$$u := \frac{\sqrt{2}(w - B) \cdot q}{\sqrt{(w - B)^4 + \max\left((w - B)^4, \varepsilon\right)}}, \tag{14}$$

where ε is a small desingularization parameter (in our numerical experiments, we have taken $\varepsilon = \min((\Delta x)^3, 10^{-4})$). Notice that this procedure will only affect the velocity computations when $h^4 < \varepsilon$. It is also important to recalculate the values of q at the points where the velocity was desingularized by setting

$$q := h \cdot u.$$

Since the flux term $\mathbf{F}$ in (6) is equivalent to that of the classical Saint-Venant system, the local propagation speeds $a^{\pm}_{j+\frac{1}{2}}$ are computed the same way using the eigenvalues of $\frac{\partial \mathbf{F}}{\partial \mathbf{U}}$:

$$\begin{aligned}
a^{+}_{j+\frac{1}{2}} &:= \max\left\{u^{+}_{j+\frac{1}{2}} + \sqrt{gh^{+}_{j+\frac{1}{2}}},\, u^{-}_{j+\frac{1}{2}} + \sqrt{gh^{-}_{j+\frac{1}{2}}},\, 0\right\},\\
a^{-}_{j+\frac{1}{2}} &:= \min\left\{u^{+}_{j+\frac{1}{2}} - \sqrt{gh^{+}_{j+\frac{1}{2}}},\, u^{-}_{j+\frac{1}{2}} - \sqrt{gh^{-}_{j+\frac{1}{2}}},\, 0\right\}.
\end{aligned}$$

Remark 1. Proof of the positivity preserving property of this reconstruction is available in [20, 24].

2.1.2. *Discretization of the Non-hydrostatic Pressure Terms.* The dispersive terms $\overline{M}_j$ and $\overline{N}_j$ are computed using the second-order midpoint rule. We first follow [5] and discretize the terms of M at x_j in the following ways:

$$\begin{aligned}
\left(\frac{1}{3}h^3u_x\right)_x(x_j) &\approx \frac{1}{3\Delta x}\left[\frac{u_{j+1}-u_j}{\Delta x}\left(h_{j+\frac{1}{2}}\right)^3 - \frac{u_j-u_{j-1}}{\Delta x}\left(h_{j-\frac{1}{2}}\right)^3\right]\\
&= \frac{1}{3(\Delta x)^2}\left[\frac{\left(h_{j+\frac{1}{2}}\right)^3}{\overline{h}_{j+1}}\overline{q}_{j+1} - \frac{\left(h_{j+\frac{1}{2}}\right)^3+\left(h_{j-\frac{1}{2}}\right)^3}{\overline{h}_j}\overline{q}_j + \frac{\left(h_{j-\frac{1}{2}}\right)^3}{\overline{h}_{j-1}}\overline{q}_{j-1}\right],
\end{aligned} \tag{15}$$

$$\begin{aligned}
\left(\frac{1}{2}h^2B_xu\right)_x(x_j) &= \left(\frac{1}{2}hB_xq\right)_x(x_j)\\
&\approx \frac{1}{2\Delta x}\left[h_{j+\frac{1}{2}}(B_x)_{j+\frac{1}{2}}q_{j+\frac{1}{2}} - h_{j-\frac{1}{2}}(B_x)_{j-\frac{1}{2}}q_{j-\frac{1}{2}}\right]\\
&= \frac{1}{4\Delta x}\Big[h_{j+\frac{1}{2}}(B_x)_{j+\frac{1}{2}}\overline{q}_{j+1}\\
&\quad + \left(h_{j+\frac{1}{2}}(B_x)_{j+\frac{1}{2}} - h_{j-\frac{1}{2}}(B_x)_{j-\frac{1}{2}}\right)\overline{q}_j - h_{j-\frac{1}{2}}(B_x)_{j-\frac{1}{2}}\overline{q}_{j-1}\Big],
\end{aligned} \tag{16}$$

$$\begin{aligned}
\left(\frac{1}{2}B_xh^2u_x\right)(x_j) &\approx \frac{1}{2}(B_x)_j\overline{h}_j^2(u_x)_j \approx \frac{1}{2}(B_x)_j\overline{h}_j^2\left[\frac{1}{\overline{h}_j}(q_x)_j - \frac{(h_x)_j}{\overline{h}_j^2}\overline{q}_j\right]\\
&= \frac{1}{4\Delta x}(B_x)_j\left[\overline{h}_j\overline{q}_{j+1} - 2\Delta x(h_x)_j\overline{q}_j - \overline{h}_j\overline{q}_{j-1}\right]
\end{aligned} \tag{17}$$

$$(B_x^2hu)(x_j) \approx (B_x)_j^2\overline{q}_j, \tag{18}$$

where $u_j := \overline{q}_j/\overline{h}_j$ and

$$\begin{aligned}
&u_{j+\frac{1}{2}} := \frac{1}{2}(u_{j+1}+u_j), \quad h_{j+\frac{1}{2}} := \frac{1}{2}(\overline{h}_{j+1}+\overline{h}_j), \quad q_{j+\frac{1}{2}} := \frac{1}{2}(\overline{q}_{j+1}+\overline{q}_j),\\
&(B_x)_j := \frac{B_{j+\frac{1}{2}}-B_{j-\frac{1}{2}}}{\Delta x}, \quad (B_x)_{j+\frac{1}{2}} := \frac{1}{2}((B_x)_{j+1}+(B_x)_j),\\
&(q_x)_j := \frac{\overline{q}_{j+1}-\overline{q}_{j-1}}{2\Delta x}.
\end{aligned} \tag{19}$$

We then replace the time derivatives h_t by its space equivalent $-q_x$ and use (19) to obtain the following discretization of N:

$$\begin{aligned} N_j = &-\frac{2}{\Delta x}\Bigg[h_{j+\frac{1}{2}}\cdot\frac{q_{j+1}-q_j}{\Delta x}\left(h_{j+\frac{1}{2}}\frac{u_{j+1}-u_j}{\Delta x}-(B_x)_{j+\frac{1}{2}}u_{j+\frac{1}{2}}\right) \\ &\qquad -h_{j-\frac{1}{2}}\cdot\frac{q_j-q_{j-1}}{\Delta x}\left(h_{j-\frac{1}{2}}\frac{u_j-u_{j-1}}{\Delta x}-(B_x)_{j-\frac{1}{2}}u_{j-\frac{1}{2}}\right)\Bigg] \\ &-2(B_x)_j(q_x)_j\Big\{(q_x)_j-[(h_x)_j+(B_x)_j]u_j\Big\} \end{aligned} \tag{20}$$

Remark 2. In equations (15)–(18), $(h_x)_j$ are obtained using the limiter as it is described in Section 2.1.1, while $(q_x)_j$ are calculated using the centered differences (see (19)). The latter is done to avoid the need to solve a nonlinear system of algebraic equations as we explain in Section 2.2.

Remark 3. We would like point out that all of the terms in (15)–(18) will be taken at either t^n or t^{n+1} depending on a particular choice of the time evolution method for the numerical integration of the system (7). The manner in which these terms are combined and treated is presented in Section 2.2.

2.1.3. *Well-Balanced Source Discretization.* Our goal is to design a numerical scheme for (4) that exactly preserves the "lake-at-rest" steady-state solution (5). This is achieved by selecting a proper discretization of the geometric source term $\overline{S}_j^{(2)}$. Such a discretization was derived for the classical Saint-Venant system in [20], and since both M_j and N_j as defined in Section 2.1.2 vanish at this steady state, we use this discretization along with an additional atmospheric pressure term for our scheme:

$$\begin{aligned} \overline{S}_j^{(2)} = &-g\,\frac{\left(w^-_{j+\frac{1}{2}}-B_{j+\frac{1}{2}}\right)+\left(w^+_{j-\frac{1}{2}}-B_{j-\frac{1}{2}}\right)}{2}\cdot\frac{\left(B_{j+\frac{1}{2}}-B_{j-\frac{1}{2}}\right)}{\Delta x} \\ &+p^a\,\frac{w^-_{j+\frac{1}{2}}-w^+_{j-\frac{1}{2}}}{\Delta x}. \end{aligned}$$

2.2. **Time Evolution.** We solve the semi-discrete system (7) by applying the third-order strong stability preserving Runge-Kutta (SSP-RK) method from [15,16], which can be written as a convex combination of three forward Euler steps. For the purpose of demonstration, we proceed by fully discretizing (7) according to the forward Euler method, and all results obtained from doing so also apply to the SSP-RK method used in all of our numerical experiments.

When fully discretized by the forward Euler method, the first component of (7) becomes

$$\overline{w}_j^{n+1} = \overline{w}_j^n - \lambda\left(H^{(1)}_{j+\frac{1}{2}} - H^{(1)}_{j-\frac{1}{2}}\right), \tag{21}$$

where $\lambda = \Delta t/\Delta x$. Notice that (21) has no contribution from $\mathcal{M}$, $\mathcal{N}$ or $\mathbf{S}$ and therefore we may advance the first component *independently* of the second one to obtain the cell averages of w at the new time level, $\{\overline{w}_j^{n+1}\}_{j=1}^N$ (and thus $\{\overline{h}_j^{n+1}\}_{j=1}^N$ since $\overline{h}_j^{n+1} := \overline{w}_j^{n+1} - B_j$, where B_j is given by (11)). The fully discretized version of the second component of (7) then becomes

$$\overline{q}_j^{n+1} + \alpha_M M_j^{n+1} = \overline{q}_j^n + \alpha_M M_j^n - \lambda\left(H^{(2)}_{j+\frac{1}{2}} - H^{(2)}_{j-\frac{1}{2}}\right) + \Delta t\overline{S}_j^{(2)} - \Delta t\alpha_N N_j^n, \tag{22}$$

where all of the terms on the RHS of (22) are taken at $t = t^n$.

Combining (15)–(18) for the discretization of M at time level t^{n+1} and inserting this into the left-hand side (LHS) of (22) leads to the tridiagonal system $\mathcal{T} = (\tau_{i,j}^{n+1})$, $j = 1, \dots, N$, $i = j-1, j, j+1$ for $\{\overline{q}_j^{n+1}\}$:

$$\overline{q}_j^{n+1} + \alpha_M M_j^{n+1} = \tau_{j-1,j}^{n+1}\overline{q}_{j-1}^{n+1} + \tau_{j,j}^{n+1}\overline{q}_j^{n+1} + \tau_{j+1,j}^{n+1}\overline{q}_{j+1}^{n+1}, \tag{23}$$

where

$$\tau_{j-1,j}^{n+1} = \alpha_M \left[\frac{\overline{h}_j^{n+1}(B_x)_j - h_{j-\frac{1}{2}}^{n+1}(B_x)_{j-\frac{1}{2}}}{4\Delta x} - \frac{\left(h_{j-\frac{1}{2}}^{n+1}\right)^3}{3\overline{h}_{j-1}^{n+1}(\Delta x)^2} \right],$$

$$\tau_{j,j}^{n+1} = 1 + \alpha_M \left[\frac{h_{j+\frac{1}{2}}^{n+1}(B_x)_{j+\frac{1}{2}} - h_{j-\frac{1}{2}}^{n+1}(B_x)_{j-\frac{1}{2}}}{4\Delta x} + \frac{\left(h_{j+\frac{1}{2}}^{n+1}\right)^3 + \left(h_{j-\frac{1}{2}}^{n+1}\right)^3}{3\overline{h}_j^{n+1}(\Delta x)^2} \right.$$
$$\left. + \frac{(B_x)_j (h_x)_j^{n+1}}{2} + (B_x)_j^2 \right],$$

$$\tau_{j,j+1}^{n+1} = \alpha_M \left[\frac{h_{j+\frac{1}{2}}^{n+1}(B_x)_{j+hf} - \overline{h}_j^{n+1}(B_x)_j}{4\Delta x} - \frac{\left(h_{j+\frac{1}{2}}^{n+1}\right)^3}{3\overline{h}_{j+1}^{n+1}(\Delta x)^2} \right].$$

Notice that the term $\overline{q}_j^n + \alpha_M M_j^n$ on the RHS of (22) is discretized in the same way, but at time level $t = t^n$.

Remark 4. The addition of the dispersive terms M and N does not affect the well-balanced property of the scheme because these terms vanish at the "lake-at-rest" steady state (5). The positivity-preserving property of the scheme is also unaffected because these terms do not appear in the first equation of (1).

Remark 5. We may write the LHS of (22) as described by (23) as $\mathcal{T}\mathbf{q}^{n+1}$, where $\mathbf{q}^{n+1}$ is the vector of the unknown cell averages $\{\overline{q}_j^{n+1}\}_{j=1}^N$. When using free boundary conditions, $\mathcal{T}$ will be strictly tridiagonal, and it is well-known that in this case, the linear algebraic system (22) can be efficiently solved using the LU decomposition; see, e.g., [8, 34] for details. In the case of periodic boundary conditions, the matrix $\mathcal{T}$ becomes circulant and one may still take advantage of the banded structure of the matrix by implementing the Sherman-Morrison algorithm proposed in [32].

3. Numerical Experiments.

In the following experiments, we will examine the role that the non-hydrostatic pressure terms play in the long-time propagation of water waves. We will use the classical Saint-Venant system for comparison, which is simply (4) with $\alpha_M = \alpha_N = p^a \equiv 0$. In all of the experiments, we take $p^a \equiv 0$, take the minmod parameter $\theta = 1.3$, and consider free boundary conditions.

Example 1 — Solitary Wave Propagation. In the first example (taken from [5]), we study propagation of the wave given by the following initial data:

$$h(x,0) = 1 + \frac{1}{10}\operatorname{sech}^2\Big(\sqrt{\frac{3}{40}}(x-70)\Big), \quad u(x,0) = \frac{\sqrt{g}}{10}\operatorname{sech}^2\Big(\sqrt{\frac{3}{40}}(x-70)\Big),$$

over a flat bottom topography with $B(x) \equiv -0.1$. We take $g = 9.81$ and divide the computational domain $[0, 400]$ into 3200 finite-volume cells. According to [5], in the case when $\alpha_M = \alpha_N = 1$, these data correspond to a solitary wave, which is a single elevation of water surface above an undisturbed surrounding, which is neither preceded nor followed by any free surface disturbances.

In our numerical experiments below, we compute the solutions until the final time $t = 50$ and demonstrate how the speed, magnitude and shape of the wave is affected by the choice of α_M and α_N. We begin with the classical Saint-Venant system ($\alpha_M = \alpha_N = 0$) and then start adding the non-hydrostatic pressure terms by gradually increasing α_M and α_N. We first observe that for a very small value of $\alpha_M = \alpha_N = 0.01$, the solutions of hydrostatic and non-hydrostatic systems are almost the same except for a small change of the shape of the wave at the top; see Figure 1. We then further increase α_M and α_N to 0.02–0.05 and observe that up to the intermediate times (around $t = 20$) the solution magnitude increases before decreasing at later times. One can also observe a substantial change in the shape of the wave as a dispersive wave structure clearly develops for $\alpha_M = \alpha_N = 0.04$ and 0.05; see Figure 2. When α_M and α_N are increased up to 0.01, the magnitude of the wave seem to increase up to about $t = 30$ and then it stabilizes; for even larger values of $\alpha_M = \alpha_N = 0.25$ and 0.5, the dispersive wave structure starts disappearing and the amplitude growth becomes less pronounced; and for $\alpha_M = \alpha_N = 1$ the expected solitary wave structure is numerically recovered; see Figure 3. Finally, in Figure 4, we show the solution obtained for larger dispersive coefficients $\alpha_M = \alpha_N = 2$ and 5. As one can see, in these two cases the magnitude of the wave decreases and a wave train is clearly formed.

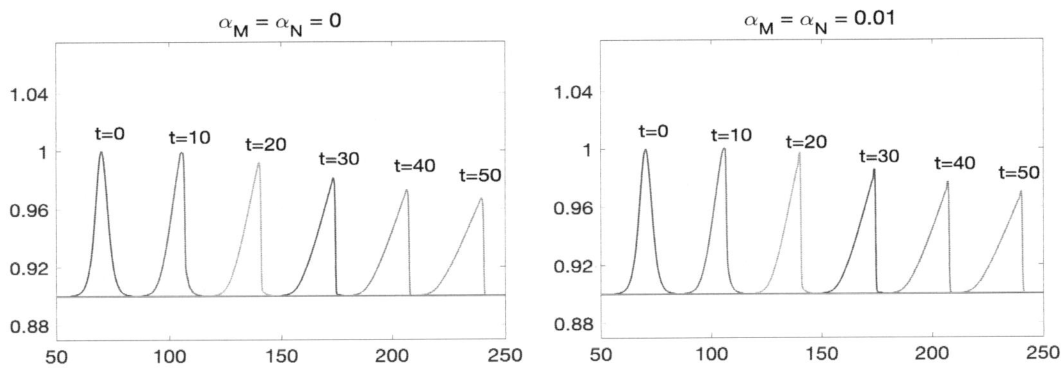

FIGURE 1. Example 1: Time evolution of the water surface for $\alpha_M = \alpha_N = 0$ (left) and 0.01 (right).

We also perform an experimental convergence study of the proposed method. To this end, we take the solution computed with $\alpha_M = \alpha_N = 1$ at time $t = 0.1$ on different grids and compare them with the reference solution obtained with 51200 finite-volume cells. The results are reported in Tables 1 and 2 for w and q, respectively. as One can observe, the expected second order of convergence is achieved in both L^∞-, L^1- and L^2-norms.

Example 2 — Large-Scale Tsunami-Like Wave Propagation. In the second example, we consider a wave that was created using a Savage-Hutter type model of submarine landslides and generated tsunami waves. This model is governed by a two-layer system in which the lower layer is considered to be a fluid-granular mixture that has a larger density than the upper layer, which is water. The lower layer slides down the slope of the solid bottom, and the through momentum exchange causes

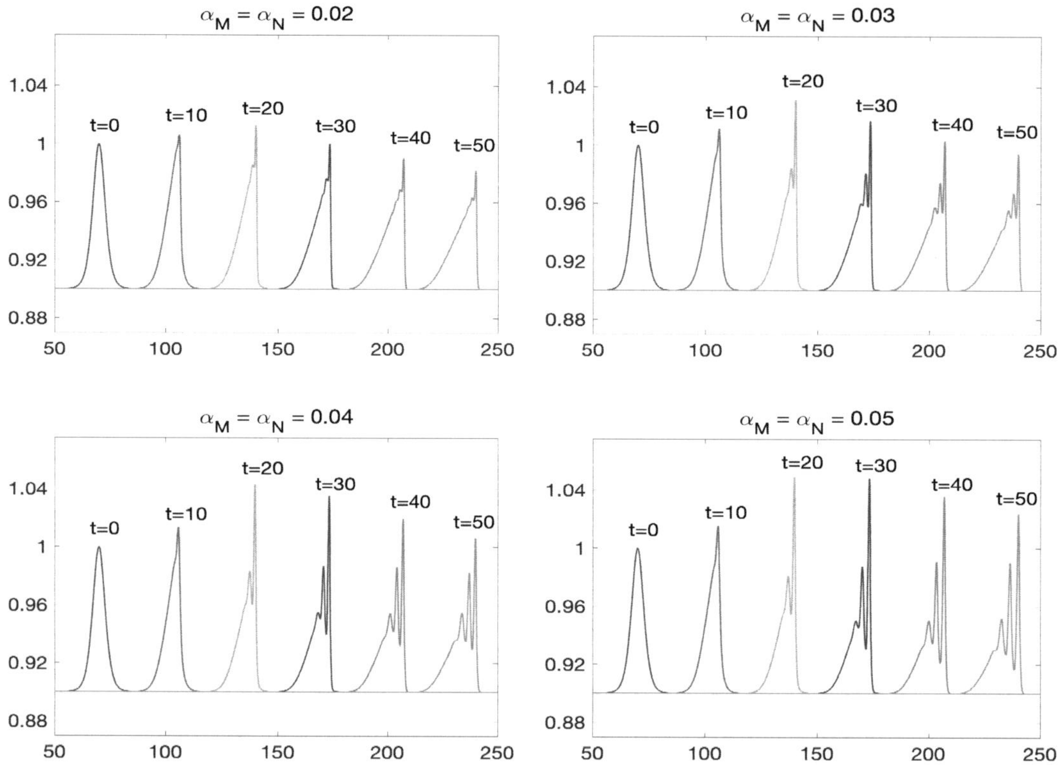

FIGURE 2. Example 1: Time evolution of the water surface for $\alpha_M = \alpha_N = 0.02$ (top left), 0.03 (top right), 0.04 (bottom left) and 0.05 (bottom right).

Number of cells	L^∞-error	Rate	L^1-error	Rate	L^2-error	Rate
400	2.94e-04	–	1.97e-04	–	1.47e-04	–
800	9.23e-05	1.67	4.46e-05	2.14	3.54e-05	2.06
1600	1.51e-05	2.61	8.99e-06	2.31	5.53e-06	2.68
3200	2.55e-06	2.56	2.04e-06	2.14	1.01e-06	2.45
6400	6.63e-07	1.94	5.13e-07	1.99	2.31e-07	2.13
12800	1.75e-07	1.92	1.49e-07	1.79	5.88e-08	1.97

TABLE 1. L^∞-, L^1- and L^2-errors in w and the corresponding experimental rates of convergence.

waves to form at the water surface. For more details of this system and associated numerical methods, see [13, 18, 22].

The initial data are obtained from [22, Section 4.5], where a submarine landslide on the ocean floor creates surface waves traveling to the left and right. We choose the right-moving wave at $t = 0.3$ as the initial condition for the non-hydrostatic system (4) and the following bottom topography function:

$$B(x) = \begin{cases} -5, & x < 0, \\ -5 + \sum_{i=1}^{5} C_i \sin(\pi(x - S_i)/L_i), & x \geq 0, \end{cases} \tag{24}$$

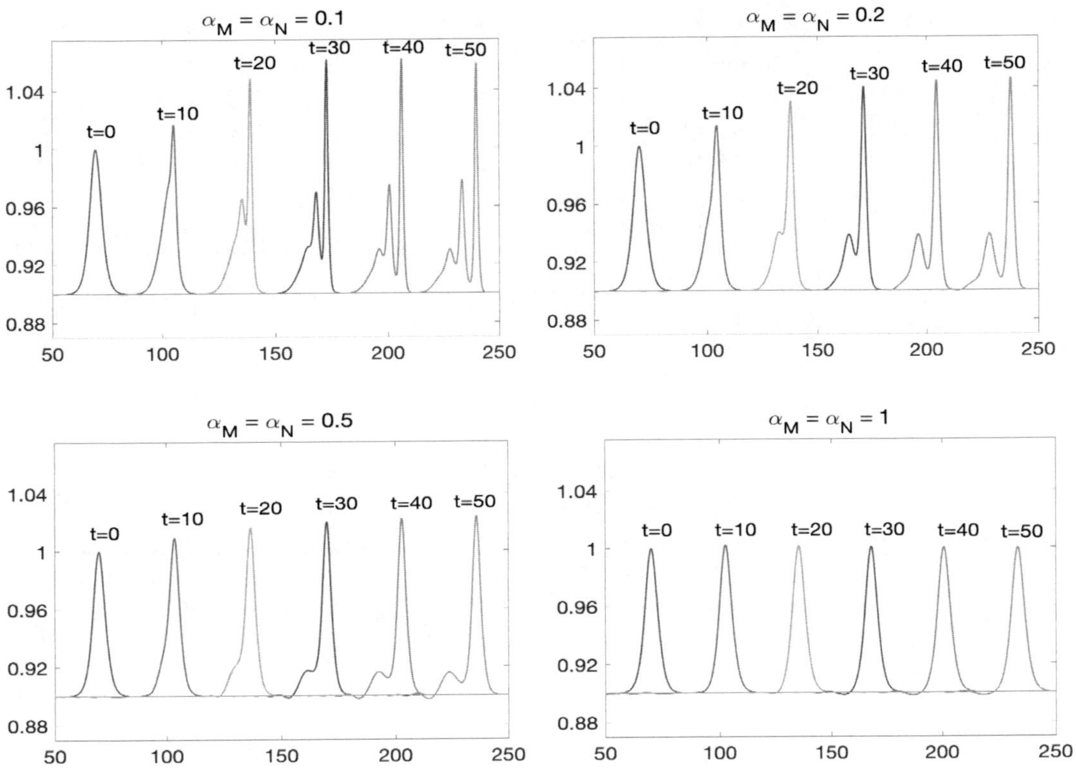

FIGURE 3. Example 1: Time evolution of the water surface for $\alpha_M = \alpha_N = 0.1$ (top left), 0.25 (top right), 0.5 (bottom left) and 1 (bottom right).

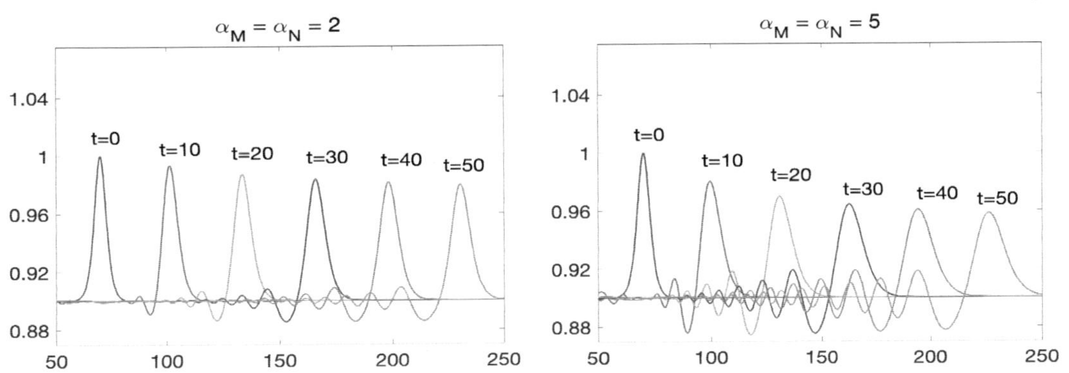

FIGURE 4. Example 1: Time evolution of the water surface for $\alpha_M = \alpha_N = 2$ (left) and 5 (right).

where the parameters C_i, S_i and L_i are given in Table 3. The initial water surface $w(x,0)$ and velocity $u(x,0)$ are plotted in Figure 5 and a nonflat part of the bottom topography is shown in Figure 6. In this example, the length scale is kilometers and the time scale is hours, so we take the corresponding gravity to be $g = 271008$ km/h^2. The computational domain, $[-150, 2200]$, is divided into 18800 finite-volume cells.

Number of cells	L^∞-error	Rate	L^1-error	Rate	L^2-error	Rate
400	2.28e-04	–	4.12e-04	–	1.90e-04	–
800	5.48e-05	2.06	1.03e-04	2.00	4.62e-05	2.04
1600	1.34e-05	2.04	2.56e-05	2.01	1.12e-05	2.05
3200	2.89e-06	2.21	6.49e-06	1.98	2.78e-06	2.01
6400	7.31e-07	1.98	1.65e-06	1.98	6.94e-07	2.00
12800	1.70e-07	2.10	4.14e-07	1.99	1.71e-07	2.02

TABLE 2. L^∞-, L^1- and L^2-errors in q and the corresponding experimental rates of convergence.

i	1	2	3	4	5
C_i	0.1	0.3	0.5	0.1	1
S_i	0	2	3	0	80
L_i	40	70	100	10	2500

TABLE 3. Parameters used in for the bottom topography functions (24) and (25).

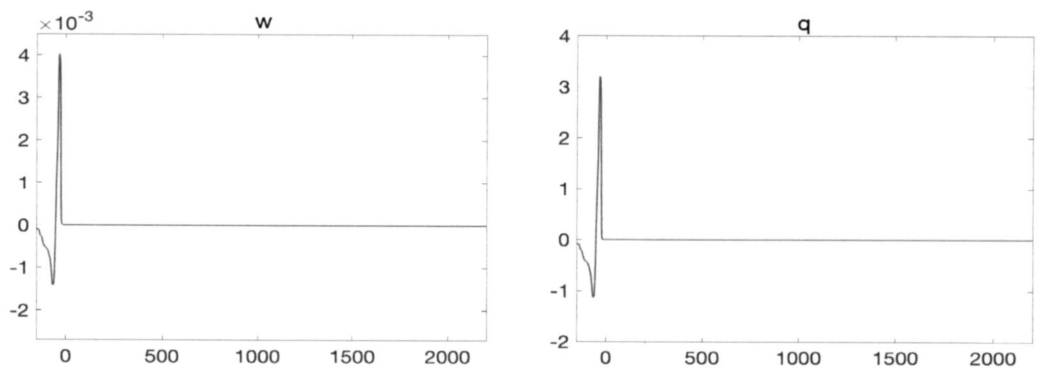

FIGURE 5. Example 2: Initial water surface (left) and discharge (right).

We compute the solutions until the final time $t = 2$ and as in Example 1 study the dependence of the computed solutions on the choice of the dispersion parameters α_M and α_N. We begin with the classical Saint-Venant system ($\alpha_M = \alpha_N = 0$) and plot the obtained results in Figure 7. As one can see, there are many small waves created behind the large wave as a result of the nonflat bottom topography, but the structure of the larger waves does not seem to be significantly affected. Figure 8 shows time snapshots of the numerical solutions of the non-hydrostatic system (4) with $\alpha_M = \alpha_N = 0.05$, 0.1, 0.15 and 0.2. As expected, dispersive wave trains start appearing and become more pronounced for larger values of α_M and α_N.

Example 3 — On-Shore Dynamics of the Large Wave. In order to further emphasize the difference between hydrostatic and non-hydrostatic solutions, we let the computed waves to approach the shore. We take the solutions at time $t = 2$ shown in Figure 7 for $\alpha_M = \alpha_N = 0$ and Figure 8 for $\alpha_M = \alpha_N = 0.2$ as initial data in the domain $[1000, 3000]$ (divided into 16000 finite-volume cells) along with following

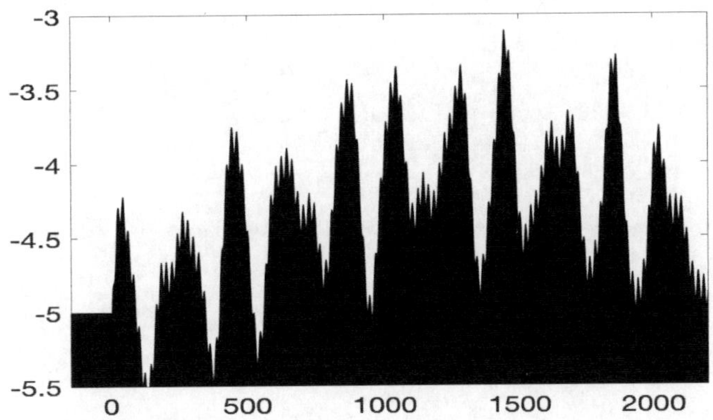

FIGURE 6. Example 2: The bottom topography function (24).

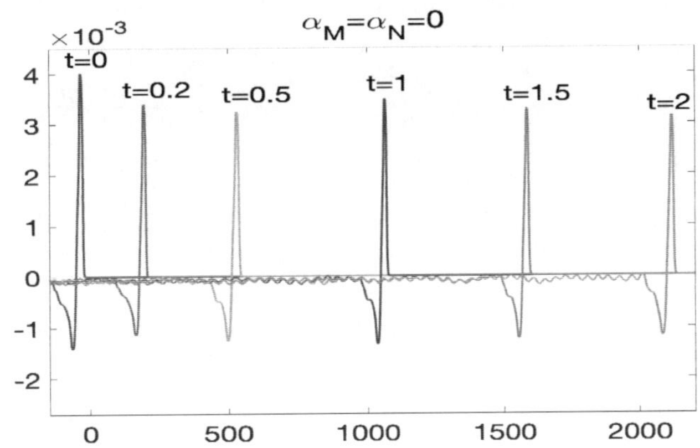

FIGURE 7. Example 2: Time evolution of the water surface for the classical Saint-Venant system ($\alpha_M = \alpha_N = 0$).

bottom topography function:

$$B(x) = \begin{cases} -5 + \sum_{i=1}^{5} C_i \sin(\pi(x - S_i)/L_i), & x < 2200, \\ -4.86 + 2.75 \exp\left[-300\left(1 - \dfrac{x}{2600}\right)\right], & 2200 < x \le 2600, \\ 10^{-10} - 2.11 \exp\left[-300\left(\dfrac{x}{2600} - 1\right)\right], & x > 2600, \end{cases} \tag{25}$$

where the coefficients C_i, S_i, and L_i are given in Table 3. We notice that near the shore, the function B is simply a smooth curve that increases from -4.86 to almost zero; see Figure 9.

In order to accurately capture the on-shore arrival of the waves, we have implemented a special well-balanced reconstruction of wet/dry fronts from [2] and computed both the hydrostatic and non-hydrostatic solutions until the final time $t = 3$. We present several time snapshots of the computed water surface in Figure 10. As one can see, both dispersive and non-dispersive waves go through the shoaling process where they slow down and increase in height, and eventually arrive on shore. If we look closer (Figure 11), we see that the trailing waves actually

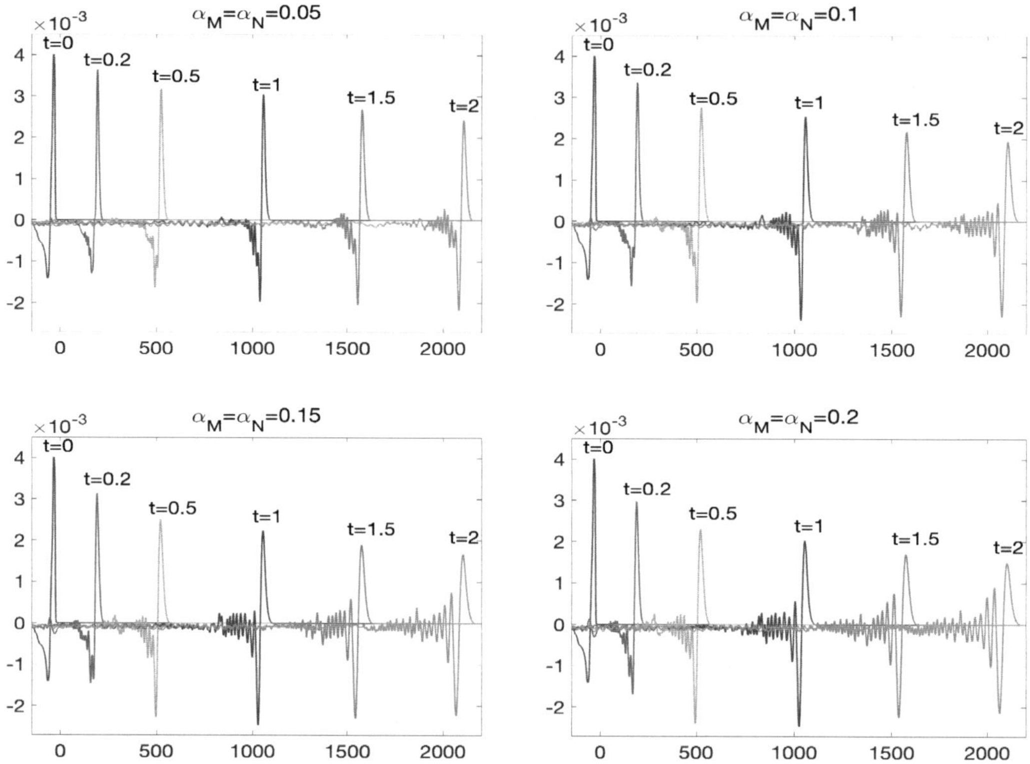

FIGURE 8. Example 2: Time evolution of the water surface for $\alpha_M = \alpha_N = 0.05$ (top left), 0.1 (top right), 0.15 (bottom left) and 0.2 (bottom right).

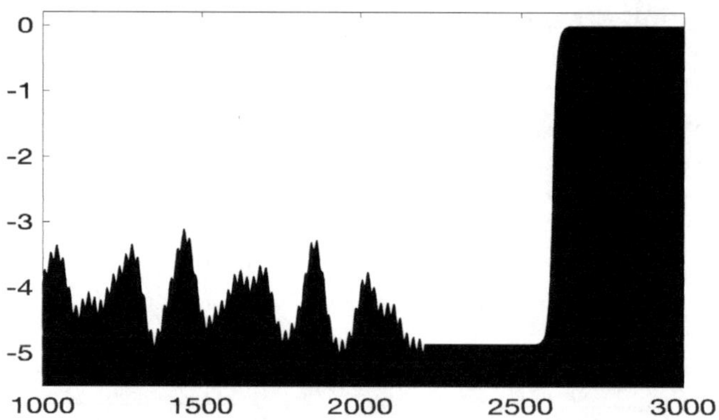

FIGURE 9. Example 3: The bottom topography function (25).

impact how the wave comes to shore: The front of the non-hydrostatic solution is about 10–20 km behind the hydrostatic one. This suggests that the non-hydrostatic terms *must* be included in a tsunami model if one wants to accurately represent the ultimate outcome of the tsunami waves.

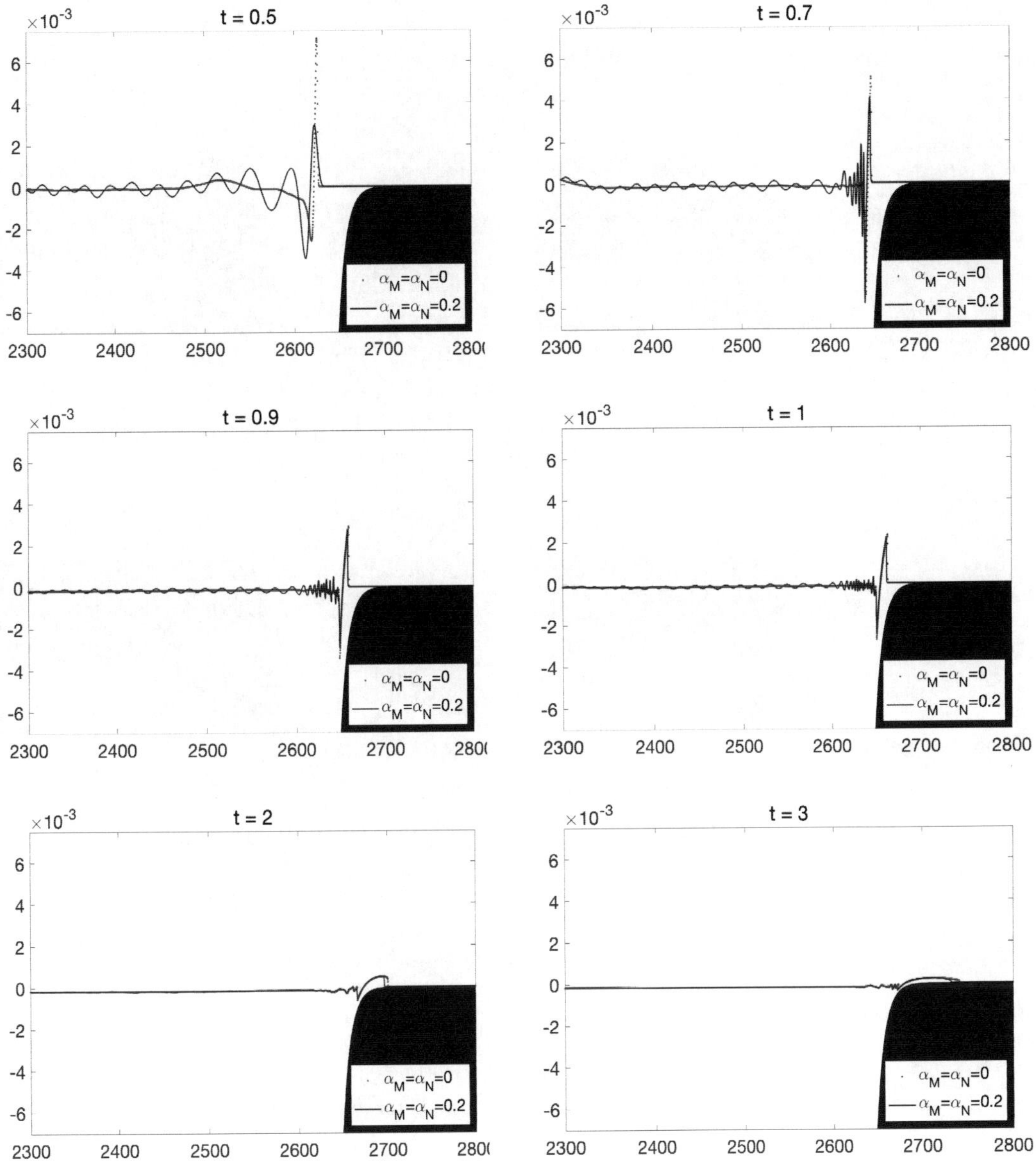

FIGURE 10. Example 3: On-shore arrival of the tsunami-like waves in the hydrostatic ($\alpha_M = \alpha_N = 0$) and non-hydrostatic with $\alpha_M = \alpha_N = 0.2$ regimes.

REFERENCES

[1] E. Barthelemy, Nonlinear shallow water theories for coastal waves, *Surv. Geophys.*, **25** (2004), 315–337.

[2] A. Bollermann, G. Chen, A. Kurganov and S. Noelle, A well-balanced reconstruction of wet/dry fronts for the shallow water equations, *J. Sci. Comput.*, **56** (2013), 267–290.

[3] J. L. Bona, M. Chen and J.-C. Saut, Boussinesq equations and other systems for small-amplitude long waves in nonlinear dispersive media. I. Derivation and linear theory, *J. Nonlinear Sci.*, **12** (2002), 283–318.

[4] J. L. Bona, M. Chen and J.-C. Saut, Boussinesq equations and other systems for small-amplitude long waves in nonlinear dispersive media. II. The nonlinear theory, *Nonlinearity*, **17** (2004), 925–952.

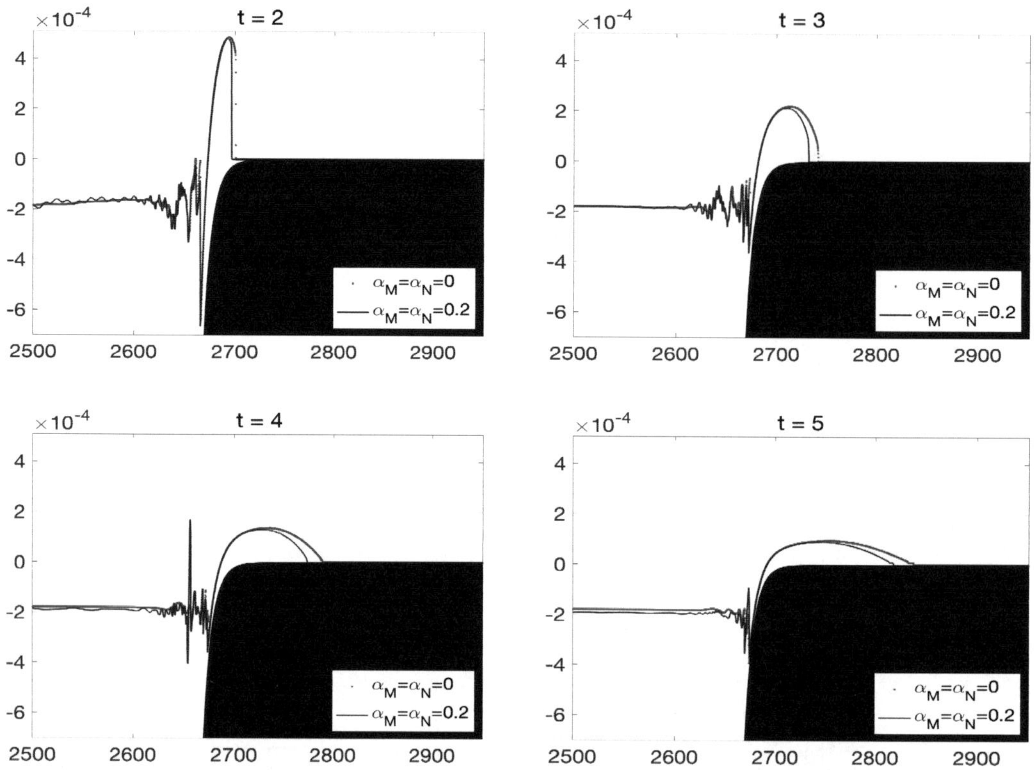

FIGURE 11. Example 3: Same as Figure 10, but zoomed in.

[5] M.-O. Bristeau, N. Goutal and J. Sainte-Marie, Numerical simulations of a non-hydrostatic shallow water model, *Comput. & Fluids*, **47** (2011), 51–64.

[6] M.-O. Bristeau and J. Sainte-Marie, Derivation of a non-hydrostatic shallow water model; comparison with Saint-Venant and Boussinesq systems, *Discrete Contin. Dyn. Syst. Ser. B*, **10** (2008), 733–759.

[7] E. Bryant, *Tsunami: the Underrated Hazard*, 2nd edition, Cambridge University Press, 2008.

[8] R. L. Burden and D. J. Faires, *Numerical Analysis*, 8th edition, Brooks Cole, 2005.

[9] M. J. Castro Díaz, A. Kurganov and T. Morales de Luna, Path-conservative central-upwind schemes for nonconservative hyperbolic systems, *ESAIM Math. Model. Numer. Anal.*, To appear.

[10] Y. Cheng and A. Kurganov, Moving-water equilibria preserving central-upwind schemes for the shallow water equations, *Commun. Math. Sci.*, **14** (2016), 1643–1663.

[11] A. Chertock, S. Cui, A. Kurganov and T. Wu, Well-balanced positivity preserving central-upwind scheme for the shallow water system with friction terms, *Internat. J. Numer. Meth. Fluids*, **78** (2015), 355–383.

[12] A. J. C. de Saint-Venant, Thèorie du mouvement non-permanent des eaux, avec application aux crues des rivière at à l'introduction des marèes dans leur lit., *C.R. Acad. Sci. Paris*, **73** (1871), 147–154.

[13] E. D. Fernández-Nieto, F. Bouchut, D. Bresch, M. J. Castro Díaz and A. Mangeney, A new Savage-Hutter type model for submarine avalanches and generated tsunami, *J. Comput. Phys.*, **227** (2008), 7720–7754.

[14] J.-F. Gerbeau and B. Perthame, Derivation of viscous Saint-Venant system for laminar shallow water; numerical validation, *Discrete Contin. Dyn. Syst. Ser. B*, **1** (2001), 89–102.

[15] S. Gottlieb, D. Ketcheson and C.-W. Shu, *Strong stability preserving Runge-Kutta and multistep time discretizations*, World Scientific Publishing Co. Pte. Ltd., Hackensack, NJ, 2011.

[16] S. Gottlieb, C.-W. Shu and E. Tadmor, Strong stability-preserving high-order time discretization methods, *SIAM Rev.*, **43** (2001), 89–112.

[17] A. Green and P. Naghdi, A derivation of equations for wave propagation in water at variable depth, *J. Fluid Mech.*, **78** (1976), 237–246.
[18] P. Heinrich, A. Piatanesi and H. Hebert, Numerical modelling of tsunami generation and propagation from submarine slumps: the 1998 Papua New Guinea event, *Geophys. J. Int.*, **145** (2001), 97–111.
[19] A. Kurganov, Finite-volume schemes for shallow-water equations, *Acta Numer.*, **27** (2018), 289–351.
[20] A. Kurganov and D. Levy, Central-upwind schemes for the saint-venant system, *M2AN Math. Model. Numer. Anal.*, **36** (2002), 397–425.
[21] A. Kurganov and C.-T. Lin, On the reduction of numerical dissipation in central-upwind schemes, *Commun. Comput. Phys.*, **2** (2007), 141–163.
[22] A. Kurganov and J. Miller, Central-upwind scheme for Savage-Hutter type model of submarine landslides and generated tsunami waves, *Comput. Methods Appl. Math.*, **14** (2014), 177–201.
[23] A. Kurganov, S. Noelle and G. Petrova, Semi-discrete central-upwind scheme for hyperbolic conservation laws and Hamilton-Jacobi equations, *SIAM J. Sci. Comput.*, **23** (2001), 707–740.
[24] A. Kurganov and G. Petrova, A second-order well-balanced positivity preserving central-upwind scheme for the saint-venant system, *Commun. Math. Sci.*, **5** (2007), 133–160.
[25] A. Kurganov and G. Petrova, Central-upwind schemes for two-layer shallow equations, *SIAM J. Sci. Comput.*, **31** (2009), 1742–1773.
[26] A. Kurganov and E. Tadmor, New high resolution central schemes for nonlinear conservation laws and convection-diffusion equations, *J. Comput. Phys.*, **160** (2000), 241–282.
[27] R. J. LeVeque, D. L. George and M. J. Berger, Tsunami modelling with adaptively refined finite volume methods, *Acta Numer.*, **20** (2011), 211–289.
[28] K.-A. Lie and S. Noelle, On the artificial compression method for second-order nonoscillatory central difference schemes for systems of conservation laws, *SIAM J. Sci. Comput.*, **24** (2003), 1157–1174.
[29] A. Mercado-Irizarry and P. L. F. Liu, *Caribbean Tsunami Hazard*, World Scientific, 2006.
[30] H. Nessyahu and E. Tadmor, Nonoscillatory central differencing for hyperbolic conservation laws, *J. Comput. Phys.*, **87** (1990), 408–463.
[31] M. Ortiz, E. Gomez-Reyes and H. Velez-Munoz, A fast preliminary estimation model for transoceanic tsunami propagation, *Geofis. Int.*, **39** (2000), 207–220.
[32] J. Sherman and W. J. Morrison, Adjustment of an inverse matrix corresponding to a change in one element of a given matrix, *Ann. Math. Statistics*, **21** (1950), 124–127.
[33] P. K. Sweby, High resolution schemes using flux limiters for hyperbolic conservation laws, *SIAM J. Numer. Anal.*, **21** (1984), 995–1011.
[34] L. Trefethen and D. Bau, *Numerical Linear Algebra*, Society for Industrial and Applied Math, 1997.

E-mail address: `chertock@math.ncsu.edu`
E-mail address: `alexander@sustech.edu.cn`
E-mail address: `jmiller8@tulane.edu`
E-mail address: `jyan9@ncsu.edu`

STABILITY OF VORTICES IN IDEAL FLUIDS: THE LEGACY OF KELVIN AND RAYLEIGH

Thierry Gallay

Institut Fourier, Université Grenoble Alpes
100 rue des Maths, 38610 Gières, France

Abstract. The mathematical theory of hydrodynamic stability started in the middle of the 19th century with the study of model examples, such as parallel flows, vortex rings, and surfaces of discontinuity. We focus here on the equally interesting case of columnar vortices, which are axisymmetric stationary flows where the velocity field only depends on the distance to the symmetry axis and has no component in the axial direction. The stability of such flows was first investigated by Kelvin in 1880 for some particular velocity profiles, and the problem benefited from important contributions by Rayleigh in 1880 and 1917. Despite further progress in the 20th century, notably by Howard and Gupta (1962), the only rigorous results so far are necessary conditions for instability under either two-dimensional or axisymmetric perturbations. This note is a non-technical introduction to a recent work in collaboration with D. Smets, where we prove under mild assumptions that columnar vortices are spectrally stable with respect to general three-dimensional perturbations, and that the linearized evolution group has a subexponential growth as $|t| \to \infty$.

1. **Introduction.** Hydrodynamic stability is the subdomain of fluid dynamics which studies the stability and the onset of instability in fluid flows. These fundamental questions were first addressed in the 19th century, with pioneering contributions by G. Stokes, H. von Helmholtz, W. Thomson (Lord Kelvin), and J. W. Strutt (Lord Rayleigh) on the theoretical side, and by O. Reynolds on the experimental side [9]. In early times the notion of stability still lacked a precise mathematical definition, but its physical meaning was already perfectly understood, as can be seen from the following quote by J. C. Maxwell [19], which dates back to 1873:

> "When the state of things is such that an infinitely small variation of the present state will alter only by an infinitely small quantity the state at some future time, the condition of the system, whether at rest or in motion, is said to be stable; but when an infinitely small variation in the present state may bring about a finite difference in the state of the system in a finite time, the system is said to be unstable."

What is exactly meant by "infinitely small" in this definition is rigorously specified, for instance, in the subsequent memoir by A. M. Lyapunov [26], which was published in 1892. The relevance of stability questions in fluid mechanics cannot

---

2000 *Mathematics Subject Classification.* Primary: 35Q31, 35B35; Secondary: 76B47, 76E07.

Key words and phrases. Hydrodynamic stability, ideal fluid, vortices, critical layers.

This work is supported by the SingFlows project, grant ANR-18-CE40-0027 of the French National Research Agency (ANR).

be overestimated. As an example, in the idealized situation where the fluid is assumed to be incompressible and inviscid, a plethora of explicit stationary solutions are known which describe shear flows, vortices, or flows past obstacles. However, depending on circumstances, these solutions may or may not be observed in real life, where experimental uncertainties, viscosity effects, and boundary conditions play an important role. To determine the relevance of a given flow, the stability analysis is certainly the first step to perform, but even in an idealized framework this often leads to difficult mathematical problems, a complete solution of which was largely out of reach in the 19th century and is still a serious challenge today.

To make the previous considerations more concrete, we analyze in this introduction three relatively simple cases, of increasing complexity, where stability can be discussed using the techniques introduced by Rayleigh [30]. These examples are classical and thoroughly studied in many textbooks [7, 8, 11, 23, 33], as well as in the excellent review article [10]. The results obtained for these model problems will serve as a guideline for the stability analysis of columnar vortices, which will be presented in Sections 2 and 3.

1.1. **The Rayleigh-Taylor Instability.** We consider the motion of an incompressible and inviscid fluid in the infinite strip $D = \mathbb{R} \times [0, L]$ with coordinates (x, z), where $x \in \mathbb{R}$ is the horizontal variable and $z \in [0, L]$ the vertical variable. The state of the fluid at time $t \in \mathbb{R}$ is defined by the density distribution $\rho(x, z, t) > 0$, the velocity field $u(x, z, t) \in \mathbb{R}^2$, and the pressure $p(x, z, t) \in \mathbb{R}$. The evolution is determined by the density-dependent incompressible Euler equations

$$\partial_t \rho + u \cdot \nabla \rho = 0\,, \quad \rho\bigl(\partial_t u + (u \cdot \nabla)u\bigr) = -\nabla p - \rho g e_z\,, \quad \operatorname{div} u = 0\,, \tag{1}$$

where g denotes the acceleration due to gravity and e_z is the unit vector in the (upward) vertical direction. Setting $u = (u_x, u_z)$, we impose the impermeability condition $u_z(x, z, t) = 0$ at the bottom and the top of the domain D, namely for $z = 0$ and $z = L$.

The PDE system (1) has a family of stationary solutions of the form $\rho = \bar\rho(z)$, $u = 0$, $p = \bar p(z)$, where the density $\bar\rho$ is an arbitrary function of the vertical coordinate z, and the associated pressure is determined (up to an irrelevant additive constant) by the hydrostatic balance $\bar p'(z) = -\bar\rho(z) g$. To study the stability of the equilibrium $(\bar\rho, 0, \bar p)$, we consider perturbed solutions of the form

$$\rho(x, z, t) = \bar\rho(z) + \tilde\rho(x, z, t)\,, \quad u(x, z, t) = \tilde u(x, z, t)\,, \quad p(x, z, t) = \bar p(z) + \tilde p(x, z, t)\,.$$

Inserting this Ansatz into (1) and neglecting all quadratic terms in $(\tilde\rho, \tilde u)$, we obtain the *linearized* equations for the perturbations $(\tilde\rho, \tilde u, \tilde p)$:

$$\begin{aligned} \bar\rho(z)\partial_t \tilde u_x &= -\partial_x \tilde p\,, & \partial_t \tilde\rho + \bar\rho'(z)\tilde u_z &= 0\,, \\ \bar\rho(z)\partial_t \tilde u_z &= -\partial_z \tilde p - \tilde\rho g\,, & \partial_x \tilde u_x + \partial_z \tilde u_z &= 0\,. \end{aligned} \tag{2}$$

Remark 1.1. It is not obvious at all that considering the linearized perturbation equations (2) is sufficient, or even appropriate, to determine the stability of stationary solutions to (1). In fact the validity of Lyapunov's linearization method in the context of fluid mechanics is a difficult question [39], which is the object of ongoing research. In particular, for ideal fluids, there is no general result asserting that a linearly stable equilibrium is actually stable in the sense of Lyapunov. However, if the linearized system is exponentially unstable, for instance due to the existence of an eigenvalue with nonzero real part, it is often possible to conclude that the equilibrium under consideration is unstable, see [5, 16, 25, 38] for a few results in this direction. To summarize, the linearization approach may be useful to detect exponential instabilities, but stability results have to be established by a different approach, for instance (in two space dimensions) using variational techniques [2, 3]

The linearized equations (2) are invariant under translations in the horizontal direction, so that we can use a Fourier transform to reduce the number of independent variables. A further simplification is made by restricting our attention to *eigenfunctions* of the linearized operator. In other words, we consider solutions of (2) of the particular form

$$\tilde{\rho}(x,z,t) = \rho(z)\, e^{ikx}\, e^{st}\,, \quad \tilde{u}(x,z,t) = u(z)\, e^{ikx}\, e^{st}\,, \quad \tilde{p}(x,z,t) = p(z)\, e^{ikx}\, e^{st}\,, \tag{3}$$

where $k \in \mathbb{R}$ is the horizontal wave number and $s \in \mathbb{C}$ is the spectral parameter. The representation (3) transforms the linearized equations (2) into an ODE system:

$$\begin{aligned} \bar{\rho}(z) s u_x &= -ikp\,, & s\rho + \bar{\rho}'(z) u_z &= 0\,, \\ \bar{\rho}(z) s u_z &= -\partial_z p - \rho g\,, & iku_x + \partial_z u_z &= 0\,, \end{aligned} \tag{4}$$

which (if $s \neq 0$) can in turn be reduced to a single equation for the vertical velocity profile u_z:

$$-\,\partial_z\big(\bar{\rho}(z)\partial_z u_z\big) + k^2 \bar{\rho}(z) u_z - \frac{k^2 g}{s^2}\,\bar{\rho}'(z) u_z = 0\,, \qquad z \in [0, L]\,. \tag{5}$$

By construction, the values of the spectral parameter $s \in \mathbb{C} \setminus \{0\}$ for which the ODE (5) has a nontrivial solution u_z satisfying the boundary conditions $u_z(0) = u_z(L) = 0$ are *eigenvalues* of the linearized operator (2) in the Fourier subspace indexed by the horizontal wavenumber $k \in \mathbb{R}$. Spectral stability is obtained if all eigenvalues are purely imaginary, whereas the existence of an eigenvalue $s \in \mathbb{C}$ with $\mathrm{Re}(s) \neq 0$ implies exponential instability of the linearized system in positive or negative times.

Remarks 1.2.

1. The Fourier transform reduces the linearized equations to a one-dimensional PDE system in the bounded domain $[0, L]$, but this does not immediately imply that the spectrum of the full linearized operator is the union of the point spectra obtained for all values of the horizontal wavenumber $k \in \mathbb{R}$. So, even if one can prove that the eigenvalues are purely imaginary for all $k \in \mathbb{R}$, an additional argument is needed to verify that the full linearized operator has indeed no spectrum outside the imaginary axis. This rather technical issue will not be discussed further in this introduction, but we shall come back to it in Section 3.

2. In the literature, the Rayleigh-Taylor equation (5) is often derived in the Boussinesq approximation, which consists in neglecting the variations of the density profile $\bar{\rho}(z)$ everywhere except in the buoyancy term. This gives the simplified eigenvalue equation

$$-\,\partial_z^2 u_z + k^2\Big(1 + \frac{N(z)^2}{s^2}\Big) u_z = 0\,, \qquad \text{where} \quad N(z)^2 = -\frac{g\bar{\rho}'(z)}{\bar{\rho}(z)}\,. \tag{6}$$

When $\bar{\rho}'(z) < 0$, the real number $N(z)$ is called the Brunt-Väisälä frequency. This is the (maximal) oscillation frequency of internal waves inside a stably stratified fluid.

Assume that, for some $k \in \mathbb{R}$ and some $s \in \mathbb{C} \setminus \{0\}$, the ODE (5) has a nontrivial solution u_z satisfying the boundary conditions $u_z(0) = u_z(L) = 0$. Multiplying both sides of (5) by the complex conjugate $\bar{u}_z$ and integrating over the vertical domain $[0,L]$, we obtain the integral identity

$$\int_0^L \bar{\rho}(z)|\partial_z u_z|^2\,\mathrm{d}z + k^2\int_0^L \bar{\rho}(z)|u_z|^2\,\mathrm{d}z - \frac{k^2 g}{s^2}\int_0^L \bar{\rho}'(z)|u_z|^2\,\mathrm{d}z \,=\, 0\,. \qquad (7)$$

The first two terms in (7) being real and positive, equality can hold only if the third term is real and negative. Thus we must have $k \neq 0$ and $\mathrm{Im}(s^2) = 0$, namely $s \in \mathbb{R}$ or $s \in i\mathbb{R}$. Now, if we assume that the fluid is *stably stratified*, in the sense that $\bar{\rho}'(z) \le 0$ for all $z \in [0,L]$, the last term in (7) is negative only if $s^2 < 0$, which means that $s \in i\mathbb{R}$. Under this assumption, we conclude that all eigenfunctions of the form (3) with $k \in \mathbb{R}$ correspond to eigenvalues s on the imaginary axis, so that the equilibrium $(\bar{\rho},0,\bar{p})$ of (1) is *spectrally stable*, up to the technical issue mentioned in Remark 1.2.1.

On the other hand, if $\bar{\rho}'(z) > 0$ for some $z \in [0,L]$, a nice argument due to Synge [36] shows that, for any $k \neq 0$, the Rayleigh equation has a nontrivial solution u_z (satisfying the boundary conditions) for a sequence of real eigenvalues $s_n \to 0$. The equilibrium $(\bar{\rho},0,\bar{p})$ of (1) is thus spectrally unstable. Summarizing, the stability of the rest state $u = 0$ in stratified ideal fluids is reasonably understood, in the sense that Rayleigh's approach provides a *necessary and sufficient* condition for spectral stability in that case.

1.2. **Shear Flows in Homogeneous Fluids.** For the same equations (1) in the domain D, we now consider a different family of equilibria, namely shear flows of the form $\rho = 1$, $u = U(z)e_x$, $p = 0$, where the horizontal velocity profile U is an arbitrary function. For the moment, we assume that the fluid is homogeneous and only allow for perturbations of the velocity field. The perturbed solutions thus take the form

$$\rho(x,z,t) \,=\, 1\,, \quad u(x,z,t) \,=\, U(z)e_x + \tilde{u}(x,z,t)\,, \quad p(x,z,t) \,=\, \tilde{p}(x,z,t)\,,$$

and the linearized equations become

$$\begin{aligned} \partial_t\tilde{u}_x + U(z)\partial_x\tilde{u}_x + U'(z)\tilde{u}_z &= -\partial_x\tilde{p}\,, \\ \partial_t\tilde{u}_z + U(z)\partial_x\tilde{u}_z &= -\partial_z\tilde{p}\,, \end{aligned} \qquad \partial_x\tilde{u}_x + \partial_z\tilde{u}_z \,=\, 0\,. \qquad (8)$$

As before, we suppose that $\tilde{u}(x,z,t) = u(z)\,e^{ikx}\,e^{st}$ and $\tilde{p}(x,z,t) = p(z)\,e^{ikx}\,e^{st}$ for some $k \in \mathbb{R}$ and some $s \in \mathbb{C}$. The functions u, p are solutions of the ODE system

$$\gamma(z)u_x + U'(z)u_z \,=\, -ikp\,, \quad \gamma(z)u_z \,=\, -\partial_z p\,, \quad iku_x + \partial_z u_z \,=\, 0\,, \qquad (9)$$

where $\gamma(z) = s + ikU(z)$ is the symbol of the material derivative $\partial_t + U(z)\partial_x$. This function plays an important role in the stability analysis, as it incorporates the spectral parameter s.

Since we are interested in detecting potential instabilities, we assume in what follows that $\mathrm{Re}(s) \neq 0$, which implies in particular that $\gamma(z) \neq 0$ for all $z \in [0,L]$. Under this hypothesis, we can reduce the ODE system (9) to the following scalar equation for the vertical velocity:

$$-\,\partial_z^2 u_z + k^2 u_z + \frac{ikU''(z)}{\gamma(z)}\,u_z \,=\, 0\,, \qquad z \in [0,L]\,. \qquad (10)$$

This equation looks simpler than (5), but is in fact substantially harder to analyze. If u_z is a nontrivial solution satisfying the boundary conditions, we have Rayleigh's identity

$$\int_0^L |\partial_z u_z|^2 \,\mathrm{d}z + k^2 \int_0^L |u_z|^2 \,\mathrm{d}z + ik \int_0^L \frac{U''(z)}{\gamma(z)} |u_z|^2 \,\mathrm{d}z \;=\; 0\,, \tag{11}$$

which can be satisfied only if $k \neq 0$ and if $U''(z)$ is not identically zero. Under these assumptions, the imaginary part of (11) gives the useful relation

$$\mathrm{Re}(s) \int_0^L \frac{U''(z)}{|\gamma(z)|^2} |u_z|^2 \,\mathrm{d}z \;=\; 0\,. \tag{12}$$

If $U''(z)$ does not change sign on $[0, L]$, the integral in (12) is nonzero, which contradicts our assumption that $\mathrm{Re}(s) \neq 0$. This gives Rayleigh's *inflection point criterion* [30] : a necessary condition for the shear flow with velocity profile $U(z)$ to be (spectrally) unstable is that the function $z \mapsto U''(z)$ changes sign on the interval $[0, L]$.

Rayleigh's inflection point criterion is not sharp, and can be improved somehow by exploiting both the real and imaginary parts of identity (11), see [15]. However, surprisingly enough, it seems difficult to formulate a necessary and sufficient stability condition for shear flows, even in the ideal case considered here. An instructive example is Kolmogorov's flow $U(z) = \sin(z - L/2)$, which is known to be stable if and only if $L \leq \pi$ [10, 24], although both Rayleigh's and Fjørtoft's criteria allow for a possible instability for any $L > 0$. In fact, the origin of inertial instabilities in shear flows seems only partially understood from a physical point of view, see [4].

1.3. **Shear Flows in Stratified Fluids.** Following the same approach as in the previous paragraphs, we now analyze the stability of shear flows in (stably) stratified fluids. We consider the Euler equations (1) in the vicinity of a stationary solution of the form $\rho = \bar\rho(z)$, $u = U(z)e_x$, $p = \bar p(z)$, where $\bar p'(z) = -\bar\rho(z)g$ (hydrostatic balance). The perturbed solutions are written in the form

$$\begin{aligned} \rho(x,z,t) &= \bar\rho(z) + \tilde\rho(x,z,t)\,, \\ u(x,z,t) &= U(z)e_x + \tilde u(x,z,t)\,, \\ p(x,z,t) &= \bar p(z) + \tilde p(x,z,t)\,, \end{aligned}$$

so that the linearized equations become

$$\begin{aligned} \bar\rho(z)\big(\partial_t \tilde u_x + U(z)\partial_x \tilde u_x + U'(z)\tilde u_z\big) &= -\partial_x \tilde p\,, & \partial_t \tilde\rho + U(z)\partial_x \tilde\rho + \bar\rho'(z)\tilde u_z &= 0\,, \\ \bar\rho(z)\big(\partial_t \tilde u_z + U(z)\partial_x \tilde u_z\big) &= -\partial_z \tilde p - \tilde\rho g\,, & \partial_x \tilde u_x + \partial_z \tilde u_z &= 0\,. \end{aligned} \tag{13}$$

For perturbations of the form (3), we arrive at the ODE system

$$\begin{aligned} \bar\rho(z)\big(\gamma(z)u_x + U'(z)u_z\big) &= -ikp\,, & \gamma(z)\rho + \bar\rho'(z)u_z &= 0\,, \\ \bar\rho(z)\gamma(z)u_z &= -\partial_z p - \rho g\,, & iku_x + \partial_z u_z &= 0\,, \end{aligned} \tag{14}$$

where $\gamma(z) = s + ikU(z)$ is the spectral function. If we assume that $\mathrm{Re}(s) \neq 0$, so that $\gamma(z) \neq 0$, we can reduce the system (14) to the *Taylor-Goldstein equation*

$$-\,\partial_z\big(\bar\rho(z)\partial_z u_z\big) + k^2\bar\rho(z)u_z + \frac{ik}{\gamma(z)}\big(\bar\rho U'\big)'(z)u_z - \frac{k^2 g}{\gamma(z)^2}\,\bar\rho'(z)u_z \;=\; 0\,, \tag{15}$$

where $z \in [0, L]$. Note that we recover the Rayleigh-Taylor equation (5) by setting $U = 0$, hence $\gamma(z) = s$, in (15). Similarly, (15) reduces to the Rayleigh stability equation (10) when $\bar\rho = 1$.

The original approach of Rayleigh does not give much information on the solutions of (15). If u_z is a nontrivial solution satisfying the boundary conditions, it is difficult to exploit the integral identity

$$\int_0^L \Big\{\bar\rho(z)|\partial_z u_z|^2 + k^2\bar\rho(z)|u_z|^2 + ik\frac{(\bar\rho U')'(z)}{\gamma(z)}\,|u_z|^2 - \frac{k^2 g}{\gamma(z)^2}\bar\rho'(z)|u_z|^2\Big\}\,\mathrm{d}z \,=\, 0\,, \quad (16)$$

because the real or imaginary parts of the last two terms in the integrand have no obvious sign. A solution to this problem was found by Miles [27] and Howard [20] in the early 60's. Following the elegant approach of [20], we perform the change of variables

$$u_z(z) \,=\, \gamma(z)^{1/2}v_z(z)\,, \qquad \text{where} \quad \gamma(z) \,=\, s + ikU(z)\,.$$

The new function v_z satisfies the modified ODE

$$-\,\partial_z\big(\bar\rho(z)\gamma(z)\partial_z v_z\big) + k^2\bar\rho(z)\gamma(z)v_z + \frac{ik}{2}\,(\bar\rho U')'(z)\,v_z + \Big(\frac{\bar\rho\gamma'^2}{4\gamma} - \frac{k^2 g\bar\rho'}{\gamma}\Big)(z)\,v_z \,=\, 0\,. \quad (17)$$

If v_z is a nontrivial solution satisfying the boundary conditions $v_z(0) = v_z(L) = 0$, we multiply both sides of (17) by the complex conjugate $\bar v_z$ and integrate over the domain $[0, L]$. After taking the real part, we obtain the useful identity

$$\mathrm{Re}(s)\int_0^L \Big\{\bar\rho(z)\big(|\partial_z v_z|^2 + k^2|v_z|^2\big) + \frac{k^2\bar\rho(z)U'(z)^2}{|\gamma(z)|^2}\Big(\mathrm{Ri}(z) - \frac{1}{4}\Big)\,|v_z|^2\Big\}\,\mathrm{d}z \,=\, 0\,, \quad (18)$$

where $\mathrm{Ri}(z)$ is the (local) *Richardson number* defined by

$$\mathrm{Ri}(z) \,=\, \frac{-\bar\rho'(z)\,g}{\bar\rho(z)}\,\frac{1}{U'(z)^2} \,=\, \Big(\frac{N(z)}{U'(z)}\Big)^2\,.$$

We assume here that $\bar\rho'(z) \le 0$ (stable stratification), so that $\mathrm{Ri}(z) \ge 0$, and we denote by $N(z)$ the Brunt-Väisälä frequency (6).

The Richardson number compares the stabilizing effect of the stratification, measured by the frequency N of the internal waves, to the potentially destabilizing effect of the shear flow, which may be proportional to the velocity gradient U' [10]. Clearly, equality (18) cannot hold if $\mathrm{Ri}(z) \ge 1/4$ for all $z \in [0, L]$, because the integrand is then positive while we assumed that $\mathrm{Re}(s) \neq 0$. This gives the celebrated *Miles-Howard criterion*: a shear flow in a stratified fluid is spectrally stable if the Richardson number is greater than or equal to $1/4$ everywhere in the fluid. The threshold value $1/4$ is known to be sharp, in the sense that it cannot be replaced by any smaller real number. However, the Miles-Howard criterion itself is by no means sharp: if $\bar\rho = 1$, any shear flow without inflection point is spectrally stable by Rayleigh's criterion, although $\mathrm{Ri}(z) \equiv 0$ in that case.

Remark 1.3. So far we concentrated on the two-dimensional case, but it is also instructive to investigate the stability of shear flows with respect to three-dimensional perturbations. In that case, we work in the domain $D' = \mathbb{R}^2 \times [0, L]$ with coordinates (x, y, z), and consider perturbations that are plane waves with horizontal wave vector $k = (k_1, k_2) \in \mathbb{R}^2$. For instance, the three-dimensional velocity field takes the form

$$u(x, y, z, t) \,=\, U(z)e_x + u(z)\,e^{i(k_1x + k_2y)}\,e^{\sigma t}\,,$$

where $\sigma \in \mathbb{C}$ is the spectral parameter. Using a similar Ansatz for the density and the pressure, it is easy to derive the 3D perturbation equations which generalize (14). Now, in the homogeneous case where $\rho \equiv 1$, a well-know result due to Squire [35] shows that, if the 3D perturbation equations have a nontrivial solution for some

$k_1, k_2 \neq 0$ and some $\sigma \in \mathbb{C}$ with $\mathrm{Re}(\sigma) \neq 0$, then the 2D perturbation equations (9) also have a nontrivial solution with $k = (k_1^2 + k_2^2)^{1/2}$ and $s = (k/k_1)\sigma$. Note that $|\,\mathrm{Re}(s)| > |\,\mathrm{Re}(\sigma)|$, so that the most unstable modes are always two-dimensional; in other words, it is sufficient to consider the 2D case to detect potential instabilities. A similar result holds in the general situation where the fluid is stratified [10], but in that case Squire's transformation also affects the acceleration due to gravity, replacing g by the larger quantity $(k^2/k_1^2)g$. This means that, to any unstable 3D mode, there corresponds a more unstable 2D mode *in a stronger gravitational field.* Therefore, unless the fluid is stably stratified, this result does not imply that the most unstable modes are necessarily two-dimensional.

2. **Classical Stability Results for Vortices in Ideal Fluids.** We now discuss our main topic, namely the stability of a family of axisymmetric stationary solutions to the three-dimensional Euler equations which describe steady vortex columns. For symmetry reasons, it is convenient to introduce cylindrical coordinates (r, θ, z), and to decompose the velocity of the fluid as $u = u_r e_r + u_\theta e_\theta + u_z e_z$, where e_r, e_θ, e_z are unit vectors in the radial, azimuthal, and vertical directions, respectively. Assuming that the fluid density is constant and equal to one, the Euler equations become

$$\begin{aligned} \partial_t u_r + (u\cdot\nabla)u_r - \frac{u_\theta^2}{r} &= -\partial_r p\,, \\ \partial_t u_\theta + (u\cdot\nabla)u_\theta + \frac{u_r u_\theta}{r} &= -\frac{1}{r}\partial_\theta p\,, \\ \partial_t u_z + (u\cdot\nabla)u_z &= -\partial_z p\,, \end{aligned} \tag{19}$$

where $u\cdot\nabla = u_r\partial_r + \frac{1}{r}u_\theta\partial_\theta + u_z\partial_z$. In addition, we have the incompressibility condition

$$\operatorname{div} u \equiv \frac{1}{r}\partial_r(ru_r) + \frac{1}{r}\partial_\theta u_\theta + \partial_z u_z = 0\,. \tag{20}$$

Columnar vortices are stationary solutions of (19), (20) of the form

$$u = V(r)\,e_\theta\,, \qquad p = P(r)\,, \tag{21}$$

where $V : \mathbb{R}_+ \to \mathbb{R}$ is an arbitrary velocity profile, and the associated pressure $P : \mathbb{R}_+ \to \mathbb{R}$ is determined, up to an irrelevant additive constant, by the centrifugal balance $rP'(r) = V(r)^2$. For the moment, we only assume that V is a piecewise differentiable function, and that the vortex (21) is localized in the sense that $V(r) \to 0$ as $r \to \infty$, but more restrictive assumptions will be added later. We introduce the angular velocity Ω and the vorticity W, which are defined as follows:

$$\Omega(r) = \frac{V(r)}{r}\,, \qquad W(r) = \frac{1}{r}\frac{\mathrm{d}}{\mathrm{d}r}(rV(r)) = r\Omega'(r) + 2\Omega(r)\,. \tag{22}$$

Without loss of generality, we normalize the vortex so that $\Omega(0) = 1$, hence $W(0) = 2$. Typical examples we have in mind are

- The *Rankine vortex*: $\Omega(r) = \begin{cases} 1 & \text{if } r \le 1\,, \\ r^{-2} & \text{if } r \ge 1\,, \end{cases} \qquad W(r) = \begin{cases} 2 & \text{if } r < 1\,, \\ 0 & \text{if } r > 1\,. \end{cases}$
- the *Lamb-Oseen vortex*: $\Omega(r) = \frac{1}{r^2}\Big(1 - e^{-r^2}\Big)\,, \qquad W(r) = 2\,e^{-r^2}\,.$
- the *Kaufmann-Scully vortex*: $\Omega(r) = \frac{1}{1+r^2}\,, \qquad W(r) = \frac{2}{(1+r^2)^2}\,.$

To study the stability of the vortex (21), we consider perturbed solutions of the form

$$u(r,\theta,z,t) \;=\; V(r)\,e_\theta + \tilde{u}(r,\theta,z,t)\,, \qquad p(r,\theta,z,t) \;=\; P(r) + \tilde{p}(r,\theta,z,t)\,.$$

This leads to the linearized evolution equations

$$\begin{aligned} \partial_t\tilde{u}_r + \Omega(r)\partial_\theta\tilde{u}_r - 2\Omega(r)\tilde{u}_\theta \;&=\; -\partial_r\tilde{p}\,, \\ \partial_t\tilde{u}_\theta + \Omega(r)\partial_\theta\tilde{u}_\theta + W(r)\tilde{u}_r \;&=\; -\frac{1}{r}\partial_\theta\tilde{p}\,, \\ \partial_t\tilde{u}_z + \Omega(r)\partial_\theta\tilde{u}_z \;&=\; -\partial_z\tilde{p}\,, \end{aligned} \tag{23}$$

where the pressure is determined so that the velocity perturbation remains divergence-free. Taking the divergence of both sides in (23), we obtain for $\tilde{p}$ the second order elliptic equation

$$-\,\partial_r^*\partial_r\tilde{p} - \frac{1}{r^2}\partial_\theta^2\tilde{p} - \partial_z^2\tilde{p} \;=\; 2\big(\partial_r^*\Omega\big)\partial_\theta\tilde{u}_r - 2\partial_r^*\big(\Omega\,\tilde{u}_\theta\big)\,, \tag{24}$$

where we introduced the shorthand notation $\partial_r^* = \partial_r + \frac{1}{r}$.

System (23) was first studied by Kelvin [37] for some particular velocity profiles. In [31], Rayleigh drew an interesting analogy between columnar vortices and shear flows in stratified fluids, on the basis of which he obtained a sufficient condition for stability with respect to axisymmetric perturbations. Further progress was made in the 20th century, notably by Howard and Gupta [21], and the state of the art is reviewed in textbooks on vortex dynamics [1, 29] or hydrodynamic stability [7, 11]. In this section we give a brief account of these classical developments, and we postpone the presentation of our own results to Section 3.

2.1. **Normal Mode Analysis.** As the coefficients in (23) only depend on the distance r to the symmetry axis, we can reduce the number of independent variables by using a Fourier series decomposition in the angular variable θ and a Fourier transform in the vertical coordinate z. Moreover, as in Sections 1.1–1.3, we focus our attention to the eigenvalues of the linearized operator. We thus consider velocities and pressures of the following form

$$\tilde{u}(r,\theta,z,t) \;=\; u(r)\,e^{i(m\theta+kz)}\,e^{st}\,, \qquad p(r,\theta,z,t) \;=\; p(r)\,e^{i(m\theta+kz)}\,e^{st}\,, \tag{25}$$

where $m \in \mathbb{Z}$ is the angular Fourier mode, $k \in \mathbb{R}$ is the vertical wave number, and $s \in \mathbb{C}$ is the spectral parameter. The velocity $u = (u_r, u_\theta, u_z)$ and the pressure p in (25) satisfy the ODE system

$$\gamma(r)u_r - 2\Omega(r)u_\theta \;=\; -\partial_r p\,, \quad \gamma(r)u_\theta + W(r)u_r \;=\; -\frac{im}{r}p\,, \quad \gamma(r)u_z \;=\; -ikp\,, \tag{26}$$

where $\gamma(r) = s + im\Omega(r)$ is the spectral function. The incompressibility condition becomes

$$\frac{1}{r}\partial_r(ru_r) + \frac{im}{r}u_\theta + iku_z \;=\; 0\,. \tag{27}$$

If $(m,k) \neq (0,0)$ it is possible to reduce the system (26), (27) to a scalar equation for the radial velocity u_r, by eliminating the pressure p and the velocity components u_θ, u_z, see [11, Section 15] or [17, Section 2]. After straightforward calculations, we obtain the second order ODE

$$-\,\partial_r\Big(\frac{r^2\partial_r^* u_r}{m^2+k^2r^2}\Big) + \Big\{1 + \frac{imr}{\gamma(r)}\partial_r\Big(\frac{W(r)}{m^2+k^2r^2}\Big) + \frac{1}{\gamma(r)^2}\frac{k^2r^2\Phi(r)}{m^2+k^2r^2}\Big\}u_r \;=\; 0\,, \tag{28}$$

where $\partial_r^* = \partial_r + \frac{1}{r}$ and $\Phi(r) = 2\Omega(r)W(r)$ is the Rayleigh function. This equation is well defined if $\gamma(r) \neq 0$ for all $r > 0$, which is the case if $\mathrm{Re}(s) \neq 0$ or, more generally, if $s \neq -imb$ for all b in the range of the angular velocity Ω. Eigenvalues of the linearized operator correspond to those values of the spectral parameter s for which equation (28) has a nontrivial solution u_r that is regular at the origin and decays to zero as $r \to \infty$.

It is instructive to notice that the stability equation (28) has a very similar structure as the Taylor-Goldstein equation (15). Both are second order Schrödinger equations involving a complex-valued potential which is a polynomial of degree two in the inverse spectral function $1/\gamma$. The coefficient of $1/\gamma(r)^2$ in (28) is proportional to the Rayleigh function Φ, and corresponds to the buoyancy term involving $-k^2 g\bar{\rho}'$ in (15). Similarly, the coefficient of $1/\gamma(r)$ in (28) is proportional to the vorticity W and its derivative, and corresponds to the inertial term involving $ik(\bar{\rho}U')'$ in (15). This analogy is grounded in deep physical reasons, which are explained in the pioneering work of Rayleigh [31]. It gives hope that the stability equation (28) can be analyzed using the techniques that were developed for shear flows, but we shall see that additional difficulties arise in the case of columnar vortices.

2.2. **Kelvin's Vibration Modes.** When the spectral parameter s is purely imaginary, the stability equation (28) has real-valued coefficients and can be studied using classical methods such as Sturm-Liouville theory. If $m \neq 0$, it is convenient to set $s = -imb$ for some $b \in \mathbb{R}$, so that $\gamma(r) = im(\Omega(r) - b)$. In that case, the equation satisfied by the radial velocity u_r becomes

$$-\partial_r\Big(\frac{r^2\partial_r^* u_r}{m^2+k^2r^2}\Big)+\Big\{1+\frac{r}{\Omega(r)-b}\partial_r\Big(\frac{W(r)}{m^2+k^2r^2}\Big)-\frac{1/m^2}{(\Omega(r)-b)^2}\frac{k^2r^2\Phi(r)}{m^2+k^2r^2}\Big\}u_r = 0\,. \tag{29}$$

This equation is well-posed if the spectral parameter b does not belong to the range of the angular velocity Ω, so that $\Omega(r) - b \neq 0$ for all $r > 0$.

In the particular case of Rankine's vortex, for which the vorticity distribution W is piecewise constant, Kelvin [37] observed that the stability equation can be explicitly solved in terms of modified Bessel functions in both regions $r < 1$ and $r > 1$. In the generic case where $k \neq 0$, matching conditions at $r = 1$ lead to the "dispersion relation"

$$\frac{I_m'(\beta)}{\beta I_m(\beta)} + \frac{2}{(1-b)\beta^2} = \frac{K_m'(k)}{kK_m(k)}\,, \qquad \text{where} \quad \beta^2 = k^2\Big(1 - \frac{4}{m^2(1-b)^2}\Big)\,. \tag{30}$$

Here I_m, K_m are modified Bessel functions of order m of the first and second kind, respectively. Those values of $b \neq 1$ for which (30) holds give purely imaginary eigenvalues of the linearized operator, which correspond to periodic oscillations of the columnar vortex. A careful analysis [37] reveals that the relation (30) is satisfied for a decreasing sequence $b_n \to 1$, and also for an increasing sequence $b_n' \to 1$, all solutions being contained in the interval $|b - 1| \leq 2/|m|$. So, for any $m \neq 0$ and $k \neq 0$, Kelvin established the existence of an infinite sequence of purely imaginary eigenvalues for the linearized operator at Rankine's vortex. He was confident that the whole spectrum could be obtained in that way [37]:

> "All possible simple harmonic vibrations are thus found: and summation, after the manner of Fourier, for different values of $[m, k]$, with different amplitudes and different epochs, gives every possible motion, deviating infinitely little from the undisturbed motion in circular orbits."

Unfortunately, the above claim is not substantiated by any argument in Kelvin's paper. Nevertheless, in the case of Rankine's vortex, one can show that the linearized operator has no eigenvalue outside the imaginary axis, so that the whole spectrum can indeed be obtained as demonstrated by Kelvin, see [17, Section 6.2].

The situation is different for a vortex with smooth angular velocity profile, as is the case for the Lamb-Oseen or the Kaufmann-Scully vortex. Assuming that $\Omega(0) = 1$ and $\Omega'(r) < 0$ for $r > 0$, it can be proved that, if $m \neq 0$ and $k \neq 0$, there exists a decreasing sequence $b_n \to 1$ of values of the spectral parameter for which the eigenvalue equation (29) has a nontrivial solution satisfying the boundary conditions. Moreover, (29) may have a solution for a finite number of negative values of b [17, Section 3.2]. So we still have an infinite number of purely imaginary eigenvalues, but in addition to these Kelvin waves there is also *continuous spectrum* filling the interval where $0 \le b \le 1$. Note that, if $0 < b < 1$, the eigenvalue equation (29) has a singularity at $r = \bar{r} := \Omega^{-1}(b)$, which is referred to as a "critical layer" in the physical literature. The interested reader is referred to [6, 14, 22, 32] for a few recent contributions to the study of Kelvin waves.

2.3. **Axisymmetric or Two-Dimensional Perturbations.** From now on we concentrate on the spectrum of the linearized operator outside the imaginary axis. The stability equation (28) is difficult to analyze in general, but important insight can be obtained by considering some particular cases.

To begin with, we restrict our attention to *axisymmetric perturbations* for which $m = 0$. In that case, we have $\gamma(r) = s$ for all $r > 0$, so that (28) reduces to the simpler equation

$$-\partial_r \partial_r^* u_r + k^2\Big(1 + \frac{\Phi(r)}{s^2}\Big) u_r = 0\,, \qquad r > 0\,. \tag{31}$$

The analogy with the Rayleigh-Taylor equation (6) is striking, and we see that the Rayleigh function Φ in (31) plays the exact role of the buoyancy term $N^2 = -g\bar{\rho}'/\bar{\rho}$ in (6). Following the same approach as in Section 1.1, we conclude that, if Φ is everywhere nonnegative, equation (31) has no nontrivial solution satisfying the boundary conditions when $\mathrm{Re}(s) \neq 0$. Moreover, if $\Phi(r) < 0$ for some $r > 0$, Synge's argument [17, 36] shows that equation (31) has a nontrivial solution for a sequence of real eigenvalues $s_n \to 0$, so that the positivity of the Rayleigh function is a necessary and sufficient condition for stability in the axisymmetric case.

Remark 2.1. The analogy between columnar vortices and shear flows in stratified fluids was already noticed by Rayleigh [31], and can be roughly explained as follows. In a stratified fluid, exchanging the positions of two fluid particles located on the same vertical line results in a gain or a loss of potential energy, depending on whether the fluid density is decreasing or increasing upwards. The first situation is thus stable, and the second unstable. A similar effect occurs in vortices, even if the fluid is homogeneous, because the centrifugal force (which plays the role of gravity) varies as a function of the distance to the vortex center. It turns out that exchanging two fluid particles on the same radial line results in a gain or a loss of energy depending on the sign of the Rayleigh function Φ, and that a stable "stratification" corresponds to $\Phi \ge 0$.

We next consider *two-dimensional perturbations*, which correspond to $k = 0$. In that case, the stability equation (28) reduces to

$$-\partial_r(r^2\partial_r^* u_r) + m^2 u_r + \frac{imrW'(r)}{\gamma(r)}\, u_r = 0\,, \qquad r > 0\,. \tag{32}$$

Here we can make a comparison with the Rayleigh stability equation (10), and we see that the vorticity derivative W' in (32) plays the role of the second order derivative U'' in (10). Thus, proceeding as in Section 1.2, we conclude that, if W' does not change sign, equation (32) has no nontrivial solution satisfying the boundary condition if $\mathrm{Re}(s) \neq 0$. The monotonicity of the vorticity profile W is thus a sufficient condition for stability with respect to two-dimensional perturbations, but as in the case of shear flows this condition is not necessary in general (and no sharp stability criterion is known).

Remarks 2.2.
1. For any localized vortex, the monotonicity of the vorticity distribution W *implies* the positivity of the Rayleigh function Φ. Indeed, if W is monotone, then $W(r) \to 0$ as $r \to \infty$ (otherwise the vortex would not be localized), hence W does not change sign, and the reconstruction formula

$$\Omega(r) = \frac{1}{r^2}\int_0^r W(s)s\,\mathrm{d}s\,, \qquad r > 0\,, \tag{33}$$

shows that Ω has the same sign as W. Thus $\Phi = 2\Omega W \geq 0$.

2. In view of the previous remark, if we extrapolate the conclusions obtained in the particular cases considered above, one may be tempted to conjecture that a columnar vortex with monotone vorticity distribution W is (spectrally) stable for all values of the Fourier parameters m, k. That daring claim has not been proved or disproved so far, and it is good to keep in mind that, in the present state of affairs, there is no analog of Squire's theorem for columnar vortices. In other words, there is no argument indicating that the most unstable modes (if any) should always correspond to axisymmetric or two-dimensional perturbations.

2.4. **Howard Identities.** We assume henceforth that $\Phi(r) > 0$ and $W'(r) < 0$ for all $r > 0$, so that the vortex under consideration is stable with respect to axisymmetric or two-dimensional perturbations. Our goal is now to study the eigenvalue equation (28) in the general case where $m \neq 0$ and $k \neq 0$. It is convenient to write the spectral parameter as $s = m(a - ib)$, where $a, b \in \mathbb{R}$, so that

$$\gamma(r) = s + im\Omega(r) = im\gamma_\star(r)\,, \qquad \text{where} \quad \gamma_\star(r) = \Omega(r) - b - ia\,. \tag{34}$$

When $a \neq 0$, we have $\gamma_\star(r) \neq 0$ for all $r > 0$, and equation (28) can be written in the condensed form

$$-\partial_r\big(\mathcal{A}(r)\partial_r^* u_r\big) + \mathcal{B}(r)u_r = 0\,, \qquad r > 0\,, \tag{35}$$

where $\partial_r^* = \partial_r + \frac{1}{r}$ and

$$\mathcal{A}(r) = \frac{r^2}{m^2 + k^2r^2}\,, \qquad \mathcal{B}(r) = 1 + \frac{r}{\gamma_\star(r)}\partial_r\Big(\frac{W(r)}{m^2 + k^2r^2}\Big) - \frac{k^2}{m^2}\,\frac{\mathcal{A}(r)\Phi(r)}{\gamma_\star(r)^2}\,. \tag{36}$$

If we assume that (35) has a nontrivial solution that is regular at the origin and decays to zero at infinity, we can multiply both sides of by $r\bar{u}_r$ and integrate over

$\mathbb{R}_+$ to arrive at the identity

$$\int_0^\infty \Big(\mathcal{A}(r)|\partial_r^* u_r|^2 + \mathcal{B}(r)|u_r|^2\Big) r\,\mathrm{d}r \,=\, 0\,. \tag{37}$$

As $|\gamma_\star(r)| \ge |a| > 0$ for all $r > 0$, we deduce from (36) that

$$|1-\mathcal{B}(r)| \,\le\, \frac{C}{m^2}\Big(\frac{1}{|a|} + \frac{1}{|a|^2}\Big)\,, \qquad r > 0\,,$$

for some constant $C > 0$ depending only on the vorticity profile W. In particular, if we suppose that $|a| > M := \max(1, 2C)$, then $\mathrm{Re}\,\mathcal{B}(r) > 0$ for all $r > 0$, and taking the real part of (37) we obtain a contradiction. Thus equation (35) has no nontrivial solution if $|a| > M$. Similarly, if we take the imaginary part of (37) and use the definitions (34), (36), we obtain the relation

$$a\int_0^\infty \Big\{\frac{r}{a^2+(\Omega{-}b)^2}\partial_r\Big(\frac{W(r)}{m^2+k^2r^2}\Big) + \frac{2(b-\Omega(r))}{(a^2+(\Omega{-}b)^2)^2}\,\frac{k^2}{m^2}\,\mathcal{A}(r)\Phi(r)\Big\}|u_r|^2 r\,\mathrm{d}r \,=\, 0\,. \tag{38}$$

If $a \neq 0$, the integral in (38) must vanish. But the first term in the integrand is negative since $W'(r) < 0$, and the second one is negative too if we suppose that $b \le 0$, because $\Omega(r) > 0$ for all $r > 0$. Thus we conclude from (38) that (35) has no nontrivial solution if $a \neq 0$ and $b \le 0$, see Figure 1.

To obtain further information on the spectrum outside the imaginary axis, we proceed as in the case of the Taylor-Goldstein equation (15), which was analyzed in Section 1.3. Following Howard's approach [20, 21], we first consider the differential equation satisfied by the new function $w_r = u_r/\gamma_\star(r)$. Straightforward calculations that are reproduced in [17, Section 3.4] show that w_r satisfies

$$-\,\partial_r\Big(\gamma_\star(r)^2\mathcal{A}(r)\partial_r^* w_r\Big) + \mathcal{D}(r) w_r \,=\, 0\,, \qquad r > 0\,, \tag{39}$$

where

$$\mathcal{D}(r) \,=\, \gamma_\star(r)^2 + 2r\gamma_\star(r)\partial_r\Big(\frac{\Omega(r)}{m^2+k^2r^2}\Big) - \frac{k^2}{m^2}\,\mathcal{A}(r)\Phi(r)\,.$$

In particular, if we multiply (39) by $r\bar{w}_r$, integrate the result over $\mathbb{R}_+$, and take the imaginary part, we obtain the relation

$$2a\int_0^\infty \Big\{(b-\Omega(r))\Big(\mathcal{A}(r)|\partial_r^* w_r|^2 + |w_r|^2\Big) - r\partial_r\Big(\frac{\Omega(r)}{m^2+k^2r^2}\Big)|w_r|^2\Big\} r\,\mathrm{d}r \,=\, 0\,. \tag{40}$$

The second term in the integrand is positive, because $\Omega'(r) < 0$, and the first one is positive too if we assume that $b \ge 1$, so that $b - \Omega(r) > 0$ for all $r > 0$. We thus conclude from (40) that equation (39), hence also equation (35), has no nontrivial solution satisfying the boundary conditions if $a \neq 0$ and $b \ge 1$, see Figure 1.

Finally, we consider the function $v_r = u_r/\gamma_\star(r)^{1/2}$ which satisfies

$$-\,\partial_r\Big(\gamma_\star(r)\mathcal{A}(r)\partial_r^* v_r\Big) + \mathcal{E}(r) v_r \,=\, 0\,, \qquad r > 0\,, \tag{41}$$

where

$$\mathcal{E}(r) \,=\, \gamma_\star(r) + \frac{r}{2}\partial_r\Big(\frac{W(r)+2\Omega(r)}{m^2+k^2r^2}\Big) + \frac{1}{4}\frac{\Omega'(r)^2}{\gamma_\star(r)}\,\mathcal{A}(r) - \frac{k^2}{m^2}\,\frac{\mathcal{A}(r)\Phi(r)}{\gamma_\star(r)}\,.$$

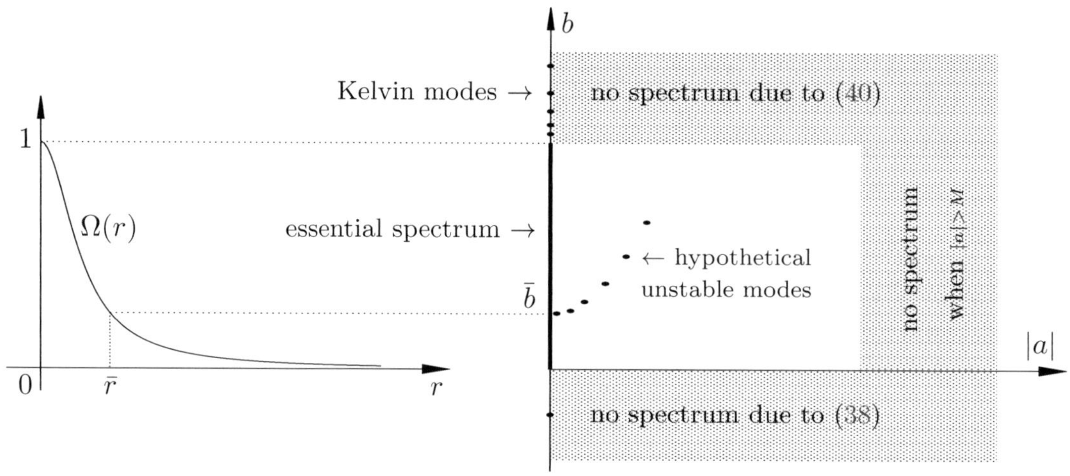

FIGURE 1. The information obtained so far on the spectrum of the linearized operator, in terms of the spectral parametrization $s = m(a - ib)$. Kelvin modes are located on the imaginary axis $a = 0$, and accumulate only at the upper edge of the essential spectrum, which fills the segment $a = 0$, $b \in [0, 1]$. The rest of the spectrum, if any, consists of isolated eigenvalues which are contained in the region $|a| \le M$, $b \in [0, 1]$, and can possibly accumulate only on the essential spectrum.

If we multiply (41) by $r\bar{v}_r$, integrate the result over $\mathbb{R}_+$, and take the imaginary part, we obtain the relation

$$-\,a \int_0^\infty \Big\{ \mathcal{A}(r)|\partial_r^* v_r|^2 + |v_r|^2 + \frac{\mathcal{A}(r)}{a^2 + (\Omega - b)^2}\Big(\frac{k^2\Phi(r)}{m^2} - \frac{\Omega'(r)^2}{4}\Big)|v_r|^2\Big\} r\,\mathrm{d}r \;=\; 0\,, \tag{42}$$

which is analogous to identity (18). Introducing the "Richardson number"

$$\mathrm{Ri}(r) \;=\; \frac{k^2}{m^2}\,\frac{\Phi(r)}{\Omega'(r)^2}\,, \tag{43}$$

we deduce from (42) that equation (41), hence also equation (35), has no nontrivial solution satisfying the boundary conditions if $a \neq 0$ and $\mathrm{Ri}(r) \ge 1/4$ for all $r > 0$. Unfortunately, unlike for the Taylor-Goldstein equation, the Richardson number (43) depends on the Fourier parameters m, k, and it is obvious that the inequality $\mathrm{Ri}(r) \ge 1/4$ cannot hold for all values of m and k. So the above approach fails to give any stability criterion that would hold for arbitrary perturbations. The situation is plainly summarized by Howard and Gupta in [21]:

> "The overall conclusion of this consideration of the non-axisymmetric case is thus essentially negative: the methods used to derive the Richardson number and semicircle results in the axisymmetric case reproduce the known results of Rayleigh for two-dimensional perturbations and pure axial flow, but seem to give very little more. In fact the present situation with regard to non-axisymmetric perturbations seems to be very unsatisfactory from a theoretical point of view."

Remark 2.3. In the spirit of Howard's semi-circle law for shear flows [10], it is possible in the case of columnar vortices to locate the (hypothetical) unstable modes in a slightly more precise way than what is depicted in Fig. 1, see e.g. [12]. We do not comment further on that, because in the next section we give conditions on the vorticity profile which entirely preclude the existence of unstable eigenvalues.

3. **Spectral Stability of Inviscid Columnar Vortices.** In this section, we present the main results that were obtained recently in collaboration with D. Smets [17, 18]. We first state our precise assumptions on the unperturbed columnar vortex.

Assumption H1: *The vorticity profile* $W : \overline{\mathbb{R}}_+ \to \mathbb{R}_+$ *is a* $\mathcal{C}^2$ *function satisfying* $W'(0) = 0$, $W'(r) < 0$ *for all* $r > 0$, $rW'(r) \to 0$ *as* $r \to \infty$, *and*

$$\Gamma := \int_0^\infty W(r) r \,\mathrm{d}r < \infty \,. \tag{44}$$

The crucial point here is the monotonicity of the vorticity distribution W, which implies stability with respect to two-dimensional perturbations, see Section 2.3. We also suppose that $W(r) \to 0$ as $r \to \infty$ fast enough so that the integral in (44) converges; in other words, the *total circulation* of the vortex is finite. It follows in particular that $W(r) > 0$ for all $r > 0$, and the expression (33) of the angular velocity shows that $\Omega(r) > 0$ and $\Omega'(r) < 0$ for all $r > 0$. As a consequence, the Rayleigh function $\Phi = 2\Omega W$ is positive everywhere, which implies stability with respect to axisymmetric perturbations too.

Assumption H2: *The "Richardson function"* $J : \mathbb{R}_+ \to \mathbb{R}_+$ *defined by*

$$J(r) = \frac{\Phi(r)}{\Omega'(r)^2}\,, \qquad r > 0\,, \tag{45}$$

satisfies $J'(r) < 0$ *for all* $r > 0$ *and* $rJ'(r) \to 0$ *as* $r \to \infty$.

This second assumption is less natural, and probably only technical in nature. The quantity $J(r)$ appears in the definition of the "Richardson number" (43), which plays an important role in the stability analysis of columnar vortices. If, for some given value of the ratio $k^2/m^2 > 0$, the Richardson number (43) is not everywhere larger than $1/4$, assumption H2 implies the existence of a unique $r_* > 0$ such that $\mathrm{Ri}(r) > 1/4$ if $r < r_*$ (stable region) and $\mathrm{Ri}(r) < 1/4$ if $r > r_*$ (possibly unstable region). If we do not suppose that the function J is monotone, more regions have to be considered, which greatly complicates the analysis. The monotonicity of J is also essential to construct simple subsolutions of equation (29) for large r, see [17, Section 4.6]. On the positive side, we emphasize that assumptions H1 and H2 are satisfied in all classical examples, such as the Lamb-Oseen vortex or the Kaufmann-Scully vortex.

The following statement is our first main result.

Theorem 3.1. [17] *Under assumptions H1, H2, the columnar vortex with vorticity profile* W *is spectrally stable in the following sense. Given any* $m \in \mathbb{Z}$ *and any* $k \in \mathbb{R}$ *with* $(m, k) \neq (0, 0)$, *the stability equation* (28) *has no nontrivial solution* $u_r \in L^2(\mathbb{R}_+, r\,\mathrm{d}r)$ *if the spectral parameter* s *has a nonzero real part.*

Theorem 3.1 asserts that, under assumptions H1, H2, the linearized operator in (23) has no unstable eigenmode of the form (25) with $\mathrm{Re}(s) \neq 0$ and $u \in L^2(\mathbb{R}_+, r\,\mathrm{d}r)^3$. In some sense, this answers a long-standing question dating back to the pioneering contributions of Kelvin and Rayleigh. This rather optimistic view has to be tempered for at least two reasons: first, the status of assumption H2 is unclear, and it is conceivable that the conclusion of Theorem 3.1 holds under the sole hypothesis that the vorticity profile is monotone, although we do not know how to prove that. Next, the proof of Theorem 3.1 given in [17] is very indirect, and does not give much insight into the physical mechanisms leading to stability. Therefore, it is not clear if our approach can be applied to more complicated problems, such as the stability analysis of columnar vortices with nonzero axial flow.

As is explained in Section 2.4, if the angular Fourier mode m and the vertical wavenumber k are both nonzero, the historical approach to hydrodynamic stability based on integral identities such as (37) does not seem sufficient to preclude the existence of unstable eigenvalues in all regions of the complex plane, see Fig. 1. However, it is easy to verify that all unstable eigenvalues (if any) are simple, isolated, and depend continuously on the vortex profile W, which can be considered as an infinite-dimensional parameter in the differential equation (28). In addition, for the rescaled Kaufmann-Scully vortex

$$W_\epsilon(r) = \frac{2}{(1+\epsilon r^2)^2}\,, \qquad \text{where} \quad 0 < \epsilon \le \frac{4k^2}{m^2}\,, \tag{46}$$

a direct calculation shows that the Richardson number (43) satisfies $\mathrm{Ri}_\epsilon(r) \ge 1/4$ for all $r > 0$. By Howard and Gupta's result [21], it follows that the associated linearized operator has no unstable eigenvalue in the Fourier subspace indexed by m, k.

These observations suggest the following contradiction argument to prove Theorem 3.1. Assume that, for some vorticity profile W satisfying assumptions H1 and H2, the linearized operator in (23) has an unstable eigenmode of the form (25) for some $s \in \mathbb{C} \setminus i\mathbb{R}$ and some Fourier parameters $m \in \mathbb{N}$, $k \in \mathbb{R}$. We know from the results of Section 2.3 that both m and k are necessarily nonzero. The idea is now to perform a *continuous homotopy* $(W_t)_{t\in[0,1]}$ between the original profile $W_0 := W$ and the reference profile $W_1 := W_\epsilon$, where W_ϵ is defined in (46). For small t, the linearized operator associated with W_t has an unstable eigenvalue $s(t)$ which depends continuously on t and satisfies $s(0) = s$. But we also know that, for $t = 1$, the linearized operator associated with the reference profile W_ϵ has no unstable eigenvalue at all. Thus we logically conclude that there exists some $t_* \in (0, 1]$ such that the unstable eigenvalue $s(t)$ merges into the continuous spectrum on the imaginary axis at $t = t_*$. The core of our contradiction argument is the claim that, under assumptions H1 and H2, such a merger is actually impossible.

The way we actually arrive at a contradiction is not easily described in a few lines, and the interested reader is referred to [17, Section 4] for full details. If t_n is an increasing sequence converging to t_*, we denote $s_n = s(t_n) = m(a_n - ib_n)$, so that $a_n \to 0$ as $n \to \infty$ by construction. Also, extracting a subsequence if needed, we can assume that $b_n \to \bar b \in [0, 1]$ as $n \to \infty$, see Fig. 1. For simplicity, we suppose here that $0 < \bar b < 1$, but of course the limiting cases $\bar b = 0$ and $\bar b = 1$ are also treated in [17]. If u_r^n denotes the (suitably normalized) eigenfunction associated with the eigenvalue s_n and the vorticity profile W_{t_n}, it is straightforward to verify that u_r^n converges as $n \to \infty$ to a solution u_r of the limiting equation (29), where $b = \bar b$ and Ω, W, Φ denote the angular velocity, vorticity, and Rayleigh function of the vortex profile at the bifurcation point $t = t_*$. That equation has a singularity at the point $\bar r := \Omega^{-1}(\bar b)$, and it is crucial to study the behavior of u_r in the vicinity of $\bar r$ (this is what is referred to as a *critical layer analysis* in the physical literature). If $\mathrm{Ri}(\bar r) > 1/4$, it is relatively easy to obtain a contradiction from identity (42), because all main terms in the integrand are positive in that case. If $\mathrm{Ri}(\bar r) < 1/4$, a contradiction can be obtained by a careful study of the solutions of (29) near the singularity, and by the construction of appropriate subsolutions in the region where $r > \bar r$, see [17].

Remark 3.2. The argument we have just sketched requires that assumption H2 be satisfied by the interpolated profile W_t for all $t \in [0, t_*]$. For that reason, we cannot use a linear interpolation of the form $W_t = (1-t)W + tW_\epsilon$, because the class of vorticity profiles satisfying H2 is not a linear space nor even a convex set. Thus an additional technical difficulty in our proof is the necessity of constructing *ad hoc* interpolation and approximation schemes in the nonlinear class of profiles satisfying assumption H2, see [17, Section 6.4].

To state our second main result, we return to the linearized system (23) which we write in condensed form $\partial_t \tilde u = L\tilde u$. The linearized operator L is given by

$$L\tilde u \,=\, \begin{pmatrix} -\Omega\partial_\theta \tilde u_r + 2\Omega \tilde u_\theta - \partial_r P[\tilde u] \\ -\Omega\partial_\theta \tilde u_\theta - W\tilde u_r - \frac{1}{r}\partial_\theta P[\tilde u] \\ -\Omega\partial_\theta \tilde u_z - \partial_z P[\tilde u] \end{pmatrix}, \tag{47}$$

where $P[\tilde u]$ denotes the solution $\tilde p$ of elliptic equation (24). Our goal is to solve the linearized system in the Hilbert space

$$X \,=\, \Big\{ u = (u_r, u_\theta, u_z) \in L^2(\mathbb{R}^3)^3 \,\Big|\, \partial_r^* u_r + \frac{1}{r}\partial_\theta u_\theta + \partial_z u_z = 0 \Big\},$$

equipped with the standard L^2 norm.

Theorem 3.3. [18] *Assume that the vorticity profile W satisfies assumptions H1, H2. Then the linear operator L defined in (47) is the generator of a strongly continuous group $(e^{tL})_{t\in\mathbb{R}}$ of bounded linear operators in the energy space X. Moreover, for any $\epsilon > 0$, there exists a constant $C_\epsilon \ge 1$ such that*

$$\|e^{tL}\|_{X\to X} \,\le\, C_\epsilon\, e^{\epsilon|t|}, \qquad \text{for all } t \in \mathbb{R}. \tag{48}$$

Estimate (48) exactly means that the spectrum of the evolution operator e^{tL} in X is contained in the unit circle of the complex plane for all $t \in \mathbb{R}$. In that sense, Theorem 3.3 is arguably the strongest way of asserting that the columnar vortex with vorticity profile W is *spectrally stable.* In view of the Hille-Yosida theorem [13], it follows the spectrum of the generator L is entirely contained in the imaginary axis of the complex plane, and we have the following resolvent bound for any $a > 0$:

$$\sup\Big\{ \|(z-L)^{-1}\|_{X\to X} \,\Big|\, z \in \mathbb{C}\,,\ |\mathrm{Re}(z)| \ge a \Big\} \,<\, \infty\,. \tag{49}$$

In fact, since X is a Hilbert space, the Gearhart-Prüss theorem [13, Section V.1] asserts that the resolvent bound (49) is *equivalent* to the group estimate (48).

Let $L_{m,k}$ denote the restriction of the linearized operator L to the Fourier subspace indexed by the angular mode $m \in \mathbb{Z}$ and the vertical wave number $k \in \mathbb{R}$. To prove Theorem 3.3, we fix some spectral parameter $s \in \mathbb{C}$ with $\mathrm{Re}(s) = a \neq 0$ and we consider the resolvent equation $(s - L_{m,k})u = f$, which is equivalent to the system

$$\begin{aligned} \gamma(r)u_r - 2\Omega(r)u_\theta \,&=\, -\partial_r p + f_r\,, \\ \gamma(r)u_\theta + W(r)u_r \,&=\, -\tfrac{im}{r}p + f_\theta\,, \\ \gamma(r)u_z \,&=\, -ikp + f_z\,, \end{aligned} \tag{50}$$

where $\gamma(r) = s + im\Omega(r)$ and the pressure $p = P_{m,k}[u]$ is chosen so as to preserve the incompressibility condition (27). Our goal is to show that the solution of (50) satisfies $\|u\| \le C(a)\|f\|$, where $C(a)$ is a positive constant depending only on the spectral abscissa a; in particular, the resolvent estimate is uniform in the Fourier parameters m, k and in the spectral parameter s on the vertical line $\mathrm{Re}(s) = a$. Such a uniform bound is essentially equivalent to (49), hence also to (48) by the Gearhart-Prüss theorem.

If $(m, k) \neq (0, 0)$, the resolvent system (50) can be reduced to a scalar equation for the radial velocity u_r, which can then be studied using the same techniques as in Section 2.4. This provides resolvent estimates with *explicit constant* $C(a)$ is some regions of the parameter space, but that approach fails in other regions where we have to invoke a contradiction argument that relies on the conclusion of Theorem 3.1. Thus our proof is again non-constructive, and does not provide any explicit expression for the constant $C(a)$ in general. In particular, we do not know if $C(a) = \mathcal{O}(|a|^{-N})$ as $a \to 0$ for some $N \in \mathbb{N}$. Such an improved estimate would indicate that the norm of the group e^{tL} grows at most polynomially as $|t| \to \infty$.

4. **Conclusion and Perspectives.** The results of the previous section apply to a large family of columnar vortices, including all classical models in atmospheric flows and engineering applications [1, 34]. They provide the first rigorous proof of spectral stability allowing for general perturbations, without any particular symmetry. In this sense, they solve an important problem that was formulated as early as 1880 by Lord Kelvin in the pioneering work [37]. However, many interesting questions remain open:

- Is assumption H2 really necessary for the conclusion of Theorem 3.1 to hold? Can one find a different proof, that does not rely on a non-constructive contradiction argument?
- Can one strengthen the conclusion of Theorem 3.3 and show that the group norm $\|e^{tL}\|$ grows at most polynomially as $|t| \to \infty$?
- Is it possible to prove some spectral stability results for more general equilibria of the form $u = V(r)e_\theta + W(r)e_z$, which include a nonzero axial flow?
- Do our results give any useful information on the stability of columnar vortices in the slightly viscous case?

REFERENCES

[1] S. V. Alekseenko, P. A. Kuibin and V. L. Okulov, *Theory of Concentrated Vortices. An Introduction*, Springer, 2007.

[2] V. I. Arnold, Conditions for the nonlinear stability of the stationary plane curvilinear flows of an ideal fluid, *Dokl. Mat. Nauk.* **162** (1965), 773–777.

[3] V. I. Arnold, Sur la géométrie différentielle des groupes de Lie de dimension infinie et ses applications à l'hydrodynamique des fluides parfaits, *Ann. Inst. Fourier* **16** (1966), 319–361.

[4] P. G. Baines and H. Mitsudera, On the mechanism of shear flow instabilities, *J. Fluid Mechanics* **276** (1994), 327–342.

[5] C. Bardos, Y. Guo, and W. Strauss, Stable and unstable ideal plane flows, *Chin. Ann. Math. Ser. B* **23**, (2002), 149–164.

[6] P. Billant and F. Gallaire, Generalized Rayleigh criterion for non-axisymmetric centrifugal instabilities, *J. Fluid Mech.* **542** (2005), 365–379.

[7] S. Chandrasekhar, *Hydrodynamic and Hydromagnetic Stability*, Clarendon Press, Oxford, 1961.

[8] F. Charru, *Hydrodynamic Instabilities*, Cambridge University Press, 2011.

[9] O. Darrigol, Stability and instability in nineteenth-century fluid mechanics, *Revue d'histoire des mathématiques* **8** (2002), 5–65.

[10] P. Drazin and L. Howard, Hydrodynamic Stability of Parallel Flow of Inviscid Fluid, *Advances in Applied Mechanics* **9** (1966), 1–89.
[11] P. Drazin and W. Reid, *Hydrodynamic stability*, Cambridge Univ. Press, 1981.
[12] C. Eckart, Extension of Howard's circle theorem to adiabatic jets, *Physics of Fluids* **6** (1963), 1042–1047.
[13] K. J. Engel and R. Nagel, *One-Parameter Semigroups for Linear Evolution Equations*, Graduate Texts in Mathematics **194**, Springer, 1999.
[14] D. Fabre, D. Sipp, and L. Jacquin, Kelvin waves and the singular modes of the Lamb-Oseen vortex, *J. Fluid Mech.* **551** (2006), 235–274.
[15] R. Fjørtoft, Application of integral theorems in deriving criteria of stability for laminar flow and for the baroclinic circular vortex, *Geofysiske Publikasjoner* **17** (1950), 2–52.
[16] S. Friedlander, W. Strauss, and M. Vishik, Nonlinear instability in an ideal fluid, *Ann. Inst. Henri Poincaré Anal. Non Linéaire* **14** (1997), 187–209.
[17] Th. Gallay and D. Smets, Spectral stability of inviscid columnar vortices, `arXiv:1805.05064`.
[18] Th. Gallay and D. Smets, On the linear stability of vortex columns in the energy space, preprint `arXiv:1811.07584`.
[19] P. M. Harman, *The Natural Philosophy of James Clerk Maxwell*, Cambridge University Press, 2001.
[20] L. N. Howard, Note on a paper of John W. Miles, *J. Fluid Mech.* **10** (1961), 509–512.
[21] L. N. Howard and A. S. Gupta, On the hydrodynamic and hydromagnetic stability of swirling flows, *J. Fluid Mechanics* **14** (1962), 463–476.
[22] S. Le Dizès and L. Lacaze, An asymptotic description of vortex Kelvin modes, *J. Fluid Mech.* **542** (2005), 69–96.
[23] Chia Chiao Lin, *The Theory of Hydrodynamic Stability*, Cambridge Univ. Press, 1955.
[24] Zhiwu Lin, Instability of Some ideal plane flows, *SIAM J. Math. Anal.* **35** (2003), 318–356.
[25] Zhiwu Lin, Nonlinear instability of ideal plane flows, *Int. Math. Res. Not.* **2004** (2004), 2147–2178.
[26] A. M. Lyapunov, *The General Problem of the Stability of Motion*, Kharkov, 1892 (in Russian). French translation: *Ann. Fac. Sci. Toulouse (2)* **9** (1907), 203–474. English translation: *Annals of Mathematics Studies* **17**, Princeton University Press, 1947.
[27] J. W. Miles, On the stability of heterogeneous shear flows, *J. Fluid Mech.* **10** (1961), 496–508.
[28] C. Marchioro and M. Pulvirenti, *Mathematical Theory of Incompressible Nonviscous Fluids*, Applied mathematical sciences **96**, Springer, 1994.
[29] P. G. Saffman, *Vortex Dynamics*, Cambridge Univ. Press, 1992.
[30] Lord Rayleigh, On the stability, or instability, of certain fluid motions, *Proceedings of the London Mathematical Society* **11** (1880), 57–72.
[31] Lord Rayleigh, On the dynamics of revolving fluids, *Proceedings of the Royal Society A* **93** (1917), 148–154.
[32] A. Roy and G. Subramanian, Linearized oscillations of a vortex column: the singular eigenfunctions, *J. Fluid Mech.* **741** (2014), 404–460.
[33] P. J. Schmid and D. S. Henningson, *Stability and Transition in Shear Flows*, Applied mathematical sciences **142**, Springer, 2001.
[34] M. N. Strasser and R. P. Selvam, Selection of a realistic viscous vortex tangential velocity profile for computer simulation of vortex-structure interaction, *J. Arkansas Academy of Science* **69** (2015), 88–97.
[35] H. B. Squire, On the stability of three-dimensional disturbances of viscous flow between parallel walls, *Proc. Roy. Soc. A* **142** (1933), 621–628.
[36] J. L. Synge, The stability of heterogeneous liquids, *Trans. Roy. Soc. Canada* **27** (1933), 1–18.
[37] Sir W. Thomson (Lord Kelvin), Vibrations of a columnar vortex, *Proceedings of the Royal Society Edinburgh* **10** (1880), 443-456. *The London, Edinburgh and Dublin Philosophical Magazine and Journal of Science* **X** (1880), 153–168.
[38] M. Vishik and S. Friedlander, Nonlinear instability in two-dimensional ideal fluids: the case of a dominant eigenvalue, *Commun. Math. Phys.* **243** (2003), 261–273.
[39] V. Yudovich, *The Linearization Method in Hydrodynamical Stability Theory*, Translations of Mathematical Monographs **74**, AMS, Providence, 1989.

E-mail address: `Thierry.Gallay@univ-grenoble-alpes.fr`

ON THE EULER-POISSON SYSTEM

YAN GUO
Division of Applied Mathematics
Brown University
Providence, RI 02912, USA

ABSTRACT. The Euler-Poisson system is a classical example of hyperbolic balance laws arising in either the two-fluid theory in plasma physics, or in the study of a self-gravitating gaseous star. This paper surveys recent mathematical progresses in the PDE study of such an Euler-Poisson system with physical importance.

The Euler-Poisson system describes compressible fluid(s) interacting with a self-consistent potential. In the classical "two-fluid" model describing plasma dynamics for ion and electron fluids, their self-consistent electrostatic potential satisfies a Poisson equation according to the *repulsive* Coulomb interaction. On the other hand, in the classical model describing a self-gravitating gaseous star, its self-consistent gravitational potential also satisfies a Poisson equation according to the *attractive* gravitational force via Newton's law.

Even though there is no general well-posedness PDE theory for hyperbolic balance laws in 3D due to possible formation of shocks, there have been important progress in the PDE study of the Euler-Poisson system for both plasma and stellar models in recent years. In the case of Euler-Poisson system for a plasma, global in time smooth solutions have been constructed *without* shock formation, thanks to enhanced dispersive effects induced by the repulsive electrostatic interaction. This is in stark contrast to the pure Euler equations for a compressible neutral gas. For the Euler-Poisson system for describing a self-gravitating star, mathematical advances have been made in local well-posedness of free boundary of the edge of the star, stability and instability of the celebrated Lane-Emden stars, *global in time dynamics* of an expanding star, as well as examples of gravitational collapse of a star. In particular, *global in time dynamics of* an expanding gas ball for the pure Euler equations can also be constructed as a consequence. These exciting new developments and mathematical techniques have opened up new lines of research in the PDE study of hyperbolic balance laws in higher dimensions.

1. **Euler-Poisson System for Plasma.** A plasma is a collection of fast-moving charged particles. It is believed that more than 90% of the matter in the universe is in the form of plasma, from sparse intergalactic plasma, to the interior of stars to neon signs. In addition, understanding of the instability formation in plasma is one of the main challenges for nuclear fusion, in which charged particles are accelerated at high speed to create energy. At high temperature and velocity, ions and electrons in a plasma tend to become two separate fluids due to their different

Yan Guo is supported by NSF grant DMS-1810868.

physical properties (inertia, charge). One of the basic fluid models for describing plasma dynamics is the so-called "two-fluid" model, the Euler-Maxwell system in which two compressible ion and electron fluids interact with their own self-consistent electromagnetic field. Similar to the classical water wave problem in fluids, such a two-fluid theory is another origin of many dispersive equations such as KdV, KP, NLS and Zaharov equations. The Euler-Poisson system is the simplified model from the Euler-Maxwell system as the speed of light $c \to \infty$:

$$\begin{aligned} \partial_t n_e + \nabla \cdot (n_e u_e) &= 0, \\ n_e m_e[\partial_t u_e + u_e \cdot \nabla u_e] + \nabla p_e &= n_e e \nabla \phi, \\ \partial_t n_i + \nabla \cdot (n_i u_i) &= 0, \\ n_i M_i[\partial_t u_i + v_i \cdot \nabla u_i] + \nabla p_i &= -Z n_i e \nabla \phi, \\ -\Delta \phi &= 4\pi e(Z n_i - n_e). \end{aligned} \tag{1}$$

This system describes a plasma composed of a compressible electron gas and a compressible ion gas. The electrons has charge $-e$, density n_e, mass m_e, velocity v_e and pressure $p_e(n_e)$, and the ions have change Ze, density n_i, mass M_i, velocity v_i, and pressure $p_i(n_i)$. These two fluids interact through the self-consistent electric flied $E = -\nabla\phi$ through the Poisson equation. For notational simplicity, we set the integer $Z = 1$.

1.1. **Electron Fluid with Constant Ion Background.** The two-fluid Euler-Poisson system (1) has a rich and complex dynamics with several distinct physical parameters. In particular, it is well-known that the ratio of the electron mass and the ion mass,

$$\frac{m_e}{M_i} \backsim \frac{1}{2000} \ll 1,$$

which can be regarded as a small parameter in a plasma. Perhaps the most simplified model of (1) is to describe an electron fluid dynamics in an ion background (Langmuir waves):

$$\begin{aligned} \partial_t n_e + \nabla \cdot (n_e u_e) &= 0, \\ n_e m_e[\partial_t u_e + u_e \cdot \nabla u_e] + \nabla p_e &= n_e e \nabla \phi, \\ -\Delta \phi &= 4\pi e(n_0 - n_e). \end{aligned} \tag{2}$$

Thanks to the fact of $\frac{m_e}{M_i} \ll 1$, the much heavier ions are treated as motionless with a constant density $n_i(t,x) \equiv n_0$, and only form a fixed charged background en_0. Such a simplified system (2) is used for describing Langmuir waves (electron waves) in the two-fluid theory. Thanks to the quasi-linear hyperbolic nature, as expected, shock waves do develop for 'large' perturbations of the constant state equilibrium of $n_e \equiv n_0, v_e \equiv 0$ (see [23]) in (2). On the other hand, in [17], the following result is established

Theorem 1.1. *[17] Assume* $\rho_0(x) \in C_c^\infty(\mathbf{R}^3, \mathbf{R})$ *and* $v_0(x) \in C_c^\infty(\mathbf{R}^3, \mathbf{R}^3)$ *such that*

$$\nabla \times v_0(x) \equiv 0 \text{ (irrotationality)}, \qquad \int_{\mathbf{R}^3} \rho_0 = 0 \text{ (neutrality)}.$$

Then there exists $\varepsilon_0 > 0$, *such that for* $0 < \varepsilon < \varepsilon_0$, *there exists a unique global smooth solution* $(n^\varepsilon(t,x), v^\varepsilon(t,x))$ *with initial condition* $(n_0 + \varepsilon\rho_0, \varepsilon v_0)$. *Moreover,* $n^\varepsilon(t,x) - n_0$ *and* $v^\varepsilon(t,x)$ *decay uniformly at any rate of* t^{-q} *for* $1 < q < \frac{3}{2}$.

There is a distinct feature of solutions to hyperbolic conservation laws in 3D: The linearized acoustic (wave) equations for irrotational compressible Euler flows are given by

$$\partial_{tt}n - \frac{p'(n_0)}{m_e}\Delta n = 0, \qquad (3)$$
$$\partial_{tt}u - \frac{p'(n_0)}{m_e}\Delta u = 0$$

whose solutions enjoy a $\frac{1}{t}$ decay rate due to dispersion. Intuitively, such a decay rate of $\frac{1}{t}$ just barely fails to be integrable in time, which can be viewed as the obstruction for persistence of smooth solutions with small amplitude. Indeed, in a remarkable classical result [44], it is shown that shock formation for compressible Euler flows as small perturbation of the equilibrium $n(t,x) \equiv n_0$ and $u(t,x) \equiv 0$.

In a stark contrast, Theorem 1.1 asserts absence of shock formation for irrotational electron flows with small amplitude to the Euler-Poisson system (2). The assumption of irrotationality is necessary to ensure time decay of the flows. This is in stark contrast to the classical result for a neutral compressible gas [44]. The key difference lies in the linearization of (2) around $n(t,x) \equiv n_0$ and $u(t,x) \equiv 0$ for irrotational flows:

$$\partial_{tt}n - \frac{p'(n_0)}{m_e}\Delta n + \frac{4\pi e^2 n_0}{m_e} n = 0, \qquad (4)$$
$$\partial_{tt}u - \frac{p'(n_0)}{m_e}\Delta u + \frac{4\pi e^2 n_0}{m_e} u = 0.$$

In comparison with the pure Euler equations for a neutral gas (3), new zeroth order terms of (n,u) arises due to electrostatic interaction, and $\frac{4\pi e^2 n_0}{m_e} = \omega_p^2$ is the plasma frequency. Such plasma oscillations create a new 'Klein-Gordon' effect in (4), which enhances linear decay rate to $t^{-\frac{3}{2}}$ (integrable) from the non-integrable decay rate of t^{-1} for (3).

To illustrate the mathematical background, consider a semi-linear Klein-Gordon equation

$$(\partial_{tt} - \Delta + 1)\alpha = f(\alpha). \qquad (5)$$

Recall the linear decay $L^\infty - L^1$ estimate in 3D:

$$||\alpha(t,\cdot)||_\infty \lesssim \frac{1}{t^{3/2}}||\alpha(0,\cdot)||_{W^{4,1}} \qquad (6)$$

for solution to the linear homogeneous Klein-Gordon equation

$$(\partial_{tt} - \Delta + 1)\alpha = 0$$

with $\alpha_t|_{t=0} = 0$. The goal is to show the high energy norm $\mathcal{E}(t) \equiv ||\alpha_t||^2_{H^k} + ||\nabla\alpha||^2_{H^k} + ||\alpha||^2_{H^k}$ $(k \gg 1)$ is bounded uniformly in time for small α for the full nonlinear Klein-Gordon equation (5). Classical energy estimate yields

$$\mathcal{E}(t) \lesssim \text{good } + \int_0^t ||\partial f(\alpha)||_2 \times ||\partial\alpha_t||_2$$
$$\lesssim \text{good } + \int_0^t ||\alpha||_{W^{k/2,\infty}} ||\partial\alpha_t||_2^2.$$

Clearly, $\sup_{0\le t\le\infty}\mathcal{E}(t)$ is bounded for small α if

$$\int_0^t ||\alpha(\tau,\cdot)||_{W^{k/2,\infty}}d\tau \lesssim \sqrt{\mathcal{E}}. \tag{7}$$

Despite a strong linear decay of $t^{-3/2}$ in (6), we remark that (7) is far from obvious, due to the mismatch between L^1 based $W^{k,1}$ norm in (6) and L^2 based energy norm $\mathcal{E}(t)$. In fact, bootstrap of the linear decay estimate (6) with Duhamel principle leads to the nonlinear decay estimate:

$$||\alpha(t,\cdot)||_{W^{k/2,\infty}} \lesssim \text{good } + \int_0^t \frac{1}{1+[t-\tau]^{3/2}}||f(\alpha(\tau))||_{W^{k/2+4,1}}d\tau.$$

If $f(\alpha)=\alpha^3$, then for $k\gg 1$,

$$||\alpha^3||_{W^{k/2+4,1}} \lesssim ||\alpha||_{W^{k/2,\infty}}||\alpha^2||_{W^{k,1}} \lesssim ||\alpha||_{W^{k/2,\infty}}\mathcal{E}. \tag{8}$$

Hence (7) is valid and global smooth solutions can be constructed.

If $f(\alpha)=\alpha^2$, then

$$||\alpha^2||_{W^{k/2+4,1}} \lesssim ||\alpha||_{W^{k/2,\infty}}||\alpha||_{W^{k,1}} \backsim \mathcal{E}$$

since $||\alpha||_{W^{k,1}}$ can not relate directly to $||\alpha||_{W^{k,2}}$ in $\mathcal{E}$.

We remark a quadratic nonlinearity is generic for small data problems. Fortunately, such a fundamental difficulty for quadratic nonlinearity can be overcome by either Klainerman's vector-field method or Shatah's normal form method, introduced in two seminal papers [34] and [43] respectively.

It is necessary to apply Shatah's normal form method in [43] to control nonlocal Poisson operator Δ^{-1} to construct global smooth solutions to (2). Theorem 1.1 demonstrates that even though there is no dissipation or relaxation effects in (2), stronger dispersive effects from the repulsive electrostatic interaction can still prevent shock formations.

There have been recent extensions of Theorem 1.1. In [25], [28], [29] and [38], global smooth irrotational flows with are constructed in 2D independently for (2). Furthermore, In [19], despite a meager $t^{-\frac{1}{2}}$ linear decay rate, global smooth flows are constructed in 1D for (2), where *nonlinear solution tends to a linear solution with phase shift*. We note that for 1D compressible flows have been studied extensively by researchers working in hyperbolic conservation/balance laws. However, it remains an outstanding question to construct global unique BV solutions to (2).

1.2. **Ion Fluid with Boltzmann Statistics.** Assume $p_e(n_e)=T_e n_e$ in (2) with a constant temperature T_e. By formally taking $\frac{m_e}{M_e}\to 0$, we deduce $\nabla p_e = T_e\nabla n_e = n_e e\nabla\phi$, or the celebrated Boltzmann statistics (relation) for the electron density (with constant n_0):

$$n_e = n_0\exp(\frac{e\phi}{T_e}).$$

We then obtain the reduced Euler-Poisson system for ion dynamics

$$\begin{aligned}\partial_t n_i + \nabla\cdot(n_i u_i) &= 0,\\ n_i M_i[\partial_t u_i + v_i\cdot\nabla u_i] + \nabla p_i &= -n_i e\nabla\phi,\\ -\Delta\phi &= 4\pi e(n_i - n_0\exp(\frac{e\phi}{T_e})).\end{aligned} \tag{9}$$

The nonlinear Poisson equation for the electric potential ϕ presents a new mathematical challenge to prevent shock formation near $n_i(t,x)\equiv n_0$, and $v_i(t,x)\equiv 0$.

Theorem 1.2. *[21] Assume $\rho_0(x) \in C_c^\infty(\mathbf{R}^3, \mathbf{R})$ and $v_0(x) \in C_c^\infty(\mathbf{R}^3, \mathbf{R}^3)$ such that*

$$\nabla \times v_0(x) \equiv 0 \text{ (irrotationality).}$$

Then there exists $\varepsilon_0 > 0$, such that for $0 < \varepsilon < \varepsilon_0$, there exists a unique global smooth solution $(n^\varepsilon(t,x), v^\varepsilon(t,x))$ with initial condition $(n_0 + \varepsilon\rho_0, \varepsilon v_0)$. Moreover, $n^\varepsilon(t,x) - n_0$ and $v^\varepsilon(t,x)$ decay uniformly at the rate of $t^{-\frac{16}{15}}$.

It is convenient to rewrite the system (9) as a nonlinear equation for a complex-valued scalar unknown $U_i(t,x)$ for the perturbation as

$$\{\partial_t + i\Lambda_i\}U_i = Q(U_i, U_i)$$

where Q is a quadratic nonlinearity (higher order perturbations are being ignored). Here the pseudo-differential operator Λ_i is defined as

$$\Lambda_i(|\xi|) = |\xi|\sqrt{\frac{2 + |\xi|^2}{1 + |\xi|^2}}. \tag{10}$$

The key is to study [16] the *profile in the phase space*

$$V_i = e^{i\Lambda_i t}\hat{U}_i$$

which satisfies

$$V_i(t,\xi) = \int_0^t \int_{\mathbf{R}^3} e^{is\Phi(\xi,\eta)} m(\xi,\eta) V_i(s, \xi - \eta) V_j(s,\eta) d\eta ds, \tag{11}$$

where $m(\xi,\eta)$ represents the multiplier from the quadratic nonlinearity Q via a convolution in the Fourier space, and the important interacting phase function is

$$\Phi(\xi,\eta) = \Lambda_i(\xi) \pm \Lambda_i(\xi - \eta) \pm \Lambda_i(\eta).$$

We remark that *if* the nonlinear dynamics behaves like the linear one, then the profile V_i should remain more or less stationary in time. The control of $\int_0^t$ in (11) therefore must come from oscillatory behavior of the interacting phase $\Phi(\xi,\eta)$ with its interaction with nonlinearity presented as multiplier $m(\xi,\eta)$. For the Klein-Gordon equation (5) with $\Lambda(\xi) = \sqrt{1 + |\xi|^2}$, $\Phi \gtrsim 1$, an integration by part in s yields

$$\begin{aligned} V_i(t,\xi) &\backsim -\int_0^t \int \frac{e^{is\Phi(\xi,\eta)}}{i\Phi(\xi,\eta)} m(\xi,\eta) V_i(s,\xi-\eta)\partial_s V_i(s,\eta) d\eta ds \\ &= -\int_0^t \int \frac{e^{is\Phi(\xi,\eta)}}{i\Phi(\xi,\eta)} m(\xi,\eta) m(\eta,\zeta) V_i(s,\xi-\eta) V_i(s,\eta-\zeta) V_i(s,\zeta) d\eta d\zeta ds \end{aligned}$$

by plugging $\partial_s V_i$ again from (11). This transform the original quadratic nonlinearity to a cubic nonlinearity as in (8), precisely Shatah's normal form transformation [43] to construct global solutions.

For (9), (10) leads to slower decay rate of $t^{-\frac{4}{3}}$ for the linearized flows, and $\Phi(\xi,\eta)$ may vanish at $|\xi| = 0$ and $|\eta| = 0$ as well:

$$\Phi(\xi,\eta) \gtrsim |\xi||\xi - \eta||\eta|,$$

A careful study of the multiplier of $m(\xi,\eta)m(\eta,\zeta) \backsim |\xi||\eta|$ is needed to control $\frac{m(\xi,\eta)m(\eta,\zeta)}{\Phi(\xi,\eta)}$ and to construct global smooth solutions.

It remains an outstanding question to determine if shock waves can develop for (9) in 2D.

1.3. **Full Euler-Poisson System.** Finally, a recent work considers the full Euler-Poisson system (1) in [18] and [20]:

Theorem 1.3. *Assume $\rho_e(x), \rho_i(x) \in C_c^\infty(\mathbf{R}^3, \mathbf{R})$ and $v_e(x), v_i(x) \in C_c^\infty(\mathbf{R}^3, \mathbf{R}^3)$ such that*

$$\nabla \times v_e(x) \equiv \nabla \times v_i(x) \equiv 0 \ (irrotationality).$$

Assume that

$$\frac{m_e}{M_e} \le 10^{-3} \text{ and } 1 \le \frac{p_e'(n_0)}{p_i'(n_0)} \le 100$$

Then there exists $\varepsilon_0 > 0$, such that for $0 < \varepsilon < \varepsilon_0$, there exists a unique global smooth solution $(n_e^\varepsilon(t,x), n_i^\varepsilon(t,x), v_e^\varepsilon(t,x), v_i^\varepsilon(t,x))$ with initial condition $(n_0 + \varepsilon\rho_e, n_0 + \varepsilon\rho_i, \varepsilon v_e, \varepsilon v_i)$. Moreover, $n_e^\varepsilon(t,x) - n_0$, $n_i^\varepsilon(t,x) - n_0$, $v_e^\varepsilon(t,x)$ and $v_i^\varepsilon(t,x)$ decay uniformly at any rate of $t^{-1-\frac{1}{200}}$.

Together with a similar result for the Euler-Maxwell system in [18], Theorem 1.3 entails a distinct dispersive feature of the two-fluid model: thanks to the self-consistent electromagnetic interaction, smooth irrotational flows with small amplitude will persist forever, without any possible shock formation.

To illustrate the method of the construction, it is convenient to rewrite (1) in terms of two complex-valued functions U_e and U_i as

$$\{\partial_t + i\Lambda_e\}U_e = Q_e(U_e, U_i)$$
$$\{\partial_t + i\Lambda_i\}U_i = Q_i(U_e, U_i)$$

(cubic nonlinearity being ignored). Here with $r :\equiv \frac{m_e}{M_e}$ and $T :\equiv \frac{p_e'(n_0)}{p_i'(n_0)}$,

$$\Lambda_e = r^{-1/2}\sqrt{\frac{(1+r) - (T+r)\Delta + \sqrt{((1-r) - (T-r)\Delta)^2 + 4r}}{2}},$$

$$\Lambda_i = r^{-1/2}\sqrt{\frac{(1+r) - (T+r)\Delta - \sqrt{((1-r) - (T-r)\Delta)^2 + 4r}}{2}}.$$

Schematically, as in (11), we may express part of V_i as

$$V_i(t,\xi) \backsim \int_0^t \int_{\mathbf{R}^3} e^{is\Phi(\xi,\eta)} m(\xi,\eta) V_i(s, \xi-\eta) V_e(s,\eta) d\eta ds, \tag{12}$$

with

$$\Phi = \Lambda_e(\xi) \pm \Lambda_i(\xi - \eta) \pm \Lambda_e(\eta).$$

As in (11), if $\Phi \neq 0$, then an integration by part in s will produce third order nonlinearity which is manageable as in (8). While if $\partial_\eta \Phi \neq 0$, it is possible to perform a different integration by parts in η via

$$e^{is\Phi(\xi,\eta)} = \frac{1}{is\partial_\eta \Phi}\partial_\eta e^{is\Phi(\xi,\eta)},$$

which produces decay of s to better control the time integral $\int_0^t$. Correspondingly, the most difficult set to control is when both $\Phi = \partial_\eta \Phi = 0$. Such a singular set

$$S = \{(\xi,\eta) : \Phi = \partial_\eta \Phi = 0\}$$

is called the *space-time resonance* set. Unfortunately, for (1), its space-resonance set S is not only non-empty, but actually contains a 2D sphere. Based on techniques introduced in [26], a systematic and robust method is developed in [18] to analyze and control such a singular 2D sphere, which leads to the construction of global smooth irrotational flows with small amplitude for (1).

1.4. **Reference for EP for Plasma.** [1] S. Alinhac, Temps de vie des solutions régulières des équations d'Euler compressibles axisymétriques en dimension deux, Invent. Math. 111 (1993), 627-670.
[2] J. A. Bittencourt, Fundamentals of Plasma Physics, 3rd ed., Springer-Verlag, New York, 2004. Zbl 1084.76001.
[3] G.-Q. Chen, J. W. Jerome, and D. Wang, Compressible Euler-Maxwell equations, in Proceedings of the Fifth International Workshop on Mathematical Aspects of Fluid and Plasma Dynamics (Maui, HI, 1998), 29 Transport Theory Statist. Phys. no. 3-5, 2000, pp. 311-331.
[4] D. Christodoulou, Global solutions of nonlinear hyperbolic equations for small initial data, Comm. Pure Appl. Math. 39 (1986), 267-282.
[5] D. Christodoulou, The formation of shocks in 3-dimensional fluids, in Recent Advances in Nonlinear Partial Differential Equations and Applications, Proc. Sympos. Appl. Math. 65, Amer. Math. Soc., Providence, RI, 2007, pp. 17-30.
[6] D. Christodoulou and S. Klainerman, The Global Nonlinear Stability of the Minkowski Space, Princeton Math. Ser. 41, Princeton Univ. Press, Princeton, NJ, 1993.
[7] S. Cordier and E. Grenier, Quasineutral limit of an Euler-Poisson system arising from plasma physics, Comm. Partial Differential Equations 25 (2000), 1099-1113.
[8] P. Degond, F. Deluzet, and D. Savelief, Numerical approximation of the Euler-Maxwell model in the quasineutral limit, J. Comput. Phys. 231 (2012), 1917-1946.
[9] J.-L. Delcroix and A. Bers, Physique des plasmas, InterEditions/CNRS Editions, Paris, 1994.
[10] J.-M. Delort and D. Fang, Almost global existence for solutions of semilinear Klein-Gordon equations with small weakly decaying Cauchy data, Comm. Partial Differential Equations 25 (2000), 2119-2169.
[11] J.-M. Delort, D. Fang, and R. Xue, Global existence of small solutions for quadratic quasilinear Klein-Gordon systems in two space dimensions, J. Fund. Anal. 211 (2004), 288-323.
[12] D. Gérard- Varet, D. Han-Kwan, and F. Rousset, Quasineutral limit of the Euler-Poisson system for ions in a domain with boundaries, Indiana Univ. Math. J. 62 (2013), 359-402.
[13] P. Germain, Global existence for coupled Klein-Gordon equations with different speeds, Ann. Inst. Fourier (Grenoble) 61 (2011), 2463-2506 (2012).
[14] P. Germain and N. Masmoudi, Global existence for the Euler-Maxwell system, Ann. Sei. Éc. Norm. Super. 47 (2014), 469-503.
[15] P. Germain, N. Masmoudi, and B. Pausader, Nonneutral global solutions for the electron Euler-Poisson system in three dimensions, SIAM J. Math. Anal. 45 (2013), 267-278.
[16] P. Germain, N. Masmoudi, and J. Shatah, Global solutions for 3D quadratic Schrödinger equations, Int. Math. Res. Not. 2009 (2009), 414-432.
[17] Y. Guo, Smooth irrotational flows in the large to the Euler-Poisson system in $\mathbf{R}^{3+1}$, Comm. Math. Phys. 195 (1998), 249-265.
[18] Y. Guo, A. Ionescu, B. Pausader, Global solutions of the Euler-Maxwell two-fluid system in 3D, Annals of Math. (2), 183 (2016), no. 2, 377{498.

[19] Y. Guo, L. Han, J. Zhang, Absence of shocks for one dimensional Euler-Poisson system, Arch. Ration. Mech. Anal. 223 (2017), no. 3, 1057-1121.
[20] Y. Guo, A. D. Ionescu, and B. Pausader, Global solutions of certain plasma fluid models in three-dimension, J. Math. Phys. 55 (2014), no. 12, 123102, 26 pp.
[21] Y. Guo and B. Pausader, Global smooth ion dynamics in the Euler-Poisson system, Comm. Math. Phys. 303 (2011), 89-125.
[22] Y. Guo and X. Pu, KdV limit of the Euler-Poisson system, KdV limit of the Euler-Poisson system, Arch. Ration. Mech. Anal. 211 (2014), 673-710.
[23] Y. Guo and A. S. Tahvildar-Zadeh, Formation of singularities in relativists fluid dynamics and in spherically symmetric plasma dynamics, in Nonlinear Partial Differential Equations (Evanston, IL, 1998), Contemp. Math. 238, Amer. Math. Soc., Providence, RI, 1999, pp. 151-161.
[24] S. Gustafson, K. Nakanishi, and T.-P. Tsai, Scattering theory for the Gross-Pitaevskii equation in three dimensions, Commun. Contemp. Math. 11 (2009), 657-707.
[25] A. D. Ionescu and B. Pausader, The Euler-Poisson system in 2D: global stability of the constant equilibrium solution, Int. Math. Res. Not. 2013 (2013), 761-826.
[26] A. D. Ionescu and B. Pausader, Global solutions of quasilinear systems of Klein-Gordon equations in 3D, J. Eur. Math. Soc. (JEMS) 16 (2014), 2355-2431.
[27] A. D. Ionescu and F. Pusateri, Global solutions for the gravity water waves system in 2d, Invent. Math. 199 (2015), 653-804. MR 3314514. Zbl 06418043.
[28] J. Jang, The two-dimensional Euler-Poisson system with spherical symmetry, J. Math. Phys. 53 (2012), 023701, 4.
[29] J. Jang, D. Li, and X. Zhang, Smooth global solutions for the two-dimensional Euler Poisson system, Forum Math. 26 (2014), 645-701.
[30] F. John, Blow-up of solutions of nonlinear wave equations in three space dimensions, Manuscripta Math. 28 (1979), 235-268.
[31] F. John and S. Klainerman, Almost global existence to nonlinear wave equations in three space dimensions, Comm. Pure Appl. Math. 37 (1984), 443-455.
[32] T. Kato, The Cauchy problem for quasi-linear symmetric hyperbolic systems, Arch. Rational Mech. Anal. 58 (1975), 181-205.
[33] S. Klainerman, Long time behaviour of solutions to nonlinear wave equations, in Proceedings of the International Congress of Mathematicians, Vol. 1, 2 (Warsaw, 1983), PWN, Warsaw, 1984, pp. 1209-1215.
[34] S. Klainerman, Global existence of small amplitude solutions to nonlinear Klein-Gordon equations in four space-time dimensions, Comm. Pure Appl. Math.
[35] S. Klainerman, Uniform decay estimates and the Lorentz invariance of the classical wave equation, Comm. Pure Appl. Math. 38 (1985)
[36] S. Klainerman, The null condition and global existence to nonlinear wave equations, in Nonlinear Systems of Partial Differential Equations in Applied Mathematics, Part 1 (Santa Fe, N.M., 1984), Lectures in Appl. Math. 23, Amer. Math. Soc., Providence, RI, 1986, pp. 293-326.
[37] D. Lannes, F. Linares, and J.-C. Saut, The Cauchy problem for the Euler-Poisson system and derivation of the Zakharov-Kuznetsov equation, in Studies in Phase Space Analysis with Applications to PDEs, Progr. Nonlinear Differential Equations Appl. 84, Springer-Verlag, New York, 2013, pp. 181-213.
[38] D. Ll and Y. Wu, The Cauchy problem for the two dimensional Euler-Poisson system, J. Eur. Math. Soc. (JEMS) 16 (2014), 2211-2266.

[39] H. Lindblad and I. Rodnianski, The weak null condition for Einstein's equations, C. R. Math. Acad. Sei. Paris 336 (2003), 901-906.
[40] H. Lindblad and I. Rodnianski, The global stability of Minkowski space time in harmonic gauge, .Arm. of Math. 171 (2010), 1401-1477.
[41] Y.-J. Peng, Global existence and long-time behavior of smooth solutions of two-fluid Euler-Maxwell equations, Ann. Inst. H. Poincaré Anal. Non Linéaire 29 (2012), 737-759.
[42] X. Pu, Dispersive limit of the Euler-Poisson system in higher dimensions, SIAM J. Math. Anal. 45 (2013), 834-878.
[43] J. Shatah, Normal forms and quadratic non-linear Klein-Gordon equations, Comm. Pure Appl. Math. 38 (1985), 685-696.
[44] T. C. Sideris, Formation of singularities in three-dimensional compressible fluids, Comm. Math. Phys. 101 (1985), 475-485.
[45] J. C. H. Simon, A wave operator for a non-linear Klein-Gordon equation, Lett. Math. Phys. 7 (1983), 387-398.
[46] B. Texier, Derivation of the Zakharov equations, Arch. Ration. Mech. Anal. 184 (2007), 121-183.

2. **Gravitational Euler-Poisson.** The Euler-Poisson system for describing a self-gravitating star takes the form of

$$\begin{aligned} &\partial_t \rho + \nabla \cdot (\rho u) = 0, \\ &\rho[\partial_t u + u \cdot \nabla u] + \nabla p = -\rho \nabla \phi, \qquad \Delta \phi = 4\pi \rho. \end{aligned} \tag{13}$$

Here $\rho, u, p(\rho)$ and ϕ denote the density, velocity, pressure and the gravitational respectively. In contrast to (1), there is an opposite sign in the Poisson equation for ϕ, thanks to the attractive Newtonian gravitational force instead of the previous repulsive Coulomb interaction in a plasma. We assume the pressure law ($\gamma > 1$):

$$p(\rho) = \rho^\gamma$$

It is important to note that, for a star with compact support, the density ρ, as well as the pressure $p(\rho)$, is usually positive inside, and vanish outside as well as at the surface of the star.

It is convenient to formulate such a free-boundary value problem by an alternative Lagrangian formulation, following the framework developed in [33]. Given a velocity field u, define the flow map η, from the reference domain of a unit ball $\Omega \to \Omega(s)$, to be the characteristic ODE for particle paths:

$$\begin{aligned} \partial_s \eta(s, y) &= u(s, \eta(s, y)) \\ \eta(0, y) &= \eta_0(y) \end{aligned} \tag{14}$$

Define the Jacobian matrix of η as

$$J :\equiv \det \nabla \eta.$$

The Abel Lemma implies that

$$\partial_s J = J \nabla \cdot u$$

along the characteristic curve (14). It then follows from the continuity equation for density ρ that along the particle path (14),

$$\frac{d}{ds}(\rho J) \equiv 0,$$
$$\rho J|_{(t,\eta(t,y))} = \rho J|_{(0,\eta(0,y)} \equiv: w^{\frac{1}{\gamma-1}}$$

for $\gamma > 1$. It is clear that the continuity equation, or ρ can be eliminated from (13), and it suffices to solve for the flow map η from the rest of (13).

We further assume *radial symmetry* with

$$\eta(s,y) = \chi(s,|y|)y,$$

where $\chi(s,|y|)$ is a scalar function. It follows in this case, the crucial quantity Jacobian J takes the form of

$$J[\chi] \equiv \det \nabla\eta = \chi^2(\chi + r\partial_r\chi), \quad r = |y|. \tag{15}$$

We note that if $\chi \equiv 1$ then the flow map $\eta \equiv y$, the identity map. Hence, $\chi - 1$ is the deviation from the identity map. In [33], (13) can be reduced to a scalar equation for χ

$$\chi_{tt} + \frac{G(r)}{\chi^2} + P[\chi] = 0, \tag{16}$$

Here, by (15), the pressure operator $P[\chi]$ is given by

$$P[\chi] \equiv: \frac{\chi^2}{w^{\frac{1}{\gamma-1}}r^2}(r\partial_r)(w^{\frac{\gamma}{\gamma-1}}J[\chi]^{-\gamma}),$$

while the gravitational force is given by $\frac{G(r)}{\chi^2}$, where

$$G(r) \equiv: \frac{1}{r^3}\int_0^r 4\pi w^{\frac{1}{\gamma-1}}(s)s^2 ds \tag{17}$$

and $\int_0^r 4\pi w^{\frac{1}{\gamma-1}}(s)s^2 ds$ is the total 'mass' of the 'initial density' $w^{\frac{1}{\gamma-1}}$ within a ball of radius r. We note that a typical profile of $w(r)$ satisfies

$$w > 0 \text{ and } w|_{r=1} = 0,$$

with $r = 1$ being the boundary of the star, giving rise to the free boundary motion via the flow map $\chi(t,|y|)y$. Physically, $w(r) \backsim 1 - r$ near $r = 1$, it is not smooth across $r = 1$.

We remark that (16) is a quasi-linear wave equations for χ with a degenerate weight $w(r)$ which vanishes at $r = 1$.

2.1. **Stability of Lane-Emden Stars.** The celebrated Lane-Emden star configurations are solutions $w(r)$ to the steady Euler-Poisson system (16) with $\chi \equiv 1$ (identity map):

$$\frac{4\pi}{r^3}\int_0^r w^{\frac{1}{\gamma-1}}(s)s^2 ds + \frac{1}{r}\partial_r(w^{\frac{\gamma}{\gamma-1}}) = 0,$$

or equivalently, the following Lane-Emden equation

$$w_{rr} + \frac{2}{r}w_r + \frac{4\pi(\gamma-1)}{\gamma}w^{\frac{\gamma}{\gamma-1}} = 0. \tag{18}$$

We note that $w_{rr} + \frac{2}{r}w_r = \Delta w$ for 3D radial functions. It is well-known that for $\frac{6}{5} < \gamma < 2$, there exists solution $w(r)$, or Lane-Emden stars, such that $w > 0$ for

$0 \leq r \leq 1$ and $w(1) = 0$. While for $\gamma = \frac{6}{5}$, there is $w(r) > 0$ for $0 \leq r < \infty$ with a finite mass.

One of the important question is the dynamical stability of these Lane-Emden stars. It has been long conjectured, via a scaling argument, that they are stable for $\gamma > \frac{4}{3}$, while unstable for $\frac{6}{5} \leq \gamma < \frac{4}{3}$. In [47], stability for $\gamma > \frac{4}{3}$ is established via a variational method for global weak solutions. In [32], nonlinear instability is established for the case $\gamma = \frac{6}{5}$.

In a recent work [33], the following result is established:

Theorem 2.1. *[33] Assume $\frac{6}{5} < \gamma < \frac{4}{3}$ for the Lane-Emden star in (18). Then there exits a $\theta > 0$, for any $\delta \ll 1$, there exist $T_\delta > 0$ and initial perturbations*

$$\{\chi^\delta - 1\}|_{t=0} = O(\delta), \qquad \partial_t \chi^\delta|_{t=0} = O(\delta),$$

but

$$\sup_{0 \leq t \leq T^\delta} |||[\chi^\delta - 1, \partial_t \chi^\delta]|||_{Z_0^\alpha} \geq \theta,$$

where the norm Z_0^α consists of a weighted H^1 norm of $\chi^\delta - 1$ and a weighted L^2 norm for $\partial_t \chi^\delta$.

The proof of this theorem is based on cumulative efforts and machineries developed in [35],[36] to study of a body of compressible Euler flows surrounded by vacuum. A general PDE framework with high-order weighted Sobolev estimates via Hardy's inequality near $r = 1$ is developed in [33] to study of dynamics for a self-gravitation star described by (13).

2.2. **Global Expanding Stars.** For constants $\delta \in [\delta^*, \infty]$ with some $\delta^* < 0$, let $w = w_\delta$ be the solution to the following *generalized* Lane-Emden equation:

$$w_{rr} + \frac{2}{r} w_r + \frac{4\pi(\gamma - 1)}{\gamma} w^{\frac{\gamma}{\gamma-1}} = -\frac{3}{4}\delta$$

For $\gamma = \frac{4}{3}$, in [23], [24] and [42], the following two families of expanding homogeneous solutions to (13) are discovered:

1) $\frac{2}{3}$ *Expanding Solutions*: For $\delta^* \leq \delta < 0$:

$$\chi_{\frac{2}{3}}(t) = (\lambda_0^{\frac{3}{2}} + \frac{3}{2}\lambda_0^{\frac{1}{2}}\lambda_1 t)^{\frac{2}{3}},$$

where (λ_0, λ_1) satisfying $(\lambda_1 > 0)$:

$$(\lambda_1^2 + \frac{2\delta}{\lambda_0}) \int_0^1 2\pi w^3 s^4 ds = 0.$$

2. *Linear Expanding Solutions*: For $\delta > 0$,

$$\chi_1 = \lambda(t)$$
$$\lambda^2 \lambda_{tt} = \delta, \qquad \lim_{t \to \infty} \lambda_t = \text{constant}.$$

Theorem 2.2. *[28] Assume $\gamma = \frac{4}{3}$, χ_1 is asymptotically stable; $\chi_{\frac{2}{3}}$ is asymptotically stable for perturbations with zero total energy.*

Due to the attractive nature of the gravitational interaction in (13), it is expected that the Euler-Poisson system for a gaseous star is even more difficult than the pure Euler equations for a neutral gas. Therefore, construction of any global in time solutions to (13) should be even more challenging.

It is striking that such an asymptotic stability result in fact leads to global well-posedness in high Sobolev norms for (13). A subtle and surprising damping effect due to these background expanding solutions $\chi_{\frac{2}{3}}$ and χ_1 is carefully captured and capitalized in a re-scaled equation, which ensures the global control of the high Sobolev norms.

Such an observation and new mathematical techniques have led to constructions of more general global expanding solutions, based on the work of [53], for Euler-Poisson system for both plasmas (2) and a gaseous star (13) in [30]. More importantly, they have led to the ground breaking construction of expanding solutions to the Euler equations in 3D [29].

2.3. **Gravitational Collapse.** The gravitational collapse of a star is one of the important and fascinating problems in astrophysics. When $\gamma = \frac{4}{3}$, self-similar blowup solutions have been constructed in [21], [24] and [42], and absence of blowup solutions is proven in [20] for $\gamma > \frac{4}{3}$.

Recall $\eta(t,y) = \chi(t,r)y$, $\rho(t,\chi(t,r)y) \equiv w^{\frac{1}{\gamma-1}}(r)J^{-1}(s,r)$ so that

$$\lim \rho(t,\chi(s,r)y) = \infty$$

if and only if the Jacobian

$$\lim J[\chi] = \lim \chi^2(\chi + \partial_r\chi) = 0.$$

In case $\chi + \partial_r\chi > 0$, then density ρ blow up if and only if $\chi = 0$ (shell focusing) for which the particle goes to the origin $r = 0$.

In [26], gravitational collapse is constructed based on the following *dust* (pressureless) model. Set $P[\chi] \equiv 0$ in (16) to obtain ODE for χ with given G in (17)

$$\chi_{tt} + \frac{G(r)}{\chi^2} = 0. \tag{19}$$

A typical solution collapsing solution takes the form of

$$\chi_{\text{dust}} = [1 - g(r)t]^{\frac{2}{3}}, \tag{20}$$

$$g(r) \equiv 3\sqrt{\frac{G(r)}{2}} = 3\sqrt{\frac{\text{mean density}}{2}}.$$

We assume $w(r)$ is decreasing ($w' < 0$), which implies that $g(r)$ is also decreasing. Clearly $\chi_{\text{dust}} \equiv 0$ along space-time blowup curve:

$$\Gamma := \{(t,r)|\ t = \frac{1}{g(r)}\}$$

which is increasing in r with the first blowup time $t_{\min}$:

$$t_{\min} = \frac{1}{g(0)} \text{ for } r = 0,\ t_{\max} = \frac{1}{g(1)} \text{ for } r = 1.$$

We note that $J[\chi_{\text{dust}}] = 0$ iff $1 - g(r)t = 0$ for $g' \le 0$:

$$J[\chi_{\text{dust}}] = (1 - g(r)t)^2 \left(1 - \frac{2}{3}\frac{trg'(r)}{1 - g(r)t}\right).$$

From $\rho_{\text{dust}}(t, \chi_{\text{dust}}(t,r)y) = w^{\frac{1}{\gamma-1}}(r)J[\chi_{\text{dust}}]^{-1}$, we deduce that

$$\lim_{t \to \frac{1}{g(r)}} \chi_{\text{dust}}(t,r) = 0, \tag{21}$$

$$\lim_{t \to \frac{1}{g(r)}} \rho_{\text{dust}}(t, \chi_{\text{dust}}(t,r)y) = \infty \tag{22}$$

Moreover, the remaining mass $M(t)$ of the star decreases continuously to zero for $t \in [t_{\min}, t_{\max}] = [\frac{1}{g(0)}, \frac{1}{g(1)}]$:

$$M(t) = 4\pi \int_{g^{-1}\circ(\frac{1}{t})}^{1} w^{\frac{1}{\gamma-1}}(\bar{r})\bar{r}^2 d\bar{r} = \int_0^{\chi_{\text{dust}}(t,1)} 4\pi\rho_{\text{dust}}(tZ)Z^2 dZ,$$

$$M(t_{\max}) = 0 \text{ (final collapse time).} \tag{23}$$

In [26], it is shown that such a pressureless collapse χ_{dust} 'persists' even in the presence of the pressure for $\gamma < \frac{4}{3}$:

Theorem 2.3. *[26] Let $\gamma < \frac{4}{3}$. Assume w smooth in $[0,1)$, decreasing, and $w(r) = w(0) - O(1)r^n$ for $r << 1$, for some $n = n(\gamma)$ sufficiently large. Then there exists a collapsing solution χ to (16) which behaves as a scaled dust solution of χ_{dust} with the same properties (21), (22) and (23).*

Setting $s = \bar{\varepsilon}^{-3/2}t$, $y = \bar{\varepsilon}^{-1}x$, thanks to $\gamma < \frac{4}{3}$, $\varepsilon \equiv: \bar{\varepsilon}^{4-3\gamma}$, we may rewrite (16) as

$$\chi_{ss} + \frac{G(r)}{\chi^2} + \varepsilon P[\chi] = 0,$$

where the pressure $\varepsilon P[\chi]$ is treated as a (unbounded!) small perturbation to the dust problem (19). To maintain the zero of $J[\chi_{\text{dust}}]$, it is important to seek a formal expansion

$$\phi = \phi_0 + \varepsilon\phi_1 + \varepsilon^2\phi_2 + ...$$

as a series solution to (16). Equating coefficient of ε^j, ϕ_j can be solved via the linearized ODE of (19), with

$$\phi_0 = \chi_{\text{dust}}$$

as expected. The key observation is that, for $\gamma < \frac{4}{3}$, there is a repeated gain of power of χ_{dust} as

$$|\phi_j| \lesssim_j |\chi_{\text{dust}}|^{1+j\delta}$$

for some $\delta > 0$. It thus follows that for $\varepsilon \ll 1$, any finite order approximation behaves like χ_{dust}

$$\phi_{\text{app}} = \phi_0 + \varepsilon\phi_1 + \varepsilon^2\phi_2 + .. + \varepsilon^j\phi_j \backsim \chi_{\text{dust}}$$
$$J[\phi_{\text{app}}] \backsim J[\chi_{\text{dust}}].$$

Furthermore, thanks $\delta > 0$, for $j \gg 1$, it is reasonable to seek a true solution

$$\chi = \phi_{\text{app}} + \frac{\chi_{\text{dust}}^m}{r} H.$$

Here the high order remainder H should satisfy a nonlinear PDE from (16), which can be controlled via energy estimates for $m \gg 1$.

2.4. **Reference of EP for Gaseous Stars.** [1] J. Binney, S. Tremaine., Galactic Dynamics. Princeton University Press, Princeton, 2008.
[2] Blottiau, P., Bouquet, S., Chi'eze, J. P., An asymptotic self-similar solution for the gravitational collapse. Astron. Astrophys. 207, (1988) 24–36.
[3] O.I Bogoyavlensky, Methods in the Qualitative Theory of Dynamical Systems in Astrophysics and Gas Dynamics, Springer, 1985.
[4] Borisov, A. V., Kilin A. A., Mamaev, I. S., The Hamiltonian dynamics of self-gravitating liquid and gas ellipsoids Regular and Chaotic Dynamics 14(2), (2008) 179–217.

[5] Bouquet, S., Feix, M. R., Fijalkow, E., Munier, A., Density bifurcation in a homogeneous isotropic collapsing star. The Astrophysical Journal 293 (1985), 494–503.
[6] Brenner, M. P., Witelski T. P., On Spherically Symmetric Gravitational Collapse. J. Stat. Phys., 93, 3-4 (1998) 863–899.
[7] Calvez, V., Carrillo, J. A., Hoffmann, F., Equilibria of homogeneous functionals in the faircompetition regime. Nonlinear Analysis TMA 159, (2017) 85–128.
[8] Calvez, V., Carrillo, J. A., Hoffmann, F., The geometry of diffusing and self-attracting particles in a one-dimensional fair-competition regime, Lecture Notes in Mathematics 2186, CIME Foundation Subseries, Springer, 2018.
[9] Carrillo, J. A., Wr ´oblewska-Kaminska, A., Zatorska, E., On long-time asymptotics for viscous hydrodynamic models of collective behavior with damping and nonlocal interactions. Preprint, arXiv:1709.09290.
[10] S. Chandrasekhar, An Introduction to the Study of Stellar Structures. Dover Publications, INC, 1967.
[11] Cheng, B., Tadmor, E., An improved local blow-up condition for Euler-Poisson equations with attractive forcing Physica D 238, (2009) 2062–2066.
[12] Chiodaroli, E., De Lellis, C., Kreml, O., Global ill-posedness of the isentropic system of gas dynamics. Comm. Pure Appl. Math. 68, no. 7, (2015) 1157–1190.
[13] Christodoulou, D., Violation of cosmic censorship in the gravitational collapse of a dust cloud. Comm. Math. Phys. 93, (1984) 171–195.
[14] D. Christodoulou, The Formation of Shocks in 3-Dimensional Fluids. EMS Monographs in Mathematics, EMS Publishing House 2007.
[15] D. Christodoulou, S. Miao, Compressible Flow and Euler's Equations, Surveys in Modern Mathematics, Volume 9, International Press, 1–602, 2014.
[16] Coutand, D., Shkoller, S., Well-posedness in smooth function spaces for the moving boundary three-dimensional compressible Euler equations in physical vacuum. Arch. Ration. Mech. Anal. 206, no. 2, (2012) 515–616.
[17] Dafermos, M., Holzegel, G., Rodnianski, I., The linear stability of the Schwarzschild solution to gravitational perturbations. Preprint, arXiv:1601.06467.
[18] Dafermos, M., Rodnianski, I., Shlapentokh-Rothman, Y., Decay for solutions of the wave equation on Kerr exterior spacetimes III: the full subextremal case |a| < M. Ann. of Math., 183 (2016), 787–913.
[19] De Lellis, C., Szkelyhidi Jr. L., High dimensionality and h-principle in PDE. Bull. Amer. Math. Soc., 54, no 2., (2017) 247–282.
[20] Deng, Y., Liu, T.P., Yang, T., Yao Z., Solutions of Euler-Poisson equations for gaseous stars. Arch. Ration. Mech. Anal. 164, no. 3, (2002) 261–285.
[21] Deng, Y., Xiang, J., Yang, T., Blowup phenomena of solutions to Euler-Poisson equations. J. Math. Anal. Appl. 286 (2003), 295–306.
[22] Dyson F. J., Dynamics of a Spinning Gas Cloud. J. Math. Mech., 18, no. 1, (1968) 91–101.
[23] Fu, C.-C.; Lin, S.-S., On the critical mass of the collapse of a gaseous star in spherically symmetric and isentropic motion. Japan J. Indust. Appl. Math., 15, no. 3 (1998) 461–469.
[24] Goldreich, P., Weber, S., Homologously collapsing stellar cores. Astrophys. J., 238 (1980), 991–997.
[25] Gu, X., Lei, Z., Local well-posedness of the three dimensional compressible Euler–Poisson equations with physical vacuum. Journal de Mathmatiques Pures et Appliques, 105, 5 (2016), 662–723.

[26] Guo, Y., Hadzic, M., Jang, J., Continued gravitational collapse for Newtonian stars. Preprint 2018: arXiv:1811.01616
[27] Gurka, P., Opic, B., Continuous and compact imbeddings of weighted Sobolev spaces. II. Czechoslovak Math. J., 39, (1989) 78–94.
[28] Hadzic, M., Jang, J., Nonlinear stability of expanding star solutions in the radially-symmetric mass-critical Euler-Poisson system. Comm. Pure Appl. Math., 71, no. 5, (2018) 827–891.
[29] Hadzic, M., Jang, J., Expanding large global solutions of the equations of compressible fluid mechanics. Invent. Math., 1–62. DOI 10.1007/s00222-018-0821-1 (2018).
[30] Hadzic, M., Jang, J., A class of global solutions to the Euler-Poisson system. Available on ArXiv at: https://arxiv.org/abs/1712.00124.
[31] Holzegel, G., Luk, J., Speck, J., Wong, W., Stable shock formation for nearly simple outgoing plane symmetric waves. Annals of PDE, 2, no. 2, (2016) 1–198.
[32] Jang, J., Nonlinear Instability in Gravitational Euler-Poisson system for $\gamma = \frac{6}{5}$. Arch. Ration. Mech. Anal. 188 (2008), 265–307.
[33] Jang, J., Nonlinear Instability Theory of Lane-Emden stars. Comm. Pure Appl. Math. 67 (2014), no. 9, 1418–1465.
[34] Jang, J., Time periodic approximations of the Euler-Poisson system near Lane-Emden stars. Anal. PDE, 9, no. 5 (2016) 1043–1078.
[35] Jang, J., Masmoudi, N., Well-posedness for compressible Euler equations with physical vacuum singularity, Comm. Pure Appl. Math. 62 (2009), 1327–1385.
[36] Jang, J., Masmoudi, N., Well-posedness of compressible Euler equations in a physical vacuum. Comm. Pure Appl. Math. 68, no. 1, (2015), 61–111.
[37] Johnson, W. P., The Curious History of Faa di Bruno's Formula. The American Mathematical Monthly, 109, (2002) 217–234.
[38] Larson, R. B., Numerical calculations of the dynamics of a collapsing protostar. Monthly Notices Roy. Astron. Soc. 145 (1969), 271–295.
[39] Luk, J., Speck, J., Shock formation in solutions to the 2D compressible Euler equations in the presence of non-zero vorticity. To appear in Invent. Math., arXiv:1610.00737.
[40] Luo, T., Smoller, J., Existence and Nonlinear Stability of Rotating Star Solutions of the Compressible Euler-Poisson Equations. Arch. Ration. Mech. Anal. 191, 3, (2009) 447–496.
[41] Luo, T., Xin, Z., Zeng, H., Well-posedness for the motion of physical vacuum of the threedimensional compressible Euler equations with or without self-gravitation. Arch. Ration. Mech. Anal. 213, no. 3, (2014) 763–831.
[42] Makino, T., Blowing up solutions of the Euler-Poisson equation for the evolution of gaseous stars. Transport Theory Statist. Phys. 21 (1992), 615–624.
[43] Makino, T., Perthame, B., Sur les Solutions a Symetrie Spherique de l'Equation d'Euler-Poisson pour l'Evolution d'Etoiles Gazeuses. Japan J. Appl. Math. 7 (1990), 165–170.
[44] Oppenheimer J. R., Snyder H., On continued gravitational contraction. Phys. Rev., 56, (1939) 455–459.
[45] Ovsiannikov, L. V., New solution of hydrodynamic equations. Dokl. Akad. Nauk SSSR, Vol lll, Nl, (1956) 47–49.
[46] Penston, M. V., Dynamics of self-gravitating gaseous spheres-III. Analytical results in the free-fall of isothermal cases. Monthly Notices Roy. Astron. Soc. 144 (1969), 425–448.

[47] Rein, G., Non-linear stability of gaseous stars. Arch. Ration. Mech. Anal. 168, no. 2, (2003) 115–130.

[48] Ringeval, C., Bouquet, S., Dynamical stability for the gravitational evolution of a homogeneous polytrope. Astron. Astrophys. 355 (2000), 564–572.

[49] Shkoller, S., Sideris, T. C., Global existence of near-affine solutions to the compressible Euler equations. Preprint, arXiv:1710.08368.

[50] Shu, F. H., Self-similar collapse of isothermal spheres and star formation. Astrophys. J., Part 1, 214, (1977), 488–497.

[51] Sideris, T. C., Formation of singularities in three-dimensional compressible fluids. Comm. Math. Phys. 101 (1985), 475–485.

[52] Sideris, T. C., The lifespan of 3D compressible flow. Centre de Mathematiques, Ecole Polytechnique. Expose No. V (1991).

[53] Sideris, T. C., Spreading of the free boundary of an ideal fluid in a vacuum. J. Differential Equations, 257(1), (2014) 1–14.

[54] Sideris, T. C., Global existence and asymptotic behavior of affine motion of 3D ideal fluids surrounded by vacuum. Arch. Ration. Mech. Anal. 225, no. 1, (2017) 141–176.

[55] J. Speck, Shock formation in small-data solutions to 3D quasilinear wave equations. Mathematical Surveys and Monographs (AMS), (2016), 1–515.

[56] Speck, J., A summary of some new results on the formation of shocks in the presence of vorticity. To appear in Harvard CMSA Series in Math.

[57] Taylor, M., The nonlinear stability of the Schwarzschild family of black holes (joint with M. Dafermos, G. Holzegel, I. Rodnianski). Oberwolfach Reports, Workshop ID 1832, Mathematical General Relativity (2018).

[58] Tohline, J. E., Hydrodynamic Collapse. Fundamentals of Cosmic Physics, 8, (1982) 1–82.

[59] Tolman, R. C., Effect of inhomogeneity on cosmological models. Proc. Nat. Acad. Sci. U. S. 20, (1934) 169–176.

[60] Yahil, A., Self-similar stellar collapse. Astrophys. J. 265 (1983), 1047–1055.

[61] Zel'dovich, Y. B., Novikov, I. D., Relativistic Astrophysics Vol. 1: Stars and Relativity. Chicago, University Press, Chicago, 1971.

E-mail address: Yan_Guo@brown.edu

THE HYDRODYNAMIC LIMIT OF THE BOLTZMANN EQUATION FOR RIEMANN SOLUTIONS

FEIMIN HUANG*

Institute of Applied Mathematics, AMSS, CAS
Beijing 100190, China

ABSTRACT. In this note, I will survey my recent works [16] on the hydydrodynamic limit of the Boltzmann equation in the setting of Riemann solution that contains the generic superposition of shock, rarefaction wave and contact discontinuity to the Euler equations..

The Boltzmann equation, as the fundamental equation in statistical mechanics, reads

$$f_t + \xi \cdot \nabla_X f = \frac{1}{\varepsilon} Q(f, f),$$

where $f(t, X, \xi)$ is the density distribution of particles at time t with location X and velocity ξ. The Knudsen number $\varepsilon > 0$ is proportional to the mean free path of the interacting particles.

It is well-known that the Boltzmann equation is closely related to the systems of fluid dynamics, in particular, the systems of Euler and Navier-Stokes equations. In fact, the first derivation of the fluid dynamical components and systems from the kinetic equations can be traced back to the dates of Maxwell and Boltzmann. Hilbert proposed a systematic expansion in 1912, and Enskog and Chapman independently proposed another expansion in 1916 and 1917 respectively.

Either the Hilbert expansion or Chapman-Enskog expansion yields the compressible Euler equations in the leading order with respect to the Knudsen number ε, and the compressible Navier-Stokes equations, Burnett equations in the subsequent orders. To justify these formal approximations in rigorous mathematics, that is, hydrodynamic limits, has been proved to be extremely challenging and most remains open, in part because the basic well-posedness and regularity questions are still mostly unsolved for these fluid equations. The justification of the fluid limits of the Boltzmann equation is also related to the Hilbert's sixth problem.

In this note, I will outline my recent works [16] on the limiting process of the Boltzmann equation to the system of the compressible Euler equations in the setting of Riemann solutions. The Riemann problem was first formulated and studied by Riemann in 1860s when he studied the one space dimensional isentropic gas dynamics with initial data being two constant states. The solution to this problem

---

2000 *Mathematics Subject Classification.* Primary: 35Q35, 35B65 Secondary: 76N10.

Key words and phrases. Hydrodynamic limit, Boltzmann equation, Euler equations, Riemann solution.

The first author is supported by National Center for Mathematics and Interdisciplinary Sciences, AMSS, CAS and NSFC Grant No. 11371349 and 11688101.

* Corresponding author: Feimin Huang.

turns out to be fundamental in the theory of hyperbolic conservation laws because it not only captures the local and global behavior of solutions, but also fully represents the effect of the nonlinearity in the structure of the solutions. It is now well known that for the system of Euler equations, there are three basic wave patterns, that is, shock wave, rarefaction wave and contact discontinuity. These three types of waves have essential differences, that is, shock is compressive, rarefaction is expansive, and contact discontinuity has some diffusive structure. Therefore, how to study the hydrodynamic limit of Boltzmann equation for the full Riemann solution that consists of the superposition of these three typical waves is still very challenging in mathematics.

By coping with the essential properties of individual wave pattern, the hydrodynamic limit for a single wave was justified in the previous works, see Yu [38] for shock wave, Xin-Zeng [37] for rarefaction wave and Huang-Wang-Yang [17] for contact discontinuity. However, up to now, how to deal with the general Riemann solution that consists of all three basic waves is still a challenging open problem. This is mainly due to the difficulty in handling the wave interactions and also unifying the different approaches in the analysis used for each single wave pattern. In order to overcome these difficulties in justifying the limit, two families of hyperbolic waves, called hyperbolic wave I and II, are introduced in [16] to capture the propagation of the extra mass created by the approximate hyperbolic rarefaction wave profile in the viscous setting and the diffusion approximation of contact discontinuity.

We now briefly explain why the two families of hyperbolic waves we introduced are essential for the proof. The approximate rarefaction wave is constructed as a hyperbolic wave profile. Therefore, we need to precisely capture the error in the second order of the approximation for the Boltzmann equation in term of the Knudsen number, that is, in the Navier-Stokes level. For this, we introduce the hyperbolic wave I as a solution to the linearized system around the approximate rarefaction wave profile with source terms given by the viscosity and heat conductivity induced by the rarefaction wave profile to recover the viscous terms. We can show that the hyperbolic wave I decays like the first-order derivative of the rarefaction wave profile so that the decay properties given in Lemma 1.2 are good enough to carry out the analysis.

The main difficulty comes from the approximation of the contact discontinuity. First of all, such an approximation, that is, 2-viscous contact wave, behaves like a diffusion wave profile as for the Navier-Stokes equations. Due to the lack of sufficient decay in ε and the non-conservative error terms when taking the anti-derivative of the perturbation, we need to remove the leading error terms and non-conservative terms in such approximation before taking the anti-derivative. The hyperbolic wave II is then constructed around the superposition of the approximate 1-rarefaction wave, the hyperbolic wave I, the 2-viscous contact wave and the 3-shock profile as a whole. Some wave interaction terms can be absorbed in the hyperbolic wave II and the other wave interaction terms can be handled by some subtle and careful calculations.

With the help of these two hyperbolic waves and the corresponding new estimates, we can justify the limiting process in [16] from the Boltzmann equation to compressible Euler equations for the generic Riemann problems by elaborate analysis after a hyperbolic scaling. Furthermore, a convergence rate is obtained in term of the Knudsen number.

We now formulate the problem. Consider the Boltzmann equation with slab symmetry

$$f_t + \xi_1 f_x = \frac{1}{\varepsilon} Q(f,f), \tag{1}$$

where $\xi = (\xi_1, \xi_2, \xi_3) \in \mathbf{R}^3$ and $x \in \mathbf{R}^1$. Here, the collision operator takes the form of

$$Q(f,g)(\xi) \equiv \frac{1}{2}\int_{\mathbf{R}^3}\int_{\mathbf{S}^2_+}\Big(f(\xi')g(\xi'_*) + f(\xi'_*)g(\xi') - f(\xi)g(\xi_*) - f(\xi_*)g(\xi)\Big)B(|\xi-\xi_*|,\hat{\theta})\, d\xi_* d\Omega,$$

where ξ', ξ'_* are the velocities after an elastic collision of two particles with velocities ξ, ξ_* before the collision. Here, $\hat{\theta}$ is the angle between the relative velocity $\xi - \xi_*$ and the unit vector Ω in $\mathbf{S}^2_+ = \{\Omega \in \mathbf{S}^2 : (\xi - \xi_*) \cdot \Omega \geq 0\}$. The conservations of momentum and energy yield the following relations between the velocities before and after collision:

$$\xi' = \xi - [(\xi - \xi_*)\cdot\Omega]\,\Omega, \qquad \xi'_* = \xi_* + [(\xi - \xi_*)\cdot\Omega]\,\Omega.$$

We will concentrate on the hard sphere model where the cross-section is

$$B(|\xi - \xi_*|, \hat{\theta}) = |(\xi - \xi_*, \Omega)| = |\xi - \xi_*| \cos\hat{\theta}.$$

On the other hand, it is noted that the analysis can be applied to at least hard potential.

Formally when the Knudsen number ε tends to zero, the limit of the Boltzmann equation (1) is the system of compressible Euler equations that consists of conservations of mass, momentum and energy:

$$\begin{cases} \rho_t + (\rho u_1)_x = 0, \\ (\rho u_1)_t + (\rho u_1^2 + p)_x = 0, \\ (\rho u_i)_t + (\rho u_1 u_i)_x = 0, \ i = 2,3, \\ [\rho(e + \frac{|u|^2}{2})]_t + [\rho u_1 (E + \frac{|u|^2}{2}) + p u_1]_x = 0, \end{cases} \tag{2}$$

where

$$\begin{cases} \rho(t,x) = \int_{\mathbf{R}^3} \varphi_0(\xi) f(t,x,\xi) d\xi, \\ \rho u_i(t,x) = \int_{\mathbf{R}^3} \varphi_i(\xi) f(t,x,\xi) d\xi, \ i = 1,2,3, \\ \rho(e + \frac{|u|^2}{2})(t,x) = \int_{\mathbf{R}^3} \varphi_4(\xi) f(t,x,\xi) d\xi. \end{cases} \tag{3}$$

Here, ρ is the density, $u = (u_1, u_2, u_3)$ is the macroscopic velocity, e is the internal energy, and $p = R\rho\theta$ with R being the gas constant is the pressure. The temperature θ is related to the internal energy by $e = \frac{3}{2}R\theta$, and $\varphi_i(\xi)(i = 0,1,2,3,4)$ are the collision invariants given by

$$\varphi_0(\xi) = 1, \ \varphi_i(\xi) = \xi_i \ (i = 1,2,3), \ \varphi_4(\xi) = \frac{1}{2}|\xi|^2, \tag{4}$$

that satisfy

$$\int_{\mathbf{R}^3} \varphi_i(\xi) Q(g_1, g_2) d\xi = 0, \quad \text{for } \ i = 0,1,2,3,4.$$

Instead of using either Hilbert expansion or Chapman-Enskog expansion, we will apply the macro-micro decomposition introduced in [26]. For a solution $f(t,x,\xi)$ of (1), set

$$f(t,x,\xi) = \mathbf{M}(t,x,\xi) + \mathbf{G}(t,x,\xi),$$

where the local Maxwellian $\mathbf{M}(t,x,\xi) = \mathbf{M}_{[\rho,u,\theta]}(\xi)$ represents the macroscopic component of the solution defined by the five conserved quantities, i.e., the mass density $\rho(t,x)$, the momentum $\rho u(t,x)$, and the total energy $\rho(e+\frac{1}{2}|u|^2)(t,x)$ given in (3), through

$$\mathbf{M} = \mathbf{M}_{[\rho,u,\theta]}(t,x,\xi) = \frac{\rho(t,x)}{\sqrt{(2\pi R\theta(t,x))^3}} e^{-\frac{|\xi-u(t,x)|^2}{2R\theta(t,x)}}. \tag{5}$$

And $\mathbf{G}(t,x,\xi)$ represents the microscopic component.

The inner product of g_1 and g_2 in $L^2_\xi(\mathbf{R}^3)$ with respect to a given Maxwellian $\tilde{\mathbf{M}}$ is denoted by:

$$\langle g_1, g_2\rangle_{\tilde{\mathbf{M}}} \equiv \int_{\mathbf{R}^3} \frac{1}{\tilde{\mathbf{M}}} g_1(\xi) g_2(\xi) d\xi. \tag{6}$$

If $\tilde{\mathbf{M}}$ is the local Maxwellian $\mathbf{M}$ defined in (5), the macroscopic space is spanned by

$$\begin{cases} \chi_0(\xi) \equiv \frac{1}{\sqrt{\rho}}\mathbf{M}, \\ \chi_i(\xi) \equiv \frac{\xi_i - u_i}{\sqrt{R\theta\rho}}\mathbf{M} \ \text{ for } \ i = 1,2,3, \\ \chi_4(\xi) \equiv \frac{1}{\sqrt{6\rho}}(\frac{|\xi-u|^2}{R\theta} - 3)\mathbf{M}, \\ \langle \chi_i, \chi_j\rangle = \delta_{ij}, \ i,j = 0,1,2,3,4. \end{cases} \tag{7}$$

By using the above base, the macroscopic projection $\mathbf{P}_0$ and microscopic projection $\mathbf{P}_1$ can be defined as

$$\mathbf{P}_0 g = \sum_{j=0}^{4} \langle g, \chi_j\rangle \chi_j, \qquad \mathbf{P}_1 g = g - \mathbf{P}_0 g.$$

Note that a function $g(\xi)$ is called microscopic if

$$\int g(\xi)\varphi_i(\xi) d\xi = 0, \ \ i = 0,1,2,3,4.$$

Notice that the solution $f(t,x,\xi)$ to the Boltzmann equation (1) satisfies

$$\mathbf{P}_0 f = \mathbf{M}, \ \ \mathbf{P}_1 f = \mathbf{G},$$

and the Boltzmann equation (1) becomes

$$(\mathbf{M}+\mathbf{G})_t + \xi_1(\mathbf{M}+\mathbf{G})_x = \frac{1}{\varepsilon}[2Q(\mathbf{M},\mathbf{G}) + Q(\mathbf{G},\mathbf{G})]. \tag{8}$$

Applying the projection operator $\mathbf{P}_1$ to (8), we have

$$\mathbf{G}_t + \mathbf{P}_1(\xi_1 \mathbf{M}_x) + \mathbf{P}_1(\xi_1 \mathbf{G}_x) = \frac{1}{\varepsilon}\left[\mathbf{L}_\mathbf{M}\mathbf{G} + Q(\mathbf{G},\mathbf{G})\right]. \tag{9}$$

Here $\mathbf{L}_\mathbf{M}$ is the linearized collision operator of $Q(f,f)$ with respect to the local Maxwellian $\mathbf{M}$ given by

$$\mathbf{L}_\mathbf{M} g = 2Q(\mathbf{M},g) = Q(\mathbf{M},g) + Q(g,\mathbf{M}).$$

Note that the null space $\mathfrak{N}$ of $\mathbf{L}_\mathbf{M}$ is spanned by the macroscopic variables $\chi_j(\xi)$, $j = 0,1,2,3,4$. Furthermore, there exists a positive constant $\widetilde{\sigma} > 0$ such that for any function $g(\xi) \in \mathfrak{N}^\perp$, cf. [14],

$$\langle g, \mathbf{L}_\mathbf{M} g\rangle \le -\widetilde{\sigma}\langle \nu(|\xi|)g, g\rangle,$$

where $\nu(|\xi|) = O(1)(1+|\xi|)$ is the collision frequency for the hard sphere model.

Consequently, the linearized collision operator $\mathbf{L_M}$ is a dissipative operator on $L^2(\mathbf{R}^3)$, and its inverse $\mathbf{L}_\mathbf{M}^{-1}$ is a bounded operator on $\mathfrak{N}^\perp$. It follows from (9) that

$$\mathbf{G} = \varepsilon \mathbf{L}_\mathbf{M}^{-1}[\mathbf{P}_1(\xi_1 \mathbf{M}_x)] + \Pi, \tag{10}$$

with

$$\Pi = \mathbf{L}_\mathbf{M}^{-1}[\varepsilon(\mathbf{G}_t + \mathbf{P}_1(\xi_1 \mathbf{G}_x)) - Q(\mathbf{G},\mathbf{G})]. \tag{11}$$

By integrating the product of the equation (8) and the collision invariants $\varphi_i(\xi)(i = 0,1,2,3,4)$ with respect to ξ over $\mathbf{R}^3$, and using (10), we have

$$\begin{cases} \rho_t + (\rho u_1)_x = 0, \\ (\rho u_1)_t + (\rho u_1^2 + p)_x = \dfrac{4\varepsilon}{3}(\mu(\theta)u_{1x})_x - \displaystyle\int \xi_1^2 \Pi_x d\xi, \\ (\rho u_i)_t + (\rho u_1 u_i)_x = \varepsilon(\mu(\theta)u_{ix})_x - \displaystyle\int \xi_1\xi_i \Pi_x d\xi, \; i = 2,3, \\ [\rho(\theta + \dfrac{|u|^2}{2})]_t + [\rho u_1(\theta + \dfrac{|u|^2}{2}) + p u_1]_x = \varepsilon(\kappa(\theta)\theta_x)_x + \dfrac{4\varepsilon}{3}(\mu(\theta)u_1 u_{1x})_x \\ \qquad + \varepsilon \displaystyle\sum_{i=2}^{3}(\mu(\theta)u_i u_{ix})_x - \int \frac{1}{2}\xi_1|\xi|^2 \Pi_x d\xi, \end{cases} \tag{12}$$

where the viscosity coefficient $\mu(\theta) > 0$ and the heat conductivity coefficient $\kappa(\theta) > 0$ are smooth functions of the temperature θ. Here, we normalize the gas constant R to be $\frac{2}{3}$ so that $e = \theta$ and $p = \frac{2}{3}\rho\theta$.

Since the problem considered is one dimensional in the space variable $x \in \mathbf{R}$, it is more convenient to rewrite the equation (1) in the *Lagrangian* coordinates. Set the coordinate transformation:

$$(t,x) \Rightarrow \Big(t, \int_{(0,0)}^{(t,x)} \rho(\tau,y)dy - (\rho u_1)(\tau,y)d\tau\Big). \tag{13}$$

We will still denote the *Lagrangian* coordinates by (t,x) for the simplicity of notations. Then (1) and (2) in the Lagrangian coordinates becomes, respectively,

$$f_t - \frac{u_1}{v} f_x + \frac{\xi_1}{v} f_x = \frac{1}{\varepsilon} Q(f,f), \tag{14}$$

and

$$\begin{cases} v_t - u_{1x} = 0, \\ u_{1t} + p_x = 0, \\ u_{it} = 0, \; i = 2,3, \\ (\theta + \dfrac{|u|^2}{2})_t + (p u_1)_x = 0. \end{cases} \tag{15}$$

Moreover,

$$\mathbf{G}_t - \frac{u_1}{v}\mathbf{G}_x + \frac{1}{v}\mathbf{P}_1(\xi_1 \mathbf{M}_x) + \frac{1}{v}\mathbf{P}_1(\xi_1 \mathbf{G}_x) = \frac{1}{\varepsilon}(\mathbf{L_M G} + Q(\mathbf{G},\mathbf{G})), \tag{16}$$

with

$$\mathbf{G} = \varepsilon \mathbf{L}_\mathbf{M}^{-1}(\frac{1}{v}\mathbf{P}_1(\xi_1 \mathbf{M}_x)) + \Pi_1, \tag{17}$$

$$\Pi_1 = \mathbf{L}_\mathbf{M}^{-1}[\varepsilon(\mathbf{G}_t - \frac{u_1}{v}\mathbf{G}_x + \frac{1}{v}\mathbf{P}_1(\xi_1 \mathbf{G}_x)) - Q(\mathbf{G},\mathbf{G})], \tag{18}$$

and

$$
\begin{cases}
v_t - u_{1x} = 0, \\
u_{1t} + p_x = \frac{4\varepsilon}{3}(\frac{\mu(\theta)}{v}u_{1x})_x - \int \xi_1^2 \Pi_{1x} d\xi, \\
u_{it} = \varepsilon(\frac{\mu(\theta)}{v}u_{ix})_x - \int \xi_1\xi_i \Pi_{1x} d\xi, \ i = 2, 3, \\
(\theta + \frac{|u|^2}{2})_t + (pu_1)_x = \varepsilon(\frac{\kappa(\theta)}{v}\theta_x)_x + \frac{4\varepsilon}{3}(\frac{\mu(\theta)}{v}u_1 u_{1x})_x \\
\qquad + \varepsilon \sum_{i=2}^{3}(\frac{\mu(\theta)}{v}u_i u_{ix})_x - \int \frac{1}{2}\xi_1 |\xi|^2 \Pi_{1x} d\xi.
\end{cases} \tag{19}
$$

The Riemann problem for the Euler system (15) is an initial value problem with initial data

$$
(v, u, \theta)(t = 0, x) = \begin{cases} (v_-, u_-, \theta_-), \ x < 0, \\ (v_+, u_+, \theta_+), \ x > 0, \end{cases}
$$

where $u = (u_1, u_2, u_3)$, $u_\pm = (u_{1\pm}, 0, 0)$ and $v_\pm > 0, u_{1\pm}, \theta_\pm > 0$ are constants. It is known that the generic solution to the Riemann problem consists of three waves that propagates at different speeds, that is, shock, rarefaction wave and contact discontinuity, cf. [10, 25]. We denote this solution by $(\widetilde{V}, \widetilde{U}, \widetilde{\Theta})(t, x)$ with $\widetilde{U} = (\widetilde{U}_1, 0, 0)$. Given the right end state (v_+, u_{1+}, θ_+), the following wave curves for the left end state (v, u_1, θ) in the phase space are defined with v and θ for the Euler equations (15).

- Contact discontinuity curve:

$$
CD(v_+, u_{1+}, \theta_+) = \{(v, u_1, \theta) | u_1 = u_{1+}, p = p_+, v \not\equiv v_+\}. \tag{20}
$$

- i-Rarefaction wave curve $(i = 1, 3)$:

$$
R_i(v_+, u_{1+}, \theta_+) := \left\{ (v, u_1, \theta) \middle| v < v_+, \ u_1 = u_{1+} - \int_{v_+}^{v} \lambda_i(\eta, s_+)\, d\eta, \ s(v, \theta) = s_+ \right\}, \tag{21}
$$

where $s_+ = s(v_+, \theta_+)$ and $\lambda_i = \lambda_i(v, s)$ is the i-th characteristic speed of (15).

- i-Shock wave curve $(i = 1, 3)$:

$$
S_i(v_+, u_{1+}, \theta_+) := \left\{ (v, u_1, \theta) \middle| \begin{array}{l} -s_i(v_+ - v) - (u_{1+} - u_1) = 0, \\ -s_i(u_{1+} - u_1) + (p_+ - p) = 0, \\ -s_i(E_+ - E) + (p_+ u_{1+} - p u_1) = 0, \end{array} \text{ and } \lambda_{i+} < s_i < \lambda_{i-} \right\}, \tag{22}
$$

where $E = \theta + \frac{|u|^2}{2}, p = \frac{2\theta}{3v}, E_+ = \theta_+ + \frac{|u_+|^2}{2}, p_+ = \frac{2\theta_+}{3v_+}$, $\lambda_{i\pm} = \lambda_i(v_\pm, \theta_\pm)$ and s_i is the $i-$shock speed.

For definiteness, we consider the case when the solution to the Riemann problem is a superposition of a 1-rarefaction and a 3-shock wave with a contact discontinuity in between, that is, $(v_-, u_{1-}, \theta_-) \in R_1\text{-}CD\text{-}S_3(v_+, u_{1+}, \theta_+)$. Then there exist uniquely two intermediate states (v_*, u_{1*}, θ_*) and (v^*, u_1^*, θ^*) such that (v_-, u_{1-}, θ_-) $\in R_1(v_*, u_{1*}, \theta_*)$, $(v_*, u_{1*}, \theta_*) \in CD(v^*, u_1^*, \theta^*)$ and $(v^*, u_1^*, \theta^*) \in S_3(v_+, u_{1+}, \theta_+)$.

Hence, the wave pattern $(\widetilde{V}, \widetilde{U}, \widetilde{\mathcal{E}})(t, x)$ can be written as

$$
\begin{pmatrix} \widetilde{V} \\ \widetilde{U}_1 \\ \widetilde{\mathcal{E}} \end{pmatrix}(t, x) = \begin{pmatrix} v^{r_1} + v^{cd} + v^{s_3} \\ u_1^{r_1} + u_1^{cd} + u_1^{s_3} \\ E^{r_1} + E^{cd} + E^{s_3} \end{pmatrix}(t, x) - \begin{pmatrix} v_* + v^* \\ u_{1*} + u_1^* \\ E_* + E^* \end{pmatrix}, \quad \widetilde{U}_2 = \widetilde{U}_3 = 0, \tag{23}
$$

where $(v^{r_1}, u_1^{r_1}, \theta^{r_1})(t,x)$ is the 1-rarefaction wave defined in (21) with the right state (v_+, u_{1+}, θ_+) given by (v_*, u_{1*}, θ_*), $(v^{cd}, u_1^{cd}, \theta^{cd})(t,x)$ is the contact discontinuity defined in (20) with the states (v_-, u_{1-}, θ_-) and (v_+, u_{1+}, θ_+) given by (v_*, u_{1*}, θ_*) and (v^*, u_1^*, θ^*) respectively, and $(v^{s_3}, u_1^{s_3}, \theta^{s_3})(t,x)$ is the 3-shock wave defined in (22) with the left state (v_-, u_{1-}, θ_-) given by (v^*, u_1^*, θ^*).

Consequently, we can define

$$\widetilde{\Theta}(t,x) = (\widetilde{\mathcal{E}}(t,x) - \frac{\widetilde{U}(t,x)^2}{2}). \tag{24}$$

Due to the singularity of the rarefaction wave at $t = 0$, in this note, we consider the problem in the time interval $[h, T]$ for any small fixed $h > 0$ up to any arbitrarily fixed time $T > 0$. To investigate the interaction between the waves and the initial layer is another interesting topic that will not be discussed here. With the above preparation, the main result can be stated as follows.

Theorem 0.1. *([16]) Let $(\widetilde{V}, \widetilde{U}, \widetilde{\Theta})(t,x)$ be a Riemann solution to the Euler equations which is a superposition of a 1-rarefaction wave, a 2-contact discontinuity and a 3-shock wave, and $\delta = |(v_+ - v_-, u_+ - u_-, \theta_+ - \theta_-)|$ be the wave strength. There exist a small positive constant δ_0, and a global Maxwellian $\mathbf{M}_\star = \mathbf{M}_{[v_\star, u_\star, \theta_\star]}$ such that if the wave strength satisfies $\delta \le \delta_0$, then in any time interval $[h, T]$ with $0 < h < T$, there exists a positive constant $\varepsilon_0 = \varepsilon_0(\delta, h, T)$, such that if the Knudsen number $\varepsilon \le \varepsilon_0$, then the Boltzmann equation admits a family of smooth solutions $f^{\varepsilon,h}(t,x,\xi)$ satisfying*

$$\sup_{(t,x)\in\Sigma_{h,T}} \|f^{\varepsilon,h}(t,x,\xi) - \mathbf{M}_{[\widetilde{V},\widetilde{U},\widetilde{\Theta}]}(t,x,\xi)\|_{L^2_\xi(\frac{1}{\sqrt{\mathbf{M}_\star}})} \le C_{h,T}\, \varepsilon^{\frac{1}{5}} |\ln \varepsilon|,$$

where $\Sigma_{h,T} = \{(t,x) | h \le t \le T, |x| \ge h, |x - s_3 t| \ge h\}$, the norm $\|\cdot\|_{L^2_\xi(\frac{1}{\sqrt{\mathbf{M}_\star}})}$ is $\|\frac{\cdot}{\sqrt{\mathbf{M}_\star}}\|_{L^2_\xi(\mathbf{R}^3)}$ and the positive constant $C_{h,T}$ depends on h and T but is independent of ε.

Remark 1. Note that this superposition of waves is the most generic case for the Riemann problem. Similar results hold for any other superpositions of waves by using the same analysis.

Remark 2. The analysis can also be applied to the vanishing viscosity limit of the one dimensional compressible Navier-Stokes equations.

Let us now review some previous works on the hydrodynamic limits to the Boltzmann equation. For the case when the Euler equations have smooth solutions, the vanishing Knudsen number limit of the Boltzmann equation has been studied even in the case with an initial layer, cf. Caflisch [6], Lachowicz [24], Nishida [31] and Ukai-Asona [34] etc. However, as well-known, solutions of the Euler equations in general develop singularities, such as shock waves and contact discontinuities. Therefore, how to verify the hydrodynamic limit from the Boltzmann equation to the Euler equations with basic wave patterns becomes a natural problem in the process to the general setting. In this direction, with slab symmetry, as mentioned earlier, there were studies on each individual wave pattern. For superposition of different types of waves, to our knowledge, there is only one result given in [18] about the superposition of two rarefaction waves and one contact discontinuity.

On the other hand, for the incompressible equations, there are works which studied direct derivations of the incompressible Navier-Stokes equations in the long time scaling, see [2, 4, 3, 13, 23, 33] and the references therein. In particular,

Golse and Saint-Raymond showed that the limits of suitably rescaled sequences of the DiPerna-Lions renormalized solutions to the Boltzmann equation are the Leray solutions to the incompressible Navier-Stokes equations. However, even in this aspect, the uniqueness and regularity of the solution are still big issues. Since we will concentrate on the compressible Euler limit, we will not go into details about the incompressible limits.

Notations: Throughout this paper, the positive generic constants which are independent of ε, T, h are denoted by $c, C, C_i (i = 1, 2, 3, \cdots)$, while $C_{h,T}$ represents a generic positive constant depending on h and T but independent of ε. And we will use $\|\cdot\|$ to denote the standard $L_2(\mathbf{R}; dy)$ norm, and $\|\cdot\|_{H^i}$ $(i = 1, 2, 3, \cdots)$ to denote the Sobolev $H^i(\mathbf{R}; dy)$ norm. Sometimes, we also use $O(1)$ to denote a uniform bounded constant which is independent of ε, T, h.

1. Approximate Wave Patterns.

1.1. Rarefaction Wave.

For the rarefaction wave, since there is no exact rarefaction wave profile for either the Navier-Stokes equations or the Boltzmann equation, the following approximate rarefaction wave profile satisfying the Euler equations was introduced in [30, 36]. If $(v_-, u_{1-}, \theta_-) \in R_1(v_+, u_{1+}, \theta_+)$, then there exists a 1-rarefaction wave $(v^{r_1}, u_1^{r_1}, E^{r_1})(x/t)$. As in [36], the approximate rarefaction wave $(V^{R_1}, U^{R_1}, \Theta^{R_1})(t, x)$ can be constructed by the solution of the Burgers equation

$$\begin{cases} w_t + ww_x = 0, \\ w(0, x) = w_\sigma(x) = w(\dfrac{x}{\sigma}) = \dfrac{w_+ + w_-}{2} + \dfrac{w_+ - w_-}{2} \tanh \dfrac{x}{\sigma}, \end{cases} \tag{25}$$

where $\sigma > 0$ is a small parameter to be determined later to be $\varepsilon^{\frac{1}{5}}$. Note that the solution $w_\sigma^r(t, x)$ of the problem (25) is given by

$$w_\sigma^r(t, x) = w_\sigma(x_0(t, x)), \qquad x = x_0(t, x) + w_\sigma(x_0(t, x))t.$$

The smooth approximate rarefaction wave profile denoted by $(V^{R_1}, U^{R_1}, \Theta^{R_1})(t, x)$ can be defined by

$$\begin{cases} S^{R_1}(t, x) = s(V^{R_1}(t, x), \Theta^{R_1}(t, x)) = s_+, \\ w_\pm = \lambda_{1\pm} := \lambda_1(v_\pm, \theta_\pm), \\ w_\sigma^r(t, x) = \lambda_1(V^{R_1}(t, x), s_+), \\ U_1^{R_1}(t, x) = u_{1+} - \displaystyle\int_{v_+}^{V^{R_1}(t,x)} \lambda_1(v, s_+) dv, \\ U_i^{R_1}(t, x) \equiv 0, \ i = 2, 3. \end{cases} \tag{26}$$

And $(V^{R_1}, U^{R_1}, \Theta^{R_1})(t, x)$ satisfies

$$\begin{cases} V_t^{R_1} - U_{1x}^{R_1} = 0, \\ U_{1t}^{R_1} + P_x^{R_1} = 0, \\ U_{it}^{R_1} = 0, \ i = 2, 3, \\ \mathcal{E}_t^{R_1} + (P^{R_1} U_1^{R_1})_x = 0, \end{cases} \tag{27}$$

where $P^{R_1} = p(V^{R_1}, \Theta^{R_1}) = \frac{2\Theta^{R_1}}{3V^{R_1}}$ and $\mathcal{E}^{R_1} = \Theta^{R_1} + \frac{|U^{R_1}|^2}{2}$. The properties of the rarefaction wave profile can be summarized as follows.

Lemma 1.1. ([36]) *The approximate rarefaction waves $(V^{R_1}, U^{R_1}, \Theta^{R_1})(t, x)$ constructed in (26) have the following properties:*

(1) $U_{1x}^{R_1}(t,x) > 0$ *for* $x \in \mathbf{R}$, $t > 0$;

(2) For any $1 \le p \le +\infty$, *the following estimates holds,*

$$\|(V^{R_1}, U_1^{R_1}, \Theta^{R_1})_x\|_{L^p(dx)} \le C \min\left\{\delta^{R_1}\sigma^{-1+1/p},\ (\delta^{R_1})^{1/p} t^{-1+1/p}\right\},$$
$$\|(V^{R_1}, U_1^{R_1}, \Theta^{R_1})_{xx}\|_{L^p(dx)} \le C \min\left\{\delta^{R_1}\sigma^{-2+1/p},\ \sigma^{-1+1/p} t^{-1}\right\},$$

where the positive constant C *depends only on* p *and the wave strength;*

(3) If $x \ge \lambda_{1+}^{R_1} t$, *then*

$$|(V^{R_1}, U^{R_1}, \Theta^{R_1})(t,x) - (v_+, u_+, \theta_+)| \le Ce^{-\frac{2|x-\lambda_{1+}t|}{\sigma}},$$
$$|\partial_x^k (V^{R_1}, U^{R_1}, \Theta^{R_1})(t,x)| \le \frac{C}{\sigma^k} e^{-\frac{2|x-\lambda_{1+}t|}{\sigma}}, \quad k = 1, 2;$$

(4) There exist positive constants C *and* σ_0 *such that for* $\sigma \in (0, \sigma_0)$ *and* $t > 0$,

$$\sup_{x \in \mathbf{R}} |(V^{R_1}, U^{R_1}, \mathcal{E}^{R_1})(t,x) - (v^{r_1}, u^{r_1}, E^{r_1})(\frac{x}{t})| \le \frac{C}{t}[\sigma \ln(1+t) + \sigma |\ln \sigma|].$$

1.2. **Hyperbolic Wave I.** Since the whole wave profile consisits of a shock wave whose rate of change in the shock region is of the order of ε^{-1}, we have to consider the anti-derivative of the perturbation in order to cope with the correct sign as in the stability analysis. From (27), we know that the approximate rarefaction wave $(V^{R_1}, U^{R_1}, \Theta^{R_1})(t,x)$ satisfies the compressible Euler equations exactly without viscous terms. Thus if we carry out the energy estimates to the anti-derivative variables, the error terms due to the viscous terms from the approximate rarefaction wave are not good enough to get the desired estimates. In order to overcome this difficulty, we introduce the hyperbolic wave I to recover these viscous terms.

This hyperbolic wave denoted by $(d_1, d_2, d_3)(t,x)$ can be defined as follows. Consider a linear system

$$\begin{cases} d_{1t} - d_{2x} = 0, \\ d_{2t} + (p_v^{R_1} d_1 + p_{u_1}^{R_1} d_2 + p_E^{R_1} d_3)_x = \frac{4}{3}\varepsilon(\frac{\mu(\Theta^{R_1})U_{1x}^{R_1}}{V^{R_1}})_x, \\ d_{3t} + [(pu_1)_v^{R_1} d_1 + (pu_1)_{u_1}^{R_1} d_2 + (pu_1)_E^{R_1} d_3]_x \\ \quad = \varepsilon(\frac{\kappa(\Theta^{R_1})\Theta_x^{R_1}}{V^{R_1}})_x + \frac{4}{3}\varepsilon(\frac{\mu(\Theta^{R_1})U_1^{R_1}U_{1x}^{R_1}}{V^{R_1}})_x, \end{cases} \tag{28}$$

where $p = \frac{R\theta}{v} = p(v,u,E) = \frac{2E-u^2}{3v}$ and $p_v^{R_1} = p_v(V^{R_1}, U^{R_1}, \mathcal{E}^{R_1})$ etc. Note that the left hand side of the above system is the linearization of the Euler equation around the rarefaction wave approximation. We want to solve this linear hyperbolic system (28) on the time interval $[h, T]$. For this, we diagonalize the above system by rewriting it as

$$\begin{pmatrix} d_1 \\ d_2 \\ d_3 \end{pmatrix}_t + \left[A^{R_1}\begin{pmatrix} d_1 \\ d_2 \\ d_3 \end{pmatrix}\right]_x = \begin{pmatrix} 0 \\ H_1^{R_1} \\ H_2^{R_1} \end{pmatrix}, \tag{29}$$

where $H_1^{R_1} = \varepsilon(\frac{\mu(\Theta^{R_1})U_{1x}^{R_1}}{V^{R_1}})_x$, $H_2^{R_1} = \varepsilon(\frac{\kappa(\Theta^{R_1})\Theta_x^{R_1}}{V^{R_1}})_x + \varepsilon(\frac{\mu(\Theta^{R_1})U_1^{R_1}U_{1x}^{R_1}}{V^{R_1}})_x$. Here, the matrix

$$A^{R_1} = \begin{pmatrix} 0 & -1 & 0 \\ p_v^{R_1} & p_{u_1}^{R_1} & p_E^{R_1} \\ (pu_1)_v^{R_1} & (pu_1)_{u_1}^{R_1} & (pu_1)_E^{R_1} \end{pmatrix}$$

has three distinct eigenvalues $\lambda_1^{R_1} := \lambda_1(V^{R_1}, s_\pm) < 0 \equiv \lambda_2^{R_1} < \lambda_3(V^{R_1}, s_\pm) := \lambda_3^{R_1}$ and the corresponding left and right eigenvectors denoted $l_j^{R_1}, r_j^{R_1}$ $(j = 1, 2, 3)$

respectively, satisfy

$$L^{R_1}A^{R_1}R^{R_1} = \mathrm{diag}(\lambda_1^{R_1}, 0, \lambda_3^{R_1}) \equiv \Lambda^{R_1}, \qquad L^{R_1}R^{R_1} = \mathrm{Id.},$$

Here $L^{R_1} = (l_1^{R_1}, l_2^{R_1}, l_3^{R_1})^t$, $R^{R_1} = (r_1^{R_1}, r_2^{R_1}, r_3^{R_1})$ with $l_i^{R_1} = l_i(V^{R_1}, U_1^{R_1}, s_+)$ and $r_i^{R_1} = r_i(V^{R_1}, U_1^{R_1}, s_+)$ $(i = 1, 2, 3)$ and Id. is the 3×3 identity matrix. Now we set

$$(D_1, D_2, D_3)^t = L^{R_1}(d_1, d_2, d_3)^t. \tag{30}$$

Then

$$(d_1, d_2, d_3)^t = R^{R_1}(D_1, D_2, D_3)^t, \tag{31}$$

and (D_1, D_2, D_3) satisfies the system

$$\begin{aligned}\begin{pmatrix} D_1 \\ D_2 \\ D_3 \end{pmatrix}_t &+ \left[\Lambda^{R_1}\begin{pmatrix} D_1 \\ D_2 \\ D_3 \end{pmatrix}\right]_x \\ &= L^{R_1}\begin{pmatrix} 0 \\ H_1^{R_1} \\ H_2^{R_1} \end{pmatrix} + L_t^{R_1}R^{R_1}\begin{pmatrix} D_1 \\ D_2 \\ D_3 \end{pmatrix} + L_x^{R_1}R^{R_1}\Lambda^{R_1}\begin{pmatrix} D_1 \\ D_2 \\ D_3 \end{pmatrix}.\end{aligned} \tag{32}$$

Now we impose the following boundary condition to the above linear hyperbolic system in the domain $(t, x) \in [h, T] \times \mathbf{R}$:

$$D_1(t = h, x) = 0, \qquad D_2(t = T, x) = D_3(t = T, x) = 0. \tag{33}$$

With this boundary condition, we can solve the linear diagonalized hyperbolic system under the conditions (33). Moreover, we have the following estimates on the solution.

Lemma 1.2. *([16]) There exists a positive constant $C_{h,T}$ independent of ε such that*

(1)

$$\|\frac{\partial^k}{\partial x^k} d_i(t, \cdot)\|^2_{L^2(dx)} \le C_{h,T}\, \frac{\varepsilon^2}{\sigma^{2k+1}}, \quad i = 1, 2, 3,\ k = 0, 1, 2, 3.$$

(2) If $x > \lambda_{1+}t$, then we have

$$|d_i(x, t)| \le C_{h,T}\, \frac{1}{\sigma} e^{-\frac{|x - \lambda_{1+}t|}{\sigma}}, \qquad |d_{ix}(x, t)| \le C_{h,T}\, \frac{1}{\sigma^2} e^{-\frac{|x - \lambda_{1+}t|}{\sigma}},\ i = 1, 2, 3.$$

1.3. **Viscous Contact Wave.** In this subsection, we construct the contact wave $(V^{CD}, U^{CD}, \Theta^{CD})(t, x)$ for the Boltzmann equation motivated by [21]. It is known (cf. [32]) that the Euler system (15) admits a contact discontinuity

$$(v^{cd}, u^{cd}, \theta^{cd})(t, x) = \begin{cases} (v_-, u_-, \theta_-), & x < 0, \\ (v_+, u_+, \theta_+), & x > 0, \end{cases} \tag{34}$$

provided that $(v_-, u_{1-}, \theta_-) \in CD(v_+, u_{1+}, \theta_+)$, that is,

$$u_{1+} = u_{1-}, \qquad p_- := \frac{2\theta_-}{3v_-} = p_+ := \frac{2\theta_+}{3v_+}. \tag{35}$$

Then for the Navier-Stokes equations, by the energy equation $(19)_4$ and the mass equation $(19)_1$ with $v \approx \frac{2\theta}{3p_+}$, cf. [18], we can obtain the following nonlinear diffusion equation

$$\theta_t = \varepsilon(a(\theta)\theta_x)_x, \ a(\theta) = \frac{9p_+\kappa(\theta)}{10\theta}. \tag{36}$$

From [1], we know that the diffusion equation (36) admits a self-similar solution $\hat{\Theta}(\eta)$, $\eta = \frac{x}{\sqrt{\varepsilon(1+t)}}$ satisfying the boundary conditions $\hat{\Theta}(\pm\infty, t) = \theta_\pm$. Let $\delta^{CD} = |\theta_+ - \theta_-|$, then $\hat{\Theta}(t, x)$ has the property that

$$\hat{\Theta}_x(t,x) = \frac{O(1)\delta^{CD}}{\sqrt{\varepsilon(1+t)}} e^{-\frac{cx^2}{\varepsilon(1+t)}}, \tag{37}$$

with some positive constant c depending only on $\theta_\pm$. We can define the Navier-Stokes profile by

$$\hat{V} = \frac{2}{3p_+}\hat{\Theta},\ \hat{U}_1 = u_{1+} + \frac{2\varepsilon a(\hat{\Theta})}{3p_+}\hat{\Theta}_x,\ \hat{U}_i = 0, i = 2, 3. \tag{38}$$

For the Boltzmann equation, if we still use the above Navier-Stokes profile $(\hat{V}, \hat{U}, \hat{\Theta})$, we can not get any decay with respect to the Knudsen number ε due to the non-fluid component. Hence, we construct a Boltzmann contact wave as follows. Set

$$\mathbf{G}^{CD}(t,x,\xi) = \frac{3\varepsilon}{2v\theta}\mathbf{L}_\mathbf{M}^{-1}\{\mathbf{P}_1[\xi_1(\frac{|\xi-u|^2}{2\theta}\Theta_x^{CD} + \xi\cdot U_x^{CD})\mathbf{M}]\}, \tag{39}$$

and

$$\Pi_{11}^{CD} = \mathbf{L}_\mathbf{M}^{-1}\left[\varepsilon(-\frac{u_1}{v}\mathbf{G}_x^{CD} + \frac{1}{v}\mathbf{P}_1(\xi_1\mathbf{G}_x^{CD})) - Q(\mathbf{G}^{CD}, \mathbf{G}^{CD})\right], \tag{40}$$

where $(V^{CD}, U^{CD}, \Theta^{CD})(t,x)$ is the viscous contact wave for the Boltzmann equation to be constructed later.

Note that for the Boltzmann equation, the leading terms in the energy equation $(19)_4$ is

$$\theta_t = \varepsilon(a(\theta)\theta_x)_x + \frac{3}{5}\Delta_{11x}, \tag{41}$$

where $a(\theta)$ is defined in (36) and

$$\Delta_{11} = \varepsilon^2\Big[g_{11}\theta_x\Theta_x^{CD} + g_{12}v_x\Theta_x^{CD} + g_{13}(\Theta_x^{CD})^2 + g_{14}\Theta_{xx}^{CD}\Big], \tag{42}$$

with $g_{1i} = g_{1i}(v,u,\theta), (i = 1,2,3,4)$ being smooth functions of (v,u,θ). To represent the microscopic effect on the wave profile, we want to define Θ^{CD} to be close to $\hat{\Theta}(\frac{x}{\sqrt{\varepsilon(1+t)}}) + \hat{\Theta}^{nf}(t,x)$ with $\hat{\Theta}$ being determined by (36), (37) and $\hat{\Theta}^{nf}$ represents the part of the nonlinear diffusion wave coming from the non-fluid component not appearing in the Navier-Stokes level. To construct $\hat{\Theta}^{nf}$, we linearize the equation (41) around the Navier-Stokes profile $\hat{\Theta}$ and drop all the higher order terms. This leads to a linear diffusion equation for $\hat{\Theta}^{nf}$

$$\hat{\Theta}_t^{nf} = \varepsilon(a(\hat{\Theta})\hat{\Theta}_x^{nf})_x + \varepsilon(a'(\hat{\Theta})\hat{\Theta}_x\hat{\Theta}^{nf})_x + \frac{3}{5}\widetilde{\Delta}_{11x}, \tag{43}$$

where $\widetilde{\Delta}_{11} = \varepsilon^2(\tilde{g}_{11} + \frac{2}{3p_+}\tilde{g}_{12} + \tilde{g}_{13})(\hat{\Theta}_x)^2 + \varepsilon^2\tilde{g}_{14}\hat{\Theta}_{xx}$ with $\tilde{g}_{1i} = \tilde{g}_{1i}(\hat{V},\hat{U},\hat{\Theta})$ $(i = 1,2,3,4)$. Integrating (43) with respect to x yields that

$$\Xi_{1t} = \varepsilon a(\hat{\Theta})\Xi_{1xx} + \varepsilon a'(\hat{\Theta})\hat{\Theta}_x\Xi_{1x} + \frac{3}{5}\widetilde{\Delta}_{11}, \tag{44}$$

where

$$\Xi_1(t,x) = \int_{-\infty}^{x} \hat{\Theta}^{nf}(t,x)dx. \tag{45}$$

Note that $\widetilde{\Delta}_{11}$ takes the form of $\frac{\varepsilon}{1+t}A^1(\frac{x}{\sqrt{\varepsilon(1+t)}})$ and satisfies that

$$|\widetilde{\Delta}_{11}| = O(\delta^{CD})\varepsilon(1+t)^{-1}e^{-\frac{x^2}{4a(\theta_\pm)\varepsilon(1+t)}}, \text{ as } x \to \pm\infty.$$

We can check that there exists a self-similar solution $\Xi_1(\frac{x}{\sqrt{\varepsilon(1+t)}})$ for (43) with the boundary conditions $\Xi_1(-\infty) = 0, \Xi_1(+\infty) = \Xi_{1+}$. Here Ξ_{1+} can be any given constant satisfying $|\Xi_{1+}| < \delta^{CD}$. It is worthy to point out that even though the function $\Xi_1(t,x)$ depends on the constant Ξ_{1+}, $\hat{\Theta}^{nf}(t,x) = \Xi_{1x}(t,x) \to 0$ as $x \to \pm\infty$. That is, the choice of the constant Ξ_{1+} has no influence on the ansantz as long as $|\Xi_{1+}| < \delta^{CD}$. From now on, we fix Ξ_{1+} so that the function $\Xi_1(t,x)$ is uniquely determined and its derivative $\Xi_{1x} = \hat{\Theta}^{nf}$ has the property

$$|\hat{\Theta}^{nf}| = |\Xi_{1x}| = O(\delta^{CD})\varepsilon^{\frac{1}{2}}(1+t)^{-\frac{1}{2}}e^{-\frac{x^2}{4a(\theta_\pm)\varepsilon(1+t)}}, \text{ as } x \to \pm\infty. \tag{46}$$

Then we apply the similar procedure to construct the second and the third components of the velocity of the contact wave denoted by U_i^{CD} $(i = 2,3)$, see [16] for details.

In summary, the viscous contact wave $(V^{CD}, U^{CD}, \Theta^{CD})(t,x)$ can be defined by

$$\begin{aligned} V^{CD} &= \frac{2}{3p_+}(\hat{\Theta} + \hat{\Theta}^{nf}), \\ U_1^{CD} &= u_{1+} + \frac{2}{3p_+}\Big[\varepsilon a(\hat{\Theta})\hat{\Theta}_x + \varepsilon(a(\hat{\Theta})\hat{\Theta}^{nf})_x + \frac{3}{5}\widetilde{\Delta}_{11}\Big], \\ \Theta^{CD} &= \hat{\Theta} + \hat{\Theta}^{nf} + H, \end{aligned} \tag{47}$$

where

$$H = O(\delta^{CD})\varepsilon(1+t)^{-2}e^{-\frac{cx^2}{\varepsilon(1+t)}}, \text{ as } x \to \pm\infty, \tag{48}$$

is a higher order correction.

Now the contact wave $(V^{CD}, U^{CD}, \Theta^{CD})(t,x)$ defined in (47) satisfies the following system

$$\begin{cases} V_t^{CD} - U_{1x}^{CD} = 0, \\ U_{1t}^{CD} + P_x^{CD} = \frac{4\varepsilon}{3}(\frac{\mu(\Theta^{CD})}{V^{CD}}U_{1x}^{CD})_x - \int \xi_1^2 \Pi_{11x}^{CD} d\xi + Q_1^{CD}, \\ U_{it}^{CD} = \varepsilon(\frac{\mu(\Theta^{CD})}{V^{CD}}U_{ix}^{CD})_x - \int \xi_1\xi_i \Pi_{11x}^{CD} d\xi + Q_i^{CD}, i = 2,3, \\ \mathcal{E}_t^{CD} + (P^{CD}U_1^{CD})_x = \varepsilon(\frac{\kappa(\Theta^{CD})}{V^{CD}}\Theta_x^{CD})_x + \frac{4\varepsilon}{3}(\frac{\mu(\Theta^{CD})U_1^{CD}U_{1x}^{CD}}{V^{CD}})_x \\ \quad + \sum_{i=2}^{3} \varepsilon(\frac{\mu(\Theta^{CD})U_i^{CD}U_{ix}^{CD}}{V^{CD}})_x - \int \xi_1 \frac{|\xi|^2}{2}\Pi_{11x}^{CD} d\xi + Q_4^{CD}, \end{cases} \tag{49}$$

where

$$Q_i^{CD} = O(1)\delta^{CD}\varepsilon(1+t)^{-2}e^{-\frac{cx^2}{\varepsilon(1+t)}}, \qquad \text{as } x \to \pm\infty,\ i = 1,2,3,4 \tag{50}$$

with some positive constant $c > 0$ depending only on $\theta_\pm$. Note that from (37), we have

$$|(V^{CD}, U^{CD}, \Theta^{CD})(t,x) - (v^{cd}, u^{cd}, \theta^{cd})(t,x)| = O(1)\delta^{CD}e^{-\frac{cx^2}{2\varepsilon(1+t)}}. \tag{51}$$

1.4. **Shock Profile.** In this subsection, we will recall the shock profile $F^{S_3}(x-\bar{s}_3t,\xi)$ of the Boltzmann equation (1) in Eulerian coordinates with its existence and properties given in the papers by Caflisch-Nicolaenko [7] and Liu-Yu [28], [29]. Then we will state the corresponding properties in the Lagrangian coordinates.

First of all, $F^{S_3}(x-\bar{s}_3t,\xi)$ satisfies

$$\begin{cases} -\bar{s}_3(F^{S_3})'+\xi_1(F^{S_3})'=\frac{1}{\varepsilon}Q(F^{S_3},F^{S_3}), \\ F^{S_3}(\pm\infty,\xi)=\mathbf{M}_\pm(\xi):=\mathbf{M}_{[\rho_\pm,u_\pm,\theta_\pm]}(\xi), \end{cases} \tag{52}$$

where $'=\frac{d}{d\vartheta}$, $\vartheta=x-\bar{s}_3t$, $u_\pm=(u_{1\pm},0,0)$ and $(\rho_\pm,u_\pm,\theta_\pm)$ satisfy Rankine-Hugoniot condition

$$\begin{cases} -\bar{s}_3(\rho_+-\rho_-)+(\rho_+u_{1+}-\rho_-u_{1-})=0, \\ -\bar{s}_3(\rho_+u_{1+}-\rho_-u_{1-})+(\rho_+u_{1+}^2+p_+-\rho_-u_{1-}^2-p_-)=0, \\ -\bar{s}_3(\rho_+E_+-\rho_-E_-)+(\rho_+u_{1+}E_++p_+u_{1+}-\rho_-u_{1-}E_--p_-u_{1-})=0, \end{cases} \tag{53}$$

and Lax entropy condition

$$\lambda_{3+}^E<\bar{s}_3<\lambda_{3-}^E, \tag{54}$$

with $\bar{s}_3$ being 3-shock wave speed and $\lambda_3^E=u_1+\frac{\sqrt{10\theta}}{3}$ being the third characteristic eigenvalue of the Euler equations in the Eulerian coordinate and $\lambda_{3\pm}^E=u_{1\pm}+\frac{\sqrt{10\theta_\pm}}{3}$.

By the macro-micro decomposition around the local Maxwellian $\mathbf{M}^{S_3}$, set

$$F^{S_3}(x,t,\xi)=\mathbf{M}^{S_3}(x,t,\xi)+\mathbf{G}^{S_3}(x,t,\xi),$$

where

$$\mathbf{M}^{S_3}(x,t,\xi)=\mathbf{M}_{[\rho^{S_3},u^{S_3},\theta^{S_3}]}(x,t,\xi)=\frac{\rho^{S_3}(x,t)}{\sqrt{(2\pi R\theta^{S_3}(x,t))^3}}e^{-\frac{|\xi-u^{S_3}(x,t)|^2}{2R\theta^{S_3}(x,t)}}.$$

Now we rewrite this shock profile in Lagrangian coordinate by using the transformation (13) and use $(\tilde{t},\tilde{x})$ for the Lagrangian coordinate to distingish it from the Eulerian coordinate (t,x) at this moment. Then the shock profile in Lagrangian coordinate can be written as $\tilde{F}^{S_3}(\tilde{x}-s_3\tilde{t},\xi)$ with s_3 determined by the 3-shock wave curve given in (22), that is,

$$s_3=\rho_\pm(\bar{s}_3-u_{1\pm}). \tag{55}$$

This shows that under the Lagrangian transformation (13), the shock profile $F^{S_3}(x-\bar{s}_3t,\xi)$ in Eulerian coordinate can be exactly transformed to the shock profile $\tilde{F}^{S_3}(\tilde{x}-s_3\tilde{t},\xi)$ in Lagrangian coordinate. Moreover, we have

$$\tilde{F}_\eta^{S_3}(\eta,\xi)=\rho^{S_3}F_\vartheta^{S_3}(\vartheta,\xi)$$

with $\eta=\tilde{x}-s_3\tilde{t}$.

For simplicity of the notations, from now on, we use (t,x) to denote the Lagrangian coordinate and $F^{S_3}(\eta,\xi)$ with $\eta=x-s_3t$ to denote the 3-shock profile of Boltzmann equation in Lagrangian coordinate. And in the Lagrangian coordinate, we have the following Lemma.

Lemma 1.3. ([16]) *Assume that $(v_-,u_-,\theta_-)\in S_3(v_+,u_+,\theta_+)$, then there exists a unique shock profile $F^{S_3}(\eta,\xi)$ with $\eta=x-s_3t$ up to a shift, to the Boltzmann* *equation (14) in Lagrangian coordinate. Moreover, there are positive constants $c_\pm$ and C such that for $\eta \in \mathbf{R}$,*

$$\begin{cases} s_3 V_\eta^{S_3} = -U_{1\eta}^{S_3} > 0, \\ U_i^{S_3} \equiv 0, \ \int \xi_1 \xi_i \Pi_1^{S_3} d\xi \equiv 0, \ i = 2, 3, \\ (|V^{S_3} - v_\pm|, |U_1^{S_3} - u_{1\pm}|, |\Theta^{S_3} - \theta_\pm|) \leq C\delta^{S_3} e^{-\frac{c_\pm \delta^{S_3}|\eta|}{\varepsilon}}, \quad \text{as } \eta \to \pm\infty, \\ \left(\int \frac{\nu(|\xi|)|\mathbf{G}^{S_3}|^2}{\mathbf{M}_0} d\xi\right)^{\frac{1}{2}} \leq C(\delta^{S_3})^2 e^{-c_\pm \frac{\delta^{S_3}|\eta|}{\varepsilon}}, \quad \text{as } \eta \to \pm\infty. \end{cases}$$

Furthermore, we have

$$V_\eta^{S_3} \sim U_{1\eta}^{S_3} \sim \Theta_\eta^{S_3} \sim \frac{1}{\varepsilon}\left(\int \frac{\nu(|\xi|)|\mathbf{G}^{S_3}|^2}{\mathbf{M}_0} d\xi\right)^{\frac{1}{2}},$$

and

$$|\partial_\eta^k (V^{S_3}, U_1^{S_3}, \Theta^{S_3})| \leq C\frac{(\delta^{S_3})^{k-1}}{\varepsilon^{k-1}} |(V_\eta^{S_3}, U_{1\eta}^{S_3}, \Theta_\eta^{S_3})|, \ k \geq 2,$$

$$\left(\int \frac{\nu(|\xi|)|\partial_\eta^k \mathbf{G}^{S_3}|^2}{\mathbf{M}_0} d\xi\right)^{\frac{1}{2}} \leq C\frac{(\delta^{S_3})^k}{\varepsilon^k}\left(\int \frac{\nu(|\xi|)|\mathbf{G}^{S_3}|^2}{\mathbf{M}_0} d\xi\right)^{\frac{1}{2}}, \ k \geq 1,$$

and

$$\left|\int \xi_1 \varphi_i(\xi) \Pi_{1\eta}^{S_3} d\xi\right| \leq C\delta^{S_3} |U_{1\eta}^{S_3}|, \ i = 1, 2, 3, 4,$$

with $\varphi_i(\xi)$ being the collision invariants.

Furthermore, we have

$$\begin{cases} V_t^{S_3} - U_{1x}^{S_3} = 0, \\ U_{1t}^{S_3} + P_x^{S_3} = \frac{4}{3}\varepsilon\left(\frac{\mu(\Theta^{S_3}) U_{1x}^{S_3}}{V^{S_3}}\right)_x - \int \xi_1^2 \Pi_{1x}^{S_3} d\xi, \\ U_{it}^{S_3} = \varepsilon\left(\frac{\mu(\Theta^{S_3}) U_{ix}^{S_3}}{V^{S_3}}\right)_x - \int \xi_1 \xi_i \Pi_{1x}^{S_3} d\xi, \ i = 2, 3, \\ \mathcal{E}_t^{S_3} + (P^{S_3} U_1^{S_3})_x = \varepsilon\left(\frac{\kappa(\Theta^{S_3}) \Theta_x^{S_3}}{V^{S_3}}\right)_x + \frac{4}{3}\varepsilon\left(\frac{\mu(\Theta^{S_3}) U_1^{S_3} U_{1x}^{S_3}}{V^{S_3}}\right)_x \\ \qquad + \varepsilon \sum_{i=2}^{3}\left(\frac{\mu(\Theta^{S_3}) U_i^{S_3} U_{ix}^{S_3}}{V^{S_3}}\right)_x - \int \xi_1 \frac{|\xi|^2}{2} \Pi_{1x}^{S_3} d\xi, \end{cases} \tag{56}$$

where $\mathcal{E}^{S_3} = \Theta^{S_3} + \frac{|U^{S_3}|^2}{2}$ and $(v_\pm, u_\pm, \theta_\pm)$ satisfy Rankine-Hugoniot condition and Lax entropy condition and s_3 is 3-shock wave speed.

Correspondingly, we have the following equation for the non-fluid part of 3-shock profile.

$$\mathbf{G}_t^{S_3} - \frac{U_1^{S_3}}{V^{S_3}} \mathbf{G}_x^{S_3} + \frac{1}{V^{S_3}} \mathbf{P}_1^{S_3}(\xi_1 \mathbf{M}_x^{S_3}) + \frac{1}{V^{S_3}} \mathbf{P}_1^{S_3}(\xi_1 \mathbf{G}_x^{S_3}) = \frac{1}{\varepsilon}\left[\mathbf{L}_{\mathbf{M}^{S_3}} \mathbf{G}^{S_3} + Q(\mathbf{G}^{S_3}, \mathbf{G}^{S_3})\right].$$

Here, $\mathbf{L}_{\mathbf{M}^{S_3}}$ is the linearized collision operator of $Q(F^{S_3}, F^{S_3})$ with respect to the local Maxwellian $\mathbf{M}^{S_3}$:

$$\mathbf{L}_{\mathbf{M}^{S_3}} g = 2Q(\mathbf{M}^{S_3}, g) = Q(\mathbf{M}^{S_3}, g) + Q(g, \mathbf{M}^{S_3}).$$

Thus

$$\mathbf{G}^{S_3} = \varepsilon \mathbf{L}^{-1}_{\mathbf{M}^{S_3}} \Big[\frac{1}{V^{S_3}} \mathbf{P}_1^{S_3}(\xi_1 \mathbf{M}_x^{S_3}) \Big] + \Pi_1^{S_3},$$
$$\Pi_1^{S_3} = \mathbf{L}^{-1}_{\mathbf{M}^{S_3}} \Big[\varepsilon \Big(\mathbf{G}_t^{S_3} - \frac{U_1^{S_3}}{V^{S_3}} \mathbf{G}_x^{S_3} + \frac{1}{V^{S_3}} \mathbf{P}_1^{S_3}(\xi_1 \mathbf{G}_x^{S_3}) \Big) - Q(\mathbf{G}^{S_3}, \mathbf{G}^{S_3}) \Big]. \tag{57}$$

1.5. **Hyperbolic Wave II.** The purpose of this subsection is to construct the second hyperbolic wave. Up to now, we can define the following approximate composite wave profile $(\bar V, \bar U, \bar{\mathcal{E}})(t,x)$

$$\begin{pmatrix} \bar V \\ \bar U_1 \\ \bar{\mathcal{E}} \end{pmatrix}(t,x) = \begin{pmatrix} V^{R_1} + d_1 + V^{CD} + V^{S_3} \\ U_1^{R_1} + d_2 + U_1^{CD} + U_1^{S_3} \\ \mathcal{E}^{R_1} + d_3 + \mathcal{E}^{CD} + \mathcal{E}^{S_3} \end{pmatrix}(t,x) - \begin{pmatrix} v_* + v^* \\ u_{1*} + u_1^* \\ E_* + E^* \end{pmatrix}, \tag{58}$$
$$\bar U_i = U_i^{CD}, i = 2,3,$$

where $\bar{\mathcal{E}} = \bar\Theta + \frac{|\bar U|^2}{2}$, $(V^{R_1}, U_1^{R_1}, \mathcal{E}^{R_1})(t,x)$ is the 1-rarefaction wave defined in (26) with the right state (v_+, u_{1+}, E_+) replaced by (v_*, u_{1*}, E_*), $(V^{CD}, U_1^{CD}, \mathcal{E}^{CD})(t,x)$ is the viscous contact wave defined in (47) with the states (v_-, u_{1-}, E_-) and (v_+, u_{1+}, E_+) replaced by (v_*, u_{1*}, E_*) and (v^*, u_1^*, E^*) respectively, and $(V^{S_3}, U_1^{S_3}, \mathcal{E}^{S_3})(t,x)$ is the fluid part of 3-shock profile of Boltzmann equation defined in (56) with the left state (v_-, u_{1-}, E_-) replaced by (v^*, u_1^*, E^*).

Moreover, we can check that this profile satisfies

$$\begin{cases} \bar V_t - \bar U_{1x} = 0, \\ \bar U_{1t} + \bar P_x = \frac{4}{3}\varepsilon(\frac{\mu(\bar\Theta)\bar U_{1x}}{\bar V})_x - \int \xi_1^2 \Pi_{11x}^{CD} d\xi - \int \xi_1^2 \Pi_{1x}^{S_3} d\xi + \bar Q_{1x} + Q_1^{CD}, \\ \bar U_{it} = \varepsilon(\frac{\mu(\bar\Theta)\bar U_{1x}}{\bar V})_x - \int \xi_1\xi_i \Pi_{11x}^{CD} d\xi - \int \xi_1\xi_i \Pi_{1x}^{S_3} d\xi + \bar Q_{ix} + Q_i^{CD}, \ i = 2,3, \\ \bar{\mathcal{E}}_t + (\bar P \bar U_1)_x = \varepsilon(\frac{\kappa(\bar\Theta)\bar\Theta_x}{\bar V})_x + \frac{4}{3}\varepsilon(\frac{\mu(\bar\Theta)\bar U_1 \bar U_{1x}}{\bar V})_x + \sum_{i=2}^{3} \varepsilon(\frac{\mu(\bar\Theta)\bar U_i \bar U_{ix}}{\bar V})_x \\ \qquad - \int \xi_1 \frac{|\xi|^2}{2} \Pi_{11x}^{CD} d\xi - \int \xi_1 \frac{|\xi|^2}{2} \Pi_{1x}^{S_3} d\xi + \bar Q_{4x} + Q_4^{CD}, \end{cases} \tag{59}$$

where $\bar P = p(\bar V, \bar\Theta)$, Q_i^{CD} $(i = 1,2,3,4)$ are defined in (50), and $\bar Q_i$ $(i = 1,2,3,4)$ represent the interaction of waves in different families, and the error terms coming from the approximate rarefaction wave and the hyperbolic wave I, which can be estimated in the stability analysis.

In order to remove the non-conservative error terms $Q_i^{CD}, (i = 1,2,3,4)$ coming from the definition of the viscous contact wave, we now introduce the following hyperbolic wave $\vec b = (b_1, b_{21}, b_{22}, b_{23}, b_3)$:

$$\begin{cases} b_{1t} - b_{21x} = 0, \\ b_{21t} + [\bar P_v b_1 + \bar P_{u_1} b_{21} + \bar P_{u_2} b_{22} + \bar P_{u_3} b_{23} + \bar P_E b_3]_x = -Q_1^{CD}, \\ b_{22t} = -Q_2^{CD}, \\ b_{23t} = -Q_3^{CD}, \\ b_{3t} + [(\bar P \bar U_1)_v b_1 + (\bar P \bar U_1)_{u_1} b_{21} + (\bar P \bar U_1)_{u_2} b_{22} + (\bar P \bar U_1)_{u_3} b_{23} + (\bar P \bar U_1)_E b_3]_x = -Q_4^{CD}, \end{cases} \tag{60}$$

where $P = \frac{2\Theta}{3V} = P(V, U, \mathcal{E})$ and $\bar P_v = P_v(\bar V, \bar U, \bar{\mathcal{E}})$, etc. Now we impose the following boundary condition to the linear hyperbolic system (60) on the domain

$(t, x) \in [h, T] \times \mathbf{R}$:

$$(B_1, B_{21}, B_{22}, B_{23}, B_3)(t = T, x) = 0. \tag{61}$$

As for the hyperbolic wave I, We can solve the linear hyperbolic system under the condition (61) to have the following lemma.

Lemma 1.4. *There exists a positive constant δ_0 such that if the wave strength $\delta \le \delta_0$, then there exists a positive constant $C_{h,T}$ which is independent of ε, such that*

$$\begin{aligned}
&\left\| \frac{\partial^k}{\partial x^k}(b_1, b_{21}, b_{22}, b_{23}, b_3)(t, \cdot) \right\|^2_{L^2(dx)} \\
&+ \int_h^T \left\| \sqrt{|U^{S_3}_{1x}|}\frac{\partial^k}{\partial x^k}(b_1, b_{21}, b_{22}, b_{23}, b_3)(t, \cdot) \right\|^2_{L^2(dx)} dt \\
&\le C_{h,T}\ \varepsilon^{\frac{5}{2}-2k},\ k = 0, 1, 2, 3.
\end{aligned} \tag{62}$$

1.6. **Superposition of Waves.** With the above preparation, finally, the approximate superposition wave $(V, U, \mathcal{E})(t, x)$ can be defined by

$$\begin{pmatrix} V \\ U_i \\ \mathcal{E} \end{pmatrix}(t, x) = \begin{pmatrix} \bar{V} + b_1 \\ \bar{U}_i + b_{2i} \\ \bar{\mathcal{E}} + b_3 \end{pmatrix}(t, x),\ i = 1, 2, 3, \tag{63}$$

where $\mathcal{E} = \Theta + \frac{|U|^2}{2}$. Thus, we have

$$\Theta = \bar{\Theta} - \sum_{i=1}^{3} \bar{U}_i b_{2i} + b_3 - \frac{|b_2|^2}{2}, \tag{64}$$

where $b_2 = (b_{21}, b_{22}, b_{23})^t$ and $|b_2|^2 = \sum_{i=1}^3 b_{2i}^2$.

From the construction of the contact wave and Lemma 1.1 and by noting that $\sigma = \varepsilon^{\frac{1}{5}}$, we have the following relation between the approximate wave pattern $(V, U, \mathcal{E}, \Theta)(t, x)$ of the Boltzmann equation and the inviscid superposition wave pattern $(\widetilde{V}, \widetilde{U}, \widetilde{\mathcal{E}}, \widetilde{\Theta})(t, x)$ to the Euler equations

$$\begin{aligned}
&|(V, U, \mathcal{E}, \Theta)(t, x) - (\widetilde{V}, \widetilde{U}, \widetilde{\mathcal{E}}, \widetilde{\Theta})(t, x)| \\
&\le C_{h,T}\Big[\varepsilon^{\frac{1}{5}}|\ln\varepsilon| + \delta^{CD}e^{-\frac{cx^2}{\varepsilon(1+t)}} + \delta^{S_3}e^{-c\frac{\delta^{S_3}|x-s_3t|}{\varepsilon}}\Big].
\end{aligned} \tag{65}$$

Moreover, the approximate wave pattern $(V, U, \mathcal{E}, \Theta)(t, x)$ satisfies

$$\begin{cases}
V_t - U_{1x} = 0, \\
U_{1t} + P_x = \frac{4}{3}\varepsilon(\frac{\mu(\Theta)U_{1x}}{V})_x - \int \xi_1^2\Pi^{CD}_{11x}d\xi - \int \xi_1^2\Pi^{S_3}_{1x}d\xi + \bar{Q}_{1x} + Q_{1x}, \\
U_{it} = \varepsilon(\frac{\mu(\Theta)U_{ix}}{V})_x - \int \xi_1\xi_i\Pi^{CD}_{11x}d\xi - \int \xi_1\xi_i\Pi^{S_3}_{1x}d\xi + \bar{Q}_{ix} + Q_{ix},\ i = 2, 3, \\
\mathcal{E}_t + (PU_1)_x = \varepsilon(\frac{\kappa(\Theta)\Theta_x}{V})_x + \frac{4}{3}\varepsilon(\frac{\mu(\Theta)U_1U_{1x}}{V})_x + \sum_{i=2}^{3}\varepsilon(\frac{\mu(\Theta)U_iU_{ix}}{V})_x \\
\qquad - \int \xi_1\frac{|\xi|^2}{2}\Pi^{CD}_{11x}d\xi - \int \xi_1\frac{|\xi|^2}{2}\Pi^{S_3}_{1x}d\xi + \bar{Q}_{4x} + Q_{4x},
\end{cases} \tag{66}$$

where $P = p(V, \Theta)$ and

$$
\begin{aligned}
Q_1 &= \Big[P - \bar{P} - (\bar{P}_v b_1 + \bar{P}_u \cdot b_2 + \bar{P}_E b_3)\Big] - \frac{4}{3}\varepsilon\Big[\frac{\mu(\Theta)U_{1x}}{V} - \frac{\mu(\bar{\Theta})\bar{U}_{1x}}{\bar{V}}\Big],\\
Q_i &= -\varepsilon\Big[\frac{\mu(\Theta)U_{ix}}{V} - \frac{\mu(\bar{\Theta})\bar{U}_{ix}}{\bar{V}}\Big], \ i = 2,3,\\
Q_4 &= \Big[PU_1 - \bar{P}\bar{U}_1 - ((\bar{P}\bar{U}_1)_v b_1 + (\bar{P}\bar{U}_1)_u \cdot b_2 + (\bar{P}\bar{U}_1)_E b_3)\Big]\\
&\quad -\varepsilon\Big[(\frac{\kappa(\Theta)\Theta_x}{V} - \frac{\kappa(\bar{\Theta})\bar{\Theta}_x}{\bar{V}}) + \frac{4}{3}(\frac{\mu(\Theta)U_1U_{1x}}{V} - \frac{\mu(\bar{\Theta})\bar{U}_1\bar{U}_{1x}}{\bar{V}})\\
&\quad + \sum_{i=2}^{3}(\frac{\mu(\Theta)U_iU_{ix}}{V} - \frac{\mu(\bar{\Theta})\bar{U}_i\bar{U}_{ix}}{\bar{V}})\Big].
\end{aligned}
\tag{67}
$$

2. Main Result.

2.1. Reformulation of the Problem.

We now reformulate the system by introducing a scaling for the independent variables. Set

$$
y = \frac{x}{\varepsilon}, \ \tau = \frac{t}{\varepsilon}. \tag{68}
$$

In the following, we will also use the notations $(v, u, \theta)(\tau, y), \mathbf{G}(\tau, y, \xi), \Pi_1(\tau, y, \xi)$ and $(V, U, \Theta)(\tau, y)$, etc., in the scaled independent variables. Set the perturbation around the superposition wave $(V, U, \Theta)(\tau, y)$ by

$$
\begin{aligned}
&(\phi, \psi, \omega, \zeta)(\tau, y) = (v - V, u - U, E - \mathcal{E}, \theta - \Theta)(\tau, y),\\
&\widetilde{\mathbf{G}}(\tau, y, \xi) = \mathbf{G}(\tau, y, \xi) - \mathbf{G}^{S_3}(\tau, y, \xi),\\
&\widetilde{f}(\tau, y, \xi) = f(\tau, y, \xi) - F^{S_3}(\tau, y, \xi).
\end{aligned}
\tag{69}
$$

Under this scaling, the hydrodynamic limit problem is reduced to a time asymptotic stability problem for the Boltzmann equation.

In particular, we can choose the initial value as

$$
(\phi, \psi, \omega)(\tau = \frac{h}{\varepsilon}, y) = (0, 0, 0), \quad \widetilde{\mathbf{G}}(\tau = \frac{h}{\varepsilon}, y, \xi) = 0. \tag{70}
$$

Introduce the anti-derivative variables

$$
(\Phi, \Psi, \bar{W})(\tau, y) = \int_{-\infty}^{y} (\phi, \psi, \omega)(\tau, y')dy'.
$$

Then $(\Phi, \Psi, \bar{W})(\tau, y)$ satisfies that

$$
\begin{cases}
\Phi_\tau - \Psi_{1y} = 0,\\
\Psi_{1\tau} + (p - P) = \frac{4}{3}\Big(\frac{\mu(\theta)u_{1y}}{v} - \frac{\mu(\Theta)U_{1y}}{V}\Big) - \int \xi_1^2(\Pi_1 - \Pi_{11}^{CD} - \Pi_1^{S_3})d\xi - \bar{Q}_1 - Q_1,\\
\Psi_{i\tau} = \Big(\frac{\mu(\theta)u_{iy}}{v} - \frac{\mu(\Theta)U_{iy}}{V}\Big) - \int \xi_1\xi_i(\Pi_1 - \Pi_{11}^{CD} - \Pi_1^{S_3})d\xi - \bar{Q}_i - Q_i, \ i = 2,3,\\
\bar{W}_\tau + (pu_1 - PU_1) = \Big(\frac{\kappa(\theta)\theta_y}{v} - \frac{\kappa(\Theta)\Theta_y}{V}\Big) + \frac{4}{3}\Big(\frac{\mu(\theta)u_1u_{1y}}{v} - \frac{\mu(\Theta)U_1U_{1y}}{V}\Big)\\
\quad + \sum_{i=2}^{3}\Big(\frac{\mu(\theta)u_iu_{iy}}{v} - \frac{\mu(\Theta)U_iU_{iy}}{V}\Big) - \int \xi_1\frac{|\xi|^2}{2}(\Pi_1 - \Pi_{11}^{CD} - \Pi_1^{S_3})d\xi - \bar{Q}_4 - Q_4.
\end{cases}
\tag{71}
$$

To precisely capture the dissiaption of heat conduction, we introduce another variable related to the absolute temperature

$$W = \bar{W} - U \cdot \Psi = \bar{W} - \sum_{i=1}^{3} U_i \Psi_i,$$

then

$$\zeta = W_y - (\frac{|\Psi_y|^2}{2} - U_y \cdot \Psi). \tag{72}$$

Linearizing the system (71) around the approximate wave pattern $(V, U, \Theta)(\tau, y)$ implies that

$$\begin{cases} \Phi_\tau - \Psi_{1y} = 0, \\ \Psi_{1\tau} - \frac{Z}{V}\Phi_y + \frac{2}{3V}W_y + \frac{2}{3V}U_y \cdot \Psi - \frac{4}{3}\frac{\mu'(\Theta)}{V}(W_y + U_y \cdot \Psi)U_{1y} = \frac{4}{3}\frac{\mu(\Theta)}{V}\Psi_{1yy} \\ \qquad - \int \xi_1^2(\Pi_1 - \Pi_{11}^{CD} - \Pi_1^{S_3})d\xi + N_1 - \bar{Q}_1 - Q_1, \\ \Psi_{i\tau} + \frac{\mu(\Theta)U_{iy}}{V^2}\Phi_y - \frac{\mu'(\Theta)}{V}(W_y + U_y \cdot \Psi)U_{iy} = \frac{\mu(\Theta)}{V}\Psi_{iyy} \\ \qquad - \int \xi_1\xi_i(\Pi_1 - \Pi_{11}^{CD} - \Pi_1^{S_3})d\xi + N_i - \bar{Q}_i - Q_i, i = 2, 3, \\ W_\tau + Z\Psi_{1y} - \sum_{i=2}^{3}\frac{\mu(\Theta)U_{iy}}{V}\Psi_{iy} + U_\tau \cdot \Psi - \frac{\kappa(\Theta)}{V}(U_y \cdot \Psi)_y + \frac{\kappa(\Theta)}{V^2}\Theta_y\Phi_y \\ -\frac{\kappa'(\Theta)}{V}(W_y + U_y \cdot \Psi)\Theta_y = \frac{\kappa(\Theta)}{V}W_{yy} - \int \xi_1\frac{|\xi|^2}{2}(\Pi_1 - \Pi_{11}^{CD} - \Pi_1^{S_3})d\xi \\ +\sum_{i=1}^{3}U_i\int \xi_1\xi_i(\Pi_1 - \Pi_{11}^{CD} - \Pi_1^{S_3})d\xi + N_4 - \bar{Q}_4 + \sum_{i=1}^{3}U_i\bar{Q}_i - Q_4 + \sum_{i=1}^{3}U_iQ_i, \end{cases} \tag{73}$$

where

$$Z = P - \frac{4}{3}\frac{\mu(\Theta)U_{1y}}{V}, \tag{74}$$

$$N_i = O(1)\Big[|\Phi_y|^2 + |\Psi_y|^2 + |\zeta|^2 + |\Psi_{yy}|^2 + |\zeta_y|^2\Big], i = 1, 2, 3, 4. \tag{75}$$

We now derive the equation for the non-fluid component $\widetilde{\mathbf{G}}(\tau, y, \xi)$ in the scaled independent variables. From (16), we have

$$\begin{aligned} \widetilde{\mathbf{G}}_\tau - \mathbf{L_M}\widetilde{\mathbf{G}} &= \frac{u_1}{v}\widetilde{\mathbf{G}}_y - \frac{1}{v}\mathbf{P}_1(\xi_1\widetilde{\mathbf{G}}_y) - \Big[\frac{1}{v}\mathbf{P}_1(\xi_1\mathbf{M}_y) - \frac{1}{V^{S_3}}\mathbf{P}_1^{S_3}(\xi_1\mathbf{M}_y^{S_3})\Big] \\ &+2Q(\widetilde{\mathbf{G}}, \mathbf{G}^{S_3}) + Q(\widetilde{\mathbf{G}}, \widetilde{\mathbf{G}}) + J_1, \end{aligned} \tag{76}$$

where

$$J_1 = (\mathbf{L_M} - \mathbf{L_{M^{S_3}}})\mathbf{G}^{S_3} + (\frac{u}{v} - \frac{U_1^{S_3}}{V^{S_3}})\mathbf{G}_y^{S_3} - \Big[\frac{1}{v}\mathbf{P}_1(\xi_1\mathbf{G}_y^{S_3}) - \frac{1}{V^{S_3}}\mathbf{P}_1^{S_3}(\xi_1\mathbf{G}_y^{S_3})\Big]. \tag{77}$$

Let

$$\mathbf{G}^{R_1}(\tau, y, \xi) = \frac{3}{2v\theta}\mathbf{L_M^{-1}}\{\mathbf{P}_1[\xi_1(\frac{|\xi - u|^2}{2\theta}\Theta_y^{R_1} + \xi \cdot U_y^{R_1})\mathbf{M}]\}, \tag{78}$$

and

$$\widetilde{\mathbf{G}}_1(\tau, y, \xi) = \widetilde{\mathbf{G}}(\tau, y, \xi) - \mathbf{G}^{R_1}(\tau, y, \xi) - \mathbf{G}^{CD}(\tau, y, \xi), \tag{79}$$

where $\mathbf{G}^{CD}(\tau, y, \xi)$ is defined in (39). Then $\mathbf{G}_1(\tau, y, \xi)$ satisfies

$$\widetilde{\mathbf{G}}_{1\tau} - \mathbf{L}_{\mathbf{M}}\widetilde{\mathbf{G}}_1 = \frac{u_1}{v}\widetilde{\mathbf{G}}_y - \frac{1}{v}\mathbf{P}_1(\xi_1\widetilde{\mathbf{G}}_y) + 2Q(\widetilde{\mathbf{G}}, \mathbf{G}^{S_3}) + Q(\widetilde{\mathbf{G}}, \widetilde{\mathbf{G}}) + J_1 + J_2 - \mathbf{G}^{R_1}_\tau - \mathbf{G}^{CD}_\tau. \tag{80}$$

with

$$\begin{aligned} J_2 &= -\Big[\frac{1}{v}\mathbf{P}_1(\xi_1\mathbf{M}_y) - \frac{1}{V^{S_3}}\mathbf{P}_1^{S_3}(\xi_1\mathbf{M}^{S_3}_y) \\ &\quad - \frac{3}{2v\theta}\mathbf{P}_1\Big(\xi_1\big(\frac{|\xi-u|^2}{2\theta}(\Theta^{R_1}_y + \Theta^{CD}_y) + \xi\cdot(U^{R_1}_y + U^{CD}_y)\big)\mathbf{M}\Big)\Big]. \end{aligned} \tag{81}$$

From (14) and the scaling transformation (68), we have

$$f_\tau - \frac{u_1}{v}f_y + \frac{\xi_1}{v}f_y = Q(f, f). \tag{82}$$

Thus, we have the equation for $\widetilde{f}$ defined in (69)

$$\widetilde{f}_\tau - \frac{u_1}{v}\widetilde{f}_y + \frac{\xi_1}{v}\widetilde{f}_y = \mathbf{L}_{\mathbf{M}}\widetilde{\mathbf{G}} + Q(\widetilde{\mathbf{G}}, \widetilde{\mathbf{G}}) + J_F, \tag{83}$$

with

$$J_F = (\frac{u_1}{v} - \frac{U^{S_3}_1}{V^{S_3}})F^{S_3}_y - (\frac{1}{v} - \frac{1}{V^{S_3}})\xi_1 F^{S_3}_y + 2Q(\mathbf{M} - \mathbf{M}^{S_3}, \mathbf{G}^{S_3}) + 2Q(\widetilde{\mathbf{G}}, \mathbf{G}^{S_3}). \tag{84}$$

Note that to prove the main theorem, it is sufficient to prove the following theorem on the Boltzmann equation (82) in the scaled independent variables based on the construction of the approximate wave pattern.

Theorem 2.1. *There exist a small positive constants δ_1 and a global Maxwellian $\mathbf{M}_\star = \mathbf{M}_{[v_\star, u_\star, \theta_\star]}$ such that if the wave strength δ satisfies $\delta \le \delta_1$, then on the time interval $[\frac{h}{\varepsilon}, \frac{T}{\varepsilon}]$ for any $0 < h < T$, there is a positive constant $\varepsilon_1(\delta, h, T)$. If the Knudsen number $\varepsilon \le \varepsilon_1$, then the problem (82) admits a family of smooth solution $f^{\varepsilon,h}(\tau, y, \xi)$ satisfying*

$$\sup_{\tau\in[\frac{h}{\varepsilon},\frac{T}{\varepsilon}]}\sup_{y\in\mathbf{R}} \|f^{\varepsilon,h}(\tau, y, \xi) - \mathbf{M}_{[V,U,\Theta]}(\tau, y, \xi)\|_{L^2_\xi(\frac{1}{\sqrt{\mathbf{M}_\star}})} \le C\varepsilon^{\frac{1}{5}}. \tag{85}$$

Consider the reformulated system (73) and (80). Since the local existence of solution to (73) and (80) is known, cf. [15] and [35], to prove the existence on the time interval $[\frac{h}{\varepsilon}, \frac{T}{\varepsilon}]$, we only need to close the following a priori estimate by the continuity argument. Set

$$\begin{aligned} \mathcal{N}(\tau) = \sup_{\frac{h}{\varepsilon}\le\tau'\le\tau} &\Big\{ \|(\Phi, \Psi, W)(\tau', \cdot)\|^2 + \|(\phi, \psi, \zeta)(\tau', \cdot)\|_1^2 + \int\int \frac{|\widetilde{\mathbf{G}}_1|^2}{\mathbf{M}_\star} d\xi dy \\ &+ \sum_{|\alpha'|=1}\int\int \frac{|\partial^{\alpha'}\widetilde{\mathbf{G}}|^2}{\mathbf{M}_\star} d\xi dy + \sum_{|\alpha|=2}\int\int \frac{|\partial^\alpha \widetilde{f}|^2}{\mathbf{M}_\star} d\xi dy \Big\} \le \chi^2 = \varepsilon^{\frac{1}{10}},\ \forall \tau \in [\frac{h}{\varepsilon}, \frac{T}{\varepsilon}], \end{aligned} \tag{86}$$

where $\partial^\alpha, \partial^{\alpha'}$ denote the derivatives with respect to y and τ, and $\mathbf{M}_\star$ is a global Maxwellian to be chosen.

2.2. **Energy Estimates.** To close the a priori estimate (86) and to prove Theorem 2.1, we need the following energy estimates given in Propositions 1 and Proposition 2. First, the lower order estimates to the system (73) and (80) are given in the following Proposition.

Proposition 1. *Under the assumptions of Theorem 2.1, there exist positive constants C and $C_{h,T}$ independent of ε such that*

$$
\begin{aligned}
&\sup_{\frac{h}{\varepsilon}\le\tau_1\le T}\Big[\|(\Phi,\Psi,W,\Phi_y)(\tau_1,\cdot)\|^2+\int\int\frac{|\widetilde{\mathbf{G}}_1|^2}{\mathbf{M}_\star}(\tau_1,y,\xi)d\xi dy\Big]\\
&\quad+\int_{\frac{h}{\varepsilon}}^{T}\Big[\|\sqrt{|U_{1y}^{S_3}|}(\Psi,W)\|^2+\|(\Phi_y,\Psi_y,W_y,\zeta,\Psi_\tau,W_\tau)\|^2\Big]d\tau+\int_{\frac{h}{\varepsilon}}^{T}\int\int\frac{\nu(|\xi|)}{\mathbf{M}_\star}|\widetilde{\mathbf{G}}_1|^2d\xi dyd\tau\\
&\le C_{h,T}\ \varepsilon\int_{\frac{h}{\varepsilon}}^{T}\|(\Psi,W)\|^2d\tau+C\sum_{|\alpha'|=1}\int_{\frac{h}{\varepsilon}}^{T}\|\partial^{\alpha'}(\phi,\psi,\zeta)\|^2d\tau\\
&\quad+C\sum_{|\alpha'|=1}\int_{\frac{h}{\varepsilon}}^{T}\int\int\frac{\nu(|\xi|)}{\mathbf{M}_\star}|\partial^{\alpha'}\widetilde{\mathbf{G}}|^2d\xi dyd\tau+C_{h,T}\ \varepsilon^{\frac{2}{5}}.
\end{aligned}
$$

The higher order estimates are given as follows,

Proposition 2. *Under the assumptions of Theorem 2.1, there exist positive constants C and $C_{h,T}$ independent of ε such that*

$$
\begin{aligned}
&\sup_{\frac{h}{\varepsilon}\le\tau_1\le T}\Big[\|(\phi,\psi,\zeta,\phi_y,\psi_y,\zeta_y)(\tau_1,\cdot)\|^2+\sum_{|\alpha'|=1}\int\int\frac{|\partial^{\alpha'}\widetilde{\mathbf{G}}|^2}{\mathbf{M}_\star}(\tau_1,y,\xi)d\xi dy\\
&\qquad+\sum_{|\alpha|=2}\int\int\frac{|\partial^{\alpha}\widetilde{f}|^2}{2\mathbf{M}_\star}(\tau_1,y,\xi)d\xi dy\Big]\\
&\qquad+\int_{\frac{h}{\varepsilon}}^{T}\sum_{1\le|\alpha|\le2}\|\partial^{\alpha}(\phi,\psi,\zeta)\|^2d\tau+\sum_{1\le|\alpha|\le2}\int_{\frac{h}{\varepsilon}}^{T}\int\int\frac{\nu(|\xi|)}{\mathbf{M}_\star}|\partial^{\alpha}\widetilde{\mathbf{G}}|^2d\xi dyd\tau\\
&\le C(\delta+C_{h,T}\chi)\int_{\frac{h}{\varepsilon}}^{T}\int\int\frac{\nu(|\xi|)}{\mathbf{M}_\star}|\widetilde{\mathbf{G}}_1|^2d\xi dyd\tau\\
&\qquad+C(\delta+C_{h,T}\chi)\int_{\frac{h}{\varepsilon}}^{T}\|(\phi,\psi,\zeta)\|^2d\tau+C_{h,T}\ \varepsilon^{\frac{1}{2}}.
\end{aligned}
$$

By combining the above lower and higher order estimates given in Propositions 1 and 2 and choosing the wave strength δ, and the Knudsen number ε to be suitably small, we obtain

$$
\begin{aligned}
&\mathcal{N}(\tau)+\int_{\frac{h}{\varepsilon}}^{T}\Big[\sum_{0\le|\alpha|\le2}\|\partial^{\alpha}(\phi,\psi,\zeta)\|^2+\|\sqrt{|U_{1y}^{S_3}|}(\Psi,W)\|^2\Big]d\tau\\
&\quad+\int_{\frac{h}{\varepsilon}}^{T}\int\int\frac{\nu(|\xi|)|\widetilde{\mathbf{G}}_1|^2}{\mathbf{M}_\star}d\xi dyd\tau+\sum_{1\le|\alpha|\le2}\int_{\frac{h}{\varepsilon}}^{T}\int\int\frac{\nu(|\xi|)|\partial^{\alpha}\widetilde{\mathbf{G}}|^2}{\mathbf{M}_\star}(\tau,y,\xi)d\xi dyd\tau\\
&\qquad\le C_{h,T}\ \varepsilon^{\frac{2}{5}}.
\end{aligned}
$$

Therefore, we close the a priori assumption (86) and then complete the proof of Theorem 2.1.

REFERENCES

[1] F. V. Atkinson and L. A. Peletier, Similarity solutions of the nonlinear diffusion equation, *Arch. Rat. Mech. Anal.*, **54** (1974), 373–392.

[2] C. Bardos, F. Golse & D. Levermore, Fluid dynamic limits of kinetic equations, I. Formal derivations, *J. Statis. Phys.*, **63**, 323-344, 1991; II. Convergence proofs for the Boltzmann equation, Comm. Pure Appl. Math., **46**, 667-753, 1993.

[3] C. Bardos, C. Levermore, S. Ukai, T. Yang, Kinetic equations: fluid dynamical limits and viscous heating, *Bull. Inst. Math. Acad. Sin.* (N.S.) **3** (2008), 1-49.

[4] C. Bardos and S. Ukai, The classical incompressible Navier-Stokes limit of the Boltzmann equation, *Math. Models Methods Appl. Sci.* **1**, (1991), 235-257.

[5] L. Boltzmann, (translated by Stephen G. Brush), "Lectures on Gas Theory," Dover Publications, Inc. New York, 1964.
[6] R. E. Caflisch, The fluid dynamical limit of the nonlinear Boltzmann equation, *Comm. Pure Appl. Math.*, **33** (1980), 491–508.
[7] R. E. Caflisch, B. Nicolaenko, Shock profile solutions of the Boltzmann equation, *Comm. Math. Phys.*, **86** (1982), 161–194.
[8] C. Cercignani, R. Illner and M. Pulvirenti, "The Mathematical Theory of Dilute Gases," Springer-Verlag, Berlin, 1994.
[9] S. Chapman and T. G. Cowling, "The Mathematical Theory of Non-Uniform Gases," 3rd edition, Cambridge University Press, 1990.
[10] R. Courant, K. O. Friedrichs, Supersonic flow and shock waves, Wiley-Interscience: New York, 1948.
[11] R. Esposito and M. Pulvirenti, From particle to fluids, in "Handbook of Mathematical Fluid Dynamics," Vol. **III**, North-Holland, Amsterdam, (2004), 1–82.
[12] F. Golse, The Boltzmann equation and its hydrodynamic limits, Evolutionary equations. Vol. II, 159-301, Handb. Differ. Equ., Elsevier/North-Holland, Amsterdam, 2005.
[13] F. Golse, L. Saint-Raymond, The incompressible Navier-Stokes limit of the Boltzmann equation for hard cutoff potentials, *J. Math. Pures Appl.*, **91** (2009), 508-552.
[14] H. Grad, "Asymptotic Theory of the Boltzmann Equation II," in "Rarefied Gas Dynamics" (J. A. Laurmann, ed.), Vol. **1**, Academic Press, New York, (1963), 26–59.
[15] Y. Guo, The Boltzmann equation in the whole space, *Indiana Univ. Math. J.*, **53** (2004), 1081–1094.
[16] F. M. Huang, Y. Wang, Y. Wang and T. Yang, The limit of the Boltzmann equation to the Euler equations for Riemann problems, *SIAM J. Math. Anal.*, **45** (2013), No.3, 1741–1811.
[17] F. M. Huang, Y. Wang and T. Yang, Hydrodynamic limit of the Boltzmann equation with contact discontinuities, *Comm. Math. Phy.*, **295** (2010), 293–326.
[18] F. M. Huang, Y. Wang and T. Yang, Fluid Dynamic Limit to the Riemann Solutions of Euler Equations: I. Superposition of rarefaction waves and contact discontinuity, *Kinet. Relat. Models*, **3** (2010), 685–728.
[19] F. M. Huang, Y. Wang and T. Yang, Vanishing Viscosity Limit of the Compressible Navier-Stokes Equations for Solutions to Riemann Problem, *Arch. Ration. Mech. Anal.*, **203** (2012), no. 2, 379-413.
[20] F. M. Huang, Z. P. Xin and T. Yang, Contact discontinuities with general perturbation for gas motion, *Adv. Math.*, **219** (2008), 1246–1297.
[21] F. M. Huang, T. Yang, Stability of contact discontinuity for the Boltzmann equation, *J. Differential Equations*, **229** (2006), 698-742.
[22] S. Kawashima, A. Matsumura, T. Nishida, On the fluid-dynamical approximation to the Boltzmann equation at the level of the Navier-Stokes equation, *Comm. Math. Phys.*, **70** (1979), 97–124.
[23] C. Levermore, N. Masmoudi, From the Boltzmann equation to an incompressible Navier-Stokes-Fourier system, *Arch. Ration. Mech. Anal.*, **196** (2010), 753-809.
[24] M. Lachowicz, On the initial layer and existence theorem for the nonlinear Boltzmann equation, *Math. Methods Appl.Sci.*, **9** (1987), 342–366.
[25] P. Lax, Hyperbolic systems of conservation laws, II, *Comm. Pure Appl. Math.*, **10** (1957), 537-566.
[26] T. Liu, T. Yang, and S. H. Yu, Energy method for the Boltzmann equation, *Physica D*, **188** (2004), 178–192.
[27] T. Liu, T. Yang, S. H. Yu and H. J. Zhao, Nonlinear stability of rarefaction waves for the Boltzmann equation, *Arch. Rat. Mech. Anal.*, **181** (2006), 333–371.
[28] T. Liu and S. H. Yu, Boltzmann equation: Micro-macro decompositions and positivity of shock profiles, *Commun. Math. Phys.*, **246** (2004), 133–179.
[29] T. Liu and S. H. Yu, Invariant manifolds for steady Boltzmann flows and applications, *Arch. Ration. Mech. Anal.*, **209** (2013), no. 3, 869-997.
[30] A. Matsumura and K. Nishihara, Asymptotics toward the rarefaction wave of the solutions of a one-dimensional model system for compressible viscous gas, *Japan J. Appl. Math.*, **3** (1986), 1–13.
[31] T. Nishida, Fluid dynamical limit of the nonlinear Boltzmann equation to the level of the compressible Euler equation, *Commun. Math. Phys.*, **61** (1978), 119–148.
[32] J. Smoller, "Shock Waves and Reaction-Diffusion Equations," New York: Springer, 1994.

[33] Y. Sone, Kinetic Theory and Fluid Dynamics, Birkhäuser, Boston, 2002.

[34] S. Ukai and K. Asano, The Euler limit and the initial layer of the nonlinear Boltzmann equation, *Hokkaido Math. J.*, **12** (1983), 303–324.

[35] S. Ukai, T. Yang and H. J. Zhao, Global solutions to the Boltzmann equation with external forces, *Analysis and Applications*, **3** (2005), 157–193.

[36] Z. P. Xin, Zero dissipation limit to rarefaction waves for the one-dimentional Navier-Stokes equations of compressible isentropic gases, *Commun. Pure Appl. Math*, **XLVI** (1993), 621–665.

[37] Z. P. Xin and H. H. Zeng, Convergence to the rarefaction waves for the nonlinear Boltzmann equation and compressible Navier-Stokes equations, *J. Differential Equations.*, **249** (2010), 827–871.

[38] S. H. Yu, Hydrodynamic limits with shock waves of the Boltzmann equations, *Commun. Pure Appl. Math.* **58** (2005), 409–443.

E-mail address: fhuang@amt.ac.cn

WELL-POSEDNESS OF BOUNDARY LAYER PROBLEM IN WIND-DRIVEN OCEANIC CIRCULATION

XIANG WANG

School of Mathematical Sciences, Shanghai Jiao Tong University,
Shanghai, P. R. China

YA-GUANG WANG*

School of Mathematical Sciences, MOE-LSC and SHL-MAC, Shanghai Jiao Tong University,
Shanghai, P. R. China

ABSTRACT. In this note we review our recent study on the boundary layer problem for the homogeneous model of the wind-driven oceanic circulation near the western coast when both of the Coriolis parameter and the Reynolds number go to infinity. When the Coriolis parameter is the square root of the Reynolds number, the inertial force, the Coriolis force and friction force have the same order near the boundary. By multi-scale analysis, we derive the boundary layer equation near the western coast, which has an additional nonlocal term arising from the Coriolis force, in contrast with the classical Prandtl equation. Under the monotonicity assumption of the tangential velocity in the normal variable, we obtain a local classical solution to this boundary layer equation by using the Crocco transformation. When this monotonicity condition is violated, a well-posedness result of this boundary layer equation is obtained by using the Littlewood-Paley theory when the velocity is analytic in the tangential variable. Finally, we show that the classical solution of the boundary layer equation blows up in a finite time in general, when the data does not satisfy the monotonic assumption, this implies that the analytic solution exists only locally in time.

1. **Introduction.** The motion of the oceanic circulation in the presence of wind in plane can be described by the following equations, cf. [6, 21, 22],

$$\begin{cases} \partial_t u + u \cdot \nabla u - (\eta_B + \beta x)u^\perp + \frac{r_0}{2}u + \nabla \Pi - Re^{-1}\Delta u = \beta\tau, \\ \text{div } u = 0, \end{cases} \tag{1}$$

in $(0,T) \times \Omega$, where $\Omega = \{x \in \mathbb{R}, 0 < y < 1\}$ and the Cartesian-like coordinates (x, y) represent latitude and longitude respectively. Here $u = (u_1, u_2)^T$, Π, η_B, Re, β and r_0 represent the velocity, pressure of the fluid, the bottom topography, the Reynolds number, the beta-plane parameter and Ekman pumping parameter due to friction on the bottom respectively, $\tau = (\tau_1, \tau_2)^T$ is the shear tensor created by wind, and $-xu^\perp$ represents the effect of the Coriolis force created by rotation with $u^\perp = (-u_2, u_1)^T$.

2000 *Mathematics Subject Classification.* Primary: 35Q30,76D10; Secondary: 74H35.
Key words and phrases. Wind-driven oceanic circulation, boundary layers, stability.
* Corresponding author: Ya-Guang Wang.

The property of flow in the oceanic circulation depends on the boundary condition sensitively. We consider the equations (1) with the classical non-slip boundary condition:

$$u\big|_{\partial\Omega} = 0. \tag{2}$$

In certain geophysical regime, the parameters β^{-1}, r_0 and Re^{-1} are very small, and it is important and interesting to understand the asymptotic behavior of flow when the parameters go to zero. Formally, setting $\beta^{-1} = Re^{-1} = r_0 = 0$ in (1), the limit equations read as

$$\begin{cases} -x(u^0)^{\perp} + \nabla\Pi^0 = \tau, \\ \text{div } u^0 = 0, \end{cases} \tag{3}$$

which implies that

$$\begin{cases} u_1^0(t,x,y) = -\text{curl}\tau(t,x,y), \\ u_2^0(t,x,y) = u_2^0(t,x,0) - \int_0^y \partial_x \text{curl}\tau(t,x,y')dy', \end{cases} \tag{4}$$

which is neither compatible with the initial data nor the boundary condition at $y = 0$ and $y = 1$ for the equations (1), in general. Therefore, there may exist initial layers and boundary layers in these limits. Especially, near the western coast, $y = 0$, the western intensification of boundary currents has been widely concerned by mathematicians and physicists, cf.[6, 21, 22, 11] and references therein. As there are different scale relations among these parameters from the different physical background, certain different kinds of boundary layers have been introduced, such as the inertial boundary layer, the Stommel boundary layer and the Munk boundary layer, cf.[6, 4, 5, 2, 21, 22].

We are interested in the study of the asymptotic behavior of solutions to the problem of the equations (1) with the non-slip boundary condition (2) in the large Reynolds number and large beta-plane parameter limit when omitting the friction on the bottom, i.e. $\eta_B = r_0 = 0$. In order to avoid the strong gap for the normal velocity between the viscous flow and the outer flow given in (1) and (3) respectively, in the following discussion, we shall assume that the vorticity generated by the shear tensor of wind satisfies

$$\frac{\partial}{\partial x}\left(\int_0^1 \text{curl}\tau(t,x,y')dy'\right) \equiv 0 \tag{5}$$

for all $t \geq 0$ and $x \in \mathbb{R}$. From (4), this assumption implies that the normal velocity field u_2^0 determined by (3) satisfies the impermeable boundary condition on $\{y = 1\}$, if it holds on $\{y = 0\}$.

To balance the inertial force, the Coriolis force and viscous friction of (1) in boundary layers near $\{y = 0\}$ and $\{y = 1\}$, we consider the case that

$$\beta = (\text{Re})^{\frac{1}{2}},$$

in (1), from which we shall know that the boundary layer thickness is

$$\varepsilon = \beta^{-1} = \text{Re}^{-\frac{1}{2}}, \tag{6}$$

which is the same as in the classical Prandtl boundary layer ([23]) as well as the Stewartson layer (see [11, Chapter 9]), which is the vertical shear layer in the rotating flow.

In the next section, by using multi-scale analysis we shall see under the assumption (5), that near the physical boundary $\{y = 0\}$, the flow given by (1) shall behave as

$$\begin{cases} u_1^\varepsilon(t, x, y) = u(t, x, \frac{y}{\varepsilon}) + o(1) \\ u_2^\varepsilon(t, x, y) = \varepsilon v(t, x, \frac{y}{\varepsilon}) + o(\varepsilon) \end{cases} \tag{7}$$

as $\varepsilon \to 0$, where the boundary layer profiles $(u(t, x, \eta), v(t, x, \eta))$ satisfy the following problem

$$\begin{cases} \partial_t u + u\partial_x u + v\partial_\eta u - \partial_\eta^2 u = \partial_t U + U\partial_x U + \int_\infty^\eta (U - u)d\eta', \\ \partial_x u + \partial_\eta v = 0, \\ (u, v)|_{\eta=0} = (0, 0), \quad \lim_{\eta \to +\infty} u(t, x, \eta) = U(t, x) \end{cases} \tag{8}$$

where $U(t, x) = u_1^0(t, x, 0)$ is the the tangential velocity on the boundary $\{y = 0\}$ of the outer flow given by (3). The boundary layer of the flow near $\{y = 1\}$ has the same behavior as above.

It is worthy to note that there is an additional integral term on the right side of the first equation given in (8), in contrast with the classical Prandtl boundary layer equation given in [23].

Till now, there are many interesting results on the well-posedness and stability of the two-dimensional classical Prandtl boundary layer equation, i.e. without the integral term in (8). Under the monotonicity assumption of the tangential velocity with respect to the normal variable, i.e., $u_y > 0$, the first well-posedness locally in time in the Hörder spaces was obtained by Oleinik et al. [19, 20] by using the Crocco transformation, this well-posedness has been re-studied recently by developing an energy method in the Sobolev spaces in [1, 18]. In addition to this monotonicity assumption, in the case of a favorite pressure gradient for the outer flow, Xin and Zhang obtained a global weak solution of the classical Prandtl equation in [26]. When the monotonicity condition of the velocity field was failed, there are interesting works on the well-posedness of the classical Prandtl equation in the space of analytic functions and Gevery functions, cf. [3, 17, 14, 16, 7, 27] and references therein, or the blow-up and instability of solutions in a finite time, cf. [8, 13, 9, 10, 15] etc.

As there is an additional integral term in the equation of the boundary layer problem (8), one needs to introduce certain weighted spaces to study the well-posedness of the problem (8). The main goal of this note is to study the well-posedness and blowup of the solution to the boundary layer problem (8). A related stationary boundary layer problem was considered recently by Dalibard and Paddick in [5] when the parameters satisfy the relation $\beta = \mathrm{Re}^2$ and $\eta_B = r_0 = 0$ in (1).

The remainder of this note is arranged as follows. In Section 2, we shall derive the boundary layer equation by using multi-scale analysis under the assumption (5) of the shear tensor of the wind. In Section 3, we study the existence and uniqueness of the classical solution in the class of the monotonicity velocity, and the existence of a solution being analytic in the tangential variable and finite order regular in the normal variable, when the monotonicity condition is failed. Finally, in Section 4, we present a blowup result under certain condition on a transversal plane for the initial velocity and outer flow, which shows that the analytic solution obtained exists locally in time in general.

2. **Asymptotic analysis.** As in (6), to balance the inertial force, the Coriolis force and viscous friction of (1) in boundary layers as both of the beta-plane parameter and the Reynolds number go to infinite, we consider the case that

$$\beta = \mathrm{Re}^{\frac{1}{2}}.$$

Setting $\varepsilon = \beta^{-1} = \mathrm{Re}^{-\frac{1}{2}}$ in (1) with $\eta_B = r_0 = 0$, consider the following problem in $(0,T)\times\Omega$ with $\Omega = \{x \in \mathbb{R}, 0 < y < 1\}$,

$$\begin{cases} \varepsilon\partial_t u^\varepsilon + \varepsilon u^\varepsilon \cdot \nabla u^\varepsilon - x(u^\varepsilon)^\perp + \nabla \Pi^\varepsilon - \varepsilon^3 \Delta u^\varepsilon = \tau, \\ \mathrm{div}\ u^\varepsilon = 0, \\ u^\varepsilon|_{y=0} = u^\varepsilon|_{y=1} = 0, \\ u^\varepsilon|_{t=0} = u_0^\varepsilon(x,y). \end{cases} \tag{9}$$

When $\epsilon \to 0$, the problem (9) formally goes to (3), with velocity u^0 being given in (4). On the other hand, as in the classical Prandtl boundary layer theory for the small viscosity limit of the incompressible viscous flow ([23, 20]), from the nonslip boundary condition given in (9), a natural condition on the velocity field u^0 given in (4) is the following impermeable condition,

$$u_2^0|_{y=0} = u_2^0|_{y=1} = 0. \tag{10}$$

Thus, we know that the shear tensor τ from wind should satisfy the constraint,

$$\frac{\partial}{\partial x}\left(\int_0^1 \mathrm{curl}\tau(t,x,y')dy'\right) \equiv 0, \quad \forall 0 \le t < T,\ x \in \mathbb{R}. \tag{11}$$

From now on, we assume that the condition (11) holds for τ in the following discussion.

By comparing the leading scale of the inertial force, the Coriolis force and friction in the boundary layer, it infers that the thickness of the boundary layer is of the order ε, therefore we assume that the solution of the problem (9) has the following ansatz

$$\begin{cases} u^\varepsilon(t,x,y) = \sum\limits_{i\ge 0} \varepsilon^i (u^{I,i}(t,x,y) + u^{B,i,0}(t,x,\frac{y}{\varepsilon}) + u^{B,i,1}(t,x,\frac{1-y}{\varepsilon})) \\ \Pi^\varepsilon(t,x,y) = \sum\limits_{i\ge 0} \varepsilon^i (\Pi^{I,i}(t,x,y) + \Pi^{B,i,0}(t,x,\frac{y}{\varepsilon}) + \Pi^{B,i,1}(t,x,\frac{1-y}{\varepsilon})) \end{cases} \tag{12}$$

where $u^{B,i,0}(t,x,\eta)$, $\Pi^{B,i,0}(t,x,\eta)$ and $u^{B,i,1}(t,x,\xi)$, $\Pi^{B,i,1}(t,x,\xi)$ are fast decay when $\eta \to +\infty$ and $\xi \to +\infty$ respectively.

In the following calculation, we denote by $\overline{u}(t,x) = u(t,x,0)$ and $\widetilde{u}(t,x) = u(t,x,1)$ the traces of $u(t,x,y)$ at boundaries $\{y=0\}$ and $\{y=1\}$ respectively.

From the nonslip boundary condition given in (9), one has

$$u^{B,k,0}(t,x,0) + \overline{u^{I,k}} = 0,\ u^{B,k,1}(t,x,0) + \widetilde{u^{I,k}} = 0, \text{ for } k \ge 0. \tag{13}$$

Plugging (12) into the divergence-free condition given in (9), we immediately get that

$$\partial_\eta u_2^{B,0,0}(t,x,\eta) \equiv 0, \quad \partial_\xi u_2^{B,0,1}(t,x,\xi) \equiv 0$$

which implies

$$u_2^{B,0,0}(t,x,\eta) \equiv 0, \quad u_2^{B,0,1}(t,x,\xi) \equiv 0 \tag{14}$$

by using $\lim\limits_{\eta\to+\infty} u^{B,0,0}(t,x,\eta) = \lim\limits_{\xi\to+\infty} u^{B,0,1}(t,x,\xi) = 0$. Therefore, from (13) we have

$$u_2^{I,0}(t,x,0) = u_2^{I,0}(t,x,1) \equiv 0. \tag{15}$$

On other hand hand, by plugging (12) into the equations given in (9), in the interior of Ω, we deduce that $(u^{I,0}, \Pi^{I,0})$ satisfies the equations

$$\begin{cases} -x(u^{I,0})^{\perp} + \nabla \Pi^{I,0} = \tau, \\ \text{div } u^{I,0} = 0, \end{cases}$$

which implies

$$\begin{cases} u_1^{I,0}(t,x,y) = -\text{curl}\tau(t,x,y), \\ u_2^{I,0}(t,x,y) = -\int_0^y \partial_x \text{curl}\tau(t,x,y')dy' \end{cases} \tag{16}$$

by using the boundary condition given in (15).

It is important to note from the constraint of τ given in (11) that, the solution $u_2^{I,0}(t,x,y)$ of (16) satisfies the boundary condition given in (15) at $\{y=1\}$, which is consistent with the phenomenon that the leading boundary layer profile $u_2^{B,0,1}(t,x,\frac{1-y}{\varepsilon})$ near $\{y=1\}$ of the normal velocity is identically equal to zero.

Successively, we can obtain that for all $k \geq 0$, $u^{I,k+1}$ can be represented as

$$\begin{cases} u_1^{I,k+1}(t,x,y) = \text{curl}(\partial_t u^{I,k} + \sum_{j=0}^k u^{I,j} \cdot \nabla u^{I,k-j} - \Delta u^{I,k-2}), \\ u_2^{I,k+1}(t,x,y) = u_2^{I,k+1}(t,x,0) - \int_0^y \partial_x u_1^{I,k+1}(t,x,z)dz, \ k \geq 0, \end{cases} \tag{17}$$

where $u^{I,-2} = u^{I,-1} \equiv 0$.

Near the boundary $y=0$, by matching the orders $O(\varepsilon^{-1})$ and $O(\varepsilon^0)$ respectively in the second component of the first equation in (9), in view of (15), one has

$$\partial_\eta \Pi^{B,0,0} = 0$$

implying

$$\Pi^{B,0,0}(t,x,\eta) \equiv 0, \tag{18}$$

and

$$-xu_1^{B,0,0} + \partial_\eta \Pi^{B,1,0} = 0. \tag{19}$$

From the order of $O(\varepsilon^1)$ in the first component of the first equation in (9), we get that

$$\begin{aligned} &\partial_t(\overline{u_1^{I,0}} + u_1^{B,0,0}) + (\overline{u_1^{I,0}} + u_1^{B,0,0})\partial_x(\overline{u_1^{I,0}} + u_1^{B,0,0}) \\ &\quad + (\overline{\eta \partial_y u_2^{I,0}} + \overline{u_2^{I,1}} + u_2^{B,1,0})\partial_\eta u_1^{B,0,0} \\ &\quad + x(u_2^{B,1,0} + \overline{u_2^{I,1}}) + \partial_x(\overline{\Pi^{I,1}} + \Pi^{B,1,0})) - \partial_\eta^2 u_1^{B,0,0} = 0. \end{aligned} \tag{20}$$

On the other hand, from the equation (19), one has

$$\partial_x \Pi^{B,1,0} = \int_{+\infty}^{\eta} u_1^{B,0,0} d\eta' - xu_2^{B,1,0}. \tag{21}$$

Denote by

$$\begin{cases} u(t,x,\eta) = u_1^{I,0}(t,x,0) + u_1^{B,0,0}(t,x,\eta), \\ v(t,x,\eta) = \eta \partial_y u_2^{I,0}(t,x,0) + u_2^{I,1}(t,x,0) + u_2^{B,1,0}(t,x,\eta), \end{cases}$$

by using (20), (21) and the boundary condition given in (13), we obtain that the boundary layer profile (u, v) satisfies the following problem

$$\begin{cases} \partial_t u + u\partial_x u + v\partial_\eta u - \partial_\eta^2 u = \partial_t U + U\partial_x U + \int_\infty^\eta (U-u)d\eta', \\ \partial_x u + \partial_\eta v = 0, \\ (u,v)|_{\eta=0} = (0,0), \quad \lim_{\eta\to+\infty} u(t,x,\eta) = U(t,x) \end{cases} \tag{22}$$

where $U(t,x) = u_1^{I,0}(t,x,0)$.

Similarly, one derives that near $y = 0$, the proceeding boundary layer profiles $(u_1^{B,k,0}, u_2^{B,k+1,0})$ statisfy

$$\begin{cases} \partial_t u_1^{B,k,0} + (\overline{\partial_x u_1^{I,0}} + \partial_x u_1^{B,0,0})u_1^{B,k,0} + (\overline{u_1^{I,0}} + u_1^{B,0,0})\partial_x u_1^{B,k,0} \\ \quad +(\overline{u_2^{I,k+1}} + u_2^{B,k+1,0})\partial_\eta u_1^{B,0,0} + (\eta\overline{\partial_y u_2^{I,0}} + \overline{u_2^{I,1}} + u_2^{B,1,0})\partial_\eta u_1^{B,k,0} \\ \quad + \int_\infty^\eta u_1^{B,k,0} d\eta - \partial_\eta^2 u_1^{B,k,0} = f_{0k}, \\ \partial_x u_1^{B,k,0} + \partial_\eta u_2^{B,k+1,0} = 0, \end{cases} \tag{23}$$

for $k \geq 1$, where f_{0k} depends on $u_1^{B,i,0}$ $(0 \leq i \leq k-1)$, $u_2^{B,i,0}$ $(0 \leq i \leq k)$, $u_1^{I,i}(0 \leq i \leq k)$ and $u_2^{I,i}(0 \leq i \leq k)$.

In the same way as above, near $y = 1$, one can derive that for all $k \geq 0$, the boundary layer profiles $(u_1^{B,k,1}, u_2^{B,k+1,1})$ satisfy the similar problems as given in (22) and (23).

Remark 2.1. As explained in (4)-(5), if the constraint

$$\int_0^1 \partial_x \text{curl}\tau(t,x,y')dy' \equiv 0, \quad \forall 0 \leq t < T, \ x \in \mathbb{R}$$

does not hold for the shear tensor of wind, then the ansatz (12) for the solution of (9) yields that the leading order profile $u^{I,0}(t,x,y)$ does not satisfy the boundary condition

$$u_2^{I,0}|_{y=0} = 0, \quad u_2^{I,0}|_{y=1} = 0.$$

To deal with this discrepancy, one could revise the ansatz (12) to be

$$u^\varepsilon(t,x,y) = \sum_{i\geq 0} \varepsilon^i u^{I,i}(t,x,y) + \sum_{i\geq -1} \varepsilon^i (u^{B,i,0}(t,x,\frac{y}{\varepsilon}) + u^{B,i,1}(t,x,\frac{1-y}{\varepsilon})) \tag{24}$$

including $O(\epsilon^{-1})$–order boundary layer, much stronger than the Prandtl one. The study of this boundary layer is a challenging problem, one may refer to [21] for some discussion.

3. **Local well-posedness of the boundary layer problem.** The first important issue is to study the well-posedness of the problem (22) for the leading order boundary layer profile. Comparing with the classical Prandtl equation studied in [20, 19, 1, 18], the additional non-local integral term in the first equation of (22) requires us to overcome certain new difficulties.

3.1. **Well-posedness under the monotonicity condition.** In this subsection, we develop the idea given in [20] to obtain the well-posedness of the problem (22) in the monotonic class.

Consider the following problem for the boundary layer equation (22) in the domain $D_T = \{0 < t < T, 0 < x < X, y > 0\}$,

$$\begin{cases} \partial_t u + u\partial_x u + v\partial_y u - \partial_y^2 u = \partial_t U + U\partial_x U + \int_\infty^y (U-u)dy', \\ \partial_x u + \partial_y v = 0, \\ u|_{t=0} = u_0(x,y), \; u|_{x=0} = u_1(t,y), \\ (u,v)|_{y=0} = (0,0), \; \lim_{y\to+\infty} u(t,x,y) = U(t,x). \end{cases} \tag{25}$$

First, we impose a monotonicity assumption on the initial and boundary data of the problem (25). Assume that there exist some positive constants C_1 and C_2 such that the initial data $u_0(x,y)$, the incoming flow $u_1(t,y)$ and the outer flow $U(t,x)$ satisfy

$$C_1(1 - \frac{u_0}{U(0,x)}) \le \frac{u_{0y}}{U(0,x)} \le C_2(1 - \frac{u_0}{U(0,x)}) \tag{26}$$

and

$$C_1(1 - \frac{u_1}{U(t,0)}) \le \frac{u_{1y}}{U(t,0)} \le C_2(1 - \frac{u_1}{U(t,0)}). \tag{27}$$

for any $0 \le t \le T$ and $0 \le x \le X$.

Obviously, from (25) we know that the convection term can be written as

$$v\partial_y u = -\partial_y u \int_0^y \partial_x u(t,x,y')dy'$$

which can not be controlled if one uses the usual energy method for the problem (25). To remove this term with a loss of regularity in x, as in [20], we introduce the following Crocco transformation

$$\tau = t, \quad \xi = x, \quad \eta = \frac{u}{U}, \quad \omega = \frac{u_y}{U}, \tag{28}$$

then we know that in the class of u satisfying $\partial_y u > 0$, the transformation from (t,x,y) to (τ,ξ,η) given in (28) is invertible, and from (25), $\omega(\tau,\xi,\eta)$ satisfies the following problem for a scalar degenerated parabolic equation in $Q_T = \{0 < \tau < T, 0 < \xi < X, 0 < \eta < 1\}$

$$\begin{cases} \omega_\tau + \eta U \omega_\xi + \widetilde{A}\omega_\eta + B\omega - \omega^2\omega_{\eta\eta} = 1 - \eta, \\ \omega|_{\tau=0} = \omega_0, \quad \omega|_{\xi=0} = \omega_1, \\ \omega|_{\eta=1} = 0, \quad \left(\omega\omega_\eta - \int_0^1 \frac{1-\eta'}{\omega} d\eta' + C\right)|_{\eta=0} = 0, \end{cases} \tag{29}$$

where

$$\omega_0(\xi,\eta) = \frac{u_{0y}(x,y)}{U(0,x)}, \; \omega_1(\tau,\eta) = \frac{u_{1y}(t,y)}{U(t,0)},$$

and

$$\widetilde{A} = A - \int_\eta^1 \frac{1-\eta'}{\omega} d\eta', \; B = \frac{U_t}{U} + \eta U_x, \quad C = U_x + \frac{U_t}{U},$$

with $A = (1-\eta)\frac{U_t}{U} + (1-\eta^2)U_x$.

Due to the degeneracy and nonlinearity structure of (29), we construct the approximation solution sequence of (29) via the following iteration scheme

$$\begin{cases} \partial_\tau \omega^n + \eta U \partial_\xi \omega^n + \widetilde{A}^{n-1}\partial_\eta \omega^n + B\omega^n - (\omega^{n-1})^2\partial_\eta^2 \omega^n = 1-\eta, \\ \omega^n|_{\tau=0} = \omega_0, \quad \omega^n|_{\xi=0} = \omega_1, \\ \omega^n|_{\eta=1} = 0, \ \left(\omega^{n-1}\partial_\eta \omega^n - \int_0^1 \frac{1-\eta'}{\omega^{n-1}} d\eta' + C\right)|_{\eta=0} = 0, \end{cases} \tag{30}$$

where $\widetilde{A}^{n-1} = A - \int_\eta^1 \frac{1-\eta'}{\omega^{n-1}} d\eta'$.

The existence of a solution $\omega^n \in C^2(Q_T)$ to (30) can be proved as long as $\frac{\widetilde{A}^{n-1}}{1-\eta}$, B and their second derivatives are bounded, the detail calculation can be found in [20]. In order to proceed the above iteration scheme, it needs to estimate $\frac{\omega^n}{1-\eta}$ and its derivatives up to the second order. This is different from the case of the classical Prandtl equation studied in [20], in which one only need to study the norm of ω^n in $C^2(Q_T)$.

To control the first and second derivatives of ω^n, we introduce that $V^n = \omega^n e^{\alpha\eta}$ with α being a positive constant, and the quantities

$$\Phi^n = \left(\frac{V_\tau^n}{1-\eta}\right)^2 + \left(\frac{V_\xi^n}{1-\eta}\right)^2 + (V_\eta^n)^2 + K_1^n\eta + K_0$$

and

$$\Psi^n = \sum_{|\beta|\le 2}\left(\partial_{tan}^\beta \frac{V^n}{1-\eta}\right)^2 + |\nabla V_\eta^n|^2 + N_1^n\eta + N_0,$$

with $\partial_{tan} = \partial_\tau$ or ∂_ξ, where K_0, K_1^n, N_0 and N_1^n are positive constants. By studying the quantities Φ^n and Ψ^n carefully as given in [20], and choosing the above parameters properly, one can obtain the following result:

Theorem 3.1. *For any given $X > 0$, assume that u_0, u_1 and U are smooth, and satisfy compatibility conditions for the problem (25) up to order two. In addition, the monotonicity conditions given in (26) and (27) hold. Then there exists $T > 0$ such that the solution ω^n of (30) in Q_T, $\frac{\omega^n}{1-\eta}$ and derivatives up to the order two are bounded uniformly in n.*

Based on Theorem 3.1, by estimating the error $\frac{\omega^{n+1}-\omega^n}{1-\eta}$, one can deduce that there is a C^2 function $\omega(\tau,\xi,\eta)$ defined in Q_T, such that

$$\omega^n \to \omega \text{ as n} \to +\infty$$

uniformly in $C^1(\overline{Q}_T)$. Meanwhile, it follows from the equation (30) that $\omega_{\eta\eta}^n$ also converges to $\omega_{\eta\eta}$ for any $\eta < 1$. Hence, ω is a classical solution to (29). Moreover, by noticing the invertibility of the Crocco transformation, it arrives at the following well-posedness result:

Theorem 3.2. *Under the same assumption as given in Theorem 3.1, the problem (25) admits a unique classical solution (u, v) in $D_T = \{0 \le t \le T, 0 \le x \le X, y \ge 0\}$ for some $T > 0$. Furthermore, the inequality*

$$M_1\left(1-\frac{u}{U}\right) \le \frac{u_y}{U} \le M_2\left(1-\frac{u}{U}\right) \tag{31}$$

holds for some $M_2 \ge M_1 > 0$ and $(t,x,y) \in D_T$.

More details of the proofs of Theorems 3.1 and 3.2 are given in [12].

3.2. **Well-posedness in the space of analytic functions.** The proposal of this subsection is to study the well-posedness of the problem (25) when the datum do not satisfy the monotonicity assumptions (26) and (27) anymore, but are analytic with respect to the variable x. We shall adapt the approach given in [27] to study this problem.

Consider the problem (25) in the domain $Q_T = \{0 < t < T, x \in \mathbb{R}, y > 0\}$,

$$\begin{cases} \partial_t u + u\partial_x u + v\partial_y u + \int_\infty^y (u-U)dy' - \partial_y^2 u = \partial_t U + U\partial_x U, \\ \partial_x u + \partial_y v = 0, \\ u|_{t=0} = u_0(x,y), \\ (u,v)|_{y=0} = (0,0), \quad \lim_{y\to+\infty} u(t,x,y) = U(t,x), \end{cases} \tag{32}$$

To study the well-posedness of this problem in the energy spaces for the y-variable, we introduce

$$\phi(t,y) = \operatorname{Erf}\left(\frac{y}{\sqrt{4(t+1)}}\right) \text{ with } \operatorname{Erf}(y) = \frac{2}{\sqrt{\pi}}\int_0^y e^{-z^2}dz,$$

to homogenize the condition of u at infinity given in (32). Obviously, $\phi(t,y)$ is a solution to the problem

$$\begin{cases} \partial_t\phi - \partial_y^2\phi = 0, \\ \phi\big|_{y=0} = 0, \quad \lim_{y\to+\infty}\phi(t,y) = 1, \\ \phi\big|_{t=0} = \operatorname{Erf}(\frac{y}{2}). \end{cases}$$

Denote by $u^s = U\phi$ and $w = u - u^s$. From (32), we know that w satisfies the following problem

$$\begin{cases} \partial_t w + (w+u^s)\partial_x w + w\partial_x u^s - \int_0^y \partial_x(w+u^s)dy'\partial_y(w+u^s) - \partial_y^2 w \\ \qquad + \int_{+\infty}^y w dy' = (1-\phi)(\partial_t U + (1+\phi)U\partial_x U) - \int_y^{+\infty} U(1-\phi)dy', \\ w\big|_{y=0} = 0, \quad \lim_{y\to+\infty} w = 1, \\ w\big|_{t=0} = w_0(x,y) = u_0(x,y) - U(0,x)\operatorname{Erf}(\frac{y}{2}). \end{cases} \tag{33}$$

To study the well-posedness of the problem (33), we recall some facts of the Littlewood-Paley theory and certain function spaces from [27, 25].

Let (φ,χ) be smooth functions such that

$$\operatorname{supp}\varphi \subset \left\{\tau\in\mathbb{R}\big|\frac{3}{4}\le|\tau|\le\frac{8}{3}\right\},\ \operatorname{supp}\chi\subset\left\{\tau\in\mathbb{R}\big||\tau|\le\frac{4}{3}\right\}$$

satisfying

$$\sum_{k\in\mathbb{N}}\varphi(2^{-k}\tau) = 1 \quad (\forall\tau\neq0), \qquad \chi(\tau) + \sum_{k\ge0}\varphi(2^{-k}\tau) = 1 \quad (\forall\tau\in\mathbb{R}).$$

Denote by $S_k f = \mathcal{F}^{-1}[\chi(2^{-k}|\xi|)\mathcal{F}[f]]$, and

$$\Delta_k f = \begin{cases} \mathcal{F}^{-1}[\varphi(2^{-k}|\xi|)\mathcal{F}[f]],\ k\ge0, \\ \mathcal{F}^{-1}[\chi(|\xi|)\mathcal{F}[f]],\ k=-1, \\ 0,\ k\le-2, \end{cases}$$

where $\mathcal{F}$ and $\mathcal{F}^{-1}$ denote the Fourier transform and Fourier inverse transform in the $x-$variable.

Introduce the following function spaces for the problem with the parameters $s > 0$, $l \in \mathbb{N}_+$ and $p \in [1, +\infty]$.

Definition 3.1. (i) The space B^s is the set of functions $u \in \mathcal{S}'(\mathbb{R})$ such that

$$\|u\|_{B^s} := \sum_{k\in\mathbb{Z}} 2^{ks}\|\Delta_k u\|_{L^2(\mathbb{R})} < +\infty.$$

(ii) The space $B^{s,l}_\psi$, with a positive function $\psi(y)$, is the space of functions $u \in \mathcal{S}'(\mathbb{R}^2_+)$ such that

$$\|u\|_{B^{s,l}_\psi} := \sum_{j=0}^{l}\sum_{k\in\mathbb{Z}} 2^{ks}\|e^{\psi(y)}\Delta_k\partial_y^j u\|_{L^2(\mathbb{R}\times\mathbb{R}_+)} < +\infty.$$

(iii) The space $\tilde{L}^p_t(B^s)$ is defined as the completion of $C([0,t];\mathcal{S}(\mathbb{R}))$ with the norm

$$\|u\|_{\tilde{L}^p_t(B^s)} := \sum_{k\in\mathbb{Z}} 2^{ks}\Big(\int_0^t \|\Delta_k u(t',\cdot)\|^p_{L^2(\mathbb{R})}dt'\Big)^{\frac{1}{p}}.$$

(iv) For any positive function $\psi(t',y)$ and nonnegative $f(t') \in L^1_{loc}(\mathbb{R}_+)$, the space $\tilde{L}^p_{t,f}(B^{s,l}_\psi)$ is defined as the completion of $C([0,t];\mathcal{S}(\mathbb{R}^2_+))$ with the norm

$$\|u\|_{\tilde{L}^p_{t,f}(B^{s,l}_\psi)} := \sum_{j=0}^{l}\sum_{k\in\mathbb{Z}} 2^{ks}\Big(\int_0^t f(t')\|e^{\psi(t',y)}\Delta_k\partial_y^j u(t',\cdot)\|^p_{L^2(\mathbb{R}\times\mathbb{R}_+)}dt'\Big)^{\frac{1}{p}}.$$

Denote $\tilde{L}^p_{t,1}(B^{s,l}_\psi)$ by $\tilde{L}^p_t(B^{s,l}_\psi)$ for simplicity when $f(t') \equiv 1$. The above notations can be properly changed when $p = +\infty$.

Similar to that given in [27], to obtain energy estimates of the solution of (33) in the normal variable, we introduce the weights

$$\psi(t,y) = \frac{1+y^2}{16(1+t)^\gamma} \text{ and } \psi_0(y) = \frac{1+y^2}{16}, \tag{34}$$

with $\gamma \geq 2$.

Denote by $\hat{w}(t,\xi,y)$ the Fourier transform of $w(t,x,y)$ in the x-variable, and

$$w_\Phi(t,x,y) = \mathcal{F}^{-1}_{\xi\to x}[e^{\Phi(t,\xi)}\hat{w}(t,\xi,y)]$$

for a given locally bounded function $\Phi(t,\xi)$. From (33), we know that w_Φ satisfies the following equation

$$\begin{aligned}
&\partial_t w_\Phi + \lambda\dot{\theta}\langle D\rangle w_\Phi + [(w+u^s)\partial_x w]_\Phi + [w\partial_x u^s]_\Phi \\
&\quad + \Big[(-\int_0^y \partial_x(w+u^s)dy')\partial_y(w+w^s)\Big]_\Phi + \Big[\int_{+\infty}^y w dy'\Big]_\Phi \\
=&\partial_y^2 w_\Phi + (1-\phi)\left[\partial_t U + (1+\phi)U\partial_x U\right]_\Phi - \int_y^{+\infty} U_\Phi(1-\phi)dy'.
\end{aligned} \tag{35}$$

To control derivatives of the solution in the x-variable for the equation (35), we take

$$\Phi(t,\xi) = (1-\lambda\theta(t))\langle\xi\rangle \tag{36}$$

with $\langle\xi\rangle = 1 + |\xi|$ and a parameter λ, in which $\theta(t)$ is mainly used to deal with energy estimates for the nonlinear terms and will be determined by the following problem

$$\begin{cases} \dot{\theta} = \langle t\rangle^{\frac{\gamma}{4}} \|\partial_y w_\Phi\|_{B_\psi^{\frac{1}{2},0}} + \langle t\rangle^{\frac{\gamma}{4}} \|U_\Phi\|_{B^{\frac{1}{2}}} + \langle t\rangle^{\frac{\gamma}{2}} \|w_\Phi\|^2_{B_\psi^{1,0}} + \|w_\Phi\|^2_{B_\psi^{\frac{1}{2},0}} \\ \quad + \langle t\rangle^{\frac{1}{2}} \|U_\Phi\|^2_{B^{\frac{1}{2}}} + \langle t\rangle^{\frac{1}{2}} \|U_\Phi\|^2_{B^1} + \langle t\rangle^{\gamma}, \\ \theta\big|_{t=0} = 0. \end{cases} \tag{37}$$

By acting the dyadic operator Δ_k on (35) and taking $L^2(Q_T)$ inner product with $e^{2\psi}\Delta_k w_\Phi$, in virtue of the Bony decomposition and the Littlewood-Paley theory, we get the following estimate for the solution of (33) by choosing parameters γ and λ properly, more details can be found in [24].

Theorem 3.3. *Suppose that $w(t,x,y)$ is a classical solution of the problem (33), then there exist $T_1 > 0$ and a positive constant G such that there holds*

$$\begin{aligned} &\|w_\Phi\|_{\tilde{L}^\infty_T(B_\psi^{\frac{1}{2},0})} + \|\sqrt{-(\psi_t + 2\psi_y^2)} w_\Phi\|_{\tilde{L}^2_T(B_\psi^{\frac{1}{2},0})} + \|\partial_y w_\Phi\|_{\tilde{L}^2_T(B_\psi^{\frac{1}{2},0})} \\ &+ \sqrt{\lambda}\Big(\|w_\Phi\|_{\tilde{L}^2_{T,\dot{\theta}}(B_\psi^{\frac{1}{2},0})} + \|w_\Phi\|_{\tilde{L}^2_{T,\dot{\theta}}(B_\psi^{1,0})}\Big) \\ \le& G\Big[\|e^{\langle D\rangle} w_0\|_{B_{\psi_0}^{\frac{1}{2},0}} + T^{\frac{1}{2}}(\|e^{\langle D\rangle}U\|_{\tilde{L}^\infty_T(B^{\frac{1}{2}})} + \|e^{\langle D\rangle}U\|_{\tilde{L}^\infty_T(B^1)}) \\ &+ (\langle T\rangle^{\frac{3}{2}} - 1)^{\frac{1}{2}} \|e^{\langle D\rangle}[\partial_t U]\|_{L^\infty_T B^{\frac{1}{2}}} + (\langle T\rangle^{\frac{5}{2}} - 1)^{\frac{1}{2}} \|e^{\langle D\rangle}U\|_{L^\infty_T(B^{\frac{1}{2}})} \\ &+ \sigma T^{\frac{1}{2}} \|e^{\langle D\rangle}U\|_{\tilde{L}^\infty_T(B^{\frac{1}{2}})} \|e^{\langle D\rangle}U\|_{\tilde{L}^\infty_T(B^{\frac{3}{2}})}\Big] \end{aligned} \tag{38}$$

for any $0 < T \le T_1$, provided that the weight $\Phi(t,\xi)$ is positive in $[0, T_1]$.

In fact, in view of (37) and (38), there exist a generic constant C_1 and a constant $C(w_0, U, t)$ depending on w_0, U and t such that

$$\begin{aligned} \theta = \int_0^t \dot{\theta} dt' =& \int_0^t [\langle t'\rangle^{\frac{\gamma}{4}} \|\partial_y w_\Phi\|_{B_\psi^{\frac{1}{2},0}} + \langle t'\rangle^{\frac{\gamma}{4}} \|U_\Phi\|_{B^{\frac{1}{2}}} + \langle t'\rangle^{\frac{\gamma}{2}} \|w_\Phi\|^2_{B_\psi^{1,0}} \\ &+ \|w_\Phi\|^2_{B_\psi^{\frac{1}{2},0}} + \langle t'\rangle^{\frac{1}{2}} \|U_\Phi\|^2_{B^{\frac{1}{2}}} + \langle t'\rangle^{\frac{1}{2}} \|U_\Phi\|^2_{B^1} + \langle t'\rangle^{\gamma}] dt' \\ \le& C_1[(\langle t\rangle^{\frac{\gamma}{2}+1} - 1)^{\frac{1}{2}} \|\partial_y w_\Phi\|_{\tilde{L}^2_t(B_\psi^{\frac{1}{2},0})} + (\langle t\rangle^{\frac{\gamma}{4}+1} - 1)\|e^{\langle D\rangle}U\|_{\tilde{L}^\infty_t(B^{\frac{1}{2}})} \\ &+ \langle t\rangle^{\frac{\gamma}{2}} \|w_\Phi\|_{\tilde{L}^2_t(B_\psi^{1,0})} + \|w_\Phi\|_{\tilde{L}^2_t(B_\psi^{\frac{1}{2},0})} + (\langle t\rangle^{\frac{3}{2}} - 1)(\|e^{\langle D\rangle}U\|^2_{\tilde{L}^\infty_t(B^{\frac{1}{2}})} \\ &+ \|e^{\langle D\rangle}U\|^2_{\tilde{L}^\infty_t(B^1)}) + (\langle t\rangle^{\gamma+1} - 1)] \\ \le& C(w_0, U, t). \end{aligned}$$

Therefore one can choose T_1 properly small such that

$$0 < T_1 \le \sup_{t>0} \left\{ \theta(t) < \frac{1}{\lambda} \right\}, \tag{39}$$

which guarantees the weight $\Phi(t,\xi)$ defined in (36) is positive on $[0, T_1]$. Thereby we obtain the a priori estimate (38) for $0 < T \le T_1$.

By using the above energy estimate, we obtain the following well-posedness result in the weighted Chemin-Lerner spaces, more details of the proof refer to [24].

Theorem 3.4. *For a given $T_0 > 0$, assume that the initial velocity $w_0(x,y)$ and the outflow velocity $U(t,x)$ are analytic in $x \in \mathbb{R}$, and*

$$e^{\langle D\rangle} w_0 \in B^{\frac{1}{2},0}_{\psi_0}$$

and

$$e^{\langle D\rangle} U \in \tilde{L}^\infty_{T_0}(B^{\frac{1}{2}}) \cap \tilde{L}^\infty_{T_0}(B^1) \cap \tilde{L}^\infty_{T_0}(B^{\frac{3}{2}}), \quad e^{\langle D\rangle} U_t \in \tilde{L}^\infty_{T_0}(B^{\frac{1}{2}}).$$

Then there exists $0 < T^* \le T_0$ such that the problem (33) has a unique solution satisfying $e^{\Phi(t,D)} w \in \tilde{L}^\infty_{T^*}(B^{\frac{1}{2},0}_{\psi})$, where D is the Fourier multiplier with respect to the x-variable.

The uniqueness of the solution to the problem (33) can be obtained by using an energy estimate similar to that one given in (38).

4. **Blowup of smooth solutions.** It is interesting to study whether the smooth solution obtained in the previous section exists globally in time for the problem (32). Consider the problem (32) under the assumption that the initial data and outer flow satisfy

$$\begin{cases} u_0(0,y) = U(0,0) = 0, \ u_{0x}(0,y) \le U_x(0,0), \\ U_x(t,0) \ge 0, \ \text{for } 0 \le t \le T. \end{cases} \tag{40}$$

Denote by $H_{t'} = \{(t,y)|\ t \in (0,t'), y > 0\}$. In what follows, the aim is to prove that for the solution of $u(t,x,y)$ of (32), $\|\partial_x u(s,0,y)\|_{L^\infty(H_t)}$ will blow up in $(0,T)$ under certain assumption. This shall be obtained by developing the idea from [15] and a contradiction argument.

Denote by $\bar{f}(t,y) = f(t,0,y)$ and $\bar{g}(t) = g(t,0)$. As in [15], by restricting the problem (32) on the plane $\{x=0\}$, we get that $\bar{u}(t,y) = u(t,0,y)$ satisfies

$$\begin{cases} \partial_t \bar{u} + \bar{u}\overline{\partial_x u} + \bar{v}\partial_y \bar{u} + \int_\infty^y \bar{u}dy' - \partial_y^2 \bar{u} = 0, \\ \overline{\partial_x u} + \partial_y \bar{v} = 0, \\ \bar{u}|_{t=0} = 0, \\ (\bar{u},\bar{v})|_{y=0} = (0,0), \quad \lim\limits_{y\to+\infty} \bar{u} = 0. \end{cases} \tag{41}$$

It can be proved that the problem (41) admits a trivial solution only by using the energy method provided that

$$\lim_{y\to+\infty} (u-U)e^{\psi} = 0 \tag{42}$$

for ψ being given in (34).

Denote by $\tilde{u} = -\partial_x u(t,0,y)$ and $\tilde{U} = -\partial_x U(t,0)$. From (32), we know that $w = \tilde{u} - \tilde{U}$ satisfies

$$\begin{cases} \partial_t w - w^2 + \partial_y^{-1}(w+\tilde{U})\partial_y w - 2\tilde{U}w + \int_\infty^y w dy' - \partial_y^2 w = 0, \\ w|_{t=0} = w_0 = \tilde{u}_0 - \tilde{U}, \\ w|_{y=0} = -\tilde{U}, \quad \lim\limits_{y\to+\infty} w = 0, \end{cases} \tag{43}$$

where $\partial_y^{-1} f(y) := \int_0^y f(y')dy'$.

One can verify that the solution of (43) is non-negative, under assumptions (40), (42) and

$$\lim_{y\to+\infty} (u_x - U_x)e^{y} = 0. \tag{44}$$

Defining a Lyapunov functional as

$$G(t) = \int_0^\infty \rho(y) w(t, y) dy$$

with

$$\rho(y) = \begin{cases} \frac{916}{7\times 405^3} y, 0 \le y < 2, \\ -\frac{415}{21\times 405^3} y^2 + \frac{4108}{21\times 405^3} y - \frac{1660}{21\times 405^3}, 2 \le y < 5, \\ \frac{1}{(y+400)^2}, y \ge 5, \end{cases}$$

we can obtain that G satisfies the following inequality

$$\frac{dG}{dt} \ge K_2 G^2 - K_1 G, \tag{45}$$

for two positive constants K_1 and K_2.

Therefore, there exists a time $0 < t^* \le T$ such that $G(t)$ goes to infinity as $t \to t^* - 0$ provided that

$$G(0) = \int_0^{+\infty} \rho(y)(U_x(0,0) - u_{0x}(0, y))dy \ge M \tag{46}$$

being properly large.

In particular, when $U \equiv 0$, the inequality (45) simplifies into

$$\frac{dG}{dt} \ge K_2 G^2, \tag{47}$$

which implies that $G(t)$ blows up in a finite time always, for any given nonzero initial value.

Therefore, we conclude the following result, the detail proof can be found in [24].

Theorem 4.1. *Under assumptions* (40), (42), (44) *and* (46), *the smooth solution* u *of* (25) *blows up in* Q_T.

Remark 4.1. When $U(t, x) \equiv 0$, from (47) we know that the smooth solution u of (25) always blows up in a finite time for any nonzero initial data satisfying

$$u_0(0, y) = 0, \; u_{0x}(0, y) \le 0, \tag{48}$$

which differs from the result obtained in [15] for the classical Prandtl equation. Therefore, in contrast with the classical Prandtl equation, the integral term in (25) triggers the formulation of singularities earlier.

Acknowledgments. This research was partially supported by the National Natural Science Foundation of China under Grant No. 11631008.

REFERENCES

[1] R. Alexandre, Y. G. Wang, C. J. Xu and T. Yang, Well-posedness of the Prandtl equation in Sobolev spaces, *J. Amer. Math. Soc.*, **339** (2015), 607–633.

[2] V. Barcilon, P. Constantin and E. S. Titi, Existence of solutions to the Stommel-Charney model of the Gulf Stream, *SIAM J. Math. Anal.*, **19** (1988), 1355–1364.

[3] R. E. Caflisch and M. Sammartino, Existence and singularities for the Prandtl boundary layer equations, *Z. Angew. Math. Mech.*, **80** (2000), 733–744.

[4] T. Colin, Remarks on a homogeneous model of ocean circulation, *Asympt. Anal.*, **12** (1996), 153–168.

[5] A. L. Dalibard and M. Paddick, An existence result for the steady rotating Prandtl equation, preprint, `arXiv:1603.05089v1`.

[6] B. Desjardins and E. Grenier, On the homogeneous model of wind-driven ocean circulation, *SIAM J. Appl. Math.*, **60** (2000), 43–60.

[7] H. Dietert and D. Gerard-Varet, Well-posedness of the Prandtl equation without any structural assumption, *Ann. Partial Diff. Equat.*, **5** (2019), Art. 8, 51pp.
[8] W. E and B. Engquist, Blowup of solutions of the unsteady Prandtl's equation, *Comm. Pure Appl. Math.*, **50** (1997), 1287–1293.
[9] D. Gérard-Varet and E. Dormy, On the ill-posedness of the Prandtl equation, *J. Amer. Math. Soc.*, **23** (2010), 591–609.
[10] Y. Guo and T. T. Nguyen, A note on Prandtl boundary layers, *Comm. Pure Appl. Math.*, **64** (2011), 1416–1438.
[11] S. Friedlander, *An Introduction to The Mathematical Theory of Geophysical Fluid Dynamics*, North-Holland Publishing Company, 1980.
[12] S. B. Gong, X. Wang and Y. G. Wang, Local well-posedness and separation of boundary layer problems for wind driven oceanic current, preprint.
[13] E. Grenier, On the nonlinear instability of Euler and Prandtl equations, *Comm. Pure Appl. Math.*, **53** (2000), 1067–1091.
[14] I. Kukavica and V. Vicol, On the local existence of analytic solutions to the Prandtl boundary layer equations, *Commun. Math. Sci.*, **11** (2013), 269–292.
[15] I. Kukavica, V. Vicol, and F. Wang, The van Dommelen and Shen singularity in the Prandtl equations, *Adv. Math.*, **307** (2017), 288–311.
[16] W. X. Li and T. Yang, Well-posedness in Gevery space for the Prandtl system with nondegenerate critical points, preprint, `arXiv:1609.08430`.
[17] M. C. Lombardo, M. Cannone and M. Sammartino, Well-posedness of the boundary layer equations, *SIAM J. Math. Anal.*, **35** (2003), 987–1004.
[18] N. Masmoudi and T. K. Wong, Local-in-time existence and uniqueness of solutions to the Prandtl equations by energy methods, *Comm. Pure Appl. Math.*, **68** (2015), 1683–1741.
[19] O. A. Oleinik, The Prandtl system of equations in boundary layer theory, *Soviet Math. Dokl.*, **4** (1963), 583–586.
[20] O. A. Oleinik and V. N. Samokhin, *Mathematical Models in Boundary Layer Theory*, Chapman & Hall/CRC, 1999.
[21] J. Pedlovsky, *Geographysical Fluid Dynamics*, 2nd edition, Springer-Verlag, New York, 1987.
[22] J. Pedlosky, *Ocean Circulation Theory*, second edition, Springer-Verlag, Berlin-Heidelberg, 1996.
[23] L. Prandtl, Über flüssigkeitsbewegungen bei sehr kleiner Reibung, in "Verh. Int. Math. Kongr., Heidelberg 1904", Teubner, 1905.
[24] X. Wang and Y. G. Wang, Well-posedness and blowup of the geophysical boundary layer problem, preprint, `arXiv:1903.07016`.
[25] Y. G. Wang and S. Y. Zhu, Mathematical analysis of boundary layers in two-dimensional incompressible viscous heat conducting flows (in Chinese), *Sci. Sin. Math.*, **49** (2019), 267–280.
[26] Z. P. Xin and L. Q. Zhang, On the global existence of solutions to the Prandtl's system, *Adv. Math.*, **181** (2004), 88–133.
[27] P. Zhang and Z. F. Zhang, Long time well-posedness of Prandtl system with small and analytic initial data, *J. Funct. Anal.*, **270** (2016), 2591–2615.

E-mail address: `xiangwangg@126.com`
E-mail address: `ygwang@sjtu.edu.cn`

Part 2

Invited Lectures

ON THE DYNAMIC OF DISSIPATIVE PARTICLES

RICARDO ALONSO

Departamento de Matemática, PUC-Rio
Rua Marquês de São Vicente 225
Rio de Janeiro, CEP 22451-900, Brazil

ABSTRACT. We present in this document a short discussion about the time asymptotic behaviour of dissipative systems with large number of particles. Two classical examples of non conservative phenomena are brought to the discussion, viscoelastic and reactive particles. Non linear techniques based on entropy and spectral analysis are used to rigorously describe the evolution of such systems toward self-similarity and equilibrium. Such techniques show universal rates of relaxation of the macroscopic quantities theorized by physicist and engineers.

1. **Introduction.** This document brings two examples of non conservative systems, viscoelastic particles and reactive particles. In the case of viscoelasticity, the temperature of the system decreases as particle collisions take place. This is because particle deformation happens as particles bear collisions producing an outflow of heat. It is natural that the energy dissipation in a particular collision depends on the impact velocity; for fast collisions the dissipation is large while for slow is small. As particles collide at all times, energy dissipation continues until particles reach zero temperature.

In the case of reactive particles that we discuss here, they follow the reaction $A+A \to \emptyset$ with certain fixed probability at each collision. In this way, as the particles collide at all times, the number of reactive particles will diminish until no particle is left. Note that while viscoelastic particles have the Dirac density as steady state due to conservation of total mass, reactive particle have the zero density as steady state. The issue that we want to understand here is precisely how such steady states are reached.

1.1. **The Boltzmann equation.** Systems with large number of particles that undergo collisions are well described by the Boltzmann equation. For spatially homogeneous gases, the Boltzmann equation is given by

$$\partial_t f(t,v) = Q(f,f)(t,v)\,, \quad v \in \mathbb{R}^d\,, \quad f(0,v) = f_0(v)\,. \tag{1}$$

A solution $f = f(t,v) \geq 0$ of this equation represent the density distribution of the particles with respect to velocity $v \in \mathbb{R}^d$ along time $t > 0$. The collisional operator

---

2000 *Mathematics Subject Classification.* Primary: 35Q20, 70F45; Secondary: 35P15.

Key words and phrases. Inelastic Boltzmann, viscoelastic particles, annihilation equation, Haff's law, entropy methods, spectral analysis.

This research has been done in collaboration with B. Lods, V. Bagland and Y. Chen. R. Alonso is partially supported by the Bolsa de Produtividade em Pesquisa CNPq.

is a bilinear form defined as

$$\begin{aligned} Q(f,g)(v) &= \int_{\mathbb{R}^d}\int_{\mathbb{S}^{d-1}} \big(f('v)g('v_*) - f(v)g(v_*)\big)\,|u|\,b(\sigma\cdot\hat{u})\,\mathrm{d}\sigma\,\mathrm{d}v_* \\ &= Q^{+}(f,g)(v) - Q^{-}(f,g)(v)\,. \end{aligned} \tag{2}$$

Here $('v,'v_*)$ are the pre collisional velocities of two particles leading, after collision, two particles with velocities (v,v_*). The relative velocity between particles is $u := v-v_*$ and the hat represents unitary vectors, so, $\hat{u} = u/|u|$. Observe that $f('v)g('v_*)$ quantify the probability of a particular collision to gain a pair with velocities (v,v_*). In the same way $f(v)g(v_*)$ quantify the probability of a pair to lose the velocities (v,v_*). Therefore, the collisional form is nothing else but the inflow (the positive part) and outflow (the negative part) of particles acquiring velocity v. This process is weighted with a collisional kernel $B(u,\sigma) = |u|\,b(\sigma\cdot\hat{u})$ representing the physics of the interaction involved. In the case of hard spheres, such as billiards, the kinetic weight $|u|$ in the collisional kernel is adequate. The reader is referred to the classical reference [14] for a complete discussion.

1.2. **Observables and entropy.** The statistical moments are defined by

$$m_k(t) = \int_{\mathbb{R}^d} f(t,v)\,|v|^k\,\mathrm{d}v\,, \qquad k\geq 0\,. \tag{3}$$

Solutions to the Boltzmann equation conserve mass, momentum, and energy (addressed also as temperature of f). That is,

$$m_0(t) = m_0(0)\,, \quad m_2(t) = m_2(0)\,, \quad \int_{\mathbb{R}^d} f(t,v)\,v\,\mathrm{d}v = \int_{\mathbb{R}^d} f_0(v)\,v\,\mathrm{d}v\,.$$

Also, the entropy decreases,

$$\int_{\mathbb{R}^d} f(t,v)\,\ln\big(f(t,v)\big)\,\mathrm{d}v =: \mathcal{H}(f(t)) \leq \mathcal{H}(f_0)\,.$$

In addition, it is known that the statistical moments of any order $k>2$ are instantaneously generated for solutions f. That is, for initial condition f_0 with finite mass and energy it follows that

$$m_k(t) \leq C(f_0)\Big(1 + t^{-k}\Big)\,, \qquad \forall\, k>2\,,\; t>0\,. \tag{4}$$

Furthermore, the emergence of exponential tails also holds,

$$\int_{\mathbb{R}^d} f(t,v)\,e^{c(f_0)\min\{t,1\}|v|}\mathrm{d}v \leq C(f_0)\,. \tag{5}$$

The constants $c(f_0)$ and $C(f_0)$ depend only on the initial datum f_0. This property is deeply related with the fact that the collisional kernel grows unbounded for large velocities. See for example [10, 11, 30, 2] for extensive discussion of such properties.

1.3. **Dissipation of entropy.** Take $f(t,v)$ a solution to the Boltzmann equation. The relative entropy is defined as

$$\mathcal{H}(f(t)|\mathcal{M}) := \int_{\mathbb{R}^d} f(t,v)\,\ln\left(\frac{f(t,v)}{\mathcal{M}(v)}\right)\mathrm{d}v\,. \tag{6}$$

Here $\mathcal{M}(v)$ is the Gaussian density with the same mass, momentum, and temperature of $f(t,v)$. A central result for the study of relaxation of the Boltzmann equation is, see [27, 29]

$$-\int_{\mathbb{R}^d} Q(f,f)(t,v)\,\ln\left(\frac{f(t,v)}{\mathcal{M}(v)}\right)\mathrm{d}v \geq C(f_0)\mathcal{H}(f(t)|\mathcal{M})^{1+\varepsilon(f_0)}\,, \tag{7}$$

where $\varepsilon(f_0) > 0$ can be explicitly computed [1]. Estimate (7) is a *functional inequality* valid under some minimal regularity requirements on $f(t,v)$ [2], see [3]. This leads to the estimate for the relative entropy

$$\frac{\mathrm{d}}{\mathrm{d}t}\mathcal{H}(f(t)|\mathcal{M}) + C(f_0)\mathcal{H}(f(t)|\mathcal{M})^{1+\varepsilon(f_0)} \leq 0\,, \quad t > 0\,.$$

Therefore,

$$\mathcal{H}(f(t)|\mathcal{M})(t) \leq C(f_0)\,(1+t)^{-\frac{1}{\varepsilon(f_0)}}\,, \tag{8}$$

and, using Cszisár - Kullback inequality

$$\|f(t)-\mathcal{M}\|_{L^1(\mathbb{R}^d)} \leq \sqrt{\mathcal{H}(f(t)|\mathcal{M})(t)} \leq \sqrt{C(f_0)}\,(1+t)^{-\frac{1}{2\varepsilon(f_0)}}\,, \tag{9}$$

which proves an algebraic relaxation toward thermal equilibrium.

1.4. **Linear theory.** The algebraic relaxation given by (8) using entropy methods can be improved to exponential relaxation by means of spectral analysis. Take f a solution to the Boltzmann equation (1). If the linearisation $f(t,v) = \mathcal{M}(v) + \sqrt{\mathcal{M}(v)}\,h(t,v)$ is used in (1), one arrives to

$$\partial_t h(t,v) = \mathcal{L}_{\mathcal{M}}(h)(t,v) + \frac{1}{\sqrt{\mathcal{M}(v)}}Q\big(\sqrt{\mathcal{M}}\,h, \sqrt{\mathcal{M}}\,h\big)(t,v)\,,$$

where

$$\mathcal{L}_{\mathcal{M}}(h) = \frac{1}{\sqrt{\mathcal{M}(v)}}\Big(Q\big(\sqrt{\mathcal{M}}\,h, \mathcal{M}\big) + Q\big(\mathcal{M}, \sqrt{\mathcal{M}}\,h\big)\Big)\,.$$

It is not difficult to check that $\mathcal{L}_{\mathcal{M}}$ is self adjoint in $L^2(\mathbb{R}^d, \mathrm{d}v)$ and $\langle\mathcal{L}_{\mathcal{M}}(h), h\rangle \leq 0$, see [13]. There are another two important observations to make. First, we can write the linearised Boltzmann operator as

$$\mathcal{L}_{\mathcal{M}} = \mathcal{K} - \mathcal{D}\,, \tag{10}$$

where $\mathcal{D}$ is a dissipative operator and $\mathcal{K}$ a relatively compact operator. In essence,

$$\mathcal{D}(h) \sim \frac{1}{\sqrt{\mathcal{M}(v)}}Q^-\big(\sqrt{\mathcal{M}}\,h, \mathcal{M}\big) \sim \nu_o\,\langle v\rangle\,h\,,$$

$$\mathcal{K}(h) \sim \frac{1}{\sqrt{\mathcal{M}(v)}}\Big(Q^+\big(\sqrt{\mathcal{M}}\,h, \mathcal{M}\big) + Q\big(\mathcal{M}, \sqrt{\mathcal{M}}\,h\big)\Big)\,.$$

Here we used the notation $\langle v\rangle = \sqrt{1+|v|^2}$. Such observations lead to the fact that $\mathcal{L}_{\mathcal{M}}$ only has real spectrum with the same essential spectrum as $\mathcal{D}$ localised in $(-\infty, \nu_0]$. Furthermore, using a compact perturbation argument, see [19], the eigenvalues are localised in $(-\nu_0, 0]$. Of course, this is equivalent to say that the operator

$$\mathcal{L}(g) = Q\big(g, \mathcal{M}\big) + Q\big(\mathcal{M}, g\big)\,, \tag{11}$$

has such spectrum as an operator in $L^2(\mathbb{R}^d, \mathcal{M}^{-1/2}\mathrm{d}v)$.

Second, it can be proved that the eigenvalue problem

$$\mathcal{L}(\psi) = \lambda\,\psi\,, \quad v \in \mathbb{R}^d\,,$$

---

[1] The smoother f is, the smaller $\varepsilon(f) > 0$ is.

[2] A lower Gaussian barrier [26] and some finite statistical moments for f and $f\ln f$ suffice [3].

lead to $\mathcal{C}^\infty$ eigenfunctions ψ, all of them having Gaussian tails. This fact is related to the splitting (10). This point is quite important in the argument used in [24] to deduce that if $\mathcal{L}$ is considered as an operator in a larger space $\mathcal{X} \supset L^2(\mathbb{R}^d, \mathcal{M}^{-1/2}\mathrm{d}v)$, then, the eigenvalues will remain invariant in such larger space. The argument proves quite useful since the Boltzmann dynamic for hard spheres take place in bigger spaces such as $L^1(\mathbb{R}^d, e^{c\langle v\rangle}\mathrm{d}v)$, recall (5). In this way, defining the perturbation h as

$$h = f - \mathcal{M}$$

we see that, by conservation of mass, momentum, and energy for f, the perturbation h has zero mass, momentum, and energy. In this way, h relaxes exponentially with the first negative eigenvalue of $\mathcal{L}$ [3].

These ideas can be formalised to prove exponential convergence of f toward the thermal equilibrium $\mathcal{M}$ in a two step approach. First, there is a non linear transitory slow relaxation guided by the entropy, then, in a second stage, an exponentially fast linear relaxation happens due to the spectral gap of $\mathcal{L}$ in, say, a space such as $\mathcal{X} = L^1(\mathbb{R}^d, e^{c\langle v\rangle}\mathrm{d}v)$. See for instance [24, 3] for technical details. This process of enlargement of the operator's space was introduced in the context of the Boltzmann equation in [24] and formalised in a more general framework in [17].

2. **Viscoelastic spheres.** Dilute granular gases can be modelled with the inelastic Boltzmann equation

$$\partial_t f(t,v) = Q_e(f,f)(t,v)\,, \quad v \in \mathbb{R}^d\,, \quad f(0,v) = f_0(v)\,. \tag{12}$$

The collision operator is given by

$$Q_e(f,g)(v) = \int_{\mathbb{R}^d}\int_{\mathbb{S}^{d-1}} \Big(\tfrac{1}{{}'e\,{}'J} f({}'v)g({}'v_*) - f(v)g(v_*)\Big)\,|u|\,b(\sigma\cdot\hat{u})\,\mathrm{d}\sigma\,\mathrm{d}v_*\,. \tag{13}$$

The operator is quite similar to the classical Boltzmann operator (2) and, in fact, many analytical properties are identical, see [4, 5, 20, 21]. However, there are essential differences, for example the Gaussian density is not longer in the kernel of the operator $Q_e(\mathcal{M},\mathcal{M}) \neq 0$. The key new parameter defining the collision between particles is the restitution coefficient $e := e(|u\cdot\hat{n}|) \in [0,1]$, which is understood as the fraction of the energy that is "restituted" after collision. Thus, $e \equiv 1$ is the classical elastic collision case and $e \equiv 0$ is the sticky particle case. In the definition of the collision operator (13), the reader notices the extra term $\frac{1}{e\,J}$. Here J is the Jacobian of the transformation $({}'v,{}'v_*) \to (v,v_*)$ which, in general, satisfies $J \leq 1$. Refer to [1] for particular examples. The term $\frac{1}{e\,J}$ is evaluated in pre collisional velocities so that conservation of mass and momentum hold. For viscoelastic particles the restitution coefficient depends on the impact velocity $|u\cdot\hat{n}|$ where $\hat{n}$ is the unit vector normal to the collision's plane. For small impact velocities deformation of particles due to collisions is minimal, thus, $e \approx 1$ in such case. In fact, it is customary to assume that

$$e(r) \approx 1 - ar^{\beta}\,, \quad r \approx 0\,,\ a \in [0,1],\ \beta > 0\,.$$

Strictly speaking, assuming linear deformation, it follows that $\beta = 1/5$, see [12]. The interested reader may go to [1, 4, 5], for additional information.

[3] $\{0\}$ belongs to the spectrum of $\mathcal{L}$. It has $d+2$ associated eigenvectors $\{\sqrt{\mathcal{M}}, v\sqrt{\mathcal{M}}, |v|^2\sqrt{\mathcal{M}}\}$ related to the $d+2$ conservation laws.

2.1. **Statistical moments and entropy.** Viscoelastic spheres conserve mass and momentum, but dissipates energy: $m_2(t) \leq m_2(0)$. The rate of energy dissipation (or cooling) is known as Haff's law [18]. In fact, it is possible to prove that, see [4, 5]

$$c(t_0, f_0)t^{-\frac{k}{1+\beta}} \leq m_k(t) \leq C(t_0, f_0)t^{-\frac{k}{1+\beta}}, \quad t \geq t_0 > 0. \tag{14}$$

Observe that the cooling rate depends uniquely on $\beta > 0$. This is the case because after many collisions most particles will have little energy left. That is, most collisions happen at low impact velocity, consequently, the behaviour of the restitution coefficient at low impact velocity is key.

An important difference with respect to elastic Boltzmann is that entropy is not monotonic. From the technical point of view, this implies that the general theory of renormalised solutions made for the Cauchy problem in the elastic case does not apply, see [15]. For the homogeneous inelastic case, however, one can prove that the entropy is uniformly bounded

$$\mathcal{H}(f)(t) \leq C\big(f_0, \mathcal{H}(f_0)\big).$$

2.2. **Self-similarity and viscoelastic relaxation.** As collisions occur in viscoelastic particles, the system evolves to a homogeneous cooling state. The reason is, again, that the granular gas is becoming more elastic as it loses energy. In order to observe such homogenisation, we search for a good scale to pose the dynamics. It turns out, a good scale is given by one that brings back the conservation of energy [5, 16]. Set

$$f(t,v) =: V^d(t) g(\tau(t), V(t)v), \quad V(t) := \sqrt{\frac{m_2(0)}{m_2(t)}}, \quad \tau(t) := \int_0^t \frac{\mathrm{d}s}{V(s)}, \tag{15}$$

then, $g(\tau, w)$ conserves mass, momentum, and energy. In fact, it solves the equation

$$\partial_\tau g(\tau, w) + V'\big(t^{-1}(\tau)\big)\nabla_w \cdot \big(w\, g(\tau, w)\big) = Q_{e_\tau}(g,g)(\tau, w), \quad g(0,w) = f_0(w). \tag{16}$$

Estimate (14) applied to $m_2(t)$ leads to

$$\tau(t) \sim C(1+t)^{\frac{\beta}{1+\beta}}, \quad V'\big(t^{-1}(\tau)\big) \sim \Big(1 + \tfrac{\beta}{1+\beta}\tau\Big)^{-1},$$

and

$$e_\tau(r) := e\Big(V'\big(t^{-1}(\tau)\big)^{1/\beta} r\Big) \sim e\Big(\Big(1 + \tfrac{\beta}{1+\beta}\tau\Big)^{-1/\beta} r\Big)$$

for sufficiently large time. Since $\lim_{\tau\to\infty} e_\tau = 1$, the self-similar scaling suggests that

$$Q_{e_\tau}(g,g) \to Q(g,g), \quad \text{as} \quad \tau \to \infty.$$

In other words, such scaling is quantifying the process at which the granular gas becomes an elastic gas.

Set $\mathcal{M}(w)$ the Gaussian with same mass, momentum, and energy as f_0. By construction, $g(\tau, w)$ and $\mathcal{M}(w)$ have the same mass, momentum, and energy. As a consequence, we can find an equation for the relative entropy using (16)

$$\frac{\mathrm{d}}{\mathrm{d}\tau}\mathcal{H}(g|\mathcal{M}) - \int_{\mathbb{R}^d} Q(g,g)(\tau, w) \ln\left(\frac{g(\tau, w)}{\mathcal{M}(w)}\right)\mathrm{d}w = \mathcal{I}_1(\tau) + \mathcal{I}_2(\tau),$$

where

$$\mathcal{I}_1(\tau) := -V'\big(t^{-1}(\tau)\big)\int_{\mathbb{R}^d}\nabla_w\cdot\big(wg(\tau,w)\big)\ln\left(\frac{g(\tau,w)}{\mathcal{M}(w)}\right)\mathrm{d}w\,,$$

$$\mathcal{I}_2(\tau) := \int_{\mathbb{R}^d}\big(Q^+_{e_\tau}(g,g)(\tau,w) - Q^+(g,g)(\tau,w)\big)\ln\left(\frac{g(\tau,w)}{\mathcal{M}(w)}\right)\mathrm{d}w\,.$$

It is possible to bound such terms as long as some suitable regularity is available for g. Refer to [5] for the technical part. More precisely,

$$\big|\mathcal{I}_1(\tau)\big| + \big|\mathcal{I}_2(\tau)\big| \le C(f_0)V'\big(t^{-1}(\tau)\big)\,.$$

This estimate and the dissipation of entropy (7) lead to

$$\frac{\mathrm{d}}{\mathrm{d}\tau}\mathcal{H}(g(\tau)|\mathcal{M}) + c(f_0)\mathcal{H}(g(\tau)|\mathcal{M})^{1+\varepsilon(f_0)} \le C(f_0)V'\big(t^{-1}(\tau)\big)\,,\quad \tau\ge 1\,.$$

We conclude with Cszisár - Kullback inequality and this estimate that

$$\|g(\tau)-\mathcal{M}\|_{L^1(\mathbb{R}^d)} \le \sqrt{\mathcal{H}(g(\tau)|\mathcal{M})} \le C(f_0)\Big(V'\big(t^{-1}(\tau)\big)\Big)^{\frac{1}{2(1+\varepsilon(f_0))}}\,. \tag{17}$$

Furthermore, we know the asymptotic behavior of $\tau(t)$. Then, scaling back to the original problem

$$\|f(t) - V^d(t)\mathcal{M}(V(t)\cdot)\|_{L^1(\mathbb{R}^d)} \le C(f_0)\,(1+t)^{-\frac{\beta}{1+\beta}\frac{1}{1+\varepsilon(f_0)}}\,. \tag{18}$$

Estimates on $\varepsilon(f_0)$ show that $\varepsilon(f_0)\sim 0$ for smooth densities f_0. Thus, the best algebraic rate of relaxation that can be expected using this method is $\frac{\beta}{1+\beta}$ which seems to be optimal.

3. **Reactive particles and ballistic annihilation.** We consider now a systems of elastic interacting particles that, at the moment of interaction, have probability α to react and produce an inert product. Such system can be modelled by the ballistic annihilation equation, see for instance [9, 28, 25]

$$\partial_t f(t,v) = (1-\alpha)Q(f,f)(t,v) - \alpha\,Q^-(f,f)(t,v)\,,\quad v\in\mathbb{R}^d\,,\quad t>0\,. \tag{19}$$

The reactive particles sustain a continuous outflow of particles, consequently, its total mass decreases. As particles leave, the system cannot conserve momentum or energy. In fact, in the long run most particles will have reacted, so, the stationary state is the zero density. The goal here is to describe the rate at which this process occurs when the probability of reaction is relatively small but nonzero $0<\alpha\ll 1$, that is, in the case where particles collide many times before a reaction occurs. In order to accomplish this, we use the technique introduced for viscoelastic particles. More precisely, use the rescaling

$$f(t,v) =: \frac{n_f(t)}{(2T_f(t))^{\frac{d}{2}}}\,g\Big(\tau(t),\frac{v-u_f(t)}{\sqrt{2T_f(t)}}\Big)\,, \tag{20}$$

where

$$n_f(t) = \int_{\mathbb{R}^d} f(t,v)\mathrm{d}v\,,\quad n_f(t)u_f(t) = \int_{\mathbb{R}^d} f(t,v)\,v\,\mathrm{d}v\,,$$

are the mass and momentum of $f(t,v)$, and

$$d\,n_f(t)\,T_f(t) = \int_{\mathbb{R}^d} f(t,v)\big|v-u_f(t)\big|^2\mathrm{d}v\,,$$

is its temperature, to construct a normalised density $g(\tau, w)$

$$\int_{\mathbb{R}^d} g(\tau,w)\begin{pmatrix}1\\ w\\ |w|^2\end{pmatrix}\mathrm{d}w = \begin{pmatrix}1\\ 0\\ \frac{d}{2}\end{pmatrix}, \qquad \tau \geq 0\,.$$

After some calculations, it is possible to find the equation for $g(\tau, w)$

$$\partial_\tau g(\tau,w) + \big(A(g)(\tau) - d\,B(g)(\tau)\big)g(\tau,w) + B(g)(\tau)\nabla_w\cdot\big((w - w_g(\tau))g(\tau,w)\big)$$
$$= (1-\alpha)Q(g,g)(\tau,w) - \alpha\, Q^-(g,g)(\tau,w)\,, \quad w\in\mathbb{R}^d\,, \quad \tau > 0\,. \quad (21)$$

The explicit definitions of the operators $A(g)(\tau)$, $B(g)(\tau)$, and $w_g(\tau)$ can be found in [6]. For the purpose of this discussion, the important fact is that they are of order α provided that minimal regularity requirements for $g(\tau, w)$ are satisfied. An important remark is that, in order to obtain (21), the time scale $\tau'(t) = n_f(t)\sqrt{2T_f(t)}$ is needed. Such time scale pops up naturally to simplify the equation for $g(\tau, w)$ as much as possible. A fine analysis of the statistical moments of the annihilation equation, see [7, 8], show that for $t > 0$

$$(1+t)^{-2} \lesssim n_f(t)\sqrt{2T_f(t)} \lesssim (1+\alpha t)^{-2}\,, \quad \text{then} \quad \tau(t) \sim C\ln(1+\tau)\,. \quad (22)$$

Of course, knowing the explicit expression of the time scale (22) is essential to find the relaxation rates in the physical scale from the relaxation rates observed for the self similar problem. It is worthwhile to remark that using nonlinear moment analysis does not render an optimal relaxation rate for $n_f(t)$ or $T_f(t)$ individually, but it does for $\tau'(t)$.

3.1. **Entropy.** Set $\mathcal{M}(w)$ the normalised Gaussian distribution, multiply equation (21) by $\ln\big(g(\tau,w)/\mathcal{M}(w)\big)$, and integrate in $w\in\mathbb{R}^d$. We are led to

$$\frac{\mathrm{d}}{\mathrm{d}\tau}\mathcal{H}(g(\tau)|\mathcal{M}) - (1-\alpha)\int_{\mathbb{R}^d} Q(g,g)(\tau,w)\ln\left(\frac{g(\tau,w)}{\mathcal{M}(w)}\right)\mathrm{d}w$$
$$= \big(d\,B(g)(\tau) - A(g)(\tau)\big) + \mathcal{I}_1(\tau)\,,$$

where

$$\mathcal{I}_1(\tau) := -\alpha\int_{\mathbb{R}^d} Q^-(g,g)(\tau,w)\ln\left(\frac{g(\tau,w)}{\mathcal{M}(w)}\right)\mathrm{d}w\,.$$

It follows, under some regularity conditions for $g(\tau, w)$ needed for the validity of (7), that

$$\frac{\mathrm{d}}{\mathrm{d}\tau}\mathcal{H}(g(\tau)|\mathcal{M}) + C(f_0)\mathcal{H}(g(\tau)|\mathcal{M})^{1+\varepsilon(f_0)} \leq C(f_0)\alpha\big(\mathcal{H}(g(\tau)|\mathcal{M}) + 1\big)\,, \quad \tau > 0\,.$$

Consequently, we conclude that

$$\mathcal{H}(g(\tau)|\mathcal{M}) \leq C(f_0)\Big((1+\tau)^{-\frac{1}{2}} + \alpha^{\frac{1}{3}}\Big)\,, \quad \tau > 0\,. \quad (23)$$

Estimate (23) shows that the normalised self similar profile $g(\tau, w)$ for the ballistic annihilation equation (19) relaxes at an algebraic rate toward a stationary state lying to a distance $\sim \alpha^{\frac{1}{3}}$, in the sense of relative entropy, of the normalised Gaussian density $\mathcal{M}$.

3.2. **Linear theory.** We want to improve the algebraic relaxation found by entropy methods to a exponential rate. Thus, we first study the location of the spectrum for the linearised annihilation operator. The stationary equation is given by

$$\begin{aligned}\big(A(G) - d\,B(G)\big)G(w) + B(G)\nabla_w \cdot \big((w - w_G)G(w)\big) \\ = (1-\alpha)Q(G,G)(w) - \alpha\, Q^-(G,G)(w)\,, \quad w \in \mathbb{R}^d\,. \quad (24)\end{aligned}$$

This equation is well posed for the regime that we are discussing $0 < \alpha \ll 1$ as shown in [7, 8]. The stationary profile $G(w) := G_\alpha(w)$ is not Gaussian, however,

$$\lim_{\alpha\to 0} G_\alpha(w) = \mathcal{M}(w)\,, \quad \text{in the } L^1 \text{ sense}\,.$$

Introducing the linearisation $g(\tau,w) = G(w) + h(\tau,w)$ one finds the equation

$$\partial_\tau g = \mathcal{L}_{ann}(h) + \mathcal{Q}_{ann}(h,h)\,.$$

More than the explicit expression of the linearised operator $\mathcal{L}_{ann}$, it is important to remark that one can write it as a perturbation of the linearised Boltzmann operator $\mathcal{L} = \mathcal{K}_{Bolt} - \mathcal{D}_{Bolt}$[4]. Indeed,

$$\mathcal{L}_{ann} = \mathcal{L} + \mathcal{L}_{ann} - \mathcal{L} = \mathcal{K}_{Bolt} - \big(\mathcal{D}_{Bolt} + \mathcal{L} - \mathcal{L}_{ann}\big) = \mathcal{K}_{Bolt} - \mathcal{D}_{ann}\,.$$

The key observation here is that $\mathcal{D}_{ann}$ is a dissipative operator in the exponentially weighted Sobolev space $W^{1,1}(\mathbb{R}^d,\, e^{c\langle v\rangle}\mathrm{d}v)$ and its spectrum lies in $\{z : \mathrm{Re}(z) \le -\nu_\alpha\}$ for a positive ν_α. In fact, $\nu_\alpha \to \nu_o$ as $\alpha \to 0$. In this sense $\mathcal{D}_{ann}$ is a perturbation of $\mathcal{D}_{Bolt}$. Compact perturbation theory, see [19], leads us to conclude that no essential spectrum, only eigenvalues, will lie in the set $\{z : \mathrm{Re}(z) > -\nu_\alpha\}$. Take β one of such eigenvalues and a corresponding eigenfunction ψ. Then

$$\mathcal{L}_{ann}(\psi) = \beta\,\psi\,, \quad \text{consequently} \quad \big(\mathcal{L}_{ann} - \mathcal{L}\big)(\psi) = \big(\beta - \mathcal{L}\big)\psi\,.$$

That is,

$$\psi = \big(\beta - \mathcal{L}\big)^{-1}\big(\mathcal{L}_{ann} - \mathcal{L}\big)(\psi)\,. \quad (25)$$

It is simple to check that

$$\big\|\big(\mathcal{L}_{ann} - \mathcal{L}\big)(\psi)\big\|_{W^{1,1}(\mathbb{R}^d,\, e^{c\langle v\rangle}\mathrm{d}v)} \le \alpha\, C\, \|\psi\|_{W^{2,1}(\mathbb{R}^d,\, \langle v\rangle e^{c\langle v\rangle}\mathrm{d}v)}\,.$$

Additionally, we commented already in relation with the eigenvalue problem for $\mathcal{L}$, the eigenvalue problem $\mathcal{L}_{ann}(\psi) = \beta\,\psi$ regularises [5]

$$\|\psi\|_{W^{2,1}(\mathbb{R}^d,\, \langle v\rangle e^{c\langle v\rangle}\mathrm{d}v)} \le \frac{C}{|\nu_\alpha - \beta|}\|\psi\|_{W^{1,1}(\mathbb{R}^d,\, e^{c\langle v\rangle}\mathrm{d}v)}\,.$$

Using these last two estimates in (25), we obtain that

$$\begin{aligned}\|\psi\|_{W^{1,1}(\mathbb{R}^d,\, e^{c\langle v\rangle}\mathrm{d}v)} &\le \big\|(\beta - \mathcal{L})^{-1}\big\|_{\mathcal{B}(W^{1,1}(\mathbb{R}^d,\, e^{c\langle v\rangle}))}\big\|\big(\mathcal{L}_{ann} - \mathcal{L}\big)(\psi)\big\|_{W^{1,1}(\mathbb{R}^d,\, e^{c\langle v\rangle}\mathrm{d}v)} \\ &\le \frac{C\,\alpha}{\mathrm{dist}(\beta, \sigma(\mathcal{L}))}\frac{\|\psi\|_{W^{1,1}(\mathbb{R}^d,\, e^{c\langle v\rangle}\mathrm{d}v)}}{|\nu_\alpha - \beta|}\,.\end{aligned}$$

[4] Recall that $\mathcal{D}_{Bolt}$ is a dissipative operator and $\mathcal{K}_{Bolt}$ is a relative compact operator.

[5] The essential difference between the elastic Boltzmann eigenvalue problem and the one of particle annihilation is that eigenvectors have Gaussian tails for the former and exponential tails, $\sim e^{-c\langle v\rangle}$, for the latter. Refer to [11] for a tail analysis in the self similar inelastic case with constant restitution.

Here $\sigma(\mathcal{L})$ is the set of eigenvalues of $\mathcal{L}$ in $W^{1,1}(\mathbb{R}^d, e^{c\langle v\rangle}dv)$, which agrees with that in $L^2(\mathbb{R}^d, \mathcal{M}^{-1/2})$. Consequently,

$$\operatorname{dist}\big(\beta, \sigma(\mathcal{L})\big) \leq \frac{C\,\alpha}{|\nu_\alpha - \beta|}\,. \tag{26}$$

Estimate (26) shows that the spectrum of $\mathcal{L}_{ann}$ is indeed a perturbation of the spectrum of $\mathcal{L}$. As a consequence, for $\alpha \ll 1$ the only eigenvalues that pose a stability problem are those near $0 \in \sigma(\mathcal{L})$. Letting P_α be the projection onto these eigenvalues, it can be proved that

$$\|\mathrm{P}_\alpha - \mathrm{P}_0\|_{\mathcal{B}(W^{1,1}(\mathbb{R}^d,\, e^{c\langle v\rangle}))} \lesssim o(\alpha) < 1\,,$$

where P_0 is the projection onto the null space of $\mathcal{L}$. Perturbation theory [19] leads us to conclude that

$$\text{Dim Range}(\mathrm{P}_\alpha) = \text{Dim Range}(\mathrm{P}_0) = d+2\,,$$

where the number $d+2$ relates to the number of conservation laws. With this information at hand, one can invoke a spectral mapping theorem [22, 23] to show that $\mathcal{L}_{ann}$ generates a semigroup $\mathcal{S}(t)$ with the property

$$\|\mathcal{S}(t)(1-\mathrm{P}_\alpha)\|_{\mathcal{B}(W^{1,1}(\mathbb{R}^d,\, e^{c\langle v\rangle}))} \leq C_\mu e^{\mu t}\,, \tag{27}$$

where $\mu < 0$ is such that $|\mu - \mu_0| \leq C\alpha$ with μ_0 the first negative eigenvalue of $\mathcal{L}$. One last important observation is needed to close the argument. Note that by construction $g(\tau, w)$ conserves mass, momentum, and energy, hence, the perturbation $h(\tau, w)$ has zero mass, momentum, and energy. Consequently, $\mathrm{P}_0 h(\tau) = 0$ and, then, if follows that

$$\psi(\tau) = \big(1-\mathrm{P}_\alpha\big)h(\tau) = \big(1-(\mathrm{P}_\alpha - \mathrm{P}_0)\big)h(\tau)\,.$$

Since $\mathrm{P}_\alpha - \mathrm{P}_0$ has norm less than 1 for sufficiently small α, it is possible to invert the right side of this equality. One concludes that

$$\begin{aligned}\|h(\tau)\|_{W^{1,1}(\mathbb{R}^d,\, e^{c\langle v\rangle}dv)} &\leq \big\|\big(1-(\mathrm{P}_\alpha - \mathrm{P}_0)\big)^{-1}\big\|_{\mathcal{B}(W^{1,1}(\mathbb{R}^d,\, e^{c\langle v\rangle}))}\|\psi\|_{W^{1,1}(\mathbb{R}^d,\, e^{c\langle v\rangle}dv)}\\ &= C_\alpha \big\|\big(1-\mathrm{P}_\alpha\big)h(\tau)\big\|_{W^{1,1}(\mathbb{R}^d,\, e^{c\langle v\rangle}dv)}\,.\end{aligned}$$

That is, the projected component of $h(\tau, w)$ controls the full norm of $h(\tau, w)$. We remark that the fact that $g(\tau, w)$ enjoys all conservation laws was essential for this last estimate. These are the main ingredients, together with the entropy analysis, to prove the exponential relaxation for the self similar profile

$$\|g(\tau) - G\|_{L^1(\mathbb{R}^d,\, e^{c\langle v\rangle}dv)} \leq C_\mu(g_0)\, e^{\mu t}\,, \quad t > 0\,. \tag{28}$$

3.3. **Physical problem and universal rates.** Recall that the time scale is known $\tau(t) \sim C\ln(1+\tau)$. Thus, we can go back to the original, physical, scaling. The result is that

$$\int_{\mathbb{R}^d} \big|f(t,v) - F(t,v)\big| \exp\left(c\,\frac{|v-u_f(t)|}{\sqrt{2T_f(t)}}\right)dv \leq C(1+t)^{-\theta}\,, \quad t > 0\,, \tag{29}$$

where the algebraic rate $\theta > 0$ is explicit. The asymptotic "cooling" profile is given by

$$F(t,v) =: \frac{n_f(t)}{(2T_f(t))^{\frac{d}{2}}}\,G\left(\frac{v-u_f(t)}{\sqrt{2T_f(t)}}\right). \tag{30}$$

Furthermore, in the limit $\alpha \to 0$, one has

$$n_f(t) \sim t^{-\frac{4d}{4d+1}}\,, \quad T_f \sim t^{-\frac{2}{4d+1}}\,, \quad \text{as} \quad t \gg 1\,. \tag{31}$$

The algebraic rates (31) were conjectured by Trizac [28, 25] after extensive numerical simulation. Such rates completely define the asymptotic cooling state (30) in the physical scale. Furthermore, estimate (29) gives theoretical information that is very hard and expensive to quantify using numerical simulation.

For a full account of the technical details, and more, we refer the interested reader to [6].

REFERENCES

[1] R. Alonso, Existence of global solutions to the Cauchy problem for the inelastic Boltzmann equation with near-vacuum data, *Indiana Univ. Math. J.*, **58** (2009), 999–1022.

[2] R. Alonso, J. A. Canizo, I. M. Gamba and C. Mouhot, A new approach to the creation and propagation of exponential moments in the Boltzmann equation, *Commun. Partial Differential Equations*, **38** (2013), 155–169.

[3] R. Alonso, I. M. Gamba and M. Tasković, Exponentially-tailed regularity and time asymptotic for the homogeneous Boltzmann equation, `https://arxiv.org/abs/1711.06596v1`, 2017.

[4] R. Alonso and B. Lods, Free cooling and high-energy tails of granular gases with variable restitution coefficient, *SIAM J. Math. Anal.*, **42** (2010), 2499–2538.

[5] R. Alonso and B. Lods, Boltzmann model for viscoelastic particles: Asymptotic behaviour, pointwise lower bounds and regularity, *Comm. Math. Phys.*, **331** (2014), 545–591.

[6] R. Alonso, V. Bagland, and B. Lods, Convergence to self-similarity for ballistic annihilation dynamics, *J. Math. Pures Appl.* (2019), https://doi.org/10.1016/j.matpur.2019.09.008

[7] V. Bagland and B. Lods, Existence of self-similar profile for a kinetic annihilation model, *J. Differential Equations*, **254** (2013), 3023–3080.

[8] V. Bagland and B. Lods, Uniqueness of the self-similar profile for a kinetic annihilation model, *J. Differential Equations*, **259** (2015), 7012–7059.

[9] E. Ben-Naim, P. Krapivsky, F. Leyvraz and S. Redner, Kinetics of ballistically controlled reactions, *J. Chem. Phys.*, **98** (1994), 7284–7288.

[10] A. Bobylev, Moment inequalities for the Boltzmann equations and application to spatially homogeneous problems, *J. Statist. Phys.*, **88** (5-6) (1997), 1183–1214

[11] A. Bobylev, I. Gamba, and V. Panferov, Moment inequalities and high-energy tails for Boltzmann equations with inelastic interactions. *J. Statist. Phys.*, **116** (5-6) (2004), 1651–1682.

[12] N. V. Brilliantov and T. Pöschel, *Kinetic Theory of Granular Gases.* Oxford University Press, 2004.

[13] T. Carleman, *Problèmes mathématiques dans la théorie cinétique des gaz*, Publ. Sci. Inst. Mittag-Löffler. 2. Almqvist & Wiksells Boktryckeri Ab, Uppsala, 1957.

[14] C. Cercignani, R. Illner and M. Pulvirenti, *The Mathematical Theory of Dilute Gases.* Springer, New York, 1994.

[15] R. J. DiPerna and P. L. Lions, On the Cauchy problem for Boltzmann equations: Global existence and weak stability, *Annals of Mathematics*, **130** (2) (1989), 321–366.

[16] F. Filbet and T. Rey, A rescaling velocity method for dissipative kinetic equations. Applications to granular media, *J. Comput. Phys.*, **248** (2013) 177–199.

[17] M. P. Gualdani, S. Mischler ,and C. Mouhot, Factorisation for non-symmetric operators and exponential H-theorem, *Mémoires de la SMF*, **153** (2017).

[18] P. K. Haff, Grain flow as a fluid-mechanical phenomenon, *J. Fluid Mech.*, **134** (1983).

[19] T. Kato, *Perturbation theory for linear operators*, Springer Verlag, Berlin, 1980.

[20] S. Mischler and C. Mouhot, Cooling process for inelastic Boltzmann equations for hard-spheres, Part II: Self-similar solution and tail behavior, *J. Statist. Phys.*, **124** (2006), 655–702.

[21] S. Mischler and C. Mouhot, Stability, convergence to self-similarity and elastic limit for the Boltzmann equation for inelastic hard-spheres, *Commun. Math. Phys.*, **288** (2009), 431–502.

[22] S. Mischler and J. Scher, Spectral analysis of semigroups and growth-fragmentation equations, *Ann. Inst. H. Poincaré Anal. Non Linéaire*, **33** (2016), 849–898.

[23] S. Mischler, Erratum: Spectral analysis of semigroups and growth-fragmentation equations, `https://hal.archives-ouvertes.fr/hal-01422273`, 2017.

[24] C. Mouhot, Rate of convergence to equilibrium for the spatially homogeneous Boltzmann equation with hard potentials, *Commun. Math. Phys.*, **261** (2006), 629–672.

[25] J. Piasecki, E. Trizac and M. Droz, Dynamics of ballistic annihilation, *Phys. Rev. E*, **66** (2002), 066111.

[26] A. Pulvirenti and B. Wennberg, A Maxwellian lower bound for solutions to the Boltzmann equation, *Commun. Math. Phys.*, **183** (1997), 145–160.
[27] G. Toscani and C. Villani, Sharp entropy dissipation bounds and explicit rate of trend to equilibrium for the spatially homogeneous Boltzmann equation, *Commun. Math. Phys.*, **203** (1999), 667–706.
[28] E. Trizac, Kinetics and scaling in ballistic annihilation, *Phys. Rev. Lett.*, **88** (2002), 160601.
[29] C. Villani, Cercignani's conjecture is sometimes true and always almost true, *Commun. Math. Phys.*, **234** (2003), 455–490.
[30] B. Wennberg, Entropy dissipation and moment production for the Boltzmann equation, *J. Stat. Phys.*, **86** (5-6) (1997), 1053–1066.

E-mail address: `ralonso@mat.puc-rio.br`

A NOTE ON 2-D DETACHED SHOCKS OF STEADY EULER SYSTEM

MYOUNGJEAN BAE*

Department of Mathematics, POSTECH
77 Cheongam-ro, Nam-gu Pohang,
Pohang, Gyungbuk, Republic of Korea 37673;
Korea Institute for Advanced Study
85 Hoegiro, Dongdaemun-gu,
Seoul 130-722, Republic of Korea

WEI XIANG

Department of Mathematics, City Univerisity of Hong Kong
Hong Kong, China

ABSTRACT. The shock polar analysis shows that if a weak solution of steady Euler system for inviscid compressible flow has a shock past a blunt body, then the shock cannot be attached to the blunt body. This observation naturally raises a question on the existence of a detached shock solution past a blunt body. The main goal of this paper is to demonstrate how a shock polar analysis is used to analyze two dimensional shocks past wedges or blunt body, and to review the recent result on the existence of detached shocks past a blunt body with a asymptotic state at far field, which is proved in [3]. And, further open questions on detached shocks are discussed.

1. **Shock polars.** The steady *Euler system*

$$
\begin{aligned}
&\partial_{x_1}(\rho u_1)+\partial_{x_2}(\rho u_2)=0\\
&\partial_{x_1}(\rho u_1^2+p)+\partial_{x_2}(\rho u_1u_2)=0\\
&\partial_{x_1}(\rho u_1u_2)+\partial_{x_2}(\rho u_2^2+p)=0\\
&\frac{1}{2}(u_1^2+u_2^2)+\frac{\gamma p}{(\gamma-1)\rho}=B_0\quad (B_0>0\text{: a constant})
\end{aligned}
\tag{1}
$$

governs two dimensional steady flow of inviscid compressible ideal polytropic gas. The constant $B_0>0$ is called the *Bernoulli constant*. And, the functions ρ, p, u_1, u_2 represent density, pressure, horizontal and vertical components of velocity, respectively. The constant $\gamma>1$ represents an adiabatic exponent. In smooth flow, if $\partial_{x_2}u_1-\partial_{x_1}u_2=0$ holds, that is, the flow is irrotational, then one can directly derive

2000 *Mathematics Subject Classification.* Primary: 35A01, 35J25, 35J62, 35M10, 35Q31, 35R35; Secondary: 76H05, 76L05, 76N10.

Key words and phrases. blunt body, detached shock, Euler system, free boundary problem, inviscid compressible flow, irrotational, shock polar, strong shock, transonic shock.

The first author is supported in part by Samsung Science and Technology Foundation under Project Number SSTF-BA1502-02. The second author is supported in part by the Research Grants Council of the HKSAR, China (Project CityU 21305215, Project CityU 11332916, Project CityU 11304817, and Project CityU 11303518).

* Corresponding author: Myoungjean Bae.

from (1) that $\frac{p}{\rho^\gamma}$ is a constant, provided that $\rho > 0$ and $p > 0$. In that case, (1) can be further simplified as

$$\begin{aligned}
&\partial_{x_1}(\rho u_1) + \partial_{x_2}(\rho u_2) = 0 \\
&\partial_{x_1} u_2 - \partial_{x_2} u_1 = 0 \\
&\frac{1}{2}(u_1^2 + u_2^2) + \frac{\rho^{\gamma-1}}{\gamma - 1} = B_0,
\end{aligned} \tag{2}$$

which is called the potential flow model of (1). Both (1) and (2) are elliptic-hyperbolic mixed type nonlinear system. In (1) and (2), *the Mach number* M are defined by $M := \frac{|\mathbf{u}|}{\sqrt{\gamma p/\rho}}$ and $M := \frac{|\mathbf{u}|}{\sqrt{\rho^{\gamma-1}}}$, respectively, for $\mathbf{u} = (u_1, u_2)$. Both (1) and (2) are hyperbolic if $M > 1$(supersonic), hyperbolic-elliptic mixed type if $M < 1$(subsonic), and hyperbolic-degenerate mixed type if $M = 1$(sonic).

Definition 1.1. (a) Let Ω be a domain in $\mathbb{R}^2$. Suppose that a non self-intersecting C^1-curve $\mathcal{S}$ divides Ω into two open and connected subsets Ω^- and Ω^+ so that $\Omega^- \cap \Omega^+ = \emptyset$ and $\Omega^- \cup \mathcal{S} \cup \Omega^+ = \Omega$. The vector valued function $(\rho, p, \mathbf{u}) \in [L^\infty(\Omega) \cap C^0(\overline{\Omega^\pm}) \cap C^1_{\rm loc}(\Omega^\pm)]^4$ is a *weak solution of* (1) with a shock $\mathcal{S}$ if the following properties are satisfied:

(i) $\rho > 0$ and $p > 0$ in Ω;

(ii) In $\Omega^\pm$, $(\rho, p, \mathbf{u})$ satisfy all the equations stated in (1) pointwisely;

(iii) For each point $\mathbf{x}_* \in \mathcal{S}$, set $(\rho^\pm, p^\pm, \mathbf{u}^\pm)(\mathbf{x}_*) := \lim_{\substack{\mathbf{x}\to\mathbf{x}_* \\ \mathbf{x}\in\Omega^\pm}} (\rho, p, \mathbf{u})(\mathbf{x})$. On $\mathcal{S}$, $(\rho, p, \mathbf{u})$ satisfy the Rankine-Hugoniot conditions

$$\begin{aligned}
&\rho^+\mathbf{u}^+ \cdot \boldsymbol{\nu} = \rho^-\mathbf{u}^- \cdot \boldsymbol{\nu}, \\
&\mathbf{u}^+ \cdot \boldsymbol{\tau} = \mathbf{u}^- \cdot \boldsymbol{\tau}, \\
&\rho^+(\mathbf{u}^+ \cdot \boldsymbol{\nu})^2 + p^+ = \rho^-(\mathbf{u}^- \cdot \boldsymbol{\nu})^2 + p^-,
\end{aligned} \tag{3}$$

where $\boldsymbol{\nu}$ is a unit normal vector field, and $\boldsymbol{\tau}$ is a unit tangential vector field on $\mathcal{S}$;

(iv) On $\mathcal{S}$, $\mathbf{u}^+ \cdot \boldsymbol{\nu} \neq 0$ holds, or equivalently $\mathbf{u}^- \cdot \boldsymbol{\nu} \neq 0$ holds;

(v) On $\partial\Omega$, the slip boundary condition $\mathbf{u} \cdot \mathbf{n} = 0$ holds for the inward unit normal vector field $\mathbf{n}$ on $\partial\Omega$.

(b) A weak solution to (2) with a shock $\mathcal{S}$ is defined to be almost same as in (a) except for the following differences:

(ii′) In $\Omega^\pm$, $(\rho, \mathbf{u})$ satisfy all the equations stated in (2) pointwisely;

(iii′) On $\mathcal{S}$, $(\rho, \mathbf{u})$ satisfy the Rankine-Hugoniot conditions

$$\rho^+\mathbf{u}^+ \cdot \boldsymbol{\nu} = \rho^-\mathbf{u}^- \cdot \boldsymbol{\nu}, \quad \text{and} \quad \mathbf{u}^+ \cdot \boldsymbol{\tau} = \mathbf{u}^- \cdot \boldsymbol{\tau}. \tag{4}$$

Definition 1.2. A weak solution $(\rho, p, \mathbf{u})$ to (1) (or (2)) is said to satisfy *the entropy solution* if

$$0 < \rho^- < \rho^+ < \infty, \quad \text{and} \quad 0 < \mathbf{u}^+ \cdot \boldsymbol{\nu} < \mathbf{u}^- \cdot \boldsymbol{\nu} < \infty \tag{5}$$

hold on $\mathcal{S}$, where the unit normal $\boldsymbol{\nu}$ on $\mathcal{S}$ points interior to Ω^+.

Note that, due to the continuity of $(\rho, p, \mathbf{u})$ up to $\mathcal{S}$ from each side, Definition 1.1(ii) implies that a weak solution of (1) with a shock $\mathcal{S}$ satisfies

$$\frac{1}{2}|\mathbf{u}|^2 + \frac{\gamma p}{(\gamma - 1)\rho} = B_0 \quad \text{on } \mathcal{S}. \tag{6}$$

Then, it follows from (6) and Definition 1.1(iii) that

$$\frac{1}{2}(\mathbf{u}^+\cdot\boldsymbol{\nu})^2+\frac{\gamma p^+}{(\gamma-1)\rho^+}=\frac{1}{2}(\mathbf{u}^-\cdot\boldsymbol{\nu})^2+\frac{\gamma p^-}{(\gamma-1)\rho^-}\quad\text{on } \mathcal{S}. \tag{7}$$

For fixed constants $\gamma>1$, $\rho_\infty>0$, $p_\infty>0$ and $u_\infty>0$ with $M_\infty:=\frac{u_\infty}{\sqrt{\gamma p_\infty/\rho_\infty}}>1$, set

$$\mathbf{u}_\infty:=(u_\infty,0),\quad u_0:=\frac{u_\infty}{\gamma+1}\left(\gamma-1+\frac{2}{M_\infty^2}\right),\quad U_0:=\frac{2u_\infty}{\gamma+1}+u_0.$$

A direct computation shows that $0<u_0<u_\infty<U_0$. Next, we define a function $\mathfrak{f}:[u_0,u_\infty]\to\mathbb{R}^+$ by

$$\mathfrak{f}(u):=(u_\infty-u)\sqrt{\frac{u-u_0}{U_0-u}}. \tag{8}$$

For each $u\in(u_0,u_\infty)$, set $\mathbf{u}^-:=\mathbf{u}_\infty$, $\rho^-:=\rho_\infty$, $p^-:=p_\infty$, $\mathbf{u}^+:=(u,\mathfrak{f}(u))$, and

$$\boldsymbol{\nu}:=\frac{\mathbf{u}_\infty-\mathbf{u}^+}{|\mathbf{u}_\infty-\mathbf{u}^+|},\quad \boldsymbol{\tau}:=\boldsymbol{\nu}^\perp.$$

Then, we define

$$\rho^+:=\frac{\rho^-\mathbf{u}^-\cdot\boldsymbol{\nu}}{\mathbf{u}^+\cdot\boldsymbol{\nu}},\quad p^+:=\rho^-(\mathbf{u}^-\cdot\boldsymbol{\nu})^2+p^--\rho^+(\mathbf{u}^+\cdot\boldsymbol{\nu})^2.$$

In [5, §121–§122], it is proved that, for each $u\in(u_0,u_\infty)$, $(\rho^\pm,p^\pm,\mathbf{u}^\pm)$ defined as above satisfies (3) and (7) on any line $\mathcal{S}$ perpendicular to the vector $\boldsymbol{\nu}$. Fix a line $\mathcal{S}$ perpendicular to $\boldsymbol{\nu}$. For a fixed point $P_0\in\mathcal{S}$, define $\Omega^+:=\{Q\in\mathbb{R}^2:\overrightarrow{P_0Q}\cdot\boldsymbol{\nu}>0\}$. Then, $(\rho^\pm,p^\pm,\mathbf{u}^\pm)$ satisfies the entropy condition in the sense of Definition 1.2. This is also proved in [5, §121–§122]. For $\theta_w=\arctan(\frac{\mathfrak{f}(u)}{u})$, if we define

$$\begin{aligned}W_0&:=\{\mathrm{x}=(x_1,x_2)\in\mathbb{R}^2_+:x_1\ge x_2\tan\theta_w\},\\ \Omega_0&:=\{Q\in\mathbb{R}^2:\overrightarrow{OQ}\cdot\boldsymbol{\nu}>0\},\end{aligned} \tag{9}$$

then $(\rho,p,\mathbf{u})$ given by

$$(\rho,p,\mathbf{u})(\mathrm{x})=(\rho^+,p^+,\mathbf{u}^+)\chi_{\Omega_0}(\mathrm{x})+(\rho^-,p^-,\mathbf{u}^-)(1-\chi_{\Omega_0})(\mathrm{x})$$

is a weak solution to (1) in $\mathbb{R}^2_+\setminus W_0$ with a shock $\mathcal{S}$ (See Fig. 1). Here, the line $\mathcal{S}$ is called an *oblique shock* past the ramp W_0. And, the function $\mathfrak{f}$ defined by (8) is called *a shock polar* of the system (1).

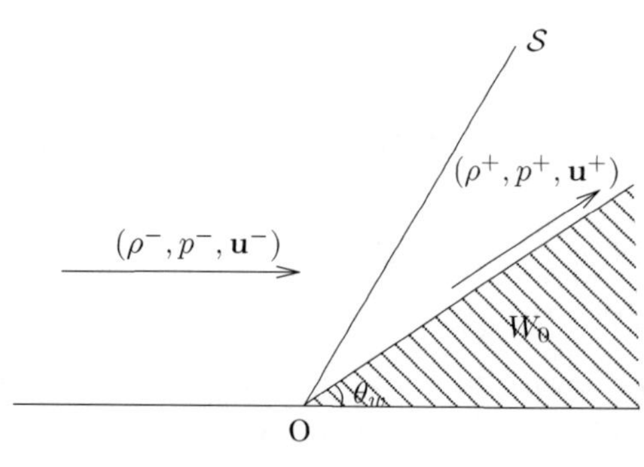

FIGURE 1. Oblique shock past a ramp W_0 in $\mathbb{R}^2_+$

In [8], the existence of a shock polar of (2) is proved. Differently from (1), it is not possible to find an explicit formula of the shock polar of (2). Instead, a more subtle approach is used in [8].

Lemma 1.3 ([8, Proposition 2.1]). *For fixed constants $\gamma > 1$, $\rho_\infty > 0$ and $u_\infty > 0$ with $M_\infty := \frac{u_\infty}{\sqrt{\rho_\infty^{\gamma-1}}} > 1$, there exist a unique constant $u_0 \in (0, u_\infty)$ and a unique function $\mathfrak{f}_{\rm polar} : [u_0, u_\infty] \to \mathbb{R}_+$ satisfying the following properties:*

(i) $\mathfrak{f}_{\rm polar} \in C^0([u_0, u_\infty]) \cap C^\infty((u_0, u_\infty))$

(ii) $\mathfrak{f}_{\rm polar}(u_0) = \mathfrak{f}_{\rm polar}(u_\infty) = 0$;

(iii) $\mathfrak{f}_{\rm polar}(u) > 0$ *for* $u_0 < u < u_\infty$;

(iv) If we set $\mathbf{u}^- := (u_\infty, 0)$, $\mathbf{u}^+ := (u, \mathfrak{f}_{\rm polar}(u))$, $\rho^\pm := \left((\gamma-1)(B_0 - \frac{1}{2}|\mathbf{u}|^2)\right)^{\frac{1}{\gamma-1}}$ *, then we have*

$$\rho^+\mathbf{u}^+ \cdot \boldsymbol{\nu} = \rho^-\mathbf{u}^- \cdot \boldsymbol{\nu} \quad \textit{for } \boldsymbol{\nu} = \frac{\mathbf{u}^- - \mathbf{u}^+}{|\mathbf{u}^- - \mathbf{u}^+|}, \tag{10}$$

$$0 < \mathbf{u}^+ \cdot \boldsymbol{\nu} < \mathbf{u}^- \cdot \boldsymbol{\nu} < \infty \quad \textit{for } u_0 < u < u_\infty. \tag{11}$$

(v) Any vector $\mathbf{u} = (u, v) \in \mathbb{R}^2$ *satisfying* (10) *and* (11) *lies either on the curve* $v = \mathfrak{f}_{\rm polar}(u)$, *or* $v = -\mathfrak{f}_{\rm polar}(u)$.

Lemma 1.4. *The shock polars* $\mathfrak{f}$ *of* (1), *given by* (8), *satisfies*

$$\mathfrak{f}''(u) < 0 \quad \textit{for } u_0 < u < u_\infty. \tag{12}$$

Also, the shock polar $\mathfrak{f}_{\rm polar}$ *of* (2), *whose existence is stated in Lemma 1.3 satisfies*

$$\mathfrak{f}_{\rm polar}''(u) < 0 \quad \textit{for } u_0 < u < u_\infty. \tag{13}$$

Proof. A direct computation with using (8) yields that

$$\mathfrak{f}\mathfrak{f}'' = \frac{-(U_0 - u_0)(u_\infty - u)}{4(U_0 - u^3)(u - u_0)} \left(4(u - u_0)(U_0 - u_\infty) - (u_\infty - u)(U_0 - u_0)\right).$$

Since $\mathfrak{f} > 0$ and $\mathfrak{f}\mathfrak{f}'' < 0$ for $u_0 < u < u_\infty$, (12) is obtained.

Since there is no explicit formula of $\mathfrak{f}_{\rm polar}$, we need a different approach to prove (13). The inequality (13) can be proved by adjusting the proof of [6, Theorem 1]. Or, one can refer to [2, Appendix A] for a detailed proof. □

Lemma 1.4 directly yields the following result.

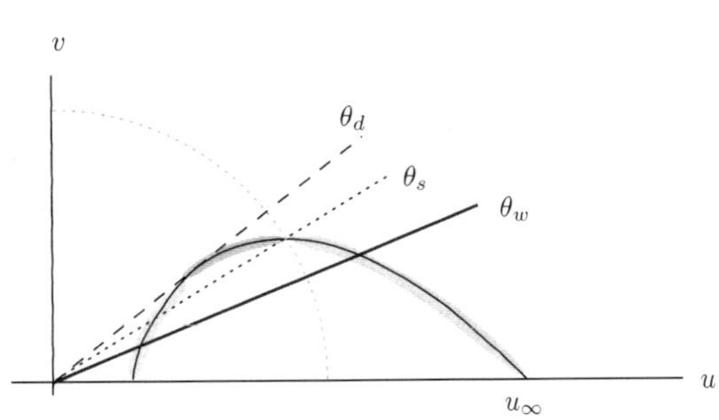

FIGURE 2. Shock polar $v = \mathfrak{f}(u)$ (or $v = \mathfrak{f}_{\rm polar}(u)$)

Corollary 1. *The shock polar $v = \mathfrak{f}(u)$ of* (1) *has a unique constant $\theta_d \in (0, \frac{\pi}{2})$ so that*

(i) if $0 \le \theta < \theta_d$, then the line $v = u \tan\theta$ intersects $v = \mathfrak{f}(u)$ at two distinct points;

(ii) if $\theta = \theta_d$, the line $v = u \tan\theta_d$ intersects $v = \mathfrak{f}(u)$ at a unique point;

(iii) if $\theta_d < \theta < \frac{\pi}{2}$, then there is no intersection of $v = u\tan\theta$ and $v = \mathfrak{f}(u)$.

Also, the shock polar $v = \mathfrak{f}_{\rm polar}(u)$ of (2) *has a unique constant $\theta_d \in (0, \frac{\pi}{2})$ that satisfies all the properties stated right above. Such a constant θ_d is called* the detachment angle.

2. **Attached shocks and detached shocks.** For the rest of the paper, we focus on the potential flow model (2) of Euler system, and its weak solutions with shocks.

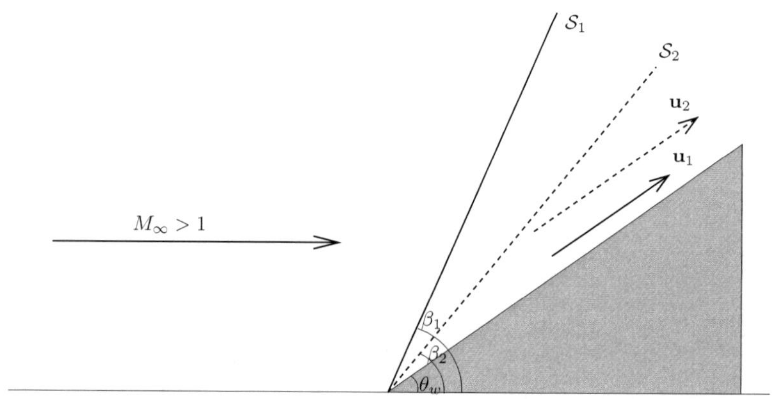

FIGURE 3. Attached oblique shocks: strong shock and weak shock

Fix constants $\gamma > 1$, $\rho_\infty > 0$ and $u_\infty > 0$ with $M_\infty := \frac{u_\infty}{\sqrt{\rho_\infty^{\gamma-1}}} > 1$. For a fixed $\theta_w \in (0, \theta_d)$, let $\mathbf{u}_1 = (u_1, \mathfrak{f}_{\rm polar}(u_1))$ and $\mathbf{u}_2 = (u_2, \mathfrak{f}_{\rm polar}(u_2))$ be two distinct intersections of $v = u\tan\theta_w$ and $v = \mathfrak{f}_{\rm polar}(u)$. Without loss of generality, assume that $u_1 < u_2$. For each $j = 1, 2$, set $\boldsymbol{\nu}_j = \frac{\mathbf{u}_\infty - \mathbf{u}_j}{|\mathbf{u}_\infty - \mathbf{u}_j|}$ for $\mathbf{u}_\infty := (u_\infty, 0)$. And, let $\mathcal{S}_j$ be the line passing through the origin, and perpendicular to $\boldsymbol{\nu}_j$. By Lemma 1.4 and the entropy condition stated in Definition 1.2, we have $\mathcal{S}_j = \{\mathrm{x} = (x_1, x_2) : x_2 = x_1 \tan\beta_j\}$ for $\beta_j \in (\theta_w, \frac{\pi}{2})$ and $\beta_1 > \beta_2$ (See Fig. 3). According to the construction of the shock polar $v = \mathfrak{f}_{\rm polar}(u)$ given in [8], we have

$$|\mathbf{u}_\infty - \mathbf{u}_2| < |\mathbf{u}_\infty - \mathbf{u}_1|. \tag{14}$$

For $\rho_j := \left((\gamma-1)(B_0 - \frac{1}{2}|\mathbf{u}_j|^2)\right)^{\frac{1}{\gamma-1}}$, $(\rho^{(j)}, \mathbf{u}^{(j)})$ given by

$$(\rho^{(j)}, \mathbf{u}^{(j)})(\mathrm{x}) := (\rho_\infty, \mathbf{u}_\infty)\chi_{\{x_1 < x_2 \tan\beta_j\}}(\mathrm{x}) + (\rho_{\rm j}, \mathbf{u}_{\rm j})\chi_{\{\mathrm{x}_1 > \mathrm{x}_2 \tan\beta_j\}}(\mathrm{x})$$

is a weak solution in $\mathbb{R}^2_+ \setminus W_0$ with a shock $\mathcal{S}_j$, and it satisfies the entropy condition. Based on the observation (14), $(\rho^{(1)}, \mathbf{u}^{(1)}, \mathcal{S}_1)$ and $(\rho^{(2)}, \mathbf{u}^{(2)}, \mathcal{S}_2)$ are called *a strong shock solution* and *a weak shock solution* of (2) in $\mathbb{R}^2_+ \setminus W_0$, respectively.

It is shown in [2, Lemma A.3(a)] that every strong shock($\mathcal{S}_1$) is *a transonic shock* in the sense that $M_1 < 1 < M_\infty$ for $M_1 := \frac{|\mathbf{u}_1|}{\sqrt{\rho_1^{\gamma-1}}}$. Weak shock($\mathcal{S}_2$) is, however,different from strong shock in that its type changes depending on θ_w. In [2, Lemma A.3(a)], it is proved that there exists a unique $\theta_s \in (0, \theta_d)$ such that

$M_2 := \frac{|\mathbf{u}_2|}{\sqrt{\rho_2^{\gamma-1}}} > 1$ for $\theta_w < \theta_s$, $M_2 = 1$ at $\theta_w = \theta_s$ and $M_2 < 1$ for $\theta_s < \theta_w < \theta_d$ (See Fig. 1). Such θ_s is called *the sonic angle.*

Even though the shocks $\mathcal{S}_1$ and $\mathcal{S}_2$ are quite different, they still share a common feature. Namely, both shocks are attached to the tip of W_0 (See Fig. 3). But what if the tip of the obstacle is not sharp, or what if the angle θ_w of W_0 is grater than the detachment angle θ_d? Would there exists a weak solution of (2) with a shock attached to the tip of the obstacle? If $\theta_w > \theta_d$, then it follows from Corollary 1 that no attached oblique shock as in Fig. 3 can be formed around W_0. Of course, the example discussed above only concerns a piecewise constant weak solution with a shock. So one may speculate that (2) can still have a weak solution with a shock $\mathcal{S}$ attached to the tip of W_0 although the solution is not a piecewise constant thus $\mathcal{S}$ is not a straight line. But, a local shock polar analysis shows that even a curved attached shock cannot be formed around W_0. And, this raises a question on the existence of a detached shock past a wedge of angle θ_w greater than the detachment angle θ_d. This question on a detached shock has been a long standing open problem, and no rigourous answer has been given to this day. There is still a remaining question on a shock solution past a blunt body.

Question 1. *For W_0 given by (9) with $\theta_w < \theta_d$, let W_b be a blunt body obtained from smoothing out the tip of W_0. Would there exists a weak solution of (2) with a shock attached to the tip of the obstacle?*

To make this question more precise, we give a definition of the blunt body.

Definition 2.1. For a fixed constant $h_0 > 0$, let a function $b : \mathbb{R} \to \mathbb{R}$ satisfy the following properties:

(b_1) $b(x_2) = b(-x_2)$ for all $x_2 \in \mathbb{R}$;
(b_2) $b \in C^3(\mathbb{R})$;
(b_3) $b'(x_2) > 0$ for all $x_2 > 0$;
(b_4) $b''(x_2) \geq 0$ for all $x_2 \geq 0$;
(b_5) $b(x_2) = x_2 \cot\theta_w$ for $x_2 \geq h_0$.

For such a function b, we define a blunt body W_b by

$$W_b := \{\mathbf{x} = (x_1, x_2) \in \mathbb{R}^2_+ : x_1 \geq b(x_2)\}. \tag{15}$$

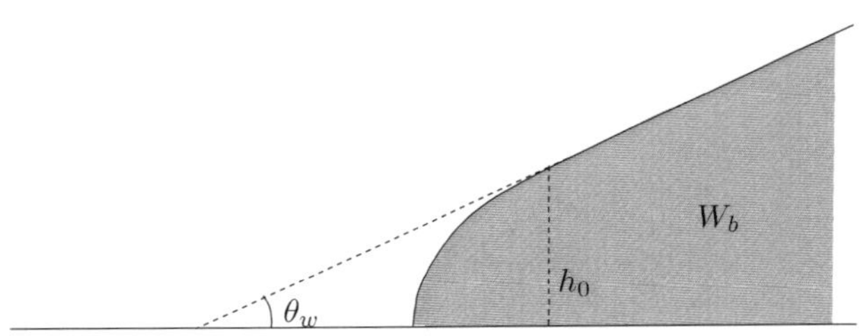

FIGURE 4.

Lemma 2.2. *The system (2) cannot have a weak solution $(\rho, \mathbf{u})$ in $\mathbb{R}^2_+ \setminus W_b$ with a shock $\mathcal{S}$ so that $(\rho, \mathbf{u})$ satisfies the entropy condition (11), and that the shock $\mathcal{S}$ is attached to the blunt body W_b.*

Proof. Let $(\rho, \mathbf{u})$ be a weak solution to (2) in $\mathbb{R}^2_+ \setminus W_b$ with a shock $\mathcal{S}$ in the sense of Definition 1.1(b). Suppose that $\mathcal{S}$ is attached to the blunt body W_b at a point P_0.

Case 1. $P_0 = (b(x_2^*), x_2^*)$ *for some* $x_2^* > 0$

Let $\boldsymbol{\nu}_0$ be the unit normal vector of $\mathcal{S}$ at P_0 pointing toward W_b. If $\boldsymbol{\nu}_0 \not\perp W_b$ at P_0, then there exists a point $Q = (b(\hat{x}_2), \hat{x}_2)$ with $\hat{x}_2 > 0$, and a small constant $r > 0$ such that

$$(\rho, \mathbf{u}) = (\rho_\infty, \mathbf{u}_\infty) \quad \text{in } B_r(Q) \setminus W_b. \text{ (Fig. 5)}$$

Then, by continuation, we have $(\rho, \mathbf{u})(Q) = (\rho_\infty, \mathbf{u}_\infty)$. We define a vector field

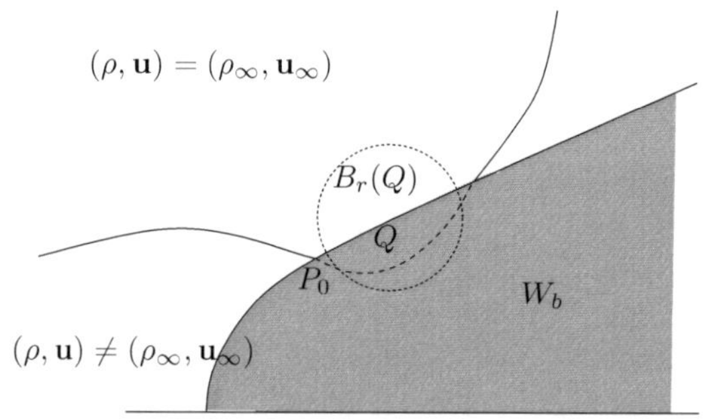

FIGURE 5.

$\mathbf{n}(x_2) := (-1, b'(x_2))$ for $x_2 > 0$. By Definition 2.1, $\mathbf{n_0} := (-1, b'(\hat{x}_2))$ is a normal vector of ∂W_b at Q, which points inward of $\mathbb{R}^2_+ \setminus W_b$. And, we have $\mathbf{u} \cdot \mathbf{n_0} = -u_\infty \neq 0$ at Q so the slip boundary condition does not hold. This contradicts to Definition 1.1(b).

Next, we suppose that $\boldsymbol{\nu}_0 \perp W_b$ at P_0. Then, we have $\boldsymbol{\nu}_0 = \frac{\mathbf{n_0}}{|\mathbf{n_0}|}$. We compute the value of $(\rho, \mathbf{u})$ at P_0 in the side of Ω^+(Fig. 6) by taking the limit

$$(\rho, \mathbf{u})(P_0) = \lim_{\substack{\mathbf{x} \to P_0 \\ \mathbf{x} \in \Omega^+}} (\rho, \mathbf{u})(\mathbf{x}) =: (\rho^+, \mathbf{u}^+). \tag{16}$$

By the slip boundary condition on the boundary of W_b, and C^3 regularity of b, we

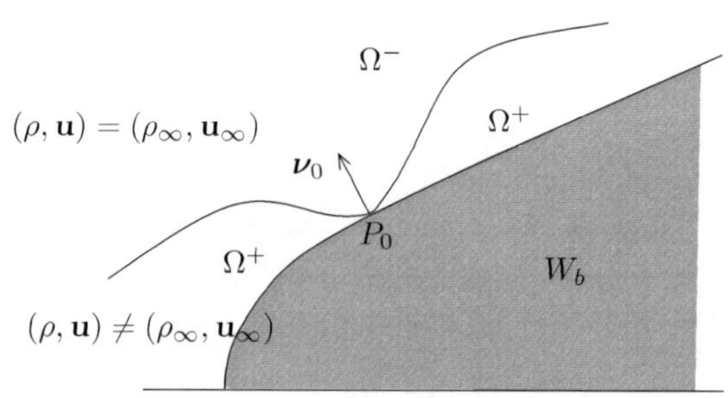

FIGURE 6.

have

$$\mathbf{u}^+ \cdot \boldsymbol{\nu}_0 = \lim_{x_2 \to \hat{x}_2} \mathbf{u}(b(x_2), x_2) \cdot \frac{\mathbf{n}(x_2)}{|\mathbf{n}(x_2)|} = 0,$$

which implies that $\rho^+\mathbf{u}^+\cdot\boldsymbol{\nu}_0 = 0$. Then the Rankine-Hugoniot condition (4) cannot not hold at P_0 because $\rho_\infty\mathbf{u}_\infty\cdot\boldsymbol{\nu}_0 \neq 0$. This is a contradiction. Therefore, $\mathcal{S}$ cannot be attached to W_b away from the tip $(b(0),0)$.

Case 2. $P_0 = (b(0),0)$

From (b_1)–(b_3) in Definition 2.1, it follows that $b'(0) = 0$. Let $\mathbf{u}^+$ be given by (16). By continuity of $\mathbf{u}$ in Ω^+(Fig. 7) up to its boundary, and slip boundary condition on the boundary of W_b, we obtain that

$$\mathbf{u}^+\cdot(-1,0) = \lim_{x_2\to 0+} \mathbf{u}(b(x_2),x_2)\cdot\frac{\mathbf{n}(x_2)}{|\mathbf{n}(x_2)|} = 0. \tag{17}$$

This implies that if the shock $\mathcal{S}$ is attached to W_b at P_0, then the horizontal

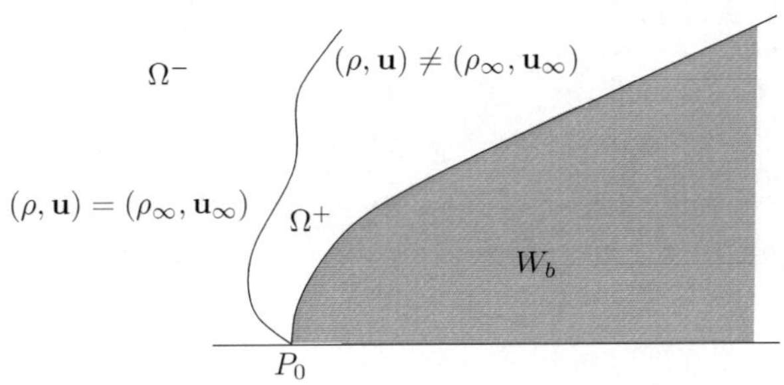

FIGURE 7.

component of $\mathbf{u}^+$ is 0. By Lemma 1.3, $\mathbf{u}^+$ must lie on either $v = \mathfrak{f}_{\rm polar}(u)$ or $v = -\mathfrak{f}_{\rm polar}(u)$. Therefore, $\mathbf{u}^+$ must have a strictly positive horizontal component but this contradicts to (17). So we conclude that $\mathcal{S}$ cannot be attached to W_b at the tip.

□

Lemma 2.2 shows that if $(\rho,\mathbf{u})$ is a weak solution of (2) in $\mathbb{R}^2_+\setminus W_b$ with a shock $\mathcal{S}$, then $\mathcal{S}$ must be detached from W_b. So our next question is as follows:

Question 2. *Does there exists a detached shock solution of* (2) *past* W_b*?*

In the next section, we present a recent result to yield an answer to this question.

3. Detached shock solutions past the blunt body W_b.

3.1. **Main result.** Fix $\gamma > 1$ and $B_0 > 0$. Given constants $\theta_w \in (0,\frac{\pi}{2})$ and $h_0 > 0$, let a function $b:\mathbb{R}\to\mathbb{R}$ be given by Definition 2.1. And, let W_b be given by (2.1). In [3], we proved the existence of detached shock solutions of (2) past W_b for an incoming state $(\rho_\infty,\mathbf{u}_\infty)$ for $M_\infty = \frac{u_\infty}{\sqrt{\rho_\infty^{\gamma-1}}}$ being sufficiently large.

Problem 1. *Find a weak solution* $(\rho,\mathbf{u})$ *of* (2) *in* $\mathbb{R}^2_+\setminus W_b$ *with a shock*

$$\Gamma_{\rm sh} = \{\mathrm{x} = (x_1,x_2) : x_1 = f_{\rm sh}(x_2),\ x_2 \geq 0\}$$

for a C^1 *function* $f_{\rm sh}:\mathbb{R}_+\to\mathbb{R}$ *satisfying* $f_{\rm sh}(x_2) < b(x_2)$ *for all* $x_2\in\mathbb{R}_+$ *so that* $(\rho,\mathbf{u})$ *uniformly converges to a piecewise constant state as* $|\mathrm{x}|\to\infty$.

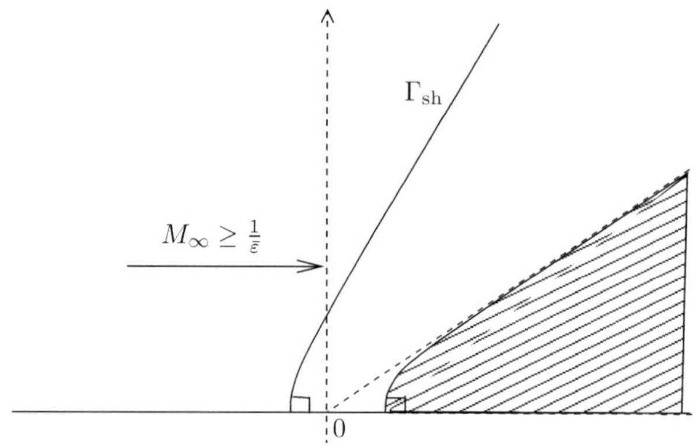

FIGURE 8. Detached shock past the blunt body W_b

In [3], it is proved that if $M_\infty = \frac{u_\infty}{\sqrt{\rho_\infty^{\gamma-1}}}$ is sufficiently large, then there exists a detached shock solution past W_b so that its asymptotic state at far field ($|\mathbf{x}| = \infty$) is given as a strong shock state, which is uniquely determined by shock polar analysis. We state the main result of [3] in a simplified form as follows:

Theorem 3.1. *For a fixed constant $d_0 > 0$, there exists a small constant $\bar{\varepsilon} > 0$ depending on $(\gamma, B_0, d_0, \theta_w, h_0)$ so that whenever the incoming supersonic state (ρ_∞, u_∞) satisfies $M_\infty = \frac{1}{\varepsilon}$ for $\varepsilon \in (0, \bar{\varepsilon}]$, the system (2) has a weak solution $(\rho, \mathbf{u})$ in $\mathbb{R}^2_+ \setminus W_b$ with a shock $\Gamma_{sh} = \{(f_{sh}(x_2), x_2) : x_2 \in \mathbb{R}_+\}$. And, the solution satisfies the following properties:*

(i) $f_{sh}(0) = b(0) - d_0$;

(ii) There exists a constant $\delta > 0$ depending only on $(\gamma, B_0, d_0, \theta_w, h_0)$ such that

$$b(x_2) - f_{sh}(x_2) \geq \delta \quad \text{for all } x_2 \geq 0;$$

(iii) Let $\mathbf{u}^\varepsilon_{st}$ be the intersection of the shock polar $v = \mathfrak{f}_{polar}(u)$ and $v = u \tan\theta_w$ corresponding to a strong shock. And, let s^ε_{st} be the slope of the corresponding strong shock. Finally, set $\rho^\varepsilon_{st} := \left((\gamma-1)(B_0 - \frac{1}{2}|\mathbf{u}^\varepsilon_{st}|^2)\right)^{\frac{1}{\gamma-1}}$. Then,

$$\lim_{\substack{|\mathbf{x}|\to\infty \\ \mathbf{x}\in\Omega_{f_{sh}}}} |(\rho, \mathbf{u})(\mathbf{x}) - (\rho^\varepsilon_{st}, \mathbf{u}^\varepsilon_{st})| = 0, \quad \text{and} \quad \lim_{x_2\to\infty} |f'_{sh}(x_2) - s^\varepsilon_{st}| = 0$$

for $\Omega_{f_{sh}} := \{\mathbf{x} = (x_1, x_2) \in \mathbb{R}^2_+ : f_{sh}(x_2) < x_1 < b(x_2),\ x_2 > 0\}$;

(iv) There exists a constant $\sigma \in (0, 1)$ depending only on $(\gamma, B_0, d_0, \theta_w, h_0)$ so that the Mach number $M(\rho, \mathbf{u}) = \frac{|\mathbf{u}|}{\sqrt{\rho^{\gamma-1}}}$ satisfy the inequality

$$M(\rho, \mathbf{u}) \leq 1 - \sigma \quad \text{in } \overline{\Omega_{f_{sh}}}.$$

To the best of our knowledge, Theorem 3.1 is the first rigorous result on the global existence of detached shock solutions past a blunt body in unbounded domain.

By Lemma 1.3 and Corollary 1, if the incoming supersonic state $(\rho_\infty, \mathbf{u}_\infty)$ is given, then the detachment angle θ_d is uniquely determined so that $\theta_w < \theta_d$ hold. Then a unique strong shock state is determined depending on θ_w. In Theorem 3.1, however, we first fix θ_w from which a wedge W_0 is given by (9), then we get a blunt body W_b by perturbing the vertex of W_0 with a C^3 function b. And, we seek a detached shock solution of past W_b with its far-field asymptotic state being determined by shock polar analysis. Therefore, to make Theorem 3.1 valid, we add a lemma given from [3]:

Lemma 3.2. *For any given $\theta_w \in (0, \frac{\pi}{2})$, there exists a small constant $\varepsilon_0 \in (0,1)$ depending on (γ, B_0, θ_w) so that whenever the incoming supersonic state (ρ_∞, u_∞) satisfies $M_\infty \geq \frac{1}{\varepsilon_0}$, the shock polar $v = \mathfrak{f}_{\rm polar}(u)$ and the line $v = u \tan\theta_w$ intersect at two distinct points. In other words, we have*

$$\theta_w < \theta_d. \tag{18}$$

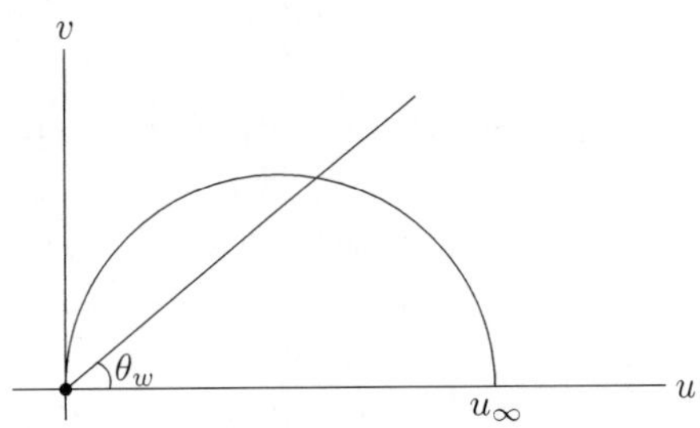

FIGURE 9. Shock polar for $M_\infty = \infty$

Proof. For fixed $\gamma > 1$ and $B_0 > 0$, direct computations with using (10) and (11) show that if $M_\infty = \infty$, then the graph of the shock polar $v = \mathfrak{f}_{\rm polar}(u)$ is a semi-circle with the center at $(\frac{u_\infty}{2}, 0)$. In this case, the detachment angle θ_d is $\frac{\pi}{2}$. Therefore, for any $\theta_w \in (0, \frac{\pi}{2})$, the line $v = u\tan\theta_w$ intersects with $v = \mathfrak{f}_{\rm polar}(u)$ at two distinct points with one of the intersection being $(0,0)$ always. Therefore, the downstream state of any strong shock solution is given as $\rho^+ = \infty$ and $\mathbf{u}^+ = (0,0)$, and the corresponding shock is a normal shock (Fig. 9). By using this observation and the implicit mapping theorem, we can prove that $\theta_w < \theta_d$ for sufficiently large M_∞. All the details are given in [3]. □

Remark 1. In [3], Theorem 3.1 is proved by using a stream function formulation, and solving a free boundary problem of stream function in a cut-off domain, then applying a limiting argument. In doing so, the asymptotic state at far field is essential to establish a compactness of approximate detached shock solutions obtained in cut-off domains.

Remark 2. The analysis given in [3] shows that if we fix $\underline{d} > 0$, then there exists a small constant $\bar{\varepsilon} > 0$ depending on $(\gamma, B_0, \underline{d}, \theta_w, h_0)$ so that whenever $d_0 \geq \underline{d}$ and $M_\infty \geq \frac{1}{\bar{\varepsilon}}$, a detached shock solution of (2) past W_b can be constructed. Furthermore, *the detached distance* $\inf\limits_{x_2 \in \mathbb{R}_+} (b(x_2) - f_{\rm sh}(x_2))$ is bounded below by a positive constant $\underline{\delta}$ depending on $(\gamma, B_0, \underline{d}, \theta_w, h_0)$.

Remark 3. Statement (iv) in Theorem 3.1 indicates that the detached shock solutions given in this theorem are transonic.

3.2. **Further questions.** While Theorem 3.1 provides an answer to Question 2 stated in Section 2, it also raises new interesting problems to be investigated in the future.

1. *For a fixed $d_0 > 0$, is the detached shock solution obtained in Theorem 3.1 unique?*

As briefly mentioned in Remark 1, each detached shock solution of Problem 1 for a given $d_0 > 0$ is obtained by a limiting argument. Therefore, it is unclear how to achieve the uniqueness of a solution. Furthermore, according to (iii) of Theorem 3.1, the tangential slope of each detached shock converges to a constant $s^{\varepsilon}_{\rm st}$, which is the slope of the strong shock corresponding to θ_w. But there is no unique asymptotic state of detached shock itself, so we cannot use the argument in [7], and this makes it even harder to achieve the uniqueness of a detached shock solution for a fixed $d_0 > 0$.

2. *For fixed incoming supersonic state* (ρ_∞, u_∞), *Theorem 3.1 yields infinitely many detached shock solutions. Does this mean that detached shock phenomenon is unstable?*

According to Remark 2, if the incoming supersonic state (ρ_∞, u_∞) has a sufficiently large Mach number M_∞, then there is a lower bound $\underline{d} > 0$ so that whenever $d_0 \geq \underline{d}$, Problem 1 has a detached shock solution past W_b. This means that there are infinitely many different detached shock solutions past W_b. This observation naturally raises question on the stability of detached shock phenomena. Or would there be any criterion to pick a *physically stable solution*?

3. *Can we seek a solution of Problem 1 so that its asymptotic state at far field is given as* weak shock solution *instead of* strong shock solution*?*

In Theorem 3.1, by fixing the asymptotic state at far field as a strong shock state, it was possible to find detached shock solutions past W_b, all of which are transonic. What would happen if we require for a detached shock solution to converge to a weak shock state at far field ($|\mathbf{x}| = \infty$)?

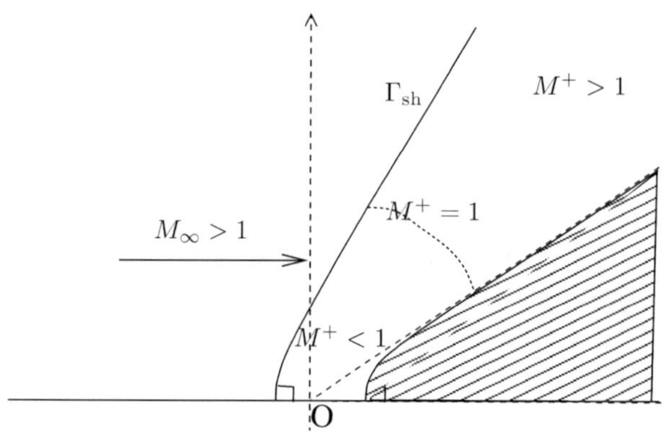

FIGURE 10.

Suppose that $(\rho, \mathbf{u})$ is a weak solution of (2) with a detached shock $\Gamma_{\rm sh}$, and that it converges to a weak shock state at far field ($|\mathbf{x}| = \infty$). As pointed out earlier in Section 2, if θ_w is greater than the sonic angle θ_s, then Mach number $M = \frac{|\mathbf{u}|}{\sqrt{\rho^{\gamma-1}}}$ of downstream state of a weak shock solution satisfies $M > 1$. Let us set $Z_0 := (b(0), 0)$, that is, Z_0 is the tip of W_b. Since $\Gamma_{\rm sh}$ is detached from W_b, one can fix a small constant $\delta > 0$ so that $(\rho, \mathbf{u})$ is continuous in $\mathcal{U}_\delta(Z_0) := B_\delta(Z_0) \cap (\mathbb{R}^2_+ \setminus W_b)$ up to the boundary. By properties (b_1)–(b_2) stated in Definition 2.1, we have $b'(0) = 0$. Then, due to the slip boundary condition on $\partial(\mathbb{R}^2_+ \setminus W_b)$ and the continuity of $\mathbf{u}$ in $\overline{\mathcal{U}_\delta(Z_0)}$. we obtain that $\mathbf{u} \cdot \mathbf{e}_1 = 0$ and $\mathbf{u} \cdot \mathbf{e}_2 = 0$ at Z_0. In other words, Z_0 is a stagnation point. Since $\rho(Z_0) = ((\gamma - 1)B_0)^{\frac{1}{\gamma-1}} \neq 0$, we have $M = \frac{|\mathbf{u}|}{\sqrt{\rho^{\gamma-1}}} = 0$ at Z_0. This observation implies that $M < 1$ near Z_0, and $M > 1$ for $|\mathrm{x}|$ sufficiently large behind the shock $\Gamma_{\rm sh}$. Perhaps, the simplest configuration of such a solution would be as in Fig. 10. Namely, the downstream state behind the shock $\Gamma_{\rm sh}$ is a smooth transonic, and the solution contains a sonic boundary on which M becomes 1. But even this case is very difficult to construct mathematically unless there is further information on the sonic boundary.

4. Can we prove Theorem 3.1 for the system (1)*?*

Even though an incoming supersonic flow is uniform state thus irrotational, the vorticity $(\nabla \times \mathbf{u})$ is generated across a shock unless the shock has a special geometric structure relative to incoming supersonic flow. The analysis given in [3] to prove Theorem 3.1 uses a stream function formulation. Namely, we introduce a scalar function ψ to satisfy $\nabla^{\perp}\psi = \rho\mathbf{u}$, then we obtain a quasi-linear second order homogeneous equation for ψ. If we apply the same approach to prove Theorem 3.1 for (1), the only difference is that the equation for the stream function ψ becomes nonhomogeneous due to variation of entropy, which is caused by a generation of vorticity across a shock. This would raise several technical difficulties in constructing a detached shock solution of (1) past W_b, especially because $\mathbb{R}^2_+ \setminus W_b$ is unbounded. But we believe that these difficulties can be overcome with more careful analysis. Further result will be given in the forthcoming work.

REFERENCES

[1] M. Bae, G.-Q. Chen and M. Feldman, Prandtl-Meyer reflection for supersonic flow past a solid ramp. *Quart. Appl. Math.,* **71** (2013), 583–600.

[2] M. Bae, G.-Q. Chen and M. Feldman, Prandtl-Meyer reflection configuration, transonic shocks and free boundary problems. Preprint

[3] M. Bae and W. Xiang, Detached shock past a blunt body, Preprint

[4] J. Chen, C. Christoforou and K. Jegdić, Existence and uniqueness analysis of a detached shock problem for the potential flow, *Nonlinear Anal.* **74**(2011), 705–720.

[5] R. Courant and K. O. Friedrichs, Supersonic Flow and Shock Waves, Springer-Verlag: New York, 1948.

[6] V. Elling, Counter examples to the sonic criterion, *Arch. Rational Mech. Anal.* **194** (2009), 987–1010.

[7] B. Fang, and W. Xiang, The uniqueness of transonic shocks in supersonic flow past a 2-D wedge. J. Math. Anal. Appl. **437**(2016), no.1, 194–213.

[8] B. L. Keyfitz and G. Warnecke, The existence of viscous profiels and admissibility for transonic shocks, *Comm. Partial Differential Equations* **16** no. 6–7, (1991) 1197–1221

E-mail address: `mjbae@postech.ac.kr`
E-mail address: `weixiang@cityu.edu.hk`

RIGIDITY IN GENERALIZED ISOTHERMAL FLUIDS

Rémi Carles*

Univ Rennes, CNRS, IRMAR - UMR 6625, F-35000 Rennes, France

Kleber Carrapatoso and Matthieu Hillairet

Institut Montpelliérain Alexander Grothendieck, CNRS, Univ. Montpellier, France

Abstract. We investigate the long-time behavior of solutions to the isothermal Euler equation. By writing the system with a suitable time-dependent scaling we prove that the densities of global solutions display universal dispersion rate and asymptotic profile. This result estends to Korteweg or quantum Navier Stokes equations, as well as generalizations of these equations where the convex pressure law is asymptotically linear near vacuum.

1. Introduction.

1.1. Isentropic Euler equation: existence of singularities.

In the isentropic case $\gamma > 1$, the Euler equation on $\mathbb{R}^d$, $d \geq 1$,

$$\begin{cases} \partial_t \rho + \operatorname{div}(\rho u) = 0, \quad \rho_{|t=0} = \rho_0, \\ \partial_t(\rho u) + \operatorname{div}(\rho u \otimes u) + \kappa \nabla(\rho^\gamma) = 0, \quad \rho u_{|t=0} = J_0, \end{cases} \tag{1}$$

enjoys the formal conservations of mass,

$$M(t) = \int_{\mathbb{R}^d} \rho(t,x) \mathrm{d}x \equiv M(0),$$

and entropy (or energy),

$$E(t) = \frac{1}{2} \int_{\mathbb{R}^d} \rho(t,x)|u(t,x)|^2 \mathrm{d}x + \frac{\kappa}{\gamma - 1} \int_{\mathbb{R}^d} \rho(t,x)^\gamma \mathrm{d}x \equiv E(0).$$

In general, smooth solutions are defined only locally in time (see [11, 7, 14]). Indeed, as first noticed in [11], considering the new unknown

$$(a, v) = \left(\rho^{\frac{\gamma-1}{2}}, v\right)$$

turns (1) into

$$\begin{cases} \partial_t a + v \cdot \nabla a + \dfrac{\gamma - 1}{2} a \operatorname{div} v = 0, \quad a_{|t=0} = \rho_0^{\frac{\gamma-1}{2}}, \\ \partial_t v + v \cdot \nabla v + \kappa \dfrac{2\gamma}{\gamma - 1} a \nabla a = 0, \quad v_{|t=0} = \dfrac{J_0}{\rho_0}. \end{cases} \tag{2}$$

2000 *Mathematics Subject Classification.* Primary: 35Q35, 35B40; Secondary: 35Q40, 76N15.

Key words and phrases. Compressible Euler equation, Korteweg equation, quantum Navier–Stokes equation, isothermal, large time.

* Corresponding author: Rémi Carles.

This system is hyperbolic symmetric (with a constant symmetrizer), so there exists a unique local solution $(a,u) \in C(0,T;H^s)^{1+d}$, provided that $s > 1 + d/2$ and $\rho_0^{\frac{\gamma-1}{2}}, v_0 = \frac{J_0}{\rho_0} \in H^s(\mathbb{R}^d)$.

It is also proven in [11] (and generalized in [14]) that if $a_{|t=0}$ and $v_{|t=0}$ are smooth and compactly supported, then no matter how small they may be (and unless both are identically zero), the solution to (2) will develop a singularity in finite time. The proof relies on two key arguments:

- As long as the solution is smooth, its speed of propagation is zero (for instance, view the equations like ODEs).
- A virial computation shows that if the solution is global, then it is dispersive:

$$\frac{d^2}{dt^2}\int_{\mathbb{R}^d} |x|^2\rho(t,x)dx \geq E(t)\inf\left(2, 3(\gamma-1)\right) = E_0 \inf\left(2, 3(\gamma-1)\right) > 0,$$

  where we have used (one more time) the assumption $\gamma > 1$ and the conservation of the energy (which is granted in the case of smooth solutions).

Suppose that the solution remains smooth for all time. Integrating the above estimate twice yields

$$\int |x|^2\rho(t,x)dx \gtrsim t^2.$$

This is incompatible with the fixed compact support of ρ and the conservation of mass, since these properties imply, for some $K > 0$ independent of time,

$$\int |x|^2\rho(t,x)dx \leq \int_{|x|<K} |x|^2\rho(t,x)dx \lesssim K^{2d}\int \rho(t,x)dx = K^{2d}M(0).$$

Therefore, a singularity appears in finite time.

1.2. Isentropic Euler equation: some global solutions and their asymptotic behavior.

A first global existence of smooth solutions was obtained by D. Serre [12], under an extra geometric assumption involving a special structure for the initial velocity. For $1 < \gamma \leq 1 + 2/d$, change the unknown functions

$$\rho(t,x) = \frac{1}{(1+t)^d}R\left(\frac{t}{1+t}, \frac{x}{1+t}\right), \quad u(t,x) = \frac{1}{1+t}U\left(\frac{t}{1+t}, \frac{x}{1+t}\right) + \frac{x}{1+t},$$

and assume that $R_0^{\frac{\gamma-1}{2}}, U_0 \in H^s$, for some $s > 1 + d/2$. This means $\rho_0^{\frac{\gamma-1}{2}} \in H^s$ (like before), and $u_0(x) - x \in H^s$ (hence $u_0 \notin L^2$).

Theorem 1.1 (D. Serre, [12]). *There exists $\eta > 0$ such that if*

$$\|(\rho_0^{(\gamma-1)/2}, U_0)\|_{H^s(\mathbb{R}^d)} \leq \eta,$$

then there is a unique global solution, in the sense that $(R,U) \in C([0,\infty);H^s(\mathbb{R}^d))^{1+d}$. In addition, there exists $R_\infty, U_\infty \in H^s(\mathbb{R}^d)$ such that

$$\left\|\left(\rho(t,x) - \frac{1}{t^d}R_\infty\left(\frac{x}{t}\right), u(t,x) - \frac{x}{1+t} - \frac{1}{1+t}U_\infty\left(\frac{x}{t}\right)\right)\right\|_{L^\infty(\mathbb{R}^d)} \underset{t\to\infty}{\longrightarrow} 0.$$

Back to the initial unknown functions, we infer (this is a rather straightforward consequence of the proof in [12], see [6]):

Corollary 1.2. *Let $1 < \gamma \leq 1+2/d$ and $s > d/2+1$. If $R_\infty, U_\infty \in H^s(\mathbb{R}^d)$ are such that $\|(R_\infty^{(\gamma-1)/2}, U_\infty)\|_{H^s(\mathbb{R}^d)} \leq \eta$, then there exists Cauchy data $\rho_0, u_0 \in H^s(\mathbb{R}^d)$ such that the solution is global in time in the same sense as above, and*

$$\left\| \left(\rho(t,x) - \frac{1}{t^d} R_\infty\left(\frac{x}{t}\right), u(t,x) - \frac{x}{1+t} - \frac{1}{1+t} U_\infty\left(\frac{x}{t}\right) \right) \right\|_{L^\infty(\mathbb{R}^d)} \underset{t\to\infty}{\longrightarrow} 0.$$

Some comments are in order:

- The assumptions on the velocity is reminiscent of the "good case" in Burgers' equation: particles spread out. Generalizations of this result can be found in [9, 8].
- In this regime, the density is dispersive, and dispersive rate is universal,
$$\|\rho(t)\|_{L^\infty(\mathbb{R}^d)} \lesssim \frac{1}{t^d}.$$
- However, the asymptotic profile R_∞ may be any smooth, small function.

1.3. **Isothermal case.** In the case $\gamma = 1$,

$$\begin{cases} \partial_t \rho + \operatorname{div}(\rho u) = 0, \\ \partial_t(\rho u) + \operatorname{div}(\rho u \otimes u) + \kappa \nabla \rho = 0, \end{cases} \tag{3}$$

with $\kappa > 0$, the mass is still conserved,

$$M(t) = \int_{\mathbb{R}^d} \rho(t,x)dx \equiv M(0),$$

as well as the entropy, which now reads

$$E(t) = \frac{1}{2}\int_{\mathbb{R}^d} \rho(t,x)|u(t,x)|^2 dx + \kappa \int_{\mathbb{R}^d} \rho(t,x) \ln \rho(t,x) dx \equiv E(0).$$

Now, the energy has no definite sign: no a priori estimate like in the argument of [11] is available. We show that rigidity results are available in this case though, involving a large time behavior in sharp contrast with the case $1 < \gamma \leq 1 + 2/d$ of [12].

2. **A large family of explicit solutions.** To simplify the presentation, we assume $d = 1$ in this section, and refer to [6] for the general case. Consider

$$\rho_0(x) = b_0 e^{-\alpha_0 x^2}, \quad u_0(x) = \beta_0 x.$$

As noticed by M. Yuen [15], the above structure is preserved by the flow:

$$\rho(t,x) = b(t)e^{-\alpha(t)x^2}, \quad u(t,x) = \beta(t)x,$$

and solving the PDE (3) becomes equivalent to solving the ODEs

$$\dot{\alpha} + 2\alpha\beta = 0, \quad \dot{\beta} + \beta^2 - 2\kappa\alpha = 0, \quad \dot{b} = -\beta b.$$

Following T. Li and D. Wang [10], seek

$$\alpha(t) = \frac{\alpha_0}{\tau(t)^2}, \quad \beta(t) = \frac{\dot{\tau}(t)}{\tau(t)}.$$

We come up with the ODE

$$\ddot{\tau} = \frac{2\kappa\alpha_0}{\tau}, \quad \tau(0) = 1, \quad \dot{\tau}(0) = \beta_0.$$

At this stage, the surprising fact is the universal behavior of solutions to the above equation, regardless of initial data.

Lemma 2.1 ([5]). *Let $a_0, \tilde{\kappa} > 0$, $\beta_0 \in \mathbb{R}$. Consider the ordinary differential equation*

$$\ddot{\tau} = \frac{2\tilde{\kappa}}{\tau}, \quad \tau(0) = a_0, \quad \dot{\tau}(0) = \beta_0.$$

It has a unique solution $\tau \in C^2(0,\infty)$, and it satisfies, as $t \to \infty$,

$$\tau(t) = 2t\sqrt{\tilde{\kappa}\ln t}\,(1 + \mathcal{O}(\ell(t))), \quad \dot{\tau}(t) = 2\sqrt{\tilde{\kappa}\ln t}\,(1 + \mathcal{O}(\ell(t))),$$

where $\ell(t) := \dfrac{\ln\ln t}{\ln t}$.

Back to (3), this yields

$$\rho(t,x) \underset{t\to\infty}{\sim} \frac{b_0}{2t\sqrt{\alpha_0\kappa\ln t}} e^{-x^2/(2t\sqrt{\kappa\ln t})^2}, \quad u(t,x) \underset{t\to\infty}{\sim} \frac{x}{t}.$$

Here, we emphasize the property

$$b_0/\sqrt{\alpha_0} \propto \|\rho_0\|_{L^1}.$$

Note that since the velocity is linear in x, this provides explicit solutions for the (Newtonian) isothermal Navier-Stokes equation.

Remark 2.2. It is possible to consider an initial Gaussian density which is not centered at the origin, or, equivalently,

$$\rho_0(x) = b_0 e^{-\alpha_0 x^2}, \quad u_0(x) = \beta_0 x + c_0.$$

Then

$$\rho(t,x) = b(t)e^{-\alpha(t)(x-\overline{x}(t))^2}, \quad u(t,x) = \beta(t)x + c(t),$$

with b, α and β like before, and

$$\overline{x}(t) = c_0 t, \quad c(t) = c_0\left(1 - \frac{\dot{\tau}(t)}{\tau(t)}t\right).$$

Remark 2.3 (Universal dynamics for the density). In this Gaussian case, the function ρ exhibits two interesting features. We get a new dispersive rate, different from the one proved in [12] in the case $1 < \gamma \leq 1 + 2/d$ (logarithmic correction). More suprisingly, the density enjoys a universal asymptotic profile: no matter what the initial variance is, the asymptotic one is always the same.

Remark 2.4 (Generalization to other equations). The same approach can be extended to

$$\begin{cases} \partial_t\rho + \operatorname{div}(\rho u) = 0, \\ \partial_t(\rho u) + \operatorname{div}(\rho u \otimes u) + \kappa\nabla\rho = \dfrac{\varepsilon^2}{2}\rho\nabla\left(\dfrac{\Delta\sqrt{\rho}}{\sqrt{\rho}}\right) + \nu\operatorname{div}(\rho D(u)), \end{cases}$$

with $\varepsilon, \nu \geq 0$ and $D(u) = \frac{1}{2}(\nabla u + {}^t\nabla u)$. The term in ε corresponds to Korteweg equation (this term models capillarity). The term in ν corresponds to quantum Navier–Stokes equation, to take dissipative effects into account (see[4] for the derivation). Essentially, we proceed like before, and get the ODE

$$\ddot{\tau}^{\varepsilon,\nu} = \frac{2\kappa\alpha_0}{\tau^{\varepsilon,\nu}} + \varepsilon^2\frac{\alpha_0^2}{(\tau^{\varepsilon,\nu})^3} - \nu\alpha_0\frac{\dot{\tau}^{\varepsilon,\nu}}{(\tau^{\varepsilon,\nu})^2}, \quad \tau^{\varepsilon,\nu}(0) = 1, \quad \dot{\tau}^{\varepsilon,\nu}(0) = \beta_0.$$

It turns out that ε and ν do not alter the large time behavior, and we observe the same universal large time dynamics as for the isothermal Euler equation; see [6] for details.

3. **Isothermal Euler equation and universal large time dynamics.** Motivated by the remarkable observations in the one-dimensional Gaussian case, we introduce another change of unknown functions, in the case of the general space dimension $d \geq 1$. Consider the universal dispersion τ,

$$\ddot{\tau} = \frac{2\kappa}{\tau}, \quad \tau(0) = 1, \quad \dot{\tau}(0) = 0, \tag{4}$$

and change the unknown functions

$$\rho(t,x) = \frac{1}{\tau(t)^d} R\left(t, \frac{x}{\tau(t)}\right) \frac{\|\rho_0\|_{L^1}}{\|\Gamma\|_{L^1}}, \quad u(t,x) = \frac{1}{\tau(t)} U\left(t, \frac{x}{\tau(t)}\right) + \frac{\dot{\tau}(t)}{\tau(t)} x,$$

where

$$\Gamma(y) = e^{-|y|^2}$$

is the Gaussian that appeared in the previous section. The system in $(\rho, \rho u)$ is equivalent to

$$\begin{cases} \partial_t R + \dfrac{1}{\tau^2} \operatorname{div}(RU) = 0, \\ \partial_t(RU) + \dfrac{1}{\tau^2} \operatorname{div}(RU \otimes U) + 2\kappa y R + \kappa \nabla R = 0. \end{cases} \tag{5}$$

Naturally, the conservation of mass remains. The good news is that we gain some positivity in the entropy. Indeed, define the pseudo-energy

$$\mathcal{E}(t) := \frac{1}{2\tau^2} \int R|U|^2 + \kappa \int (R|y|^2 + R \ln R).$$

Formally, it satisfies

$$\dot{\mathcal{E}}(t) = -\mathcal{D}(t) =: -\frac{\dot{\tau}}{\tau^3} \int R|U|^2.$$

We also have

$$\mathcal{E}(t) := \frac{1}{2\tau^2} \int R|U|^2 + \kappa \int R \ln \frac{R}{\Gamma},$$

and since $\int R = \int \Gamma$, the Csiszár-Kullback inequality (see e.g. [1])

$$\|f - g\|^2_{L^1(\mathbb{R}^d)} \leq 2\|f\|_{L^1(\mathbb{R}^d)} \int f(x) \ln\left(\frac{f(x)}{g(x)}\right) dx$$

shows that $\mathcal{E}$ is the sum of two non-negative terms. As a matter of fact, Csiszár-Kullback inequality is not used: it is just a hint that more estimates are available in terms of (R, U) than in terms of (ρ, u).

Lemma 3.1. *Suppose that $\int_{\mathbb{R}^d} R(t,y)dy$ is bounded and $\mathcal{E}(t) \leq \Lambda$ for all $t \geq 0$. There exists $C_0 > 0$ such that*

$$\frac{1}{\tau^2} \int_{\mathbb{R}^d} R|U|^2 + \int_{\mathbb{R}^d} R(1 + |y|^2 + |\ln R|) \leq C_0, \quad \forall t \geq 0.$$

Proof. We decompose $\mathcal{E}$ in order to introduce a sum of positive terms,

$$\mathcal{E}_+(t) := \frac{1}{2\tau^2} \int R|U|^2 + \kappa \left(\int R|y|^2 + \int_{R \geq 1} R \ln R \right).$$

Since $\mathcal{E}$ is non-increasing, we have

$$\mathcal{E}_+(t) \leq \Lambda + \kappa \int_{R<1} R \ln \frac{1}{R} \lesssim 1 + \int_{\mathbb{R}^d} R^{1-\eta}.$$

By interpolation,

$$\int_{\mathbb{R}^d} R^{1-\eta} \leq C_\eta \|R\|_{L^1(\mathbb{R}^d)}^{1-\eta-d\eta/2} \||y|^2 R\|_{L^1(\mathbb{R}^d)}^{d\eta/2}, \quad 0<\eta<\frac{2}{d+2}.$$

Therefore, since the mass is conserved,

$$\mathcal{E}_+(t) \leq \Lambda + C\mathcal{E}_+(t)^{d\eta/4}$$

and since $d\eta/4 < 1$, $\mathcal{E}_+(t)$ is uniformly bounded for $t \geq 0$. As $\mathcal{E}_+$ is the sum of three non-negative terms, each one is uniformly bounded, and the only remaining term in $\mathcal{E}$ is also bounded. □

Recall that

$$\dot{\mathcal{E}}(t) = -\mathcal{D}(t) =: -\frac{\dot{\tau}}{\tau^3}\int R|U|^2 \leq 0.$$

Therefore, $\mathcal{E}$ is naturally bounded from above. The previous lemma shows that $\mathcal{E}$ is bounded from below, hence

$$\int_0^\infty \mathcal{D}(t)dt < \infty.$$

We can now state our main result.

Theorem 3.2. *Let (R,U) be a global weak solution, with constant mass.*

1. *If $\sup_{t\geq 0} \mathcal{E}(t) < \infty$, then*

$$\int_{\mathbb{R}^d} yR(t,y)dy \underset{t\to\infty}{\longrightarrow} 0 \quad \textit{and} \quad \left|\int_{\mathbb{R}^d} (RU)(t,y)dy\right| \underset{t\to\infty}{\longrightarrow} \infty,$$

   unless $\int yR(0,y)dy = \int (RU)(0,y)dy = 0$ (a case where each of these quantites remains identically zero).

2. *If $\sup_{t\geq 0} \mathcal{E}(t) < \infty$ and the energy E satisfies $E(t) = o(\ln t)$ as $t\to\infty$, then*

$$\int_{\mathbb{R}^d} |y|^2 R(t,y)dy \underset{t\to\infty}{\longrightarrow} \int_{\mathbb{R}^d} |y|^2\Gamma(y)dy.$$

3. *If $\sup_{t\geq 0} \mathcal{E}(t) + \int_0^\infty \mathcal{D}(t)dt < \infty$, then $R(t,\cdot) \rightharpoonup \Gamma$ weakly in $L^1(\mathbb{R}^d)$ as $t\to\infty$.*

Remark 3.3 (Wasserstein distance). Theorem 3.2 implies the large time convergence of R to Γ in the Wasserstein distance W_2, defined, for ν_1 and ν_2 probability measures, by

$$W_p(\nu_1,\nu_2) = \inf\left\{\left(\int_{\mathbb{R}^d\times\mathbb{R}^d} |x-y|^p \mathrm{d}\mu(x,y)\right)^{1/p}; \quad (\pi_j)_\sharp \mu = \nu_j\right\},$$

where μ varies among all probability measures on $\mathbb{R}^d\times\mathbb{R}^d$, and $\pi_j : \mathbb{R}^d\times\mathbb{R}^d \to \mathbb{R}^d$ denotes the canonical projection onto the j-th factor. This implies, for instance, the convergence of fractional momenta (see e.g. [13, Theorem 7.12])

$$\int |y|^{2s} R(t,y)\mathrm{d}y \underset{t\to\infty}{\longrightarrow} \int |y|^{2s}\Gamma(y)\mathrm{d}y, \quad 0\leq s\leq 1.$$

In [6], we prove the above result in the following generalized framework:

- The result remains valid in the presence of capillarity (Korteweg equation) and quantum dissipation (quantum Navier-Stokes).

- The result remains valid also with a generalized pressure law:
$$\begin{cases} \partial_t \rho + \operatorname{div}(\rho u) = 0, \\ \partial_t(\rho u) + \operatorname{div}(\rho u \otimes u) + \nabla P(\rho) = \dfrac{\varepsilon^2}{2}\rho \nabla \left(\dfrac{\Delta \sqrt{\rho}}{\sqrt{\rho}} \right) + \nu \operatorname{div}(\rho D u), \end{cases}$$
with P convex and $P'(0) = \kappa > 0$.

Remark 3.4. In the case where $0 \leq \varepsilon \leq \nu$ and $\nu > 0$, with $P(\rho) \equiv \kappa\rho$, we have another rigidity result: up to an extraction, sequences of solutions on $(0,T)$ enjoying uniformly the conservation of mass, and "natural" inequalities (energy dissipation, BD-entropy dissipation, Mellet-Vasseur inequality in (R,U)), converge to a weak solution on $(0,T)$. The proof, presented in [6], follows the same argument as in [2].

4. **Elements of proof.** We present the main arguments to prove Theorem 3.2. Details can be found in [6]. Many features of the proof are similar to the arguments given in [5] in the context of a Schrödinger equation with logarithmic nonlinearity.

4.1. **Main Theorem: proof of the first point.** Define
$$\mathcal{I}_1(t) = \int_{\mathbb{R}^d} (RU)(t,y)dy, \quad \mathcal{I}_2(t) = \int_{\mathbb{R}^d} yR(t,y)dy.$$
We compute
$$\dot{\mathcal{I}}_1 = -\frac{1}{\tau^2}\int_{\mathbb{R}^d} \operatorname{div}(RU \otimes U) - 2\kappa \mathcal{I}_2 - \kappa \int_{\mathbb{R}^d} \nabla R = -2\kappa \mathcal{I}_2,$$
thanks to Lemma 3.1. Similarly, we compute
$$\dot{\mathcal{I}}_2 = -\frac{1}{\tau^2}\int_{\mathbb{R}^d} y \operatorname{div}(RU) = \frac{1}{\tau^2}\int_{\mathbb{R}^d} RU \equiv \frac{1}{\tau^2}\mathcal{I}_1,$$
since, from Lemma 3.1 and Cauchy-Schwarz inequality,
$$R|y||U| \in L^\infty_{\mathrm{loc}}(0,\infty; L^1(\mathbb{R}^d)).$$
Then $J_2 := \tau \mathcal{I}_2$ satisfies $\ddot{J}_2 = 0$, hence
$$\mathcal{I}_2(t) = \frac{-\mathcal{I}_1(0)t + \mathcal{I}_2(0)}{\tau(t)}, \quad \mathcal{I}_1(t) = \mathcal{I}_1(0) - 2\kappa \int_0^t \mathcal{I}_2(s)ds.$$
The first point is then an easy consequence of Lemma 2.1.

4.2. **Main Theorem: proof of the second point.** Recall that the energy (entropy) E is formally conserved,
$$E(t) = \frac{1}{2}\int_{\mathbb{R}^d} \rho(t,x)|u(t,x)|^2 dx + \kappa \int_{\mathbb{R}^d} \rho(t,x)\ln \rho(t,x)dx.$$
In view of the change of unknown functions $(\rho, u) \mapsto (R, U)$, rewrite E:
$$E(t) = \frac{1}{2\tau^2}\int R|U|^2 dy + \frac{(\dot{\tau})^2}{2}\int R|y|^2 dy + \frac{\dot{\tau}}{\tau}\int Ry\cdot U dy + \kappa \int R \ln R dy$$
$$- \kappa \ln\left(\tau^d\right)\int R dy.$$

We already know that the first and fourth terms are uniformly bounded (Lemma 3.1), and that the third term is $\mathcal{O}(\sqrt{\ln t}))$ (Lemma 3.1, Cauchy-Schwarz inequality, and

Lemma 2.1), while each of the second and last term is potentially of order $\ln t$ (Lemma 2.1). Therefore, if $E(t) = o(\ln t)$,

$$2\int R|y|^2 dy - d\int R dy = o(1), \quad \text{hence} \quad \int R|y|^2 dy \underset{t\to\infty}{\longrightarrow} \frac{d}{2}\int \Gamma dy = \int |y|^2\Gamma(y)dy.$$

4.3. **Main Theorem: proof of the last point.** Discarding terms which seem negligible for large time in (5), we get

$$\begin{cases} \partial_t R + \dfrac{1}{\tau^2}\operatorname{div}(RU) = 0, \\ \partial_t(RU) + 2\kappa y R + \kappa\nabla R = 0, \end{cases}$$

hence

$$\partial_t\left(\tau^2\partial_t R\right) = \kappa\mathcal{L}R, \quad \text{where} \quad \mathcal{L}f = \Delta f + 2\operatorname{div}(yf)$$

is a Fokker–Planck operator. Since $\tau^2 \ll (\dot\tau\tau)^2$ as $t\to\infty$ (Lemma 2.1), we expect

$$\partial_t\left(\tau^2\partial_t R\right) = \tau^2\partial_t^2 R + 2\dot\tau\tau\partial_t R \approx 2\dot\tau\tau\partial_t R,$$

hence, for large time, $\partial_s R \approx \mathcal{L}R$, for the new time variable

$$s(t) = \kappa\int\frac{1}{\tau\dot\tau} = \frac{1}{2}\int\frac{\ddot\tau}{\dot\tau} = \frac{1}{2}\ln\dot\tau(t) \underset{t\to\infty}{\sim} \frac{1}{4}\ln\ln t.$$

The large time behavior is thus expected to be dictated by the Fokker–Planck equation

$$\partial_s R_\infty = \mathcal{L}R_\infty, \quad \mathcal{L}f = \Delta f + 2\operatorname{div}(yf).$$

It was established in [3] that any solution to this equation, obeying the bounds given by Lemma 3.1, satisfies

$$\|R_\infty(t) - \Gamma\|_{L^1(\mathbb{R}^d)} \underset{t\to\infty}{\longrightarrow} 0.$$

To make the argument rigorous, set $s(t) = \frac{1}{2}\ln\dot\tau(t)$. At this stage, we emphasize that this rescaled time turns out to be rather natural: in view of Lemma 2.1,

$$s(t) \underset{t\to\infty}{\sim} \frac{1}{4}\ln\ln t.$$

This property conciles the fact that R_∞ converges to Γ exponentially fast in s (due to a spectral gap), and the fact that the convergence of the above quadratic quantities involved a logarithmic convergence in t.

Now denote by $\alpha : s \mapsto \alpha(s) = t$ its inverse mapping. Set $\bar R(s,y) = R(t,y)$, $\bar U(s,y) = U(t,y)$:

$$\partial_s\bar R - \frac{2\kappa}{(\dot\tau\circ\alpha)^2}\partial_s\bar R + \frac{\kappa}{(\dot\tau\circ\alpha)^2}\partial_s^2\bar R = \mathcal{L}\bar R + \frac{1}{(\tau\circ\alpha)^2}\nabla^2 : (\bar R\bar U\otimes\bar U).$$

As a consequence of Lemma 3.1, $\int_0^\infty \mathcal{D}(t)dt < \infty$, which now reads

$$\int_0^\infty\left(\frac{\dot\tau\circ\alpha}{\tau\circ\alpha}\right)^2\left\|\sqrt{\bar R}\bar U\right\|_{L^2(\mathbb{R}^d)}^2 ds < \infty.$$

For $s\in[0,1]$ and $s_n\to\infty$, let $\bar R_n(s,y) = \bar R(s+s_n,y)$, $\bar U_n(s,y) = \bar U(s+s_n,y)$:

$$\sup_{n\in\mathbb{N}}\sup_{s\in[0,1]}\int_{\mathbb{R}^d}\bar R_n(1+|y|^2+|\ln\bar R_n|)dy \le C,$$

$$\lim_{n\to\infty}\int_0^1\left(\frac{\dot\tau\circ\alpha_n}{\tau\circ\alpha_n}\right)^2\left\|\sqrt{\bar R_n}\bar U_n\right\|_{L^2(\mathbb{R}^d)}^2 ds = 0.$$

Dunford–Pettis criterion implies that there exists $R_\infty \in L^1((0,1)\times\mathbb{R}^d)$, such that, up to extracting a subsequence,

$$\bar{R}_n \rightharpoonup R_\infty \text{ weakly in } L^1((0,1)\times\mathbb{R}^d) \text{ as } n\to\infty,$$

and $\int_{\mathbb{R}^d} R_\infty = \int_{\mathbb{R}^d} \bar{R}_n = \int_{\mathbb{R}^d} \Gamma$ (tightness).

$$\lim_{n\to\infty} \int_0^1 \left(\frac{\dot{\tau}\circ\alpha_n}{\tau\circ\alpha_n}\right)^2 \left\|\sqrt{\bar{R}_n}\bar{U}_n\right\|^2_{L^2(\mathbb{R}^d)} ds = 0$$

yields

$$\frac{1}{(\tau\circ\alpha_n)^2}\nabla^2 : (\bar{R}_n\bar{U}_n\otimes\bar{U}_n) \rightharpoonup 0, \quad \text{hence} \quad \partial_s R_\infty = \mathcal{L}R_\infty.$$

On the other hand, we can show that R_∞ is stationary, $\partial_s R_\infty = 0$, and we conclude thanks to the result of Arnold, Markowich, Toscani and Unterreiter [3]. The limit is unique, so no extraction is actually needed.

REFERENCES

[1] C.Ané, S.Blachère, D.Chafaï, P.Fougères, I.Gentil, F.Malrieu, C.Roberto, and G.Scheffer, Sur les inégalités de Sobolev logarithmiques, *Panoramas et Synthèses*, vol. 10, Société Mathématique de France, Paris, 2000.

[2] P. Antonelli and S. Spirito, On the compactness of finite energy weak solutions to the quantum Navier-Stokes equations, *J. Hyperbolic Differ. Equ.* **15** (2018), 133–147.

[3] A. Arnold, P. Markowich, G. Toscani, and A. Unterreiter, On convex Sobolev inequalities and the rate of convergence to equilibrium for Fokker-Planck type equations, *Comm. Partial Differential Equations* **26** (2001), no. 1-2, 43–100.

[4] S. Brull and F. Méhats, Derivation of viscous correction terms for the isothermal quantum Euler model, *ZAMM, Z. Angew. Math. Mech.* **90** (2010), no. 3, 219–230.

[5] R. Carles and I. Gallagher, Universal dynamics for the defocusing logarithmic Schrödinger equation, *Duke Math. J.* **167** (2018), 1761–1801.

[6] R. Carles, K. Carrapatoso, and M. Hillairet, Rigidity results in generalized isothermal fluids, *Annales Henri Lebesgue* **1** (2018), 47–85.

[7] J.-Y. Chemin, Dynamique des gaz à masse totale finie, *Asymptotic Anal.* **3** (1990), 215–220.

[8] M. Grassin, Global smooth solutions to Euler equations for a perfect gas, *Indiana Univ. Math. J.* **47** (1998), 1397–1432.

[9] M. Grassin and D. Serre, Existence de solutions globales et régulières aux équations d'Euler pour un gaz parfait isentropique, *C. R. Acad. Sci. Paris Sér. I Math.* **325** (1997), 721–726.

[10] T. Li and D. Wang, Blowup phenomena of solutions to the Euler equations for compressible fluid flow, *J. Differential Equations* **221** (2006), 91–101.

[11] T. Makino, S. Ukai, and S. Kawashima, Sur la solution à support compact de l'équation d'Euler compressible, *Japan J. Appl. Math.* **3** (1986), 249–257.

[12] D. Serre, Solutions classiques globales des équations d'Euler pour un fluide parfait compressible, *Ann. Inst. Fourier* **47** (1997), 139–153.

[13] C. Villani, *Topics in optimal transportation*, American Mathematical Society, Providence, RI, 2003.

[14] Z. Xin, Blowup of smooth solutions of the compressible Navier-Stokes equation with compact density, *Comm. Pure Appl. Math.* **51** (1998), 229–240.

[15] M. Yuen, Self-similar solutions with elliptic symmetry for the compressible Euler and Navier-Stokes equations in R^N, *Commun. Nonlinear Sci. Numer. Simul.* **17** (2012), 4524–4528.

E-mail address: `Remi.Carles@math.cnrs.fr`
E-mail address: `Kleber.Carrapatoso@umontpellier.fr`
E-mail address: `Matthieu.Hillairet@umontpellier.fr`

ASYMPTOTIC ANALYSIS FOR VLASOV-FOKKER-PLANCK/COMPRESSIBLE NAVIER-STOKES EQUATIONS WITH A DENSITY-DEPENDENT VISCOSITY

YOUNG-PIL CHOI*

Department of Mathematics,
Yonsei University,
Seoul 03722, Korea (Republic of)

JINWOOK JUNG

Department of Mathematical Sciences,
Seoul National University,
Seoul 08826, Korea (Republic of)

ABSTRACT. We study a hydrodynamic limit of a system of coupled kinetic and fluid equations under a strong local alignment force and a strong Brownian motion. More precisely, we consider the Vlasov-Fokker-Planck-type equation coupled with compressible Navier-Stokes equations with a density-dependent viscosity. Based on a relative entropy argument, by assuming the existence of weak solutions to that kinetic-fluid system, we rigorously derive a two-phase fluid model consisting of isothermal Euler equations and compressible Navier-Stokes equations.

1. **Introduction.** The present work is devoted to the asymptotic analysis of a system of kinetic-fluid equations, namely Vlasov-Fokker-Planck equation with a local alignment force coupled with compressible Navier-Stokes equations through the drag force. This system describes the time evolution of dispersed particles immersed in a compressible fluid, in which particles interact with each other via local alignment forces, and particles and fluid are interacting through a drag force. To be more precise, let $f = f(x, \xi, t)$ be the number density function on the phase point $(x, \xi) \in \mathbb{R}^d \times \mathbb{R}^d$ at time $t \in \mathbb{R}_+$, and $n = n(x, t)$ and $v = v(x, t)$ be the local mass density and the bulk velocity of the compressible fluid, respectively. Then, our main system is governed by

$$\begin{aligned}
&\partial_t f + \xi \cdot \nabla_x f + \nabla_\xi \cdot ((v - \xi) f) = \nabla_\xi \cdot (\nabla_\xi f - (u - \xi) f), \ (x, \xi) \in \mathbb{R}^d \times \mathbb{R}^d, \ t > 0, \\
&\partial_t n + \nabla_x \cdot (nv) = 0, \\
&\partial_t (nv) + \nabla_x \cdot (nv \otimes v) + \nabla_x p - 2\nabla_x \cdot (\nu(n) \mathbb{D} v) = -\int_{\mathbb{R}^d} (v - \xi) f \, d\xi,
\end{aligned} \tag{1}$$

2000 *Mathematics Subject Classification.* Primary: 35Q70, 35Q83; Secondary: 35B25.

Key words and phrases. Vlasov/Navier-Stokes equations, asymptotic analysis, hydrodynamic limit, two-phase fluid system, relative entropy method.

* Corresponding author: Young-Pil Choi.

subject to initial data:

$$f(x,\xi,0) = f_0(x,\xi),\ n(x,0) = n_0(x),\ v(x,0) = v_0(x), \quad (x,\xi) \in \mathbb{R}^d \times \mathbb{R}^d$$

and the boundary conditions:

$$f(x,\xi,t) \to 0, \quad n(x,t) \to n_\infty \in \mathbb{R}_+, \quad \text{and} \quad v(x,t) \to 0,$$

sufficiently fast as $|x|$, $|\xi| \to \infty$, where $\mathbb{D}v$ is the deformation tensor given by $\mathbb{D}v := (\nabla_x v + (\nabla_x v)^t)/2$, ν is the viscosity coefficient which is a function of the fluid density n, $p = p(n) := n^\gamma$ $(\gamma > 0)$ is the pressure law, and ρ and u denote the average local densitiy and velocity of f, respectively:

$$\rho(x,t) := \int_{\mathbb{R}^d} f(x,\xi,t)\,d\xi \quad \text{and} \quad (\rho u)(x,t) := \int_{\mathbb{R}^d} \xi f(x,\xi,t)\,d\xi.$$

Those types of kinetic-fluid systems have been extensively studied. The global well-posedness of weak and strong solutions for the Vlasov-type kinetic equations coupled with the incompressible Navier-Stokes equations are discussed in [2, 3, 6, 7, 14, 20, 29, 31] and coupled with compressible Navier-Stokes [1, 8, 15, 27, 28]. The local-in-time existence of classical solutions for the Vlasov-Boltzmann/compressible Euler equations is obtained in [26], and more recently, the global-in-time existence of weak solutions for the BGK/incompressible Navier-Stokes is also discussed in [16]. We refer to [12] and [13] for a priori estimate of large-time behavior of solutions and the finite-time blow-up phenomena of Vlasov-type/Navier-Stokes equations, respectively.

In the current work, we are interested in the asymptotic regime corresponding to a strong drag force and a strong Brownian motion. More specifically, we consider the following system:

$$\begin{aligned}
&\partial_t f^\varepsilon + \xi\cdot\nabla_x f^\varepsilon + \nabla_\xi\cdot((v^\varepsilon - \xi)f^\varepsilon) = \frac{1}{\varepsilon}\nabla_\xi\cdot(\nabla_\xi f^\varepsilon - (u^\varepsilon - \xi)f^\varepsilon),\\
&\partial_t n^\varepsilon + \nabla_x\cdot(n^\varepsilon v^\varepsilon) = 0,\\
&\partial_t(n^\varepsilon v^\varepsilon) + \nabla_x\cdot(n^\varepsilon v^\varepsilon\otimes v^\varepsilon) + \nabla_x p(n^\varepsilon) - 2\nabla_x\cdot(\nu(n^\varepsilon)\mathbb{D}v^\varepsilon) = -\rho^\varepsilon(v^\varepsilon - u^\varepsilon),
\end{aligned} \tag{2}$$

where

$$\rho^\varepsilon(x,t) := \int_{\mathbb{R}^d} f^\varepsilon(x,\xi,t)\,d\xi \quad \text{and} \quad (\rho^\varepsilon u^\varepsilon)(x,t) := \int_{\mathbb{R}^d} \xi f^\varepsilon(x,\xi,t)\,d\xi.$$

Here, since we are concerned with unbounded domain, we assumed the far-field behavior $n^\varepsilon \to n_\infty$ as $|x| \to \infty$ for all $\varepsilon \geq 0$. Note that the global-in-time strong solutions to the kinetic equation in (2) around the global Maxwellian is studied in [11] and the global-in-time existence of weak solutions to the Vlasov-Fokker-Planck/compressible Navier-Stokes equations with a constant viscosity coefficient in a bounded domain is established in [27]. Our main purpose is to investigate the convergence of weak solutions $(f^\varepsilon, n^\varepsilon, v^\varepsilon)$ of the above system (2) to the strong solutions (ρ, u, n, v) to the following system of fluid equations:

$$\begin{aligned}
&\partial_t\rho + \nabla_x\cdot(\rho u) = 0,\\
&\partial_t(\rho u) + \nabla_x\cdot(\rho u\otimes u) + \nabla_x\rho = \rho(v - u),\\
&\partial_t n + \nabla_x\cdot(nv) = 0,\\
&\partial_t(nv) + \nabla_x\cdot(nv\otimes v) + \nabla_x p(n) - 2\nabla_x\cdot(\nu(n)\mathbb{D}v) = -\rho(v - u).
\end{aligned} \tag{3}$$

The hydrodynamic limit of kinetic equation appeared in (2) coupled with the incompressible Navier-Stokes equations is addressed in [6] based on the relative entropy method which relies on the "weak-strong" uniqueness principle [17]. The hydrodynamic limit found in [6] holds as long as there exists a unique strong solution to the limiting system, which is a system of Euler/incompressible Navier-Stokes equations. Later, in [9], the global-in-time existence and uniqueness of strong solutions to that limiting system is obtained. We refer to [21, 22, 28] for other kind of hydrodynamical limits.

Our main strategy relies on the relative entropy argument, which is widely used to analyze hydrodynamic limits of kinetic equations [19, 24, 30], together with some entropy inequalities. In order to establish the hydrodynamic limit, we first need to show the existence of weak and strong solutions to the systems (2) and (3) at least locally in time, and estimate the error between them by means of relative entropy method. However, in the current work, we focus on the relative entropy estimates by assuming the existence of weak solutions to the kinetic-fluid system (2). Since local existence theories for the types of balance laws have been well developed, the local-in-time existence and uniqueness of solutions for the limiting system (3) can be obtained under suitable assumption on the viscosity coefficient ν, see [25] for the readers who are interested in it. We also refer to [10] where the global-in-time existence of a unique strong solution under suitable smallness and regularity assumptions on the initial data is discussed. This yields that once we obtain the existence of weak solutions to the system (2), our analysis becomes fully rigorous. We emphasize that the asymptotic regime we considered for the system (2) has not been studied so far, to the best of our knowledge.

1.1. **Formal derivation of the asymptotic system.** The right-hand side of the kinetic equation in (2) reads

$$\nabla_\xi \cdot [\nabla_\xi f^\varepsilon - (u^\varepsilon - \xi) f^\varepsilon] = \nabla_\xi \cdot \left(M_{f^\varepsilon} \nabla_\xi \left(\frac{f^\varepsilon}{M_{f^\varepsilon}} \right) \right),$$

where $M_{f^\varepsilon} = M_{f^\varepsilon}(x, \xi, t)$ is the Maxwellian given by

$$M_{f^\varepsilon}(x, \xi, t) := \frac{1}{(2\pi)^{d/2}} e^{-\frac{|\xi - u^\varepsilon(x,t)|^2}{2}}.$$

Thus, once we have $\rho^\varepsilon \to \rho$ and $u^\varepsilon \to u$ as $\varepsilon \to 0$, we find

$$f^\varepsilon \to M_{\rho,u} := \frac{\rho(x,t)}{(2\pi)^{d/2}} e^{-\frac{|\xi - u(x,t)|^2}{2}} \quad \text{as} \quad \varepsilon \to 0.$$

This enables us to close the momentum equations derived from the kinetic equation (2) and the limiting solutions (ρ, u, n, v), where $(n^\varepsilon, v^\varepsilon) \to (n, v)$ as $\varepsilon \to 0$, satisfy the two-phase fluid system presented in (3). See [6, 10] for more detailed discussion.

Without loss of generality, throughout this paper, we may assume $\|f_0^\varepsilon\|_{L^1} = 1$ for all $\varepsilon > 0$. This together with the conservation of mass yields $\|f^\varepsilon(\cdot, \cdot, t)\|_{L^1} = \|f_0^\varepsilon\|_{L^1}$ for $\varepsilon > 0$ and $t \geq 0$. In fact, we only need to assume $\|f_0^\varepsilon\|_{L^1} \leq C$ for all $\varepsilon > 0$, where $C > 0$ is independent of ε.

1.2. **Main result.** For the hydrodynamic limit, we will use the following notion of weak solutions to the system (1) and strong solutions to the system (3).

Definition 1.1. For $T \in (0, \infty)$, we say a triplet (f, n, v) is a weak solution to the system (1) if the following conditions are satisfied:

1. $f \in L^\infty(0, T; (L^1_+ \cap L^\infty)(\mathbb{R}^d \times \mathbb{R}^d))$, $\quad (|x|^2 + |\xi|^2) f \in L^\infty(0, T; L^1(\mathbb{R}^d \times \mathbb{R}^d))$.

2. $n - n_\infty \in L^\infty(0,T;(L^1_+ \cap L^\gamma)(\mathbb{R}^d))$, $\quad n|v|^2 \in L^\infty(0,T;L^1(\mathbb{R}^d))$, $\sqrt{\nu(n)}\nabla_x v \in L^2(0,T;L^2(\mathbb{R}^d))$.
3. (f,n,v) satisfies (1) in a distributional sense.

Definition 1.2. Let $s > d/2+2$. For $T \in (0,\infty)$, (ρ,u,n,v) is called a strong solution of (3) on the time interval $[0,T]$ if it satisfies the system (3) in the sense of distributions, and it also satisfies the following regularity conditions:

$$(\rho,u,n,v) \in \mathcal{C}([0,T];H^s(\mathbb{R}^d))\times\mathcal{C}([0,T];H^s(\mathbb{R}^d))\times\mathcal{C}([0,T];H^s(\mathbb{R}^d))\times\mathcal{C}([0,T];H^s(\mathbb{R}^d)).$$

Remark 1. As discussed before, the local-in-time existence and uniqueness of strong solutions in the sense of Definition 1.2 can be obtained under suitable assumptions on the initial data and the viscosity coefficient ν.

We now state our main result on the hydrodynamic limit of (2).

Theorem 1.3. *Let $d > 2$, $\gamma \in [1,2]$, and $(f^\varepsilon,n^\varepsilon,v^\varepsilon)$ be a weak solution to the system (2) up to time $T>0$ in the sense of Definition 1.1 with the initial data $(f_0^\varepsilon,n_0^\varepsilon,v_0^\varepsilon)$ satisfying*

$$\begin{aligned} &f_0^\varepsilon \in (L^1_+ \cap L^\infty)(\mathbb{R}^d\times\mathbb{R}^d),\ (|x|^2+|\xi|^2)f_0^\varepsilon \in L^1(\mathbb{R}^d\times\mathbb{R}^d),\\ &n_0^\varepsilon - n_\infty \in (L^1_+\cap L^\gamma)(\mathbb{R}^d),\ n_0^\varepsilon|v_0^\varepsilon|^2 \in L^1(\mathbb{R}^d),\ \textit{and}\ \sqrt{\nu(n_0^\varepsilon)}\nabla_x v_0^\varepsilon \in L^2(\mathbb{R}^d). \end{aligned} \tag{4}$$

Let $s > d/2+2$ and (ρ,u,n,v) be a strong solution to the system (3) up to time $T>0$ in the sense of Definition 1.2 with the initial data (ρ_0,u_0,n_0,v_0) satisfying

$$\rho_0 > 0 \ \textit{in}\ \mathbb{R}^d, \qquad \inf_{x\in\mathbb{R}^d} n_0(x) > 0, \qquad \textit{and}$$

$$(\rho_0,u_0,n_0,v_0) \in H^s(\mathbb{R}^d)\times H^s(\mathbb{R}^d)\times H^s(\mathbb{R}^d)\times H^s(\mathbb{R}^d).$$

Suppose that the viscosity coefficient $\nu \in \mathcal{C}^1(\mathbb{R}_+)$ is Lipschitz continuous satisfying

$$|\nu(x)-\nu(y)| \le \nu_{Lip}|x-y|, \quad \nu(x) \ge \nu_* > 0, \quad \textit{and} \quad x^2 \le c_0\nu(x)p(x), \tag{5}$$

for all $x,y\in\mathbb{R}_+$, where ν_{Lip}, ν_*, and c_0 are positive constants. Moreover, the initial data $(f_0^\varepsilon,n_0^\varepsilon,v_0^\varepsilon)$ and (ρ_0,u_0,n_0,v_0) are well-prepared such that

(H1):

$$\begin{aligned} &\int_{\mathbb{R}^d}\left(\int_{\mathbb{R}^d} f_0^\varepsilon\left(1+\log f_0^\varepsilon + \frac12(|\xi|^2+|x|^2)\right)d\xi + \frac12 n_0^\varepsilon|v_0^\varepsilon|^2 + H(n_0^\varepsilon)\right)dx\\ &\quad - \int_{\mathbb{R}^d}\left(\rho_0\left(1+\log\rho_0+\frac12(|u_0|^2+|x|^2)\right)+\frac12 n_0|v_0|^2 + H(n_0)\right)dx\\ &= \mathcal{O}(\sqrt{\varepsilon}), \end{aligned}$$

where

$$H(x) := x\int_{n_\infty}^x \frac{p(z)}{z^2}\,dz - \frac{p(n_\infty)}{n_\infty}(x-n_\infty).$$

(H2):

$$\begin{aligned} &\int_{\mathbb{R}^d}\rho_0^\varepsilon|u_0^\varepsilon-u_0|^2\,dx + \int_{\mathbb{R}^d} n_0^\varepsilon|v_0^\varepsilon - v_0|^2\,dx\\ &\quad + \int_{\mathbb{R}^d}\int_{\rho_0}^{\rho_0^\varepsilon}\frac{\rho_0^\varepsilon - z}{z}\,dzdx + \int_{\mathbb{R}^d}\left(n_0^\varepsilon\int_{n_0}^{n_0^\varepsilon}\frac{p(z)}{z^2}\,dz - \frac{p(n_0)}{n_0}(n_0^\varepsilon - n_0)\right)dx\\ &= \mathcal{O}(\sqrt{\varepsilon}). \end{aligned}$$

Then we have

$$\begin{aligned}
&\int_{\mathbb{R}^d} \rho^\varepsilon |u^\varepsilon - u|^2\,dx + \int_{\mathbb{R}^d} (n^\varepsilon)|v^\varepsilon - v|^2\,dx + \int_{\mathbb{R}^d}\int_\rho^{\rho^\varepsilon} \frac{\rho^\varepsilon - z}{z}\,dzdx \\
&\quad + \int_{\mathbb{R}^d}\left(n^\varepsilon \int_n^{n^\varepsilon} \frac{p(z)}{z^2}\,dz - \frac{p(n)}{n}(n^\varepsilon - n)\right)dx \\
&\quad + \int_0^t\int_{\mathbb{R}^d} \nu(n^\varepsilon)|\mathbb{D}(v - v^\varepsilon)|^2\,dxds + \int_0^t\int_{\mathbb{R}^d} \rho^\varepsilon|(u^\varepsilon - v^\varepsilon) - (u - v)|^2\,dxds \\
&\quad \le C\sqrt{\varepsilon},
\end{aligned} \tag{6}$$

where C is a positive constant independent of ε.

As a consequence, we have the following strong convergences of weak solutions $(f^\varepsilon, n^\varepsilon, v^\varepsilon)$ to the system (2) towards the strong solutions (ρ, u, n, v) to the system (3):

$f^\varepsilon \to M_{\rho,u}$ *a.e. and in* $L^1_{loc}(0,T; L^1(\mathbb{R}^d \times \mathbb{R}^d))$,

$(\rho^\varepsilon, n^\varepsilon) \to (\rho, n)$ *a.e. and in* $L^1_{loc}(0,T;L^1(\mathbb{R}^d)) \times L^1_{loc}(0,T;L^p_{loc}(\mathbb{R}^d))$ $\forall p \in [1,\gamma]$,

$(\rho^\varepsilon u^\varepsilon, n^\varepsilon v^\varepsilon) \to (\rho u, nv)$ *a.e. and in* $L^1_{loc}(0,T;L^1(\mathbb{R}^d)) \times L^1_{loc}(0,T;L^1_{loc}(\mathbb{R}^d))$, *and*

$(\rho^\varepsilon|u^\varepsilon|^2, n^\varepsilon|v^\varepsilon|^2) \to (\rho|u|^2, n|v|^2)$ *a.e. and in* $L^1_{loc}(0,T;L^1(\mathbb{R}^d)) \times L^1_{loc}(0,T;L^1_{loc}(\mathbb{R}^d))$,

as $\varepsilon \to 0$.

Remark 2. Since n^ε is not integrable in $\mathbb{R}^d$, we only provide the convergences related to the compressible Navier-Stokes system in (2) locally in $\mathbb{R}^d$.

Remark 3. The technical condition $\gamma \in [1,2]$ is also used in [28], where the asymptotic analysis of the Vlasov-Fokker-Planck equations coupled with the compressible Navier-Stokes equaiton with the constant viscosity coefficient in a bounded domain under strong drag force and strong Brownian motion is studied.

Remark 4. By Young's inequality, we find

$$r^{2-\gamma} \le (\gamma - 1) + (2 - \gamma)r \quad \text{for} \quad \gamma \in [1,2].$$

This yields that $\nu(r) = 1 + r$ satisfies the assumption (5) with $\nu_{\text{Lip}} = \nu_* = 1$ and $c_0 = \max(\gamma - 1, 2 - \gamma) > 0$. It looks that the assumptions on ν (5) do not allow us to consider the constant viscosity coefficient. However, if $\nu \equiv \nu_*$ for an example, the third assumption in (5) is not needed in our estimate. To be more specific, the term K_7 in Section 3 vanishes. Thus our strategy can be directly applied to the constant viscosity coefficient case.

Remark 5. Recently, a non-trivial relative entropy for compressible Navier-Stokes equations with density-dependent viscosities is introduced, and some applications, for examples, weak-strong uniqueness, inviscid limit or low Mach number limit, are discussed in [4, 5, 18].

The next section is devoted to derive an evolution equation for the integrated relative entropy. Finally, in Section 3, we provide the details of proof of Theorem 1.3.

Before closing this section, we introduce several notations used throughout the paper. For a function $f = f(x,\xi)$ defined on $(x,\xi) \in \mathbb{R}^d \times \mathbb{R}^d$, $u = u(x,t)$ on $x \in \mathbb{R}^d$ and $p \in [1,\infty)$, we denote $\|f\|_{L^p}$ and $\|u\|_{L^p}$ by the usual $L^p(\mathbb{R}^d \times \mathbb{R}^d)$- and $L^p(\mathbb{R}^d)$-norm, respectively. $H^k(\mathbb{R}^d)$ is the k-th order L^2-Sobolev space. We also denote by

C a generic positive constant which may differ from line to line; $C = C(\alpha, \beta, \cdots)$ represents the positive constants depending on $\alpha, \beta, \cdots$.

2. **Relative entropy estimate.** In this section, we present some entropy inequalities and relative entropy estimates which will be crucially used for the hydrodynamic limit of the system (2).

2.1. **Entropy inequalities.** In this part, we show that the weak solutions to the system (2) in the sense of Definition 1.1 satisfy several entropy inequalities. Similarly to [6, Section 5], let us set

$$\mathcal{F}(f^\varepsilon, n^\varepsilon, v^\varepsilon) := \int_{\mathbb{R}^d\times\mathbb{R}^d} f^\varepsilon\left(\log f^\varepsilon + \frac{|\xi|^2}{2}\right) dxd\xi + \int_{\mathbb{R}^d} \frac{1}{2} n^\varepsilon |v^\varepsilon|^2\,dx + \int_{\mathbb{R}^d} H(n^\varepsilon)\,dx,$$

$$D_1(f^\varepsilon) := \int_{\mathbb{R}^d\times\mathbb{R}^d} \frac{1}{f^\varepsilon}|\nabla_\xi f^\varepsilon - (u^\varepsilon - \xi)f^\varepsilon|^2\,dxd\xi, \quad \text{and}$$

$$D_2(f^\varepsilon, n^\varepsilon, v^\varepsilon) := \int_{\mathbb{R}^d\times\mathbb{R}^d} |v^\varepsilon - \xi|^2 f^\varepsilon\,dxd\xi + \int_{\mathbb{R}^d} \nu(n^\varepsilon)|\mathbb{D}v^\varepsilon|^2\,dx,$$

where $H = H(n)$ is given by

$$H(n) := K(n) - K'(n_\infty)(n - n_\infty), \quad K(n) := n\int_{n_\infty}^n \frac{p(z)}{z^2}\,dz.$$

Then we can easily find

$$\mathcal{F}(f^\varepsilon, n^\varepsilon, v^\varepsilon) + \frac{1}{\varepsilon}\int_0^t D_1(f^\varepsilon)\,ds + \int_0^t D_2(f^\varepsilon, n^\varepsilon, v^\varepsilon)\,ds \le \mathcal{F}(f_0^\varepsilon, n_0^\varepsilon, v_0^\varepsilon) + dt \qquad (7)$$

for $t \ge 0$. Note that the term $\int_{\mathbb{R}^d\times\mathbb{R}^d} f^\varepsilon \log f^\varepsilon\,dxd\xi$ has an indefinite sign, however, in the lemma below, we show that it can be controlled by $\mathcal{F}(f_0^\varepsilon, n_0^\varepsilon, v_0^\varepsilon)$ and second spatial moment of f_0^ε.

Lemma 2.1. *Let $T > 0$ and suppose that $(f^\varepsilon, n^\varepsilon, v^\varepsilon)$ is a weak solution to the system (2) on the time interval $[0, T)$ in the sense of Definition 1.1 with the initial data $(f_0^\varepsilon, n_0^\varepsilon, v_0^\varepsilon)$ satisfying (4). Then we have*

$$\int_{\mathbb{R}^d\times\mathbb{R}^d} f^\varepsilon\left(1 + |\log f^\varepsilon| + \frac{1}{4}(|x|^2 + |\xi|^2)\right) dxd\xi + \frac{1}{2}\int_{\mathbb{R}^d} n^\varepsilon|v^\varepsilon|^2\,dx + \int_{\mathbb{R}^d} H(n^\varepsilon)\,dx$$
$$+ \frac{1}{\varepsilon}\int_0^t D_1(f^\varepsilon)\,ds + \int_0^t D_2(f^\varepsilon, n^\varepsilon, v^\varepsilon)\,ds \le C(T) + \mathcal{O}(\sqrt{\varepsilon}),$$

for $t \in (0, T)$, where $C = C(T)$ is a positive constant independent of ε.

Proof. It follows from (7) that

$$\begin{aligned}\frac{d}{dt}&\left(\mathcal{F}(f^\varepsilon, n^\varepsilon, v^\varepsilon) + \int_{\mathbb{R}^d\times\mathbb{R}^d} f^\varepsilon\frac{|x|^2}{2}\,dxd\xi\right) + \frac{1}{\varepsilon}D_1(f^\varepsilon) + D_2(f^\varepsilon, n^\varepsilon, v^\varepsilon)\\
&\le \int_{\mathbb{R}^d\times\mathbb{R}^d} f^\varepsilon(x\cdot\xi)\,dxd\xi + d\\
&\le \int_{\mathbb{R}^d\times\mathbb{R}^d}\left(f^\varepsilon\left(\frac{|\xi|^2}{2} + \frac{|x|^2}{2}\right) + 2f^\varepsilon\log_- f^\varepsilon - 2f^\varepsilon\log_- f^\varepsilon\right) dxd\xi + d,\end{aligned}$$

where $\log_- g(x) := \max\{0, -\log g(x)\}$. On the other hand, we get

$$2\int_{\mathbb{R}^d\times\mathbb{R}^d} f^\varepsilon\log_- f^\varepsilon\,dxd\xi$$

$$\le \int_{\mathbb{R}^d\times\mathbb{R}^d} f^\varepsilon \left(\frac{|x|^2}{2} + \frac{|\xi|^2}{2}\right) dxd\xi + \frac{1}{e}\int_{\mathbb{R}^d\times\mathbb{R}^d} e^{-\frac{|\xi|^2}{4}-\frac{|x|^2}{4}}dxd\xi,$$

and this implies

$$\begin{aligned}\frac{d}{dt}&\left(\mathcal{F}(f^\varepsilon,n^\varepsilon,v^\varepsilon) + \int_{\mathbb{R}^d\times\mathbb{R}^d} f^\varepsilon\frac{|x|^2}{2}\,dxd\xi\right)\\ &\le 2\left(\mathcal{F}(f^\varepsilon,n^\varepsilon,v^\varepsilon) + \int_{\mathbb{R}^d\times\mathbb{R}^d} f^\varepsilon\frac{|x|^2}{2}\,dxd\xi\right) + C,\end{aligned}$$

Thus we obtain

$$\mathcal{F}(f^\varepsilon,n^\varepsilon,v^\varepsilon) + \int_{\mathbb{R}^d\times\mathbb{R}^d} f^\varepsilon\frac{|x|^2}{2}\,dxd\xi \le \left(\mathcal{F}(f_0^\varepsilon,n_0^\varepsilon,v_0^\varepsilon) + \int_{\mathbb{R}^d\times\mathbb{R}^d} f_0^\varepsilon\frac{|x|^2}{2}\,dxd\xi\right)e^{C(T)}.$$

Finally, we combine the above inequality with (7) and **(H1)** to conclude the desired result. □

We now present an uniform-in-ε estimate of a modified entropy inequality which can be obtained by using almost the same argument as in [6, Section 5.1].

Lemma 2.2. *Let $T > 0$ and suppose that $(f^\varepsilon, n^\varepsilon, v^\varepsilon)$ is a weak solution to the system (2) on the time interval $[0,T)$ in the sense of Definition 1.1 with the initial data $(f_0^\varepsilon, n_0^\varepsilon, v_0^\varepsilon)$ satisfying (4). Then we have*

$$\begin{aligned}\mathcal{F}(f^\varepsilon,n^\varepsilon,v^\varepsilon) + \frac{1}{2\varepsilon}\int_0^t D_1(f^\varepsilon)\,ds + \int_0^t\int_{\mathbb{R}^d}\rho^\varepsilon|u^\varepsilon - v^\varepsilon|^2dxds&\\ + \int_0^t\int_{\mathbb{R}^d}\nu(n^\varepsilon)|\mathbb{D}v^\varepsilon|^2\,dxds \le \mathcal{F}(f_0^\varepsilon,n_0^\varepsilon,v_0^\varepsilon) + C(T)\varepsilon.&\end{aligned}\tag{8}$$

2.2. **Relative entropy estimate.** In this subsection, we provide the relative entropy estimates. For this purpose, we introduce

$$U = \begin{pmatrix}\rho\\ m\\ n\\ w\end{pmatrix},\quad A(U) := \begin{pmatrix} m & 0 & 0 & 0\\ (m\otimes m)/\rho & \rho\mathbb{I}_d & 0 & 0\\ w & 0 & 0 & 0\\ (w\otimes w)/n & n^\gamma\mathbb{I}_d & 0 & 0\end{pmatrix},$$

and

$$F(U) = \begin{pmatrix} 0\\ \rho(v-u)\\ 0\\ -\rho(v-u) + 2\nabla_x\cdot(\nu(n)\mathbb{D}v)\end{pmatrix},$$

where $\mathbb{I}_d$ denotes the $d\times d$ identity matrix, $m := \rho u$, and $w := nv$, and then we rewrite the system (3) in the form of conservation of laws:

$$U_t + \nabla_x\cdot A(U) = F(U).$$

For notational simplicity, we drop x-dependence of differential operators, i.e., $\nabla f := \nabla_x f$ and $\Delta f = \Delta_x f$ for the rest of this paper. The corresponding macroscopic entropy $E(U)$ to above system is given by

$$E(U) := \frac{m^2}{2\rho} + \frac{w^2}{2n} + \rho\log\rho + H(n),$$

and the relative entropy functional $\mathcal{H}$ is defined as

$$\mathcal{H}(V|U) := E(V) - E(U) - DE(U)(V-U), \qquad V = \begin{pmatrix} \bar\rho \\ \bar m \\ \bar n \\ \bar w \end{pmatrix}.$$

A straightforward computation yields

$$\mathcal{H}(V|U) = \frac{\bar\rho}{2}|u-\bar u|^2 + \frac{\bar n}{2}|v-\bar v|^2 + P(\bar\rho|\rho) + \tilde P(\bar n|n),$$

where $P(x|y)$ and $\tilde P(x|y)$ are relative pressures given by

$$P(x|y) := x\log x - y\log y + (y-x)(1+\log y) = \int_y^x \frac{x-z}{z}\,dz \ge \frac{1}{2}\min\left\{\frac{1}{x},\frac{1}{y}\right\}|x-y|^2$$

and

$$\tilde P(x|y) := \begin{cases} P(x|y) & \text{if } \gamma = 1, \\ \dfrac{1}{\gamma-1}(x^\gamma - y^\gamma) + \dfrac{\gamma}{\gamma-1}(y-x)y^{\gamma-1} & \text{if } \gamma > 1, \end{cases}$$

respectively. Note that

$$\begin{aligned} \tilde P(x|y) &= K(x) - K(y) - K'(y)(x-y) \\ &\ge \gamma\min\left\{x^{\gamma-2}, y^{\gamma-2}\right\}|x-y|^2 \ge \frac{\gamma}{2}\max\{x^{2-\gamma}, y^{2-\gamma}\}^{-1}|x-y|^2, \end{aligned}$$

for $\gamma > 1$. Using those newly defined notations, we derive an evolution equation for the relative entropy functional $\mathcal{H}$.

Lemma 2.3. *The relative entropy $\mathcal{H}$ satisfies the following equation:*

$$\begin{aligned} &\int_{\mathbb{R}^d} \mathcal{H}(V|U)\,dx + \int_0^t\int_{\mathbb{R}^d} \nu(\bar n)|\mathbb{D}(v-\bar v)|^2\,dxds + \int_0^t\int_{\mathbb{R}^d} \bar\rho|(\bar u-\bar v)-(u-v)|^2\,dxds \\ &\quad = \int_{\mathbb{R}^d} \mathcal{H}(V_0|U_0)\,dx + \int_0^t\int_{\mathbb{R}^d} \partial_s E(V)\,dxds + \int_0^t\int_{\mathbb{R}^d} \nu(\bar n)|\mathbb{D}\bar v|^2\,dxds \\ &\qquad + \int_0^t\int_{\mathbb{R}^d} \bar\rho|\bar u-\bar v|^2\,dxds - \int_0^t\int_{\mathbb{R}^d} DE(U)(\partial_s V + \nabla\cdot A(V) - F(V))\,dxds \\ &\qquad - \int_0^t\int_{\mathbb{R}^d} (\nabla DE(U)) : A(V|U)\,dxds + \int_0^t\int_{\mathbb{R}^d} \left(\frac{\bar n}{n}\rho - \bar\rho\right)(v-\bar v)(u-v)\,dxds \\ &\qquad + 2\int_{\mathbb{R}^d} \left(\frac{\bar n}{n} - 1\right)(\nabla\cdot(\nu(n)\mathbb{D}v))\cdot(v-\bar v)\,dx \\ &\qquad + 2\int_{\mathbb{R}^d} (\nabla\cdot((\nu(n)-\nu(\bar n))\mathbb{D}v))\cdot(v-\bar v)\,dx, \end{aligned}$$

where $A : B = \sum_{i=1}^m\sum_{j=1}^n a_{ij}b_{ij}$ for $A = (a_{ij}), B = (b_{ij}) \in \mathbb{R}^{mn}$ and $A(V|U)$ is the relative flux functional defined by

$$A(V|U) := A(V) - A(U) - DA(U)(V-U).$$

Proof. A straightforward computation gives

$$\begin{aligned}\frac{d}{dt}\int_{\mathbb{R}^d}\mathcal{H}(V|U)\,dx &= \int_{\mathbb{R}^d}\partial_t E(V)\,dx - \int_{\mathbb{R}^d} DE(U)(\partial_t V + \nabla\cdot A(V) - F(V))\,dx \\ &\quad + \int_{\mathbb{R}^d} D^2E(U)\nabla\cdot A(U)(V-U) + DE(U)\nabla\cdot A(V)\,dx \\ &\quad - \int_{\mathbb{R}^d} D^2E(U)F(U)(V-U) + DE(U)F(V)\,dx \\ &=: \sum_{i=1}^4 I_i.\end{aligned}$$

In order to get the desired result, it suffices to estimate I_3 and I_4 only. For the estimate of I_3, we directly use the idea of [6, Appendix A] to get

$$I_3 = -\int_{\mathbb{R}^d}(\nabla DE(U)) : A(V|U)\,dx.$$

For the estimate of I_4, we first notice that

$$\begin{aligned}&D^2E(U)F(U)(V-U)\\ &= \begin{pmatrix} * & -m/\rho^2 & * & 0 \\ * & 1/\rho & * & 0 \\ * & 0 & * & -w/n^2 \\ * & 0 & * & 1/n \end{pmatrix}\begin{pmatrix} 0 \\ \rho(v-u) \\ 0 \\ -\rho(v-u) + 2\nabla\cdot(\nu(n)\mathbb{D}v)\end{pmatrix}\begin{pmatrix}\bar\rho-\rho \\ \bar m - m \\ \bar n - n \\ \bar w - w\end{pmatrix} \\ &= -(v-u)\cdot u(\bar\rho-\rho) + (v-u)\cdot(\bar m - m) + \frac{v}{n}\cdot(\rho(v-u) - 2\nabla\cdot(\nu(n)\mathbb{D}v))(\bar n - n) \\ &\quad - \frac{1}{n}(\rho(v-u) - 2\nabla\cdot(\nu(n)\mathbb{D}v))\cdot(\bar w - w),\end{aligned}$$

and

$$\begin{aligned}DE(U)F(V) &= \bar\rho(\bar v - \bar u)\cdot u - (\bar\rho(\bar v - \bar u) - 2\nabla\cdot(\nu(\bar n)\mathbb{D}\bar v))\cdot v \\ &= \bar\rho(\bar v - \bar u)\cdot(u-v) + 2\,(\nabla\cdot(\nu(\bar n)\mathbb{D}\bar v))\cdot v.\end{aligned}$$

Combining the above inequalities, we find

$$\begin{aligned}&D^2E(U)F(U)(V-U) + DE(U)F(V)\\ &= (u-v)\cdot u(\bar\rho-\rho) - (u-v)\cdot(\bar\rho\bar u - \rho u) \\ &\quad - \frac{\rho v}{n}\cdot(u-v)(\bar n - n) + \frac{\rho}{n}(u-v)\cdot((\bar n)\bar v - (n)v) \\ &\quad + \bar\rho(\bar v\cdot u - \bar u\cdot u - \bar v\cdot v + \bar u\cdot v) - \frac{2}{n}(\bar n - n)\,(\nabla\cdot(\nu(n)\mathbb{D}v))\cdot v \\ &\quad + \frac{2}{n}\,(\nabla\cdot(\nu(n)\mathbb{D}v))\cdot(\bar w - w) + 2\,(\nabla\cdot(\nu(\bar n)\mathbb{D}\bar v))\cdot v \\ &= \bar\rho(u-\bar u)\cdot(u-v) - \frac{\bar n}{n}\rho(v-\bar v)\cdot(u-v) \\ &\quad + \bar\rho(\bar v\cdot u - \bar u\cdot u - \bar v\cdot v + \bar u\cdot v) \\ &\quad - \frac{2}{n}(\bar n - n)(\nabla\cdot(\nu(n)\mathbb{D}v))\cdot v + \frac{2}{n}(\nabla\cdot(\nu(n)\mathbb{D}v))\cdot(\bar w - w) + 2(\nabla\cdot(\nu(\bar n)\mathbb{D}\bar v))\cdot v \\ &=: \sum_{i=1}^3 J_i,\end{aligned}$$

where J_2 can be rewritten as

$$\begin{aligned}
J_2 &= \bar\rho(\bar v\cdot u - \bar u\cdot u - \bar v\cdot v + \bar u\cdot v)\\
&= \bar\rho(\bar v\cdot u - \bar u\cdot u - \bar v\cdot v + \bar u\cdot v + |u|^2 - u\cdot v + \bar u\cdot v - \bar u\cdot u)\\
&\quad - \bar\rho(|u|^2 - u\cdot v + \bar u\cdot v - \bar u\cdot u)\\
&= \bar\rho(-2(\bar u - \bar v)\cdot(u-v) - \bar v\cdot(u-v) - u\cdot v + |u|^2)\\
&\quad - \bar\rho(|u|^2 - u\cdot v + \bar u\cdot v - \bar u\cdot u)\\
&= \bar\rho(-2(\bar u - \bar v)\cdot(u-v) + |u-v|^2) + \bar\rho(-|u-v|^2 - \bar v\cdot(u-v) - u\cdot v + |u|^2)\\
&\quad - \bar\rho(|u|^2 - u\cdot v + \bar u\cdot v - \bar u\cdot u)\\
&= \bar\rho|(\bar u - \bar v) - (u-v)|^2 - \bar\rho|\bar u - \bar v|^2 - \bar\rho(u-v)\cdot((u-\bar u) - (v-\bar v)).
\end{aligned}$$

Thus we obtain

$$J_1 + J_2 = \bar\rho|(\bar u - \bar v) - (u-v)|^2 - \bar\rho|\bar u - \bar v|^2 + \left(\frac{\bar n}{n}\rho - \bar\rho\right)(v-\bar v)\cdot(v-u). \tag{9}$$

For J_3, we estimate

$$\begin{aligned}
J_3 &= -2\left(\nabla\cdot(\nu(n)\mathbb{D}v)\right)\cdot(v-\bar v) + 2\left(\nabla\cdot(\nu(\bar n)\mathbb{D}\bar v)\right)\cdot v\\
&\quad - 2\left(\frac{\bar n}{n} - 1\right)\left(\nabla\cdot(\nu(n)\mathbb{D}v)\right)\cdot(v-\bar v)\\
&= -2\left(\nabla\cdot\left((\nu(n)-\nu(\bar n))\,\mathbb{D}v\right)\right)\cdot(v-\bar v) - 2\left(\nabla\cdot(\nu(\bar n)\mathbb{D}(v-\bar v))\right)\cdot(v-\bar v)\\
&\quad + 2\left(\nabla\cdot(\nu(\bar n)\mathbb{D}\bar v)\right)\cdot\bar v - 2\left(\frac{\bar n}{n} - 1\right)\left(\nabla\cdot(\nu(n)\mathbb{D}v)\right)\cdot(v-\bar v),
\end{aligned}$$

which together with (9) gives

$$\begin{aligned}
I_4 &= -\int_{\mathbb{R}^d}\bar\rho|(\bar u - \bar v) - (u-v)|^2\,dx + \int_{\mathbb{R}^d}\bar\rho|\bar u - \bar v|^2 dx\\
&\quad + \int_{\mathbb{R}^d}\left(\frac{\bar n}{n}\rho - \bar\rho\right)(v-\bar v)\cdot(u-v)\,dx + \int_{\mathbb{R}^d}\nu(\bar n)|\mathbb{D}\bar v|^2 dx\\
&\quad - \int_{\mathbb{R}^d}\nu(\bar n)|\mathbb{D}(v-\bar v)|^2 dx + 2\int_{\mathbb{R}^d}\left(\frac{\bar n}{n} - 1\right)\left(\nabla\cdot(\nu(n)\mathbb{D}v)\right)\cdot(v-\bar v)\,dx\\
&\quad + 2\int_{\mathbb{R}^d}\left(\nabla\cdot\left((\nu(n)-\nu(\bar n))\,\mathbb{D}v\right)\right)\cdot(v-\bar v)\,dx.
\end{aligned}$$

This completes the proof. □

3. **Proof of Theorem 1.3.** In this section, we provide the details of proof of Theorem 1.3. Let

$$U := \begin{pmatrix}\rho\\ \rho u\\ n\\ nv\end{pmatrix} \quad\text{and}\quad U^\varepsilon := \begin{pmatrix}\rho^\varepsilon\\ \rho^\varepsilon u^\varepsilon\\ n^\varepsilon\\ n^\varepsilon v^\varepsilon\end{pmatrix},$$

where $(f^\varepsilon, n^\varepsilon, v^\varepsilon)$ and (ρ, u, n, v) are weak solutions to the system (2) and a unique strong solution to the system (3), respectively. Then it follows from Lemma 2.3 that

$$\begin{aligned}
&\int_{\mathbb{R}^d}\mathcal{H}(U^\varepsilon|U)\,dx + \int_0^t\int_{\mathbb{R}^d}\nu(n^\varepsilon)|\mathbb{D}(v-v^\varepsilon)|^2\,dxds\\
&\quad + \int_0^t\int_{\mathbb{R}^d}\rho^\varepsilon|(u^\varepsilon - v^\varepsilon) - (u-v)|^2\,dxds
\end{aligned}$$

$$
\begin{aligned}
&= \int_{\mathbb{R}^d} \mathcal{H}(U_0^\varepsilon|U_0)\,dx \\
&\quad + \int_0^t\int_{\mathbb{R}^d} \partial_s E(U^\varepsilon)\,dxds + \int_0^t\int_{\mathbb{R}^d} \nu(n^\varepsilon)|\mathbb{D}v^\varepsilon|^2\,dxds + \int_0^t\int_{\mathbb{R}^d} \rho^\varepsilon|u^\varepsilon - v^\varepsilon|^2\,dxds \\
&\quad - \int_0^t\int_{\mathbb{R}^d} DE(U)(\partial_s U^\varepsilon + \nabla\cdot A(U^\varepsilon) - F(U^\varepsilon))\,dxds \\
&\quad - \int_0^t\int_{\mathbb{R}^d} (\nabla DE(U)) : A(U^\varepsilon|U)\,dxds + \int_0^t\int_{\mathbb{R}^d}\left(\frac{n^\varepsilon}{n}\rho - \rho^\varepsilon\right)(v - v^\varepsilon)\cdot(u-v)\,dxds \\
&\quad + 2\int_0^t\int_{\mathbb{R}^d}\left(\frac{n^\varepsilon - n}{n}\right)(\nabla\cdot(\nu(n)\mathbb{D}v))\cdot(v - v^\varepsilon)\,dxds \\
&\quad + 2\int_0^t\int_{\mathbb{R}^d}(\nabla\cdot((\nu(n) - \nu(n^\varepsilon))\,\mathbb{D}v))\cdot(v - v^\varepsilon)\,dxds \\
&=: \sum_{i=1}^7 K_i.
\end{aligned}
$$

We separately estimate $K_i, i = 1, \ldots, 7$ as follows.

◇ (Estimates for K_1): It follows from **(H2)** that

$$K_1 = \mathcal{O}(\sqrt{\varepsilon}).$$

◇ (Estimates for K_2): Similar to [6, Proposition 5.2], we estimate

$$
\begin{aligned}
K_2 &= \int_{\mathbb{R}^d} E(U^\varepsilon)\,dx - \mathcal{F}(f^\varepsilon, n^\varepsilon, v^\varepsilon) + \mathcal{F}(f^\varepsilon, n^\varepsilon, v^\varepsilon) + \int_0^t\int_{\mathbb{R}^d} \nu(n^\varepsilon)|\mathbb{D}v^\varepsilon|^2\,dxds \\
&\quad + \int_0^t\int_{\mathbb{R}^d} \rho^\varepsilon|u^\varepsilon - v^\varepsilon|^2\,dxds - \mathcal{F}(f_0^\varepsilon, n_0^\varepsilon, v_0^\varepsilon) + \mathcal{F}(f_0^\varepsilon, n_0^\varepsilon, v_0^\varepsilon) - \int_{\mathbb{R}^d} E(U_0)\,dx \\
&\le C(T)\varepsilon + \mathcal{F}(f_0^\varepsilon, n_0^\varepsilon, v_0^\varepsilon) - \int_{\mathbb{R}^d} E(U_0)\,dx,
\end{aligned}
$$

where we used the entropy inequality (8) and the fact that

$$\int_{\mathbb{R}^d} E(U^\varepsilon)\,dx \le \mathcal{F}(f^\varepsilon, n^\varepsilon, v^\varepsilon).$$

We then use the assumption **(H1)** on the well-prepared initial data to obtain

$$K_2 \le C\sqrt{\varepsilon},$$

for some $C > 0$ independent of ε.

◇ (Estimates for K_3): It follows from (2) that

$$
\begin{aligned}
&\partial_t\rho^\varepsilon + \nabla\cdot(\rho^\varepsilon u^\varepsilon) = 0, \\
&\partial_t(\rho^\varepsilon u^\varepsilon) + \nabla\cdot(\rho^\varepsilon u^\varepsilon\otimes u^\varepsilon) + \nabla\rho^\varepsilon - \rho^\varepsilon(v^\varepsilon - u^\varepsilon) = \nabla\cdot\left(\int_{\mathbb{R}^d}(u^\varepsilon\otimes u^\varepsilon - \xi\otimes\xi + \mathbb{I}_d)f^\varepsilon\,d\xi\right), \\
&\partial_t n^\varepsilon + \nabla\cdot(n^\varepsilon v^\varepsilon) = 0, \\
&\partial_t(n^\varepsilon v^\varepsilon) + \nabla\cdot(n^\varepsilon v^\varepsilon\otimes v^\varepsilon) + \nabla p(n^\varepsilon) - 2\nabla\cdot(\nu(n^\varepsilon)\mathbb{D}v^\varepsilon) + \rho^\varepsilon(v^\varepsilon - u^\varepsilon) = 0,
\end{aligned}
$$

in the sense of distributions. This gives

$$- \int_0^t\int_{\mathbb{R}^d} DE(U)(\partial_s U^\varepsilon + \nabla\cdot A(U^\varepsilon) - F(U^\varepsilon))\,dxds$$

$$
\begin{aligned}
&= -\int_0^t \int_{\mathbb{R}^d} D_m E(U) \cdot \left(\nabla \cdot \left(\int_{\mathbb{R}^d} (u^\varepsilon \otimes u^\varepsilon - \xi \otimes \xi + \mathbb{I}_d) f^\varepsilon \, d\xi\right)\right) dxds \\
&= \int_0^t \int_{\mathbb{R}^d} \nabla u : \left(\int_{\mathbb{R}^d} (u^\varepsilon \otimes u^\varepsilon - \xi \otimes \xi + \mathbb{I}_d) f^\varepsilon \, d\xi\right) dxds
\end{aligned}
$$

due to $D_m E(U) = u$. We then follow the proof of [23, Lemma 4.4] to get

$$
K_3 \le C\sqrt{\varepsilon},
$$

where $C = C(\|\nabla u\|_{L^\infty})$ is a positive constant independent of ε.

$\diamond$ (Estimates for K_4): Note that

$$
\begin{aligned}
A(U^\varepsilon|U) &= A(U^\varepsilon) - A(U) - DA(U)(U^\varepsilon - U) \\
&= \begin{pmatrix} 0 & 0 & 0 & 0 \\ \rho^\varepsilon(u^\varepsilon - u) \otimes (u^\varepsilon - u) & 0 & 0 & 0 \\ 0 & 0 & 0 & 0 \\ n^\varepsilon(v^\varepsilon - v) \otimes (v^\varepsilon - v) & (\gamma - 1)\tilde{P}(n^\varepsilon|n)\mathbb{I}_d & 0 & 0 \end{pmatrix}.
\end{aligned}
$$

This implies

$$
\begin{aligned}
\int_{\mathbb{R}^d} |A(U^\varepsilon|U)| \, dx &\le \int_{\mathbb{R}^d} \rho^\varepsilon |u^\varepsilon - u|^2 + n^\varepsilon |v^\varepsilon - v|^2 + d(\gamma - 1)\tilde{P}(n^\varepsilon|n) \, dx \\
&\le C \int_{\mathbb{R}^d} \mathcal{H}(U^\varepsilon|U) \, dx,
\end{aligned}
$$

where $C > 0$ only depends on d and γ. Thus we obtain

$$
K_4 \le C \int_0^t \int_{\mathbb{R}^d} \mathcal{H}(U^\varepsilon|U) \, dxds.
$$

$\diamond$ (Estimates for K_5): We divide K_5 into two terms:

$$
\begin{aligned}
K_5 &= \int_0^t \int_{\mathbb{R}^d} (\rho - \rho^\varepsilon)(v - v^\varepsilon) \cdot (u - v) \, dxds \\
&\quad + \int_0^t \int_{\mathbb{R}^d} \rho \left(\frac{n^\varepsilon - n}{n}\right) (v - v^\varepsilon) \cdot (u - v) \, dxds \\
&=: K_5^1 + K_5^2.
\end{aligned}
$$

For the estimate of K_5^1, we use the following elementary inequality

$$
1 = \min\left\{x^{-1}, y^{-1}\right\} \max\{x, y\} \le \min\left\{x^{-1}, y^{-1}\right\}(x + y) \quad \text{for} \quad x, y > 0, \tag{10}
$$

to get

$$
\begin{aligned}
&\left|\int_{\mathbb{R}^d} (\rho - \rho^\varepsilon)(v - v^\varepsilon) \cdot (u - v) \, dx\right| \\
&\quad \le \left(\int_{\mathbb{R}^d} \min\left\{\frac{1}{\rho^\varepsilon}, \frac{1}{\rho}\right\} (\rho - \rho^\varepsilon)^2 \, dx\right)^{1/2} \left(\int_{\mathbb{R}^d} (\rho + \rho^\varepsilon)|v - v^\varepsilon|^2 |u - v|^2 \, dx\right)^{1/2} \\
&\quad \le C \left(\int_{\mathbb{R}^d} \mathcal{H}(U^\varepsilon|U) \, dx\right)^{1/2} \left(\int_{\mathbb{R}^d} (\rho + \rho^\varepsilon)|v - v^\varepsilon|^2 |u - v|^2 \, dx\right)^{1/2}.
\end{aligned}
$$

On the other hand, the second term on the above inequality can be estimated as

$$
\int_{\mathbb{R}^d} (\rho + \rho^\varepsilon)|v - v^\varepsilon|^2 |u - v|^2 \, dx
$$

$$
\begin{aligned}
&\leq \|\rho\|_{L^\infty}\|v-v^\varepsilon\|_{L^{p*}}^2\|u-v\|_{L^d}^2 \\
&\quad + 2\int_{\mathbb{R}^d}\left(\rho^\varepsilon|(u-u^\varepsilon)-(v-v^\varepsilon)|^2+\rho^\varepsilon|u-u^\varepsilon|^2\right)|u-v|^2\,dx \\
&\leq C\|\nabla(v-v^\varepsilon)\|_{L^2}^2\|u-v\|_{L^d}^2 \\
&\quad + 2\|u-v\|_{L^\infty}^2\left(\int_{\mathbb{R}^d}\rho^\varepsilon|(u-u^\varepsilon)-(v-v^\varepsilon)|^2\,dx+\int_{\mathbb{R}^d}\rho^\varepsilon|u-u^\varepsilon|^2\,dx\right),
\end{aligned}
$$

where $1/p^* = 1/2 - 1/d$ and we used Gagliardo-Nirenberg-Sobolev inequality. Note that

$$
\|\rho\|_{L^\infty} \leq C\|\rho\|_{H^s}, \quad \|u-v\|_{L^d} \leq \|u-v\|_{L^\infty}^{(d-2)/d}\|u-v\|_{L^2}^{2/d} \leq C\|u-v\|_{H^s},
$$

due to $s > d/2+2$, and

$$
\begin{aligned}
\frac{1}{2}\int_{\mathbb{R}^d}|\nabla(v-v^\varepsilon)|^2\,dx &\leq \frac{1}{2}\int_{\mathbb{R}^d}|\nabla(v-v^\varepsilon)|^2\,dx+\frac{1}{2}\int_{\mathbb{R}^d}|\nabla\cdot(v-v^\varepsilon)|^2\,dx \\
&\leq \int_{\mathbb{R}^d}|\mathbb{D}(v-v^\varepsilon)|^2\,dx \leq \frac{1}{\nu_*}\int_{\mathbb{R}^d}\nu(n^\varepsilon)|\mathbb{D}(v-v^\varepsilon)|^2\,dx.
\end{aligned}
$$

These together with using Young's inequality give

$$
\begin{aligned}
K_5^1 &\leq C\int_0^t\int_{\mathbb{R}^d}\mathcal{H}(U^\varepsilon|U)\,dxds+\frac{1}{8}\int_0^t\int_{\mathbb{R}^d}\nu(n^\varepsilon)|\mathbb{D}(v-v^\varepsilon)|^2\,dxds \\
&\quad +\frac{1}{2}\int_0^t\int_{\mathbb{R}^d}\rho^\varepsilon|(u-u^\varepsilon)-(v-v^\varepsilon)|^2\,dxds,
\end{aligned}
$$

where $C = C(\|\rho\|_{L^\infty}, \|u-v\|_{L^\infty(0,T;L^d\cap L^\infty)}, \nu_*)$ is a positive constant. For the term K_5^2, we let $n_* := \inf_{x\in\mathbb{R}^d} n(x) > 0$ and use the inequality (10) to get

$$
\begin{aligned}
&\left|\int_{\mathbb{R}^d}\rho\left(\frac{n^\varepsilon-n}{n}\right)(v-v^\varepsilon)\cdot(u-v)\,dx\right| \\
&\leq \frac{\|\rho\|_{L^\infty}}{n_*}\int_{\mathbb{R}^d}|n^\varepsilon-n||v^\varepsilon-v||u-v|\,dx \\
&\leq C\left(\int_{\mathbb{R}^d}\min\left\{(n^\varepsilon)^{\gamma-2},n^{\gamma-2}\right\}(n-n^\varepsilon)^2\,dx\right)^{1/2} \\
&\qquad\times\left(\int_{\mathbb{R}^d}\left(n^{2-\gamma}+(n^\varepsilon)^{2-\gamma}\right)|v-v^\varepsilon|^2|u-v|^2\,dx\right)^{1/2} \\
&\leq C\left(\int_{\mathbb{R}^d}\mathcal{H}(U^\varepsilon|U)\,dx\right)^{1/2}\left(\int_{\mathbb{R}^d}\left(n^{2-\gamma}+(n^\varepsilon)^{2-\gamma}\right)|v-v^\varepsilon|^2|u-v|^2\,dx\right)^{1/2},
\end{aligned}
$$

where $C = C(\|\rho\|_{L^\infty}, n_*, \gamma)$ is a positive constant. We further estimate

$$
\begin{aligned}
&\int_{\mathbb{R}^d}\left(n^{2-\gamma}+(n^\varepsilon)^{2-\gamma}\right)|v-v^\varepsilon|^2|u-v|^2\,dx \\
&\quad\leq \|n\|_{L^\infty}^{2-\gamma}\|v-v^\varepsilon\|_{L^{p^*}}^2\|u-v\|_{L^d}^2+\int_{\mathbb{R}^d}(n^\varepsilon)^{2-\gamma}|v-v^\varepsilon|^2|u-v|^2\,dx \\
&\quad\leq C\|n\|_{L^\infty}^{2-\gamma}\|u-v\|_{L^d}^2\|\nabla(v-v^\varepsilon)\|_{L^2}^2+\int_{\mathbb{R}^d}(n^\varepsilon)^{2-\gamma}|v-v^\varepsilon|^2|u-v|^2\,dx.
\end{aligned}
$$

For $\gamma = 1$ or 2, we easily get

$$\int_{\mathbb{R}^d} (n^\varepsilon)^{2-\gamma}|v-v^\varepsilon|^2|u-v|^2\,dx \le \begin{cases} \|u-v\|_{L^\infty}^2 \displaystyle\int_{\mathbb{R}^d} \mathcal{H}(U^\varepsilon|U)\,dx & \text{for } \gamma = 1, \\ \|u-v\|_{L^d}^2 \|\nabla(v-v^\varepsilon)\|_{L^2}^2 & \text{for } \gamma = 2. \end{cases}$$

For $\gamma \in (1,2)$, we first use Young's inequality to obtain

$$\begin{aligned}
&\int_{\mathbb{R}^d} (n^\varepsilon)^{2-\gamma}|v-v^\varepsilon|^2|u-v|^2\,dx \\
&\quad \le \int_{\mathbb{R}^d} (n^\varepsilon)^{2-\gamma}|v-v^\varepsilon|^{(4-2\gamma)+(2\gamma-2)}|u-v|^2\,dx \\
&\quad \le (2-\gamma)\int_{\mathbb{R}^d} n^\varepsilon|v-v^\varepsilon|^2\,dx + (\gamma-1)\int_{\mathbb{R}^d} |v-v^\varepsilon|^2|u-v|^{2/(\gamma-1)}\,dx \\
&\quad \le (2-\gamma)\int_{\mathbb{R}^d} n^\varepsilon|v-v^\varepsilon|^2\,dx + (\gamma-1)\|v-v^\varepsilon\|_{L^{p^*}}^2 \|u-v\|_{L^{\frac{d}{\gamma-1}}}^{\frac{2}{\gamma-1}} \\
&\quad \le C\int_{\mathbb{R}^d} \mathcal{H}(U^\varepsilon|U)\,dx + C\int_{\mathbb{R}^d} |\nabla(v-v^\varepsilon)|^2\,dx,
\end{aligned}$$

where $C = C(\gamma, \|u-v\|_{L^\infty(0,T;L^d\cap L^\infty)})$ is a positive constant. Note that $d/(\gamma-1) > d > 2$. Using the similar argument as in the estimate of K_5^1, we find

$$K_5^2 \le C\int_0^t\int_{\mathbb{R}^d} \mathcal{H}(U^\varepsilon|U)\,dxds + \frac{1}{8}\int_0^t\int_{\mathbb{R}^d} \nu(n^\varepsilon)|\mathbb{D}(v-v^\varepsilon)|^2\,dxds,$$

for any $\gamma \in [1,2]$. Thus, we collect the estimates for K_5^1 and K_5^2 to yield

$$\begin{aligned}
K_5 &\le C\int_0^t\int_{\mathbb{R}^d} \mathcal{H}(U^\varepsilon|U)\,dxds + \frac{1}{4}\int_0^t\int_{\mathbb{R}^d} \nu(n^\varepsilon)|\mathbb{D}(v-v^\varepsilon)|^2\,dxds \\
&\quad + \frac{1}{2}\int_0^t\int_{\mathbb{R}^d} \rho^\varepsilon|(u-u^\varepsilon)-(v-v^\varepsilon)|^2\,dxds,
\end{aligned}$$

where $C = C(\gamma, n_*, \nu_*, \|\rho\|_{L^\infty}, \|u-v\|_{L^\infty(0,T;L^d\cap L^\infty)})$ is a positive constant.

◇ (Estimates for K_6): By using almost the same argument as in the estimate of K_5^2, we have

$$\begin{aligned}
&2\left|\int_{\mathbb{R}^d} \left(\frac{n^\varepsilon - n}{n}\right)(\nabla\cdot(\nu(n)\mathbb{D}v))\cdot(v-v^\varepsilon)\,dx\right| \\
&\quad \le \frac{2}{n_*}\int_{\mathbb{R}^d} |n^\varepsilon - n||v-v^\varepsilon||\nabla\cdot(\nu(n)\mathbb{D}v)|\,dx \\
&\quad \le C\left(\int_{\mathbb{R}^d} \min\left\{(n^\varepsilon)^{\gamma-2}, n^{\gamma-2}\right\}(n-n^\varepsilon)^2\,dx\right)^{1/2} \\
&\qquad\quad \times \left(\int_{\mathbb{R}^d} (n^{2-\gamma} + (n^\varepsilon)^{2-\gamma})|v-v^\varepsilon|^2|\nabla\cdot(\nu(n)\mathbb{D}v)|^2\,dx\right)^{1/2} \\
&\quad \le C\int_{\mathbb{R}^d} \mathcal{H}(U^\varepsilon|U)\,dx + \frac{1}{8}\int_{\mathbb{R}^d} \nu(n^\varepsilon)|\mathbb{D}(v-v^\varepsilon)|^2\,dx.
\end{aligned}$$

This asserts

$$K_6 \le C\int_0^t\int_{\mathbb{R}^d} \mathcal{H}(U^\varepsilon|U)\,dxds + \frac{1}{8}\int_0^t\int_{\mathbb{R}^d} \nu(n^\varepsilon)|\mathbb{D}(v-v^\varepsilon)|^2\,dxds,$$

where $C = C(\gamma, n_*, \nu_*, \|\rho\|_{L^\infty}, \|\nabla\cdot(\nu(n)\mathbb{D}v)\|_{L^\infty(0,T;L^d\cap L^\infty)})$ is a positive constant. We notice that

$$\begin{aligned}\|\nabla\cdot(\nu(n)\mathbb{D}v)\|_{L^d\cap L^\infty} &\le C\|\nabla n\|_{L^\infty}\|\mathbb{D}v\|_{L^d\cap L^\infty} + C\|\nabla\mathbb{D}v\|_{L^d\cap L^\infty}\\ &\le C\|\nabla n\|_{L^\infty}\|\mathbb{D}v\|_{H^{s-2}} + C\|\nabla\mathbb{D}v\|_{H^{s-2}}\\ &\le C(1+\|\nabla n\|_{H^{s-1}})\|\nabla v\|_{H^{s-1}},\end{aligned}$$

due to $\nu\in\mathcal{C}^1(\mathbb{R}_+)$, $d>2$, and $s-2>d/2$.

$\diamond$ (Estimates for K_7): Using the integration by parts and symmetry of $\mathbb{D}v$, we find

$$2\int_{\mathbb{R}^d}(\nabla\cdot((\nu(n)-\nu(n^\varepsilon))\,\mathbb{D}v))\cdot(v-v^\varepsilon)\,dx = -\int_{\mathbb{R}^d}(\nu(n)-\nu(n^\varepsilon))\,\mathbb{D}v:\mathbb{D}(v-v^\varepsilon)\,dx.$$

Then we estimate

$$\begin{aligned}&\left|\int_{\mathbb{R}^d}(\nu(n)-\nu(n^\varepsilon))\,\mathbb{D}v:\mathbb{D}(v-v^\varepsilon)\,dx\right|\\ &\quad\le \nu_{\mathrm{Lip}}\|\mathbb{D}v\|_{L^\infty}\int_{\mathbb{R}^d}|\mathbb{D}(v-v^\varepsilon)||n-n^\varepsilon|\,dx\\ &\quad\le \nu_{\mathrm{Lip}}\|\mathbb{D}v\|_{L^\infty}\left(\int_{\mathbb{R}^d}\min\left\{(n^\varepsilon)^{\gamma-2},n^{\gamma-2}\right\}(n-n^\varepsilon)^2\,dx\right)^{1/2}\\ &\qquad\times\left(\int_{\mathbb{R}^d}\left(n^{2-\gamma}+(n^\varepsilon)^{2-\gamma}\right)|\mathbb{D}(v-v^\varepsilon)|^2\,dx\right)^{1/2}\\ &\quad\le C\nu_{\mathrm{Lip}}\|\mathbb{D}v\|_{L^\infty}\left(\int_{\mathbb{R}^d}\mathcal{H}(U^\varepsilon|U)\,dx\right)^{1/2}\left(\int_{\mathbb{R}^d}(n^{2-\gamma}+(n^\varepsilon)^{2-\gamma})|\mathbb{D}(v-v^\varepsilon)|^2\,dx\right)^{1/2}.\end{aligned}$$

On the other hand, by using the assumption on ν (5), we get $(n^\varepsilon)^{2-\gamma}\le c_0\nu(n^\varepsilon)$, and this gives

$$\begin{aligned}&\int_{\mathbb{R}^d}(n^{2-\gamma}+(n^\varepsilon)^{2-\gamma})|\mathbb{D}(v-v^\varepsilon)|^2\,dx\\ &\quad\le\frac{\|n\|_{L^\infty}^{2-\gamma}}{\nu_*}\int_{\mathbb{R}^d}\nu(n^\varepsilon)|\mathbb{D}(v-v^\varepsilon)|^2\,dx + c_0\int_{\mathbb{R}^d}\nu(n^\varepsilon)|\mathbb{D}(v-v^\varepsilon)|^2\,dx.\end{aligned}$$

This together with using Young's inequality provides

$$K_7\le\frac{1}{8}\int_{\mathbb{R}^d}\nu(n^\varepsilon)|\mathbb{D}(v-v^\varepsilon)|^2\,dx + C\int_{\mathbb{R}^d}\mathcal{H}(U^\varepsilon|U)\,dx,$$

where $C = C(\nu_{\mathrm{Lip}},\nu_*,c_0,\|\mathbb{D}v\|_{L^\infty},\|n\|_{L^\infty},\gamma)$ is a positive constant independent of ε.

By combining all of the above estimates, we have

$$\begin{aligned}&\int_{\mathbb{R}^d}\mathcal{H}(U^\varepsilon|U)\,dx + \frac{1}{2}\int_0^t\int_{\mathbb{R}^d}\nu(n^\varepsilon)|\mathbb{D}(v-v^\varepsilon)|^2\,dxds\\ &\quad+\frac{1}{2}\int_0^t\int_{\mathbb{R}^d}\rho^\varepsilon|(u^\varepsilon-v^\varepsilon)-(u-v)|^2\,dxds\le C\left(\int_0^t\int_{\mathbb{R}^d}\mathcal{H}(U^\varepsilon|U)\,dxds+\sqrt{\varepsilon}\right),\end{aligned}$$

where C is a positive constant depeding on $\nu_{\mathrm{Lip}}, c_0, \gamma, n_*, \nu_*, \|\rho\|_{L^\infty}, \|u-v\|_{L^d\cap L^\infty}$, $\|n\|_{L^\infty}, \|\mathbb{D}v\|_{L^\infty}$, $\|\nabla\cdot(\nu(n)\mathbb{D}v)\|_{L^d\cap L^\infty}$ and $\|\nabla u\|_{L^\infty}$. Finally, we apply Grönwall's inequality to the above to conclude the desired result.

We next provide the strong convergence appeared in Theorem 1.3 by using the relative entropy inequality (6). Since the convergence of ρ^ε, $\rho^\varepsilon u^\varepsilon$, and $\rho^\varepsilon|u^\varepsilon|^2$ can be obtained by the same argument as in [23], we only show the strong convergence of n^ε, $n^\varepsilon v^\varepsilon$ and $n^\varepsilon|v^\varepsilon|^2$ below.

$\diamond$ (Convergence of n^ε to n): Before proceding, we claim that the following inequality holds: if $x, y > 0$ and $0 < y_{min} \le y \le y_{max} < \infty$, then

$$\tilde{P}(x|y) = K(x) - K(y) - K'(y)(x-y) \ge \begin{cases} \gamma(2y_{max})^{\gamma-2}|x-y|^2 & \text{if } y/2 \le x \le 2y, \\ \dfrac{\gamma y_{min}^\gamma}{4(1+y_{min}^\gamma)}(1+x^\gamma) & \text{otherwise.} \end{cases} \tag{11}$$

If $y/2 \le x \le 2y$, we easily find

$$\begin{aligned} K(x) - K(y) - K'(y)(x-y) &\ge \gamma \min\{x^{\gamma-2}, y^{\gamma-2}\}|x-y|^2 \\ &\ge \gamma(2y)^{\gamma-2}|x-y|^2 \ge \gamma(2y_{max})^{\gamma-2}|x-y|^2. \end{aligned}$$

If $x > 2y > y$ $(> y_{min})$, i.e., $y/x < 1/2$, we get

$$\begin{aligned} &K(x) - K(y) - K'(y)(x-y) \\ &\ge \gamma \min\{x^{\gamma-2}, y^{\gamma-2}\}|x-y|^2 = \gamma x^{\gamma-2}|x-y|^2 = \gamma x^\gamma \left|1 - \frac{y}{x}\right|^2 \\ &\ge \frac{\gamma x^\gamma}{4} = \frac{\gamma}{4}(1+x^\gamma)\left(1 - \frac{1}{1+x^\gamma}\right) \ge \frac{\gamma}{4}(1+x^\gamma)\left(1 - \frac{1}{1+y_{min}^\gamma}\right). \end{aligned}$$

On the other hand, if $x < y/2$, i.e., $x/y < 1/2$, we obtain

$$\begin{aligned} K(x) - K(y) - K'(y)(x-y) &\ge \gamma y^{\gamma-2}|x-y|^2 = \gamma y^\gamma \left|1 - \frac{x}{y}\right|^2 \\ &\ge \frac{\gamma y^\gamma}{4} = \frac{\gamma}{4}(1+y^\gamma)\left(1 - \frac{1}{1+y^\gamma}\right) \\ &\ge \frac{\gamma}{4}(1+x^\gamma)\left(1 - \frac{1}{1+y_{min}^\gamma}\right). \end{aligned}$$

Thus we have the inequality (11). We now use that inequality (11) to show the convergence of n^ε to n. For $\Omega \subset \mathbb{R}^d$ with $|\Omega| < \infty$, we estimate

$$\begin{aligned} \int_\Omega |n^\varepsilon - n|^\gamma\,dx &= \int_{\Omega \cap \{n/2 \le n^\varepsilon \le 2n\}} |n^\varepsilon - n|^\gamma\,dx + \int_{\Omega \cap \{n/2 \le n^\varepsilon \le 2n\}^c} |n^\varepsilon - n|^\gamma\,dx \\ &=: L_1^\varepsilon + L_2^\varepsilon. \end{aligned}$$

For L_1^ε, we find

$$\begin{aligned} L_1^\varepsilon &\le \left(\int_{\Omega \cap \{n/2 \le n^\varepsilon \le 2n\}} \min\{(n^\varepsilon)^{\gamma-2}, n^{\gamma-2}\}|n^\varepsilon - n|^2\,dx\right)^{\frac{\gamma}{2}} \\ &\quad \times \left(\int_{\Omega \cap \{n/2 \le n^\varepsilon \le 2n\}} \max\{(n^\varepsilon)^\gamma, n^\gamma\}\,dx\right)^{\frac{2-\gamma}{2}} \\ &\le C\left(\int_{\Omega \cap \{n/2 \le n^\varepsilon \le 2n\}} \mathcal{H}(U^\varepsilon|U)\,dx\right)^{\frac{\gamma}{2}} \left(\left(2\|n\|_{L^\infty}\right)^\gamma |\Omega|\right)^{\frac{2-\gamma}{2}} \longrightarrow 0, \end{aligned}$$

as $\varepsilon \to 0$, where $C = C(\gamma)$ is a positive constant independent of ε. For L_2^ε, we use (11) to get

$$
\begin{aligned}
L_2^\varepsilon &\le \int_{\Omega\cap\{n/2\le n^\varepsilon\le 2n\}^c} \|n\|_{L^\infty}^\gamma \left|\frac{n^\varepsilon}{n}+1\right|^\gamma dx\\
&\le \int_{\Omega\cap\{n/2\le n^\varepsilon\le 2n\}^c} (2\|n\|_{L^\infty})^\gamma \left(\left(\frac{n^\varepsilon}{n}\right)^\gamma+1\right) dx\\
&\le \int_{\Omega\cap\{n/2\le n^\varepsilon\le 2n\}^c} (2\|n\|_{L^\infty})^\gamma \left(\left(\frac{n^\varepsilon}{n_*}\right)^\gamma+1\right) dx\\
&\le C\int_{\Omega\cap\{n/2\le n^\varepsilon\le 2n\}^c} (1+(n^\varepsilon)^\gamma)\, dx\\
&\le C\int_{\Omega\cap\{n/2\le n^\varepsilon\le 2n\}^c} H(U^\varepsilon|U)\, dx \longrightarrow 0,
\end{aligned}
$$

as $\varepsilon \to 0$, where $C = C(\|n\|_{L^\infty}, n_*, \gamma)$ is a positive constant independent of ε. Thus we have the convergence $n^\varepsilon \to n$ in $L^1_{loc}(0,T;L^\gamma_{loc}(\mathbb{R}^d))$, and this together with the integrability condition yields that it also holds in $L^1_{loc}(0,T;L^p_{loc}(\mathbb{R}^d))$ with $p\in[1,\gamma]$.

$\diamond$ (Convergence of $n^\varepsilon v^\varepsilon$ to nv): For $\Omega\subseteq\mathbb{R}^d$ with $|\Omega|<\infty$, similarly as before, we estimate

$$
\int_\Omega |n^\varepsilon v^\varepsilon - nv|\,dx \le \int_\Omega (n^\varepsilon|v^\varepsilon - v| + |n^\varepsilon - n||v|)\,dx =: L_3^\varepsilon + L_4^\varepsilon,
$$

where L_3^ε can be bounded by

$$
L_3^\varepsilon \le \left(\int_\Omega n^\varepsilon|v^\varepsilon - v|^2\,dx\right)^{1/2}\left(\int_\Omega n^\varepsilon\,dx\right)^{1/2}.
$$

Note that n^ε is locally integrable in $\mathbb{R}^d$, and furthermore, we find

$$
\begin{aligned}
\int_\Omega n^\varepsilon\,dx &= \int_{\Omega\cap\{n/2\le n^\varepsilon\le 2n\}} n^\varepsilon\,dx + \int_{\Omega\cap\{n/2\le n^\varepsilon\le 2n\}^c} n^\varepsilon\,dx\\
&\le 2\|n\|_{L^\infty}|\Omega| + |\Omega|^{\frac{\gamma-1}{\gamma}}\left(\int_{\Omega\cap\{n/2\le n^\varepsilon\le 2n\}^c}(n^\varepsilon)^\gamma\,dx\right)^{\frac{1}{\gamma}}\\
&\le 2\|n\|_{L^\infty}|\Omega| + |\Omega|^{\frac{\gamma-1}{\gamma}}\left(\int_{\Omega\cap\{n/2\le n^\varepsilon\le 2n\}^c}\mathcal{H}(U^\varepsilon|U)\,dx\right)^{\frac{1}{\gamma}}.
\end{aligned}
$$

This gives $L_3^\varepsilon \to 0$ as $\varepsilon\to 0$. For the estimate of L_4^ε, we obtain

$$
L_4^\varepsilon \le \|v\|_{L^\infty}|\Omega|^{\frac{\gamma-1}{\gamma}}\left(\int_\Omega |n^\varepsilon - n|^\gamma\,dx\right)^{1/\gamma} \longrightarrow 0,
$$

as $\varepsilon\to 0$. This gives the desired result for the convergence of $n^\varepsilon v^\varepsilon$.

$\diamond$ (Convergence of $n^\varepsilon|v^\varepsilon|^2$ to $n|v|^2$): Note that the following identity holds:

$$
n^\varepsilon|v^\varepsilon|^2 - n|v|^2 = n^\varepsilon|v^\varepsilon - v|^2 + 2v\cdot(n^\varepsilon v^\varepsilon - nv) + |v|^2(n - n^\varepsilon).
$$

This relation together with the previous convergence results yields the desired strong convergence of $n^\varepsilon|v^\varepsilon|^2$. This completes the proof.

Acknowledgments. The work of Y.-P. Choi is supported by National Research Foundation of Korea(NRF) grant funded by the Korea government(MSIP) (No. 2017R1C1B2012918 and 2017R1A4A1014735) and POSCO Science Fellowship of POSCO TJ Park Foundation. The work of J. Jung is supported by the German Research Foundation (DFG) under the project number IRTG2235.

REFERENCES

[1] H.-O. Bae, Y.-P. Choi, S.-Y. Ha, and M.-J. Kang, Global existence of strong solution for the Cucker-Smale-Navier-Stokes system, *J. Differential Equations*, **257** (2014), 2225-2255.

[2] H.-O. Bae, Y.-P. Choi, S.-Y. Ha, and M.-J. Kang, Asymptotic flocking dynamics of Cucker-Smale particles immersed in compressible fluids, *Discrete Contin. Dyn. Syst., Ser. A*, **34** (2014), 4419-4458.

[3] L. Boudin, L. Desvillettes, C. Grandmont, and A. Moussa, Global existence of solution for the coupled Vlasov and Naiver-Stokes equations, *Differ. Integral Equ.*, **22** (2009), 1247-1271.

[4] D. Bresch, P. Noble, and J.-P. Vila, Relative entropy for compressible Navier-Stokes equations with density-dependent viscosities and applications, *C. R. Acad. Sci. Paris, Ser. I*, **354** (2016), 45-49.

[5] D. Bresch, P. Noble, and J.-P. Vila, Relative entropy for compressible Navier-Stokes equations with density dependent viscosities and various applications, *ESAIM: Proc.*, **58** (2017), 40-57.

[6] J. A. Carrillo, Y.-P. Choi, and T. K. Karper, On the analysis of a coupled kinetic-fluid model with local alignment forces, *Ann. I. H. Poincaré - AN.*, **33** (2016), 273-307.

[7] J.A. Carrillo, R. Duan, and A. Moussa, Global classical solutions close to the equilibrium to the Vlasov-Fokker-Planck-Euler system, *Kinet. Relat. Models*, **4** (2011), 227-258.

[8] J.A. Carrillo and T. Goudon, Stability and asymptotic analysis of a fluid-particle interaction model, *Commun. Partial Differ. Equ.*, **31** (2006), 1349-1379.

[9] Y.-P. Choi, Compressible Euler equations interacting with incompressible flow, *Kinet. Relat. Models*, **8** (2015), 335-358.

[10] Y.-P. Choi, Global classical solutions and large-time behavior of the two-phase fluid model, *SIAM J. Math. Anal.*, **48** (2016), 3090-3122.

[11] Y.-P. Choi, Global classical solutions of the Vlasov-Fokker-Planck equation with local alignment forces, *Nonlinearity*, **29** (2016), 1887-1916.

[12] Y.-P. Choi, Large-time behavior for the Vlasov/compressible Navier-Stokes equations, *J. Math. Phys.*, **57** (2016), 071501.

[13] Y.-P. Choi, Finite-time blow-up phenomena of Vlasov/Navier-Stokes equations and related systems, *J. Math. Pures Appl.*, **108** (2017), 991-1021.

[14] Y.-P. Choi, S.-Y. Ha, J. Jung, and J. Kim, Global dynamics of the thermodynamic Cucker-Smale ensemble immersed in incompressible viscous fluid, *Nonlinearity*, **32** (2019), 1579-1640.

[15] Y.-P. Choi, S.-Y. Ha, J. Jung, and J. Kim, On the coupling of kinetic thermomechanical Cucker-Smale equation and compressible viscous fluid system, preprint.

[16] Y.-P. Choi and S.-B. Yun, Global existence of weak solutions for Navier-Stokes-BGK system, preprint.

[17] C. M. Dafermos, The second law of thermodynamics and stability, *Arch. Ration. Mech. Anal.*, **70** (1979), 167-179.

[18] E. Feireisl, B. J. Jin, and A. Novotný, Relative entropies, suitable weak solutions, and weak-strong uniqueness for the compressible Navier-Stokes system, *J. Math. Fluid Mech.*, **14** (2012), 717-730.

[19] P. Goncalves, C. Landim, and C. Toninelli, Hydrodynamic limit for a particle system with degenerate rates, *Ann. Inst. Henri Poincaré Probab. Stat.*, **45** (2009), 887-909.

[20] T. Goudon, L. He, A. Moussa, and P. Zhang, The Navier-Stokes-Vlasov-Fokker-Planck system near equilibrium, *SIAM J. Math. Anal.*, **42** (2010), 2177-2202.

[21] T. Goudon, P.-E. Jabin, and A. Vasseur, Hydrodynamic limit for the Vlasov-Navier-Stokes equations: I. Light particles regime, *Indiana Univ. Math. J.*, **53** (2004), 1495-1515.

[22] T. Goudon, P.-E. Jabin, and A. Vasseur, Hydrodynamic limit for the Vlasov-Navier-Stokes equations: II. Fine particles regime, *Indiana Univ. Math. J.*, **53** (2004), 1517-1536.

[23] T.K. Karper, A. Mellet and K. Trivisa, Hydrodynamic limit of the kinetic Cucker-Smale flocking model, *Math. Models Methods Appl. Sci.*, **25** (2014), 131-163.

[24] C. Landim, Hydrodynamic limit of interacting particle systems, in *School and Conference on Probability Theory*, ICTP Lect. Notes, vol. XVII, Abdus Salam Int. Cent. Theoret. Phys., Trieste, 2004, 57100 (electronic).
[25] A. Majda, *Compressible fluid flow and systems of conservation laws in several space variables*, Applied Mathematical Sciences, vol. 53, Springer-Verlag, New York, 1984.
[26] J. Mathiaud, Local smooth solutions of a thin spray model with collisions, *Math. Models Methods Appl. Sci.*, **20** (2010), 191-221.
[27] A. Mellet and A. Vasseur, Global weak solutions for a Vlasov-Fokker-Planck/Navier-Stokes system of equations, *Math. Models Methods Appl. Sci.*, **17** (2007), 1039-1063.
[28] A. Mellet and A. Vasseur, Asymptotic analysis for a Vlasov-Fokker-Planck/compressible Navier-Stokes equations, *Comm. Math. Phys.*, **281** (2008), 573-596.
[29] D. Wang and C. Yu, Global weak solutions to the inhomogeneous Navier-Stokes-Vlasov equations, *J. Differential Equations*, **259** (2014), 3976-4008.
[30] H. T. Yau, Relative entropy and hydrodynamics of Ginzburg-Landau models, *Lett. Math. Phys.*, **22** (1991), 6380.
[31] C. Yu, Global weak solutions to the incompressible Navier-Stokes-Vlasov equations, *J. Math. Pures Appl.*, **100** (2013), 275-293.

E-mail address: `ypchoi@yonsei.ac.kr`
E-mail address: `warp100@snu.ac.kr`

ON NON-UNIQUENESS BELOW ONSAGER'S CRITICAL EXPONENT

SARA DANERI

Gran Sasso Science Institute
L'Aquila, 67100 Italy

ERIS RUNA

Quant Institute, Deutsche Bank AG
Berlin, 10585 Germany

ABSTRACT. In these notes we review and we announce some results concerning the non-uniqueness of solutions of the Euler equations in Hölder spaces $C^{0,\beta}$ with $\beta < 1/3$, obtained also in collaboration with L. Székelyhidi. In particular, we can show the existence of dense sets of *wild initial data*, namely data for which non-uniqueness of energy dissipating solutions occur, up to Onsager's critical exponent.

1. **Introduction.** In these notes we consider the following initial value problem for the Euler equations on the three-dimensional torus $\mathbb{T}^3$

$$\begin{cases} \partial_t v + \operatorname{div}(v \otimes v) + \nabla p = 0 & \text{in } (0,T) \times \mathbb{T}^3 \\ \operatorname{div} v = 0 & \text{in } (0,T) \times \mathbb{T}^3 \\ v(\cdot, 0) = v_0 & \text{on } \mathbb{T}^3 \end{cases} \tag{1}$$

In 1, $v : [0,T) \times \mathbb{T}^3 \to \mathbb{R}^3$ is the velocity field of the fluid, $p : [0,T) \times \mathbb{T}^3 \to \mathbb{R}$ the pressure field and $v_0 : \mathbb{T}^3 \to \mathbb{R}^3$ is a given divergence free velocity field, the prescribed initial datum for the Cauchy problem.

While for initial data in $C^{1,\alpha}$ one has short time existence and uniqueness of classical solutions [16], a completely different picture appears for weak L^∞ solutions. In the seminal paper [9], De Lellis and Székelyhidi showed the existence of infinitely many bounded solutions of the Euler equations with compact support in space and time, in any dimension greater than or equal to two. A feature of these non-physical solutions is that the total kinetic energy of the fluid, namely the map

$$[0,T) \ni t \mapsto \int_{\mathbb{T}^3} |v(t,x)|^2 \, dx.$$

increases at time $t = 0$. Therefore, in [10] they considered solutions satisfying an additional *admissibility condition*, which among the possible formulations takes the form

$$\int_{\mathbb{T}^3} |v(t,x)|^2 \, dx \le \int_{\mathbb{T}^3} |v_0|^2 \, dx, \qquad \forall t \ge 0, \tag{2}$$

2000 *Mathematics Subject Classification.* Primary: 35-06, 35Q31; Secondary: 76D03.

Key words and phrases. Euler equations, initial value problem, nonuniqueness, convex integration, Onsager's conjecture.

and asked themselves if in this class one can prevent non-uniqueness. The answer turns out to be negative: there are initial data v_0 in L^∞, called by the authors *wild initial data*, which give rise to infinitely many bounded and admissible weak solutions of 1. Moreover, in [24] they were shown to be dense in the solenoidal fields in L^2. Notice that, due to the weak-strong uniqueness result by Brenier, De Lellis and Székelyhdidi [1], not any initial datum can be a wild initial datum, since whenever a classical solution exists, this is the unique solution in the class of weak admissible solutions with the same initial datum. As shown in [23], the vortex sheet is an L^∞-wild initial datum.

Therefore a natural question arises, namely whether there exists a regularity threshold above which solutions are unique, for all initial data, and below which non-uniqueness may happen.

The aim of these notes is to show, by reviewing some already published results and announcing a result obtained by the two authors in collaboration with L. Székelyhidi, that such a threshold must be bigger than Hölder continuity in space of order $\beta = 1/3$ (with Hölder constant uniformly bounded in time).

In order to state the main results obtained in answer to this question, which are reviewed or announced in these notes, we need the following definitions.

Definition 1.1. Let $\beta \in (0,1)$. We say that $v : \mathbb{T}^3 \times [0,T) \to \mathbb{R}^3$ is in $C^{0,\beta}$ if $\exists C > 0$ s.t. $\forall x, y \in \mathbb{T}^3$ and $\forall t \in [0,T)$

$$|v(t,x) - v(t,y)| \leq C|x-y|^\beta. \tag{3}$$

Definition 1.2. Let $0 < \beta \leq \beta_0 < 1$. We say that a divergence free vector field $v_0 \in C^{0,\beta_0}(\mathbb{T}^3)$ is a wild initial datum in $C^{0,\beta}$ if there exist infinitely many solutions v to 1 on $\mathbb{T}^3 \times [0,T)$ satisfying 2 and 3 for all $t \in (0,T)$.

Notice that in the last definition we allow the solutions of the Euler equation to possibly lose a bit of regularity for $t > 0$.

The results that have been obtained about the existence of wild initial data are the following.

The first existence result for wild initial data in Hölder spaces was given by the first author in [6].

Theorem 1.3. *For every $\varepsilon > 0$, there exist vector fields in $C^{0,1/10-\varepsilon}(\mathbb{T}^3)$ which are wild initial data in $C^{0,1/16-\varepsilon}$.*

Moreover, they are infinitely many.

Then, by the first author in collaboration with L. Székelyhidi in [8], the result was improved to larger Hölder exponents without having loss of regularity for positive times and showing the density in L^2 of the wild initial data.

Theorem 1.4. *Let $\theta < \frac{1}{5}$. Then, there exist vector fields $v_0 \in C^{0,\theta}(\mathbb{T}^3)$ which are wild initial data in $C^{0,\theta}$.*

Moreover, the set of such initial data is dense in $L^2(\mathbb{T}^3)$.

Finally, the two authors of these notes in collaboration with L. Székelyhidi show, in a work in preparation [7], that such an existence and density result for wild initial data can be extended up to Hölder regularity less than 1/3.

Theorem 1.5. *Let $\theta < \frac{1}{3}$. Then, there exist vector fields $v_0 \in C^{0,\theta}(\mathbb{T}^3)$ which are wild initial data in $C^{0,\theta}$.*

Moreover, the set of such initial data is dense in $L^2(\mathbb{T}^3)$.

Such a regularity class is connected to the celebrated Onsager's conjecture [19], according to which solutions of 1 in the class of $C^{0,\beta}$-functions conserve the total kinetic energy $\int_{\mathbb{T}^3} |v(t,x)|^2\, dx$ as soon as $\beta > 1/3$ and may dissipate it if $\beta < 1/3$. While the first part of the conjecture was proven in [5] after a partial result in [13], the proof of the second part of the conjecture required a much longer time, expressed through a series of results by different authors (see Section 2), starting with [9] and culminating with [15]. In particular, in the proof of Theorem 1.4 the first author in collaboration with L. Székelyhidi introduced a set of flows, called *Mikado flows*, which turned out to be fundamental for the proof of the Onsager's conjecture given in [15].

The method adopted both to show the existence of dissipative solutions to 1 (Onsager's conjecture) and to show non-uniqueness is the so-called *convex integration*. Such a technique, introduced for the first time by Nash [18] in order to prove the existence of infinitely many C^1 isometric embeddings of n-dimensioanl Riemannian manifolds in $\mathbb{R}^{n+2}$, has found application in different areas of analysis and geometry. It was applied in the context of differential inclusions for the first time by Müller and Sverak [17] and to the non-uniqueness problem for the Euler equations by De Lellis and Szekelyhidi in [9].

The plan of these notes is the following: in Section 2, after reviewing the known results about the existence of dissipative solutions to 1, we explain the general strategy and the ideas of the proof of the Onsager's conjecture, in the form presented in [3]; in Section 4 we explain the additional difficulties in the non-uniqueness problem (existence and density of wild initial data) and how they have been overcome (for solutions in $C^{0,\beta}$ with $\beta < 1/5$) in [8].

2. **Existence of dissipative solutions and Onsager's conjecture.** In his famous paper [19] in 1949, Onsager conjectured that the regularity threshold for energy conservation of solutions of the Euler equations is $C^{0,1/3}$. In particular, for $\beta > 1/3$ solutions must conserve the total kinetic energy, while for $\beta < 1/3$ they might not.

Energy conservation for $\beta > 1/3$ was proved by Constantin, E and Titi in [5], after a partial result by Eyink in [13] (see [4] for a sharper result in L^3 spaces).

Concerning the existence of non-conservative solutions of the Euler equations, in a pioneering paper Scheffer [20] constructed an example of compactly supported weak solutions in L^2 to the Euler-equations in dimension two. Later, Shnirelman [21] gave a different proof of the same result. In [22], Shnirelman proved the existence of L^2 weak solutions with energy decreasing in time. In [9] De Lellis and Szekelyhidi were the first to understand that Nash's convex integration method could be adapted to the Euler framework to prove a much stronger result, namely the existence of infinitely many non-conservative weak solutions in the space $L^\infty_t L^\infty_x$, in any space dimension. Their techniques opened the way for all the subsequent results on the negative part of the Onsager's conjecture. In [10], they were able to show that none of the available admissibility criteria, in the spirit of 2, is able to single out a unique solution for 1, for some initial data. In [11], they introduced new techniques (in particular, convex integration using perturbations of stationary solutions of the Euler equations called Beltrami flows) and they proved that, given a smooth and positive function $e : [0,T] \to \mathbb{R}$, there exist infinitely many solutions of the Euler equations with kinetic energy e which are continuous in space and time. In particular, if e is nonincreasing, such solutions must be dissipative. In [12], they show that actually the regularity of such solutions can be increased to Hölder 1/10. After that, in his PhD thesis [14], Isett introduced new ideas and managed to prove the existence of non-conservative solutions with Hölder regularity exponent up to 1/5. In [2], the result was improved to provide with $1/5 - \varepsilon$ solutions with prescribed smooth and positive kinetic energy e. Then, in [8], the first author together with Székelyhidi proved the existence of infinitely many $C^{0,1/5-\varepsilon}$ wild initial data in $C^{0,1/5-\varepsilon}$, namely Theorem 1.4, and their density in L^2 solenoidal initial data. While proving the density of wild initial data in L^2, the authors used in this process a new class of stationary solutions of the Euler equations, that were called by the authors *Mikado flows*. In [15], Isett was able to substitute the Beltrami flows with the Mikado flows in the whole convex integration scheme, thanks to a procedure that he called the *gluing*. A convex integration scheme based on Mikado flows turned out to have better error estimates, which lead to non-conservative solutions in the regularity class $C^{0,1/3-\varepsilon}$, namely the second part of Onsager's conjecture. In [3], the result was improved to get solutions in $C^{0,1/3-\varepsilon}$ with preassigned smooth and positive kinetic energy e.

3. **Some ideas of the proof of the Onsager's conjecture.** We give here a rough review of some parts of the proof of Onsager's conjecture, with the formalism used in [3]. Our aim is to give an idea of some of the basic estimates which are required in the process, giving up precision in return for heuristics and motivation.

The general strategy of a proof via *convex integration*, starting from [18], consists in the following:

- Start from a *subsolution* of the problem, namely a solution of a "relaxed" version of the original problem. The error from being an exact solution provides the room for perturbing a subsolution and approaching gradually the space of solutions;
- Add iteratively a sequence of nonlinear perturbations to the original subsolution, in such a way that after each iteration one obtains still a subsolution but with a smaller and smaller gap from being a solution;
- At each iteration check that the perturbed subsolution enjoys estimates which guarantee in the limit the convergence to a solution of the problem in the desired regularity class.

Given that a solution of the problem is a solution of the Euler equations in the regularity class $C^{0,\beta}$ for $0 < \beta < 1/3$ and with prescribed kinetic energy e, a subsolution at step $q \in \mathbb{N}$ is a triple $(v_q, p_q, R_q) : [0,T] \times \mathbb{T}^3 \to \mathbb{R}^3 \times \mathbb{R} \times \mathcal{S}^{3\times 3}_+$, where $\mathcal{S}^{3\times 3}_+$ denotes the space of symmetric positive definite matrices, which satisfies the following Euler-Reynolds system

$$\begin{cases} \partial_t v_q + \operatorname{div}(v_q \otimes v_q) + \nabla p_q = -\operatorname{div} R_q \\ \operatorname{div} v_q = 0 \end{cases} \tag{4}$$

on $[0,T] \times \mathbb{T}^3$ and

$$\int_{\mathbb{T}^3} |v_q(t,x)|^2 \, dx < e(t), \quad \forall t \in [0,T].$$

The Euler-Reynolds system appears naturally also in turbulence theory if one considers averages of the Euler flow, where the appearance of the *Reynolds stress* R_q is due to the fact that the tensor product $v_q \otimes v_q$ does not commute with averaging. We ask also that

$$R_q(t,x) = \rho(t)\mathrm{Id} + \mathring{R}_q(t,x),$$

where $\mathring{R}_q(t,x)$ is a traceless symmetric 3×3 matrix. Moreover, the subsolution has to satisfy quantitative estimates on the energy gap $e(t) - \int_{\mathbb{T}^3} |v_q(t,x)|^2\, dx$, on the C^0 and C^1 norms of v_q and on the C^0 norm of the Reynolds stress R_q. In particular, if

$$\|v_q - v_{q-1}\|_0 \le \delta_q^{1/2}, \tag{5}$$

$$\|v_q\|_1 \le \delta_q^{1/2}\lambda_q, \tag{6}$$

where $\{\delta_q\}_{q\in\mathbb{N}}$ and $\{\lambda_q\}_{q\in\mathbb{N}}$ are two sequences of parameters linked by the relation

$$\delta_q = \lambda_q^{-2\beta'}, \quad \beta' > \beta, \quad \lambda_q \to +\infty, \tag{7}$$

then by interpolation

$$\|v_q - v_{q-1}\|_\beta \le \|v_q - v_{q-1}\|_0^{(1-\beta)} \|v_q - v_{q-1}\|_1^\beta \le \delta_q^{1/2}\lambda_q^\beta,$$

which implies the convergence of v_q in $C^{0,\beta}$ as $q \to +\infty$. Therefore, 5 and 6 are required to be a subsolution. More precisely, one chooses double exponential sequences

$$\lambda_q = [a^{b^q}], \quad a \gg 1, \quad 1 < b < 1 + \varepsilon. \tag{8}$$

Moreover, so that (v_q, p_q, R_q) converges to a solution of the problem, $\|R_q\|_0$ and $e(t) - \int_{\mathbb{T}^3} |v_q(t,x)|^2\, dx$ have also to converge to 0. In order to understand how small such quantities should be, in term of the sequences $\{\delta_q\}$ and $\{\lambda_q\}$, let us set

$$v_{q+1} = v_q + w_{q+1}, \quad p_{q+1} = p_q + p$$

where w_{q+1} is a suitable divergence free perturbation and let us see how large errors $\|R_q\|_0$ and $e(t) - \int_{\mathbb{T}^3} |v_q(t,x)|^2\, dx$ can be reduced by w_{q+1}.

One has that

$$\begin{aligned}
&\partial_t v_{q+1} + \mathrm{div}\,(v_{q+1} \otimes v_{q+1}) + \nabla p_{q+1} = \\
&= -\mathrm{div}\, \mathring{R}_{q+1} \\
&= \mathrm{div}\,(w_{q+1} \otimes w_{q+1} + p\mathrm{Id} - \mathring{R}_q) && (9)\\
&+ \partial_t w_{q+1} + v_q \cdot \nabla w_{q+1} && (10)\\
&+ w_{q+1} \cdot \nabla v_q. && (11)
\end{aligned}$$

One calls 9 the *oscillation error*, 10 the *transport error* and 11 the *Nash error*. In first approximation (here we are not precise, our aim is just to heuristically and gradually motivate the choice of the perturbation as a mean to decrease the errors), the perturbation w_{q+1} is an highly oscillating flow of the form

$$w_{q+1}(t,x) = \sum_{k\in\mathbb{N}} a_k(t,x) e^{i\lambda_{q+1}k\cdot x} = W_{q+1}(t,x,\lambda_{q+1}x), \tag{12}$$

$$W_{q+1}(t,x,\xi) = \sum_{k\in\mathbb{N}} a_k(t,x) e^{ik\cdot\xi}. \tag{13}$$

The fact that an highly oscillating flow can produce a smaller new error $\mathring{R}_{q+1}$ is expressed by the following stationary phase lemma.

Lemma 3.1. *If* $\operatorname{div} w = 0$, *there exists* $\mathring{R}_{q+1} = \mathcal{R}($*"oscillation error"* $+$ $+$ *"transport error"* $+$ *"Nash error"*$)$, *with* $\mathcal{R}(f)^{ij} = \mathcal{R}^{ijk} f^k$

$$\mathcal{R}^{ijk} = -\frac{1}{2}\Delta^{-2}\partial_1\partial_j\partial_k + \frac{1}{2}\Delta^{-1}\partial_k\delta_{ij} - \Delta^{-1}\partial_i\delta_{jk} - \Delta^{-1}\partial_j\delta_{ik}.$$

Moreover, for every $m \in \mathbb{N}$, $\theta \in (0,1)$, $\exists C = C(m,\theta)$ *s.t.* $\forall F(x) = a(x)e^{i\lambda k\cdot x}$, $k \neq 0$

$$\|\mathcal{R}(F)\|_\theta \leq C\Big(\frac{\|a\|_0}{\lambda^{1-\theta}} + \frac{\|a\|_m}{\lambda^{m-\theta}} + \frac{\|a\|_{m+\theta}}{\lambda^m}\Big). \tag{14}$$

So, if the terms of order λ_q or higher in the errors 9, 10 and 11 vanish, choosing λ_q large enough the error $\|\mathring{R}_{q+1}\|_0$ will be smaller. At this aim, looking at the oscillation error it turns out that the perturbation w_{q+1} has to be built on stationary solutions of the Euler equations, namely

$$\begin{cases} \operatorname{div}_\xi W \otimes W + \nabla p = 0 \\ \operatorname{div}_\xi W = 0 \end{cases} \tag{15}$$

and moreover

$$\int_{\mathbb{T}^3} W \otimes W \, d\xi = f(t)\mathrm{Id} + \mathring{R}_q. \tag{16}$$

Since by 5 $\|w_{q+1}\|_0 = \|v_{q+1} - v_q\|_0 \sim \delta_{q+1}^{1/2}$, then by 16 one asks that a subsolution (v_q, p_q, R_q) fulfils

$$\|\mathring{R}_q\|_0 \leq \delta_{q+1}.$$

In the papers [11, 12, 14, 2] the perturbations w_{q+1} fulfilling 15 and 16 were the so-called *Beltrami flows*. In [8], in order to prove the density in L^2 of the wild initial data, a new class of flows satisfying 15 and 16 was introduced, namely the *Mikado flows*. One has the following

Lemma 3.2. *For any compact subset* $\mathcal{N} \subset\subset \mathcal{S}^{3\times 3}_+$, *there exists a smooth vector field* $W : \mathcal{N} \times \mathbb{T}^3 \to \mathbb{R}^3$ *such that,* $\forall R \in \mathcal{N}$

$$\begin{cases} \operatorname{div}_\xi W(R,\xi) \otimes W(R,\xi) = 0 \\ \operatorname{div}_\xi W(R,\xi) = 0 \end{cases} \tag{17}$$

$$\int_{\mathbb{T}^3} W(R,\xi) \otimes W(R,\xi)\, d\xi = R,$$

$$\int_{\mathbb{T}^3} W(R,\xi) = 0. \tag{18}$$

In order to explain how the vector field W is constructed, we need the following lemma.

Lemma 3.3. *For every compact subset* $\mathcal{N} \subset\subset \mathcal{S}^{3\times 3}_+$ *there exist* $\lambda_0 \geq 1$ *and smooth functions* $\Gamma_k \in C^\infty(\mathcal{N};[0,1])$ *for every* $k \in \mathbb{Z}^3$ *with* $|k| \leq \lambda_0$ *such that*

$$R = \sum_{k\in\mathbb{Z}^3, |k|\leq\lambda_0} \Gamma_k^2(R) k \otimes k, \qquad \forall R \in \mathcal{N}.$$

The choice of W in Lemma 3.2 is then of the form

$$W(R,\xi) = \sum_{k\in\mathbb{Z}^3, |k|\leq\lambda_0} \Gamma_k(R)\psi_k(\xi)k,$$

where $\psi_k(\xi) = g_k(\operatorname{dist}(\xi, \ell_{p_k,k}))$, with $g_k \in C_c^\infty((0, r_k))$, $r_k > 0$, and $\ell_{p_k,k}$ is the $\mathbb{T}^3$-periodic extension of the line $\{p_k + tk : t \in \mathbb{R}\}$ passing through the point p_k and pointing in direction k. The points p_k and the radii r_k are chosen in such a way that the supports of the functions ψ_k are disjoint as k varies. Moreover, g_k is such that $\int_{\mathbb{T}^3} \psi_k^2(\xi)\, d\xi = 1$. It is easy to see that 17 and 18 are satisfied. Moreover,

$$\int_{\mathbb{T}^3} W(R,\xi) \otimes W(R,\xi)\, d\xi = \sum_k \Gamma_k^2(R) \int_{\mathbb{T}^3} \psi_k^2(\xi)\, d\xi k \otimes k = R.$$

The presence of the transport error, where a spatial gradient of the perturbation w_{q+1} appears, suggests however that the first ansatz 12 for w_{q+1} is still rough: indeed one has to find a way to eliminate the appearance of such a term of order λ_{q+1}. The idea, introduced already in [14], is to choose a nonlinear phase governed by the flow of the underlying vector field v_q, namely a w_{q+1} of the form

$$w_{q+1}(t,x) = \sum_{j,k} \chi_j(t) e^{i\lambda_{q+1} k\cdot\phi_j(t,x)} a_{j,k}(t,x),$$

where

$$\begin{cases} \partial_t \phi_j + v_q \cdot \nabla \phi_j = 0 \\ \phi_j(t_j, x) = x, \end{cases}$$

is the flow of v_q, $t_j = j\tau_q$, $0 < \tau_q \ll 1$, χ_j are cut-off functions with support of order τ_q centered at t_j. The reason for such a choice is that in the transport error the terms of order λ_{q+1} coming from a differentiation of the phase now disappear, and the resason why the nonlinear phase starting from t_j is taken only for a time interval of order τ_q is that, provided

$$\tau_q \le \frac{1}{2} \|v_q\|_1^{-1},$$

then

$$\|\nabla \phi_j - \mathrm{Id}\|_0 \le \tau_q \|v_q\|_1 \le \frac{1}{2},$$

namely one is close to a linear flow for short times. On one hand this is an advantage and it is indeed necessary to carry on the estimates. On the other hand, the supports of χ_j and χ_{j+1} intersect and therefore one has to estimate the Reynolds stresses produced by the tensor products

$$\chi_j \chi_{j+1} W_j \otimes W_{j+1}.$$

While these can be estimated when W_j are perturbed Beltrami flows (even though leading to regularity $1/5 - \varepsilon$), if W_j and W_{j+1} are two different Mikado flows, this leads to problems in the estimation of the error. Indeed, in [8] Mikado flows where used only in the first step of the convex integration scheme, where rougher estimates were needed and in particular just one flow ϕ was used (no time dependent cut-off functions χ_j). In order to substitute Mikado flows to Beltrami flows in the whole scheme, this issue had to be overcome, and this was done by Isett in [15] (and later in [3]) introducing an intermediate step preliminary to the perturbation called "gluing".

The aim of the gluing is to produce from v_q a flow $\bar{v}_q$ whose associated Reynolds stress $\mathring{\bar{R}}_q$ has support in pairwise disjoint temporal regions of width and distance τ_q. In this way, one needs to perturb only on these regions, and the associated cut-off functions χ_j are now disjoint in time. The name gluing comes from the fact that such $\bar{v}_q$ is obtained by gluing with a partition of unity exact classical short-time solutions of the Euler equations with initial datum $v_q(t_j)$.

As for reducing the energy gap, one has that

$$\begin{aligned}\int_{\mathbb{T}^3} |v_{q+1}|^2(t,x)\,dx &= \int_{\mathbb{T}^3} |v_q(t,x)|^2\,dx + \int_{\mathbb{T}^3} |w_{q+1}(t,x)|^2\,dx \\ &\quad + 2\int_{\mathbb{T}^3} v_q \cdot w_{q+1}(t,x)\,dx \\ &\sim \int_{\mathbb{T}^3} |v_q(t,x)|^2\,dx + \int_{\mathbb{T}^3} |w_{q+1}(t,x)|^2\,dx \\ &\sim \int_{\mathbb{T}^3} |v_q(t,x)|^2\,dx + 3(2\pi)^3 f(t).\end{aligned}$$

Therefore, setting

$$f(t) = \frac{1}{3(2\pi)^3}\left(e(t) - \int_{\mathbb{T}^3} |v_q(t,x)|^2\,dx - \delta_{q+2}\right)$$

and asking that

$$e(t) - \int_{\mathbb{T}^3} |v_q(t,x)|^2\,dx \sim \delta_{q+1},$$

the new energy gap $e(t) - \int_{\mathbb{T}^3} |v_{q+1}(t,x)|^2\,dx$ is of the order δ_{q+2}, thus smaller.

The Nash error 11 tells us that the best regularity one can hope to obtain with such method is Hölder 1/3. Indeed, looking at 5, 6 and 14, one has that

$$\|\mathcal{R}(w_{q+1}\cdot\nabla v_q)\|_0 \sim \frac{\delta_{q+1}^{1/2}\delta_q^{1/2}\lambda_q}{\lambda_{q+1}}. \tag{19}$$

In order to have that 19 is smaller than δ_{q+2}, which is the expected order of magnitude for $\|\mathring{R}_{q+1}\|_0$, by the choice of parameters δ_q and λ_q made in 7 and 8 one finds

$$a^{b^q(-\beta b-\beta+1-b)} \le a^{b^q(-2\beta b^2)},$$

which implies that

$$\beta < \frac{1}{2b+1}, \frac{1}{3},$$

since $b > 1$.

4. **The initial value problem.** The aim of Theorems 1.3, 1.4 and 1.5 is twofold:

1. to show that if some initial data satisfy suitable conditions, they generate infinitely many admissible weak solutions in the appropriate regularity space;

2. to show that such wild initial data exist and are infinitely many.

Then, there is the issue of the density of wild initial data, that we do not pursue here.

The idea from [6] is to solve both points 1 and 2 with a convex integration scheme, which though in comparison with the one described in the last section has to satisfy some additional requirements. Indeed, if the perturbation (and in Theorem 1.5 the gluing) stages are applied uniformly in time, then the solutions so obtained will be infinitely many but in general different at time $t = 0$. Hence, if we want to use a convex integration scheme leading to solutions with the same initial datum, we have to start from a concept of subsolution (*adapted subsolution*) that at time 0 is already a solution with energy $e(0)$ and then apply perturbations that at time $t = 0$ must all be null, in order not to change the initial datum. This will answer point 1 above. In order to show that such adapted subsolutions exist, and in particular that wild initial data exist (the fact that their initial data are wild is given by point 1), one performs another convex integration scheme starting from a classical (*strong*) subsolution (namely, a subsolution as in Section 3) and adding perturbations which are nonzero in smaller and smaller neighbourhoods of $t = 0$. The difficulties, as we will see, arise from the fact that the estimates on the C^0 and C^1 norms are not uniform in time any more as it was the case in Section 3, due to the presence of time cut-offs.

Let us first consider the convex integration scheme leading to an adapted subsolution, namely point 2 above. We will give a rough idea of the construction present in the paper [8], leaving comments on [7] at the end of this section. One starts from (v_0, p_0, R_0) classical (or strong) subsolution, for example the identically zero subsolution. Then, at step $q+1$, one takes

$$v_{q+1} \sim (1-\varphi_{q+1})v_q + \varphi_{q+1}(v_q + w_{q+1})$$

with $\varphi \in C_c^\infty([0,T];[0,1])$ cut-off in time such that

$$\varphi_{q+1} = \begin{cases} 1 & \text{on } [0, 2^{-q}T] \\ 0 & \text{on } [2^{-(q-1)}T, T] \end{cases}$$

Since $\text{supp}\varphi_{q+2} \subset \{\varphi_{q+1} = 1\}$, in the next step the perturbation will be supported in a region where one has uniform estimates of the 1/5-scheme (1/3 in [7]) of the previous Section.

One has to show that on the remaining regions, namely where $\varphi_{q+2} = 0$, one still has a quantitative control on the decay/growth of the C^0/C^1 norms which will now depend on t. Indeed, in the estimates for the Reynolds stress generated by the transport error, also a derivative of the cut-off function φ_q will appear. By performing careful estimates, one can show convergence of (v_q, p_q, R_q) to an *adapted subsolution*, which is defined by the following properties: defining

$$\rho(t) := e(t) - \int_{\mathbb{T}^3} |v(t,x)|^2 \, dx,$$

an adapted subsolution is a triple $(v, p, \mathring{R}) \in C^\infty((0,T]) \cap C^0([0,T])$ solving 4 on $(0,T]$ with

$$\int_{\mathbb{T}^3} |v(0,x)|^2 \, dx = e(0), \qquad \mathring{R}(0,x) \equiv 0$$

and satisfying among others the following (non-uniform) estimates

$$\|\mathring{R}\|_0 \leq \sigma\rho \tag{20}$$

$$\|v\|_1 \leq \rho^{-2-\alpha} \tag{21}$$

$$|\partial_t \rho| \leq \rho^{-1-\alpha}. \tag{22}$$

for some sufficiently small and positive σ and $0 < \alpha \ll 1$. Of course, in order to have convergence of the flow in Hölder spaces with exponent $1/3 - \varepsilon$ instead of $1/5 - \varepsilon$ one needs better exponents in 20-22. Notice that, since at time $t = 0$ the vector field v will be only in $C^{0,1/5-\varepsilon}$, the corresponding C^1 norms blow up at $t = 0$, where $\rho = 0$. Moreover, the energy gap ρ is the quantity dictating the bounds on the decay/growth of norms.

In order now to perform a convex integration scheme to solve point (1) at the beginning of this section, one has to start from an adapted subsolution, which carries naturally estimates which are non-uniform in time. Now the cut-off functions have to be chosen null at $t = 0$, so that the obtained solutions are all equal to $v(0, \cdot)$ at time 0. In [6], the cut-off functions where chosen of the form $1 - \varphi_q$, where φ_q are the cut-off functions used in the convex integration scheme for point (2). However, the choice of cut-off functions supported on dyadic intervals leads to quantitative estimates which are worse (therefore a loss in the exponent from 1/10 to 1/16) since at step q on the support of $1 - \varphi_q$ the best available uniform estimates of the adapted subsolution are the uniform estimates of point (2) for step $q-1$, instead of step q. The idea of [8] is instead to localize the perturbations using cut-off functions which are adapted to regions where the energy gap is bounded from below, which implies by 20-22 that one has uniform bounds on the decay/growth of the C^0/C^1 norms.

An additional difficulty in [7] is that a careful localized gluing step has also to be implemented.

REFERENCES

[1] Y. Brenier, C. De Lellis and L. Székelyhidi Jr. Weak-strong uniqueness for measure-valued solutions *Comm. Math. Phys.*, **305** (2011), 351–361.

[2] T. Buckmaster, C. De Lellis, P. Isett and L. Székelyhidi Jr. Anomalous dissipation for 1/5-Hölder Euler flows *Ann. of Math.*, **182** 1 (2015), 127–172.

[3] T. Buckmaster, C. De Lellis, L. Székelyhidi Jr. and V. Vicol. Onsager's conjecture for admissible weak solutions *Comm. Pure Appl. Math.*, **72** 2 (2019), 229-274.

[4] A. Cheskidov, P. Constantin, S. Friedlander and R. Shvydkoy. Energy conservation and Onsager's conjecture for the Euler equations *Nonlinearity*, **21** 6 (2008), 1233–1252.

[5] P. Constantin, W. E and E. S. Titi. Onsager's conjecture on the energy conservation for solutions of Euler's equation *Comm. Math. Phys.*, **165** 1 (1994), 207–209.

[6] S. Daneri Cauchy problem for dissipative Hölder solutions to the incompressible Euler equations *Comm. Math. Phys.* **329** 2 (2014), 745–786.

[7] S. Daneri, E. Runa and L. Székelyhidi Jr. Non-uniqueness for the Euler equations up to Onsager's critical exponent *In preparation.*

[8] S. Daneri and L. Székelyhidi Jr. Non-uniqueness and h-principle for Hölder-continuous weak solutions of the Euler equations *Arch. Rat. Mech. Anal.* **224** (2017), 471–514.

[9] C. De Lellis and L. Székelyhidi Jr. The Euler equations as a differential inclusion *Ann. Math.* **170** 3 (2009), 1417–1436.

[10] C. De Lellis and L. Székelyhidi Jr. On admissibility criteria for weak solutions of the Euler equations *Arch. Rat. Mech. Anal.* **195** 1 (2010), 225–260.

[11] C. De Lellis and L. Székelyhidi Jr. Dissipative continuous Euler flows *Invent. Math.* **193** 2 (2013), 377–407.

[12] C. De Lellis and L. Székelyhidi Jr. Dissipative Euler flows and Onsager's conjecture *J. Eur. Math. Soc.* **16** 7 (2014), 1467–1505.

[13] G. L. Eyink Energy dissipation without viscosity in ideal hydrodynamics. I. Fourier analysis and local energy transfer *Phys. D*, **78** 3-4 (1994), 222–240.

[14] P. Isett Hölder Continuous Euler Flows in Three Dimensions with Compact Support in Time *Princeton University Press* (2017).

[15] P. Isett A proof of Onsager's conjecture *Ann. Math.* **188** (2018), 871–963.

[16] L. Lichtenstein Über einige Existenzprobleme der Hydrodynamik homogener unzusammendrückbarer, reibungloser Flüssigkeiten und die Helmholtzschen Wirbelsatze *Mat. Zeit. Phys.* **23** (1925), 89–154; **26** (1927), 193–323; **32** (1930), 608.

[17] S. Müller and V. Sverak Convex integration for Lipschitz mappings and counterexamples to regularity *Ann. Math.* **157** 3 (2003), 715–742.

[18] J. Nash C^1 isometric embeddings *Ann. Math.* **60** (1954), 383–396.

[19] L. Onsager Statistical hydrodynamics *Il Nuovo Cimento* **6** (1949), 279–287.

[20] V. Scheffer An inviscid flow with compact support in space-time *J. Geom. Anal.* **3** (1993), 343–401.

[21] A. Shnirelman On the nonuniqueness of weak solution of the Euler equation *Comm. Pure Appl. Math.* **50** (1997), 1261–1286.

[22] A. Shnirelman Weak solutions with decreasing energy of incompressible Euler equations *Comm. Math. Phys.* **210** (2000), 541–603.

[23] L. Székelyhidi Jr. Weak solutions to the incompressible Euler equations with vortex sheet initial data *Comptes Rendus Mathematique* **349** 19-20 (2011), 1063–1066.
[24] L. Székelyhidi Jr. and E. Wiedemann Young measures generated by ideal incompressible fluid flows *Arch. Rat. Mech. Anal.* **206** 1 (2012), 333–366.

E-mail address: sara.daneri@gssi.it
E-mail address: eris.runa@gmail.com

MODELLING, NUMERICAL METHOD AND ANALYSIS OF THE COLLAPSE OF CYLINDRICAL SUBMARINES GRANULAR MASS

E.D. FERNÁNDEZ-NIETO
Departamento de Matemática Aplicada I, Universidad de Sevilla. E.T.S. Arquitectura, Avda, Reina Mercedes, s/n. 41012 Sevilla, Spain.

MANUEL J. CASTRO
Departamento de Análisis Matemático, Universidad de Málaga, F. Matemáticas, Campus Teatinos S/N, Spain.

ANNE MANGENEY
Equipe de Sismologie, IPGP, 4, pl. Jussieu 75232, Paris cedex 05, France.

ABSTRACT. In this work we focus on the numerical study of shallow submarine avalanches. Submarine avalanches could be modeled by a two-layer shallow-water Savage-Hutter type model (see [9]). The system is discretized by a finite volume solver named as IFCPH, that results form a combination of IFCP solver (see [11]) and the standard HLL solver (see [13]). Concerning the applications, we focus on the collapse of an initially cylindrical submarine granular mass along an horizontal plane. It is well stablished by laboratory studies ([14]) and the dimensional analysis of the Savage-Hutter model and numerical simulations ([16]), that the final profile of the landslide depends on the aspect ratio $a = H_i/R_i$, where H_i and R_i are the initial height and radius, respectively, and the effective friction angle. In this work, a similar behavior, for the two-layer model, and the final profile of the landslide only depends on the two aspect ratios: $a_H = H_1/H_2$ and $a_2 = H_2/R$, with R and H_2 the initial radius and height of the sediment column, respectively, and H_1 the initial height of the water above the sediment column. The sensitivity of the granular dynamics and of the associated water perturbation to these two aspect ratios is investigated.

1. **Introduction.** Submarine avalanches may occur when a sediment layer lying on the ocean bottom suddenly becomes unstable. These avalanches may generate tsunami waves that carry the signature of their characteristics and dynamics. These processes are however difficult to simulate because of the complex interaction between the granular and the fluid phases [2] and because of the accurate derivation of the shallow approximation for both the sediment and fluid layers. In [9] a two-layer Shallow Water Equation (SWE) system has been proposed to simulate submarine avalanches and the potentially generated tsunami waves. The first layer corresponds to the fluid and the second one to the sediment layer.

2000 *Mathematics Subject Classification.* Primary: 35L45, 35L65, 65Z05; Secondary: 65M99.
Key words and phrases. Submarine Avalanches, Bilayer Shallow Water, Well-Balanced, Finite Volume Method.
This research has been partially supported by the Spanish Government under grants MTM 2015-70490-C2-1-R and MTM 2015-70490-C2-2-R with the participation of FEDER, by the ANR contract ANR-11-BS01-0016 LANDQUAKES, the USPC PEGES project and the ERC contract ERC-CG-2013-PE10-617472 SLIDEQUAKES.
* Corresponding author: Enrique D. Fernández-Nieto.

For the sediment layer, a Savage-Hutter type model is considered. The pioneering work of Savage-Hutter [24] derives a model to describe granular flows over a sloping plane based on a Coulomb friction law that describes the avalanche/bottom interaction.

One of the characteristics of the model proposed in [9] is that the definition of the Coulomb friction term takes into account bouyancy effects involved in submarine avalanches. Another characteristics is that, depending on the ratio between the water density and the sediment density, the motion of the sediment avalanche can be more or less influenced by the presence of the fluid.

In this work we present a two-dimensional two-layer model that is a generalization of the 1D model presented in [9] in cartesian coordinates. One of the question arising in the deduction of the model is the choice of the coordinate system in which the model is deduced. Let us remember that the Saint-Venant equations are set up in cartesian coordinates, but it is valid only for almost flat topography, thus not relevant for debris avalanches in particular. On the other hand, the Savage-Hutter model uses the curvilinear coordinate along a sloping plane. New Savage-Hutter models over a general bottom have been proposed by Bouchut et al. in [1], taking into account the curvature of the topography. In [3], Bouchut and Westdickenberg generalize the previous models for small or for general slope variation in two dimensions. The 1D model introduced in [9] for submarine avalanches has been also deduced on local coordinate along the topography, by taking into account the curvature of the bottom. Here, we only focus on the spreading of an initially cylindrical submarine granular mass on a flat bottom, therefore, cartesian coordinates could be used. The resulting model has non-conservative terms, that come from the pressure terms, and can be written under the general formulation

$$\frac{\partial W}{\partial t} + \frac{\partial F_1}{\partial x_1}(W) + \frac{\partial F_2}{\partial x_2}(W) + B_1(W)\frac{\partial W}{\partial x_1} + B_2(W)\frac{\partial W}{\partial x_2} = S(W), \qquad (1)$$

where the unknown $W(\mathbf{x}, t)$ is defined in the domain $D \times (0, T)$, where D is a subset of $\mathbb{R}^2$, with values in an open subset Ω of $\mathbb{R}^N$; F_i, $i = 1, 2$ are regular functions from Ω to $\mathbb{R}^N$; B_i, $i = 1, 2$ are regular function matrices from Ω to $\mathcal{M}_{N\times N}(\mathbb{R})$ and S, is defined from Ω to $\mathbb{R}^N$.

Finite volume path-conservative schemes ([19]) are well-adapted to approximate non-conservative hyperbolic system (1). Here, we propose to combine two particular path-conservative schemes: the IFCP solver (Intermediate Field Capturing Parabola method, see [11]), that is very well-adapted to approximate two-layer shallow-water type systems, with the robust extension of HLL solver to the non-conservative framework (see [6]). IFCP solver provides accurate results, similar to the standard path-conservative Roe scheme ([20]), being IFCP more efficient, from the computational point of view, but may present disturbances in wet/dry fronts, while HLL solver is more robust in such situations. Therefore, the main objective is to naturally combine both solvers, and this can be easily done in the framework of PVM schemes. Both solvers, IFCP and HLL could be re-written as PVM methods with a similar structure, that allows to combine them in a very natural way, obtaining the IFCPH solver.

As the two-dimensional landslide model is rotationally invariant, IFCPH solver could be extended as the HLL solver to deal with the contact discontinuities associated to the tangential velocities (see [26])

This work is organized as follows. In Section 2 we present a 2D extension of the model proposed in [9] for submarine avalanches. Section 3 is devoted to the presentation of the IFCPH finite volume solver. Finally, in Section 4, we analyze the dependency of the model on the parameters involved both in terms of avalanche dynamics and water wave generation.

2. **2D two-layer Submarine landslide model.** In [9] a two-layer 1D model is presented to study submarine avalanches. The first layer corresponds to the water and it is modeled by the standard shallow-water system and the submerged sediment layer is modeled by a Savage-Hutter type system (see [24]).

Savage-Hutter model is characterized by the presence of a Coulomb friction term. This term opposes the avalanche motion and depends on the pressure at the bottom and on a friction coefficient. When the driving forces are higher than a threshold, the avalanche is moving and Coulomb friction applies to the flow [15]. When the driving forces are smaller, the material stops. The Coulomb friction term in the model proposed by [9] also includes the bouyancy effect.

In this section we present a 2D simplified extension of the model proposed in [9] with flat bottom topography. With subindex 1 we denote the unknowns corresponding to the fluid layer: h_1 is the height of the fluid layer and $\vec{q}_1 = (q_{11}, q_{12}) = (h_1 U_1, h_1 V_1)$ is the fluid flux, with $\vec{u}_1 = (U_1, V_1)$ the fluid velocity vector. Index 2 corresponds to the sediment layer: by h_2 we denote the height of the sediment layer and $\vec{q}_2 = (q_{21}, q_{22}) = (h_2 U_2, h_2 V_2)$ is the flux of the granular material, with $\vec{u}_2 = (U_2, V_2)$, the granular velocity vector

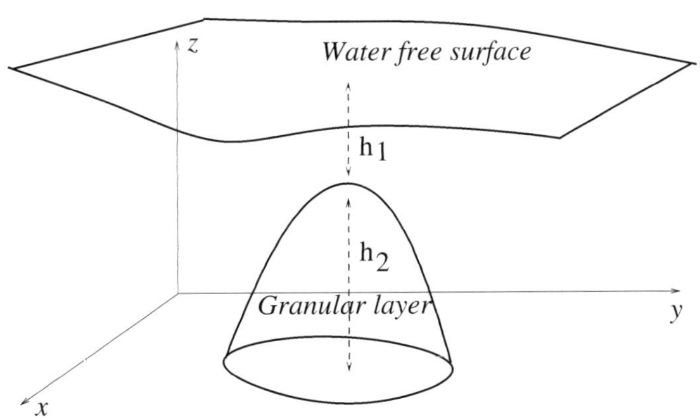

FIGURE 1. Notation: 2D submarine avalanche on a flat bottom

$$\left\{\begin{array}{l}\partial_t(h_1)+\partial_x(h_1U_1)+\partial_y(h_1V_1)=0,\\[2mm] \partial_t(h_1U_1)+\partial_x(h_1U_1^2)+\partial_y(h_1U_1V_1)+gh_1\partial_x(h_1+h_2)=0\\[2mm] \partial_t(h_1V_1)+\partial_x(h_1U_1V_1)+\partial_y(h_1V_1^2)+gh_1\partial_y(h_1+h_2)=0\\[2mm] \partial_t(h_2)+\partial_x(h_2U_2)+\partial_y(h_2V_2)=0,\\[2mm] \partial_t(h_2U_2)+\partial_x(h_2U_2^2)+\partial_y(h_2U_2V_2)+gh_2\partial_x(rh_1+h_2)=\mathcal{T}_x\\[2mm] \partial_t(h_2V_2)+\partial_x(h_2U_2V_2)+\partial_y(h_2V_2^2)+gh_2\partial_y(rh_1+h_2)=\mathcal{T}_y\end{array}\right. \qquad (2)$$

where g is the gravity acceleration, $r=\rho_f/\rho_s$ is the ratio between the fluid density, that it is supposed to be $\rho_f=1000$ kg.m^{-3} and the density of the granular material, ρ_s. Typical values of ρ_s are between 1200 to 2500 kg.m^{-3} depending on the solid volume fraction and on the material involved. Note that we consider here quite dense granular material consistent with our model. As a result, the density ratio is $0.4<r<0.8$. $\mathcal{T}=(\mathcal{T}_x,\mathcal{T}_y)$ denotes the Coulomb friction term:

$$\mathcal{T}=-\frac{g(1-r)h_2\mu}{\sqrt{U_2^2+V_2^2}}\begin{pmatrix}U_2\\ V_2\end{pmatrix}.$$

Note that this term is multi-valuated when $|\vec{u}_2|=0$.

The simplest friction law corresponds to a constant friction coefficient:

$$\mu=\tan(\delta), \qquad (3)$$

where δ is the friction angle, although more complex friction terms have been used to simulate natural subaerial or submarine landslides (see [17], [22]). For example, in order to incorporate turbulence effects, McDougall and Hungr [18] proposed to add a turbulent friction term proportional to $(U_2^2+V_2^2)$. Other definitions, deduced from experimental data, have been proposed by Pouliquen (see [23]) where the friction coefficient depends on the velocity and thickness of the granular layer. This law is widely used in the literature but involves at least three parameters that are difficult to calibrate for natural landslides (see e. g. [4]).

Model (2) can be written in the same form as (1), by setting:

$$W=\begin{pmatrix}h_1\\ h_1U_1\\ h_1V_1\\ h_2\\ h_2U_2\\ h_2V_2\end{pmatrix},\quad F_1(W)=\begin{pmatrix}h_1U_1\\ h_1U_1^2\\ h_1U_1V_1\\ h_2U_2\\ h_2U_2^2\\ h_2U_2V_2\end{pmatrix},\quad F_2(W)=\begin{pmatrix}h_1V_1\\ h_1U_1V_1\\ h_1V_1^2\\ h_2V_2\\ h_2U_2V_2\\ h_2V_2^2\end{pmatrix},$$

$$B_1(W)=\begin{pmatrix}0&0&0&0&0&0\\ gh_1&0&0&gh_1&0&0\\ 0&0&0&0&0&0\\ 0&0&0&0&0&0\\ rgh_2&0&0&gh_2&0&0\\ 0&0&0&0&0&0\end{pmatrix},$$

$$B_2(W) = \begin{pmatrix} 0 & 0 & 0 & 0 & 0 & 0 \\ 0 & 0 & 0 & 0 & 0 & 0 \\ gh_1 & 0 & 0 & gh_1 & 0 & 0 \\ 0 & 0 & 0 & 0 & 0 & 0 \\ 0 & 0 & 0 & 0 & 0 & 0 \\ rgh_2 & 0 & 0 & gh_2 & 0 & 0 \end{pmatrix},$$

$S(W) = S_{1,\mathcal{T}}(W) + S_{2,\mathcal{T}}(W)$ is defined by the Coulomb Friction terms,

$$S_{1,\mathcal{T}}(W) = -\frac{g(1-r)h_2\mu}{\sqrt{U_2^2+V_2^2}}\begin{pmatrix} 0 \\ 0 \\ U_2 \\ 0 \end{pmatrix}, \quad S_{2,\mathcal{T}}(W) = -\frac{g(1-r)h_2\mu}{\sqrt{U_2^2+V_2^2}}\begin{pmatrix} 0 \\ 0 \\ 0 \\ V_2 \end{pmatrix}.$$

Note that $F_i(W)$, $i = 1, 2$ represent the convective terms and $B_i(W)\partial_x W$ are the pressure terms.

System (2) is rotationally invariant. Thus, if we denote by $\eta = (\eta_1, \eta_2)$ an unit vector, and

$$R_\eta = \begin{pmatrix} \eta_1 & \eta_2 \\ -\eta_2 & \eta_1 \end{pmatrix}, \quad T_\eta = \left(\begin{array}{cc|cc} 1 & 0 & 0 & 0 \\ 0 & R_\eta & 0 & 0 \\ \hline 0 & 0 & 1 & 0 \\ 0 & 0 & 0 & R_\eta \end{array}\right).$$

and if we also denote

$$F_\eta(W) = F_1(W)\eta_1 + F_2(W)\eta_2, \quad B_\eta(W, \mathbf{s}) = B_1(W)\eta_1 + B_2(W)\eta_2,$$

$$S_\eta(W) = S_{1,\mathcal{T}}(W)\eta_1 + S_{2,\mathcal{T}}(W)\eta_2.$$

Then, the following properties follows:

$$T_\eta F_\eta(W) = F_1(T_\eta W), \; T_\eta B_\eta(W) = B_1(T_\eta W) \text{ and } T_\eta S_\eta(W) = S_{1,\mathcal{T}}(T_\eta W). \quad (4)$$

Moreover, for any unit vector η, system (2) can be rewritten as follows:

$$\partial_t W + \partial_\eta F_\eta + \partial_{\eta^\perp} F_{\eta^\perp} + B_\eta(W)\partial_\eta W + B_{\eta^\perp}(W)\partial_{\eta^\perp} W = S_\eta(W) + S_{\eta^\perp}(W).$$

Multiplying previous system by T_η and using (4) we obtain

$$\partial_t(T_\eta W) + \partial_\eta F_1(T_\eta W) + B_1(T_\eta W)\partial_\eta T_\eta W = S_1(T_\eta W,) + \mathcal{R}_{\eta^\perp} \quad (5)$$

where

$$\mathcal{R}_{\eta^\perp} = -T_\eta\Big(\partial_{\eta^\perp} F_{\eta^\perp} + B_{\eta^\perp}(W)\partial_{\eta^\perp} W - S_{\eta^\perp}(W)\Big).$$

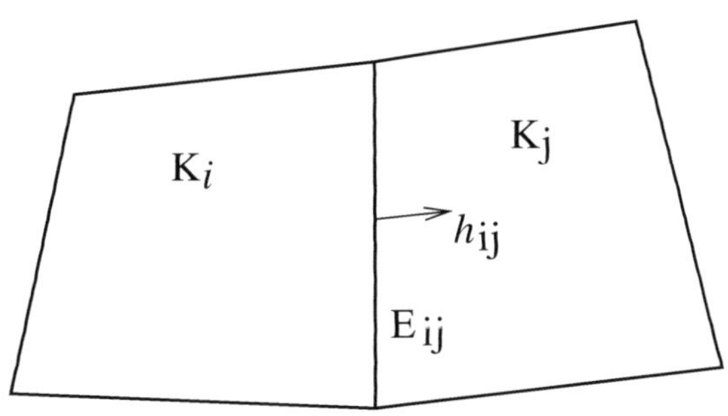

FIGURE 2. Notation, control volumes

Up to our knowledge, there is not in the bibliography results on the existence and uniqueness of solution of this model. It can be seen as a bilayer Shallow Water system with a specific definition of the friction terms. In this sense we can remark that in [25] an existence theorem of global weak solutions is presented for a bilayer Shallow Water system with other friction terms and capillary effects.

3. **Numerical scheme.** First we set a partition of the domain Ω into control volumes. We denote the volumes that define the mesh by K_i. Here, quadrilateral finite volume meshes are considered. In any case, the description of the numerical scheme is also valid for arbitrary meshes. Let also denote by $|K_i|$ the are of the volume K_i and by $E_{i\,j}$ the common edge between the volumes K_i and K_j. d_{ij} is the distance between the center of mass of both volumes, and $\eta_{i\,j}$ is the unitary normal vector to $E_{i\,j}$ outward to K_i (see Figure 2). We will also denote by W_i^n the approximation computed by the numerical scheme of the cell average of the solution at every volume K_i at time t^n:

$$W_i^n \approx \frac{1}{|K_i|}\int_{K_i} W(x,t^n)\,dx.$$

Here a two step method is used to discretize (2). In the first step, the Coulomb friction term is neglected and the non-conservative hyperbolic system is discretized by means of the IFCPH path-conservative finite volume solver, to obtain the value $W_i^{n+1/2}$ given by:

$$W_i^{n+1/2} = W_i^n - \frac{\Delta t}{|K_i|}\sum_{j\in K_i} |E_{i\,j}|\mathcal{D}^-(W_i^n, W_j^n, \eta_{i\,j}). \tag{6}$$

In the second step, that corresponds to the discretization of the Coulomb friction term, we obtain W_i^{n+1} from $W_i^{n+1/2}$ as follows:

We set $h_{1,i}^{n+1} = h_{1,i}^{n+1/2}$, $\vec{q}_{1,i}^{\,n+1} = \vec{q}_{1,i}^{\,n+1/2}$ and $h_{2,i}^{n+1} = h_{2,i}^{n+1/2}$. In order to compute $\vec{q}_{2,i}^{\,n+1}$, let us first define $\vec{u}_{2,i}^{\,*}$ as follows,

$$\vec{u}_{2,i}^{\,*} = \vec{u}_{2,i}^{\,n+1/2} - \Delta t\frac{g(1-r)\mu}{|\vec{u_{2,i}}^{\,n+1/2}|}\vec{u}_{2,i}^{\,n+1/2}.$$

Then we proceed as follows: if $|\vec{u}_{2,i}^{\,*}| \leq g(1-r)\mu$ then we set $\vec{u}_{2,i}^{\,n+1} = \vec{u}_{2,i}^{\,*}$. Otherwise $\vec{u}_{2,i}^{\,n+1} = 0$. Finally, $\vec{q}_{2,i}^{\,n+1}$ is obtained multiplying $\vec{u}_{2,i}^{\,n+1}$ by $h_{2,i}^{n+1}$.

In order to define $\mathcal{D}^-(W_i^n, W_j^n, \eta_{i\,j})$, we consider at each edge E_{ij} of the finite volume mesh the following 1D projected Riemann problem (see [5], [7] and [10]):

$$\begin{cases} \partial_t W + \partial_{\eta_{ij}} F_\eta + B_{\eta_{ij}}(W)\partial_{\eta_{ij}} W = 0, \\ \\ W(x,y,t=0) = \begin{cases} W_i & \text{if } (x,y)\in K_i, \\ W_j & \text{if } (x,y)\in K_j. \end{cases} \end{cases}$$

Notice that taking into account the invariance by rotation property (4)

$$\partial_t W + \partial_{\eta_{ij}} F_\eta + B_{\eta_{ij}}(W)\partial_{\eta_{ij}} W =$$

$$= T_{\eta_{ij}}^{-1}\Big(\partial_t T_{\eta_{ij}} W + \partial_{\eta_{ij}} F_1(T_{\eta_{ij}} W) + B_1(T_{\eta_{ij}} W)\partial_{\eta_{ij}} T_{\eta_{ij}} W\Big).$$

Then, we propose to define

$$\mathcal{D}^-(W_i^n, W_j^n, \eta_{i\,j}) = T_{\eta_{ij}}^{-1} D^-(T_{\eta_{ij}} W_i^n, T_{\eta_{ij}} W_j^n)$$

being $D^-(T_{\eta_{ij}}W_i^n, T_{\eta_{ij}}W_j^n)$ the path-conservative fluctuation associated to the following 1D problem:

$$\begin{cases} \partial_t w(\xi,t) + \partial_\xi F_1(w) + B_1(w)\partial_\xi w = 0, \\ w(\xi, t=0) = \begin{cases} T_{\eta_{ij}}W_i & \text{if } \xi < 0, \\ T_{\eta_{ij}}W_j & \text{if } \xi > 0. \end{cases} \end{cases} \tag{7}$$

System (7) has non-conservative products. The presence of the nonconservative product implies that the notion of weak solution in the sense of distributions cannot be used. The theory introduced by Dal Maso, LeFloch, and Murat [8] is followed here to define weak solutions of (7). This theory allows one to define the nonconservative product $B_1(w)\partial_\xi w$ as a bounded measure provided a family of Lipschitz continuous paths $\varphi : [0,1] \times \Omega \times \Omega \to \Omega$ is prescribed, which must satisfy certain natural regularity conditions. Here, the family of straight segments is considered:

$$\varphi(s; w_L, w_R) = w_L + s(w_R - w_L).$$

Moreover, the chosen path will play also an important role in the discretization of the system. As mentioned before, here path-conservative finite volume framework will be used.

Moreover, system (7) has two linearly degenerated fields associated to the tangential velocities of each layer with respect to the normal vector η_{ij}, that act as two passive scalars. In this way, the definition of $D^-(T_{\eta_{ij}}W_i, T_{\eta_{ij}}W_j)$ is blocked based: the first block corresponds to the non passive scalar unknowns and the second block to the passive scalar unknowns.

To define $D^-(T_{\eta_{ij}}W_i, T_{\eta_{ij}}W_j)$, we introduce the following notation. Let $\mathcal{N}$ denote the set of index associated to the non passive scalar unknowns, that for that (7) is $\mathcal{N} = \{1,2,4,5\}$. We also denote by $[D^-]_\mathcal{N}$ the vector defined by the components of D^- with index in $\mathcal{N}$.

The definition of the numerical scheme is done in the following two steps:

- *Step 1:* Definition of $[D^-(w_i, w_j)]_\mathcal{N}$.

We consider here path-conservative numerical schemes, corresponding to the following definition:

$$\begin{aligned} [D^-(w_i,w_j)]_\mathcal{N} = \; & \frac{1}{2}\Big([F_1(w_j) - F_1(w_i)]_\mathcal{N} + B_{1,ij}[w_j - w_i]_\mathcal{N} \\ & -Q_{ij}([w_j - w_i]_\mathcal{N} + A_{ij}^{-1}[S_{1,\mathcal{T},ij}]_\mathcal{N})\Big), \end{aligned} \tag{8}$$

In the previous equation A_{ij} is a generalized Roe matrix (see [20, 19]) associated to (7) for the equations of the set $\mathcal{N}$, that is

$$A_{ij}[w_j - w_i]_\mathcal{N} = [F_1(w_j) - F_1(w_i)]_\mathcal{N} + B_{1,ij}[w_j - w_i]_\mathcal{N},$$

where

$$B_{1,ij}[w_j - w_i]_\mathcal{N} = \int_0^1 [B_1(\varphi(s,w_i,w_j))\partial_s\varphi(s,w_i,w_j)]_\mathcal{N} ds$$

and

$$[S_{1,\mathcal{T},ij}]_\mathcal{N} = [S_{1,\mathcal{T}}(w_{ij})]_\mathcal{N},$$

being w_{ij} an intermediate state computed from w_i and w_j.

Note that the numerical diffusion term $Q_{ij}([w_j - w_i]_{\mathcal{N}} + A_{ij}^{-1}[S_{1,\mathcal{T},ij}]_{\mathcal{N}})$ depends on the Coulomb friction term. This correction is critical in order to preserve accurately the stationary solutions of the form:

$$\vec{u}_1 = \vec{0},\ \vec{u}_2 = \vec{0},\ h_1 + h_2 = cst,\ \text{and } \partial_x h_2 \le \mu\ \partial_y h_2 \le \mu.$$

The viscosity matrix Q_{ij} is defined by considering the IFCP method (see [11]):

$$Q_{ij} = \alpha_0^{ij} Id + \alpha_1^{ij} A_{ij} + \alpha_2^{ij} A_{ij}^2 \tag{9}$$

where α_l^{ij}, $l = 0,\ 1\,2$ are defined in terms of the wave speed of the system:

$$\begin{aligned} \alpha_0^{ij} &= \delta_L\, S_R^{ij}\, S_{int}^{ij} + \delta_R\, S_L^{ij}\, S_{int}^{ij} + \delta_{int}\, S_L^{ij}\, S_R^{ij}, \\ \alpha_1^{ij} &= -S_L^{ij}(\delta_R + \delta_{int}) - S_R^{ij}(\delta_L + \delta_{int}) - S_{int}^{ij}(\delta_L + \delta_R), \\ \alpha_2^{ij} &= \delta_L + \delta_R + \delta_{int} \end{aligned} \tag{10}$$

with

$$\delta_L = \frac{|S_L^{ij}|}{(S_L^{ij} - S_R^{ij})(S_L^{ij} - S_{int}^{ij})}, \qquad \delta_R = \frac{|S_R^{ij}|}{(S_R^{ij} - S_L^{ij})(S_R^{ij} - S_{int}^{ij})},$$

$$\delta_{int} = \frac{|S_{int}^{ij}|}{(S_{int}^{ij} - S_L^{ij})(S_{int}^{ij} - S_R^{ij})}.$$

Here, S_L^{ij} and S_R^{ij} are approximations of the slowest and fastest waves (respectively) of the Riemann problem associated to intercell E_{ij}. Here, the following expressions are used:

$$S_L^{ij} = \min(\lambda_{ext,i}^-, \lambda_{ext,ij}^-), \quad S_R^{ij} = \max(\lambda_{ext,j}^+, \lambda_{ext,ij}^+).$$

S_{int}^{ij} is defined by

$$S_{int}^{ij} = s_{ij} \max(|\lambda_{int,ij}^-|, |\lambda_{int,ij}^+|) \tag{11}$$

with

$$s_{ij} = \begin{cases} \text{sign}(S_L^{ij}) & \text{if } |S_L^{ij}| \ge |S_R^{ij}|, \\ \text{sign}(S_R^{ij}) & \text{otherwise.} \end{cases} \tag{12}$$

where $\lambda_{ext,ij}^- < \lambda_{int,ij}^- < \lambda_{int,ij}^+ < \lambda_{ext,ij}^+$ are the eigenvalues of A_{ij}. Moreover, for the case of wet/dry fronts, that is if $h_{k,i}$ or $h_{k,j}$ is zero for $k = 1, 2$, we consider the following definition of the coefficients α_l^{ij}, $l = 0, 1, 2$:

$$\alpha_0^{ij} = \frac{S_R^{ij}|S_L^{ij}| - S_L^{ij}|S_R^{ij}|}{S_R^{ij} - S_L^{ij}}, \quad \alpha_1^{ij} = \frac{|S_R^{ij}| - |S_L^{ij}|}{S_R^{ij} - S_L^{ij}}, \quad \alpha_2^{ij} = 0.$$

With this choice, it is straightforward to check that the Riemann solver reduces to HLL solver (see [6]), which is more robust when wet/dry fronts appear.

Therefore, the resulting numerical scheme reduces to HLL solver in wet/dry areas and in other case reduces to IFCP solver.

∘ *Step 2:* In this second step, we define the components of the numerical fluctuation, D^-, corresponding to the passive scalar unknowns. In this particular system, they correspond to equations 3 and 6. Taking into account the relation between the passive scalar and the other variables, and that their associated wave speeds only depends on the normal velocities of both layers, we propose the following definition:

$$[D^-(w_i, w_j)]_3 = \Big([D^-(w_i, w_j)]_1 + [F_1(w_i)]_1\Big) C_{1,\eta_{ij}^\pm}^* - [F_1(w_i)]_3$$

$$[D^-(w_i,w_j)]_6 = \Big([D^-(w_i,w_j)]_4 + [F_1(w_i)]_4\Big)\mathcal{C}^*_{2,\eta^\perp_{ij}} - [F_1(w_i)]_6.$$

where, $\mathcal{C}^*_{l,\eta^\perp_{ij}}$, $l=1,2$ is an uncentered approximation of the tangential velocities of each layer through edges E_{ij}:

$$\mathcal{C}^*_{1,\eta^\perp_{ij}} = \begin{cases} [T_{\eta_{ij}}W_i]_3/h_{1,i} & \text{if } S^*_{1,ij}<0,\\ [T_{\eta_{ij}}W_j]_3/h_{1,j} & \text{if } S^*_{1,ij}>0,\end{cases} \qquad \mathcal{C}^*_{2,\eta^\perp_{ij}} = \begin{cases} [T_{\eta_{ij}}W_i]_6/h_{2,i} & \text{if } S^*_{2,ij}<0,\\ [T_{\eta_{ij}}W_j]_6/h_{2,i} & \text{if } S^*_{2,ij}>0,\end{cases} \tag{13}$$

The values $S^*_{l,ij}$, $l=1,2$ are an approximation of the normal velocities through edges E_{ij}. We can use for example $S^*_{1,ij} = ([D^-(w_i,w_j)]_1 + [F_1(w_i)]_1)/h_{1,ij}$ and $S^*_{2,ij} = ([D^-(w_i,w_j)]_4 + [F_1(w_i)]_4)/h_{2,ij}$, respectively, being $h_{l,ij} = \frac{h_{l,i}+h_{l,j}}{2}$. Some other definitions are possible, as the one proposed in [10].

Theorem 3.1. *The previous numerical scheme exactly preserves the water at rest solutions given by*

$$\vec{u}_{1,i}=0, \quad \vec{u}_{2,i}=0, \quad h_{1,i}+h_{2,i}=cst, \quad \frac{1}{|K_i|}\sqrt{\sum_{j\in K_i}\left(\frac{h_{2,j}-h_{2,i}}{\Delta x\eta_{i,j,1}+\Delta y\eta_{i,j,2}}\right)^2} \le \mu.$$

The proof is similar to the one performed in [9].

4. **Numerical tests: submarine collapse of initially cylindrical granular masses.** In this section we simulate a set of submarine circular dam-break problems and we also compare them with some existing laboratory data for the case of aerial avalanches. Let us denote by R the radius of the initial granular column and by H_2 its initial height. H_1 designs the initial height of the water layer above the sediment column (See Figure 3). We introduce two aspect ratios : $a_H = H_1/H_2$ and $a_2 = H_2/R$. By scaling the equations as proposed below, we observe that the dimensionless equations only depend on a_H and a_2 and not on the granular mass or on the gravity acceleration g.

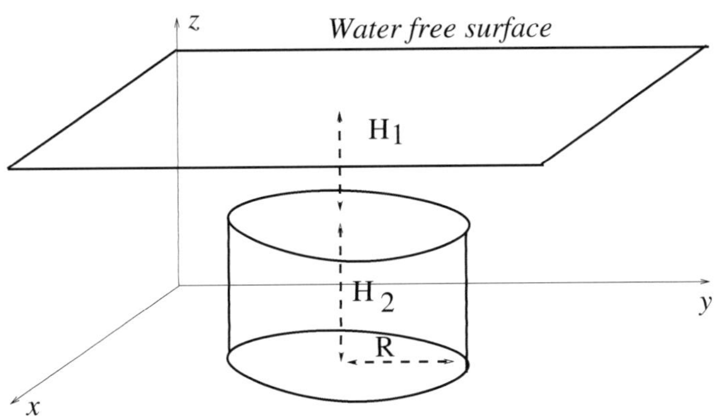

FIGURE 3. Initial condition and notation

The following change of variable is done:

$$(x,y) = (R\tilde{x}, R\tilde{y}), \quad t = \sqrt{\frac{R}{g}}\,\tilde{t},$$

$$U_l = \sqrt{gH_l}\tilde{U}_l, \quad V_l = \sqrt{gH_l}\tilde{V}_l, \quad h_l = H_l\tilde{h}_l, \quad l=1,2.$$

By omitting the tildes, we obtain the following system

$$\begin{cases} \sqrt{\dfrac{a_H}{a_2}}\,\partial_t\,(h_1) + a_H\partial_x(h_1U_1) + a_H\partial_y(h_1V_1) = 0, \\[2ex] \sqrt{\dfrac{a_H}{a_2}}\,\partial_t\,(h_1U_1) + a_H\partial_x(h_1U_1^2) + a_H\partial_y(h_1U_1V_1) + h_1\partial_x(a_Hh_1 + h_2) = 0 \\[2ex] \sqrt{\dfrac{a_H}{a_2}}\,\partial_t\,(h_1V_1) + a_H\partial_x(h_1U_1V_1) + a_H\partial_y(h_1V_1^2) + h_1\partial_y(a_Hh_1 + h_2) = 0 \\[2ex] \dfrac{1}{\sqrt{a_2}}\partial_t\,(h_2) + \partial_x(h_2U_2) + \partial_y(h_2V_2) = 0, \\[2ex] \dfrac{1}{\sqrt{a_2}}\partial_t\,(h_2U_2) + \partial_x(h_2U_2^2) + \partial_y(h_2U_2V_2) + h_2\partial_x(ra_Hh_1 + h_2) = \mathcal{T}_x \\[2ex] \dfrac{1}{\sqrt{a_2}}\partial_t\,(h_2V_2) + \partial_x(h_2U_2V_2) + \partial_y(h_2V_2^2) + h_2\partial_y(ra_Hh_1 + h_2) = \mathcal{T}_y \end{cases}$$

where

$$\mathcal{T} = -\frac{a_2(1-r)h_2\mu}{\sqrt{U_2^2 + V_2^2}} \begin{pmatrix} U_2 \\ V_2 \end{pmatrix}.$$

and $\mu = \tan\delta$. So, the solutions are mainly governed by the values of a_H, a_2, r and δ.

In what follows, we initially check that the previous numerical scheme is able to recover the stationary profiles of aerial avalanches. Aerial avalanches (not submerged) can be described here by setting $r = 0$. Next, we will consider fully submerges landslides and we perform some sensitivity analysis with respect to the parameters a_H, a_2, r and δ.

The initial condition is $\vec{q_1} = \vec{0}$, $\vec{q_2} = \vec{0}$,

$$h_1(\mathbf{x}, 0) = a_Ha_2R + a_2R - h_2(\mathbf{x}, 0),$$

$$h_2(\mathbf{x}, 0) = \begin{cases} a_2R & \text{if } (x - x_0)^2 + (y - y_0)^2 \le (R)^2, \\ 0 & \text{otherwise.} \end{cases}$$

We set the domain $[0, 0.6]\text{m}\times[0, 0.6]\text{m}$, the center of the cylinder is $(x_0, y_0) = (0.3\text{m}, 0.3\text{m})$ and $R = 0.0705\text{m}$. The domain is decomposed in 200×200 square finite volumes.

We compare the numerical solutions for $r = 0$, $r = 0.2$, $r = 0.4$, $r = 0.6$ and $r = 0.8$ with laboratory data of dry granular flows, i. e. corresponding to $r = 0$. The case $r = 0.2$ is presented to show the transition between submarine and aerial avalanches. The test is done for $a_2 = 0.56$ as in the experiments (1) in the sub-aerial case and (2) for different values of the relative height $a_H = 1, 2, 10$ to assess the sensitivity of the flow dynamics and generated tsunami to the water depth.

We also denote

$$\eta(\mathbf{x}, t) = h_1(\mathbf{x}, t) + h_2(\mathbf{x}, t) - A_{ref}.$$

where A_{ref} is the reference water surface. For this test $A_{ref} = h_1(\mathbf{x}, 0) + h_2(\mathbf{x}, 0)$.

We begin with the experiments corresponding to $a_2 = 0.56$. First, in Figure 4 the comparison between the experiment with $r = 0$ and experimental data is presented. A good agreement of the numerical results and the laboratory data can be observed.

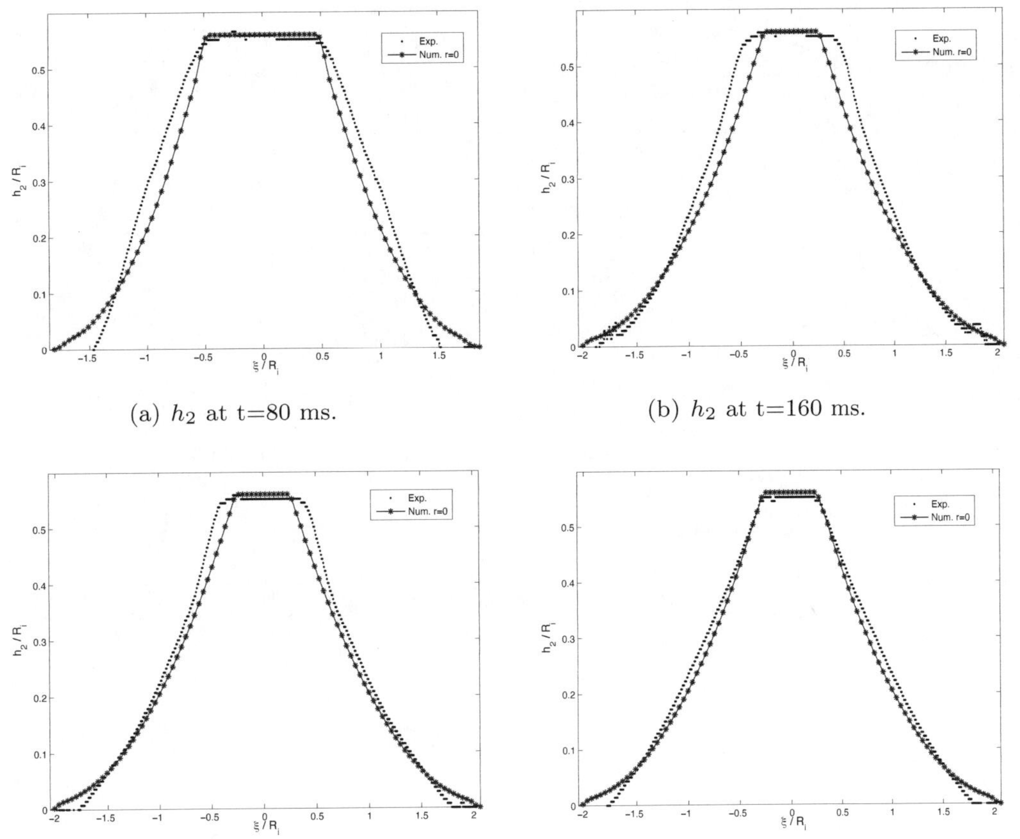

(a) h_2 at t=80 ms. (b) h_2 at t=160 ms.

(c) h_2 at t=240 ms. (d) h_2 at t=580 ms.

FIGURE 4. Granular mass profiles. h_2 evolution for $r = 0$.

In Figure 5 a comparison between the evolution of the sediment layer for different values of r is presented, where a_H is set to 1. For other values of a_H, similar behaviour is obtained. The evolution of the submarine avalanche depends on r. For smaller values of r, the final deposit is quickly reached.

In Figure 6 we present the evolution of the front of the avalanche denoted by x_{front}. Figure 6(c) correspond to L_f, the final length of the deposit. We observe that the final length of the avalanche is smaller for bigger values of r and bigger values of a_H. The main difference of the final length between the aerial avalanche (corresponding to $r = 0$) and submarine avalanches corresponds to $r = 0.8$ and $a_H = 10$.

In Figure 7 we present the evolution of x_{front}, for the values $a_H = 1$ and $a_H = 10$. We can observe how, effectively the final deposit is reached previously for smaller values of r. , i. e. the propagation time is smaller. The front position is more sensitive to r for $a_H = 1$ than for $a_H = 10$.

In Figure 8 we represent the evolution of $\max_x |\eta(\mathbf{x}, t)|$, for $t = 80$, 160, 240 and 540 ms. We observe that the perturbation of the water surface are bigger from smaller values of r and also smaller values of the aspect ratio a_H. Indeed, the spreading has been shown to be faster for small values of r and the water surface is more sensitive to the granular flow if it is closer to it. For $a_H = 10$ there is almost no perturbation of the water surface, during the submarine avalanche. In Figure

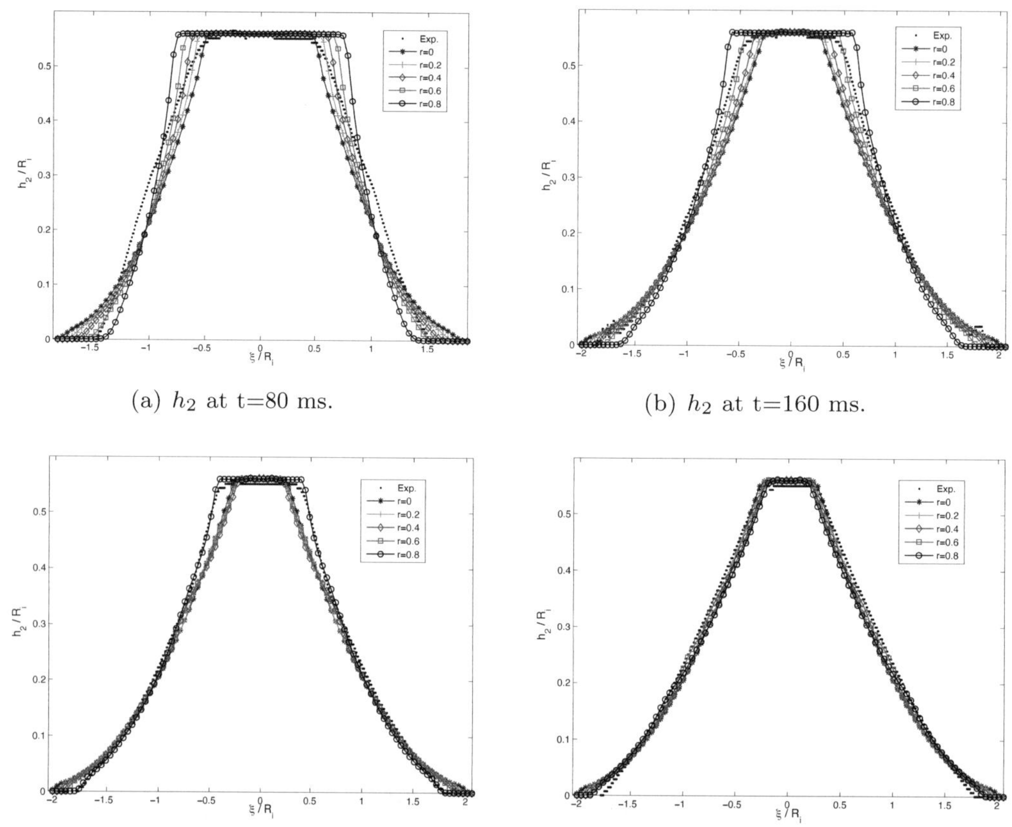

(a) h_2 at t=80 ms.

(b) h_2 at t=160 ms.

(c) h_2 at t=240 ms.

(d) h_2 at t=580 ms.

FIGURE 5. Granular mass profiles. h_2 evolution for $a_H = 1$, $r \in \{0, 0.2, 0.4, 0.6, 0.8\}$.

9 the evolution of the water surface for $a_H = 1$ and $a_H = 2$ is presented. We can observe the different behaviour of the water surface, depending on the initial aspect ratios.

In Figure 10 the tridimensional evolution of the sediment avalanche and water surface, for $r = 0.4$ and $a_H = 1$ is presented.

5. **Conclusion.** In this work we present a preliminary study of the influence of the ratio of densities and the characteristic dimensions of a cylindrical submarine landslide over a flat bottom topography. This is done by considering a 2D generalization of the model presented in [9] where the bottom topography is supposed to be flat. The 2D system is discretized by a first order Riemann solver that results of the combination of the IFCP and HLL Riemann solvers. In particular, the solver reduces to the HLL in wet/dry regions, while the IFCP solver is used in the other regions. Finally, the model has been written in non-dimensional variables in terms of the aspect ratio between the initial height of the avalanche and the initial height of the fluid above the granular mass, and the aspect ratio of the initial sediment mass. The evolution of the maximum amplitude of the free surface and the front position has been studied in terms of the aspect ratios, the ratio of densities r and the friction angle. A comparison with experimental data for the limit case when $r = 0$, corresponding to aerial avalanches has been also presented. We observe that

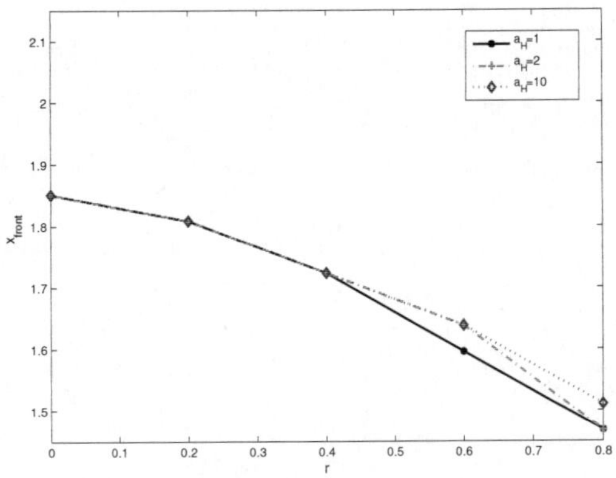

(a) x_{front} at t=80 ms.

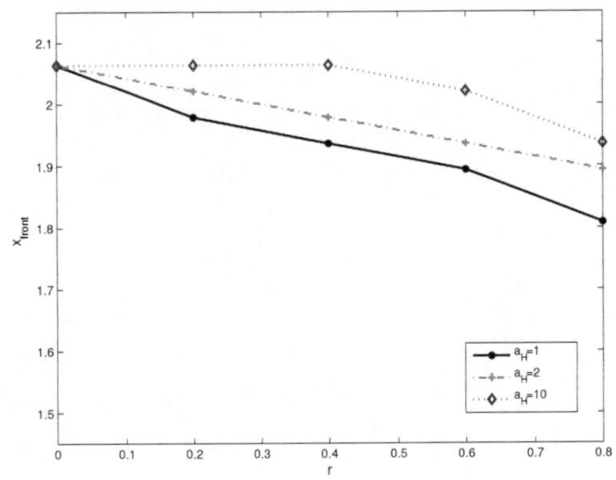

(b) x_{front} at t=240 ms.

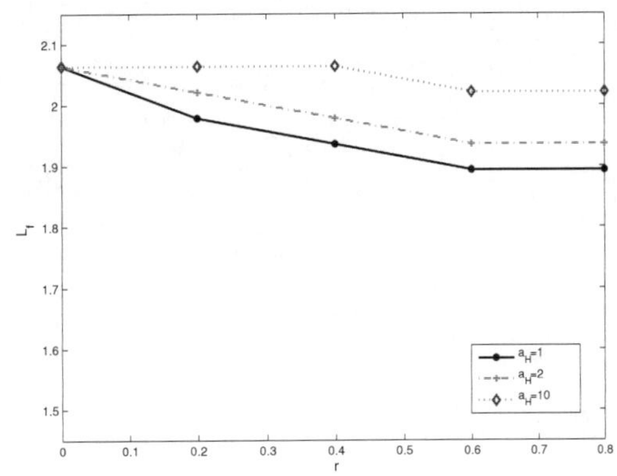

(c) L_f

FIGURE 6. x_{front} evolution and L_f for $r \in \{0, 0.2, 0.4, 0.6, 0.8\}$, $a_H \in \{1, 2, 10\}$.

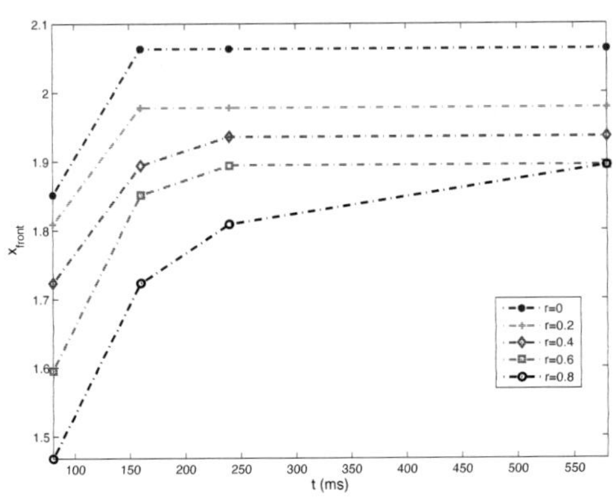

(a) x_{front} evolution $a_H = 1$

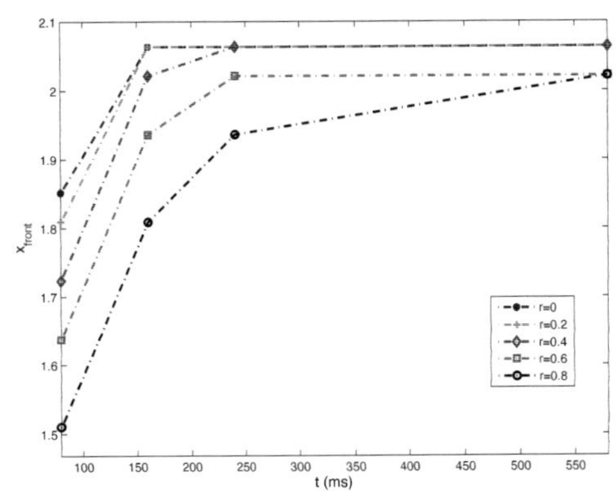

(b) x_{front} evolution $a_H = 10$

FIGURE 7. x_{front} for $a_H = 1$ and $a_H = 10$ for $r \in \{0, 0.2, 0.4, 0.6, 0.8\}$.

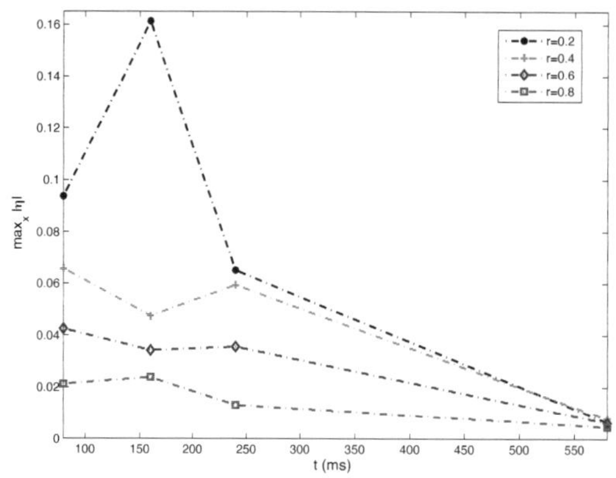

(a) $a_H = 1$

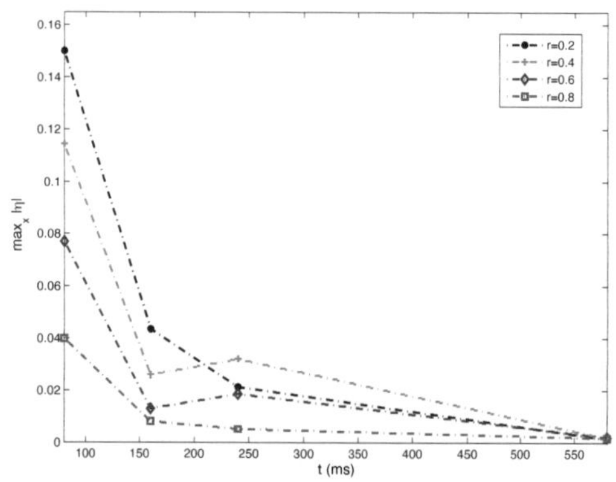

(b) $a_H = 2$

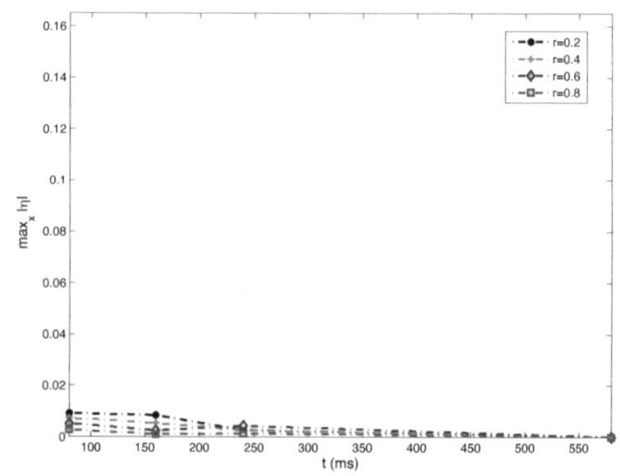

(c) $a_H = 10$

FIGURE 8. $\max_x |\eta|$ for $a_H = 1$, $a_H = 2$ and $a_H = 10$, $r \in \{0.2, 0.4, 0.6, 0.8\}$.

the mass spreading takes more time and lead to smaller runout distance for granular flows of smaller density, leading to a smaller amplitude of the generated water wave. When the granular mass is closer to the water free surface, the runout of the granular flow is smaller but the generated water wave is bigger than when it is 10 times deeper. For intermediate values of the water depth, the behavior is more complex.

REFERENCES

[1] F. Bouchut, A. Mangeney-Castelnau, B. Perthame, J.P Vilotte, A new model of Saint Venant and Savage-Hutter type for gravity driven shallow flows. *C.R. Acad. Sci. Paris, Ser I*, **336** (2003), 531–536.

[2] F. Bouchut, E.D. Fernández-Nieto, A. Mangeney, G. Narbona-Reina. A two-phase two-layer model for fluidized granular flows with dilatancy effects. *Journal of Fluid Mechanics*, **801** (2016), 166–221.

[3] F. Bouchut, M. Westdickenberg. Gravity driven shallow water model for arbitrary topography. *Comm. Math. Sci.*, **2(3)** (2004), 359–389.

[4] M. Brunet, L. Moretti, A. Le Friand, A. Mangeney, E. Fernandez-Nieto, F. Bouchut. Numerical simulation of the 30-45 ka debris avalanche flow of Montagne Pelée volcano, Martinique: from volcano flank-collapse to submarine emplacement comparison between measurements and numerical modelling. *Natural Hazards*, **87** (2017), 1189–1222.

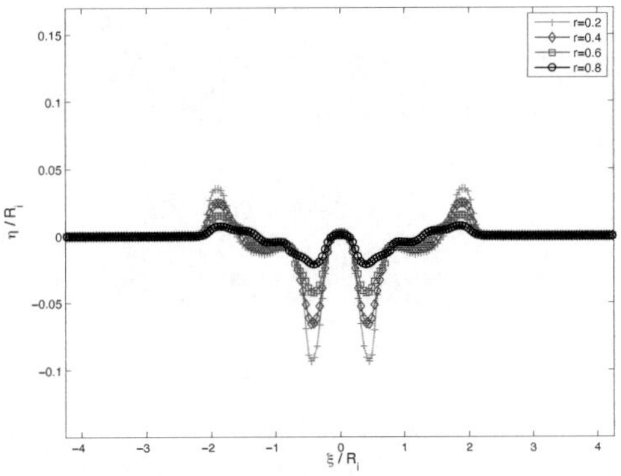

(a) $a_H = 1$, t=80 ms.

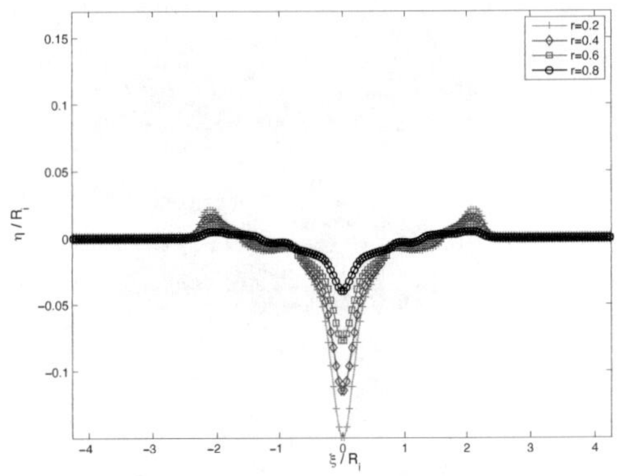

(b) $a_H = 2$ t=80 ms.

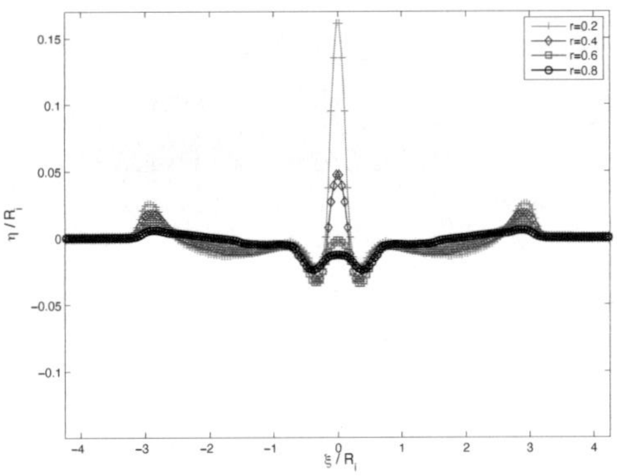

(c) $a_H = 1$ t=160 ms.

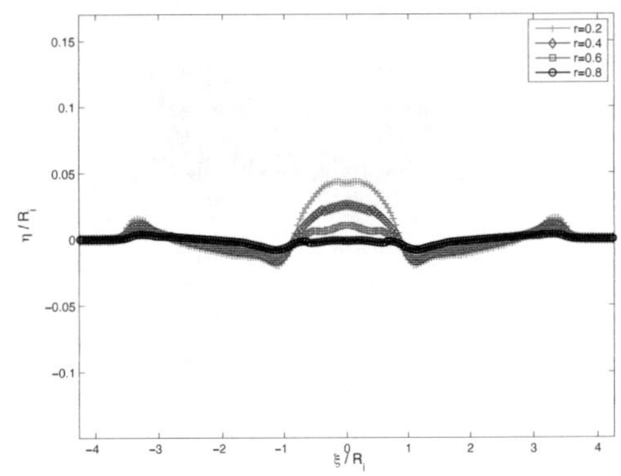

(d) $a_H = 2$ t=160 ms.

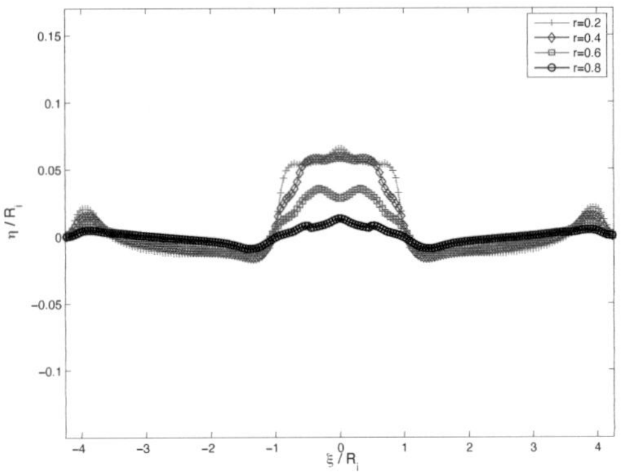

(e) $a_H = 1$ t=240 ms.

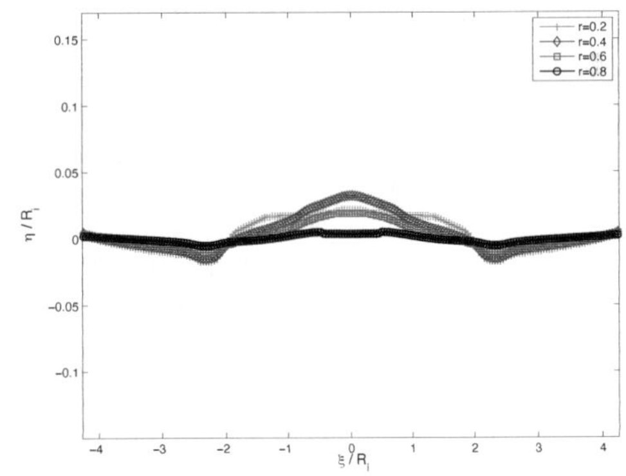

(f) $a_H = 2$ t=240 ms.

FIGURE 9. Free surface evolution for $a_H = 1$ and $a_H = 2$, $r \in \{0.2, 0.4, 0.6, 0.8\}$.

[5] M.J. Castro Díaz, T. Chacón Rebollo, E.D. Fernández-Nieto, J.M. González Vida. C. Parés. Well-balanced finite volume schemes for 2D non-homogeneous hyperbolic systems. Application to the dam break of Aznalcóllar. *Comput. Methods Appl. Mech. Engrg.* **197(45-48)** (2008), 3932–3950.

[6] M. J. Castro-Díaz and E. D. Fernández-Nieto. A class of computationally fast first order finite volume solvers: PVM methods. *SIAM Journal on Scientific Computing*, **34(4)** (2012), A2173–A2196.

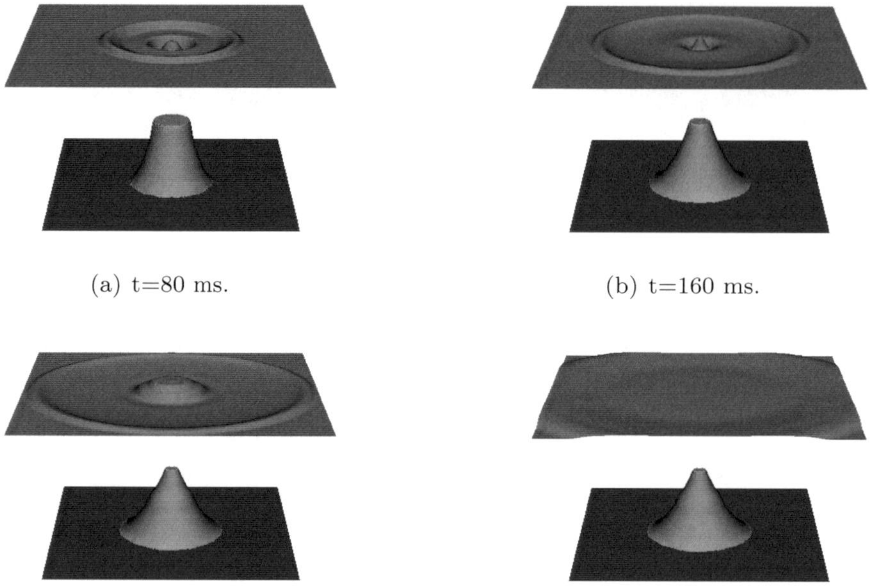

(a) t=80 ms. (b) t=160 ms.

(c) h_2 at t=240 ms. (d) t=580 ms.

FIGURE 10. h_2 and surface evolution for $a_H = 1$, $r = 0.4$.

[7] M.J. Castro, E.D. Fernández-Nieto, A. Ferreiro, J.A. García-Rodríguez, C. Parés. High order extensions of Roe schemes for two-dimensional nonconservative hyperbolic systems. *J. Sci. Comput.* **39(1)** (2009), 67–114.

[8] G. Dal Maso, P.G. LeFloch, F. Murat. Definition and weak stability of nonconservative products. *J. Math. Pures Appl.* **74** (1995), 483–548.

[9] E. D. Fernández-Nieto, F. Bouchut, D. Bresch, M.J. Castro, A. Mangeney. A new Savage-Hutter type model for submarine avalanches and generated tsunami. *J. Comput. Phys.* **227(16)** (2008), 7720–7754.

[10] E.D. Fernández-Nieto, D. Bresch, J. Monnier. A consistent intermediate wave speed for a well-balanced HLLC solver. C. R. Math. Acad. Sci. Paris **346(13-14)** (2008), 795–800.

[11] E.D. Fernández-Nieto, M.J. Castro-Díaz, C. Parés, C. On an Intermediate Field Capturing Riemann Solver Based on a Parabolic Viscosity Matrix for the Two-Layer Shallow Water System. Journal of Scientific Computing, **48(1)** (2011), 117–140.

[12] E. Godlewski E, P.A. Raviart. Numerical approximation of hyperbolic systems of conservation laws. New York: Springer-Verlag, (1996).

[13] A. Harten, P. D. Lax, and B. van Leer. On upstream differencing and Godunov-type schemes for hyperbolic conservation laws. *SIAM Review*, **25(1)** (1983), 35–61.

[14] Lajeunesse, E., A. Mangeney-Castelnau, J. P. Vilotte. Spreading of a granular mass on a horizontal plane. *Phys. Fluids*, **16** (2004), 2371–2381.

[15] A. Mangeney-Castelnau, J.P. Vilotte, M.O. Bristeau, B. Perthame, F. Bouchut, C. Simeoni, S. Yernini. Numerical modeling of avalanches based on Saint-Venant equations using a kinetic scheme. *J. Geophys. Res.* (2003) 108 (B11), 2527.

[16] A. Mangeney-Castelnau, F. Bouchut, J. P. Vilotte, E. Lajeunesse, A. Aubertin, M. Pirulli. On the use of Saint Venant equations to simulate the spreading of a granular mass. *J. Geoph. Res.* **110** (2005), B09103.

[17] A. Mangeney, F. Bouchut, N. Thomas, J.P. Vilotte, M.O. Bristeau, M.O. Numerical modeling of self-channeling granular flows and of their levee-channel deposits. *J. Geophys. Res.* (2007) **112**, F02017.

[18] S. McDougall, O. Hungr, O. Dynamic modelling of entrainment in rapid landslides. *Can. Geotech. J.* **42** (2005), 1437–1448.

[19] C. Parés. Numerical methods for nonconservative hyperbolic systems: a theoretical framework. *SIAM J. Numer. Anal.* **44(1)** (2004), 300–321.
[20] C. Parés, M.J. Castro. On the well-balance property of Roe's method for nonconservative hyperbolic systems. Applications to shallow-water systems. *ESAIM: M2AN*, **38(5)** (2004), 821–852.
[21] C. Parés, M.L. Muñoz-Ruiz. On some difficulties of the numerical approximation of nonconservative hyperbolic systems. *Bol. Soc. Esp. Mat. Apl. (SEMA)*, **47** (2009), 23–52.
[22] M. Pirulli, A. Mangeney. Result of Back-Analysis of the Propagation of Rock Avalanches as a Function of the Assumed Rheology. *Rock Mech. Rock Engng.*, **41(1)** (2008), 59–84.
[23] O. Pouliquen. Scaling laws in granular ßows down rough inclined planes. *Phys. Fluid.* **11** (1999), 542–548.
[24] S.B. Savage, K. Hutter. The dynamics of avalanches of granular materials from initiation to run-out. *Acta Mech.* **86** (1991) 201–223.
[25] J.D. Zabsonré, G. Narbona-Reina. Existence of global weak solution for a 2D viscous bilayer Shallow-Water model. *Nonlin. Anal. Real World Appl.* **10(5)** (2009) 2971–2984.
[26] E.F. Toro. *Shock-Capturing Methods for Free-Surface Shallow Flows*, Wiley, England (2001).

E-mail address: edofer@us.es
E-mail address: castro@anamat.cie.uma.es
E-mail address: mangeney@ipgp.jussieu.fr

ON STATIONARY BIFURCATION PROBLEM FOR THE COMPRESSIBLE NAVIER-STOKES EQUATIONS

Yoshiyuki Kagei

Faculty of Mathematics *
Kyushu University
Fukuoka 819-0395, Japan

Abstract. Bifurcation of wave trains from the Poiseuille flow of the compressible Navier-Stokes equations is studied. Results on instability of the Poiseuille flow and the bifurcation of wave trains are summarized. A sketch of the proof of the bifurcation is given to illustrate a scheme to prove stationary bifurcation in the compressible Navier-Stokes equations.

1. **Introduction.** This article is concerned with stationary bifurcation problem for the compressible Navier-Stokes equations. To discuss it we will review the result in [7] on the bifurcation of wave trains (spatio-temporal periodic traveling waves) from the Poiseuille flow. Let us consider the following compressible Navier-Stokes system for a barotropic motion:

$$\partial_t \rho + \mathrm{div}\,(\rho \boldsymbol{v}) = 0, \tag{1}$$

$$\rho(\partial_t \boldsymbol{v} + \boldsymbol{v}\cdot\nabla \boldsymbol{v}) - \mu\Delta \boldsymbol{v} - (\mu+\mu')\nabla \mathrm{div}\, \boldsymbol{v} + \nabla P(\rho) = \rho \boldsymbol{g}, \tag{2}$$

where $\rho = \rho(x,t)$ and $\boldsymbol{v} = {}^\top(v^1(x,t), v^2(x,t))$ are the unknown density and velocity, respectively, at time $t \geq 0$ and position $x \in \mathbb{R}^2$; $P = P(\rho)$ is the pressure that is assumed to be a smooth function of ρ; μ and μ' are the viscosity constants; and $\boldsymbol{g}$ is a given external force. Here and in what follows ${}^\top\cdot$ stands for the transposition.

We assume that $P'(\rho_*) > 0$ for a given positive constant ρ_* and $\mu > 0$, $\mu+\mu' \geq 0$.

The system 1–2 is considered in a two-dimensional infinite layer $\Omega = \mathbb{R}\times(0,\ell)$:

$$\Omega = \{x = (x_1, x_2) \,:\, x_1 \in \mathbb{R},\, 0 < x_2 < \ell\}.$$

The external force $\boldsymbol{g}$ is assumed to have the form

$$\boldsymbol{g} = g\boldsymbol{e}_1,$$

where g is a positive constant and $\boldsymbol{e}_1 = {}^\top(1,0) \in \mathbb{R}^2$.

We consider the system 1–2 under the boundary condition

$$\boldsymbol{v}|_{x_2=0,\ell} = \boldsymbol{0}. \tag{3}$$

2000 *Mathematics Subject Classification.* Primary: 35Q30 ; Secondary: 76N15.

Key words and phrases. Compressible Navier-Stokes equations, Poiseuille flow, bifurcation, wave trains.

The author is partly supported by JSPS KAKENHI Grant Numbers 16H03947 and 16H06339.

* The present address: Department of Mathematics, Tokyo Institute of Technology, Tokyo 152-8551, JAPAN.

We impose the periodic boundary condition on ρ and $\boldsymbol{v}$ in x_1:

$$\rho(x_1 + \tfrac{2\pi k}{\alpha}, x_2, t) = \rho(x_1, x_2, t), \quad \boldsymbol{v}(x_1 + \tfrac{2\pi k}{\alpha}, x_2, t) = \boldsymbol{v}(x_1, x_2, t), \tag{4}$$

where $\alpha > 0$ is a given wave number and k is any integer.

One can easily verify that 1–4 has a stationary solution $\overline{u}_s = {}^\top(\overline{\rho}_s, \overline{\boldsymbol{v}}_s)$ of the form

$$\overline{\rho}_s = \rho_*, \quad \overline{\boldsymbol{v}}_s = \frac{\rho_* g}{2\mu} x_2(\ell - x_2)\boldsymbol{e}_1.$$

This stationary solution is called the plane Poiseuille flow.

In this article we will summarize the result in [7] on the bifurcation of wave trains from the plane Poiseuille flow and will give a sketch of its proof to illustrate a scheme to prove stationary bifurcation in the compressible Navier-Stokes equations.

Bifurcation problems for equations describing fluid motions has been extensively studied for the incompressible Navier-Stokes equations since 1960's; see, e.g., [5, 9, 13, 14], and so on. Classical bifurcation theories such as the one by Crandall and Rabinowitz [1] is directly applicable to bifurcation problems for the incompressible Navier-Stokes equations. This is because the incompressible Navier-Stokes equations can be classified in semilinear parabolic systems. In contrast to the incompressible case, bifurcation problems for the multi-dimensional compressible Navier-Stokes equations which are classified in quasilinear hyperbolic-parabolic systems have begun to be studied recently. The first result for compressible bifurcation problems was given by Nishida, Padula and Teramoto [11] (cf., [10]) who proved the existence of bifurcating convection solutions for thermal convection problem. The main difficulty in the proof of the bifurcation for the compressible system arises from the convection term $\boldsymbol{v} \cdot \nabla\rho$ in 1; this term may cause the derivative-loss; in other words, it is not Frechét differentiable in a standard setting in classical bifurcation analysis. In [11], the effective viscous flux is used to overcome this difficulty and establish the necessary estimates for the proof of the bifurcation of stationary convective patterns. In [7], a bifurcation problem of wave trains from the plane Poiseuille flow in viscous compressible fluids was studied. The effective viscous flux is not employed in the proof in [7]. Instead, an iterative argument based on the method of characteristics is employed, that is, the convection term $\boldsymbol{v} \cdot \nabla\rho$ in 1 is regarded as a part of the principal part as in the proof of the local solvability of the time evolution problem. To prove the existence of bifurcating wave trains, the time evolution problem is rewritten to a stationary problem in a moving coordinates. The Lyapunov-Schmidt reduction then applies to decompose the stationary problem into the parts on the null space of the linearized operator and its complementary subspace. One of the points of the proof is to establish the solvability in the complementary subspace. The complementary part is solved based on the estimates obtained by the Matsumura-Nishida energy method [12] and the results by Heywood and Padula [3] on the resolvent problem for transport equation including the convective term $\boldsymbol{v} \cdot \nabla\rho$ as in 1 with a given velocity $\boldsymbol{v}$. The method in [7] will be widely applicable to stationary bifurcation problems for certain classes of quasilinear hyperbolic-parabolic systems.

We mention one more remark. The bifurcation theory in [1, 2] also provides the stability of bifurcating solutions and smooth dependence of bifurcating solutions on the bifurcation parameter, which is applicable to incompressible problems. The situation in compressible cases is different. In fact, it is not straightforward to conclude the stability and smooth dependence because of the derivative-loss in $\boldsymbol{v} \cdot \nabla\rho$. These issues will be discussed in [8].

The remaining of this article is devoted to surveying the bifurcation result and a sketch of its proof given in [7]. In section 2, a non-dimensional form of the system 1–2 is firstly derived and then it is rewritten into the system of equations for the perturbation. The instability result of the plane Poiseuille flow obtained in [6] is given in section 3. We state in section 4 the result on the existence of bifurcating wave trains obtained in [7]; and we give a sketch of the proof of the bifurcation result in section 5.

2. **Preliminaries.** In this section we first derive a non-dimensional form of the system 1–2 and then give the system of equations for the perturbation. In the second part of this section we introduce notations used in this article.

2.1. **Non-dimensionalization.** We transform the problem into the non-dimensional form under the following variable transformations: $x = \ell\tilde{x}$, $t = \frac{\ell}{V}\tilde{t}$, $\boldsymbol{v} = V\tilde{\boldsymbol{v}}$, $\rho = \rho_*\tilde{\rho}$, $P = \rho_* P'(\rho_*)p$, where $V = \frac{\rho_* g\ell^2}{\mu}$.

Using these new non-dimensional variables, we arrive at the system of equations, after omitting tildes,

$$\partial_t\rho + \mathrm{div}\,(\rho\boldsymbol{v}) = 0, \tag{5}$$

$$\rho(\partial_t\boldsymbol{v} + \boldsymbol{v}\cdot\nabla\boldsymbol{v}) - \nu\Delta\boldsymbol{v} - (\nu+\nu')\nabla\mathrm{div}\,\boldsymbol{v} + \gamma^2\nabla p(\rho) = \nu\rho\boldsymbol{e}_1. \tag{6}$$

Here ν, ν' and γ are the non-dimensional parameters given by $\nu = \frac{\mu}{\rho_*\ell V}$, $\nu' = \frac{\mu'}{\rho_*\ell V}$ and $\gamma = \frac{\sqrt{P'(\rho_*)}}{V}$. The assumption $P'(\rho_*) > 0$ is reduced to the form $p'(1) = 1$. To derive 6 we have used the relation $\frac{\ell g}{V^2} = \nu$.

The system 5–6 is then considered on the two-dimensional infinite layer:

$$\{x = (x_1, x_2)\,;\, x_1 \in \mathbb{R},\, 0 < x_2 < 1\}.$$

Our purpose is to show the existence of wave trains of 5–6 bifurcating from the plane Poiseuille flow. Under the above non-dimensionalization, the plane Poiseuille flow is transformed into $u_s = {}^\top(\rho_s, \boldsymbol{v}_s)$, where

$$\rho_s = 1, \quad \boldsymbol{v}_s = {}^\top(v_s^1(x_2), 0), \quad v_s^1(x_2) = \frac{1}{2}(-x_2^2 + x_2).$$

We substitute $u(t) = {}^\top(\phi(t), \boldsymbol{w}(t))$ with $\phi(t) = \gamma^2(\rho(t) - \rho_s)$ and $\boldsymbol{w}(t) = \boldsymbol{v}(t) - \boldsymbol{v}_s$ into 5 and 6 to obtain the equations for the perturbation. Since $\rho_s = 1$, $\boldsymbol{v}_s = {}^\top(v_s^1(x_2), 0)$ and $-\Delta\boldsymbol{v}_s = \boldsymbol{e}_1$, we have

$$\partial_t\phi + v_s^1\partial_{x_1}\phi + \gamma^2\mathrm{div}\,\boldsymbol{w} = f^0, \tag{7}$$

$$\partial_t\boldsymbol{w} - \nu\Delta\boldsymbol{w} - \tilde{\nu}\nabla\mathrm{div}\,\boldsymbol{w} + \nabla\phi - \frac{\nu}{\gamma^2}\phi\boldsymbol{e}_1 + v_s^1\partial_{x_1}\boldsymbol{w} + (\partial_{x_2}v_s^1)w^2\boldsymbol{e}_1 = \boldsymbol{f}. \tag{8}$$

Here $\tilde{\nu} = \nu + \nu'$; and f^0 and $\boldsymbol{f} = {}^\top(f^1, f^2)$ are the nonlinear terms:

$$f^0 = -\mathrm{div}\,(\phi\boldsymbol{w}),$$

$$\boldsymbol{f} = -\boldsymbol{w}\cdot\nabla\boldsymbol{w} - \frac{\phi}{\gamma^2+\phi}\left(\nu\Delta\boldsymbol{w} + \frac{\nu}{\gamma^2}\phi\boldsymbol{e}_1 + \tilde{\nu}\nabla\mathrm{div}\,\boldsymbol{w}\right) + P^{(1)}(\phi)\phi\nabla\phi$$

with

$$P^{(1)}(\phi) = \frac{1}{\gamma^2+\phi}\left(1 - \int_0^1 p''(1+\theta\gamma^{-2}\phi)\,d\theta\right).$$

The boundary conditions are written as

$$\boldsymbol{w}|_{x_2=0,1} = \boldsymbol{0}, \quad \phi,\ \boldsymbol{w}\colon\ \tfrac{2\pi}{\alpha}\text{-periodic in } x_1, \tag{9}$$

where α is a given positive number.

2.2. **Notation.** For a given $\alpha > 0$, we denote the basic period cell by $\Omega_\alpha = \mathbb{T}_\alpha \times (0,1)$, where $\mathbb{T}_\alpha = \mathbb{R}/\frac{2\pi}{\alpha}\mathbb{Z}$.

We denote by $L^2(\Omega_\alpha)$ the usual L^2 space on Ω_α with norm $\|\cdot\|_2$, and likewise, by $H^k(\Omega_\alpha)$ the k th order L^2 Sobolev space on Ω_α with norm $\|\cdot\|_{H^k}$. We also denote by $C_0^\infty(\Omega_\alpha)$ the space of functions in $C^\infty(\Omega_\alpha)$ which vanish near $x_2 = 0, 1$. We define $H_0^1(\Omega_\alpha)$ by the $H^1(\Omega_\alpha)$ -closure of $C_0^\infty(\Omega_\alpha)$.

The inner product of $f_j \in L^2(\Omega_\alpha)$ $(j = 1, 2)$ is denoted by

$$(f_1, f_2) = \int_{\Omega_\alpha} f_1(x)\overline{f_2(x)}\,dx.$$

Here $\overline{z}$ denotes the complex conjugate of z.

We define $\langle\phi\rangle$ by

$$\langle\phi\rangle = \frac{1}{|\Omega_\alpha|}\int_{\Omega_\alpha} \phi(x)\,dx.$$

We also define $L_*^2(\Omega_\alpha)$ by

$$L_*^2(\Omega_\alpha) = \{\phi \in L^2(\Omega_\alpha);\ \langle\phi\rangle = 0\}.$$

Furthermore, we set

$$H_*^k(\Omega_\alpha) = H^k(\Omega_\alpha) \cap L_*^2(\Omega_\alpha).$$

The inner product of $u_j = {}^\top(\phi_j, \boldsymbol{w}_j) \in L^2(\Omega_\alpha)$ $(j = 1, 2)$ is defined by

$$\langle u_1, u_2\rangle = \frac{1}{\gamma^2}\int_{\Omega_\alpha} \phi_1(x)\overline{\phi_2(x)}\,dx + \int_{\Omega_\alpha} \boldsymbol{w}_1(x)\cdot\overline{\boldsymbol{w}_2(x)}\,dx.$$

In the following we omit Ω_α in $L^2(\Omega_\alpha)$, $H^k(\Omega_\alpha)$, $\cdots$, and etc., and simply write them as L^2, H^k, $\cdots$, and etc.

The resolvent set of a closed operator A is denoted by $\rho(A)$ and the spectrum of A by $\sigma(A)$. We denote the null space and the range of A by $N(A)$ and $R(A)$, respectively.

3. **Instability of plane Poiseuille flow.** In this section we briefly state the instability result on the plane Poiseuille flow obtained in [6].

We consider the linearized problem

$$\partial_t\phi + v_s^1\partial_{x_1}\phi + \gamma^2\mathrm{div}\,\boldsymbol{w} = 0, \tag{10}$$

$$\partial_t\boldsymbol{w} - \nu\Delta\boldsymbol{w} - \tilde{\nu}\nabla\mathrm{div}\,\boldsymbol{w} + \nabla\phi - \frac{\nu}{\gamma^2}\phi\boldsymbol{e}_1 + v_s^1\partial_{x_1}\boldsymbol{w} + (\partial_{x_2}v_s^1)w^2\boldsymbol{e}_1 = \boldsymbol{0}, \tag{11}$$

$$\boldsymbol{w}|_{x_2=0,1} = \boldsymbol{0}, \quad \phi, \boldsymbol{w}:\ \tfrac{2\pi}{\alpha}\text{-periodic in } x_1, \tag{12}$$

$$u|_{t=0} = u_0 = {}^\top(\phi_0, \boldsymbol{w}_0). \tag{13}$$

We define the space X by

$$X = L_*^2 \times (L^2)^2$$

and the operator L on X by

$$D(L) = \left\{u = {}^\top(\phi, \boldsymbol{w}) \in X;\ \boldsymbol{w} \in (H_0^1)^2,\ Lu \in X\right\},$$

$$L = \begin{pmatrix} v_s^1\partial_{x_1} & \gamma^2\mathrm{div} \\ \nabla & -\nu\Delta - \tilde{\nu}\nabla\mathrm{div} \end{pmatrix} + \begin{pmatrix} 0 & 0 \\ -\frac{\nu}{\gamma^2}\boldsymbol{e}_1 & v_s^1\partial_{x_1} + (\partial_{x_2}v_s^1)\boldsymbol{e}_1{}^\top\boldsymbol{e}_2 \end{pmatrix}.$$

Here $\boldsymbol{e}_2 = {}^\top(0, 1)$. Following the argument in [4] one can show that $-L$ generates a C_0-semigroup in X.

We have the following instability criterion for the plane Poiseuille flow.

Theorem 3.1. ([6]) *There exist positive constants r_0 and η_0 such that if $\alpha \le r_0$, then*

$$\sigma(-L) \cap \{\lambda \in \mathbb{C}; |\lambda| \le \eta_0\} = \{\lambda_{\alpha k};\ |k| = 1, \cdots, n_0\}$$

for some $n_0 \in \mathbb{N}$, where $\lambda_{\alpha k}$ are simple eigenvalues of $-L$ that satisfy

$$\lambda_{\alpha k} = -\frac{i}{6}(\alpha k) + \kappa_0(\alpha k)^2 + O(|\alpha k|^3)$$

as $\alpha k \to 0$. Here κ_0 is the number given by

$$\kappa_0 = \frac{1}{12\nu}\left[\left(\frac{1}{280} - \gamma^2\right) - \frac{\nu}{30\gamma^2}(3\nu + \nu')\right].$$

As a consequence, if $\gamma^2 < \frac{1}{280}$ and $\nu(3\nu + \nu') < 30\gamma^2\left(\frac{1}{280} - \gamma^2\right)$, then $\kappa_0 > 0$ and the plane Poiseuille flow $u_s = {}^\top(\phi_s, \boldsymbol{v}_s)$ is linearly unstable.

We note that the eigenspace for $\lambda_{\alpha k}$ is spanned by a function of the form $u(x_2)e^{i\alpha k x_1}$ with an eigenfunction $u(x_2)$ for the eigenvalue $\lambda_{\alpha k}$ of $-L_{\eta,k}$, where $L_{\eta,k}$ is an operator given in 18 below. See [6, Sections 4–6].

4. **Bifurcation of wave trains.** In this section we state the result on the existence of wave trains bifurcating from the plane Poiseuille flow after its becoming unstable.

We fix γ in such a way that $\frac{1}{280} - \gamma^2 > 0$; and we regard ν as a bifurcation parameter. We denote the eigenvalue $\lambda_{\alpha k}$ by $\lambda_{\alpha k}(\nu)$:

$$\lambda_{\alpha k} = \lambda_{\alpha k}(\nu),$$

and the linearized operator L by L_ν:

$$L = L_\nu.$$

We take $\tilde{\nu}_0 > 0$ in such a way that $\kappa_0 = 0$, where κ_0 is the coefficient of $(\alpha k)^2$ of $\lambda_{\alpha k}(\nu)$ described in Theorem 3.1. Then, a perturbation argument shows that for each $0 < \alpha \ll 1$, there exists $\nu_0 > 0$ such that $\operatorname{Re}\lambda_{\pm\alpha}(\nu_0) = 0$, $\operatorname{Re}\lambda_{\pm\alpha}(\nu) < 0$ iff $\nu > \nu_0$ and $\operatorname{Re}\lambda_{\pm\alpha}(\nu) > 0$ iff $\nu < \nu_0$; if $\alpha \ll 1$, then $\lambda_{\pm\alpha}(\nu)$ cross the imaginary axis from left to right at $\nu = \nu_0$ when ν is decreased. See [7, Section 6].

On the spectrum, we assume the following:

$$\sigma(-L_{\nu_0}) \cap \{\lambda; \operatorname{Re}\lambda = 0\} = \{\lambda_\alpha(\nu_0), \lambda_{-\alpha}(\nu_0)\}. \tag{14}$$

The result on the bifurcation of wave trains is now stated as follows.

Theorem 4.1. ([7]) *Assume that 14 holds true. Then there is a solution branch $\{\nu, u\} = \{\nu_\varepsilon, u_\varepsilon\}$ ($|\varepsilon| \ll 1$) such that*

$$\begin{aligned}
&\nu_\varepsilon = \nu_0 + O(\varepsilon),\\
&u_\varepsilon = u_\varepsilon(x_1 - c_\varepsilon t, x_2), \quad u_\varepsilon(x_1 + \tfrac{2\pi}{\alpha}, x_2) = u_\varepsilon(x_1, x_2),\\
&u_\varepsilon(x_1, x_2) = \varepsilon \begin{pmatrix} 1 \\ \frac{1}{2\gamma^2}(-x_2^2 + x_2) \\ 0 \end{pmatrix} \tfrac{\sqrt{2}}{2}\cos\alpha x_1(1 + O(\alpha)) + O(\varepsilon^2),\\
&c_\varepsilon = \tfrac{1}{6} + O(\varepsilon).
\end{aligned}$$

5. **Sketch of Proof of Theorem 4.1.** In this section we give a sketch of proof of Theorem 4.1. Details can be found in [7, Sections 5–6].

We introduce a new bifurcation parameter $\eta = \nu - \nu_0$. We then write $L_{\eta+\nu_0}$ as L_η for simplicity in notation.

Let us consider the nonlinear problem which is written in the form:

$$\partial_t \tilde{u} + L_\eta \tilde{u} = F(\eta, \tilde{u}), \tag{15}$$

where $F(\eta, \tilde{u})$ denotes the nonlinear term.

Our task is to find a nontrivial solution in the form $\tilde{u}(x_1, x_2, t) = u(x_1 - ct, x_2)$. Substituting this into 15, we have

$$\mathcal{L}_{c,\eta} u = F(\eta, u), \tag{16}$$

where $\mathcal{L}_{c,\eta} = L_\eta - c\partial_{x_1}$.

To solve 16, we investigate the spectrum of $\mathcal{L}_{c,\eta}$.

5.1. **Spectrum of $-L_0$.** We consider the resolvent problem for $-L_\eta$:

$$\lambda u + L_\eta u = F. \tag{17}$$

To investigate this problem, u and F are expanded into the Fourier series in x_1:

$$u = \sqrt{\frac{\alpha}{2\pi}} \sum_{k\in\mathbb{Z}} u_k(x_2) e^{i\alpha k x_1}, \quad u_k = {}^\top(\phi_k, \boldsymbol{w}_k),$$

$$F = \sqrt{\frac{\alpha}{2\pi}} \sum_{k\in\mathbb{Z}} F_k(x_2) e^{i\alpha k x_1}, \quad F_k = {}^\top(f_k^0, \boldsymbol{f}_k)$$

with $\int_0^1 \phi_0(x_2)\, dx_2 = \int_0^1 f_0^0(x_2)\, dx_2 = 0$. The problem is then reduced to the following problem for each $k \in \mathbb{Z}$:

$$(\lambda + L_{\eta,k}) u_k = F_k. \tag{18}$$

Here $L_{\eta,k}$ is the operator on $L_k^2(0,1) \times L^2(0,1)^2$ obtained by replacing ∂_{x_1} in L_η by $i\alpha k$ with domain $D(L_{\eta,k}) = \{u_k = {}^\top(\phi_k, \boldsymbol{w}_k) \in L_k^2(0,1) \times L^2(0,1)^2; \boldsymbol{w}_k \in H_0^1(0,1)^2, L_{\eta,k} u_k \in L_k^2(0,1) \times L^2(0,1)^2\}$, where

$$L_k^2(0,1) = \begin{cases} L^2(0,1) & (k \neq 0) \\ L^2(0,1) \cap \{\phi; \int_0^1 \phi(x_2)\, dx_2 = 0\} & (k = 0). \end{cases}$$

We denote by $\tilde{L}_\eta$ the extension of L_η to $\tilde{X} = L^2 \times (L^2)^2$, and likewise, we define an operator $\tilde{L}_{\eta,k}$ on $L^2(0,1) \times L^2(0,1)^2$ by the extension of $L_{\eta,k}$ to $L^2(0,1) \times L^2(0,1)^2$. It follows that $\tilde{L}_{\eta,k} = L_{\eta,k}$ when $k \neq 0$ and $L_{\eta,0}$ is the restriction of $\tilde{L}_{0,\eta}$ to $L_0^2(0,1) \times L^2(0,1)^2$. The adjoint operator $\tilde{L}_\eta^*$ with respect to the inner product $\langle \cdot, \cdot \rangle$ is given by

$$\tilde{L}_\eta^* = \begin{pmatrix} -v_s^1 \partial_{x_1} & -\nu^\top \boldsymbol{e}_1 - \gamma^2 \mathrm{div} \\ -\nabla & -\nu\Delta - \tilde{\nu}\nabla \mathrm{div} - v_s^1 \partial_{x_1} + (\partial_{x_2} v_s^1) \boldsymbol{e}_2 {}^\top \boldsymbol{e}_1 \end{pmatrix}.$$

The adjoint operators $\tilde{L}_{\eta,k}^*$ of $\tilde{L}_{\eta,k}$ are similarly given.

Since X is an invariant space of $\tilde{L}_\eta$, if λ is an eigenvalue of $-L_\eta$, then the eigenprojection for λ of $-L_\eta$ is the restriction of the eigenprojection for λ of $-\tilde{L}_\eta$. The same assertions also hold for eigenprojections of $L_{\eta,0}$ and $\tilde{L}_{\eta,0}$.

Under the assumption 14, we have the following information on the spectrum. In the following we set

$$\lambda_{\pm\alpha}(\nu_0) = \pm ia,$$

where $a = -\frac{\alpha}{6}(1+O(\alpha^2)) \in \mathbb{R}\setminus\{0\}$.

Proposition 1. *There holds* $\sigma(-L_{0,\pm 1}) \cap \{\lambda; \mathrm{Re}\,\lambda = 0\} = \{\pm ia\}$, *where* $\pm ia$ *are isolated simple eigenvalues of* $-L_{0,\pm1}$, $N(\pm ia + L_{0,\pm1}) = \mathrm{span}\{v_{\pm1}\}$, *and* $v_{-1} = \overline{v_{+1}}$. *Furthermore, there exists a positive constant* β *such that* $\sigma(-L_{0,k}) \subset \{\lambda; |\mathrm{Re}\,\lambda| \geq \beta\}$ *for all* $k \in \mathbb{Z}$ *with* $k \neq \pm 1$.

As for the eigenprojections for the eigenvalues $\pm ia$, we have the following

Proposition 2. *The eigenprojections* $\Pi_\pm$ *for the eigenvalues* $\pm ia$ *are given by* $\Pi_\pm u = \langle\langle u, v^*_{\pm1}\rangle\rangle v_{\pm1}$, *where* $N(\mp ia + \tilde{L}^*_{0,\pm1}) = \mathrm{span}\{v^*_{\pm1}\}$, $\langle\langle v_{\pm1}, v^*_{\pm1}\rangle\rangle = 1$. *Here, for* $u_j = {}^\top(\phi_j, \boldsymbol{w}_j)$
$\in L^2(0,1)$ $(j = 1,2)$, *the symbol* $\langle\langle u_1, u_2\rangle\rangle$ *denotes*

$$\langle\langle u_1, u_2\rangle\rangle = \frac{1}{\gamma^2}\int_0^1 \phi_1(x_2)\overline{\phi_2(x_2)}\,dx_2 + \int_0^1 \boldsymbol{w}_1(x_2)\cdot\overline{\boldsymbol{w}_2(x_2)}\,dx_2.$$

We thus conclude that $\sigma(-L_0)$ has the following properties.

Proposition 3. *There holds* $\sigma(-L_0) \cap \{\lambda; \mathrm{Re}\,\lambda = 0\} = \{\pm ia\}$. *Here* $\pm ia$ *are isolated simple eigenvalues of* $-L_0$ *and* $N(\pm ia + L_0) = \mathrm{span}\{V_\pm\}$, *where* $V_\pm = v_{\pm1}(x_2)e^{\pm i\alpha x_1}$.

We set $V^*_\pm = \frac{\alpha}{2\pi}v^*_{\pm1}(x_2)e^{\pm i\alpha x_1}$. It then follows that $-\tilde{L}^*_0 V^*_\pm = \mp ia V^*_\pm$, $\langle V_\pm, V^*_\pm\rangle = 1$, $\langle V_\pm, V^*_\mp\rangle = 0$, and the eigenprojections $P_\pm$ for $\pm ia$ of $-L_0$ are given by

$$P_\pm V = \langle V, V^*_\pm\rangle V_\pm.$$

5.2. **Spectrum of** $-\mathcal{L}_{c_0,0}$**.** We next investigate the spectrum of the critical operator $-\mathcal{L}_{c_0,0}$.

Proposition 4. *Set* $c_0 = -\frac{a}{\alpha}$. *Then* $\sigma(-\mathcal{L}_{c_0,0}) \cap \{\lambda; \mathrm{Re}\,\lambda = 0\} = \{0\}$, *where* 0 *is an isolated semisimple eigenvalue of* $-\mathcal{L}_{c_0,0}$ *and* $N(-\mathcal{L}_{c_0,0}) = \mathrm{span}\{V_+, V_-\}$ *with* $V_- = \overline{V_+}$.

Let us consider the eigenprojection for the eigenvalue 0 of $-\mathcal{L}_{c_0,0}$. We have

$$N(-\mathcal{L}_{c_0,0}) = \mathrm{span}\{V_1, V_2\}, \quad \langle V_j, V_k^*\rangle = \delta_{jk}, \quad j,k = 1,2.$$

Here

$$V_1 = \sqrt{2}\mathrm{Re}\,V_+, \quad V_2 = \sqrt{2}\mathrm{Im}\,V_+, \quad V_1^* = \sqrt{2}\mathrm{Re}\,V_+^*, \quad V_2^* = \sqrt{2}\mathrm{Im}\,V_+^*.$$

We define the symbol $[\![u]\!]_j$ $(j=1,2)$ by $[\![u]\!]_j = \langle u, V_j^*\rangle$. We set P, P_1 and P_2 as

$$Pu = P_1 u + P_2 u, \quad P_j u = [\![u]\!]_j V_j \quad (j = 1,2).$$

We have the following properties of P_j.

Proposition 5. P *is the eigenprojection for the eigenvalue* 0 *of* $-\mathcal{L}_{c_0,0}$; *and*

$$R(P_j) = \mathrm{span}\{V_j\}, \quad P_j^2 = P_j, \quad P_jP_k = O\ (j \neq k).$$

For each nonnegative integer k, P_j *are bounded from* $L^2_* \times (L^2)^2$ *to* $H^k_* \times (H^k)^2$:

$$\|P_j u\|_{H^k\times(H^k)^2} \leq C\|u\|_2.$$

Furthermore, $u \in R(I - P_j)$ *if and only if* $[\![u]\!]_j = 0$.

5.3. **Lyapunov-Schmidt reduction.** To look for nontrivial solutions of 16, we employ the Lyapunov-Schmidt reduction. We wet

$$c = c_0 + \varepsilon\sigma, \quad u = \varepsilon(V_1 + \varepsilon V), \quad V \in R(Q), \quad Q = I - P,$$

Here ε is a small parameter.

We write

$$L_\eta = L_0 + \eta K_0,$$

where

$$K_0 = \frac{1}{\eta}(L_\eta - L_0) = \begin{pmatrix} 0 & 0 \\ -\frac{1}{\gamma^2}\boldsymbol{e}_1 & -\Delta - \nabla\mathrm{div} \end{pmatrix}.$$

It follows that

$$\mathcal{L}_{c,\eta} = \mathcal{L}_{c_0,0} - \varepsilon\sigma\partial_{x_1} + \eta K_0.$$

Setting $\eta = \varepsilon\omega$, we rewrite the problem 16 as

$$\mathcal{L}_{c_0,0}V - \sigma\partial_{x_1}(V_1 + \varepsilon V) + \omega K_0(V_1 + \varepsilon V) = \frac{1}{\varepsilon^2}F(\varepsilon\omega, \varepsilon(V_1 + \varepsilon V)). \tag{19}$$

We write the right-hand side as

$$\frac{1}{\varepsilon^2}F(\varepsilon\omega, \varepsilon(V_1 + \varepsilon V)) = -N[V_1 + \varepsilon V](V_1 + \varepsilon V) + G(\varepsilon, \varepsilon\omega, V_1 + \varepsilon V),$$

where

$$N[\tilde{u}]u = {}^\top(\mathrm{div}\,(\phi\tilde{\boldsymbol{w}}), \boldsymbol{0})$$

for $\tilde{u} = {}^\top(\tilde{\phi}, \tilde{\boldsymbol{w}})$ and $u = {}^\top(\phi, \boldsymbol{w})$, and

$$G(\varepsilon, \omega, u) = {}^\top(0, \boldsymbol{g}(\varepsilon, \omega, u))$$

with

$$\begin{aligned} \boldsymbol{g}(\varepsilon, \omega, u) &= -\boldsymbol{w}\cdot\nabla\boldsymbol{w} - \tfrac{\phi}{\gamma^2+\varepsilon\phi}\Big((\nu_0+\omega)\Delta\boldsymbol{w} + \tfrac{(\nu_0+\omega)}{\gamma^2}\phi\boldsymbol{e}_1 + (\hat{\nu}_0+\omega)\nabla\mathrm{div}\,\boldsymbol{w}\Big) \\ &\quad + P^{(1)}(\varepsilon\phi)\phi\nabla\phi \end{aligned}$$

for $u = {}^\top(\phi, \boldsymbol{w})$, where $\hat{\nu}_0 = \nu_0 + \nu'$.

We now decompose 19 into a finite-dimensional part and an infinite-dimensional part.

Taking the inner product of 19 with V_j^* $(j = 1, 2)$ and applying Q to 19 we have

$$\begin{aligned} \omega[\![K_0V_1]\!]_1 &= -\varepsilon\omega[\![K_0V]\!]_1 - [\![N[V_1+\varepsilon V](V_1+\varepsilon V)]\!]_1 \\ &\quad + [\![G(\varepsilon, \varepsilon\omega, V_1+\varepsilon V)]\!]_1, \end{aligned} \tag{20}$$

$$\begin{aligned} \omega[\![K_0V_1]\!]_2 + \alpha\sigma &= -\varepsilon\omega[\![K_0V]\!]_2 - [\![N[V_1+\varepsilon V](V_1+\varepsilon V)]\!]_2 \\ &\quad + [\![G(\varepsilon, \varepsilon\omega, V_1+\varepsilon V)]\!]_2, \end{aligned} \tag{21}$$

$$\begin{aligned} &\omega QK_0V_1 + (\mathcal{L}_{c_0,0} - \varepsilon\sigma Q\partial_{x_1} + \varepsilon QN[V_1+\varepsilon V])\,V \\ &= -\varepsilon\omega QK_0V - QN[V_1+\varepsilon V]V_1 + QG(\varepsilon, \varepsilon\omega, V_1+\varepsilon V). \end{aligned} \tag{22}$$

Here we have used $[\![\partial_{x_1}(V_1+\varepsilon V)]\!]_1 = 0$ and $[\![\partial_{x_1}(V_1+\varepsilon V)]\!]_2 = -\alpha$.

In the classical bifurcation theory, the nonlinearity is regarded as a perturbation of the linearized part. This does not work well for the problem under consideration, since the term $\varepsilon QN[V_1+\varepsilon V]V$ causes derivative loss in a standard setting of the classical bifurcation theory. We thus put $\varepsilon QN[V_1+\varepsilon V]V$ on the left-hand side of 22 to regard it as a part of the principal part as in the proof of the local solvability of the time quasilinear evolution problem. This is the main difference to the case of the incompressible problem.

The problem 20–22 is now formulated in the form

$$T(\varepsilon,\sigma,V)U = \mathcal{F}(\varepsilon,U), \tag{23}$$

where $U = {}^\top(\omega,\sigma,V) \in \mathbb{R}\times\mathbb{R}\times X^2$. Here and in what follows we define the function space X^ℓ by $X^\ell = H^\ell_* \times (H^{\ell+1}\cap H^1_0)^2$, $\ell = 1,2$. The map $T(\varepsilon,\sigma,V)U$ is defined as follows; for a given $(\tilde\sigma,\tilde V) \in \mathbb{R}\times X^2$, we define the linear map $T(\varepsilon,\tilde\sigma,\tilde V)$ by

$$T(\varepsilon,\tilde\sigma,\tilde V) : \mathbb{R}\times\mathbb{R}\times QX^\ell \ \to\ \mathbb{R}\times\mathbb{R}\times Q(H^\ell\times(H^{\ell-1})^2),\ \ell = 1,2,$$

$$T(\varepsilon,\tilde\sigma,\tilde V) \ = \ \begin{pmatrix} [\![K_0V_1]\!]_1 & 0 & 0 \\ [\![K_0V_1]\!]_2 & \alpha & 0 \\ QK_0V_1 & 0 & \mathcal{L}_{c_0,0} - \varepsilon\tilde\sigma Q\partial_{x_1} + \varepsilon QN[V_1+\varepsilon\tilde V] \end{pmatrix}.$$

The map $\mathcal{F}(\varepsilon,U)$ on the right-hand side of 23 is defined in such a way that $\mathcal{F}(\varepsilon,U) = {}^\top(\mathcal{F}_1(\varepsilon,U),\mathcal{F}_2(\varepsilon,U),\mathcal{F}_3(\varepsilon,U))$ with $\mathcal{F}_j(\varepsilon,U)$ $(j = 1,2,3)$ given by the right-hand side of 20, 21, 22, respectively.

If we would have a suitable invertibility of the map $T(\varepsilon,\tilde\sigma,\tilde V)$ we could solve the problem 23. One can show the following

Proposition 6. *If $0 < \alpha \ll 1$, then the following assertions hold true.*

(i) $[\![K_0V_1]\!]_1 > 0$.

(ii) *For a given positive number M, there exists a positive constant ε_1 such that if $|\varepsilon| \le \varepsilon_1$ and $|\tilde\sigma| + \|\tilde V\|_{X^2} \le M$, then $\mathcal{L}_{c_0,0} - \varepsilon\tilde\sigma Q\partial_{x_1} + \varepsilon QN[V_1+\varepsilon\tilde V]$ has a bounded inverse from $Q(H^\ell_*\times(H^{\ell-1})^2)$ to QX^ℓ $(\ell = 1,2)$.*

Proposition 6 (i) can be proved by a perturbation argument for $0 < \alpha \ll 1$. To prove Proposition 6 (ii), we apply the Matsumura-Nishida energy method [12] and the results on the resolvent problem for transport equation by Heywood and Padula [3]; see [7, Section 6].

Proposition 6 implies the invertibility of $T(\varepsilon,\tilde\sigma,\tilde V)$ as follows.

Proposition 7. *Under the assumption of* Proposition 6, *the operator $T(\varepsilon,\tilde\sigma,\tilde V)$ has a bounded inverse from $\mathbb{R}\times\mathbb{R}\times Q(H^\ell_*\times(H^{\ell-1})^2)$ to $\mathbb{R}\times\mathbb{R}\times QX^\ell$ $(\ell = 1,2)$, and the estimates*

$$\|T(\varepsilon,\tilde\sigma,\tilde V)^{-1}U\|_{\mathbb{R}\times\mathbb{R}\times X^\ell} \le C_1\|U\|_{\mathbb{R}\times\mathbb{R}\times H^\ell\times(H^{\ell-1})^2}, \quad \ell = 1,2,$$

hold uniformly for $U = {}^\top(\omega,\sigma,V)$.

The nonlinear map $\mathcal{F}(\varepsilon,U)$ satisfies the following estimates. Let C_S be the positive constant appearing in the Sobolev inequality: $\|\phi\|_{L^\infty} \le C_S\|\phi\|_{H^2}$.

Proposition 8. *For given $M \in (0,\frac{\gamma^2}{2C_S}]$, there exists a positive constant ε_2 such that if $|\varepsilon| \le \varepsilon_2$, $\|U\|_{\mathbb{R}\times\mathbb{R}\times X^2} \le M$ and $\|U^{(j)}\|_{\mathbb{R}\times\mathbb{R}\times X^2} \le M$ $(j = 1,2)$, then the estimates*

$$\|\mathcal{F}(\varepsilon,U) - \mathcal{F}(0,0)\|_{\mathbb{R}\times\mathbb{R}\times H^2\times(H^1)^2} \le C(M)M|\varepsilon|,$$

$$\|\mathcal{F}(\varepsilon,U^{(1)}) - \mathcal{F}(\varepsilon,U^{(2)})\|_{\mathbb{R}\times\mathbb{R}\times H^1\times(L^2)^2} \le C(M)|\varepsilon|\|U^{(1)} - U^{(2)}\|_{\mathbb{R}\times\mathbb{R}\times X^1},$$

hold true, where $C(M) > 0$ is a nondecreasing continuous function of M.

5.4. **Iteration argument.** To obtain a solution branch of wave trains, we employ an iteration argument based on Propositions 7 and 8.

We construct approximate solutions $U^{(n)} = {}^\top(\omega^{(n)}, \sigma^{(n)}, V^{(n)})$ $(n \geq 1)$ as follows. Let $U^{(1)}$ be the solution of

$$\begin{aligned} T(0,0,0)U^{(1)} &= \mathcal{F}(0,0) \\ &= {}^\top([\![F(0,V_1)]\!]_1, [\![F(0,V_1)]\!]_2, QF(0,V_1)). \end{aligned}$$

Here $F(0,V_1)$ is given by $F(0,V_1) = -N[V_1]V_1 + G(0,0,V_1)$. Proposition 7 shows

$$\|U^{(1)}\|_{\mathbb{R}\times\mathbb{R}\times X^2} \leq C_1\|\mathcal{F}(0,0)\|_{\mathbb{R}\times\mathbb{R}\times H^2\times(H^1)^2} < \infty. \tag{24}$$

Let $M = 2C_1\|\mathcal{F}(0,0)\|_{\mathbb{R}\times\mathbb{R}\times H^2\times(H^1)^2}$ and assume that $|\varepsilon| \leq \min\{\varepsilon_1, \varepsilon_2, \frac{1}{2C_1C(M)}\}$. Then $U^{(n)}$ $(n \geq 2)$ can be defined by the solution of

$$T(\varepsilon, \sigma^{(n-1)}, V^{(n-1)})U^{(n)} = \mathcal{F}(\varepsilon, U^{(n-1)}). \tag{25}$$

By using Propositions 7 and 8, one can show, with an inductive argument, that

$$\|U^{(n)}\|_{\mathbb{R}\times\mathbb{R}\times X^2} \leq M$$

for all $n \geq 1$, and that $\{U^{(n)}\}$ is a Cauchy sequence in $\mathbb{R}\times\mathbb{R}\times X^1$ for sufficiently small ε. It then follows that if $|\varepsilon| \leq \varepsilon_0$ for some small positive constant ε_0, there exists $U = {}^\top(\omega, \sigma, V) \in \mathbb{R}\times\mathbb{R}\times X^2$ satisfying $T(\varepsilon,\sigma,V)U = \mathcal{F}(\varepsilon, U)$. The desired branch of wave trains is now obtained as $\nu = \nu_0 + \varepsilon\omega$, $u = \varepsilon V_1(x_1 - ct, x_2) + \varepsilon^2 V(x_1 - ct, x_2)$, $c = c_0 + \varepsilon\sigma$.

REFERENCES

[1] M. Crandall and P. Rabinowitz, Bifurcation from simple eigenvalues, *J. Functional Analysis*, **8** (1971), 321–340.

[2] M. Crandall and P. Rabinowitz, Bifurcation, perturbation of simple eigenvalues and linearized stability, *Arch. Rational Mech. Anal.*, **52** (1973), 161–180.

[3] J. G. Heywood and M. Padula, On the steady transport equation, in *Fundamental Directions in Mathematical Fluid Mechanics* (ed. by G. P. Galdi, J. G. Heywood, R, Rannacher), Birkhäuser, Basel, (2000), 149–170.

[4] G. Iooss and M. Padula, Structure of the linearized problem for compressible parallel fluid flows, *Ann. Univ. Ferrara, Sez. VII*, **43** (1997), 157–171.

[5] V. I. Iudovich, Secondary flows and fluid instability between rotating cylinders, *Prikl. Mat. Meh.*, **30** (1966), 688–698 (Russian); translated as *J. Appl. Math. Mech.*, **30** (1966), 822–833.

[6] Y. Kagei and T. Nishida, Instability of plane Poiseuille flow in viscous compressible gas, *J. Math. Fluid Mech.*, **17** (2015), 129–143.

[7] Y. Kagei and T. Nishida, Traveling waves bifurcating from plane Poiseuille flow of the compressible Navier-Stokes equation, *Arch. Rational Mech. Anal.*, **231** (2019), 1–44.

[8] Y. Kagei, T. Nishida and Y. Teramoto, Bifurcation of the compressible Taylor vortex, in preparation.

[9] K. Kirchgässner and H. Kielhöfer, Stability and bifurcation in fluid dynamics, Rocky Mountain Consortium Symposium on Nonlinear Eigenvalue Problems (Santa Fe, N.M., 1971), *Rocky Mountain J. Math.*, **3** (1973), 275–318.

[10] T. Nishida, M. Padula and Y. Teramoto, Heat convection of compressible viscous fluids: I, *J. Math. Fluid Mech.*, **15** (2013), 525–536.

[11] T. Nishida, M. Padula and Y. Teramoto, Heat convection of compressible viscous fluids. II, *J. Math. Fluid Mech.*, **15** (2013), 689–700.

[12] A. Matsumura and T. Nishida, Initial boundary value problems for the equations of motion of compressible viscous and heat-conductive fluids. *Comm. Math. Phys.*, **89** (1983) , 445–464.

[13] P. H. Rabinowitz, Existence and nonuniqueness of rectangular solutions of the Bénard problem, *Arch. Rational Mech. Anal.*, **29** (1968), 32–57.

[14] W. Velte, Stabilität und Verzweigung stationärer Lösungen der Navier-Stokesschen Gleichugen beim Taylorproblem, (German), *Arch. Rational Mech. Anal.*, **22** (1966), 1–14.

E-mail address: kagei@math.titech.ac.jp

ON STRUCTURE-PRESERVING HIGH ORDER METHODS FOR CONSERVATION LAWS

Hailiang Liu

Department of Mathematics
Iowa State University
Ames, IA 50010, USA

Abstract. In this paper we review the algorithm development in high order methods for some conservation laws. The emphasis is on our recent contribution in the study of two model classes: Fokker-Planck-type equations and hyperbolic conservation law systems. For the former we will review free-energy-satisfying and positivity-preserving schemes. For the later we will review the general invariant-region-preserving (IRP) limiter, and its application to high order methods for multi-dimensional hyperbolic systems of conservation laws.

1. **Introduction.** Systems of conservation laws for field quantities arise in diverse applications. Their solutions may be visualized as evolving observables or propagating waves. When the system is nonlinear, solution profiles can become steeper as shocks or even concentrated as measures, propagation of these profiles cause mathematical and numerical challenges in solving systems of conservation laws.

We are interested in building structure-preserving high order numerical methods for time-dependent conservation laws through model refinement. In this paper we restrict to two model classes: Fokker-Planck-type equations and hyperbolic conservation law systems. By structure preserving algorithms we mean algorithms that can preserve certain intrinsic solution properties at the discrete level.

For Fokker-Planck-type equations, the three main solution properties are mass conservation, non-negativity, and the free energy/entropy dissipation law. We present a second order explicit-implicit scheme that satisfies all three properties at the discrete level, without a strict time step restriction [15], and discuss how to incorporate these solution properties into a high order discontinuous Galerkin (DG) method of arbitrary order [19]. For multi-dimensional hyperbolic conservation law systems endowed with a convex invariant region in the phase space, main solution properties are also in three aspects: solution conservation, invariant region preservation, and the entropy dissipation law. Here we only review the invariant-region-preserving (IRP) limiter designed in [10], and has been tested in [9, 11] for systems of Euler equations.

The organization of this paper is as follows. Section 2 is devoted to two models and their main mathematical properties. Section 3 gives a short account of the direct DG discretization techniques. Section 4 contains a review of the entropy satisfying methods for Fokker-Planck type equations. In section 5 we address the

2000 *Mathematics Subject Classification.* Primary: 65M60, 35L65; Secondary: 35L45.

Key words and phrases. Structure-preserving, High order methods, Conservation laws.

The author is supported by NSF grant DMS1812666.

invariant-region-preserving limiter and its applications to multi-D hyperbolic systems of conservation laws, and finally in section 6 we give some concluding remarks.

2. **PDE models and solution properties.** We begin with the fundamental transport equation

$$\partial_t \rho(t,x) + \nabla_x \cdot (\rho(t,x)u) = 0, \tag{1}$$

for which the probability density space

$$\mathbb{P} = \{\rho, \quad \rho \geq 0, \quad \int \rho = 1\}$$

is invariant. This transport equation alone is not closed, unless u can be related to ρ or governed by further equations.

In dynamics driven by an entropy/ free energy functional $E = E[\rho]$, a direct verification (assuming zero-flux boundary condition) shows that fast decay of E along the transport dynamics (1) can be ensured if $u = -\nabla_x (\delta_\rho E)$, where δ_ρ denotes the usual L^2 variational derivative. We are led to the Fokker-Planck type equation

$$\partial_t \rho = \nabla_x \cdot (\rho \nabla_x \delta_\rho E[\rho]). \tag{2}$$

Dictated by different forms of E, this class includes many equations such as the heat equation, the Fokker-Planck equation [28], the aggregation equation [12, 33] with

$$E = \int \left[\rho \log \rho + V(x)\rho + \frac{1}{2} W * \rho\rho\right] dx,$$

as well as drift-diffusion models such as the Poisson-Nernst-Planck equation [7] and the Keller-Segel system [26]. Equation (2) is a natural gradient flow generated by functionals $E[\rho]$ in Wasserstein distance, directly linked to the minimization problem $\min_{\rho\in\mathbb{P}} E[\rho]$, and has received ample attention in multiple contexts. Solutions to (2) are usually not sensitive to initial distributions, but often to the critical mass, some patterns will emerge as time evolves leading to rich solution structures when coupled with notrivial forces. In order for a numerical method to generate solutions with satisfying long time behavior, it is crucial to preserve some intrinsic solution properties. The main solution properties are

- nonnegativity principle, $\rho_0 \geq 0 \Longrightarrow \rho(t,x) \geq 0 \quad \forall t > 0.$
- mass conservation $\int \rho(t,x)\, dx = \int \rho_0(x)\, dx \quad \forall t > 0.$
- the entropy/energy dissipation inequality

$$\frac{d}{dt} E = -\int \rho |\nabla_x \delta_\rho E|^2 dx \leq 0.$$

These properties are naturally desired for high order numerical schemes.

In Eulerian dynamics of 'fluids', velocity field is governed by the moment equation

$$\partial_t u + u \cdot \nabla u = F.$$

Dictated by different forcing F, examples of such system include the Euler equation, the Navier-Stokes equation, the Euler-Poisson equation, etc. For such Eulerian balance laws the solution is often sensitive to the initial velocity field, leading to the so called critical threshold (CT) phenomena! [16]. We note that gradient flows (2) can be seen as describing the long-time response of an Euler equation with friction [4, system (2.1)].

The simplest hyperbolic balance laws is the system of compressible Euler equations, which belongs to the following model class:

$$\partial_t \mathbf{w} + \sum_{j=1}^{d} \partial_{x_j} F_j(\mathbf{w}) = 0, \quad x \in \mathbb{R}^d, \quad t > 0; \quad \mathbf{w}(0, x) = \mathbf{w}_0, \tag{3}$$

where $\mathbf{w} \in \mathbb{R}^l$ with $l > 1$, and the flux function $F_j(\mathbf{w}) \in \mathbb{R}^l$. It is known that discontinuities can develop at finite time even for smooth initial data [13], hence entropy inequalities should be used to single out the physically relevant solution among many weak solutions. In application problems, the pointwise range of solutions (invariant region) may be known from physical considerations.

The main solution properties, also desired at discrete level, are

- Invariant region $\quad \mathbf{w}_0 \in \Sigma \Longrightarrow \mathbf{w}(t, \cdot) \in \Sigma \quad \forall t > 0.$
- Conservation $\quad \int \mathbf{w}(t, x) dx = \int \mathbf{w}_0(x)\, dx \quad \forall t > 0.$
- Entropy inequality. $\quad \partial_t \eta(\mathbf{w}(t, x)) + \nabla_x \cdot \Psi(\mathbf{w}(t, x)) \leq 0, a.e,$ where (η, Ψ) is an admissible entropy-pair.

In the construction of structure-preserving algorithms for the above two model classes, we have adopted the following strategy:

- *Direct DG (DDG) discretization* of the PDE weak formulation, *choosing proper numerical fluxes* to preserve solution conservation and certain entropy dissipation law, together with Runge–Kutta methods [3] for time discretizaiton.
- *Limiting numerical solutions* to weakly enforce the point-wise solution bounds.

3. **Discretization techniques.** For solutions with either concentrations or discontinuities, the finite volume method as a natural choice can lead to the conservation form of a scheme which is a main ingredient of shock capturing methods for hyperbolic conservation laws. Its high order extension is the Discontinuous Galerkin (DG) method, which is also a class of finite element methods, using a completely discontinuous piecewise polynomial space for the numerical solution and the test functions [8, 29, 30].

For DG methods to conservative PDEs, the key is to design suitable numerical fluxes so that the resulting scheme satisfies the desired properties.

Taking $\partial_t u + \partial_x \cdot J = 0$ as an example, a simple integration by parts over any computational cell I gives

$$\int_I \partial_t u v dx - \int_I J v_x dx + Jv|_{\partial I} = 0.$$

Here ∂I denotes the boundary of I. To complete the DG method, a single valued numerical flux $\hat{J}$ is needed to replace J, and values from inside I for the test function v. For first order scalar conservation laws $J = f(u)$, it is simple to take a monotone flux

$$\hat{J} = \hat{f}(u^-, u^+),$$

including the celebrated Lax-Friedrichs flux and Godunov flux, see [30].

However, for high order PDEs, it is subtle to define $\hat{J}$. For example, for $J = -\partial_x u$, there is a need to define a flux for $\partial_x u$. The average of $\partial_x u$ from traces of derivatives of two neighboring polynomials is known to give a wrong solution for P^1 polynomials! Indeed, various ideas have appeared in the literature to overcome such difficulty, see e.g. [29].

The solution of the heat equation $\partial_t u = \partial_x^2 u$ with initial data g which has only one discontinuity at $x = 0$ gives

$$u_x(t,0) = \frac{1}{\sqrt{4\pi t}}[g] + \{\partial_x g\} + \sqrt{\frac{t}{\pi}}[\partial_x^2 g] + \cdots,$$

where $[\cdot]$ denotes the jump and $\{\cdot\}$ the average. This led us to the flux formula in [23]

$$\hat{u}_x = \beta_0 \frac{[u]}{\Delta x} + \{u_x\} + \sum_{m=1}^{\lfloor k/2 \rfloor} \beta_m (\Delta x)^{2m-1} [\partial_x^{2m} u].$$

Motivated by such formula, we design a refined DDG for diffusion in [24] as

$$\int_{I_j} \partial_t u v dx + \int_{I_j} \partial_x u \partial_x v dx - \widehat{u_x} v \Big|_{\partial I_j} + (\{u\} - u) v_x \Big|_{\partial I_j} = 0,$$

where

$$\hat{u}_x = \beta_0 \frac{[u]}{\Delta x} + \{u_x\} + \beta_1 \Delta x [u_{xx}].$$

In [14], the DDG method is shown L^2 stable in the sense that

$$\int u^2(t,x)dx + \{\cdots\} \le \int u^2(0,x)dx,$$

with $\{\cdots\} \ge 0$ if

$$\beta_0 > \Gamma(\beta_1) := k^2 \left(1 - \beta_1(k^2 - 1) + \frac{\beta_1^2}{3}(k^2 - 1)^2\right).$$

The use of β_0, β_1 provides extra room for incorporating more desired solution properties. Sharp L^2 error estimates are obtained in [14] as

$$\|u_{exact}(t,\cdot) - u(t,\cdot)\|_{L^2_x} \le Ch^{(k+1)},$$

when using polynomial elements of degree k for $\partial_t u + \nabla_x \cdot f(u) = \Delta u$. Moreover, 3rd order maximum-principle-preserving DG scheme (P_2 polynomials) is possible, if

$$\frac{1}{8} < \beta_1 < \frac{1}{4}, \quad \beta_0 \ge 1;$$

as shown for linear Fokker-Planck equations [21], and for a class of convection-diffusion equations [34]. In addition, super-convergence rate of h^{2k} at nodes has been proved by Cao, Liu and Zhang [2] if

$$\beta_1 = \frac{1}{2k(k+1)}; \quad \beta_1 = \frac{1}{12} \quad \text{if } k = 2.$$

The results also include rate h^{k+1} for solution derivatives at Gauss points, h^{k+2} at Lobatto points, and h^{2k} at nodes.

4. **Fokker-Planck-type equations.** We begin with the aggregation model

$$\partial_t \rho = \nabla \cdot (\nabla \rho + \rho \nabla (V(\mathbf{x}) + W * \rho)),$$

where $V(x)$ is a given potential, and W is a symmetric, positive kernal with integral 1. Based on the reformulation of the form

$$\partial_t \rho = \nabla \cdot \left(M \nabla \left(\frac{\rho}{M}\right)\right), \quad M = e^{-V(x) - W*\rho},$$

we introduced in [15] an explicit-implicit scheme:

$$h_j \frac{\rho_j^{n+1} - \rho_j^n}{\Delta t} = h_{j+1/2}^{-1} M_{j+1/2}^n \left(\frac{\rho_{j+1}^{n+1}}{M_{j+1}^n} - \frac{\rho_j^{n+1}}{M_j^n} \right) - h_{j-1/2}^{-1} M_{j-1/2}^n \left(\frac{\rho_j^{n+1}}{M_j^n} - \frac{\rho_{j-1}^{n+1}}{M_{j-1}^n} \right),$$

where ρ_j^n approximates $\rho(t, x_j)$ at time $t = n\Delta t$. This scheme is easy to implement, and is shown to preserve all three desired properties without a strict time step restriction. This has extended and improved upon our earlier works [20, 17]. Extensions to multi-dimensional settings and/or the case when $W * \rho$ is replaced by Ψ solved by a Poisson equation are doable as shown in [15].

It is more challenging to design a high order scheme (3rd or higher order) while three properties remain preserved at the discrete level. Next we show how this can be achieved through a drift-diffusion system. A detailed account can be found in [19], also earlier works [18, 22].

In a mean field approximation of diffusive molecules or ions, one finds the Poisson–Nernst–Planck (PNP) system, $i = 1, \cdots, m$,

$$\partial_t c_i = \nabla \cdot (\nabla c_i + q_i c_i \nabla \psi) \quad x \in \Omega,\ t > 0 \tag{4a}$$

$$-\nabla \cdot (\epsilon(x) \nabla \psi) = \sum_{i=1}^{m} q_i c_i + \rho_0(x), \quad x \in \Omega,\ t > 0, \tag{4b}$$

$$c_i(0, x) = c_i^{\text{in}}(x), \quad x \in \Omega, \tag{4c}$$

$$\frac{\partial \psi}{\partial \mathbf{n}} = \sigma, \quad \frac{\partial c_i}{\partial \mathbf{n}} + q_i c_i \frac{\partial \psi}{\partial \mathbf{n}} = 0, \qquad x \in \partial\Omega,\ t > 0. \tag{4d}$$

Here $c_i = c_i(t, x)$ denotes density of i-th charged particle with charge q_i, at time t and position x, and $\psi = \psi(t, x)$ the electro-static potential. The PNP system has been widely accepted in applications in electrical engineering and electrokinetics, electrochemistry, and biophysics: for example in biological channels [7] or semiconductor devices [25].

Main mathematical features of the system include the conservation of ions, positivity of concentration, and dissipation of the free energy

$$\frac{d}{dt} F = -\sum_i^m \int_\Omega c_i^{-1} |\nabla c_i + c_i \nabla \psi|^2 dx \le 0$$

where

$$F = \int_\Omega \sum_{i=1}^m c_i \log c_i dx + \frac{1}{2} \int_\Omega |\nabla_x \psi|^2 dx.$$

In order to construct a DG scheme to incorporate these solution properties, we reformulate the PNP system (one dimensional case and $\epsilon = 1$, for notational simplicity) as

$$\begin{aligned} \partial_t c_i &= \partial_x (c_i \partial_x p_i),\ i = 1, \cdots, m, \\ p_i &= q_i \psi + \log c_i, \\ -\partial_x^2 \psi &= \sum_{i=1}^m q_i c_i + \rho_0(x). \end{aligned}$$

Let V_h denote a DG solution space (piecewise polynomials), then the DDG spatial discretization when coupled with the Euler forward time discretization gives us the scheme: find $c_{ih}^n, p_{ih}^n, \psi_h^n \in V_h$, $\forall\ v_i, r_i, \eta \in V_h$, $i = 1, \cdots, m$,

$$\int_{I_j} \frac{c_{ih}^{n+1} - c_{ih}^n}{\Delta t} v_i dx = -\int_{I_j} c_{ih}^n \partial_x p_{ih}^n \partial_x v_i dx + \{c_{ih}^n\} \left(\widehat{\partial_x p_{ih}^n} v_i + (p_{ih}^n - \{p_{ih}^n\})\partial_x v_i\right)\Big|_{\partial I_j},$$

$$\int_{I_j} p_{ih}^n r_i dx = \int_{I_j} (q_i \psi_h^n + \log c_{ih}^n) r_i dx,$$

$$\int_{I_j} \partial_x \psi_h^n \partial_x \eta dx - \left(\widehat{\partial_x \psi_h^n} \eta + (\psi_h^n - \{\psi_h^n\})\partial_x \eta\right)\Big|_{\partial I_j} = \int_{I_j} \left[\sum_{i=1}^m q_i c_{ih}^n + \rho_0\right] \eta dx,$$

with flux $\widehat{\partial_x p_{ih}} = Fl(p_{ih})$ and $\widehat{\partial_x \psi_h} = Fl(\psi_h)$, and

$$Fl(w) := \beta_0 \frac{[w]}{h} + \{\partial_x w\} + \beta_1 h [\partial_x^2 w].$$

The numerical solution is shown to have following properties.

Theorem 4.1. [19]

1. *The fully discrete scheme is conservative*

$$\sum_{j=1}^N \int_{I_j} c_{ih}^n dx = \sum_{j=1}^N \int_{I_j} c_{ih}^{n+1} dx, \quad i = 1, \cdots, m, \qquad n \in \mathbb{N}.$$

2. *Assuming* $c_{ih}^n(x) > 0$, *there exists* $\mu^* > 0$ *such that if the mesh ratio* $\mu = \frac{\Delta t}{\Delta x^2} \in (0, \mu^*)$, *then the fully discrete free energy*

$$F^n = \sum_{j=1}^N \int_{I_j} \left[\sum_{i=1}^m c_{ih}^n \log c_{ih}^n + \frac{1}{2}\left(\sum_{i=1}^m q_i c_{ih}^n + \rho_0\right) \psi_h^n\right] dx + \frac{1}{2}\int_{\partial\Omega} \sigma \psi_h^n ds,$$

$$F^{n+1} \le F^n - \frac{\Delta t}{2} \sum_{i=1}^m A_{c_{ih}^n}(p_{ih}^n, p_{ih}^n).$$

where $A_c(\cdot,\cdot)$ *is a weighted bilinear operator, which is coercive if* β_0 *is suitably large, and* $\beta_1 = 0$ *in* $Fl(\psi_h)$.

The free energy dissipation law is also proved for high order strong stability preserving Runge-Kutta methods [3], which are convex combinations of several formal forward Euler steps.

As a result, steady states can well be preserved: if initial data c_{ih}^0 is already at steady states, i.e., $\log c_{ih}^0 + q_i \psi_h^0(x) = C_i$. By induction, it can be shown that the following holds:

$$\log c_{ih}^n + q_i \psi_h^n(x) = C_i \quad \forall n \in \mathbb{N}.$$

The scheme requires c_{ih} be positive, which is difficult to achieve for high order approximations. Inspired by the Zhang-Shu limiter [35] for scalar conservation laws, we impose a limiter. For approximation $w_h \in P^k(I_j)$ with cell averages $\bar{w}_j > \delta$, we reconstruct

$$w_h^\delta(x) = \bar{w}_j + \frac{\bar{w}_j - \delta}{\bar{w}_j - \min_{I_j} w_h(x)} (w_h(x) - \bar{w}_j), \quad \text{if } \min_{I_j} w_h(x) < \delta.$$

This reconstruction maintains same cell averages, satisfies $\min_{I_j} w^\delta(x) \ge \delta$, and does not destroy accuracy when $\delta < h^{k+1}$.

The algorithm in [19] can be summarized in following steps.

1. (Initialization) Project $c_i^{\text{in}}(x)$ onto V_h to obtain $c_{ih}^0(x)$.

2. (Reconstruction) From $c_{ih}^n(x)$, apply, if necessary, the reconstruction limiter to update c_{ih}^n so that $c_{ih}^n > \delta$.
3. (Poisson solver) Solve the Poisson equation to obtain ψ_h^n.
4. (Projection) Obtain $p_{ih}^n \in V_h$ by projection of $q_i\psi_h^n + \log c_{ih}^n$.
5. (Update) Solve the NP equations to obtain c_{ih}^{n+1} with some Runge-Kutta solver.

   Repeat steps 2-5 until final time T.

5. **IRP limiter for hyperbolic systems.** An invariant region to (3) is an open set in phase space $\mathbb{R}^l$ such that if the initial data is in this set, then the solution will remain in this set. It was proved by Hoff [6] that an invariant region for one dimensional hyperbolic conservation laws must be convex. For 2×2 systems such as the isentropic Euler system, an invariant region can be described by two Riemann invariants [31]. For general hyperbolic conservation law systems, it is a challenging task to identify a useful invariant region.

Shock capturing numerical methods have seen revolutionary developments over the past 40 years, with both conservation and entropy stability as two main ingredients in each scheme construction. However, it remains a difficult task to preserve an invariant region by a high order numerical method unless some nonlinear limiter is frequently imposed (Refs [1, 5] for first order IRP schemes). Indeed, recent efforts using limiting techniques have been made to construct high order maximum-principle-preserving schemes for scalar conservation laws (see [35]) and positivity-preserving schemes for hyperbolic systems including compressible Euler equations (see e.g. [27, 36, 38]). The work by Zhang and Shu in [37] introduced a limiter to preserve the minimum-entropy-principle [32] for high order schemes to the compressible Euler equation.

We now discuss the general explicit limiter introduced in [10]. Assume the multi-dimensional system of conservation laws admits an invariant region Σ, characterized by

$$\Sigma = \{\mathbf{w} \mid \quad U(\mathbf{w}) \le 0\},$$

where U is convex. Denote the interior of Σ by Σ_0. A key fact we have used is that for any bounded domain K, the averaging defined by

$$\bar{\mathbf{w}} = \frac{1}{|K|}\int_K \mathbf{w}(x)dx$$

is a contraction operator.

Lemma 5.1. *Let $\mathbf{w}(x)$ be non-trivial piecewise continuous vector functions. If $\mathbf{w}(x) \in \Sigma$ for all $x \in K \subset \mathbb{R}^d$ and U is strictly convex, then $\bar{\mathbf{w}} \in \Sigma_0$ for any bounded domain K.*

This lemma sets the foundation for using the domain average as a reference to limit the existing polynomials, through a linear convex combination as in [35, 37]. In the system case, the question of particular interest is whether the limited approximation is still high order accurate.

Let $\mathbf{w}_h(x)$ be a sequence of vector polynomials over K, a high order accurate approximation to the function $\mathbf{w}(x) \in \Sigma$. Assume $\bar{\mathbf{w}}_h \in \Sigma_0$, but $\mathbf{w}_h(x)$ is not entirely located in Σ. We construct

$$\tilde{\mathbf{w}}_h(x) = \theta \mathbf{w}_h(x) + (1-\theta)\bar{\mathbf{w}}_h,$$

where $\theta \in (0,1]$ is defined by $\theta = \min\{1, \theta_1\}$, where

$$\theta_1 = \frac{U(\bar{\mathbf{w}}_h)}{U(\bar{\mathbf{w}}_h) - U_h^{\max}}, \quad U_h^{\max} = \max_{x \in K} U(\mathbf{w}_h(x)) > 0.$$

If $\Sigma = \bigcap_{i=1}^{M} \{\mathbf{w} | \quad U_i(\mathbf{w}) \le 0\}$, then the limiter parameter needs to be modified as

$$\theta = \min\{1, \theta_1, \cdots, \theta_M\}.$$

This reconstruction has been shown to satisfy three desired properties.

Theorem 5.2. [10] *The reconstructed polynomial $\tilde{\mathbf{w}}_h(x)$ satisfies the following three properties:*

(i) *the average is preserved, i.e.* $\bar{\mathbf{w}}_h = \bar{\tilde{\mathbf{w}}}_h$;
(ii) $\tilde{\mathbf{w}}_h(x)$ *lies entirely within invariant region* $\Sigma, \forall x \in K$;
(iii) *order of accuracy is maintained, i.e., if* $\|\mathbf{w}_h - \mathbf{w}\|_\infty \le 1$, *then*

$$\|\tilde{\mathbf{w}}_h - \mathbf{w}\|_\infty \le \frac{C}{|U(\bar{\mathbf{w}}_h)|} \|\mathbf{w}_h - \mathbf{w}\|_\infty,$$

where $C > 0$ *depends on* $\mathbf{w}$ *and* Σ.

Let $\mathbf{w}_h^n$ be the numerical solution at n-th time step generated from a high order finite-volume-type scheme of an abstract form

$$\mathbf{w}_h^{n+1} = \mathcal{L}(\mathbf{w}_h^n), \quad \mathbf{w}_h^n = \mathbf{w}_h^n(x) \in V_h.$$

Provided that the scheme has the following property: there exists λ_0, and a test set S such that if

$$\lambda := \frac{\Delta t}{\Delta x} \le \lambda_0 \quad \text{and} \quad \mathbf{w}_h^n(x) \in \Sigma \text{ for } x \in S,$$

then

$$\bar{\mathbf{w}}_h^{n+1} \in \Sigma_0;$$

the limiter can then be applied with K replaced by $S_K : S \cap K$, i.e.,

$$U_h^{\max} = \max_{x \in S_K} U(\mathbf{w}_h(x)),$$

through the following algorithm:
Step 1. *Initialization:* take the piecewise L^2 projection of $\mathbf{w}_0$ onto V_h, such that

$$\langle \mathbf{w}_h^0 - \mathbf{w}_0, \phi \rangle = 0, \quad \forall \phi \in V_h.$$

Step 2. *Limiting:* Impose the modified limiter on $\mathbf{w}_h^n$ for $n = 0, 1, \cdots$ to obtain $\tilde{\mathbf{w}}_h^n$.
Step 3. *Update* by the scheme:

$$\mathbf{w}_h^{n+1} = \mathcal{L}(\tilde{\mathbf{w}}_h^n).$$

Return to Step 2.

A limiter as such was first reported in [11] for one-dimensional Euler equations, and in [9] for the isentropic gas dynamics. The limiter in [11] is explicit and simultaneously preserves the positivity of density and pressure and also a minimum principle for the specific entropy [32].

For multi-dimensional systems of conservation laws, there is a need to check whether the projected system shares the same invariant region as that for the full multi-D system. For 2D compressible Euler equations with $\mathbf{w} = (\rho, m, n, E)^\top$, $\mathbf{F}(\mathbf{w}) = (F_1(\mathbf{w}), F_2(\mathbf{w}))$, where

$$F_1(\mathbf{w}) = (m, \rho u^2 + p, \rho uv, (E+p)u)^\top,$$
$$F_2(\mathbf{w}) = (n, \rho uv, \rho v^2 + p, (E+p)v)^\top$$
$$m = \rho u, \quad n = \rho v, \quad E = \frac{1}{2}\rho u^2 + \frac{1}{2}\rho v^2 + \frac{p}{\gamma - 1}, \ \gamma > 1,$$

this as been shown true with the invariant region expressed as

$$\Sigma = \{\mathbf{w} | \quad \rho > 0, \ p > 0, \ q < 0\},$$

where $s = \log\left(\frac{p}{\rho^\gamma}\right)$ and $s_0 = \inf_x \log\left(\frac{p_0(x)}{\rho_0^\gamma(x)}\right)$, and $q = (s_0 - s)\rho$ is convex in $\mathbf{w}$. Hence, a corresponding IRP algorithm can be well established, and has been tested in [10].

6. **Conclusions and outlook.** In this paper, we have reviewed some of our contributions to the development of structure-preserving algorithms for two model classes. It is clear from the works we have reviewed, and from related references in the literature, that these techniques are not limited to these model equations, it is interesting to check the algorithmic improvement with more complex systems. Interesting directions worth further investigation include: (1) Design of explicit-implicit structure-preserving algorithms for nonlinear Fokker-Planck-type equations so to enhance computational efficiency; (2) Design of local IRP algorithms for multi-dimensional systems of hyperbolic conservation laws, with more realistic applications.

Acknowledgments. This is part of an ongoing project with several collaborators, including Yi Jiang, Wumaier Maimaitiyiming, Zhongming Wang, Nianyu Yi, and Hui Yu. The author benefitted a great deal from discussions with them.

REFERENCES

[1] H. Frid. *Maps of Convex Sets and Invariant Regions for Finite-Difference Systems of Conservation Laws.* Archive for rational mechanics and analysis, 160(3): 245-269, 2001.
[2] W.-X. Cao, H. Liu and Z.-M. Zhang. *Superconvergence of the direct discontinuous Galerkin method for convection-diffusion equations.* Numerical Methods for Partial Differential Equations, 33(1): 290–317, 2017.
[3] S. Gottlieb, D.I. Ketcheson and C.-W. Shu. *Strong Stability Preserving Runge-Kutta and Multistep Time Discretizations.* World Scientific, 2011.
[4] J. Giesselmann, C. Lattanzio and A. E. Tzavaras. *Relative energy for the Korteweg Theory and related Hamiltonian flows in gas dynamics.* Arch. Rational Mech. Anal. 223: 1427–1484, 2017.
[5] J.-L. Guermond and B. Popov. *Invariant domains and first-order continuous finite element approximation for hyperbolic systems*, SIAM J. Numer. Anal., 54(4): 2466–2489, 2016.
[6] D. Hoff. *Invariant regions for systems of conservation laws.* Trans. Amer. Math. Soc., 289(2):591–610, 1985.
[7] B. Hille. *Ionic Channels and Excitable Membranes.* Sinauer Sunderland, MA, 1992.
[8] Jan S. Hesthaven and Tim Warburton. *Nodal Discontinuous Galerkin Methods: Algorithms, Analysis, and Applications.* Springer Publishing Company, Incorporated, 1st edition, 2007.
[9] Yi Jiang and H. Liu. *An invariant region preserving limiter for DG schemes to isentropic Euler equations.* Numerical methods for PDEs, 35(1):5–33, 2019.
[10] Yi Jiang and H. Liu, *Invariant-region-preserving DG methods for multi-dimensional hyperbolic conservation law systems, with an application to compressible Euler equations.* Journal of Comput. Phys. 373(15), 385–409, 2018.

[11] Yi Jiang and H. Liu. *An Invariant-region-preserving (IRP) limiter to DG methods for compressible Euler equations.* HYP 2016: Theory, Numerics and Applications of Hyperbolic Problems II, 71–83, 2018. The Springer Proceedings in Mathematics and Statistics book series (PROMS, volume 237).

[12] K. Kawasaki. *Diffusion and the formation of spatial distributions.* Math. Sci., 16 (183): 47–52,1978.

[13] P.D. Lax. *Development of singularities of solutions of nonlinear hyperbolic partial differential equations.* J. Math. Phys., 5: 611–613, 1964.

[14] H. Liu. *Optimal error estimates of the direct discontinuous Galerkin method for convection-diffusion equations.* Math. Comp. 84 (2015), 2263–2295.

[15] H. Liu and W. Maimaitiyiming. *Positive and free energy satisfying schemes for diffusion with interaction potentials.* submitted to SINUM (2018).

[16] H. Liu and E. Tadmor. *Spectral dynamics of the velocity gradient field in restricted flows.* Commun. Math. Phys., 228:435–466, 2002.

[17] H. Liu and Z.-M. Wang. *A free energy satisfying finite difference method for Poisson–Nernst–Planck equations.* J. Comput. Phys. 268 (2014), 363–376.

[18] H. Liu and Z.-M. Wang. *An entropy satisfying discontinuous Galerkin method for nonlinear Fokker-Planck equations.* J Sci Comput., 68:1217–1240, 2016.

[19] H. Liu and Z.-M. Wang. *A free energy satisfying discontinuous Galerkin method for Poisson-Nernst-Planck systems.* J. Comput. Phys. 238: 413–437, 2017.

[20] H. Liu and H. Yu. *An entropy satisfying conservative method for the Fokker Planck equation of FENE dumbbell model.* SIAM J. Numer. Anal. 50(3) (2012), 1207–1239.

[21] H. Liu and H. Yu. *Maximum-principle-satisfying third order discontinuous Galerkin schemes for Fokker-Planck equations.* SIAM J. Sci. Comput. 36(5) (2014), A2296–A2325.

[22] H. Liu and H. Yu. *The entropy satisfying discontinuous Galerkin method for Fokker–Planck equations.* J. Sci. Comput. 62 (2015), 803–830.

[23] H. Liu and J. Yan. *The Direct Discontinuous Galerkin (DDG) methods for diffusion problems.* SIAM Journal on Numerical Analysis, 47(1): 475–698, 2009.

[24] H. Liu and J. Yan. *The Direct Discontinuous Galerkin (DDG) method for diffusion with interface corrections.* Commun. Comput. Phys. 8(3): 541–564, 2010.

[25] P.A. Markowich, C.A. Ringhofer and C. Schmeiser. *Semiconductor Equations*, Springer-Verlag Inc., New York, 1990.

[26] B. Perthame. *Transport Equations in Biology.* Frontiers in Mathematics, Birkauser Verlag, Basel, 2007.

[27] B. Perthame and C.-W. Shu. *On positivity preserving finite volume schemes for Euler equations.* Numerische Mathematik, 73: 119–130, 1996.

[28] H. Risken. *The Fokker-Planck Equation: Methods of Solution and Applications.* Lecture Notes in Mathematics. Springer London, Limited, 1996.

[29] Bèatrice Riviére. *Discontinuous Galerkin Methods for Solving Elliptic and Parabolic Equations.* Society for Industrial and Applied Mathematics, 2008.

[30] Chi-Wang Shu. Discontinuous Galerkin methods: General approach and stability. *Numerical Solutions of Partial Differential Equations, S. Bertoluzza, S. Falletta, G. Russo and C.-W. Shu, Advanced Courses in Mathematics CRM Barcelona*, pages 149–201, 2009. Birkhauser, Basel.

[31] J. Smoller. *Shock waves and reaction-diffusion equations.* volume 258 of Grundlehren der Mathematischen Wissenschaften. Springer-Verlag, New York-Berlin, 1983.

[32] E. Tadmor. *A minimum entropy principle in the gas dynamics equations.* Applied Numerical Mathematics, 2, 211–219,1986.

[33] C. M. Topaz, A. L. Bertozzi and M. A. Lewis. *A nonlocal continuum model for biological aggregation.* Bull. Math. Bio., 68: 1601–1623, 2006.

[34] H. Yu and H. Liu. *Third order maximum-principle-satisfying DG schemes for convection-diffusion problems with anisotropic diffusivity.* submitted to JCP (2018).

[35] X. Zhang and C.-W. Shu. *On maximum-principle-satisfying high order schemes for scalar conservation laws.* Journal of Computational Physics, 229: 3091–3120, 2010.

[36] X. Zhang and C.-W. Shu. *On positivity preserving high order discontinuous Galerkin schemes for compressible Euler equations on rectangular meshes.* Journal of Computational Physics, 229: 8918–8934, 2010.

[37] X. Zhang and C.-W. Shu. *A minimum entropy principle of high order schemes for gas dynamics equations.* Numerische Mathematik, 121:545–563, 2012.

[38] X. Zhang, Y. Xia and C.-W. Shu. *Maximum-principle-satisfying and positivity-preserving high order discontinuous Galerkin schemes for conservation laws on triangular meshes.* Journal of Scientific Computing, 50: 29–62, 2012.

[39] X. Zhang. *On positivity-preserving high order discontinuous Galerkin schemes for compressible Navier-Stokes equations.* Journal of Computational Physics, 328: 301D343, 2017.

E-mail address: hliu@iastate.edu

Part 3

Contributed Lectures

ERROR BOUNDEDNESS OF CORRECTION PROCEDURE VIA RECONSTRUCTION / FLUX RECONSTRUCTION AND THE CONNECTION TO RESIDUAL DISTRIBUTION SCHEMES

RÉMI ABGRALL, ELISE LE MÉLÉDO, PHILIPP ÖFFNER* AND HENDRIK RANOCHA

Institute of Mathematics, University of Zurich
Winterthurerstrasse 190
CH-8057 Zürich, Switzerland

HENDRIK RANOCHA

Institute Computational Mathematics, TU Braunschweig
Universitätsplatz 2
38106 Braunschweig, Germany

ABSTRACT. We focus on the correction procedure via reconstruction (CPR) / flux reconstruction (FR) methods for hyperbolic conservation laws. Their long time error behavior is investigated and their connection with the Residual Distribution schemes is pointed out. Considering a model problem, we start by deriving an error equation that will be investigated in detail. There, we show that the choice between upwinding and central numerical fluxes affects the growth rate and asymptotic value of the error. Furthermore, the selection of the bases themselves (Gauß-Lobatto-Legendre or Gauß-Legendre) highly impacts the solution. In particular, using Gauß-Legendre basis, the error reaches the asymptotic value faster than using Gauß-Lobatto-Legendre basis [8, 9] which also appears to be smaller. In the second part of this contribution, we demonstrate that FR schemes can be transformed into the Residual Distribution (RD) framework and vice versa. As a consequence, we can directly apply the known results from RD schemes to CPR/FR methods [2].

1. **Introduction.** Various physical processes are modeled with hyperbolic conservation laws, including fluid mechanics and electromagnetism. Since the existence of analytical solutions is still unknown, especially for non-linear equations, numerical methods have to be applied. So far, many numerical methods are based either on finite element (FE) or finite difference (FD) approaches. However, one can transform and reformulate numerical schemes from one to another. Thus, techniques which are originally used in some framework can be transferred to the other ones. Here, summation-by-parts (SBP) operators are a good example [3]. In this contribution, we mainly focus on two numerical methods by considering two different topics. First, we are considering the correction procedure via reconstruction (CPR) / flux

2000 *Mathematics Subject Classification.* Primary: 65N15; 65N12 Secondary: 65M06; 65M60.

Key words and phrases. Flux Reconstruction, Residual Distribution, SBP Operators, Error boundedness, Stability.

The second and third authors have been funded in by the the SNF project (Number 175784). H. Ranocha was supported by the German Research Foundation (DFG) under Grant SO 363/14-1.

* Corresponding author: Philipp Öffner, philipp.oeffner@math.uzh.ch.

reconstruction (FR). These methods unify several high-order methods like discontinuous Galerkin (DG), spectral volume (SV) and spectral difference (SD) schemes in a common framework [5]. We investigate their long time error behavior. In the literature several examples demonstrating a linear error growth can be found, even though stability issues should exclude this. In the same time, other examples show a bounded temporal error growth. It was also shown in [7] that when considering one-block FD schemes the error behavior depends only on the choice of boundary procedure, whereas in the DG framework the internal approximation has indeed an influence [6]. The selection of the numerical fluxes is therefore essential. We extend the investigation from [6] to the CPR/FR framework, consider different bases (Gauss-Lobatto-Legendre, Gauss-Legendre) and include variable coefficients in the model problem [8, 9]. We also focus on the residual distribution (RD) approach that leads to a general framework containing several numerical methods including DG [1]. As there is a close relation between RD and CPR/FR to DG, it seems natural to analyze the link between RD and CPR/FR. In the second part of this contribution we demonstrate how to embed the CPR/FR approach into the RD framework, and derive two conditions for the construction of the CPR/FR correction functions to guarantee that the remaining schemes have favorable properties (e.g. conservation). This builds the foundation of further developments in the context of FR/CPR methods [2].

2. **Correction Procedure via Reconstruction / Flux reconstruction using Summation-by-parts Operators.** Flux reconstruction schemes were introduced by Huynh [4] as an alternative to other high order methods. Rather than using a weak/variational formulation or integral form in the spirit of common DG methods, the semidiscretisation of FR uses a differential formulation. This approach has been extended to unstructured grids in [12], and in [5] the authors suggested the common name correction procedure via reconstruction (CPR). Today the literature mostly refers to those methods by the term flux reconstruction, term that we adopt here under the short-name FR. Let us start now by explaining the main idea of FR. For simplicity, we consider a one-dimensional scalar conservation law

$$\partial_t u + \partial_x f(u) = 0 \tag{1}$$

in the domain $\Omega \subset \mathbb{R}$. FR performs a semidiscretisation by using a polynomial approximation within elements. The domain $\Omega \subset \mathbb{R}$ is thus split into disjoint intervals $\Omega_i \subset \Omega$, and each element Ω_i is transferred onto a standard element, in our case $[-1, 1]$, where all the calculations are done. The solution $u(t) = u(t, \cdot)$ is approximated by a polynomial $U \in \mathbb{P}^p$ of degree smaller or equal than p. To this end, in the basic formulation of FR a nodal Lagrange basis is employed. The coefficients of $\underline{u}$ are given by the nodal values $\underline{u}_i = u(\zeta_i), i \in \{0, \dots, p\}$, where $-1 \le \zeta_i \le 1$ are interpolation points in $[-1, 1]$. The numerical solution is obtained by $U(\xi) = \sum_{i=0}^{p} \underline{u}_i l_i(\xi)$, where $l_i(\xi)$ is the i-th Lagrange interpolation polynomial that satisfies $l_j(\xi_j) = \delta_{ij}$. Besides the solution u, the flux is also approximated by a polynomial $\hat{f}$ with coefficients $\underline{f}_i = f(\underline{u}_i) = f(u(\zeta_i))$. Note that the possible discontinuities of the numerical solution will also appear in the discrete flux. There, instead of using only a numerical flux f^{num} to avoid this issue, correction terms/functions working at the boundaries between the elements are also applied. This is the main idea of the FR approach. More precisely, a FR semidiscretisation of (1) is given by

$$\partial_t U + \partial_\xi \left(\hat{f}(\xi) + \left(f_L^{\text{num}} - \hat{f}_L \right) g_L(\xi) + \left(f_L^{\text{num}} - \hat{f}_R \right) g_R(\xi) \right) = 0, \tag{2}$$

where g_i, $i = L, R$ denote the left and right correction functions. The properties of the FR methods highly rely on their definitions. In [11], linear stability of the FR schemes is demonstrated using the following corrections

$$g_L = \frac{(-1)^p}{2}\left[L_p - \left(\frac{\eta_p L_{p-1} + L_{p+1}}{1+\eta_p}\right)\right], \quad g_R = \frac{1}{2}\left[L_p + \left(\frac{\eta_p L_{p-1} + L_{p+1}}{1+\eta_p}\right)\right],$$

where L_p are the Legendre polynomials, $\eta_p = \frac{c(2p+1)((2p)!)^2}{2^{2p+1}(p!)^2}$ and c is a free parameter bounded from below which finally specifies completely the methods, c.f [11] for details. In [10], a reformulation of FR in the general framework of summation-by-parts (SBP) operators using simultaneous approximation terms (SATs) is provided, yielding proofs of conservation and stability in discrete norms. Classically, a nodal basis associated with a quadrature rule (Gauß-Lobatto-Legendre or Gauß-Legendre) is given and the mass matrix $\underline{\underline{M}} = \operatorname{diag}(\omega_0, \dots, \omega_p)$ corresponds to a SBP operator. We denote by $\underline{\underline{B}} = \operatorname{diag}(-1, 1)$ the boundary matrix, $\underline{\underline{D}}$ the discrete derivative matrix and $\underline{\underline{R}}$ the restriction term. The main idea of SBP is to mimic integration by parts on a discrete level, meaningly

$$\underline{u}^T \underline{\underline{M}}\,\underline{\underline{D}}\,\underline{v} + \underline{u}^T \underline{\underline{D}}^T \underline{\underline{M}}\,\underline{v} \approx \int_{-1}^{1} u\,\partial_x v + \int_{-1}^{1} \partial_x u\, v = u\,v\big|_{-1}^{1} \approx \underline{u}^T \underline{\underline{R}}^T \underline{\underline{B}}\,\underline{\underline{R}}\,\underline{v}.$$

The SBP property reads $\underline{\underline{M}}\,\underline{\underline{D}} + \underline{\underline{D}}^T \underline{\underline{M}} = \underline{\underline{R}}^T \underline{\underline{B}}\,\underline{\underline{R}}$. Therefore, the semidiscretisation of (2) is given by

$$\partial_t \underline{u} = -\underline{\underline{D}}\,\underline{f} - \underline{\underline{C}}\left(\underline{f}^{\text{num}} - \underline{\underline{R}}\,\underline{f}\right),$$

where $\underline{\underline{C}}$ is the correction matrix and $\underline{f}^{\text{num}}$ are the coefficients of the numerical flux. Again, the choice of the correction matrices $\underline{\underline{C}}$ will determine the numerical methods. Considering $\underline{\underline{C}} = \underline{\underline{M}}^{-1}\underline{\underline{R}}^T\underline{\underline{B}}$ is the canonical choice and is equivalent to the DGSEM of [3]. Choosing $\underline{\underline{C}} = (\underline{\underline{M}} + \underline{\underline{K}})^{-1}\underline{\underline{R}}^T\underline{\underline{B}}$ where $\underline{\underline{K}} = c(\underline{\underline{D}}^p)^T\underline{\underline{M}}\,\underline{\underline{D}}^p$ is a symmetric matrix, $\underline{\underline{M}} + \underline{\underline{K}} > 0$ (i.e. positive definite) and $\underline{\underline{K}}\,\underline{\underline{D}} = 0$ leads to the above mentioned schemes. In particular, the SBP-FR semidiscretisation of a linear advection equation with constant coefficient one (i.e. $f(u) = u$ in (1)) reads

$$\partial_t \underline{u} = -\underline{\underline{D}}\,\underline{u} - (\underline{\underline{M}} + \underline{\underline{K}})^{-1}\underline{\underline{R}}^T\underline{\underline{B}}\left(\underline{f}^{\text{num}} - \underline{\underline{R}}\,\underline{u}\right). \tag{3}$$

3. **Long-Time Error Behavior of Flux Reconstruction Schemes.** We study the long time error behavior of SBP-FR methods for a scalar linear advection equation with non-periodic boundary conditions. We consider the following model problem.

$$\begin{aligned} &\partial_t u(t,x) + \partial_x(a(x)u(t,x)) = 0, \quad x \in [0, L], \quad t \geq 0 \\ &u(t,0) = g(t), \qquad u(0,x) = u_0(x). \end{aligned} \tag{4}$$

Theory of constant coefficients. In the first part of this contribution, we set $a(x) \equiv 1$. Then, (4) is similar to the problems investigated in [7, 6]. We further assume that $u(t,x) \in H_c^m(0,L)$ for $m > 1$ and that $||u||_{H_c^m}$ is uniformly bounded in time. H_c^m denotes the function space equipped with a broken Sobolev norm which is used in [11] to demonstrate linear stability for their methods. For the commonly used methods, c tends to (or is) zero. We presume the latter, see [8] for details. The entire interval $[0, L]$ is divided into several elements $e^k = [x^{k-1}, x^k]$, $k = 1, \dots, K$, where the x^k are the element boundaries and where $x^0 = 0$ and $x^K = L$. We set $\frac{\Delta x_k}{2} = \frac{x^k - x^{k-1}}{2}$ a transformation factor from our standard element $[-1, 1]$. The change of the discrete norm for the total energy $||\underline{u}^k||^2 = \underline{u}^{k,T}(\underline{\underline{M}} + \underline{\underline{K}})\underline{u}^k$ is investigated. Here, ^{T} denotes the transposed vector. By multiplying $\underline{u}^{k,T}\left(\underline{\underline{M}} + \underline{\underline{K}}\right)$ to equation (3) we obtain

$$\frac{\Delta x_k}{2}\underline{u}^{k,T}\left(\underline{\underline{M}} + \underline{\underline{K}}\right)\partial_t \underline{u}^k = -\underline{u}^{k,T}\left(\underline{\underline{M}} + \underline{\underline{K}}\right)\underline{\underline{D}}\,\underline{u}^k - \underline{u}^{k,T}\underline{\underline{R}}^T\underline{\underline{B}}\left(\underline{f}^{\text{num},k} - \underline{\underline{R}}\,\underline{u}^k\right). \tag{5}$$

With the SBP property, the rate of change of the total energy is derived as follows (see [8] for more details).

$$\frac{1}{2}\frac{\mathrm{d}}{\mathrm{d}t}\sum_{k=1}^{K}\frac{\Delta x_k}{2}||\underline{u}^k||^2_{M+\mathrm{K}} + \sum_{k=1}^{K}\underline{u}^{k,T}\underline{\underline{R}}^T\underline{\underline{B}}\left(\underline{f}^{\mathrm{num},k} - \frac{1}{2}\underline{\underline{R}}\,\underline{u}^k\right) = 0$$

The error in every element $E^k = u(x(\xi,t)) - U^k(\xi,t)$ can be split into two parts:

$$E^k = u(x(\xi,t)) - U^k(\xi,t) = \underbrace{(\mathbb{I}^N(u)^k - U^k)}_{=:\epsilon_1^k\in\mathbb{P}^N} + \underbrace{(u - \mathbb{I}^N(u)^k)}_{=:\epsilon_p^k},$$

where $\mathbb{I}^N$ denotes the interpolation operator. Then, ϵ_p^k is the interpolation error which is the sum of the series truncation error and the aliasing error. The error can be bounded by the triangle inequality in the discrete norm $||E^k||_{\mathrm{K}} \le ||\epsilon_1^k||_{\mathrm{K}} + ||\epsilon_p^k||_{\mathrm{K}}$. As $||\epsilon_p^k||_{\mathrm{K}}$ decays spectrally fast, fulfilling the above mentioned assumption and using either Gauß-Lobatto-Legendre or Gauß-Legendre nodes allow us to neglect the interpolation error in our investigation (see [8]). As it can be seen in [8], applying E^k in the continuous equation together with (5) allow to derive the error equation for ϵ_1^k

$$\begin{aligned}\frac{1}{2}\frac{\mathrm{d}}{\mathrm{d}t}\sum_{k=1}^{K}\frac{\Delta x_k}{2}||\epsilon_1^k||^2_{M+\mathrm{K}} &+ \sum_{k=1}^{K}\underline{\epsilon}_1^{T,k}\underline{\underline{R}}^T\underline{\underline{B}}\left(\underline{\epsilon}_1^{\mathrm{num},k} - \frac{1}{2}\underline{\underline{R}}\,\underline{\epsilon_1}^k\right)\\ &= \sum_{k=1}^{K}\left(\frac{\Delta x_k}{2}\left(\left(T^k(u),\epsilon_1^k\right) + \left(\underline{Q(u)}^k,\epsilon_1^k\right)_{M+\mathrm{K}}\right) - \tilde{\epsilon}_2^k\right),\end{aligned} \tag{6}$$

with $\tilde{\epsilon}_2^k = \epsilon_1^k\mathbb{I}^N(u)^k|_{-1}^{1} - \underline{\epsilon}_1^{T,k}\underline{\underline{R}}^T\underline{\underline{B}}\,\underline{f}^{\mathrm{num},k}(\mathbb{I}^N(u)^{k,-},\mathbb{I}^N(u)^{k,+})$, $\underline{\epsilon}_1^{\mathrm{num},k} = \underline{f}^{\mathrm{num},k}\left((\epsilon_1^k)^-,(\epsilon_1^k)^+\right)$, $T^k(u) = -\left\{\partial_t\epsilon_p^k + \partial_x\epsilon_p^k + Q(u)^k\right\}$, and where Q measures the projection error of a polynomial of degree N to a polynomial of degree $N-1$. Note that $\tilde{\epsilon}_2^k$ is zero if the Gauß-Lobatto-Legendre nodes are applied. All above terms are well-defined under the given conditions. By fundamental estimations (Cauchy-Schwarz) the right side of (6) can be estimated from above by $C_1||\epsilon_1||_{\mathrm{K}}$. By splitting the sum on the left side in (6) into three parts (one for the left physical boundary, one for the right physical boundary and a sum over the internal element endpoints), we get

$$\begin{aligned}&\sum_{k=1}^{K}\underline{\epsilon}_1^{T,k}\underline{\underline{R}}^T\underline{\underline{B}}\left(\underline{\epsilon}_1^{\mathrm{num},k} - \frac{1}{2}\underline{\underline{R}}\,\underline{\epsilon_1}^k\right) = \sum_{k=1}^{K}\underline{\epsilon}_1^{T,k}\underline{\underline{R}}^T\underline{\underline{B}}\left(\underline{f}^{\mathrm{num},k}\left(\left(\epsilon_1^k\right)^-,\left(\epsilon_1^k\right)^+\right) - \frac{1}{2}\underline{\underline{R}}\,\underline{\epsilon}_1^k\right)\\ &= -\mathbf{E}_L^1\left(f_L^{\mathrm{num},1} - \frac{\mathbf{E}_L^1}{2}\right) + \sum_{k=2}^{K}\left(f_L^{\mathrm{num},k} - \frac{1}{2}\left(\mathbf{E}_R^{k-1} + \mathbf{E}_L^k\right)\right)\left(\mathbf{E}_R^{k-1} - \mathbf{E}_L^k\right) + \mathbf{E}_R^K\left(f_R^{\mathrm{num},K} - \frac{\mathbf{E}_R^K}{2}\right).\end{aligned}$$

Here, $\mathbf{E}_i$ $(i = L, R)$ is the approximated error of ϵ_1, the indices give the position in the elements, $f_L^{\mathrm{num},k} := f^{\mathrm{num},k}\left(\mathbf{E}_R^{k-1},\mathbf{E}_L^k\right)$, $f_L^{\mathrm{num},1} := f^{\mathrm{num},1}\left(0,\mathbf{E}_L^1\right)$ and $f_R^{\mathrm{num},K} := f^{\mathrm{num},1}\left(\mathbf{E}_R^K,0\right)$. We set on the left physical boundary U^1 to g and the external states to zero. At the right boundary, an upwind numerical flux is used and there is no need to prescribe the external state since its coefficient in the numerical solution is zero. Thus, using $[\![\mathbf{E}^k]\!] = \mathbf{E}_R^{k-1} - \mathbf{E}_L^k$ we obtain in the internal elements

$$\sum_{k=2}^{K}\left(f_L^{\mathrm{num},k} - \frac{1}{2}\left(\mathbf{E}_R^{k-1} + \mathbf{E}_L^k\right)\right)\left(\mathbf{E}_R^{k-1} - \mathbf{E}_L^k\right) = \sum_{k=2}^{K}\frac{\sigma}{2}\left([\![\mathbf{E}^k]\!]\right)^2 \ge 0,$$

with $\sigma = 0$ central flux and $\sigma = 1$ upwind flux. Finally, we conclude from (6) that the energy growth rate is bounded by

$$\frac{1}{2}\frac{\mathrm{d}}{\mathrm{d}t}||\epsilon_1||_{\mathrm{K}}^2 + \underbrace{\frac{\sigma}{2}\left(\left(\mathbf{E}_R^K\right)^2 + \left(\mathbf{E}_L^1\right)^2\right) + \frac{\sigma}{2}\sum_{k=2}^{K}\left(\left[\!\left[\mathbf{E}^k\right]\!\right]\right)^2}_{BTs} \leq C_1||\epsilon_1||_{\mathrm{K}}. \tag{7}$$

There, $BTs \geq 0$. Defining $\eta(t) := \frac{BTs}{||\epsilon_1||_{\mathrm{K}}^2}$, we obtain from (7)

$$\frac{\partial}{\partial t}||\epsilon_1||_{\mathrm{K}} + \eta(t)||\epsilon_1||_{\mathrm{K}} \leq C_1. \tag{8}$$

Let us now assume that the mean value of $\eta(t)$ over any finite time interval is bounded by a positive constant δ_0 from below (i.e. $\overline{\eta} \geq \delta_0 > 0$). This together with integration over time yields

$$||\epsilon_1(t)||_{\mathrm{K}} \leq \frac{1 - \exp(-\delta_0 t)}{\delta_0} C_1. \tag{9}$$

Remark 1. Our investigation demonstrates that both the selection of bases and numerical fluxes have an essential influence on the error behaviors. Furthermore, from (9) we predict that applying Gauß-Legendre nodes leads in general to lower total errors as using Gauß-Lobatto-Legendre nodes. Indeed, the error ϵ_1 is smaller and δ_0 can therefore be chosen bigger. Also, the use of the upwind flux should be preferred, fact that was already seen for the DGSEM in [6]. Taking the terms in (9) into account, we expect that the selection of bases is even more important.

Numerical Simulations. Let us now support our theoretical investigation by some numerical simulations. First, we consider the interval $[0, 2\pi]$ together with the initial condition $u_0 = \sin(12(x - 0.1))$. The boundary function $g(t)$ is chosen to match with the exact solution $u(x,t) = \sin(12(x - t - 0.1))$. In the Figure 1, we represented the long-time error behaviors obtained by the use of polynomial of order $p = 4$, $K = 50$ elements and of the correction term $\underline{\underline{C}} = \underline{\underline{M}}^{-1}\underline{\underline{R}}^T\underline{\underline{B}}$. In contrast,

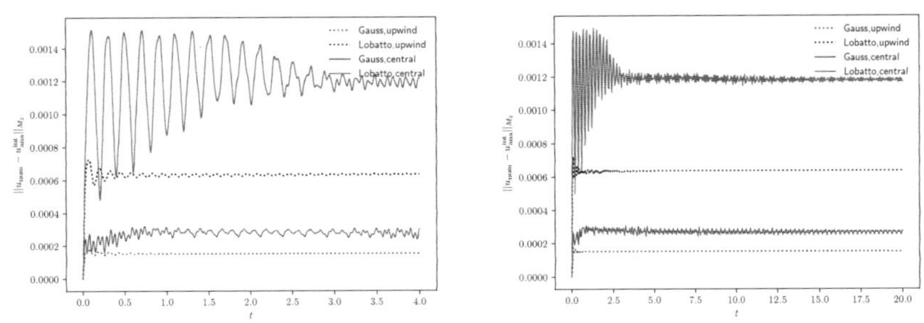

FIGURE 1. Error behaviors for $t = 4$ and $t = 20$

in the Figure 2 we show the error behaviors using $p = 3$, $K = 20$ and correction terms $\underline{\underline{C}} = (\underline{\underline{M}} + \underline{\underline{K}})^{-1}\underline{\underline{R}}^T\underline{\underline{B}}$ with correction parameters $c_{SD} = 1/1050$ (SD methods) and $c_{Hu} = 8/4725$ (Huynh scheme [4]). These simulations clearly support our conclusion drawn in the Remark 1. The use of Gauß-Legendre nodes together with the upwind fluxes yields always the smallest error. For more numerical tests and some discussion we recommend again [8].

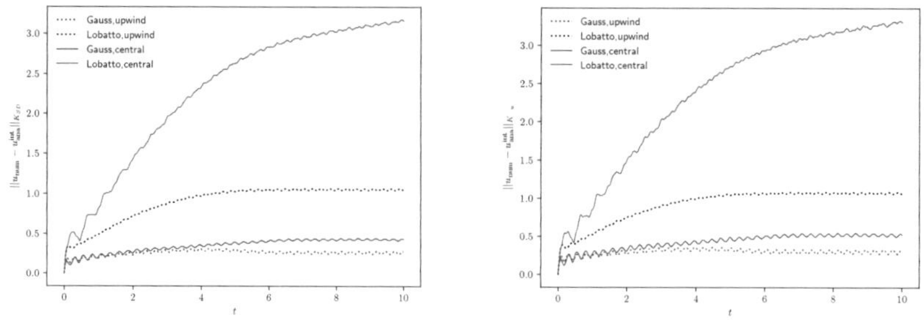

FIGURE 2. SD and Huynh

Extension to variable coefficients. Instead of using $a(x) \equiv 1$ variable coefficients $a(x)$ can also be applied in (4). This makes the investigation quite more difficult and demonstrating stability already requires a splitting approach (see [9]). Using the correction term $\underline{\underline{C}} = \underline{\underline{M}}^{-1}\underline{\underline{R}}^T\underline{\underline{B}}$ the long-time error behavior for the model problem (4) with variable coefficients is finally described in [9]. One obtains an estimation for the energy growth rate similar to (7). It reads

$$\frac{1}{2}\frac{\mathrm{d}\|\epsilon_1\|_N^2}{\mathrm{d}t} + \frac{\sigma}{2}\left(a_R^K\left(\mathbf{E}_R^K\right)^2 + a_L^1\left(\mathbf{E}_L^1\right)^2\right) + \frac{\sigma}{2}\sum_{k=2}^{K} a_R^{k-1}\left(\llbracket \mathbf{E}^k \rrbracket\right)^2 + \sum_{k=1}^{K}\frac{\Delta x_k}{4}\left(\underline{\epsilon}_1^k\underline{\epsilon}_1^k, \partial_x \underline{a}^k\right)_N \le C_2\|\epsilon_1\|_N,$$

where the values of the coefficients and their derivatives at the boundaries also play a fundamental role. If both are strictly non-negative, we obtain following the same steps as before an estimation similar to (9). However, if for example $a'(x) < 0$ for some values, then the behaviors highly depend on the numerical dissipation. To show it, we consider (4) with $a(x) = \cos(x)$ and the initial condition $u_0(x) = \sin(5x)$. Its solution is given by $u(t,x) = u_0\big(x_0(t,x)\big)\frac{\cos\left(x_0(t,x)\right)}{\cos(x)}$, with $x_0(t,x) = -2\arctan\big(\tanh\left(t/2 - \operatorname{artanh}(\tan(x/2))\right)\big)$. We apply the SBP-CPR/FR method with the correction term $\underline{\underline{C}} = \underline{\underline{M}}^{-1}\underline{\underline{R}}^T\underline{\underline{B}}$ in the interval $I = [0.1, \pi/3]$. In the Figure 3, the left picture shows the error behaviors using different bases and numerical fluxes until $t = 40$. One can realize that the errors using central flux increase whereas with upwind fluxes the errors remain bounded. It can be better seen in the right picture where a logarithmic scale is used, and the errors are plotted up to $t = 100$. This boundedness results from the introduction of dissipation through the upwind flux definition. More examples along with a general discussion can be found in [9].

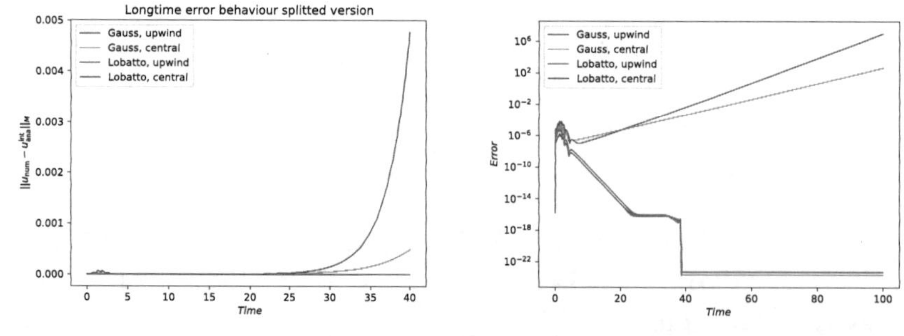

FIGURE 3. $p = 3$, $K = 30$,$t = 40$ and logarithmic scale up to $t = 100$

4. Connection between Flux Reconstruction and Residual Distribution. In this section, we shortly demonstrate the connection between the residual distribution schemes and the flux reconstruction approach. We show how FR fits in the RD framework and *vice versa*. This knowledge can be used to study the properties of these methods from a different perspective and to construct new methods with favorable properties as done in [2]. Before explaining the relation between these two frameworks, we shortly introduce the residual distribution methods from [1]. We strongly recommend this paper and references therein for a more detailed introduction. We are considering the steady state problem with initial condition

$$\operatorname{div} \boldsymbol{f}(\boldsymbol{u}) = \sum_{j=1}^{d} \frac{\partial \boldsymbol{f}_j}{\partial x_j}(\boldsymbol{u}) = 0 \text{ for } \boldsymbol{x} \in \Omega \subset \mathbb{R}^d,\ (\nabla_{\boldsymbol{u}} \boldsymbol{f}(\boldsymbol{u}) \cdot \boldsymbol{n}(\boldsymbol{x}))^- (\boldsymbol{u} - \boldsymbol{u}_b) = 0 \text{ on } \partial\Omega. \tag{10}$$

There, $\boldsymbol{u}_b$ is a regular enough function and $\boldsymbol{n}$ denotes the outward normal vector at $\mathbf{x} \in \partial\Omega$. The flux is given by $\boldsymbol{f}_j = (f_{1,j}, \cdots, f_{p,j})^T \subset \mathbb{R}^p$ and $\boldsymbol{u} = (u_1, \cdots, u_p)^T \in D \subset \mathbb{R}^p$ is the conserved variable. Again, the domain Ω is split into a partition K (triangles or general polygons), and the solution in each element is approximated by a polynomial of degree k. The term $\boldsymbol{u}^h$ indicates the numerical solution. Let us also introduce the notations related to the RD formulation and denote by S the set of degrees of freedom (DOF), $\sum_K$ the set of DOF of linear forms acting on the set $\mathbb{P}^k$ and $\{\phi_\sigma\}_{\sigma \in \sum_K}$ the set of basis functions with which for all $\boldsymbol{x} \in K$ the relation $\sum_{\sigma \in K} \phi_\sigma(\boldsymbol{x}) = 1$ is fulfilled. The main idea of the RD schemes is to define residuals Φ_σ^K on every element K, satisfying the following conservation relations.

$$\sum_{\sigma \in K} \Phi_\sigma^K(\boldsymbol{u}^h) = \oint_{\partial K} \boldsymbol{f}^{\text{num}}(\boldsymbol{u}^h, \boldsymbol{u}^{h,-}) \cdot \mathbf{n} \mathrm{d}\gamma,\ \sum_{\sigma \in \Gamma} \Phi_\sigma^\Gamma(\boldsymbol{u}^h) = \oint_{\partial \Gamma} \boldsymbol{f}^{\text{num}}(\boldsymbol{u}^h, \boldsymbol{u}_b) \cdot \mathbf{n} - \boldsymbol{f}(\boldsymbol{u}^h) \cdot \boldsymbol{n} \mathrm{d}\gamma \tag{11}$$

There, $\mathbf{u}^{h,-}$ describes the approximated solution on the other side of the local edge of K, $\boldsymbol{f}^{\text{num}}$ is a consistent numerical flux (i.e. $\boldsymbol{f}^{\text{num}}(\mathbf{u}, \mathbf{u}) = \mathbf{f}(\mathbf{u}) \cdot \mathbf{n}$), $\oint_K$ is the boundary integral evaluated by a numerical quadrature rule and Γ term the boundary elements. The formula for the discretisation of (10) reads: for any $\sigma \in S$,

$$\sum_{K \subset \Omega,\ \sigma \in K} \Phi_\sigma^K(\boldsymbol{u}^h) + \sum_{\Gamma \subset \partial\Omega,\ \sigma \in \Gamma} \Phi_\sigma^\Gamma(\boldsymbol{u}^h) = 0. \tag{12}$$

This relation shows the advantage of RD having a general formulation. Thus, depending on the solution space V^h and the exact definition of the residuals, we can embed several numerical methods like finite element or DG into this framework ([1]). Let us now demonstrate the connection between RD and FR. With $\mathbf{f}^h$ being the approximated flux function the discretisation of (10) in the FR framework reads

$$\operatorname{div}(\mathbf{f}^h + \boldsymbol{\alpha} \nabla \boldsymbol{\psi}) = 0 \Longleftrightarrow \operatorname{div}\left(\mathbf{f}^h + \left(\boldsymbol{f}^{\text{num}} \cdot \mathbf{n} - \mathbf{f}^h \cdot \mathbf{n}\right) \nabla \boldsymbol{\psi}\right) = 0, \tag{13}$$

where $\boldsymbol{\alpha} \nabla \boldsymbol{\psi}$ are our correction functions with the scaling term $\boldsymbol{\alpha} = \mathbf{f}^{\text{num}} \cdot \mathbf{n} - \mathbf{f}^h \cdot \mathbf{n}$. Using an Galerkin approach with $\boldsymbol{v}^h \in V^h$ and the Gauß theorem, we obtain

$$-\int_K \boldsymbol{\nabla} v^h \cdot \left(\mathbf{f}^h + \boldsymbol{\alpha} \nabla \boldsymbol{\psi}\right) \mathrm{d}\,\mathbf{x} + \int_{\partial K} \boldsymbol{v}^h \cdot \left(\mathbf{f}^h \cdot \mathbf{n} + \left(\left(\boldsymbol{f}^{\text{num}} \cdot \mathbf{n} - \mathbf{f}^h \cdot \mathbf{n}\right) \nabla \boldsymbol{\psi} \cdot \mathbf{n}\right)\right) \mathrm{d}\,\gamma = 0. \tag{14}$$

Because of the conservation relation, the flux over the element boundaries should be expressed only by the numerical flux of elements sharing this boundary. Therefore, we demand $\nabla \boldsymbol{\psi} \cdot \mathbf{n} \equiv 1$ on the boundary and obtain

$$\left(\mathbf{f}^h \cdot \mathbf{n} + \left(\boldsymbol{f}^{\text{num}}(\mathbf{u}^h, \mathbf{u}^{h,-}) \cdot \mathbf{n} - \mathbf{f}^h \cdot \mathbf{n}\right) \nabla \boldsymbol{\psi} \cdot \mathbf{n}\right) = \boldsymbol{f}^{\text{num}}(\mathbf{u}^h, \mathbf{u}^{h,-}) \cdot \mathbf{n}.$$

This is the first property on our correction function $\nabla\boldsymbol{\psi}$. The key of the RD schemes is a proper definition of the residuals. When considering (14), by passing from integrals to quadrature formulas, splitting $\boldsymbol{v}^h$ along $\{\phi_\sigma\}_{\sigma\in S}$ and $\nabla\boldsymbol{\psi}\cdot\mathbf{n}\equiv 1$, we can define the residuals in the following manner.

$$\Phi_\sigma^{K,FR}(\mathbf{u}^h) = -\oint_K \nabla\phi_\sigma\cdot\mathbf{f}^h \mathrm{d}\mathbf{x} + \oint_{\partial K}\phi_\sigma \boldsymbol{f}^{\mathrm{num}}(\mathbf{u}^h,\mathbf{u}^{h,-})\cdot\mathbf{n}\mathrm{d}\gamma \overbrace{-\oint_K \nabla\phi_\sigma\cdot\boldsymbol{\alpha}\nabla\boldsymbol{\psi}\mathrm{d}\mathbf{x}}^{:=r_\sigma} = \Phi_\sigma^{K,DG}(\mathbf{u}^h) + r_\sigma \tag{15}$$

There, $\Phi_\sigma^{K,DG}(\mathbf{u}^h)$ denotes the residuals of the DG scheme, see [1]. A second condition on $\nabla\psi$ is provided by the conservation relation (11)

$$\sum_{\sigma\in K} r_\sigma = -\sum_{\sigma\in K}\oint_K \nabla\phi_\sigma\cdot\boldsymbol{\alpha}\nabla\boldsymbol{\psi}\,\mathrm{d}\,\mathbf{x} = 0. \tag{16}$$

In summary, using the residuals (15) in (12), we embed the flux reconstruction within the RD framework. By ensuring that conditions (16) and $\nabla\boldsymbol{\psi}\cdot\mathbf{n}\equiv 1$ hold, the conservation relation (11) is guaranteed. The theoretical results of RD can be now applied for the FR schemes under consideration. This opens up new possibilities to construct new FR schemes with favourable properties on arbitrary meshes [2] and we are looking forward to do this.

REFERENCES

[1] R. Abgrall. A general framework to construct schemes satisfying additional conservation relations. application to entropy conservative and entropy dissipative schemes. *Journal of Computational Physics*, **372** (2018), 640–666.

[2] R. Abgrall, E. l. Meledo, and P. Öffner. On the connection between residual distribution schemes and flux reconstruction, preprint `arXiv:1807.01261` (2018).

[3] G. J. Gassner. A skew-symmetric discontinuous Galerkin spectral element discretization and its relation to SBP-SAT finite difference methods. *SIAM Journal on Scientific Computing*, **35(3)** (2013), 1233–1253.

[4] H. Huynh. A flux reconstruction approach to high-order schemes including discontinuous Galerkin methods. *AIAA paper*, **4079:2007** (2007).

[5] H. Huynh, Z. J. Wang, and P. E. Vincent. High-order methods for computational fluid dynamics: A brief review of compact differential formulations on unstructured grids. *Computers & Fluids*, **98** (2014), 209–220.

[6] D. A. Kopriva, J. Nordström, and G. J. Gassner. Error boundedness of discontinuous Galerkin spectral element approximations of hyperbolic problems. *Journal of Scientific Computing*, **72(1)** (2017), 314–330.

[7] J. Nordström. Error bounded schemes for time-dependent hyperbolic problems. *SIAM Journal on Scientific Computing*, **30(1)** (2007), 46–59.

[8] P. Öffner. Error Boundedness of Correction Procedure via Reconstruction / Flux Reconstruction, preprint, `arXiv:1806.01575` (2018).

[9] P. Öffner and H. Ranocha. Error boundedness of discontinuous Galerkin methods with variable coefficients. *Journal of Scientific Computing*, **79(3)** (2019), 1572–1607.

[10] H. Ranocha, P. Öffner, and T. Sonar. Summation-by-parts operators for correction procedure via reconstruction. *Journal of Computational Physics*, **311** (2016), 299–328.

[11] P. E. Vincent, P. Castonguay, and A. Jameson. A new class of high-order energy stable flux reconstruction schemes. *Journal of Scientific Computing*, **47(1)** (2011), 50–72.

[12] Z. J. Wang and H. Gao. A unifying lifting collocation penalty formulation including the discontinuous Galerkin, spectral volume/difference methods for conservation laws on mixed grids. *Journal of Computational Physics*, **228** (2009), 8161–8186.

E-mail address: `remi.abgrall@math.uzh.ch,elise.lemeledo@math.uzh.ch`
E-mail address: `philipp.oeffner@math.uzh.ch, h.ranocha@tu-bs.de`

A WEAK ASYMPTOTIC SOLUTION ANALYSIS FOR A LAGRANGIAN-EULERIAN SCHEME FOR SCALAR HYPERBOLIC CONSERVATION LAWS

Eduardo Abreu

University of Campinas - IMECC/UNICAMP
Campinas, SP - Brazil

Wanderson Lambert

Federal University of Alfenas
Poços de Caldas, MG - Brazil

John Pérez

ITM Instituicion Universitaria
Medellín - Colombia

Arthur Santo*

Oil Research Center - CEPETRO - University of Campinas
Campinas, SP - Brazil

Abstract. In this work, we aim to investigate the theoretical and numerical properties of weak asymptotic solutions within the Lagrangian-Eulerian framework for hyperbolic conservation laws. Recent successful applications of the locally conservative Lagrangian-Eulerian framework has been achieved in situations where the dynamic forward tracking must preserve the delicate well-balancing between the first-order hyperbolic flux and the source term. In this framework, no approximate or exact Riemann solvers and no upwind source term discretization are used. Numerical solutions for the shallow water equations on an horizontal bed with topography are presented to illustrate the significant potential of the novel approach.

1. **Introduction.** The Lagrangian-Eulerian approach is a promising tool for numerically solving partial differential equations of several types. This framework has been used for solving hyperbolic conservation laws [1, 4], balance laws problems [2, 4, 9]. In the work [11], it was identified the region in the space-time domain where the mass conservation takes place, but linked to a scalar convection-dominated nonlinear parabolic problem (see also [8]). More recently in [1, 4, 5, 6], such ideas were extended to a wide range of nonlinear purely hyperbolic conservation laws and balance laws – scalar and systems – with applications to various physical models. Similar developments based on Lagrangian-Eulerian ideas, focusing on increasing order and accuracy can be found in [2].

2000 *Mathematics Subject Classification.* Primary: 35L65, 35Q99, 34E18, 65M12; Secondary: 76.

Key words and phrases. Balance Laws, Weak Asymptotic Method, Lagrangian-Eulerian Finite Volume.

* Corresponding author: Arthur Santo.

In this work, we present the formal construction of an Lagrangian-Eulerian scheme for hyperbolic conservation laws in a non-staggered form, and our goal in the current work is to investigate its numerical properties of weak asymptotic solutions, as seen from [7].

Weak asymptotic methods have been introduced in [10] in the framework of the Maslov-Whitham asymptotic analysis; see [7] and references cited therein. They have proved to be an efficient mathematical tool to study creation and superposition of singular solutions to various nonlinear PDEs, such as δ-waves and the more general $\delta(n)$-waves. The weak asymptotic methods presented in this paper are constructed by transforming each scalar PDE into a family of ODEs of the Lipschitz type in the Banach spaces of continuous functions and essentially bounded functions. Numerical experiments illustrating the explicit calculation of the weak asymptotic approximations for concrete conservation law equations are also presented. Qualitatively correct numerical approximations for the shallow water equations on an horizontal bed with topography are presented to illustrate the significant potential of the novel approach.

2. The Lagrangian-Eulerian finite volume method.

We present a non-staggered form of the Lagrangian-Eulerian framework for the following first-order scalar conservation law $x \in \mathbb{R}, t \in \mathbb{R}^+, u = u(x,t) : \mathbb{R} \times \mathbb{R}^+ \to \Omega \subset \mathbb{R}, H : \Omega \to \mathbb{R}$.

$$\frac{\partial u}{\partial t} + \frac{\partial H(u)}{\partial x} = 0, \quad x \in \mathbb{R}, \quad t > 0, \qquad u(x,0) = u_0(x). \tag{1}$$

As in the Lagrangian-Eulerian schemes [1, 2, 3, 4, 5, 6, 9, 11], local conservation is obtained by integrating the conservation law over the region in the space-time domain where the conservation of the mass flux takes place. Consider the Lagrangian-Eulerian finite-volume cell centers

$$D_j^n = \{(t,x) \ / \ t^n \leq t \leq t^{n+1}, \ \sigma_{j-\frac{1}{2}}(t) \leq x \leq \sigma_{j+\frac{1}{2}}(t)\}, \tag{2}$$

where $\sigma_{j-\frac{1}{2}}^n(t)$ is the parameterized integral curve such that $\sigma_{j-\frac{1}{2}}^n(t^n) = x_{j-\frac{1}{2}}^n$. These curves are the lateral boundaries of the domain D_j^n in (2) and we define $\bar{x}_{j-\frac{1}{2}}^n := \sigma_{j-\frac{1}{2}}^n(t^{n+1})$ and $\bar{x}_{j+\frac{1}{2}}^n := \sigma_{j+\frac{1}{2}}^n(t^{n+1})$ as their endpoints in time t^{n+1}. The numerical scheme is expected to satisfy some type of mass conservation (due to the inherent nature of the conservation law) from time t^n in the space domain $\left[x_{j-\frac{1}{2}}^n, x_{j+\frac{1}{2}}^n\right]$ to time t^{n+1} in the space domain $\left[\bar{x}_{j-\frac{1}{2}}^{n+1}, \bar{x}_{j+\frac{1}{2}}^{n+1}\right]$. With this, we must have the flux through curves $\sigma_{j-\frac{1}{2}}^n(t)$ to be zero. From the integration of (1) and the divergence theorem, using the fact that the line integrals over curves $\sigma_j^n(t)$ vanish,

$$\int_{\bar{x}_{j-\frac{1}{2}}^{n+1}}^{\bar{x}_{j+\frac{1}{2}}^{n+1}} u(x,t^{n+1})dx = \int_{x_{j-\frac{1}{2}}^n}^{x_{j+\frac{1}{2}}^n} u(x,t^n)dx. \tag{3}$$

The linear case from [9] is essentially imitated, but here the curves $\sigma_{j-1/2}^n(t)$ are not straight lines in general, but rather solutions of the set of local nonlinear differential equations [1, 9]: $\frac{d\sigma_{j-1/2}^n(t)}{dt} = \frac{H(u)}{u}$, for $t^n < t \leq t^{n+1}$, with the initial condition $\sigma_{j-1/2}^n(t^n) = x_{j-1/2}^n$, assuming $u \neq 0$ (for the sake of presentation). This construction follows naturally from the finite volume formulation of the linear Lagrangian-Eulerian scheme as building block to construct *local* approximations such as $f_{j-1/2}^n = \frac{H(U_{j-1/2}^n)}{U_{j-1/2}^n} \approx \frac{H(u)}{u}$ with the initial condition $\sigma_{j-1/2}^n(t^n) = x_{j-1/2}^n$.

Indeed, distinct and high-order approximations are also acceptable for $\frac{d\sigma^n_{j-1/2}(t)}{dt}$ and can be viewed as ingredients to improve accuracy of the new family of Lagrangian-Eulerian methods. Equation (3) defines mass conservation but in a different mesh cell-centered in points $\bar{x}^n_{j+\frac{1}{2}}$ of width h^{n+1}_j. Along the linear approximation for $f^n_{j-1/2}$, we find out that $\bar{x}_{j-\frac{1}{2}} = x_{j-\frac{1}{2}} + f_{j-1/2}\Delta t$ and $\bar{x}_{j+\frac{1}{2}} = x_{j+\frac{1}{2}} + f_{j+1/2}\Delta t$. Equation (3) defines a local mass balance between space intervals at time t^n and intervals at time t^{n+1}. We will later address how to project these volumes back to the original mesh.

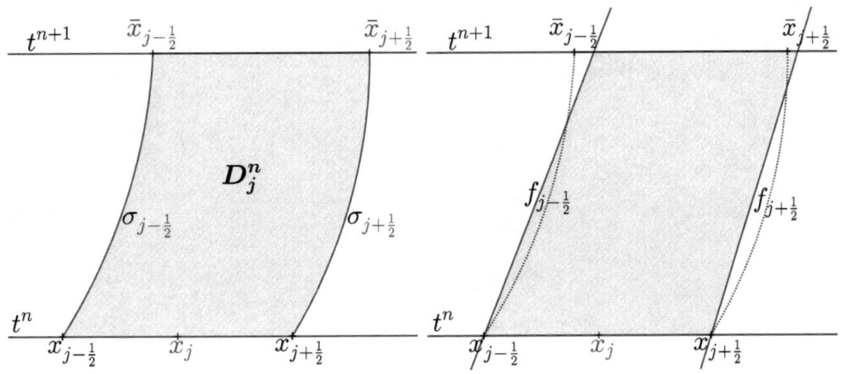

FIGURE 1. The Integral tube. Left: nonlinear, right: linear approximation.

Using the approximations[1]

$$\overline{U}^{n+1}_j := \frac{1}{h^{n+1}_j}\int_{\bar{x}^{n+1}_{j-\frac{1}{2}}}^{\bar{x}^{n+1}_{j+\frac{1}{2}}} u(x,t^{n+1})dx, \quad \text{and} \quad U^n_j := \frac{1}{h}\int_{x^n_{j-\frac{1}{2}}}^{x^n_{j+\frac{1}{2}}} u(x,t^n)dx,$$

the discrete version of equation (3) is

$$\overline{U}^{n+1}_j = \frac{1}{h^{n+1}_j}\int_{\bar{x}^{n+1}_{j-\frac{1}{2}}}^{\bar{x}^{n+1}_{j+\frac{1}{2}}} u(x,t^{n+1})dx = \frac{1}{h^{n+1}_j}\int_{x^n_{j-\frac{1}{2}}}^{x^n_{j+\frac{1}{2}}} u(x,t^n)dx = \frac{h}{h^{n+1}_j}U^n_j, \tag{4}$$

Solutions $\sigma^n_{j-1/2}(t)$ of the differential system are obtained also using the linear approximations $L(x,t)$. As in [6], the piecewise constant numerical data is reconstructed into a piecewise linear approximation (but high-order reconstructions are acceptable), through the use of MUSCL-type interpolants $L_j(x,t) = u_j(t) + (x - x_j)\frac{1}{\Delta x}u'_j$. For the numerical derivative $\frac{1}{\Delta x}u'_j$, there are several choices of slope limiters (see, e.g., [8, 13]). A priori choice of such slope limiters is quite hard, but they are chosen upon the underlying model problem under investigation.

[1]We must notice that the approximation of $f^n_{j-1/2}$ may cause spurious oscillation in Riemann problems, specially in shocks and discontinuity regions. For that, we use a polynomial reconstruction of second degree to smooth out the approximation and also slope limiters approximation of the form (see, e.g., [8, 13]). The numerical solutions have shown qualitatively correct behavior for nonlinear hyperbolic conservation laws. The convergence order remains unchanged even with the reconstruction, being a first-order approximation. In the reconstruction we may use the nonlinear Lagrange polynomial in U_{j-1}, U_j and U_{j+1}.

The approximation of $U_{j-\frac{1}{2}}$ is given by:

$$\begin{aligned} U_{j-\frac{1}{2}} &= \frac{1}{h}\int_{x_{j-1}^n}^{x_j^n} L(x,t)dx = \frac{1}{h}\left(\int_{x_{j-1}^n}^{x_{j-\frac{1}{2}}^n} L_{j-1}(x,t)dx + \int_{x_{j-\frac{1}{2}}^n}^{x_j^n} L_j(x,t)dx\right) \\ &= \frac{1}{2}(U_{j-1}+U_j) + \frac{1}{8}(U_j' - U_{j-1}'). \end{aligned} \tag{5}$$

Next, we obtain the resulting projection formula as follows

$$U_j^{n+1} = \frac{1}{h}\left(c_{-1,j}\overline{U}_{j-1}^n + c_{0,j}\overline{U}_j^n + c_{1,j}\overline{U}_{j+1}^n\right), \quad \text{where } h = \Delta x, \tag{6}$$

where the projection coefficients are

$$c_{-1,j} = \frac{1}{2}\left(1 + \text{sgn}(f_{j-\frac{1}{2}})\right)|f_{j-\frac{1}{2}}|\Delta t =: f^+(U_{j-\frac{1}{2}})\Delta t, \tag{7}$$

$$c_{+1,j} = \frac{1}{2}\left(1 - \text{sgn}(f_{j+\frac{1}{2}})\right)|f_{j+\frac{1}{2}}|\Delta t =: f^-(U_{j+\frac{1}{2}})\Delta t, \tag{8}$$

$$c_{0,j} = (h - c_{-1,j} - c_{+1,j}). \tag{9}$$

Here Δt is obtained under CFL-condition

$$\max_j\left\{|f_{j-\frac{1}{2}}\Delta t|\right\} \leq \frac{h}{2}, \tag{10}$$

which is taken by construction of method. We note that in the linear case, when $a(x,t) = a > 0$ (or $a < 0$), the numerical scheme (4)-(6) is a generalization of the Upwind scheme, but our scheme can approximate solution in both cases $a > 0$ and $a < 0$. The CFL-condition in this case is $|a\,\Delta t| \leq h$ as in the Upwind scheme. We now investigate the theoretical properties of the Lagrangian-Eulerian scheme via weak asymptotic solutions.

2.1. **Sketch of a convergence proof from the weak asymptotic solution.** The weak asymptotic solution method (see [7, 10] and references therein), is used to study the existence of solutions of scalar and system of hyperbolic equations, giving a new sense of definition for the solution. An interesting characteristic of this theory is the possibility of proving the existence of a solution from numerical methods. In the current work, we give a definition of the weak asymptotic solution for a scalar equation (1) and a sketch of proof of stability of the numerical method; in an article in preparation, [3], we describe the complete proof of convergence.

To define the weak asymptotic method, we consider the one-dimensional scalar equation (1). Here, to avoid boundary conditions in the bounded domain from numerical purposes, we consider $x \in \mathbb{S}^1 = \mathbb{R}/\mathbb{Z}$, $t \in \mathbb{R}^+$, $u = u(x,t) : \mathbb{S}^1 \times \mathbb{R}^+ \to \Omega \subset \mathbb{R}$ and the flux function $H(u) : \Omega \to \mathbb{R}$. The weak asymptotic solution is a sequence of solution $(u_\epsilon)_\epsilon = (u(x,t,\epsilon))_\epsilon$ of class $\mathcal{C}^1$ in t and of class L^∞ and piecewise continuous in x such that for all $\psi \in \mathcal{C}_c^\infty$ and for all t:

$$\lim_{\epsilon\to 0}\int_R ((u_\epsilon)_t\psi - H(u_\epsilon)\psi_x)dx = 0 \quad \text{and} \quad u_\epsilon(x,0) = u_0(x). \tag{11}$$

The weak asymptotic solution consists on first proposing a PDE with a special flux (using the parameter ϵ); then, for each fixed x, we obtain an ordinary differential equation (ODE). From the theory of ODEs, we prove existence and stability of the solution. Finally, we prove that when taking $\epsilon \to 0$, the limit satisfies (11). The idea is that the flux represents the numerical method, thus the existence and stability of the PDE of special class can represent an extension of the numerical method.

For our method, we propose the PDE:

$$\partial_t(u_\epsilon) = \frac{1}{\epsilon}\left[u_{\epsilon,-}f^+(\hat{u}_{\epsilon,-}) - u_\epsilon f^+(\hat{u}_{\epsilon,-}) - u_\epsilon f^-(\hat{u}_{\epsilon,+}) + u_{\epsilon,+}f^-(\hat{u}_{\epsilon,+})\right], \quad (12)$$

with initial condition $u_\epsilon(x,0) = u_0(x)$, where, we define the $u_{\epsilon,-}$, $u_{\epsilon,+}$ and $u_{\epsilon,\cdot}$ as:

$$u_{\epsilon,-} = u(x-\epsilon,t,\epsilon), \quad u_{\epsilon,+} = u(x+\epsilon,t,\epsilon) \quad \text{and} \quad u_\epsilon = u(x,t,\epsilon) \quad (13)$$

We remember that $f(u) = H(u)/u$. In our sketch of proof, we assume that $u \neq 0$ to avoid technical details. We remark that $H(u)$ (and then $f(u)$) can explicitly depend on x and t, however, here we describe the sketch only for which there is no this explicit dependence.

The states $\hat{u}_{\epsilon,-}$ and $\hat{u}_{\epsilon,+}$ are obtained from a combination of known states, i.e., there is a function $L(\cdot,\cdot) : \Omega\times\Omega \to \mathbb{R}$ assumed to be Lipschitzian (in both variables), such that:

$$\hat{u}_{\epsilon,-} = L(u_{\epsilon,-}, u_\epsilon) \text{ and if } u_{\epsilon,-} \text{ is continuous then } \lim_{\epsilon\to 0} L(u_{\epsilon,-}, u_\epsilon) = \lim_{\epsilon\to 0} u_\epsilon. \quad (14)$$

To prove our result, we assume that $H(u)$ is Lipschitzian and $u \neq 0$, such that $f(u)$ is also Lipschitzian. Notice that, since we assume that the reconstruction L, Eq. (14), is also Lipschitzian, thus f applied to $\hat{u}_{\epsilon,-}$ is Lipschitzian.

We state the existence and stability result:

Proposition 1. *We construct as solution of* (12) *a family of functions* $(x,t) \to u(x,t,\epsilon) : \mathbb{S}^1 \times \mathbb{R} \to \mathbb{R}$ *for* ϵ *small enough, which for a fixed* ϵ*, are of class* $\mathbb{C}^1$ *and class* $\mathbb{L}^\infty$ *for* $x \in \mathbb{S}^1$ *and satisfy* (11). *The family* $\{u(\cdot,t,\epsilon)\}_\epsilon$ *is bounded in* $\mathbb{L}^1(\mathbb{S}^1)$ *uniformly in* ϵ*; in fact,* $||u(t,\epsilon)||_{\mathbb{L}^1(\mathbb{S}^1)} \leq ||u_0||_{\mathbb{L}^1(\mathbb{S}^1)}$ *for all* t*. Moreover, if the initial condition* $u_0(x)$ *and* $H(u)$ *are continuous, then* $u(x,t,\epsilon)$ *is also continuous in* x*.*

Sketch of proof. First, we fix x and ϵ and we obtain a ODE from (12). Since $f(\cdot)$ and the reconstruction $L(\cdot,\cdot)$ are Lipschitzian functions, we obtain that the flux is also Lipschitzian. Thus, from classical theory for ODEs in Banach spaces in the Lipschitizian case, there is a local solution for $t \in [0, \delta(\epsilon)]$ for some $\delta(\epsilon)$ that depends on ϵ. For the global solution, since f is bounded, we can prove that we can extend the solution for $\delta(\epsilon) \to \infty$. To prove that the solutions of ODEs provide a weak asymptotic solution for (1), we will prove L^1 is bounded uniformly with respect to ϵ. To do so, let $T > 0$, for $t + dt \leq T$ and $dt > 0$. It follows from the mean value theorem that we can write (12) as:

$$\begin{aligned} u(x,t+dt,\epsilon) =& u_\epsilon + \frac{dt}{\epsilon}\left[u_{\epsilon,-}f^+(\hat{u}_{\epsilon,-}) - u_\epsilon f^+(\hat{u}_{\epsilon,-}) - u_\epsilon f^-(\hat{u}_{\epsilon,+}) + u_{\epsilon,+}f^-(\hat{u}_{\epsilon,+})\right] \\ &+ dt r(x,t,dt), \end{aligned} \quad (15)$$

where $||r(\cdot,t,dt)|| \to 0$ when $dt \to 0$. Since we are interested in obtaining the L^1 bound, we take the absolute value:

$$\begin{aligned} |u(x,t+dt,\epsilon)| \leq& |u_\epsilon|\left(1 - \frac{dt}{\epsilon}(f^+(\hat{u}_{\epsilon,-}) + f^-(\hat{u}_{\epsilon,+})\right) + \frac{dt}{\epsilon}\left[|u_{\epsilon,-}|f^+(\hat{u}_{\epsilon,-})\right] \\ &+ \frac{dt}{\epsilon}\left[|u_{\epsilon,+}|f^-(\hat{u}_{\epsilon,+})\right] + dt|r(x,t,dt)|, \end{aligned} \quad (16)$$

Eq. (16) is satisfied if $1 - \frac{dt}{\epsilon}(f^+(\hat{u}_{\epsilon,-}) + f^-(\hat{u}_{\epsilon,+}) \geq 0$, and from definition of f^+ and f^-, Eqs. (7)-(8), we obtain that if the CFL condition (10) is satisfied, then

(16) is true. This prove that the CFL condition provides stability for the method, since by integrating (16) and due translations of $\pm\epsilon$, one can prove the L^1 bound as:

$$\int_{\mathbb{S}}^{1} |u(x,T,\epsilon)|dx \leq \int_{\mathbb{S}}^{1} |u_0(x)|dx \tag{17}$$

To finish the prove of proposition, we define the integral I:

$$I = \int_{\mathbb{S}^1} \left(\frac{1}{\epsilon} \left[u_{\epsilon,-} f^+(\hat{u}_{\epsilon,-}) - u_\epsilon f^+(\hat{u}_{\epsilon,-}) - u_\epsilon f^-(\hat{u}_{\epsilon,+}) + u_{\epsilon,+} f^-(\hat{u}_{\epsilon,+}) \right] \psi(x) \right. \tag{18}$$
$$\left. - H(u_\epsilon)\partial_x \psi(x) \right) dx$$

and we prove that $I \to 0$ when $\epsilon \to 0$, obtaining (11). □

To finish the convergence of numerical method, we show that (6) can be written as a particular case of (12) taking $\epsilon = \Delta t$.

Proposition 2. *The numerical scheme* (6) *is compatible with the ODE:*

$$U_t = U_{j-1}^n f_{j-\frac{1}{2}}^+ - U_j^n f_{j-\frac{1}{2}}^- - U_j^n f_{j+\frac{1}{2}}^+ + U_{j+1}^n f_{j+\frac{1}{2}}^-. \tag{19}$$

Sketch of proof. We substitute Eqs. (4), (7)-(9) in Eq. (6) and we obtain:

$$U_j^{n+1} = U_j^n \frac{h}{h_j^{n+1}} + \Delta t \left(f_{j-\frac{1}{2}}^+ \frac{U_{j-1}^n}{h_{j-1}^{n+1}} - (f_{j-\frac{1}{2}}^+ + f_{j+\frac{1}{2}}^-) \frac{U_j^n}{h_j^{n+1}} + f_{j+\frac{1}{2}}^- \frac{U_{j+1}^n}{h_{j+1}^{n+1}} \right). \tag{20}$$

Using that $h_j^{n+1} = h + (f_{j+\frac{1}{2}} - f_{j-\frac{1}{2}})\Delta t$ that gives $\frac{h}{h_j^{n+1}} = 1 - \frac{f_{j+\frac{1}{2}} - f_{j-\frac{1}{2}}}{h_j^{n+1}} \Delta t$ and substituting this result in the Eq. (20), one can prove that:

$$U_j^{n+1} - U_j^n = -U_j^n \Delta t \left(\frac{f_{j+\frac{1}{2}} - f_{j-\frac{1}{2}}}{h_j^{n+1}} \right) +$$
$$+ \frac{\Delta t}{h} \left(f_{j-\frac{1}{2}}^+ U_{j-1}^n - (f_{j-\frac{1}{2}}^+ + f_{j+\frac{1}{2}}^-) U_j^n + f_{j+\frac{1}{2}}^- U_{j+1}^n \right) + o(\Delta t^2). \tag{21}$$

Using that $f = f^+ - f^-$ and taking the limit of $\Delta t \to 0$ in Eq. (21), we prove that the numerical method is compatible with the (19). □

Notice that Proposition 2 shows that the numerical method is compatible with the ODE (12) constructed in Proposition 1. This technique is very powerful because we can construct the ODE (12) to represent the proposed numerical scheme. The complete proofs of Propositions 1 and 2 are obtained in [3]. Moreover, we extend the result for n-dimensional spatial domain and we prove that the numerical scheme satisfies the Kruzhkov entropy.

2.2. **Numerical Experiments.** We consider, as in [12], a 2×2 nonlinear system of balance laws modeling the flow of water downing in a channel having a rectangular cross section. This is a prototype model for shallow-water flow (see [1]) on an horizontal bed with topography:

$$\frac{\partial h}{\partial t} + \frac{\partial (hu)}{\partial x} = 0, \qquad \frac{\partial (hu)}{\partial t} + \frac{\partial \left(hu^2 + \frac{1}{2} g h^2 \right)}{\partial x} = -gh \frac{\partial z}{\partial x}, \tag{22}$$

where h is the height of the free surface and u is the averaged horizontal velocity. Precisely, as in [12], z is the elevation of the bed above a reference level. Details for discretization strategies of the source term, see [1, 2, 3, 4]. The calculations were performed in the order of seconds with Matlab on a standard laptop with 2.60 GHz Intel Core i7-4510U CPU and 8.0 GB of RAM memory. On physical grounds, in this model problem it was assumed the hydrostatic balance in the vertical direction and surface tension was ignored. The first case (RP1) corresponds to a dam break over wet bed, i.e. initial conditions with left and right velocities equal to zero and different water depths ($h_L > h_R$). The solution of the Riemann Problem, is constituted by a left moving 1-Rarefaction, the bottom step discontinuity and a right-moving 2-Shock (assuming that the left rarefaction does not span across the x-axis). Numerical approximations are shown in Figure 3 with a clearly qualitatively correct approximations at $t = 8$. The second case (RP2) we have two rarefactions moving away from the step, one to the left and one to the right. Thus, the solution of the Riemann Problem is given by a left-propagating 1-Rarefaction, the bottom step discontinuity and a right-propagating 2-Rarefaction.

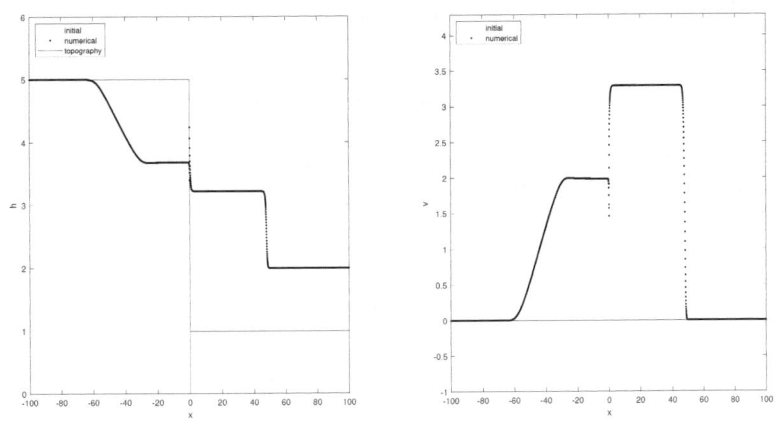

FIGURE 2. Numerical solutions to shallow water system (RP1) (22) with 2000 cells, $h + Z$ (height) on the left and v (velocity) on the right, at time t=8.0. The elapsed computer time is 10 sec.

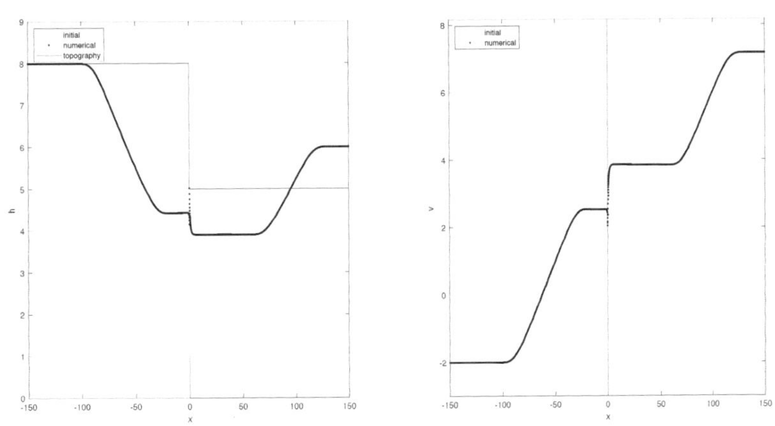

FIGURE 3. Numerical solutions to shallow water system (RP2) (22) with 2000 cells, $h + Z$ (height) on the left and v (velocity) on the right, at time t=8.0. The elapsed computer time is 10 sec.

3. **Conclusions.** We discussed the formal construction of a Lagrangian-Eulerian scheme for solving hyperbolic conservation laws with source terms. By the application of weak asymptotic solutions theory (see [7]) we investigated theoretical properties of the scheme. Numerical solutions for the shallow water equations were presented to illustrate the significant potential of the novel analysis approach.

Acknowledgments. E. Abreu thanks in part by FAPESP 2016/23374-1 (São Paulo, Brazil). A. Santo would like to thank the XVII HYP2018 committee for funding the opportunity to present at the conference.

REFERENCES

[1] E. Abreu and J. Perez. A fast, robust, and simple Lagrangian-Eulerian solver for balance laws and applications. Computers & Mathematics with Applications Available online 18 December 2018 https://doi.org/10.1016/j.camwa.2018.12.019.

[2] E. Abreu, V. Matos, J. Pérez and P. Rodríguez-Bermúdez, A shock-capturing and high-resolution Lagrangian-Eulerian method for first order hyperbolic problems with forcing terms, submitted.

[3] E. Abreu, W. Lambert, J. Perez and A. Santo, Convergence of a Lagrangian-Eulerian method via weak asymptotic method, in preparation (2019).

[4] E. Abreu, W. Lambert, J. Perez and A. Santo, A new finite volume approach for transport models and related applications with balancing source terms. Mathematics and Computers in Simulation, 137 (2017) 2-28.

[5] E. Abreu, J. Perez and A. Santo, A conservative Lagrangian-Eulerian finite volume approximation method for balance law problems. Proceeding Series of the Brazilian Society of Computational and Applied Mathematics, 6(1) (2018) 010296-1–010296-7, DOI: 10.5540/03.2018.006.01.0296

[6] E. Abreu, J. Perez and A. Santo, Solving hyperbolic conservation laws by using Lagrangian-Eulerian approach. Proceeding Series of the Brazilian Society of Computational and Applied Mathematics, 5(1) (2017) 010329-1–010329-7, DOI: 10.5540/03.2017.005.01.0329

[7] E. Abreu, M. Colombeau, and E. Panov. Weak asymptotic methods for scalar equations and systems. Journal of Mathematical Analysis and Applications 444(2) (2016) 1203-1232.

[8] E. Abreu, Numerical modelling of three-phase immiscible flow in heterogeneous porous media with gravitational effects. Mathematics and Computers in Simulation 97 (2014) 234–259.

[9] J. Aquino, A. S. Francisco, F. Pereira, T. Jordem Pereira and H.P. Amaral Souto, A lagrangian strategy for the numerical simulation of radionuclide transport problems. Progress in Nuclear Energy, 52(3) (2010) 282–291.

[10] V. Danilov and V. Shelkovich, Dynamics of propagation and interaction of δ-shock waves in conservation law systems. J. Differential Equations 211 (2005) 333–381.

[11] J. Douglas Jr., F. Pereira and L.-M. Yeh, A locally conservative Eulerian-Lagrangian numerical method and its application to nonlinear transport in porous media. Computational Geosciences, 4(1) (2000) 1–40.

[12] G. Rosatti and L. Begnudelli, The Riemann Problem for the one-dimensional, free-surface Shallow Water Equations with a bed step: Theoretical analysis and numerical simulations. Journal of Computational Physics, 229(3) (2010) 760–787.

[13] E. Tadmor, Entropy Stable Schemes, Handbook of Numerical Analysis, Vol. 17. http://dx.doi.org/10.1016/bs.hna.2016.09.006 (2016).

E-mail address: `eabreu@ime.unicamp.br`
E-mail address: `wanderson.lambert@gmail.com`
E-mail address: `jhonperez@itm.edu.co`
E-mail address: `arthurm@ime.unicamp.br`

DECAY IN L^∞ FOR THE DAMPED SEMILINEAR WAVE EQUATION ON A BOUNDED 1D DOMAIN

DEBORA AMADORI, FATIMA AL-ZAHRA' AQEL AND EDDA DAL SANTO

DISIM, University of L'Aquila
Via Vetoio, Coppito, 67100 L'Aquila, Italy

ABSTRACT. In this paper we study the long time behavior for a semilinear wave equation with space-dependent and nonlinear damping term, rewritten as a first order system. Under appropriate assumptions on the nonlinearity, we prove the exponential convergence in L^∞, as $t \to +\infty$, of the solution towards a stationary solution.

1. **Introduction.** In this paper we study the initial–boundary value problem for the 2×2 system in one space dimension

$$\begin{cases} \partial_t \rho + \partial_x J = 0, \\ \partial_t J + \partial_x \rho = -2k(x) g(J) \end{cases} \tag{1}$$

where $x \in I = [0,1]$ and $t \geq 0$, and

$$(\rho, J)(x,0) = (\rho_0, J_0)(x)\,, \qquad J(0,t) = J(1,t) = J_b \tag{2}$$

for $(\rho_0, J_0) \in BV(I)$ and for a constant $J_b \in \mathbb{R}$. Assume that

$$0 < k_1 \leq k(x) \leq k_2 \quad \forall x\,, \qquad k_1,\ k_2 > 0 \tag{3}$$

and that

$$g \in C^1(\mathbb{R})\,, \qquad g(0) = 0\,, \qquad g'(J) > 0 \quad \forall J\,. \tag{4}$$

The long time behavior of the solutions to (1), (2) is addressed by means of the stationary equation

$$\partial_x J = 0\,, \qquad \partial_x \rho = -2k(x) g(J)\,.$$

The initial and boundary conditions (2) lead to a stationary solution $(\widetilde{\rho}, \widetilde{J})$:

$$\widetilde{\rho}(x) = -2g(J_b) \int_0^x k(y)\,dy + C\,, \qquad \widetilde{J}(x) = J_b\,, \tag{5}$$

the constant C being uniquely identified by

$$\int_0^1 \widetilde{\rho}(x)\,dx = \int_0^1 \rho_0(x)\,dx\,. \tag{6}$$

2000 *Mathematics Subject Classification.* 35L50, 35B40, 35L20.

Key words and phrases. Space-dependent relaxation model, L^∞-error estimate, damped wave equation, initial-boundary value problem in one dimension.

Partially supported by 2018 INdAM-GNAMPA Project 'Equazioni iperboliche e applicazioni'.

By the change of variable $(\rho, J) \mapsto (\rho - \widetilde{\rho}, J - J_b)$ and $g(J) \mapsto g(J + J_b) - g(J)$, we can reduce to the case

$$J_b = 0\,, \qquad \int_0^1 \rho_0(x)\,dx = 0\,. \tag{7}$$

Problem (1), (2), (7) is related to the one-dimensional damped semilinear wave equation on a bounded interval: indeed the function

$$u(x,t) = -\int_0^x \rho(y,t)\,dy$$

satisfies $u_x = -\rho$, $u_t = J$ and

$$\partial_{tt}u - \partial_{xx}u + 2k(x)g(\partial_t u) = 0\,. \tag{8}$$

The equation (8) has been considered in several papers, see [9, 6, 7, 11], the recent monograph [8] and references therein. It is well known that the initial-boundary value problem for (8) is well-posed for initial data $(u_0, \partial_t u_0) \in H_0^1(I) \times L^2(I)$, for $k(x) \in L^\infty(I)$ with $k(x) \geq 0$, and decay estimates for the energy are obtained, either exponential or polynomial.

Moreover, in [7], L^p decay estimates with $2 \leq p \leq \infty$ are studied for the 1-dimensional problem. These estimates are obtained under the assumption that g' vanishes at 0, and using the hypotheses of sufficiently regular data, $(u_0, \partial_t u_0) \in W^{2,\infty}(I) \times W^{1,\infty}(I)$.

In this paper we study the decay in L^∞ for a very similar problem, assuming that the damping is space-dependent and that $g' > 0$, 4. Our main contribution is to develop an alternative approach that originates from the point of view of the hyperbolic systems of balance laws. In particular, we construct approximate solutions that allow us to get an accurate description of the solution, whose evolution is recast as a discrete time system. Then we provide a strategy for the analysis of this system, that makes use of a discrete representation formula. This eventually leads to the decay in L^∞ of the solution in terms of (u_x, u_t).

Here $u_x(\cdot,t)$, $u_t(\cdot,t))$ belong to $BV(I) \subset L^\infty(I)$ so that $(u(\cdot,t), u_t(\cdot,t))$ are in $W^{1,\infty}(I) \times L^\infty(I)$.

The main result of this paper here follows.

Theorem 1.1. *Let k satisfy (3) and g satisfy (4). Define*

$$d_1 = k_1 \min_{J \in D_J} g'(J) > 0\,, \qquad d_2 = k_2 \max_{J \in D_J} g'(J) \tag{9}$$

where D_J is a closed bounded interval depending on the data, which is invariant for J. Finally assume that

$$e^{d_2} - d_2 < e^{d_1}\,. \tag{10}$$

Let $(\rho, J)(x,t)$ be the solution of the problem (1), (2), (7) with $(\rho_0, J_0) \in BV(I)$.

Then there exist constant values $C_1 > 0$ and $C_2 > 0$, that depend only on the coefficients of the equation and on the initial and boundary data, such that

$$\begin{aligned} \|J(\cdot,t)\|_\infty &\leq C_1 e^{-C_3 t}\,, \\ \|\rho(\cdot,t)\|_\infty &\leq C_2 e^{-C_3 t}\,. \end{aligned} \tag{11}$$

where

$$C_3 = |\log C(d_1,d_2)|\,, \qquad C(d_1,d_2) = e^{-d_1}(e^{d_2} - d_2) < 1\,.$$

2. **Approximate solutions.** In this section we present our approach for the definition of approximate solutions. It consists of an adaptation of the scheme for the Cauchy problem developed in [3]. Our approach is based on the formulation of system (1) that is obtained by adding an equation for the antiderivative of $k(x)$

$$a(x) = \int_0^x k(s)\,ds\,. \tag{12}$$

More precisely, we introduce the non-conservative 3×3 system

$$\begin{cases} \partial_t \rho + \partial_x J & = 0\,, \\ \partial_t J + \partial_x \rho + 2g(J)\partial_x a & = 0\,, \\ \partial_t a & = 0\,, \end{cases} \tag{13}$$

and the piecewise constant initial data

$$\begin{aligned} &\left(\rho_0^{\Delta x}, J_0^{\Delta x}, a^{\Delta x}\right)(x) = \left(\rho_0(x_j+), J_0(x_j+), a(x_j)\right) \qquad x \in (x_j, x_{j+1}) \\ &x_j = j\Delta x \qquad\qquad j = 0, \ldots, N, \qquad\qquad \Delta x = \frac{1}{N}, \end{aligned} \tag{14}$$

where $N \in 2\mathbb{N}$ is a fixed positive even number determining the size of the space mesh. In this way, we can set up a so-called Well-Balanced algorithm to construct approximate *wave-front tracking* solutions [5], with discontinuities uniformly distributed on a grid in the (x,t)-plane. We define an approximate solution as follows.

> An approximate solution $(\rho^{\Delta x}, J^{\Delta x}, a^{\Delta x})(x,t)$ is an **exact** solution to the initial-boundary value problem (13)–(14) with boundary condition $J^{\Delta x}(0,t) = J^{\Delta x}(1,t) = 0$. In particular, $a^{\Delta x}(x)$ is piecewise constant with discontinuities located at each x_j and $(\rho^{\Delta x}, J^{\Delta x})$ is a piecewise constant function, w.r.t. (x,t), with discontinuities traveling along segments in the (x,t)-plane with slopes $\in \{\pm 1, 0\}$.

As $\Delta x \to 0$, the approximate solutions converge in L^1_{loc} (up to a subsequence) to a weak solution of (13).

The characterization of such approximate solution is based on the Riemann problem for (13), that is the initial-value problem for (13) with unknown $U = (\rho, J, a)$ and data

$$U(x,0) = \begin{cases} U_\ell & x < 0, \\ U_r & x > 0, \end{cases} \tag{15}$$

for a given *left state* $U_\ell = (\rho_\ell, J_\ell, a_\ell)$ and *right state* $U_r = (\rho_r, J_r, a_r)$. By assuming (4) and that $a_\ell \leq a_r$, this problem is uniquely solved by

$$U(x,t) = \begin{cases} U_\ell & x/t < -1, \\ U_* = (\rho_{*,\ell}, J_*, a_\ell) & -1 < x/t < 0, \\ U_{**} = (\rho_{*,r}, J_*, a_r) & 0 < x/t < 1, \\ U_r & x/t > 1, \end{cases} \tag{16}$$

where $\rho_{*,\ell}, \rho_{*,r}, J_*$ satisfy suitable conditions. See Figure (1) for a diagram of (16) in the (x,t)-plane, where the discontinuities travel along lines separating the couples (U_ℓ, U_*), (U_*, U_{**}) and (U_{**}, U_r), which stand for a -1-*wave*, a 0-*wave* and a $+1$-*wave*, respectively. In general, we call *i-wave* a couple of states (U_ℓ, U_r) separated

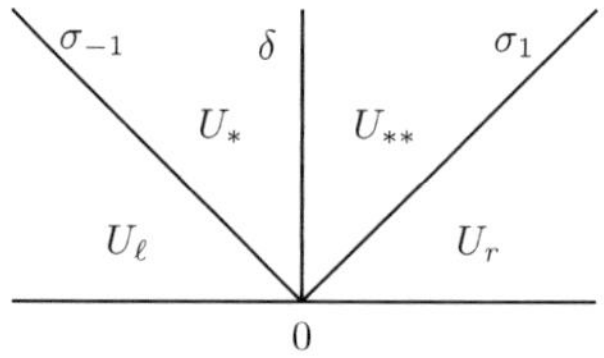

FIGURE 1. The solution to the Riemann problem (15).

by a discontinuity with *speed* (i.e. slope) $i \in \{0, \pm 1\}$ and we denote its size by

$$\begin{aligned} \sigma_{\pm 1} &= J_r - J_\ell = \pm(\rho_r - \rho_\ell) && \text{if } i = \pm 1, \\ \delta &= a_r - a_\ell && \text{if } i = 0. \end{aligned} \tag{17}$$

In the following we describe the approximate solutions in more detail; the procedure can be also regarded as a Well-Balanced scheme. See Figure (3) for a picture of the scheme in the case $N = 4$.

Step 1. The initial data is approximated as in (14); the 0-waves are located at each $0 < x_j < 1$, with size given by

$$\delta_j = a(x_j) - a(x_{j-1}) = \int_{x_{j-1}}^{x_j} k(x)dx \tag{18}$$

for $j = 1, \ldots, N-1$. Since $k \in L^\infty(I)$, we assume $\Delta x = 1/N$ to be sufficiently small so that

$$(\sup g') \cdot \delta_j < \frac{1}{2}. \tag{19}$$

Step 2. At time $t = 0+$ the solution is constructed by piecing together the solutions to the local Riemann problems at each $0 < x_j < 1$ (see (16)) and at the boundaries $x = 0$ and $x = 1$. Remark that at the boundaries the solution consists of a single $+1$-wave at $x = 0$ and of a single -1-wave at $x = 1$, respectively.

Step 3. At time $t = t^n = n\Delta t$ with $n \geq 1$ and $\Delta t = \Delta x$, multiple interactions of waves occur at $0 < x_j < 1$ (i.e. multiple segments intersect at each (x_j, t)) and the newly generated Riemann problems are solved according to

$$\begin{pmatrix} \sigma_{-1}^+ \\ \sigma_1^+ \end{pmatrix} = \begin{pmatrix} 1 - c_j & c_j \\ c_j & 1 - c_j \end{pmatrix} \begin{pmatrix} \sigma_{-1}^- \\ \sigma_1^- \end{pmatrix}, \qquad c_j := \frac{g'(s_j^n)\delta_j}{g'(s_j^n)\delta_j + 1}, \tag{20}$$

where $s_j^n \in D_J$, σ_{-1}^-, σ_1^- are the sizes of the incoming waves, σ_{-1}^+, σ_1^+ are the sizes of the outgoing ones and c is *transition coefficient*. The size of the 0-wave involved in the interaction remains constantly equal to δ_j (see (18)) across time t. Moreover, the waves hitting the boundaries $x = 0$ and $x = 1$ are both reflected and bounce back with the same size they had before the interaction. See Figure (2) for a picture of these two situations. We remark that a key property is that approximating $a(x)$ by a piecewise constant function implies that the source term is concentrated at the points x_j and results in the discontinuities with 0-slope in the solutions to the Riemann problems.

3. The iteration matrix.

The semilinear character of system (1) and the presence of the (reflecting) boundary conditions allow us to view the problem as the time evolution of the solutions to a finite dimensional linear system of the form

$$\boldsymbol{\sigma}(t^n+) = B(t^n)\,\boldsymbol{\sigma}(t^{n-1}+) = \cdots = B(t^n)B(t^{n-1})\cdots B(0+)\boldsymbol{\sigma}(0+). \tag{21}$$

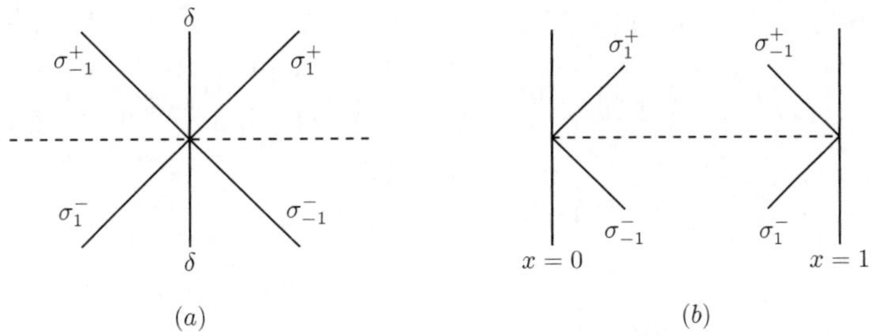

FIGURE 2. Interactions at $t = t_n > 0$: an example of multiple interaction at $0 < x_j < 1$ in (a); an example of interaction at the boundaries in (b).

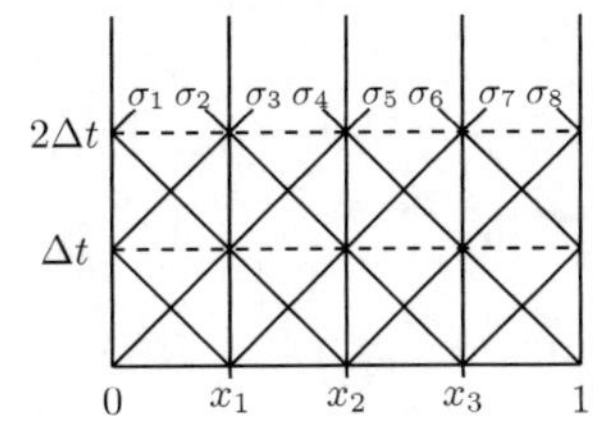

FIGURE 3. Well-balanced scheme for $N = 4$.

The components of the vector

$$\boldsymbol{\sigma}(t) = (\sigma_1, \ldots, \sigma_{2N}) \in \mathbb{R}^{2N}, \qquad N \in 2\mathbb{N},$$

are the wave sizes, see (17), that occur in the approximate solution to (13)–(14) at time t^n, ordered according to increasing space position; while the matrix $B \in \mathbb{R}^{2N\times 2N}$ is a doubly stochastic matrix (i.e. a nonnegative matrix for which the sum of all the elements by row is 1, as well as by column) given by

$$B(\boldsymbol{c}) = \begin{bmatrix} 0 & 1 & 0 & 0 & \cdots & 0 & 0 & 0 & 0 \\ c_1 & 0 & 0 & 1-c_1 & \cdots & 0 & 0 & 0 & 0 \\ 1-c_1 & 0 & 0 & c_1 & & \vdots & \vdots & & \\ \vdots & \vdots & & \vdots & \vdots & \vdots & \vdots & & \\ 0 & 0 & 0 & 0 & \cdots & c_{N-1} & 0 & 0 & 1-c_{N-1} \\ 0 & 0 & 0 & 0 & \cdots & 1-c_{N-1} & 0 & 0 & c_{N-1} \\ 0 & 0 & 0 & 0 & \cdots & 0 & 0 & 1 & 0 \end{bmatrix},$$

where $\boldsymbol{c} = (c_1, \cdots, c_{N-1}) \in \mathbb{R}^{N-1}$ and by the smallness of δ_j (see (18), (19)) we have that

$$\frac{\inf g'}{2}\delta_j \le c_j \le (\sup g')\delta_j\,, \qquad j = 1, \ldots, N-1\,. \tag{22}$$

In general the vector $\boldsymbol{c}$ depends on n, which is the index for the time: $t = t^n = n\Delta t$. The eigenvalues λ_i of B satisfy $|\lambda_i| \le 1$ for all $i = 1, \ldots, 2N$. In particular, $\lambda = \pm 1$ are eigenvalues with corresponding (left and right) eigenvectors

$$\begin{aligned} \lambda_- &= -1\,, & v_- &= (1, -1, -1, 1, \ldots, 1, -1, -1, 1)\,, \\ \lambda_+ &= 1\,, & e &= (1, 1, \ldots, 1, 1)\,. \end{aligned} \tag{23}$$

Denote by E_- the $(2N-2)$–dim eigenspace related to λ_i with $|\lambda_i| < 1$.

It is well known (Birkhoff Theorem, [10, Theorem 8.7.2]) that doubly stochastic matrices can be written as a convex combination of permutations.

In case of $\boldsymbol{c} = c(1, \cdots, 1) \in \mathbb{R}^{N-1}$ for $c \in [0, 1/2)$, the decomposition is obtained with two terms:

$$B(\boldsymbol{c}) = (1-c)B(\mathbf{0}) + cB_1 = (1-c)\left[B(\mathbf{0}) + \gamma B_1\right], \tag{24}$$

where

$$\gamma = \frac{c}{1-c} = \frac{(\sup g')\bar{k}}{N} := \frac{d}{N},$$

$B(\mathbf{0})$ is the matrix $B(\boldsymbol{c})$ with $\boldsymbol{c} = \mathbf{0}$ and B_1 is a permutation matrix that switches two consecutive rows $(2k-1)$ and $2k$. We can rewrite (24) as

$$B(\boldsymbol{c}) = \left(1 + \frac{d}{N}\right)^{-1}\left[B(\mathbf{0}) + \frac{d}{N}B_1\right].$$

On the other hand, if $\boldsymbol{c}$ is not constant (that is the case for nonlinear damping), we can bound each matrix $B = B(\boldsymbol{c}^n)$, $n \in \mathbb{N}$ with a term-by-term inequality by

$$B(\boldsymbol{c}^n) \le \left(1 + \frac{d_1}{N}\right)^{-1}\left[B(\mathbf{0}) + \frac{d_2}{N}B_1\right] \tag{25}$$

where d_1, d_2 are defined in (9).

3.1. **Total variation estimates.** Here we give a proof of the fact that

$$L_{\pm}(t) = \sum_{(\pm 1)-waves} |\Delta f^{\pm}| = \mathrm{TV}\, J^{\Delta x}(\cdot, t)$$

is not increasing in time, by means of the properties of doubly stochastic matrices. We recall here some results from [4, pp.149–153].

Definition 3.1 (Majorization of vectors)**.** Let $v, u \in \mathbb{R}^n$ and denote

$$v_{[1]} \ge v_{[2]} \ge \cdots \ge v_{[n]}, \qquad u_{[1]} \ge u_{[2]} \ge \cdots \ge u_{[n]},$$

the components rearranged in non-increasing order. We say that v is *majorized* by u if the following conditions hold:

$$\sum_{i=1}^{n} v_i = \sum_{i=1}^{n} u_i,$$

$$\sum_{i=1}^{h} v_{[i]} \le \sum_{i=1}^{h} u_{[i]} \qquad h = 1, \ldots, n-1.$$

The following theorem is a useful characterization of majorization.

Theorem 3.2 (Hardy-Littlewood-Polya)**.** *Let $v, u \in \mathbb{R}^n$. Then, v is majorized by u if and only if there exists a doubly stochastic matrix A such that $v = Au$.*

Lemma 3.3. *Let $v, u \in \mathbb{R}^n$. If v is majorized by u and $\phi : \mathbb{R} \to \mathbb{R}$ is a convex function, then*

$$\sum_{i=1}^{n} \phi(v_i) \le \sum_{j=1}^{n} \phi(u_i). \tag{26}$$

The following corollary is an easy consequence of these results, and it proves that $L_{\pm}$ is non-increasing in time.

Corollary 1. *Denote σ_j^n the jth component of $\boldsymbol{\sigma}(t^n+)$. Then,*

$$\sum_{j=1}^{2N} |\sigma_j^{n+1}| \le \sum_{j=1}^{2N} |\sigma_j^n|, \qquad n \ge 0.$$

Proof. Since $\boldsymbol{\sigma}(t^{n+1}+) = B^{(n)}\boldsymbol{\sigma}(t^n+)$ and $B^{(n)}$ is doubly stochastic, we have that $\boldsymbol{\sigma}(t^{n+1}+)$ is majorized by $\boldsymbol{\sigma}(t^n+)$. Then, we can conclude by applying the previous lemma to $\phi(\cdot) = |\cdot|$ and $v = \boldsymbol{\sigma}(t^{n+1}+)$, $u = \boldsymbol{\sigma}(t^n+)$. □

4. **A discrete representation formula.** The proof of Theorem (1.1) is given in [1] (with a slight improvement in the condition (10) given in [2]). Here we provide some key points.

First, a proposition which relates the L^∞-norm of $J(\cdot, t^n)$, $\rho(\cdot, t^n)$ as $n \to \infty$ to the evolution of the ℓ_1–norm of the operator $\mathcal{B}_n$:

$$\mathcal{B}_n \doteq \left[B^{(n)} B^{(n-1)} \cdots B^{(2)} B^{(1)}\right], \quad B^{(n)} = B(\boldsymbol{c}^n) \in M_{2N}, \qquad n \ge 1 \tag{27}$$

on the eigenspace $E_- \doteq < e, v_- >^\perp$, see (23).

Proposition 1. *For some constant values $\widetilde{C}_j > 0$, $j = 1, 2, 3$, independent on Δx one has that for every $t \in (t^n, t^{n+1})$*

$$\|J^{\Delta x}(\cdot, t)\|_\infty \le \widetilde{C}_1 \Delta x + \|\mathcal{B}_n \widetilde{\boldsymbol{\sigma}}(0+)\|_{\ell^1}$$
$$\|\rho^{\Delta x}(\cdot, t)\|_\infty \le \widetilde{C}_2 \Delta x + \widetilde{C}_3 \|\mathcal{B}_n \widetilde{\boldsymbol{\sigma}}(0+)\|_{\ell^1}$$

where $\widetilde{\boldsymbol{\sigma}}(0+)$ is the projection of $\boldsymbol{\sigma}(0+)$ onto E_-.

Next, the goal is to prove that $\|\mathcal{B}_n \widetilde{\boldsymbol{\sigma}}(0+)\|_{\ell^1}$ decays exponentially fast as $n \to \infty$, uniformly as $\Delta x = N^{-1} \to 0$. We focus our analysis on the iteration of the matrices $B = B(\boldsymbol{c}^n)$ up to time

$$t^N = N\Delta t = N\Delta x = 1.$$

Recalling (25), we get the following inequality:

$$\mathcal{B}_N \le \left(1 + \frac{d_1}{N}\right)^{-N} \left[B_0 + \frac{d_2}{N} B_1\right]^N, \qquad B_0 \doteq B(\mathbf{0}). \tag{28}$$

It is clear that

$$\left(1 + \frac{d_1}{N}\right)^{-N} \to \mathrm{e}^{-d_1} \qquad \text{as } N \to \infty,$$

while it takes a bigger effort to estimate the second factor

$$[B_0 + \gamma B_1]^N = \sum_{k=0}^{N} \gamma^k S_k(B_0, B_1), \qquad \gamma = \frac{d_2}{N} \tag{29}$$

since the matrices B_0, $B_1 \in M_{2N}$ **do not commute**. Each term $S_k(B_0, B_1)$ is the sum of all possible products of $2N$ matrices of size $2N$ equal to either B_1 or B_0 (and in which B_1 appears exactly k times). In particular,

$$S_0 = B_0^N, \qquad S_1 = \sum_{j=0}^{N-1} B_0^{2j} \doteq \widehat{P}.$$

In the following theorem we provide an estimate of the sum in (29) for the terms with $k \ge 2$.

Theorem 4.1. *Let $N \in 2\mathbb{N}$. Then,*

$$\left[B_0 + \frac{d}{N} B_1\right]^N = B_0^N + \frac{d}{N}\widehat{P} + \sum_{j=0}^{N-1} \zeta_{j,N} B_0^{2j+1} B_1 + \sum_{j=1}^{N-1} \eta_{j,N} B_0^{2j}\,, \tag{30}$$

where the scalar coefficients in the sums are bounded by:

$$0 \le \sum_{j=0}^{N-1} \zeta_{j,N} \le \sinh(d) - d + \frac{f_0(d)}{N}\,,$$

$$0 \le \sum_{j=1}^{N-1} \eta_{j,N} \le \cosh(d) - 1 + \frac{f_1(d)}{N}\,,$$

with terms $f_0(d)$ and $f_1(d)$ containing modified Bessel functions of the first type.

Thanks to (30) we can prove the following contraction property:

$$\|\mathcal{B}_N \widetilde{\boldsymbol{\sigma}}(0+)\|_{\ell^1} \le C_N(d_1, d_2) \|\widetilde{\boldsymbol{\sigma}}(0+)\|_{\ell^1} \tag{31}$$

where

$$C_N(d_1, d_2) \to \mathrm{e}^{-d_1}(\mathrm{e}^{d_2} - d_2) \doteq C(d_1, d_2) < 1\,, \qquad N \to \infty\,.$$

The last inequality follows from the assumption (10). By iterating the estimate (31), recalling Prop. (1) and sending $N \to \infty$, it is possible to prove the L^∞ decay stated in (11).

REFERENCES

[1] D. Amadori, E. Dal Santo and F. Aqel. Decay of approximate solutions for the damped semilinear wave equation on a bounded 1d domain. *J. Math. Pures Appl.* (2019), to appear

[2] D. Amadori, F. Aqel. Decay in L^∞ for the 1D semilinear damped wave equation on a bounded domain. In preparation.

[3] D. Amadori, L. Gosse. Error Estimates for Well-Balanced and Time-Split Schemes on a locally Damped Semilinear Wave Equation. *Math. Comp.* **85** (2016), 601–633

[4] R.B. Bapat, T. E. S. Raghavan. Nonnegative Matrices and Applications, Encyclopedia of Mathematics and Its Applications **64** Cambridge University Press, 1997

[5] A. Bressan. Hyperbolic Systems of Conservation Laws – The one-dimensional Cauchy problem, Oxford Lecture Series in Mathematics and its Applications **20**, Oxford University Press, 2000

[6] S. Cox, E. Zuazua. The rate at which energy decays in a damped string. *Comm. Partial Differential Equations* **19** (1994), no. 1-2, 213–243

[7] A. Haraux. L^p estimates of solutions to some non-linear wave equations in one space dimension. Int. J. of Mathematical Modelling and Numerical Optimisation **1** (2009), 146–152

[8] A. Haraux. Nonlinear vibrations and the wave equation. *SpringerBriefs in Mathematics*, BCAM SpringerBriefs, 2018

[9] A. Haraux, E. Zuazua. Decay estimates for some semilinear damped hyperbolic problems. *Arch. Rational Mech. Anal.* **100** (1988), no. 2, 191–206

[10] R.A. Horn, C.R. Johnson. Matrix Analysis, Cambridge University Press, 2nd edition, 2013

[11] E. Zuazua. Propagation, Observation, and Control of Waves Approximated by Finite Difference Methods SIAM Rev. **47** (2005), no. 2, 197–243

E-mail address: `debora.amadori@univaq.it`
E-mail address: `fatimaalzahraan.aqel@graduate.univaq.it`
E-mail address: `dalsantoedda@gmail.com`

ON L^1-STABILITY OF BV SOLUTIONS FOR A MODEL OF GRANULAR FLOW

Fabio Ancona

Dipartimento di Matematica, Università di Padova, Via Trieste 63, 35121 Padova, Italy

Laura Caravenna

Dipartimento di Matematica, Università di Padova, Via Trieste 63, 35121 Padova, Italy

Cleopatra Christoforou*

Department of Mathematics and Statistics, Univer. of Cyprus, 1678 Nicosia, Cyprus

Abstract. We are concerned with the well-posedness of a model of granular flow that consists of a hyperbolic system of two balance laws in one-space dimension, which is linearly degenerate along two straight lines in the phase plane and genuinely nonlinear in the subdomains confined by such lines. This note provides a survey of recent results [3] on the Lipschitz $\mathbf{L^1}$-continuous dependence of the entropy weak solutions on the initial data, with a Lipschitz constant that grows exponentially in time. Our analysis relies on the extension of a Lyapunov like functional and provide the first construction of a Lipschitz semigroup of entropy weak solutions to the regime of hyperbolic systems of balance laws (i) with characteristic families that are neither genuinely nonlinear nor linearly degenerate and (ii) initial data of arbitrarily large total variation.

1. **Introduction.** We consider the system of balance laws

$$\begin{aligned} h_t - (hp)_x &= (p-1)h, \\ p_t + ((p-1)h)_x &= 0, \end{aligned} \tag{1}$$

with $h \geq 0$ and $p \geq 0$. System (1) represents the model in the one space dimensional setting proposed by Hadeler and Kuttler [12] for the flow of granular material and describes he evolution of a moving layer on top and of a resting layer at the bottom. Here, the unknown $h = h(x,t)$ and $p(x,t)$ represent, respectively, the thickness of the rolling layer and the slope of the standing layer, while $t \geq 0$ and $x \in \mathbb{R}$ are the time and space variables. The evolution equations (1) show that the moving layer slides downhill with speed proportional to the slope of the standing layer in the direction of steepest descent. The model (1) is written in normalised form, assuming that the critical slope is $p = 1$. This means that, if $p > 1$, then grains initially at rest are hit by rolling matter of the moving layer and hence they start moving too; thus, the moving layer gets thicker. On the other hand, if $p < 1$,

2000 *Mathematics Subject Classification.* Primary: 35L65, 76T25; Secondary: 35L45, 35B35.

Key words and phrases. balance laws, stability, large BV solutions, granular flow, Lipschitz semigroup.

The first and second authors were partially supported by the Istituto Nazionale di Alta Matematica "F.Severi" (INdAM), through GNAMPA.

* Corresponding author was supported by the Internal grant (SBLawsMechGeom) #21036 from University of Cyprus .

then rolling grains can be deposited on the standing bed and, hence, the moving layer becomes thinner. Typical examples of granular material whose dynamics are described by such models are dry sand and gravel in dunes and heaps, or snow in avalanches.

This article serves as a survey of the analysis in [3] on the well-posedness of the Cauchy problem for (1). More precisely, in [3], the authors obtain a Lipschitz continuous semigroup of entropy weak solutions to the nonlinear system of balance laws (1) via a Lyapunov type functional with large initial data. Besides the motivation of this analysis in the setting of the granular flow model, the results provide the first construction of a semigroup for

(i) systems with characteristic families that are neither genuine nonlinear (GNL) nor linear degenerate (LD) (nor of Temple class), and
(ii) initial data with arbitrary large total variation.

The aim here is to provide a short exposition on the analysis of [3] pointing out the challenges that arise by these features and comparing the Lyapunov functional introduced in [3] with the classical one of Bressan et al [9].

Since, in general, global smooth solutions to hyperbolic systems do not exist, we consider weak solutions in the sense of distributions and in particular, an *entropy-admissible weak solution* of (1), that means a weak solution, admissible in the sense of Lax. Global existence of classical smooth solutions to (1) were established for a special class of initial data by Shen [14]. In the case of more general initial data with bounded but possibly large total variation, the existence of entropy weak solutions globally defined in time was proved by Amadori and Shen [2].

For systems without source term and small BV data, the Lipschitz $\mathbf{L}^1$-continuous dependence of solutions on the initial data, was first established by Bressan and collaborators in [7, 8] under the assumptions that all characteristic families are genuinely nonlinear (GNL) or linearly degenerate (LD), relying on a homotopy method that is lengthy and involves several technical points. An extension of these results is established in [4] to a class of 2×2 systems with non GNL characteristic fields that does not comprise the convective part of system (1). A different proof of the $\mathbf{L}^1$-stability of solutions for conservation laws with GNL or LD characteristic fields that is less technical and more transparent was later achieved by a technique introduced by Liu and Yang in [13] and then developed by Bressan et al [9]. Extensions of $\mathbf{L}^1$-stability results to the setting of large BV data was obtained for systems of conservation laws with Temple type characteristic fields and other special systems and also for balance laws with small data. A rich bibliography on these references can be found in [3] as well as further ones on other models of granular flow.

However, our system (1) does not fulfill these classical assumptions and in addition, its special source terms do not belong within a class for which L^1 stability results are available in the literature. The heart of the matter in [3] is to construct a Lyapunov-like nonlinear functional Φ, equivalent to the $\mathbf{L}^1$-distance, which is decreasing in time along any pair of solutions. In this review article, we state some preliminary results in Section 2, and then present the stability functional Φ in Section 3 comparing it with the classical one of Bressan et al [9] and providing the motivation of our construction. In Section 4, we conclude stating our main theorems and refering to [3] for the proofs and further analysis.

2. **Preliminaries.** It is easy to verify that system (1) is strictly hyperbolic on the domain

$$\Omega \doteq \{(h,p) : h \geq 0,\ p > 0\} \tag{2}$$

and weakly linearly degenerate at the point $(h,p) = (0,1)$. We observe that the line $p = 1$ separates the domain Ω into two invariant regions for solutions of the Riemann problem: the quarter $\{h \geq 0,\ p > 1\}$ and the half-strip $\{h \geq 0,\ 0 < p < 1\}$. Indeed, the rarefaction and Hugoniot curves of the first family through a point (h_ℓ, p_ℓ), with $p_\ell \neq 1$, never meets the line $p = 1$, while the rarefaction and Hugoniot curves of the second family through a point (h_ℓ, p_ℓ), with $h_\ell > 0$, never meets the line $h = 0$. On the the other hand, the lines $p = 1$ and $h = 0$ are also invariant regions for solutions of the Riemann problem since they coincide with the rarefaction and Hugoniot curves of the first and second family, respectively, passing through any of their points. Notice that, although the characteristic field of the first family does not satisfy the classical GNL assumption, no composite waves are present in the solution of a Riemann problem for

$$\begin{aligned} h_t - (hp)_x &= 0, \\ p_t + ((p-1)h)_x &= 0, \end{aligned} \tag{3}$$

since in each invariant region $\{p > 1\}$, $\{p < 1\}$ the field is GNL. In fact, the general solution of a Riemann problem for (3) consists of at most one simple wave for each family which can be either a rarefaction or a compressive shock or a contact discontinuity.

Let $u = u(x,t) \doteq (h^{s,\varepsilon}, p^{s,\varepsilon})(x,t)$ be a piecewise constant s-ε-approximate solution converging to an entropy weak solutions to (1) with initial data

$$h(x,0) = \overline{h}(x), \qquad p(x,0) = \overline{p}(x) \qquad \text{for a.e. } x \in \mathbb{R}. \tag{4}$$

constructed as in [2] by the usual operator splitting scheme as $\varepsilon \to 0+$ and $s \to 0+$. Here, $s = \Delta t > 0$ stands for the time step and a parameter $\varepsilon > 0$ a small positive parameter of the front tracking algorithm. We refer to [11] and [1] for the early works on this subject and also point out that the source term (1) does not belong in the class of the so-called "dissipative" terms exploited in [11, 1, 10]. As usual, a-priori bounds on the total variation of $u(t) \doteq u(\cdot, t)$ outside the time steps are obtained in [2] by analyzing suitable wave strength and wave interaction potential that are defined as follows.

First, the sizes of wave fronts of approximate solutions of (1) are defined as the jumps between the left and right states either measured with the original variables (h,p) or with the corresponding Riemann coordinates (H,P) associated to system (1). So given a wave front with left and right states (h_ℓ, p_ℓ) and (h_r, p_r), respectively, let (H_ℓ, P_ℓ) and (H_r, P_r) be the corresponding Riemann coordinates. Then, the wave size of the jump $\big((h_\ell, p_\ell), (h_r, p_r)\big)$ can be defined in two coordinate systems as follows:

- the size of a 1-wave (h-wave) is measured by $\rho_h = H_r - H_\ell$ or $\gamma_h = h_r - h_\ell$ in Riemann or original coordinates, respectively.
- the size of a 2-wave (p-wave) is measured by $\rho_p = P_r - P_\ell$ or $\gamma_p = p_r - p_\ell$ in Riemann or original coordinates, respectively.

Next, at any time $t > 0$ where no interaction occurs and away from time steps, let $\mathcal{J}_i\big(u(t)\big)$ denote a set of indexes α associated to the jumps of the i-th family of $u(t)$ located at x_α and let $p_\alpha^\ell \doteq P(x_\alpha -)$. Also, set $\mathcal{J}\big(u(t)\big) \doteq \mathcal{J}_1\big(u(t)\big) \bigcup \mathcal{J}_2\big(u(t)\big)$ to denote the collection of indexes associated to all jumps of $u(t)$ and $k_\alpha \in \{1, 2\}$

the characteristic family of the jump $\alpha \in \mathcal{J}(u(t))$, so that, in particular, one has $\alpha \in \mathcal{J}_{k_\alpha}(u(t))$. Then, we define the *total strength* of waves in $u(t)$ as:

$$
\begin{aligned}
V_i(u(t)) &\doteq \sum_{\alpha \in \mathcal{J}_i((u(t))} |\rho_\alpha|, \quad i = 1, 2, \\
V(u(t)) &= V_1(u(t)) + V_2(u(t)) \doteq \sum_{\alpha \in \mathcal{J}(u(t))} |\rho_\alpha|,
\end{aligned} \tag{5}
$$

and the *interaction potential* as:

$$
\mathcal{Q}(u(t)) \doteq \mathcal{Q}_{hh} + \mathcal{Q}_{hp} + \mathcal{Q}_{pp}\,. \tag{6}
$$

where

$$
\mathcal{Q}_{hh} \doteq \sum_{\substack{k_\alpha = k_\beta = 1 \\ x_\alpha < x_\beta}} \omega_{\alpha,\beta} |\rho_\alpha||\rho_\beta|, \quad \mathcal{Q}_{hp} \doteq \sum_{\substack{k_\alpha = 2, k_\beta = 1 \\ x_\alpha < x_\beta}} |\rho_\alpha \rho_\beta|, \quad \mathcal{Q}_{pp} \doteq \sum_{(\alpha,\beta) \in Appr_2} |\rho_\alpha \rho_\beta| \tag{7}
$$

with the weights $\omega_{\alpha,\beta} := \overline{\delta} \cdot \min\{|p_\alpha^\ell - 1|, |p_\beta^\ell - 1|\}$ if ρ_α, ρ_β are 1-shocks lying on the same side of $p = 1$, otherwise $\omega_{\alpha,\beta} := 0$, for a suitable constant $\overline{\delta} > 0$ sufficiently small. Also, $Appr_2$ denotes the set of pairs of indexes of approaching p-waves. Note that $\mathcal{Q}_{hh}$ is the modified interaction potential of waves of the first family (h-waves) introduced in [2] and the others are defined the usual way. Relying on the interaction estimates established in [2], the *Glimm functional*

$$
\mathcal{G}(u(t)) \doteq V(u(t)) + \mathcal{Q}(u(t)) \tag{8}
$$

is nonincreasing in any time interval $]t_k, t_{k+1}[$ between two consecutive time steps. Instead, the estimates derived in [2] on the variation of the strength of waves when the solution is updated with the source term, imply that at any time step $t_k = k\Delta t = k\,s$ there holds

$$
\mathcal{G}(u(t_k+)) \le (1 + \mathcal{O}(1)\Delta t) \cdot \mathcal{G}^-(u(t_k-))\,, \tag{9}
$$

i.e. $\mathcal{G}$ is increasing across t_k.

3. **Stability Functional.** Let u and $v : \mathbb{R}_+ \times \mathbb{R} \to \mathbb{R}^n$ be two approximate solutions to (1) and consider any piecewice constant function z with the property that for fixed t, $z(t,\cdot) : \mathbb{R} \to \mathbb{R}^2$ is a L^1 function of small total variation. In addition, $z(t,x)$ has finitely many discontinuities that are polygonal lines and the slope of such a line is bounded in absolute value by a fixed number $\hat{\lambda}$. Also, there exists a constant $\sigma > 0$ such that $Tot.Var.z(t) \le \sigma$, for all $t > 0$. We clarify that z is an arbitrary function with the aforementioned properties and is not related to the system (1). Next, consider the i-shock curve $\mathbf{S}_i(\cdot;\cdot)$ and the scalar functions η_i $i = 1, 2$ defined implicitly by

$$
w(t,x) = \mathbf{S}_2(\eta_2(t,x);\cdot) \circ \mathbf{S}_1(\eta_1(t,x); u(t,x))\,, \tag{10}
$$

where $w \doteq v + z$. According to this definition, the parameter η_i denotes the strength in the original coordinates along the i-shock curves connecting u and $w = v + z$. We clearly have

$$
\frac{1}{C_0}|u(x) - w(x)| \le \sum_i |\eta_i(x)| \le C_0 |u(x) - w(x)| \tag{11}
$$

for some constant $C_0 > 0$. We can now define the *stability functional*

$$\Phi_z(u(t), v(t)) \doteq \sum_{i=1}^{2} \int_{-\infty}^{\infty} |\eta_i(x,t)| W_i(x,t)\, dx \tag{12}$$

with weights W_i of be

$$W_i(x,t) \doteq 1 + \kappa_1 \mathcal{A}_i(x,t) + \kappa_2 \big[\mathcal{G}(u(t)) + \mathcal{G}(v(t))\big], \tag{13}$$

for suitable positive constants $\kappa_1 < \kappa_2$ to be specified. Here $\mathcal{G}$ is the Glimm functional defined in (5)-(8), and $\mathcal{A}_i(t;x)$ measures the total amount of waves in $u(t)$ and $v(t)$ which approach the i-wave η_i located at x defined as follows:

$$\mathcal{A}_1(t;x) \doteq \sum_{\substack{\alpha \in \mathcal{J}(u) \cup \mathcal{J}(v) \\ k_\alpha = 2,\, x_\alpha < x}} |\rho_\alpha| + \begin{cases} \Big[\sum_{\alpha \in \mathcal{Z}_{neg}(u)} + \sum_{\alpha \in \mathcal{Z}_{neg}(v)} \Big] |p_\alpha^\ell - 1||\rho_\alpha| & \text{if } \eta_1(t,x) < 0 \\ \Big[\sum_{\alpha \in \mathcal{Z}_{pos}(u)} + \sum_{\alpha \in \mathcal{Z}_{pos}(v)} \Big] |p_\alpha^\ell - 1||\rho_\alpha| & \text{if } \eta_1(t,x) > 0 \end{cases} \tag{14}$$

and

$$\mathcal{A}_2(t;x) \doteq \sum_{\substack{\alpha \in \mathcal{J}(u) \cup \mathcal{J}(v) \\ k_\alpha = 1,\, x_\alpha > x}} |\rho_\alpha| + \begin{cases} \Big[\sum_{\substack{\alpha \in \mathcal{J}(u), k_\alpha = 2 \\ x_\alpha > x}} + \sum_{\substack{\alpha \in \mathcal{J}(v), k_\alpha = 2 \\ x_\alpha < x}} \Big] |\rho_\alpha| & \text{if } \eta_2(t,x) < 0 \\ \Big[\sum_{\substack{\alpha \in \mathcal{J}(v), k_\alpha = 2 \\ x_\alpha > x}} + \sum_{\substack{\alpha \in \mathcal{J}(u), k_\alpha = 2 \\ x_\alpha < x}} \Big] |\rho_\alpha| & \text{if } \eta_2(t,x) > 0 \end{cases} \tag{15}$$

where $\mathcal{Z}$ denotes the set of selected 1-waves α chosen as follows

$$\mathcal{Z}_{neg}(u) := \{\alpha \in \mathcal{J}_1(u) : \text{ either } [u_2(x_\alpha-) > 1,\ x_\alpha < x] \text{ or } [u_2(x_\alpha-) < 1,\ x_\alpha > x]\}$$
$$\mathcal{Z}_{neg}(v) := \{\alpha \in \mathcal{J}_1(v) : \text{ either } [v_2(x_\alpha-) > 1,\ x_\alpha > x] \text{ or } [v_2(x_\alpha-) < 1,\ x_\alpha < x]\}$$

for $\eta_1 < 0$, and

$$\mathcal{Z}_{pos}(v) := \{\alpha \in \mathcal{J}_1(v) : \text{ either } [v_2(x_\alpha-) > 1,\ x_\alpha < x] \text{ or } [v_2(x_\alpha-) < 1,\ x_\alpha > x]\}$$
$$\mathcal{Z}_{pos}(u) := \{\alpha \in \mathcal{J}_1(u) : \text{ either } [u_2(x_\alpha-) > 1,\ x_\alpha > x] \text{ or } [u_2(x_\alpha-) < 1,\ x_\alpha < x]\}$$

for $\eta_1 > 0$ and p_α^ℓ denotes the left state of the jump located at x_α and by ρ_α the corresponding strength of the jump in Riemann coordinates.

Notice that the main novelty of our functional is encoded in the weight W_1 and in particular in $\mathcal{A}_1$, whereas W_2 has almost the same expression of the weight given in [9] for GNL and LD characteristic fields. In fact, the only difference between the definition of the weight W_2 here and the one given in [9] relies in the presence of the whole Glimm functional $\mathcal{G}$ of u and v in W_i, instead of their interaction potential $\mathcal{Q}$. Indeed, in comparison to the weights W_i used in [6, § 8], here the terms of the *Glimm functionals* $\mathcal{G}$ and not only the *interaction potential* $\mathcal{Q}$ are needed in the definition of W_i to control the change $\mathcal{A}_i$ across an interaction time. This is due to the fact that, since the first characteristic family is not GNL, we adopt as in [2] a wave interaction potential $\mathcal{Q}$, suited to (1), that is in general not decreasing in presence of interactions of 1-waves of different sign (1-shocks with 1-rarefaction waves). Therefore, one needs to exploit the decrease of the total strength V of waves due to cancellation in order to control the possible increase of the potential interaction $\mathcal{Q}$ occurring at such interactions.

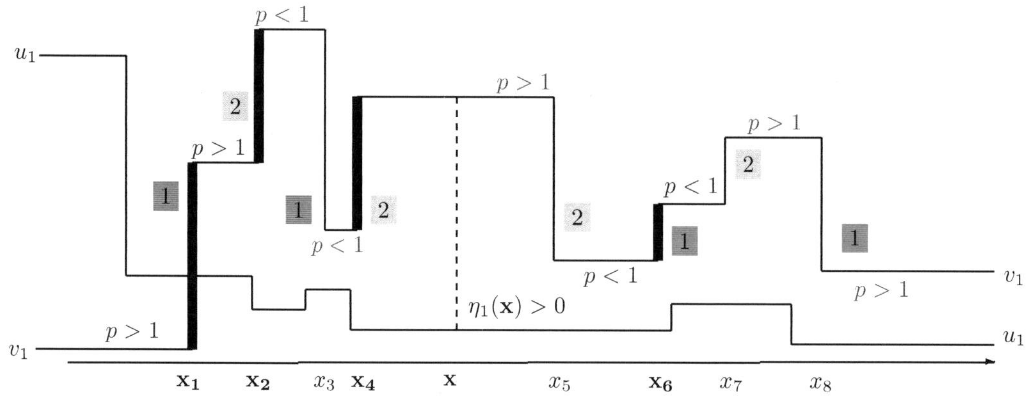

FIGURE 1. *Approaching waves* in v towards $\eta_1(\mathbf{x}) > 0$ are indicated by the jumps marked with bolded lines. Also, regions $p < 1$, $p > 1$ can only be connected by 2−waves crossing the line $p = 1$. The selected 1−waves that are located at $\mathbf{x}_\alpha$ with $\mathbf{x}_\alpha < \mathbf{x}$ correspond to $\gamma \to \lambda_1(\gamma; \cdot)$ strictly increasing, i.e. $\{p > 1\}$. On the other hand, the selected 1−waves that are located at $\mathbf{x}_\alpha$ with $\mathbf{x}_\alpha > \mathbf{x}$ correspond to $\gamma \to \lambda_1(\gamma; \cdot)$ strictly decreasing, i.e. $\{p < 1\}$.

Instead, because of the properties of the non GNL first characteristic family, the definition of 1-waves approaching η_1 varies if the left state of such waves lies on the left or on the right of $\{p = 1\}$ (see Figure 1). The key ingredient in the definition of $\mathcal{A}_1$ is the appropriate formulation of *approaching wave* of the first family for a given wave $\eta_1(x)$ in the jump $(u(x), v(x))$, which extends to our case the definition given in [9] for GNL characteristic fields. Observe that, letting $\gamma \mapsto \mathbf{S}_1(\gamma; h_0, p_0)$ be the Rankine-Hugoniot curve of right states of the first family issuing from a given state $(h_0, p_0) \in \Omega$, and denoting $\lambda_1(\gamma; h_0, p_0)$ the Rankine-Hugoniot speed of the jump connecting (h_0, p_0) with $\mathbf{S}_1(\gamma; h_0, p_0)$, by the properties of system (1) it follows that $\gamma \mapsto \lambda_1(\gamma; h_0, p_0)$ is strictly increasing on $\{p > 1\}$, strictly decreasing on $\{0 < p < 1\}$, and constant along $\{p = 1\}$. Therefore, if the size $\eta_1(x)$ is positive, we shall regard as approaching all the 1-waves present in v which either have left state in the region $\{p > 1\}$ and are located on the left of $\eta_1(x)$, or have left state in the region $\{0 < p < 1\}$ and are located on the right of $\eta_1(x)$. On the contrary, we regard as approaching to $\eta_1(x) > 0$ all the 1-waves present in u which either have left state in the region $\{p > 1\}$ and are located on the right of $\eta_1(x)$, or have left state in the region $\{0 < p < 1\}$ and are located on the left of $\eta_1(x)$. Similar definition is given in the case where $\eta_1(x) < 0$.

Moreover, in [9], the weights W_i are expressed only in terms of the strength of the approaching waves. Instead here the terms of $\mathcal{A}_1$ related to the approaching waves of the first family have the form of the product of the strength of the waves $|\rho_\alpha|$ times the distance from $\{p = 1\}$ of the left state of the waves $|p_\alpha - 1|$. The presence of the factor $|p_\alpha - 1|$ is crucial to guarantee the decreasing property of $\Phi_z(u(t, \cdot), v(t, \cdot))$ at times of interactions involving a 1-wave, say of strength $|\rho_\alpha|$, and a 2-wave crossing $\{p = 1\}$ (i.e. connecting two states lying on opposite sides of $\{p = 1\}$), say of strength $|\rho_\beta|$. In fact, in this case the possible increase of $\mathcal{A}_1$ turns out to be of order $|p_\beta - 1||\rho_\alpha| \approx |\rho_\alpha \rho_\beta|$, and thus it can be controlled by the decrease of $\mathcal{G}$ determined by the corresponding decrease of the interaction potential. Unfortunately, because of the presence of these quadratic terms in the weight W_1, we are forced to establish sharp fourth order interaction estimates in order to carry on the analysis of the variation of $\Phi_z(u(t,\cdot),v(t,\cdot))$. This is achieved deriving accurate Taylor expansions of the Hugoniot and rarefaction curves of each famiy, which rely on the specific geometric features of system (1). Namely, the rarefaction and Hugoniot curves through the same point are "almost" straight lines and have "almost" third order tangency at their issuing point near $\{p=1\}$ for the first family and near $\{h=0\}$ for the second family. We say that the characteristic fields of (1) are "almost Temple class".

4. **Main Theorems.** It should be noted that for fixed κ_1 and κ_2, the functional W_i is locally bounded. Hence, the functional Φ_z is equivalent to the $\mathbf{L^1}$ distance between $u(t)$ and $w(t)=v(t)+z(t)$:

$$\frac{1}{C_0}\big\|u(t)-w(t)\big\|_{L^1}\leq\Phi_z(u(t),v(t))\leq C_0\cdot W^*\cdot\big\|u(t)-w(t)\big\|_{L^1}\qquad\forall\,t>0. \tag{16}$$

In the same spirit of [9], we prove that Φ_z is "almost decreasing" in time if the only effect of the convective part of (1), otherwise it is exponentially increasing in time with the increase to be estimated using the operator splitting scheme. To prove this, we clarify that the functional $\Phi_z(u,v)$ in (12) is employed in two ways: either when both u and v are approximate solutions to the non-homogeneous system (1) and $z\equiv 0$ or when u and v are approximate solutions to the homogeneous system (3) and $z\neq 0$ is arbitrary.

First, consider domains $\mathcal{D}$ of the form

$$\begin{aligned}\mathcal{D}(M_0,p_0,\delta_0)=&cl\big\{(h,p)\in\mathbf{L}^1(\mathbb{R};\mathbb{R}^2):\ \ h,p\ \text{ are piecewise constant,}\\ &0\leq h(x)\leq\delta_0,\ p(x)\geq p_0\ \text{ for a.e. } x,\\ &\text{and }\ \text{TotVar}\{(h,p)\}\leq M_0,\quad\|h\|_{\mathbf{L}^1}+\|p-1\|_{\mathbf{L}^1}\leq M_0\big\},\end{aligned}\tag{17}$$

where cl denotes the $\mathbf{L}^1$-closure, $\text{TotVar}\{(h,p)\}\doteq\text{TotVar}\{h\}+\text{TotVar}\{p\}$, and M_0,p_0,δ_0 are positive constants. Given $M_0,p_0>0$, we prove in [3] that there exist constants $\delta_0,\delta_0^*,p_0^*,p_1^*,\kappa_1,\kappa_2,\sigma,C_1,C_2>0$, so that, letting Φ_z be the functional defined in (12)-(15), the followings hold true.

(i) Let u and $v:\mathbb{R}\times\mathbb{R}_+\to\mathbb{R}^2$ be two *ε-front tracking approximate solution* to (3) with initial data $u(\cdot,0),v(\cdot,0)\in\mathcal{D}(M_0,p_0,\delta_0)$ and with values in $[0,\delta_0^*]\times[p_0^*,p_1^*]$. Let z be a piecewise constant function as in Section 3, then

$$\Phi_z\big(u(\tau_2),v(\tau_2)\big)\leq\Phi_z\big(u(\tau_1),v(\tau_1)\big)+C_1\cdot(\varepsilon+\sigma)(\tau_2-\tau_1)\qquad\forall\,\tau_2>\tau_1>0. \tag{18}$$

(ii) Let u and $v:\mathbb{R}\times\mathbb{R}_+\to\mathbb{R}^2$ be two s-ε-approximate solution of (1) with initial data $u(\cdot,0),v(\cdot,0)\in\mathcal{D}(M_0,p_0,\delta_0)$ and with values in $[0,\delta_0^*]\times[p_0^*,p_1^*]$. Then, letting $t_k\doteq k\Delta t=k\,s$, $(k\in\mathbb{N})$ be the time steps, there holds

$$\Phi_0\big(u(\tau_2),v(\tau_2)\big)\leq\Phi_0\big(u(\tau_1),v(\tau_1)\big)+C_1\cdot\varepsilon(\tau_2-\tau_1)\qquad\forall\,t_k<\tau_1<\tau_2<t_{k+1}\,, \tag{19}$$

and

$$\begin{aligned}\Phi_0\big(u(t_k+),v(t_k+)\big)\leq\Phi_0\big(u(t_h+),v(t_h+)\big)\big(1+C_2\cdot\Delta t\big)^{(k-h)}+&\\ +C_1\cdot\varepsilon\,\Delta t\sum_{i=1}^{k-h}\big(1+C_2\cdot\Delta t\big)^{i}&\qquad\forall\,0\leq h<k,\end{aligned}\tag{20}$$

for all $k\in\mathbb{N}$.

The proofs of (i) and (ii) above can be found in [3, §4]. By estimate (18), the front tracking approximate solutions to the homogeneous system (3) converge to a unique limit, depending Lipschitz continuously on the initial data in the L^1 norm, that defines a semigroup solution operator $\mathcal{S}_t$, $t \geq 0$, on the domains $\mathcal{D}$ defined above. In other words, for any given initial data $\overline{u} \doteq (\overline{h}, \overline{p}) \in \mathcal{D}(M_0, p_0, \delta_0)$, the map $u(t,x) \doteq \mathcal{S}_t \overline{u}(x)$ provides an entropy weak solution of the Cauchy problem for (3)–(4). The statement is the following:

Theorem 4.1. *Given $M_0, p_0 > 0$, there exist $\delta_0, \delta_0^*, M_0^*, p_0^*, L > 0$ and a unique (up to the domain) semigroup map*

$$\mathcal{S} : [0, +\infty) \times \mathcal{D}_0 \to \mathcal{D}_0^*, \qquad (\tau, \overline{u}) \mapsto \mathcal{S}_\tau \overline{u}\,, \tag{21}$$

with $\mathcal{D}_0 \doteq \mathcal{D}(M_0, p_0, \delta_0)$, $\mathcal{D}_0^* \doteq \mathcal{D}(M_0^*, p_0^*, \delta_0^*)$ domains defined as in (17), which enjoys the following properties:

(i) $\mathcal{S}_{\tau_2}(\mathcal{S}_{\tau_1} \overline{u}) \in \mathcal{D}_0^* \quad \forall\, \overline{u} \in \mathcal{D}_0, \ \ \forall\, \tau_1, \tau_2 \geq 0$;

(ii) $\mathcal{S}_0 \overline{u} = \overline{u}, \quad \mathcal{S}_{\tau_1+\tau_2} u = \mathcal{S}_{\tau_2}(\mathcal{S}_{\tau_1} \overline{u}) \quad \forall\, \overline{u} \in \mathcal{D}_0, \ \ \forall\, \tau_1, \tau_2 \geq 0$;

(iii) $\left\|\mathcal{S}_{\tau_2}\overline{u} - \mathcal{S}_{\tau_1}\overline{v}\right\|_{\mathbf{L}^1} \leq L \cdot (|\tau_1 - \tau_2| + \|\overline{u} - \overline{v}\|_{\mathbf{L}^1}) \quad \forall\, \overline{u}, \overline{v} \in \mathcal{D}_0, \ \ \forall\, \tau_1, \tau_2 \geq 0$;

(iv) For any $\overline{u} \doteq (\overline{h}, \overline{p}) \in \mathcal{D}_0$, the map $\big(h(x,\tau), p(x,\tau)\big) \doteq \mathcal{S}_\tau \overline{u}(x)$ provides an entropy weak solution of the Cauchy problem (3), (4). Moreover, $\mathcal{S}_\tau \overline{u}(x)$ coincides with the unique limit of front tracking approximations.

(v) If $\overline{u} \in \mathcal{D}_0$ is piecewise constant, then for τ sufficiently small $u(\,\cdot\,, \tau) \doteq \mathcal{S}_\tau \overline{u}$ coincides wtth the solution of the Cauchy problem (3), (4) obtained by piecing together the entropy solutions of the Riemann problems determined by the jumps of $\overline{u}$.

It should be noted that the image of the map $\mathcal{S}_t$ in (21) is the same for every $t > 0$, but the domain $\mathcal{D}_0$ is not positively invariant under the action of $\mathcal{S}$. Indeed, it turns out that the $\mathbf{L}^\infty$, $\mathbf{L}^1$- norms as well as the total variation of the solution (that appear in the definition of the domain (17)) may well increase in presence of interactions (see the analysis in [2, Section 5]).

Moreover, relying on (19)–(20) and on Theorem 4.1, we prove that approximate solutions of (1) generated by a front-tracking algorithm combined with an operator splitting scheme, in turn, converge to a map that defines a Lipschitz continuous semigroup operator $\mathcal{P}_t$, $t \geq 0$, on domains as (17), with a Lipschitz constant that grows exponentially in time and the trajectories $u(t) = \mathcal{P}_t \bar{u}$ are entropy weak solution of the Cauchy problem (1), (4). Let us point out that, although the source term of system (1) is not dissipative, relying on the global existence result established in [2], we construct a semigroup map whose image $\mathcal{D}_0^*$ is the same for every time $t > 0$. Also, the uniqueness of the limit of approximate solutions to (1) and of the semigroup operator $\mathcal{P}$, is achieved as in [1] deriving the key estimate

$$\left\|\mathcal{P}_\theta \overline{u} - \mathcal{S}_\theta \overline{u} - \theta \cdot \big((\overline{p} - 1)\overline{h}\big)\right\|_{\mathbf{L}^1} = \mathcal{O}(1) \cdot \theta^2 \qquad \text{as} \qquad \theta \to 0\,, \tag{22}$$

relating the solutions operators of the homogeneous and nonhomogeneous systems, and invoking a general uniqueness result for quasidifferential equations in metric spaces [5]. Here is our theorem:

Theorem 4.2. *Given $M_0, p_0 > 0$, there exist $\delta_0, \delta_0^*, M_0^*, p_0^*, L', C > 0$ so that the conclusions of Theorem 4.1 hold together with the following. There exist a map*

$$\mathcal{P} : [0, +\infty) \times \mathcal{D}_0 \to \mathcal{D}_0^*, \qquad (\tau, \overline{u}) \mapsto \mathcal{P}_\tau \overline{u}\,, \tag{23}$$

(with $\mathcal{D}_0, \mathcal{D}_0^*$ domains as in (17)), which enjoys the properties:

(i) $\mathcal{P}_{\tau_1}\big(\mathcal{P}_{\tau_2}\overline{u}\big) \in \mathcal{D}_0^* \qquad \forall\, \overline{u} \in \mathcal{D}_0, \quad \forall\, \tau_1, \tau_2 \geq 0;$

(ii) $\mathcal{P}_0\overline{u} = \overline{u}, \quad \mathcal{P}_{\tau_1+\tau_2}u = \mathcal{P}_{\tau_2}\big(\mathcal{P}_{\tau_1}\overline{u}\big) \qquad \forall\, \overline{u} \in \mathcal{D}_0, \quad \forall \tau_1, \tau_2 \geq 0;$

(iii) $\big\|\mathcal{P}_{\tau_1}\overline{u} - \mathcal{P}_{\tau_2}\overline{v}\big\|_{\mathbf{L}^1} \leq L'\big(e^{C_4\cdot\tau_2}\cdot\|\overline{u}-\overline{v}\|_{\mathbf{L}^1} + (\tau_2-\tau_1)\big) \qquad \forall\, \overline{u},\, \overline{v} \in \mathcal{D}_0, \quad \forall\, \tau_2 > \tau_1 > 0\,,$

(iv) *For any* $\overline{u} \doteq (\overline{h}, \overline{p}) \in \mathcal{D}_0$*, the map* $\big(h(x,\tau), p(x,\tau)\big) \doteq \mathcal{P}_\tau\overline{u}(x)$ *provides an entropy weak solution of the Cauchy problem* (1)*,* (4)*.*

Acknowledgments. The authors would like to thank the organizers of *XVII International Conference on Hyperbolic Problems Theory, Numerics, Applications* (Hyp2018) that took place at PennState from June 25th until 29th of 2018 for the invitation and the warm hospitality.

REFERENCES

[1] D. Amadori and G. Guerra, Uniqueness and continuous dependence for systems of balance laws with dissipation, *Nonlinear Anal.* **49** (7) (2002), 987–1014.

[2] D. Amadori and W. Shen, Global existence of large *BV* solutions in a model of granular flow, *Comm. Part. Diff. Equations* **34** (2009), 1003–1040.

[3] F. Ancona, L. Caravenna and C. Christoforou, Exponential stability of large BV solutions in a model of granular flow, *Preprint* (2019).

[4] F. Ancona and A. Marson, Well-posedness for general 2×2 systems of conservation laws, *Mem. Amer. Math. Soc.*, **169** (2004), (801).

[5] A. Bressan, On the Cauchy problem for systems of conservation laws, Actes du 29ème Congrès d'Analyse Numérique: CANum'97 (Larnas, 1997) Soc. Math. Appl. Indust., Paris, 1998, *ESAIM Proc.*, 3, 23–36 (electronic).

[6] A. Bressan, *Hyperbolic systems of conservation laws. The one-dimensional Cauchy problem.* Oxford Lecture Series in Mathematics and its Applications, 20. Oxford University Press, 2000.

[7] A. Bressan, R.M. Colombo, The semigroup generated by 2×2 systems of conservation laws, *Arch. rational Mech. Anal.*, **133** (1996), 1-75.

[8] A. Bressan, G. Crasta, B. Piccoli, Well-posedness of the Cauchy problem for $n\times n$ conservation laws, *Amer. Math. Soc. Memoir*, **146** (2000), (694).

[9] A. Bressan, T.P. Liu and T. Yang, L^1 stability estimates for $n \times n$ conservation laws, *Arch. Rational Mech. Anal.*, **149**, (1999) 1-22.

[10] C. Christoforou, Hyperbolic systems of balance laws via vanishing viscosity, *J. Differential Equations* **221** (2006), 470–541.

[11] C. M. Dafermos and L. Hsiao, Hyperbolic systems of balance laws with inhomogeneity and dissipation, *Indiana U. Math. J.* **31** (1982), 471– 491.

[12] K. P., Hadeler, and C. Kuttler, Dynamical models for granular matter. *Granular Matter* **2** (1999), 9–18.

[13] T.P. Liu and T. Yang, L^1-stability for 2×2 systems of hyperbolic conservation laws, *J. Amer. Math. Soc.*, **12** (1999), (3), 729–774.

[14] W. Shen, On the shape of avalanches, *J. Math. Anal. Appl.* **339** (2008), 828–838.

E-mail address: ancona@math.unipd.it
E-mail address: laura.caravenna@unipd.it
E-mail address: christoforou.cleopatra@ucy.ac.cy

QUANTITATIVE COMPACTNESS ESTIMATE FOR SCALAR CONSERVATION LAWS WITH NONCONVEX FLUXES

FABIO ANCONA

Dipartimento di Matematica "Tullio Levi-Civita"
Università degli Studi di Padova
Via Trieste 63, 35121 Padova, Italy

OLIVIER GLASS

Ceremade, Université Paris-Dauphine, CNRS UMR 7534
Place du Maréchal de Lattre de Tassigny
75775 Paris Cedex 16, France

KHAI T. NGUYEN

Department of Mathematics
North Carolina State University
2108 SAS Hall, Raleigh, NC 27695, USA

ABSTRACT. This note provides a survey of recent results establishing upper and lower estimates for the Kolmogorov ε-entropy of the image through the mapping S_t of bounded sets in $\mathbf{L}^1 \cap \mathbf{L}^\infty$ for scalar conservation laws with non-convex fluxes in one space dimension. As suggested by Lax [25], these quantitative compactness estimates could provide a measure of the order of "resolution" of the numerical methods implemented for these equations.

1. Introduction.

Consider a scalar conservation law in one dimensional space

$$u_t + f(u)_x = 0, \tag{1}$$

where $u = u(t,x)$ is the state variable, and $f : \mathbb{R} \to \mathbb{R}$ is the twice continuously differentiable flux. In the classical setting, the problem is well-posed only locally in time, therefore one considers solutions in the sense of distributions. For sake of uniqueness, the weak solution is required to satisfy an entropy admissibility criterion [16] equivalent to the celebrated Oleĭnik E-condition [30] which generalizes the classical stability conditions introduced by Lax [24]:

Oleĭnik E-condition. *A shock discontinuity located at x and connecting a left state $u^L \doteq \lim_{y\to x-} u(t,y)$ with a right state $u^R \doteq \lim_{y\to x+} u(t,y)$ is entropy admissible if and only if there holds*

$$\frac{f(u^L) - f(u)}{u^L - u} \geq \frac{f(u^R) - f(u)}{u^R - u} \tag{2}$$

2000 *Mathematics Subject Classification.* Primary: 35L65, 47B06; Secondary: 94A17.

Key words and phrases. Hyperbolic equations, conservation laws, characteristics, compactness estimates, Kolmogorov entropy.

The third author was partially supported by a grant from the Simons Foundation/SFARI (521811, NTK).

for every u between u^L and u^R.

A celebrated theorem of Kruzkov establishes a $\mathbf{L}^1$-contractive semigroup of solutions $(S_t)_{t\geq 0}$ of (CL) that associates, to every given initial data $u_0 \in \mathbf{L}^1(\mathbb{R})\cap\mathbf{L}^\infty(\mathbb{R})$, the unique entropy admissible weak solution $S_t u_0 \doteq u(t,\cdot)$ of the corresponding Cauchy problem. In the case of strongly convex fluxes, say $f''(u) \geq c > 0$, P. D. Lax proved in [23] that the map $(S_t)_{t\geq 0}$ is compact in $\mathbf{L}^1_{loc}$ for $t > 0$. Following a suggestion by Lax [25, 26], De Lellis and Golse [17] used the concept of Kolmogorov ε-entropy, recalled below, to provide a quantitative estimate of this compactness effect.

Definition 1.1. *Let (X, d) be a metric space and K a totally bounded subset of X. For $\varepsilon > 0$, let $\mathcal{N}_\varepsilon(K)$ be the minimal number of sets in a cover of K by subsets of X having diameter no larger than 2ε. Then the ε-entropy of K is defined as*

$$\mathcal{H}_\varepsilon(K \mid X) \doteq \log_2 \mathcal{N}_\varepsilon(K).$$

In other words, this is the minimum number of binary digits (bits) needed to represent a point in a given subset E with accuracy ε w.r.t. the metric d.

Basing on the classical Oleĭnik inequality, $D_x u(t,\cdot) \leq \frac{1}{ct}$, they proved an upper bound on the number of bits needed to represent an entropy solution u of (1) at any given time $t > 0$, with accuracy ε w.r.t. the $\mathbf{L^1}$-distance. In [3], we established a lower bound on such ε-entropy which is of the same order of magnitude as the upper bound given in [17]. This result was also extended to balance laws with strictly convex flux and to strictly hyperbolic systems of conservation laws in [5, 4]. Similar results in the context of vanishing viscosity solutions of Hamilton-Jacobi equations have been establishd in [1, 2]. Notice that, when one removes the assumption of uniform convexity of the flux function of (1), the above Oleĭnik inequality does not hold and the weak entropy solution may have unbounded variation (see [12]). In the case of $\mathcal{C}^2$ strictly convex fluxes, exploiting the one side Lipschitz property of the derivative of the flux (see in [14, 20]), i.e.,

$$f'(u(t,x)) - f'(u(t,y)) \leq \frac{1}{t}\cdot(x-y) \qquad \forall t > 0, x \geq y,$$

we provided in [6] upper and lower estimates on $\mathcal{H}_\varepsilon\left(S_t([C_{L,M}])\middle|\mathbf{L}^1(\mathbb{R})\right)$ with

$$\mathcal{C}_{[L,M]} \doteq \left\{u_0 \in \mathbf{L}^\infty(\mathbb{R}) \mid \mathrm{Supp}\,(u_0) \subset [-L, L]\ ,\ \|u_0\|_{\mathbf{L}^\infty} \leq M\right\} \tag{3}$$

the set of bounded, compactly supported initial data.

Aim of this note is to discuss some recent extended results on this topic to the scalar conservation law (1) with the non-convex fluxes. More precisely, in the next section we will present a sharp estimate on $\mathcal{H}_\varepsilon\left(S_t([C_{L,M}])\middle|\mathbf{L}^1(\mathbb{R})\right)$ in the case of fluxes with a single inflection point having polynomial degeneracy (see Theorem 2.2). Notice that for fluxes having one inflection point where all derivatives vanishes, the composition of the derivative of the flux with the solution of (1) fails in general to belong to the BV space (see [28]). In the section 3, combining results on the generalized BV regularity of weak entropy solution in [28] and the ε-entropy for a class of generalized BV functions in [19], we obtain in Theorem 3.1 an upper estimate on $\mathcal{H}_\varepsilon\left(S_t([C_{L,M}])\middle|\mathbf{L}^1(\mathbb{R})\right)$ for weakly genuinely nonlinear fluxes, i.e., fluxes with no flat parts.

2. **Flux with one inflection point.** Without loss of generality, we suppose that

$$f'(0) = 0, \tag{4}$$

since one may always reduce the general case to this one by performing the space-variable and flux transformations $x \to x + tf'(0)$ and $f(u) \to f(u) - uf'(0)$. In this section, we assume that

(A1) *the flux $f : \mathbb{R} \to \mathbb{R}$ is a smooth, non convex function with a single inflection point at 0 having polynomial degeneracy, i.e. such that*

$$\begin{aligned} f^{(j)}(0) &= 0 \quad \text{for all} \quad j = 2, \dots, m, \qquad f^{(m+1)}(0) \neq 0\,, \\ f''(u) \cdot u \cdot \mathrm{sign}\big(f^{(m+1)}(0)\big) &> 0 \qquad \forall\, u \in \mathbb{R} \setminus \{0\}\,, \end{aligned} \tag{5}$$

for some even integer $m \in \mathbb{Z}^+$.

By the monotonicity of the solution operator S_t, and recalling that $S_t u_0$ can be obtained as limit of piecewise constant front tracking approximations [10, Chapter 6], one can show that

$$\big\|S_T u_0\big\|_{\mathbf{L}^\infty(\mathbb{R})} \leq M \quad \text{and} \quad Supp(S_T u_0) \subseteq \big[-l_{[L,M,T]}, l_{[L,M,T]}\big] \tag{6}$$

where

$$f'_M \doteq \sup_{|v| \leq M} |f'(v)| \qquad \text{and} \qquad l_{[L,M,T]} \doteq L + T \cdot f'_M\,.$$

Under the assumption **(A1)**, the uniform upper bounds on the total variation of the flux of an entropy weak solutions have been established in [13, Theorem 3.4, Theorem 4.9] (see also [28, Theorem 2]) [6, Lemma 2.3]).

Lemma 2.1. *For any $L, M, T > 0$ and for every $u_0 \in \mathcal{C}_{[L,M]}$, there holds*

$$TV\big\{f' \circ S_T u_0 \mid \mathbb{R}\big\} \leq C_1 \left(1 + \frac{L}{T}\right), \tag{7}$$

where $C_1 = \frac{2C_M \cdot l_{[L,M,T]}}{T} + \tilde{C}_M$, and the positive constants $C_M, \tilde{C}_M > 0$ depends only on the flux f and M.

Exploiting this BV bound and establishing a controllability result for (1), we obtain our main result.

Theorem 2.2. *Assume that f satisfies (5). For any given $L, M, T > 0$, and for every $\varepsilon > 0$ sufficiently small, the following estimates hold:*

$$\Gamma^- \cdot \frac{1}{\varepsilon^m} \leq \mathcal{H}_\varepsilon\Big(S_T(\mathcal{C}_{[L,M]}) \mid \mathbf{L}^1(\mathbb{R})\Big) \leq \Gamma^+ \cdot \frac{1}{\varepsilon^m} \tag{8}$$

where

$$\Gamma^- = c_2 \cdot \frac{L^{m+1}}{T} \qquad \text{and} \qquad \Gamma^+ = c_2 \cdot \left(1 + L + T + \frac{L^2}{T}\right)^{m+1}$$

for some constant $c_2 > 0$ depending only on f and M.

Sketch of proof.

Upper estimate. We shall provide here an outline of the proof of the upper estimate for $\mathcal{H}_\varepsilon\Big(S_T(\mathcal{C}_{[L,M]}) \mid \mathbf{L}^1(\mathbb{R})\Big)$.

Step 1. Let's consider the set

$$\mathcal{L}_{[L,M,T]} := \big\{f' \circ u \;\big|\; u \in S_T\left(\mathcal{C}_{[L,M]}\right)\big\}$$

From (6) and Lemma 2.1, it holds

$$\mathcal{L}_{[L,M,T]} \subseteq \Big\{ w \in \mathbf{L}^\infty(\mathbb{R}) \cap \mathbf{L}^1(\mathbb{R}) \;\Big|\; supp(w) \in \big[-l_{[L,M,T]},\, l_{[L,M,T]}\big], \\ \|w\|_{\mathbf{L}^\infty} \leq f'_M \quad \text{and} \quad TV\{w \mid \mathbb{R}\} \leq C_1\left(1+\frac{L}{T}\right)\Big\}.$$

Thus, thanks to a result on an upper estimate of ε-entropy for class of uniformly bounded BV function (see ([7, Theorem 1] or [18, Lemma 2.3]), one obtains that for any $\varepsilon' > 0$ sufficiently small, there exists a set of piecewise nonnegative constant functions, $\{g_1, \dots, g_p\} \subset \mathcal{L}_{[L,M,T]}$, with

$$p \;\leq\; \left\lfloor 2^{\left(\frac{\Gamma_1^+}{2\,\varepsilon'}\right)} \right\rfloor + 1, \qquad \Gamma_1^+ = c_1\left(L+T+\frac{L^2}{T}\right)$$

for some constant $c_1 > 0$ depending only on f and M, such that for all $i \in 1, \dots, p$ one has

$$g_i(x) \;=\; g_i(x_\nu) \qquad \forall\, x \in [x_\nu,\, x_{\nu+1}), \qquad \nu \in \{0, 1, \dots, N-1\},$$

with

$$x_\nu \;\doteq\; -l_{[L,M,T]} + \frac{2\,l_{[L,M,T]}}{N}\cdot \nu, \qquad \nu \in \{0, 1, ..., N\},$$

$$N \;\geq\; \left\lfloor \frac{8\,l_{[L,M,T]}\cdot V_{[L,M,T]}}{\varepsilon'} \right\rfloor, \qquad V_{[L,M,T]} \doteq \max\left\{\frac{C_1}{2}\cdot\left(1+\frac{L}{T}\right),\, f'_M\right\},$$

and

$$\mathcal{L}_{[L,M,T]} \;\subseteq\; \bigcup_{i=1}^{p} B(g_i,\, \varepsilon') \tag{9}$$

where $B(g_i,\, \varepsilon')$ denotes the $\mathbf{L}^1(\mathbb{R})$-ball centered at g_i of radius ε'.

Step 2. For every g_i, $i = 1, \dots, p$, and in connection with any N-tuple $\iota = (\iota_0, \dots, \iota_{N-1}) \in \{-1, 1\}^N$, we now define a piecewise constant map $\mathcal{T}_\iota^N(g_i)$ as follows. Let f'_{-1}, f'_1 denote the restrictions of f' to the semilines $(-\infty, 0]$ and $[0, +\infty)$, respectively. Then, set

$$\mathcal{T}_\iota^N(g_i)(x) \;\doteq\; \begin{cases} \left(f'_{\iota_\nu}\right)^{-1}\big(g_i(x_\nu)\big) \quad \forall\, x \in [x_\nu, x_{\nu+1}) & \text{if} \quad x \in \big[-l_{[L,M,T]},\, l_{[L,M,T]}\big), \\ 0 & \text{otherwise.} \end{cases} \tag{10}$$

Using the assumption **(A1)**, for any $u \in \mathcal{S}_T(C_{[L,M]})$, one can find $i_u \in \overline{1,p}$ and $\iota_u \in \{-1, 1\}^N$ such that

$$\big\|u - \mathcal{T}_{\iota_u}^N(g_{i_u})\big\|_{\mathbf{L}^1(\mathbb{R})} \;\leq\; \big(2 + 4\,l_{[L,M,T]}\big)\cdot \Delta_{f,M}^{-1}(2\varepsilon') \tag{11}$$

where a map $\Delta_{f,M} : (0, +\infty) \to \mathbb{R}$ measuring the oscillation of f', defined by setting

$$\Delta_{f,M}(s) \;\doteq\; s \cdot \inf_{\substack{|u|,|v|\leq M \;\; u\cdot v \,\geq\, 0 \\ |v-u|\geq s}} \left|\frac{f'(v) - f'(u)}{v-u}\right| \qquad \forall\, s > 0\,.$$

Step 3. Given $\varepsilon > 0$ sufficiently small, choosing $\varepsilon' = \frac{1}{2} \cdot \Delta_{f,M}\left(\frac{\varepsilon}{2+4\,l_{[L,M,T]}}\right)$, we obtain from (9) and (11) that

$$S_T(C_{[L,M]}) \subseteq \bigcup_{\iota \in \{-1,1\}^N} \bigcup_{i=1}^{p} B(\mathcal{T}_\iota^N(g_i), \varepsilon).$$

From the assumption **(A1)**, it holds

$$\frac{s^m}{\beta_M} \leq \Delta_{f,M}(s) \leq \beta_M \cdot s^m$$

for some constant β_M depending only on f and M. Thus,

$$\mathcal{H}_\varepsilon\Big(S_T(\mathcal{C}_{[L,M]}) \mid \mathbf{L}^1(\mathbb{R})\Big) \leq N + \log_2(p)$$

and it yields the second inequality in (8).

Lower estimate. The main steps of the proof of the lower bound in (8) are the following:
1. *(A controllability result).* Given $L, h, T > 0$, setting

$$b_h^- := \frac{1}{2T \cdot \max_{z\in[-h,0]} |f''(z)|}, \qquad b_h^+ := \frac{1}{2T \cdot \max_{z\in[0,h]} |f''(z)|}, \tag{12}$$

we introduce two classes of functions

$$\mathcal{A}^+_{[L,h]} := \left\{v \in \mathcal{C}_{[L/2,\,h]} \cap BV(\mathbb{R}) \;\middle|\; v(x) \geq 0 \;\; \forall\, x \in \mathbb{R}, \quad \mathrm{sign}(f''(h)) \cdot Dv \leq b_h^+\right\},$$

$$\mathcal{A}^-_{[L,h]} := \left\{v \in \mathcal{C}_{[L/2,\,h]} \cap BV(\mathbb{R}) \;\middle|\; v(x) \leq 0 \;\; \forall\, x \in \mathbb{R}, \quad \mathrm{sign}(f''(-h)) \cdot Dv \leq b_h^-\right\}.$$

Using the method of backward characteristics, one show that

$$\mathcal{A}^+_{[L,\,h]} \bigcup \mathcal{A}^-_{[L,\,h]} \subseteq S_T(\mathcal{C}_{[L,h]}) \tag{13}$$

for all $h > 0$ such that $\max_{|z|\leq h} f'(z) \leq \frac{L}{2T}$.

2. From ([3, Proposition 2.2]), one can derive that for $0 < \varepsilon \leq \frac{Lh}{6}$, it holds

$$\mathcal{H}_\varepsilon\left(\mathcal{A}^\pm_{[L,\,h]} \;\middle|\; \mathbf{L}^1(\mathbb{R})\right) \geq \frac{L^2}{54 \ln 2 \cdot b_h^\pm} \cdot \frac{1}{\varepsilon}.$$

Thus, (13) and (12) imply that

$$\mathcal{H}_\varepsilon\left(S_T(\mathcal{C}_{[L,h]}) \;\middle|\; \mathbf{L}^1(\mathbb{R})\right) \geq \frac{L^2}{108 \ln 2 \cdot T} \cdot \frac{1}{\min\left\{\max\limits_{z\in[0,h]} |f''(z)|,\; \max\limits_{z\in[-h,0]} |f''(z)|\right\}} \cdot \frac{1}{\varepsilon}.$$

From the assumption **(A1)**, there exists a constant $\bar{\alpha} > 0$ depending only on f such that

$$\min\left\{\max_{z\in[0,h]} |f''(z)|,\; \max_{z\in[-h,0]} |f''(z)|\right\} \leq \bar{\alpha} \cdot h^{m-1}$$

for all $h > 0$ sufficiently small. Therefore, for every $\varepsilon > 0$ sufficiently small, choosing $h = \frac{6\varepsilon}{L}$, we obtain

$$\mathcal{H}_\varepsilon\left(S_T(\mathcal{C}_{[L,h]}) \;\middle|\; \mathbf{L}^1(\mathbb{R})\right) \geq \frac{L^{m+1}}{108 \ln 2 \cdot 6^{m-1} \cdot \overline{\alpha} \cdot T} \cdot \frac{1}{\varepsilon^m}$$

which yields the first inequality in (8). □

3. **Weakly genuinely nonlinear flux.** In this section, we will provide an upper estimate on $\mathcal{H}_\varepsilon\left(S_T(\mathcal{C}_{[L,M]}) \,\middle|\, \mathbf{L}^1(\mathbb{R})\right)$ for weakly nonlinear flux $f \in \mathcal{C}^2(\mathbb{R})$, i.e.,

$$\{u \in \mathbb{R} \mid f''(u) \neq 0\} \text{ is dense in } \mathbb{R}. \tag{14}$$

Introduce the function $\mathfrak{d} : [0,+\infty) \to [0,+\infty)$ such that

$$\mathfrak{d}(h) \;=\; \min_{a\in[-M,M-h]}\left(\inf_{g\in\mathcal{A}_{[a,a+h]}} \|f-g\|_{\mathbf{L}^\infty([a,a+h])}\right)$$

where $\mathcal{A}_{[a,a+h]}$ is the set of affine functions defined on $[a,a+h]$. Let Φ be the convex envelop of $\mathfrak{d}$, i.e.,

$$\Phi \;=\; \sup_{\varphi\in\mathcal{G}} \varphi \quad \text{with} \quad \mathcal{G} \;=\; \{\varphi : [0,+\infty) \to [0,+\infty) \text{ convex} \mid \varphi(0)=0,\ \varphi \le \mathfrak{d}\},$$

and denote by

$$\Psi(x) := \Phi(x/2)\cdot x \qquad \forall x \in [0+\infty).$$

It is clear that Ψ is a convex, strictly increasing function on $[0,+\infty)$ with $\Psi(0)=0$. Relying on this function, we obtain the following result.

Theorem 3.1. *Assume that $f \in \mathcal{C}^2$ satisfies (14). Given constants $L, M, T > 0$, for every $\varepsilon > 0$ sufficiently small, it holds*

$$\mathcal{H}_\varepsilon\left(S_T(\mathcal{C}_{[L,M]}) \,\middle|\, \mathbf{L}^1(\mathbb{R})\right) \;\le\; 32(L+Tf'_M)\cdot\left(\frac{2M}{\varepsilon} + \frac{\gamma_{[L,M]}(1+T)}{T\cdot\Psi\left(\frac{\varepsilon}{4(L+Tf'_M)}\right)}\right) \tag{15}$$

for a constant $\gamma_{[L,M]}$ depends only on L, M and f.

To prove the above theorem, let us recall a class of generalized bounded total variation functions which was introduced in [29].

Definition 3.2. *Given an open interval $]a,b[$, we say that a function $g : \mathbb{R} \to \mathbb{R}$ has a Ψ-bounded total variation on $]a,b[$, and we denote $g \in BV^\Psi(]a,b[)$, if*

$$TV^\Psi\{g \mid]a,b[\} \;\doteq\; \sup_{n\in\mathbb{N}, a<x_1<\ldots<x_n<b} \sum_{i=1}^{n-1} \Psi\left(\left|g(x_{i+1}) - g(x_i)\right|\right) \;<\; +\infty.$$

As a consequence of [28, Theorem 1], the following holds

Lemma 3.3. *Given $L, M, T > 0$, for any $u_0 \in \mathcal{C}_{[L,M]}$, the function $S_T(u_0)$ has a Ψ-bounded total variation on $\mathbb{R}$ and*

$$TV^\Psi\{S_t u_0 \mid \mathbb{R}\} \;\le\; \gamma_{[L,M,T]} := \gamma_{[L,M]}\cdot\left(1+\frac{1}{T}\right) \tag{16}$$

for a constant $\gamma_{[L,M]}$ depending only on L, M and f.

For any $R, M, V > 0$, let us introduce a class of uniformly bounded generalized variation functions on $\mathbb{R}$ with compact supports

$$\mathcal{G}_{[R,M,V]} \;\doteq\; \left\{g \in BV^\Psi(\mathbb{R}) \mid supp(g) \subset [-R,R],\ \|f\|_{\mathbf{L}^\infty} \le M, TV^\Psi\{g \mid \mathbb{R}\} \le V\right\}.$$

Thanks to a forthcoming result in [19], one has that

Lemma 3.4. *For $\varepsilon > 0$ sufficiently small, it holds*

$$\mathcal{H}_\varepsilon\left(\mathcal{G}_{[R,M,V]} \,\middle|\, \mathbf{L}^1(\mathbb{R})\right) \;\le\; \frac{64RM}{\varepsilon} + \frac{32RV}{\Psi\left(\frac{\varepsilon}{4R}\right)}. \tag{17}$$

To complete this section, let us give a short proof of theorem 3.1 relying on lemma 3.3 and lemma 3.4.

Proof of theorem 3.1. Recalling that

$$f'_M = \max_{z\in[-M,M]} |f'(z)| \qquad \text{and} \qquad l_{[L,M,T]} = L + T\cdot f'_M,$$

it holds

$$\|S_T u_0\|_{\mathbf{L}^\infty(\mathbb{R})} \leq M \quad \text{and} \quad supp(S_T u_0) \subseteq \left[-l_{[L,M,T]}, -l_{[L,M,T]}\right] \quad \forall u_0 \in \mathcal{C}_{[L,M]}.$$

Recalling (16), we then have $S_T\left(\mathcal{C}_{[L,M]}\right) \subseteq \mathcal{G}_{[l_{[L,M,T]},M,\gamma_{[L,M,T]}]}$. Thus, from (17), it holds

$$\begin{aligned}\mathcal{H}_\varepsilon\left(S_T(\mathcal{C}_{[L,M]}) \,\Big|\, \mathbf{L}^1(\mathbb{R})\right) &\leq \mathcal{H}_\varepsilon\left(\mathcal{G}_{[l_{[L,M,T]},M,\gamma_{[L,M,T]}]} \,\Big|\, \mathbf{L}^1(\mathbb{R})\right)] \\ &\leq \frac{64\, l_{[L,M,T]} M}{\varepsilon} + \frac{32\, l_{[L,M,T]}\cdot\gamma_{[L,M,T]}}{\Psi\left(\frac{\varepsilon}{4\, l_{[L,M,T]}}\right)}\end{aligned}$$

and a direct computation yields (15). □

Remark 1. *The upper estimate on* $\mathcal{H}_\varepsilon\left(S_T(\mathcal{C}_{[L,M]}) \,\Big|\, \mathbf{L}^1(\mathbb{R})\right)$ *in Thererem 3.1 is in general not optimal. It remain an open problem to see if it is possible to improve the estimate (15) for weakly genuinely nonlinear fluxes and to obtain a lower estimate of* $\mathcal{H}_\varepsilon\left(S_T(\mathcal{C}_{[L,M]}) \,\Big|\, \mathbf{L}^1(\mathbb{R})\right)$ *of the same order.*

Acknowledgments. The first author was partially supported by the Istituto Nazionale di Alta Matematica "F.Severi" (INdAM), through GNAMPA. The second author was partially supported by the Agence Nationale de la Recherche, Project DYFICOLTI (ANR-13-BS01-0003). The research by K. T. Nguyen was partially supported by a grant from the Simons Foundation/SFARI (521811, NTK).

REFERENCES

[1] F. Ancona, P. Cannarsa, K. T. Nguyen, Quantitative compactness estimates for Hamilton-Jacobi equations, *Arch. Ration. Mech. Anal*, **219** (2016), 793–828.

[2] F. Ancona, P. Cannarsa, K. T. Nguyen, Compactness estimates for Hamilton-Jacobi equations depending on space, *Bull. Inst. Math. Acad. Sinica* **11** (2016), 63–113.

[3] F. Ancona, O. Glass and K. T. Nguyen, Lower compactness estimates for scalar balance laws, *Comm. Pure Appl. Math.* **65** (2012), 1303-1329.

[4] F. Ancona, O. Glass and K. T. Nguyen, On quantitative compactness estimates for hyperbolic conservation laws, in "Hyperbolic problems: theory, numerics and applications". Proceedings of the 14-th International Conference (HYP2012), AIMS, Springfield, MO, 2014, pp. 249-257.

[5] F. Ancona, O. Glass and K. T. Nguyen, On compactness estimates for hyperbolic systems of conservation laws, *Ann. Inst. H. Poincaré Anal. Non Linaire*, **32** (2015), 1229-1257.

[6] F. Ancona, O. Glass and K. T. Nguyen, On Kolmogorov entropy compactness estimates for scalar conservation laws without uniform convexity, to appear in *SIAM J. Math. Anal.*. Preprint 2018, arXiv:1806.07758.

[7] P. L. Bartlett, S. R. Kulkarni, S. E. Posner, Covering numbers for real-valued function classes, *IEEE Trans. Inform. Theory*, **43** (1997), 1721–1724.

[8] S. Bianchini, Stability of L^∞ solutions for hyperbolic systems with coinciding shocks and rarefactions, *SIAM Journal Math. Anal.* **33** (2001), 959–981.

[9] S. Bianchini, A. Bressan, Vanishing viscosity solutions to nonlinear hyperbolic systems, *Annals of Mathematics*, **161** (2005), 223–342.

[10] A. Bressan, *Hyperbolic Hystems of Conservation Laws. The One Dimensional Cauchy Problem.* Oxford University Press, Oxford, 2000.

[11] A. Bressan, P. Goatin, Stability of L^∞ solutions of Temple class systems, *Differ. Integral Equat.* **13** (2000), 1503–1528.

[12] K.-S. Cheng, The space BV is not enough for hyperbolic conservation laws *J. Math. Anal. Appl.*, **91** (1983), 559–561.

[13] K.-S. Cheng, A regularity theorem for a nonconvex scalar conservation law. *J. Differential Equations* **61** (1986), 79–127.

[14] C. M. Dafermos, Characteristics in hyperbolic conservation laws. A study of the structure and the asymptotic behaviour of solutions, in "Nonlinear analysis and mechanics: Heriot-Watt Symposium (Edinburgh, 1976)", 1–58.

[15] C. M. Dafermos, Generalized characteristics and the structure of solutions of hyperbolic conservation laws *Indiana Univ. Math. J.* , **26** (1977), 1097–1119.

[16] C. M. Dafermos, *Hyperbolic conservation laws in continuum physics.* Grundlehren Math. Wissenschaften Series, Vol. 325. Second Edition. Springer Verlag, 2005.

[17] C. De Lellis, F. Golse, A Quantitative Compactness Estimate for Scalar Conservation Laws, *Comm. Pure Appl. Math.* **58**, (2005), 989–998.

[18] P. Dutta and K. T. Nguyen, Covering numbers for bounded variation functions, *J. Math. Anal. Appl.* **468** (2018), 1131–1143.

[19] P. Dutta and K. T. Nguyen, Covering numbers for a class of bounded generalized variation functions, to appear.

[20] Hoff D., The sharp form of Oleinik's entropy condition in several space variables, *Trans. Amer. Math. Soc.*, **276** (1983), 707–714.

[21] L. Hörmander, Lectures on nonlinear hyperbolic differential equations. *Mathematiques & Applications*, Vol. 26. Springer Verlag, Berlin, 1997.

[22] S. N. Kružkov, First order quasilinear equations with several independent variables, *Math. USSR Sbornik*, **123** (1970), 217–243.

[23] P. D. Lax, Weak solutions of nonlinear hyperbolic equations and their numerical computation, *Comm. Pure Appl. Math.*, **7** (1954), 159–193.

[24] P. D. Lax, Hyperbolic systems of conservation laws II, *Comm. Pure Appl. Math.*, **10** (1957), 537–566.

[25] P. D. Lax, Accuracy and resolution in the computation of solutions of linear and nonlinear equations. Recent advances in numerical analysis, in "Proc. Sympos., Math. Res. Center, Univ. Wisconsin, Madison, Wis.", (1978).

[26] P. D. Lax, *Course on hyperbolic systems of conservation laws*, in "XXVII Scuola Estiva di Fisica Matematica, Ravello, 2002."

[27] T. P. Liu, The Riemann problem for general systems of conservation laws, *J. Differential Equations*, **18** (1975), 218–234.

[28] E. Marconi, Regularity estimates for scalar conservation laws in one space dimension, *J. Hyperbolic Differ. Equat.*, **15** (2018), 623–691.

[29] J. Musielak and W. Orlicz, On generalized variations, *Studia Math.*, **18** (1959), 11-41.

[30] O. Oleinik, Discontinuous solutions of non-linear differential equations, *Ann. Math. Soc. Trans.*, Ser. 2, **26**, (1957), 95–172.

E-mail address: `ancona@math.unipd.it`
E-mail address: `glass@ceremade.dauphine.fr`
E-mail address: `khai@math.ncsu.edu`

THE INCOMPRESSIBLE LIMIT FOR FINITE ENERGY WEAK SOLUTIONS OF QUANTUM NAVIER-STOKES EQUATIONS

PAOLO ANTONELLI

Gran Sasso Science Institute, viale Francesco Crispi, 7,
L'Aquila, 67100, Italy

LARS ERIC HIENTZSCH*

Gran Sasso Science Institute, viale Francesco Crispi, 7,
L'Aquila, 67100, Italy

PIERANGELO MARCATI

Gran Sasso Science Institute, viale Francesco Crispi, 7,
L'Aquila, 67100, Italy

ABSTRACT. This paper is devoted to the analysis of the incompressible limit for Quantum Navier-Stokes equations on $\mathbf{R}^3$. We present the main result of [1] where we show that for general ill-prepared data, finite energy weak solutions of the Quantum Navier-Stokes equations strongly converge to weak solutions of incompressible Navier-Stokes equations. The strong convergence result is achieved by introducing refined Strichartz estimates that analyse accurately the dispersion of acoustic waves given by the Bogoliubov dispersion relation.

1. **Introduction.**

We investigate the low Mach number limit for finite energy weak solutions to the Quantum Navier-Stokes equations on $(0,T)\times\mathbf{R}^3$ given by

$$\begin{cases} \partial_t\rho + \operatorname{div}(\rho u) = 0, \\ \partial_t(\rho u) + \operatorname{div}(\rho u\otimes u) + \nabla P(\rho) = 2\nu\operatorname{div}(\rho\mathbf{D}u) + 2\kappa^2\rho\nabla\left(\frac{\Delta\sqrt{\rho}}{\sqrt{\rho}}\right), \end{cases} \tag{1}$$

where the physical unknowns are the mass density ρ and the velocity field u. We equip (1) with non-zero conditions at infinity,

$$\rho\to 1 \qquad \text{as} \qquad |x|\to\infty. \tag{2}$$

The energy functional associated to (4) reads

$$E(t) = \int_{\mathbf{R}^3}\frac{1}{2}\rho|u|^2 + 2\kappa^2|\nabla\sqrt{\rho}|^2 + \pi(\rho)\mathrm{d}x, \tag{3}$$

with internal energy

$$\pi = \pi(\rho) = \frac{\rho^\gamma - 1 - \gamma(\rho-1)}{\gamma(\gamma-1)}.$$

The choice of the internal energy encodes the boundary condition (2) for finite energy weak solutions. System (1) presents a viscous stress tensor whose viscosity coefficient is degenerate, namely it vanishes in the vacuum region. Further, the

2000 *Mathematics Subject Classification.* Primary: 35Q35; Secondary: 35Q30, 76Y99.

Key words and phrases. Compressible and Incompressible Navier-Stokes equation, Quantum fluids, Low Mach number limit, Acoustic Waves, Strichartz estimates, Energy estimates.

third order tensor is referred to as quantum pressure and takes in account for capillarity effects in the fluid. The system (1) enters the class of Navier-Stokes-Korteweg equations [12] describing e.g. fluid flow including capillarity effects. Moreover, in the inviscid case, namely for $\nu = 0$, system (1) reduces to the Quantum Hydrodynamical system (QHD) [3, 4] arising for instance in Bose-Einstein condensation and superfluidity [17]. In this paper, we are concerned with the low Mach number regime, for that purpose we denote the scaled Mach number $\varepsilon \ll 1$ and after an appropriate rescaling system (1) reads

$$\begin{cases} \partial_t\rho_\varepsilon + \operatorname{div}(\rho_\varepsilon u_\varepsilon) = 0, \\ \partial_t(\rho_\varepsilon u_\varepsilon) + \operatorname{div}(\rho_\varepsilon u_\varepsilon \otimes u_\varepsilon) + \frac{1}{\varepsilon^2}\nabla P(\rho_\varepsilon) = 2\nu \operatorname{div}(\rho_\varepsilon \mathbf{D} u_\varepsilon) + 2\kappa^2\rho_\varepsilon\nabla\left(\frac{\Delta\sqrt{\rho_\varepsilon}}{\sqrt{\rho_\varepsilon}}\right), \end{cases} \tag{4}$$

with initial data

$$\rho_\varepsilon(0,x) = \rho_{\varepsilon,0},$$
$$(\rho_\varepsilon u_\varepsilon)(0,x) = \rho_{\varepsilon,0},$$

The scaled internal energy is given by

$$\pi_\varepsilon = \pi(\rho_\varepsilon) = \frac{\rho_\varepsilon^\gamma - 1 - \gamma(\rho_\varepsilon - 1)}{\varepsilon^2\gamma(\gamma-1)}. \tag{5}$$

The main result we present shows that the dynamics is asymptotically governed by the incompressible Navier-Stokes equation,

$$\partial_t u + u\cdot\nabla u + \nabla p = \nu\Delta u, \qquad \operatorname{div} u = 0. \tag{6}$$

Heuristically, one expects that ρ_ε tends to 1 as ε goes to 0 and consequently also $\operatorname{div}(\rho_\varepsilon u_\varepsilon)$ is expected to converge to 0; yet the system propagates rapidly oscillating acoustic waves that require a suitable control. Here, we rigorously prove that any sequence of finite energy weak solutions of (4) converges strongly to a weak solution of (6) without requiring any further assumptions on the initial data such as regularity, smallness or well-preparedness. Our method is based on refined Strichartz estimates that allow to analyse accurately the dispersion of acoustic waves that is described by the Bogoliubov dispersion relation [8], see (21) below. It takes into account the quantum pressure and differs significantly from the dispersion relation observed for compressible fluids. Different augmented dispersion relations for the acoustic waves also appear in other contexts e.g. in the study of the quasineutral limit in Navier-Stokes-Korteweg system [10], while the Bogolibuov dispersion is typical for quantum fluids. Capturing precisely the dispersion phenomena allows to obtain strong convergence of the acoustic waves at improved convergence rates and to simplify the method compared to previous works [15, 18]. Moreover, we consider general ill-prepared data giving rise to finite energy solutions to the system (4) without damping for which in particular no control on the velocity field is available. Here, we retrieve global weak solution to the limiting system, while [15, 18] achieve convergence to local strong solutions.

The low Mach number limit for compressible fluids has been extensively studied in literature, we refer the reader to the monograph [13]. Let us mention that our method is somehow inspired by [9] where the authors use dispersive effects by means of Strichartz estimates to infer the strong convergence of the irrotational part of the momentum $\rho_\varepsilon u_\varepsilon$. The low Mach number limit for the inviscid counterpart, i.e. the QHD system, has been studied in [11] on the d-dimensional torus. Due to absence of dispersion for periodic solutions, the method is completely different. The low

Mach number analysis on the whole space $\mathbf{R}^d$ will be addressed in the forthcoming paper [2].

This note provides a brief overview of the low Mach number analysis for (4) discussing in section 2 recent existence results and uniform estimates on finite energy weak solutions before introducing the main result in Section 3. In Section 4, we introduce the analysis of the acoustic waves based on refined Strichartz estimates. Finally, we sketch the ideas of the proof of Theorem 3.1 in Section 5. The interested reader can find more details in [1].

2. **The Cauchy Problem and uniform estimates.** In our study we deal with finite energy weak solutions to (4). In the two and three dimensional torus the global existence of such solutions are proved in [5, 16]. To our knowledge, no such result exists for the whole space with condition (2). Here we postulate the global existence of those solutions and we postpone this question to future research. As it is clear from [5] the problem is best studied by using the variables $\sqrt{\rho_\varepsilon}$ and $\Lambda_\varepsilon = \sqrt{\rho_\varepsilon}u_\varepsilon$, see also [3] where it is done similarly for the QHD system. In particular, at at no moment neither the velocity field u_ε nor its gradient ∇u_ε are defined. For the same reason - see also [6] for a similar problem in the context of a Navier-Stokes-Korteweg system - the viscous tensor should be rather thought of as

$$\rho_\varepsilon \mathbf{D} u_\varepsilon = \sqrt{\rho_\varepsilon}\mathbf{S}_\varepsilon, \tag{7}$$

where $\mathbf{S}_\varepsilon$ is the symmetric part of the tensor $\mathbf{T}_\varepsilon$ defined by the distributional identity

$$\sqrt{\rho_\varepsilon}\mathbf{T}_\varepsilon = \nabla m_\varepsilon - 2\nabla\sqrt{\rho_\varepsilon}\otimes\Lambda_\varepsilon. \tag{8}$$

Indeed, it is not clear whether finite energy weak solutions satisfy the energy inequality

$$E(t) + 2\nu\int_0^t\int_{\mathbf{R}^3}\rho_\varepsilon|\mathbf{D}u_\varepsilon|^2 dxdt \le E(0).$$

To circumvent this issue we introduce a weaker energy inequality in Definition 2.1, see also [1, 5, 16, 6]. Recalling that the dispersive tensor can be rewritten alternatively in $\mathcal{D}'$ as

$$2\rho\nabla\left(\frac{\Delta\sqrt{\rho}}{\sqrt{\rho}}\right) = \operatorname{div}\left(\rho\nabla^2\log\rho\right) = \nabla\Delta\rho - 4\operatorname{div}(\nabla\sqrt{\rho}\otimes\nabla\sqrt{\rho}), \tag{9}$$

the equation for the moment density in (4) then reads

$$\partial_t m_\varepsilon + \operatorname{div}\left(\Lambda_\varepsilon\otimes\Lambda_\varepsilon + 4\kappa^2\nabla\sqrt{\rho_\varepsilon}\otimes\nabla\sqrt{\rho_\varepsilon}\right) + \frac{1}{\varepsilon^2}\nabla P(\rho_\varepsilon) = 2\nu\operatorname{div}(\sqrt{\rho_\varepsilon}\mathbf{S}_\varepsilon) + \kappa^2\nabla\Delta\rho_\varepsilon.$$

This motivates the following.

Definition 2.1. A pair $(\rho_\varepsilon, u_\varepsilon)$ with $\rho_\varepsilon \ge 0$ is said to be a finite energy weak solution of the Cauchy Problem (4) if

(i) integrability conditions

$$\sqrt{\rho_\varepsilon}\in L^2_{loc}((0,T)\times\mathbf{R}^3); \qquad \sqrt{\rho_\varepsilon}u_\varepsilon\in L^2_{loc}((0,T)\times\mathbf{R}^3);$$
$$\nabla\sqrt{\rho_\varepsilon}\in L^2_{loc}((0,T)\times\mathbf{R}^3);$$

(ii) continuity equation: for any $\phi\in C_c^\infty([0,T)\times\mathbf{R}^3)$,

$$\int_{\mathbf{R}^3}\rho_{\varepsilon,0}\phi(0) + \int_0^T\int_{\mathbf{R}^3}\rho_\varepsilon\phi_t + \sqrt{\rho_\varepsilon}\sqrt{\rho_\varepsilon}u_\varepsilon\nabla\phi = 0.$$

(iii) momentum equation: for any $\psi \in C_c^\infty([0,T)\times \mathbf{R}^3;\mathbf{R}^3)$,

$$
\begin{aligned}
&\int_{\mathbf{R}^d} \rho_{\varepsilon,0} u_{\varepsilon,0}\psi(0) + \int_0^T\int_{\mathbf{R}^d} \sqrt{\rho_\varepsilon}\sqrt{\rho_\varepsilon}u_\varepsilon \psi_t + (\sqrt{\rho_\varepsilon}u_\varepsilon \otimes \sqrt{\rho_\varepsilon}u_\varepsilon)\nabla\psi + \frac{1}{\varepsilon^2}\rho_\varepsilon^\gamma \operatorname{div}\psi \\
&- 2\nu \int_0^T\int_{\mathbf{R}^d} (\sqrt{\rho_\varepsilon}u_\varepsilon \otimes \nabla\sqrt{\rho_\varepsilon})\nabla\psi - 2\nu\int_0^T\int_{\mathbf{R}^d}(\nabla\sqrt{\rho_\varepsilon}\otimes\sqrt{\rho_\varepsilon}u_\varepsilon)\nabla\psi \\
&+ \nu\int_0^T\int_{\mathbf{R}^d}\sqrt{\rho_\varepsilon}\sqrt{\rho_\varepsilon}u_\varepsilon \Delta\psi + \nu\int_0^T\int_{\mathbf{R}^d}\sqrt{\rho_\varepsilon}\sqrt{\rho_\varepsilon}u_\varepsilon\nabla\operatorname{div}\psi \\
&- 4\kappa^2\int_0^T\int_{\mathbf{R}^d}(\nabla\sqrt{\rho_\varepsilon}\otimes\nabla\sqrt{\rho_\varepsilon})\nabla\psi + 2\kappa^2\int_0^T\int_{\mathbf{R}^d}\sqrt{\rho_\varepsilon}\nabla\sqrt{\rho_\varepsilon}\nabla\operatorname{div}\psi = 0.
\end{aligned}
$$

(iv) there exists $\mathbf{T}_\varepsilon \in L^2((0,T)\times\mathbf{R}^3)$ satisfying (8), such that for a.e. $t\in[0,T]$,

$$
E(t) + 2\nu\int_0^t\int_{\mathbf{R}^3}|\mathbf{S}_\varepsilon|^2 \mathrm{d}x\mathrm{d}t \le E(0), \tag{10}
$$

where $\mathbf{S}_\varepsilon = \mathbf{T}_\varepsilon^{sym}$.

(v) Let $\mu = \nu - \sqrt{\nu^2-\kappa^2}$ and for $0<c<\mu$ define

$$
B_\varepsilon(t) = \int_{\mathbf{R}^3}\frac{1}{2}\left|\sqrt{\rho_\varepsilon}u_\varepsilon + 2c\nabla\sqrt{\rho_\varepsilon}\right|^2 + \pi_\varepsilon + \tilde{\kappa}^2\left|\nabla\sqrt{\rho_\varepsilon}\right|^2\mathrm{d}x,
$$

then the Bresch-Desjardins entropy inequality holds for a.e. $t\in[0,T]$,

$$
\begin{aligned}
&B_\varepsilon(t) + c\int_0^t\int_{\mathbf{R}^3}\frac{1}{2}\left|\mathbf{A}_\varepsilon\right|^2\mathrm{d}x\mathrm{d}s \\
&+ C\int_0^t\int_{\mathbf{R}^3}\left|\nabla^2\sqrt{\rho_\varepsilon}\right|^2\mathrm{d}x\mathrm{d}s + \frac{c\gamma}{2\varepsilon^2}\int_0^t\int_{\mathbf{R}^3}\left|\nabla\rho_\varepsilon^{\frac{\gamma}{2}}\right|^2\mathrm{d}x\mathrm{d}s \\
&\le \int_{\mathbf{R}^3}\frac{1}{2}\left|\sqrt{\rho_{\varepsilon,0}}u_{\varepsilon,0} + 2c\nabla\sqrt{\rho_{\varepsilon,0}}\right|^2 + \pi_{\varepsilon,0} + \tilde{\kappa}^2\left|\nabla\sqrt{\rho_{\varepsilon,0}}\right|^2\mathrm{d}x,
\end{aligned} \tag{11}
$$

where $\mathbf{A}_\varepsilon = \mathbf{T}_\varepsilon^{asym}$.

2.1. **Uniform estimates.** We summarize the most relevant uniform estimates for finite energy weak solutions of (4). The lack of integrability of $\sqrt{\rho_\varepsilon}$ is compensated by regularity properties of $\sqrt{\rho_\varepsilon}-1$ and control on $\mathbf{T}_\varepsilon$ provided by (10) and (11). This leads to a new uniform bound at Sobolev regularity for the momentum m_ε that is crucial for our method.

Lemma 2.2. *If the initial data $(\rho_\varepsilon^0, u_\varepsilon^0)$ is of finite energy, then there exists $C>0$ independent from $\varepsilon>0$ such that*

(i) *$\sqrt{\rho_\varepsilon^0}-1 \in H^1(\mathbf{R}^3)$ and in particular for $2\le p<6$ and $\frac{2(6-p)}{p(6-\gamma)} \le \alpha(p,\gamma) \le \frac{6-p}{2p}$, the following bound holds true $\|\sqrt{\rho_\varepsilon^0}-1\|_{L^p} \le C\varepsilon^{\alpha(p,\gamma)}$.*

(ii) *$\rho_\varepsilon^0 u_\varepsilon^0 \in L^2(\mathbf{R}^3) + L^{\frac{3}{2}}(\mathbf{R}^3)$. In particular $\rho_\varepsilon^0 u_\varepsilon^0 \in H^{-s}(\mathbf{R}^3)$ with $s>\frac{1}{2}$.*

Lemma 2.3. *If $(\rho_\varepsilon, u_\varepsilon)$ is a finite energy weak solution of (4), then there exists $C>0$ independent from $\varepsilon>0$ such that*

(i) *such that $\|\rho_\varepsilon - 1\|_{L^\infty(\mathbf{R}_+;L^2(\mathbf{R}^3))} \le C\varepsilon^\beta$, where $\beta=\beta(\gamma)$ satisfies $\frac{2}{5}\le\beta(\gamma)\le 1$*

(ii) *$\sqrt{\rho_\varepsilon}-1 \in L^\infty(\mathbf{R}_+;H^1(\mathbf{R}^3))$ and in particular for $2\le p<6$ and for $\frac{2(6-p)}{p(6-\gamma)} \le \alpha(p,\gamma) \le \frac{(6-p)}{2p}$, it holds $\|\sqrt{\rho_\varepsilon}-1\|_{L^\infty(\mathbf{R}_+;L^p(\mathbf{R}^3))} \le C\varepsilon^{\alpha(p)}$.*

(iii) *for any* $0 \leq s < 2$ *and* $2 \leq p < \frac{4}{s}$, *there exists* $0 < \beta(p,s) < 2$ *such that* $\|\sqrt{\rho_\varepsilon} - 1\|_{L^p(\mathbf{R}_+;H^s(\mathbf{R}^3))} \leq C\varepsilon^\beta$. *Moreover, for* $1 < s \leq 2$, $\|\sqrt{\rho_\varepsilon} - 1\|_{L^{\frac{2}{s-1}}(\mathbf{R}_+;H^s(\mathbf{R}^3))} \leq C$. *In particular,* $\sqrt{\rho_\varepsilon} - 1 \in L^2(\mathbf{R}_+; L^\infty(\mathbf{R}^3))$.

(iv) $\|\sqrt{\rho_\varepsilon} u_\varepsilon\|_{L^\infty(\mathbf{R}_+;L^2(\mathbf{R}^3))} \leq C$,

(v) *if* $\mathbf{T}_\varepsilon$ *is defined as in* (8) *then* $\|\mathbf{T}_\varepsilon\|_{L^2(\mathbf{R}_+;L^2(\mathbf{R}^3))} \leq C$.

(vi) *for any* $0 \leq s \leq \frac{1}{2}$ *and* $1 \leq p < \frac{4}{1+4s}$, *it holds*

$$\rho_\varepsilon u_\varepsilon \in L^p(0,T; H^s(\mathbf{R}^3)), \tag{12}$$

where the bound is uniform in $\varepsilon > 0$. *In particular for any* $0 \leq s_1 < \frac{1}{4}$, *one has* $\rho_\varepsilon u_\varepsilon \in L^2(0,T; H^{s_1}(\mathbf{R}^3))$.

3. **Statement of the main result.** We consider initial data $(\rho_{\varepsilon,0}, u_{\varepsilon,0})$ of finite energy, namely such that

$$\|\nabla\sqrt{\rho_\varepsilon^0}\|_{L^2(\mathbf{R}^3)} \leq C, \quad \|\sqrt{\rho_\varepsilon^0} u_\varepsilon^0\|_{L^2(\mathbf{R}^3)} \leq C, \quad \|\pi_\varepsilon(\rho_\varepsilon^0)\|_{L^1(\mathbf{R}^3)} \leq C, \tag{13}$$

where is C independent on $\varepsilon > 0$. In addition, we assume that

$$\sqrt{\rho_\varepsilon^0} u_\varepsilon^0 \rightharpoonup u_0 \quad \text{in } L^2(\mathbf{R}^3). \tag{14}$$

No further regularity or smallness assumptions are required, in particular the initial data is ill-prepared, i.e. $\pi_\varepsilon(\rho_\varepsilon^0)$ is only bounded in $L^1(\mathbf{R}^3)$. We now state the main Theorem characterising the low Mach number regime for (4).

Theorem 3.1. *Let* $1 < \gamma < 3$, *let* $(\rho_\varepsilon, u_\varepsilon)$ *be a finite energy weak solution of* (4) *with initial data satisfying* (13) *and* (14) *and let* $0 < T < \infty$ *be an arbitrary time. Then* $\rho_\varepsilon - 1$ *converges strongly to* 0 *in* $L^\infty(0,T; L^2(\mathbf{R}^3)) \cap L^4(0,T; H^s(\mathbf{R}^3))$ *for any* $0 \leq s < 1$. *For any subsequence (not relabeled)* $\sqrt{\rho_\varepsilon} u_\varepsilon$ *converging weakly to* u *in* $L^\infty(0,T, L^2(\mathbf{R}^3))$, *then* $u \in L^\infty(0,T; L^2(\mathbf{R}^3)) \cap L^2(0,T; \dot{H}^1(\mathbf{R}^3))$ *is a global weak solution to the incompressible Navier-Stokes equation* (6) *with initial data* $u_{|t=0} = \mathbf{P}(u_0)$ *and* $\sqrt{\rho_\varepsilon} u_\varepsilon$ *converges strongly to* u *in* $L^2(0,T; L^2_{loc}(\mathbf{R}^3))$. *Moreover,* $\mathbf{Q}(\rho_\varepsilon u_\varepsilon)$ *converges strongly to* 0 *in* $L^2(0,T, L^q(\mathbf{R}^3))$ *for any* $2 < q < \frac{9}{4}$.

We emphasize that the whole sequence $\mathbf{Q}m_\varepsilon$ converges strongly to 0, no extraction of subsequences is required. The sequence m_ε is strongly compact in $L^2(0,T; L^2_{loc}(\mathbf{R}^3))$. While the compressible system only satisfies the energy inequality in its weak form (10), we obtain that $u \in L^\infty(0,T; L^2(\mathbf{R}^3)) \cap L^2(0,T; \dot{H}^1(\mathbf{R}^3))$. Moreover, the limit function u satisfies $u \in L^p(0,T; H^s(\mathbf{R}^3))$ with $0 \leq s \leq \frac{1}{2}$ and $1 \leq p < \frac{4}{1+4s}$ provided by (viii) in Lemma 2.3. If the formation of an initial layer is ruled out by stronger assumption on the preparation of the initial data, then the limiting function u satisfies the energy inequality, i.e. is a Leray weak solution. We require

$$\begin{aligned} &\sqrt{\rho_\varepsilon^0} u_\varepsilon^0 \to u_0 = \mathbf{P}(u_0) \quad \text{strongly in} \quad L^2(\mathbf{R}^3), \\ &\pi_\varepsilon(\rho_\varepsilon^0) \to 0 \quad \text{strongly in} \quad L^1(\mathbf{R}^3), \\ &\nabla\sqrt{\rho_\varepsilon^0} \to 0 \quad \text{strongly in} \quad L^2(\mathbf{R}^3). \end{aligned} \tag{15}$$

Proposition 1. *Under the same assumptions of Theorem 3.1, let* $(\rho_\varepsilon^0, u_\varepsilon^0)$ *further satisfy* (15). *Then the limiting solution* u *to* (6) *satisfies the energy inequality*

$$\int_{\mathbf{R}^3} |u(t)|^2 dx + \nu \int_0^t \int_{\mathbf{R}^3} |\nabla u|^2 dx dt' \leq \int_{\mathbf{R}^3} |u_0|^2 dx, \tag{16}$$

for almost every $t \in [0,T]$.

To prove Theorem 3.1, we decompose the momentum m_ε by means of the Leray-Helmholtz projections on a divergence free field $\mathbf{P}m_\varepsilon$ and a irrotational field $\mathbf{Q}m_\varepsilon$. The refined analysis of acoustic waves allows to conclude the strong convergence of $\mathbf{Q}m_\varepsilon$ to 0 while the dynamics is governed in the limit by the limit of $\mathbf{P}m_\varepsilon$ for which strong convergence is achieved by a Aubin-Lions compactness argument based on the Sobolev regularity of m_ε from Lemma 2.3.

4. **Analysis of acoustic waves.** This Section is devoted to the convergence to 0 of $\sigma_\varepsilon = \frac{\rho_\varepsilon - 1}{\varepsilon}$ and $\mathbf{Q}m_\varepsilon$ in suitable space-time norms. When considering ill-prepared data rapid oscillations in time occur and only weak convergence can be expected. However, by means of refined Strichartz estimates capturing accurately the dispersion on the whole space we obtain the following statement.

Theorem 4.1. *Let $(\rho_\varepsilon, u_\varepsilon)$ be a finite energy weak solution of (4). Then, for any $0 < T < \infty$,*

(i) the density fluctuations $\rho_\varepsilon - 1$ converge strongly to 0 in $C^0(0,T;L^2(\mathbf{R}^3))$ and in $L^4(0,T;H^s(\mathbf{R}^3))$ for any $s \in (-\frac{3}{2}, 1)$,
(ii) If $\gamma = 2$, then σ_ε converges strongly to 0 in $L^2(0,T;L^q(\mathbf{R}^3))$ for any $2 < q < 6$,
(iii) and for any $2 < q < \frac{9}{4}$ there exists $\delta > 0$ such that $\mathbf{Q}(m_\varepsilon)$ converges strongly to 0 in $L^2(0,T;B^\delta_{q,2}(\mathbf{R}^3))$.

We remark that this implies in particular that $\mathbf{Q}m_\varepsilon$ converges strongly to 0 in $L^2(0,T;L^2_{loc}(\mathbf{R}^3))$. Theorem 4.1 is proven by observing that upon using (9) the linearized system for $(\sigma_\varepsilon, m_\varepsilon)$ reads

$$\begin{cases} \partial_t \sigma_\varepsilon + \frac{1}{\varepsilon}\operatorname{div}(m_\varepsilon) = 0, \\ \partial_t m_\varepsilon + \frac{1}{\varepsilon}\nabla\left(1 - \kappa^2\varepsilon^2\Delta\right)\sigma_\varepsilon = F_\varepsilon, \end{cases} \tag{17}$$

where

$$F_\varepsilon = \operatorname{div}\left(-\Lambda_\varepsilon \otimes \Lambda_\varepsilon - 4\kappa^2 \nabla\sqrt{\rho_\varepsilon} \otimes \nabla\sqrt{\rho_\varepsilon} + 2\nu\sqrt{\rho_\varepsilon}\mathbf{S}_\varepsilon\right) - (\gamma - 1)\nabla\pi_\varepsilon. \tag{18}$$

The initial datum for (17) satisfies

$$\sigma_\varepsilon^0 = \frac{\rho_\varepsilon^0 - 1}{\varepsilon} \in H^{-\frac{3}{2}}(\mathbf{R}^3), \qquad m_\varepsilon^0 = \rho_\varepsilon^0 u_\varepsilon^0 \in H^{-\frac{1}{2}}(\mathbf{R}^3),$$

in virtue of Lemma 2.2. The desired control of $(\sigma_\varepsilon, \mathbf{Q}m_\varepsilon)$ in suitable space-time norms in terms of the scaled Mach number ε are consequence of Strichartz estimates of a symmetrization of system (17). Namely, we introduce

$$\tilde{\sigma_\varepsilon} := (1 - \varepsilon^2\kappa^2\Delta)^{\frac{1}{2}}\sigma_\varepsilon, \qquad \tilde{m_\varepsilon} := (-\Delta)^{-\frac{1}{2}}\operatorname{div} m_\varepsilon,$$

and observe that $(\tilde{\sigma_\varepsilon}, \tilde{m_\varepsilon})$ satisfies

$$\begin{cases} \partial_t \tilde{\sigma_\varepsilon} + \frac{1}{\varepsilon}(-\Delta)^{\frac{1}{2}}(1 - \kappa^2\varepsilon^2\Delta)^{\frac{1}{2}}\tilde{m_\varepsilon} = 0, \\ \partial_t \tilde{m_\varepsilon} - \frac{1}{\varepsilon}(-\Delta)^{\frac{1}{2}}(1 - \kappa^2\varepsilon^2\Delta)^{\frac{1}{2}}\tilde{\sigma_\varepsilon} = \tilde{F_\varepsilon}, \end{cases} \tag{19}$$

where $\tilde{F_\varepsilon} = (-\Delta)^{-\frac{1}{2}}\operatorname{div} F_\varepsilon$. The evolution of (19) is characterised by the unitary semigroup e^{-itH_ε}, where

$$H_\varepsilon = \frac{1}{\varepsilon}\sqrt{(-\Delta)(1 - (\varepsilon\kappa)^2\Delta)} \tag{20}$$

is a self-adjoint operator with Fourier multiplier given by

$$\omega(\xi) = \frac{1}{\varepsilon}|\sqrt{|\xi|^2 + \varepsilon^2\kappa^2|\xi|^4}. \tag{21}$$

Unlike the case of compressible fluid where the dispersion relation is linear, here the dispersion relation (21) corresponds to the Bogoliubov dispersion relation arising in the excitation spectrum for Bose-Einstein condensates [8]. The dispersion relation (21) behaves linearly with slope $\frac{1}{\varepsilon}$ for frequencies below the threshold $\frac{1}{\varepsilon}$ and Schrödinger like for frequencies above the threshold. For $\varepsilon = 1$, the Strichartz estimates for the semigroup operator e^{itH_1} have been introduced in [14]. For the analysis of the low Mach number limit, we need to track the ε-dependence in the estimates. Let us mention that this is not trivial given the non-homogeneity of (21). Here, the bound of $(\tilde{\sigma_\varepsilon}, \tilde{m_\varepsilon})$ in terms of the scaled Mach number ε is rather due to the observation that the Strichartz estimates are Schrödinger like rather than wave-like for the whole frequency spectrum and behave slightly better than the one for the free Schrödinger evolution around the Fourier origin. This improved behavior is particularly relevant for the low-frequency regime while for high-frequencies one may always gain a factor ε to a small power in the estimates by Sobolev embedding. Being related to the curvature of the hyper-surface $\tau = \frac{1}{\varepsilon}|\sqrt{|\xi|^2 + \varepsilon^2\kappa^2|\xi|^4}$ our argument provides the desired decay in dimension $d \geq 3$. In dimension $d \geq 2$, the evolution of the semigroup operator e^{itH_ε} has been addressed in [7]. The authors distinguish the low and high frequency regime and approximate H_ε by the linear wave operator for low frequencies and by the Schrödinger operator for high frequencies respectively. The approximation of the low frequency regime by the wave equation leads to a higher loss of regularity in the estimates compared to the ones introduced in [1].

5. **Sketch of the Proof of the main result.** We present the argument that allows to conclude that $\mathbf{P}m_\varepsilon$ is strongly compact in $L^2(0,T; L^2_{loc}(\mathbf{R}^3))$. Together with Theorem 4.1 this is enough in order to infer that the whole sequence m_ε converges strongly to u in $L^2(0,T; L^2_{loc}(\mathbf{R}^3))$ being weak solution of the incompressible Navier-Stokes equation.

Proposition 2. *The sequence $\mathbf{P}(m_\varepsilon)$ converges strongly to u in $L^2(0,T; L^2_{loc}(\mathbf{R}^3))$ as ε goes to 0. Further,*

1. *m_ε converges strongly to u in $L^2(0,T; L^2_{loc}(\mathbf{R}^3))$,*
2. *Λ_ε converges strongly to u in $L^2(0,T; L^2_{loc}(\mathbf{R}^3))$.*

Moreover, the limit function u is weak solution of (6) with initial data $u\big|_{t=0} = \mathbf{P}(u_0)$ defined in (14).

The first part of the Proposition is proven by noticing that from Lemma 2.3 we have that $\mathbf{P}m_\varepsilon \in L^2(0,T; H^{\frac{1}{8}}(\mathbf{R}^3)$ and $\partial_t \mathbf{P}m_\varepsilon \in L^2(0,T; H^{-s}(\mathbf{R}^3)$ for some $s > \frac{5}{2}$. Hence, in the virtue of the Aubin-Lions compactness Lemma we conclude that $\mathbf{P}m_\varepsilon$ converges locally strongly to u and thus Theorem 4.1 then yields that m_ε converges strongly to u in $L^2(0,T; L^2_{loc}(\mathbf{R}^3))$. Writing $\Lambda_\varepsilon = m_\varepsilon - (\sqrt{\rho_\varepsilon} - 1)\Lambda_\varepsilon$ we infer the desired convergence for Λ_ε. The strong compactness of Λ_ε is sufficient to pass to the limit in the weak formulation of (4). Finally, we remark that the weak L^2-limit of tensor $\mathbf{S}_\varepsilon$ can be identified with the velocity gradient, i.e.

$$\mathbf{S}_\varepsilon \rightharpoonup \mathbf{D}u,$$

from (7) combined with the convergences of m_ε, $\sqrt{\rho_\varepsilon} - 1$ and Λ_ε. By the lower-semicontinuity of the norms, we conclude that $\mathbf{D}u \in L^2(0,T, L^2_{loc}(\mathbf{R}^3)$ and since $\operatorname{div} u = 0$ this implies $\nabla u \in L^2(0,T, L^2_{loc}(\mathbf{R}^3)$. If we further assume that the initial data is well-prepared, namely (15), then we may pass to the limit in the energy inequality (10) and conclude that u is a Leray weak solution of (6).

REFERENCES

[1] P. Antonelli, L.E. Hientzsch and P. Marcati, On the low Mach number limit for Quantum Navier-Stokes equations arXiv:1902.00402.

[2] P. Antonelli, L.E. Hientzsch and P. Marcati, On the Cauchy problem for the QHD system with infinite mass and energy: applications to quantum vortex dynamics, to appear.

[3] P. Antonelli and P. Marcati On the finite energy weak solutions to a system in Quantum Fluid Dynamics, *Comm. Math. Phys.*, **287** (2009), no 2, 657–686.

[4] P. Antonelli and P. Marcati, Some results on systems for quantum fluids, *Recent Advances in Partial Differential Equations and Application, Cont. Math.* **666** (2016), 41–54.

[5] P. Antonelli and S. Spirito, Global existence of finite energy weak solutions of quantum Navier-Stokes equations, *Arch. Rat. Mech. Anal.* **225**, no.3 (2017), 1161–1199.

[6] P. Antonelli and S. Spirito, On the compactness of weak solutions to the Navier-Stokes-Korteweg equations for capillary fluids, *Nonlinear Analysis*, **187** (2019), 110–124.

[7] F. Béthuel, R. Danchin and D. Smets, On the linear wave regime of the Gross-Pitaevskii equation, *Journal d'Analyse Mathématique*, **110**, no. 1 (2010), 297–338.

[8] C. Boccato, C. Brennecke, S. Cenatiempo and B. Schlein, Bogoliubov Theory in the Gross-Pitaevskii Limit, *accepted for publication on Acta Mathematica* (2019), arXiv:1801.01389.

[9] B. Desjardins and E. Grenier, Low Mach number limit of viscous compressible flows in the whole space, *Proceedings of the Royal Society of London. Series A: Mathematical, Physical and Engineering Sciences*, **455**, no.1986 (1999), 2271–2279.

[10] D. Donatelli and P. Marcati, Quasineutral limit, dispersion and oscillations for Korteweg type fluids, *SIAM J. Math. Anal.*, **47** (2015), 2265–2282.

[11] D. Donatelli and P. Marcati, Low Mach number limit for the quantum hydrodynamics system, *Res. in Math. Sci.*, **3** (2016), 3–13.

[12] E. Dunn and J. Serrin, On the thermomechanics of interstitial working, *Arch. Rational Mech. Anal.*, **88**, no. 2 (1985), 95–133.

[13] E. Feireisl and A. Novotný, *Singular Limits in Thermodynamics of Viscous Fluids*, Advances in Mathematical Fluid Mechanics, Birkhäuser, Basel, 2017.[10.1007/978-3-319-63781-5]

[14] S. Gustafson, K. Nakanishi and T.-P. Tsai, Scattering for the Gross–Pitaevskii equation, *Mathematical Research Letters*, **13** (2005), 273–285.

[15] Y.-S. Kwon and F. Li, Incompressible limit of the degenerate quantum compressible Navier–Stokes equations with general initial data, *J. Differ. Eqns.*, **264**, no. 5 (2018), 3253–3284.

[16] I. Lacroix-Violet and A. Vasseur, Global weak solutions to the compressible quantum Navier-Stokes and its semi-classical limit, *J. Math. Pures Appl.*, **114** (2018), 191–210.

[17] L. Pitaevskii and S. Stringari, *Bose-Einstein Condensation and Superfluidity*, The Clarendon Press, Oxford University Press, 2016.[10.1093/acprof:oso/9780198758884.001.0001]

[18] J. Yang, Q. Ju and Y.-F. Yang, Asymptotic limits of Navier–Stokes equations with quantum effects, *Z. Angew. Math. Phys.* **66**, no. 5 (2015), 2271–2283.

E-mail address: paolo.antonelli@gssi.it
E-mail address: larseric.hientzsch@gssi.it
E-mail address: pierangelo.marcati@gssi.it

1D QUANTUM HYDRODYNAMIC SYSTEM: GLOBAL EXISTENCE, STABILITY AND DISPERSION

HAO ZHENG*

Gran Sasso Science Institute
Viale Francesco Crispi, 7
L'Aquila, AQ 67100, Italy

PAOLO ANTONELLI AND PIERANGELO MARCATI

Gran Sasso Science Institute
Viale Francesco Crispi, 7
L'Aquila, AQ 67100, Italy

ABSTRACT. In this paper we consider the Cauchy problem for the one-dimensional quantum hydrodynamic (QHD) system. We show global existence of weak solutions. Moreover, by introducing a novel functional which is uniformly bounded in time along the flow of solutions and controls some higher order norms of the unknowns, we provide a stability result for sequence of weak solutions satisfying those bounds. Finally, we present some dispersive properties of solutions to the QHD system.

1. **Introduction.** This work is concerned about the following one dimensional quantum hydrodynamic (QHD) system

$$\begin{cases} \partial_t \rho + \partial_x J = 0 \\ \partial_t J + \partial_x \left(\frac{J^2}{\rho} \right) + \partial_x P(\rho) = \frac{1}{2} \rho \partial_x \left(\frac{\partial_x^2 \sqrt{\rho}}{\sqrt{\rho}} \right). \end{cases} \tag{1}$$

This system describes a compressible, inviscid fluid with quantum effects described by the third order dispersive term on the right hand side of the equation for the momentum density. This model is used in the description of physical phenomena in superfluidity and BEC [15] or in the modeling of semiconductor devices at nanoscales [12]. The unknowns ρ and J in (1) represent the mass and momentum densities of the fluid, respectively, $P(\rho) = \frac{\gamma-1}{\gamma}\rho^\gamma$ is the pressure term, with $1 < \gamma < \infty$. Under some suitable regularity assumptions the quantum term can also be written in different ways, like

$$\frac{1}{2}\rho\partial_x \left(\frac{\partial_x^2 \sqrt{\rho}}{\sqrt{\rho}} \right) = \frac{1}{4}\partial_x^3 \rho - \partial_x(\partial_x \sqrt{\rho})^2 = \frac{1}{4}\partial_x \left(\rho \partial_x^2 \log \rho \right). \tag{2}$$

2000 *Mathematics Subject Classification.* Primary: 58F15, 58F17; Secondary: 53C35.
Key words and phrases. Quantum hydrodynamics, QHD, global existence, stability, dispersion.
The first author is supported by Gran Sasso Science Institute.
* Corresponding author: Hao Zheng.

The system (1) is Hamiltonian and its total energy

$$E = \int_{\mathbf{R}^d} \frac{1}{2}(\partial_x \sqrt{\rho})^2 + \frac{1}{2}\frac{J^2}{\rho} + f(\rho)\, dx \tag{3}$$

is formally conserved along the flow of solutions. The internal energy in (3) is defined from the pressure by the formula $f(\rho) = \rho \int_0^\rho p(s)/s^2\, ds = \frac{1}{\gamma}\rho^\gamma$.

System (1) is intimately related to the following nonlinear Schrödinger equation

$$\begin{cases} i\partial_t \psi = -\frac{1}{2}\partial_x^2 \psi + f'(|\psi|^2)\psi \\ \psi(0) = \psi_0 \in H^1(\mathbf{R}). \end{cases} \tag{4}$$

This can be formally seen through by expressing the wave function in terms of its amplitude and its phase, $\psi = \sqrt{\rho}e^{iS}$. By plugging this ansatz inside the NLS equation (4), by separating the real and imaginary parts and after some algebra, we find that (ρ, S) solves the following system

$$\begin{cases} \partial_t \rho + \partial_x(\rho \partial_x S) = 0 \\ \partial_t S + \frac{1}{2}(\partial_x S)^2 + f'(\rho) = \frac{1}{2}\frac{\partial_x^2 \sqrt{\rho}}{\sqrt{\rho}}. \end{cases} \tag{5}$$

Given (ρ, S) satisfying system (5) we see it is possible to define the velocity field $v = \partial_x S$ and we have that (ρ, J), with $J = \rho v = \rho \partial_x S$, satisfy the QHD system. However this approach fails in the nodal region, namely the set where the wave function vanishes $\{\rho = 0\}$, since there the phase is not well-defined.

Alternatively the hydrodynamical quantities (ρ, J) associated to a wave function ψ are defined by means of the Madelung transformations, i.e. $\rho = |\psi|^2$, $J = \mathrm{Im}(\bar{\psi}\partial_x \psi)$. In [3, 4] the authors set up a polar factorisation approach for finite energy wave functions in order to show the existence of global in time finite energy weak solutions to (1) in three and two space dimensions, respectively. The main advantage of this approach is that the polar factorisation technique allows to define the hydrodynamical quantities $\sqrt{\rho}$ and $\Lambda = J/\sqrt{\rho}$ (see Lemma 3.1), thus overcoming the problem of defining the velocity field in the nodal region. In this way they show the existence of finite energy weak solutions to (1) by considering the Madelung transform of a wave function, solution to (4), see also Theorem 3.2 below.

Conversely, it is not clear whether it is possible to give an existence result for weak solutions to (1) without passing through the analogue wave function dynamics given by (4). This problem is also linked to the more general question in quantum mechanics, the so called Pauli problem, which asks whether it is possible to determine a quantum state given a set of its observables. In this paper we shall present some partial answers to those questions. First of all, we show that in 1D it is possible to invert the Madelung transform. More precisely, given a set of finite energy hydrodynamical quantities $(\sqrt{\rho}, \Lambda)$ such that Λ vanishes in the vacuum region, it is indeed possible to define an associated wave function. As a consequence we obtain the global existence of weak solutions to (1) without assuming that the initial data are generated by a wave function. Furthermore by requiring some further integrability/regularity hypotheses on the initial data we can also show compactness properties for solutions. Finally, if we further assume that the initial data ρ_0 has finite variance then we can prove that the solutions obtain some dispersive properties.

The results announced here will be proved and discussed more extensively in the forthcoming paper [6].

2. **Preliminaries.** Let us first introduce the concept of finite energy weak solutions to (1), for more details we address the reader to [5]. Following [3, 4] we consider the quantities $(\sqrt{\rho}, \Lambda)$ which define the hydrodynamic variables by $\rho = (\sqrt{\rho})^2$, $J = \sqrt{\rho}\Lambda$. Using this definition and by exploiting the identity (2), we rewrite system (1) in the following way

$$\begin{cases} \partial_t \rho + \partial_x J = 0 \\ \partial_t J + \partial_x(\Lambda^2 + p(\rho) + (\partial_x\sqrt{\rho})^2) = \frac{1}{4}\partial_x^3\rho. \end{cases} \tag{6}$$

Thus we say that $(\sqrt{\rho}, \Lambda)$ is a finite energy weak solution to (1) if $\sqrt{\rho} \in L^\infty([0,T], H^1(\mathbf{R}))$, $\Lambda \in L^\infty([0,T], L^2(\mathbf{R}))$ and they solve (6) in the sense of distribution for some $T > 0$.

To precisely characterise the regularity condition of the initial data and solutions, we define the following conditions:

$$\|\sqrt{\rho_0}\|_{H^1} + \|\Lambda_0\|_{L^2} \leq M_1 \tag{7}$$

and

$$\|\frac{\Lambda_0^2}{\sqrt{\rho_0}}\|_{L^2} + \|\partial_x^2\sqrt{\rho_0}\|_{L^2} + \|\frac{\partial_x J_0}{\sqrt{\rho_0}}\|_{L^2} \leq M_2. \tag{8}$$

Condition (7) is equivalent to require the initial mass and energy are finite. On the other hand, assumptions (8) are related to the definition of a novel functional, see (11) below, which is introduced in order to study the compactness issue. From the physical point of view this functional formally gives a control on L^2-norm of the chemical potential

$$\mu = -\frac{1}{2}\frac{\partial_x^2\sqrt{\rho}}{\sqrt{\rho}} + \frac{1}{2}v^2 + f'(\rho), \tag{9}$$

in $\rho\, dx$, where formally $v = J/\rho$ is the velocity field. More rigorously, we shall consider the following quantity

$$\lambda = -\frac{1}{2}\partial_x^2\sqrt{\rho} + \frac{\Lambda^2}{\sqrt{\rho}} + f'(\rho)\sqrt{\rho}, \tag{10}$$

which formally equals $\lambda = \sqrt{\rho}\mu$. The functional we are going to study is then defined by

$$I(t) = \int \lambda^2 + (\partial_t\sqrt{\rho})^2\, dx, \tag{11}$$

so that the bounds in (8) yield $I(0) \leq M_2^2$. Unfortunately proving a uniform estimate on $I(t)$ will not guarantee that the bounds in (8) are preserved along the evolution. Nevertheless it will provide some a priori estimates which will yield the compactness for solutions to (1).

As it will be clear through a direct computation, for Schrödinger-generated hydrodynamical momenta, say $\rho = |\psi|^2$ and $J = \mathrm{Im}(\bar{\psi}\partial_x\psi)$, the functional (11) can be written as

$$I(t) = \int |\partial_t\psi|^2\, dx, \tag{12}$$

so that intuitively $I(t)$ controls the H^2 norm of the solution to (4).

Let us first recall some basic facts on (4) which will be used later. The reader will find more details and proofs in [8].

Theorem 2.1. *Let $\psi_0 \in H^1(\mathbf{R})$ then there exists a unique global solution $\psi \in \mathcal{C}(\mathbf{R}; H^1(\mathbf{R}))$ to (4) such that the total mass and energy are conserved at all times. If moreover $\psi_0 \in H^2(\mathbf{R})$, then we also have $\psi \in \mathcal{C}(\mathbf{R}; H^2(\mathbf{R})) \cap \mathcal{C}^1(\mathbf{R} : L^2(\mathbf{R}))$ and for any $0 < T < \infty$ we have*

$$\|\psi\|_{L^\infty(0,T;H^2(\mathbf{R}))} + \|\partial_t\psi\|_{L^\infty(0,T;L^2(\mathbf{R}))} \leq C(T, \|\psi_0\|_{H^2(\mathbf{R})}). \tag{13}$$

In what follows we shall also use the following fact, see Theorem 6.19 in [14].

Lemma 2.2. *Let $f : \Omega \to \mathbb{R}$ be in $H^1(\Omega)$, and*

$$B = f^{-1}(\{0\}) = \{x \in \Omega \ : \ f(x) = 0\}.$$

Then $\nabla f(x) = 0$ for almost every $x \in B$.

3. **Wave Function Lifting and Global Existence of Weak Solutions.** In this Section we first review some known facts about the polar factorization, then we introduce the wave function lifting in order to invert the Madelung transform.

The polar factorization, developed in [3, 4], allows to define the hydrodynamic quantities $(\sqrt{\rho}, \Lambda)$ and sets up a correspondence between the wave function dynamics and the hydrodynamical system. The main advantage of this approach with respect to the usual method for instance is that vacuum regions are allowed in the theory. For a more detailed presentation we address to Section 3 in [2]. Given any function $\psi \in H^1(\mathbf{R})$ we can define the set of polar factors as

$$P(\psi) := \{\phi \in L^\infty(\mathbf{R}) \mid \|\phi\|_{L^\infty_x} \leq 1, \ \psi = \phi|\psi| \ a.e.\}.$$

Lemma 3.1. *Let $\psi \in H^1(\mathbb{R})$, $\sqrt{\rho} := |\psi|$ and $\phi \in P(\psi)$. Then $\partial_x\sqrt{\rho} = \mathrm{Re}(\bar{\phi}\partial_x\psi) \in H^1(\mathbb{R})$ and by setting $\Lambda := \mathrm{Im}(\bar{\phi}\partial_x\psi)$, we have*

$$|\partial_x\psi|^2 = (\partial_x\sqrt{\rho})^2 + \Lambda^2, \quad a.e.\ x \in \mathbf{R}. \tag{14}$$

Furthermore if $\{\psi_n\} \subset H^1(\mathbf{R})$ is such that $\|\psi_n - \psi\|_{H^1} \to 0$, then

$$\partial_x\sqrt{\rho_n} \to \partial_x\sqrt{\rho}, \quad \Lambda_n \to \Lambda, \quad in\ L^2(\mathbf{R}). \tag{15}$$

By using the previous Lemma and Theorem 2.1 it is possible to prove the following result on global existence of finite energy weak solutions to (1).

Theorem 3.2. *Let $\psi_0 \in H^1(\mathbf{R})$ and let us define $\rho_0 = |\psi_0|^2, J_0 = \mathrm{Im}(\bar{\psi}_0\partial_x\psi_0)$. Then there exists a global in time finite energy weak solution to (1) such that the total mass and total energy are conserved at all times.*

The above Theorem was first proved in [3, 4] in the three and two dimensional case, then alternative proofs appeared also in [9, 1]. We point out that the main results in [3, 4] in fact concern the existence of global solutions for a dissipative version of the QHD system, where the equation for the momentum density in (1) is augmented by a linear damping term which destroys the analogy with (4), see [3, 4] for more details. The study of that system then requires a more delicate analysis which passes through the construction of a sequence of approximating solutions by means of an operator splitting argument and the analysis of suitable compactness estimates given by the dispersive effects encoded in the system. Here we focus only on the Hamiltonian system (1).

Lemma 3.1 allows us to determine suitable hydrodynamical quantities $(\sqrt{\rho}, \Lambda)$ from a given finite energy wave function $\psi \in H^1$. The opposite result is given by the following wave function lifting proposition.

Proposition 1. *Let* $(\sqrt{\rho}, \Lambda)$ *be satisfying* (7) *and let us further assume that* $\Lambda = 0$ *a.e. on* $\{\sqrt{\rho} = 0\}$. *Then there exists a wave function* $\psi \in H^1(\mathbf{R})$ *such that*

$$\sqrt{\rho} = |\psi|, \quad \Lambda = \operatorname{Im}(\bar{\phi}\partial_x\psi),$$

where $\phi \in P(\psi)$. *If we furthermore assume that* $(\sqrt{\rho}, \Lambda)$ *satisfy also the bounds* (8), *then* $\psi \in H^2(\mathbf{R})$ *and we have*

$$\|\psi\|_{H^2(\mathbf{R})} \leq C(M_1, M_2). \tag{16}$$

Proof. Here we briefly sketch the proof of the proposition, for more details we refer to [6]. Let $\delta(x) = e^{-x^2}$ and let $\sqrt{\rho_n}(x) = \sqrt{\rho}(x) + \frac{1}{n}\delta(x)$, $\Lambda_n = J/\sqrt{\rho_n}$. By definition $\sqrt{\rho_n}$ converges to $\sqrt{\rho}$ in $H^1(\mathbf{R})$ and pointwise. Moreover, by assumption we have $\Lambda(x) = 0$ a.e. in $\{\sqrt{\rho} = 0\}$ and by construction the same holds also for Λ_n. Hence the pointwise convergence of $\sqrt{\rho_n}(x)$ also implies $\Lambda_n(x) \to \Lambda(x)$ a.e. As a consequence $\Lambda_n(x) \to \Lambda(x)$ in $L^2(\mathbf{R})$ by dominant convergence theorem. Since $\sqrt{\rho_n}$ is positive everywhere, we can apply the inverse of Madelung transformation to hydrodynamic data $(\sqrt{\rho_n}, \Lambda_n)$ to define a phase function $S_n(x) = \int_0^x v_n(y)ds$ and a wave function $\psi_n = \sqrt{\rho_n}e^{iS_n} \in H^1(\mathbf{R})$. Direct computation shows $\|\psi_n\|_{H^1} \leq \|\sqrt{\rho}\|_{H^1} + \|\Lambda\|_{L^2}$, which implies upto a subsequence ψ_n converges weakly to a $\psi \in H^1(\mathbf{R})$. Using the strong convergence of $\sqrt{\rho_n}$ and Λ_n, we can show ψ is exactly the wave function we want. The additional condition (8) implies that the sequence $\psi_n \subset H^2(\mathbf{R})$, with $H^2(\mathbf{R})$ norm is bounded by $C(M_1, M_2)$. □

We remark that the assumption on Λ vanishing on the vacuum is quite reasonable in view of the polar factorization and of Lemma 2.2. Indeed for $\psi \in H^1$, we have $\partial_x\psi = 0$ a.e. on $\{\sqrt{\rho} = 0\}$ and consequently Λ constructed in Lemma 3.1 satisfies $\Lambda = 0$ a.e. in $\{\sqrt{\rho} = 0\}$. By using Proposition 1 we can show a global existence result for finite energy initial data.

Theorem 3.3 (Global Existence). *Let* $d = 1$. *Consider a pair* $(\sqrt{\rho_0}, \Lambda_0)$ *of initial data with finite energy, i.e. satisfying bounds* (7) *and let us further assume that* $\Lambda_0 = 0$ *a.e. on the set* $\{\sqrt{\rho_0} = 0\}$. *Then there exists a global in time finite energy weak solution to the Cauchy problem* (1) *which conserves the total energy for all times. Moreover, if we also assume that the initial data satisfy the estimate in* (8), *then for any* $0 < T < \infty$ *we have*

$$\|\rho\|_{L^\infty(0,T;H^2(\mathbf{R}))} + \|J\|_{L^\infty(0,T;H^1(\mathbf{R}))} + \|\sqrt{e}\|_{L^\infty(0,T;H^1(\mathbf{R}))} \leq C(T, M_1, M_2), \tag{17}$$

where e *is the kinetic energy density defined by*

$$e = \frac{1}{2}(\partial_x\sqrt{\rho})^2 + \frac{1}{2}\Lambda^2. \tag{18}$$

The proof of the global existence theorem follows by applying Proposition 1 to $(\sqrt{\rho_0}, \Lambda_0)$. This gives $\psi_0 \in H^1(\mathbf{R})$ and by Theorem 2.1 we obtain $\psi \in L^\infty(I, H^1(\mathbf{R}))$ solution to (4) which preserves the mass and energy. Then using the polar decomposition Lemma 3.1 we define $\sqrt{\rho} = |\psi|$, $\Lambda = \operatorname{Im}(\bar{\phi}\partial_x\psi)$ and show $(\sqrt{\rho}, \Lambda)$ is a weak solution to (6). If further assume (8), then $\psi_0 \in H^2(\mathbf{R})$ and again by Theorem 2.1 $\psi \in \mathcal{C}(\mathbf{R}; H^2(\mathbf{R}))$ satisfies(13). By using (12), we see $I(t)$ is uniformly bounded. The higher order bounds of ρ, J and $\sqrt{e}$ are consequence of the bound for $I(t)$.

4. **Stability.** After the existence of global solutions, we can show that the framework determined by the existence theorem ensures suitable compactness properties for sequences of solutions to (1).

Theorem 4.1 (Stability). *Let us assume $\{(\sqrt{\rho_n}, \Lambda_n)\}_{n\geq 1}$ is a sequence of solutions to (1) with uniform bounded total mass, energy and functional $I(t)$. Then upto subsequence we have*

$$\begin{aligned} \sqrt{\rho_n} &\to \sqrt{\rho} \quad \text{in } L^\infty(0,T;H^1_{loc}(\mathbf{R})) \\ \Lambda_n &\to \Lambda \quad \text{in } L^\infty(0,T;L^2_{loc}(\mathbf{R})), \end{aligned}$$

for any $0 < T < \infty$, where $(\sqrt{\rho}, \Lambda)$ is a finite energy weak solution to (1). Furthermore, we have weak convergence

$$\left.\begin{aligned} \partial_x^2\rho_n &\rightharpoonup \partial_x^2\rho \\ \partial_x J_n &\rightharpoonup \partial_x J \\ \partial_x\sqrt{e_n(\cdot)} &\rightharpoonup \partial_x\sqrt{e(\cdot)} \\ \lambda_n &\rightharpoonup \lambda \end{aligned}\right. \quad \text{in } L^\infty_t L^2_x,$$

where $e_n(t)$ is the kinetic energy density of $(\rho_n, J_n)(t,x)$, and λ_n defined as (8).

The weak convergence is given by the uniform bound of $I(t)$ and (17). We first denote $\sqrt{\nu}$ the weak limit of $\sqrt{e}$, which is also the weak limit of the non-linearity of (6), then to prove the compactness it is sufficient to show the following proposition:

Proposition 2. *We have the following identity*

$$\nu^2 = \frac{1}{2}(\partial_x\sqrt{\rho})^2 + \frac{1}{2}\Lambda^2 \tag{19}$$

is satisfied a.e. $x \in \mathbf{R}$, and consequently we have

$$\begin{aligned} \partial_x\sqrt{\rho_n} &\to \sqrt{\rho}, \quad L^\infty(0,T;L^2_{loc}(\mathbf{R})), \\ \Lambda_n &\to \Lambda, \quad L^\infty(0,T;L^2_{loc}(\mathbf{R})). \end{aligned}$$

The idea of the proof is to consider ν away from and inside the vacuum region separately. When away from the vacuum, since $\rho(x) > 0$ it is sufficient to consider $\rho\nu$, for which the local strong convergence is given by the higher order bound and Sobolev embedding. For the vacuum region, it is important to notice that e_n and ν vanish almost everywhere by the L^2 boundedness of λ_n and λ. Combining this fact with Lemma 2.2, we show that in the vacuum region both sides of (19) vanish a.e..

5. **Dispersion.** In this last Section we provide some results about the asymptotic behaviour of finite energy weak solutions to the QHD system (1).

Theorem 5.1 (Dispersion). *Let (ρ, J) be a finite energy weak solution to system (1) such that the energy is conserved and let us further assume that $\int |x|^2\rho_0(x)\,dx < \infty$. Then we have*

$$\|\partial_x\sqrt{\rho}(t)\|_{L^2} + \|\Lambda - \frac{x}{t}\sqrt{\rho}\|_{L^2} \lesssim t^{-\sigma}, \tag{20}$$

where $\sigma = \min\{1, \frac{1}{2}(\gamma - 1)\}$.

The main idea stems from writing the hydrodynamical analogue of the pseudo-conformal energy for the NLS equation. Similar functionals are also studied in classical fluid dynamics [10]. We consider the functional

$$V(t) = \int_{\mathbf{R}} \frac{x^2}{2}\rho(t,x)\,dx - t\int_{\mathbf{R}} x \cdot J(t,x)\,dx + t^2E(t), \tag{21}$$

where the energy $E(t)$ is defined in (3). Theorem 5.1 is proved by using the result in Proposition below and an argument similar to the one given in [7].

Proposition 3. *Let* (ρ, J) *be a finite energy weak solution to* (1) *such that* $||x^2\rho_0||_{L^1} < \infty$ *and the energy is conserved for all times. Then we have*

$$V(t) + (1 - \frac{3}{\gamma}) \int_0^t \int_{\mathbf{R}} \rho^\gamma(s,x)\, dxds = \int_{\mathbf{R}} \frac{x^2}{2} \rho_0(x)\, dx.$$

REFERENCES

[1] P. Antonelli, Remarks on the derivation of finite energy weak solutions to the QHD system, to appear on *Proc. AMS*, (2019).

[2] P. Antonelli, L.E. Hientzsch, P. Marcati and H. Zheng On some results for quantum hydrodynamical models, `http://www.kurims.kyoto-u.ac.jp/~kyodo/kokyuroku/contents/2070.html` RIMS Kôkyûroku **2070** (2018), 107–129.

[3] P. Antonelli and P. Marcati, On the finite energy weak solutions to a system in Quantum Fluid Dynamics, *Comm. Math. Phys.*, **287** (2009), no 2, 657–686.

[4] P. Antonelli and P. Marcati, The Quantum Hydrodynamics system in two space dimensions, *Arch. Rat. Mech. Anal.*, **203** (2012), 499–527.

[5] P. Antonelli and P. Marcati, Some results on systems for quantum fluids, Recent Advances in Partial Differential Equations and Application, *Cont. Math.*, **666** (2016), 41–54.

[6] P. Antonelli, P. Marcati and H. Zheng, Global existence, stability and scattering for weak solutions to the one dimensional QHD system, preprint.

[7] J. Barab, Nonexistence of asymptotically free solutions for a nonlinear Schrödinger equation, *J. Math. Phys.*, **25** (1984), 3270.

[8] T. Cazenave, *Semilinear Schrödinger Equations*, Courant Lecture Notes in Mathematics vol. 10, New York University, Courant Institute of Mathematical Sciences, AMS, 2003.

[9] R. Carles, R. Danchin and J.-C. Saut, Madelung, Gross-Pitaevskii and Korteweg, *Nonlinearity*, **25** (2012), 2843–2873.

[10] J.Y. Chemin, Dynamique des gas à masse totale finie, *Asympt. Anal.*, **3** (1990), 215–220.

[11] R.P. Feynman, Application of quantum mechanics to liquid helium, *Progr. Low Temp. Phys.*, **1** (1955), 17–53.

[12] C. Gardner, The quantum hydrodynamic model for semincoductor devices, *SIAM J. Appl. Math.*, **54** (1994), 409–427.

[13] J. Ginibre and G. Velo, On a class of nonlinear Schrödinger equations. II. Scattering theory, general case, *J. Funct. Anal.*, **32** (1979), 33–71.

[14] E. Lieb and M. Loss, Analysis, Graduate Studies in Mathematics, vol. 14, AMS, 2001.

[15] L. Pitaevskii and S. Stringari, Bose-Einstein condensation and superfluidity, Clarendon Press, Oxford, 2016.

E-mail address: `hao.zheng@gssi.it`
E-mail address: `paolo.antonelli@gssi.it`
E-mail address: `pierangelo.marcati@gssi.it`

ABOUT VISCOUS APPROXIMATIONS OF THE BITEMPERATURE EULER SYSTEM

Denise Aregba-Driollet*

Univ. Bordeaux, CNRS, Bordeaux INP, IMB, UMR 5251
F-33400, Talence, France

Stéphane Brull

Univ. Bordeaux, CNRS, Bordeaux INP, IMB, UMR 5251
F-33400, Talence, France

Abstract. This paper is devoted to the study of the construction of a viscous approximation of the nonconservative bitemperature Euler system. Starting from a BGK model coupled with Ampère and Poisson equations proposed in [1], we perform a Chapman-Enskog expansion up to order 1 leading to a Navier-Stokes system. Next, we prove that this system is compatible with the entropy of the bitemperature Euler system.

1. **Introduction.** This paper is devoted to a viscous approximation of the bitemperature Euler system that has been studied in [1]. This fluid model describes the interaction of a mixture of one species of ions and one species of electrons in thermal nonequilibrium, with applications in the field of Inertial Confinement Fusion where solutions with shocks occur. Quasineutrality being assumed, the electronic and ionic mass fractions are constant: subscripts e and i standing for electron and ions respectively,

$$\rho_e = m_e n_e = c_e \rho, \quad \rho_i = m_i n_i = c_i \rho, \quad c_e + c_i = 1$$

and the model consists of two conservation equations for mass and momentum and two nonconservative equations for each energy.

Moreover the pressure of each species is supposed to satisfy a gamma-law with its own γ constant:

$$p_e = (\gamma_e - 1)\rho_e \varepsilon_e = n_e k_B T_e, \quad p_i = (\gamma_i - 1)\rho_i \varepsilon_i = n_i k_B T_i, \tag{1}$$

where k_B is the Boltzmann constant, ε_α and T_α represent respectively the internal specific energy and the temperature of species α, $\alpha \in \{e, i\}$.

The total energies are given by $\mathcal{E}_\alpha = \rho_\alpha \varepsilon_\alpha + \frac{1}{2}\rho_\alpha u^2$, $\alpha \in \{e, i\}$. We denote $\nu_{ei} \geq 0$ the interaction coefficient between electronic and ionic temperatures. The

2000 *Mathematics Subject Classification.* Primary: 35L60 ; Secondary: 82D10, 76X05.

Key words and phrases. Bitemperature Euler, Chapman-Enskog expansion, BGK model, Dissipative entropy, Plasmas.

* Corresponding author: D. Aregba-Driollet.

bitemperature Euler system is the following:

$$\begin{cases} \partial_t \rho + \partial_x(\rho u) = 0, \\ \partial_t(\rho u) + \partial_x(\rho u^2 + p_e + p_i) = 0, \\ \partial_t \mathcal{E}_e + \partial_x(u(\mathcal{E}_e + p_e)) - u(c_i \partial_x p_e - c_e \partial_x p_i) = \nu_{ei}(T_i - T_e), \\ \partial_t \mathcal{E}_i + \partial_x(u(\mathcal{E}_i + p_i)) + u(c_i \partial_x p_e - c_e \partial_x p_i) = -\nu_{ei}(T_i - T_e). \end{cases} \tag{2}$$

A first step in the comprehension of this system is to suppose that $\gamma_e = \gamma_i = \gamma$. In this case, one can define a global internal energy $\varepsilon = c_e \varepsilon_e + c_i \varepsilon_i$ which satisfies $p_e + p_i = (\gamma - 1)\rho\varepsilon$. Denoting $\mathcal{E} = \mathcal{E}_e + \mathcal{E}_i$ the total energy one has

$$\mathcal{E} = \rho\varepsilon + \frac{1}{2}\rho u^2$$

and

$$\partial_t \mathcal{E} + \partial_x(u(\mathcal{E} + p)) = 0$$

so that $(\rho, \rho u, \mathcal{E})$ satisfies the usual Euler 3×3 system with γ law. Nevertheless, even in this case one needs to solve a nonconservative equation in order to get T_e and T_i separately. In our context, the nonconservativity is not only due to source terms but especially to terms multiplying u by pressure gradients, making delicate the definition of admissible shocks. In order to define nonconservative products, Dal Maso, Le Floch and Murat proposed in [5] a new theory based on the definition of family of paths. In [4], the authors consider the bitemperature Euler system with diffusive terms. By assuming that the electrons are isentropic, the system is transformed into a conservative model. In [7], the authors consider a kinetic system for sprays and derive a nonconservative hyperbolic system that is studied in [6].

In [1], the Euler bitemperature system has been derived by hydrodynamic limit of an underlying kinetic model which consists of a BGK model coupled with Poisson equation in the quasi-neutral regime. Moreover the obtained fluid system has been proved to be entropy dissipative by a direct approach and also by using the Boltzmann entropy. In particular, the nonconservative terms are obtained from the definition of the electric field according to a generalized Ohm's law.

In the present paper, we perform a Chapman-Enskog expansion of our kinetic model up to order one in order to get rigorously a viscous, Navier-Stokes type approximation of the bitemperature Euler system in the case $\gamma_e = \gamma_i$. As a result, we obtain conservative and nonconservative second order terms. To go into details, let us denote $\mathcal{U} = (\rho, \rho u, \mathcal{E}_e, \mathcal{E}_i)$. The Euler bitemperature system (2) being written in condensed form as

$$\partial_t \mathcal{U} + A(\mathcal{U})\partial_x \mathcal{U} = S(\mathcal{U}),$$

for a fixed relaxation parameter $\tau > 0$ the obtained second order system can be written under the form

$$\partial_t \mathcal{U}^\tau + A(\mathcal{U}^\tau)\partial_x \mathcal{U}^\tau = S(\mathcal{U}^\tau) + \tau\left(u^\tau \partial_x\left(J(\mathcal{U}^\tau)\partial_x \mathcal{U}^\tau\right) + \partial_x\left(D(\mathcal{U}^\tau)\partial_x \mathcal{U}^\tau\right)\right). \tag{3}$$

Here $J(\mathcal{U}^\tau)$ and $D(\mathcal{U}^\tau)$ are 4×4 matrices, while u^τ is the velocity. This result completes known models such as the one studied by C. Chalons and F. Coquel in [3], by constructing rigorously some second order terms to their system.

Next we prove the compatibility of the entropy of the bitemperature Euler system with the diffusive terms. We recall that a dissipative entropy η exists for (2), namely

([1])

$$\begin{aligned} \eta(\mathcal{U}) &= \overline{\eta}_e(\rho c_e, \varepsilon_e) + \overline{\eta}_i(\rho c_i, \varepsilon_i), \\ \overline{\eta}_\alpha(\rho_\alpha, \varepsilon_\alpha) &= -\frac{\rho_\alpha}{m_\alpha(\gamma_\alpha - 1)} \ln\left(\frac{p_\alpha}{\rho_\alpha^{\gamma_\alpha}}\right), \quad \alpha \in \{e, i\}. \end{aligned} \tag{4}$$

We prove here that the solutions $\mathcal{U}^\tau$ of (3) formally satisfy the following inequality:

$$\partial_t \eta(\mathcal{U}^\tau) + \partial_x(u\eta(\mathcal{U}^\tau)) \leq -\frac{\nu_{ei}}{k_B T_i T_e}(T_i - T_e)^2 - \tau \frac{5k_B}{2m_\alpha} \sum_{\alpha=e,i} \partial_x \left(n_\alpha \partial_x T_\alpha\right).$$

This is a first step to prove that $\mathcal{U}^\tau$ owns a limit $\mathcal{U}$ which is a weak entropy solution of the Euler bitemperature system.

The paper is organized as follows. The Section 2 deals with the derivation of a Navier-Stokes system starting from the kinetic system proposed in [1]. In section 3, the diffusive terms are shown to be dissipative w.r.t. the entropy of the bitemperature Euler system. Finally, section 4 gives conclusions to this work.

2. Derivation of the Navier-Stokes system.

2.1. Notations. Kinetic models are described by the distribution function f_α of each species depending on the time variable $t \in \mathbb{R}_+$, on the position $x \in \mathbb{R}^3$ and on the velocity $v \in \mathbb{R}^3$. The macroscopic quantities can be obtained by extracting moments on these distribution functions w.r.t the velocity variable. Indeed density, velocity and total energy of the species α can be defined as

$$n_\alpha = \int_{\mathbb{R}^3} f_\alpha dv, \quad u_\alpha = \frac{1}{n_\alpha} \int_{\mathbb{R}^3} v_1 f_\alpha dv, \mathcal{E}_\alpha = \frac{3}{2}\rho_\alpha \frac{k_B}{m_\alpha} T_\alpha + \frac{1}{2}\rho_\alpha u_\alpha^2 = \int_{\mathbb{R}^3} m_\alpha \frac{v^2}{2} f_\alpha dv. \tag{5}$$

The present model is monoatomic ($\gamma = \frac{5}{3}$). Hence, the internal specific energy of species α writes

$$\varepsilon_\alpha = \frac{3}{2m_\alpha} k_B T_\alpha.$$

In the following, we denote U_α the moments of f_α

$$U_\alpha = \begin{pmatrix} \rho_\alpha \\ \rho_\alpha u_\alpha \\ \mathcal{E}_\alpha \end{pmatrix} = m_\alpha \int_{\mathbb{R}^3} \begin{pmatrix} 1 \\ v_1 \\ \frac{v^2}{2} \end{pmatrix} f_\alpha dv. \tag{6}$$

Usually the velocity and the temperature of the mixture are defined by

$$u = \frac{\rho_e u_e + \rho_i u_i}{\rho_e + \rho_i}, \quad nk_B T = \sum_\alpha (\frac{1}{2}\rho_\alpha(u_\alpha^2 - u^2)) + \sum_\alpha (n_\alpha k_B T_\alpha), \tag{7}$$

where $n = n_e + n_i$.

Moreover, the current of the plasma j and the total charge $\bar{\rho}$ are defined by

$$\begin{aligned} \overline{\rho} &= \int_{\mathbb{R}^3} (q_e f_e + q_i f_i)\, dv = n_e q_e + n_i q_i, \\ j &= \int_{\mathbb{R}^3} v_1 (q_e f_e + q_i f_i)\, dv = n_e q_e u_e + n_i q_i u_i, \end{aligned} \tag{8}$$

where $q_e = -e$, $q_i = Ze$ are the particle charges.

2.2. **Chapman-Enskog expansion.** Consider the following kinetic model in the quasi-neutral regime

$$\begin{cases} \partial_t f_\alpha + v_1 \partial_x f_\alpha + \dfrac{q_\alpha}{m_\alpha} E \partial_{v_1} f_\alpha = \dfrac{1}{\tau}(\mathcal{M}_\alpha - f_\alpha) + \dfrac{1}{\tau_{ei}}(\overline{\mathcal{M}_\alpha} - f_\alpha), \\ \partial_t E = -\dfrac{j}{\tau^2}, \\ \partial_x E = \dfrac{\overline{\rho}}{\tau^2}, \end{cases} \tag{9}$$

where τ is a positive parameter proportional to the Knudsen number and $1/\tau_{ei}$ corresponds to the collision frequency for the ion/electron interaction.

$\mathcal{M}_\alpha$ and $\overline{\mathcal{M}_\alpha}$ are the two Maxwellian distribution functions

$$\mathcal{M}_\alpha(f_\alpha) = \frac{n_\alpha}{(2\pi k_B T_\alpha/m_\alpha)^{3/2}} exp(-\frac{|v-u_\alpha|^2}{2k_B T_\alpha/m_\alpha}),\ \alpha = e, i, \tag{10}$$

$$\overline{\mathcal{M}_\alpha}(f_e, f_i) = \frac{n_\alpha}{(2\pi k_B T/m_\alpha)^{3/2}} exp(-\frac{|v-u|^2}{2k_B T/m_\alpha}),\ \alpha = e, i. \tag{11}$$

Next we perform a first order Chapman-Engskog expansion up to order 1. Hence the solution of the system (9, 10, 11) f_α is researched as the expansion

$$f_\alpha = \mathcal{M}_\alpha + \tau g_\alpha, \quad \alpha \in \{e, i\}, \tag{12}$$

with the constraints

$$\begin{aligned} &\int_{\mathbb{R}^3} f_\alpha dv = \int_{\mathbb{R}^3} \mathcal{M}_\alpha dv, \quad \int_{\mathbb{R}^3} v_1 f_\alpha\, dv = \int_{\mathbb{R}^3} v_1\, \mathcal{M}_\alpha\, dv, \\ &\int_{\mathbb{R}^3} v^2 f_\alpha\, dv = \int_{\mathbb{R}^3} v^2\, \mathcal{M}_\alpha\, dv. \end{aligned} \tag{13}$$

By neglecting the terms g_α, one obtains the bitemperature Euler system as an hydrodynamic limit as in ([1]). Our goal here is to compute explicitly the first order term g_α to get the related Navier-Stokes system.

2.3. **Obtention of the viscous fluid system.** We expand f_α as in (12, 13) and we extract the moments *w.r.t.* 1, v_1, v^2.

One important point to determine the viscous terms of the Navier-Stokes is to compute the term g_α of the expansion (12, 13). The calculus is performed in the following proposition.

Proposition 1. *The first order terms g_e and g_i of the expansion (12, 13) write*

$$\begin{aligned} g_e \;=\; & -\Bigg(\Bigg((v_1-u)\left(\frac{\partial_x n_e}{n_e} - \frac{3}{2}\frac{\partial_x T_e}{T_e}\right) + \partial_x u \left(\frac{(v_1-u)^2}{\frac{k_B}{m_e}T_e} - \frac{1}{3}\frac{(v-u)^2}{\frac{k_B}{m_e}T_e}\right) \\ & + \frac{\nu_{ei}}{n_e k_B T_e}(T_i - T_e)\left(\frac{(v-u)^2}{3\frac{k_B}{m_e}T_e} - 1\right) \\ & - (v_1-u)\frac{(v-u)^2}{2\frac{k_B}{m_e}}\partial_x(\frac{1}{T_e}) - \frac{(v_1-u)}{\frac{k_B}{m_e}T_e\rho}\partial_x(p_e+p_i)\Bigg)\mathcal{M}_e \\ & + \frac{q_e}{m_e}E\partial_{v_1}\mathcal{M}_e - \frac{1}{\tau_{ei}}\left(\overline{\mathcal{M}_e} - \mathcal{M}_e\right)\Bigg), \end{aligned} \tag{14}$$

$$
\begin{aligned}
g_i &= -\Bigg(\Bigg((v_1-u)\left(\frac{\partial_x n_i}{n_i}-\frac{3}{2}\frac{\partial_x T_i}{T_i}\right)+\partial_x u\left(\frac{(v_1-u)^2}{\frac{k_B}{m_i}T_i}-\frac{1}{3}\frac{(v-u)^2}{\frac{k_B}{m_i}T_i}\right)\\
&+ \frac{\nu_{ei}}{n_i k_B T_i}(T_e-T_i)\left(\frac{(v-u)^2}{3\frac{k_B}{m_i}T_i}-1\right)\\
&- (v_1-u)\frac{(v-u)^2}{2\frac{k_B}{m_i}}\partial_x(\frac{1}{T_i})-\frac{(v_1-u)}{\frac{k_B}{m_i}T_i\rho}\partial_x(p_e+p_i)\Bigg)\mathcal{M}_i\\
&+ \frac{q_i}{m_i}E\partial_{v_1}\mathcal{M}_i-\frac{1}{\tau_{ei}}\left(\overline{\mathcal{M}}_i-\mathcal{M}_i\right)\Bigg). \qquad (15)
\end{aligned}
$$

Proof. g_α is given by the relation

$$
g_\alpha = -\left(\partial_t\mathcal{M}_\alpha + v_1\partial_x\mathcal{M}_\alpha + \frac{q_\alpha}{m_\alpha}E\partial_{v_1}\mathcal{M}_\alpha - \frac{1}{\tau_{ei}}(\overline{\mathcal{M}_\alpha}-\mathcal{M}_\alpha)\right).
$$

A direct computation gives

$$
\partial_t\mathcal{M}_e = \left((\frac{\partial_t n_e}{n_e}-\frac{3}{2}\frac{\partial_t T_e}{T_e})+(v_1-u)\frac{\partial_t u}{\frac{k_B}{m_e}T_e}-\frac{(v-u)^2}{2\frac{k_B}{m_e}T_e}\partial_t\left(\frac{1}{T_e}\right)\right)\mathcal{M}_e \qquad (16)
$$

and

$$
v_1\partial_x\mathcal{M}_e = \left((\frac{v_1\partial_x n_e}{n_e}-\frac{3}{2}\frac{v_1\partial_x T_e}{T_e})+(v_1-u)\frac{v_1\partial_x u}{\frac{k_B}{m_e}T_e}-\frac{(v-u)^2}{2\frac{k_B}{m_e}T_e}v_1\partial_x\left(\frac{1}{T_e}\right)\right)\mathcal{M}_e.
$$

By using the non-conservative Euler system (2), the time derivatives of (16) are computed in function of the space derivatives up to $\mathcal{O}(\tau)$ terms, as follows

$$
\begin{aligned}
\frac{\partial_t n_e}{n_e}-\frac{3}{2}\frac{\partial_t T_e}{T_e} &= -u\frac{\partial_x n_e}{n_e}+\frac{3}{2}u\frac{\partial_x T_e}{T_e}-\frac{\nu_{ei}}{n_e k_B T_e}(T_i-T_e)+\mathcal{O}(\tau),\\
\partial_t u &= -u\partial_x u-\frac{1}{\rho}\partial_x(p_e+p_i)+\mathcal{O}(\tau),\\
\frac{\partial_t n_i}{n_i}-\frac{3}{2}\frac{\partial_t T_i}{T_i} &= -u\frac{\partial_x n_i}{n_i}+\frac{3}{2}u\frac{\partial_x T_i}{T_i}-\frac{\nu_{ei}}{n_i k_B T_i}(T_e-T_i)+\mathcal{O}(\tau),\\
\frac{\partial_t T_e}{T_e} &= -u\frac{\partial_x T_e}{T_e}-\frac{2}{3}\partial_x u+\frac{2}{3}\frac{\nu_{ei}}{n_e k_B T_e}(T_i-T_e)+\mathcal{O}(\tau)\\
\frac{\partial_t T_i}{T_i} &= -u\frac{\partial_x T_i}{T_i}-\frac{2}{3}\partial_x u+\frac{2}{3}\frac{\nu_{ei}}{n_i k_B T_i}(T_e-T_i)+\mathcal{O}(\tau).
\end{aligned}
$$

Hence up to $\mathcal{O}(\tau)$ order terms, we get

$$
\begin{aligned}
\partial_t\mathcal{M}_e+v_1\partial_x\mathcal{M}_e &= \Bigg[(v_1-u)\Big(\frac{\partial_x n_e}{n_e}-\frac{3}{2}\frac{\partial_x T_e}{T_e}\Big)+\partial_x u\left(\frac{(v_1-u)^2}{\frac{k_B}{m_e}T_e}-\frac{(v-u)^2}{3\frac{k_B}{m_e}T_e}\right)\\
&+ \frac{\nu_{ei}}{n_e k_B T_e}(T_i-T_e)\left(\frac{(v-u)^2}{3\frac{k_B}{m_e}T_e}-1\right)\\
&- (v_1-u)\frac{(v-u)^2}{2\frac{k_B}{m_e}}\partial_x\left(\frac{1}{T_e}\right)-\frac{(v_1-u)}{\frac{k_B}{m_e}\rho T_e}\partial_x(p_e+p_i)\Bigg]\mathcal{M}_e
\end{aligned}
$$

and we recover (14). The same result holds for (15). □

The proposition 1 is the cornerstone of this paper. It allows us to obtain the following proposition after some more computations.

Proposition 2. *The viscous approximation of the kinetic system (9, 10, 11) writes*

$$\partial_t \rho + \partial_x(\rho u) = 0, \quad (17)$$

$$\partial_t(\rho u) + \partial_x(\rho u^2 + p_e + p_i) - \frac{4}{3}\tau\partial_x(p\,\partial_x u) = 0, \quad (18)$$

$$\begin{aligned}\partial_t(\rho_e\varepsilon_e + \frac{1}{2}\rho_e u^2) + \partial_x(u(\rho_e\varepsilon_e + \frac{1}{2}\rho_e u^2 + p_e)) - u(c_i\partial_x p_e - c_e\partial_x p_i)&\\ -u\left(\frac{4}{3}\tau c_e\partial_x(p_i\partial_x u) - \frac{4}{3}\tau c_i\partial_x(p_e\partial_x u)\right)&\\ -\frac{4}{3}\tau\partial_x\left(u p_e\,\partial_x u\right) - \frac{5}{2}\tau\partial_x(\frac{k_B}{m_e}p_e\,\partial_x T_e) &= \nu_{ei}(T_e - T_i),\end{aligned} \quad (19)$$

$$\begin{aligned}\partial_t(\rho_i\varepsilon_i + \frac{1}{2}\rho_i u^2) + \partial_x(u(\rho_i\varepsilon_i + \frac{1}{2}\rho_i u^2 + p_i)) + u(c_i\partial_x p_e - c_e\partial_x p_i)&\\ +u\left(\frac{4}{3}\tau c_e\partial_x(p_i\partial_x u) - \frac{4}{3}\tau c_i\partial_x(p_e\partial_x u)\right)&\\ -\frac{4}{3}\tau\partial_x\left(u p_i\,\partial_x u\right) - \frac{5}{2}\tau\partial_x(\frac{k_B}{m_i}p_i\,\partial_x T_i) &= \nu_{ei}(T_i - T_e),\end{aligned} \quad (20)$$

and the electric field E is given by

$$\begin{aligned}\left(\frac{n_e q_e}{\rho_e} - \frac{n_i q_i}{\rho_i}\right)E &= \frac{\rho}{\rho_e\rho_i}n_e q_e E = -\frac{\rho}{\rho_e\rho_i}n_i q_i E\\ &= \frac{\partial_x p_e}{\rho_e} - \frac{\partial_x p_i}{\rho_i} - \frac{4}{3}\frac{\tau}{\rho_e}\partial_x\left(p_e\,\partial_x u\right) + \frac{4}{3}\frac{\tau}{\rho_i}\partial_x\left(p_i\,\partial_x u\right).\end{aligned} \quad (21)$$

Remark 1. The relation (21) is an approximation at order τ of the Ohm law given in [1]. The nonconservative terms of the system (17, 18, 19, 20) are shown to appear from this relation defining E.

Remark 2. By using the internal energy variable, our result can be compared to the model studied in ([2], [3]). However, we obtain additional terms which are not present in those papers.

3. Dissipativity of the second order terms with respect to the entropy.

This section is devoted to the proof of the entropy dissipativity of the viscous system (17-20).

Proposition 3. *We assume that $\gamma_e = \gamma_i = 5/3$. Let $\mathcal{U}^\tau$ be a solution of the second order system (17-20). Then $\mathcal{U}^\tau$ satisfies the following entropy inequality:*

$$\partial_t\eta(\mathcal{U}^\tau) + \partial_x(u^\tau\eta(\mathcal{U}^\tau)) \leq -\frac{\nu_{ei}}{k_B T_i^\tau T_e^\tau}(T_i^\tau - T_e^\tau)^2 - \tau\frac{5k_B}{2}\sum_{\alpha=e,i}\frac{1}{m_\alpha}\partial_x\left(n_\alpha^\tau\partial_x T_\alpha^\tau\right) \quad (22)$$

where η is defined by (4).

Proof. The result is obtained by multiplying (17)–(20) by $\eta'(\mathcal{U}^\tau)$. The system (17)–(20) being written in the synthetic form (3), we denote W the viscous terms

$$W = u^\tau\partial_x\left(J(\mathcal{U}^\tau)\partial_x\mathcal{U}^\tau\right) + \partial_x\left(D(\mathcal{U}^\tau)\partial_x\mathcal{U}^\tau\right).$$

In [1] we have shown that

$$\eta'(\mathcal{U}^\tau)\left[\partial_t\mathcal{U}^\tau + A(\mathcal{U}^\tau)\partial_x\mathcal{U}^\tau - S(\mathcal{U}^\tau)\right] = \frac{\nu_{ei}}{k_B T_i^\tau T_e^\tau}(T_i^\tau - T_e^\tau)^2.$$

It remains to prove that

$$\eta'(\mathcal{U}^\tau)W \leq -\frac{5k_B}{2}\sum_{\alpha=e,i}\frac{1}{m_\alpha}\partial_x\left(n_\alpha^\tau\partial_x T_\alpha^\tau\right).$$

A straightforward computation gives

$$\begin{aligned}\eta'(\mathcal{U})W &= -\frac{4}{3k_B}\left(\partial_x u\right)^2\left(\frac{p_e}{T_e}+\frac{p_i}{T_i}\right) - \frac{5}{2k_BT_e}\partial_x\left(\frac{k_B}{m_e}p_e\partial_x T_e\right)\\ &\quad - \frac{5}{2k_BT_i}\partial_x\left(\frac{k_B}{m_i}p_i\partial_x T_i\right)\\ &= -\frac{4n}{3}\left(\partial_x u\right)^2 - \sum_{\alpha=e,i}\left(\frac{5}{2m_\alpha T_\alpha}\partial_x\left(n_\alpha k_B T_\alpha\partial_x T_\alpha\right)\right).\end{aligned}$$

Using the fact that

$$T_\alpha^{-1}\partial_x\left(n_\alpha T_\alpha\partial_x T_\alpha\right) = \partial_x\left(n_\alpha\partial_x T_\alpha\right) + \frac{n_\alpha\left(\partial_x T_\alpha\right)^2}{T_\alpha},$$

we thus have

$$\eta'(\mathcal{U})W = -\frac{4n}{3}\left(\partial_x u\right)^2 - \frac{5k_B}{2}\sum_{\alpha=e,i}\frac{1}{m_\alpha}\left(\partial_x\left(n_\alpha\partial_x T_\alpha\right) + \frac{n_\alpha\left(\partial_x T_\alpha\right)^2}{T_\alpha}\right).$$

□

4. **Conclusion.** In this paper, starting from a kinetic model, we have derived a viscous approximation of the bitemperature Euler system from a Chapman-Enskog expansion. We have been able to compute explicitly all the viscous terms and we have obtained a generalization of the model proposed in [3]. Then we have proved an entropy inequality. These results support the approach taken in [1] where the same kinetic model is the basis of the numerical approximation of the system (2). The case $\gamma_e \neq \gamma_i$ can be handled by using a kinetic model with internal energy variable. In a future work we plan to study the shocks obtained by limits of travelling waves constructed from this viscous model and to compare them to the ones numerically computed in our previous work [1].

REFERENCES

[1] D. Aregba-Driollet, J. Breil, S. Brull, B. Dubroca and E. Estibals, Modelling and numerical approximation for the nonconservative bitemperature Euler model, *ESAIM Math. Model. Numer. Anal.*, **52** (2018), 1353–1383.

[2] C. Berthon and F. Coquel, Nonlinear projection methods for multi-entropies Navier-Stokes systems, *Math of Comp.*, **76** (2007), 1163-1194.

[3] C. Chalons and F. Coquel, Navier-Stokes equations with several independent pressure laws and explicit predictor-corrector schemes, *Numerisch Math.*, **101** (2005), 451-478.

[4] F. Coquel and C. Marmignon, Numerical methods for weakly ionized gas, *Astrophys. Space Sci*, **260** (1998), 15-27.

[5] G. Dal Maso, P. G. Le Floch and F. Murat, Definition and weak stability of nonconservative products, *J. Math. Pures Appl*, **74** (1995), 483-548.

[6] P. A. Raviart and L. Sainsaulieu, A nonconservative hyperbolic system modeling spray dynamics. I. Solution of the Riemann problem, *Math. Models Methods Appl. Sci.*, **5** (1995), 297–333.

[7] L. Sainsaulieu, Equilibrium velocity distribution functions for a kinetic model of two-phase flows, *Math. Models Methods Appl. Sci.*, **5** (1995), 191–211.

E-mail address: aregba@math.u-bordeaux.fr
E-mail address: brull@math.u-bordeaux.fr

AN ASYMPTOTIC PRESERVING TIME INTEGRATOR FOR LOW MACH NUMBER LIMITS OF THE EULER EQUATIONS WITH GRAVITY

K. R. ARUN AND S. SAMANTARAY*

School of Mathematics
Indian Institute of Science Education and Research Thiruvananthapuram
Thiruvananthapuram - 695551, India

ABSTRACT. We consider two distinguished asymptotic limits of the Euler equations in a gravitational field, namely the incompressible and Boussinesq limits. Both these limits can be obtained as singular limits of the Euler equations under appropriate scaling of the Mach and Froude numbers. We propose and analyse an asymptotic preserving (AP) time discretisation for the numerical approximation of the Euler system in these asymptotic regimes. A key step in the construction of the AP scheme is a semi-implicit discretisation of the fluxes and the source term. The non-stiff convective terms are treated explicitly whereas the stiff pressure-gradient and source term are implicit. The implicit terms are combined to get a nonlinear elliptic equation. We show that the overall scheme is consistent with the respective limit system when the Mach number goes to zero. A linearised stability analysis confirms the L^2-stability of the proposed scheme. The results of numerical experiments validate the theoretical findings.

1. **Introduction.** The presence of sound/acoustic waves poses a major challenge in atmospheric and meteorological flow computations due to their fast characteristic time scales. Hence, in most of the practical computations, one relies on the so-called 'sound-proof' models in which the sound waves are eliminated. The incompressible equations, Boussinesq equations, pseudo-incompressible equations, anelastic equations etc. are sound-proof models frequently used in the literature, to name but a few. The derivation and analysis of sound-proof models, study of their regimes of validity etc. are topics of active research even today; see, e.g., [2] and the references cited therein for more details.

A powerful and systematic method to derive a sound-proof model is an asymptotic analysis of the Euler equations in which one or more of the non-dimensional quantities, such as the Mach, Froude or Rossby numbers, assume the role of limiting parameters [4]. However, from a mathematical point of view, a sound-proof model is often recognised as a singular limit of the Euler equations under appropriate scalings. In addition, sound-proof equation systems are typically of hyperbolic-elliptic

---

2000 *Mathematics Subject Classification.* Primary: 35L45, 35L65, 35L67; Secondary: 65M06, 65M08, 65M20.

Key words and phrases. Asymptotic preserving, Low Mach number limit, Boussinesq limit, IMEX-RK scheme, L^2-stability.

* Corresponding author: S. Samantaray.

in nature, as opposed to the purely hyperbolic compressible Euler equations. On the other hand, from a numerical point of view, approximation of singular limits poses several challenges: stiffness arising from stringent stability requirements, reduction of order of accuracy due to the presence of limiting parameters and so on.

The goal of the present work is to obtain the incompressible and Boussinesq equations as two distinguished singular limits of the Euler equations in a gravitational field under appropriate scalings of the Mach and Froude numbers. We present their numerical resolution via the so-called asymptotic preserving (AP) methodology. An AP discretisation for a singularly perturbed problem in general is a one which reduces to a consistent discretisation of the limit model when the limits of perturbation parameters are taken. In addition, the stability requirements of the discretisation should remain independent of the perturbation parameters; see [6]. A key step in the construction of our AP scheme is a semi-implicit time discretisation based on a splitting of the flux and source terms into stiff and non-stiff terms. We show the asymptotic consistency of the scheme with the incompressible and Boussinesq limits as the Mach number approaches zero. As a first step towards the stability of the scheme in the asymptotic regime, we perform an L^2-stability analysis of the proposed scheme on a linearised model, namely the wave equation system. The results of our numerical experiments presented here clearly validate the AP nature of the proposed scheme.

2. Isentropic Euler System with Gravity and Its Asymptotic Limits.

We consider the scaled, isentropic compressible Euler equations with gravity:

$$\partial_t \rho + \nabla \cdot (\rho u) = 0, \tag{1}$$

$$\partial_t(\rho u) + \nabla \cdot (\rho u \otimes u) + \frac{\nabla p}{\mathrm{Ma}^2} = -\frac{\rho e_3}{\mathrm{Fr}^2}, \tag{2}$$

where $\rho > 0$ is the density and $u \in \mathbb{R}^3$ is the velocity vector. Here, ∇, $\nabla\cdot$ and $\otimes$ are respectively the gradient, divergence and tensor product operators and e_3 is the unit vector in the x_3-direction. We assume a simplified equation of state of an isentropic process, therein the pressure is related to density via $p = P(\rho) = \rho^\gamma$, where γ is a constant. In (1)-(2), the non-dimensional parameters Ma and Fr are respectively, the reference Mach and Froude numbers.

The goal of the present work is the numerical approximation of some distinguished asymptotic limits of the Euler system (1)-(2) which models slow convection in a highly stratified medium; see, e.g. [2, 4] for more details. In order to describe these asymptotic regimes, in the following, we consider two important scalings of Ma and Fr in terms of an infinitesimal parameter ε.

- $\mathrm{Ma} = \varepsilon$ and $\mathrm{Fr} = 1$. In this case, the pressure gradient term dominates the gravity term and we obtain the low Mach number limit.
- $\mathrm{Ma} = \varepsilon$ and $\mathrm{Fr} = \sqrt{\varepsilon}$. In this case, the gravitational term is also significant, and we derive the Boussinesq limit.

As a first step towards the derivation of the low Mach and Boussinesq limits, we expand all the dependent variables using the following three-term ansatz:

$$f(t, x) = f_{(0)}(t, x) + \varepsilon f_{(1)}(t, x) + \varepsilon^2 f_{(2)}(t, x). \tag{3}$$

We do not intent to provide the details of the derivation, but refer the interested reader to [4] for more details.

2.1. **Zero Mach Number Limit.** We set $\mathrm{Ma} = \varepsilon$ and $\mathrm{Fr} = 1$ in (1)-(2) and let $\varepsilon \to 0$ to obtain the zero Mach number limit model:

$$\partial_t u_{(0)} + \nabla \cdot \left(u_{(0)} \otimes u_{(0)}\right) + \nabla p_{(2)} = -e_3, \tag{4}$$

$$\nabla \cdot u_{(0)} = 0. \tag{5}$$

The above system (4)-(5) is the standard incompressible Euler system for the unknowns $u_{(0)}$ and $p_{(2)}$.

Remark 1. Throughout our analysis and the numerical experiments presented in this paper, we assume either periodic or wall boundary conditions. As a consequence, the leading order density $\rho_{(0)}$ is a constant and the leading order velocity $u_{(0)}$ is divergence-free. Therefore, both the zero Mach and Boussinesq limits fall in the category of 'sound-proof' models.

2.2. **Boussinesq Limit.** Now we set $\mathrm{Ma} = \varepsilon$ and $\mathrm{Fr} = \sqrt{\varepsilon}$ in (1)-(2). Letting $\varepsilon \to 0$ yields the Boussinesq model:

$$\partial_t u_{(0)} + \nabla \cdot \left(u_{(0)} \otimes u_{(0)}\right) + \nabla p_{(2)} = -\rho_{(1)} e_3, \tag{6}$$

$$\nabla \cdot u_{(0)} = 0. \tag{7}$$

Since the first order density $\rho_{(1)}$ appears in (6)-(7), we need a closure relation. Using the multiscale ansatz (3) in the equation of state $p = \rho^\gamma$ and using the hydrostatic balance $\nabla p_{(1)} = -\rho_{(0)} e_3$ gives

$$\rho_{(1)} = 1 - \frac{x_3}{\gamma}. \tag{8}$$

Remark 2. It has be noted that both zero Mach and the Boussinesq limit systems are hyperbolic-elliptic in nature.

3. **Semi-implicit Time Discretisation.** In this section we present the time discretisation of the Euler system (1)-(2) based on implicit-explicit (IMEX) Runge Kutta (RK) schemes. These schemes were originally designed for stiff ordinary differential equations; see .e.g. [5] and the references therein.

Let $0 = t^0 < t^1 < \cdots < t^n < t^{n+1} < \cdots$ be an increasing sequence of times and let Δt be the uniform time-step. Let us denote by $f^n(x)$, the approximation to the value of any function f at time t^n, i.e. $f^n(x) \sim f(t^n, x)$.

A first order accurate semi-discrete scheme for the Euler equations (1)-(2) is defined as

$$\frac{\rho^{n+1} - \rho^n}{\Delta t} + \nabla \cdot q^{n+1} = 0, \tag{9}$$

$$\frac{q^{n+1} - q^n}{\Delta t} + \nabla \cdot \left(\frac{q \otimes q}{\rho}\right)^n + \frac{\nabla p(\rho^{n+1})}{\varepsilon^2} = -\frac{\rho^{n+1}}{\varepsilon^\alpha} e_3. \tag{10}$$

Here, $q = \rho u$ denotes the momentum and $\alpha \in \{0, 1\}$ is a parameter so that $\alpha = 0$ corresponds to the low Mach limit and $\alpha = 1$ corresponds to the Boussinesq limit. Though the scheme (9)-(10) consists of a fully implicit step (9) and a semi-implicit step (10), its numerical resolution is fairly simple. Eliminating q^{n+1} between (9) and (10) yields the nonlinear elliptic equation:

$$-\frac{\Delta t^2}{\varepsilon^2} \Delta P(\rho^{n+1}) - \frac{\Delta t^2}{\varepsilon^\alpha} \nabla \cdot \left(\rho^{n+1} e_3\right) + \rho^{n+1} = \rho^n - \Phi(\rho^n, u^n), \tag{11}$$

where the known expression Φ is given by

$$\Phi(\rho^n, u^n) := \Delta t \nabla \cdot q^n + \Delta t^2 \nabla^2 : \left(\frac{q \otimes q}{\rho}\right)^n \tag{12}$$

with : denoting the contracted product. Solving the elliptic equation (11) yields the updated density ρ^{n+1}. The velocity u^{n+1} can then be updated using (10), which is now an explicit evaluation. Hence, the scheme (9)-(10) consists of solving the elliptic equation (11), followed by an explicit evaluation of (10).

4. **Asymptotic Preserving Property.** A numerical scheme for a singular perturbation problem, such as the Euler system (1)-(2), may not resolve the existing multiple scales in space and time. In addition, when the perturbation parameter goes to zero, the scheme may approximate a completely different set of equations than the actual limiting systems. An asymptotic preserving (AP) scheme is the one which is consistent with the limiting set of equations in the singular limit; see [6] for a review of AP schemes.

Theorem 4.1. *The time semi-discrete scheme* (9)-(10) *for* $\alpha = 0$ *is asymptotically consistent with the low Mach number model as* $\varepsilon \to 0$.

Proof. First, we apply the same ansatz (3) for all the dependent variables at times t^n and t^{n+1} in the semi-discrete scheme (9)-(10) and balance the like-powers of ε. The lowest order terms gives $\nabla P(\rho_{(0)}^{n+1}) = 0$ and the equation of state $P(\rho) = \rho^\gamma$ then yields that $\rho_{(0)}^{n+1}$ is constant. Therefore, from the mass update (9) we get

$$-\nabla \cdot u_{(0)}^{n+1} = \frac{\rho_{(0)}^{n+1} - \rho_{(0)}^n}{\rho_{(0)}^{n+1} \Delta t}. \tag{13}$$

We integrate the above equation (13) over a domain Ω and use Gauss' divergence theorem to obtain:

$$-\frac{1}{|\Omega|} \int_{\partial\Omega} u_{(0)}^{n+1} \cdot \nu d\sigma = \frac{\rho_{(0)}^{n+1} - \rho_{(0)}^n}{\rho_{(0)}^{n+1} \Delta t}. \tag{14}$$

Hence, the leading order density $\rho_{(0)}$ rises or falls only due to compressions or expansions at the boundary. The temporal variations in $\rho_{(0)}$ can produce nonzero divergences in the leading order velocity $u_{(0)}$. It can be proved that the integral on the left hand side of (14) vanishes under most of the physically relevant boundary conditions. In this case, we obtain $\rho_{(0)}^{n+1} = \rho_{(0)}^n$ and this in turn enforces the divergence constraint at t^{n+1} as

$$\nabla \cdot u_{(0)}^{n+1} = 0. \tag{15}$$

Combining (15) and the $\mathcal{O}(1)$ terms in (10), we have the following limiting system:

$$\frac{u_{(0)}^{n+1} - u_{(0)}^n}{\Delta t} + \nabla \cdot (u_{(0)}^n \otimes u_{(0)}^n) + p_{(2)}^{n+1} = -e_3, \tag{16}$$

$$\nabla \cdot u_{(0)}^{n+1} = 0. \tag{17}$$

The above system (16)-(17) is clearly a consistent discretisation of the low Mach number limit system (4)-(5). □

Theorem 4.2. *The time semi-discrete scheme* (9)-(10) *for* $\alpha = 1$ *is asymptotically consistent with the Boussinesq model.*

Proof. The proof is similar to that of Theorem 4.1 and hence omitted. □

5. L^2 Stability Analysis of the Semi-discrete Scheme.

The aim of this section is to present the results of an L^2-stability analysis of the semi-discrete scheme (9)-(10). To this end, we consider the homogeneous linear wave equation system:

$$\partial_t \rho + (\bar{u} \cdot \nabla)\rho + \bar{\rho}\nabla \cdot u = 0, \tag{18}$$

$$\partial_t u + (\bar{u} \cdot \nabla)u + \frac{\bar{a}^2}{\bar{\rho}\varepsilon^2}\nabla \rho = 0 \tag{19}$$

as a simplified model of the Euler system (1)-(2). Here, $(\bar{\rho}, \bar{u})$ is a linearisation state and $\bar{a}$ is a linearisation state for the sound velocity. Applying the AP methodology introduced in (9)-(10) to (18)-(19) yields the semi-discrete scheme:

$$\frac{\rho^{n+1} - \rho^n}{\Delta t} + (\bar{u} \cdot \nabla)\rho^n + \bar{\rho}\nabla \cdot u^{n+1} = 0, \tag{20}$$

$$\frac{u^{n+1} - u^n}{\Delta t} + (\bar{u} \cdot \nabla)u^n + \frac{\bar{a}^2}{\bar{\rho}\varepsilon^2}\nabla \rho^{n+1} = 0. \tag{21}$$

In the following, we use a stability result due to Richtmyer; see e.g. [7, 8] for details. Note that any difference scheme of the form $B_1 U^{n+1} = B_2 U^n$, where B_1, B_2 are $p \times p$ matrices, independent of t and x, and $U^n \in \mathbb{R}^p$ is the approximation to the original solution at time t^n, can be reduced to $\hat{U}^{n+1} = G(\Delta t, \xi)\hat{U}^n$ in the Fourier variable ξ. Here, $G(\Delta t, \xi)$ is the Fourier transform of the matrix $(B_1)^{-1}B_2$ and is called the amplification matrix. The stability result due to Richtmyer states that

Theorem 5.1. *A difference scheme given by $B_1 U^{n+1} = B_2 U^n$ is stable if*

(i) the elements of $G(0, \xi)$ are bounded for all $\xi \in \mathbb{L}$, where $\mathbb{L}$ is a lattice where ξ varies,
(ii) $\|G(0, \xi)\| \le 1$ and
(iii) $G(\Delta t, \xi)$ is Lipschitz continuous at $\Delta t = 0$ in the sense that

$$G(\Delta t, \xi) = G(0, \xi) + \mathcal{O}(\Delta t) \text{ as } \Delta t \to 0.$$

Using the above theorem, we have the following stability result.

Theorem 5.2. *The semi-discrete scheme (20)-(21) is L^2-stable.*

Proof. Taking the Fourier transform of (20)-(21) and re-arranging the terms gives

$$\hat{U}^{n+1} = G(\Delta t, \xi)\hat{U}^n, \tag{22}$$

where

$$G(\Delta t, \xi) = \gamma \begin{pmatrix} 1 & -i\Delta t\bar{\rho}\xi_1 & -i\Delta t\bar{\rho}\xi_2 \\ -i\Delta t\lambda\xi_1 & 1 + \Delta t^2\bar{\rho}\lambda\xi_2^2 & -\Delta t^2\bar{\rho}\lambda\xi_1\xi_2 \\ -i\Delta t\lambda\xi_2 & -\Delta t^2\bar{\rho}\lambda\xi_1\xi_2 & 1 + \Delta t^2\bar{\rho}\lambda\xi_1^2 \end{pmatrix}, \tag{23}$$

$$\lambda = \frac{\bar{a}^2}{\bar{\rho}\varepsilon^2} \quad \text{and} \quad \gamma = \frac{1 - i\Delta t(\bar{u} \cdot \xi)}{1 + \Delta t^2\bar{\rho}\frac{\bar{a}^2}{\varepsilon^2}|\xi|^2}. \tag{24}$$

Now, $G(0,\xi)$ reduces to the 3×3 identity matrix and hence conditions (i) and (ii) of Theorem 5.1 are automatically satisfied. Further,

$$G(\Delta t,\xi)-G(0,\xi)$$

$$= \Delta t \begin{pmatrix} -\frac{\Delta t\bar{\rho}\lambda|\xi|^2+i(\bar{u}\cdot\xi)}{1+\Delta t^2\bar{\rho}\lambda|\xi|^2} & -i\bar{\rho}\xi_1\gamma & -i\bar{\rho}\xi_2\gamma \\ -i\lambda\xi_1\gamma & -\frac{\Delta t\bar{\rho}\lambda\xi_1^2+i(\bar{u}\cdot\xi)(1+\Delta t^2\bar{\rho}\lambda\xi_2^2)}{1+\Delta t^2\bar{\rho}\lambda|\xi|^2} & -\Delta t^2\bar{\rho}\lambda\xi_1\xi_2\gamma \\ -i\lambda\xi_2\gamma & -\Delta t^2\bar{\rho}\lambda\xi_1\xi_2\gamma & -\frac{\Delta t\bar{\rho}\lambda\xi_2^2+i(\bar{u}\cdot\xi)(1+\Delta t^2\bar{\rho}\lambda\xi_1^2)}{1+\Delta t^2\bar{\rho}\lambda|\xi|^2} \end{pmatrix}. \tag{25}$$

Note that the matrix on the right hand side in (25) is bounded for every bounded lattice $\mathbb{L}$. Hence, by Theorem 5.1, the semi-discrete scheme (20)-(21) is L^2-stable. □

6. **Numerical Experiments.** We do not intend to discuss the space discretisation in detail as we use employ standard techniques. We use a finite volume approach to approximate the semi-discrete scheme (9)-(10). The explicit flux terms are approximated by a Rusanov-type flux whereas the implicit terms by simple central differences. The nonlinear elliptic equation (11) is solved iteratively after discretisation of the derivatives by central differences.

In the following, we consider a test problem in two dimensions to demonstrate the AP property of the scheme. We take the well-prepared initial data given in [1] which reads

$$\rho(0,x_1,x_2)=1+\varepsilon^2\sin^2(2\pi(x_1+x_2)), \tag{26}$$

$$q_1(0,x_1,x_2)=\sin(2\pi(x_1-x_2))+\varepsilon^2\sin(2\pi(x_1+x_2)), \tag{27}$$

$$q_2(0,x_1,x_2)=\sin(2\pi(x_1-x_2))+\varepsilon^2\cos(2\pi(x_1+x_2)). \tag{28}$$

The computational domain $[0,1]\times[0,1]$ is divided into 50×50 mesh points and we apply periodic boundary conditions on all four sides. The CFL number is set to 0.45 and we perform the computations up to a final time $T=1.0$. The parameter ε is set to 0.1. Note that our CFL condition is independent of ε.

In Figures 1 and 2 we plot the density, x_1-velocity and the divergence of the velocity at times $t=0$ and $t=1$, for the low Mach and Boussinesq cases, respectively. It can be noted from the figures that in both the cases the density converges to the constant value 1 and the divergence approach 0. This is in conformity with the AP nature of the scheme in both the cases.

7. **Conclusion.** An AP semi-implicit time discretisation is proposed for the numerical approximation of the isentropic Euler equations with gravity in the low Mach number and Boussinesq limits. The schemes are theoretically shown to be asymptotically consistent as well as linearly stable. The results of numerical experiments provide a justification to AP nature of the scheme.

Acknowledgement. The authors thank Arnab Das Gupta for several useful discussions on the topic.

REFERENCES

[1] P. Degond and M. Tang, All speed scheme for the low Mach number limit of the isentropic Euler equations, *Commun. Comput. Phys.*, **10** (2011), 1–31.

[2] Dale R. Durran, *Numerical Methods for Fluid Dynamics*, 2nd edition, Springer, New York, 2010.

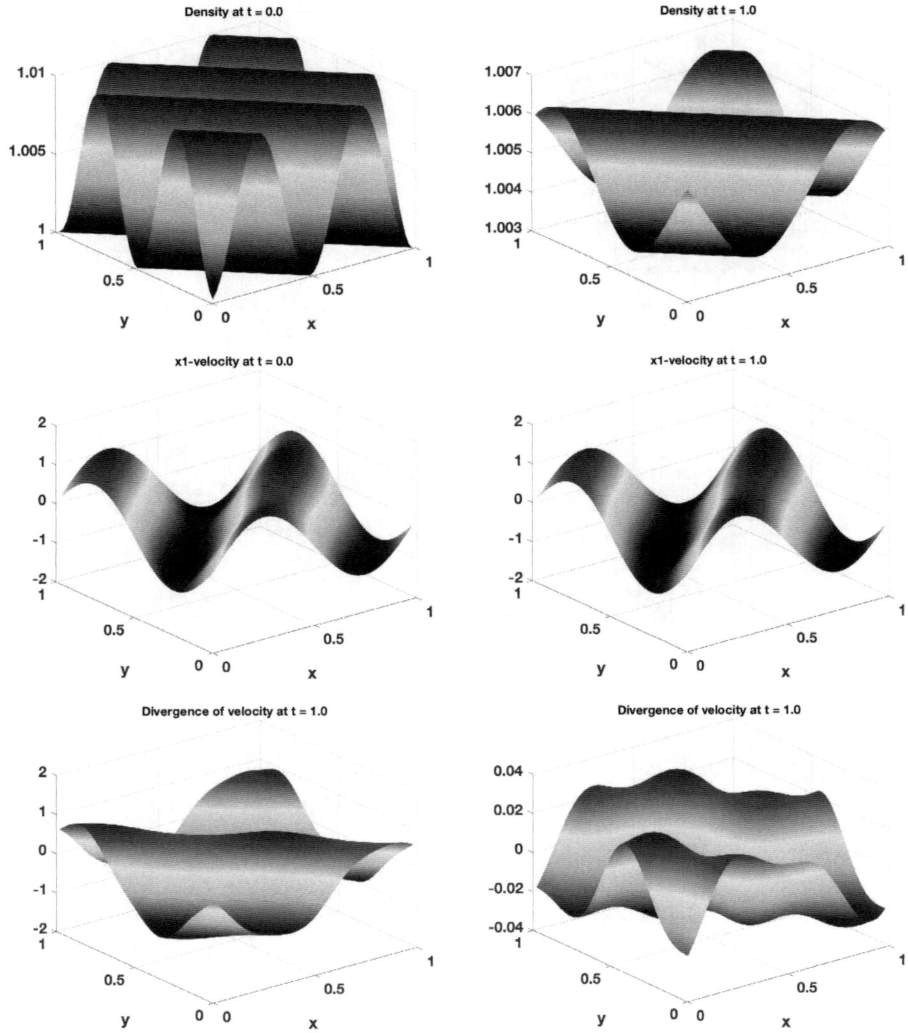

FIGURE 1. For $\varepsilon = 0.1$, the density, x_1-velocity and velocity divergence at $t = 0$ (left) the density, x_1-velocity and velocity divergence at $t = 1$. The low Mach number limit.

[3] A. Meister, Asymptotic single and multiple scale expansions in the low Mach number limit, *SIAM J. Appl. Math.*, **60** (2000), no. 1, 256–271.

[4] R. Klein, Asymptotic analyses for atmospheric flows and the construction of asymptotically adaptive numerical methods *Z. Angew. Math. Mech.*, **80** (2000), no. 11-12, 765–777.

[5] L. Pareschi and G. Russo Implicit-Explicit Runge-Kutta schemes and applications to hyperbolic systems with relaxation *J. Sci. Comput.*, **25** (2005), no. 1-2, 129–155.

[6] S. Jin, Asymptotic preserving (AP) schemes for multiscale kinetic and hyperbolic equations: a review, *Riv. Math. Univ. Parma (N.S.)*, **3** (2012), no. 2, 177–216.

[7] R. D. Richtmyer, Difference methods for initial-value problems, Interscience tracts in pure and applied mathematics. Iract 4, Interscience Publishers, Inc., New. York, 1957.

[8] M. L. Buchanan, A necessary and sufficient condition for stability of difference schemes for second-order initial value problems, *J. Soc. Indust. Appl. Math.*, **11** (1963), 474–501.

E-mail address: arun@iisertvm.ac.in
E-mail address: sauravsam13@iisertvm.ac.in

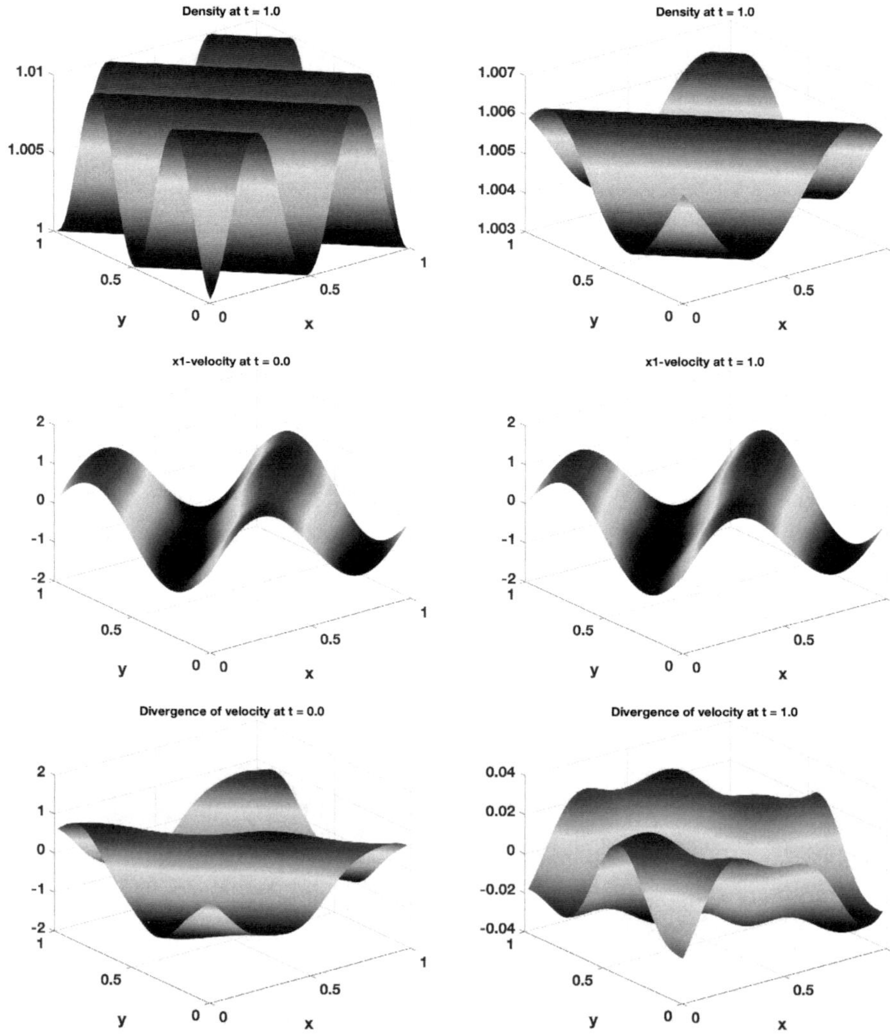

FIGURE 2. For $\varepsilon = 0.1$, the density, x_1-velocity and velocity divergence at $t = 0$ (left) the density, x_1-velocity and velocity divergence at $t = 1$. The Boussinesq limit.

ON THE CHAPMAN-ENSKOG ASYMPTOTICS FOR A MIXTURE OF MONOATOMIC AND POLYATOMIC RAREFIED GASES

Céline Baranger

CEA-CESTA, 15 avenue des sablières - CS 60001 33116 Le Barp Cedex, France

Marzia Bisi

University of Parma, Dept. of Mathematics, Physics and Computer Sciences,
Parco Area delle Scienze 53/A, I-43124, Parma

Stéphane Brull*

Bordeaux INP, IMB, UMR 5251, F-33400

Laurent Desvillettes

Univ. Paris Diderot, Sorbonne Paris Cité, Institut de Mathématiques de Jussieu,
Paris Rive Gauche, UMR 7586, CNRS, Sorbonne Universités
UPMC Univ. Paris 06, F-75013, Paris, France

Abstract. In this paper, we propose a formal Chapman-Enskog expansion in the context of mixtures of monoatomic and polyatomic gases. We start from a Boltzmann model that is based on the Borgnakke-Larsen procedure ([4]) and we derive a compressible Navier-Stokes system. In a last part, we perform some explicit computations of the transport coefficients in the case of Maxwell molecules for diatomic gases.

1. **Introduction.** This paper presents the perform a Chapman-Enskog expansion developped in ([2], [3]) in a polyatomic setting starting from a collisional model ([5], [9]). The kinetic model uses the unknown $f^{(i)}(t,x,v,I)$ as the number density of the $i-th$ species at time t, position x, velocity v and a one-dimensional internal energy parameter $I>0$. In particular, this modelling is necessary in order to treat some physical situations ([14], [11], [15], [13]). For example, in [13], the authors study different types of shock profiles by using the model proposed in ([1], [6]). Moreover, the introduction of such a parameter allows to get a general energy law at the fluid limit ([7], [9]). Remark that in some polyatomic models, the energy can be described by a discrete energy variable ([10], [12]).

The paper is organised as follows. In section 2, the collision operators are detailed. In section 3, the Chapman-Enskog expansion is performed and in section 4 the Navier-Sokes system is presented. Section 5 is devoted to the case when all the cross sections are equal in the diatomic case.

2000 *Mathematics Subject Classification.* Primary: 82B40; Secondary: 35Q80.

Key words and phrases. Boltzmann equation equation, polyatomic, Chapman-Enskog expansion.

2. **Boltzmann kernels.** We define in this section the collision operators of the kinetic model that is used to construct the Navier-Stokes system.
In the case of collisions between monoatomic molecules, we define (for $f := f(v) \geq 0, g := g(v) \geq 0$):

$$Q_{ij}(f,g)(v) = \int_{\mathbb{R}^3}\int_{S^2} \Big\{ f(v')\,g(v'_*) - f(v)\,g(v_*) \Big\}\, B_{ij}\Big(|v-v_*|, \frac{v-v_*}{|v-v_*|}\cdot\sigma \Big)\, d\sigma dv_*, \tag{1}$$

with

$$v' = \frac{m_i\,v + m_j\,v_*}{m_i+m_j} + \frac{m_j}{m_i+m_j}\,|v-v_*|\,\sigma, \qquad v'_* = \frac{m_i\,v + m_j\,v_*}{m_i+m_j} - \frac{m_i}{m_i+m_j}\,|v-v_*|\,\sigma. \tag{2}$$

In the case of collisions between polyatomic molecules, we define ([9], [2], [3]) (for $f := f(v,I) \geq 0, g := g(v,I) \geq 0$):

$$Q_{ij}(f,g)(v,I) = \int_{\mathbb{R}^3}\int_0^\infty \int_{S^2}\int_0^1\int_0^1 \Big\{ f(v',I')\,g(v'_*,I'_*) - f(v,I)\,g(v_*,I_*) \Big\} \tag{3}$$

$$\times\, B_{ij}\Big(\sqrt{E}, R^{1/2}\,|v-v_*|, \frac{v-v_*}{|v-v_*|}\cdot\sigma \Big)\,(1-R)\,R^{1/2}\,\varphi_i(I)^{-1} dr dR d\sigma dI_* dv_*,$$

with

$$v' = \frac{m_i\,v + m_j\,v_*}{m_i+m_j} + \frac{m_j}{m_i+m_j}\sqrt{\frac{2R\,E}{\mu_{ij}}}\,\sigma, \quad v'_* = \frac{m_i\,v + m_j\,v_*}{m_i+m_j} - \frac{m_i}{m_i+m_j}\sqrt{\frac{2R\,E}{\mu_{ij}}}\,\sigma, \tag{4}$$

$$I' = r\,(1-R)\,E, \qquad I'_* = (1-r)\,(1-R)\,E, \tag{5}$$

where $\mu_{ij} = \frac{m_i\,m_j}{m_i+m_j}$ is the reduced mass, $E = \frac{1}{2}\,\mu_{ij}\,|v-v_*|^2 + I + I_*$ is the total energy of the two molecules in the center of mass reference frame, and r, R lie in $[0,1]$.

In the case of collisions between polyatomic and monoatomic molecules, we define ([2], [3]) (for $f := f(v,I)$ and $g := g(v)$):

$$Q_{ij}(f,g)(v,I) = \int_{\mathbb{R}^3}\int_{S^2}\int_0^1 \Big\{ f(v',I')\,g(v'_*) - f(v,I)\,g(v_*) \Big\}$$

$$B_{ij}\Big(\sqrt{E}, R^{1/2}\,|v-v_*|, \frac{v-v_*}{|v-v_*|}\cdot\sigma \Big)\,R^{1/2}\,\varphi_i(I)^{-1} dR d\sigma dv_*, \tag{6}$$

with

$$v' = \frac{m_i\,v + m_j\,v_*}{m_i+m_j} + \frac{m_j}{m_i+m_j}\sqrt{\frac{2R\,E}{\mu_{ij}}}\,\sigma, \quad v'_* = \frac{m_i\,v + m_j\,v_*}{m_i+m_j} - \frac{m_i}{m_i+m_j}\sqrt{\frac{2R\,E}{\mu_{ij}}}\,\sigma, \tag{7}$$

$$I' = (1-R)\,E, \tag{8}$$

where $\mu_{ij} = \frac{m_i\,m_j}{m_i+m_j}$ is the reduced mass, $E = \frac{1}{2}\,\mu_{ij}\,|v-v_*|^2 + I$ is the total energy of the two molecules in the center of mass reference frame, and the parameter R lies in $[0,1]$.

We also define the symmetric operator (with the same cross section)

$$Q_{ji}(g,f)(v) = \int_{\mathbb{R}^3}\int_0^\infty \int_{S^2}\int_0^1 \Big\{ g(v')\,f(v'_*,I'_*) - g(v)\,f(v_*,I_*) \Big\}$$

$$\times B_{ij}\left(\sqrt{E}, R^{1/2}\,|v-v_*|, \frac{v-v_*}{|v-v_*|}\cdot\sigma\right) R^{1/2}\, dR d\sigma dv_* dI_*,$$

with

$$v' = \frac{m_j\,v + m_i\,v_*}{m_i+m_j} + \frac{m_i}{m_i+m_j}\sqrt{\frac{2R\,E}{\mu_{ij}}}\,\sigma, \quad v'_* = \frac{m_j\,v + m_i\,v_*}{m_i+m_j} - \frac{m_j}{m_i+m_j}\sqrt{\frac{2R\,E}{\mu_{ij}}}\,\sigma, \tag{9}$$

$$I'_* = (1-R)\,E, \tag{10}$$

where $\mu_{ij} = \frac{m_i\,m_j}{m_i+m_j}$ and $E = \frac{1}{2}\,\mu_{ij}\,|v-v_*|^2 + I_*$.

3. Chapman-Enskog expansion for a mixture of a mono-and polyatomic gases.

3.1. Macroscopic quantities.

Firstly we define the mass m_i of a molecule of species i, and recall the definition of macroscopic quantities:

The (macroscopic) mass of monoatomic species $i \in \{1, ..., A\}$:

$$\rho^{(i)} = m_i\, n^{(i)}(t,x) := \int_{\mathbb{R}^3} f^{(i)}(t,x,v)\, m_i\, dv.$$

The (macroscopic) mass of polyatomic species $i \in \{A+1, ..., A+B\}$:

$$\rho^{(i)} = m_i\, n^{(i)}(t,x) := \int_{\mathbb{R}^3}\int_0^\infty f^{(i)}(t,x,v,I)\, m_i\, \varphi_i(I)\, dI dv.$$

The momentum of monoatomic species $i \in \{1, ..., A\}$:

$$m_i\, n^{(i)}(t,x)\, u^{(i)}(t,x) := \int_{\mathbb{R}^3} f^{(i)}(t,x,v)\, m_i\, v\, dv.$$

The momentum of polyatomic species $i \in \{A+1, ..., A+B\}$:

$$m_i\, n^{(i)}(t,x)\, u^{(i)}(t,x) := \int_{\mathbb{R}^3}\int_0^\infty f^{(i)}(t,x,v,I)\, m_i\, v\, \varphi_i(I)\, dI dv.$$

The (macroscopic, internal) energy of monoatomic species $i \in \{1, ..., A\}$:

$$m_i\, n^{(i)}(t,x)\, e^{(i)}(t,x) := \int_{\mathbb{R}^3} f^{(i)}(t,x,v)\, m_i\, \frac{|v-u^{(i)}(t,x)|^2}{2}\, dv.$$

The (macroscopic, internal) energy of polyatomic species $i \in \{A+1, ..., A+B\}$:

$$m_i\, n^{(i)}(t,x)\, e^{(i)}(t,x) := \int_{\mathbb{R}^3}\int_0^\infty f^{(i)}(t,x,v,I)\left(m_i\,\frac{|v-u^{(i)}(t,x)|^2}{2} + I\right)\varphi_i(I)\, dI dv.$$

3.2. Linearized Boltzmann operator.

We first introduce the scalar product that will be used throughout the paper. Given two vectors $\underline{k} = (k^{(1)}, ..., k^{(A+B)})$ and $\underline{l} = (l^{(1)}, ..., l^{(A+B)})$, with $k^{(1)}, ..., k^{(A)}$, $l^{(1)}, ..., l^{(A)}$ functions of V, and $k^{(A+1)}, ..., k^{(A+B)}$, $l^{(A+1)}, ..., l^{(A+B)}$ functions of V, J, we define

$$\langle \underline{k} |\, \underline{l}\rangle := \sum_{i=1}^{A} n^{(i)} \int_{\mathbb{R}^3} \frac{e^{-m_i\frac{|V|^2}{2}}}{(2\pi/m_i)^{3/2}}\, k^{(i)}(V)\, l^{(i)}(V)\, dV$$

$$+ \sum_{i=A+1}^{A+B} n^{(i)} \int_0^\infty\int_{\mathbb{R}^3} \frac{e^{-(m_i\frac{|V|^2}{2}+J)}}{(2\pi/m_i)^{3/2}}\, k^{(i)}(V,J)\, l^{(i)}(V,J)\, \frac{T\,\varphi_i(J\,T)}{q_i(T)}\, dV dJ. \tag{11}$$

Next, we consider the linearized Boltzmann operator $\mathcal{K}$ that is obtained by linearizing the Boltzmann operator described in section 2 around its equilibrium states given in (17). It can also be proved that $\mathcal{K}$ is symmetric for the scalar product (3.2).

The kernel $\mathbb{K}$ of $\mathcal{K}$ can easily be found (provided that all cross sections B_{ij} are strictly positive). It is constituted by the vectors $\underline{l}^{\Delta,j}$ ($j = 1, ..., A+B$), $\underline{l}^{U,z}$ ($z = 1,2,3$) and $\underline{l}^{E}$ ([9]), defined as

$$\underline{l}^{\Delta,j} = \begin{pmatrix} l^{(1),\Delta,j} \\ . \\ . \\ . \\ l^{(j),\Delta,j} \\ . \\ . \\ . \\ l^{(A+B),\Delta,j} \end{pmatrix} = \begin{pmatrix} 0 \\ 0 \\ . \\ 0 \\ 1 \\ 0 \\ . \\ 0 \\ 0 \end{pmatrix}, \tag{12}$$

$$\underline{l}^{U,z} = \begin{pmatrix} l^{(1),U,z} \\ . \\ . \\ l^{(A+B),U,z} \end{pmatrix} = \begin{pmatrix} m_1 V_z \\ . \\ . \\ m_{A+B} V_z \end{pmatrix}, \tag{13}$$

$$\underline{l}^{E} = \begin{pmatrix} l^{(1),E} \\ . \\ . \\ l^{(A+B),E} \end{pmatrix} = \begin{pmatrix} m_1 \frac{V^2}{2} + r_1 J \\ . \\ . \\ m_{A+B} \frac{V^2}{2} + r_{A+B} J \end{pmatrix}. \tag{14}$$

3.3. **Principle of the expansion.** We first define the rescaled (w.r.t the Knudsen number) system of Boltzmann equations

$$\partial_t f^{(i)} + v \cdot \nabla_x f^{(i)} = \frac{1}{\varepsilon} \sum_{j=1}^{A+B} Q_{ij}(f^{(i)}, f^{(j)}), \quad i = 1, ..., A+B, \tag{15}$$

where the operators Q_{ij} are defined by formulas (1), (6), (3).

We then look for solutions of the Boltzmann equation (15) under the form

$$f^{(i)} = M_\varepsilon^{(i)} \, (1 + \varepsilon \, g_\varepsilon^{(i)}), \tag{16}$$

where $M_\varepsilon^{(i)}$ is a Maxwellian distribution of (number) density $n_\varepsilon^{(i)} := n_\varepsilon^{(i)}(t,x) \geq 0$, macroscopic velocity $u_\varepsilon := u_\varepsilon(t,x) \in \mathbb{R}^3$, and temperature $T_\varepsilon := T_\varepsilon(t,x) \geq 0$, that is

$$M_\varepsilon^{(i)} = \frac{n_\varepsilon^{(i)}}{(2\pi \, T_\varepsilon/m_i)^{3/2} \, q_i(T_\varepsilon)} \exp\left(- \frac{m_i \, |v - u_\varepsilon|^2 + 2 \, r_i \, I}{2 \, T_\varepsilon} \right), \tag{17}$$

with $r_i = 0$ for $i = 1, ..., A$ and $r_i = 1$ for $i = A+1, ..., A+B$.
In formula (17), $q_i(T) = 1$ for $i = 1, ..., A$ and for $i = A+1, ..., A+B$,

$$q_i(T) = \int_0^{+\infty} \varphi_i(I) e^{-\frac{I}{T}} \, dI.$$

We also assume that the vector of perturbed distributions $\underline{g} = (g_\varepsilon^{(1)}, ..., g_\varepsilon^{(A+B)})$, with functions $g_\varepsilon^{(i)} := g_\varepsilon^{(i)}(t,x,v) \in \mathbb{R}$ for $i = 1, ..., A$, and $g_\varepsilon^{(i)} := g_\varepsilon^{(i)}(t,x,v,I) \in \mathbb{R}$ for $i = A+1, ..., A+B$, satisfies

$$\forall i = 1, ..., A+B, \quad \langle \underline{g} \, | \, \underline{l}^{\Delta,i} \rangle = 0, \tag{18}$$

$$\forall z = 1, ..., 3, \quad \langle \underline{g} \,|\, \underline{l}^{U,z} \rangle = 0, \tag{19}$$

$$\langle \underline{g} \,|\, \underline{l}^{E} \rangle = 0, \tag{20}$$

where $\langle \cdot \,|\, \cdot \rangle$ is the scalar product defined in (3.2) and vectors $\underline{l}^{\Delta,i}$, $\underline{l}^{U,z}$, $\underline{l}^{E}$ are provided in (12, 13, 14). Taking $f^{(i)}$ as the Maxwellian distribution (16, 17) leads to the compressible Euler system ([9]).

4. Navier-Stokes system.

4.1. Computation of the l.h.s of the linear equation.

A straightforward computation gives ([2], [3])

$$\begin{aligned}(M^{(i)})^{-1}\,[\partial_t M^{(i)} + v \cdot \nabla_x M^{(i)}] \;&=\; k^{(i),W} + k^{(i),P} : \left(\frac{\nabla_x u + \nabla_x u^T}{2} \right) \\ &+\; k^{(i),D}\,(\nabla_x \cdot u) + k^{(i),Q} \cdot \frac{\nabla_x T}{\sqrt{T}},\end{aligned} \tag{21}$$

where $k^{(i),W}$, $k^{(i),D}$, $k^{(i),Q} = (k^{(i),Q,p})_{p \in \{1;3\}}$, $k^{(i),P} = (k^{(i),P,p,q})_{p,q \in \{1;3\}}$ write

$$\underline{k}^{P,p,q} = \begin{pmatrix} k^{(1),P,p,q} \\ \cdot \\ \cdot \\ k^{(A+B),P,p,q} \end{pmatrix} = \begin{pmatrix} P_{pq}(V)\, m_1 \\ \cdot \\ \cdot \\ P_{pq}(V)\, m_{A+B} \end{pmatrix},$$

$$\underline{k}^{W} = \begin{pmatrix} k^{(1),W} \\ \cdot \\ \cdot \\ k^{(A+B),W} \end{pmatrix} = \begin{pmatrix} \sqrt{T} n \frac{d^{(1)}}{n^{(1)}} \cdot V \\ \cdot \\ \cdot \\ \sqrt{T} n \frac{d^{(A+B)}}{n^{(A+B)}} \cdot V \end{pmatrix},$$

$$\underline{k}^{Q,p} = \begin{pmatrix} k^{(1),Q,p} \\ \cdot \\ \cdot \\ k^{(A+B),Q,p} \end{pmatrix} = \begin{pmatrix} V_p \left(\frac{m_1}{2} V^2 + r_1 J - (\frac{5}{2} + r_1 \frac{\overline{E}_1}{T}) \right) \\ \cdot \\ \cdot \\ V_p \left(\frac{m_{A+B}}{2} V^2 + r_{A+B} J - (\frac{5}{2} + r_{A+B} \frac{\overline{E}_{A+B}}{T}) \right) \end{pmatrix},$$

and

$$\underline{k}^{D} = \begin{pmatrix} k^{(1),D} \\ \cdot \\ \cdot \\ k^{(A+B),D} \end{pmatrix} = \begin{pmatrix} 2 r_1 \Lambda(T) \left(\frac{\overline{E}_1}{T} - J \right) + 2 \left(\frac{1}{3} - \Lambda(T) \right) \left(m_1 \frac{V^2}{2} - \frac{3}{2} \right) . \\ \cdot \\ \cdot \\ 2 r_{A+B} \Lambda(T) \left(\frac{\overline{E}_{A+B}}{T} - J \right) + 2 \left(\frac{1}{3} - \Lambda(T) \right) \left(m_{A+B} \frac{V^2}{2} - \frac{3}{2} \right) \end{pmatrix},$$

with

$$P(v) = v \otimes v - \frac{1}{3}\,|v|^2\, Id, \quad d^{(i)} = \nabla_x \left(\frac{p_i}{p}\right) + \left(\frac{p_i}{p} - \frac{\rho^{(i)}}{\sum_{j=1}^{A+B} \rho^{(j)}}\right) \frac{\nabla_x p}{p}$$

and

$$\overline{E}_i = \frac{\eta_i(T)}{q_i(T)}, \; \eta_i(T) = \int_0^\infty I\, \varphi_i(I)\, e^{-I/T}\, dI, \; i \in \{A+1, ..., A+B\},$$

$$\Lambda(T) = \frac{\sum_{j=1}^{A+B} n^{(j)}}{3 \sum_{j=1}^{A+B} n^{(j)} + 2 \sum_{j=A+1}^{A+B} n^{(j)} \left(\frac{\eta_j}{q_j} \right)' (T)}.$$

Next, thanks to the Galilean invariance ([8]), we can write for $i = 1, ..., A$:

$$h^{(i),P,p,q}(V) = \tilde{h}^{(i),P}(|V|)\, P_{pq}(V),$$

$$h^{(i),Q,p}(V) = \tilde{h}^{(i),Q}(|V|)\, V_p,\ h^{(i),D}(V) = \tilde{h}^{(i),D}(|V|),$$

and for $i = A+1, ..., A+B$:

$$h^{(i),P,p,q}(V,J) = \tilde{h}^{(i),P}(|V|,J)\, P_{pq}(V), \qquad h^{(i),Q,p}(V,J) = \tilde{h}^{(i),Q}(|V|,J)\, V_p,$$
$$h^{(i),D}(V,J) = \tilde{h}^{(i),D}(|V|,J).$$

Thanks to the computations (21), the previous definitions give

$$i = 1, ..., A \qquad g^{(i)}(V\sqrt{T}+u) = \tilde{h}^{(i),P}(|V|)\, P(V) : \left(\frac{\nabla_x u + \nabla_x u^T}{2}\right) \tag{22}$$
$$+\tilde{h}^{(i),D}(|V|)\, \nabla_x \cdot u + \tilde{h}^{(i),Q}(|V|)\, V \cdot \frac{\nabla_x T}{\sqrt{T}} + \sqrt{T}\, h^{(i),W}(V),$$

$$i = A+1, ..., A+B \qquad g^{(i)}(V\sqrt{T}+u, JT) = \tilde{h}^{(i),P}(|V|,J)\, P(V) : \left(\frac{\nabla_x u + \nabla_x u^T}{2}\right) \tag{23}$$
$$+\tilde{h}^{(i),D}(|V|,J)\, \nabla_x \cdot u + \tilde{h}^{(i),Q}(|V|,J)\, V \cdot \frac{\nabla_x T}{\sqrt{T}} + \sqrt{T}\, h^{(i),W}(V,J).$$

4.2. **Navier-Sokes system.** We begin by considering, for $i = 1, ..., A$ and $k = 1, ..., 3$ the mass flux

$$D_k^{(i)} := \int_{\mathbb{R}^3} M^{(i)}\, g^{(i)}\, m_i\, v_k\, dv. \tag{24}$$

In the same way, for $i = A+1, ..., A+B$ and $k = 1, ..., 3$, we get

$$D_k^{(i)} := \int_0^{+\infty} \int_{\mathbb{R}^3} M^{(i)}\, g^{(i)}\, m_i\, v_k\, \varphi_i(I)\, dv dI.$$

We then consider the stress tensor F_{kl}, for $k, l = 1, ..., 3$,

$$F_{kl} := \sum_{i=1}^{A} \int_{\mathbb{R}^3} M^{(i)}\, g^{(i)}\, m_i\, v_k\, v_l\, dv + \sum_{i=A+1}^{A+B} \int_0^{\infty} \int_{\mathbb{R}^3} M^{(i)}\, g^{(i)}\, m_i\, v_k\, v_l\, \varphi_i(I)\, dv dI. \tag{25}$$

We finally consider (for $k = 1, ..., 3$)

$$\begin{aligned} G_k &= \sum_{i=1}^{A} \int_{\mathbb{R}^3} M^{(i)}\, g^{(i)}\, m_i\, \frac{|v|^2}{2}\, v_k\, dv \\ &+ \sum_{i=A+1}^{A+B} \int_0^{\infty} \int_{\mathbb{R}^3} M^{(i)}\, g^{(i)} \left(m_i\, \frac{|v|^2}{2} + I\right) v_k\, \varphi_i(I)\, dv dI. \end{aligned} \tag{26}$$

We finally write down the Navier-Stokes system in the following semi-explicit form:

$$i = 1, ..., A+B \qquad \partial_t(m_i\, n^{(i)}) + \nabla_x \cdot (m_i\, n^{(i)} u) = -\varepsilon \nabla_x \cdot D^{(i)}, \tag{27}$$

while, for $k = 1, 2, 3$,

$$\partial_t\Big(\sum_{i=1}^{A+B} m_i\, n^{(i)}\, u_k\Big) + \sum_l \partial_{x_l}\Big(\sum_{i=1}^{A+B} [m_i\, n^{(i)}\, u_k\, u_l + n^{(i)}\, T\, \delta_{kl}]\Big) = -\varepsilon \sum_l \partial_{x_l} F_{kl}, \tag{28}$$

$$\partial_t\Big(\sum_{i=1}^{A}[m_i\,n^{(i)}\,\frac{|u|^2}{2}+\frac{3}{2}n^{(i)}\,T]+\sum_{i=A+1}^{A+B}[m_i\,n^{(i)}\,\frac{|u|^2}{2}+n^{(i)}\left[\frac{3}{2}T+\frac{\eta_i(T)}{q_i(T)}\right]]\Big) \quad (29)$$

$$+\sum_l \partial_{x_l}\Big(\sum_{i=1}^{A}[m_i\,n^{(i)}\,\frac{|u|^2}{2}\,u_l+\frac{5}{2}\,n^{(i)}\,T\,u_l]+\sum_{i=A+1}^{A+B}[m_i\,n^{(i)}\,\frac{|u|^2}{2}\,u_l+n^{(i)}\,u_l\left[\frac{5}{2}\,T+\frac{\eta_i(T)}{q_i(T)}\right]]\Big)$$
$$=-\varepsilon\nabla_x\cdot G.$$

Introduce the specific enthalpy of the ith species h_i by

$$h_i=(\frac{5}{2}T+r_i\overline{E}_i)\frac{1}{m_i}. \quad (30)$$

In that case, G_k writes

$$G_k=\sum_{l=1}^{3}F_{kl}\,u_l-\lambda\partial_{x_k}T-p\sum_{i=1}^{A+B}\theta_i d^{(i)}+\sum_{i=1}^{A+B}h_i D_k^{(i)}, \quad (31)$$

with

$$\theta_i:=-\frac{T}{n^{(i)}}\langle\mathcal{K}^{-1}(\underline{k}^{Q,k}),\psi^{D_i}\rangle,\quad \lambda:=-T\langle\mathcal{K}^{-1}(\underline{k}^{Q,k}),\underline{k}^{Q,k}\rangle. \quad (32)$$

Next by using the definition $\underline{h}^W$ and the symmetry of $\mathcal{K}^{-1}$, $D_k^{(i)}$, it comes that

$$D_k^{(i)}=-\rho_i\theta_i\partial_{x_k}\ln(T)-\sum_{j=1}^{A+B}C_{ji}d^{(j)}, \quad (33)$$

with

$$C_{ji}=-m_iT\frac{n}{n_j}\langle\psi^{D_j},\mathcal{K}^{-1}(\psi^{D_i})\rangle.$$

The coefficients (θ_j) are the thermal diffusion coefficients whereas the terms C_{ji} correspond to the multicomponent flux diffusion coefficients.

5. **Computation in the case of constant cross sections.** Moreover, in order to be coherent with the fact that in the air, the main polyatomic species (that is, O_2 and N_2) are in fact diatomic, we have for $i=A+1,...,A+B$, $\varphi_i(I)=1$, and $q_i(T)=T$. Moreover, the quantities $\underline{k}^W$, $\underline{k}^P$, $\underline{k}^D$, and $\underline{k}^Q$ can be written in the following way:

$$k^{(i),W}=s^{(i)}\,V,\quad k^{(i),P,p,q}=m_i\,P_{pq}(V),\quad k^{(i),Q,p}=\left(m_i\frac{|V|^2}{2}-\frac{5}{2}\right)V_p+r_i\,(J-1)\,V_p,$$

$$k^{(i),D}=(m_i|V|^2-3)\left(\frac{1}{3}-\Lambda\right)-2r_i\,\Lambda\,(J-1).$$

When the cross sections B_{ij} are constant, the functions $h^{(i),W}$, $\tilde{h}^{(i),P}$, $\tilde{h}^{(i),D}$ and $\tilde{h}^{(i),Q}$ may be cast in compact form, for $i=1,...,A+B$, as

$$\begin{aligned} h^{(i),W}&=m_i\,W^{(i)}\cdot V,\quad \tilde{h}^{(i),P}=m_i\,\Pi^{(i)},\\ \tilde{h}^{(i),D}&=\Delta^{(i)}\big(m_i|V|^2-3\big)+r_i\,\widetilde{\Delta}^{(i)}(J-1),\\ \tilde{h}^{(i),Q}&=Q^{(i)}\big(m_i|V|^2-5\big)+r_i\,\widetilde{Q}^{(i)}(J-1),\end{aligned} \quad (34)$$

where the constant coefficients $W^{(i)}$, $\Pi^{(i)}$, $\Delta^{(i)}$, $\widetilde{\Delta}^{(i)}$, $Q^{(i)}$, $\widetilde{Q}^{(i)}$ fulfil suitable linear systems. Hence we get

$$D_k^{(i)}=m_i\,W_k^{(i)}\,n^{(i)}\,T,\qquad i=1,...,A+B. \quad (35)$$

F_{kl} may be cast as

$$F_{kl} = -\mu\left[\frac{\nabla_x u + \nabla_x u^T}{2} - \frac{1}{3}\nabla_x \cdot u\, Id\right]_{kl} - \kappa\,\nabla_x \cdot u\,\delta_{kl}\,,$$

where

$$\mu = -\,2\,T\sum_{i=1}^{A+B} n^{(i)}\,\Pi^{(i)}$$

represents the shear viscosity and the bulk viscosity κ is provided by the formula

$$\kappa = -\,2\,T\sum_{i=1}^{A+B} n^{(i)}\,\Delta^{(i)}\,.$$

Finally, G_k can be written as

$$G_k = \sum_{l=1}^{3} F_{kl}\,u_l - \lambda\,\partial_{x_k} T + \sum_{i=1}^{A+B} h_i D_k^{(i)}, \qquad \lambda = -\,T\sum_{i=1}^{A+B}\frac{n^{(i)}}{m_i}\left(5\,Q^{(i)} + r_i\,\widetilde{Q}^{(i)}\right).$$

Moreover, by comparison with (31), the Dufour and the Soret terms yield zero.

REFERENCES

[1] P. Andries, P. Le Tallec, J.P. Perlat, B. Perthame, Entropy condition for the ES BGK model of Boltzmann equation for mono and polyatomic gases, *Eur. J. Mech. B/fluids*, **19**, (2000), 813-830

[2] C.Baranger, M.Bisi, S.Brull, L.Desvillettes, The Chapman-Enskog asymptotics of mixture of monoatomic and polyatomic rarefied gases, *Kin. Rel. Mod.*, **11**, (2018), 821-858

[3] C.Baranger, M.Bisi, S.Brull, L.Desvillettes, The Chapman-Enskog asymptotics of mixture of monoatomic and polyatomic rarefied gases, *to appear in the proceeding of RGD*

[4] C. Borgnakke, P.S. Larsen, Statistical collision model for Monte-Carlo simulation of polyatomic mixtures, *Journ. Comput. Phys.*, **18**, (1975), 405-420

[5] J.F. Bourgat, L. Desvillettes, P. Le Tallec B. Perthame, Microreversible collisions for polyatomic gases and Boltzmann's theorem, *Eur. J. Mech. B/ Fluids*, **13**, (1994), 237-254

[6] S.Brull, J.Schneider, On the Ellipsoidal Statistical Model for polyatomic gases, *Cont. Mech. Thermodyn.*, **20**, (2009), 489-508

[7] L. Desvillettes, Sur un modèle de type Borgnakke-Larsen conduisant à des lois d'énergie non-linéaires en température pour les gaz parfaits polyatomiques, *Ann. Fac. Sci. Toulouse Math.*, **6**, (1997), 257-262

[8] L. Desvillettes, F. Golse, Advances in Kinetic Theory and Computing, Series on Advances in Mathematics for Applied Sciences, *World Scientific Publications*, Singapour, (1994)

[9] L. Desvillettes, R. Monaco, F. Salvarani, A kinetic model allowing to obtain the energy law of polytropic gases in the presence of chemical reactions, *Eur. J. Mech. B/Fluids*, **24**, (2005), 219-236

[10] A. Ern, V. Giovangigli, The kinetic equilibrium regime, *Physica A*, **260**, (1998), 49-72

[11] H. Funagane, S.Takata, K.Aoki, K.Kugimoto, Poiseuille flow and thermal transpiration of a rarefied polyatomic gas through a circular tube with applications to microflows, *Boll. Unione Mat. Ital.*, **9**, (2011), 19-46

[12] V. Giovangigli, *Multicomponent flow modeling*, MESST Series, Birkhauser Boston, (1999)

[13] S. Kosuge, K. Aoki, Shock-wave structure for a polyatomic gas with large bulk viscosity, *Phys. review fluids*, **3**, 1-42, (2018)

[14] S. Kosuge, K. Aoki, T. Goto, *Shock wave structure in polyatomic gases: Numerical analysis using a model Boltzmann equation*, AIP Conf.Proc., (2016).

[15] S. Takata, H. Funagane, K.Aoki, Fluid modeling for the Knudsen compressor: case of polyatomic gases, *Kinet. Relat. Models*, **3**, (2010), 353-372

E-mail address: celine.baranger@cea.fr
E-mail address: marzia.bisi@unipr.it
E-mail address: Stephane.Brull@math.u-bordeaux1.fr
E-mail address: desvillettes@math.univ-paris-diderot.fr

STATIONARY STATES OF FINITE VOLUME DISCRETIZATIONS OF MULTI-DIMENSIONAL LINEAR HYPERBOLIC SYSTEMS

WASILIJ BARSUKOW

Institut für Mathematik, Zürich University
Winterthurer Strasse 190
8057 Zürich, Switzerland *and*
Institut für Mathematik, Würzburg University
Emil-Fischer-Straße 40
97074 Würzburg, Germany

ABSTRACT. In multiple spatial dimensions linear hyperbolic systems have stationary states given by differential constraints. This paper shows that finite volume discretizations of such systems typically introduce diffusion even if the setup should remain stationary. Recently characterized schemes, called *stationarity preserving*, on the other hand keep stationary a discretization of all the analytic stationary states. The behaviour of the two classes of schemes is discussed in detail and their abilities compared in numerical simulations.

1. **Introduction.** This paper considers the initial value problem for linear hyperbolic $n \times n$ systems in d spatial dimensions in the following general form:

$$\partial_t q + (\mathbf{J} \cdot \nabla) q = 0 \qquad q : \mathbb{R}^+_0 \times \mathbb{R}^d \to \mathbb{R}^n \tag{1}$$

$$q(0, \mathbf{x}) = q_0(\mathbf{x}) \tag{2}$$

Each entry of the vector $\mathbf{J}$ is the Jacobian matrix into the corresponding direction; e.g. in 3 spatial dimensions[1]:

$$\mathbf{J} \cdot \nabla = J^x \partial_x + J^y \partial_y + J^z \partial_z \tag{3}$$

The system is hyperbolic if the linear combination $\mathbf{k} \cdot \mathbf{J}$ of the Jacobians is diagonalizable with real eigenvalues for all $\mathbf{k} \in \mathbb{R}^d$. One or several eigenvalues can vanish; such hyperbolic systems are of special interest here, for reasons explained below.

Even linear hyperbolic systems (1) can have complicated properties in multiple spatial dimensions that are hard to capture numerically. As an example consider

---

2000 *Mathematics Subject Classification.* MSC 35L40, MSC 65M06, MSC 65M08, MSC 39A70.

Key words and phrases. Stationarity preserving, linear acoustics, system wave equation, vorticity preserving, involution.

The author acknowledges support of the German Academic Exchange Service (DAAD) with funds from the German Federal Ministry of Education and Research (BMBF) and the European Union (FP7-PEOPLE-2013-COFUND – grant agreement no. 605728).

[1]In this paper, boldface letters denote vectors with d components. Unless stated differently, the components of the vector are given the same letter endowed with an index: e.g. in 3 spatial dimensions $\mathbf{k} = (k_x, k_y, k_z)$. Sometimes upper indices are used; nowhere in the paper does an index denote a derivative. $\mathbb{i}$ is the imaginary unit.

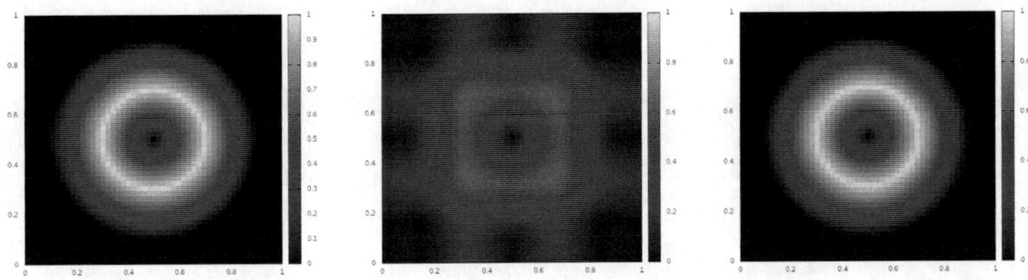

FIGURE 1. Stationary vortex setup for the acoustic equations (4)–(5). Color coded is the absolute value of the velocity. *Left*: exact stationary solution: constant pressure and divergencefree velocity. *Center*: Solution at time $t = 2$ with the upwind/Roe scheme. *Right*: Solution at time $t = 2$ using the stationarity preserving scheme (16) (see below).

linear acoustics

$$\partial_t \mathbf{v} + \nabla p = 0 \qquad \mathbf{v} : \mathbb{R}_0^+ \times \mathbb{R}^d \to \mathbb{R}^d \qquad q = (\mathbf{v}, p) \tag{4}$$

$$\partial_t p + c^2 \nabla \cdot \mathbf{v} = 0 \qquad p : \mathbb{R}_0^+ \times \mathbb{R}^d \to \mathbb{R} \tag{5}$$

This system arises upon linearization of the Euler equations around a constant and static state of the fluid.

If the initial data for (4)–(5) fulfill $p = \text{const}$ and $\operatorname{div} \mathbf{v} = 0$, then they remain stationary. This poses challenges to numerical methods. Numerical analysis seeks discretizations of the differential operators which operate on only a finite set of values of q. The discretization thus has access to only a reduced amount of information. $p = \text{const}$ can be easily represented on a numerical grid, $\operatorname{div} \mathbf{v}$ cannot. Considering a numerical scheme for (4)–(5) the question is not whether it keeps stationary all vector fields whose divergence vanishes, but what "divergence" means in the discrete. There are, for instance, many different *discretizations* of the divergence. In [2] it has been shown that for many schemes *no* discrete divergence exists that would lead to a discrete stationary state. Certain schemes, however, are able to keep the initial data exactly stationary if some discrete divergence vanishes. Such schemes are called *stationarity preserving*. In a sense, keeping stationary *one* discretization of the divergence is the best one can hope for: stationary states of (1) are governed by differential operators that cannot be evaluated exactly with the limited information available in the discrete setting.

Figure 1 shows a vortex setup whose exact solution is stationary. Simulation results of a stationarity preserving scheme and of a not stationarity preserving scheme are shown. One clearly observes the superiority of the former.

There exist numerical schemes for (4)–(5) which do not keep *any* discrete divergence stationary. It would be wrong to think, however, that they do not have stationary states *at all*. The constant state $q = \text{const}$, for example, is easily captured by virtually any kind of numerical scheme. This is why in [2] *trivial* and *non-trivial* stationary states are defined. The former are generally found not to present particular numerical difficulties. To capture non-trivial stationary states is more challenging. For the acoustic system for example, they are the ones related to $\operatorname{div} \mathbf{v} = 0$.

This paper aims at providing a detailed study of numerical stationary states for (1) and giving insight into the difference between trivial and non-trivial stationary states. The stationary states of numerical schemes are analyzed with techniques from [2]. The theoretical statements are illustrated by careful measurements in actual simulations. It is shown that the framework is able to explain details of the behaviour of numerical simulations of (1).

The paper is organized as follows: in section 2 stationary states and their relation to involutions of (1) are discussed in the continuous setting. In section 3 the behaviour of numerical schemes in the context of stationary states is discussed and 4 provides numerical experiments that illustrate the theory.

2. The relation between involutions and nontrivial stationary states.

In order to study (1) and also linear numerical schemes for (1) the Fourier transform is of great help, as both the differential and the finite difference operators turn into algebraic factors. One expresses q as

$$q(t, \mathbf{x}) = \int \mathrm{d}\mathbf{k}\, \hat{q}(t, \mathbf{k}) \exp(\mathbb{i}\mathbf{k} \cdot \mathbf{x}) \tag{6}$$

and $\hat{q} : \mathbb{R}_0^+ \times \mathbb{R}^d \to \mathbb{R}^n$ is referred to as the Fourier transform[2] of q.

It is instructive to restrict the analysis first on just one Fourier mode $\hat{q}(t, \mathbf{k}) \exp(\mathbb{i}\mathbf{k} \cdot \mathbf{x})$ corresponding to some wave vector $\mathbf{k}$. Such a Fourier mode is a solution of (1) if it evolves according to

$$\partial_t \hat{q}(t, \mathbf{k}) + \mathbb{i}\mathbf{J} \cdot \mathbf{k}\hat{q}(t, \mathbf{k}) = 0 \tag{7}$$

A Fourier mode $\hat{q}(t, \mathbf{k}) \exp(\mathbb{i}\mathbf{k} \cdot \mathbf{x})$ is a stationary solution of (1) if $(\mathbf{J} \cdot \mathbf{k})\hat{q} = 0$. (Recall that $\mathbf{J} \cdot \mathbf{k}$ is a $n \times n$ matrix and $\hat{q}(t, \mathbf{k}) \in \mathbb{R}^n$.)

Definition 2.1. The system (1) possesses *non-trivial stationary states* if for all $\mathbf{k}$ there exists $\hat{q}(\mathbf{k}) \neq 0$ such that the Fourier mode $\hat{q}(\mathbf{k}) \exp(\mathbb{i}\mathbf{k} \cdot \mathbf{x})$ is a stationary solution of (1). If stationary modes can only be obtained by restricting $\mathbf{k}$, then (1) has only *trivial* stationary states.

Theorem 2.2. *Non-trivial stationary states exist if* $\det(\mathbf{k} \cdot \mathbf{J}) = 0 \quad \forall \mathbf{k}$.

The existence of non-trivial stationary states turns out to single out hyperbolic systems with particular properties. For example, by taking the curl of (4) it follows that the *vorticity* $\nabla \times \mathbf{v}$ is always stationary, even if the solution is not:

$$\partial_t(\nabla \times \mathbf{v}) = 0 \tag{8}$$

Such a function generally is called *involution* (see e.g. [4]):

Theorem 2.3 (Involution). *Iff* (1) *possesses non-trivial stationary states, then it possesses an involution* Ωq *such that* $\partial_t(\Omega q) = 0$ *for any initial data, even if the solution itself is not stationary.*

Proof. See [2]. Note that in general Ω is a differential operator. □

[2]This is to be understood in the sense of distributions (as in [3]). However, such an approach would unnecessarily obscure the presentation. Here one thus has to restrict oneself to functions $q(t, \mathbf{x})$ which have a Fourier transform in the sense of functions.

Example 1. *In case of the acoustic equations (4)–(5) in two spatial dimensions and writing* $\hat{q} = (\hat{u}, \hat{v}, \hat{p})^T$ *one finds*

$$(\mathbf{J} \cdot \mathbf{k})\hat{q} = \begin{pmatrix} ck_x\hat{p} \\ ck_y\hat{p} \\ c(k_x\hat{u} + k_y\hat{v}) \end{pmatrix} \tag{9}$$

One can put conditions on $\mathbf{k}$ *for this vector to vanish. For example, if* $k_x = 0 \wedge \hat{v} = 0 \wedge \hat{p} = 0$*, i.e. the Fourier mode is stationary if* u *only depends on* y *and the remaining components of* q *vanish.*

However, if $\hat{p} = 0$ *and* $(\hat{u}, \hat{v})$ *are chosen such that* $k_x\hat{u} + k_y\hat{v} = 0$ *(divergencefree) then the mode is stationary* ***for all*** $\mathbf{k}$*. Observe that* $\mathbf{J} \cdot \mathbf{k}$ *has a zero eigenvalue, and this choice is such that* $\hat{q}$ *is parallel to the corresponding eigenvector*

$$e_0 = (k_y, -k_x, 0)^T \tag{10}$$

On the other hand, the left eigenvector $(k_y, -k_x, 0)(\mathbf{J}\cdot\mathbf{k}) = 0$ *implies* $\partial_t(k_y\hat{u} - k_x\hat{v}) = 0$ *which is the Fourier transform of* $\partial_t(\nabla \times \mathbf{v}) = 0$*. The involution of the acoustic system is a stationary vorticity* $\nabla \times \mathbf{v}$.

3. Discrete stationary states and involutions.

Having discussed the stationary states of hyperbolic systems, this section now focuses on their discretization. Consider an equidistant Cartesian computational grid in two spatial dimensions[3] with cell spacings Δx, Δy. The cells are indexed by integers i, j; $q_{ij}(t)$ denotes the value of q in cell (i, j) at time t, i.e. $q_{ij} : \mathbb{R}^+_0 \to \mathbb{R}^n$.

On an equidistant grid one can study the evolution of a discrete Fourier mode, in 2-d given by

$$\hat{q}(t, \mathbf{k}) \exp\Big(\mathbb{i}(k_x i\Delta x + k_y j\Delta y)\Big) \tag{11}$$

Similarly to the continuous Fourier transform, the complete solution can be constructed by considering a linear combination of discrete Fourier modes with $\mathbf{k}$ chosen from a countable set. For example the behaviour of the vortex shown in figure 1 can be understood as the combined evolution of individual discrete Fourier modes.

A semidiscrete (time-continuous) linear finite-difference scheme describing the evolution of q_{ij} replaces the spatial derivatives by linear functions of the variables $\{(r, s) \in \mathbb{Z}^2 | q_{i+r,j+s}\}$ (in two spatial dimensions). Just as differential operators become algebraic factors upon the (continuous) Fourier transform, shifts $q_{i+r,j+s}$ of a Fourier mode q_{ij} can now be rewritten as $\exp(\mathbb{i}r\Delta x + \mathbb{i}s\Delta x)q_{ij}$. It is helpful to introduce the shift operators

$$t_x = \exp(\mathbb{i}k_x\Delta x) \qquad\qquad t_y = \exp(\mathbb{i}k_y\Delta y) \qquad \text{for } d = 2 \tag{12}$$

Any linear scheme solving (1) consists of linear combinations of such shifts. As upon the Fourier transform they become factors, for any such scheme there exists a matrix $\mathcal{E}$ (called *evolution matrix*) describing the evolution of a Fourier mode (11) as

$$\partial_t\hat{q}(t, \mathbf{k}) + \mathcal{E}\hat{q} = 0 \tag{13}$$

This matrix is constructed explicitly in [2]. In general, it depends on t_x, t_y and thus on $\mathbf{k}$. It is the counterpart of $\mathbb{i}\mathbf{k} \cdot \mathbf{J}$ in (7).

[3] The concepts are detailed in [2] for any number of dimensions but are expressed here in two spatial dimensions for the ease of presentation.

Example 2. *Consider a directionally split semi-discrete upwind scheme for* (1) *in two spatial dimensions*

$$\begin{aligned}\partial_t q_{ij} &+ \frac{1}{2\Delta x}\Big(J^x(q_{i+1,j}-q_{i-1,j}) - D^x(q_{i+1,j}-2q_{ij}+q_{i-1,j})\Big)\\ &+ \frac{1}{2\Delta y}\Big(J^y(q_{i,j+1}-q_{i,j-1}) - D^y(q_{i,j+1}-2q_{ij}+q_{i,j-1})\Big) = 0\end{aligned} \tag{14}$$

with $D^x = |J^x|$, $D^y = |J^y|$. *If scheme* (14) *is applied to the acoustic equations* (4)–(5), *the Fourier mode* (11) *is evolving according to* $\partial_t \hat{q} + \mathcal{E}\hat{q} = 0$ *with*

$$\mathcal{E} = \begin{pmatrix} -\frac{c(t_x-1)^2}{2\Delta x t_x} & 0 & \frac{(t_x-1)(t_x+1)}{2\Delta x t_x} \\ 0 & -\frac{c(t_y-1)^2}{2\Delta y t_y} & \frac{(t_y-1)(t_y+1)}{2\Delta y t_y} \\ \frac{c^2(t_x-1)(t_x+1)}{2\Delta x t_x} & \frac{c^2(t_y-1)(t_y+1)}{2\Delta y t_y} & -\frac{c(t_x-1)^2}{2\Delta x t_x} - \frac{c(t_y-1)^2}{2\Delta y t_y} \end{pmatrix} \tag{15}$$

Example 3. *Consider a multi-dimensional extension of* (14)

$$\begin{aligned}\partial_t q_{ij} &+ \frac{1}{2\Delta x}\Big(J^x\langle q_{i+1,\cdot} - q_{i-1,\cdot}\rangle_j - D^x\langle q_{i+1,\cdot} - 2q_{i,\cdot} + q_{i-1,\cdot}\rangle_j\Big)\\ &\qquad - \frac{1}{4\Delta y}S^x J^y\Big(q_{i+1,j+1} - q_{i-1,j+1} - q_{i-1,j+1} + q_{i-1,j-1}\Big)\\ &+ \frac{1}{2\Delta y}\Big(J^y\langle q_{\cdot,j+1} - q_{\cdot,j-1}\rangle_i - D^y\langle q_{\cdot,j+1} - 2q_{\cdot,j} + q_{\cdot,j-1}\rangle_i\Big)\\ &\qquad - \frac{1}{4\Delta x}S^y J^x\Big(q_{i+1,j+1} - q_{i-1,j+1} - q_{i-1,j+1} + q_{i-1,j-1}\Big) = 0\end{aligned} \tag{16}$$

with D^x, D^y *as in example 2 and*

$$\langle q_{\cdot,j}\rangle_i := \frac{1}{4}(q_{i+1,j} + 2q_{ij} + q_{i-1,j}) \qquad J^x S^x = |J^x| \qquad J^y S^y = |J^y| \tag{17}$$

For more details see [1]; *applied to the acoustic equations* (4)–(5), *this scheme appears in* [6, 9, 7].

The Fourier mode (11) *is evolving according to* $\partial_t \hat{q} + \mathcal{E}\hat{q} = 0$ *with*

$$\begin{aligned}\mathcal{E} &= \left(\mathbb{1} - S^x\frac{t_x-1}{t_x+1} - S^y\frac{t_y-1}{t_y+1}\right) \times \\ &\times \left(J_x\frac{(t_x-1)(t_x+1)}{2t_x\Delta x}\cdot\frac{(t_y+1)^2}{4t_y} + J_y\frac{(t_y-1)(t_y+1)}{2t_y\Delta y}\cdot\frac{(t_x+1)^2}{4t_x}\right)\end{aligned} \tag{18}$$

whose determinant vanishes identically whenever $\det(\mathbf{J}\cdot\mathbf{k}) = 0 \quad \forall\mathbf{k}$.

From here on, the argumentation concerning both the discrete involutions and the discrete stationary states is exactly following that of section 2. Instead of $\mathbb{i}\mathbf{k}\cdot\mathbf{J}$ the analysis focuses on studying the zero eigenvalues of the evolution matrix $\mathcal{E}$ of the scheme. This, e.g. leads to the necessary condition $\det\mathcal{E} = 0$ governing the existence of non-trivial *discrete* stationary states.

Example 4. *Consider scheme* (14) *solving* (4)–(5). *The determinant of its evolution matrix* (15) *is*

$$\det\mathcal{E} = \frac{c^3(\Delta x + \Delta y)(t_x-1)^2(t_y-1)^2}{2\Delta x^2\Delta y^2 t_x t_y} \tag{19}$$

It is thus only possible to find non-vanishing vectors $\hat{q}$ *with* $\mathcal{E}\hat{q} = 0$ *if either* $t_x = 1$ *or* $t_y = 1$. *This restriction of* $\mathbf{k}$ *implies that all stationary states of* (14) *are trivial.*

In the context of the acoustic system, numerical schemes that keep stationary a discretization of the vorticity are called *vorticity preserving* ([8, 6, 9]).

Section 4 compares numerical discretizations which have non-trivial stationary states with those which do not. It is shown how in the latter case the evolution is governed by a transition towards a trivial numerical stationary state.

4. **Numerical evolution towards a discrete stationary state.** Under explicit time integration, for a numerical scheme for (1) stability in the von Neumann sense means that no discrete Fourier mode is growing in time. For instance, scheme (14) is stable for a CFL number $\frac{c\Delta t}{\min(\Delta x, \Delta y)} < \frac{1}{2}$ (e.g. [5]).

In order to demonstrate that the above theory leads to measurable predictions and allows to understand in detail the behaviour of a numerical scheme, the following setup shall be studied in more detail:

Setup 1. *The acoustic system* (4)–(5) *shall be solved numerically on an equidistant Cartesian grid covering* $[0,1]^2$ *with periodic boundaries. Forward Euler is used to integrate forward in time. Numerical results are shown at the grid center as functions of time.*

The first example illustrates a trivial discrete stationary state.

Example 5 (Trivial stationary state). *Consider the setup 1 with* $t_y = 1$, *i.e.* $k_y = 0$ *solved with scheme* (14). *Then its evolution matrix* $\mathcal{E}$ *has an eigenvalue zero and the corresponding eigenvector is* $(0,1,0)^T$. *The discrete time evolution of* $(0,1,0)^T \cos(2\pi x)$ *is stationary, because the numerical fluxes vanish identically.*

Turn now to the numerical evolution of a Fourier mode which, at PDE level, is a stationary state of (1). It can be discretized by either a scheme which only has trivial stationary states, or by a scheme with non-trivial stationary states (stationarity preserving scheme).

Example 6 (Nontrivial stationary state). *Consider the (numerical) time evolution of*

$$\left(1, -\frac{1}{10}, 0\right)^T \cos(2\pi x + 10 \cdot 2\pi y) \tag{20}$$

using setup 1. Here thus $k_x = 2\pi$ *and* $k_y = 20\pi$. *From* (10) *one deduces that* (20) *is stationary at PDE level.*

(i) First, the mode (20) *is evolved with scheme* (14). *The Fourier mode* (20) *is not a stationary state for this scheme; moreover the only stationary states of scheme* (14) *are trivial (as shown in Example 4). In Figure 2 the corresponding simulations show a rapid decay of the numerical solution (von Neumann stability). The computations have been performed on two grids:* 50×50 *and* 100×100, *which shows that the decay can be slowed down by choosing a finer mesh, but the qualitative behaviour remains the same.*

Note that the decay rate of the semi-discrete scheme can be computed from the non-zero eigenvalues of $\mathcal{E}$. *The observed decay rate in an actual simulation also contains the diffusion introduced by the time integration method.*

(ii) Now, the mode (20) *is evolved with scheme* (16). *Again,* (20) *is not a stationary mode of the scheme! However,* (16) *possesses non-trivial stationary*

states. The mode (20) can be decomposed into the basis of eigenvectors of the evolution matrix $\mathcal{E}$ (18) of scheme (16):

$$(1, -\frac{1}{10}, 0)^T \simeq -0.0867e_0 + 0.0067e_1 - 0.0067e_2 \qquad (50 \times 50) \qquad (21)$$

$$(1, -\frac{1}{10}, 0)^T \simeq -0.0968e_0 + 0.0016e_1 - 0.0016e_2 \qquad (100 \times 100) \qquad (22)$$

with $e_0 = \left(-\frac{\Delta x(t_x+1)(t_y-1)}{\Delta y(t_x-1)(t_y+1)}, 1, 0 \right)$ *the stationary eigenvector of $\mathcal{E}$. On finer grids (20) thus is moving closer to the stationary mode e_0 of the numerical scheme. This is observed in Figure 2: The stationarity preserving scheme (16) settles on the stationary state of the numerical scheme. The discrete stationary state is approximating the exact stationary state the more closely, the finer the grid is.*

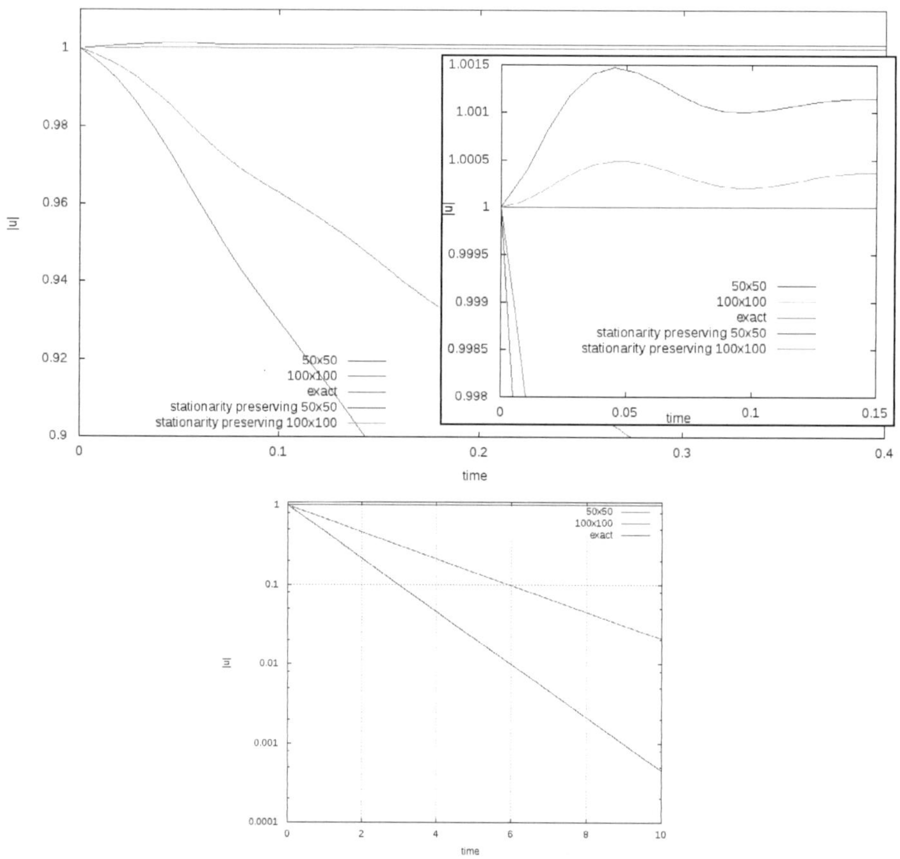

FIGURE 2. Time evolution of the first component of mode (20). *Blue* curve: exact solution (stationary). *Red* and *green* curves: numerical evolution using scheme (14). One observes a rapid decay (*top* figure; *bottom* figure shows the decay in a logarithmic plot over longer times). *Cyan* and *purple* curves: numerical evolution using scheme (16). The numerical solution settles down on a discrete stationary state after a short transition phase (*inset*).

5. **Conclusion and outlook.** This paper presented the behaviour of numerical schemes for linear hyperbolic systems in multiple spatial dimensions with a focus on stationary states. Such systems possess stationary states characterized by differential constraints. It is thus impossible for a numerical scheme to capture them *exactly* with the limited information available on a computational grid.

On the other hand, virtually all numerical schemes possess stationary states, e.g. the spatially constant state. Such states might be insufficient representatives of the analytical stationary states though. It has been shown that this can be made precise through the definition of trivial and non-trivial stationary states. When solving a hyperbolic system with non-trivial stationary states, the numerical scheme should also possess non-trivial stationary states (*stationarity preserving*). In this case its discrete stationary states are a discretization of all the analytic stationary states.

In stationarity preserving schemes the stabilizing diffusion is added to certain modes only, leaving enough room for discrete representations of all analytic stationary states, which are saved from decaying.

Already the case of linear systems is numerically challenging in multiple spatial dimensions. Although Fourier transform methods are not applicable to nonlinear conservation laws, the results presented here hopefully are a stepping stone towards an understanding of discrete stationary states in more complex contexts.

REFERENCES

[1] W.Barsukow. Low Mach number finite volume methods for the acoustic and Euler equations. Doctoral thesis, Universität Würzburg, 2018.

[2] W.Barsukow. Stationarity preserving schemes for multi-dimensional linear systems. *Mathematics of Computation*, **88** (2019),1621–1645, .

[3] W.Barsukow and C.Klingenberg. Exact solution and a truly multidimensional godunov scheme for the acoustic equations. submitted, 2017.

[4] C.M.Dafermos. Quasilinear hyperbolic systems with involutions. *Archive for Rational Mechanics and Analysis*, **94** (1986), 373–389.

[5] S.K.Godunov, A.V.Zabrodin, M.Ivanov, A.N.Kraiko, and G.P.Prokopov. Numerical solution of multidimensional problems of gas dynamics. *Moscow Izdatel Nauka*, **1**, 1976.

[6] R.Jeltsch and M.Torrilhon. On curl-preserving finite volume discretisations for shallow water equations. *BIT Numerical Mathematics*, **46** (2006), 35–53.

[7] T.B.Lung and P.L.Roe. Toward a reduction of mesh imprinting. *International J. Numerical Meth. Fluids*, **76** (2014), 450–470.

[8] K.W.Morton and P.L.Roe. Vorticity-preserving Lax–Wendroff-type schemes for the system wave equation. *SIAM J. Sci. Comp.*, **23** (2001), 170–192.

[9] S.Mishra and E.Tadmor. Constraint preserving schemes using potential-based fluxes ii. genuinely multi-dimensional central schemes for systems of conservation laws. *ETH preprint*, (2009-32), 2009.

E-mail address: wasilij.barsukow@math.uzh.ch

SMOOTH SOLUTIONS FOR NONLINEAR ELASTIC WAVES WITH SOFTENING

Harold Berjamin

Aix-Marseille Univ., CNRS, Centrale Marseille, LMA, Marseille, France

Stéphane Junca*

Université Côte d'Azur, Inria & CNRS, LJAD

Bruno Lombard

Aix-Marseille Univ., CNRS, Centrale Marseille, LMA, Marseille, France

Abstract. A new hyperbolic softening model has been proposed for wave propagation in damaged solids [*Proc. R. Soc. A*, **473** (2017), 20170024]. The linear elasticity becomes nonlinear through an additional internal variable. This thermodynamically relevant model yields a dissipative energy. The 3×3 nonlinear hyperbolic system so-obtained is totally linearly degenerate like the well-known Kerr-Debye system. Existence of global smooth solutions is studied here thanks to the Kawashima condition. Moreover, shocks never appear with smooth initial data. Thus, the only possible blow-up of smooth solutions is the blow-up in L^∞ as for ODEs.

1. **Introduction.** The system of interest has been introduced in [2] to model nonlinear wave propagation in solids:

$$\partial_t \varepsilon - \partial_x v = 0, \tag{1}$$

$$\rho_0 \partial_t v - \partial_x \sigma = 0, \tag{2}$$

$$\partial_t g = \frac{1}{\tau}(W(\varepsilon) - \phi'(g)), \tag{3}$$

where the constants are $\rho_0 > 0$ the density, $E > 0$ the Young modulus, $\tau > 0$ the relaxation time. The variables are $\varepsilon > -1$ the strain, v the velocity, $\sigma = (1-g)E\,\varepsilon$ the stress, $W(\varepsilon) = \frac{1}{2}E\varepsilon^2$ the strain energy, g an internal variable representing the damage. The system is completed with three initial data at time $t = 0$: $\varepsilon_0(x)$, $v_0(x)$, $0 \le g_0(x) < 1$.

The storage function $\phi(g)$ has to satisfy $\phi'(0) = 0$ (to preserve the equilibrium $(\varepsilon, g) = (0, 0)$ and to keep $g \ge 0$), $\phi' \ge 0$ and $\phi'' > 0$ (to ensure the stability of constant equilibrium). Moreover, $g < 1$ is required since for $g = 1$ the solid is broken. An example of function ϕ to ensure these constraints is $\phi(g) = -\frac{1}{2}\gamma \ln(1 - g^2)$ with $\gamma > 0$.

2000 *Mathematics Subject Classification.* Primary: 35L45, 35B65; Secondary: 74D10, 74J30.

Key words and phrases. Damaged solids; nonlinear balance laws; linearly degenerate flux; conservation law; finite-volume method.

* Corresponding author: Stéphane Junca.

The initial-value problem for the system of balance laws (1)-(3) can be rewritten in vectorial form with $c = c(g) = \bar{c}\sqrt{1-g}$ and $\bar{c} = \sqrt{E/\rho_0}$:

$$\partial_t U + \partial_x F(U) = G(U), \tag{4}$$

$$U = (\varepsilon, v, g)^\top, \quad F(U) = (-v, -c^2\varepsilon, 0)^\top, \quad G(U) = \frac{1}{\tau}\left(0, 0, W(\varepsilon) - \phi'(g)\right)^\top.$$

The nonincreasing total energy $\mathcal{E}$ and the internal energy e are:

$$\mathcal{E} := \rho_0(v^2/2 + e) = \rho_0\left(v^2/2 + (1-g)W(\varepsilon) + \phi(g)\right). \tag{5}$$

For smooth solutions the dissipation of the energy is

$$\frac{d}{dt}\int_{\mathbb{R}} \mathcal{E}dx = -\frac{\rho_0}{\tau}\int_{\mathbb{R}} \left(W(\varepsilon) - \phi'(g)\right)^2 dx = -\rho_0\tau \int_{\mathbb{R}} (\partial_t g)^2 dx \leq 0. \tag{6}$$

It is a partially dissipative system [8, 1] which may ensure existence of global smooth solutions [16] under the Kawashima condition [15].

In Section 2, the totally linearly degenerate 3×3 homogeneous system deduced from (4) is studied. The Kawashima condition for the full system with the source term are directly related to the function ϕ in Section 3. The comparison with the Kerr-Debye system and the non existence of shock wave for the system (4) are in Section 4. Finally, numerical simulations of smooth solutions for the system (4) conclude the paper in Section 5.

2. **The linearly degenerate homogeneous system.** Consider the system (4) with no source: $G = 0$. The Jacobian matrix of the flux F has three eigenvectors r_-, r_0, r_+ associated to 3 linearly degenerate eigenvalues: $-c, 0, +c$ in the hyperbolic region $g < 1$:

$$A = DF(U) = \begin{pmatrix} 0 & -1 & 0 \\ -c^2 & 0 & \varepsilon\,\bar{c}^2 \\ 0 & 0 & 0 \end{pmatrix}, \qquad r_\pm = \begin{pmatrix} 1 \\ \mp c \\ 0 \end{pmatrix}, \; r_0 = \begin{pmatrix} \varepsilon\,\bar{c}^2 \\ 0 \\ c^2 \end{pmatrix}. \tag{7}$$

Many things are known for 2×2 totally linearly degenerate system [13, 14]. Less is known for 3×3 system except under special conditions as in [11].

The homogeneous version of (3) means simply $g \equiv g_0$ thus the nonlinear system (1)-(2) gives a linear wave equation with the variable sound speed $c_0(x) = \bar{c}\sqrt{1-g_0(x)}$:

$$\partial_t^2 \varepsilon - \partial_x^2(c_0^2(x)\varepsilon) = 0. \tag{8}$$

Physically, it corresponds simply to the linear elasticity with a varying Young modulus depending only on the space variable x. Then a proof using the Riemann invariants [9] of the elastodynamics yields the existence of global smooth solutions:

Proposition 1 (Global smooth solution for the homogeneous system).
Let us assume that the initial data at time $t = 0$: $\varepsilon_0(x), v_0(x)$ belongs to the space $Lip_{loc}(\mathbb{R},\mathbb{R})$ of locally Lipschitz-functions, $g_0(x) \in Lip(\mathbb{R},\mathbb{R})$, $\sup_{\mathbb{R}} g_0(x) < 1$ and $\partial_x g_0 \in Lip_{loc}(\mathbb{R},\mathbb{R})$. Then the homogeneous hyperbolic system admits a unique global smooth solution (ε, v) with the same regularity in space as the initial data:

$$\varepsilon, v \in L^\infty_{loc}([0,+\infty[, Lip_{loc}(\mathbb{R},\mathbb{R})) \cap C^1([0,+\infty[, L^\infty_{loc}(\mathbb{R},\mathbb{R})).$$

A proof using the Riemann invariants of the 2×2 system of linear elastodynamics is proposed. They are not Riemann invariants for the full 3×3 system, nevertheless some computations are possible involving $\partial_x g$. Since $g(t,x) = g_0(x)$ the term $\partial_x g$

is easily controlled by the Lipschitz initial datum $g_0(x)$. For the system with the source term, the following proof fails because $\partial_x g$ cannot be estimated so easily.

Proof. The dimensionless 3×3 nonlinear system is simply rewritten as a linear 2×2 system with a variable coefficient:

$$\partial_t \varepsilon - \partial_x v = 0, \tag{9}$$
$$\partial_t v - \partial_x[(1-g_0(x))\varepsilon] = 0, \tag{10}$$

or in a short way:

$$\partial_t U + \partial_x F(x,U) = 0, \qquad U = (\varepsilon, v)^\top, \qquad F(x,U) = (-v, (g_0(x)-1)\varepsilon)^\top.$$

Let A and G be the 2×2 variable matrices which depend on the space variable x:

$$A(x) = \partial_U F = \begin{pmatrix} 0 & -1 \\ g_0(x)-1 & 0 \end{pmatrix}, \qquad G(x,U) = \partial_x F = \begin{pmatrix} 0 \\ (\partial_x g_0(x))\ \varepsilon \end{pmatrix}.$$

The gradient of the Riemann invariants $\nabla_U z_\pm = (\mp c_0, 1)$ are the left eigenvectors of the matrix A. Thus, the Riemann invariants are $Z = (z_+, z_-)^\top$. The system (9)-(10) reads:

$$\partial_t U + A(x)\,\partial_x U = -G(x,U), \tag{11}$$
$$\begin{aligned}\partial_t z_\pm \pm c_0\,\partial_x z_\pm &= -\nabla_U z_\pm \cdot G(x,U) \\ &= -(\partial_x g_0)\ \varepsilon \\ &= -(\partial_x g_0)\ \frac{z_- - z_+}{2c_0\,g_0}.\end{aligned} \tag{12}$$

Notice that $\nabla_U z_\pm$, c_0 and g_0 are only functions of x. The function σ is linear with respect to ε, thus Lipschitz with respect to Z.

The system (11) with the variable U has been rewritten with the vector Z for the 2×2 system (12), which variable coefficients are smooth while $g_0 < 1$. Using the characteristics $X_\pm(t,x)$ and the Riemann invariants evaluated along the characteristics $z_\pm(X_\pm(t,x),t)$, the 2×2 linear PDE system (12) becomes a family of the 4×4 nonlinear ODE system parametrized by $x\in\mathbb{R}$ and involving changes of variables:

$$\frac{dX_\pm}{dt} = \pm c_0(X_\pm) = \pm\sqrt{1-g_0(X_\pm)}, \qquad X_\pm(0,x) = x, \tag{13}$$
$$\frac{dz_\pm}{dt} = -\left((\partial_x g_0)\ \frac{z_- - z_+}{2c_0\,g_0}\right)(X_\pm, t), \qquad Z(x,0) = Z_0(x). \tag{14}$$

This system is block triangular. The first two equations (13) are decoupled: the characteristics are global smooth functions since g_0 is globally Lipschitz and $\sup g_0 < 1$, so that the function c_0 is also globally Lipschitz. The characteristics have at most an exponential growth with respect to the time t.

Let us turn to the two coupled last equations (14). Notice that the coupling involves change of variables between X_- and X_+: $z_\mp(X_\pm, t)$ instead of $z_\pm(X_\pm)$. This is classical [6, 7] and can be managed by a fixed point strategy. Then the global existence follows. □

For the full system with a source term the situation is more intricate.

3. The (SK) condition for the complete system. The famous (SK) condition, defined by Shizuta and Kawashima in [15], yields existence of global smooth solutions near an equilibrium [16].

Consider an equilibrium U_e of the system (4), that is a constant solution: $G(U_e) = 0$. The (SK) condition writes at U_e:

$$\textbf{Ker}\ DG(U_e) \cap \{\text{eigenvectors of } DF(U_e)\} = \{0\}. \tag{15}$$

The equilibrium $U_e = (\varepsilon_e, v_e, g_e)$ for the system of interest is given by the equation $W(\varepsilon_e) = \phi'(g_e)$. When $g_e > 0$, there are two equilibrium $(\varepsilon_e, v_e, g_e)$, with

$$\varepsilon_e = \pm\sqrt{2\phi'(g_e)/E} \tag{16}$$

and without restriction on v_e. When $g_e = 0$ the equilibrium is $(0, v_e, 0)$.

The linearized source term is a rank one matrix with the equation of the kernel,

$$\tau\, DG(U_e) = \begin{pmatrix} 0 & 0 & 0 \\ 0 & 0 & 0 \\ E\varepsilon_e & 0 & -\phi''(g_e) \end{pmatrix}, \qquad E\,\varepsilon_e\,\varepsilon = \phi''(g_e)\,g.$$

The eigenvectors $r_\pm$ of $DF(U_e)$ associated to the eigenvalues $\pm c_e$ do not belong to **Ker** $DG(U_e)$. The only problem remains for the eigenvector r_0 in the kernel of $DF(U_e)$. It also belongs to **Ker** $DG(U_e)$ if and only if $E\,\varepsilon_e^2 = (1-g_e)\phi''(g_e)$. Replacing the left hand side thanks to (16) yields only one equation to check when the (SK) condition is not fulfilled, $(0 < g_e < 1)$:

$$2\phi'(g_e) = (1-g_e)\phi''(g_e). \tag{17}$$

Since $\phi'' > 0$ the case $g_e = 0$ is excluded.

Let us consider the example $\phi(g) = -\frac{1}{2}\gamma\ln(1-g^2)$. A simple computation yields

$$\phi'(g) = \gamma\frac{g}{1-g^2}, \qquad \phi''(g) = \gamma\frac{1+g^2}{(1-g^2)^2} \geq \gamma.$$

Thus, the (SK) condition is fulfilled except when $g_e = \sqrt{2} - 1 \simeq 0.414$.

As a consequence of the previous study, with the condition $\phi'(0) = 0$ and $\phi'' > 0$, the (SK) condition is always fulfilled if and only if, $\forall g \in]0,1[$,

$$2\phi'(g) < (1-g)\phi''(g). \tag{18}$$

Lemma 3.1 (Loss of (SK) condition).
If $\lim_{g\to 1}(1-g)\phi(g) = 0$ *then the (SK) condition is not always satisfied.*

Proof. Notice that inequality (18) is always satisfied near $g = 0$ since $\phi'(0) = 0$ and $\phi''(0) > 0$. Let g_0 belong to $]0,1[$. It suffices to integrate the differential inequality (18) to get for all $g \in]g_0, 1[$,

$$\phi'(g) > \phi'(g_0)\left(\frac{1-g_0}{1-g}\right)^2, \quad \text{then} \quad \phi(g) > \phi(g_0) + \phi'(g_0)\frac{1-g_0}{1-g}(g-g_0).$$

Thus $\liminf_{g\to 1}(1-g)\phi(g) > \phi'(g_0)(1-g_0)^2 > 0$ and the lemma follows by contradiction. □

A family of examples satisfying always the (SK) condition is given by $\phi(g) = \frac{1}{2}\gamma g^2(1-g)^{-\alpha}$, with $\alpha > 1$. The condition $\alpha > 1$ is necessary and sufficient. The proof is direct and needs only to check: $\phi'(0) = 0$, $\phi'' > 0$ on $[0,1[$ and (18).

As a direct consequence of (18) and the proof of Lemma 3.1, one gets the following lemma.

Lemma 3.2 (Not (SK) fulfilled on a continuum).
If the (SK) condition is not satisfied on an interval $[g_1, g_2]$ *then*

$$\phi(g) = \phi(g_1) + \phi'(g_1)\frac{1-g_1}{1-g}(g-g_1).$$

4. **Comparison with the Kerr-Debye model.** In this section we compare the system (1)-(3) to the Kerr-Debye system well-known in nonlinear optics. In the latter system smooth initial data in the Sobolev space $H^2(\mathbb{R})$ yield global smooth solutions. The system (1)-(3) can be rewritten in a similar form as the Kerre-Debye system, except the source term which is modified. This modification prevents from transfering all known results on Kerr-Debye system to our case. In particular, we cannot deduce existence of global solutions. However, other results are known for the modified Kerr-Debye system [5] ensuring that no discontinuity can appear in finite time. The only catastrophe which can occur is a L^∞ blow-up as for the solutions of ODEs [12].

The Kerr-Debye model.

$$\partial_t d + \partial_x h = 0, \tag{19}$$

$$\partial_t h + \partial_x e = 0, \tag{20}$$

$$\partial_t \chi = \frac{1}{\tau}\left(e^2 - \chi\right), \tag{21}$$

where $d = (1+\chi)e$ and the initial condition $(d, h, \chi)(x, 0) = (d_0, h_0, \chi_0)(x)$. With $\chi_0 \geq 0$ it follows immediatly that $\chi \geq 0$. The semilinear behavior of solutions of the system is proven in [4]. That means that a smooth solution is not global only if the solution blows up in sup-norm [12]. The global existence of all smooth solution is proven in [5]. This system is also endowed with a strictly convex partially dissipative energy,

$$\widetilde{\mathcal{E}} = \frac{d^2}{1+\chi} + h^2 + \frac{\chi^2}{2}, \qquad \frac{d}{dt}\int_{\mathbb{R}} \widetilde{\mathcal{E}} dx = -\tau \int_{\mathbb{R}} \left(\partial_t \chi\right)^2 dx \leq 0 \tag{22}$$

To prove (22) the system (19)-(21) is rewritten in variables $W = (e, h, \chi)$, to obtain a symmetric system. The semilinear behavior is proven by energy estimates. More precisely, if W is bounded in $L^\infty([0, T*[\times\mathbb{R})$ then W is also bounded in $L^\infty([0, T*[, H^2(\mathbb{R}))$ which is enough to prevent the blow up of the gradient i.e. shock-wave.

Our system rewritten in Kerr-Debye variables. Motivated by the previous results on the Kerr-Debye system, the system (1)-(3) is rewritten in Kerr-Debye variables:

$$d = \varepsilon, \quad h = -v, \quad \frac{d}{1+\chi} = e = \sigma = (1-g)\varepsilon \implies 1+\chi = \frac{1}{1-g}. \tag{23}$$

Thus, $\chi_t = g_t(1-g)^{-2} = (1+\chi)^2 g_t = (1+\chi)^2(d^2/2 - \phi'(g))$ and our system becomes with $\rho_0 = 1$, $E = 1$ and $\tau = 1$:

$$\partial_t d + \partial_x h = 0, \tag{24}$$

$$\partial_t h + \partial_x e = 0, \tag{25}$$

$$\partial_t \chi = (1+\chi)^2\left(\frac{d^2}{2} - \phi'(g)\right) = (1+\chi)^4\frac{e^2}{2} - \psi'(\chi), \tag{26}$$

where ψ' is the increasing function defined by, $\psi'(\chi) = (1+\chi)^2\phi'(1-(1+\chi)^{-1})$. Comparing (21) with (26), there appears only two changes, the weight $(1+\chi)^4$ and the function ψ. Notice that ψ' linear – as for the Kerr-Debye system – corresponds to the following choice for ϕ': $\phi'(g) = g(1-g)$. For this choice, $g < 1$ for all time since $\chi > 0$ for all time. Unfortunately, this choice is not consistent with the requirement on ϕ: $\phi'' > 0$.

Our model can be seen as a nonlinear generalization of the Kerr-Debye system. The nonlinear generalization consists in the nonlinear relaxation with respect to the variable χ. This additional nonlinearity prevents to obtain a global energy estimate for the derivatives of the solutions of (1)-(3) as in [5].

The mapping, $g \mapsto \chi$ is increasing, $g = 0 \Leftrightarrow \chi = 0$, $g = 1 \Leftrightarrow \chi = +\infty$. Thus, the constraint on g becomes a condition of no blow up for χ. The equation (26) yields automatically the positivity of $\chi > 0$ and then the constraint $g < 1$ required by our model.

Moreover, the semilinear behavior for generalized Kerr-Debye systems is known, Theorem 4.1 in [5]. Thus, our system enjoys a semilinear behavior. It means that no schock can occur with smooth initial data:

Corollary 1 (No shock). *Let ε_0, v_0, g_0 belong to $H^2(\mathbb{R})$ and $\sup_{\mathbb{R}} g_0 < 1$ then the solution of the system* (1)-(3) *remains in H^2 as soon as it remains in L^∞.*

The solution is then global smooth or blows up. The blow up means that ε or v blow up in L^∞ or $g = 1$ in finite time.

5. **Numerical solution for the complete system.** The system of balance laws (4) is solved numerically. Following Sec. 4.2 of [3], an explicit time-stepping formula is used, which involves the numerical flux of a finite-volume scheme (a flux-limiter method based on the Roe scheme). The initial data is chosen as follows: $v_0(x)$ is zero, $g_0(x) = g_e$ is constant, while the strain $\varepsilon_0(x) = \varepsilon_e - 2VF(kx)$ has a smooth waveform $F(x) = \frac{4}{3\sqrt{3}}\left(\sin(x) - \frac{1}{2}\sin(2x)\right)\mathbf{1}_{0\leq x\leq 2\pi}$ with fundamental wavelength $2\pi/k = 0.2$ and amplitude V. The domain $x \in [-5, 5]$ is discretized with 20 000 points and the Courant number is 0.95. Outflow conditions are implemented at the boundaries of the domain, as presented in Sec. 7.2.1 of [10]. In this section, the physical constants ρ_0, E equal one, and τ, γ equal 10^{-4} (SI).

Numerical results at the time $t = 4.5$ are shown in Fig. 1 for $\varepsilon_e = g_e = 0$. In the small amplitude limit, the solution converges towards the solution obtained for linear elasticity ($g \equiv 0$), where the initial data is transported at constant speed. As amplitudes are increased, wavefront steepening is observed, along with a diminution of the wave amplitude and of the speed of sound (delay). Nevertheless, the solution keeps smooth.

At $\varepsilon_e = \pm\sqrt{\gamma/E}$ and $g_e = \sqrt{2} - 1$, the (SK) condition is no longer satisfied. However, the stability of the equilibrium is verified numerically. This is illustrated in Fig. 2, which displays the numerical solution for $V = 0.001$ at various times. The dynamics of the system seems to be driven by its stable equilibrium points.

Acknowledgments. This work was supported by the interdisciplinary mission of CNRS (INFINITI). The project leading to this publication has received funding from Excellence Initiative of Aix-Marseille University - A*MIDEX, a French "Investissements d'Avenir" programme. It has been carried out in the framework of the Labex MEC.

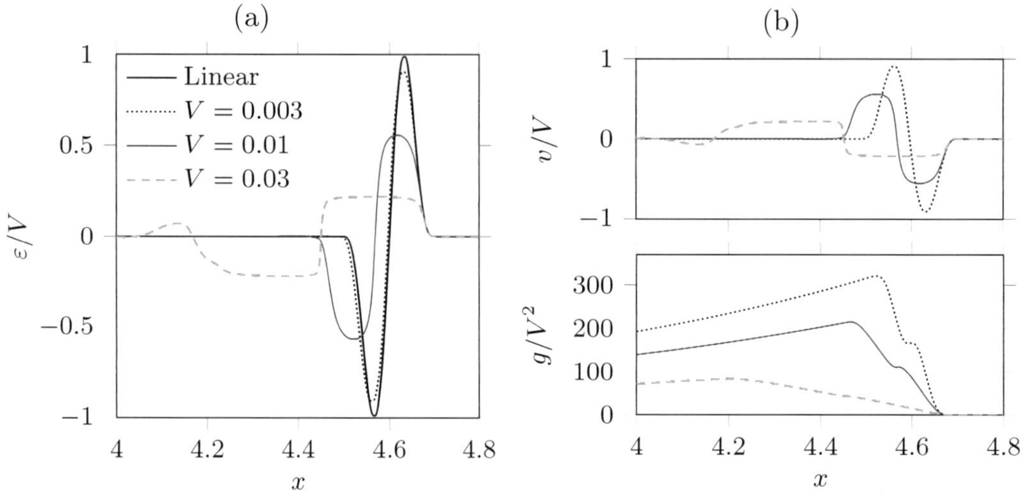

FIGURE 1. Equilibrium $\varepsilon_e = g_e = 0$. Numerical solution at $t = 4.5$ for several amplitudes V. (a) Normalized strain ε/V; (b) normalized velocity v/V and softening g/V^2.

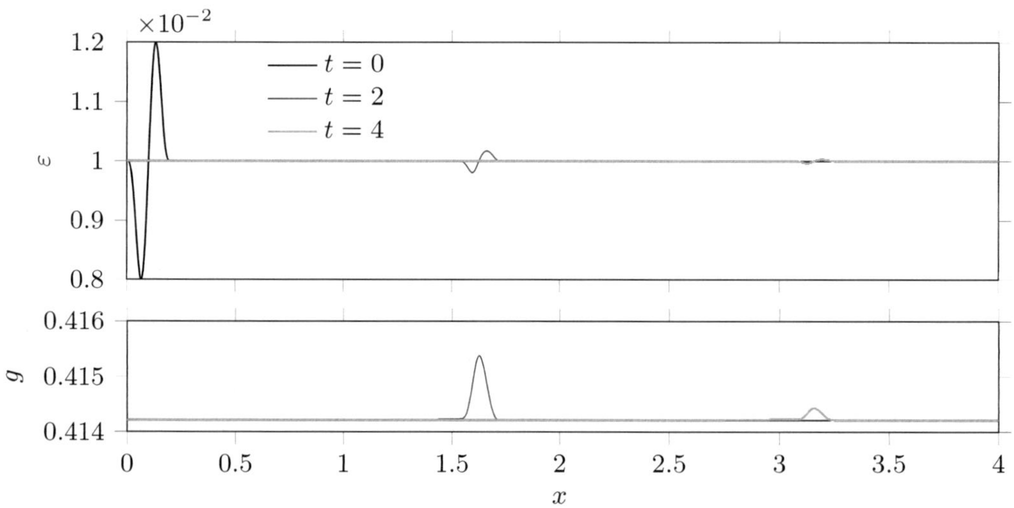

FIGURE 2. Equilibrium $\varepsilon_e = \sqrt{\gamma/E}$, $g_e = \sqrt{2} - 1$. Numerical solution for $V = 0.001$ at several times. Strain ε (top); softening g (bottom).

REFERENCES

[1] K. Beauchard and E. Zuazua, Large time asymptotics for partially dissipative hyperbolic systems, *Arch. Ration. Mech. Anal.*, **199** (2011), 177–227.

[2] H. Berjamin, N. Favrie, B. Lombard and G. Chiavassa, Nonlinear waves in solids with slow dynamics: an internal variable model, *Proc. R. Soc. A*, **473** (2017), 20170024.

[3] H. Berjamin, B. Lombard, G. Chiavassa and N. Favrie, A finite-volume approach to 1D nonlinear elastic waves: Application to slow dynamics, *Acta Acust. united Ac.*, **104** (2018), 561–570.

[4] G. Carbou and B. Hanouzet, Comportement semi-linéaire d'un système hyperbolique quasi-linéaire: le modèle de Kerr Debye, (French) [Semilinear behaviour for a quasilinear hyperbolic system: the Kerr Debye model], *C. R. Acad. Sci. Paris, Ser. I*, **343** (2006), 243–247.

[5] G. Carbou, B. Hanouzet and R. Natalini, Semilinear behavior for totally linearly degenerate hyperbolic systems with relaxation, *J. Differ. Equ.*, **246** (2009), 291–319.

[6] R. Courant and D. Hilbert, *Methods of Mathematical Physics*, vol. II, Interscience Publishers, N.Y., 1962.

[7] T. Goudon and S. Junca, Vanishing pressure in gas dynamics equations, *Z. Angew. Math. Phys.*, **51** (2000), 143–148.

[8] B. Hanouzet and R. Natalini, Global existence of smooth solutions for partially dissipative hyperbolic systems with a convex entropy, *Arch. Ration. Mech. Anal.*, **169** (2003), 89–117.

[9] P.-D. Lax, Hyperbolic partial differential equations, *Courant Lecture Notes in Mathematics*, vol. 14, American Mathematical Society, Providence, RI, 2006.

[10] R.J. LeVeque, *Finite Volume Methods for Hyperbolic Problems*, Cambridge University Press, 2002.

[11] T.-T. Li, Y.-J. Peng and J. Ruiz, Entropy solutions for linearly degenerate hyperbolic systems of rich type, *J. Math. Pures Appl.*, **91** (2009), 553–568.

[12] A. Majda, *Compressible Fluid Flow and Systems of Conservation Laws in Several Space Variables*, Springer-Verlag New York, 1984.

[13] V. Neves and D. Serre, Ill-posedness of the cauchy problem for totally degenerate system of conservation laws, *Electron. J. Differ. Equ.*, Paper No. 124, 25 p. (2005).

[14] Y.-J. Peng, Explicit solutions for 2×2 linearly degenerate systems, *Appl. Math. Lett.*, **11** (1998), 75–78.

[15] Y. Shizuta and S. Kawashima, Systems of equations of hyperbolic-parabolic type with applications to the discrete Boltzmann equation, *Hokkaido Math. J.*, **14** (1985), 249–275.

[16] W.-A. Yong, Entropy and global existence for hyperbolic balance laws, *Arch. Ration. Mech. Anal.*, **172** (2004), 247–266.

E-mail address: berjamin@lma.cnrs-mrs.fr
E-mail address: junca@unice.fr
E-mail address: lombard@lma.cnrs-mrs.fr

UNTANGLING OF TRAJECTORIES FOR NON-SMOOTH VECTOR FIELDS AND BRESSAN'S COMPACTNESS CONJECTURE

STEFANO BIANCHINI

SISSA, via Bonomea 265, 34136 Trieste, Italy

PAOLO BONICATTO*

Universität Basel, Departement Mathematik und Informatik,
Spiegelgasse 1, 4051 Basel, Switzerland

ABSTRACT. Given $d \geq 1$, $T > 0$ and a vector field $\boldsymbol{b}\colon [0,T] \times \mathbb{R}^d \to \mathbb{R}^d$, we study the problem of uniqueness of weak solutions to the associated transport equation $\partial_t u + \boldsymbol{b} \cdot \nabla u = 0$ where $u\colon [0,T] \times \mathbb{R}^d \to \mathbb{R}$ is a scalar function. In the classical setting, the method of characteristics provides an explicit formula for the solution of the PDE, in terms of the flow of $\boldsymbol{b}$. However, when we drop regularity assumptions on the velocity field, uniqueness is in general lost. We present an approach to the problem of uniqueness based on the concept of Lagrangian representation. This tool allows to represent a suitable class of vector fields as superposition of trajectories: we then give local conditions to ensure that this representation induces a partition of the space-time made up of disjoint trajectories, along which the PDE can be disintegrated into a family of 1-dimensional equations. We finally show that, if $\boldsymbol{b}$ is locally of class BV in the space variable, the decomposition satisfies this structural assumption, yielding a positive answer to the (weak) Bressan's Compactness Conjecture.

1. **Introduction.** We present some recent advances (obtained in [12]) in the study of two partial differential equations of the first order, namely the *continuity equation*

$$\begin{cases} \partial_t u + \operatorname{div}(u\boldsymbol{b}) = 0, & \text{in } [0,T] \times \mathbb{R}^d \\ u(0,\cdot) = \overline{u}(\cdot) \end{cases} \tag{1}$$

and the *transport equation*

$$\begin{cases} \partial_t u + \boldsymbol{b} \cdot \nabla u = 0, & \text{in } [0,T] \times \mathbb{R}^d \\ u(0,\cdot) = \overline{u}(\cdot) \end{cases} \tag{2}$$

where $\boldsymbol{b}\colon [0,T] \times \mathbb{R}^d \to \mathbb{R}^d$ is a given vector field, $u\colon [0,T] \times \mathbb{R}^d \to \mathbb{R}$ is a scalar function and $\overline{u}\colon \mathbb{R}^d \to \mathbb{R}$ is the initial datum.

The continuity and the transport equations are among the *cardinal equations* of Mathematical Physics: for instance, the conservation of mass in Euler's equations of fluid-mechanics has the form of (1). In that case, a solution u to (1) can be

---

2000 *Mathematics Subject Classification.* 35F10, 35L03, 28A50, 35D30.

Key words and phrases. Transport equation, continuity equation, renormalization, uniqueness, Superposition Principle.

The second author is supported by ERC Starting Grant 676675 FLIRT.

* Corresponding author: P. Bonicatto.

thought as the density of a continuous distribution of particles moving according to the velocity field $\boldsymbol{b}$; in other terms, the quantity $u(t,x)$ represents the number of particles per unit volume at time $t \in [0,T]$ and position $x \in \mathbb{R}^d$. Notice, moreover, that (1) and (2) are equivalent when $\operatorname{div} \boldsymbol{b} = 0$.

When $\boldsymbol{b}$ is sufficiently regular, existence and uniqueness results for (classical) solutions to Problems (1) and (2) are well known. They rely on the so called *method of characteristics* which establishes a deep connection between the "Eulerian" problems (1), (2) and their "Lagrangian" counterpart, given by the ordinary differential equation driven by $\boldsymbol{b}$:

$$\begin{cases} \partial_t \boldsymbol{X}(t,x) = \boldsymbol{b}(t, \boldsymbol{X}(t,x)), & (t,x) \in [0,T] \times \mathbb{R}^d \\ \boldsymbol{X}(0,x) = x. \end{cases} \tag{3}$$

Under suitable regularity assumptions on $\boldsymbol{b}$, it is well known (and goes under the name of *Cauchy-Lipschitz theory*) that a *flow* exists, i.e. there is a smooth map $\boldsymbol{X}$ solving (3). A simple observation yields that, if u is a solution to (2), then the function $t \mapsto u(t, \boldsymbol{X}(t,x))$ has to be constant: this allows to conclude that the unique solution u of (2) is the *transport* of the initial data $\overline{u}$ along the *characteristics* of (3), i.e. along the curves $[0,T] \ni t \mapsto \boldsymbol{X}(t,x)$. Thus we end up with an explicit formula for the solution u to (2):

$$u(t,x) = \overline{u}\left(\boldsymbol{X}(t,\cdot)^{-1}(x)\right).$$

Similarly one can obtain an explicit formula for solutions to (1).

However, in view of the applications to fluid-mechanics, one would like to deal with velocity fields or densities which are not necessarily smooth. For instance, continuity equation and transport equation with non-smooth vector fields are related to Boltzmann [23, 25] and Vlasov-Poisson equations [22], and also to *hyperbolic conservation laws.* In particular the *Keyfitz and Kranzer system* (introduced in [27]) is a system of conservation laws that reads as

$$\partial_t u + \operatorname{div}\left(\boldsymbol{f}(|u|)u\right) = 0 \qquad \text{in } [0,T] \times \mathbb{R}^d, \tag{4}$$

where the map $\boldsymbol{f}\colon \mathbb{R}^+ \to \mathbb{R}^d$ is assumed to be smooth. It has been shown in [5] that (4) can be formally decoupled in a scalar conservation law for the modulus $r = |u|$ and a transport equation (with field $\boldsymbol{f}(r)$) for the angular part $\vartheta = u/|u|$:

$$\begin{cases} \partial_t r + \operatorname{div}\left(\boldsymbol{f}(r)r\right) = 0, \\ \partial_t \vartheta + \boldsymbol{f}(r) \cdot \nabla\vartheta = 0. \end{cases}$$

As it is well known, solutions to systems of conservation laws are in general non-smooth, hence the vector field $\boldsymbol{f}(r)$ appearing in the transport equation is not regular enough to apply the method of characteristics: we thus have to go beyond the Cauchy-Lipschitz setting.

1.1. **The classical approach: renormalized solutions.** The exploration of the non-smooth framework started with the paper of DiPerna and Lions [24]. They realized that an interplay between Eulerian and Lagrangian coordinates could be exploited to deduce well-posedness results for the ODE (3) from analogous results on PDEs (1) and (2).

On the one hand, due to the linearity of the PDEs, the *existence* of weak solutions to (1), (2) is always guaranteed under reasonable summability assumptions on the vector field $\boldsymbol{b}$ and its spatial divergence; on the other hand, the problem of uniqueness turns out to be much more delicate. A possible strategy, introduced by [24], to recover uniqueness, is based on the notions of *renormalized solution* and of *renormalization property*.

Roughly speaking, a bounded function $u \in L^\infty([0,T] \times \mathbb{R}^d)$ is said to be a *renormalized solution* to (2) if for all $\beta \in C^1(\mathbb{R})$ the function $\beta(u)$ is a solution to the corresponding Cauchy problem:

$$\begin{cases} \partial_t u + \boldsymbol{b} \cdot \nabla u = 0, \\ u(0,\cdot) = \bar{u} \end{cases} \implies \begin{cases} \partial_t(\beta(u)) + \boldsymbol{b} \cdot \nabla(\beta(u)) = 0 \\ \beta(u(0,\cdot)) = \beta(\bar{u}(\cdot)) \end{cases} \quad \text{for every } \beta \in C^1(\mathbb{R}).$$

This can be interpreted as a sort of weak "Chain Rule" for the function u, saying that u is differentiable along the flow generated by $\boldsymbol{b}$. In [24] it is shown that the validity of this property for every $\beta \in C^1(\mathbb{R})$ implies, under general assumptions, uniqueness of weak solutions for (2). Moreover, when this property is satisfied by all solutions, this can be transferred into a property of the vector field itself, which will be said to have the *renormalization property.*

The problem of uniqueness of solutions is thus shifted to prove the renormalization property for $\boldsymbol{b}$: this seems to require some regularity of vector field (tipically in terms of spatial weak differentiability), as counterexamples by Depauw [21] and Bressan [17] show. With an approximation scheme, in [24] the authors proved that renormalization property holds under Sobolev regularity assumptions on the vector field; some years later, Ambrosio [4] improved upon this result, showing that renormalization holds for vector fields which are of class BV (locally in space) with absolutely continuous divergence.

From the Lagrangian point of view, the uniqueness of the solution to the transport equation (2) translates into well-posedness results of the so-called *Regular Lagrangian Flow* of $\boldsymbol{b}$, which is the by-now standard notion of flow in the non-smooth setting. This concept was introduced by Ambrosio in [4]: in a sense, among all possible integral curves of $\boldsymbol{b}$ passing through a point, the Regular Lagrangian Flow selects the ones that do not allow for concentration, in a quantitative way with respect to some reference measure (usually the Lebesgue measure $\mathscr{L}^d$ in $\mathbb{R}^d$). It is worth pointing out that a number of recent papers are devoted to the study of its properties, in particular we mention [6] where a purely *local* theory of Regular Lagrangian Flows has been proposed, thus establishing a complete analogy with the Cauchy-Lipschitz theory.

1.2. **Bressan's Compactness Conjecture.** As we have seen, the theory developed by DiPerna-Lions-Ambrosio settles the Sobolev and the BV case, when the divergence of $\boldsymbol{b}$ does not contain singular terms (with respect to $\mathscr{L}^d$). However, in connections with applications to conservation laws, it would be interesting to cover also the case in which $\boldsymbol{b}$ is of bounded variation in the space, but its divergence may contain non-trivial singular terms: indeed the natural assumption at the level of the divergence of $\boldsymbol{b}$ seems to be not really absolute continuity with bounded density, as considered in Ambrosio [4], but rather the existence of a nonnegative density ρ transported by $\boldsymbol{b}$, with ρ uniformly bounded from above and from below away from zero. Such vector fields are called *nearly incompressible*, according to the following definition.

Definition 1.1. A locally integrable vector field $\boldsymbol{b}\colon (0,T) \times \mathbb{R}^d \to \mathbb{R}^d$ is called *nearly incompressible* if there exists a function $\rho\colon (0,T) \times \mathbb{R}^d \to \mathbb{R}$ (called *density* of $\boldsymbol{b}$) and a constant $C > 0$ such that $0 < C^{-1} \leq \rho(t,x) \leq C$ for Lebesgue almost every $(t,x) \in (0,T) \times \mathbb{R}^d$ and

$$\partial_t \rho + \operatorname{div}_x(\rho \boldsymbol{b}) = 0 \qquad \text{in the sense of distributions on } (0,T) \times \mathbb{R}^d.$$

Notice that no assumption is made on the divergence of $\boldsymbol{b}$; on the other hand, it is rather easy to see (for instance, by mollifications) that if $\operatorname{div} \boldsymbol{b}$ is bounded then $\boldsymbol{b}$ is nearly incompressible.

Nearly incompressible vector fields are strictly related to a conjecture, raised by A. Bressan (studying the well-posedness of the Keyfitz and Kranzer system (4)), predicting the strong compactness of a family of flows associated to smooth vector fields:

Conjecture 1 (Bressan's Compactness Conjecture - Lagrangian formulation)**.** *Let $\boldsymbol{b}_k \colon [0,T] \times \mathbb{R}^d \to \mathbb{R}^d$, $k \in \mathbb{N}$, be a sequence of smooth vector fields and denote by $\boldsymbol{X}_k$ the associated flows, i.e. the solutions of*

$$\begin{cases} \partial_t \boldsymbol{X}_k(t,x) = \boldsymbol{b}_k(t, \boldsymbol{X}_k(t,x)) \\ \boldsymbol{X}_k(0,x) = x. \end{cases}$$

Assume that the quantity $\|\boldsymbol{b}_k\|_\infty + \|\nabla \boldsymbol{b}_k\|_{L^1}$ is uniformly bounded and assume furthermore that there exists $C > 0$ such that for every $k \in \mathbb{N}$ it holds

$$\frac{1}{C} \leq \det\left(\nabla_x \boldsymbol{X}_k(t,x)\right) \leq C, \qquad \forall (t,x) \in [0,T] \times \mathbb{R}^d.$$

Then the sequence $\{\boldsymbol{X}_k\}_{k\in\mathbb{N}}$ is strongly precompact in $L^1_{\mathrm{loc}}([0,T] \times \mathbb{R}^d)$.

By standard compactness arguments, it is readily seen that Conjecture 1 deals essentially with an ordinary differential equation, driven by a nearly incompressible, BV vector field. From the Eulerian point of view, one can thus expect that Conjecture 1 is proved as soon as one can show well posedness at the PDE level for a vector field of class BV and nearly incompressible, extending the well-posedness result of Ambrosio [4]. This is indeed the case: as it has been proved in [5], Conjecture 1 would follow from the following one:

Conjecture 2 (Bressan's Compactness Conjecture - Eulerian formulation)**.** *Any nearly incompressible vector field $\boldsymbol{b} \in L^1([0,T]; \mathrm{BV}_{\mathrm{loc}}(\mathbb{R}^d))$ has the renormalization property.*

The main result is the following Theorem, which answers affirmatively to the conjectures above.

Main Theorem. *Bressan's Compactness Conjecture holds true.*

More precisely, we prove Conjecture 2. It is important to mention various approaches that have been tried in the recent years, also at a purely Lagrangian level: for instance, explicit compactness estimates have been proposed in [10, 19] (and further developed in [16]; see also [26, 18]).

Before presenting the techniques we use to prove the Main Theorem we briefly discuss a particular setting, namely the two-dimensional one, where finer results are availble in view of the Hamiltonian structure.

2. **The two-dimensional case.** The problem of uniqueness of weak solutions to the transport equation (2) in the two dimensional (autonomous) case is addressed in the papers [3], [2] and [15]. In two dimensions and for divergence-free autonomous vector fields, renormalization theorems are available under quite mild assumptions, because of the underlying Hamiltonian structure. Indeed, if $\operatorname{div} \boldsymbol{b} = 0$ in $\mathbb{R}^2$, then there exists a Lipschitz Hamiltonian $H\colon \mathbb{R}^2 \to \mathbb{R}$ such that $\boldsymbol{b} = \nabla^\perp H$, where $\nabla^\perp = (-\partial_2, \partial_1)$. Heuristically it is readily seens that level sets of H are invariant under the flow of $\boldsymbol{b}$, since

$$\frac{d}{dt} H(\gamma(t)) = \nabla H(\gamma(t)) \cdot \dot{\gamma}(t) = \nabla H(\gamma(t)) \cdot \boldsymbol{b}(\gamma(t)) = 0$$

as $\boldsymbol{b}$ and ∇H are orthogonal. This suggests the possibility of decomposing the two-dimensional transport equation into a family of one-dimensional equations, along the level sets of H. By means of this strategy, and building on a fine description of the structure of level sets of Lipschitz maps (obtained in the paper [2]), in [3], the authors characterize the autonomous, divergence-free vector fields $\boldsymbol{b}$ on the plane for which uniqueness holds, within the class of bounded (or even merely integrable) solutions. The characterization they present relies on the so called *Weak Sard Property*, which is a (weaker) measure theoretic version of Sard's Lemma and is used to separate the dynamic where $\boldsymbol{b} \neq 0$ from the regions in which $\boldsymbol{b} = 0$. An extension of these Hamiltonian techniques to the two-dimensional nearly incompressible case was obtained in [14], whose main result is the following:

Theorem 2.1 ([14]). *Every bounded, autonomous, compactly supported, nearly incompressible* BV *vector field on* $\mathbb{R}^2$ *has the renormalization property.*

However, that in the general d-dimensional case, with $d > 2$, the Hamiltonian approach cannot be applied, as there are not enough first integrals of the ODE (which is to say, bounded divergence-free vector fields in $\mathbb{R}^d$ do not admit in general a Lipschitz potential).

3. **The chain rule approach.** We now come back to the general d-dimensional setting and we briefly discuss an approach towards Bressan's Conjecture 2 that has been tried.

In [9], the authors proposed to face the conjecture by establishing a *Chain rule formula* for the divergence operator. Given a bounded, Borel vector field $\boldsymbol{b}\colon \mathbb{R}^d \to \mathbb{R}^d$, a bounded, scalar function $\rho\colon \mathbb{R}^d \to \mathbb{R}$, one would like to characterize (compute) the distribution $\operatorname{div}(\beta(\rho)\boldsymbol{b})$, for $\beta \in C^1(\mathbb{R};\mathbb{R})$, in terms of the quantities $\operatorname{div} \boldsymbol{b}$ and $\operatorname{div}(\rho\boldsymbol{b})$. In the smooth setting one can use the standard chain rule formula to get

$$\operatorname{div}(\beta(\rho)\boldsymbol{b}) = \beta'(\rho)\operatorname{div}(\rho\boldsymbol{b}) + (\beta(\rho) - \rho\beta'(\rho))\operatorname{div}\boldsymbol{b} \tag{5}$$

In the general case, however, the r.h.s. of (5) cannot be written in that form, being only a distribution. In the case the vector field $\boldsymbol{b} \in \mathrm{BV}(\mathbb{R}^d)$, it can be shown that $\operatorname{div}(\beta(\rho)\boldsymbol{b})$ is a measure, controlled by $\operatorname{div}\boldsymbol{b}$ but, as noted in [9], the main problem is to give a meaning to the r.h.s. of (5) when the measure $\operatorname{div}\boldsymbol{b}$ is singular and ρ is only defined almost everywhere with respect to Lebesgue measure. To overcome this difficulty, in the BV setting, the authors split the measure $\operatorname{div}\boldsymbol{b}$ into its absolutely continuous part, jump part and Cantor part and treat the cases separately.

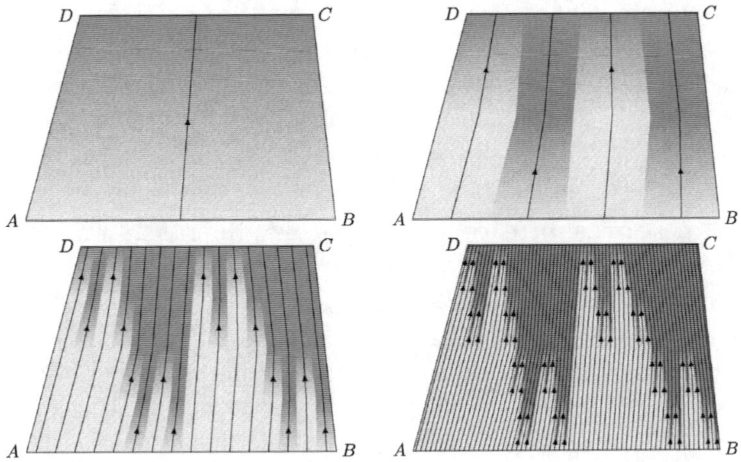

FIGURE 1. Example of [15]: the tangential set of the vector field $\boldsymbol{b}$ (only the integral curves have been drawn here) is a Cantor like set of dimension 3/2. Notice that each trajectory γ meets the tangential set in exactly one point, at time t_γ: the density ρ, computed along the curve, is piecewise constant, having a unique jump of size 1 in t_γ.

The absolutely continuous part. Their first result ([9, Thm. 3]) is that in all Lebesgue points of ρ the formula (5) holds (possibly being $\operatorname{div} \boldsymbol{b}$ singular), where ρ is replaced by its Lebsgue value $\tilde{\rho}$. This is achieved along the same techniques of [4], which are in turn a (non-trivial) extension of the ones employed in [24]: essentially, an approximation argument via convolution is performed (leading to the study of the so called *commutators*). One can control the singular terms by taking suitable convolution kernels which look more elongated in some directions.

The jump part. By exploiting properties of Anzellotti's weak *normal traces* for measure divergence vector fields (see [11]), Ambrosio, De Lellis and Malý managed to settle also the jump part: they obtain an explicit formula (in the spirit of (5)), involving the traces of $\boldsymbol{b}$ and $\rho\boldsymbol{b}$ along a $\mathscr{H}^{d-1}$-rectifiable set (see also [8] for an extension of these results to the BD case).

The Cantor part. In order to tackle the Cantor part, a "transversality condition" between the vector field and its derivative is assumed in [9]: it is shown that, if in a point $(\bar{t},\bar{x})$ one has $(D\boldsymbol{b}\cdot\boldsymbol{b})(\bar{t},\bar{x}) \neq 0$ (where $\boldsymbol{b}(\bar{t},\bar{x})$ is the Lebesgue value of $\boldsymbol{b}$ in $(\bar{t},\bar{x})$) then the point $(\bar{t},\bar{x})$ is a Lebesgue point for ρ.

From the analysis of [9], it thus remains open the case of tangential points, i.e. the set of points at which $D\boldsymbol{b}\cdot\boldsymbol{b}$ vanishes, which make up the so called *tangential set*. This is actually relevant, as shown in [15]: answering negatively to one of the questions in [9], in [15] the authors exhibited an example of BV, nearly incompressible vector field with non empty tangential set. Even worse, the tangential set is a Cantor-like set of non integer dimension but, at level of the density ρ, one sees a pure jump. This severe pathology is depicted in Figure 1 and we refer the reader to [15] for a detailed construction.

4. **A new approach.** We now want to present in more details our main contribution, discussing briefly the theorems we obtained in [12] and the strategy leading to their proofs. The starting point of our approach is the notion of *Lagrangian representation* η of the $\mathbb{R}^{d+1}$-valued vector field $\rho(1, \boldsymbol{b})$, defined in the subsequent paragraph.

4.1. **Lagrangian representations.** In the general non-smooth setting, one could recover a link between the continuity equation (1) and the ODE (3) thanks to the so called *Superposition Principle*, which has been established by Ambrosio in [4] (see also [28]). Roughly speaking, it asserts that, if the vector field is globally bounded, every non-negative (possibly measure-valued) solution to the PDE (1) can be written as a superposition of solutions obtained via propagation along integral curves of $\boldsymbol{b}$, i.e. solutions to the ODE (3).

More generally, let us consider a locally integrable vector field $\boldsymbol{b} \in L^1_{\mathrm{loc}}((0,T) \times \mathbb{R}^d)$ and let ρ be a non-negative solution to the balance law

$$\partial_t \rho + \operatorname{div}(\rho \boldsymbol{b}) = \mu, \qquad \mu \in \mathscr{M}((0,T) \times \mathbb{R}^d). \tag{6}$$

with $\rho \in L^1_{\mathrm{loc}}((1 + |\boldsymbol{b}|)\mathscr{L}^{d+1})$ (so that a distributional meaning can be given). For simplicity, we will often write (6) in the shorter form

$$\operatorname{div}_{t,x} \big(\rho(1, \boldsymbol{b})\big) = \mu. \tag{7}$$

Let us denote the space of continuous curves by

$$\Upsilon := \big\{(t_1, t_2, \gamma) \in \mathbb{R}^+ \times \mathbb{R}^+ \times C(\mathbb{R}^+, \mathbb{R}^d),\, t_1 < t_2\big\}$$

and let us tacitly identify the triplet $(t^-_\gamma, t^+_\gamma, \gamma) \in \Upsilon$ with γ, so that we will simply write $\gamma \in \Gamma$. We say that a finite, non negative measure η over the set Υ is a *Lagrangian representation* of the vector field $\rho(1, \boldsymbol{b})$ if the following conditions hold:

1. η is concentrated on the set of characteristics Γ, defined as

   $$\Gamma := \{(t_1, t_2, \gamma) \in \Upsilon : \gamma \text{ characteristic of } \boldsymbol{b} \text{ in } (t_1, t_2)\}\,;$$

   we explicitly recall that a curve γ is said to be a characteristic of the vector field $\boldsymbol{b}$ in the interval I_γ if it is an absolutely continuous solutions to the ODE

   $$\dot\gamma(t) = \boldsymbol{b}(t, \gamma(t)),$$

   in I_γ, which means that for every $(s,t) \subset I_\gamma$ we have

   $$\int_\Gamma \left|\gamma(t) - \gamma(s) - \int_s^t \boldsymbol{b}(\tau, \gamma(\tau))\, d\tau\right| \eta(d\gamma) = 0.$$

2. The solution ρ can be seen as a superposition of the curves selected by η, i.e. if $(\mathbb{I}, \gamma) \colon I_\gamma \to I_\gamma \times \mathbb{R}^d$ denotes the map defined by $t \mapsto (t, \gamma(t))$, we ask that

   $$\rho \mathscr{L}^{d+1} = \int_\Gamma (\mathbb{I}, \gamma)_\sharp \mathscr{L}^1\, \eta(d\gamma);$$

3. we can decompose μ, the divergence of $\rho(1, \boldsymbol{b})$, as a local superposition of Dirac masses without cancellation, i.e.

   $$\mu = \int_\Gamma \Big[\delta_{t^-_\gamma, \gamma(t^-_\gamma)}(dt\, dx) - \delta_{t^+_\gamma, \gamma(t^+_\gamma)}(dt\, dx)\Big]\, \eta(d\gamma),$$

   $$|\mu| = \int_\Gamma \Big[\delta_{t^-_\gamma, \gamma(t^-_\gamma)}(dt\, dx) + \delta_{t^+_\gamma, \gamma(t^+_\gamma)}(dt\, dx)\Big]\, \eta(d\gamma).$$

The existence of such a decomposition into curves is a consequence of general structural results of 1-dimensional normal currents (see [28] and, for the case $\mu = 0$, [7, Thm. 12]). The non-negativity assumption on $\rho \geq 0$ (i.e. the *a-cyclicity* of $\rho(1, \boldsymbol{b})$ in the language of currents) plays here a role, allowing to reparametrize the curves in such a way they become characteristic of $\boldsymbol{b}$, i.e. they satisfy Point (1).

4.2. **Restriction of Lagrangian representations and proper sets.** One problem we face immediately lies in the fact that η is a *global* object, thus it is not immediate to relate suitable *local estimates* with η: in other words, in general, η cannot be restricted to a set, without losing the property of being a Lagrangian representation. If we are given an open set $\Omega \subset \mathbb{R}^{d+1}$ and a curve γ, we can write

$$\gamma^{-1}(\Omega) = \bigcup_{i=1}^{\infty} (t_\gamma^{i,-}, t_\gamma^{i,+})$$

and then consider the family of curves

$$\mathtt{R}_\Omega^i \gamma := \gamma \llcorner_{(t_\gamma^{i,-}, t_\gamma^{i,+})}.$$

We can now define

$$\eta_\Omega := \sum_{i=1}^{\infty} (\mathtt{R}_\Omega^i)_\sharp \eta. \tag{8}$$

In general, the series in (8) does not converge. Moreover, even if the quantity in (8) is well defined as a measure, since η satisfies Points (1) and (2) of the definition of Lagrangian representation given above, it certainly holds

$$\rho(1, \boldsymbol{b}) \mathscr{L}^{d+1} \llcorner_\Omega = \int_\Gamma (\mathbb{I}, \gamma)_\sharp \big((1, \dot{\gamma}) \mathscr{L}^1 \big) \, \eta_\Omega(d\gamma).$$

but, in general, Point (3) is not satisfied by η_Ω (more precisely the second formula): in other words, η_Ω might not be a Lagrangian representation of $\rho(1, \boldsymbol{b}) \mathscr{L}^{d+1} \llcorner_\Omega$: the key point is that the sets of γ which are exiting from or entering in Ω are not disjoint.

Thus the first question we have to answer to is to characterize the open sets $\Omega \subset \mathbb{R}^{d+1}$ for which η_Ω is a Lagrangian representation of $\rho(1, \boldsymbol{b}) \mathscr{L}^{d+1} \llcorner_\Omega$. It turns out that there are sufficiently many open sets Ω with this property: apart from having a piecewise C^1-regular boundary and assuming that $\mathscr{H}^d \llcorner_{\partial\Omega}$-a.e. point is a Lebesgue point for $\rho(1, \boldsymbol{b})$, the fundamental fact is that there are two Lipschitz functions $\phi^{\delta,\pm}$ such that

$$\mathbb{1}_\Omega \leq \phi^{\delta,+} \leq \mathbb{1}_{\Omega + B_\delta^{d+1}(0)}, \qquad \mathbb{1}_{\mathbb{R}^{d+1} \setminus \Omega} \leq \phi^{\delta,-} \leq \mathbb{1}_{\mathbb{R}^{d+1} \setminus \Omega + B_\delta^{d+1}(0)}$$

and

$$\lim_{\delta \to 0} \rho |(1, \boldsymbol{b}) \cdot \nabla \phi^{\delta,\pm}| \, \mathscr{L}^{d+1} = \rho |(1, \boldsymbol{b}) \cdot \mathbf{n}| \, \mathscr{H}^d \llcorner_{\partial\Omega} \qquad \text{in the sense of measures on } \mathbb{R}^{d+1},$$

which essentially mean that $\rho(1, \boldsymbol{b}) \mathscr{H}^d \llcorner_{\partial\Omega}$ is measuring the flux of $\rho(1, \boldsymbol{b})$ across $\partial\Omega$. We call these set $\rho(1, \boldsymbol{b})$-*proper* (or just *proper* for shortness) and we study carefully their properties: we show that there are sufficiently many proper sets and that they can be perturbed in order to adapt to the vector field under study.

4.3. **Cylinders of approximate flow.** Once we are able to localize the problem in a proper set, we can start studying which are the pieces of information on the local behavior of the vector field that one needs in order to deduce global uniqueness results.

Given a proper set $\Omega \subset \mathbb{R}^{d+1}$, we assume we can construct locally *cylinders of approximate flow* as follows:

Assumption 4.1. *There are constants* $\mathtt{M}, \varpi > 0$ *and a family of functions* $\{\phi^\ell_\gamma\}_{\ell>0,\gamma\in\Gamma}$ *such that:*

1. *for every* $\gamma \in \Gamma, \ell \in \mathbb{R}^+$*, the function* $\phi^\ell_\gamma \colon [t^-_\gamma, t^+_\gamma] \times \mathbb{R}^d \to [0,1]$ *is Lipschitz, so that it can be used as a test function;*
2. *the* shrinking ratio *of the cylinder* ϕ^ℓ_γ *is controlled in time, preventing it collapses to a point: more precisely, for* $t \in [t^-_\gamma, t^+_\gamma]$ *and* $x \in \mathbb{R}^d$*,*
$$\mathbb{1}_{\gamma(t)+B^d_{\ell/\mathtt{M}}(0)}(x) \leq \phi^\ell_\gamma(t,x) \leq \mathbb{1}_{\gamma(t)+B^d_{\mathtt{M}\ell}(0)}(x);$$
3. *we control in a quantitative way the flux through the "lateral boundary of the cylinder" (compared to the total amount of curves starting from the "base of the cylinder") with the quantity* ϖ*: more precisely, denoting by*
$$\begin{aligned}\mathsf{Flux}^\ell(\gamma) &:= \begin{array}{c}\textit{flux of the the vector field } \rho(1,\boldsymbol{b})\\ \textit{across the "boundary of the cylinder" } \phi^\ell_\gamma\end{array}\\ &= \iint_{(t^-_\gamma,t^+_\gamma)\times\mathbb{R}^d} \rho(t,x)\big|(1,\boldsymbol{b})\cdot\nabla\phi^\ell_\gamma(t,x)\big|\,\mathscr{L}^{d+1}(dx\,dt),\end{aligned}$$
$$\sigma^\ell(\gamma) := \textit{amount of curves starting from the base of the cylinder } \phi^\ell_\gamma$$
and
$$\eta^{\mathrm{in}}_\Omega := \eta_\Omega \llcorner \{\textit{curves entering in } \Omega\}$$
we ask that
$$\int_\Gamma \frac{1}{\sigma^\ell(\gamma)} \mathsf{Flux}^\ell(\gamma)\,\eta^{\mathrm{in}}_\Omega(d\gamma) \leq \varpi. \tag{9}$$

We decided to call *cylinders of approximate flow* the family of functions $\{\phi^\ell_\gamma\}_{\ell>0,\gamma\in\Gamma}$: indeed, if γ is a characteristic of the vector field $\boldsymbol{b}$, the function ϕ^ℓ_γ can be thought as generalized, smoothed cylinder centered at γ. Notice that the measure $\eta^{\mathrm{in}}_\Omega$ makes sense if Ω is a proper set, in view of the above analysis. Thus the ultimate meaning of the assumption is that one controls the ratio between the flux of $\rho(1,\boldsymbol{b})$ across the lateral boundary of the cylinders and the total amount of curves entering through its base in a uniform way (w.r.t. ℓ), as the cylinder shrinks to a trajectory γ. A completely similar computation can be performed backward in time, by considering η_Ω restricted to the exiting trajectories and adopting suitable modifications.

4.4. **Passing to the limit via transport plans.** At this point, one would like to determine what the cylinder estimate (9) yields in the limit $\ell \to 0$. In order to perform this passage to the limit, we borrow some tools from the Optimal Transportation Theory. The language of transference plans is particularly suited for our purposes: we define
$$\Gamma^{\mathrm{cr}}(\Omega) := \big\{\gamma \in \Gamma : \gamma(t^\pm_\gamma) \in \partial\Omega\big\}, \qquad \Gamma^{\mathrm{in}}(\Omega) := \big\{\gamma \in \Gamma : \gamma(t^-_\gamma) \in \partial\Omega\big\}$$
and we consider plans between $\eta^{\mathrm{cr}}_\Omega := \eta_\Omega \llcorner_{\Gamma^{\mathrm{cr}}(\Omega)}$ and the entering trajectory measure $\eta^{\mathrm{in}}_\Omega$. Notice that $\eta^{\mathrm{cr}}_\Omega$ is concentrated, by definition, on the set of trajectories entering in and exiting from Ω (*crossing* trajectories).

In the correct estimate one has to take into account also of trajectories which end inside the set Ω and this, in view of Point 3 of the definition of Lagrangian representation, is estimated by the negative part μ^- of the divergence μ, defined in (7). Thus one obtains the following

Proposition 1. *Let $\Omega \subset \mathbb{R}^{d+1}$ be a proper set and η be a Lagrangian representation of $\rho(1, \boldsymbol{b})$. If Assumption 4.1 holds then there exist $N_1 \subset \Gamma^{\mathrm{cr}}(\Omega), N_2 \subset \Gamma^{\mathrm{in}}(\Omega)$ such that*

$$\eta_\Omega^{\mathrm{cr}}(N_1) + \eta_\Omega^{\mathrm{in}}(N_2) \leq \inf_{C>1} \left\{ 2\varpi + C\varpi + \frac{\mu^-(\Omega)}{C-1} \right\}$$

and for every $(\gamma, \gamma') \in (\Gamma^{\mathrm{cr}} \setminus N_1) \times (\Gamma^{\mathrm{in}} \setminus N_2)$

$$\textit{either } \operatorname{clos Graph} \gamma' \subset \operatorname{clos Graph} \gamma \textit{ or } \operatorname{clos Graph} \gamma, \operatorname{clos Graph} \gamma' \textit{ are disjoint.} \tag{$\star$}$$

Proposition 1 gives essentially a uniqueness result (from the Lagrangian point of view) at a *local* level, namely inside a proper set Ω: it says that, under Assumption 4.1, up to removing a set of trajectories whose measure is controlled, one gets a family of essentially disjoint trajectories (meaning that are either disjoint or one contained in the other).

4.5. ***Untangling* of trajectories.** It seems at this point natural to try to perform some "local-to-global" argument, seeking a global analog of Proposition 1. In order to do this, we introduce the following *untangling functional for η^{in}*, defined on the class of proper sets as

$$\boldsymbol{f}^{\,\mathrm{in}}(\Omega) := \inf \left\{ \eta_\Omega^{\mathrm{cr}}(N_1) + \eta_\Omega^{\mathrm{in}}(N_2) : \forall (\gamma, \gamma') \in (\Gamma \setminus N_1) \times (\Gamma \setminus N_2) \text{ condition } (\star) \text{ holds} \right\}$$

and, in a similar fashion, one can define an untangling functional for the trajectories that are exiting from the domain Ω. In a sense, these functionals are measuring the minimum amount of curves one has to remove so that the remaning ones are essentially disjoint, i.e. they satisfy condition $(\star)$. The main property of these functionals is that they are subadditive with respect to the domain Ω, meaning that

$$\boldsymbol{f}^{\,\mathrm{in}}(\Omega) \leq \boldsymbol{f}^{\,\mathrm{in}}(\mathrm{U}) + \boldsymbol{f}^{\,\mathrm{in}}(\mathrm{V}),$$

whenever $\mathrm{U}, \mathrm{V} \subset \mathbb{R}^{d+1}$ are proper sets whose union $\Omega := \mathrm{U} \cup \mathrm{V}$ is proper. The subadditivity suggests the possibility of having a local control in terms of a measure ϖ^τ, whose mass is $\tau > 0$, replacing the constant ϖ in Proposition 1 with $\varpi^\tau(\Omega)$. In view of Proposition 1 one has to combine ϖ^τ with the divergence and this can be done by introducing a suitable measure $\zeta_C^\tau \approx C\varpi^\tau + \frac{|\mu|}{C}$ on $\mathbb{R}^{d+1}$. If Assumption 4.1 is satisfied locally by a suitable family of balls, then one can show, by means of a non-trivial covering argument, the following fundamental proposition, which is the global analog of Proposition 1.

Proposition 2. *There exists a set of trajectories $N \subset \Gamma$ such that*

$$\eta(N) \leq C_d \zeta_C^\tau(\mathbb{R}^{d+1})$$

and for every $(\gamma, \gamma') \in (\Gamma \setminus N)^2$ it holds

$$\begin{aligned} &\textit{either } \operatorname{Graph} \gamma \subset \operatorname{Graph} \gamma' \textit{ or } \operatorname{Graph} \gamma' \subset \operatorname{Graph} \gamma \\ &\textit{or } \operatorname{Graph} \gamma, \operatorname{Graph} \gamma' \textit{ are disjoint (up to the end points).} \end{aligned} \tag{$\star\star$}$$

The interesting situation is when the measure ζ_τ^C can be taken arbitrarily small, i.e. when $\tau \to 0$: in that case η is said to be *untangled*, i.e. it is concentrated on a set Δ such that for every $(\gamma, \gamma') \in \Delta \times \Delta$ the condition $(\star\star)$ holds (see also Figure 2).

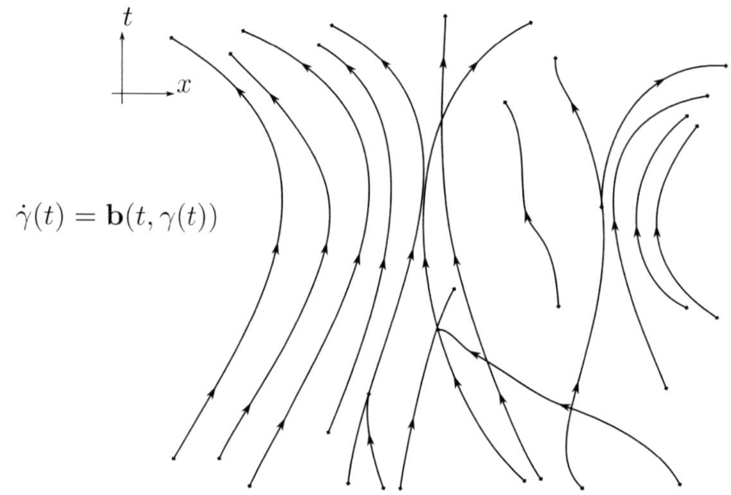

(A) Initial configuration: the curves ay intersect several times, overlap and bifurcate.

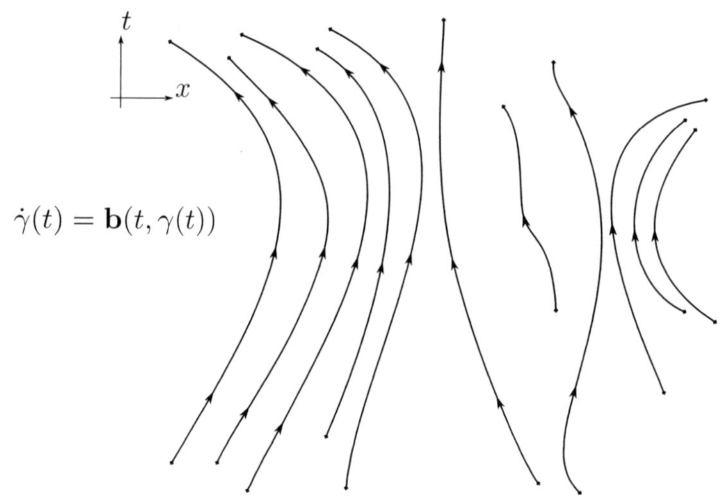

(B) Final configuration: after the untangling, the curves are disjoint, thus forming a partition $\{\wp_\mathfrak{a}\}_\mathfrak{a}$ of $\mathbb{R}^{d+1}$ up to a $\rho\mathcal{L}^{d+1}$-negligible set.

FIGURE 2. Visual effect of the *untangling* of trajectories: we start by removing locally a set of curves, whose η measure is controlled, in such a way that the curves are disjoint in a small ball. Iterating this step - thanks to subadditivity - we end up with a family of disjoint, untangled trajectories.

4.6. **Partition via characteristics and Lagrangian uniqueness.** The *untangling* of trajectories is the core of our approach and it encodes, in our language, the uniqueness issues and the computation of the chain rule. Indeed, once the untangled set $\varDelta$ is selected, we can construct an equivalence relation on it, identifying trajectories whenever they coincide in some time interval: this gives a partition of $\varDelta$ into equivalence classes $E_{\mathfrak{a}} := \{\wp_{\mathfrak{a}}\}_{\mathfrak{a}}$, being $\mathfrak{A}$ a suitable set of indexes. This, in turn, induces a partition of $\mathbb{R}^{d+1}$ (up to a set $\rho\mathscr{L}^{d+1}$-negligible) into disjoint trajectories (that we still denote by $\wp_{\mathfrak{a}}$): both partitions admit a Borel section (i.e. there exist Borel functions $\mathtt{f}\colon \mathbb{R}^{d+1} \to \mathfrak{A}$ and $\hat{\mathtt{f}}\colon \varDelta \to \mathfrak{A}$ such that $\wp_{\mathfrak{a}} = \mathtt{f}^{-1}(\mathfrak{a})$ and $\hat{\mathtt{f}}^{-1}(\mathfrak{a}) = E_{\mathfrak{a}}$ for every $\mathfrak{a} \in \mathfrak{A}$): hence a disintegration approach can be adopted, like in the two-dimensional setting. One reduces the PDE (7) into a family of one-dimensional ODEs along the trajectories $\{\wp_{\mathfrak{a}}\}_{\mathfrak{a}\in\mathfrak{A}}$: we are thus recovering a sort of method of the characteristic in the weak setting.

To formalize this disintegration issue, we propose to call a Borel map $\mathtt{g}\colon \mathbb{R}^{d+1} \to \mathfrak{A}$ a *partition via characteristics* of the vector field $\rho(1, \boldsymbol{b})$ if:

- for every $\mathfrak{a} \in \mathfrak{A}$, $\mathtt{g}^{-1}(\mathfrak{a})$ coincides with $\operatorname{Graph}\gamma_{\mathfrak{a}}$, where $\gamma_{\mathfrak{a}}\colon I_{\mathfrak{a}} \to \mathbb{R}^{d+1}$ is a characteristic of $\boldsymbol{b}$ in some open domain $I_{\mathfrak{a}} \subset \mathbb{R}$;
- if $\hat{\mathtt{g}}$ denotes the corresponding map $\hat{\mathtt{g}}\colon \varDelta \to \mathfrak{A}$, $\hat{\mathtt{g}}(\gamma) := \mathtt{g}(\operatorname{Graph}\gamma)$, setting $m := \hat{\mathtt{g}}_\sharp \eta$ and letting $w_{\mathfrak{a}}$ be the disintegration
$$\rho\,\mathscr{L}^{d+1} = \int_{\mathfrak{A}} (\mathbb{I}, \gamma_{\mathfrak{a}})_\sharp (w_{\mathfrak{a}} \mathscr{L}^1)\, m(d\mathfrak{a})$$
then
$$\frac{d}{dt} w_{\mathfrak{a}} = \mu_{\mathfrak{a}} \in \mathscr{M}(\mathbb{R}), \tag{10}$$
where $w_{\mathfrak{a}}$ is considered extended to 0 outside the domain of $\gamma_{\mathfrak{a}}$;
- it holds
$$\mu = \int (\mathbb{I}, \gamma_{\mathfrak{a}})_\sharp \mu_{\mathfrak{a}}\, m(d\mathfrak{a}) \qquad \text{and} \qquad |\mu| = \int (\mathbb{I}, \gamma_{\mathfrak{a}})_\sharp |\mu_{\mathfrak{a}}|\, m(d\mathfrak{a}).$$

We will say the partition is *minimal* if moreover
$$\lim_{t\to\bar{t}\pm} w_{\mathfrak{a}}(t) > 0 \qquad \forall \bar{t} \in I_{\mathfrak{a}}.$$

In view of the discussion above, the family of equivalence classes $\{\wp_{\mathfrak{a}}\}_{\mathfrak{a}\in\mathfrak{A}}$ arising from the untangled set $\varDelta$ constitutes a partition via characteristics. Since the function $w_{\mathfrak{a}}$ is a BV function on $\mathbb{R}$, in view of (10), we can further split the equivalence classes so that it becomes a minimal partition via characteristics of $\rho(1, \boldsymbol{b})$. Furthermore, if we take $u \in L^\infty((0,T) \times \mathbb{R}^d)$ such that $\operatorname{div}(u\rho(1, \boldsymbol{b})) = \mu'$ is a measure, we can repeat the computations for the vector field $(2\|u\|_\infty + u)\rho(1, \boldsymbol{b})$ obtaining that the same partition via characteristics works also for $u\rho(1, \boldsymbol{b})$. This yields the following uniqueness result, which is the core of our work:

Theorem 4.1 ([12])**.** *If η is untangled, then there exists a minimal partition via characteristics $\mathtt{f}$ of $\rho(1, \boldsymbol{b})$. Furthermore, if $u \in L^\infty((0,T) \times \mathbb{R}^d)$ is a solution to $\operatorname{div}(u\rho(1, \boldsymbol{b})) = \mu'$, then map $\mathtt{f}$ is a partition via characteristics of $u\rho(1, \boldsymbol{b})$ as well.*

In particular, by disintegrating the PDE $\operatorname{div}(u\rho(1, \boldsymbol{b})) = \mu'$ along the characteristics $\wp_{\mathfrak{a}} = \mathtt{f}^{-1}(\mathfrak{a})$, we obtain the one-dimensional equation
$$\frac{d}{dt}\Big(u(t, \wp_{\mathfrak{a}}(t)) w_{\mathfrak{a}}(t) \Big) = \mu'_{\mathfrak{a}}.$$

At this point, an application of Volpert's formula for one-dimensional BV functions allows an explicit computation of $\frac{d}{dt}(\beta(u \circ \wp_a) w_{\mathfrak{a}})$, i.e. of $\operatorname{div}(\beta(u)\rho(1, \boldsymbol{b}))$ thus establishing the Chain rule in the general setting.

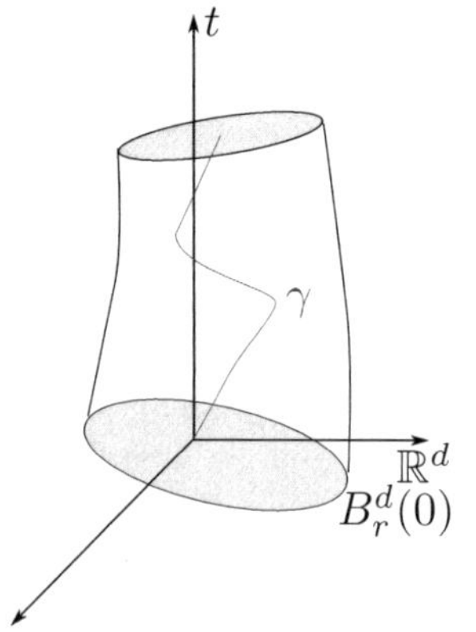

(A) In the absolutely continuous part of $D\boldsymbol{b}$ the cylinders evolve under a constant matrix A, which will be taken close to $D^a\boldsymbol{b}$.

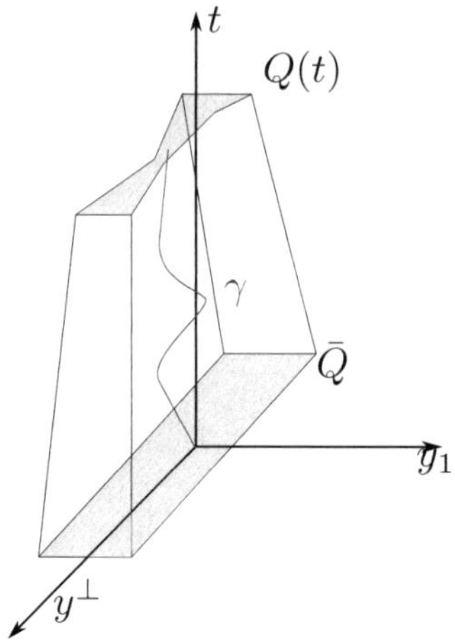

(B) The singular case: the cylinders shrink (if $\operatorname{div}\boldsymbol{b} < 0$) in a controlled way, their sides being graph of monotone Lipschitz functions which solve suitable differential equations.

FIGURE 3. Approximate cylinders of flow in the BV (nearly incompressible) case.

4.7. **The** BV **nearly incompressible case and Bressan's Compactness Conjecture.** To conclude the proof of the Main Theorem, establishing Bressan's Compactness Conjecture, it remains to show how we can construct cylinders of approximate flow satisfying Assumption 4.1, for a vector field of the form $\rho(1, \boldsymbol{b})$, with $\rho \in (C^{-1}, C)$ and $\boldsymbol{b} \in L^1((0,T); \mathrm{BV}_{\mathrm{loc}}(\mathbb{R}^d))$. In view of Theorem 4.1, without loss of generality, we can assume $\rho = 1$ so that the vector field under consideration is exactly $(1, \boldsymbol{b})$: as usual, we denote by $D\boldsymbol{b}$ the derivative of $\boldsymbol{b}$ and we split it into the absolutely continuous part and the singular part.

In a Lebesgue point $(\bar{t}, \bar{x})$ of the absolutely continuous part, the construction of the cylinders is rather easy: essentially, one replaces the real evolution under the flow of $\boldsymbol{b}$ of a ball $B^d_\ell(0)$ with an ellipsoid, obtained by letting everything evolve under a fixed matrix A (compare with Figure 3a). Some standard computations show that the difference between the two evolutions can be made arbitrarily small, when compared to the volume of $B^d_\ell(0)$, by taking A to be the Lebesgue value of $D\boldsymbol{b}$ in the point $(\bar{t}, \bar{x})$.

The estimates for the singular part are more delicate and depend heavily on the shape of the approximate cylinders of flow. Here the geometric structure of BV functions (Alberti's Rank-One Theorem [1, 20]) plays a role, as in the original proof of [4]. The main idea is to choose properly the (non-transversal) sides' lenghts of the cylinders, in such a way to cancel the effect of the divergence. Indeed, by Rank One Theorem, we can find a suitable (local) coordinate system $\boldsymbol{y} = (y_1, y^\perp) \in \mathbb{R}^d$ in which the derivative $D\boldsymbol{b}$ is essentially directed toward a fixed direction (without loss of generality, the one given by $\boldsymbol{e}_1$). Accordingly, we define the (section at time

t of the) cylinder

$$Q = Q_{\ell^{\pm}_{1,\gamma},\ell}(t) := \gamma(t) + \Big\{ \boldsymbol{y} = (y_1, y^{\perp}) : -\ell_1^-(t, y^{\perp}) \le y_1 \le \ell_1^+(t, y^{\perp}),\ |y^{\perp}| \le \ell \Big\}, \tag{11}$$

where $\ell > 0$ is a real number and $\ell^{\pm}_{1,\gamma}$ are suitable functions to be chosen, Lipschitz in $y^{\perp}$ and monotone in t. This is indeed a crucial step: we show it is possible to adapt locally the cylinders of approximate flows, by imposing that the sides' lengths $\ell^{\pm}_{1,\gamma}(t)$ are monotone functions satisfying suitable differential equations (see Figure 3b). In a simplified setting, i.e. if the level set of $b_1(t)$ were exactly of the form $y_1 =$ constant, then we would impose

$$\frac{d}{dt}\ell^+_{1,\gamma}(t) = (Db_1)\big(\gamma(t), \gamma(t) + \ell^+_{1,\gamma}(t)\big) \tag{12}$$

(and an analogous relation for $\ell^-_{1,\gamma}$). Plugging the solution of (12) into the definition of the cylinder (11), we can show that the flux of $\boldsymbol{b}$ through the lateral boundary of Q is under control. Actually, a technical variation of this is needed in order to take into account the fact that the level sets are not of the form $y_1 =$ constant: to do this we exploit Coarea Formula and a classical decomposition of finite perimeter sets into rectifiable parts (relying ultimately on De Giorgi's Rectifiability Theorem). We show that, up to a $|D^{\mathrm{sing}}\boldsymbol{b}|$-small set, one can find Lipschitz functions $y_1 = L_{t,h}(y^{\perp})$ in a fixed set of coordinates $(y_1, y^{\perp}) \in \mathbb{R} \times \mathbb{R}^{d+1}$, whose graphs cover a large fraction of the singular part $D^{\mathrm{sing}}\boldsymbol{b}\llcorner_{B^{d+1}_r(\bar{t},\bar{x})}$. We can at this point reverse the procedure, i.e. we construct a vector field starting from the level sets: this yields a BV vector field $\mathcal{U}(t)$ whose component $\mathcal{U}_1$ can be put into the right hand side of (12) and we can now perform the precise estimate of the flux of $\boldsymbol{b}$ through the lateral boundary of Q.

By an application of the Radon-Nikodym Theorem, it follows that on large compact set it holds that the flow integral (9) is controlled by $\tau|D^{\mathrm{sing}}\boldsymbol{b}|(B^{d+1}_r(\bar{t},\bar{x}))$. Finally a covering argument implies that the measure ζ^C_τ can be taken, in the BV case, to be $\tau|D\boldsymbol{b}|$: in view of the discussion above this is enough to conclude finally the proof of the Main Theorem.

5. **Further developments of the *untangling*.** In a work in progress (that will appear in a forthcoming paper [13]) we study some possible refinements of the concept of *untangling*. In particular, by imposing a control on the intersection of the curves only *forward in time* some estimates and propositions of the approach presented above simplify. More precisely, we define a Lagrangian representation η of $\rho(1,\boldsymbol{b})$, with $\mathrm{div}(\rho(1,\boldsymbol{b})) = \mu \in \mathscr{M}([0,T] \times \mathbb{R}^d)$, to be *forward untangled* when it is concentrated on a set $\Delta^{\mathrm{forward}}$ of curves which may intersect, but if they do then they remain the same curve in the future. In a sense, this means that trajectories can bifurcate only in the past.

This formulation arises naturally when one translates well-posedness of the ODEs in terms of Lagrangian representations: restricting for simplicity to the case in which $\mu = 0$ one would like to replace Assumption 4.1 with the following one:

Assumption 5.1. *Let η be a Lagrangian representation of $\rho(1,\boldsymbol{b})$ in $(0,T) \times \mathbb{R}^d$. Let $\varpi > 0$ and assume that for all $R > 0$ there exists $r = r(R) > 0$ such that*

$$\int_{\Gamma} \frac{1}{\sigma^r(\gamma)} \eta\Big(\Big\{\gamma' \in \Gamma : |\gamma(0) - \gamma'(0)| \le r, |\gamma(T) - \gamma'(T)| \ge R\Big\}\Big)\eta(d\gamma) \le \varpi.$$

where now

$$\sigma^r(\gamma) := \textit{amount of curves starting from the ball or radius } r > 0 \textit{ around } \gamma(0).$$

Assumption 5.1 has the advantage of making more transparent and easier some of the proofs used in the approach presented above. One can repeat the general scheme presented above: first one formulates Assumption 5.1 locally, in a proper set and shows that - up to a set of curves whose measure is controlled - the (restricted) Lagrangian representation η is forward untangled. In this way, one obtains a simpler proof of Theorem 1, avoiding the introduction of the crossing trajectories. Then one introduces the *forward untangling functional*, which turns out to be subadditive as well, exactly as in the setting above, allowing the usual local-to-global argument. Using this formulation of the untangling, we are able to recover in our setting the results of [16], where the authors considered vector fields whose derivative can be written as convolution between a singular kernel and a L^1 function and we also derive a quantitative stability estimate for a class of vector fields satisfying a suitable weak L^p bound on the gradient.

REFERENCES

[1] G. Alberti. Rank one property for derivatives of functions with bounded variation. *Proc. Roy. Soc. Edinburgh Sect. A*, 123(2):239–274, 1993.

[2] G. Alberti, S. Bianchini, and G. Crippa. Structure of level sets and Sard-type properties of Lipschitz maps. *Ann. Sc. Norm. Super. Pisa Cl. Sci. (5)*, 12(4):863–902, 2013.

[3] G. Alberti, S. Bianchini, and G. Crippa. A uniqueness result for the continuity equation in two dimensions. *J. Eur. Math. Soc. (JEMS)*, 16(2):201–234, 2014.

[4] L. Ambrosio. Transport equation and Cauchy problem for BV vector fields. *Inventiones mathematicae*, 158(2):227–260, 2004.

[5] L. Ambrosio, F. Bouchut, and C. De Lellis. Well-posedness for a class of hyperbolic systems of conservation laws in several space dimensions. *Comm. Partial Differential Equations*, 29(9-10):1635–1651, 2004.

[6] L. Ambrosio, M. Colombo, and A. Figalli. Existence and uniqueness of maximal regular flows for non-smooth vector fields. *Archive for Rational Mechanics and Analysis*, 218(2):1043–1081, 2015.

[7] L. Ambrosio and G. Crippa. Existence, uniqueness, stability and differentiability properties of the flow associated to weakly differentiable vector fields. In *Transport equations and multi-D hyperbolic conservation laws*, volume 5 of *Lect. Notes Unione Mat. Ital.*, pages 3–57. Springer, Berlin, 2008.

[8] L. Ambrosio, G. Crippa, and S. Maniglia. Traces and fine properties of a BD class of vector fields and applications. *Ann. Fac. Sci. Toulouse Math. (6)*, 14(4):527–561, 2005.

[9] L. Ambrosio, C. De Lellis, and J. Malý. On the chain rule for the divergence of BV-like vector fields: applications, partial results, open problems. In *Perspectives in nonlinear partial differential equations*, volume 446 of *Contemp. Math.*, pages 31–67. Amer. Math. Soc., Providence, RI, 2007.

[10] L. Ambrosio, M. Lecumberry, and S. Maniglia. Lipschitz regularity and approximate differentiability of the DiPerna-Lions flow. *Rend. Sem. Mat. Univ. Padova*, 114:29–50 (2006), 2005.

[11] G. Anzellotti. Traces of bounded vectorfields and the divergence theorem. 1983.

[12] S. Bianchini and P. Bonicatto. A uniqueness result for the decomposition of vector fields in $\mathbb{R}^d$. *Preprint SISSA 15/2017/MATE*, 2017.

[13] S. Bianchini and P. Bonicatto. Work in progress. 2017.

[14] S. Bianchini, P. Bonicatto, and N. A. Gusev. Renormalization for autonomous nearly incompressible BV vector fields in two dimensions. *SIAM J. Math. Anal.*, 48(1):1–33, 2016.

[15] S. Bianchini and N. A. Gusev. Steady nearly incompressible vector fields in two-dimension: chain rule and renormalization. *Arch. Ration. Mech. Anal.*, 222(2):451–505, 2016.

[16] F. Bouchut and G. Crippa. Lagrangian flows for vector fields with gradient given by a singular integral. *J. Hyperbolic Differ. Equ.*, 10(2):235–282, 2013.

[17] A. Bressan. An ill posed Cauchy problem for a hyperbolic system in two space dimensions. *Rend. Sem. Mat. Univ. Padova*, 110:103–117, 2003.

[18] N. Champagnat and P.-E. Jabin. Well posedness in any dimension for Hamiltonian flows with non BV force terms. *Comm. Partial Differential Equations*, 35(5):786–816, 2010.

[19] G. Crippa and C. De Lellis. Estimates and regularity results for the DiPerna-Lions flow. *J. Reine Angew. Math.*, 616:15–46, 2008.

[20] C. De Lellis. Notes on hyperbolic systems of conservation laws and transport equations. In *Handbook of differential equations: evolutionary equations. Vol. III*, Handb. Differ. Equ., pages 277–382. Elsevier/North-Holland, Amsterdam, 2007.

[21] N. Depauw. Non unicité des solutions bornées pour un champ de vecteurs BV en dehors d'un hyperplan. *C. R. Math. Acad. Sci. Paris*, 337(4):249–252, 2003.

[22] R. J. DiPerna and P.-L. Lions. Global weak solutions of Vlasov-Maxwell systems. *Comm. Pure Appl. Math.*, 42(6):729–757, 1989.

[23] R. J. DiPerna and P.-L. Lions. On the Cauchy problem for Boltzmann equations: global existence and weak stability. *Ann. of Math. (2)*, 130(2):321–366, 1989.

[24] R. J. DiPerna and P.-L. Lions. Ordinary differential equations, transport theory and Sobolev spaces. *Invent. Math.*, 98(3):511–547, 1989.

[25] R. J. DiPerna and P.-L. Lions. Global solutions of Boltzmann's equation and the entropy inequality. *Arch. Rational Mech. Anal.*, 114(1):47–55, 1991.

[26] P.-E. Jabin. Differential equations with singular fields. *Journal de MathÃ©matiques Pures et AppliquÃ©es*, 94(6):597 – 621, 2010.

[27] B. L. Keyfitz and H. C. Kranzer. A system of nonstrictly hyperbolic conservation laws arising in elasticity theory. *Arch. Rational Mech. Anal.*, 72(3):219–241, 1979/80.

[28] S. K. Smirnov. Decomposition of solenoidal vector charges into elementary solenoids and the structure of normal one-dimensional currents. *St. Petersburg Math. J.*, 5(4):841–867, 1994.

E-mail address: bianchin@sissa.it
E-mail address: paolo.bonicatto@unibas.ch

CONSERVATION LAWS WITH REGULATED FLUXES

Alberto Bressan*, Graziano Guerra† and Wen Shen*

*Department of Mathematics, Penn State University, University Park, PA 16802, U.S.A.

† Department of Mathematics and its Applications, University of Milano - Bicocca.

Abstract. Scalar conservation laws $\partial_t u + \partial_x f(t, x, u) = 0$ where the flux f is discontinuous w.r.t. the time and space variables t, x arise in many applications, related to physical models in rough media. Typical examples include traffic flow with variable road conditions and polymer flooding in porous media. An extensive body of recent literature has dealt with fluxes that are discontinuous along a finite number of curves in the t-x plane. Here we are interested in the existence and uniqueness of solutions obtained via vanishing viscosity approximations i.e. solutions to $\partial_t u + \partial_x f(t, x, u) = \varepsilon \partial_{xx} u$ when $\varepsilon \to 0^+$, for more general discontinuous fluxes.

We first give a definition of regulated functions in two variables. After recalling some results about parabolic equations with discontinuous coefficients, we show how the knowledge of the existence and uniqueness of the vanishing viscosity limit for fluxes with a single discontinuity at $x = 0$ can be used as a building block to prove the existence and uniqueness of the vanishing viscosity limit for regulated fluxes.

1. **Introduction.** We consider the Cauchy problem for a scalar conservation law of the form

$$\begin{cases} u_t + f(t, x, u)_x &= 0, \\ u(0, x) &= \bar{u}(x) \in \mathbf{L}^1(\mathbb{R}), \end{cases} \tag{1}$$

where the flux function f is smooth w.r.t. the unknown u but can be discontinuous w.r.t. both variables t and x. Our main concern is the convergence of the viscous approximations u^ε, which solve

$$\begin{cases} u_t + f(t, x, u)_x &= \varepsilon\, u_{xx}, \\ u(0, x) &= \bar{u}(x) \in \mathbf{L}^1(\mathbb{R}), \end{cases} \tag{2}$$

to a unique weak solution u to (1), as the viscosity parameter $\varepsilon \to 0^+$.

Starting with the works by N. Risebro and collaborators (see [2, 9, 10, 14] and references therein) scalar conservation laws with discontinuous coefficients have now become the subject of an extensive literature also including some multi-dimensional cases (see [1, 2, 7, 12, 13, 17] and references therein).

Results on the uniqueness and stability of vanishing viscosity solutions have been obtained mainly in the case where the flux f is piecewise smooth with discontinuities located on finitely many smooth curves on the (t, x) plane. Aim of this note is to describe an alternative approach, introduced in [3, 11], based on comparison

2000 *Mathematics Subject Classification.* 35L65, 35R05.

Key words and phrases. Nonlinear semigroups of contractions. Conservation law with discontinuous flux, regulated flux function, vanishing viscosity, Hamilton-Jacobi equation, existence and uniqueness of solutions.

estimates for solutions to the corresponding Hamilton–Jacobi equation. This yields the uniqueness of the vanishing viscosity limit under the more general assumption that $f(t,x,\omega) = F(v(t,x),\omega)$ where F is a smooth function and $v(t,x)$ is a *regulated* function of the two variables t and x, as in Definition 1.1 below.

We recall that a function of a single variable $v : \mathbb{R} \mapsto \mathbb{R}$ is *regulated* if it admits left and right limits at every point. This is true if and only if, for every interval $[x_1, x_2]$ and every $\varepsilon > 0$, there exists a piecewise constant function χ such that $\|\chi - v\|_{\mathbf{L}^\infty([x_1,x_2])} \leq \varepsilon$. We extend this concept to functions of two variables, as follows.

Definition 1.1. *(see Fig. 1) We say that a bounded function $v = v(t,x)$ is* **regulated** *if, for every intervals $[x_1, x_2]$ and $[0,T]$, and any $\varepsilon > 0$, the following holds.*

There exist finitely many disjoint subintervals $[a_i, b_i] \subseteq [0,T]$, Lipschitz continuous curves $\gamma_{i,1}(t) < \cdots < \gamma_{i,N_i}(t)$, $t \in [a_i, b_i]$, and constants $\alpha_{i,0}, \ldots, \alpha_{i,N_i}$ such that

(i) For every $t \in [a_i, b_i]$, the step function

$$\chi_i(t,x) \doteq \begin{cases} \alpha_{i,0}, & \text{if } x < \gamma_{i,1}(t), \\ \alpha_{i,k}, & \text{if } \gamma_{i,k}(t) < x < \gamma_{i,k+1}(t), \qquad k = 1,2,\ldots,N_i - 1, \\ \alpha_{i,N_i}, & \text{if } \gamma_{i,N_i}(t) < x, \end{cases} \tag{3}$$

satisfies $\|\chi_i(t,\cdot) - v(t,\cdot)\|_{\mathbf{L}^\infty([x_1,x_2])} \leq \varepsilon$.

(ii) For every i,k, the time derivative $\dot\gamma_{i,k}(t) = \frac{d}{dt}\gamma_{i,k}(t)$ coincides a.e. with a regulated function.

(iii) The intervals $[a_i, b_i]$ cover most of $[0,T]$, namely $T - \sum_i (b_i - a_i) \leq \varepsilon$.

We remark that, if $v = v(x)$ is independent of time, then it satisfies Definition 1.1 if and only if it is a regulated function in the usual sense.

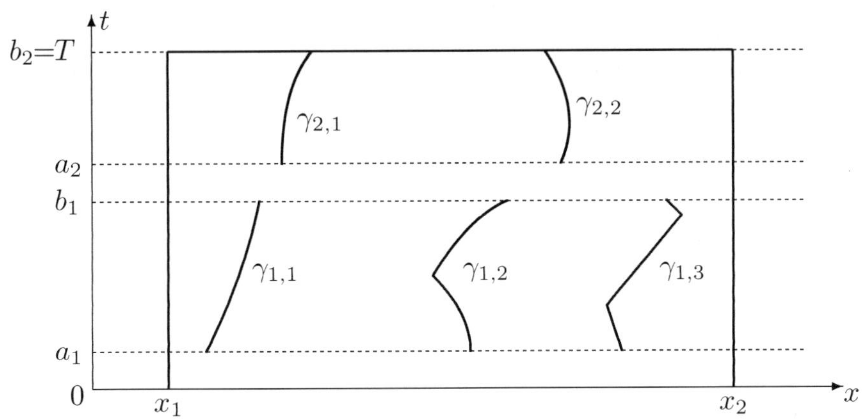

FIGURE 1. According to Definition 1.1, a **regulated** function of two variables $v = v(t,x)$ can be approximated by a piecewise constant function, with jumps along finitely many Lipschitz curves $\gamma_{i,k}$. The time derivatives $\dot\gamma_{i,k}$ are regulated functions.

Let $T > 0$ be given and consider the open domain $\Omega \doteq]0,T[\times \mathbb{R}$. For future use, we collect here some assumptions that will be imposed on the flux function $f : \Omega \times \mathbb{R} \mapsto \mathbb{R}$, at various stages of the analysis.

(F1): The function f satisfies:

(i) For each fixed $\omega \in \mathbb{R}$, the map $(t, x) \mapsto f(t, x, \omega)$ is in $\mathbf{L}^\infty(\Omega)$.
(ii) The map $\omega \mapsto f(t, x, \omega)$ is twice continuously differentiable for any $(t, x) \in \Omega$ and there exists a constant $L \geq 0$ such that

$$\big|f(t,x,\omega_1) - f(t,x,\omega_2)\big| \;\leq\; L\,|\omega_1 - \omega_2| \qquad \forall \omega_1, \omega_2 \in \mathbb{R},\ (t,x) \in \Omega. \tag{4}$$

(iii) There exists a constant $L_1 \geq 0$ such that, $\int_{\mathbb{R}} \big|f(t,x,0)\big|\,dx \;\leq\; L_1$, $\forall t \in\,]0, T[$.

(F2): For every $(t, x) \in \Omega$, the function f satisfies $f(t, x, 0) = 0$ and $f(t, x, 1) = h(t)$ for some $h \in \mathbf{L}^\infty\,(]0, T[\,, \mathbb{R})$.

(F3): The flux f has the form $f(t, x, \omega) = F(v(t, x), \omega)$, where $F(\alpha, \omega)$ is Lipschitz continuous w.r.t. α and twice continuously differentiable w.r.t. ω satisfying $F(\alpha, 0) = 0$ and $F(\alpha, 1) = h_1 \in \mathbb{R}$ for any $\alpha \in \mathbb{R}$, moreover v is a regulated function.

(F4): The flux f has the following form

$$f(x,\omega) = \begin{cases} f_l(\omega) & \text{if } x \leq 0, \\ f_r(\omega) & \text{if } x > 0, \end{cases}$$

where f_l and f_r are smooth functions satisfying $f_l(0) = f_r(0) = 0$ and $f_l(1) = f_r(1)$.

2. **Parabolic equations with discontinuous coefficients.** If f is smooth, under mild hypotheses on the growth of the solution, the Cauchy problem (2) is equivalent to the integral equation $u = \mathcal{P}^\varepsilon u$, where the transformation $\mathcal{P}^\varepsilon$ is defined by

$$(\mathcal{P}^\varepsilon u)(t,x) \;\doteq\; \int_{\mathbb{R}} G^\varepsilon(t, x-y)\,\bar{u}(y)\,dy - \int_0^t \int_{\mathbb{R}} G_x^\varepsilon(t-s,\ x-y) f\big(s, y, u(s,y)\big)\,dy\,ds. \tag{5}$$

For $t > 0$, the functions $G(t,x) \;\doteq\; \frac{1}{\sqrt{4\pi t}}\,e^{-x^2/4t}$ and $G^\varepsilon(t,x) \;\doteq\; \frac{1}{\sqrt{4\varepsilon\pi t}}\,e^{-x^2/4\varepsilon t}$ are the standard Gauss kernels. Observe that the equation $u = \mathcal{P}^\varepsilon u$ is meaningful even when f is discontinuous. Following [16], we say that $u = u(t, x)$ is a **mild solution** to the Cauchy problem (2) if it is a fixed point for of the transformation $\mathcal{P}^\varepsilon$. The following facts about mild solutions to (2) are proved in [3].

Theorem 2.1. *Consider the Banach space $Y_T \doteq \mathcal{C}^0([0,T], \mathbf{L}^1(\mathbb{R}))$ endowed with the supremum norm $\|u\|_T \doteq \sup_{t\in[0,T]} \|u(t)\|_{\mathbf{L}^1(\mathbb{R})}$. Let the flux function f satisfy* **(F1)**. *Then there exists a unique mild solution $u \in Y_T$ to the Cauchy problem (2). If u and v are two mild solutions of the parabolic equation in (2), with initial data $\bar{u}, \bar{v} \in \mathbf{L}^1(\mathbb{R})$. Then the following properties hold.*

(i) The total mass is conserved in time: $\int_{\mathbb{R}} u(t,x)\,dx = \int_{\mathbb{R}} \bar{u}(x)\,dx$, $\forall t \in [0,T]$.
(ii) A comparison principle holds: $\bar{u} \leq \bar{v} \implies u(t,\cdot) \;\leq\; v(t,\cdot)$, $\forall t \in [0,T]$.
(iii) The $\mathbf{L}^1$ distance between the two solutions is non-increasing in time:

$$\int_{\mathbb{R}} |u(t,x) - v(t,x)|\,dx \;\leq\; \int_{\mathbb{R}} |\bar{u}(x) - \bar{v}(x)|\,dx \qquad \textit{for all } t \geq 0. \tag{6}$$

We now consider a second Cauchy problem with different flux and initial data:

$$\begin{cases} u_t + f^\sharp(t,x,u)_x \;=\; \varepsilon u_{xx}\,, \\ u(0,x) \;=\; \bar{u}^\sharp(x) \in \mathbf{L}^1(\mathbb{R})\,. \end{cases} \tag{7}$$

The following theorem is based on comparison estimates for solutions to the related Hamilton–Jacobi equation. It provides a comparison between two solutions corresponding to not only different initial data, but also possibly different fluxes.

Theorem 2.2. [3, Theorem 2.3] *Let u and $u^\sharp$ be solutions to (2) and (7), respectively. Assume that both fluxes f and $f^\sharp$ satisfy* **(F1)**. *Let U and $U^\sharp$ be the integrated functions:*

$$U(t,x) = \int_{-\infty}^{x} u\,(t,\xi)\; d\xi, \qquad U^\sharp(t,x) = \int_{-\infty}^{x} u^\sharp\,(t,\xi)\; d\xi. \tag{8}$$

Then the following comparison property holds.

Let I be an interval containing the range of $u^\sharp(t,x)$ and assume that, for some $\eta \in \mathbf{L}^\infty\big([0,T]\big)$ and some constant $\bar\eta \geq 0$, one has

$$\begin{cases} f^\sharp(t,x,\omega) \leq f(t,x,\omega) + \eta(t) & \text{for all } (t,x,\omega) \in]0,T[\times\mathbb{R}\times I,\\ U(0,x) \leq U^\sharp(0,x) + \bar\eta & \text{for all } x \in \mathbb{R}. \end{cases} \tag{9}$$

Then, for all $t \in [0,T]$ and $x \in \mathbb{R}$, one has

$$U(t,x) \;\leq\; U^\sharp(t,x) + \bar\eta + \int_0^t \eta(s)\,ds\,. \tag{10}$$

3. The unique weak vanishing viscosity limit.

Without further hypotheses on the flux f, the solution to (2) could blow up as $\varepsilon \to 0^+$. Indeed consider the following linear example,

$$\begin{cases} u^\varepsilon_t + [\Theta(x)]_x = \varepsilon u^\varepsilon_{xx},\\ u\,(0,\cdot) = 0, \end{cases} \qquad \text{where} \qquad \Theta(x) = \begin{cases} 0 & \text{for } x \leq 0,\\ 1 & \text{for } x > 0. \end{cases}$$

Its mild solution is given by

$$u^\varepsilon(t,x) = -t\frac{1}{\sqrt{\varepsilon t}}\Phi\left(\frac{x}{\sqrt{\varepsilon t}}\right), \qquad \text{where} \quad \Phi\,(y) = 2G\,(1,y) - |y|\int_{|y|}^{+\infty} G\,(1,\xi)\,d\xi.$$

Since $u^\varepsilon \overset{*}{\rightharpoonup} -t\delta_0(x)$ as $\varepsilon \to 0^+$, it does not converge to any function even in a weak sense.

This motivates the introduction of Hypothesis **(F2)** that allows us to apply the maximum principle (namely, *(ii)* in Theorem 2.1) to the mild solutions to the parabolic equation (2). Indeed, let $f = f(t,x,\omega)$ be a flux function satisfying **(F1)**, **(F2)**, and consider the domain

$$\mathcal{D} \doteq \big\{u \in \mathbf{L}^1(\mathbb{R})\,;\; u(x) \in [0,1] \quad \text{for all} \quad x\big\}\,. \tag{11}$$

Let an initial data $\bar u \in \mathcal{D}$ be given. Since the constant functions $u^*(t,x) = 1$ and $u_*(t,x) = 0$ are solutions to the parabolic equation in (2) for any $\varepsilon > 0$, by a standard comparison argument the solution $u^\varepsilon(t,x)$ to (2) satisfies $u(t,\cdot) \in \mathcal{D}$ for all $t \in [0,T]$.

The bound in $\mathbf{L}^\infty$ gives weak* compactness of the sequence of functions, but the uniqueness of the limit as $\varepsilon \to 0^+$ requires additional analysis. Our main goal is to find a general class $\mathcal{F}$ of flux functions for which the vanishing viscosity limits are unique, for any fixed initial data in $\mathcal{D}$. As a starting point, by Theorem 5.2 in [11] we know that this class contains all fluxes $f = f(x,u)$ having one single discontinuity at $x = 0$. Next, we prove that this class is closed under certain elementary operations and suitable limits. By repeatedly applying these operations and taking limits, we conclude that all flux functions of the form $f(t,x,u) = F(v(t,x),u)$, with F

Lipschitz and v regulated, as in **(F3)**, lie in this class. Hence, for these fluxes the weak solutions obtained as vanishing viscosity limits are unique.

Definition 3.1. *We denote by $\mathcal{F}_{[a,b]}$ the family of all fluxes $f = f(t,x,u)$ that satisfy* **(F1)**, **(F2)** *for $t \in [a,b]$ (instead of $[0,T]$), and for which the following property holds. For any initial data $\bar{u} \in \mathcal{D}$, calling u^ε the solutions to the viscous Cauchy problem*

$$\begin{cases} u_t + f(t,x,u)_x &= \varepsilon\, u_{xx}, \\ u(a,x) &= \bar{u}(x) \in \mathbf{L}^1(\mathbb{R}), \end{cases} \tag{12}$$

as $\varepsilon \to 0^+$ the integrated functions

$$U^\varepsilon(t,x) = \int_{-\infty}^{x} u^\varepsilon(t,y)\, dy$$

converge uniformly in $[a,b] \times \mathbb{R}$ to a unique limit.

Since uniform convergence of the integrated function U^ε corresponds to weak convergence of u^ε (see [3, Lemma 3.1]), if $f \in \mathcal{F}_{[0,T]}$, then as $\varepsilon \to 0^+$ the solutions $u^\varepsilon(t,\cdot)$ to (2) converge weakly to a unique limit $u(t,\cdot)$ in the weak topology of $\mathbf{L}^1(\mathbb{R})$ for any fixed $t \in [0,T]$. Our eventual goal is to show that $\mathcal{F}_{[0,T]}$ contains all the flux functions satisfying **(F3)**. The following result, proved in [3] with the help of Theorem 2.2, describes the uniform limit under which $\mathcal{F}_{[a,b]}$ is closed.

Theorem 3.2. *Consider a flux $f = f(t,x,\omega)$ defined in $[0,T]\times\mathbb{R}\times[0,1]$, satisfying* **(F1)** *and* **(F2)**. *Assume that, for any $\delta > 0$, there exists times $0 < a_1 < b_1 < \cdots < a_N < b_N < T$ and flux functions $f_i \in \mathcal{F}_{[a_i,b_i]}$ such that $T - \sum_{i=1}^{N}(b_i - a_i) < \delta$,*

$$|f(t,x,\omega) - f_i(t,x,\omega)| < \delta, \quad \forall (t,x,\omega) \in [a_i,b_i] \times \mathbb{R} \times [0,1],\ i = 1,\dots,N. \tag{13}$$

Then $f \in \mathcal{F}_{[0,T]}$.

The classical result by Kruzhkov [15] implies that the vanishing viscosity limit exists and is unique for conservation law with smooth flux. Consequently, smooth fluxes belong to $\mathcal{F}_{[0,T]}$. An extensive body of more recent literature has dealt with fluxes satisfying hypothesis **(F4)**. In this case, one can again conclude that $f \in \mathcal{F}_{[0,T]}$, for every $T > 0$.

A detailed proof, based on the theory of nonlinear semigroups [4, 6, 5], can be found in [11]. Our approach avoids the technicalities in previous literature such as traces, Riemann problems, interface conditions, compensated compactness and entropy inequalities etc. , which generally require some additional hypotheses. Consequently the results in [11] holds under the general assumption **(F4)**. Theorems 3.4 and 5.2 in [11] can be restated in the following form.

Theorem 3.3. *Under hypothesis **(F4)**, the parabolic equation in (2) generates a unique continuous semigroup of contractions $S_t^\varepsilon : \mathcal{D} \to \mathcal{D}$ whose trajectories $S_t^\varepsilon \bar{u}$ are the unique mild solutions to (2). Moreover, as $\varepsilon \to 0^+$, for any $\bar{u} \in \mathcal{D}$, $S_t^\varepsilon \bar{u}$ converges in $\mathbf{L}^1(\mathbb{R})$ to $S_t \bar{u}$ uniformly on bounded t intervals, where $S_t : \mathcal{D} \to \mathcal{D}$ is a continuous semigroup of contractions whose trajectories are weak solutions to the Cauchy problem (1). Consequently if the flux f satisfies hypotheses **(F4)**, then $f \in \mathcal{F}_{[0,T]}$.*

By a change of variables it can be proved that the existence and uniqueness of the weak limit also holds when the interface between the two fluxes varies in time, under mild regularity assumptions.

Lemma 3.4. ([3, Lemma 3.5]) *Let f satisfy* ***(F4)****. Let $\gamma : [0,T] \mapsto \mathbb{R}$ be a Lipschitz function whose derivative $\dot\gamma$ coincides a.e. with a regulated function. Then the flux function $\tilde f$ defined by $\tilde f(t,x) = \tilde f(x - \gamma(t))$ belongs to $\mathcal{F}_{[0,T]}$.*

Thanks to the finite speed of propagation, the functions in $\mathcal{F}_{[0,T]}$ can be patched together horizontally, provided that they coincide on an intermediate domain.

Lemma 3.5. ([3, Lemma 3.6]) *Consider two flux functions f_1, f_2, both satisfying* **(F1)** *and* **(F2)**. *Assume that f_1, $f_2 \in \mathcal{F}_{[0,T]}$ and that there exists $\alpha < \beta$ such that $f_1(t,x,\omega) = f_2(t,x,\omega)$ for all $t \in [0,T]$, $x \in]\alpha,\beta[$, and $\omega \in [0,1]$. Then the flux f defined by*

$$f(t,x,\omega) \doteq \begin{cases} f_1(t,x,\omega) & \text{if } x < \beta \\ f_2(t,x,\omega) & \text{if } x > \alpha \end{cases} \tag{14}$$

belongs to $\mathcal{F}_{[0,T]}$.

Lemma 3.6. ([3, Lemma 3.8]) *Let $f = f(t,x,\omega)$ be a flux function satisfying* ***(F1)****,* ***(F2)****. Assume that, for every bounded interval $[x_1,x_2]$ the function*

$$\hat f(t,x,\omega) = \begin{cases} f(t,x_1,\omega) & \text{if } x < x_1, \\ f(t,x,\omega) & \text{if } x \in [x_1,x_2], \\ f(t,x_2,\omega) & \text{if } x > x_2, \end{cases} \tag{15}$$

lies in $\mathcal{F}_{[0,T]}$. Then $f \in \mathcal{F}_{[0,T]}$ as well.

Combining the previous results, the main theorem can be proved.

Theorem 3.7. *Let $f = f(t,x,\omega)$ be a flux function satisfying* ***(F3)****. Then $f \in \mathcal{F}_{[0,T]}$.*

Proof. By the assumption **(F3)**, the flux function f satisfies **(F1)** and **(F2)**.

Fix an interval $[x_1,x_2]$. Let $\delta > 0$ be given. Since v is regulated we can find disjoint intervals $[a_i,b_i]$, Lipschitz continuous curves $\gamma_{i,k}$ and constants $\alpha_{i,k}$ such that all conditions *(i)–(iii)* in Definition 1.1 hold.

For each i, let the piecewise constant function $\chi_i(t,x)$ be as in (3). Applying Lemma 3.5 and Lemma 3.4, by induction we show that the flux function

$$\begin{aligned} f_i(t,x,\omega) \doteq F(\chi_i(t,x),\omega) &= F(\alpha_{i,0},\omega)\,\chi_{\{x<\gamma_{i,1}(t)\}} \\ &\quad + \sum_{k=1}^{N_i-1} F(\alpha_{i,k},\omega)\,\chi_{\{\gamma_{i,k}(t)<x<\gamma_{k+1,1}(t)\}} \\ &\quad + F(\alpha_{i,N_i},\omega)\,\chi_{\{x>\gamma_{i,N_i}(t)\}} \end{aligned}$$

lies in $\mathcal{F}_{[a_i,b_i]}$. In turn, an application of Theorem 3.2 shows that the function $\hat f$ in (15) lies in $\mathcal{F}_{[0,T]}$. Since the interval $[x_1,x_2]$ is arbitrary, by Lemma 3.6, the flux function f lies in $\mathcal{F}_{[0,T]}$ as well. □

4. **The strong vanishing viscosity limit.** In this section, we assume **(F3)**. Moreover we consider the following additional hypotheses.

(V1): $v(t,x)$ is a bounded measurable function whose total variation w.r.t. x is integrable. More precisely, for every rectangular domain of the form $[0,T] \times [x_1,x_2]$ one has

$$\int_0^T (\text{Tot.Var.}\{v(t,\cdot);[x_1,x_2]\})\,dt < +\infty. \tag{16}$$

(V2): For each $\alpha \in \mathbb{R}$ the partial derivative $\omega \mapsto F_\omega(\alpha, \omega)$ is not constant on any open interval.

Under **(V1)**, the *unique* weak limit found in the previous section is a solution to the conservation law

$$u_t + f(t, x, u)_x = 0. \tag{17}$$

Moreover, if we assume **(V2)** as well, the convergence of u^ε is in $\mathbf{L}^1([0,T] \times \mathbb{R})$. These results can be obtained using a well established compensated compactness argument [8, 18].

Theorem 4.1. ([3, Theorem 4.2]) *Let the flux f satisfy* **(F1)**, **(F2)**, **(F3)** *and* ***(V1)***, *and choose an initial data $\bar{u} \in \mathcal{D}$. Let u^ε be the solution to the Cauchy problem* (2). *Then the unique weak viscosity limit $u(t, \cdot) = \lim_{\varepsilon \to 0} u^\varepsilon(t, \cdot)$ is a weak solution to the conservation law* (2).

Moreover if the flux satisfies ***(V2)*** *as well, then the convergence $u^\varepsilon \to u$ is in* $\mathbf{L}^1(\Omega)$ *endowed with its strong topology.*

REFERENCES

[1] B. Andreianov, K. H. Karlsen and N. H. Risebro, On vanishing viscosity approximation of conservation laws with discontinuous flux, *Netw. Heterog. Media*, **5** (2010), 617–633,

[2] B. Andreianov, K. H. Karlsen and N. H. Risebro, A theory of L^1-dissipative solvers for scalar conservation laws with discontinuous flux, *Arch. Ration. Mech. Anal.*, **201** (2011), 27–86,

[3] A. Bressan, G. Guerra and W. Shen, Vanishing viscosity solutions for conservation laws with regulated flux, *Journal of Differential Equations*, **266** (2019), 312 – 351.

[4] H. Brézis and A. Pazy, Convergence and approximation of semigroups of nonlinear operators in Banach spaces, *J. Functional Analysis*, **9** (1972), 63–74.

[5] M. G. Crandall and T. M. Liggett, Generation of semi-groups of nonlinear transformations on general Banach spaces, *Amer. J. Math.*, **93** (1971), 265–298,

[6] M. G. Crandall, The semigroup approach to first order quasilinear equations in several space variables, *Israel J. Math.*, **12** (1972), 108–132,

[7] G. Crasta, V. De Cicco and G. De Philippis, Kinetic formulation and uniqueness for scalar conservation laws with discontinuous flux, *Comm. Partial Differential Equations*, **40** (2015), 694–726,

[8] C. M. Dafermos, *Hyperbolic Conservation Laws in Continuum Physics*, vol. 325 of Grundlehren der Mathematischen Wissenschaften [Fundamental Principles of Mathematical Sciences], 3rd edition, Springer-Verlag, Berlin, 2010,

[9] T. Gimse and N. H. Risebro, Riemann problems with a discontinuous flux function, in *Third International Conference on Hyperbolic Problems, Vol. I, II (Uppsala, 1990)*, Studentlitteratur, Lund, 1991, 488–502.

[10] T. Gimse and N. H. Risebro, Solution of the Cauchy problem for a conservation law with a discontinuous flux function, *SIAM J. Math. Anal.*, **23** (1992), 635–648,

[11] G. Guerra and W. Shen, Vanishing viscosity and backward Euler approximations for conservation laws with discontinuous flux. *SIAM J. Math. Anal.* **51** (2019), 3112–3144.

[12] P. Gwiazda, A. Świerczewska-Gwiazda, P. Wittbold and A. Zimmermann, Multi-dimensional scalar balance laws with discontinuous flux, *J. Funct. Anal.*, **267** (2014), 2846–2883,

[13] K. H. Karlsen, M. Rascle and E. Tadmor, On the existence and compactness of a two-dimensional resonant system of conservation laws, *Commun. Math. Sci.*, **5** (2007), 253–265,

[14] R. A. Klausen and N. H. Risebro, Stability of conservation laws with discontinuous coefficients, *J. Differential Equations*, **157** (1999), 41–60,

[15] S. N. Kružkov, First order quasilinear equations with several independent variables, *Mat. Sb. (N.S.)*, **81 (123)** (1970), 228–255.

[16] R. H. Martin Jr., *Nonlinear Operators and Differential Equations in Banach spaces*, Wiley-Interscience [John Wiley & Sons], New York-London-Sydney, 1976, Pure and Applied Mathematics.

[17] E. Y. Panov, Existence and strong pre-compactness properties for entropy solutions of a first-order quasilinear equation with discontinuous flux, *Arch. Ration. Mech. Anal.*, **195** (2010), 643–673,

[18] D. Serre, *Systems of Conservation Laws. 1, 2*, Cambridge University Press, Cambridge, 2000.

E-mail address: axb62@psu.edu
E-mail address: graziano.guerra@unimib.it
E-mail address: wxs27@psu.edu

INITIAL DATA AND BLACK HOLES FOR MATTER MODELS

ANNEGRET Y. BURTSCHER
Department of Mathematics, IMAPP, Radboud University
PO Box 9010, Postvak 59
6500 GL Nijmegen, The Netherlands

ABSTRACT. To observe the dynamic formation of black holes in general relativity, one essentially needs to prove that closed trapped surfaces form during evolution from initial data that do not already contain trapped surfaces. We discuss the recent development of the construction of such admissible initial data for matter models. In addition, we extend known results for the Einstein equations coupled to perfect fluids in spherical symmetry and with linear equation of state to unbounded domains. Polytropic equations of state and regularity issues with the direct application of the singularity theorems in general relativity are discussed briefly.

1. **Introduction.** The Einstein equations in general relativity, with speed of light and Newton's gravitational constant normalized to 1, read

$$G_{\mu\nu} = 8\pi T_{\mu\nu}, \tag{1}$$

where the left hand side, the so-called Einstein tensor, is given in terms of the Ricci curvature and scalar curvature, $G_{\mu\nu} = R_{\mu\nu} + \frac{1}{2}R\, g_{\mu\nu}$, and the right hand side is the energy-momentum tensor of a particular matter model. Solutions to this equation are four-dimensional manifolds M with Lorentzian metric tensors $\mathbf{g}$, describing how light and particles travel in our universe. The first results on the local existence and uniqueness of solutions (for the vacuum equations) have been obtained in 1952 by Choquet–Bruhat [18]. Ever since, the global behavior of solutions is in the focus of attention.

Singular solutions are known since the discovery of the Schwarzschild solution in 1916, however, only several decades later the systematic study of singularities and black holes has taken off. According to Penrose's Singularity Theorem from 1965 (see, e.g., [20, 28]), a spacetime (M, g) is null geodesically incomplete if the following three conditions are met:

(i) $R_{\mu\nu}X^\mu X^\nu \geq 0$ for all null vectors X^μ,
(ii) there is a non-compact Cauchy surface in M, and
(iii) there is a closed trapped surface in M.

The first two conditions are met by any reasonable matter model, as the first condition is tied to the strong energy condition. The third condition, although tangible, is difficult to verify in general circumstances. It is therefore necessary

---

2000 *Mathematics Subject Classification.* Primary: 83C05; Secondary: 83C57, 83C75, 35A01, 35D30, 35L60, 76N10.

Key words and phrases. General relativity, perfect fluid, trapped surface, singularity, black hole, initial data, static solution, spherical symmetry.

to have some control over the parameters that illustrate trapping throughout the spacetime, initially as well as during evolution. We briefly examine how this was achieved for certain matter models. We will not discuss the vacuum case here, as it differs significantly from the treatment for matter models (spherically symmetry cannot be employed due to Birkhoff's Theorem) and an excellent review has already been written by Bieri [6].

The first time gravitational collapse was observed in the homogeneous spherically symmetric dust model by Oppenheimer and Snyder in 1939. Initially, the dust collapses into a region $r < 2M$, and then the scalar curvature at the singularity in the center blows up. Only later, however, singularities came into the picture and the term *black hole* was coined by Wheeler.

In a series of papers in the 1990s, Christodoulou considered global existence, uniqueness and regularity of solutions to the Einstein equations coupled to a massless scalar field in spherical symmetry. In [11] he provided conditions on the initial data that guaranteed the formation of trapped surfaces during evolution, and also proved the weak cosmic censorship conjecture in this case [15]. The large accumulation of mass in a controlled annular region guaranteed the existence of a trapped surface, even if the initial conditions were far from containing a trapped surface (though more time is required in case the initial data are close to flat). The initial conditions are specified on a future null cone and stated in terms of the mass function m and radius r.

In a more realistic setting, Rendall [27] and more explicitly Andréasson, Kunze and Rein [3] considered the gravitational collapse of collision gas modeled by the Vlasov equation in Schwarzschild coordinates. In astrophysics this model is used to describe galaxies and globular clusters. The solutions of the Einstein–Vlasov system are smooth, and no singularities occur for small initial data. Andréasson and Rein proved the formation of trapped surfaces in generalized Eddington–Finkelstein coordinates later in [4]. The benefit of the latter coordinates is that they can be used to cover the whole spacetime and do not break down at the event horizon. With the advanced null coordinate v and area radius r, dynamical spherically symmetric spacetimes are of the form

$$\mathbf{g} = -a(v,r)b^2(v,r)dv^2 + 2b(v,r)dvdr + r^2(d\theta^2 + \sin^2\theta d\varphi^2). \tag{2}$$

Asymptotic flatness is tied to the condition

$$\lim_{r\to\infty} a(v,r) = \lim_{r\to\infty} b(v,r) = 1. \tag{3}$$

A trapped surface $\{v_\bullet\} \times \mathbb{S}^2(r_\bullet)$ is present if $a(v_\bullet, r_\bullet) < 0$. Overall similar to the work of Christodoulou, the authors constructed suitable initial data leading to the formation of trapped surfaces out of spherically symmetric steady states at the center that are surrounded by a shell of matter moving inwards. The particle density was the key property that was adjusted to achieve this. Weak cosmic censorship holds for these data due to the work of Dafermos and Rendall [16, 17].

In the universe, black holes are expected to form when very massive stars collapse. In general relativity, stellar objects are described by a perfect fluid and modelled by the Einstein–Euler equations (1), where $T_{\mu\nu}$ is the energy-momentum tensor of a perfect fluid, given in terms of the pressure p, density ρ and velocity vector field u_μ, i.e.,

$$T_{\mu\nu} = (\rho + p)u_\mu u_\nu + p\, g_{\mu\nu}.$$

One of the major difficulties still is to describe the matter-vacuum boundary during evolution [7, 24]. In order to avoid this difficulty at first, LeFloch and the author studied the gravitational collapse of (spherically symmetric) perfect fluids with a priori infinite extent [10]. More precisely, the linear equation of state,

$$p = k^2 \rho,$$

for $k \in (0,1)$ representing the (normalized) speed of sound, was employed. In [10] it was shown that spherically symmetric steady states can be perturbed in an annular region by manipulating the normalized velocity in a way that while the initial data did not contain trapped surfaces, during the evolution trapped surfaces form. This approach also made use of the generalized Eddington–Finkelstein coordinates (2), however, (3) was not (could not) be used due to the unknown asymptotic behavior of static solutions. Thus rather than integrating from spatial infinity the analysis had to be restricted to an (albeit arbitrarily large but nevertheless) compact region. Only in recent work of Andersson and the author [2] on spherically symmetric static solutions of the Einstein–Euler equations, it became ultimately clear that perfect fluid solutions with linear equation of state are not asymptotically flat and how (3) needed to be modified in order to describe common perfect fluid solutions with infinite extent globally. The situation is different for equations of state that are only piecewise linear, e.g., as studied in the work of Christodoulou [12, 13, 14] and Fournodavlos and Schlue [19], however, no results on the formation of trapped surfaces are known in this setting and we will not discuss it further.

On the following pages, we employ the geometric description derived in [2] to extend the trapping results for perfect fluids of [10] to unbounded domains. We focus here on constructing admissible initial data, since the remaining local existence and trapping analysis based on a generalized random choice scheme and control during evolution can be carried over directly from [10].

2. **Construction of admissible initial data.** The crucial step in [10] is the construction of admissible initial data, that is, initial data that do not contain trapped surfaces but will evolve into solutions that do contain trapped surfaces during their time of existence. The nonexistence of trapped surfaces in the initial data is—in theory—easy to achieve, since it only requires to check that $a(v_0, r) > 0$ at the initial time v_0 for all r in question. In general, a can be computed using the integral representation

$$a(v,r) = 1 - \tfrac{4\pi(1+k^2)}{r} \int_0^r \tfrac{b(v,r')}{b(v,r)} M(v,r')(2k^2|V(v,r')| + 1) r'^2 \, dr', \qquad (4)$$

where $M = b^2 \rho u^0 u^0$ is a normalized mass and $V = \frac{u^1}{bu^0} - \frac{a}{2}$ is a normalized velocity [10, Sec. 2.3]. In practice, however, obtaining this positivity control on a is nontrivial. We investigate this problem in detail.

As mentioned in the Introduction, the idea to obtain admissible initial data is to construct static solutions and then introduce a large but localized perturbation to initiate trapped surface formation. Static solutions satisfy

$$V_{\text{static}} = -\tfrac{a_{\text{static}}}{2},$$

and do not contain trapped surfaces. The latter property should be preserved, to some extend, even with a large perturbation. Around the center $r = 0$ the sign of a is clearly positive due to the integral representation in (4), however, this property may not hold for large r. This problem did not occur in the work of Andréasson and Rein [4, Sec. 5], because due to the asymptotically flat model they used, the ADM mass M was finite and they could simply integrate a from spatial infinity. For

$$a(v,r) = 1 - \tfrac{2m(r)}{r},$$

their integral representation [4, Eq. (5.2)] from infinity is determined by

$$m(v,r) = \tfrac{M}{b(v,r)} - \tfrac{1}{2}\int_r^\infty 4\pi\eta^2(T_{11}+S)e^{-\int_\eta^r 4\pi\rho T_{11}\,d\sigma}\,d\eta, \tag{5}$$

where S depends on the density, the conserved angular momentum and canonical momenta corresponding to the coordinates (v, r, θ, φ).

In the setting of perfect fluids with linear equation of state an analogous integral representation of a is not possible due to the infiniteness of the ADM mass of the static solution. Therefore, in [10], we restricted our attention to solutions on a bounded domain. Recently, Andersson and the author investigated the asymptotic behavior of the static solutions to perfect fluids models with linear and polytropic-type equations of state in more detail. In [2, Thm. 1.2] was established that the solutions for linear equations of state are, in fact, asymptotically conical with deficit angle[1] $\alpha = \frac{4k^2}{(1+k^2)^2+4k^2}$ depending solely on the normalized speed of sound k. In a spacetime version, this behavior fits into the quasi-asymptotically set-up of Nucamendi and Sudarsky [23] (see also [5]), for which an alternative notion of ADM mass has been defined. This so-called ADMα mass is coordinate invariant and thus represents a geometric invariant, however, neither an analogue of the Positive Mass Theorem nor the fact that is constant over time have yet been established. A reasonable premise when dealing with perfect fluids with linear equation of state in general relativity would be to simply *assume* that the solutions are quasi-asymptotically flat. For the kind of initial data we are interested in, this assumption is satisfied due to [2, Thm. 1.2] (compact perturbations do not change the asymptotic behavior) and we can replace the use of the integral representation (5) in the Vlasov case involving the ADM mass M by employing the deficit angle α in a suitable way.

2.1. **Asymptotic behavior for static solutions revisited.** In order to understand quasi-asymptotic flatness in terms of the metric representation in coordinates (v, r, θ, φ) we rewrite the static solution in these coordinates.

Lemma 2.1 (Static solutions in generalized Eddington–Finkelstein coordinates). *Static spherically symmetric solutions of the Einstein–Euler equations for linear equations of state $p = k^2\rho$ are of the form*

$$\mathbf{g} = -a(r)b^2(r)dv^2 + 2b(r)dvdr + r^2(d\theta^2 + \sin^2\theta d\varphi^2) \tag{6}$$

with $a(r) = 1 - \frac{2m(r)}{r}$ for the mass function m, conical angle $\alpha = \frac{4k^2}{(1+k^2)^2+4k^2}$ and decay

$$\lim_{r\to\infty} a(r) = 1 - \alpha, \tag{7}$$

$$\lim_{r\to\infty} r^{-\frac{2k^2}{1+k^2}} b(r) = \left(\tfrac{2\rho_0}{\pi\alpha}\right)^{\frac{k^2}{1+k^2}} \tfrac{1}{\sqrt{1-\alpha}}, \tag{8}$$

[1]Note that in the notation of [2] the squared (normalized) speed of sound is denoted by $K = k^2$.

Proof. According to [2, Cor. 2.6] and [2, Cor. 3.6] solutions are of the form

$$\mathbf{g} = -e^{2\nu(r)}dt^2 + e^{2\lambda(r)}dr^2 + r^2(d\theta^2 \sin^2\theta d\varphi^2)$$

with

$$e^{2\Lambda} := \lim_{r\to\infty} e^{2\lambda(r)} = \lim_{r\to\infty}\left(1-\tfrac{2m(r)}{r}\right)^{-1} = \tfrac{(1+k^2)^2+4k^2}{(1+k^2)^2} = (1-\alpha)^{-1}$$

and $\nu'(r) = O(r^{-\frac{1}{2}})$ as $r\to\infty$. We set

$$v := t + \int_0^r e^{\lambda(s)-\nu(s)}\,ds.$$

Note that the integral converges because, as $r\to 0$ the asymptotic behavior is the metric coefficients is $e^{\lambda(r)} = \left(1-\frac{2m(r)}{r}\right)^{-\frac{1}{2}} \sim \frac{1}{\sqrt{1-r^2}} \to 1$ and $e^{\nu(r)} = \left(\frac{\rho_0}{\rho(r)}\right)^{\frac{k^2}{1+k^2}} \sim \left(\frac{\rho_0}{\rho_0}\right)^{\frac{k^2}{1+k^2}} = 1$ (cf. [2, (3.3) and Sec. 3.1]). Thus

$$dv = dt + e^{\lambda(r)-\nu(r)}dr,$$

and therefore

$$e^{2\nu(r)}dt^2 = e^{2\nu(r)}dv^2 - 2e^{\lambda(r)+\nu(r)}dvdr + e^{2\lambda(r)}dr^2.$$

The metric $\mathbf{g}$ in coordinates (v,r,θ,φ) thus is of the form

$$\mathbf{g} = -e^{2\nu(r)}dv^2 + 2e^{\lambda(r)+\nu(r)}dvdr + r^2(d\theta^2\sin^2\theta d\varphi^2),$$

which for

$$b(r) = e^{\lambda(r)+\nu(r)} \quad\text{and}\quad a(r) = e^{-2\lambda(r)} = 1-\tfrac{2m(r)}{r} \tag{9}$$

yields the desired form (6). By the above and by [2, Cor. 3.6] we obtain

$$\lim_{r\to\infty} a(r) = \lim_{r\to\infty} 1-\frac{2m(r)}{r} = 1-\alpha,$$

$$\lim_{r\to\infty} r^{-\frac{2k^2}{1+k^2}}b(r) = \lim_{r\to\infty} r^{-\frac{2k^2}{1+k^2}}\left(\tfrac{\rho_0}{\rho(r)}\right)^{\frac{k^2}{1+k^2}}\left(1-\tfrac{2m(r)}{r}\right)^{-\frac{1}{2}} = \left(\tfrac{2\rho_0}{\pi\alpha}\right)^{\frac{k^2}{1+k^2}}\tfrac{1}{\sqrt{1-\alpha}}.\ \square$$

Remark 1. The proof of the asymptotic behavior as $r\to\infty$ is based on the analysis in [2]. An explicit, so-called singular, solution of the static Einstein–Euler equations in spherical symmetry exists, to which all other solutions are asymptotic as $r\to\infty$. The density of this solution blows up at the center, hence the name "singular solution". In [9] we have shown that this solution is, although singular, still surprisingly well-behaved in a way that it satisfies the second Bianchi identity weakly. The stability of this solution may be studied using metric convergence, e.g., in the sense of Gromov–Hausdorff convergence or Sormani–Wenger intrinsic flat convergence [1, 6, 22, 29, 30].

In a general dynamic setting for the spherically symmetric Einstein–Euler equations with linear equation of state, one can reasonably assume that the initial data have the same asymptotic behavior as that obtained for static solutions in Lemma 2.1. Since we ore only interested in initial data based on static solutions with a compact perturbation, this is not a restriction for our set-up in the next Section.

2.2. **Construction of admissible initial data.** The idea is to construct admissible initial data for trapped surface formation on an *unbounded* domain. The presentation is inspired by [10, Sec. 6.2], where an analogous result has been obtained for arbitrarily large but *bounded* domains.

Let us recall the set-up of [10] for constructing admissible initial data for the spherically symmetric Einstein–Euler equations. The main goal was to observe the dynamic formation of trapped surface from untrapped initial data. The property that initial data do *not* contain trapped surfaces requires that

$$a(v_0, r) > 0 \qquad \text{for all } r \geq 0 \tag{10}$$

initially. In order to observe the formation of trapped surfaces, which corresponds to a sign change, i.e.,

$$a(v_\bullet, r_\bullet) < 0 \qquad \text{for some } v_\bullet, r_\bullet > 0,$$

we need to make sure that the initial data, in addition to (10), also satisfy

$$a_v(v_0, r) \ll 0 \qquad \text{for } r \in [r_* - \delta, r_* + \delta] \subseteq [0, \infty),$$

meaning that the derivative is large and negative in a small region. In [10] we proved (10) for arbitrarily large domains $[0, r_* + \Delta]$. The following result, based on on the asymptotic analysis of static solutions in Section 2.1, generalizes it to all of $[0, \infty)$. We start with a definition.

Definition 2.2. Let $(M^{(0)}, V^{(0)}, a^{(0)}, b^{(0)})$ be a static solution of the spherically symmetric Einstein–Euler equations with linear equation of state and central density $\rho_0 > 0$. Let $r_* > 0$, $\Delta \in (0, r_*)$ and $\delta \in (0, \Delta)$ and $h > 0$ be given. We consider a perturbation of the normalized fluid velocity, defined by a step function

$$V^{(1)}(r) = \begin{cases} 0 & r < r_* - \delta, \\ \frac{V^{(0)}(r)}{h} & r_* - \delta \leq r \leq r_* + \delta, \\ 0 & r > r_* + \delta. \end{cases}$$

We call (M_0, V_0, a_0, b_0) the *(r_*, δ, h)-perturbed initial data* if

$$M_0 = M^{(0)}, \qquad V_0 = V^{(0)} + V^{(1)}, \qquad b_0 = b^{(0)}, \tag{11}$$

and a_0 is given by the integral (cf. [10, Eq. (6.9)])

$$\begin{aligned} a_0(r) &= 1 - \tfrac{4\pi(1+k^2)}{r} \int_0^r \tfrac{b_0(s)}{b_0(r)} M_0(s) \left(2\tfrac{1-k^2}{1+k^2} |V_0(s)| + 1 \right) s^2 \, ds \\ &= 1 - \tfrac{4\pi(1+k^2)}{r} \int_0^r \tfrac{b^{(0)}(s)}{b^{(0)}(r)} M^{(0)}(s) \left(1 + \tfrac{1-k^2}{1+k^2} \left(1 + \tfrac{1}{h} \chi_{[r_*-\delta, r_*+\delta]}\right) a^{(0)}(s) \right) s^2 \, ds. \end{aligned} \tag{12}$$

Theorem 2.3. *Let (M_0, V_0, a_0, b_0) be a (r_*, δ, h)-perturbed initial data set to the spherically symmetric Einstein–Euler equations with linear equation of state $p = k^2\rho$, $k \in (0, 1)$ and central density $\rho_0 > 0$. Then there exist constants $C_1, C_2, C_3, C_4 > 0$ depending on $r_* > 0$ and a fixed*[2] *$\Delta \in (0, r_*)$ such that for all $\delta, h > 0$ with*

[2] We can also simply choose, for instance, $\Delta = \frac{r_*}{2}$ in order to avoid another parameter.

$\frac{\delta}{h} \le \frac{1}{C_1}$ *the following holds:*

$$0 < a_0(r) \le a^{(0)}(r), \quad r \ge 0,$$
$$\partial_v a_0(r) \begin{cases} = 0 & 0 \le r < r_* - \delta, \\ < 0 & r > r_* - \delta, \\ \le -C_2 \frac{\delta}{h^3} & r_* - \delta \le r \le r_* + \delta, \\ \le -C_4 \frac{1}{h^2} & r_* - \delta \le r \le r_* + \delta, \\ \le -C_3 \frac{\delta}{h} & r \in (r_* + \delta, r_* + \Delta]. \end{cases}$$

In particular, this initial data set does not contain trapped surfaces and $\partial_v a_0 \ll 0$ for suitably chosen δ and h.

Proof. We proceed as in the proof of [10, Prop. 6.1]. The major difference is Step 1, and we also generalize Step 2 and add an additional Step 5. Steps 3 and 4 can be obtained in the same fashion for a fixed $\Delta \in (0, r_*)$ (or simply $\Delta := \frac{r_*}{2}$).

Step 1. Positivity of a_0. Static solutions do not contain trapped surfaces, and thus $a^{(0)}$ is positive throughout. Due to (12), this immediately implies that

$$a_0(r) = a^{(0)}(r) > 0 \qquad \text{for all } r < r_* - \delta.$$

Let $r \ge r_* - \delta$. Then, by (12) and for $a^{(1)} := a_0 - a^{(0)}$,

$$\begin{aligned} a_0(r) &= a^{(0)}(r) + a^{(1)}(r) \\ &= a^{(0)}(r) - \frac{4\pi(1-k^2)}{rh} \int_{r_*-\delta}^{\min(r, r_*+\delta)} \frac{b^{(0)}(s)}{b^{(0)}(r)} M^{(0)}(s) a^{(0)}(s) s^2 \, ds \\ &\ge a^{(0)}(r) - \frac{4\pi(1-k^2)}{rh} \int_{r_*-\delta}^{r_*+\delta} \frac{b^{(0)}(s)}{b^{(0)}(r)} M^{(0)}(s) a^{(0)}(s) s^2 \, ds, \end{aligned} \tag{13}$$

since $M^{(0)}, b^{(0)}, a^{(0)} > 0$. By Lemma 2.1, and the fact that a is monotonically decreasing (cf. [10, Sec. 4.]) we know that

$$a^{(0)}(r) > 1 - \alpha > 0, \qquad \text{for all } r \ge 0,$$

where $\alpha = \frac{4k^2}{(1+k^2)^2 + 4k^2}$ is a constant strictly less than 1 for all $k \in [0, 1]$. It thus remains to be shown that the integral term in (13) is less than $1 - \alpha$. We show that this can be achieved for certain ratios of δ and h. Since, as $r \to \infty$, $b^{(0)} \ge 1$ is increasing and $\rho_0 \ge \rho^{(0)} = a^{(0)} M^{(0)} > 0$ (cf. [10, Eq. (4.5)]) is monotonically decreasing by [10, Thm. 4.3]) we obtain that

$$\begin{aligned} 0 < -a^{(1)}(r) &\le \frac{4\pi(1-k^2)}{rh} b^{(0)}(r_* + \delta) a^{(0)}(r_* - \delta) M^{(0)}(r_* - \delta) \left[\frac{r^3}{3}\right]_{r_*-\delta}^{r_*+\delta} \\ &\le \frac{8\pi(1-k^2)}{3} \frac{\delta}{h} \frac{\delta^2 + 3r_*^2}{r_* - \delta} b^{(0)}(r_* + \delta) \rho_0. \end{aligned}$$

Without loss of generality we may assume that $\delta \le \min\{\frac{r_*}{2}, \Delta\}$, hence $\frac{\delta^2 + 3r_*^2}{r_* - \delta} \le \frac{13 r_*}{2}$, so that we obtain

$$0 < -a^{(1)}(r) \le \frac{52\pi(1-k^2)}{3} \rho_0 \, r_* b^{(0)} \left(\tfrac{3r_*}{2}\right) \frac{\delta}{h}$$

Thus for $\frac{\delta}{h}$ sufficiently small, more precisely, for $\frac{\delta}{h} \le \frac{1}{C_1}$ with $C_1(r_*, \rho_0, k) := \frac{52\pi(1-k^2)}{3} \rho_0 \, r_* b^{(0)} \left(\frac{3r_*}{2}\right) (1 - \alpha)^{-1}$, we thus obtain that

$$-a^{(1)}(r) \le 1 - \alpha.$$

Therefore, for any $r \geq r_* - \delta$, we have

$$a_0(r) = a^{(0)}(r) + a^{(1)}(r) > 1 - \alpha - (1-\alpha) = 0.$$

Thus

$$a_0(r) > 0 \qquad \text{for all } r \geq 0,$$

and hence the initial datum does not contain trapped surfaces.

Step 2. Negativity of $\partial_v a_0$. By [10, Eq. (3.3)] we know that a must satisfy

$$a_v(v_0, r) = 2\pi r b^{(0)}(r) M^{(0)}(r)(a_0^2(r) - 4V_0^2(r)).$$

By [10, Thm. 4.3], static solutions satisfy $0 < a^{(0)} = -2V^{(0)} \leq 1$. Then (12) implies[3], for any $r \geq 0$,

$$a_v(v_0,r) = 2\pi r b^{(0)}(r)M^{(0)}(r)\big((a^{(0)}(r)+a^{(1)}(r))^2 - \big[a^{(0)}(r)(1+\tfrac{1}{h}\chi_{[r_*-\delta,r_*+\delta]}(r))\big]^2\big)$$

$$=2\pi r b^{(0)}(r)M^{(0)}(r)\big(a^{(1)}(r)(a_0(r)+a^{(0)}(r)) - \chi_{[r_*-\delta,r_*+\delta]}(a^{(0)}(r))^2\tfrac{2h+1}{h^2}\big). \tag{14}$$

Since $a^{(0)}$ and a_0 are positive for all $r > 0$ by Step 1, and $a^{(1)}$ is negative for $r > r_* - \delta$ by construction, we have that

$$a_v(v_0, r) < 0, \qquad \text{for all } r > r_* - \delta.$$

Step 3 and 4. Bounds for $\partial_v a_0$. One can proceed as in [10, Prop. 6.1] to obtain these bounds.

Step 5. Additional bound for $\partial_v a_0$ on $[r_* - \delta, r_* + \delta]$. As in Step 4 of [10] one obtains

$$a_v(v_0, r) \leq -2\pi r b^{(0)}(r) M^{(0)}(r)(a^{(0)}(r))^2 \tfrac{2h+1}{h^2}.$$

Since $b^{(0)} \geq 1$ is increasing and $\rho^{(0)} = M^{(0)}a^{(0)}$ is decreasing, and $a^{(0)} \geq 1 - \alpha$,

$$\begin{aligned} a_v(v_0, r) &\leq -2\pi(r_* - \delta)\rho^{(0)}(r_* + \delta)(1-\alpha)\tfrac{2h+1}{h^2} \\ &\leq -\tfrac{C_4}{h^2}, \qquad \text{for all } r \in [r_* - \delta, r_* + \delta], \end{aligned}$$

where C_4 depends on r_*, δ (or Δ, r_*),k, and ρ_0. □

Compared to [10, Prop. 6.1], the above Theorem 2.3 establishes three additional properties. We have shown that

(i) a_0 is positive for *all* $r \geq 0$ (and not just up to some $r_* + \Delta$),
(ii) $a_v < 0$ for *all* $r > r_* - \delta$ (and not just up to some $r_* + \Delta$),
(iii) $a_v \leq -C_4 \frac{1}{h^2}$ for $r \in [r_* - \delta, r_* + \delta]$ holds (in addition to $a_v \leq -C_3 \frac{\delta}{h^3}$).

Property (i), in particular, shows that admissible initial data can be constructed that do *not* contain trapped surfaces on the *unbounded domain* $\mathbb{R}^3$. All other properties of [10, Prop. 6.1] are preserved, so that the same procedure as in [10, Sec. 6 and 7] establishes the dynamic formation of trapped surfaces. The above Theorem 2.3 thus generalizes [10] to unbounded domains. For an exact formulation with all assumptions we refer the reader to [10, Thm. 6.4].

Corollary 1. *The initial value problem for the spherically symmetric Einstein–Euler equations with linear equation of state for a class of (r_*, δ, h)-perturbed initial data sets, prescribed on an unbounded Cauchy surface, leads to solutions with bounded variation with the following properties:*

(i) The spacetime is a spherically symmetric, future development of the initial data set.

[3] Note that the calculation [10, Eq. (6.13)] contains two minor typos.

(ii) The initial hypersurface does not contain trapped surfaces.
(iii) The spacetime does contain trapped surfaces.

Remark 2 (Generalization to other equations of state)**.** While no analysis on the formation of trapped surfaces for perfect fluid models have yet been performed for equations of state other than the linear one (even in spherical symmetry), the asymptotic behavior of static solutions w.r.t. polytropic-type equations of state, that is, equations of state of the form $p = K\rho^{\frac{n+1}{n}}$ with polytropic index $n > 5$, has also been described by Andersson and the author in [2]. These static solutions also have infinite extend and are also not asymptotically flat. Eventually, of course, one would be interested to study the formation of trapped surfaces for *bounded* fluid balls (models of stars). At the moment, this seems out of reach, as no suitable setting is yet available to study such evolution problems with a fluid–vacuum boundary, but may become available in the future [24].

3. **From trapped surfaces to black holes.** While the Penrose Singularity Theorem discussed in the Introduction would yield a singularity based on the existence of a closed trapped surface, this result requires a metric regularity of C^2 (and also a generalization requires at least $C^{1,1}$ [21]). In [10] solutions of bounded variation have been obtained which do not guarantee this regularity for all available derivatives. As such, the Singularity Theorems known today are not directly applicable. It may be possible to either extend the Singularity Theorems or to improve the regularity along the lines of [25, 26] of the solutions obtained in [10].

REFERENCES

[1] B. Allen and A.Y. Burtscher, Properties of the null distance and spacetime convergence, preprint, arXiv:1909.04483.
[2] L. Andersson and A.Y. Burtscher, On the Asymptotic Behavior of Static Perfect Fluids, *Ann. Henri Poincaré* **20** (2019), no. 3, 813–857.
[3] H. Andréasson, M. Kunze and G. Rein, The formation of black holes in spherically symmetric gravitational collapse, *Math. Ann.* **350** (2011), no. 3, 683–705.
[4] H. Andréasson and G. Rein, Formation of trapped surfaces for the spherically symmetric Einstein-Vlasov system, *J. Hyperbolic Differ. Equ.* **7** (2010), no. 4, 707–731.
[5] M. Barriola and A. Vilenkin, Gravitational field of a global monopole, *Physical Review Letters* **63** (1989), no. 4, 341–343.
[6] L. Bieri, Black hole formation and stability: a mathematical investigation, *Bull. Amer. Math. Soc. (N.S.)* **55** (2018), no. 1, 1–30.
[7] I. Brito and F.C. Mena, Initial boundary-value problem for the spherically symmetric Einstein equations with fluids with tangential pressure, *Proc. A.* **473** (2017), no. 2204, 20170113, 14.
[8] A.Y. Burtscher, Length structures on manifolds with continuous Riemannian metrics, *New York J. Math.* **21** (2015), 273–296.
[9] A.Y. Burtscher, M.K-H. Kiessling and A.S. Tahvildar-Zadeh, Weak second Bianchi identity for spacetimes with timelike singularities, preprint, arXiv:1901.00813.
[10] A.Y. Burtscher and P.G. LeFloch, The formation of trapped surfaces in spherically-symmetric Einstein-Euler spacetimes with bounded variation, *J. Math. Pures Appl. (9)* **102** (2014), no. 6, 1164–1217.
[11] D. Christodoulou, The formation of black holes and singularities in spherically symmetric gravitational collapse, *Comm. Pure Appl. Math.* **44** (1991), no. 3, 339–373.
[12] D. Christodoulou, Self-gravitating relativistic fluids: a two-phase model, *Arch. Rational Mech. Anal.* **130** (1995), no. 4, 343–400.
[13] D. Christodoulou, Self-gravitating relativistic fluids: the continuation and termination of a free phase boundary, *Arch. Rational Mech. Anal.* **133** (1996), no. 4, 333–398.
[14] D. Christodoulou, Self-gravitating relativistic fluids: the formation of a free phase boundary in the phase transition from soft to hard, *Arch. Rational Mech. Anal.* **134** (1996), no. 2, 97–154.

[15] D. Christodoulou, The instability of naked singularities in the gravitational collapse of a scalar field, *Ann. of Math. (2)* **149** (1999), no. 1, 183–217.

[16] M. Dafermos, Spherically symmetric spacetimes with a trapped surface, *Classical Quantum Gravity* **22** (2005), no. 11, 2221–2232.

[17] M. Dafermos and A.D. Rendall, An extension principle for the Einstein–Vlasov system in spherical symmetry, *Ann. Henri Poincaré* **6** (2005), no. 6, 1137–1155.

[18] Y. Fourès-Bruhat, Théorème d'existence pour certains systèmes d'équations aux dérivées partielles non linéaires,*Acta Math.* **88** (1952), 141–225.

[19] G. Fournodavlos and V. Schlue, On 'hard stars' in general relativity, *Ann. Henri Poincaré* (2019), DOI 10.1007/s00023-019-00793-4.

[20] S.W. Hawking and G.F.R. Ellis, *The large scale structure of space-time*, Cambridge Monographs on Mathematical Physics, No. 1, Cambridge University Press, London-New York, 1973.

[21] M. Kunzinger, R. Steinbauer and J.A. Vickers, The Penrose singularity theorem in regularity $C^{1,1}$, *Classical Quantum Gravity* **32** (2015), no. 15, 155010, 12.

[22] P.G. LeFloch and C. Sormani, The nonlinear stability of rotationally symmetric spaces with low regularity, *J. Funct. Anal.* **268** (2015), no. 7, 2005–2065.

[23] U. Nucamendi and D. Sudarsky, Quasi-asymptotically flat spacetimes and their ADM mass, *Classical Quantum Gravity* **14** (1997), no. 5, 1309–1327.

[24] T.A. Oliynyk, Dynamical relativistic liquid bodies, preprint, arXiv:1907.08192.

[25] M. Reintjes and B. Temple, No regularity singularities exist at points of general relativistic shock wave interaction between shocks from different characteristic families, *Proc. A.* **471** (2015), no. 2177, 20140834, 17.

[26] M. Reintjes, Spacetime is locally inertial at points of general relativistic shock wave interaction between shocks from different characteristic families, *Adv. Theor. Math. Phys.* **21** (2017), no. 6, 1525–1612.

[27] A.D. Rendall, Cosmic censorship and the Vlasov equation, *Classical Quantum Gravity* **9** (1992), no. 8, L99–L104.

[28] J.M.M. Senovilla and D. Garfinkle, The 1965 Penrose singularity theorem, *Classical Quantum Gravity* **32** (2015), no. 12, 124008, 45.

[29] C. Sormani and C. Vega, Null distance on a spacetime, *Classical Quantum Gravity* **33** (2016), no. 8, 085001, 29.

[30] C. Sormani and S. Wenger, The intrinsic flat distance between Riemannian manifolds and other integral current spaces, *J. Differential Geom.* **87** (2011), no. 1, 117–199.

E-mail address: burtscher@math.ru.nl

DISPERSIVE DYNAMICS OF THE DIRAC EQUATION ON CURVED SPACES

Federico Cacciafesta*

Dipartimento di Matematica, Università degli studi di Padova,
Via Trieste, 63, 35131
Padova PD, Italy.

Abstract. We discuss some results concerning dispersive properties of the Dirac dynamics on non-flat geometries: in particular we present some *local smoothing estimates* for asymptotically flat and warped products manifolds, and *local Strichartz estimates* for spherically symmetric manifolds. These results are obtained in collaboration with Anne-Sophie de Suzzoni and are contained in the papers [7], [8].

1. **Introduction.** The study of the dynamics of dispersive PDEs on curved manifolds is a subject that has attracted increasing interest in the last years, and has seen several striking contributions and breakthroughs. It is indeed now quite well understood how several parameters of a manifold, as e.g. compactness, presence of "trapping components", spherical symmetry, asymptotic behaviour of the coefficients, can affect the dispersion of dispersive flows. Attempting a detailed discussion of the state-of-the art of the theory is out of the scope of this note; we limit to mention the fact that the vast majority of the available results deal with the most celebrated models of the Schrödinger and wave equations in various geometrical contexts (compact, asymptotically flat, asymptotically conic manifolds...). The purpose of this note is to review some recent results in this direction concerning the *Dirac equation* for which, despite its huge interest in relativistic quantum mechanics and its relevance in several fields of applications, to the best of our knowledge nothing is known.

First of all, we recall that the *Dirac equation on* $\mathbb{R}^{1+3}$ is written as

$$iu_t + \mathcal{D}u + m\beta u = 0 \tag{1}$$

where $u : \mathbb{R}_t \times \mathbb{R}^3_x \to \mathbb{C}^4$, $m \geq 0$ is called the *mass*, the Dirac operator is defined as

$$\mathcal{D} = i^{-1} \sum_{k=1}^{n} \alpha_k \frac{\partial}{\partial x_k} = i^{-1}(\alpha \cdot \nabla),$$

and the 4×4 Dirac matrices can be written as

$$\alpha_k = \begin{pmatrix} 0 & \sigma_k \\ \sigma_k & 0 \end{pmatrix}, \quad k = 1, 2, 3, \qquad \beta = \begin{pmatrix} I_2 & 0 \\ 0 & -I_2 \end{pmatrix} \tag{2}$$

2000 *Mathematics Subject Classification.* Primary: 35Q41; Secondary: 37L50.

Key words and phrases. Dispersive PDEs, Dirac equation, Smoothing estimates, Strichartz estimates, curved spacetimes.

in terms of the Pauli matrices

$$\sigma_1 = \begin{pmatrix} 0 & 1 \\ 1 & 0 \end{pmatrix}, \quad \sigma_2 = \begin{pmatrix} 0 & -i \\ i & 0 \end{pmatrix}, \quad \sigma_3 = \begin{pmatrix} 1 & 0 \\ 0 & -1 \end{pmatrix}. \tag{3}$$

The first problem one has to face is the definition of the Dirac operator on a manifold. This is somehow a classical procedure, and requires the use of the so called *vierbein*, that roughly speaking are a set of matrices that "connect" the curved spacetime to the Minkowski one. It turns out that the Dirac operator on a manifold with given metric g_{jk} takes the form

$$\mathcal{D} = i\gamma^a e_a^\mu D_\mu \tag{4}$$

where the matrices $\gamma_0 = \beta$ and $\gamma^j = \gamma^0 \alpha_j$ for $j = 1, 2, 3$, e^j is a vierbein and D_j is the covariant derivative for fermionic fields. We refer to [14] and [7] for all the details. We stress the fact that such a construction is considerably more delicate than the one of the Laplace-Beltrami operator, that represents the counterpart of the generalization of the Laplacian to curved spaces: this fact is ultimately due to the rich algebraic structure of the Dirac operator.

In order to deal with more familiar objects, we consider metrics g_{jk} that decouple space and time, i.e. that have the following structure

$$g_{jk} = \begin{cases} \phi^{-2}(t) & \text{if } j = k = 0 \\ 0 & \text{if } jk = 0 \text{ and } j \neq k \\ -h_{jk}(\overrightarrow{x}) & \text{otherwise.} \end{cases} \tag{5}$$

The function ϕ is assumed to be strictly positive for all t and can in fact be assumed to be 1 after a change of variables in time. Within this setting, the Dirac equation can be written in the convenient form

$$i\phi\partial_t u - Hu = 0 \tag{6}$$

where H is an operator such that $H^2 = -\Delta_h + \frac{1}{4}\mathcal{R}_h + m^2$, and Δ_h and $\mathcal{R}_h$ are respectively the Laplace-Beltrami operator and the scalar curvature associated to the metrics h.

The rest of this note will be devoted to a brief presentation of the results contained in [7] and [8], and in particular to the discussion of the *linear estimates* associated to the flow of equation (6). More precisely, in section 2 we will present *local smoothing estimates* in the case of asymptotically flat and warped products manifolds, in section 3 *local Strichartz estimates* in the case of spherically symmetric mainfolds.

2. **Local smoothing estimates: the multiplier method.** The multiplier method has been widely used to prove local smoothing (or Morawetz) estimates for linear perturbations of dispersive flows in many different frameworks. Indeed, the main advantage of this technique is in its flexibility, which allows to deal with somehow rough objects and domains, and the "simple" nature of the calculations, that ultimately rely on integration by parts, that typically provide explicit conditions (smallness, positivity...) on the perturbations that may directly be checked. In the framework of the Dirac equation, we should mention at least [2] and [4] in which this strategy is developed to prove dispersive estimates for the electromagnetic Dirac equation in respectively 3D and higher space dimensions.

The first result we prove in [7] is a local smoothing estimate for *asymptotically flat manifolds*, that is manifolds (M, g) with g having the form (5) and the symmetric matrix $h(x) = [h_{jk}(x)]_{j,k=1}^n$ satisfying the following standard assumptions:

Assumptions (A1).

- $\nu|\xi|^2 \leq h_{jk}(x)\xi_j\xi_k \leq N|\xi|^2$ for some $\nu, N \in \mathbb{R}^+$ and for every $x,\ \xi \in \mathbb{R}^3$;
- $|h(x) - I| + |x||h'(x)| + |x|^2|h''(x)| + |x|^3|h'''(x)| \leq C_h\langle x\rangle^{-\sigma}$ for some C_h small enough and $\sigma \in (0,1)$.

With this set of assumptions, we have the following result.

Theorem 2.1. *Let u be a solution to* (6) *with initial condition u_0, with g satisfying* (5), *and assume that h satisfies Assumptions* **(A1)** *with the constants involved small enough. Then for $\eta_1, \eta_2 > 0$, there exists $C_{\eta_1,\eta_2} > 0$ independent from u such that*

$$\|\langle x\rangle^{-3/2-\eta_1}u\|_{L^2_\phi L^2_x} + \|\langle x\rangle^{-1/2-\eta_2}\nabla u\|_{L^2_\phi L^2_x} \leq C_{\eta_1,\eta_2}\|Hu_0\|^2_{L^2(\mathcal{M}_h)}. \tag{7}$$

Remark 1. The spaces L^2_ϕ and $L^2(\mathcal{M}_h)$ are defined in a standard way as

$$\|u\|^2_{L^2_\phi} = \int_0^{+\infty}\frac{|u(t)|^2}{\phi(t)}dt, \qquad \|f\|^2_{L^2(\mathcal{M}_h)} = \int_{D(h)}|f(x)|^2\sqrt{\det(h(x))}dx.$$

where $D(h)$ is the set where h is defined. Also, we will need the norm

$$\|f\|^2_{L^2(\mathcal{M}_g)} = \int_{\mathbb{R}\times D(h)}|f(t,x)|^2\sqrt{\det(g(t,x))}dxdt$$

Remark 2. We should mention that this result is close in spirit to [5] and [6], in which similar calculations are developed in a different contest, i.e. the one of the *Helmholtz equation*, to obtain weighted estimates and the limiting absorption principle in the same setting.

The second main result proved in [7] deals with the setting of *warped products manifolds*: beyond (5), we require on h the additional structure

$$h_{11} = 1, h_{1i} = h_{i1} = 0 \text{ if } i \neq 1, h_{ij} = \varphi(x^1)\omega_{ij}(x^2, x^3) \tag{8}$$

where ω is a 2×2 metric. In what follows we will use the more intuitive notation $r = x^1$. Notice that the choice $\varphi(r) = r^2$ and ω the metrics on the sphere $\mathbb{S}^2$ retrieves the flat case. With this condition, we get the following result.

Theorem 2.2. *Let u be a solution to* (6) *with initial condition u_0, with g satisfying* (5) *and h as in* (8). *Then the following results hold.*

- *(Hyperbolic-type metrics). Take $\varphi(r) = e^{r/2}$ in* (8) *and assume that for all* (x^2, x^3)

  $$\mathcal{R}_\omega(x^2, x^3) > 0, \qquad m^2 > \frac{3}{32}.$$

  Let $\eta_1, \eta_2 > 0$. There exists $C_{\eta_1,\eta_2} > 0$ such that for all u solution of the linear Dirac equation, we have

$$\|e^{-r/4}\langle r\rangle^{-(1+\eta_1)}u\|^2_{L^2(\mathcal{M}_g)} + \|e^{-r/4}\langle r\rangle^{-(1/2+\eta_2)}\nabla_h u\|^2_{L^2(\mathcal{M}_g)} \leq C_{\eta_1,\eta_2}\|Hu_0\|_{L^2(\mathcal{M}_h)}. \tag{9}$$

- *(Flat-type metrics). Take $\varphi(r) = r^2$ in* (8) *and assume that for all* (x^2, x^3),

  $$\mathcal{R}_\omega \geq 2, \qquad m > 0.$$

  Let $\eta_1, \eta_2 > 0$. There exists $C_{\eta_1,\eta_2} > 0$ such that for all u solution of the linear Dirac equation, we have

$$\|\langle r\rangle^{-(3/2+\eta_1)}u\|^2_{L^2(\mathcal{M}_g)} + \|\langle r\rangle^{-(1/2+\eta_2)}\nabla_h u\|^2_{L^2(\mathcal{M}_g)} \leq C_{\eta_1,\eta_2}\|Hu_0\|_{L^2(\mathcal{M}_h)}. \tag{10}$$

- *(Sub-flat type metrics) Take* $\varphi(r) = r^n$ *in* (8) *with* $n \in]2-\sqrt{2}, 4/3]$. *There exists* $C_n > 0$ *such that if for all* (x^2, x^3), $\mathcal{R}_\omega \geq C_n$, *then for all* $\eta_1, \eta_2 > 0$, *there exists* $C_{\eta_1,\eta_2,n} > 0$ *such that for all* u *solution of the linear Dirac equation, we have*

$$\|\langle r\rangle^{-(3/2+\eta_1)} u\|^2_{L^2(\mathcal{M}_g)} + \|\langle r\rangle^{-(1/2+\eta_2)} \nabla_h u\|^2_{L^2(\mathcal{M}_g)} \leq C_{\eta_1,\eta_2,n} \|Hu_0\|_{L^2(\mathcal{M}_h)}. \quad (11)$$

Remark 3. The various assumptions of positivity on the curvature and the mass are somehow technical, and are due to the nature of the mutiplier method, that ultimately requires to rely on the positivity of the various terms.

3. **Weighted Strichartz estimates in the spherically symmetric case.** The problem of proving Strichartz estimates, which typically represent the main tool in nonlinear applications and thus represent the ultimate goal to be proved in linear analysis, is considerably more complicated: indeed, as mentioned, the standard trick of relying on the available estimates on $\mathbb{R}^3$, Duhamel formula and local smoothing can not be applied as we are not dealing with zero-order perturbations, even in the asymptotically flat case. Anyway, it is possible to rely on some "radial structure" of the Dirac operator and on the so called *partial wave decomposition* to obtain some results in the case of spherically symmetric manifolds. We mention that our strategy is strongly inspired by [1], in which the authors develop the same argument for the Schrödinger equation. We assume indeed to have a manifold (M, g) defined by $M = \mathbb{R}_t \times \Sigma$ where $\Sigma = \mathbb{R}_x \times \mathbb{S}^2_{\theta,\phi}$ equipped with the Riemannian metrics

$$d\sigma = dr^2 + \varphi(r)^2 d\omega^2_{\mathbb{S}^2} \quad (12)$$

where $d\omega^2_{\mathbb{S}^2} = (d\theta + \sin^2\theta d\phi)$ is the Euclidean metrics on the 2D sphere $\mathbb{S}^2$. This of course is a special case of (8). Notice that taking $\varphi(r) = r$ reduces Σ to the standard 3D euclidean space, and therefore M to be the standard Minkowski space. We assume the following set of hypothesis on $\varphi(r)$, that are fairly natural in this contest.

Assumptions (A2) Take $\varphi(r) \in C^\infty(\mathbb{R}^+)$ strictly positive on $(0, +\infty)$, such that

$$\varphi(0) = \varphi^{(2n)}(0) = 0, \qquad \varphi'(0) = 1, \qquad \frac{|\varphi'(r)|}{\varphi(r)} \leq C \text{ for} |x| \geq 1. \quad (13)$$

With these assumptions, we are able to rely on some radial version of the Dirac operator: a crucial role in our analysis will be played indeed by the so called partial wave decomposition, for the details of which we refer to [15]. For the purpose of this note, we only limit to recall the existence of an isomorphism

$$L^2(\mathbb{R}^3)^4 \cong \bigoplus_{j,m_j,k_j} L^2((0,+\infty), \varphi^2(r)dr) \otimes \mathcal{H}_{j,m_j,k_j}$$

where the spaces $\mathcal{H}_{j,m_j,k_j}$, the so called partial wave subspaces, are 2-dimensional Hilbert spaces.

The main result we get is the following

Theorem 3.1. *Let* g *be as in* (5), h *having the structure* (12) *and satisfying Assumptions* **(A2)**. *Then for all bounded intervals* $I = (0, T)$, $T > 0$, *there exists a constant* C_T *such that the solutions* u *to* (6) *with initial condition* $u_0 \in$

$H^s((0,+\infty)\varphi(r)^2dr) \otimes \mathcal{H}_{j,m_j,k_j}$ *for any admissible triple* (j, m_j, k_j)*, with* $s = 2/p$ *or* $s = 1/p$ *if* $m > 0$*, for a fixed triple* n *satisfy estimates*

$$\left\| u \left(\frac{\varphi(r)}{r} \right)^{1-\frac{2}{q}} \right\|_{L^p_t(I)L^q(M)} \leq C_T \langle k_j \rangle \|u_0\|_{H^s(M)} \tag{14}$$

provided that $\frac{2}{p} + \frac{2}{q} = 1$ *and* $p \in]2, \infty]$ *but also*

$$\left\| u \left(\frac{\varphi(r)}{r} \right)^{1-\frac{2}{q}} \right\|_{L^p_t(I)L^q(M)} \leq C_T \langle k_j \rangle \|u_0\|_{H^s(M)} \tag{15}$$

provided that $m \neq 0$*,* $\frac{2}{p} + \frac{3}{q} = \frac{3}{2}$ *and* $p \in [2, \infty]$*.*

Remark 4. It is interesting to compare this result with the one obtained for the Schrödinger equation in the same setting. Indeed, in [1] the authors were able to obtain *global* Strichartz estimates for the dynamics, while we are here only able to obtain local ones. The reason for this difference is the following. The main idea of the proof in [1] relies on using radial coordinates to reduce the problem to a second order ODE (for radial initial data), then introduce a weighted function that roughly speaking transforms the equation in a Schrödinger equation on $\mathbb{R}^N$, with $N > 3$, perturbed by an electric potential $V(r)$. Such a potential, in general, shows a *scaling critical* decay at infinity, i.e. it behaves as r^{-2} for r large. Then, the existing theory on the Schrödinger equation with potentials (see [3]) can be exploited directly to obtain Strichartz estimates for the "weighted function", and so to obtain, after a re-change of variables, weighted Strichartz estimates for the original dynamics. Such a strategy can be transferred to the Dirac equation, but has some severe problems: first of all, the Dirac operator does not preserve radiality, and therefore one needs to rely on this much more sophisticated 2-dimensional decomposition mentioned above. Then, and this is the major difficulty, Strichartz estimates for the Dirac equation with scaling critical perturbations are *not* known, and therefore one can not rely on existing theory as in the Schrödinger case. Indeed, the problem of proving Strichartz estimates for the Dirac equation perturbed with critical potentials (e.g. the Coulomb potential) in the Euclidean setting is a major open problem of independent interest. To the best of our knowledge, the only results available are some local smoothing estimates obtained in [11] for the Dirac-Coulomb model and in [10] for the Dirac equation in Aharonov-Bohm field.

Theorem 3.1 can be extended to generic initial conditions, provided one introduces Sobolev spaces with angular regularity. The spaces $H^{a,b}$ for $a, b \in \mathbb{R}$ are defined by the norms

$$\|f\|_{H^{a,b}} = \Big[\sum_{j,m_j,k_j,\pm} \Big(\langle k_j \rangle^{2b} \|f_{j,m_j,k_k,\pm}\|^2_{H^a(\varphi^2(r)dr)} + \|f_{j,m_j,k_k,\pm}\|^2_{H^a(\varphi^2(r)dr)} \Big) \Big]^{1/2}$$

where

$$f_{j,m_j,k_j,\pm} = \langle f, \Phi^{\pm}_{m_j,k_j} \rangle_{L^2(\mathbb{S}^2)}$$

with $\{\Phi^+_{m_j,k_j}, \Phi^-_{m_j,k_j}\}$ an orthonormal basis of $\mathcal{H}_{j,m_j,k_j}$. In other words, we are taking a derivatives in radial coordinates, b derivatives in angular coordinates, and the L^2 norm on the whole manifold. We mention that spaces of this form are widely used in the contest of nonlinear dispersive PDEs. Then, one can deduce the following

Corollary 1. *Let g be as in (5), h having the structure (12) and satisfying Assumptions* **(A2)**. *Let $p, q \in [2,\infty]$ and $a, b \geq 0$. Assume either $p > 2$, $b > \frac{4}{p}$, $\frac{1}{p} + \frac{1}{q} = \frac{1}{2}$ and $\frac{2}{pa} + \frac{2}{pb} < 1$ or $m \neq 0$, $b \geq \frac{3}{p}$, $\frac{2}{p} + \frac{3}{q} = \frac{3}{2}$ and $\frac{1}{pa} + \frac{2}{pb} \leq 1$. Then for all bounded intervals $I = (0,T)$, $T > 0$, there exists a constant C_T such that the solutions u to (6) with initial condition u_0 such that $u_0 \in H^{a,b}$ satisfy the estimates*

$$\left\| u \left(\frac{\varphi(r)}{r} \right)^{1-\frac{2}{q}} \right\|_{L^p_t(I,L^q)} \leq C_T \|u_0\|_{H^{a,b}}. \tag{16}$$

Corollary 1 can be now exploited, in a more or less standard way, to obtain a well-posedness result for some nonlinear equations. The result is the following

Theorem 3.2. *Let g be as in (5), h having the structure (12) and satisfying Assumptions* **(A2)**. *Let $r > 0$, $r' = \max(r,2)$ and let $s_1 = \frac{3}{2} - \frac{3}{r'}$. Let $a, b > s_1$ such that $a < 2$ and*

$$\frac{2}{r'}\left(\frac{1}{a-s_1} + \frac{1}{b-s_1}\right) < 1 \text{ and } b > \frac{1}{r'} + \frac{3}{2}.$$

Then, for all $R \geq 0$ there exists $T(R) > 0$ such that for all $u_0 \in H^{a,b}$ with $\|u_0\|_{H^{a,b}} \leq R$, the Cauchy problem

$$\begin{cases} i\partial_t u - Hu = |\langle \beta u, u \rangle|^{\frac{r}{2}} u, \\ u(0,x) = u_0(x) \in H^{a,b} \end{cases} \tag{17}$$

has a unique solution in $\mathcal{C}([-T,T], H^{a,b})$ and the flow hence defined is continuous in the initial datum.

Perspectives. The results presented above represent only the first steps in the understanding of the (rich) dynamics of the Dirac equation on manifolds, and many questions and open problems naturally arise. A first one, which is of independent interest, is the study of dispersive dynamics of the Dirac equation with scaling-critical potential perturbations, and in particular of the Dirac-Coulomb equation, which naturally appears in many physical applications: a good understanding of the long time behaviour of it would allow to describe the dynamics of several interesting nonlinear models (see e.g. [9]). Then, it would be important to understand whether global Strichartz estimates for the Dirac equation on symmetric manifolds under assumptions **(A2)** hold or not; this might be done first for metrics that coincide with the flat one in some ball large enough, and then is asymptotically flat, or hyperbolic, or of some prescribed polynomial growth. This would allow to give also a global analog of Theorem (3.2), at least for small initial data.

REFERENCES

[1] V. Banica and T. Duyckaerts, Weighted Strichartz estimates for radial Schroedinger equation on noncompact manifolds. *Dyn. Partial Differ. Equ.* **4** (2007), no. 4, 335-359.

[2] N. Boussaid, P. D'Ancona, and L. Fanelli, Virial identity and weak dispersion for the magnetic dirac equation. *Journal de Mathématiques Pures et Appliquées*, **95** (2011) 137–150.

[3] N. Burq, F. Planchon, J.G. Stalker and A. Tahvildar-Zadeh Shadi, Strichartz estimates for the wave and Schrödinger equations with the inverse-square potential. *J. Funct. Anal.* **203** (2) (2003), 519–549.

[4] F. Cacciafesta, Virial identity and dispersive estimates for the n-dimensional Dirac equation. *J. Math. Sci. Univ. Tokyo* **18** (2011), 1-23.

[5] F. Cacciafesta, P. D'Ancona, and R. Lucá, Helmholtz and dispersive equations with variable coefficients on exterior domains. *SIAM J. Math. Anal.* **48**, (2016) no.3 1798-1832.

[6] F. Cacciafesta, P. D'Ancona, and R. Lucá, A limiting absorption principle with variable coefficients. *J. Spectral Theory* **4** (2018) 1349-1392.
[7] F. Cacciafesta and A.S. de Suzzoni, Weak dispersion for the Dirac equation on asymptotically flat and warped products spaces, *Discrete Contin. Dyn. Syst* **39** (2019), 4359-4398.
[8] F. Cacciafesta and A.S. de Suzzoni, Strichartz estimates for the Dirac equation on spherically symmetric spaces, arxiv: `https://arxiv.org/abs/1902.07572`.
[9] F. Cacciafesta, A. S. de Suzzoni, D. Noja, A Dirac field interacting with point nuclear dynamics. arxiv: `https://arxiv.org/abs/1709.05317`, to appear on *Math Ann.*.
[10] F. Cacciafesta and L. Fanelli, Dispersive estimates for the Dirac equation in an Aharonov-Bohm field. *J. Differential Equations*, **263** 7 (2017), 4382-4399.
[11] F. Cacciafesta and Eric Séré, Local smoothing estimates for the Dirac Coulomb equation in 2 and 3 dimensions. *J. Funct. Anal.* **271** 8 (2016), 2339-2358.
[12] L.D. Landau, L.M. Lifshitz, Quantum mechanics - Relativistic quantum theory.
[13] S. Machihara, M. Nakamura, K. Nakanishi, T. Ozawa, Endpoint Strichartz estimates and global solutions for the nonlinear Dirac equation. *J. Funct. Anal.*, **219** (2005), 1-20.
[14] L. E. Parker, D. J. Toms, Quantum field theory in curved spacetime. Cambridge university press.
[15] B. Thaller, *The Dirac Equation*, Springer-Verlag, Texts and Monographs in Physics (1992).

E-mail address: `cacciafe@math.unipd.it`

HIGH-ORDER FINITE VOLUME WENO SCHEMES FOR NON-LOCAL MULTI-CLASS TRAFFIC FLOW MODELS

Felisia Angela Chiarello and Paola Goatin

Inria Sophia Antipolis - Méditerranée, Université Côte d'Azur, Inria, CNRS, LJAD
2004 route des Lucioles - BP 93
06902 Sophia Antipolis Cedex, France.

Luis Miguel Villada*

GIMNAP-DMAT, Universidad del Bío-Bío, Casilla 5-C, Concepción, Chile
and
CI^{2}MA, Universidad de Concepción, Casilla 160-C, Concepción, Chile.

Abstract. This paper focuses on the numerical approximation of a class of non-local systems of conservation laws in one space dimension, arising in traffic modeling, proposed by [F. A. Chiarello and P. Goatin. Non-local multi-class traffic flow models. Networks and Heteroge-neous Media, 14(2), 371-387, 2019]. We present the multi-class version of the Finite Volume WENO (FV-WENO) schemes [C. Chalons, P. Goatin, and L. M. Villada. High-order numerical schemes for one-dimensional non-local conservation laws. SIAM Journal on Scientific Computing, 40(1), A288–A305, 2018], with quadratic polynomial reconstruction in each cell to evaluate the non-local terms in order to obtain high-order of accuracy. Simulations using FV-WENO schemes for a multi-class model for autonomous and human-driven traffic flow are presented for $M = 3$.

1. **Introduction.** We consider the following class of non-local systems of M conservation laws in one space dimension, introduced in [5] to model multi-class traffic dynamics:

$$\partial_t \rho_i(t,x) + \partial_x \left(\rho_i(t,x) v_i((r * \omega_i)(t,x))\right) = 0, \qquad i = 1, ..., M, \tag{1}$$

where

$$r(t,x) := \sum_{i=1}^{M} \rho_i(t,x), \tag{2}$$

$$v_i(\xi) := v_i^{\max} \psi(\xi), \tag{3}$$

$$(r * \omega_i)(t,x) := \int_x^{x+\eta_i} r(t,y)\omega_i(y-x)\,\mathrm{d}y\,, \tag{4}$$

where ρ_i is the density of vehicles belonging to the i-th class, v_i is the class-specific mean velocity and η_i is proportional to the look-ahead distance.

2000 *Mathematics Subject Classification.* Primary: 35L65, 90B20; Secondary: 65M08.

Key words and phrases. System of conservation laws, non-local flux, macroscopic traffic flow models, multi-class model, finite volume schemes, weighted essentially non-oscillatory scheme.

* Corresponding author: Luis Miguel Villada.

We assume that the following hypotheses hold:

(H1) The convolution kernels $\omega_i \in \mathbf{C}^1([0,\eta_i];\mathbb{R}^+)$, $\eta_i > 0$, are non-increasing functions with interaction strength $J_i := \int_0^{\eta_i} \omega_i(y)\, dy$.
We set $W_0 := \max_{i=1,\ldots,M} \omega_i(0)$.

(H2) $v_i^{\max}$ are the maximal velocities, with $0 < v_1^{\max} \leq v_2^{\max} \leq \ldots \leq v_M^{\max}$.

(H3) $\psi : \mathbb{R}^+ \to \mathbb{R}^+$ is a smooth non-increasing function such that $\psi(0) = 1$ and $\psi(r) = 0$ for $r \geq 1$.

We couple (1) with an initial datum

$$\rho_i(0,x) = \rho_i^0(x), \qquad i = 1,\ldots,M. \tag{5}$$

Model (1) is a generalization of the n-population model for traffic flow described in [1] and it is a multi-class version of the one dimensional scalar conservation law with non-local flux proposed in [2]. The term "non-local" refers to the speed functions v_i evaluated on a neighborhood of $x \in \mathbb{R}$ defined by the downstream convolution between the weight functions ω_i and the sum of the densities r. This is intended to describe the reaction of drivers that adapt their velocity to the downstream traffic, assigning greater importance to closer vehicles, see also [7, 8]. We consider different anisotropic discontinuous kernels for each equation of the system.

The model takes into account the distribution of heterogeneous drivers and vehicles characterized by their maximal speeds and look-ahead visibility in a traffic stream. It is worth to point out that in multi-class dynamic faster vehicles can overtake slower ones and slower vehicles slow down the faster ones, avoiding one of the biggest limitations of the standard LWR traffic flow model [9, 10], i.e. the first-in-first-out rule.

The computation of numerical solutions for (1) is challenging due to the high non-linearity of the system and the dependence of the flux function on integral terms. First and second order finite volume schemes for (1) were proposed and analyzed in [5, 6]. In this paper, a high-order finite-volume WENO (FV-WENO) scheme is proposed to solve the non-local multi-class system (1). The procedure proposed in [4] is used and extended to the multi-class cases in order to evaluate the non-local term that appears in the flux functions.

The paper is organized as follows. First, in Section 2, we describe the implementation of the high-order FV-WENO scheme for the non-local system (1). In Section 3, we provide a couple of numerical test in the case of three populations ($M = 3$) and convergence studies for third, fifth and seventh accuracy order.

2. Finite Volume WENO schemes.

In this section, we solve the non-local system of conservation laws (1) by using a high-order finite volume WENO scheme [11, 12]. First we consider $\{I_j\}_{j=1}^M$ as a partition of $[-L, L]$ and the points x_j are the center of the cells $I_j = [x_{j-\frac{1}{2}}, x_{j+\frac{1}{2}}]$, with length $|I_j| = \Delta x = \frac{2}{M}$. We denote the unknowns by $\rho_{i,j}(t)$, the cell average of the exact solution $\rho_i(t,\cdot)$ in the cell I_j :

$$\rho_{i,j}(t) := \frac{1}{\Delta x} \int_{I_j} \rho_i(t,x)dx.$$

We extend $\omega_i(x) = 0$ for $x > \eta_i$, and set

$$\omega_i^k := \frac{1}{\Delta x} \int_{(k-1)\,\Delta x}^{k\,\Delta x} \omega_i(x)\,\mathrm{d}x\,, \qquad k \in \mathbb{N}^*, \tag{6}$$

so that $\Delta x \sum_{k=1}^{+\infty} \omega_i^k = \int_0^{\eta_i} \omega_i(x)\,\mathrm{d}x = J_i$ (the sum is indeed finite since $\omega_i^k = 0$ for $k \geq N_i$ sufficiently large). Moreover, we set $r_j(t) := \sum_{i=1}^m \rho_{i,j}(t)$ and define the convolution term in the form $R_i(t,x) := (r * \omega_i)(t,x)$. Integrating (1) over I_j we obtain

$$\frac{d}{dt}\rho_{i,j}(t) = -\frac{1}{\Delta x}\left(f_i(t, x_{j+1/2}) - f_i(t, x_{j-1/2})\right), \quad i = 1,\dots,M, \quad \forall j \in \mathbb{Z},$$

where $f_i(t,x_{j+1/2}) := \rho_i(t, x_{j+\frac{1}{2}}) v_i(R_i(t, x_{j+\frac{1}{2}}))$. This equation is approximated by the semi-discrete conservative scheme

$$\frac{d}{dt}\rho_{i,j}(t) = -\frac{1}{\Delta x}\left(f_{i,j+\frac{1}{2}} - f_{i,j-\frac{1}{2}}\right), \quad i = 1,\dots,M, \quad \forall j \in \mathbb{Z}, \tag{7}$$

where $f_{i,j+\frac{1}{2}}$ is a consistent approximation of flux $\rho_i v_i(R_i)$ at interface $x_{j+1/2}$. Here, we consider the multi-class version of the Godunov scheme [5]

$$f_{i,j+\frac{1}{2}} := f(\rho^l_{i,j+\frac{1}{2}}, \rho^r_{i,j+\frac{1}{2}}) = \rho^l_{i,j+\frac{1}{2}} v_i(R^r_{i,j+1/2}), \tag{8}$$

where $\rho^l_{i,j+\frac{1}{2}}$ and $\rho^r_{i,j+\frac{1}{2}}$ are some left and right high-order WENO reconstructions of $\rho_i(t, x_{j+\frac{1}{2}})$ obtained from the cell averages $\{\rho_{i,j}(t)\}_{j\in\mathbb{Z}}$. In this work, we consider the classical WENO scheme proposed in [11, 12]. $R^r_{i,j+1/2}$ is the right approximation of $R_i(t,x)$ at the interface $x_{j+1/2}$. Since R_i is defined by a convolution, we naturally set $R^r_{i,j+1/2} = R_i(t, x_{j+1/2}) := R_{i,j+1/2}(t)$.

In order to compute the integral $R_{i,j+1/2}$, we use the technique proposed in [4], i.e., we consider a reconstruction of $\rho_i(x,t)$ on I_j by taking advantage of the high-order WENO reconstructions $\rho^r_{i,j-\frac{1}{2}}$ and $\rho^l_{i,j+\frac{1}{2}}$ at the boundaries of I_j, as well as the approximation of the cell average $\rho^n_{i,j}$. We consider a quadratic polynomial $p_{i,j}(x)$ defined on I_j such that

$$p_{i,j}(x_{j-\frac{1}{2}}) = \rho^r_{i,j-\frac{1}{2}}, \quad p_{i,j}(x_{j+\frac{1}{2}}) = \rho^l_{i,j+\frac{1}{2}}, \quad \frac{1}{\Delta x}\int_{I_j} p_{i,j}(x)dx = \rho^n_{i,j}.$$

In particular, we take

$$p_{i,j}(x) := a_{i,j,0}v^{(0)}(\xi_j(x)) + a_{i,j,1}v^{(1)}(\xi_j(x)) + a_{i,j,2}v^{(2)}(\xi_j(x)), \qquad x \in I_j, \tag{9}$$

with

$$v^{(0)}(y) = 1, \qquad v^{(1)}(y) = y, \qquad v^{(2)}(y) = \frac{1}{2}\left(3y^2 - 1\right), \qquad \xi_j(x) = \frac{x - x_j}{\Delta x/2}.$$

Coefficients in (9) can be easily computed as

$$a_{i,j,0} = \rho^n_{i,j}, \quad a_{i,j,1} = \frac{1}{2}\left(\rho^l_{i,j+\frac{1}{2}} - \rho^r_{i,j-\frac{1}{2}}\right), \quad a_{i,j,2} = \frac{1}{2}\left(\rho^l_{i,j+\frac{1}{2}} + \rho^r_{i,j-\frac{1}{2}}\right) - \rho^n_{i,j}.$$

Now, summing for $i = 1,\dots,M$, we have

$$P_j(x) := \sum_{i=1}^M p_{i,j}(x) = \hat{a}_{j,0}v^{(0)}(\xi_j(x)) + \hat{a}_{j,1}v^{(1)}(\xi_j(x)) + \hat{a}_{j,2}v^{(2)}(\xi_j(x)), \quad x \in I_j,$$

with

$$\hat{a}_{j,0} := \sum_{i=1}^M a_{i,j,0} = r_j, \qquad \hat{a}_{j,1} := \sum_{i=1}^M a_{i,j,1}, \qquad \hat{a}_{j,2} := \sum_{i=1}^M a_{i,j,2}.$$

With this polynomial $P_j(x)$, we can compute $R_{i,j+\frac{1}{2}}$ as

$$\begin{aligned} R_{i,j+\frac{1}{2}} &= \sum_{k=1}^{M} \int_{I_{j+k}} P_{j+k}(y)\omega_i(y - x_{j+\frac{1}{2}})dy \\ &= \sum_{k=1}^{N} \int_{I_{j+k}} \omega_i(y - x_{j+\frac{1}{2}}) \sum_{l=0}^{2} \hat{a}_{j+k,l} v^{(l)}(\zeta_{j+k}(y))dy \\ &= \sum_{k=1}^{M}\sum_{l=0}^{2} \hat{a}_{j+k,l} \int_{I_{j+k}} \omega_i(y - x_{j+\frac{1}{2}}) v^{(l)}(\zeta_{j+k}(y))dy \\ &= \sum_{k=1}^{M}\sum_{l=0}^{2} \hat{a}_{j+k,l} \underbrace{\frac{\Delta x}{2} \int_{-1}^{1} \omega_i\left(\frac{\Delta x}{2} y + (k - \frac{1}{2})\Delta x\right) v^{(l)}(y)dy} \\ &= \sum_{k=1}^{M}\sum_{l=0}^{2} \hat{a}_{j+k,l}\Gamma_{i,k,l}, \end{aligned} \tag{10}$$

where the coefficients $\Gamma_{i,k,l}$ are computed exactly or using a high-order quadrature approximation.

The utilization of the quadratic polynomial on each cell to evaluate the convolution term suggests the following algorithm to approach the solution of non-local system (1):

Algorithm: FV-WENO scheme for non-local multi-class traffic models. Given $\rho_{i,j}^n$ for $j \in \mathbb{Z}$, $i = 1, \ldots, M$, approximation of the cell averages of $\rho_i(x,t)$ at t^n.

1. Compute $\rho_{i,j+\frac{1}{2}}^l$ and $\rho_{i,j+\frac{1}{2}}^r$, the left and right high-order WENO approximations for $j \in \mathbb{Z}$ and $i = 1, \ldots, M$;
2. Calculate $R_{i,j+\frac{1}{2}}$ for $j \in \mathbb{Z}$ and $i = 1, \ldots, M$;
3. Calculate the Godunov numerical flux (8) for $j \in \mathbb{Z}$ and $i = 1, \ldots, M$;
4. Use a high-order accurate Runge-Kutta method to solve the semi-discrete system (7), with the CFL condition

$$\frac{\Delta t}{\Delta x} v_M^{\max} \|\psi\|_\infty \leq \frac{1}{2}. \tag{11}$$

In this paper, we use the WENO method of third (WENO3), fifth (WENO5) and seventh (WENO7) accuracy order proposed by [11, 12]. For the temporal discretization, in order to match the order of spatial accuracy, fifth or seventh explicit Runge-Kutta schemes are used [3].

3. **Numerical tests.** In the following numerical tests, we solve (1) numerically in the intervals $x \in [-1, 1]$ and $t \in [0, 2]$. We propose two tests in order to illustrate the dynamics of the model (1) for autonomous and human-driven vehicles, using FV-WENO5 scheme with $1/\Delta x = 400$. For each integration, we set Δt to satisfy the CFL condition (11).

To test the accuracy order of the proposed method, since we cannot compute the exact solution explicitly, we use a reference solution $\overline{\rho}^{ref}$ obtained using FV-WENO7 on a refined mesh ($1/\Delta x = 6400$). The $\mathbf{L^1}$-error for the cell average is given by

$$L^1(\Delta x) = \sum_{i=1}^{M} \left(\frac{1}{N} \sum_{j=1}^{N} |\overline{\rho}_{i,j} - \overline{\rho}_{i,j}^{ref}| \right),$$

where $\overline{\rho}_{i,j}$ and $\overline{\rho}_{i,j}^{ref}$ are the cell averages of the numerical approximation and the reference solution respectively. The Experimental Order of Accuracy (E.O.A.) is naturally defined by

$$\gamma(\Delta x) = \log_2 \left(L^1(\Delta x)/L^1(\Delta x/2) \right).$$

3.1. **Test 1, circular road.** The aim of this test is to study the possible impact of the presence of Connected Autonomous Vehicles (CAVs) on road traffic performances, as proposed in [7, Section 4.2]. Let us consider a circular road modeled by the space interval $[-1,1]$ with periodic boundary conditions at $x = \pm 1$. The interaction radius of CAVs is much grater than the one of human-driven cars. Moreover, we can assign a constant convolution kernel to CAVs, since we assume that the information they get about surrounding traffic is transmitted through wireless connections and its degree of accuracy does not depend on distance. We consider the following initial data and parameters

$$\rho_1(0,x) = \alpha p(x), \qquad \omega_1(x) = \frac{1}{\eta_1}, \qquad \eta_1 = 0.3, \qquad v_1^{\max} = 0.8, \tag{12}$$

$$\rho_2(0,x) = \beta\, p(x), \qquad \omega_2(x) = \frac{1}{\eta_2}, \qquad \eta_2 = 0.3, \qquad v_2^{\max} = 1.2, \tag{13}$$

$$\rho_3(0,x) = \gamma\, p(x), \qquad \omega_3(x) = \frac{2}{\eta_3}\left(1 - \frac{x}{\eta_3}\right), \qquad \eta_3 = 0.05, \qquad v_3^{\max} = 1.2, \tag{14}$$

where $p(x) = 0.5 + 0.3\sin(5\pi x)$ is the total initial density, $\alpha, \beta, \gamma \geq 0$ and $\alpha + \beta + \gamma = 1$. Above, ρ_1 represents the density of autonomous trucks, ρ_2 is the density of autonomous cars and ρ_3 is the density of human-driven cars. In Figure 1(a) we consider the penetration rates

$$\alpha = 0.5, \quad \beta = 0.3, \quad \gamma = 0.2,$$

and we can compare the total density $r = \rho_1 + \rho_2 + \rho_3$ with that one in Figure 1(b) where we have no human-driven cars:

$$\alpha = 0.5, \quad \beta = 0.5, \quad \gamma = 0.$$

We observe that oscillations are reduced if only autonomous vehicles are present.

Finally, we compute the E.O.A. for the FV-WENO schemes. We consider parameters $\alpha = 0.5$, $\beta = 0.3$, $\gamma = 0.2$, and compute the $\mathbf{L^1}$-error at $T = 0.2$ in Table 1. As expected, we obtain the correct order.

3.2. **Test 2, stretch of straight road.** In this test case, we consider a stretch of road populated by cars and trucks as in the example proposed in [5, Section 4.2]. The space domain is given by the interval $[-1,1]$ and we impose absorbing conditions at the boundaries. The dynamics is described by the equation (1) with

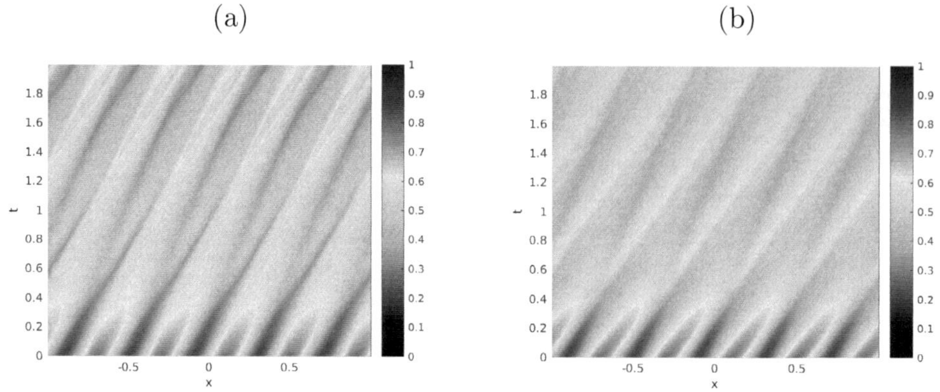

FIGURE 1. (t,x)−plots of the total density $r(t,x) = \rho_1(t,x) + \rho_2(t,x) + \rho_3(t,x)$ computed with the FV-WENO5 scheme, corresponding to different penetration rates of autonomous and non-autonomous vehicles: (a) $\alpha = 0.5$, $\beta = 0.3$, $\gamma = 0.2$, mixed autonomous / human-driven traffic, (b) $\alpha = 0.5$, $\beta = 0.5$, $\gamma = 0$. fully autonomous traffic.

	FV-WENO3		FV-WENO5		FV-WENO7	
$1/\Delta x$	L^1-err	$\gamma(\Delta x)$	L^1-err	$\gamma(\Delta x)$	L^1-err	$\gamma(\Delta x)$
100	1.51e-03	–	1.09e-04	–	5.64e-05	–
200	1.38e-04	3.44	9.44e-06	3.53	1.54e-06	5.19
400	1.20e-05	3.53	4.01e-07	4.56	1.58e-08	6.61
800	1.27e-06	3.24	1.26e-08	4.99	1.68e-10	6.55
1600	1.05e-07	3.01	3.60e-10	5.12	4.71e-12	5.15

TABLE 1. E.O.A. Test 1, initial condition (12)-(14), with $\alpha = 0.5$, $\beta = 0.3$, $\gamma = 0.2$ and final time $T = 0.2$. The reference solution is computed with FV-WENO7 scheme for $1/\Delta x = 6400$.

$M = 3$, and the following initial conditions and parameter values

$$\rho_1(0,x) = 0.5\chi_{[-0.6,-0.1]}(x), \quad \omega_1(x) = \frac{2}{\eta_1}\left(1 - \frac{x}{\eta_1}\right), \quad \eta_1 = 0.1, \quad v_1^{\max} = 0.8, \tag{15}$$

$$\rho_2(0,x) = \alpha_1\chi_{[-0.9,-0.6]}(x), \quad \omega_2(x) = \frac{1}{\eta_2}, \quad \eta_2 = 0.5, \quad v_2^{\max} = 1.3, \tag{16}$$

$$\rho_3(0,x) = \beta_1\chi_{[-0.9,-0.6]}(x), \quad \omega_3(x) = \frac{2}{\eta_3}\left(1 - \frac{x}{\eta_3}\right), \quad \eta_3 = 0.05, \quad v_2^{\max} = 1.3. \tag{17}$$

In this setting, $\rho_1(t,x)$ describes the density of human-driven trucks, $\rho_2(x,t)$ the density of autonomous cars and $\rho_3(x,t)$ is density of human driven cars. We have a red traffic light located at $x = -0.1$, which turns green at the initial time $t = 0$.

In Figure 2(a) we consider the rates

$$\alpha_1 = 0.25, \quad \beta_1 = 0.25,$$

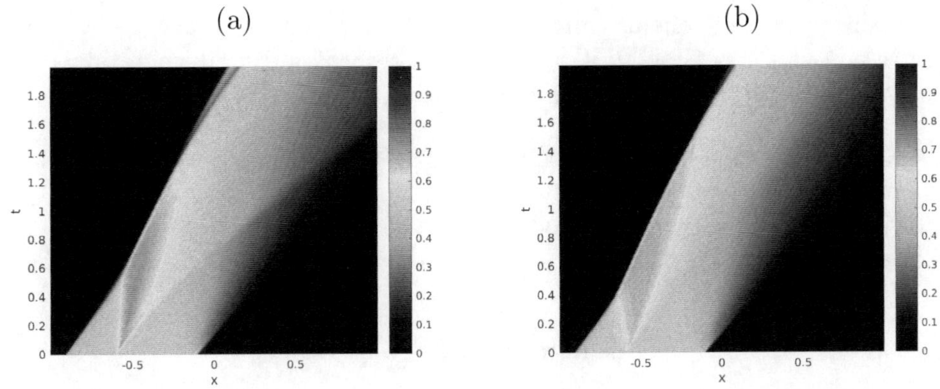

FIGURE 2. (t, x)−plots of the total density $r(t, x) = \rho_1(t, x) + \rho_2(t, x) + \rho_3(t, x)$ computed with FV-WENO5 scheme, corresponding to different penetration rates of autonomous and non-autonomous vehicles. (a) $\alpha_1 = 0.25$, $\beta_1 = 0.25$, (b) $\alpha_1 = 0$, $\beta_1 = 0.5$.

and we can compare the space-time evolution of the total density $r = \rho_1 + \rho_2 + \rho_3$ with the one in Figure 2(b), where

$$\alpha_1 = 0, \quad \beta_1 = 0.5.$$

In this case, the presence of autonomous cars in a heterogeneous traffic of human-driven vehicles induces higher vehicle densities during the overtaking phase, but for shorter time. In Figure 3 we display the density profiles of ρ_1, ρ_2 and ρ_3 computed with different FV-WENO schemes at time $t = 0.5$ in the same setting of Test 2(a). We can appreciate the efficiency of FV-WENO schemes in presence of discontinuities in comparison with the finite volume Godunov type scheme.

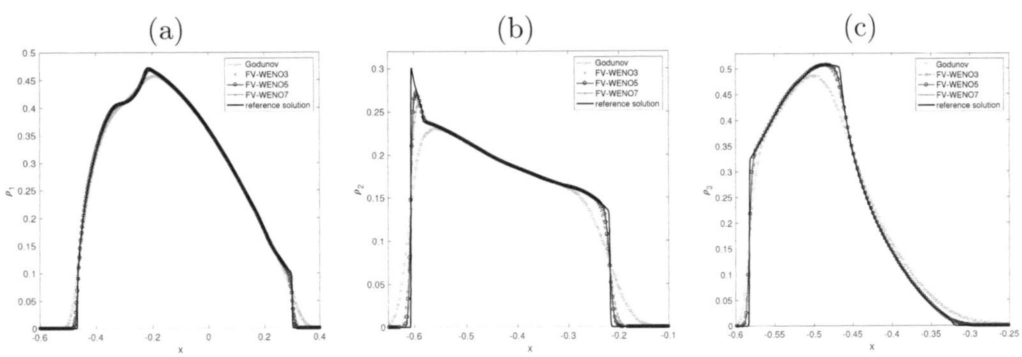

FIGURE 3. Test 2(a). (a) Profile of ρ_1, (b) profile of ρ_2, (c) profile of ρ_3, computed with different numerical schemes at time=0.5 and $1/\Delta x = 400$. The reference solution is computed with $1/\Delta x = 3200$.

4. **Conclusions.** In this paper, we applied high-order finite volume WENO schemes to the non-local multi-class traffic flow model proposed in [5]. We used quadratic polynomial reconstructions in each cell to evaluate the non-local terms in order to obtain high order of accuracy. The numerical results of the accuracy test

show that the proposed schemes maintain the correct order of accuracy. Besides, the considered examples allow to illustrate the interaction dynamics of mixed traffic consisting of both autonomous and human-driven vehicles.

Acknowledgements. This research was supported by the Inria Associated Team *Efficient numerical schemes for non-local transport phenomena (NOLOCO; 2018-2020)*. LMV is supported by Fondecyt project 1181511 and CONICYT/PIA/Concurso Apoyo a Centros Científicos y Tecnológicos de Excelencia con Financiamiento Basal AFB170001.

REFERENCES

[1] S. Benzoni-Gavage and R. M. Colombo, An n-populations model for traffic flow, *European J. Appl. Math.*, **14** (2003), 587–612.

[2] S. Blandin and P. Goatin, Well-posedness of a conservation law with non-local flux arising in traffic flow modeling, *Numer. Math.*, **132** (2016), 217–241.

[3] P. Buchmüller and C. Helzel, Improved accuracy of high-order weno finite volume methods on Cartesian grids, *Journal of Scientific Computing*, **61** (2014), 343–368.

[4] C. Chalons, P. Goatin and L. M. Villada, High-order numerical schemes for one-dimensional nonlocal conservation laws, *SIAM Journal on Scientific Computing*, **40** (2018), A288–A305.

[5] F. A. Chiarello and P. Goatin, Non-local multi-class traffic flow models, *Netw. Heterog. Media*, **14** (2019), 371–387.

[6] F. A. Chiarello, P. Goatin and L. M. Villada, Lagrangian-Antidiffusive Remap schemes for non-local multi-class traffic flow models, preprint.

[7] F. A. Chiarello and P. Goatin, Global entropy weak solutions for general non-local traffic flow models with anisotropic kernel, *ESAIM: M2AN*, **52** (2018), 163–180.

[8] P. Goatin and S. Scialanga, Well-posedness and finite volume approximations of the LWR traffic flow model with non-local velocity, *Netw. Heterog. Media*, **11** (2016), 107–121.

[9] M. J. Lighthill and G. B. Whitham, On kinematic waves. II. A theory of traffic flow on long crowded roads, *Proc. Roy. Soc. London. Ser. A.*, **229** (1955), 317–345.

[10] P. I. Richards, Shock waves on the highway, *Operations Res.*, **4** (1956), 42–51.

[11] C. W. Shu, Essentially non-oscillatory and weighted essentially non-oscillatory schemes for hyperbolic conservation laws, in *Advanced Numerical Approximation of Nonlinear Hyperbolic Equations*, Springer, (1998), 325–432.

[12] C. W. Shu and S. Osher, Efficient implementation of essentially non-oscillatory shock-capturing schemes, *Journal of Computational Physics*, **77** (1988), 439–471.

E-mail address: felisia.chiarello@inria.fr
E-mail address: paola.goatin@inria.fr
E-mail address: lvillada@ubiobio.cl

ON SMOOTH APPROXIMATIONS OF ROUGH VECTOR FIELDS AND THE SELECTION OF FLOWS

GENNARO CIAMPA

GSSI - Gran Sasso Science Institute
Viale Francesco Crispi 7, 67100, L'Aquila, Italy
and
Department Mathematik und Informatik Universität Basel
Spiegelgasse 1, CH-4051, Basel, Switzerland

GIANLUCA CRIPPA

Department Mathematik und Informatik Universität Basel
Spiegelgasse 1, CH-4051, Basel, Switzerland

STEFANO SPIRITO

DISIM, Università degli Studi dell'Aquila
Via Vetoio, 67100, L'Aquila, Italy

ABSTRACT. In this work we deal with the selection problem of flows of an irregular vector field. We first summarize an example from [4] of a vector field b and a smooth approximation b_ε for which the sequence X^ε of flows of b_ε has subsequences converging to different flows of the limit vector field b. Furthermore, we give some heuristic ideas on the selection of a subclass of flows in our specific case.

1. **Introduction and notations.** Consider the system of ordinary differential equations

$$\begin{cases} \dfrac{\mathrm{d}}{\mathrm{d}t}X(t,x) = b(t,X(t,x)), \\ X(0,x) = x, \end{cases} \tag{1.1}$$

where $(t,x) \in (0,T) \times \mathbb{R}^d$ are the independent variables, with $T < \infty$, $b : (0,T) \times \mathbb{R}^d \to \mathbb{R}^d$ is a given vector field and $X : (0,T) \times \mathbb{R}^d \to \mathbb{R}^d$ is the unknown. A solution X of (1.1) is called *flow* of b. The well posedness of (1.1) is a well known result when the vector field b is globally Lipschitz in space uniformly in time. The system (1.1) is strictly connected to the Cauchy problem for the linear transport equation

$$\begin{cases} \partial_t u + b \cdot \nabla u = 0, \\ u|_{t=0} = u_0, \end{cases} \tag{1.2}$$

since in a smooth setting, the unique solution of (1.2) is given by the formula $u(t,x) = u_0((X(t,\cdot))^{-1}(x))$, where X is the unique flow of b.

2000 *Mathematics Subject Classification.* 34A12, 34A36, 34A45.

Key words and phrases. Ordinary differential equations with non smooth vector fields; transport and continuity equations; regular Lagrangian flow; selection problem; smooth approximation.

This research has been supported by the ERC Starting Grant 676675 FLIRT.

Besides the theoretical interest, due to applications to several equations from mathematical physics the setting of smooth vector fields is too restrictive and a theory under assumptions of lower regularity was developed in the last years. Exploiting the connection between (1.2) and (1.1), DiPerna and Lions in [8] proved the well posedness of (1.2) under hypotesis of Sobolev regularity for the vector field and bounded divergence. As a consequence of their result, they proved well posedness of (1.1) under the same hypothesis. Similarly, Ambrosio in [1] improved the result of [8] to the case of BV regularity and bounded divergence for b. On the other hand, a well posedness theory based only on *a priori* estimates of the flow was developed in [5] for $W^{1,p}$ vector field with $p > 1$ and in [3, 6] for the case $p = 1$ and vector fields which gradient is given by a singular integral of a L^1 function. This latter is a class of interest in the context of 2D Euler equations. More recently Nguyen in [9] improved the result to vector fields which can be represented as singular integral of a function in BV.

Various counterexamples show that weak differentiability assumptions on the vector field are in general necessary in order to obtain well posedness, see for instance [7, 8]. For a general survey on this topic, we refer to [2]. The aim of this note is to discuss the selection problem for solutions of (1.1) in a low regularity setting. To better explain what we mean by selection, let us first recall some preliminary notations and definitions. We denote by $\mathscr{L}^d$ the Lebesgue measure on $\mathbb{R}^d$. If $f : \mathbb{R}^d \to \mathbb{R}$ is a Borel map we denote by $f\#\mathscr{L}^d$ the *push forward*, that is, the measure defined by the following relation

$$f\#\mathscr{L}^d(E) = \mathscr{L}^d(f^{-1}(E)) \qquad \text{for every Borel set } E \subset \mathbb{R}^d.$$

The definition of flow of a vector field b, when b is not smooth, is the following:

Definition 1.1. Let $b \in L^1((0,T); L^1_{\mathrm{loc}}(\mathbb{R}^d; \mathbb{R}^d))$ be given. We say that $X : (0,T) \times \mathbb{R}^d \to \mathbb{R}^d$ is a regular Lagrangian flow associated to b if

1. for a.e. $x \in \mathbb{R}^d$ the map $t \mapsto X(t,x)$ is an absolutely continuous integral solution of the ordinary differential equation

$$\begin{cases} \dfrac{\mathrm{d}}{\mathrm{d}t} X(t,x) = b(t, X(t,x)), \\ X(0,x) = x, \end{cases} \tag{1.3}$$

2. there exists a constant L indipendent of t such that

$$X(t,\cdot)\#\mathscr{L}^d \leq L\mathscr{L}^d. \tag{1.4}$$

If the vector field is divergence-free, L can be taken to be 1 and (1.4) is an equality. This means that the flow X is measure preserving. Condition (1.4) is a first selection: we only consider among all solutions of (1.1) those that do not "compress" too much the Lebesgue measure. This selection is necessary in the theory since it is not known if there is uniqueness in the class of flows that can compress the Lebesgue measure, even under assumptions of weak differentiability and zero divergence for the vector field. The paper is divided as follows. In Section 2 we give a precise statement for the selection problem and we give an example of a vector field and of an approximation for which the selection is not true. In Section 3 we characterize a class of flows through measure preserving maps of the unit circle. Finally in Section 4 we introduce a new question about the selection of a subset of the set of all flows and we give some ideas and heuristics about what we can expect.

2. **The problem of selection.** Let us consider a weakly differentiable vector field b which falls into the class of well-posedness like those discussed in the introduction. To prove the existence of solutions of (1.1), the natural approach is to rely on a compactness argument for an approximating sequence X^ε. This latter is usually constructed as the (unique) flow of a smooth approximation b_ε of b, see [2]. Consider, instead, a vector field b that has more than one regular Lagrangian flow and let b_ε be a smooth approximation of b. Consider the solution X^ε of the ODE relative to b_ε and assume that X^ε converges to a regular Lagrangian flow X of b. We wonder if for every approximation b_ε the corresponding flows X^ε can converge to only one regular Lagrangian flow: if this were true, this procedure could be considered as a selection principle for the flows of an irregular vector field. We can summarize the previous discussion in the following: question

(Q1) *Does the approximation procedure obtained by smoothing the vector field select a unique solution of* (1.1)*?*

In [4] we give a negative answer to the previous question showing a counterexample. Precisely, we consider this vector field, which is a 3D analogous of an example of DiPerna and Lions [8]:

$$b(x,y,z)=\begin{cases}\left(-\operatorname{sgn}(z)\dfrac{x}{|z|^2},-\operatorname{sgn}(z)\dfrac{y}{|z|^2},-\dfrac{2}{|z|}\right) & \text{if } x\in P,\\ (0,0,0) & \text{otherwise,}\end{cases}\tag{2.1}$$

where $P\subset\mathbb{R}^3$ denotes the set

$$P=P^+\cup P^-=\{(x,y,z)\in\mathbb{R}^3 : x^2+y^2\le z\}\cup\{\,(x,y,z)\in\mathbb{R}^3 : x^2+y^2\le -z\},$$

the union of two symmetric paraboloids.

The vector field b is divergence-free and it is out of the class of uniqueness of solutions of (1.1). In particular, observe that b is not in any Sobolev space $W^{1,p}(\mathbb{R}^3)$ or in $BV(\mathbb{R}^3)$.

We want to define two different regular Lagrangian flows $\bar X,\tilde X$ of b and, since we are considering flows defined almost everywhere, we need to define $\bar X,\tilde X$ only on $\mathbb{R}^3\setminus\{0\}$. We start for $\mathbf{x}\in\mathbb{R}^3\setminus P$: in this region the vector field is identically 0 so that we define $\bar X,\tilde X$ simply as

$$\bar X(t,\mathbf{x})=\mathbf{x}=\tilde X(t,\mathbf{x})\qquad \forall t\ge 0.$$

If $\mathbf{x}=(x,y,z)\in P^-$ we define $\bar X,\tilde X$ as

$$\begin{cases}\bar X_1(t,x,z)=\tilde X_1(t,x,z)=\dfrac{x}{\sqrt{-z}}\sqrt[4]{z^2+4t}\\ \bar X_2(t,y,z)=\tilde X_2(t,y,z)=\dfrac{y}{\sqrt{-z}}\sqrt[4]{z^2+4t}\\ \bar X_3(t,z)=\tilde X_3(t,z)=-\sqrt{z^2+4t}\end{cases}\qquad \forall\, t\ge 0.\tag{2.2}$$

Finally, when $\mathbf{x}=(x,y,z)\in P^+$ define the flows as

$$\begin{cases}\bar X_1(t,x,z)=\tilde X_1(t,x,z)=\dfrac{x}{\sqrt{z}}\sqrt[4]{z^2-4t}\\ \bar X_2(t,y,z)=\tilde X_2(t,y,z)=\dfrac{y}{\sqrt{z}}\sqrt[4]{z^2-4t}\\ \bar X_3(t,z)=\tilde X_3(t,z)=\sqrt{z^2-4t}\end{cases}\qquad \text{for } t\in\left[0,\frac{z^2}{4}\right].\tag{2.3}$$

At time $t = \frac{z^2}{4}$ the trajectories reach the origin and then one possible way to extend them for later times is

$$\begin{cases} \bar{X}_1(t,x,z) = \frac{x}{\sqrt{z}}\sqrt[4]{4t-z^2}\cos\Theta - \frac{y}{\sqrt{z}}\sqrt[4]{4t-z^2}\sin\Theta \\ \bar{X}_2(t,y,z) = \frac{x}{\sqrt{z}}\sqrt[4]{4t-z^2}\sin\Theta + \frac{y}{\sqrt{z}}\sqrt[4]{4t-z^2}\cos\Theta \qquad t \geq \frac{z^2}{4}, \\ \bar{X}_3(t,z) = -\sqrt{4t-z^2} \end{cases} \tag{2.4}$$

while

$$\begin{cases} \tilde{X}_1(t,x,z) = \frac{x}{\sqrt{z}}\sqrt[4]{4t-z^2}\cos\Phi - \frac{y}{\sqrt{z}}\sqrt[4]{4t-z^2}\sin\Phi \\ \tilde{X}_2(t,y,z) = \frac{x}{\sqrt{z}}\sqrt[4]{4t-z^2}\sin\Phi + \frac{y}{\sqrt{z}}\sqrt[4]{4t-z^2}\cos\Phi \qquad t \geq \frac{z^2}{4}, \\ \tilde{X}_3(t,z) = -\sqrt{4t-z^2} \end{cases} \tag{2.5}$$

where $\Theta, \Phi \in (0, 2\pi]$ and $\Theta \neq \Phi$. An easy computation shows that $\bar{X}, \tilde{X}$ are two different regular Lagrangian flows of b. We call those kind of solutions respectively X^Θ, X^Φ, where Θ and Φ represent a rotation in the xy plane. Heuristically, we can define this kind of flows as a consequence of the fact that the trajectories once they reach the origin can come out arbitrarily. In [4] one of our main results is the following:

Theorem 2.1. *There exists a sequence of vector fields $b_n \in C^\infty(\mathbb{R}^3)$ such that:*

- *b_n is divergence-free;*
- *$b_n \to b$ in $L^1_{\text{loc}}(\mathbb{R}^3)$;*
- *the sequence X^n of regular Lagrangian flows of b_n has two different subsequences converging in $L^\infty((0,T); L^1_{\text{loc}}(\mathbb{R}^3))$ to two different regular Lagrangian flows of b.*

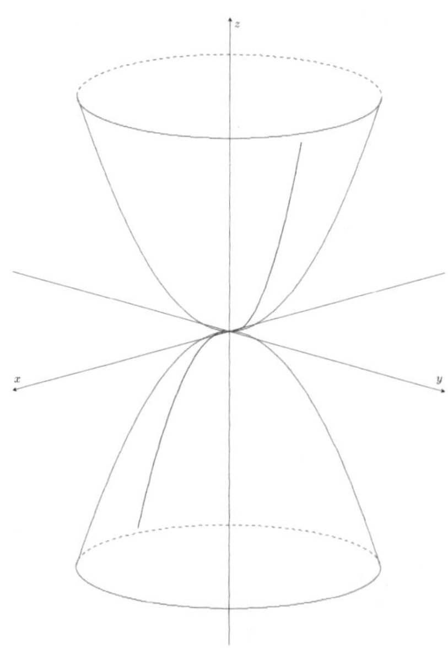

FIGURE 1. An example of solution X^Θ

In the proof of Theorem 2.1, given $\Theta, \Phi \in (0, 2\pi]$ with $\Theta \neq \Phi$, we construct an explicit approximation which has two different subsequences converging respectively to X^{Θ} and X^{Φ}. The strategy of the approximation is based on smoothing b nearby the origin and forcing the trajectories to rotate very fast along a given helix. We basically modify b in a small region with contains the singularity and leave the rest unchanged.

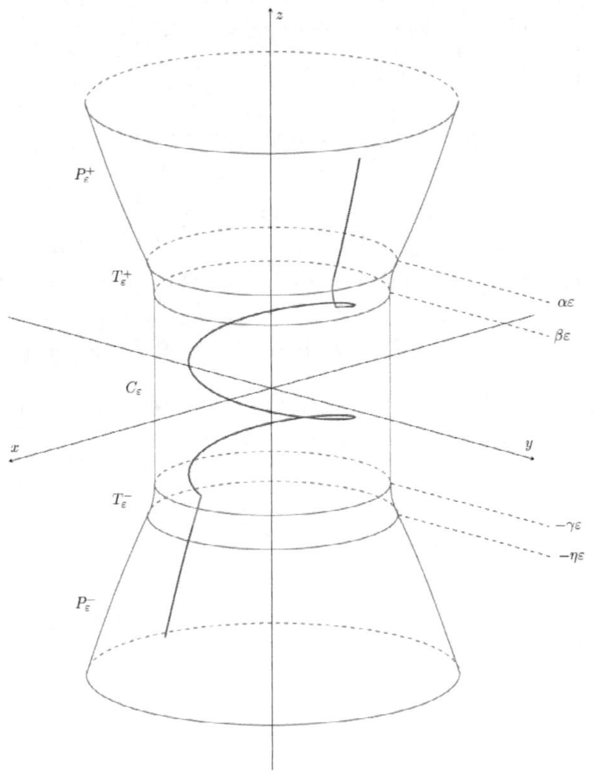

FIGURE 2. The figure represents an approximated trajectory in the construction of the proof of Theorem 2.1

The theorem answers question (Q1) in the negative. However with our approach we are able to obtain only solutions of the form X^{Θ}. Indeed, note that another possible way to define a flow for $\mathbf{x} \in P^+$ is the following:

$$\begin{cases} X_1(t,r,\theta,z) = \dfrac{r}{\sqrt{z}} \sqrt[4]{z^2 - 4t} \, \cos\theta \\ X_2(t,r,\theta,z) = \dfrac{r}{\sqrt{z}} \sqrt[4]{z^2 - 4t} \, \sin\theta \\ X_3(t,z) = \sqrt{z^2 - 4t} \end{cases} \quad \text{for } t \in \left[0, \frac{z^2}{4}\right], \tag{2.6}$$

and

$$\begin{cases} X_1(t,r,\theta,z) = \dfrac{r}{\sqrt{z}} \sqrt[4]{4t - z^2} \, \cos\psi(\theta) \\ X_2(t,r,\theta,z) = \dfrac{r}{\sqrt{z}} \sqrt[4]{4t - z^2} \, \sin\psi(\theta) \\ X_3(t,z) = -\sqrt{4t - z^2} \end{cases} \quad \text{for } t \geq \frac{z^2}{4}, \tag{2.7}$$

where the map $\psi : [0, 2\pi] \to [0, 2\pi]$ is arbitrary and (r, θ, z) denote the cylindrical coordinates in $\mathbb{R}^3$. It is easy to check that the map in (2.6),(2.7) is a solution of the ODE relative to b; we call X_ψ such a map. It will turn out to be useful the flow on $P \setminus \{0\}$ and not only in $\mathring{P}$ although we deal with functions defined almost everywhere with respect to the 3D Lebesgue measure. The reason for that lies in the fact that for our purpose we will compute X on $\partial P \setminus \{0\}$; this would not make sense without a suitable definition of X on the boundary of P. Such definition is made accordingly to the everywhere definition of b. In the next section we will discuss the conditions that the map ψ has to satisfied in order for X_ψ to be a regular Lagrangian flow of b.

3. Regular Lagrangian flows and measure preserving map on the circle.

In this section we prove that solutions of the form X_ψ are regular Lagrangian flows of b when ψ is a measure preserving map. Before doing this, note that the map X_ψ associated to $\psi(\theta) = \alpha$, where $\alpha \in (0, 2\pi]$ is fixed, is a solution of the ODE but it does not preserve the 3D Lebesgue measure and then it is not a regular Lagrangian flow.

Now we recall the definition of a mesure preserving map on the unit circle.

Definition 3.1. Let $\psi : \mathbb{S}^1 \to \mathbb{S}^1$ be a measurable map, where $\mathbb{S}^1 = \mathbb{R}/2\pi\mathbb{Z}$ is the unit circle with the 1D Lebesgue measure. The map ψ is called measure preserving if

$$\psi \# \mathscr{L}^1 = \mathscr{L}^1.$$

We identify $\mathbb{S}^1 \sim [0, 2\pi]$ and we define the set $\mathscr{M}$ as

$$\mathscr{M} := \{\psi : [0, 2\pi] \to [0, 2\pi] : \psi \text{ satisfies Definition 3.1}\}.$$

Moreover, define the maps

$$I_\pm : \theta \in [0, 2\pi] \to (\cos\theta, \sin\theta, \pm 1) \in \mathbb{R}^3.$$

Proposition 3.2. *Given a regular Lagrangian flow X there exists $\psi \in \mathscr{M}$ such that $X = X_\psi$. Viceversa given $\psi \in \mathscr{M}$ there exists a unique regular Lagrangian flow X such that $X = X_\psi$.*

Proof. Consider a regular Lagrangian flow $X(t, \mathbf{x})$ and define

$$\psi(\theta) = I_-^{-1}\left(X\left(\frac{1}{2}, I_+(\theta)\right)\right) \qquad \theta \in [0, 2\pi].$$

We need to show that such a map preserves the 1D Lebesgue measure: consider a Borel set $E \subseteq [0, 2\pi]$ and define $\mathbf{E}$ as the set

$$\mathbf{E} = \{(\rho, \theta, z) : \theta \in E, \rho \in [0, \sqrt{z}], z \in [-1, 0]\}.$$

A straightforward computation shows that

$$X^{-1}\left(\frac{1}{2}, \cdot\right)(\mathbf{E}) = \{(\rho, \theta, z) : \theta \in \psi^{-1}(E), \rho \in [0, \sqrt{z}], z \in [1, \sqrt{2}]\}$$

and

$$\mathscr{L}^1(\psi^{-1}(E)) = 4\mathscr{L}^3\left(X^{-1}\left(\frac{1}{2}, \cdot\right)(\mathbf{E})\right) = 4\mathscr{L}^3(\mathbf{E}) = \mathscr{L}^1(E), \tag{3.1}$$

hence ψ is measure preserving.

We now prove the other implication. Consider a measure preserving map ψ, a point $\mathbf{x} \in \mathbb{S}^1 \times \{1\}$ and solve the system

$$\begin{cases} \dot{X}(t,\mathbf{x}) = b(X(t,\mathbf{x})), \\ X(0,\mathbf{x}) = \mathbf{x}, \\ X\left(\frac{1}{2},\mathbf{x}\right) = I_-\left(\psi(I_+^{-1}(\mathbf{x}))\right). \end{cases} \tag{3.2}$$

It is easy to see that (3.2) admits a unique solution X_ψ. We have to prove that X_ψ is measure preserving. A computation like (3.1) shows that $X_\psi(t,\mathbf{E})\#\mathscr{L}^3 = \mathscr{L}^3(\mathbf{E})$ for all sets $\mathbf{E}$ of the form

$$\mathbf{E} = \{(r,\theta,z) : \theta \in E_1, r \in [0,\sqrt{z}], z \in E_2\}, \tag{3.3}$$

where $E_1 \subset [0,2\pi], E_2 \subset \mathbb{R}$. Sets of the form (3.3) are a basis for the Borel σ-algebra, hence X_ψ preserve the 3D Lebesgue measure on Borel sets. Since X_ψ maps null sets into null sets, it follows that it is a regular Lagrangian flow. □

4. **Some ideas and heuristics on possible extensions.** Consider the maps

$$\psi_1(\theta) = \begin{cases} \theta & \text{if } \theta \in [0,\pi), \\ 3\pi - \theta & \text{if } \theta \in [\pi, 2\pi], \end{cases}$$

and

$$\psi_2(\theta) = \begin{cases} 2\theta & \text{if } \theta \in [0,\pi), \\ 2(\theta - \pi) & \text{if } \theta \in [\pi, 2\pi]. \end{cases}$$

The map ψ_1 leaves half a circle fixed and flips the other half, while the map ψ_2 rotates twice around $\mathbb{S}^1$. Since the strategy of the proof of Theorem 2.1 produces in the limit only solutions of the form X^Θ, we wonder if it is possible to obtain, as limit of a suitable approximation, the flows X_{ψ_1}, X_{ψ_2} associated to ψ_1, ψ_2 as in the proof of Proposition 3.2. This is a concrete example of the following general question:

(Q2) *Does the approximation procedure obtained by smoothing the vector field select a subset of the flows of b?*

The strategy of [4] selects the regular Lagrangian flows corresponding to measure preserving map of the form $\psi(\theta) = \theta + \Theta \mod 2\pi$. These flows are in a sense "better" than the others for the following reasons:

- the flows X^Θ self intersect only in the origin, while this is not true for X_{ψ_2}, which is not even a.e. invertible;
- the Jacobian of X^Θ does not change sign, while this is the case for X_{ψ_1}.

Consider a general smooth approximation b_ε of the vector field b; the corresponding Cauchy problem admits a uniquely defined sequence of flows X^ε and one can ask to which X_ψ the sequence X^ε may converge. It is not clear to us if it is possible to construct an approximation of b in such a way that the approximated flow converge to X_{ψ_1} or X_{ψ_2}, especially if we want to approximate b only close to the singularity at the origin. We can however provide some heuristics motivating why it is not trivial to exclude the possibility of getting X_{ψ_1} in the limit just by arguing on the base of "topological obstructions". In fact, we can approximate the flow X_{ψ_1} with maps X^ε of the form:

$$X^\varepsilon(t,\mathbf{x}) = \begin{cases} X(t,\mathbf{x}) & \text{for } 0 \le t \le t_1^\varepsilon := \frac{z^2-\varepsilon^2}{4}, \\ \frac{t-t_1^\varepsilon}{t_2^\varepsilon-t_1^\varepsilon} I_-\left(\psi_1(I_+^{-1}(\mathbf{x}))\right) + \frac{t_2^\varepsilon-t}{t_2^\varepsilon-t_1^\varepsilon} X(t_1^\varepsilon,\mathbf{x}) & \text{for } t_1^\varepsilon \le t \le t_2^\varepsilon := \frac{z^2}{4} + \frac{\varepsilon^2}{4}, \\ X\left(t - t_2^\varepsilon, I_-\left(\psi_1(I_+^{-1}(\mathbf{x}))\right)\right) & \text{for } t_2^\varepsilon \le t < \infty, \end{cases}$$

where $\mathbf{x} \in P^+$. Each X^ε is a well-defined map, which is however not a flow a vector field. Therefore, this does not answer our question. However, this example tells us that an answer in the positive to our question could not just rely on topological properties of the approximating flows.

References

[1] L. Ambrosio, Transport equation and Cauchy problem for BV vector fields. *Inventiones Mathematicae*, **158** (2004), 227-260.

[2] L. Ambrosio, G. Crippa, Continuity equations and ODE flows with non-smooth velocities. *Proc. Roy. Soc. Edinburgh Sect. A*, **144** (2014), 1191-1244.

[3] F. Bouchut, G. Crippa, Lagrangian flows for vector fileds with gradient given by a singular integral. *J. Hyperbolic Diff. Equ.*, **10** (2013), 235-282.

[4] G. Ciampa, G. Crippa, S. Spirito, Smooth approximation is not a selection principle for the transport equation. arXiv:1902.08084

[5] G. Crippa, C. De Lellis, Estimates and regularity results for the DiPerna-Lions flow. *J. Reine Angew. Math.*, **616** (2008), 15-46.

[6] G. Crippa, C. Nobili, C. Seis, S. Spirito, Eulerian and Lagrangian solutions to the continuity and Euler equations with L^1 vorticity. *SIAM J. Math. Anal.*, **49** (2017), no.5, 3973-3998.

[7] N. Depauw, Non unicité des solutions bornées pour un champ de vecteurs BV en dehors d'un hyperplan. *C.R. Math. Sci. Acad. Paris*, **337** (2003), 249-252.

[8] R.J. DiPerna, P.-L. Lions, Ordinary differential equations, transport theory and Sobolev spaces. *Invententiones Mathematicae*, **98** (1989), 511-547.

[9] Q.-H. Nguyen, Quantitative estimates for regular Lagrangian flows with *BV* vector fields. Available from: http://cvgmt.sns.it/media/doc/paper/3848

E-mail address: gennaro.ciampa@gssi.it
E-mail address: gianluca.crippa@unibas.ch
E-mail address: stefano.spirito@univaq.it

RECENT RESULTS ON THE SINGULAR LOCAL LIMIT FOR NONLOCAL CONSERVATION LAWS

MARIA COLOMBO

EPFL SB, Station 8, CH-1015 Lausanne, Switzerland

GIANLUCA CRIPPA

Departement Mathematik und Informatik, Universität Basel,
Spiegelgasse 1, CH-4051 Basel, Switzerland

MARIE GRAFF

Department of Mathematics, University of Auckland,
Private Bag 92019, Auckland 1142, New Zealand

LAURA V. SPINOLO*

IMATI-CNR, via Ferrata 5, I-27100 Pavia, Italy

ABSTRACT. We provide an informal overview of recent developments concerning the singular local limit of nonlocal conservation laws. In particular, we discuss some counterexamples to convergence and we highlight the role of numerical viscosity in the numerical investigation of the nonlocal-to-local limit. We also state some open questions and describe recent related progress.

1. **Introduction.** We consider the nonlocal conservation law

$$\partial_t u + \partial_x\big[uV(u * \eta)\big] = 0. \tag{1}$$

In the previous expression, the unknown is the function $u : \mathbb{R}_+ \times \mathbb{R} \to \mathbb{R}$, the function $V : \mathbb{R} \to \mathbb{R}$ is Lipschitz continuous and the term $u * \eta$ is the convolution, computed with respect to the space variable x only, of the solution u with the convolution kernel $\eta : \mathbb{R} \to \mathbb{R}$. For the time being we assume that η satisfies the following assumptions:

$$\eta \in C^1_c(\mathbb{R}), \qquad \eta \geq 0, \qquad \int_{\mathbb{R}} \eta(x)dx = 1, \tag{2}$$

but actually the regularity assumptions on η can be relaxed, as we will see in §4. Nonlocal equations in the form (1) have been extensively studied in recent years owing to the applications to (among others) models of sedimentation and pedestrian and vehicular traffic, see for instance [2, 3, 5, 9, 10] and the references therein.

Consider the Cauchy problem posed by coupling (1) with the initial datum

$$u(0, x) = \bar{u}(x). \tag{3}$$

2000 *Mathematics Subject Classification.* Primary: 35L65, 65M08.

Key words and phrases. nonlocal conservation law, traffic model, downstream traffic density, numerical viscosity, singular limit, local limit.

* Corresponding author.

Existence and uniqueness results have been obtained in various frameworks by several authors, see among others [3, 10, 11, 14].

In this note we review some recent progress in the analysis of the singular local limit of (1), which is defined as follows. Fix a parameter $\varepsilon > 0$, consider the rescaled function $\eta_\varepsilon(x) := \eta(x/\varepsilon)/\varepsilon$ and note that, owing to the third condition in (2), when $\varepsilon \to 0^+$ the family η_ε converges weakly* in the sense of measures to the Dirac delta. By plugging η_ε into (1),(3) we arrive at the family of Cauchy problems

$$\begin{cases} \partial_t u_\varepsilon + \partial_x \big[u_\varepsilon V(u_\varepsilon * \eta_\varepsilon)\big] = 0 \\ u_\varepsilon(0, x) = \bar{u}(x). \end{cases} \tag{4}$$

We now consider the limit $\varepsilon \to 0^+$: since η_ε converges to the Dirac delta, from the equation at the first line of (4) we formally recover the nonlinear conservation law

$$\partial_t u + \partial_x \big[uV(u)\big] = 0. \tag{5}$$

The above derivation is completely formal, and whether or not it can be rigorously justified is the object of the following question, which was originally posed in [1].

Question 1. *Does u_ε, solution of* (4)*, converge (in some suitable topology) to the entropy admissible solution of* (3),(5) *as $\varepsilon \to 0^+$?*

We refer to [12] for the definition of entropy admissible solution of (3),(5). In this work we overview some recent developments concerning Question 1. The exposition is organized as follows: in §2 we show that, notwithstanding numerical evidence suggesting the opposite, the answer to Question 1 is in general negative. In §3 we discuss a possible explanation of the reason why the numerical evidence provides the wrong intuition. Finally, in §4 we introduce Question 3, which is a refinement of Question 1 in a more specific setting motivated by the applications to vehicular traffic models. Question 3 is still open, but recent progress has been recently achieved and we discuss it in §4.

2. The nonlocal-to-local limit.

Question 1 was originally motivated by numerical evidence. More precisely, in [1] the authors exhibit numerical experiments where the solution of the nonlocal Cauchy problem (4) gets closer and closer to the entropy admissible solution of (3),(5) as $\varepsilon \to 0^+$, thus suggesting a positive answer to Question 1. This was later confirmed by other numerical experiments, see for instance [3].

Another positive partial answer to Question 1 is provided by [19, Proposition 4.1], which loosely speaking states that the answer to Question 1 is positive provided that the convolution kernel η is even (i.e. $\eta(x) = \eta(-x)$, for every x) and the limit solution u is smooth. The rationale underpinning [19, Proposition 4.1] is basically the following. Assume that the initial datum $\bar{u}$ is smooth and say compactly supported, then there is a time interval $[0, T]$ where the entropy admissible solution of (3),(5) is smooth, i.e. it is a classical solution. Proposition 4.1 in [19] states that on the interval $[0, T]$ the family u_ε converges to u, in the uniform C^0 norm.

Despite the above mentioned results, the answer to Question 1 is, in general, negative. More precisely, in [8] we exhibit three counterexamples that rule out the possibility that the family u_ε solving (4) converge to the entropy admissible solution of (3),(5). The counterexamples are completely explicit and rule out not only strong convergence, but also i) weak convergence and ii) the possibility of extracting from u_ε a (strongly or weakly) converging subsequence. In one case we even manage to rule out the possibility that u_ε converges to a distributional solution of (3),(5), i.e. we do not need to require that the limit u is entropy admissible to rule out convergence. The counterexamples are constructed in [8, §5.1,§5.2,§5.3] and at the beginning of each of §5.1, §5.2 and §5.3 the basic ideas underpinning the construction of the counterexample are overviewed. Loosely speaking the very basic mechanism is that in each of the counterexamples we manage to single out a property that i) is satisfied by the solution u_ε of (4), for every $\varepsilon > 0$; ii) is stable under weak or strong convergence, i.e. it passes to the weak or strong limit; iii) is *not* satisfied by the entropy admissible solution of (3),(5). The exact property verifying conditions i), ii) and iii) is different in each counterexample: in the first one it is the fact that the integral over $\mathbb{R}_-$ is constant in time, in the second one the fact that u_ε is identically 0 at positive values of x. Finally, in the third counterexample we single out a functional that is constant in time when evaluated at $u_\varepsilon(t, \cdot)$ and strictly decreasing when evaluated at $u(t, \cdot)$.

3. Numerical experiments and viscosity.

We now go back to the numerical experiments in [1], which as we have seen provide the wrong intuition concerning Question 1. A possible explanation of the reason why the numerical evidence is not reliable is given by the following argument.

The numerical results in [1] have been obtained by relying on a Lax-Friedrichs type scheme. The Lax-Friedrichs scheme is a finite volume scheme which is very commonly used to construct numerical solutions of conservation laws, see [17] for an exhaustive discussion. The Lax-Friedrichs scheme contains a large amount of what is called *numerical viscosity*: very loosely speaking, the numerical viscosity is a collection of finite difference terms which are the numerical counterpart of some analytical viscosity, i.e. of some second order term. In other words, the presence of the numerical viscosity implies that the model equation for the Lax-Friedrichs scheme for the conservation law (5) is actually the *viscous* conservation law

$$\partial_t u + \partial_x \big[u V(u) \big] = \nu \partial^2_{xx} u, \tag{6}$$

where the viscosity coefficent $\nu > 0$ is of the same order of the space mesh, see [17]. When the Lax-Friedrichs scheme is applied to the nonlocal conservation law (1), the presence of the numerical viscosity implies that the model equation is

$$\partial_t u + \partial_x \big[u V(u * \eta) \big] = \nu \partial^2_{xx} u. \tag{7}$$

This in turn implies that in order to get some insight on the discrepancy between the numerical evidence in [1] and the analytic counterexamples in [8] it might be useful to consider the family of Cauchy problems[1]

$$\begin{cases} \partial_t u_\varepsilon + \partial_x \big[u_\varepsilon V(u_\varepsilon * \eta_\varepsilon) \big] = \nu \, \partial^2_{xx} u_\varepsilon \\ u_\varepsilon(0, \cdot) = \bar{u} \end{cases} \tag{8}$$

and pose the following "viscous counterpart" of Question 1.

Question 2. *Does u_ε, solution of (8), converge to the solution of (3),(6) as $\varepsilon \to 0^+$?*

The answer to Question 2 is largely positive, and it is given by the following result.

[1]Existence and uniqueness results for the Cauchy problem (8) can be obtained by combining a fixed point argument with fairly standard parabolic estimates, see [8, §2.1]

Theorem 3.1. *Assume (2), fix $\nu > 0$ and $T > 0$ and assume that the function $V : \mathbb{R} \to \mathbb{R}$ is Lipschitz continuous. If $\bar{u} \in L^1(\mathbb{R}) \cap L^\infty(\mathbb{R})$, then the solution of (8) converge to the solution of (3),(6) strongly in $L^2([0,T] \times \mathbb{R})$ as $\varepsilon \to 0^+$.*

The proof of Theorem 3.1 is provided in [8], actually yields a slightly stronger result and applies in greater generality to the case of several space dimensions: we refer to [8, Theorem 1.1] for the precise statement. Note furthermore that Theorem 3.1 was established in [4] under the additional assumptions that the initial datum $\bar{u}$ is regular and that $V(u) = u$.

We can now consider the family of Cauchy problems (8), keep the nonlocal parameter $\varepsilon > 0$ fixed, vary the viscosity parameter ν and consider the inviscid limit $\nu \to 0^+$. In this way we recover the inviscid nonlocal problem (4): more precisely, [8, Proposition 1.2] states that the solutions of (8) converge to the solution of (4) when $\nu \to 0^+$. Finally, we recall that a celebrated result by Kružkov [16] states that the solutions of (3),(6) converge to the entropy admissible solution of (3),(5) when $\nu \to 0^+$.

We now put together all the previous convergence results and we combine them with the counterexamples mentioned in §2. We denote by $u_{\varepsilon\nu}$ the solution of the viscous nonlocal equation at the first line of (8) to stress that it depends on both the nonlocal parameter ε and the viscosity parameter ν. We arrive at the following diagram:

$$\begin{array}{ccc}
\partial_t u_{\varepsilon\nu} + \partial_x \big[u_{\varepsilon\nu} V(u_{\varepsilon\nu} * \eta_\varepsilon)\big] = \nu\, \partial^2_{xx} u_{\varepsilon\nu} & \xrightarrow[\text{Theorem 3.1}]{\varepsilon \to 0^+} & \partial_t u_\nu + \partial_x \big[u_\nu V(u_\nu)\big] = \nu\, \partial^2_{xx} u_\nu \\
\nu \to 0^+ \Big\downarrow \text{[8, Proposition 1.2]} & & \nu \to 0^+ \Big\downarrow \text{Kružkov [16]} \\
\partial_t u_\varepsilon + \partial_x \big[u_\varepsilon V(u_\varepsilon * \eta_\varepsilon)\big] = 0 & \xrightarrow[\text{False in general}]{\varepsilon \to 0^+} & \partial_t u + \partial_x \big[u V(u)\big] = 0
\end{array}$$

We can now go back to the numerical evidence erroneously suggesting a positive answer to Question 1. A possible explanation is the following: the numerical experiments were designed to test the convergence of u_ε to the entropy admissible solution u. However, owing to the numerical viscosity, what the numerical experiments were actually testing was the convergence of $u_{\varepsilon\nu}$ to u_ν, which holds true owing to Theorem 3.1. In other words, the numerical schemes were designed to provide an answer to Question 1, but as a matter of fact they provide an answer to Question 2. Since the two questions have opposite answers, the numerical schemes provide the wrong intuition. This explanation is validated by recent numerical experiments collected in [6]. In particular, in [6], we have used the Lax-Frierichs type scheme to test the nonlocal-to-local limit from (4) to (3),(5) in the case of the counterexamples mentioned in §2. More precisely, we have computed the numerical solution of (4) in the case where (4) is the same as in the counterexamples. Next, we have computed the L^1 norm (evaluated at a given positive time $t > 0$) of the difference between the numerical solution of (4) and the numerical entropy admissible solution of (3),(5). In Figure 1 we display some of the results concerning one of the counterexamples, more precisely the one discussed in [8, §5.1]. The blue line refers to the L^1 error between the numerical solutions obtained by the Lax-Frierichs type scheme and strongly suggests that the L^1 error is converging to 0 as $\varepsilon \to 0^+$, i.e. it erroneously suggests a positive answer to Question 1. The red line refers to the L^1 error between the numerical solutions obtained by a Godunov type scheme. Godunov type schemes for the nonlocal conservation law (1) were introduced in [5, 13] and the reason why we used them to test the nonlocal-to-local limit is because the Godunov scheme

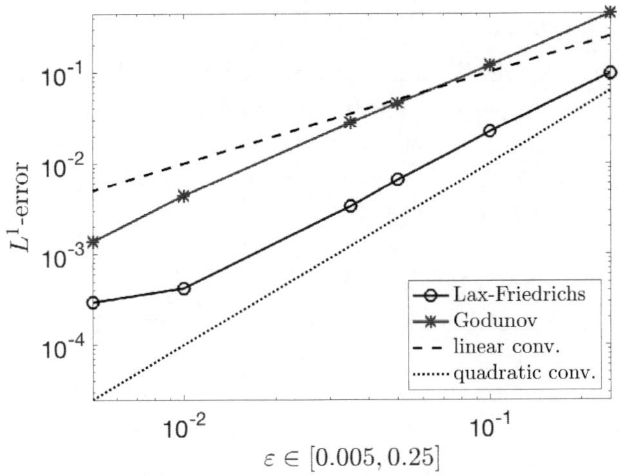

FIGURE 1. L^1-error at $t = 2$, for different values of ε, comparing the solution of (4) to the entropy admissible solution of (3),(5) computed with Lax-Friedrichs and Godunov type schemes in the case where (4) is the same as in [8, §5.1]. The space mesh is fixed and it is $h = 0.001$.

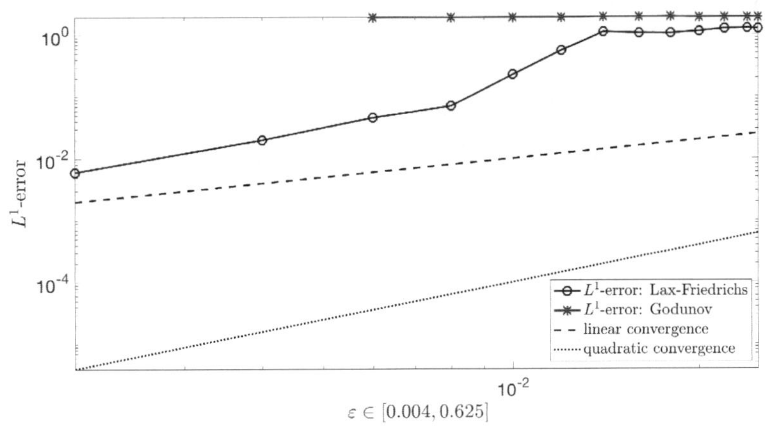

FIGURE 2. L^1-error at $t = 2$, for different values of ε, between the solutions of the nonlocal equations (4) and the entropy solution of (3),(5) computed with Godunov and Lax Friedrichs schemes in the case where (4) is the same as in [8, §5.2]. The space mesh h depends on ε and the relation is $\varepsilon = 1000h^2$. The time step k decreases linearly with the mesh size, satisfying the CFL condition $k/h = 1/6$. The L^1 error of the Godunov scheme is much larger for small values of ε.

is known to have a smaller amount of numerical viscosity than the Lax-Friedrichs scheme, see [18]. In the example studied in Figure 1, the numerical results obtained with both the Godunov and the Lax-Friedrichs scheme erroneously suggest convergence in the nonlocal-to-local limit. However, in other cases there is a difference between the two schemes. For instance, Figure 2 displays some of the numerical results concerning the counterexample discussed in [8, §5.2]: there is a remarkable difference between the Lax-Friedrichs and the Godunov scheme. Indeed, the numerical results obtained with the Lax-Friedrichs type scheme erroneously suggest convergence, whereas the numerical results obtained with the Godunov type scheme are more consistent with the analytic results, which rule out convergence. This is consistent with the fact that the Godunov scheme contains less numerical viscosity than the Lax-Friedrichs scheme, see [18].

4. **Anisotropic traffic models: total variation blow up and open questions.** In recent years, several authors have been focusing on (1) in the case where the function V is decreasing, $V' < 0$, the initial datum $\bar{u}$ is nonnegative and the convolution kernel η in equation (5) is supported on $]-\infty, 0]$. This case is extremely relevant for the applications to vehicular traffic models. Indeed, in these models u represents the density of cars (and is therefore nonnegative) and V their speed. The function V is evaluated at $u * \eta$ because the model postulates that drivers regulate their speed based on the density of cars ahead of them. The fact that the function V is decreasing is a classical assumption in traffic models and takes into account the fact that drivers tend to slow down when the traffic is congested, and conversely to speed up when the traffic is light. If the convolution kernel is supported on the interval $]-\infty, 0]$, then the convolution kernel $u * \eta$ evaluated at the point x only depends on the value of u on the interval $[x, +\infty[$. In other words, choosing this kind of convolution kernels aims at modeling the fact that drivers only look forward, not backward, and hence their speed only depends on the downstream traffic density.

To avoid some technicalities, in the following we focus on the case

$$V(u) = 1 - u, \quad \eta = \mathbb{1}_{[-1,0]}, \quad 0 \le \bar{u} \le 1, \tag{9}$$

but as a matter of fact the following discussion applies to more general cases than (9). In the previous formula, $\mathbb{1}_{[-1,0]}$ denotes the characteristic function of the interval $[-1, 0]$. Note that, strictly speaking, the regularity assumptions on the function η given in (2) are violated when $\eta = \mathbb{1}_{[-1,0]}$. Notwithstanding the lack of regularity, in [3, 14] the authors established existence and uniqueness results for the Cauchy problem (1),(3). By exploiting the anisotropy of the kernel, the analysis in [3] establishes better a-priori estimates on the solution than those available in the smooth case (2). In particular, they established a maximum principle: under (9), the solution of (1),(3) satisfies $0 \le u \le 1$. To complete the picture, we point out that the counterexamples exhibited in [8] *do not* apply in the case (9).

Summing up, the case (9) is very relevant from the modeling viewpoint, stronger analytic results apply and the counterexamples do not work. This yields the following refinement of Question 1.

Question 3. *Does u_ε, solution of (4), converge to the entropy admissible solution of (3),(5) as $\varepsilon \to 0^+$, provided (9) holds true?*

Question 1 is presently open and it is the object of current investigation. However, some progress have been recently achieved in [7]. Before discussing the results in [7], we need some preliminary considerations.

Assume (9), then, owing to the maximum principle, the solution of the Cauchy problem (4) satisfies the uniform bound

$$\|u_\varepsilon\|_{L^\infty} \le 1, \quad \text{for every } \varepsilon > 0.$$

This yields compactness in the weak-* topology and implies that we can extract a subsequence that converges to some limit function w weakly-* in $L^\infty(\mathbb{R}^+ \times \mathbb{R})$. Note however that, owing to the nonlinear nature of the problem, nothing a priori tells us that the limit w is a distributional solution (let alone entropy admissible) of the conservation law (3),(5). A natural strategy to establish a positive answer to Question 2 is hence to look for compactness in some *strong* topology. A fairly classical argument to establish strong L^1 compactness combines the Helly-Kolmogorov Compactness Theorem with a uniform bound on the total variation, i.e. an estimate like

$$\operatorname{TotVar} u_\varepsilon(t,\cdot) \le C, \quad \text{for every } t>0,\ \varepsilon>0 \text{ and for some constant } C>0. \tag{10}$$

This yields the following question:

Question 4. *Assume* (9) *and that* TotVar $\bar{u}$ *is finite. Does* u_ε*, solution of* (4)*, satisfy the uniform bound* (10)*?*

Before addressing Question 4 we make some preliminary remarks. First, the semigroup of entropy admissible solutions of (3),(5) is total variation decreasing, i.e.

$$\operatorname{TotVar} u(t,\cdot) \le \operatorname{TotVar} \bar{u}, \quad \text{for every } t>0, \tag{11}$$

provided TotVar $\bar{u}$ is finite. In other words, the entropy admissible solution of (3),(5) satisfies estimate (10) with $C = \operatorname{TotVar} \bar{u}$. Second, numerical experiments discussed in [3] suggest that, under (9), the semigroup of solutions of (4) is also total variation decreasing, and hence in particular that the answer to Question 4 is positive. Third, by combining the maximum principle with the monotonicity preserving property established in [3] one can show that, under (9), if the initial datum $\bar{u}$ is *monotone*, then the total variation does not increase in time, i.e. (10) is satisfied with $C = \operatorname{TotVar} \bar{u}$. In other words, we know that the answer to Question 4 is positive provided the initial datum is monotone, see [15].

Notwithstanding the numerical evidence and the positive answer in the case of monotone data, a counterexample constructed in [7] shows that the answer to Question 4 is in general negative. More precisely, there is an initial datum $\bar{u}$ such that TotVar $\bar{u}$ is finite and the solution of the Cauchy problem (4) satisfies

$$\sup_{\varepsilon>0} \operatorname{TotVar} u_\varepsilon(t,\cdot) = +\infty, \text{ for every } t>0,$$

which in particular implies that (10) cannot be true.

The fact that the answer to Question 4 is negative does not by any mean imply that the answer to Question 3 is also negative. However, it rules out the most classical and natural strategy to achieve an hypothetical positive answer to Question 3. Note, furthermore, that the initial datum $\bar{u}$ in [7] has finite total variation and attains values in the physical range $0 \le \bar{u} \le 1$, but it is also highly oscillating and it is unlikely to describe a realistic initial density of vehicles in some real-word applications. In principle it might be possible that, under (9), the uniform bound (10) holds true provided $\bar{u}$ is an initial datum with finite total variation which satisfies some further condition making it more "realistic". Even if this were true, however, the counterexample in [7] would provide some useful information because it implies that (10) cannot be established by relying only on the maximum principle and on the boundedness of TotVar $\bar{u}$. To establish (10) in the case of "realistic" initial data one should likely rely on some more refined information on the structure of the solution, which is in general harder to obtain.

Acknowledgments. GC is partially supported by the Swiss National Science Foundation grant 200020-156112 and by the ERC Starting Grant 676675 FLIRT. MG was partially supported by the Swiss National Science Foundation grant P300P2-167681. LVS is a member of the GNAMPA group of INDAM and of the PRIN National Project "Hyperbolic Systems of Conservation Laws and Fluid Dynamics: Analysis and Applications".

REFERENCES

[1] Amorim, P., Colombo, R. M. and Teixeira, A., On the numerical integration of scalar nonlocal conservation laws, *ESAIM Math. Model. Numer. Anal.*, **49** (2015), 19–37.

[2] Betancourt, F., Bürger, R. and Karlsen, K. H. and Tory, E. M., On nonlocal conservation laws modelling sedimentation, *Nonlinearity*, **24** (2011), 855–885.

[3] Blandin, S. and Goatin, P., Well-posedness of a conservation law with non-local flux arising in traffic flow modeling, *Numer. Math.*, **132** (2016), 217–241.

[4] Calderoni, P. and Pulvirenti, M., Propagation of chaos for Burgers' equation, *Ann. Inst. H. Poincaré Sect. A (N.S.)*, **39** (1983), 85–97.

[5] Chiarello, F.A. and Goatin, P., Non-local multi-class traffic flow models, *Netw. Heterog. Media*, **14** (2019), 371–387.

[6] Colombo, M., Crippa, G., Graff, M. and Spinolo, L. V., On the role of numerical viscosity in the study of the local limit of nonlocal conservation laws, *ArXiv:1902.07513.*

[7] Colombo, M., Crippa, G. and Spinolo, L. V., Blow-up of the total variation in the local limit of a nonlocal traffic model, *ArXiv:1808.03529.*

[8] Colombo, M., Crippa, G. and Spinolo, L. V., On the singular local limit for conservation laws with nonlocal fluxes, *Arch. Ration. Mech. Anal.*, **233** (2019), 1131–1167.

[9] Colombo, R. M., Garavello, M. and Lécureux-Mercier, M., A class of nonlocal models for pedestrian traffic, *Math. Models Methods Appl. Sci.*, **22** (2012), 1150023, 34 p.

[10] Colombo, R. M., Herty, M. and Mercier, M., Control of the continuity equation with a non local flow, *ESAIM Control Optim. Calc. Var.*, **17** (2011), 353–379.

[11] Crippa, G. and Lécureux-Mercier, M., Existence and uniqueness of measure solutions for a system of continuity equations with non-local flow, *NoDEA Nonlinear Differential Equations Appl.*, **20** (2013), 523–537.

[12] Dafermos, C. M., *Hyperbolic Conservation Laws in Continuum Physics*, 4th edition, Springer-Verlag, Berlin, 2016.

[13] Friedrich, J., Kolb, O. and Göttlich, S., A Godunov type scheme for a class of LWR traffic flow models with non-local flux, *Netw. Heterog. Media*, **13** (2018), 531-547.

[14] Keimer, A. and Pflug, L., Existence, uniqueness and regularity results on nonlocal balance laws, *J. Differential Equations*, **263** (2017), 4023–4069.

[15] Keimer, A. and Pflug, L., On approximation of local conservation laws by nonlocal conservation laws, *J. Math. Anal. Appl.*, **475** (2019), 1927–1955.

[16] Kružkov, S. N., First order quasilinear equations with several independent variables, *Mat. Sb. (N.S.)*, **81(123)** (1970), 228–255.

[17] LeVeque, R. J., *Finite Volume Methods for Hyperbolic Problems*, Cambridge University Press, Cambridge, 2002.

[18] Tadmor, E., Numerical viscosity and the entropy condition for conservative difference schemes, *Math. Comp.*, **43** (1984), 369–381.

[19] Zumbrun, K., On a nonlocal dispersive equation modeling particle suspensions, *Quart. Appl. Math.*, **57** (1999), 573–600.

E-mail address: maria.colombo@epfl.ch
E-mail address: gianluca.crippa@unibas.ch
E-mail address: marie.graff@auckland.ac.nz
E-mail address: spinolo@imati.cnr.it

A FEEDBACK STRATEGY IN HYPERBOLIC CONTROL PROBLEMS

RINALDO M. COLOMBO

INdAM Unit, c/o DII
University of Brescia
Via Branze 38, 25123 Brescia, Italy

MAURO GARAVELLO*

Department of Mathematics and its Applications
University of Milano – Bicocca
Via R. Cozzi 55, 20125 Milano, Italy

ABSTRACT. We consider a control problem where some controllers aim at confining or directing a multitude of individuals in a given target region. A non–anticipative feedback strategy devoted to this task is defined and tested by means of numerical integrations.

1. **Introduction.**

A group of k leaders aims at gathering a multitude of individuals towards a target region $\mathcal{T}$, which is a given subset of $\mathbb{R}^n$. The leaders are described through their positions, say $P_1, \ldots, P_k$ in $\mathbb{R}^n$, while the many individuals are determined through their density $\rho = \rho(t, x)$, t being the time variable and x the space coordinate. For simplicity, we assume throughout that $n = 2$, i.e., that leaders and individuals move in $\mathbb{R}^2$, although the case of higher dimensional spaces also fit in the same theoretical framework.

A natural tool to describe the evolution of individuals in the present setting is given by the continuity equation

$$\partial_t \rho + \operatorname{div}_x \left(\rho\, v\left(t, x, P_1(t), \ldots, P_k(t)\right)\right) = 0\,,$$

where the speed v describes the individuals' movement and, in particular, how the leaders exert their influence on the individuals. The common goal of all the leaders is formalized through the minimization of the quantity

$$\mathcal{J} = \int_{\mathbb{R}^2} \rho(T, x)\, \psi(x)\, \mathrm{d}x \tag{1}$$

where T is the terminal time and ψ is a cost function. Typically, $\psi(x)$ is $d(x, \mathcal{T}) = \inf_{y \in \mathcal{T}} \|x - y\|$, i.e. the Euclidean distance between the position x and the target $\mathcal{T}$, with $\mathcal{T} \subset \mathbb{R}^2$. The leader P_i aims at minimizing $\mathcal{J}$ through a careful choice of its speed, say u_i, reasonably constrained to $\|u_i\| \leq U_i$, for given positive $U_1, \ldots, U_k$.

2000 *Mathematics Subject Classification.* Primary: 35L65; Secondary: 91A23; 93B52.

Key words and phrases. Hyperbolic Consensus Game; Multi-agent Consensus Strategies; Differential Games; Instantaneous Control; Feedback Strategy.

The authors are supported by the GNAMPA 2017 project *Conservation Laws: from Theory to Technology.* The *IBM Power Systems Academic Initiative* substantially contributed to the numerical integrations.

* Corresponding author: Mauro Garavello.

The resulting framework consists of (1) together with the following mixed ODE – PDE system

$$\left\{\begin{array}{l}\partial_t \rho + \mathrm{div}_x \left(\rho\, v(t,x,P_1(t),\ldots,P_k(t))\right) = 0\\ \rho(0,x) = \bar\rho(x)\end{array}\right. \quad \text{where} \left\{\begin{array}{l}\dot P_i = u_i(t)\\ P_i(0) = \bar P_i\end{array}\right. \quad i = 1,\ldots,k\,, \qquad (2)$$

so that $\mathcal{J}$ in (1) has to understood as a function of the controls $u_1, \ldots, u_k$. The function $\bar\rho$ and the points $\bar P_i$ respectively describe the initial distribution of the individuals and the initial positions of the agents.

Here, we only mention that the strategy introduced below can be exploited also when different controllers have different targets, so that (1)–(2) turns into a *game* rather being a *control problem.* The current literature offers a variety of results to related problems, see for instance [1, 3, 5, 6, 7, 10, 11].

The next section presents an analytic framework where (1)–(2) can be effectively formalized and studied. Then, the feedback strategy introduced in [4] is presented and in Section 3 some of its properties are shown by means of numerical integrations. The last section contains the conclusions.

2. **The Feedback Strategy.** Throughout, we keep the terminal time T fixed. For $x_o \in \mathbb{R}^2$ and $r > 0$, $B(x_o, r)$ stands for the open ball centered at x_o with radius r.

Under natural assumptions on the map v, the well posedness of (2) is well known, see for instance [2, 9]. The existence of a strategy $(u_1, \ldots, u_k) \in \mathbf{L}^\infty([0,T]; B(0,U))$ that minimizes (1) then follows by a standard compactness argument, see [4, Proposition 2.1]. However, such a strategy requires the controller P_i to have at any time $t \in [0,T]$ a complete knowledge of the evolution described by (2) also beyond time t, a perfect coordination among the controllers is also necessary. Not always these features are acceptable or realistic: the behavior of individuals might be unpredictable to the leaders and communications among leaders might be impossible or difficult.

To this aim, seeking the strategy of the i-th controller, we set $P = P_i$, $u = u_i$, $\bar P = \bar P_i$ and comprise within the time dependence of v all other strategies u_j, for $j \neq i$, obtaining the problem

$$\left\{\begin{array}{l}\partial_t \rho + \mathrm{div}_x \left(\rho\, v(t,x,P(t))\right) = 0\\ \rho(0,x) = \bar\rho(x)\end{array}\right. \quad \text{where} \quad \left\{\begin{array}{l}\dot P = u(t)\\ P(0) = \bar P\,.\end{array}\right. \qquad (3)$$

For a positive (suitably small) Δt, we seek the best choice of a speed $w \in \overline{B(0,U)}$ on the interval $[t, t+\Delta t]$ such that the solution $\rho_w = \rho_w(\tau, x)$ to

$$\left\{\begin{array}{l}\partial_\tau \rho_w + \mathrm{div}_x \left(\rho_w\, v(t,x,P(t) + (\tau - t)w)\right) = 0\\ \rho_w(t,x) = \rho(t,x)\end{array}\right. \qquad \tau \in [t, t+\Delta t] \qquad (4)$$

is likely to best contribute to decrease the value of $\mathcal{J}$. Remark that the dependence of v on t in (4) is frozen at time t. It is this choice that will later lead to a non anticipative strategy. A proof that (4) is well posed is provided in [4, Lemma 4.6]. Note also that this auxiliary control problem reduces to the original one, choosing t as the initial time, $t+\Delta t$ as the final time, and considering only one leader. Letting $\Delta t \to 0$, this construction produces the so called *myopic strategy.*

In the case of the functional (1), a natural choice for the agent P at time t is then to choose on the time interval $[t, t+\Delta t]$ a speed w, with $\|w\| \leq U$, to minimize the quantity

$$\begin{array}{rccl} \mathcal{J}_{t,\Delta t} & : & \mathbb{R}^2 & \to & \mathbb{R} \\ & & w & \to & \displaystyle\int_{\mathbb{R}^2} \rho_w(t+\Delta t, x)\, \psi(x)\, \mathrm{d}x\,. \end{array} \tag{5}$$

The following result, proved in [4], provides an effective hint on a non anticipative optimal choice of w.

Theorem 2.1. *Fix $T > 0$ and $U > 0$. Let $v \in \mathbf{C}^{0,1}([0,T] \times \mathbb{R}^2 \times \mathbb{R}^2; \mathbb{R}^2)$ and $\psi \in \mathbf{L}^\infty(\mathbb{R}^2; \mathbb{R})$. As initial data in (3), choose a boundedly supported $\bar\rho \in \mathbf{L}^1(\mathbb{R}^2; \mathbb{R})$ and a $\bar P \in \mathbb{R}^2$. Define ρ as the solution to (3) and ρ_w as the solution to (4), for a $w \in \overline{B(0,U)}$.*

Then, for any $t \in [0,T[$ and $\Delta t \in\,]0, T-t]$ the map (5) is well defined and Lipschitz continuous.

Moreover, if $v \in \mathbf{C}^2([0,T] \times \mathbb{R}^2 \times \mathbb{R}^2; \mathbb{R}^2)$, the map (5) admits the expansion

$$\mathcal{J}_{t,\Delta t}(w + \delta_w) = \mathcal{J}_{t,\Delta t}(w) + \operatorname{grad}_w \mathcal{J}_{t,\Delta t}(w) \cdot \delta_w + o(\delta_w) \qquad \text{as } \delta_w \to 0 \tag{6}$$

where, as $\Delta t \to 0$,

$$\begin{aligned} &\operatorname{grad}_w \mathcal{J}_{t,\Delta t}(w) = \\ &= \frac{(\Delta t)^2}{2} \int_{\mathbb{R}^2} \Big[\operatorname{grad}_x \rho(t,x) D_P v(t,x,P(t)) - \rho(t,x) \operatorname{grad}_P \operatorname{div}_x v(t,x,P(t))\Big] \psi(x)\, \mathrm{d}x \\ &\quad + o(\Delta t)^2. \end{aligned} \tag{7}$$

On the basis of Theorem 2.1, the definition of an effective non anticipative strategy for P_i can be easily achieved as follows. Split the interval $[0,T]$ in smaller portions $[t_\ell, t_{\ell+1}[$, where $t_\ell = \ell\, \Delta t$. On each of these intervals, define $u_i(t) = w_\ell$, where w_ℓ minimizes on $\overline{B(0,U)}$ the cost $\mathcal{J}_{t_\ell,\Delta t}$ defined in (5). The leading term in the right hand side of (7) is independent of w, so that for Δt small it is reasonable to choose $-w_\ell$ as

$$\frac{U \displaystyle\int_{\mathbb{R}^2} \Big[\operatorname{grad}_x \rho(t_\ell, x)\; D_P v\,(t_\ell, x, P_i(t_\ell)) - \rho(t_\ell, x)\; \operatorname{grad}_P \operatorname{div}_x v\,(t_\ell, x, P_i(t_\ell))\Big] \psi(x)\, \mathrm{d}x}{\left\| \displaystyle\int_{\mathbb{R}^2} \Big[\operatorname{grad}_x \rho(t_\ell, x)\; D_P v(t_\ell, x, P_i(t_\ell)) - \rho(t_\ell, x)\; \operatorname{grad}_P \operatorname{div}_x v(t_\ell, x, P_i(t_\ell))\Big] \psi(x)\, \mathrm{d}x \right\|}$$

as long as the denominator above does not vanish, in which case we set $w_\ell = 0$. Remark that, through the term ρ_ℓ, the right hand side above depends on all the past values $w_0, \ldots, w_{\ell-1}$ attained by u_i.

3. Numerical Simulations.

Below we consider a few sample integrations of (2) where the controllers $P_1, \ldots, P_k$ use the strategy based on Theorem 2.1. The different cases considered here fit in the well posedness result in [4, Lemma 4.6], which requires the Lipschitz continuous dependence of the speed v on its arguments. To this aim, we use the following regularized, that is Lipschitz continuous, normalization in $\mathbb{R}^2$:

$$\mathcal{N}(x) = \frac{x}{\max\{\varepsilon, \|x\|\}} \qquad \text{with} \qquad \varepsilon = 0.01\,. \tag{8}$$

The numerical algorithm employed is the usual Lax–Friedrichs method, see [8, § 3.1], with uniform mesh, in a numerical domain which is, typically, a square. This algorithm, as is well known, is conservative so that the total mass of the numerical solutions is conserved. Nevertheless, in the integrations below the support of ρ can be significantly both shrunk or enlarged, due to the lack of an *a priori* upper bound on ρ independent of the initial datum, see [4] for further details.

3.1. **A Single Attractive Controller Coping with an Obstacle.** We first show that, in spite of its myopic nature, a leader following the strategy based on Theorem 2.1 is able to drive individuals around an obstacle.

Consider (3) in the numerical domain $\Omega = [-10,10] \times [-10,10]$, with

$$k=1, \quad \begin{array}{c} v(t,x,P) = -e^{-0.05\,\|x-P\|^2}\mathcal{N}(x-P)\,, \\ U=4\,, \end{array} \quad \begin{array}{cc} \bar\rho = \chi_{B((-7,0),1)}, & \mathcal{T}=\{(7,0)\}\,, \\ \bar P_1 \equiv (-5,-5), & T=35\,, \end{array} \tag{9}$$

where $\mathcal{N}$ is as in (8). The attraction of the leader has unbounded support, but decreases exponentially with the square of the distance between the leader and the individuals. Initially, the individuals are uniformly distributed in the ball centered

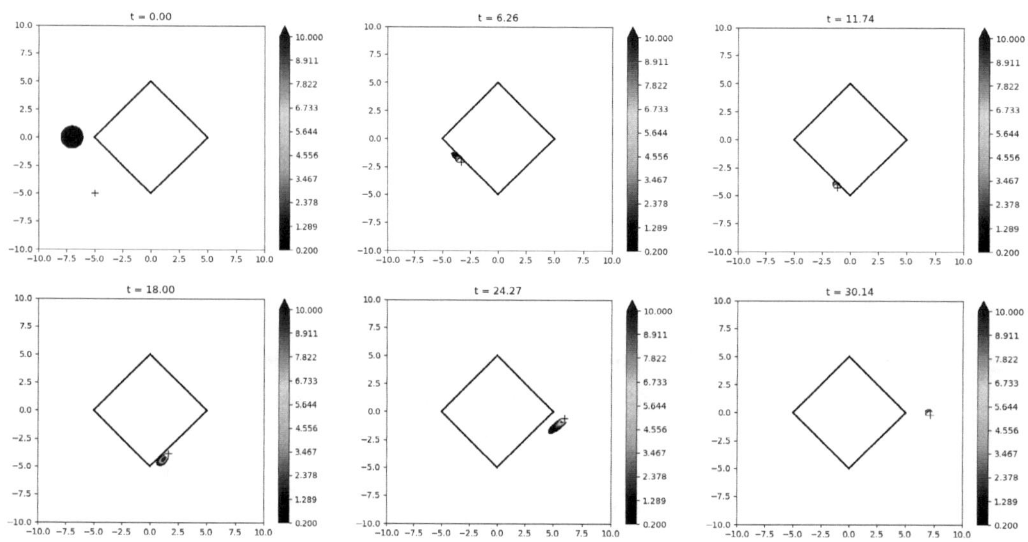

FIGURE 1. Integration of (2) with the choices (9). Note that, in spite of the myopic nature of the strategy suggested by Theorem 2.1, the leader first heads towards the mass of individuals, then succeeds in bypassing the obstacle.

at $(-7,0)$ with radius 1 and the target where the leader has to concentrate ρ is the point $(7,0)$.

In the central part of the domain, the square $\mathcal{S}$ with vertices at points $(-5,0)$, $(0,5)$, $(0,-5)$ and $(5,0)$ is forbidden to all individuals. This feature is accomplished through an *ad hoc* penalization of the distance function appearing in (1), in the sense that we use the cost

$$\mathcal{J} = \int_{\mathbb{R}^2} \rho(T,x)\,\psi(x)\,\mathrm{d}x \qquad \text{where} \qquad \psi(x) = 50\,\chi_{\mathcal{S}}(x) + (x_1-7)^2 + {x_2}^2\,. \tag{10}$$

Note that the analytical results in [4] comprehend also this setting.

We now compute the solution to (2) with u piecewise constant given by the above strategy based on (7), piecewise constant on the intervals $[j\,\Delta t, (j+1)\,\Delta t]$, where $\Delta t = 0.1$. The resulting solution, obtained by means of the Lax–Friedrichs [8] algorithm on a numerical grid of $n_x \times n_y = 2000 \times 2000$ cells, is displayed in Figure 1.

The strategy relying on Theorem 2.1 is myopic by its very definition, in the sense that it is based on an optimization over a short time interval, namely from t

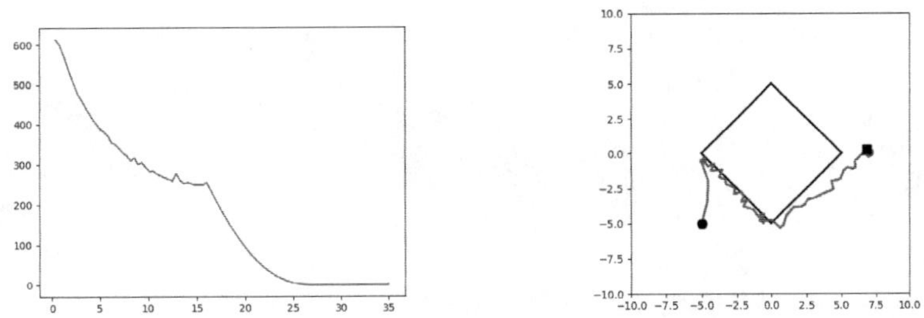

FIGURE 2. Integration of (2) with the choices (9). Left, the value of the cost (1) as a function of time: the lower corner of the obstacle is bypassed at about $t \approx 16.5$. Right, the trajectory followed by the leader: first it moves towards the individuals, then it drives them around the obstacle to the target.

to $t + \Delta t$. However, remarkably, in the present case the leader P_1 first approaches the individuals and then moves along a side of the obstacle. Note also that the speed of the leader increases after passing the lower vertex of the obstacle, see Figure 2, right. The resulting cost (1) vanishes before the final time, see Figure 2, left. Here, we also remark the sharp decrease in the cost at about time $t \approx 16.5$, corresponding to the controller passing the lower corner of the obstacle.

3.2. **Three Cooperating Repulsive Controllers.** We now use (3) to describe three agents whose aim is to push a multitude of individuals out of a given region. More precisely, in the numerical domain $\Omega = [-8, 8] \times [-8, 8]$, we consider $k = 3$ controllers, initially located at the positions $\bar{P}_i$ and with maximal speeds U_i, where

$$\begin{array}{lclllcllcl} \bar{P}_1 & \equiv & (0,-5)\,, & \bar{P}_2 & \equiv & (5,0)\,, & \bar{P}_3 & \equiv & (-1,3)\,, \\ U_1 & = & 5\,, & U_2 & = & 3\,, & U_3 & = & 1\,. \end{array} \tag{11}$$

The initial individuals' density is

$$\bar{\rho} = 0.25 * \chi_{B((1,1),1)} + 0.5 * \chi_{B((1,2),1.5)} + 0.75 * \chi_{B((-1,-1),0.75)}. \tag{12}$$

Each controller is repulsive, since the individuals' speed v is given by

$$v(t, x, P_1, P_2, P_3) = -e^{-0.1\,\|x\|^2}\,\mathcal{N}(x) + \sum_{i=1}^{3} e^{-0.1\,\|x-P_i\|^2}\,\mathcal{N}(x - P_i)\,, \tag{13}$$

where $\mathcal{N}$ is as in (8). The first summand in v describes how the individuals tend to move towards the origin while, at the same time, the latter summand models a repulsive interaction between individuals and each of the leaders P_1, P_2, and P_3. The target for each agent is the complement in $\mathbb{R}^2$ of the ball centered at the origin with radius 4, i.e., in (1) we have

$$\mathcal{T} = \mathbb{R}^2 \setminus B(0, 4) \qquad \text{so that} \qquad \psi(x) = \max\left\{0, 16 - \|x\|^2\right\}, \tag{14}$$

and the time at which this goal has to be obtained is $T = 20$.

The resulting solution, obtained on a numerical grid of $n_x \times n_y = 2000 \times 2000$ cells, is displayed in Figure 3. In this computation, the solution to (2) is obtained with u piecewise constant given by the strategy (7), constant on intervals $[j\,\Delta t, (j+1)\,\Delta t]$, where $\Delta t = 0.1$.

At about $t = 12$, the cost vanishes and the controllers succeed in keeping the individuals in $\mathcal{T}$, i.e. outside $B(0, 4)$, up to the final time $T = 20$. The controllers'

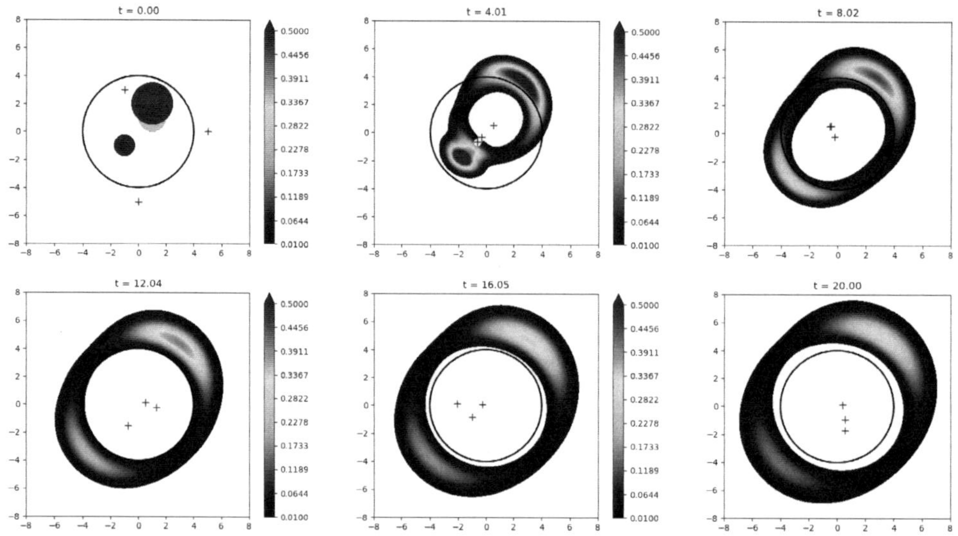

FIGURE 3. Integration of (2) with the choices (11)–(12)–(13). Already at time $t = 12.04$ (bottom left picture), the individuals' density ρ is supported almost completely in the target, i.e., outside $B(0,4)$. The black circle represents the boundary of $B(0,4)$.

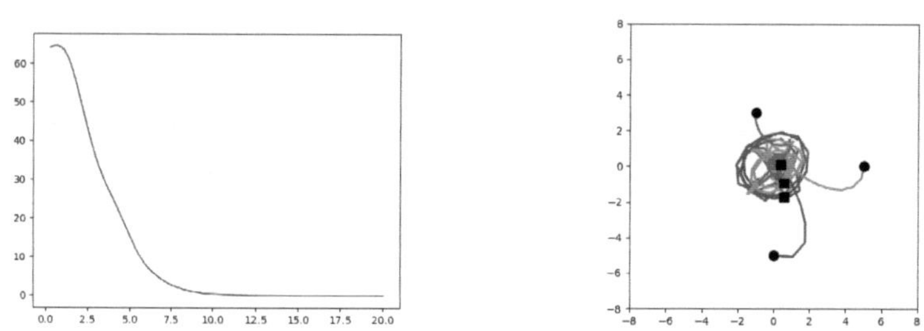

FIGURE 4. Integration of (2) with the choices (11)–(12)–(13). Left, the value of the cost (1)–(14) as a function of time: coherently with Figure 3, $B(0,4)$ is emptied approximately by time $t = 12$. Right, the trajectories followed by the controllers: first they move towards the origin, then they move repelling the individuals towards the target $\mathbb{R}^2 \setminus B(0,4)$.

trajectories first are direct towards the origin, then they apparently cover the sphere $B(0,4)$ to maintain the individuals within the target.

3.3. **Grouping Many Individuals.** We now display an example where repulsive and attracting leaders cooperate. Three controllers ($k = 3$) are initially located at $\bar{P}_i$, for $i = 1,2,3$ and have the same maximal speed U, where

$$\bar{P}_1 \equiv (5,-5)\,, \quad \bar{P}_2 \equiv (-9,-9)\,, \quad \bar{P}_3 \equiv (9,9) \quad \text{and} \quad U = 3\,, \tag{15}$$

while the speed in (2), which describes the individual–leader interactions, is

$$\begin{aligned} v(t,x,P_1,P_2,P_3) = & -e^{-0.05\,\|x-P_1\|^2}\,\mathcal{N}(x-P_1) + e^{-0.05\,\|x-P_2\|^2}\,\mathcal{N}(x-P_2) \\ & + e^{-0.05\,\|x-P_3\|^2}\,\mathcal{N}(x-P_3) \end{aligned} \tag{16}$$

with $\mathcal{N}$ is as in (8). P_1 is attractive while P_2 and P_3 are repulsive. All controllers

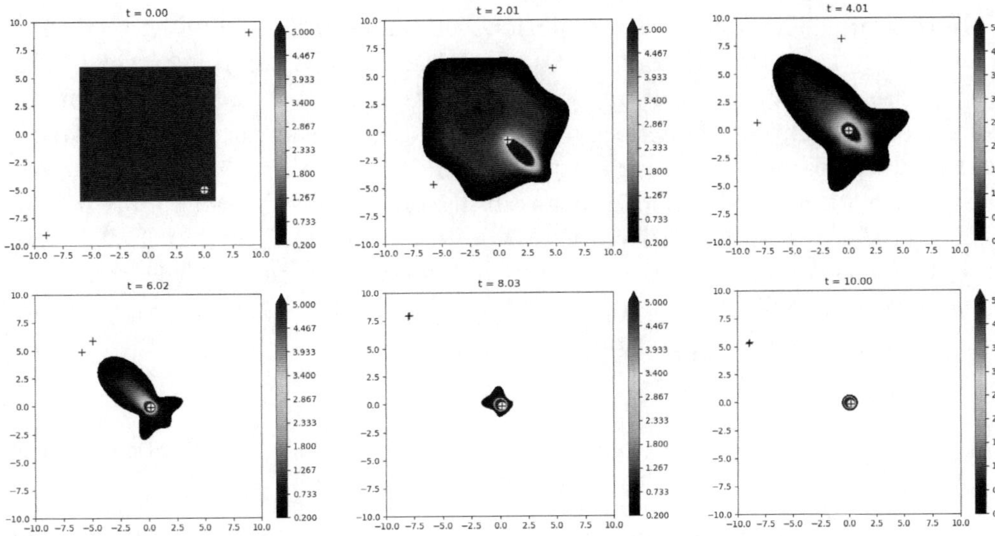

FIGURE 5. Integration of (2) with the choices (15)–(16)–(17)–(18). The two repulsive controllers, initially near to opposite corners of the numerical domain, surround the individuals, while the attractive controller moves directly to the origin.

have the same target, i.e., to bring the individuals near to the origin, so that

$$\mathcal{T} = \{(0,0)\} \qquad \text{and} \qquad \psi(x) = \|x\|^2 . \tag{17}$$

The initial individuals' distribution $\bar{\rho}$ is uniform over $[-6,6] \times [-6,6]$, so that

$$\bar{\rho} = \chi_{[-6,6]\times[-6,6]} . \tag{18}$$

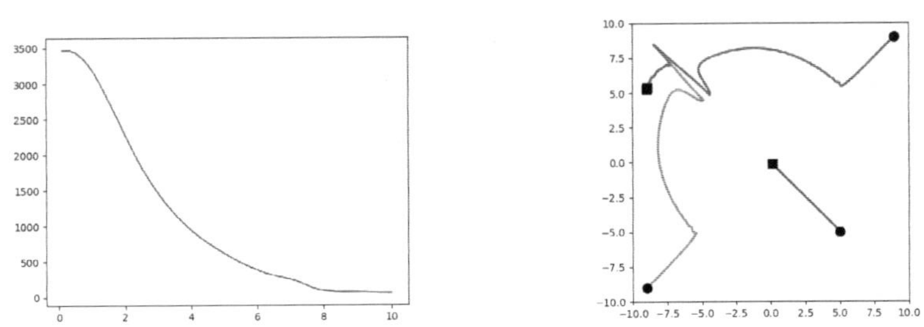

FIGURE 6. Integration of (2) with the choices (15)–(16)–(17)–(18). Left, the value of the cost (1)–(17) as a function of time. The combined strategy of the three controllers is rather effective, see the diagram of the cost on the left. Right: the different trajectories of the 3 controllers.

The numerical domain is $\Omega = [-10,10] \times [-10,10]$. The resulting integration, see Figure 5, shows that the three controllers successfully cooperate. Indeed, P_1, the attractive leader, heads directly towards the target $\mathcal{T}$, reaches it at about $t \approx 2.5$ and remains there attracting the individuals, see also Figure 6, right. In the mean time, the two repulsive controllers encircle the individuals, cut their escape route towards the top left corner and help getting the full confinement. The resulting cost at time $t = 10$, although positive, is about $\mathcal{J} = 52.8$, which shows a remarkable improvement with respect to its value at $t = 0$, see Figure 6, left.

4. **Conclusions.** The numerical simulations presented in Section 3 show that the myopic strategy is, at least in some cases, effective. However, the theoretical framework currently available, essentially based on [4], is not yet sufficient to provide rigorous results ensuring the efficacy of this strategy. Quantitative estimates on how far this strategy is from optimality are also, to our knowledge, unavailable.

The framework presented above naturally leads to consider also *competing* controllers. Again, the strategy above was proved to be effective in [4], at least in some cases, but no rigorous result has ever been provided. Currently, in the game theoretic framework, completely open natural questions concern the existence of Nash equilibria and their characterization.

REFERENCES

[1] R. Borsche, R.M. Colombo, M. Garavello, and A. Meurer. Differential equations modeling crowd interactions. *J. Nonlinear Sci.*, 25(4):827–859, 2015.

[2] A. Bressan. *Hyperbolic systems of conservation laws*, volume 20 of *Oxford Lecture Series in Mathematics and its Applications*. Oxford University Press, Oxford, 2000. The one-dimensional Cauchy problem.

[3] M. Caponigro, M. Fornasier, B. Piccoli, and E. Trélat. Sparse stabilization and control of alignment models. *Math. Models Methods Appl. Sci.*, 25(3):521–564, 2015.

[4] R.M. Colombo and M. Garavello. Hyperbolic consensus games. *Comm. Math. Sci.*, to appear.

[5] R.M. Colombo and M. Mercier. An analytical framework to describe the interactions between individuals and a continuum. *Journal of Nonlinear Science*, 22(1):39–61, 2012.

[6] R.M. Colombo and N. Pogodaev. Confinement strategies in a model for the interaction between individuals and a continuum. *SIAM J. Appl. Dyn. Syst.*, 11(2):741–770, 2012.

[7] R. Hegselmann and U. Krause. Opinion dynamics under the influence of radical groups, charismatic leaders, and other constant signals: a simple unifying model. *Netw. Heterog. Media*, 10(3):477–509, 2015.

[8] H. Holden and N.H. Risebro. *Front tracking for hyperbolic conservation laws*, volume 152 of *Applied Mathematical Sciences*. Springer, Heidelberg, second edition, 2015.

[9] S.N. Kružhkov. First order quasilinear equations with several independent variables. *Mat. Sb. (N.S.)*, 81 (123):228–255, 1970.

[10] R. Olfati-Saber, J.A. Fax, and R.M. Murray. Consensus and cooperation in networked multi-agent systems. *Proceedings of the IEEE*, 95(1):215–233, Jan 2007.

[11] B. Piccoli, N. Pouradier Duteil, and B. Scharf. Optimal control of a collective migration model. *Math. Models Methods Appl. Sci.*, 26(2):383–417, 2016.

E-mail address: rinaldo.colombo@unibs.it
E-mail address: mauro.garavello@unimib.it

ADJOINT APPROXIMATION OF NONLINEAR HYPERBOLIC SYSTEMS WITH NON-CONSERVATIVE PRODUCTS

Frédéric Coquel

CMAP, École Polytechnique, 91128 Palaiseau Cedex, France

Claude Marmignon, Pratik Rai and Florent Renac*

DAAA, ONERA, Université Paris Saclay, F-92322 Châtillon, France

Abstract. We consider the approximation of adjoint-based derivatives for discontinuous solutions of the Cauchy problem associated to one-dimensional nonlinear non-conservative hyperbolic systems. We first derive the adjoint equations in strong form with a discontinuous primal solution together with the associated jump relations across the discontinuity. The adjoint solution may be discontinuous at the discontinuity in contrast to the case of conservative systems. Then, we consider first-order finite volume (FV) approximations to the primal problem and show that, using the Volpert path family of schemes, the discrete adjoint solution is consistent with the strong form adjoint solution. Numerical experiments are shown for a nonlinear 2×2 system with a genuinely nonlinear (GNL) field and a linearly degenerate (LD) field associated to the non-conservative product.

1. **Introduction.** The discussion in this paper focuses on the adjoint analysis of the Cauchy problem for nonlinear hyperbolic systems in non-conservative form:

$$\begin{aligned} \partial_t \mathbf{u} + \mathbf{A}(\mathbf{u})\partial_x \mathbf{u} &= 0 \quad \text{in } \Omega := \mathbb{R} \times (0,T), && \text{(1a)} \\ \mathbf{u}(\cdot, 0) &= \mathbf{u}_0(\cdot) \quad \text{in } \mathbb{R}, && \text{(1b)} \end{aligned}$$

where $\mathbf{u}(x,t)$ is the vector of unknowns with values in the set of states $\Omega^a \subset \mathbb{R}^m$ and $\mathbf{A} : \Omega^a \ni \mathbf{u} \mapsto \mathbf{A}(\mathbf{u}) \in \mathbb{R}^{m\times m}$ is a smooth matrix-valued function with entries $a_{ij}(\mathbf{u})$, $1 \leq i,j \leq m$. We assume that (1a) is strictly hyperbolic over Ω^a. In the general case where $\mathbf{A}$ is not the Jacobian of a flux function, the works in [11, 4] generalize the notion of weak solutions from conservation laws to (1) and allow to define the non-conservative product $\mathbf{A}(\mathbf{u})\partial_x\mathbf{u}$ at a point of discontinuity of the solution for functions of bounded variations. The definition is based on a family of consistent and Lipschitz paths $\phi : [0,1] \times \Omega^a \times \Omega^a \to \Omega^a$. Across a discontinuity of speed σ, the non-conservative product is thus defined as the unique Borel measure defined by the so-called generalized Rankine-Hugoniot (RH) relations on Σ:

$$\sigma[\![\mathbf{u}]\!] = \boldsymbol{\mathcal{A}}_\phi(\mathbf{u}^-, \mathbf{u}^+) := \int_0^1 \mathbf{A}\big(\phi(s,\mathbf{u}^-,\mathbf{u}^+)\big)\partial_s\phi(s,\mathbf{u}^-,\mathbf{u}^+)ds, \tag{2}$$

where $[\![\mathbf{u}]\!] = \mathbf{u}^+ - \mathbf{u}^-$, and $\mathbf{u}^\pm$ are the limits of $\mathbf{u}$ at Σ (see section 2).

2000 *Mathematics Subject Classification.* Primary: 65M12, 65M60.

Key words and phrases. Non-conservative hyperbolic systems, adjoint equations, finite volume method.

* Corresponding author: florent.renac@onera.fr.

In this work, we consider the adjoint equations of (1). Methods based on adjoint equations are widely used for shape optimization, control, receptivity-sensitivity-stability analyses, data assimilation, error analysis, etc. These methods are often used for the linear analysis of nonlinear conservation laws where the adjoint is defined as the dual to the linearized equations around a given primal solution, $\mathbf{u}$. In the case of hyperbolic equations, this raises the question of the validity of this linearization around discontinuities in $\mathbf{u}$ because the adjoint equations are linear with discontinuous coefficients for which the Cauchy problem is not well posed in general. The analysis must include the linearization of the jump relations at the discontinuity [6] which leads to a so-called interior boundary condition for the adjoint variables [10]. Existence, uniqueness and stability of backward solutions to scalar equations have been established in [1] with Lipschitz initial condition and OSLC coefficients [9]. The interior condition at the shock has been shown to be satisfied by such backward solutions [10]. In the case of systems of conservation laws, well-posedness of the adjoint problem with GNL and LD fields has been shown in [2], while the interior boundary condition is satisfied at the discrete level providing that the primal and adjoint solutions are vanishing viscosity limits of regularized problems [8].

In § 2, we first derive the adjoint equations associated to the primal equations in strong form and then derive the adjoint equations associated to a first-order FV approximation in § 3. We prove consistency of the discrete adjoint equations for the Volpert path family of schemes for which the consistency condition can be expressed in closed form. An example of a 2×2 system with GNL and LD fields is provided in § 4 and numerical experiments are given in § 5.

2. **Adjoint formulation of linearized perturbations.** We are interested in Fréchet differentiable tracking-type output functionals of the form

$$J(\mathbf{u}) = \int_{\mathbb{R}} j\big(\mathbf{u}(x,T)\big)dx, \tag{3}$$

where $j : \Omega^a \to \mathbb{R}$ is a smooth function. We assume that the solution admits one isolated discontinuity along the curve $\Sigma := \big\{(x_s(t),t) : \ 0 < t < T\big\}$ in Ω (see figure 1) and consider infinitesimal perturbations imposed on the solution, $\mathbf{u}(x,t)+\boldsymbol{\psi}(x,t)$, and on the location of the discontinuity, $x_s(t)+\zeta_s(t)$ with $\zeta_s(0)=0$, (see figure 1), which may follow from perturbations of the initial condition (1b). Setting $\mathbf{j}'(\mathbf{u}) = \big(\partial_{u_i} j(\mathbf{u})\big)_{1\leq i\leq m}$, linear perturbations on $J(\mathbf{u})$ read [10]

$$J'(\mathbf{u};\boldsymbol{\psi},\zeta_s) = \int_{\mathbb{R}\backslash x_s(T)} \mathbf{j}'\big(\mathbf{u}(x,T)\big)\cdot\boldsymbol{\psi}(x,T)dx - \zeta_s(T)[\![j\big(\mathbf{u}(x_s(T),T)\big)]\!].$$

The speed of the discontinuity in (2), $\sigma = x_s'(t)$, may be expressed in terms of components of the normal to Σ, $\mathbf{n} = (n_x,n_t)^\top = \big(1+x_s'(t)^2\big)^{-1/2}\big(1,-x_s'(t)\big)^\top$: $\sigma = -\frac{n_t}{n_x}$. Linearized perturbations in the speed of the discontinuity read

$$\frac{\zeta_s'(t)}{x_s'(t)} = \frac{\delta n_t}{n_t} - \frac{\delta n_x}{n_x}, \tag{4}$$

so we get $n_x ds = n_x\sqrt{dx_s(t)^2+dt^2} = dt$ with $s(t)$ the curvilinear coordinate. The traces at Σ in the direction $\mathbf{n}$ are $\mathbf{u}^\pm = \lim_{\epsilon\downarrow 0}\mathbf{u}(x_s(t)\pm\epsilon n_x, t\pm\epsilon n_t)$, and the jump relations (2) now read

$$n_t[\![\mathbf{u}]\!] + n_x\boldsymbol{\mathcal{A}}_\phi(\mathbf{u}^-,\mathbf{u}^+) = 0 \quad \text{on } \Sigma. \tag{5}$$

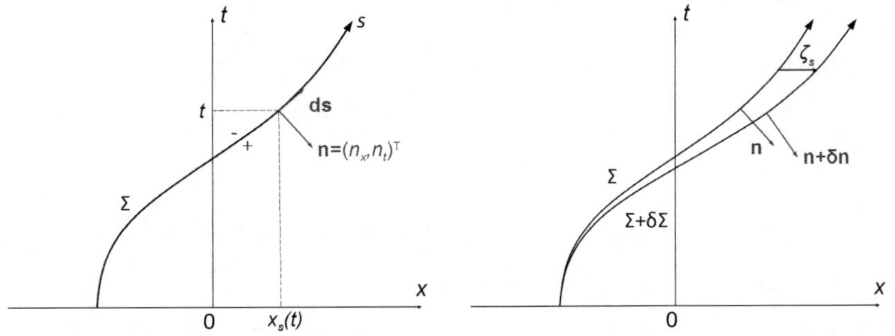

FIGURE 1. Discontinuity curve Σ in the space-time domain Ω: (left) definitions of traces and normal to Σ, (right) perturbation of Σ.

The adjoint formulation of linearized perturbations is obtained by introducing the following Lagrangian functional

$$\begin{aligned}\mathcal{L}(\mathbf{u};\mathbf{z},\mathbf{z}^s,\mathbf{z}^0) &= J(\mathbf{u}) - \int_{\Omega\setminus\Sigma} \mathbf{z}\cdot\big(\partial_t\mathbf{u}+\mathbf{A}(\mathbf{u})\partial_x\mathbf{u}\big)dxdt \\ &- \int_\Sigma \mathbf{z}^s\cdot\big(n_t[\![\mathbf{u}]\!]+n_x\boldsymbol{\mathcal{A}}_{\boldsymbol{\phi}}(\mathbf{u}^-,\mathbf{u}^+)\big)ds - \int_{\mathbb{R}} \mathbf{z}^0\cdot\big(\mathbf{u}(\cdot,0)-\mathbf{u}_0\big)dx, \end{aligned} \tag{6}$$

where the adjoint variables $\mathbf{z}(x,t):\Omega\setminus\Sigma\to\mathbb{R}^m$, $\mathbf{z}^s(x,t):\Sigma\to\mathbb{R}^m$, and $\mathbf{z}^0(x):\mathbb{R}\to\mathbb{R}^m$ are Lagrange multipliers associated to constraints (1a), (5), and (1b). Linearizing formally $\mathcal{L}$ in the perturbation direction around a state $\mathbf{u}$, we obtain

$$\mathcal{L}'(\mathbf{u};\boldsymbol{\psi},\zeta_s,\mathbf{z},\mathbf{z}^s,\mathbf{z}^0) = J'(\mathbf{u};\boldsymbol{\psi},\zeta_s) \tag{7a}$$

$$- \int_{\Omega\setminus\Sigma} \mathbf{z}\cdot\Big(\partial_t\boldsymbol{\psi}+\big(\mathbf{A}'(\mathbf{u})\boldsymbol{\psi}\big)\partial_x\mathbf{u}+\mathbf{A}(\mathbf{u})\partial_x\boldsymbol{\psi}\Big)dxdt \tag{7b}$$

$$+ \int_\Sigma n_x\zeta_s[\![\mathbf{z}\cdot\big(\partial_t\mathbf{u}+\mathbf{A}(\mathbf{u})\partial_x\mathbf{u}\big)]\!]ds \tag{7c}$$

$$- \int_\Sigma \mathbf{z}^s\cdot\Big(n_t[\![\boldsymbol{\psi}]\!]+n_x\big(\partial_{\mathbf{u}^-}\boldsymbol{\mathcal{A}}_{\boldsymbol{\phi}}(\mathbf{u}^-,\mathbf{u}^+)\boldsymbol{\psi}^- + \partial_{\mathbf{u}^+}\boldsymbol{\mathcal{A}}_{\boldsymbol{\phi}}(\mathbf{u}^-,\mathbf{u}^+)\boldsymbol{\psi}^+\big)\Big)ds \tag{7d}$$

$$- \int_\Sigma \zeta_s\,\mathbf{z}^s\cdot\big(n_t[\![\partial_x\mathbf{u}]\!]+n_x\partial_x\boldsymbol{\mathcal{A}}_{\boldsymbol{\phi}}(\mathbf{u}^-,\mathbf{u}^+)\big)ds \tag{7e}$$

$$- \int_\Sigma \mathbf{z}^s\cdot\big(\delta n_t[\![\mathbf{u}]\!]+\delta n_x\boldsymbol{\mathcal{A}}_{\boldsymbol{\phi}}(\mathbf{u}^-,\mathbf{u}^+)\big)ds \tag{7f}$$

$$- \int_{\mathbb{R}} \mathbf{z}^0\cdot\boldsymbol{\psi}(\cdot,0)dx, \tag{7g}$$

and the adjoint variables are defined as stationary points of $\mathcal{L}$ in (6):

$$\mathbf{z},\mathbf{z}^s,\mathbf{z}^0:\quad \mathcal{L}'(\mathbf{u};\boldsymbol{\psi},\zeta_s,\mathbf{z},\mathbf{z}^s,\mathbf{z}^0)=0\quad \forall\boldsymbol{\psi},\ \zeta_s. \tag{8}$$

Theorem 2.1 (Adjoint problem). *Let* $\mathbf{u}$ *be the solution of the nonlinear Cauchy problem (1) satisfying the generalized RH relations (5) at an isolated discontinuity*

$\Sigma \subset \Omega$. *Then, the adjoint solutions to (8) satisfy the following problem*

$$\begin{aligned} \partial_t \mathbf{z} + \mathbf{A}(\mathbf{u})^\top \partial_x \mathbf{z} + \Big(\big(\mathbf{A}'(\mathbf{u})^\top - \mathbf{B}(\mathbf{u})\big) \partial_x \mathbf{u} \Big) \mathbf{z} &= 0 \quad \text{in } \Omega \backslash \Sigma, \qquad (9a) \\ \mathbf{z}(\cdot, T) &= \mathbf{j}'(\mathbf{u}(\cdot, T)) \quad \text{in } \mathbb{R}, \qquad (9b) \end{aligned}$$

together with the jump relations across Σ*:*

$$\big(n_t \mathbf{I} + n_x \mathbf{A}(\mathbf{u}^\pm)^\top\big) \mathbf{z}^\pm = \big(n_t \mathbf{I} \pm n_x \partial_{\mathbf{u}^\pm} \boldsymbol{\mathcal{A}}_\phi(\mathbf{u}^-, \mathbf{u}^+)\big) \mathbf{z}^s, \qquad (10)$$

$\mathbf{z}^0(\cdot) = \mathbf{z}(\cdot, 0)$ *in* $\mathbb{R}$*, and the equation for* $\mathbf{z}^s$*:*

$$\begin{aligned} [\![\mathbf{u}]\!] \cdot d_t \mathbf{z}^s + \big(\partial_x \boldsymbol{\mathcal{A}}_\phi(\mathbf{u}^-, \mathbf{u}^+) - [\![\mathbf{A}(\mathbf{u}) \partial_x \mathbf{u}]\!]\big) \cdot \mathbf{z}^s &= 0 \quad \text{on } \Sigma, \qquad (11a) \\ \big([\![\mathbf{u}]\!] \cdot \mathbf{z}^s(T) - [\![j(\mathbf{u})]\!]\big)_{x_s(T)} &= 0. \qquad (11b) \end{aligned}$$

The tensor operators in (9a) are defined by

$$\mathbf{A}'(\mathbf{u})^\top_{ijk} = \partial_{u_k} a_{ji}(\mathbf{u}), \quad \mathbf{B}(\mathbf{u})_{ijk} = \partial_{u_i} a_{jk}(\mathbf{u}), \quad 1 \le i, j, k \le m, \qquad (12)$$

and $\mathbf{B}$ *satisfies* $\boldsymbol{\psi}^\top (\mathbf{B}(\mathbf{u}) \partial_x \mathbf{u}) \mathbf{z} = \mathbf{z}^\top (\mathbf{A}'(\mathbf{u}) \boldsymbol{\psi}) \partial_x \mathbf{u}$*, for all* $\boldsymbol{\psi}$ *in* $\mathbb{R}^m$ *and* $\mathbf{u}$ *in* Ω^a*.*

Proof. First, (7c) vanishes due to (1a). Integration by parts in (7b) gives

$$\begin{aligned} (7b) \quad = \quad & \int_{\Omega \backslash \Sigma} \boldsymbol{\psi} \cdot \Big(\partial_t \mathbf{z} + \partial_x \big(\mathbf{A}(\mathbf{u})^\top \mathbf{z}\big) \Big) - \mathbf{z} \big(\mathbf{A}'(\mathbf{u}) \boldsymbol{\psi}\big) \partial_x \mathbf{u} dx dt \\ + \quad & \int_\Sigma [\![\boldsymbol{\psi} (n_t + n_x \mathbf{A}(\mathbf{u})^\top) \mathbf{z}]\!] ds + \int_{\mathbb{R}} \boldsymbol{\psi}(x, 0) \cdot \mathbf{z}(x, 0) - \boldsymbol{\psi}(x, T) \cdot \mathbf{z}(x, T) dx. \end{aligned}$$

Then, using (5) and (4), we get $\delta n_t [\![\mathbf{u}]\!] + \delta n_x \boldsymbol{\mathcal{A}}_\phi(\mathbf{u}^-, \mathbf{u}^+) = -n_x \zeta_s'(t) [\![\mathbf{u}]\!]$. Using again (1a), the term in (7e) may be be recast into

$$\begin{aligned} n_t [\![\partial_x \mathbf{u}]\!] + n_x \partial_x \boldsymbol{\mathcal{A}}_\phi(\mathbf{u}^-, \mathbf{u}^+) &= [\![n_t \partial_x \mathbf{u} - n_x (\partial_t \mathbf{u} + \mathbf{A}(\mathbf{u}) \partial_x \mathbf{u})]\!] + n_x \partial_x \boldsymbol{\mathcal{A}}_\phi(\mathbf{u}^-, \mathbf{u}^+) \\ &= -n_x d_t [\![\mathbf{u}]\!] + n_x \big(\partial_x \boldsymbol{\mathcal{A}}_\phi - [\![\mathbf{A}(\mathbf{u}) \partial_x \mathbf{u}]\!]\big), \end{aligned}$$

where $d_t \equiv \partial_t + x_s'(t) \partial_x$ and using integration by parts we get

$$\begin{aligned} (7e) + (7f) &= \int_\Sigma \mathbf{z}^s \cdot \Big(d_t (\zeta_s [\![\mathbf{u}]\!]) - \zeta_s \big(\partial_x \boldsymbol{\mathcal{A}}_\phi - [\![\mathbf{A}(\mathbf{u}) \partial_x \mathbf{u}]\!]\big) \Big) n_x ds \\ &= -\int_0^T \zeta_s \Big([\![\mathbf{u}]\!] \cdot d_t \mathbf{z}^s + \big(\partial_x \boldsymbol{\mathcal{A}}_\phi - [\![\mathbf{A}(\mathbf{u}) \partial_x \mathbf{u}]\!]\big) \cdot \mathbf{z}^s \Big) dt + \zeta_s(T) \mathbf{z}^s(T) \cdot [\![\mathbf{u}]\!]_{x_s(T)}, \end{aligned}$$

where we have used $dt = n_x ds$ and $\zeta_s(0) = 0$. We thus obtain

$$\begin{aligned} \mathcal{L}'(\mathbf{u}; \boldsymbol{\psi}, \zeta_s, \mathbf{z}, \mathbf{z}^s, \mathbf{z}^0) \quad = \quad & \int_{\mathbb{R} \backslash x_s(T)} \mathbf{j}'\big(\mathbf{u}(x, T)\big) \cdot \boldsymbol{\psi}(x, T) dx - \zeta_s(T) [\![j\big(\mathbf{u}(x_s(T), T)\big)]\!] \\ + \quad & \int_{\Omega \backslash \Sigma} \boldsymbol{\psi} \cdot \Big(\partial_t \mathbf{z} + \partial_x \big(\mathbf{A}(\mathbf{u})^\top \mathbf{z}\big) \Big) - \mathbf{z} \big(\mathbf{A}'(\mathbf{u}) \boldsymbol{\psi}\big) \partial_x \mathbf{u} dx dt \\ + \quad & \int_\Sigma [\![\boldsymbol{\psi} (n_t + n_x \mathbf{A}(\mathbf{u})^\top) \mathbf{z}]\!] ds - \int_{\mathbb{R}} \big[\boldsymbol{\psi}(x, t) \cdot \mathbf{z}(x, t)\big]_{t=0}^T dx \\ - \quad & \int_\Sigma \mathbf{z}^s \cdot \Big(n_t [\![\boldsymbol{\psi}]\!] + n_x \big(\partial_{\mathbf{u}^-} \boldsymbol{\mathcal{A}}_\phi \boldsymbol{\psi}^- + \partial_{\mathbf{u}^+} \boldsymbol{\mathcal{A}}_\phi \boldsymbol{\psi}^+\big) \Big) ds \\ - \quad & \int_0^T \zeta_s \Big([\![\mathbf{u}]\!] \cdot d_t \mathbf{z}^s + \big(\partial_x \boldsymbol{\mathcal{A}}_\phi - [\![\mathbf{A}(\mathbf{u}) \partial_x \mathbf{u}]\!]\big) \cdot \mathbf{z}^s \Big) dt \\ + \quad & \zeta_s(T) \mathbf{z}^s(T) \cdot [\![\mathbf{u}]\!]_{x_s(T)} - \int_{\mathbb{R}} \mathbf{z}^0 \cdot \boldsymbol{\psi}(\cdot, 0) dx. \end{aligned}$$

Then, collecting terms against $\boldsymbol{\psi}$ and ζ_s, we obtain the desired results. □

2.1. **The case of conservative systems.** We now consider the particular case where (1a) reduces to a conservation law, i.e., $\mathbf{A}(\mathbf{u}) = \mathbf{f}'(\mathbf{u})$. Assuming that the flux $\mathbf{f}$ is a $\mathcal{C}^2$ function, its Hessian is symmetric. Hence $\partial_{\mathbf{u}_k} a_{ij} = \partial_{\mathbf{u}_j} a_{ik}$, $1 \leq i, j, k \leq m$, which is equivalent to $\mathbf{A}'(\mathbf{u})^\top = \mathbf{B}(\mathbf{u})$ from (12), so (9a) reduces to the classical adjoint equation of first-order conservation laws:

$$\partial_t \mathbf{z} + \mathbf{A}(\mathbf{u})^\top \partial_x \mathbf{z} = 0 \quad \text{in } \Omega \backslash \Sigma.$$

Then, for all paths in (2), we have $\boldsymbol{\mathcal{A}}_\phi(\mathbf{u}^-, \mathbf{u}^+) = [\![\mathbf{f}(\mathbf{u})]\!]$, so we obtain

$$\partial_{\mathbf{u}^\pm} \boldsymbol{\mathcal{A}}_\phi(\mathbf{u}^-, \mathbf{u}^+) = \pm \mathbf{A}(\mathbf{u}^\pm), \quad \partial_x \boldsymbol{\mathcal{A}}_\phi(\mathbf{u}^-, \mathbf{u}^+) = [\![\mathbf{A}(\mathbf{u})\partial_x \mathbf{u}]\!],$$

and the jump relations (10) now read

$$\left(n_t \mathbf{I} + n_x \mathbf{A}(\mathbf{u}^\pm)^\top\right)(\mathbf{z}^\pm - \mathbf{z}^s) = 0 \quad \text{on } \Sigma.$$

For a non-characteristic discontinuity, the matrices $n_t \mathbf{I} + n_x \mathbf{A}(\mathbf{u}^\pm)$ are nonsingular and we obtain the so-called interior boundary condition on Σ [10]: $\mathbf{z}^\pm = \mathbf{z}^s$.

3. Space-time discretization.

3.1. **Finite volume method.** The nonlinear problem (1) is discretized with a first-order FV method and explicit time stepping. The degrees-of-freedom are

$$\mathbf{u}_h(x, t^{(n)}) = \mathbf{U}_i^n, \quad \forall x \in \kappa_i = (x_{i-\frac{1}{2}}, x_{i+\frac{1}{2}}), \quad i \in \mathbb{Z}, \quad 0 \leq n \leq N,$$

with $x_{i+\frac{1}{2}} = i\Delta x$, $t^{(n)} = n\Delta t$, $\Delta x > 0$ and $\Delta t = \frac{T}{N} > 0$ the space and time steps. The numerical scheme reads (see [7] and references therein)

$$\mathbf{U}_i^{n+1} - \mathbf{U}_i^n + \tfrac{\Delta t}{\Delta x}\left(\mathbf{D}_{i+\frac{1}{2}}^- + \mathbf{D}_{i-\frac{1}{2}}^+\right) = 0, \quad i \in \mathbb{Z}, \quad 0 \leq n < N, \tag{13}$$

with smooth fluctuation fluxes $\mathbf{D}_{i+\frac{1}{2}}^\pm = \mathbf{D}^\pm(\mathbf{U}_i^n, \mathbf{U}_{i+1}^n)$ satisfying consistency: $\mathbf{D}^\pm(\mathbf{u}, \mathbf{u}) = 0$ for all $\mathbf{u}$ in Ω^a. The initial condition is projected onto the space grid:

$$\mathbf{U}_i^0 = \langle \mathbf{u}_0 \rangle_i := \frac{1}{\Delta x} \int_{\kappa_i} \mathbf{u}_0(x) dx, \quad i \in \mathbb{Z}. \tag{14}$$

3.2. **Discrete adjoint solution.** We now consider the adjoint solution to the discrete nonlinear problem and look again for a piecewise constant discrete solution:

$$\mathbf{z}_h(x, t^{(n)}) = \mathbf{Z}_i^n, \quad \forall x \in \kappa_i, \quad i \in \mathbb{Z}, \quad 0 \leq n \leq N.$$

We introduce the discrete Lagrangian functional containing an approximation of the output functional (3) and the multipliers to the constraints (13) and (14):

$$\begin{aligned} \mathcal{L}_h(\mathbf{u}_h; \mathbf{z}_h) = & \sum_{i \in \mathbb{Z}} \Delta x \, j(\mathbf{U}_i^N) - \sum_{i \in \mathbb{Z}} \sum_{n=1}^{N-1} \Delta x \, \mathbf{Z}_i^n \cdot \left(\mathbf{U}_i^{n+1} - \mathbf{U}_i^n + \tfrac{\Delta t}{\Delta x}(\mathbf{D}_{i+\frac{1}{2}}^- + \mathbf{D}_{i-\frac{1}{2}}^+)\right) \\ & - \sum_{i \in \mathbb{Z}} \Delta x \, \mathbf{Z}_i^0 \cdot (\mathbf{U}_i^0 - \langle \mathbf{u}_0 \rangle_i). \end{aligned} \tag{15}$$

Linearizing (15) around $\mathbf{u}_h$ and looking for stationary solutions give the discrete adjoint equations which again constitute a backward problem in time:

$$\begin{aligned} \mathbf{Z}_i^{n-1} - \mathbf{Z}_i^n + \tfrac{\Delta t}{\Delta x}\Big(\partial_{\mathbf{u}^-} \mathbf{D}_{i+\frac{1}{2}}^{-\top} \mathbf{Z}_i^n + \partial_{\mathbf{u}^+} \mathbf{D}_{i-\frac{1}{2}}^{-\top} \mathbf{Z}_{i-1}^n & \\ + \partial_{\mathbf{u}^-} \mathbf{D}_{i+\frac{1}{2}}^{+\top} \mathbf{Z}_{i+1}^n + \partial_{\mathbf{u}^+} \mathbf{D}_{i-\frac{1}{2}}^{+\top} \mathbf{Z}_i^n\Big) & = 0, \quad 0 < n \leq N, \end{aligned} \tag{16a}$$

$$\mathbf{Z}_i^N = \mathbf{j}'(\mathbf{U}_i^N). \tag{16b}$$

3.3. **The Volpert path family of schemes.** Let us consider fluctuation fluxes based on linear paths [11] of the form $\mathbf{D}^{\pm}(\mathbf{u}^-,\mathbf{u}^+) = \mathbf{A}^{\pm}(\mathbf{u}^-,\mathbf{u}^+)[\![\mathbf{u}]\!]$, where the consistency relation now reads

$$\mathbf{A}^{+}(\mathbf{u},\mathbf{u}) + \mathbf{A}^{-}(\mathbf{u},\mathbf{u}) = \mathbf{A}(\mathbf{u}), \quad \forall \mathbf{u}\in\Omega^a. \tag{17}$$

Theorem 3.1 (Adjoint consistency). *The discrete adjoint scheme (16) with the Volpert path family of fluxes is a consistent approximation of the adjoint problem (9) at points* $(x_i, t^{(n)})$ *with* $x_i = \frac{1}{2}(x_{i-\frac{1}{2}} + x_{i+\frac{1}{2}})$.

Proof. We use an usual finite difference analysis. Let $\mathbf{u}$ and $\mathbf{z}$ be smooth solutions to the primal and adjoint equations. Use Taylor expansions in time to show that $\frac{1}{\Delta t}(\mathbf{Z}_i^{n-1} - \mathbf{Z}_i^n) = -\partial_t\mathbf{z}(x_i,t^{(n)}) + \mathcal{O}(\Delta t)$. Setting $a_{ij}^{\pm}(\mathbf{u}^-,\mathbf{u}^+) = \mathbf{A}^{\pm}(\mathbf{u}^-,\mathbf{u}^+)_{ij}$ and differentiating (17), we get for $1\leq i,j,k\leq m$:

$$\partial_{u_k^-}a_{ij}^+(\mathbf{u},\mathbf{u}) + \partial_{u_k^+}a_{ij}^+(\mathbf{u},\mathbf{u}) + \partial_{u_k^-}a_{ij}^-(\mathbf{u},\mathbf{u}) + \partial_{u_k^+}a_{ij}^-(\mathbf{u},\mathbf{u}) = \partial_{u_k}a_{ij}(\mathbf{u}), \;\; \mathbf{u}\in\Omega^a. \tag{18}$$

Now, we decompose the space terms in (16) into $\mathcal{R}_1 + \mathcal{R}_2$ with

$$\begin{aligned}
\mathcal{R}_1(x_i,t^{(n)}) &= -\frac{1}{\Delta x}\Big(\mathbf{A}_{i+\frac{1}{2}}^{-\top}\mathbf{Z}_i^n - \mathbf{A}_{i-\frac{1}{2}}^{-\top}\mathbf{Z}_{i-1}^n + \mathbf{A}_{i+\frac{1}{2}}^{+\top}\mathbf{Z}_{i+1}^n - \mathbf{A}_{i-\frac{1}{2}}^{+\top}\mathbf{Z}_i^n\Big),\\
\mathcal{R}_2(x_i,t^{(n)}) &= \frac{1}{\Delta x}\Big(\big(\partial_{\mathbf{u}^-}\mathbf{A}_{i+\frac{1}{2}}^{-}[\![\mathbf{u}]\!]_{i+\frac{1}{2}}\big)^\top\mathbf{Z}_i^n + \big(\partial_{\mathbf{u}^+}\mathbf{A}_{i-\frac{1}{2}}^{-}[\![\mathbf{u}]\!]_{i-\frac{1}{2}}\big)^\top\mathbf{Z}_{i-1}^n\\
&+ \big(\partial_{\mathbf{u}^-}\mathbf{A}_{i+\frac{1}{2}}^{+}[\![\mathbf{u}]\!]_{i+\frac{1}{2}}\big)^\top\mathbf{Z}_{i+1}^n + \big(\partial_{\mathbf{u}^+}\mathbf{A}_{i-\frac{1}{2}}^{+}[\![\mathbf{u}]\!]_{i-\frac{1}{2}}\big)^\top\mathbf{Z}_i^n\Big).
\end{aligned}$$

We thus obtain from (17) and (18)

$$\begin{aligned}
\mathcal{R}_1(x_i,t^{(n)})_k &= -\frac{1}{\Delta x}\sum_l\Big(a_{lk}^-(\mathbf{U}_i^n,\mathbf{U}_{i+1}^n)Z_i^{l,n} - a_{lk}^-(\mathbf{U}_{i-1}^n,\mathbf{U}_i^n)Z_{i-1}^{l,n}\\
&+ a_{lk}^+(\mathbf{U}_i^n,\mathbf{U}_{i+1}^n)Z_{i+1}^{l,n} - a_{lk}^+(\mathbf{U}_{i-1}^n,\mathbf{U}_i^n)Z_i^{l,n}\Big)\\
&= -\sum_{l,m}\big((\partial_{u_m^+}a_{lk}^- + \partial_{u_m^-}a_{lk}^- + \partial_{u_m^+}a_{lk}^+ + \partial_{u_m^-}a_{lk}^+)\partial_x u_m\big)_i^n Z_i^{l,n}\\
&- \sum_l(a_{lk}^- + a_{lk}^+)\partial_x Z_i^{l,n} + \mathcal{O}(\Delta x)\\
&= -\Big(\mathbf{A}(\mathbf{u})^\top\partial_x\mathbf{z} + \big(\mathbf{A}'(\mathbf{u})^\top\partial_x\mathbf{u}\big)\mathbf{z}\Big)_k(x_i,t^{(n)}) + \mathcal{O}(\Delta x),
\end{aligned}$$

and similarly $\mathcal{R}_2(x_i,t^{(n)})_k = \big((\mathbf{B}(\mathbf{u})\partial_x\mathbf{u})\mathbf{z}\big)_k(x_i,t^{(n)}) + \mathcal{O}(\Delta x)$ from (18). □

4. **Non-conservative product associated to a LD field.** Let us introduce the following nonlinear hyperbolic system [7] typical of two-phase flow models where the characteristic LD field plays the role of an interface velocity [3]:

$$\partial_t u + g(\mathbf{u})\partial_x u = 0, \quad \partial_t v + \partial_x f(\mathbf{u}) = 0, \tag{19}$$

with $g(\mathbf{u}) = u+v$ and $f(\mathbf{u}) = \frac{v^2-u^2}{2}$. The eigenvalues are $g(\mathbf{u})$ associated to the LD field and v associated to a GNL field so the system is strictly hyperbolic over $\Omega^a = \{(u,v)^\top\in\mathbb{R}^2 : u>0\}$. The generalized RH relations (5) read

$$n_t[\![u]\!]_{x_s(t)} + n_x\mathcal{G}_\phi(\mathbf{u}^-,\mathbf{u}^+) = 0, \quad [\![n_t v + n_x f(\mathbf{u})]\!]_{x_s(t)} = 0 \quad \text{on } \Sigma,$$

where $\mathcal{G}_\phi(\mathbf{u}^-,\mathbf{u}^+) := \int_0^1 g(\phi(\theta;\mathbf{u}^-,\mathbf{u}^+))\partial_\theta\phi_u(\theta;\mathbf{u}^-,\mathbf{u}^+)d\theta$, so for a linear path $\mathcal{G}_\phi(\mathbf{u}^-,\mathbf{u}^+) = \overline{u+v}[\![u]\!]$, where $\overline{a} = \frac{a^-+a^+}{2}$ denotes the average operator. Let us stress that the LD field $g(\cdot)$ is continuous across a contact discontinuity, so the generalized RH relations are independent of the choice of path which motivates the choice of a linear path.

The adjoint equations for $\mathbf{z} = (y, z)^\top$ read

$$\partial_t y - u\partial_x z + (u+v)\partial_x y + (\partial_x u)y = 0, \quad \partial_t z + v\partial_x z - (\partial_x u)y = 0 \quad \text{in } \Omega\backslash\Sigma, \tag{20}$$

together with the jump relations on Σ

$$\begin{aligned} \big(n_t + n_x(u^\pm + v^\pm)\big)y^\pm - \big(n_t + n_x(u^\pm + \overline{v})\big)y^s - n_x u^\pm(z^\pm - z^s) &= 0, \quad (21a)\\ -\tfrac{n_x}{2}[\![u]\!]y^s \pm \big(n_t + n_x v^\pm\big)(z^\pm - z^s) &= 0, \quad (21b) \end{aligned}$$

and the equation for $\mathbf{z}^s$:

$$[\![\mathbf{u}]\!]_{x_s(t)} \cdot d_t\mathbf{z}^s + \Big([\![u]\!]\overline{\partial_x(u+v)} - [\![u+v]\!]\overline{\partial_x u}\Big)_{x_s(t)} y^s(t) = 0 \quad \text{in } (0,T). \tag{22}$$

The above relations at Σ may be simplified in the following two cases:

- isolated non-characteristic shock ($[\![u]\!] = 0$ and $n_t + \overline{v}n_x = 0$):
$$y^\pm = \tfrac{2u}{2u\pm[\![v]\!]}y^s, \; z^\pm = z^s, \; z^s(\cdot) \equiv \tfrac{[\![j(\mathbf{u})]\!]_{x_s(T)}}{[\![v]\!]_{x_s(T)}};$$
- isolated characteristic contact ($[\![u+v]\!] = 0$ and $n_t + (u^\pm + v^\pm)n_x = 0$):
$$z^\pm = z^s \pm \tfrac{[\![v]\!]y^s}{2u^\pm}, \; z^s = \tfrac{\overline{uz}}{\overline{u}}, \; z^s(\cdot) - y^s(\cdot) \equiv \tfrac{[\![j(\mathbf{u})]\!]_{x_s(T)}}{[\![v]\!]_{x_s(T)}}.$$

5. Numerical experiments.

We consider Riemann problems for (19) with initial conditions $\mathbf{u}_0(x) = \mathbf{u}_L$ if $x < 0$, and $\mathbf{u}_0(x) = \mathbf{u}_R$ if $x > 0$:

test	problem	left state $\mathbf{u}_L$	right state $\mathbf{u}_R$	T
RP1	shock	$(\frac{3}{2}, 3)^\top$	$(\frac{3}{2}, 1)^\top$	0.1
RP2	shock	$(\frac{1}{2}, 3)^\top$	$(\frac{1}{2}, 1)^\top$	0.1
RP3	contact	$(1, 3)^\top$	$(2, 2)^\top$	0.05

The output functional reads $J(\mathbf{u}) = \frac{1}{2}\int_{\mathbb{R}} \mathbf{u}(x,T)^2 dx$ which imposes $\mathbf{z}(\cdot, T) = \mathbf{u}(\cdot, T)$ as final condition. We compute approximate solutions with a numerical flux described in [7] that falls into the family of Volpert schemes. Figure 2 compares the numerical solution in dashed lines with the exact solution in continuous lines and displays the characteristics of both primal and adjoint problems in Ω. The exact solutions are obtained from the method of generalized characteristics [5, 1, 10], the adjoint equations (20) and jump relations (21). Results are obtained on a very fine mesh to check the consistency of the discrete adjoint method. In particular, it may be checked that the adjoint solutions satisfy the RH relations derived in § 4.

REFERENCES

[1] F. Bouchut and F. James, One-dimensional transport equations with discontinuous coefficients, *Nonlinear Anal.*, **32** (1998), 891–933.

[2] A. Bressan and A. Marson, A maximum principle for optimally controlled systems of conservation laws, *Rend. Sem. Mat. Univ. Padova*, **94** (1995), 79–94.

[3] F. Coquel, T. Gallouët, J.-M. Hérard and N. Seguin, Closure laws for a two-fluid two-pressure model, *C. R. Acad. Sci. Paris*, **334** (2002), 927–932.

[4] G. Dal Maso, P. G. LeFloch and F. Murat, Definition and weak stability of nonconservative products, *J. Math. Pures Appl.*, **74** (1995), 483–548.

[5] A. F. Filippov, Differential equations with discontinuous right-hand side, *Amer. Math. Soc. Transl.*, **42** (1964), 199–231.

[6] A. Majda, The stability of multidimensional shock fronts, Memoirs of the AMS, 275, Amer. Math. Soc., Providence, RI, 1983.

[7] F. Renac, Entropy stable DGSEM for nonlinear hyperbolic systems in nonconservative form with application to two-phase flows, *J. Comput. Phys.*, **382** (2019), 1–26.

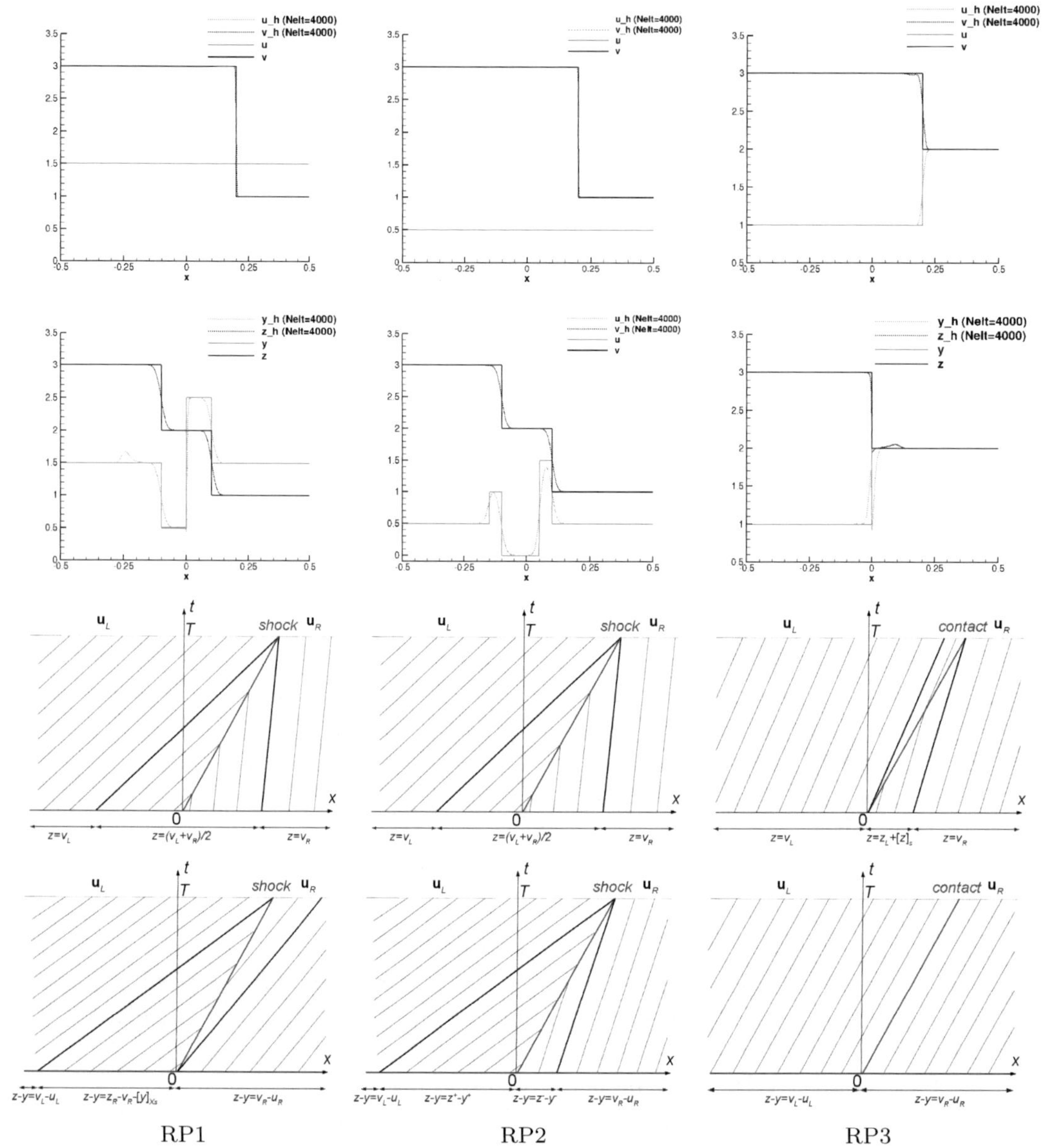

FIGURE 2. Riemann problems discretized with $N = 4000$ cells. From top raw to bottom raw: primal solutions at $t = T$, adjoint solutions at $t = 0$, v-characteristics, $(u + v)$-characteristics.

[8] J. Schütz, S. Noelle, C. Steiner and G. May, A note on adjoint error estimation for one-dimensional stationary balance laws with shocks, *SIAM J. Numer. Anal.*, **51** (2013), 126–136.

[9] E. Tadmor, Local error estimates for discontinuous solutions of nonlinear hyperbolic equations, *SIAM J. Numer. Anal.*, **28** (1991), 891–906.

[10] S. Ulbrich, A sensitivity and adjoint calculus for discontinuous solutions of hyperbolic conservation laws with source terms, *SIAM J. Control Optim.*, **41** (2002), 740–797.

[11] A. L. Volpert, The space BV and quasilinear equations, *Math. USSR Sbornik*, **73** (1967), 225–267.

E-mail address: `frederic.coquel@cmap.polytechnique.fr`
E-mail address: `{claude.marmignon,pratik.rai,florent.renac}@onera.fr`

MODELS OF COLLECTIVE MOVEMENTS WITH NEGATIVE DEGENERATE DIFFUSIVITIES

ANDREA CORLI *

Department of Mathematics and Computer Science
University of Ferrara
I-44121 Italy

LUISA MALAGUTI

Department of Sciences and Methods for Engineering
University of Modena and Reggio Emilia
I-42122 Italy

ABSTRACT. We consider an advection-diffusion equation whose diffusivity can be negative. This equation arises in the modeling of collective movements, where the negative diffusivity simulates an aggregation behavior. Under suitable conditions we prove the existence, uniqueness and qualitative properties of traveling-wave solutions connecting states where the diffusivity has opposite signs. These results are extended to end states where the diffusivity is positive but is negative in between. The vanishing-viscosity limit is also considered. Examples from real-world models are provided.

1. **Introduction.** We consider the advection-diffusion equation

$$\rho_t + f(\rho)_x = (D(\rho)\rho_x)_x\,, \qquad t \geq 0,\, x \in \mathbb{R}, \tag{1}$$

where $\rho \in [0,1]$. Our main assumptions are that, for some $\alpha \in (0,1)$,

(f) $f \in C^1[0,1]$, $f(0) = 0$;

(D1) $D \in C^1[0,1]$, $D(\rho) > 0$ for $\rho \in (0,\alpha)$ and $D(\rho) < 0$ for $\rho \in (\alpha, 1)$.

We refer to Figure 1 for two possible plots of D and f. Even if in the following examples, discussed in Section 3, we have $f \geq 0$ and $D(0) = D(1) = 0$, as depicted for simplicity, we emphasize that these conditions are not required in our general results.

Condition (D1) makes (1) a forward-backward parabolic equation. These equations, which arise in a natural way in several physical [11, 18] and biological [17] models, are unstable in the backward regime, where also uniqueness is lost for non-smooth solutions [16].

Equation (1) also arises in the modeling of vehicular traffic flows or crowds dynamics, where $\rho(x,t)$ represents the normalized density at time t and place x of vehicules or pedestrians. In this case the corresponding flow is $q(\rho) = f(\rho) = \rho v(\rho)$, where the velocity v is an assigned function. In the famous inviscid LWR model proposed in [22, 28] the density-flow pairs lie on a curve (the graph of q) in the (ρ, q)-plane. However, experimental data clearly show that such pairs usually cover

2000 *Mathematics Subject Classification.* Primary: 35K65, 35C07; Secondary: 35K55, 35K57.

Key words and phrases. Degenerate parabolic equations, negative diffusivity, traveling-wave solutions, collective movements.

The authors are supported by GNAMPA of INdAM.

* Corresponding author: Andrea Corli.

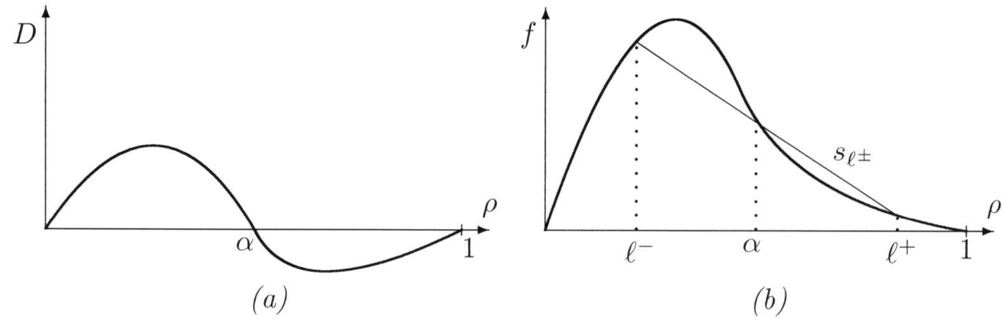

FIGURE 1. *(a)*: a diffusivity D satisfying assumption (D1); *(b)*: the flux function f.

a two-dimensional region. To reproduce this effect, either one considers second-order models [1, 34] or, as in this paper, introduces a diffusive term. In the latter case the physical (or parabolic) flow is $q_p = f(\rho) - D(\rho)\rho_x$; if we assume that ρ_x varies in an interval $[-a, a]$, then q_p can be understood as a perturbation of q and the density-flow pairs cover a full two-dimensional region around the curve $q = q(\rho)$ in the (ρ, q)-plane. The introduction of D also avoids the appearance of shock waves and then the occurrence of an infinite acceleration. We refer to [4, 6] for several models where the diffusivity D can vanish but otherwise remains positive. The *negativity* of D simulates an aggregative behavior; it occurs, for instance, in vehicular flows for high car densities and limited sight distance ahead [26].

We recall that the introduction of viscosity in traffic flow models has been criticized since the famous paper [10], because it could yield negative velocities of the cars. It is not difficult to reject this objection. The contribution to the equation of viscosity, either deduced from asymptotic expansions or by physical motivations, is indeed much smaller than the corresponding contribution due to the velocity v, see [8]. Moreover, in almost every model known in the literature [8], the diffusivity degenerates and is required to vanish where v does, namely, at 1. This means that even at the maximum density where v vanishes, cars do not move backward because of the parabolic term.

A general framework for the study of equation (1), *in the case* $f = 0$, has been proposed in [12, 27]. The "correct" solutions to (1), in the sense of Young measures, are characterized as limits for $\to_{\varepsilon} 0$ of the solutions of a pseudo-parabolic third-order equation $\rho^\varepsilon_t = (D(\rho^\varepsilon)u^\varepsilon_x)_x + \varepsilon\psi(\rho^\varepsilon)_{xxt}$. In this case, solutions of (1) satisfy some *entropy conditions*, in analogy with the hyperbolic setting. We refer to [25, 29, 30] and references there. We do not follow this approach: first, the case $f \neq 0$ is still an outstanding open problem; second, the above third-order approximation has no clear meaning for collective movements; third, we are interested in traveling-wave solutions and in the vanishing-viscosity limit to the conservation law.

More precisely, we are concerned with *traveling-wave solutions* $\rho(x,t) = \phi(x-ct)$. Then the profile ϕ satisfies the differential equation

$$(D(\phi)\phi')' + (c\phi - f(\phi))' = 0. \tag{2}$$

We require that ϕ connects a state ℓ^- with $D(\ell^-) > 0$ to a state ℓ^+ with $D(\ell^-) < 0$:

$$\phi(-\infty) = \ell^-, \qquad \phi(+\infty) = \ell^+. \tag{3}$$

More general cases are also studied. Traveling-wave solutions in the case $f = 0$, but then equation (1) is endowed of a source term g, have been considered in [2, 3, 13, 20, 23, 24] for different nonpositive D. Traveling-wave solutions in the case $f \neq 0$ have also been considered in [11, 33] for a model of infiltration through porous media; in these latter papers the unstable region is bypassed by inserting in the solution a shock wave, which is uniquely determined by a higher-order regularization (either of pseudo-parabolic or of Cahn-Hilliard type), analogously to the aforemontioned approach in [12, 27]. The case of solutions of the form $u(x,t) = \phi(x/\sqrt{t})$ that enter the unstable region has been considered in [15] in the case $f = 0$.

Our results give necessary and sufficient conditions for the existence of wavefronts when D changes sign once or twice. We also study the smoothness of the profiles and the vanishing-viscosity limit to discontinuous (nonentropic) solutions to the hyperbolic conservation law

$$\rho_t + f(\rho)_x = 0. \tag{4}$$

As a byproduct, we show that some nonclassical shock waves [21], considered in [7] in the modeling of panic situation in crowds dynamics, admit a viscous profile. Full details are provided in [8].

2. **Main results.** Traveling-wave solutions are meant in the weak sense [14]; in particular $\varphi \in C(I)$ and $D(\varphi)\varphi' \in L^1_{\mathrm{loc}}(I)$ for some interval $I \subset \mathbb{R}$. A traveling-wave solution is *global* if $I = \mathbb{R}$, *classical* if φ is differentiable and $D(\varphi)\varphi'$ is absolutely continuous, *sharp at* ℓ if there exists $\xi_\ell \in I$, $\phi(\xi_\ell) = \ell$, with ϕ classical in $I \setminus \{\xi_\ell\}$ and not differentiable at ξ_ℓ. A global, bounded traveling-wave solution with a monotonic, non-constant profile ϕ satisfying (3) with $\ell^-, \ell^+ \in [0,1]$ is said to be a *wavefront solution* from ℓ^- to ℓ^+. The line joining $(\ell^-, f(\ell^-))$ with $(\ell^+, f(\ell^+))$ is denoted by $s_{\ell^\pm} = s_{\ell^\pm}(\rho)$, see Figure 1*(b)*.

If $D > 0$ in $(0,1)$, then profiles are uniquely determined up to a shift [14]; the loss of uniqueness is more severe under (D1). Indeed, assume that (1) admits a wavefront solution with profile ϕ connecting $\ell^- \in [0,\alpha)$ with $\ell^+ \in (\alpha, 1]$. Then, after a suitable shift, there is a unique $\xi_1 \geq 0$ such that

$$\phi(0) = \alpha \text{ and } \phi(\xi) < \alpha \text{ for } \xi < 0, \qquad \phi(\xi_1) = \alpha \text{ and } \phi(\xi) > \alpha \text{ for } \xi > \xi_1. \tag{1}$$

We refer to Figure 2 for the case $\xi_1 > 0$. It is then clear that a linear change of the parameter ξ_1 provides another profile. For simplicity, we focus on the case

$$\xi_1 = 0. \tag{2}$$

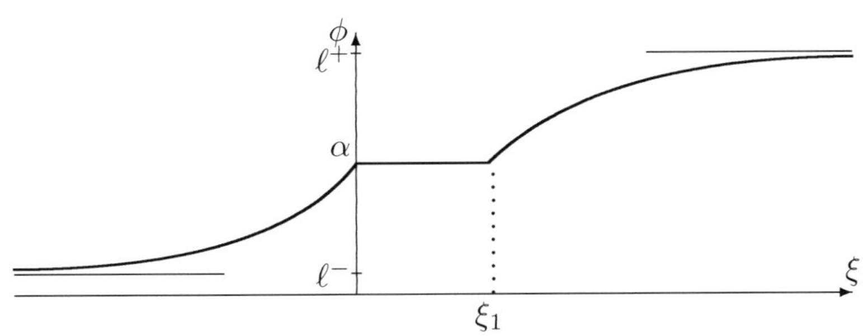

FIGURE 2. Under (D1), a profile ϕ in the case $\xi_1 > 0$.

Theorem 2.1. *Assume* (f), (D1), $\ell^- \in [0,\alpha)$, $\ell^+ \in (\alpha,1]$. *Equation* (1) *has a wavefront solution whose profile* ϕ *satisfies* (3) *if and only if the following three conditions are satisfied:*

$$\frac{f(\alpha)-f(\ell^-)}{\alpha-\ell^-} = \frac{f(\ell^+)-f(\alpha)}{\ell^+-\alpha} =: c_{\ell^\pm}, \tag{3}$$

$$f(\rho) > s_{\ell^\pm}(\rho) \ \text{for all} \ \rho \in (\ell^-,\alpha), \qquad f(\rho) < s_{\ell^\pm}(\rho) \ \text{for all} \ \rho \in (\alpha,\ell^+), \tag{4}$$

$$\frac{D(\rho)}{f(\rho)-s_{\ell^\pm}(\rho)} \in L^1(I_\alpha), \tag{5}$$

for some neighborhood I_α *of* α. *The wave speed is* $c_{\ell^\pm}$ *and* $f'(\alpha) \le c_{\ell^\pm}$. *Under* (1)–(2) *the profile* ϕ *is unique; moreover,* $\phi'(\xi) > 0$ *when* $\ell^- < \phi(\xi) < \ell^+$, $\xi \neq 0$, *while*

$$\lim_{\xi\to 0} \phi'(\xi) = \begin{cases} \frac{f'(\alpha)-c_{\ell^\pm}}{D'(\alpha)} & \text{if } D'(\alpha) < 0, \\ \infty & \text{if } D'(\alpha) = 0 \text{ and } f'(\alpha)-c_{\ell^\pm} < 0. \end{cases} \tag{6}$$

We refer to Figure 1*(b)* for the geometric meaning of conditions (3), (4) and to [8, Th. 2.1] for the proof of the theorem. Condition (5), see [14, Th. 9.1], guarantees the existence of profiles that reach α for a finite value of ξ; it is needed only in case $f'(\alpha) = c_{\ell^\pm}$, i.e., when the line $s_{\ell^\pm}$ is tangent to the graph of f at $(\alpha, f(\alpha))$.

The results of Theorem 2.1 can be extended either to the case when the sign of D is opposite to the one considered in (D1), or to the case when D satisfies

(D2) $D \in C^1[0,1]$, $D(\rho) > 0$ for $\rho \in (0,\alpha) \cup (\beta,1)$ and $D(\rho) < 0$ for $\rho \in (\alpha,\beta)$.

We refer to Figure 3*(a)* for a possible plot of a diffusivity D satisfying (D2) and to [8] for a precise statement.

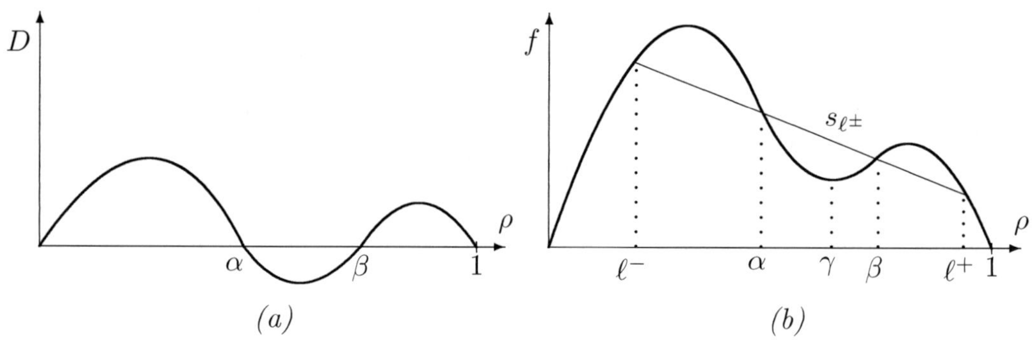

FIGURE 3. *(a)*: the diffusivity D in case (D2); *(b)*: the flux function f.

At last, we consider the family of equations

$$\rho_t + f(\rho)_x = (\varepsilon D(\rho)\rho_x)_x, \qquad t \ge 0, \ x \in \mathbb{R}, \tag{7}$$

for $\varepsilon \in (0,1]$. About the convergence of solutions ρ_ε of (7) to a solution ρ of (4), if $D > 0$, a positive answer is provided in [19]; see [9, §6] for more information. The case $D \ge 0$ was first considered in [32]. We provide now a convergence result concerning wavefronts when D changes sign. For sake of simplicity, we suppose

$$D'(\alpha) < 0. \tag{8}$$

Theorem 2.2. *Assume* (f), (D1) *and* $\ell^- \in [0,\alpha)$, $\ell^+ \in (\alpha, 1]$; *also assume* (3)–(5) *and* (8). *Let* ϕ_ε *be the unique profiles of wavefront solutions to* (7) *satisfying* (3) *and* (2). *Then*

$$\lim_{\varepsilon\to 0+} \phi_\varepsilon(\xi) =: \phi_0(\xi) = \begin{cases} \ell^- & \text{if } \xi < 0, \\ \ell^+ & \text{if } \xi > 0. \end{cases} \tag{9}$$

The convergence is uniform in every interval $(-\infty, -\delta)$ *and* $(\delta, +\infty)$ *with* $\delta > 0$.

We refer to [8, Th. 2.3] for the proof. Because of Theorem 2.2 we can briefly comment the previous results from the hyperbolic point of view. First of all, notice that the function $\rho_0(x,t) = \phi_0(x - c_{\ell\pm}t)$ is a weak solution to equation (4) because the Rankine-Hugoniot conditions are satisfied. Conditions analogous to (4) are well known in the hyperbolic setting [5, Thm. 4.4]. However the discontinuous solution ρ_0 is not entropic: referring to the case depicted in Figure 1*(b)*, the Lax inequality $f'(\ell^-) > c_{\ell\pm}$ is satisfied while $c_{\ell\pm} > f'(\ell^+)$ fails: the shock is compressive on the left and undercompressive on the right. However, even if ρ_0 is not entropic, Theorem 2.1 shows that it has a viscous profile, where "viscous" refers to a negative diffusivity in the nonentropic part of the solution; such a wave is unstable in the sense of [5, Rem. 4.7]. Notice that the one-sided sonic case $c_{\ell\pm} = f'(\ell^+) \neq f'(\ell^-)$ (or $c_{\ell\pm} = f'(\ell^-) \neq f'(\ell^+)$) has been considered in [7] in the framework of nonclassical shocks.

The proof of Theorem 2.1 exploits, and extends to the case of negative diffusivities, some results of [14]; then, a suitable pasting of the profiles thus obtained leads to the above result. The same strategy is used with case (D2). The proof of Theorem 2.2 makes use of an approximation technique to cope with the degeneracy of the diffusivity at α.

3. Applications to collective movements.

In the case of collective movements, assumption (f) specializes to [22, 28]

(fcm) $f(\rho) = \rho v(\rho)$, with $v \in C^1[0,1]$, $v(\rho) \geq 0$ for $\rho \in [0,1)$ and $v(1) = 0$.

From a modeling point of view, the velocity v vanishes at 1 and is decreasing at least in a right neighborhood of 0. About D, we focus on case (D1) and the properties $D(0) = D(1) = 0$ would be desirable [4, 6]. We refer to [8] for a list of the diffusivities proposed in the literature; here, we only consider two cases. The case

$$D(\rho) = -\rho v'(\rho)\left(hv^2(\rho) + \tau\rho v'(\rho)\right) \tag{1}$$

has been proposed in [26] for vehicular flows. In the case of pedestrian flows one may consider [6, Figure 4]

$$D(\rho) = -\rho v'(\rho)\left(hv(\rho) + \tau\rho v'(\rho)\right). \tag{2}$$

Here $\tau > 0$ is a reaction time and $h > 0$ a proportionality parameter.

We first consider case (1). In order that $D(1) = 0$ holds, we need v vanishes at second order at $\rho = 1$; then, we consider

$$v(\rho) = \overline{v}(1-\rho)^2, \tag{3}$$

for $\overline{v} > 0$. We have

$$D(\rho) = 2h\overline{v}^3\rho(1-\rho)^2\left[(1-\rho)^3 - \sigma\rho\right], \qquad \sigma := 2\tau/(h\overline{v}) > 0. \tag{4}$$

Lemma 3.1. *Let v be given by (3) and D by (1) with $\tau > 0$, see (4). Then D satisfies* (D1) *for any positive $\overline{v}, h, \tau$ such that $\alpha = \alpha(\overline{v}, h, \tau)$ is the unique root in $(0,1)$ of*

$$(1-\alpha)^3 = \sigma\alpha. \tag{5}$$

The function $\alpha(\overline{v}, h, \tau)$ covers the interval $(0,1)$ for $\tau, h, \overline{v} \in (0, \infty)$. If $\tau, h, \overline{v}$ are such that $\alpha(\overline{v}, h, \tau) \in (1/2, 1)$, then there are infinitely many pairs (ℓ^-, ℓ^+) such that (3)–(5) hold.

For the proof, see [8, Lemma 3.3]. We refer to Figure 4 for an illustration of the example in the case of real-world data [26]. There, we use dimensional variables.

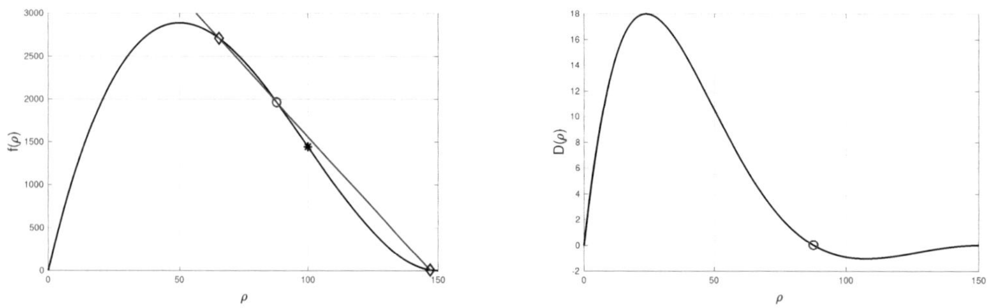

FIGURE 4. Plots of flows and diffusivities; the end states are depicted with diamonds. Here $v(\rho) = \overline{v}(1 - \frac{\rho}{\overline{\rho}})^2$, D as in (1), $\overline{\rho} = 150$ cars/km, $\overline{v} = 130$ km/h, $\tau = 2$ s, $h = 1/15800$ h^2/km, see [26]. An empty circle localizes $\alpha \sim 88$, an asterisk the inflection point of f, which is 100. For $\ell^+ = 147$ we find $\ell^- \sim 65$.

Second, we consider case (2) together with the velocity [31]:

$$v(\rho) = \begin{cases} \overline{v} & \text{if } \rho \le a, \\ \overline{v} e^{\gamma \frac{a-\rho}{1-\rho}} & \text{if } \rho > a, \end{cases} \tag{6}$$

where $\gamma > 0$, $\overline{v} > 0$ and $0 \le a < 1$ is a critical density that separates free from congested flow, $a\gamma < 2$. Then f is strictly concave (convex) in $[a, \tilde{\rho})$ (in $(\tilde{\rho}, 1]$) for a suitable $\tilde{\rho} \in (a, 1)$. In this case, for real-world data, conditions (3), (4) are satisfied (in $[a, 1]$) if a and τ are sufficiently small; plots analogous to those in Figure 4 can be shown [8]. About (5), it can be easily shown to be generically satisfied.

REFERENCES

[1] A. Aw and M. Rascle, Resurrection of "second order" models of traffic flow. *SIAM J. Appl. Math.*, **60**(3) (2000), 916–938.

[2] L. Bao and Z. Zhou, Traveling wave in backward and forward parabolic equations from population dynamics. *Discrete Contin. Dyn. Syst. Ser. B*, **19**(6) (2014),1507–1522.

[3] L. Bao and Z. Zhou, Traveling wave solutions for a one dimensional model of cell-to-cell adhesion and diffusion with monostable reaction term. *Discrete Contin. Dyn. Syst. Ser. S*, **10**(3) (2017), 395–412.

[4] N. Bellomo, M. Delitala, and V. Coscia, On the mathematical theory of vehicular traffic flow. I. Fluid dynamic and kinetic modelling. *Math. Models Methods Appl. Sci.*, **12**(12) (2002),1801–1843.

[5] A. Bressan, *Hyperbolic Systems of Conservation Laws.* Oxford University Press, 2000.

[6] L. Bruno, A. Tosin, P. Tricerri, and F. Venuti, Non-local first-order modelling of crowd dynamics: a multidimensional framework with applications. *Appl. Math. Model.*, **35**(1) (2011), 426–445.

[7] R. M. Colombo and M. D. Rosini, Pedestrian flows and non-classical shocks. *Math. Methods Appl. Sci.*, **28**(13) (2005), 1553–1567.

[8] A. Corli and L. Malaguti, Viscous profiles in models of collective movement with negative diffusivity. *Zeit. Ang. Math. Phys.*, **70** (2019).

[9] C. M. Dafermos, *Hyperbolic Conservation Laws in Continuum Physics*. Springer-Verlag, Berlin, 3nd edition, 2010.

[10] C. Daganzo, Requiem for second-order fluid approximation of traffic flow. *Transportation Res. B*, **29**(4) (1995), 277–286.

[11] D. A. DiCarlo, R. Juanes, L. Tara, and T. P. Witelski, Nonmonotonic traveling wave solutions of infiltration into porous media. *Water Resources Res.*, **44** (2008), 1–12.

[12] L. C. Evans and M. Portilheiro, Irreversibility and hysteresis for a forward-backward diffusion equation. *Math. Models Methods Appl. Sci.*, **14**(11) (2004), 1599–1620.

[13] L. Ferracuti, C. Marcelli, and F. Papalini, Travelling waves in some reaction-diffusion-aggregation models. *Adv. Dyn. Syst. Appl.*, **4**(1) (2009), 19–33.

[14] B. H. Gilding and R. Kersner, *Travelling Waves in Nonlinear Diffusion-Convection Reaction*. Birkhäuser Verlag, Basel, 2004.

[15] B. H. Gilding and A. Tesei, The Riemann problem for a forward-backward parabolic equation. *Phys. D*, **239**(6) (2010), 291–311.

[16] K. Höllig, Existence of infinitely many solutions for a forward backward heat equation. *Trans. Amer. Math. Soc.*, **278**(1) (1983), 299–316.

[17] D. Horstmann, K. J. Painter, and H. G. Othmer, Aggregation under local reinforcement: from lattice to continuum. *European J. Appl. Math.*, **15**(5) (2004), 546–576.

[18] B. S. Kerner and V. V. Osipov, *Autosolitons*. Kluwer Academic Publishers Group, Dordrecht, 1994.

[19] S. N. Kružkov, First order quasilinear equations with several independent variables. *Mat. Sb. (N.S.)*, **81** (123) (1970), 228–255.

[20] M. Kuzmin and S. Ruggerini. Front propagation in diffusion-aggregation models with bi-stable reaction. *Discrete Contin. Dyn. Syst. Ser. B*, **16**(3) (2011), 819–833.

[21] P. G. LeFloch. *Hyperbolic systems of conservation laws*. Lectures in Mathematics ETH Zürich. Birkhäuser Verlag, Basel, 2002.

[22] M. J. Lighthill and G. B. Whitham. On kinematic waves. II. A theory of traffic flow on long crowded roads. *Proc. Roy. Soc. London. Ser. A.*, **229** (1955), 317–345.

[23] P. K. Maini, L. Malaguti, C. Marcelli, and S. Matucci. Diffusion-aggregation processes with mono-stable reaction terms. *Discrete Contin. Dyn. Syst. Ser. B*, **6**(5) (2006), 1175–1189.

[24] P. K. Maini, L. Malaguti, C. Marcelli, and S. Matucci. Aggregative movement and front propagation for bi-stable population models. *Math. Models Methods Appl. Sci.*, **17**(9) (2007), 1351–1368.

[25] C. Mascia, A. Terracina, and A. Tesei. Two-phase entropy solutions of a forward-backward parabolic equation. *Arch. Ration. Mech. Anal.*, **194**(3) (2009), 887–925.

[26] P. Nelson. Synchronized traffic flow from a modified Lighthill-Whitham model. *Phys. Review E*, **61** (2000), R6052–R6055.

[27] P. I. Plotnikov. Passage to the limit with respect to viscosity in an equation with a variable direction of parabolicity. *Diff. Uravn.*, **30**(4) (1994), 665–674.

[28] P. I. Richards. Shock waves on the highway. *Oper. Res.*, **4** (1956), 42–51.

[29] F. Smarrazzo and A. Tesei. Degenerate regularization of forward-backward parabolic equations: the regularized problem. *Arch. Ration. Mech. Anal.*, **204**(1) (2012), 85–139.

[30] F. Smarrazzo and A. Tesei. Degenerate regularization of forward-backward parabolic equations: the vanishing viscosity limit. *Math. Ann.*, **355**(2) (2013), 551–584.

[31] F. Venuti and L. Bruno. An interpretative model of the pedestrian fundamental relation. *C. R. Mech.*, **335** (2007), 194–200.

[32] A. I. Volpert and S. I. Hudjaev. The Cauchy problem for second order quasilinear degenerate parabolic equations. *Mat. Sb. (N.S.)*, **78** (120) (1969), 374–396.

[33] T. P. Witelski. The structure of internal layers for unstable nonlinear diffusion equations. *Stud. Appl. Math.*, **97**(3) (1996), 277–300.

[34] H. M. Zhang. A non-equilibrium traffic model devoid of gas-like behavior. *Transp. Res. B*, **36** (2002), 275–290.

E-mail address: andrea.corli@unife.it
E-mail address: luisa.malaguti@unimore.it

LINEAR STABILITY OF A VECTORIAL KINETIC RELAXATION SCHEME WITH A CENTRAL VELOCITY

CLÉMENTINE COURTÈS*

Institut de Mathématiques de Toulouse
UMR CNRS 5219, Université de Toulouse, INSA
F-31077 Toulouse, France

EMMANUEL FRANCK

INRIA Nancy-Grand Est, TONUS team and IRMA
UMR CNRS 7501, Université de Strasbourg
F-67000 Strasbourg, France

ABSTRACT. This article deals with the linear stability of an implicit vectorial kinetic relaxation scheme with a central velocity used to solve numerically some multi-scale hyperbolic systems.

1. **Introduction.** Hyperbolic systems are often used to model complex physical phenomena such as multi-scale problems. In such problems, characteristic waves do not propagate with the same speed: fast waves interact with slower waves.
Discretizing such physical phenomena is still an open issue. Explicit methods are prohibited due to their very restrictive CFL condition imposed by fastest scales and implicit methods are computational time-consuming and memory cost-consuming due to the inversion of ill-conditioned nonlinear systems. In order to better grasp numerically these multi-scale problems, an alternative is to use kinetic relaxation methods.
The key idea of these kinetic relaxation methods is to consider the unknown of the hyperbolic system as the macroscopic moment of a kinetic distribution function. The main advantage is that the distribution function satisfies a mesoscopic kinetic equation, which is easier to process because it is composed of an advection equation (at constant speeds) combined with a relaxation term, often chosen of Bhatnagar-Gross-Krook type (in short BGK) [2]. The relaxation term enables kinetic equation to tend toward hyperbolic system for an asymptotically small relaxation parameter.

An important degree of freedom in the kinetic relaxation methods is the choice of the number and the values of the constant advection speeds for the distribution function. We follow here the vectorial kinetic relaxation method, introduced in [7, 1], which consists of fixing the same (small) set of advection speeds for each component of the unknown of the hyperbolic system. A suitable choice for multi-scale problems is the one introduced in [4] and mainly developed in [5], where

2000 *Mathematics Subject Classification.* Primary: 65M12, 35L65, 35L02; Secondary: 65F15.

Key words and phrases. linear stability, von Neumann analysis, kinetic relaxation, semi-Lagrangian method, implicit splitting scheme, eigenvalues.

The authors are supported by a PEPS INSMI CNRS 2018 (Projet Exploratoire Premier Soutien) "Jeunes chercheuses et jeunes chercheurs".

* Corresponding author: clementine.courtes@math.univ-toulouse.fr.

three advection speeds $\lambda_-, \lambda_0, \lambda_+$ are associated to each of the components of the unknown of the hyperbolic system. The central speed λ_0 is added to treat the slowest scale of the physical phenomenon.

From numerical point of view, vectorial kinetic relaxation models are often discretized with numerical schemes which split the advection part from the relaxation term. There seems to be widespread agreement that the relaxation term is treated numerically as a source term. The noticeable difference between numerical schemes mainly comes from the numerical processing of the advection part. It may be treated, for example, by an Exact discrete Transport, as in [6] or by a Semi-Lagrangian method as in [5], which has the advantage to avoid matrices storage and CFL condition. The properties of such a numerical scheme are detailed in [5], particularly for the consistency. The stability analysis is much more difficult and is rather sketchy.
The aim of this current paper is precisely to review all results on that stability property. For simplicity, we restrict our study to the notion of linear stability (or L^2-stability). Note that other notions of stability such as entropic one is briefly discussed in [5]. The outline of the current paper is constructed as follow. Section 2 gathers the notations of the splitting scheme associated to the vectorial kinetic relaxation model with a central velocity. Section 3 is a brief reminder of the notion of L^2-stability. This linear stability issue is raised in Section 4 for a Semi-Lagrangian method for the advection part and in Section 5 for an Exact discrete Transport method.

2. The vectorial kinetic relaxation scheme.

Let us consider a 1D linear hyperbolic system $\partial_t \boldsymbol{U} + \partial_x \boldsymbol{F}(\boldsymbol{U}) = 0$, with $\boldsymbol{U}(t,x) \in \mathbb{R}^N$. The flux $\boldsymbol{F}$ is assumed to be linear : $\boldsymbol{F}(\boldsymbol{U}) = A\boldsymbol{U}$ with the square matrix $A \in \mathcal{M}_N(\mathbb{R})$.
The kinetic relaxation representation. By following the notations introduced in [5], a fixed set of velocities $\{\lambda_-, \lambda_0, \lambda_+\}$ with $\lambda_- < \lambda_0 < \lambda_+$ is associated to each of the N components of $\boldsymbol{U}$. Then, $\boldsymbol{U}$ is considered as a macroscopic moment of a kinetic distribution function $\boldsymbol{f} \in \mathbb{R}^{3N}$, which satisfies the following kinetic relaxation equation

$$\partial_t \boldsymbol{f} + \Lambda \partial_x \boldsymbol{f} = \frac{1}{\varepsilon}\left(\boldsymbol{f}^{eq}(\boldsymbol{U}) - \boldsymbol{f}\right). \tag{1}$$

According to the choice of the advection speeds set, we decompose $\boldsymbol{f}$ such as $\boldsymbol{f} = (\boldsymbol{f}_-, \boldsymbol{f}_0, \boldsymbol{f}_+)^t$, with $\boldsymbol{f}_j = (f_{j,k})_{k\in\{1,\ldots,N\}} \in \mathbb{R}^N$ for $j \in \{-,0,+\}$. The left hand side of (1) consists on the advection part with the diagonal matrix $\Lambda = \mathrm{diag}(\lambda_-\mathrm{Id}, \lambda_0\mathrm{Id}, \lambda_+\mathrm{Id})$, which contains all the advection speeds (Id is the N-identity matrix). The right hand side of (1) consists on the BGK relaxation part with $\varepsilon > 0$ the relaxation parameter and $\boldsymbol{f}^{eq} = (\boldsymbol{f}_-^{eq}, \boldsymbol{f}_0^{eq}, \boldsymbol{f}_+^{eq})^t$ the equilibrium vector, which is a function of $\boldsymbol{U}$ and which satisfies some consistency properties.
In order to determine $\boldsymbol{f}_-^{eq}$, $\boldsymbol{f}_0^{eq}$ and $\boldsymbol{f}_+^{eq}$, we perform the decentered flux vector splitting detailed in [5]. It consists to decompose the hyperbolic flux $\boldsymbol{F}$ into three parts, which commute each other: $\boldsymbol{F}(\boldsymbol{U}) = \boldsymbol{F}_0^-(\boldsymbol{U}) + \boldsymbol{F}_0^+(\boldsymbol{U}) + \lambda_0\boldsymbol{U}$. In the linear case, the hyperbolic flux $\boldsymbol{F}$ writes $\boldsymbol{F}(\boldsymbol{U}) = A\boldsymbol{U}$ with A a diagonalizable square matrix (because $\boldsymbol{F}$ is hyperbolic) and the previous decomposition is also linear: there exist two commuting diagonalizable square matrices $A_0^{\pm}$ such that

$$A\boldsymbol{U} = A_0^-\boldsymbol{U} + A_0^+\boldsymbol{U} + \lambda_0\boldsymbol{U}. \tag{2}$$

Decomposition (2) together with $\boldsymbol{U} = \sum_{j\in\{-,0,+\}} \boldsymbol{f}_j$ (since $\boldsymbol{U}$ is the macroscopic moment of $\boldsymbol{f}$) enable to define each $\boldsymbol{f}_j^{eq}$ for $j \in \{-,0,+\}$ as follow, where Id is the identity matrix, (for more details, see [5])

$$\begin{cases} \boldsymbol{f}_-^{eq}(\boldsymbol{U}) = -\dfrac{1}{\lambda_0-\lambda_-} A_0^- \left(\boldsymbol{f}_- + \boldsymbol{f}_0 + \boldsymbol{f}_+\right), \\ \boldsymbol{f}_0^{eq}(\boldsymbol{U}) = \left[\mathrm{Id} - \left(\dfrac{1}{\lambda_+-\lambda_0} A_0^+ - \dfrac{1}{\lambda_0-\lambda_-} A_0^-\right)\right] \left(\boldsymbol{f}_- + \boldsymbol{f}_0 + \boldsymbol{f}_+\right), \\ \boldsymbol{f}_+^{eq}(\boldsymbol{U}) = \dfrac{1}{\lambda_+-\lambda_0} A_0^+ \left(\boldsymbol{f}_- + \boldsymbol{f}_0 + \boldsymbol{f}_+\right). \end{cases} \tag{3}$$

The numerical scheme. As in [5], we fix $\Delta t > 0$ and $\Delta x > 0$ the time and space steps and denote $\boldsymbol{f}^n = \left(f_{j,k}^n\right)_{j\in\{-,0,+\},k\in\{1,\ldots,N\}}$ the distribution vector at time $t^n = n\Delta t \in [0,T]$. The numerical scheme chosen to discretize (1) is the following splitting scheme: $\boldsymbol{f}^{n+1} = \mathcal{R}_\varepsilon(\Delta t, \Delta x, \theta) \circ \mathcal{T}(\Delta t, \Delta x)\boldsymbol{f}^n$.
For convenient, we denote $\boldsymbol{f}^* = \mathcal{T}(\Delta t, \Delta x)\boldsymbol{f}^n$.

• The transport step, named $\mathcal{T}(\Delta t, \Delta x)$, may be either a Semi-Lagrangian scheme (SL hereafter), defined by

$$f_{j,k}^*(x) = \mathrm{I}_{\Delta x}\left(f_{j,k}^n\right)(x - \lambda_j \Delta t), \quad \forall j \in \{-,0,+\} \text{ and } \forall k \in \{1,..,N\}, \tag{4}$$

where, for any $g : \mathbb{R} \mapsto \mathbb{R}$, $\mathrm{I}_{\Delta x}(g)$ is a piecewise polynomial interpolation of the values taken by g on the mesh points, or an Exact discrete Transport scheme (ET hereafter), defined by $\mathrm{I}_{\Delta x} = \mathrm{Id}$ (the identity map)

$$f_{j,k}^*(x) = f_{j,k}^n(x - \lambda_j \Delta t), \quad \forall j \in \{-,0,+\} \text{ and } \forall k \in \{1,..,N\}. \tag{5}$$

Remark 1. Assuming an Exact discrete Transport scheme, as Relation (5), leads to a CFL condition since it makes $\frac{\lambda_j \Delta t}{\Delta x}$ be an integer, for all $j \in \{-,0,+\}$.

• The relaxation step, named $\mathcal{R}_\varepsilon(\Delta t, \Delta x, \theta)$, consists on a θ-scheme, with $\theta \in [\frac{1}{2}, 1]$, defined by $\frac{\boldsymbol{f}^{n+1}-\boldsymbol{f}^*}{\Delta t} = \theta \frac{\boldsymbol{f}^{eq}(\boldsymbol{U}^{n+1})-\boldsymbol{f}^{n+1}}{\varepsilon} + (1-\theta)\frac{\boldsymbol{f}^{eq}(\boldsymbol{U}^*)-\boldsymbol{f}^*}{\varepsilon}$.
Since $\boldsymbol{U}^{n+1} = \boldsymbol{U}^*$ during the relaxation step, cf [5], it may be rewritten in the form:

$$\boldsymbol{f}^{n+1} = \boldsymbol{f}^* + \omega\left(\boldsymbol{f}^{eq}(\boldsymbol{U}^*) - \boldsymbol{f}^*\right) \quad \text{with } \omega = \frac{\Delta t}{\varepsilon + \theta \Delta t} \in [0,2]. \tag{6}$$

The final numerical scheme is thus obtained by combining (4)-(6) for the Semi-Lagrangian choice (or (5)-(6) for the Exact discrete Transport choice).

3. Linear stability.

We restrict our study to a linear (or L^2) stability, by a von Neumann analysis.

3.1. A review of L^2-stability.

Let G be the amplification matrix of a one-step linear scheme (S) : $\boldsymbol{f}^{n+1} = G(\Delta t, \Delta x)\boldsymbol{f}^n$. We recall the notion of L^2-stability in the following definition.

Definition 3.1. The scheme (S) is L^2-stable if there exists a constant $K > 0$ such that, for all Δt and Δx small enough (and possibly satisfying a CFL condition), for all $n \geq 0$ such that $n\Delta t \leq T$, one has $||\boldsymbol{f}^{n+1}||_{\ell^2} \leq (1 + K\Delta t)||\boldsymbol{f}^n||_{\ell^2}$.

In terms of amplification matrix, the L^2-stability notion translates into the following necessary and sufficient condition : $\sqrt{\rho\left([G(\Delta t, \Delta x)]^* G(\Delta t, \Delta x)\right)} \leq 1 + K\Delta t$, with, for a square matrix G, $\rho(G)$ the spectral radius of G and G^* the adjoint matrix of G. This necessary and sufficient condition is not always easy to verify so we focus only to the sufficient condition of the following proposition.

Proposition 1. ***Sufficient condition :*** *In space Fourier variables* $\widehat{\boldsymbol{f}^n}(\xi)$ *with* $\xi \in [0, \frac{2\pi}{\Delta x}]$*, a sufficient condition to ensure the* L^2*-stability is as follow:*
$\sup_{\xi\in[0,\frac{2\pi}{\Delta x}]} \rho(G(\Delta t,\xi)) < 1$*, or* $\sup_{\xi\in[0,\frac{2\pi}{\Delta x}]} \rho(G(\Delta t,\xi)) = 1$ *and the eigenvalues of* $G(\Delta t,\xi)$ *with modulus equal to 1 are simple.*

3.2. **Amplification matrix for the vectorial kinetic relaxation scheme.** To deal with the L^2-stability, we have to first compute the amplification matrix.
Reformulation of the relaxation step in the linear case. Since A_0^- and A_0^+ commute and are both diagonalizable, they are diagonalizable in the same basis to obtain $A_0^\pm = B_0 D_0^\pm B_0^{-1}$ with $D_0^\pm$ the diagonal matrices
$D_0^\pm = \text{diag}\left(\lambda_k(A_0^\pm)\right)_{k\in\{1,..,N\}}$ and B_0 an invertible matrix. The diagonal term $\lambda_k(A_0^\pm)$ corresponds to the kth eigenvalue of $A_0^\pm$.

Relaxation step (6) is thus rewritten under the following bloc matrices form:

$$\begin{pmatrix} \boldsymbol{f}_-^{n+1} \\ \boldsymbol{f}_0^{n+1} \\ \boldsymbol{f}_+^{n+1} \end{pmatrix} = \mathcal{B}_0 R_\varepsilon(\Delta t,\omega)\mathcal{B}_0^{-1} \begin{pmatrix} \boldsymbol{f}_-^{*} \\ \boldsymbol{f}_0^{*} \\ \boldsymbol{f}_+^{*} \end{pmatrix} \tag{7}$$

with $\mathcal{B}_0 = \text{diag}(B_0, B_0, B_0)$, $\mathcal{B}_0^{-1} = \text{diag}(B_0^{-1}, B_0^{-1}, B_0^{-1})$ and $R_\varepsilon(\Delta t,\omega)$ the relaxation amplification matrix defined by blocs by

$$R_\varepsilon(\Delta t,\omega) = \text{diag}(\text{Id},\text{Id},\text{Id}) + \omega\tilde{R}_\varepsilon(\Delta t),$$

with $\tilde{R}_\varepsilon(\Delta t) =$

$$\begin{pmatrix} -\frac{1}{\lambda_0-\lambda_-}D_0^- - \text{Id} & -\frac{1}{\lambda_0-\lambda_-}D_0^- & -\frac{1}{\lambda_0-\lambda_-}D_0^- \\ \text{Id} - \left(\frac{1}{\lambda_+-\lambda_0}D_0^+ - \frac{1}{\lambda_0-\lambda_-}D_0^-\right) & -\frac{1}{\lambda_+-\lambda_0}D_0^+ + \frac{1}{\lambda_0-\lambda_-}D_0^- & \text{Id} - \left(\frac{1}{\lambda_+-\lambda_0}D_0^+ - \frac{1}{\lambda_0-\lambda_-}D_0^-\right) \\ \frac{1}{\lambda_+-\lambda_0}D_0^+ & \frac{1}{\lambda_+-\lambda_0}D_0^+ & \frac{1}{\lambda_+-\lambda_0}D_0^+ - \text{Id} \end{pmatrix}.$$

Fourier analysis. We introduce the Fourier variable (in space) $\widehat{\boldsymbol{f}^n}$ in $L^2([0,\frac{2\pi}{\Delta x}])$ defined by $\widehat{\boldsymbol{f}^n}(\xi) = \sum_{x\in\{\text{mesh points}\}} \boldsymbol{f}^n(x)e^{ix\xi}$, $\forall \xi \in [0,\frac{2\pi}{\Delta x}]$.

With this Fourier decomposition, transport step of the scheme rewrites :

$$\begin{pmatrix} \widehat{\boldsymbol{f}_-^*}(\xi) \\ \widehat{\boldsymbol{f}_0^*}(\xi) \\ \widehat{\boldsymbol{f}_+^*}(\xi) \end{pmatrix} = \begin{pmatrix} T_-(\Delta t,\xi) & & 0 \\ & T_0(\Delta t,\xi) & \\ 0 & & T_+(\Delta t,\xi) \end{pmatrix} \begin{pmatrix} \widehat{\boldsymbol{f}_-^n}(\xi) \\ \widehat{\boldsymbol{f}_0^n}(\xi) \\ \widehat{\boldsymbol{f}_+^n}(\xi) \end{pmatrix},$$

with
- $T_j(\Delta t,\xi) = T_j^{SL}(\Delta t,\xi)$ the amplification factor of the Semi-Lagrangian scheme (4),
- or $T_j(\Delta t,\xi) = e^{i\lambda_j\Delta t\xi}\text{Id}$ in the case of an Exact discrete Transport scheme (5).

Since relaxation step does not depend on the space, Relation (7) is also true with $\widehat{\boldsymbol{f}^{n+1}}$ (resp. $\widehat{\boldsymbol{f}^*}$) instead of $\boldsymbol{f}^{n+1}$ (resp. $\boldsymbol{f}^*$).

Eventually, the total amplification matrix is equal to

$$G(\Delta t,\xi,\varepsilon,\omega) = \mathcal{B}_0 R_\varepsilon(\Delta t,\omega)\mathcal{B}_0^{-1} \begin{pmatrix} T_-(\Delta t,\xi) & & 0 \\ & T_0(\Delta t,\xi) & \\ 0 & & T_+(\Delta t,\xi) \end{pmatrix}, \tag{8}$$

with $T_j(\Delta t, \xi) = T_j^{SL}(\Delta t, \xi)$ for the scheme (4)-(6), or with $T_j(\Delta t, \xi) = e^{i\lambda_j \Delta t \xi}\mathrm{Id}$ for the scheme (5)-(6), for $j \in \{-, 0, +\}$.

4. **Linear stability for the Semi-Lagrangian step.** Let us ensure our first linear stability result.

Proposition 2. *Let the hyperbolic flux* $\boldsymbol{F}$ *be linear and be decomposed into* $\boldsymbol{F}(\boldsymbol{U}) = A\boldsymbol{U} = A_0^-\boldsymbol{U} + A_0^+\boldsymbol{U} + \lambda_0\boldsymbol{U}$, *with* A_0^+ *and* A_0^- *two commuting and diagonalizable matrices.*
The numerical scheme (4)-(6) *with the advection speeds set* $\{\lambda_-, \lambda_0, \lambda_+\}^N$, *with* $\lambda_- < \lambda_0 < \lambda_+$ *and equilibrium* (3) *is* L^2*-stable on the following conditions :*

- $\omega \in [0, 1]$,
- $\mathrm{Id} - \left(\frac{1}{\lambda_+ - \lambda_0}A_0^+ - \frac{1}{\lambda_0 - \lambda_-}A_0^-\right)$ *is a positive semidefinite matrix,*
- A_0^+ *(resp.* A_0^-*) is a positive (resp. negative) semidefinite matrix.*

Remark 2. Note that here a matrix is said to be positive semidefinite (resp. negative semidefinite) if all its eigenvalues are nonnegative (resp. nonpositive).

Remark 3. Sufficient conditions which involve in Proposition 2 are exactly the same as the ones required for the entropy stability property, proved in [5].

To prove Proposition 2, we follow the main guidelines of [6] which suggest to study the Gershgorin discs of the amplification matrix, for more details see [9].

Definition 4.1. Let $G = (g_{ij})_{i,j} \in \mathcal{M}_N(\mathbb{C})$ be a complex square matrix. The kth-Gershgorin disc corresponds to the disc $\mathcal{D}_k = \{z \in \mathbb{C}, |g_{kk} - z| \leq \sum_{j\neq k}|g_{jk}|\}$, for $k \in \{1, ..., N\}$.

Theorem 4.2 (Gershgorin's theorem). *Let* $G = (g_{ij})_{i,j} \in \mathcal{M}_N(\mathbb{C})$ *be a complex square matrix. Every eigenvalue of* G *belongs at least to one Gershgorin disc of* G.

Proof of Proposition 2. As the Semi-Lagrangian step is unconditionally stable [3], we may omit the Semi-Lagrangian amplification factors $T_j^{SL}(\Delta t, \xi)$ for $j \in \{-, 0, +\}$ in our study. It only remains to consider eigenvalues of $R_\varepsilon(\Delta t, \omega)$.

Let λ be an eigenvalue of $R_\varepsilon(\Delta t, \omega) = (r_{\varepsilon,ij})_{i,j\in\{1,...,3N\}}$. According to Theorem 4.2, there exists $\bar{k} \in \{1, ..., 3N\}$ such that $|r_{\varepsilon,\bar{k}\bar{k}} - \lambda| \leq \sum_{j\neq\bar{k}}|r_{\varepsilon,j\bar{k}}|$. Then by a triangular inequality, $|\lambda| \leq |r_{\varepsilon,\bar{k}\bar{k}} - \lambda| + |r_{\varepsilon,\bar{k}\bar{k}}| \leq \sum_{j=1}^{3N}|r_{\varepsilon,j\bar{k}}|$. However, one has

- for $\bar{k} \in \{1, ..., N\}$

$$\sum_{j=1}^{3N}|r_{\varepsilon,j\bar{k}}| = \left|1 - \omega\frac{\lambda_{\bar{k}}(A_0^-)}{\lambda_0 - \lambda_-} - \omega\right| + \left|\omega - \omega\left(\frac{\lambda_{\bar{k}}(A_0^+)}{\lambda_+ - \lambda_0} - \frac{\lambda_{\bar{k}}(A_0^-)}{\lambda_0 - \lambda_-}\right)\right| + \left|\omega\frac{\lambda_{\bar{k}}(A_0^+)}{\lambda_+ - \lambda_0}\right|,$$

- for $\bar{k} \in \{N+1, ..., 2N\}$

$$\sum_{j=1}^{3N}|r_{\varepsilon,j\bar{k}}| = \left|-\omega\frac{\lambda_{\bar{k}}(A_0^-)}{\lambda_0 - \lambda_-}\right| + \left|1 - \omega + \omega\left(1 - \left(\frac{\lambda_{\bar{k}}(A_0^+)}{\lambda_+ - \lambda_0} - \frac{\lambda_{\bar{k}}(A_0^-)}{\lambda_0 - \lambda_-}\right)\right)\right| + \left|\omega\frac{\lambda_{\bar{k}}(A_0^+)}{\lambda_+ - \lambda_0}\right|,$$

- for $\bar{k} \in \{2N+1, ..., 3N\}$

$$\sum_{j=1}^{3N}|r_{\varepsilon,j\bar{k}}| = \left|-\omega\frac{\lambda_{\bar{k}}(A_0^-)}{\lambda_0 - \lambda_-}\right| + \left|\omega - \omega\left(\frac{\lambda_{\bar{k}}(A_0^+)}{\lambda_+ - \lambda_0} - \frac{\lambda_{\bar{k}}(A_0^-)}{\lambda_0 - \lambda_-}\right)\right| + \left|1 + \omega\frac{\lambda_{\bar{k}}(A_0^+)}{\lambda_+ - \lambda_0} - \omega\right|.$$

Hypotheses of Proposition 2 enable to remove modulus in the previous relations and to obtain for all $\bar{k} \in \{1, ..., 3N\}$, $\sum_{j=1}^{3N} |r_{\varepsilon,j\bar{k}}| = 1$.

Hence, all eigenvalues of $R_\varepsilon(\Delta t, \omega)$ have a modulus smaller than 1 which implies the L^2- stability of the numerical scheme (4)-(6).

□

Example 1. Here is a non-exhaustive list of flux decompositions which enable to obtain a L^2-stable scheme for scalar case when $F(u) = au = a_0^- u + a_0^+ u + \lambda_0 u$:
• (Rusanov) We choose $\lambda_- \leq \min(0, a)$, $\lambda_0 = 0$ and $\max(0, a) \leq \lambda_+$ and we define $a_0^+ = \lambda_+ \left(\frac{a-\lambda_-}{\lambda_+-\lambda_-}\right)$ and $a_0^- = -\lambda_- \left(\frac{a-\lambda_+}{\lambda_+-\lambda_-}\right)$. A particular Rusanov decomposition consists to choose $-\lambda_- = \lambda_+ = |a|$.
• (Upwind) We choose $\lambda_- \leq \min(\lambda_0, a)$ and $\max(\lambda_0, a) \leq \lambda_+$ and we define $a_0^+ = \mathbb{1}_{a>\lambda_0}(a - \lambda_0)$ and $a_0^- = \mathbb{1}_{a<\lambda_0}(a - \lambda_0)$, with $\mathbb{1}$ the indicator function.
• (Lax-Wendroff) We choose $-\lambda_- = \lambda_+ = \lambda > 0$ with $\sqrt{\alpha}|a| \leq \lambda \leq \alpha|a|$ and $\lambda_0 = 0$ and we define $a_0^\pm = \frac{1}{2}\left(a \pm \alpha\frac{a^2}{\lambda}\right)$ with $\alpha \in [1, 2]$.
For more details about these flux decompositions, we refer the readers to [5].

Remark 4. Proposition 2 is also valid for the scheme (5)-(6) (with an Exact discrete Transport step). However, the following section improves those results.

5. Linear stability for the Exact discrete Transport step.

For simplicity, we focus only on the scalar linear case: $\partial_t u + a\partial_x u = 0$ with $u(t,x) \in \mathbb{R}$ and $a \in \mathbb{R}$, which implies $\mathcal{B}_0 = 1$ in (8). The Exact discrete Transport step enables to improve sufficient conditions of Proposition 2, in particular in the range of admissible ω.

Proposition 3. *Let the scalar hyperbolic flux F be linear and be decomposed into $F(u) = au = a_0^- u + a_0^+ u + \lambda_0 u$.*
The numerical scheme (5)-(6) with the advection speeds set $\{\lambda_-, \lambda_0, \lambda_+\}$, with $\lambda_- < \lambda_0 < \lambda_+$ and equilibrium (3) is L^2-stable on the following conditions :
- $\omega \in [0, 2]$,
- $1 - \left(\frac{a_0^+}{\lambda_+-\lambda_0} - \frac{a_0^-}{\lambda_0-\lambda_-}\right) \geq 0$,
- $a_0^+ \geq 0$ *and* $a_0^- \leq 0$,
- *One of the three following equalities is satisfied :* $a_0^+ = 0$ *or* $a_0^- = 0$ *or* $1 - \left(\frac{a_0^+}{\lambda_+-\lambda_0} - \frac{a_0^-}{\lambda_0-\lambda_-}\right) = 0$.

Remark 5. As the scalar case is considered here, condition $1-\left(\frac{a_0^+}{\lambda_+-\lambda_0} - \frac{a_0^-}{\lambda_0-\lambda_-}\right) \geq 0$ may be written in the simplest form: $\lambda_- \leq a \leq \lambda_+$. In deed, in the scalar case,

$$1-\left(\frac{a_0^+}{\lambda_+ - \lambda_0} - \frac{a_0^-}{\lambda_0 - \lambda_-}\right) = \frac{\lambda_+ - a}{\lambda_+ - \lambda_0} + \frac{a_0^-(\lambda_+ - \lambda_-)}{(\lambda_+ - \lambda_0)(\lambda_0 - \lambda_-)} = \frac{a - \lambda_-}{\lambda_0 - \lambda_-} - \frac{a_0^+(\lambda_+ - \lambda_-)}{(\lambda_+ - \lambda_0)(\lambda_0 - \lambda_-)}.$$

The nonnegativity of this quantity together with the hypotheses $\lambda_- < \lambda_0 < \lambda_+$, $a_0^+ \geq 0$ and $a_0^- \leq 0$ implie that $\lambda_- \leq a \leq \lambda_+$.

Instead of using Gershgorin discs to prove Proposition 3, we need a more specific tool to localize the eigenvalues and we use Rouché's theorem, as suggested in [8].

Theorem 5.1 (Rouché's theorem). *Let γ be a closed simple path in $\Omega \subset \mathbb{C}$ and assume that γ has an interior. Let f and g be holomorphic (analytic) on Ω and $|f(\zeta) - g(\zeta)| < |f(\zeta)|$ for all ζ on γ. Then f and g have the same number of zeros in the interior of γ.*

Proof of Proposition 3. Let us define $\mathcal{A}_0^+ = \frac{a_0^+}{\lambda_+ - \lambda_0}$ and $\mathcal{A}_0^- = \frac{a_0^-}{\lambda_0 - \lambda_-}$. In the scalar case, the amplification matrix writes $G(\Delta t, \xi, \varepsilon, \omega) =$

$$\begin{pmatrix} (1-\omega\mathcal{A}_0^- - \omega)\, e^{i\lambda_-\Delta t\xi} & -\omega\mathcal{A}_0^- e^{i\lambda_0\Delta t\xi} & -\omega\mathcal{A}_0^- e^{i\lambda_+\Delta t\xi} \\ \omega\left(1-(\mathcal{A}_0^+ - \mathcal{A}_0^-)\right) e^{i\lambda_-\Delta t\xi} & \left(1-\omega\left(\mathcal{A}_0^+ - \mathcal{A}_0^-\right)\right) e^{i\lambda_0\Delta t\xi} & \omega\left(1-(\mathcal{A}_0^+ - \mathcal{A}_0^-)\right) e^{i\lambda_+\Delta t\xi} \\ \omega\mathcal{A}_0^+ e^{i\lambda_-\Delta t\xi} & \omega\mathcal{A}_0^+ e^{i\lambda_0\Delta t\xi} & (1+\omega\mathcal{A}_0^+ - \omega)\, e^{i\lambda_+\Delta t\xi} \end{pmatrix}.$$

The characteristic polynomial of G is as follow: $\chi(X) = \mu_0 + \mu_1 X + \mu_2 X^2 - X^3$, with

$$\mu_0 = (1-\omega)^2 e^{i(\lambda_+ + \lambda_0 + \lambda_-)\Delta t\xi},$$

$$\mu_1 = (1-\omega)\Big[(-1-\omega\mathcal{A}_0^-)e^{i(\lambda_0+\lambda_+)\Delta t\xi} + (-1+\omega\mathcal{A}_0^+)e^{i(\lambda_0+\lambda_-)\Delta t\xi}$$
$$+(-1+\omega(1-\mathcal{A}_0^+ + \mathcal{A}_0^-))e^{i(\lambda_+ + \lambda_-)\Delta t\xi}\Big],$$

$$\mu_2 = (1-\omega+\omega\mathcal{A}_0^+)e^{i\lambda_+\Delta t\xi} + (1-\omega+\omega(1-\mathcal{A}_0^+ + \mathcal{A}_0^-))e^{i\lambda_0\Delta t\xi}$$
$$+ (1-\omega-\omega\mathcal{A}_0^-)e^{i\lambda_-\Delta t\xi}.$$

In the three particular studied cases, the previous characteristic polynomial writes

$$\chi(X) = \left\{(1-\omega)e^{i\nu_1\Delta t\xi} - X\right\}\tilde{\chi}(X), \tag{9}$$

where $\tilde{\chi}(X) = (1-\omega)e^{i(\nu_2+\nu_3)\Delta t\xi} - Xe^{i\left(\frac{\nu_2+\nu_3}{2}\right)\Delta t\xi}\left[(2-\omega)\cos(\Xi) + i\omega\eta\sin(\Xi)\right] + X^2$, with

- **Case $a_0^+ = 0$:** $\nu_1 = \lambda_+$, $\nu_2 = \lambda_0$, $\nu_3 = \lambda_-$, $\Xi = \frac{\lambda_0 - \lambda_-}{2}\Delta t\xi$, $\eta = 1 + 2\frac{a_0^-}{\lambda_0 - \lambda_-}$,
- **Case $a_0^- = 0$:** $\nu_1 = \lambda_-$, $\nu_2 = \lambda_0$, $\nu_3 = \lambda_+$, $\Xi = \frac{\lambda_0 - \lambda_+}{2}\Delta t\xi$, $\eta = 1 - 2\frac{a_0^+}{\lambda_+ - \lambda_0}$,
- **Case $1 - \frac{a_0^+}{\lambda_+ - \lambda_0} + \frac{a_0^-}{\lambda_0 - \lambda_-} = 0$:** $\nu_1 = \lambda_0$, $\nu_2 = \lambda_+$, $\nu_3 = \lambda_-$, $\Xi = \frac{\lambda_+ - \lambda_-}{2}\Delta t\xi$, $\eta = \frac{a_0^+}{\lambda_+ - \lambda_0} + \frac{a_0^-}{\lambda_0 - \lambda_-}$.

Particular cases $\omega = 0$ or $\omega = 2$: In these two cases, the three roots are transparent: $\left\{e^{i\nu_1\Delta t\xi}, [\cos(\Xi) \pm i\sin(\Xi)]\, e^{i\left(\frac{\nu_2+\nu_3}{2}\right)\Delta t\xi}\right\} = \left\{e^{i\nu_1\Delta t\xi}, e^{i\nu_2\Delta t\xi}, e^{i\nu_3\Delta t\xi}\right\}$ for $\omega = 0$ and $\left\{-e^{i\nu_1\Delta t\xi}, \left[\pm\sqrt{1-\eta^2\sin^2(\Xi)} + i\eta\sin(\Xi)\right] e^{i\left(\frac{\nu_2+\nu_3}{2}\right)\Delta t\xi}\right\}$ for $\omega = 2$.

Equality (2) in the scalar case enables to simplify η: $\eta = -\frac{\nu_2 - 2a + \nu_3}{\nu_2 - \nu_3}$. Conditions $\lambda_- < \lambda_0 < \lambda_+$, $\lambda_- \le a \le \lambda_+$, $a_0^- \le 0$ and $a_0^+ \ge 0$ imply that $\eta \in [-1,1]$. Thus, the square root $\sqrt{1-\eta^2\sin^2(\Xi)}$ is well defined. If $-1 < \eta < 1$, the roots are simple since $1-\eta^2\sin^2(\Xi) \neq 0$. Otherwise, the roots for $\omega = 2$ simplify into $\left\{-e^{i\nu_1\Delta t\xi}, e^{i\nu_2\Delta t\xi}, -e^{i\nu_3\Delta t\xi}\right\}$ if $\eta = 1$ and $\left\{-e^{i\nu_1\Delta t\xi}, -e^{i\nu_2\Delta t\xi}, e^{i\nu_3\Delta t\xi}\right\}$ if $\eta = -1$.

To conclude with $\omega \in \{0,2\}$, all these roots have a modulus equal to 1 and are simple.

General case $\omega \in]0,2[$: Obviously, according to (9), one of the three roots of χ is $(1-\omega)e^{i\nu_1\Delta t\xi}$ which has a modulus strictly smaller than 1 if $\omega \in]0,2[$. We have to determine the two other roots.

• If $\Xi \equiv 0[\pi]$, $\tilde{\chi}$ writes $\tilde{\chi}_\pm(X) := (1-\omega)e^{i(\nu_2+\nu_3)\Delta t\xi} \pm [2-\omega]e^{i\left(\frac{\nu_2+\nu_3}{2}\right)\Delta t\xi}X + X^2$. The roots of $\tilde{\chi}_-$ are $\frac{(2-\omega)\pm\omega}{2}e^{i\left(\frac{\nu_2+\nu_3}{2}\right)\Delta t\xi}$ and those of $\tilde{\chi}_+$ are $\frac{-(2-\omega)\pm\omega}{2}e^{i\left(\frac{\nu_2+\nu_3}{2}\right)\Delta t\xi}$. They are all simple and their modulus are equal to 1 or to $|1-\omega| < 1$ since $\omega \in]0,2[$.

• If $\Xi \not\equiv 0[\pi]$, we define ψ such as $\psi(X) := \tilde{\chi}(X)$ for $\omega = 1$, which gives

$$\psi(X) = X\left[-e^{i\left(\frac{\nu_2+\nu_3}{2}\right)\Delta t\xi}\left[\cos(\Xi) + i\eta\sin(\Xi)\right] + X\right].$$

The key point is to compare the zeros of both $\tilde{\chi}$ and ψ in the unit ball, as in [8].

Roots of ψ. The roots of ψ are $X_1 = 0$ and $X_2 = e^{i\left(\frac{\nu_2+\nu_3}{2}\right)\Delta t\xi}\left[\cos(\Xi) + i\eta\sin(\Xi)\right]$. The modulus of X_2 is $|X_2|^2 = \cos^2(\Xi)\left[1-\eta^2\right] + \eta^2 \in \left[\eta^2, 1\right[$, since $\eta \in [-1,1]$. Since $\Xi \not\equiv 0[\pi]$, $|X_2|$ does not be equal to 1. The function ψ has thus two roots strictly contained in the open unit ball.

Comparison between $\tilde{\chi}$ and ψ on the unit circle. For $\theta \in \mathbb{R}$, one has

$$|\tilde{\chi}(e^{i\theta}) - \psi(e^{i\theta})| = |1-\omega|\left|e^{i(\nu_2+\nu_3)\Delta t\xi} - e^{i(\theta + (\frac{\nu_2+\nu_3}{2})\Delta t\xi)}\left[\cos(\Xi) - i\eta\sin(\Xi)\right]\right|.$$

Multiplying by $|-e^{-i(\theta+(\frac{\nu_2+\nu_3}{2})\Delta t\xi)}|$ and taking the complex conjugate give

$$|\tilde{\chi}(e^{i\theta}) - \psi(e^{i\theta})| = |1-\omega|\left|-e^{-i(\frac{\nu_2+\nu_3}{2})\Delta t\xi + i\theta} + \left[\cos(\Xi) + i\eta\sin(\Xi)\right]\right|.$$

Computation of ψ on the unit circle. One has $|\psi(e^{i\theta})| = \left|-e^{i(\frac{\nu_2+\nu_3}{2})\Delta t\xi}\left[\cos(\Xi) + i\eta\sin(\Xi)\right] + e^{i\theta}\right| = \left|\cos(\Xi) + i\eta\sin(\Xi) - e^{-i(\frac{\nu_2+\nu_3}{2})\Delta t\xi + i\theta}\right|$.

The latest equality is obtained by a multiplication by $|-e^{-i(\frac{\nu_2+\nu_3}{2})\Delta t\xi}|$.

Use of Rouché's theorem 5.1. One chooses the closed simple path γ be equal to the unit circle. Since $\omega \in]0,2[$, one has $|\tilde{\chi}(e^{i\theta}) - \psi(e^{i\theta})| = |1-\omega||\psi(e^{i\theta})| < |\psi(e^{i\theta})|$ for all $\theta \in \mathbb{R}$. By Rouché's theorem 5.1, $\tilde{\chi}$ has the same number of roots in the open unit ball than ψ.

All in all, each case of ω leads to three roots of χ with modulus strictly less than 1 or equal to 1 and simple. The L^2-stability is thus a consequence of Proposition 1. □

Example 2. The three first flux decompositions of Example 1 satisfy hypotheses of Proposition 3. They are also satisfied by the Lax-Wendroff decomposition only with the extremal choice $|a| = \lambda/\alpha$ or $|a| = \lambda/\sqrt{\alpha}$. (Note that the L^2-stability may be proved by a directe computation with $\alpha = 1$ and $\lambda \geq |a|$ in the particular choice $\omega = 1$).

REFERENCES

[1] D. Aregba-Driollet and R. Natalini, Discrete kinetic schemes for systems of conservation laws, in *Hyperbolic Problems: Theory, Numerics, Applications. International Series of Numerical Mathematics* (vol 129), Birkhäuser, Basel, (1999), 1–10.

[2] P. L. Bhatnagar, E. P. Gross and M. Krook, A model for collision processes in gases. I. Small amplitude processes in charged and neutral one-component systems, *Physical review*, **94** (1954), 511–525.

[3] F. Charles, B. Després and M. Mehrenberger, Enhanced convergence estimates for semi-Lagrangian schemes. Application to the Vlasov–Poisson equation, *SIAM Journal on Numerical Analysis*, **51** (2013), 840–863.

[4] D. Coulette, E. Franck, P. Helluy, M. Mehrenberger and L. Navoret, High-order implicit palindromic discontinuous Galerkin method for kinetic-relaxation approximation, preprint, (2018).

[5] D. Coulette, C. Courtès, E. Franck and L. Navoret, Vectorial kinetic relaxation model with central velocity. Application to implicit relaxation schemes, preprint, (2018).

[6] B. Graille, Approximation of mono-dimensional hyperbolic systems: A lattice Boltzmann scheme as a relaxation method, *Journal of Computational Physics*, **266** (2014), 74–88.

[7] R. Natalini, A discrete kinetic approximation of entropy solutions to multidimensional scalar conservation laws, *Journal of Differential Equations*, **148** (1998), 292 – 317.

[8] M. Rheinländer, Stability and multiscale analysis of an advective lattice Boltzmann scheme, *Progress in Computational Fluid Dynamics*, **8** (2008), 56–68.

[9] D. Serre, *Matrices: Theory and Applications*, Graduate Texts in Mathematics, vol. 216, Springer New York, 2010.

E-mail address: `clementine.courtes@math.univ-toulouse.fr`

E-mail address: `emmanuel.franck@inria.fr`

A POSTERIORI ERROR ANALYSIS FOR PATCH-WISE LOCAL PROJECTION STABILIZED FEM FOR CONVECTION-DIFFUSION PROBLEMS

ASHA K. DOND*

Department of Mathematics, Indian Institute of Science, Bangalore - 560012

THIRUPATHI GUDI

Department of Mathematics, Indian Institute of Science, Bangalore - 560012

ABSTRACT. In this article, we derive a posteriori error estimator for the patch-wise local projection (PLP) stabilized conforming finite element method for convection-diffusion-reaction problems. The present a posteriori error analysis builds on the general approach of Verfürth which exploits the generic equivalence of error and residual, and get explicit and computable bounds for errors. The considered PLP stabilized method is a composition of the standard Galerkin finite element method, the patch-wise local projection stabilization, and weakly imposed Dirichlet boundary conditions on the discrete solution. Therefore in a posteriori estimates, the main issues are to deal with the consistency error arising due to non-consistent PLP stabilization term and weakly imposed boundary conditions. We present numerical results to validate the performance of the adaptive estimator.

1. **Introduction.** Convection-dominated diffusion equations arise in many applications like pollutant transport, Navier-Stokes equations and oil recovery models. The standard Galerkin finite element method (FEM) fails to provide a stable and non-oscillatory solution for these cases. Therefore, there is a wide range of stabilized FEMs has been proposed and analyzed in the literature (see [5, 6, 7] and the reference therein). It is well understood that the standard Galerkin method fails due to the presence of interior or boundary layers in the solution and these are not adequately resolved by this method. These layers cause high gradient of the solution in some small subregions of the domain. Although stabilized methods provide non-oscillatory solutions, it is essential to identify the critical regions in the domain where these layers occur and then to adopt a local refinement strategy to increase the accuracy of the finite element solution. Robust reliable and efficient adaptive error estimator and adaptive algorithms can play a vital role in this scenario. Therefore, investigating robust a posteriori error estimator for the stabilized FEMs is an active research area (see [7, 9, 8] and the references given there).

2000 *Mathematics Subject Classification.* Primary: 65N30, 65N12.

Key words and phrases. A posteriori error analysis, local projection stabilized, Convection-diffusion problems.

The first author is supported by National Board for Higher Mathematics, Government of India.

* Corresponding author: Asha K. Dond.

In this article, our aim is to derive a posteriori error estimator for the conforming patch-wise local projection (PLP) stabilized finite element method for convection-diffusion equations, described in [3]. To obtain a posteriori error estimator for convection-diffusion problems, Verfürth [7, 9] has proposed a general abstract framework which is based on the equivalence of the error and the associated residual. This equivalence is completely independent of the discretization. The PLP stabilized discrete formulation comprises the standard finite element method, PLP stabilization, and the Nitsche's technique for boundary condition (see [1, 3, 4] for more details). This formulation lacks the Galerkin orthogonality and also has weakly imposed boundary conditions in the discrete solution. Due to this, while deriving a posteriori error estimator for PLP method, in addition to Verfürth approach, we need to control two consistency errors by some computable estimators.

A posteriori estimator derived in this paper has four sub-estimators: first two are the standard residuals over the triangles and the jump of normal gradient over the edges which are obtained by the standard adaptive estimator approach, the third estimator is a patch-wise estimator arising from consistency error due to PLP stabilization term and the last estimator is a boundary integral due to weakly imposed boundary conditions in the formulation.

The paper is organized as follows. Section 2 recalls the convection-diffusion problem and the conforming PLP stabilized finite element method. Section 3 investigates a posteriori error estimator for PLP stabilization method. Finally, Section 4 presents numerical results of the adaptive algorithm using the proposed estimator in Section 3.

We conclude the introductory section with some notation used throughout this paper. Standard notation applies to Lebesgue and Sobolev spaces. The $L_2(\Omega)$ and $L_\infty(\Omega)$ norms are respectively denoted by $\|\cdot\|$ and $\|\cdot\|_\infty$, and $\|\cdot\|_k (k \geq 1)$ denotes the standard norm on the Sobolev space $H^k(\Omega)$. The $L_2(\Omega)$ inner-product is denoted by $(\cdot,\cdot)$. The notation $a \lesssim b$ abbreviates $a \leq Cb$, where C denotes a generic constant, which may depend on the shape-regularity of the triangulation but is independent of the mesh-size and diffusion parameter ϵ.

2. Convection-diffusion equations.

2.1. Variational Problem.

Consider the following convection-diffusion-reaction problem with homogeneous Dirichlet boundary conditions

$$-\epsilon\Delta u + b\cdot\nabla u + a_0 u = f \quad \text{in} \quad \Omega, \qquad u = 0 \quad \text{on} \quad \Gamma, \tag{1}$$

where $\Omega \subset \mathbb{R}^2$ is a bounded polygonal domain with boundary Γ. We make the following assumptions on the data: (1) The force function $f \in L_2(\Omega)$, (2) The coefficient ϵ is a positive constant, $b \in [W^1_\infty(\Omega)]^2$ and $a_0 \in L_\infty(\Omega)$ are given. The problem is convection dominated, that is, $\epsilon \ll |b|$, (3) There exist two constants $\beta > 0$ and $c_b \geq 0$ such that $(a_0 - \operatorname{div} b/2) \geq \beta > 0$ and $\|a_0\|_\infty \leq c_b\beta$.

Let V be the standard $H^1_0(\Omega)$ space which is defined by $H^1_0(\Omega) := \{v \in H^1(\Omega) : \gamma(v) = 0 \text{ on } \Gamma\}$, where $\gamma : H^1(\Omega) \to H^{1/2}(\Gamma)$ is the trace operator. A weak formulation of the model problem (1) consists of finding $u \in V$ such that

$$a(u, v) = \langle \ell, v\rangle \quad \text{for all } v \in V, \tag{2}$$

where $a(u, v) := \epsilon(\nabla u, \nabla v) + (b\cdot\nabla u, v) + (a_0 u, v)$ and $\langle \ell, v\rangle := (f, v)$.

A use of integration by parts yields that

$$a(v,v) \geq \|v\|^2 \quad \forall\, v \in V, \quad \text{where} \quad \|v\|^2 := \epsilon\|\nabla v\|^2 + \beta\|v\|^2. \tag{3}$$

The corresponding dual norm on $H^{-1}(\Omega) = (H_0^1(\Omega))^*$ is defined by

$$\|\phi\|_* = \sup_{v \in H_0^1(\Omega)/\{0\}} \frac{\langle \phi, v\rangle}{\|v\|}, \tag{4}$$

where $\langle \cdot, \cdot \rangle$ denotes the corresponding duality pairing. We have for all $v,\ w \in H_0^1(\Omega)$ from [7, Proposition 4.17],

$$a(v,w) \leq \max\{c_b, 1\}\{\|v\| + \|b\cdot\nabla v\|_*\}\,\|w\|\,, \tag{5}$$

$$\inf_{v \in H_0^1(\Omega)/\{0\}} \sup_{w \in H_0^1(\Omega)/\{0\}} \frac{a(v,w)}{(\|v\| + \|b\cdot\nabla v\|_*)\,\|w\|} \geq \frac{1}{2+\max\{c_b,1\}}. \tag{6}$$

Hence, the problem (2) admits a unique solution $u \in H_0^1(\Omega)$ for every right-hand side $\ell \in H^{-1}(\Omega)$ and

$$\|\ell\|_* \lesssim \|u\| + \|b\cdot\nabla u\|_* \lesssim \|\ell\|_*\,. \tag{7}$$

Here, the constants are independent of ϵ and β.

2.2. **Discrete Problem with Conforming Elements.** Let $\mathcal{T}_h$ be a locally quasi-uniform and regular triangulation of Ω into triangles. The set of all interior (resp. boundary) vertices of $\mathcal{T}_h$ are denoted by $\mathcal{N}_h^i$ (resp. $\mathcal{N}_h^b$). Set $\mathcal{N}_h := \mathcal{N}_h^i \cup \mathcal{N}_h^b$. For any $z \in \mathcal{N}_h$, we denote by ω_z (patch of z) the union of all the triangles that share the vertex z. Let $\mathcal{E}_h := \mathcal{E}_h^i \cup \mathcal{E}_h^b$, where $\mathcal{E}_h^i$ (resp. $\mathcal{E}_h^b$) is the set of all interior (resp. boundary) edges in $\mathcal{T}_h$. Let n_e be the unit outward normal vector along the edge e. Let $h_T := |T|^{1/2}$, where $|T|$ is the area of the element T and $h := \max\{h_T : T \in \mathcal{T}_h\}$. The length of any edge e is denoted by h_e. Also let h_z be the diameter of ω_z and $|\omega_z|$ be the area of ω_z. Let $|v|$ be the modulus function of v and define $v^{\ominus} := \frac{1}{2}(|v| - v)$.

We introduce the mesh-functions $h_{\mathcal{T}}$ defined by $h_{\mathcal{T}}|_T = h_T$ and $h_{\mathcal{E}}$ defined by $h_{\mathcal{E}}|_T = h_e$. We also use following notation: for $s \in \mathbb{R}$ and $k \geq 0$

$$\|h_{\mathcal{T}}^s v\|_k = \Big(\sum_{T\in\mathcal{T}_h} h_T^{2s}\,\|v\|^2_{H^k(T)}\Big)^{1/2} \quad \text{for all } v \in H^k(\mathcal{T}_h).$$

Let $k \geq 0$ be an integer and $\mathbb{P}_k(T)$ be the space of polynomials of degree at most k over the element T. The broken polynomial space $\mathbb{P}_k(\mathcal{T}_h)$ is defined as

$$\mathbb{P}_k(\mathcal{T}_h) := \{v \in L_2(\Omega) : \forall T \in \mathcal{T}_h, v|_T \in \mathbb{P}_k(T)\}.$$

Define the discrete space for the conforming finite element approximation by

$$V_h := \{v \in H^1(\Omega) : \forall T \in \mathcal{T}_h,\ v|_T \in \mathbb{P}_1(T)\}.$$

For any $z \in \mathcal{N}_h$, define the fluctuation operator $S_z : H^1(\omega_z) \to L_2(\omega_z)$ by

$$S_z(v) := b\cdot\nabla v - \fint_{\omega_z} b\cdot\nabla v\,dx, \ \text{where} \ \fint_{\omega_z} v\,dx = \frac{1}{|\omega_z|}\int_{\omega_z} v\,dx.$$

For each $z \in \mathcal{N}_h$, let $\delta_z := \delta h_z$ and $\delta \geq 0$. Define the bilinear form F_h of fluctuations by

$$F_h(w,v) := \sum_{z\in\mathcal{N}_h} \delta_z \int_{\omega_z} S_z(w) S_z(v)\,dx. \tag{8}$$

The patch-wise local projection stabilized conforming finite element method is defined as follows: find $u_h \in V_h$ such that

$$A_h(u_h, v_h) = \langle \ell, v_h \rangle \quad \text{for all } v_h \in V_h, \tag{9}$$

where the bilinear form A_h is defined by

$$A_h(w, v) := a_h(w, v) + d_h(w, v) + F_h(w, v) \qquad \text{and } \langle \ell, v_h \rangle := (f, v_h), \tag{10}$$

the discrete bilinear forms a_h and d_h are defined by

$$a_h(w, v) := \epsilon(\nabla w, \nabla v) - \sum_{e \in \mathcal{E}_h^b} \int_e \epsilon \frac{\partial w}{\partial n_e} v \, ds - \sum_{e \in \mathcal{E}_h^b} \int_e \epsilon \frac{\partial v}{\partial n_e} w \, ds + \sum_{e \in \mathcal{E}_h^b} \int_e \frac{\epsilon \sigma}{h_e} wv \, ds, \tag{11}$$

$$d_h(w, v) := (b \cdot \nabla w, v) + (a_0 w, v) + \sum_{e \in \mathcal{E}_h^b} \int_e (b \cdot n_e)^{\ominus} wv \, ds, \tag{12}$$

where σ is a positive constant, it is the penalty parameter for weakly imposed boundary condition. The choice of σ comes through the coercivity results [3, Lemma 3.5]. The well-posedness of the discrete stabilized formulation (9) has been discussed in [3].

3. **A posteriori error analysis.**

In this section, we derive a posteriori error estimator for the considered problem. The current analysis utilizes the approach from Verfürth [9] which is based on the equivalence of the error to the associated residual. Let $u \in V$ and $u_h \in V_h$ be the unique solutions of problems (2) and (9), respectively. Here, note that $u_h \in V_h \not\subset H_0^1(\Omega)$, hence in order to use equivalence estimates (7) for the error equation $a(u - u_h, v) = (f, v) - a(u_h, v)$, we need to find some companion $\tilde{u}_h \in V_h \cap V$ of u_h. Then, a posteriori error estimator can be found in the two steps: first step will be the estimation of error $u - \tilde{u}_h$ using the equivalence and second step will be the estimation of $u_h - \tilde{u}_h$ in the corresponding norms. This kind of approach can be found in [2, Section 3] and references therein.

The error $u - \tilde{u}_h$ solves the variational problem (2), that is, for every $v \in H_0^1(\Omega)$

$$a(u - \tilde{u}_h, v) = (f, v) - a(\tilde{u}_h, v), \tag{13}$$

with ℓ replaced by the residual $\tilde{R}$ which is defined by $\langle \tilde{R}, v \rangle = (f, v) - a(\tilde{u}_h, v)$. We have the following equivalence of error and residual from (7)

$$\left|\left|\left| \tilde{R} \right|\right|\right|_* \lesssim |||u - \tilde{u}_h||| + |||b \cdot \nabla(u - \tilde{u}_h)|||_* \lesssim \left|\left|\left| \tilde{R} \right|\right|\right|_* . \tag{14}$$

The approximation $\tilde{u}_h \in V_h \cap V$ is defined as $\tilde{u}_h(z) := \frac{1}{\#\mathcal{T}_h(z)} \sum_{T \in \omega_z} u_h|_T$ for all interior nodes $z \in \mathcal{N}_h^i$ and $\tilde{u}_h(z) = 0$ for the boundary nodes $z \in \mathcal{N}_h^b$, here $\#\mathcal{T}_h(z)$ denotes the cardinality of triangles that sharing node z. This approximation has following properties (for more details see [2, Theorem 5.1])

$$\left\| h_{\mathcal{T}}^{-1}(u_h - \tilde{u}_h) \right\|_{L_2(\Omega)} \lesssim \left\| h_{\mathcal{E}}^{-1/2} [u_h] \,|_{\mathcal{E}} \right\|_{L_2(\mathcal{E}_h)} \lesssim \min_{v \in V} \|\nabla u_h - \nabla u\| . \tag{15}$$

Since $u_h \in \mathbb{P}_1(\mathcal{T}_h) \cap H^1(\Omega)$ is a continuous function, $\tilde{u}_h(z) = u_h(z)$, for all $z \in \mathcal{N}_h^i$ and the jumps of u_h over the interior edges $\mathcal{E}_h^i$ are zero. Hence (15) become

$$\left\| h_{\mathcal{T}}^{-1}(u_h - \tilde{u}_h) \right\|_{L_2(\Omega)} \lesssim \left\| h_{\mathcal{E}}^{-1/2} u_h \right\|_{L_2(\mathcal{E}_h^b)} . \tag{16}$$

Let $I_h : H_0^1(\Omega) \to \mathbb{P}_1(\mathcal{T}_h) \cap H_0^1(\Omega)$ be the quasi-interpolation operator [9, (3.22)] which satisfies for $T \in \mathcal{T}_h$ and $e \in \mathcal{E}_h$ the following local approximation properties ([9, proposition 3.33], [8, Lemma 3.2])

$$\begin{aligned}
&\|v - I_h v\|_{L_2(T)} \lesssim h_T \|\nabla v\|_{L_2(\omega_T)} \lesssim \alpha_T \|\!|v|\!\|_{L_2(\omega_T)}, \\
&\|v - I_h v\|_{L_2(T)} \lesssim h_T \|\nabla v\|_{L_2(\omega_T)}, \ \|\nabla(v - I_h v)\|_{L_2(T)} \lesssim \|\nabla v\|_{L_2(\omega_T)}, \\
&\|v - I_h v\|_{L_2(e)} \lesssim h_e^{1/2} \|\nabla v\|_{L_2(\omega_e)} \lesssim \epsilon^{-1/4} \alpha_e^{1/2} \|\!|v|\!\|_{L_2(\omega_T)},
\end{aligned} \tag{17}$$

where $\alpha_s := \min\{\epsilon^{-1/2}\mathrm{diam}(s), \ \beta^{-1/2}\}$, and ω_T and ω_e denote the union of all elements in $\mathcal{T}_h$ sharing at least a point with T and e, respectively.

3.1. A reliable a posteriori error estimator.

This subsection discovers a posteriori error estimator. The complete adaptive estimator has four parts: first two are the standard residual over the triangles and the jump of normal gradient over the edges

$$\begin{aligned}
\eta_T^2 :=& \alpha_T^2 \|f + \epsilon \Delta u_h - b \cdot \nabla u_h - a_o u_h\|_{L_2(T)}^2, \\
\eta_E^2 :=& \frac{1}{2} \sum_{e \in \mathcal{E}_T} \alpha_e \epsilon^{-1/2} \|[\epsilon \nabla u_h \cdot n_e]\|_{L_2(e)}^2,
\end{aligned}$$

where $\mathcal{E}_T$ is the set of edges of triangle T and $\alpha_s := \min\{\epsilon^{-1/2}\mathrm{diam}(s), \ \beta^{-1/2}\}$. Let $\mathcal{N}_T$ be the set of vertices of the triangle T. Since the PLP stabilized discrete formulation is not consistent, there is also the following PLP estimator term:

$$\eta_{\mathrm{PLP}}^2 = \sum_{z \in \mathcal{N}_T} \alpha_z \inf_{q_z \in \mathbb{P}_0(\omega_z)} \|b \cdot \nabla u_h - q_z\|_{L_2(T)}^2.$$

The estimator due to use of Nitsche's or weakly imposed boundary conditions

$$\eta_B^2 = \sum_{e \in \mathcal{E}_h^b} \max\{\epsilon, \alpha_e^2\} \frac{1}{h_e} \|u_h\|_{L_2(e)}^2.$$

The complete adaptive estimator η_h is defined by

$$\eta_h^2 := \sum_{T \in \mathcal{T}_h} \{\eta_T^2 + \eta_E^2 + \eta_{\mathrm{PLP}}^2 + \eta_B^2\}. \tag{18}$$

The following theorem derives the reliable estimator η_h.

Theorem 3.1. *Let $u \in V$ and $u_h \in V_h$ be the unique solutions of problems (2) and (9), respectively. Then, it holds*

$$\|\!|u - u_h|\!\| + \|\!|b \cdot \nabla(u - u_h)|\!\|_* \le C \eta_h.$$

Proof. Let $\tilde{u}_h$ be the connecting companion as defined in (15). We rewrite error as

$$\begin{aligned}
\|\!|u - u_h|\!\| + \|\!|b \cdot \nabla(u - u_h)|\!\|_* \le& \|\!|u - \tilde{u}_h|\!\| + \|\!|b \cdot \nabla(u - \tilde{u}_h)|\!\|_* \\
&+ \|\!|\tilde{u}_h - u_h|\!\| + \|\!|b \cdot \nabla(\tilde{u}_h - u_h)|\!\|_*.
\end{aligned} \tag{19}$$

The first two error terms on the right-hand sides of (19) can be estimated using the equivalence relation (14). In that, to estimate the residual $\|\!|\tilde{R}|\!\|_*$ we need

$$\langle \tilde{R}, v \rangle = (f, v) - a(\tilde{u}_h, v) = (f, v) - a(u_h, v) + a(u_h - \tilde{u}_h, v) \text{ for all } v \in H_0^1(\Omega).$$

Let $\langle R, v\rangle := (f, v) - a(u_h, v)$. The boundedness property (5) of the bilinear form $a(\cdot,\cdot)$ implies

$$a(u_h - \tilde{u}_h, v) \leq \max\{c_b, 1\}\{\|\!|u_h - \tilde{u}_h|\!\| + \|\!|b\cdot\nabla(u_h - \tilde{u}_h)|\!\|_*\}\,\|\!|v|\!\|\,. \tag{20}$$

From the combination of (19)-(20) and the definition (4), it is clear that in the proof we have to estimate $\langle R, v\rangle$ and $\|\!|\tilde{u}_h - u_h|\!\| + \|\!|b\cdot\nabla(\tilde{u}_h - u_h)|\!\|_*$.

Step I. Let us start with $\langle R, v\rangle$. Introduction of the quasi-interpolation I_h as defined in (17) yields,

$$\langle R, v\rangle \leq \langle R, v - I_h v\rangle + \langle R, I_h v\rangle \text{ for all } v \in H_0^1(\Omega). \tag{21}$$

The standard procedure of deriving a posteriori error estimates [8, Section 4], [9, Theorem 3.57] yields the following upper bound for the dual norm of the residual

$$\frac{\langle R, v - I_h v\rangle}{\|\!|v|\!\|} \lesssim \Big(\sum_{T\in\mathcal{T}_h}(\eta_T^2 + \eta_E^2)\Big)^{1/2}.$$

In the second term on the right-hand side of (21), use of the weak formulation (2) and (9), and the fact $I_h v \in H_0^1(\Omega)$ result in

$$\begin{aligned}\langle R, I_h v\rangle &= (f, I_h v) - a(u_h, I_h v)\\ &= (f, I_h v) - \epsilon(\nabla u_h, \nabla I_h v) - (b\cdot\nabla u_h, I_h v) - (a_0 u_h, I_h v)\\ &= -\sum_{e\in\mathcal{E}_h^b}\int_e \epsilon\frac{\partial I_h v}{\partial n_e} u_h\, ds + F_h(u_h, I_h v).\end{aligned} \tag{22}$$

Consider the first term on the right-hand side of (22) over an edge $e \in \mathcal{E}_h^b$. A use the Cauchy-Schwarz inequality and the trace inequality result in

$$\begin{aligned}\int_e \epsilon\frac{\partial I_h v}{\partial n_e} u_h\, ds &\leq \epsilon\left\|\frac{\partial I_h v}{\partial n_e}\right\|_{L_2(e)}\|u_h\|_{L_2(e)} \leq \epsilon\,\|\nabla I_h v\|_{L_2(e)}\|u_h\|_{L_2(e)}\\ &\leq \frac{\epsilon^{1/2}}{h_e^{1/2}}\|u_h\|_{L_2(e)}\,\epsilon^{1/2}\,\|\nabla I_h v\|_{L_2(T)}\,.\end{aligned}$$

The sum over all boundary edges and the fact $\epsilon^{1/2}\|\nabla I_h v\| \leq \epsilon^{1/2}\|\nabla v\| \leq \|\!|v|\!\|$ result in

$$\begin{aligned}\sum_{e\in\mathcal{E}_h^b}\int_e \epsilon\frac{\partial I_h v}{\partial n} u_h\, ds &\leq \Big(\sum_{e\in\mathcal{E}_h^b}\frac{\epsilon}{h_e}\|u_h\|_{L_2(e)}^2\Big)^{1/2}\epsilon^{1/2}\,\|\nabla I_h v\|\\ &\leq \Big(\sum_{e\in\mathcal{E}_h^b}\frac{\epsilon}{h_e}\|u_h\|_{L_2(e)}^2\Big)^{1/2}\|\!|v|\!\|\,.\end{aligned} \tag{23}$$

The second term on the right-hand side of (22) can be bounded as

$$\begin{aligned}F_h(u_h, I_h v) &= \sum_{z\in\mathcal{N}_h}\delta_z\int_{\omega_z}\Big(b\cdot\nabla u_h - ⨍_{\omega_z} b\cdot\nabla u_h\, dx\Big)\Big(b\cdot\nabla I_h v - ⨍_{\omega_z} b\cdot\nabla I_h v\, dx\Big)\, dx\\ &\leq \sum_{z\in\mathcal{N}_h}\delta_z\left\|b\cdot\nabla u_h - ⨍_{\omega_z} b\cdot\nabla u_h\, dx\right\|_{L_2(\omega_z)}\left\|b\cdot\nabla I_h v - ⨍_{\omega_z} b\cdot\nabla I_h v\, dx\right\|_{L_2(\omega_z)}.\end{aligned} \tag{24}$$

For any $q_z \in \mathbb{P}_0(\omega_z)$, $q_z - \fint_{\omega_z} q_z = 0$. The boundedness of the averaging operator implies

$$\left\| b \cdot \nabla u_h - \fint_{\omega_z} b \cdot \nabla u_h \, dx \right\|_{L_2(\omega_z)} = \left\| (b \cdot \nabla u_h - q_z) - \fint_{\omega_z} (b \cdot \nabla u_h - q_z) \, dx \right\|_{L_2(\omega_z)} \lesssim \| b \cdot \nabla u_h - q_z \|_{L_2(\omega_z)} . \tag{25}$$

The boundedness of operator and the inequality $\|\nabla I_h v\|_{L_2(\omega_z)} \leq h_z^{-1} \alpha_z \|v\|_{L_2(\omega_z)}$([9, Section 3.8]) show

$$\delta_z \left\| b \cdot \nabla I_h v - \fint_{\omega_z} b \cdot \nabla I_h v \, dx \right\|_{L_2(\omega_z)} \lesssim \delta_z \| b \cdot \nabla I_h v \|_{L_2(\omega_z)} \lesssim h_z \|b\|_\infty \|\nabla I_h v\|_{L_2(\omega_z)} \lesssim h_z h_z^{-1} \alpha_z \|v\|_{L_2(\omega_z)} . \tag{26}$$

The substitution of (25) and (26) in (24) implies

$$F_h(u_h, I_h v) \lesssim \sum_{z \in \mathcal{N}_h} \alpha_z \| b \cdot \nabla u_h - q_z \|_{L_2(\omega_z)} \|v\|_{L_2(\omega_z)} \lesssim \Big(\sum_{z \in \mathcal{N}_h} \alpha_z^2 \| b \cdot \nabla u_h - q_z \|_{L_2(\omega_z)}^2 \Big)^{1/2} \|v\| . \tag{27}$$

Since q_z is arbitrary, we have

$$F_h(u_h, I_h v) \lesssim \Big(\sum_{z \in \mathcal{N}_h} \inf_{q_z \in \mathbb{P}_0(\omega_z)} \alpha_z^2 \| b \cdot \nabla u_h - q_z \|_{L_2(\omega_z)}^2 \Big)^{1/2} \|v\| .$$

The final step is the rearrangement of summations, that is,

$$\sum_{z \in \mathcal{N}_h} \alpha_z \inf_{q_z \in \mathbb{P}_0(\omega_z)} \| b \cdot \nabla u_h - q_z \|_{L_2(\omega_z)}^2 = \sum_{z \in \mathcal{N}_h} \sum_{T \in \omega_z} \alpha_z \inf_{q_z \in \mathbb{P}_0(\omega_z)} \| b \cdot \nabla u_h - q_z \|_{L_2(T)}^2 = \sum_{T \in \mathcal{T}_h} \Big(\sum_{z \in \mathcal{N}_T} \alpha_z \inf_{q_z \in \mathbb{P}_0(\omega_z)} \| b \cdot \nabla u_h - q_z \|_{L_2(T)}^2 \Big). \tag{28}$$

The substitution of consistent error (23) and (27) in (22), and (28) result in

$$\frac{\langle R, I_h v \rangle}{\|v\|} \lesssim \Big(\sum_{e \in \mathcal{E}_h^b} \frac{\epsilon}{h_e} \|u_h\|_{L_2(e)}^2 \Big)^{1/2} + \Big(\sum_{z \in \mathcal{N}_h} \alpha_z^2 \inf_{q_z \in \mathbb{P}_0(\omega_z)} \| b \cdot \nabla u_h - q_z \|_{L_2(\omega_z)}^2 \Big)^{1/2} .$$

Step II In the second part, we have to estimate $\|u_h - \tilde{u}_h\| + \|b \cdot \nabla (u_h - \tilde{u}_h)\|_*$. In the first term, a use of (16) and the inverse inequality result in

$$\|u_h - \tilde{u}_h\|^2 = \epsilon \|\nabla (u_h - \tilde{u}_h)\|^2 + \beta \|u_h - \tilde{u}_h\|^2 \lesssim \epsilon \left\| h_{\mathcal{T}}^{-1} (u_h - \tilde{u}_h) \right\|^2 + \beta \|u_h - \tilde{u}_h\|^2 \lesssim \epsilon \left\| h_{\mathcal{E}}^{-1/2} u_h \right\|_{L_2(\mathcal{E}_h^b)}^2 + \beta \left\| h_{\mathcal{T}} h_{\mathcal{E}}^{-1/2} u_h \right\|_{L_2(\mathcal{E}_h^b)}^2 \lesssim 2 \max\{\epsilon, \beta h_{\mathcal{T}}^2\} \left\| h_{\mathcal{E}}^{-1/2} u_h \right\|_{L_2(\mathcal{E}_h^b)}^2 .$$

Note that, $\epsilon \leq h_T$ and $h_e \leq \alpha_e$, so the final coefficient of the edge estimator will be α_e^2. The second term

$$\|b \cdot \nabla (u_h - \tilde{u}_h)\|_* = \sup_{v \in H_0^1(\Omega)/\{0\}} \frac{\langle b \cdot \nabla (u_h - \tilde{u}_h), v \rangle}{\|v\|} . \tag{29}$$

The Cauchy-Schwarz inequality, (16) and $\|v\| \leq \beta^{-1/2} \|v\|$ imply

$$\langle b \cdot \nabla (u_h - \tilde{u}_h), v \rangle \leq \|b \cdot \nabla (u_h - \tilde{u}_h)\| \|v\| \lesssim \beta^{-1/2} \left\| h_{\mathcal{E}}^{-1/2} u_h \right\|_{L_2(\mathcal{E}_h^b)} \|v\| . \tag{30}$$

Also, by an integration by parts and using data assumption (2), (16) and the fact that $\epsilon < \beta$ we have

$$\begin{aligned}\langle b\cdot\nabla(u_h-\tilde{u}_h),v\rangle &= -(u_h-\tilde{u}_h,\nabla\cdot(bv)) = -(u_h-\tilde{u}_h, b\nabla v + \mathrm{div} b\ v)\\ &\lesssim \|u_h-\tilde{u}_h\|\,(\|v\|+\|\nabla v\|) = \|u_h-\tilde{u}_h\|\,\big(\frac{1}{\beta^{1/2}}\beta^{1/2}\,\|v\| + \frac{1}{\epsilon^{1/2}}\epsilon^{1/2}\,\|\nabla v\|\,\big)\\ &\lesssim \epsilon^{-1/2}\left\|h_{\mathcal{T}}h_{\mathcal{E}}^{-1/2}u_h\right\|_{L_2(\mathcal{E}_h^b)}\|\!|v|\!\|. \qquad (31)\end{aligned}$$

From (30) and (31), we obtain

$$\begin{aligned}\langle b\cdot\nabla(u_h-\tilde{u}_h),v\rangle &\lesssim \min\{\epsilon^{-1/2}h_{\mathcal{T}},\beta^{-1/2}\}\left\|h_{\mathcal{E}}^{-1/2}u_h\right\|_{L_2(\mathcal{E}_h^b)}\|\!|v|\!\|\\ &\lesssim \alpha_e\left\|h_{\mathcal{E}}^{-1/2}u_h\right\|_{L_2(\mathcal{E}_h^b)}\|\!|v|\!\|.\end{aligned}$$

Altogether, (19), Step I and Step II lead to the final estimator (18), this concludes the proof.

4. **Computational Results.** This section presents some numerical experiments to substantiate the above theoretical results. The standard adaptive algorithm has used in the numerical experiments, which contains the loop of the four-steps **Solve-Estimates-Mark-Refine** (see [9, Algorithm 1.1]).

Example 4.1. *Consider the PDE (1) with coefficients* $\epsilon = 10^{-8}$, $b = (2,3)$ *and* $a_0 = 1$ *with homogeneous Dirichlet boundary condition on the domain* $\Omega = (0,1)^2$, *and the exact solution*

$$u(x,y) = x^3(1-x)y(1-y^4).$$

This example has convection-dominated coefficients and a smooth polynomial solution. The numerical simulation initializes on the criss-cross mesh. After applying several successive adaptive iterations, resulting adaptive mesh and discrete solution are shown in Figure 1. It can be seen that the mesh is more refined near the boundary $x = 1$ and $y = 1$. This is because the solution has large variation near the boundary and this is nicely captured by the estimator. Figure 2 depicts the

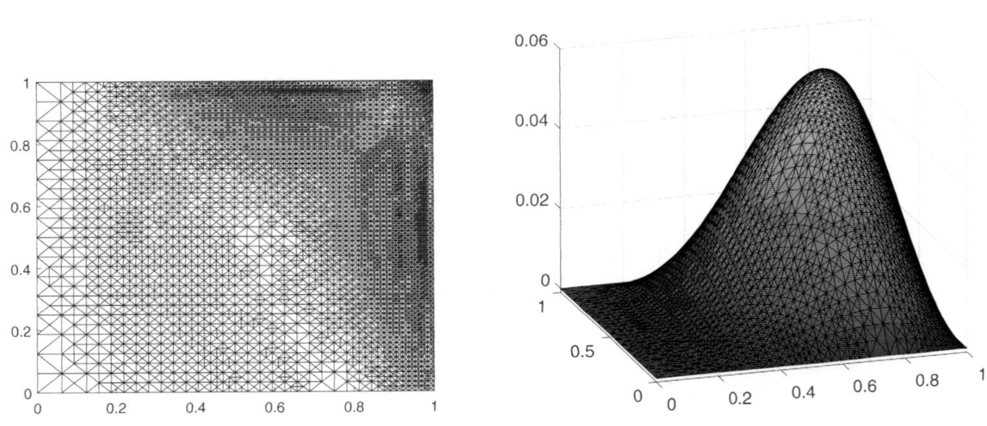

FIGURE 1. Adaptive mesh and discrete solution of Example 4.1

convergence results for errors and estimator with respect to the number of degrees

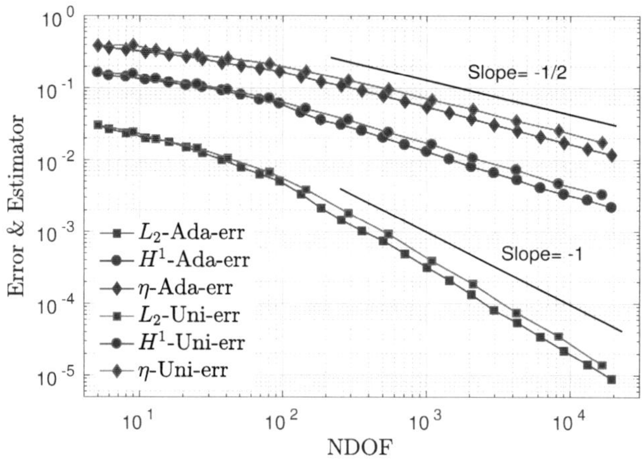

FIGURE 2. Convergence plot: NDOF Vs Error and estimator of Example 4.1

of freedom (NDOF). In the error plots, we have considered the L_2 and H^1-norms of the error $(u - u_h)$ and the complete estimator. The observations are that the H^1-error and estimator have convergence rate $\mathrm{NDOF}^{1/2}$ which is the almost linear in terms of mesh-size. For the L_2-error, we have observed the optimal convergence rate, that is, order one for both uniform and adaptive refinement with respect to NDOF.

Example 4.2. (Circular internal layer) *Consider the PDE (1) with coefficients* $\epsilon = 10^{-8}$, $b = (2,3)$ *and* $a_0 = 2$ *with the domain* $\Omega = (0,1)^2$. *The exact solution is*

$$u(x,y) = 16x(1-x)y(1-y)\Big(\frac{1}{2} + \frac{\tan^{-1}(200(0.25^2 - (x-0.5)^2 - (y-0.5)^2)}{\pi}\Big).$$

The right-hand side function f is chosen according to the exact solution u. This problem is a homogeneous Dirichlet boundary problem. The solution possesses a circular internal layer on the circumference of the circle, centered at (0.5,0.5) with radius 0.25, in the unit square domain. In the numerical experiment, we observed that the adaptive algorithm with the derived adaptive estimator has successfully refined the right elements which are the part of the interior layer of the solution. Figure 3 shows the adaptive-refined meshes, which are generated through the adaptive refinements. Figure 4 (left) shows the discrete solution with the adaptive refinements. Figure 4 (right) depicts the convergence rates. Since the adaptive refinement are more concentrated at the circular region, the adaptive method achieves optimal convergence rate in a few iterations, while in the uniform case, it requires more numbers of refinements. From the error plots, it is visible that the adaptive refinements give more accurate solution compared to uniform refinements.

5. **conclusions.** We have derived a reliable a posteriori error estimator for PLP stabilized conforming FEM. In the numerical experiment, we have shown that the derived a posteriori estimator captures inner-layer. Further, the adaptive algorithm achieves the optimal convergences rates. In the case of boundary layer problems, the PLP stabilization provides a non-oscillatory solution, but do not resolve the boundary layer (see [3, Example 5.2]). It will be interesting to look at the performance of the adaptive algorithm with the current estimator over boundary layer

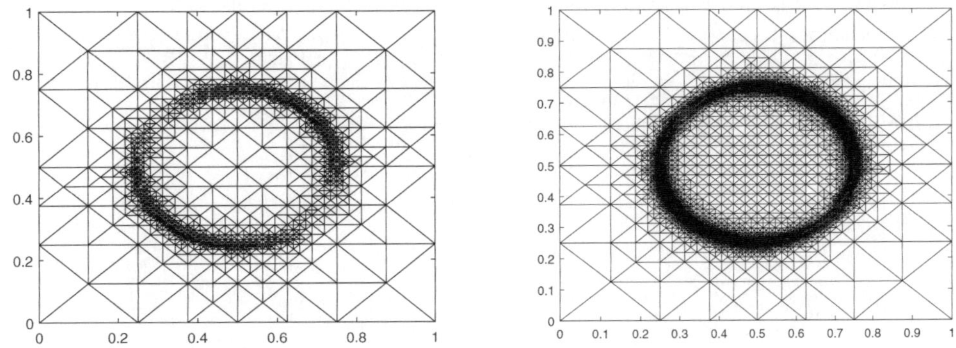

FIGURE 3. Adaptive meshes at 1314 and 5312 NDOF of Example 4.2

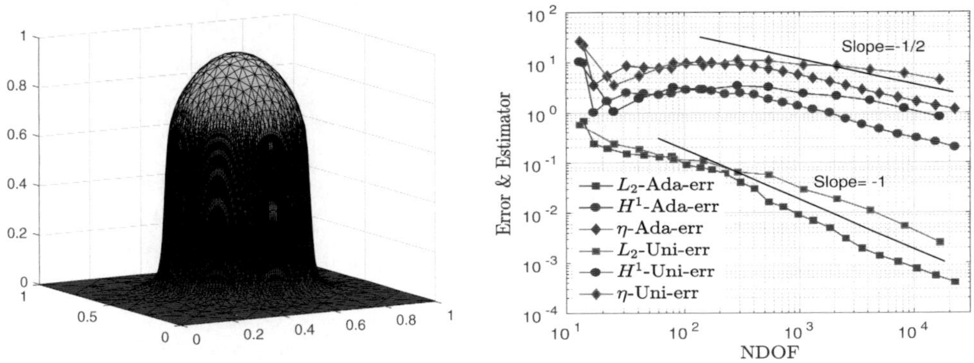

FIGURE 4. Discrete solution and convergence rates of Example 4.2

problems. Further, proving the efficiency of the presented adaptive estimators will be part of the interest of future work.

Acknowledgments. The first author gratefully acknowledges financial support from **the National Board for Higher Mathematics**, Government of India.

REFERENCES

[1] R. Biswas, A.K. Dond and T. Gudi, Edge patch-wise local projection stabilized nonconforming FEM for the Oseen problem *Comput. Meth. Appl. Math.*, **19** (2019), 189-214.

[2] C. Carstensen, M. Eigel, R. H. W. Hoppe, and C. Löbhard, A review of unified a posteriori finite element error control, *Numer. Math. Theory Methods Appl.*, **5** (2012), 509-558.

[3] A.K. Dond and T. Gudi., Patch-wise local projection stabilized finite element methods for convection-diffusion-reaction problems, *Numer Methods Partial Differential Eq.*, **35** (2019), 638-663.

[4] P. Knobloch, A generalization of the local projection stabilization for convection-diffusion-reaction equations, *SIAM J. Numer. Anal.*, **48** (2010), 659–680.

[5] C. Johnson. *Numerical Solution of Partial Differential Equations by the Finite Element Method*, Cambridge Univ. Press. 1987.

[6] H.-G.Roos, M. Stynes and L. Tobiska. *Robust numerical methods for singularly perturbed differential equations*. Springer-Verlag, Berlin., 2008.

[7] Tobiska, L. and Verfürth, R., Robust *a posteriori* error estimates for stabilized finite element methods, *IMA J. Numer. Anal.*, **35** (2015), 1652-1671.

[8] R. Verfürth, A posteriori error estimators for convection-diffusion equations, *Numer. Math.*, **80** (1998), 641-663.

[9] R. Verfürth. *A posteriori error estimation techniques for finite element methods*, Numerical Mathematics and Scientific Computation, Oxford University Press, Oxford, 2013.

E-mail address: aasha29@gmail.com
E-mail address: gudi@math.iisc.ernet.in

MODELING MOVING BOTTLENECKS ON ROAD NETWORKS

Nikodem Dymski*
Inria Sophia Antipolis - Méditerranée
Université Côte d'Azur, Inria, CNRS, LJAD
06902 Sophia-Antipolis, France
and
Instytut Matematyki, Uniwersytet Marii Curie-Skłodowskiej
pl. Marii Curie-Skłodowskiej 1
20-031 Lublin, Poland

Paola Goatin
Inria Sophia Antipolis - Méditerranée
Université Côte d'Azur, Inria, CNRS, LJAD
06902 Sophia-Antipolis, France

Massimiliano D. Rosini
Instytut Matematyki, Uniwersytet Marii Curie-Skłodowskiej
pl. Marii Curie-Skłodowskiej 1
20-031 Lublin, Poland

Abstract. In this paper we generalize the Lighthill-Witham-Richards model for vehicular traffic coupled with moving bottlenecks to the case of road networks. Such models can be applied to study the traffic evolution in the presence of a slow-moving vehicle, like a bus. At last, a numerical experiment is shown.

1. **Introduction.** In this paper we study the Lighthill-Witham-Richards (LWR) model for vehicular traffic coupled with moving bottlenecks on road networks. We recall that the LWR model was introduced in [16, 17] and gave rise to macroscopic modelling of traffic flow. A moving bottleneck models the presence of a slow vehicle, like a bus or a truck, which causes the reduction of the road capacity at its position and thus generates a moving constraint for the traffic flow. From the analytical point of view our model is the natural generalization to the case of a network of the LWR model with moving constraint on a single road developed in [8], which in turn can be considered as a generalization of the fixed in space point constraint on the flow theory, see [2, 5], [18, Chapter 6]. For completeness we mention that a 2×2 system of conservation laws coupled with a fixed in space point constraint on the flow has been studied in [1, 11, 12] in the case of a single road, while the case of a phase transition model coupled with a fixed in space point constraint on the flow has been studied in [6].

We describe the evolution of the traffic in presence of a slow-moving vehicle by the strongly coupled PDE-ODE system (1) introduced in [8, 14], where the PDE

2000 *Mathematics Subject Classification.* Primary: 35L65 ; Secondary: 90B20 .
Key words and phrases. scalar conservation laws, traffic flow, networks, Riemann solver, moving bottleneck.
* Corresponding author: N. Dymski.

(1a) consists of a scalar conservation law which models the evolution of traffic, while the ODE (1b) describes the trajectory of the slow-moving vehicle. The study of coupled PDE-ODE systems is not new in the conservation laws framework, we refer the reader to [3, 7, 8, 9, 11, 15].

The paper is organized as follows. In the next section we consider the case of a single unidirectional road. The case of a network is then considered in Section 3. Finally, in Section 4, we compare the solutions of the standard model with that with moving constraint for the same Riemann problem.

2. A single unidirectional road.

We consider a single road parametrized by the coordinate $x \in \mathbb{R}$ and assume that the vehicles move in the direction of increasing x with maximal speed $V > 0$. Let $y = y(t) \in \mathbb{R}$ be the position of the bus and $\rho = \rho(t,x) \in [0,1]$ be the mean (normalized) density of cars at time $t \geq 0$ and position $x \in \mathbb{R}$. The resulting model is expressed by the following system

$$\partial_t \rho + \partial_x f(\rho) = 0 \qquad (t,x) \in \mathbb{R}_+ \times \mathbb{R}, \tag{1a}$$

$$\dot{y}(t) = \omega\big(\rho(t, y(t)+)\big) \qquad t \in \mathbb{R}_+, \tag{1b}$$

$$f\big(\rho(t,y(t))\big) - \dot{y}(t)\rho(t,y(t)) \leq \frac{\alpha}{4V}\big(V - \dot{y}(t)\big)^2 \qquad t \in \mathbb{R}_+. \tag{1c}$$

Above, the flux $f \geq 0$ is defined by $f(\rho) := \rho v(\rho)$, where $v(\rho)$ is the mean velocity of the cars. We let $v\colon [0,1] \to \mathbb{R}_+$ be the strictly decreasing function defined by $v(\rho) := V(1-\rho)$. Clearly $f\colon [0,1] \to \mathbb{R}_+$ is a strictly concave function such that $f(0) = f(1) = 0$, $f(1/2) = \max_{\rho\in[0,1]} f(\rho)$ and $\operatorname{sign}(\rho - 1/2)f'(\rho) < 0$ $\forall \rho \in [0,1] \setminus \{1/2\}$. We stress that $v(1) = 0$ and $v(0) = V$. If the bus has maximal speed $V_b \in [0, V[$, then it moves with velocity $\omega(\rho) := \min\{v(\rho), V_b\}$. As a result, the trajectory of the bus is given by the function $y : \mathbb{R}_+ \to \mathbb{R}$ satisfying (1b). Notice that if the road is sufficiently congested, then $v(\rho) < V_b$ and the speed of the bus coincides with the speed of the cars. Condition (1c) can be derived as follows. By setting $X = x - y(t)$ we obtain the bus reference frame, where the velocity of the bus is zero and the conservation law (1a) becomes

$$\partial_t \rho + \partial_X\big(f(\rho) - \dot{y}\rho\big) = 0.$$

The presence of the bus hinders the maximum flow at $X = 0$ according to the rule

$$f(\rho) - \dot{y}\rho \leq \frac{\alpha}{4V}\big(V - \dot{y}\big)^2,$$

where the constant coefficient $\alpha \in]0,1[$ is the reduction rate of the road capacity due to the presence of the bus. Notice that a higher velocity of the bus $\dot{y}$ corresponds to a lower capacity of the road at its position and that

$$\frac{\alpha}{4V}\big(V - \dot{y}\big)^2 \in \left[\frac{\alpha}{4V}\big(V - V_b\big)^2, \frac{\alpha V}{4}\right].$$

We augment system (1) with an initial datum for the density of the form of a Heaviside function with a jump at the initial bus position, which is assumed to be $x = 0$, that is

$$\rho(0,x) = \begin{cases} \rho_\ell & \text{if } x < 0, \\ \rho_r & \text{if } x \geq 0, \end{cases} \tag{2a}$$

$$y(0) = 0, \tag{2b}$$

where $\rho_\ell, \rho_r \in [0,1]$ are fixed constants. We consider solutions of the problem (1), (2) that are self-similar, hence the bus velocity $\dot{y}(t)$ is assumed to be constant.

Let $\mathcal{RS}\colon [0,1]^2 \to \mathbf{L}^1_{loc}(\mathbb{R};[0,1])$ be the standard Riemann solver for (1a), (2a), which is described for instance in [4]. We point out that the associated self-similar weak solution $(t,x) \mapsto \mathcal{RS}(\rho_\ell,\rho_r)(x/t)$ does not always satisfy the constraint condition (1c). For this reason we define below the constrained Riemann solver $\mathcal{RS}_\alpha\colon [0,1]^2 \to \mathbf{L}^1_{loc}(\mathbb{R};[0,1])$ for the Riemann problem (1), (2), see [8]. First, see Figure 1, we need to introduce the density values $\check\rho_\alpha$ and $\hat\rho_\alpha$ defined by

$$\check\rho_\alpha = \min\{\rho \in [0,1]\colon f(\rho) = \rho V_b + F_\alpha\}, \qquad \hat\rho_\alpha = \max\{\rho \in [0,1]\colon f(\rho) = \rho V_b + F_\alpha\},$$

where $F_\alpha := \frac{\alpha}{4V}\left(V - V_b\right)^2$.

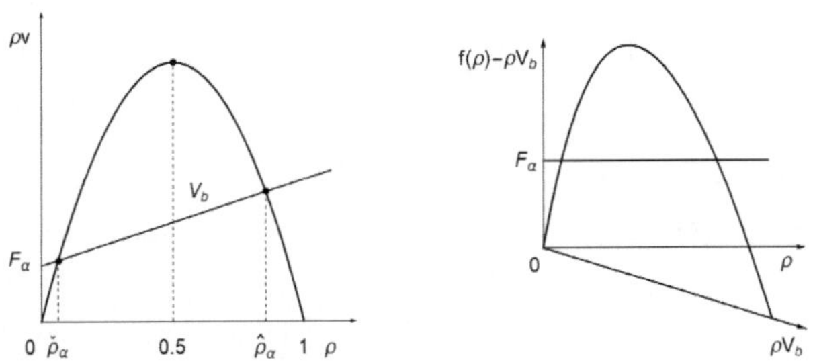

FIGURE 1. Fundamental diagram with constraint. Left: Fixed reference frame. Right: Bus reference frame.

Definition 2.1. The constrained Riemann solver $\mathcal{RS}_\alpha\colon [0,1]^2 \to \mathbf{L}^1_{loc}(\mathbb{R};[0,1])$ is defined as follows:

1. If $f\big((\mathcal{RS}(\rho_\ell,\rho_r)(V_b)\big) \le V_b \mathcal{RS}(\rho_\ell,\rho_r)(V_b) + F_\alpha$, then
$$\mathcal{RS}_\alpha(\rho_\ell,\rho_r)(x/t) = \mathcal{RS}(\rho_\ell,\rho_r)(x/t) \qquad \text{and} \qquad y(t) = \omega(\rho_r)t.$$
2. If $f\big((\mathcal{RS}(\rho_\ell,\rho_r)(V_b)\big) > V_b \mathcal{RS}(\rho_\ell,\rho_r)(V_b) + F_\alpha$, then
$$\mathcal{RS}_\alpha(\rho_\ell,\rho_r)(x/t) = \begin{cases} \mathcal{RS}(\rho_\ell,\hat\rho_\alpha)(x/t) \text{ if } x < V_b t, \\ \mathcal{RS}(\check\rho_\alpha,\rho_r)(x/t) \text{ if } x \ge V_b t, \end{cases} \qquad \text{and} \qquad y(t) = V_b t.$$

Notice that if constraint condition (1c) is not satisfied by the standard weak solution $(t,x) \mapsto \mathcal{RS}(\rho_\ell,\rho_r)(x/t)$, then the weak solution $(t,x) \mapsto \mathcal{RS}_\alpha(\rho_\ell,\rho_r)(x/t)$ has a single undercompressive shock $(\hat\rho_\alpha,\check\rho_\alpha)$ moving with speed of propagation equal to V_b, according to the Rankine-Hugoniot condition.

3. **Networks.** In this section we introduce the LWR model with moving constraint on road networks. As in [13], we define a network as a directed graph $(\mathcal{I},\mathcal{J})$, that is a pair consisting of a finite set $\mathcal{I}$ of unidirectional roads and a finite set $\mathcal{J}$ of junctions. For the rest of the work, if it is not stated differently, a junction is placed at $x = 0$.

Below we consider a node $J \in \mathcal{J}$ having n incoming roads $I_i =]-\infty,0[\in \mathcal{I}$ for $i \in \mathsf{I} := \{1,\ldots,n\}$ and m outgoing roads $I_j =]0,\infty[\in \mathcal{I}$ for $j \in \mathsf{J} := \{n+1,\ldots,n+m\}$. Let $f_h(\rho) := V_h\rho(1-\rho)$ be the flux corresponding to the road I_h, for $h \in \mathsf{H} = \mathsf{I} \cup \mathsf{J}$. Assume that at time $t = 0$ the bus is at the junction, that is $y(0) = 0$. Let I_k, $k \in \mathsf{J}$, be the road corresponding to the route of the bus. A constrained Riemann problem at the node J is the following system of scalar conservation laws with constant initial datum on every road, augmented by the ODE for the bus trajectory and the constraint inequality:

$$\begin{aligned}
&\begin{cases}\partial_t \rho_i + \partial_x f_i(\rho_i) = 0\\ \rho_i(0,x) = \rho_i^0\end{cases} && (t,x) \in \mathbb{R}_+ \times I_i,\ i \in \mathsf{I},\\
&\begin{cases}\partial_t \rho_j + \partial_x f_j(\rho_j) = 0\\ \rho_j(0,x) = \rho_j^0\end{cases} && (t,x) \in \mathbb{R}_+ \times I_j,\ j \in \mathsf{J},\\
&\begin{cases}\dot{y}(t) = \omega(\rho_k(t, y(t)+)),\\ y(0) = 0,\end{cases}\\
&f_k\big(\rho_k(t,y(t))\big) - \dot{y}(t)\rho_k(t,y(t)) \le \frac{\alpha_k}{4V_k}\big(V_k - \dot{y}(t)\big)^2 && t \in \mathbb{R}_+,
\end{aligned} \tag{3}$$

for some α_k depending on the k-th road characteristics. Before stating the constrained Riemann solver at the junction for (3), we define the admissible solutions.

Definition 3.1. An admissible constrained Riemann solver at the junction $J \in \mathcal{J}$ for (3) is a map $\mathcal{RS}^J_{\alpha_k} \colon [0,1]^{n+m} \to [0,1]^{n+m}$ such that for any $(\rho_1^0, \ldots, \rho_{n+m}^0) \in [0,1]^{n+m}$ we have that $(\bar{\rho}_1, \ldots, \bar{\rho}_{n+m}) := \mathcal{RS}^J_{\alpha_k}(\rho_1^0, \ldots, \rho_{n+m}^0)$ satisfies the following properties:

- For every $i \in \mathsf{I}$, $\mathcal{RS}_i(\rho_i^0, \bar{\rho}_i)$ has only waves with negative speed.
- For every $j \in \mathsf{J} \setminus \{k\}$, $\mathcal{RS}_j(\bar{\rho}_j, \rho_j^0)$ and $\mathcal{RS}_{\alpha_k}(\bar{\rho}_k, \rho_k^0)$ have only waves with positive speed.
- The mass through the junction is conserved, that is: $\sum_{i=1}^{n} f_i(\bar{\rho}_i) = \sum_{j=n+1}^{n+m} f_j(\bar{\rho}_j)$.
- $\mathcal{RS}^J_{\alpha_k}$ is consistent, that is: $\mathcal{RS}^J_{\alpha_k}(\bar{\rho}_1, \ldots, \bar{\rho}_{n+m}) = (\bar{\rho}_1, \ldots, \bar{\rho}_{n+m})$.

The last condition above says that the vector of traces at junction of an admissible solution is a fixed point for $\mathcal{RS}^J_{\alpha_k}$. We propose below the possible traces and their maximal fluxes. To reach this goal, we need first to define the following function.

Definition 3.2. For any $h \in \mathsf{H}$, the function $\tau_h \colon [0,1] \to [0,1]$ is such that

- $f_h(\tau_h(\rho)) = f_h(\rho)$ for every $\rho \in [0,1]$;
- $\tau_h(\rho) \neq \rho$ for every $\rho \in [0,1] \setminus \{1/2\}$.

The function τ_h is well defined, continuous and satisfies

$$0 \le \rho \le 1/2 \iff 1/2 \le \tau_h(\rho) \le 1, \qquad 1/2 \le \rho \le 1 \iff 0 \le \tau_h(\rho) \le 1/2.$$

In next propositions, we show the range of admissible fluxes for a given initial datum.

Proposition 1. *Let $i \in \mathsf{I}$ and ρ_i^0 be the initial datum on the incoming road I_i. The set of reachable fluxes $f_i(\bar{\rho}_i)$ is*

$$\Omega_i(\rho_i^0) = \begin{cases}[0, f_i(\rho_i^0)] & \text{if } \rho_i^0 \in [0,1/2],\\ [0, f_i(1/2)] & \text{if } \rho_i^0 \in \,]1/2, 1].\end{cases}$$

Proof. Since the constraint does not affect an incoming road, we can apply the construction done in [13, Proposition 4.3.3]. For definiteness, we consider the case $\rho_i^0 \in [0,1/2]$; the case $\rho_i^0 \in \,]1/2,1]$ is analogous. We stress that $\mathcal{RS}_i(\rho_i^0, \bar{\rho}_i)$ must have only waves with negative speed. If $\bar{\rho}_i \in \{\rho_i^0\} \cup \,]\tau_i(\rho_i^0), 1]$ then $\mathcal{RS}_i(\rho_i^0, \bar{\rho}_i)$ is either constant or has a single shock with negative speed. On the other hand, if $\bar{\rho}_i \in [0, \tau_i(\rho_i^0)] \setminus \{\rho_i^0\}$ then $\mathcal{RS}_i(\rho_i^0, \bar{\rho}_i)$ is either a rarefaction or a single shock, but in both cases with non negative speed, which concludes the proof. □

A direct consequence of the above proposition is the following

Corollary 1. *The maximal flow of the incoming road I_i at the junction J is*

$$\gamma_i^{\max}(\rho_i^0) = \begin{cases} f_i(\rho_i^0) & \text{if } \rho_i^0 \in [0,1/2], \\ f_i(1/2) & \text{if } \rho_i^0 \in \,]1/2,1]. \end{cases}$$

Additionally, there exists a unique $\bar{\rho}_i \in [0,1]$ such that the admissible solution of the Riemann problem with initial datum $(\rho_i^0, \bar{\rho}_i)$ consists of waves with only negative speed and the condition $f_i(\bar{\rho}_i) = \gamma_i^{\max}(\rho_i^0)$ holds.

Proposition 2. *Let $j \in \mathrm{J}$ and ρ_j^0 be the initial datum on the outgoing road I_j. The set of reachable fluxes $f_j(\bar{\rho}_j)$ is*

$$\Omega_j(\rho_j^0) = \begin{cases} \begin{cases} [0, f_j(1/2)] & \text{if } \rho_j^0 \in [0,1/2] \text{ and } j \neq k, \\ [0, f_j(\rho_j^0)] & \text{if } \rho_j^0 \in \,]1/2,1] \text{ and } j \neq k, \end{cases} \\ \begin{cases} [0, f_k(\hat{\rho}_{\alpha_k})] & \text{if } \rho_k^0 \in [0, \hat{\rho}_{\alpha_k}], \\ [0, f_k(\rho_k^0)] & \text{if } \rho_k^0 \in \,]\hat{\rho}_{\alpha_k}, 1]. \end{cases} \end{cases}$$

Proof. The proof for $j \neq k$ is analogous to proof of Proposition 1. The only difference is that $\mathcal{RS}_j(\bar{\rho}_j, \rho_j^0)$ must have only waves with positive speed. Let $j = k$ and $\rho_k^0 \in [0, \hat{\rho}_{\alpha_k}]$. We observe that $\bar{\rho}_k \in [0, \check{\rho}_{\alpha_k}]$ can be connected with ρ_k^0 by a classical waves. For $\bar{\rho}_k \in \,]\check{\rho}_{\alpha_k}, \tau_k(\hat{\rho}_{\alpha_k})[\, \cup \{\hat{\rho}_{\alpha_k}\}$ the $\mathcal{RS}_k(\bar{\rho}_k, \rho_k^0)$ consists of a possibly null shock $(\bar{\rho}_k, \hat{\rho}_{\alpha_k})$, followed by a non-classical shock $(\hat{\rho}_{\alpha_k}, \check{\rho}_{\alpha_k})$ and a shock $(\check{\rho}_{\alpha_k}, \rho_k^0)$. Notice that $\bar{\rho}_k \in [\tau(\hat{\rho}_{\alpha_k}), \hat{\rho}_{\alpha_k}[$ cannot be joined with $\hat{\rho}_{\alpha_k}$ by a wave with positive speed. The case $j = k$ and $\rho_k^0 \in \,]\hat{\rho}_{\alpha_k}, 1]$ is analogous to the situation when $j \neq k$ and $\rho_j^0 \in \,]1/2, 1]$. □

Corollary 2. *The maximal flow of the outgoing road I_j at the junction J is*

$$\gamma_j^{\max}(\rho_j^0) = \begin{cases} \begin{cases} f_j(1/2) & \text{if } \rho_j^0 \in [0,1/2] \text{ and } j \neq k, \\ f_j(\rho_j^0) & \text{if } \rho_j^0 \in \,]1/2,1] \text{ and } j \neq k, \end{cases} \\ \begin{cases} f_k(\hat{\rho}_{\alpha_k}) & \text{if } \rho_k^0 \in [0, \hat{\rho}_{\alpha_k}], \\ f_k(\rho_k^0) & \text{if } \rho_k^0 \in \,]\hat{\rho}_{\alpha_k}, 1]. \end{cases} \end{cases}$$

Additionally, there exists a unique $\bar{\rho}_j \in [0,1]$ such that the admissible solution of the Riemann problem with initial datum $(\bar{\rho}_j, \rho_j^0)$ consists of waves with only positive speed and the condition $f_j(\bar{\rho}_j) = \gamma_j^{\max}(\rho_j^0)$ holds.

For each junction we consider a traffic distribution matrix, i.e. a matrix representing the distribution of cars among the roads.

Definition 3.3. A distribution matrix $A_{m\times n}$ for the junction $J \in \mathcal{J}$ is given by

$$A_{m\times n} = \begin{pmatrix} \alpha_{n+1,1} & \cdots & \alpha_{n+1,n} \\ \vdots & \ddots & \vdots \\ \alpha_{n+m,1} & \cdots & \alpha_{n+m,n} \end{pmatrix},$$

where $\alpha_{j,i} \geq 0$ for every i, j and $\sum_{j=n+1}^{n+m} \alpha_{j,i} = 1$ for every i.

A distribution matrix $A_{m\times n}$ gives the percentage of cars from each incoming road I_i choosing the outgoing road I_j. In other words, if C is the amount of cars coming from road I_i, then $C\alpha_{j,i}$ is the amount of cars moving towards road I_j from I_i.

The construction of the admissible solution at the junction J corresponding to the initial datum $(\rho_1^0, \ldots, \rho_{n+m}^0) \in [0,1]^{n+m}$ is the following:

1. Fix a distribution matrix $A_{m\times n}$ by choosing $m \times n$ non-negative constants $\alpha_{j,i}$ such that $\sum_{j=n+1}^{n+m} \alpha_{j,i} = 1$ for every $i \in \mathsf{I}$.
2. Define the closed, convex and non-empty sets of admissible fluxes

$$\Omega = \{(\gamma_1, \ldots, \gamma_n) \in \Omega_1 \times \ldots \times \Omega_n \colon A \cdot (\gamma_{n+1}, \ldots \gamma_{n+m})^T \in \Omega_{n+1} \times \ldots \times \Omega_{n+m}\},$$

   where $\Omega_i(\rho_i^0) = [0, \gamma_i^{\max}(\rho_i^0)]$ and $\Omega_j(\rho_j^0) = [0, \gamma_j^{\max}(\rho_j^0)]$ are respectively defined in Propositions 1 and 2, see also Corollaries 1 and 2.
3. Compute a vector $(\bar\gamma_1, \ldots, \bar\gamma_n) \in \Omega$ such that

$$\sum_{i=1}^{n} \bar\gamma_i = \max_{(\gamma_1,\ldots,\gamma_n)\in\Omega} \sum_{i=1}^{n} \gamma_i. \tag{4}$$

   Then by Corollary 1 there exists unique $\bar\rho_i \in [0,1]$ such that $f_i(\bar\rho_i) = \bar\gamma_i$.
4. Compute the vector $(\bar\gamma_{n+1}, \ldots, \bar\gamma_{n+m})$ such that

$$\bar\gamma_j = \sum_{i=1}^{n} \alpha_{j,i}\bar\gamma_i.$$

   Then by Corollary 2 there exists unique $\bar\rho_j \in [0,1]$ such that $f_j(\bar\rho_j) = \bar\gamma_j$.
5. Finally, let $\mathcal{RS}_{\alpha_k}^{J}(\rho_1^0, \ldots, \rho_{n+m}^0) = (\bar\rho_1, \ldots, \bar\rho_{n+m})$.

Remark 1. The maximization problem (4) may admit more than one solution. Additional assumptions are in general required to get uniqueness of the Riemann solver. This can be obtained either imposing further conditions on the distribution matrix A, see [13, Section 5.1], or introducing a priority vector as in [10].

4. **A case study.** We consider a junction with two incoming ($n = 2$) and two outgoing ($m = 2$) roads. Let $V_h = 4$, namely $f_h(\rho) = 4\rho(1-\rho)$, $h \in \{1, \ldots, 4\}$. Fix constant initial density $(\rho_1^0, \ldots, \rho_4^0) \in [0,1]^4$, see Figure 2, center, such that

$$\begin{array}{llll} 0 < \rho_1^0 < 1/2, & 1/2 < \rho_2^0 < 1, & 1/2 < \rho_3^0 < 1, & 1/2 < \rho_4^0 < 1, \\ f(\rho_1^0) = 1/2, & f(\rho_2^0) = 2/5, & f(\rho_3^0) = 7/10, & f(\rho_4^0) = 1/2. \end{array}$$

The parameter α_3 is suitably chosen to obtain $f(\hat\rho_{\alpha_3}) = 7/20$ and we take the distribution matrix

$$A = \begin{pmatrix} 1/2 & 1/3 \\ 1/2 & 2/3 \end{pmatrix}.$$

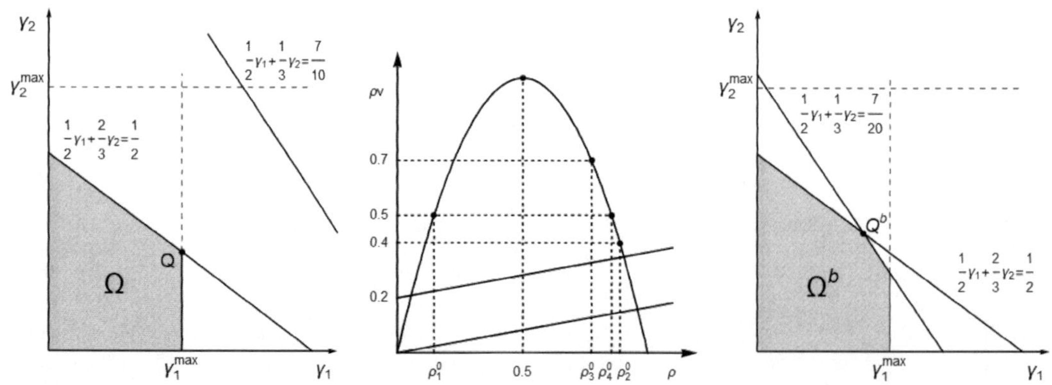

FIGURE 2. Left: the set Ω. Center: the fundamental diagram with initial datum. Right: the set Ω^b.

We consider two cases: the case a slow vehicle with maximal velocity $V_b = 1/6$ enters road I_3 and the case there is no slow vehicle at the junction. According to Propositions 1 and 2, the sets of admissible fluxes at the junction are

$$\Omega_1 = \Omega_4 = [0, 1/2], \qquad \Omega_2 = [0, 1], \qquad \Omega_3^b = [0, 7/20], \qquad \Omega_3 = [0, 7/10],$$

where Ω_3^b and Ω_3 are the sets of admissible fluxes on I_3 in case the bus is present or not, respectively. In the case without the bus we let

$$\Omega = \{(\gamma_1, \gamma_2) \in \Omega_1 \times \Omega_2 \colon A \cdot (\gamma_1, \gamma_2)^T \in \Omega_3 \times \Omega_4\},$$

and find that the maximal admissible flow through junction $\max_{(\gamma_1,\gamma_2)\in\Omega}(\gamma_1 + \gamma_2)$ is reached at the point $Q = (1/2, 3/8)$, see Figure 2, left, hence the solution for the fluxes of this problem is $(\bar{\gamma}_1, \dots, \bar{\gamma}_4) = (1/2, 3/8, 3/8, 1/2)$. In the case with the bus, we let

$$\Omega^b = \{(\gamma_1, \gamma_2) \in \Omega_1 \times \Omega_2 \colon A \cdot (\gamma_1, \gamma_2)^T \in \Omega_3^b \times \Omega_4\},$$

and find that the maximal admissible flow through the junction $\max_{(\gamma_1,\gamma_2)\in\hat{\Omega}^b}(\gamma_1 + \gamma_2)$ is reached at the point $Q^b = (2/5, 9/20)$, see Figure 2, right, therefore

$$1/2 < \bar{\rho}_1 < 1, \qquad 1/2 < \bar{\rho}_2 < 1, \qquad \bar{\rho}_3 = \hat{\rho}_{\alpha_3}, \qquad \bar{\rho}_4 = \rho_{4,0},$$
$$f(\bar{\rho}_1) = 2/5, \qquad f(\bar{\rho}_2) = 9/20, \qquad f(\bar{\rho}_3) = 7/20, \qquad f(\bar{\rho}_4) = 1/2.$$

The solution of the Riemann problem at the junction is completely determined. For better understanding the solution behavior, we display in Figure 3 the two solutions at time $t = 1/5$. The blue line describes the density profile without the bus, while the red line represents the solution in the presence of the bus. We notice that a shock wave arises on road I_1, on road I_2 we observe a rarefaction wave instead of a shock wave, on road I_3 the undercompressive shock is visible in the situation with the bus. Only the solution on road I_4 is the same in both cases.

Acknowledgments. ND was partially supported by the French Government Scholarship program for joint PhD thesis of the French Embassy in Poland. MDR acknowledges the support of the National Science Centre, Poland, Project "Mathematics of multi-scale approaches in life and social sciences" No. 2017/25/B/ST1/00051.

REFERENCES

[1] B. Andreianov, C. Donadello, and M. D. Rosini. A second-order model for vehicular traffics with local point constraints on the flow. *Math. Models Methods Appl. Sci.*, 26(4):751–802, 2016.

[2] B. Andreianov, P. Goatin, and N. Seguin. Finite volume schemes for locally constrained conservation laws. *Numer. Math.*, 115(4):609–645, 2010.

[3] R. Borsche, R. M. Colombo, and M. Garavello. On the coupling of systems of hyperbolic conservation laws with ordinary differential equations. *Nonlinearity*, 23(11):2749–2770, 2010.

[4] A. Bressan. *Hyperbolic systems of conservation laws*, volume 20 of *Oxford Lecture Series in Mathematics and its Applications*. Oxford University Press, Oxford, 2000.

[5] R. M. Colombo and P. Goatin. A well posed conservation law with a variable unilateral constraint. *J. Differential Equations*, 234(2):654–675, 2007.

[6] E. Dal Santo, M. D. Rosini, N. Dymski, and M. Benyahia. General phase transition models for vehicular traffic with point constraints on the flow. *Mathematical Methods in the Applied Sciences*, 40(18):6623–6641, 2017.

[7] M. L. Delle Monache and P. Goatin. A front tracking method for a strongly coupled PDE-ODE system with moving density constraints in traffic flow. *Discrete Contin. Dyn. Syst. Ser. S*, 7(3):435–447, 2014.

[8] M. L. Delle Monache and P. Goatin. Scalar conservation laws with moving constraints arising in traffic flow modeling: an existence result. *J. Differential Equations*, 257(11):4015–4029, 2014.

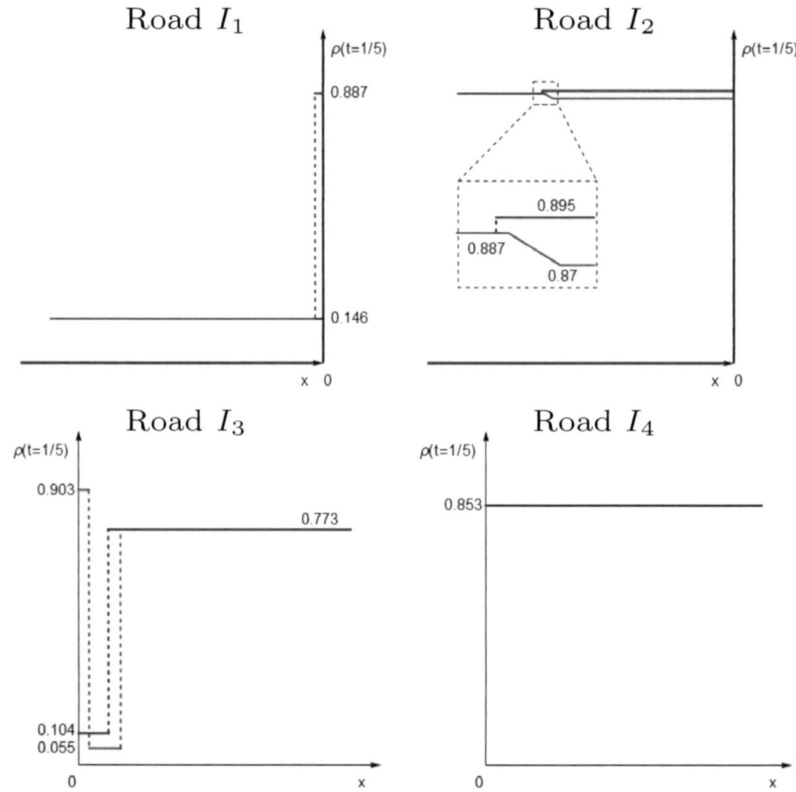

FIGURE 3. Blue line indicates the situation without bus and red with the bus.

[9] M. L. Delle Monache and P. Goatin. A numerical scheme for moving bottlenecks in traffic flow. *Bull. Braz. Math. Soc. (N.S.)*, 47(2):605–617, 2016. Joint work with C. Chalons.

[10] M. L. Delle Monache, P. Goatin, and B. Piccoli. Priority-based Riemann solver for traffic flow on networks. *Commun. Math. Sci.*, 16(1):185–211, 2018.

[11] N. S. Dymski, P. Goatin, and M. D. Rosini. Existence of **BV** solutions for a non-conservative constrained Aw-Rascle-Zhang model for vehicular traffic. *J. Math. Anal. Appl.*, 467(1):45–66, 2018.

[12] M. Garavello and P. Goatin. The Aw-Rascle traffic model with locally constrained flow. *Journal of Mathematical Analysis and Applications*, 378(2):634 – 648, 2011.

[13] M. Garavello and B. Piccoli. *Traffic flow on networks*, volume 1 of *AIMS Series on Applied Mathematics*. American Institute of Mathematical Sciences (AIMS), Springfield, MO, 2006.

[14] F. Giorgi. Prise en compte des transports en commun de surface dans la modélisation macroscopique de l'écoulement du trafic. *Institut National des Sciences Appliques de Lyon*, 2002.

[15] C. Lattanzio, A. Maurizi, and B. Piccoli. Moving bottlenecks in car traffic flow: a PDE-ODE coupled model. *SIAM J. Math. Anal.*, 43(1):50–67, 2011.

[16] M. J. Lighthill and G. B. Whitham. On kinematic waves. II. A theory of traffic flow on long crowded roads. *Proc. Roy. Soc. London. Ser. A.*, 229:317–345, 1955.

[17] P. I. Richards. Shock waves on the highway. *Operations Res.*, 4:42–51, 1956.

[18] M. D. Rosini. *Macroscopic models for vehicular flows and crowd dynamics: theory and applications.* Understanding Complex Systems. Springer, Heidelberg, 2013.

E-mail address: `nikodem.dymski@inria.fr`
E-mail address: `paola.goatin@inria.fr`
E-mail address: `mrosini@umcs.lublin.pl`

STABILITY PRESERVING APPROXIMATIONS OF A SEMILINEAR HYPERBOLIC GAS TRANSPORT MODEL

HERBERT EGGER AND THOMAS KUGLER*

Technische Universität Darmstadt, Dolivostr. 15
64293 Darmstadt, Germany

BJÖRN LILJEGREN-SAILER

Universität Trier, Universitätsring 15
54296 Trier, Germany

ABSTRACT. We consider the discretization of a semilinear damped wave equation arising, for instance, in the modeling of gas transport in pipeline networks. For time invariant boundary data, the solutions of the problem are shown to converge exponentially fast to steady states. We further prove that this decay behavior is inherited uniformly by a class of Galerkin approximations, including finite element, spectral, and structure preserving model reduction methods. These theoretical findings are illustrated by numerical tests.

1. **Introduction.** The propagation of pressure waves through a network of pipes can be described by a semilinear hyperbolic system on each pipe together with appropriate coupling conditions [4, 10]. Due to friction at the pipe walls, the kinetic energy of the gas flow gets damped resulting in a dissipative behavior and, as a consequence, the system relaxes to steady states exponentially fast; see [8]. While structure preserving model reduction methods [9] allow to guarantee the dissipative nature also after discretization, the rates of the exponential decay in the discretized models may in general degenerate with the discretization parameter; see e.g. [13].

In this work we extend our previous results [7] to problems with nonlinear damping and make the following contributions: First, we prove the exponential decay for the infinite dimensional problem in a form that can be extended to pipe networks. Second, we analyze a class of Galerkin discretizations which inherit the exponential decay behavior uniformly in the discretization parameter.

2. **Analytical results.** We consider the semilinear instationary wave propagation problem

$$\partial_t p(x,t) + \partial_x m(x,t) = \bar{f}(x), \qquad x \in (0,1),\ t \in [0,T], \tag{1}$$

$$\partial_t m(x,t) + \partial_x p(x,t) + d(m(x,t)) = \bar{g}(x), \qquad x \in (0,1),\ t \in [0,T], \tag{2}$$

$$p(x,t) = \bar{h}(x), \qquad x \in \{0,1\},\ t \in [0,T], \tag{3}$$

2000 *Mathematics Subject Classification.* Primary: 35L50, 65M60; Secondary: 35L05, 65L60, 65L20.

Key words and phrases. damped wave equation, partial differential equations on networks, Galerkin approximations, exponential stability, uniform estimates.

The authors are supported by DFG grants GSC 233, TRR 146, TRR 154, Eg-331/1-1.

* Corresponding author: Thomas Kugler.

with nonlinear damping function d satisfying the following assumptions.

Assumption 2.1. $d \in C^1(\mathbb{R})$ with $d(0) = 0$, $d'(m) > d_0$, and $|d'(m)| \le d_1 + d_2|m|^p$ for some constants $d_0 > 0$ and $d_1, d_2, p \ge 0$.

These conditions allow us to prove the well-posedness of the above problem. As a preparatory step, let us consider corresponding stationary problems of the form

$$\partial_x \tilde m(x) = \tilde f(x), \qquad x \in (0,1), \tag{4}$$

$$\partial_x \tilde p(x) + \tilde d(\tilde m(x)) = \tilde g(x), \qquad x \in (0,1), \tag{5}$$

$$\tilde p(x) = \tilde h(x), \qquad x \in \{0,1\}. \tag{6}$$

Note that solutions of 4–6 with $(\tilde d, \tilde f, \tilde g, \tilde h) = (d, \bar f, \bar g, \bar h)$ are steady states $(\bar p, \bar m)$ for the system 1–3. Using Assumption 2.1 and results about nonlinear variational problems under constraints [12, Proposition 2.3], we obtain the following.

Lemma 2.2. *Let Assumption 2.1 hold. Then for any $\tilde f, \tilde g \in L^2(0,1)$ and $\tilde h \in \mathbb{R}^2$ the system 4–6 has a unique solution $(\tilde p, \tilde m) \in H^1(0,1) \times H^1(0,1)$ and there exists a constant $c > 0$ independent of $\tilde d$ and of $\tilde f, \tilde g, \tilde h$, such that*

$$\|\tilde m\|_{H^1} \le \frac{c}{d_0}(\|\tilde g\|_{L^2} + |\tilde h|_1 + d_1\|\tilde f\|_{L^2} + d_2\|\tilde f\|_{L^2}^{p+1}) + c\|\tilde f\|_{L^2} := M$$

$$\|\tilde p\|_{H^1} \le c\,(\|\tilde g\|_{L^2} + |\tilde h|_1 + d_1 M + d_2 M^{p+1}).$$

Let us now return to the instationary problem. Using the previous result and energy estimates, we can show the following a-priori bounds.

Lemma 2.3. *Let (p, m) be a smooth solution of 1–3. Then*

$$\|\partial_t p(t)\|_{L^2} + \|\partial_t m(t)\|_{L^2} + \|m(t)\|_{H^1} \le c\,(\|\bar f\|_{L^2}, \|\bar g\|_{L^2}, |\bar h|_1, \|p(0)\|_{H^1}, \|m(0)\|_{H^1})$$

with a constant c depending only on $\bar f, \bar g, \bar h, p(0), m(0)$ but not on times t and T.

Here and below, we interpret p and m as functions of time with values in a Hilbert space, and write $p(t)$ and $m(t)$ for the corresponding functions of x at time t.

Proof. Subtracting equations 4–6 for $(\tilde f, \tilde g, \tilde h) = (\bar f, \bar g, \bar h)$ from 1–3 yields a problem of the form 1–3 for the functions $(p(t) - \bar p, m(t) - \bar m)$ with $\bar f, \bar g, \bar h = 0$ and damping term $d(m)$ replaced by $d(m(t)) - d(\bar m)$. By testing this problem with the functions $(p(t) - \bar p, m(t) - \bar m)$ and noting that $(d(m) - d(\bar m), m - \bar m) \ge 0$ due to Assumption 2.1, one can see that

$$\|p(t) - \bar p\|_{L^2}^2 + \|m(t) - \bar m\|_{L^2}^2 \le \|p_0 - \bar p\|_{L^2}^2 + \|m_0 - \bar m\|_{L^2}^2.$$

Differentiation of 1–3 with respect to time, testing with $(\partial_t p(t), \partial_t m(t))$, and using that $d'(m) > 0$ by Assumption 2.1, further shows that

$$\|\partial_t p(t)\|_{L^2}^2 + \|\partial_t m(t)\|_{L^2}^2 \le \|\partial_t p(0)\|_{L^2}^2 + \|\partial_t m(0)\|_{L^2}^2.$$

The right-hand side in this estimate can be bounded using 1–3 for $t = 0$. Then the splitting $\|m(t)\| \le \|m(t) - \bar m\| + \|\bar m\|$ and $\|\partial_x m(t) - \partial_x \bar m\| = \|\partial_t p(t)\|$ together with the bounds of Lemma 2.2 and the previous estimates implies the result. □

We are now in the position to show well-posedness of the instationary problem.

Lemma 2.4. *Let Assumption 2.1 hold. Then for any $\bar f, \bar g \in L^2(0,1)$, any $\bar h \in \mathbb{R}^2$, and any $p_0, m_0 \in H^1(0,1)$ there exists a unique solution $(p, m) \in C(0,T; H^1 \times H^1) \cap C^1(0,T; L^2 \times L^2)$ of the system 1–3 with initial value $p(0) = p_0$ and $m(0) = m_0$.*

Proof. By Assumption 2.1, the nonlinear damping term $d(m)$ in equation 2 is locally Lipschitz continuous, and existence of a unique solution (p, m) local in time thus follows by semigroup theory; cf. [11, Theorem 6.1.4]. The uniform a-priori estimates of Lemma 2.3 allow to extend the solution globally in time. □

We can now state our first main result, i.e. the exponential decay of the energies

$$E(q, v) := \tfrac{1}{2}\|q\|_{L^2}^2 + \tfrac{1}{2}\|v\|_{L^2}^2$$

for the two choices $(q, v) = (p(t) - \bar{p}, m(t) - \bar{m})$ and $(q, v) = (\partial_t p(t), \partial_t m(t))$.

Theorem 2.5. *Let (p, m) be a solution of 1–3 provided by Lemma 2.4. Then*

$$E(p(t) - \bar{p}, m(t) - \bar{m}) \le ce^{-\gamma t} \quad \text{and} \quad E(\partial_t p(t), \partial_t m(t)) \le c'e^{-\gamma t},$$

for $0 \le t \le T$ with $c, c', \gamma > 0$ only depending on $\|\bar{f}\|_{L^2}, \|\bar{g}\|_{L^2}, |\bar{h}|_2, \|p_0\|_{H^1}, \|m_0\|_{H^1}$.

The proof follows in the same way as that of Theorem 3.5 given below, and is therefore omitted. Similar results can also be found in [2, 8, 14].

3. Galerkin discretization in space.

Let $Q_h \subset L^2(0,1)$ and $V_h \subset H^1(0,1)$ and consider the following Galerkin approximation of the stationary problem 4–6: Find $(\tilde{p}_h, \tilde{m}_h) \in Q_h \times V_h$ such that

$$(\partial_x \tilde{m}_h, q_h) = (\tilde{f}, q_h), \tag{7}$$

$$-(\tilde{p}_h, \partial_x v_h) + (\tilde{d}(\tilde{m}_h), v_h) = (\tilde{g}, v_h) - \tilde{h}v_h|_0^1, \tag{8}$$

for all $q_h \in Q_h$ and $v_h \in V_h$. For convenience we write $(\cdot,\cdot) := (\cdot,\cdot)_{L^2}$ in the sequel. We will assume that the spaces Q_h, V_h satisfy the following compatibility conditions.

Assumption 3.1. $Q_h \subset L^2(0,1)$ and $V_h \subset H^1(0,1)$ are finite dimensional and

$$Q_h = \partial_x V_h \quad \text{and} \quad ker(\partial_x) \subset V_h. \tag{9}$$

Well-posedness of the discretized stationary problem 7–8 now follows with the same arguments as used in Lemma 2.2 for the analysis on the continuous level.

Lemma 3.2. *Let Assumptions 2.1 and 3.1 hold. Then for any $\tilde{f}, \tilde{g} \in L^2(0,1)$ and $\tilde{h} \in \mathbb{R}^2$ there exists a unique solution $(\tilde{p}_h, \tilde{m}_h) \in Q_h \times V_h$ of the system 7–8 and a constant $c > 0$ independent of $\tilde{d}$, of $\tilde{f}, \tilde{g}, \tilde{h}$ and of the space Q_h, V_h, such that*

$$\|\tilde{m}_h\|_{H^1} \le \frac{c}{d_0}(\|\tilde{g}\|_{L^2} + |\tilde{h}|_1 + d_1\|\tilde{f}\|_{L^2} + d_2\|\tilde{f}\|_{L^2}^{p+1}) + c\|\tilde{f}\|_{L^2} := M$$

$$\|\tilde{p}_h\|_{H^1} \le c\,(\|\tilde{g}\|_{L^2} + |\tilde{h}|_1 + d_1 M + d_2 M^{p+1}).$$

The corresponding discretization of the instationary problem 1–3 reads as follows: Find $(p_h, m_h) \in H^1(0, T; Q_h \times V_h)$ such that

$$(\partial_t p_h(t), q_h) + (\partial_x m_h(t), q_h) = (\bar{f}, q_h), \tag{10}$$

$$(\partial_t m_h(t), v_h) - (p_h(t), \partial_x v_h) + (d(m_h(t)), v_h) = (\bar{g}, v_h) - \bar{h}v_h|_0^1, \tag{11}$$

for all $q_h \in Q_h$ and $v_h \in V_h$, and for $0 \le t \le T$. In addition, we require that

$$p_h(0) = p_{h,0} \quad \text{and} \quad m_h(0) = m_{h,0}, \tag{12}$$

where $(p_{h,0}, m_{h,0})$ solves problem 7–8 with $(\tilde{f}, q_h) = (\partial_x m_0, q_h)$ and $(\tilde{g}, v_h) = (d(m_0), v_h) - (p_0, \partial_x v_h)$. By Lemma 3.2, $p_{h,0}, m_{h,0}$ and $\partial_t p_{h,0}, \partial_t m_{h,0}$ can be bounded in terms of the data of the continuous problem. In order to prove the existence of a global solution, we proceed similarly as on the continuous level. We denote by $(\bar p_h, \bar m_h)$ the steady states of the system 10–11, which correspond to the solution of 7–8 with $(\tilde d, \tilde f, \tilde g, \tilde h) = (d, \bar f, \bar g, \bar h)$, and obtain the following a-priori bounds.

Lemma 3.3. *Any solution $(p_h, m_h) \in H^1(0,T; Q_h \times V_h)$ of 10–12 with initial values $p_{h,0}$ and $m_{h,0}$ as described above, satisfies*

$$\|\partial_t p_h(t)\|_{L^2} + \|\partial_t m_h(t)\|_{L^2} + \|m_h(t)\|_{H^1} \le c\,(\|\bar f\|_{L^2}, \|\bar g\|_{L^2}, |\bar h|_1, \|p_0\|_{H^1}, \|m_0\|_{H^1})$$

with a constant $c > 0$ depending only on $\bar f, \bar g, \bar h, p_0, m_0$ but not on t, T, or Q_h, V_h.

By the Picard-Lindelöf theorem, one then obtains the existence of a unique solution.

Lemma 3.4. *Let the conditions of Lemma 2.4 and Assumption 3.1 hold. Then there exists a unique solution $(p_h, m_h) \in H^1(0,T; Q_h \times V_h)$ of problem 10–12.*

We are now in the position to prove the main result of our paper.

Theorem 3.5. *Under the assumptions of Lemma 3.2 and 3.4, there holds*

$$E(p_h(t) - \bar p_h, m_h(t) - \bar m_h) \le c e^{-\gamma t} \quad \text{and} \quad E(\partial_t p_h(t), \partial_t m_h(t)) \le c' e^{-\gamma t},$$

for all $0 \le t \le T$ with constants $c, c', \gamma > 0$ depending only on the data.

In particular, the estimate is independent of T and the choice of the spaces Q_h, V_h.

Proof. For any $t \in [0,T]$ the difference $(\tilde p_h, \tilde m_h) := (p_h(t) - \bar p_h, m_h(t) - \bar m_h)$ satisfies 7–8 with $\tilde f = \partial_t p_h(t)$, $\tilde g = \partial_t m_h(t)$, and damping $\tilde d(\tilde m_h) := d(\tilde m_h + \bar m_h) - d(\bar m_h)$. From Assumption 2.1, one can deduce that $\tilde d'(m) \ge d_0 > 0$ and

$$|\tilde d'(m)| = |d'(m + \bar m_h)| \le d_1 + d_2 |m + \bar m_h|^p \le \tilde d_1 + \tilde d_2 |m|^p,$$

for some constants $\tilde d_1$, $\tilde d_2$ depending only on d_1, d_2, p and the norm of the steady state $\bar m_h$, which is bounded uniformly by Lemma 3.2 in terms of the data. Therefore, the a-priori estimates of Lemma 3.2 apply and we can further estimate the terms $\|\tilde f\|_{L^2}^p$ and M^p appearing in the estimate of Lemma 3.2 by Lemma 3.3. As a consequence

$$\|p_h(t) - \bar p_h\|_{L^2} + \|m_h(t) - \bar m_h\|_{L^2} \le c\,(\|\partial_t p_h(t)\|_{L^2} + \|\partial_t m_h(t)\|_{L^2}) \tag{13}$$

with some constant c independent of t, T, and of the spaces Q_h, V_h. Let us define a modified energy $E_{h,\varepsilon}^1 := E(\partial_t p_h, \partial_t m_h) + \varepsilon(\partial_t m_h, m_h - \bar m_h)_{L^2}$ and note that

$$\tfrac{1}{2} E(\partial_t p_h, \partial_t m_h) \le E_{h,\varepsilon}^1 \le \tfrac{3}{2} E(\partial_t p_h, \partial_t m_h) \tag{14}$$

for all parameters $0 \le \varepsilon \le \varepsilon^*$ sufficiently small, i.e., the two energies are equivalent. From 10–11, one can further deduce that

$$\begin{aligned}\frac{d}{dt} E_{h,\varepsilon}^1 &= \frac{d}{dt} E(\partial_t p_h, \partial_t m_h) + \varepsilon \|\partial_t m_h\|_{L^2}^2 + \varepsilon(\partial_{tt} m_h, m_h - \bar m_h)_{L^2} \\ &\le -(d_0 - \varepsilon)\|\partial_t m_h\|_{L^2}^2 + \varepsilon(\partial_{tt} m_h, m_h - \bar m_h)_{L^2}.\end{aligned}$$

The second term in this estimate can be bounded by

$$\begin{aligned}(\partial_{tt} m_h, m_h - \bar m_h)_{L^2} &= (\partial_t p_h, \partial_x(m_h - \bar m_h))_{L^2} - (d'(m_h)\partial_t m_h, m_h - \bar m_h)_{L^2} \\ &\le -\|\partial_t p_h\|_{L^2}^2 + c\|\partial_t m_h\|_{L^2}\|m_h - \bar m_h\|_{L^2} \\ &\le -\tfrac{1}{2}\|\partial_t p_h\|_{L^2}^2 + \tilde c\|\partial_t m_h\|_{L^2}^2,\end{aligned}$$

where the global a-priori bounds in Lemma 3.3, equation 13, and the assumptions on d were used. By choosing $\varepsilon^* > 0$ sufficiently small, we can conclude that

$$\frac{d}{dt} E^1_{h,\varepsilon} \leq -\varepsilon E(\partial_t p_h, \partial_t m_h) \leq -\tfrac{2\varepsilon}{3} E^1_{h,\varepsilon}, \qquad \text{for all } 0 < \varepsilon \leq \varepsilon^*.$$

By integration in time, this yields $E^1_{h,\varepsilon}(t) \leq e^{-\frac{2\varepsilon}{3}} E^1_{h,\varepsilon}(0)$, and using the equivalence of the two energies 14, we obtain the second estimate of the theorem. With the help of inequality 13, we also obtain the first estimate. □

Remark 1. In the next section, we will make use of the following simple observation: Let $(\cdot,\cdot)_h$ be a semi inner product which is equivalent to $(\cdot,\cdot)_{L^2}$ on V_h, i.e.,

$$\tfrac{1}{2}\|v_h\|_{L^2} \leq \|v_h\|_h \leq \tfrac{3}{2}\|v_h\|_{L^2} \quad \text{for all } v_h \in V_h. \tag{15}$$

Then the assertions of Theorem 3.5 remain valid when replacing $(\partial_t m_h(t), v_h)$ and $(d(m_h(t)), v_h)$ in problem 10–12 by the approximations $(\partial_t m_h(t), v_h)_h$ and $(d(m_h(t)), v_h)_h$, which can be verified by a close inspection of the previous proof. This modification may substantially simplify the numerical solution.

4. **Approximation schemes.** After a basis is chosen for Q_h and V_h, the discretized system 10–11 reads

$$\mathbf{M}_p \partial_t \mathbf{p}(t) + \mathbf{G}\mathbf{m}(t) = \mathbf{f}(t),$$
$$\mathbf{M}_m \partial_t \mathbf{m}(t) - \mathbf{G}^T \mathbf{p}(t) + \mathbf{D}(\mathbf{m}(t))\mathbf{m}(t) = \mathbf{g}(t) - \mathbf{B}\mathbf{h}(t).$$

Here $\mathbf{p}, \mathbf{m}$ are the coordinate vectors for the functions p_h, m_h. Following Remark 1, we define quadrature points ξ_n and weights ω_n, and we set

$$(v, \tilde{v})_h := \sum_{n=0}^{N} \omega_n v(\xi_n) \tilde{v}(\xi_n), \qquad \text{for } v, \tilde{v} \in H_1. \tag{16}$$

We now discuss some typical choices for the subspaces Q_h, V_h for method 10–11.

Example 4.1 (Finite element method). Let T_h be a uniform mesh with nodes $x_n = nh$, $h = 1/N$, and let $P_p(T_h)$ be the space of piecewise polynomials of order p. We set $Q_h = P_0(T_h)$ and $V_h = P_1(T_h) \cap H^1$ and note that Assumption 3.1 is satisfied. We further choose $\xi_n = x_n$ and $\omega_0 = \omega_N = h/2$ and $\omega_n = h$ for $0 < n < N$ for 16, which corresponds to numerical quadrature with the trapezoidal rule, and note that 15 is fulfilled. Moreover, the matrices $\mathbf{M}_p, \mathbf{M}_m$, and $\mathbf{D}(\mathbf{m})$ are all diagonal and approximation order h^2 can be expected for sufficiently smooth solutions.

Example 4.2 (Spectral method). For $Q_h = P_{p-1}(0,1)$ and $V_h = P_p(0,1) \cap H^1$, Assumption 3.1 holds as well. Now let ξ_n and ω_n, $0 \leq n \leq p$, be the quadrature points and weights for the Gauss-Lobatto quadrature rule on $[0,1]$, then also norm equivalence 15 is valid; cf. [5], When choosing the Lagrange polynomials for the points $\{\xi_n\}_n$ as basis for V_h and the Legendre polynomials as basis for Q_h, the matrices $\mathbf{M}_p, \mathbf{M}_m$, and $\mathbf{D}(\mathbf{m})$ are again diagonal. Here exponential convergence in p can expected for smooth solutions [5].

Example 4.3 (Projection based model reduction). Let $Q_h, V_h, \omega_n, \xi_n$ be chosen as in Example 1 for small h and let $Q_H \subset Q_h$, $V_H \subset V_h$ be constructed by a structure preserving model reduction approach [3], together with the modifications proposed in [7]. Then Assumption 3.1 holds and $\mathbf{M}_p$, $\mathbf{M}_m$ are diagonal for an appropriate choice of basis. Note that the evaluation of the nonlinear term $\mathbf{D}(\mathbf{m}(t))$ via 16 still has the complexity of the high dimensional space V_h. Replacing 16 by a quadrature rule with fewer quadrature points may be used to further reduce the complexity [1]. Let us note that uniform exponential stability can still be guaranteed for this complexity-reduction approach, as long as 15 is valid.

5. **Numerical illustration.** Let us note that our results and methods of proof can be generalized almost verbatim to networks; see [6]. This will be illustrated now by some numerical tests, for which we utilize the network in Fig. 1. The topology of

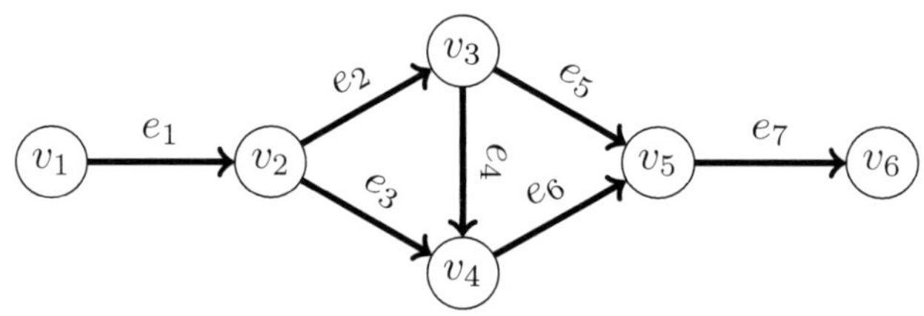

FIGURE 1. Network used for numerical tests.

the network is represented by a directed graph $(\mathcal{V}, \mathcal{E})$ with vertices $\mathcal{V} = \{v_1, \ldots, v_6\}$, divided into interior and exterior vertices $\mathcal{V}_0 = \{v_2, \ldots, v_5\}$ and $\mathcal{V}_\partial = \{v_1, v_6\}$, and edges $\mathcal{E} = \{e_1, \ldots, e_7\} \subset \mathcal{V} \times \mathcal{V}$. We denote by $\mathcal{E}(v) = \{e = (v, \cdot) \text{ or } e = (\cdot, v)\}$ the set of edges adjacent to the vertex v and define $n^e(v) = -1$ for ingoing and $n^e(v) = 1$ for the outgoing pipes.

We then consider the following problem on the network: For every pipe $e \in \mathcal{E}$, the solution (p^e, m^e) restricted to the pipe should satisfy 1–2 with data $\bar{f}, \bar{g} \equiv 0$, and damping function $d(m^e) = |m^e| m^e$. At the interior vertices of the network, the solution is required to satisfy the coupling conditions

$$p^e(v,t) = p^{e'}(v,t), \qquad \text{for all } e, e' \in \mathcal{E}(v),\ v \in \mathcal{V}_0,\ t > 0,$$
$$\sum_{e \in \mathcal{E}(v)} n^e(v) m^e(v,t) = 0, \qquad \text{for all } v \in \mathcal{V}_0,\ t > 0,$$

and we prescribe time dependent boundary conditions

$$p(v_1, t) = 90 + 10 \max\{(1-t), 0\} \quad \text{and} \quad p(v_6, t) = 70.$$

As initial conditions $p(0)$, $m(0)$, we choose the stationary solutions for the boundary data at time $t = 0$. The time discretization is chosen sufficiently accurate such that time integration errors can be neglected. For $T = 50$, we depict in Table 1 the exponential convergence of all methods. The POD method with n_{sv} singular values is trained by an h-FEM method with $h = 10^{-3}$ and the correct boundary data. We choose N Gauss-Lobatto points on each pipe such that 15 is satisfied. As predicted in Theorem 3.5 the exponential decay is uniform in the discretization parameters.

Acknowledgments. The authors would like to gratefully acknowledge financial support by the German Research Foundation (DFG) via grants GSC 233, TRR 146, TRR 154, and Eg-331/1-1.

method\t^n	0	10	20	30	40	50	γ
Ex. 4.1; $h = 0.2$	99.136	23.693	6.943	2.051	0.607	0.180	0.122
Ex. 4.1; $h = 0.05$	99.192	23.709	6.947	2.052	0.607	0.180	0.122
Ex. 4.2; $p = 3$	99.196	23.904	7.005	2.069	0.613	0.182	0.122
Ex. 4.2; $p = 10$	99.196	23.710	6.947	2.052	0.607	0.180	0.122
Ex. 4.3; $n_{sv} = 2$	99.196	23.850	6.984	2.062	0.610	0.181	0.122
Ex. 4.3; $n_{sv} = 10$	99.196	23.710	6.947	2.052	0.607	0.180	0.122

TABLE 1. Exponential convergence of $E(p_h(t) - \bar{p}_h, m_h(t) - \bar{m}_h)$ for the methods in Example 4.1-4.3.

REFERENCES

[1] H. Antil, S. E. Field, F. Herrmann, R. H. Nochetto and M. Tiglio, Two-step greedy algorithm for reduced order quadratures, *J. Sci. Comput.*, **57** (2013), 604–637.

[2] A. V. Babin and M. I. Vishik, Regular attractors of semigroups and evolution equations, *J. Math. Pures Appl.* **62** (1983), 441–491.

[3] P. Benner, V. Mehrmann and D. C. Sorensen, Dimension Reduction of Large-Scale Systems, Springer (2005).

[4] J. Brouwer, I. Gasser and M. Herty, Gas pipeline models revisited: Model hierarchies, non-isothermal models and simulations of networks, *Multiscale Model. Simul.* **9** (2011), 601–623.

[5] C. Canuto, M. Y. Hussaini, A. Quarteroni and T. A. Zang, Spectral methods: Fundamentals in single domains, Springer-Verlag, Berlin (2006).

[6] H. Egger and T. Kugler, Damped wave systems on networks: Exponential stability and uniform approximations, *Numerische Mathematik* **138** (2016), 839–867.

[7] H. Egger, T. Kugler, B. Liljegren-Sailer, N. Marheineke and V. Mehrmann, On structure-preserving model reduction for damped wave propagation in transport networks, *SIAM J. Sci. Comput.* **40** (2018), A331–A365.

[8] S. Gatti and V. Pata, A one-dimensional wave equation with nonlinear damping, *Glasgow Math. J.* **48** (2000), 419–430.

[9] S. Gugercin, R. V. Polyuga, C. Beattie and A. van der Schaft, Structure-preserving tangential interpolation for model reduction of port-Hamiltonian systems, *Automatica J. IFAC* **48** (2012), 1963-1974.

[10] D. Mugnolo, Semigroup methods for evolution equations on networks, Springer, (2014).

[11] A. Pazy, Semigroups of linear operators and applications to partial differential equations, Springer-Verlag, New York (1983).

[12] B. Scheurer, Existence et approximation de points selles pour certains problèmes non linéaires, *ESAIM: Math. Model. and Num. Anal.* **11** (1977), 369–400.

[13] L. R. T. Tebou and E. Zuazua, Uniform exponential long time decay for the space semi-discretization of a locally damped wave equation via an artificial numerical viscosity, *Num. Math.* **95** (2003), 563–598.

[14] E. Zuazua, Stability and decay for a class of nonlinear hyperbolic problems, *Asymptotic Anal.* **1** (1988), 161–185.

E-mail address: egger@mathematik.tu-darmstadt.de
E-mail address: kugler@mathematik.tu-darmstadt.de
E-mail address: bjoern.sailer@uni-trier.de

MOTION OF INTERFACES FOR HYPERBOLIC VARIATIONS OF THE ALLEN–CAHN EQUATION

Raffaele Folino*

DISIM, Università degli Studi dell'Aquila
Via Vetoio, 67100, L'Aquila, Italy

Corrado Lattanzio

DISIM, Università degli Studi dell'Aquila
Via Vetoio, 67100, L'Aquila, Italy

Corrado Mascia

Dipartimento di Matematica "Guido Castelnuovo", Sapienza, Università di Roma
P.le Aldo Moro, 2 - 00185, Roma, Italy

Abstract. The Allen–Cahn equation is a (parabolic) reaction-diffusion equation with a balanced bistable reaction term, which describes phase transition processes. It is well-known that when the diffusion coefficient is very small, the solutions exhibit very interesting phenomena. In the one-dimensional case, we have an example of *metastable dynamics*, while in the multi-dimensional case the Allen–Cahn equation is strictly related to the *mean curvature flow*. In this paper we discuss such phenomena in the case of some hyperbolic variations of the Allen–Cahn equation. In particular, in the one-dimensional case we focus the attention on the assumptions needed to have metastability and we show some numerical solutions in the case such assumptions are not satisfied.

1. **Introduction.** In this paper we are interested in the limiting behavior as $\varepsilon \to 0^+$ of the solutions to the following *hyperbolic Allen–Cahn equation*

$$\tau u_{tt} + g(u)u_t = \varepsilon^2 \Delta u - F'(u), \qquad \boldsymbol{x} \in \Omega,\ t > 0, \tag{1}$$

where $u(\boldsymbol{x}, t) \in \mathbb{R}$, $\Omega \subset \mathbb{R}^n$, with $n = 1, 2$ or 3, the diffusion coefficient ε and the parameter τ are positive, and F is a double well potential with wells of equal depth (we will specify later the precise assumptions on F). Regarding the function $g : \mathbb{R} \to \mathbb{R}$, the main examples we have in mind correspond to the choices $g \equiv 1$ and $g = 1 + \tau F''$: in the first case one has the *damped nonlinear wave equation with bistable nonlinearity*

$$\tau u_{tt} + u_t = \varepsilon^2 \Delta u - F'(u), \qquad \boldsymbol{x} \in \Omega,\ t > 0, \tag{2}$$

in the second case, we have the *Allen–Cahn equation with relaxation*

$$\tau u_{tt} + (1 + \tau F''(u))u_t = \varepsilon^2 \Delta u - F'(u), \qquad \boldsymbol{x} \in \Omega,\ t > 0. \tag{3}$$

2000 *Mathematics Subject Classification.* 35L20, 35B25, 35B36, 35K57.

Key words and phrases. Allen–Cahn equation, slow motion, metastability, motion by mean curvature, singular perturbations.

* Corresponding author: Raffaele Folino.

Both equations (2) and (3) are hyperbolic variations of the classic *Allen–Cahn equation*

$$u_t = \varepsilon^2 \Delta u - F'(u), \qquad \boldsymbol{x} \in \Omega,\, t > 0, \tag{4}$$

proposed in [1] to describe the motion of antiphase boundaries in iron alloys. Formally, we obtain eq. (4) by passing to the limit as $\tau \to 0$ in (2) or (3).

Before presenting our results on the limiting behavior as $\varepsilon \to 0^+$ of the solutions to (1), let us recall that there are different interpretations of the equation (3) with a generic potential F. From the physical point of view, such equation describes heat propagation by conduction with finite speed and it has been obtained by substituting the Fourier's law with a relaxation law of Maxwell-Cattaneo type [5]. In this case, the parameter $\tau > 0$ represents a relaxation time. Moreover, equation (3) can be seen as a reaction-diffusion equation with memory. Finally, in the one-dimensional case, equation (3) has also a probabilistic interpretation and it describes a *correlated random walk*. In both the last two interpretations (reaction-diffusion with memory and correlated random walk) the parameter $1/\tau$ is the rate of a Poisson process. Details for derivation and interpretations of the equation (3) can be found in [15, 9, 12] and references therein.

The solutions to the hyperbolic variations (1) with a generic positive (smooth) function g exhibit the same phenomena of the ones to the classic Allen–Cahn equation (4) when the diffusion coefficient $\varepsilon \to 0^+$. In the one-dimensional case, i.e. when $n = 1$ and $\Omega = [a, b]$ in (1) and (4), we have an example of *metastable dynamics* and there exist *metastable patterns* which maintain an unstable structure for an exponentially long time as $\varepsilon \to 0^+$, that is for a time $T_\varepsilon = \mathcal{O}(\exp(C/\varepsilon))$, where $C > 0$. It is impossible to mention all the papers devoted to the study of the metastable dynamics for the Allen–Cahn equation (4); here we only cite the pioneering works [2, 4, 7, 14]. Metastability for the hyperbolic version (1) has been investigated in detail in [9, 10, 11, 12]. We will briefly review these results in Section (2), where we also discuss the role of the assumptions on the functions F, g and present some numerical solutions in the case such assumptions are not satisfied.

On the other hand, in the multi-dimensional case, the Allen–Cahn equation is strictly related to the *mean curvature flow*. The link between the equation (4) and the motion by mean curvature was firstly observed by Allen and Cahn in [1] on the basis of a formal analysis. Among others, rigorous proofs can be found in [3, 6, 8]. In Section (3) we present a result contained in [13], where we study in detail the case of radially symmetric solutions to the damped version (2).

2. The one-dimensional case: metastability.

In this section we consider the one-dimensional version of (1) in a bounded interval $[a, b]$, subject to homogeneous Neumann boundary conditions

$$u_x(a, t) = u_x(b, t) = 0, \qquad t > 0, \tag{5}$$

and initial data

$$u(x, 0) = u_0(x), \qquad u_t(x, 0) = u_1(x), \qquad x \in [a, b]. \tag{6}$$

In [9, 11, 12] it has been proved that if the potential $F \in C^3(\mathbb{R})$ satisfies

$$F(\pm 1) = F'(\pm 1) = 0, \qquad F''(\pm 1) > 0, \qquad F(u) > 0 \;\; \forall u \neq \pm 1, \tag{7}$$

and the damping coefficient $g \in C^1(\mathbb{R})$ is a strictly positive function, namely

$$g(u) \geq \kappa > 0, \tag{8}$$

then there exist metastable states for the IBVP (1)-(5)-(6), which maintain a *transition layer structure* for a time $T_\varepsilon = \mathcal{O}(\exp(C/\varepsilon))$ as $\varepsilon \to 0^+$, where $C > 0$. In other words, if the initial datum u_0 has a particular structure with N transitions between -1 and $+1$ (the global minimal points of the potential F), and the initial velocity u_1 is sufficiently small as $\varepsilon \to 0^+$, then the solution maintains the same transition layer structure of the initial datum for an exponentially long time. A construction of a function with a *transition layer structure* can be found in [10, pag. 553]; roughly speaking, a metastable state with N transitions between $+1$ and -1 located at $\boldsymbol{h} = (h_1, \dots, h_N)$ is a function that is approximately ± 1 except in an $\mathcal{O}(\varepsilon)$-neighborhood of $h_1, \dots, h_N$. In formula, given $\boldsymbol{h} \in \mathbb{R}^N$ and $\beta \in \{-1, +1\}$, the metastable state $U^{\boldsymbol{h}} = U^{\boldsymbol{h}}(x)$ satisfies for ε small

$$U^{\boldsymbol{h}}(x) \approx \begin{cases} \beta, & x \in [0, h_1 - \mathcal{O}(\varepsilon)], \\ (-1)^i \beta, & x \in [h_i + \mathcal{O}(\varepsilon), h_{i+1} - \mathcal{O}(\varepsilon)], \qquad i = 1, \dots, N-1, \\ (-1)^N \beta, & x \in [h_N + \mathcal{O}(\varepsilon), 1], \end{cases}$$

and $U^{\boldsymbol{h}}(h_i) = 0$, for $i = 1, \dots, N$. In [12, Section 4], we proved that the layer dynamics is described by the ODE

$$\tau h_i'' + \gamma h_i' = \frac{\varepsilon}{c_0} \mathcal{P}_i(\boldsymbol{h}), \qquad i = 1, \dots, N, \tag{9}$$

where $h_i := h_i(t)$ denotes the position of the i-th transition point at time t and the constants c_0, γ are defined by

$$c_0 := \int_{-1}^{+1} \sqrt{2F(s)}\, ds, \qquad \gamma := \frac{1}{c_0} \int_{-1}^{+1} g(s) \sqrt{2F(s)}\, ds.$$

Regarding $\mathcal{P}_i(\boldsymbol{h})$, the precise formula can be found in [4] or [12]; here we only recall that, when F is an even function, one has

$$\mathcal{P}_i(\boldsymbol{h}) := \frac{1}{2} A^2 K^2 \left\{ \exp\left(-\frac{A(h_{i+1} - h_i)}{\varepsilon} \right) - \exp\left(-\frac{A(h_i - h_{i-1})}{\varepsilon} \right) \right\}, \tag{10}$$

for $i = 1, \dots, N$, where we used the notations $h_0 := 2a - h_1$, $h_{N+1} := 2b - h_N$ and the constants A, K are given by

$$A := \sqrt{F''(\pm 1)}, \qquad K = 2 \exp\left\{ \int_0^1 \left(\frac{A}{\sqrt{2F(t-1)}} - \frac{1}{t} \right) dt \right\}.$$

Taking $\tau = 0$ and $\gamma = 1$ in the ODE (9), we obtain the equation describing the layer dynamics in the classical Allen–Cahn equation (4). Therefore, the term $\frac{\varepsilon}{c_0} \mathcal{P}_i(\boldsymbol{h})$ represents the speed of the i-th transition point in the case of equation (4); it follows that such speed depends only on the distance between h_i and the *neighbours* h_{i-1} and h_{i+1}. In particular, h_i is attracted by the closest transition point and moves with an exponentially small velocity (provided $A > 0$).

In Figure (1) we show numerical solutions of the classical Allen–Cahn equation (left picture) and the Allen–Cahn equation with relaxation (3) (right). In both cases, we see that at time $T = 1.5 * 10^4$ the solution has the same transition layer structure of the initial profile and only the positions of the two closest transition layers slightly change; the other points appear to be stationary. Based on the ODE (9) with $\mathcal{P}_i(\boldsymbol{h})$ given by (10), the two closest transition points collapse after a time $T \approx e^{1/\varepsilon} = e^{10} \approx 2.2 * 10^4$.

The assumptions (7)-(8) are fundamental to prove the metastable dynamics of the solutions. Here, we will show what happens when they are not satisfied. First,

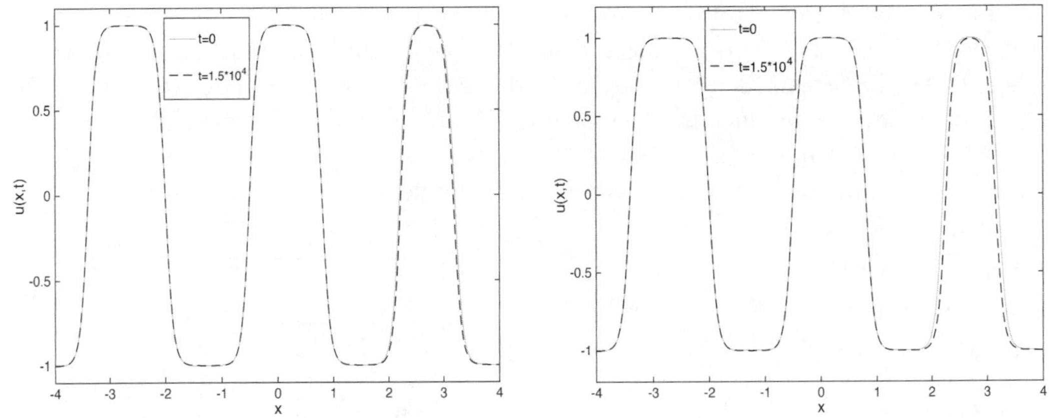

FIGURE 1. Evolution of an initial profile with 6 transitions, located at $\boldsymbol{h} = (-3.4, -2, -0.5, 0.8, 2.2, 3.2)$, in the case of the Allen–Cahn equation (left) and the Allen–Cahn equation with relaxation (right). In both cases, we choose $\varepsilon = 0.1$ and $F(u) = \frac{1}{4}(u^2 - 1)^2$. In the relaxation case, we choose $\tau = 0.8$ and the initial velocity $u_1 = 0$.

we consider the case when the assumption (8) on g is not satisfied. Fix $g = 1+\tau F''$, with $F(u) = \frac{1}{4}(u^2 - 1)^2$. Hence, if $\tau \leq \delta < 1$ then $g(u) \geq 1 - \delta > 0$; otherwise we have

$$g(u) = 3u^2, \qquad \text{for } \tau = 1, \qquad g(u) \leq 0, \text{ if } |u| \leq \sqrt{\frac{\tau - 1}{3\tau}} \text{ for } \tau > 1.$$

In Figure (2) we show the numerical solutions with the same initial data and the same value of ε of Figure (1), but different values of τ. In the left picture we choose $\tau = 1$ and see that there are small differences with the case $\tau = 0.8$; in the right picture, $\tau = 2$ and we see that at time $T = 380$ the two closest transitions collapse.

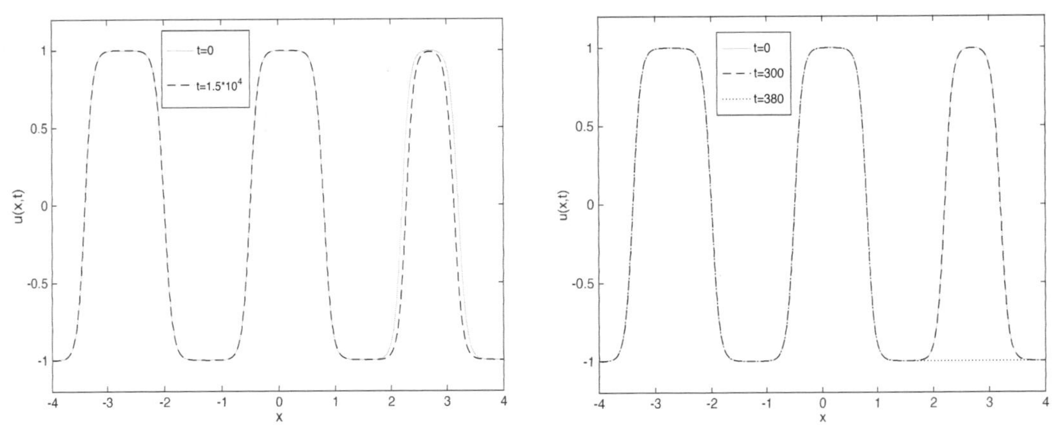

FIGURE 2. Numerical solutions to the Allen–Cahn equation with relaxation. The initial data and the value of ε are the same of Figure (1); we choose $\tau = 1$ in the left picture and $\tau = 2$ in the right picture.

Now, we focus the attention on the assumptions on F (7) and show that the condition $F''(\pm 1) > 0$ is fundamental for the metastability. Indeed, in Figure (3) we show the numerical solutions in the case of the Allen–Cahn equation (left) and the Allen–Cahn equation with relaxation (right), when the potential F has two global minimal points at ± 1 , but $F''(\pm 1) = 0$, and we see that the two closest transitions points collapse after a time much smaller than Figure (1).

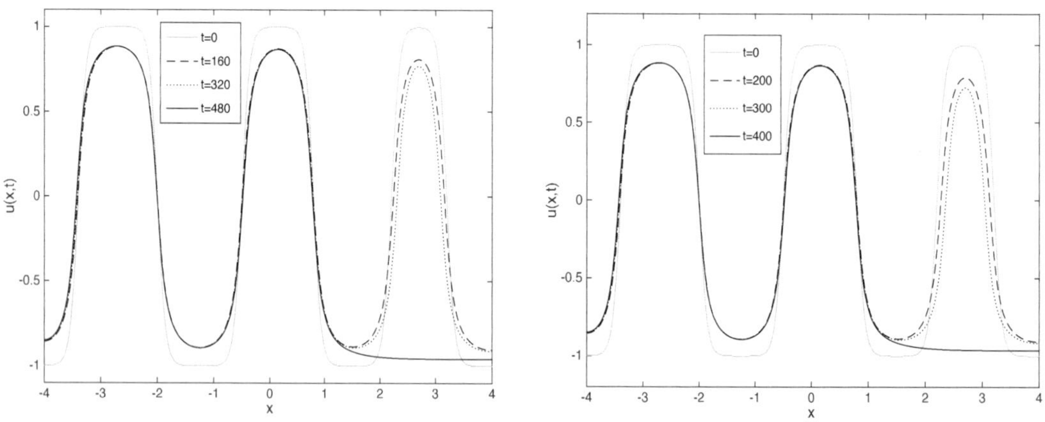

FIGURE 3. Numerical solutions to the Allen–Cahn equation (left picture) and Allen–Cahn equation with relaxation (right). The values of the parameters ε, τ and the initial data are the same of Figure (1); we change the double well potential and choose $F(u) = \frac{1}{8}(u^2-1)^4$.

3. **The multi-dimensional case: mean curvature flow.** In this section we consider radially symmetric solutions to (2) and present the main result of [13], where we rigorously proved the connection between equation (2) and the motion by mean curvature in the radial case. As in the one-dimensional case, we consider well-prepared initial data: we assume that u_0 has a particular *transition layer structure* and that u_1 is sufficiently small as $\varepsilon \to 0^+$. In general, we can divide the domain $\Omega \subset \mathbb{R}^n$, where we consider equation (2) in three regions: two regions Ω_+, Ω_- where $u_0 > 0$ and $u_0 < 0$, respectively, and the *interface* Γ_0, where $u_0 = 0$. Therefore, if $u^\varepsilon = u^\varepsilon(\boldsymbol{x}, t)$ is a solution to (2), we are interested in studying the propagation of the *interface*

$$\Gamma^\varepsilon(t) := \{\boldsymbol{x} \in \Omega \,:\, u^\varepsilon(\boldsymbol{x}, t) = 0\}.$$

In the one-dimensional case, Γ^ε consists of a finite number of points: their dynamics is described by the ODE (9) and they move with an exponentially small velocity. In the multi-dimensional case, for (1) a formal computation (see [13, Section 2.4]) shows that Γ^ε moves by mean curvature flow and its velocity is of order $\mathcal{O}(\varepsilon^2)$. Hence, we can rescale equation (1) and in the new scale the solution reaches its asymptotic limit in a time which does not depend on ε. To the best of our knowledge, the rigorous description of the interface motion for (1) with a generic positive damping coefficient g and a generic domain $\Omega \subset \mathbb{R}^n$ is an open problem. In [13] we consider the case with $g \equiv 1$ and Ω the ball of center 0 and of radius 1, i.e. $\Omega = B(0,1) = \{\boldsymbol{x} \in \mathbb{R}^n \,:\, |\boldsymbol{x}| \leq 1\}$, $n = 2, 3$. If $u^\varepsilon(\boldsymbol{x}, t)$ is a solution to (2), then $w^\varepsilon(\boldsymbol{x}, t) = u^\varepsilon(\boldsymbol{x}, \varepsilon^{-2}t)$ solves

$$\varepsilon^2 \tau w^\varepsilon_{tt} + w^\varepsilon_t = \Delta w^\varepsilon - \varepsilon^{-2} F'(w^\varepsilon), \qquad \boldsymbol{x} \in B(0,1),\, t > 0. \tag{11}$$

Consider equation (11) in radial coordinates

$$\varepsilon^2\tau w^\varepsilon_{tt} + w^\varepsilon_t = w^\varepsilon_{rr} + \frac{n-1}{r}w^\varepsilon_r - \varepsilon^{-2}F'(w^\varepsilon), \qquad r\in(0,1),\ t>0, \tag{12}$$

subject to initial conditions

$$w^\varepsilon(r,0) = w^\varepsilon_0(r), \qquad w^\varepsilon_t(r,0) = w^\varepsilon_1(r), \qquad r\in(0,1), \tag{13}$$

and Dirichlet boundary condition

$$w^\varepsilon(1,t) = 1, \qquad \forall\, t\geq 0; \tag{14}$$

moreover, at $r=0$ w^ε must satisfy $w^\varepsilon_r(0,t)=0$ for any $t\geq 0$. Let us assume that F satisfies (7), that w^ε_1 is sufficiently small as $\varepsilon\to 0^+$ and that w^ε_0 has a *single transition sphere*; the precise assumptions on the initial data can be found in [13, Section 3]. In particular, we assume that w^ε_0 satisfies

$$\lim_{\varepsilon\to 0}\int_0^1 |w^\varepsilon_0(r)-\bar w(r)|\, r^{n-1}\,dr = 0, \qquad \text{where} \qquad \bar w(r) := \begin{cases} -1, & r<\rho_0, \\ +1, & r>\rho_0. \end{cases} \tag{15}$$

Hence, ρ_0 is the radius of the initial transition sphere and the goal is to prove that $\Gamma_0 := \{\boldsymbol{x}\in\Omega : |\boldsymbol{x}| = \rho_0\}$ moves by mean curvature flow in the singular limit $\varepsilon\to 0^+$. It is well-known that if Γ_0 is a sphere in $\mathbb{R}^n$ of radius ρ_0 which evolves by mean curvature, then it remains a sphere and shrinks into a point in a finite time; precisely, at time t we have

$$\Gamma_t := \{\boldsymbol{x}\in\Omega : |\boldsymbol{x}| = \rho(t)\},$$

where ρ satisfies

$$\rho' = -\frac{n-1}{\rho}, \qquad \rho(0)=\rho_0, \tag{16}$$

and, as a consequence,

$$\rho(t) = \sqrt{\rho_0^2 - 2(n-1)t}, \qquad t\in[0,\rho_0^2/2(n-1)].$$

In [13] we prove that the motion of the interface is governed by the law (16) in the case of the IBVP (12)-(13)-(14) for well-prepared initial data (in particular, w^ε_0 satisfies (15)) by using an energy approach introduced in [3] to study the propagation of a transition sphere in the case of the classic Allen–Cahn equation (4). To do this, we must require that the parameter τ depends on ε and goes to 0 as $\varepsilon\to 0^+$; precisely, we assume that there exists a positive number $\mu\ll 1$ such that

$$\tau(\varepsilon) = o(\varepsilon^\mu). \tag{17}$$

However, we believe that such condition on the smallness of τ is indeed technical, as confirmed by numerical evidence in [13, Section 1].

Introduce the new variable $R = r-\rho(t)$ and define

$$v^\varepsilon(R,t) := w^\varepsilon(R+\rho(t),t), \qquad \text{or, equivalently} \qquad w^\varepsilon(r,t) = v^\varepsilon(r-\rho(t),t). \tag{18}$$

The function v^ε is defined in $[-\rho(t), 1-\rho(t)]\times[0,T]$ for some $T>0$, and we want to choose $\rho=\rho(t)$ in a way such that v^ε has a transition at $R=0$ as $\varepsilon\to 0^+$, namely

$$\lim_{\varepsilon\to 0^+} v^\varepsilon(R,t) = \begin{cases} -1, & R<0, \\ +1, & R>0. \end{cases} \tag{19}$$

If (19) is satisfied, then the function $\rho = \rho(t)$ describes the propagation of the interface for w^ε. Using the change of variables (18), we deduce that v^ε satisfies

$$\varepsilon^2\tau v^\varepsilon_{tt} - 2\varepsilon^2\tau\rho' v^\varepsilon_{tR} + v^\varepsilon_t = \left(1-\varepsilon^2\tau(\rho')^2\right) v^\varepsilon_{RR} + \left(\varepsilon^2\tau\rho'' + \rho' + \frac{n-1}{R+\rho}\right) v^\varepsilon_R - \varepsilon^{-2}F'(v^\varepsilon). \tag{20}$$

Inspired by [3], we shall rewrite the first two terms of the right-hand side of (20) in weigthed divergence form, by introducing an appropriate integrating factor. To this aim, the ODE (16) for ρ in our hyperbolic model is replaced by

$$\varepsilon^2\tau\rho'' + \rho' = -\frac{n-1}{\rho}. \tag{21}$$

Equation (20) with ρ satisfying (21) can be rewritten in the form

$$\varepsilon^2\tau v^\varepsilon_{tt} - 2\varepsilon^2\tau\rho' v^\varepsilon_{tR} + v^\varepsilon_t = \left(1-\varepsilon^2\tau(\rho')^2\right)\frac{(\phi^\varepsilon v^\varepsilon_R)_R}{\phi^\varepsilon} - \varepsilon^{-2}F'(v^\varepsilon), \tag{22}$$

where ϕ^ε is the aforementioned integrating factor. In [13], we study the dynamics of the solutions to (22), for well-prepared initial data and τ satisfying (17), and prove that v^ε satisfies (19). Coming back to the original variables, we have that

$$\lim_{\varepsilon\to 0}\int_0^T\int_0^1 \left|w^\varepsilon(r,t) - w^0(r,t)\right| r^{n-1}\,dr\,dt = 0,$$

for any $T \in (0, \rho_0^2/2(n-1))$, where

$$w^0(r,t) := \begin{cases} -1, & r < \rho(t), \\ +1, & r > \rho(t), \end{cases} \qquad \text{with } \rho(t) = \sqrt{\rho_0^2 - 2(n-1)t}\,.$$

Therefore, concerning equation (2), we rigorously prove that the motion of a transition sphere is governed by the law (16) in the singular limit $\varepsilon \to 0^+$, and so, in the radial case the interface moves by mean curvature flow. However, in order to rigorously describe the dynamics of the solutions to (11), we used the ODE (21), which takes into account the inertial term $\varepsilon^2\tau\rho''$, and it is different from (16) as long as ε is (small, but) strictly positive.

REFERENCES

[1] S. Allen and J. Cahn, A microscopic theory for antiphase boundary motion and its application to antiphase domain coarsening, *Acta Metall.*, **27** (1979), 1085–1095.

[2] L. Bronsard and R. Kohn, On the slowness of phase boundary motion in one space dimension, *Comm. Pure Appl. Math.*, **43** (1990), 983–997.

[3] L. Bronsard and R. Kohn, Motion by mean curvature as the singular limit of Ginzburg–Landau dynamics, *J. Differential Equations*, **90** (1991), 211–237.

[4] J. Carr and R. L. Pego, Metastable patterns in solutions of $u_t = \varepsilon^2 u_{xx} - f(u)$, *Comm. Pure Appl. Math.*, **42** (1989), 523–576.

[5] C. Cattaneo, Sulla conduzione del calore, *Atti del Semin. Mat. e Fis. Univ. Modena*, **3** (1948), 83–101.

[6] X. Chen, Generation and propagation of interfaces for reaction-diffusion equations, *J. Differential Equations*, **96** (1992), 116–141.

[7] X. Chen, Generation, propagation, and annihilation of metastable patterns, *J. Differential Equations*, **206** (2004), 399–437.

[8] P. de Mottoni and M. Schatzman, Geometrical evolution of developed interfaces, *Trans. Amer. Math. Soc.*, **347** (1995), 1533–1589.

[9] R. Folino, Slow motion for a hyperbolic variation of Allen–Cahn equation in one space dimension, *J. Hyperbolic Differ. Equ.*, **14** (2017), 1–26.

[10] R. Folino, Metastability for hyperbolic variations of Allen–Cahn equation, in *Theory, Numerics and Applications of Hyperbolic Problems I. HYP 2016. Springer Proceedings in Mathematics & Statistics* (eds. C. Klingenberg and M. Westdickenberg), vol 236. Springer, Cham, (2018), 551–563.

[11] R. Folino, Slow motion for one-dimensional nonlinear damped hyperbolic Allen–Cahn systems, preprint, `arXiv:1612.03203`.

[12] R. Folino, C. Lattanzio and C. Mascia, Metastable dynamics for hyperbolic variations of the Allen–Cahn equation, *Commun. Math. Sci.*, **15** (2017), 2055–2085.

[13] R. Folino, C. Lattanzio and C. Mascia, Motion of interfaces for a damped hyperbolic Allen–Cahn equation, preprint, `arXiv:1802.05038`.

[14] G. Fusco and J. Hale, Slow-motion manifolds, dormant instability, and singular perturbations, *J. Dynamics Differential Equations*, **1** (1989), 75–94.

[15] T. Hillen, Qualitative analysis of semilinear Cattaneo equations, *Math. Models and Methods Appl. Sci.*, **8** (1998), 507–519.

E-mail address: `raffaele.folino@univaq.it`
E-mail address: `corrado@univaq.it`
E-mail address: `mascia@mat.uniroma1.it`

MODEL ADAPTATION OF CHEMICALLY REACTING FLOWS BASED ON A POSTERIORI ERROR ESTIMATES

Jan Giesselmann and Hrishikesh Joshi*

Numerical Analysis and Scientific Computing
TU-Darmstadt,
Dolivostraße 15, Darmstadt, 64293, Germany

Abstract. Many physical phenomena can be described by using models of various levels of complexity. The more complex the model the higher the level of details it accounts for, but as a result it is more expensive to simulate. The computational expenses can be saved by decomposing the computational domain and solving the simple model where sufficient and the complex model everywhere else. This paper is concerned with model adaptation based on domain decomposition for systems of hyperbolic partial differential equations with stiff source terms. To this end, we derive a posteriori estimates, i.e an upper bound for the L_2 distance between the numerical solution to the simple system and the exact solution to the complex system. We also account for discretization errors, which enables mesh and model adaptation.

1. **Introduction.** Chemically reacting flows are of interest in many industrial applications such as simulation of combustion in engines, electrochemistry in batteries and manufacturing processes in pharmaceutical and chemical industry. Chemically reacting flows are extremely costly to simulate due to the interaction between various mechanisms like convection and reaction. Chemically reacting flows can also have large system sizes, due to the large number of constituents being present. At chemical equilibrium, the reaction terms vanish and the partial densities depend on each other by algebraic relations, simplifying the governing equations and hence significantly reducing the computational costs. The a posteriori error analysis presented in this paper provides computable bounds for the L_2 distance between the numerical solution to the chemical-equilibrium system and the exact solution to the chemical non-equilibrium system. This is crucial in devising model adaptive schemes.

Previously proposed model adaptive algorithms were based on dual-weighted residuals (see [3]) and Chapman-Enskog expansions (see [9]). We present error estimators based on the relative entropy framework which uses the weak-strong stability of entropy admissible weak solutions to systems of hyperbolic balance laws. The error estimates are inspired from the analysis of hyperbolic relaxation systems in [12] and [10].

The paper is organised as follows: firstly, the modelling of chemically reacting flows is discussed in Section 2, followed by its abstract form in Section 3. Finally,

2000 *Mathematics Subject Classification.* Primary: 58F15, 58F17; Secondary: 53C35.

Key words and phrases. A posteriori error estimates, Hyperbolic balance laws, Relative entropy, Model adaptation.

* Corresponding author: Hrishikesh Joshi.

the a posteriori error analysis is described in Section 4, in which first a brief description of the reconstruction methodology is given, then the error estimates and the coupling between the two models is discussed.

2. **Chemically reacting flows.**

In this section, a class of models describing chemically reacting flows and its properties are discussed.
Consider the following set of $M \in \mathbb{N}$ chemical reactions for $N_c \in \mathbb{N}$ constituents

$$\sum_{i=1}^{N_c} \alpha_i^j A_i \rightleftharpoons \sum_{i=1}^{N_c} \beta_i^j A_i \quad j = 1, \dots, M\,,$$

where $\alpha_i^j, \beta_i^j \in \mathbb{N}$ are the stoichiometric coefficients and A_i the constituents.
Neglecting heat conduction and viscosity, chemically reacting flows can be modelled by

$$\begin{aligned} \partial_t \rho_i + \nabla \cdot (\rho_i \mathbf{v}) &= M_i \left(\beta_i - \alpha_i\right) R_i, \ i = 1, \dots, N_c \\ \partial_t(\rho \mathbf{v}) + \nabla \cdot (\rho \mathbf{v} \otimes \mathbf{v}) + \nabla p &= 0, \\ \partial_t(\rho e) + \nabla \cdot ((p + \rho e)\mathbf{v}) &= 0, \end{aligned} \tag{1}$$

where $\mathbf{v}$ is the velocity, ρ_j are the partial densities, ρe is the total energy and the total density is defined as $\rho = \sum_{i=1}^{N_c} \rho_i$. Further, we assume ideal gas mixtures, hence the pressure is given by $p = RT \sum_{i=1}^{N_c} \frac{\rho_i}{M_i}$, where R is the universal gas constant, M_i are the molecular masses and T is the temperature.
The temperature can be calculated from the total energy, which is defined as

$$\rho e = \sum_{i=1}^{N_c} \rho_i c_{v,i} T + \sum_{i=1}^{N_c} \rho_i h_{f,i} + \frac{1}{2}\rho |\mathbf{v}|^2. \tag{2}$$

Here, $c_{v,i} > 0$ are the specific heats at constant volume and $h_{f,i} \in \mathbb{R}$ are the enthalpies of formation. The reader is referred to Section 9 in [2] for more details.

Chemically reacting flows may have various dissipative mechanisms such as heat conduction, radiation, diffusion and reaction. These dissipative mechanisms drive the system to chemical and mechanical equilibrium and as a result to thermodynamic equilibrium. Entropy of the system tells us how far this process has advanced and each of these mechanisms leads to a positive entropy production as known from the second law of thermodynamics, see [11]. This entropic structure enables us to derive computable a posteriori error estimates to carry out model adaptation. For the sake of simplicity we only consider reactions. Other dissipative mechanisms can be incorporated in the presented analysis in a similar fashion.

3. **Abstract form.**

3.1. **Balance laws.**

The system of governing equations describing chemically reacting flows (1) can be cast in an abstract form as

$$\partial_t \mathbf{U} + \sum_{\alpha} \partial_{x_\alpha} \mathbf{F}_\alpha(U) = \frac{1}{\varepsilon} \mathbf{R}(\mathbf{U}), \mathbf{U} : \mathbb{R}^d \times \mathbb{R}^+ \to \mathbb{R}^N, \tag{3}$$

where $\varepsilon > 0$. From hereon system (3) is referred to as the complex system.
Projecting the complex system using some projection matrix $\mathbb{P} : \mathbb{R}^N \to \mathbb{R}^n$ such that

$$\mathbb{P}\mathbf{R}(\mathbf{U}) = 0 \tag{4}$$

and $\mathbf{u} := \mathbb{P}\mathbf{U}$, we get

$$\partial_t \mathbf{u} + \sum_\alpha \partial_{x_\alpha} \mathbb{P}\mathbf{F}_\alpha(\mathbf{U}) = 0, \ \ \mathbf{u} : \mathbb{R}^d \times \mathbb{R}^+ \to \mathbb{R}^n. \tag{5}$$

Further, the equilibrium solutions are parametrized by so called Maxwellian, $M(\mathbf{u})$, such that

$$\mathbf{U}_{eq} = M(\mathbf{u}). \tag{6}$$

Hence, in the limit $\varepsilon \to 0$, (5) reduces to

$$\partial_t \mathbf{u} + \sum_\alpha \partial_{x_\alpha} \mathbb{P}\mathbf{F}_\alpha(M(\mathbf{u})) = 0. \tag{7}$$

From hereon system (7) is referred to as the simple system.
Our objective is to carry out model adaptation between the simple system (7) and the complex system (3).

The relaxation system is equipped with a convex entropy-entropy flux pair $(H(\mathbf{U}), Q(\mathbf{U}))$ satisfying,

$$\mathrm{D}\, H(\mathbf{U})\, \mathrm{D}\, \mathbf{F}(\mathbf{U}) = \mathrm{D}\, Q(\mathbf{U}), \tag{8}$$

so that smooth solutions of (3) satisfy

$$\partial_t H(\mathbf{U}) + \sum_\alpha \partial_{x_\alpha} Q_\alpha(\mathbf{U}) = \frac{1}{\varepsilon} \frac{\partial H(\mathbf{U})}{\partial \mathbf{U}} \cdot \mathbf{R}(\mathbf{U}) \leq 0. \tag{9}$$

The above inequality implies that the system dissipates entropy.

Furthermore, the Maxwellian induces an entropy-entropy flux pair for the equilibrium system via $\eta(\mathbf{u}) := H(M(\mathbf{u}))$, $q_\alpha := Q_\alpha(M(\mathbf{u}))$, smooth solutions of which satisfy

$$\partial_t \eta(\mathbf{u}) + \sum_\alpha \partial_{x_\alpha} q_\alpha(\mathbf{u}) = 0. \tag{10}$$

The error estimates we present are applicable to all systems having the structure described above. For the application under consideration the model adaptation will be carried out between the complex system (3), i.e. the chemical non-equilibrium system and the simple system (7), i.e the chemical equilibrium system. In this case, ε corresponds to the ratio of time scales of reaction and convection, and the system size is $N = N_c + d + 1$, where d is the number of spatial dimensions. As the reaction rates increase ε tends to zero, the reaction terms vanish and the system reaches equilibrium. The Maxwellian can be calculated from the conditions required for chemical equilibrium.

3.2. **A primer on the relative entropy framework.** In this section we briefly discuss the nature of the solutions of hyperbolic systems of balance laws.

3.2.1. *Entropy admissible weak solutions.*
It is well known that weak solutions to hyperbolic systems of conservation laws are non-unique. Hence, we look for solutions that satisfy an additional entropy admissibility condition. Scalar problems have unique entropy solutions, whereas entropy solutions to systems of conservation laws in two or more dimensions are not unique, see [5]. But, even in this case, a weak-strong uniqueness principle holds, which can be proven based on the relative entropy framework.

Definition 3.1. The relative entropy and relative entropy flux between states $\mathbf{u}$ and $\mathbf{v}$ is defined as

$$H(\mathbf{u}|\mathbf{v}) = H(\mathbf{u}) - H(\mathbf{v}) - \frac{\partial H}{\partial \mathbf{u}}(\mathbf{v})(\mathbf{u}-\mathbf{v}), \tag{11}$$

$$Q(\mathbf{u}|\mathbf{v}) = Q(\mathbf{u}) - Q(\mathbf{v}) - \frac{\partial H}{\partial \mathbf{u}}(\mathbf{v})(\mathbf{F}(\mathbf{u}) - \mathbf{F}(\mathbf{v})) \tag{12}$$

respectively.

3.2.2. *Weak-strong stability.*
Weak-strong stability implies uniqueness of entropy admissible weak solution as long as a Lipschitz continuous solution to the system exists. Next, weak-strong stability is described, which forms the basis of the relative entropy framework employed to derive the error estimates. For general background on hyperbolic conservation laws the reader is referred to [4].

Theorem 3.2. *Let $\bar{\mathbf{u}} : \mathbb{R}^d \times \mathbb{R}^+ \to \mathbb{R}^n$ be a smooth solution of the equilibrium system (7), with initial data $\bar{\mathbf{u}}_0$. Let $\mathbf{U}^\varepsilon : \mathbb{R}^d \times \mathbb{R}^+ \to \mathbb{R}^N$ be a family of entropy admissible weak solutions of the complex system (3) with initial data $\mathbf{U}_0^\varepsilon$, and the relative entropy $H_r(x,t) := H(\mathbf{U}^\epsilon(x,t)|M(\bar{\mathbf{u}}(x,t)))$, then there exist constants $C > 0$ and $s > 0$ independent of ε so that for any $\mathcal{R} > 0$*

$$\int_{|x|<\mathcal{R}} H_r(x,t)\,\mathrm{d}x \le C\left(\int_{|x|\le\mathcal{R}+st} H_r(x,0)\,\mathrm{d}x + \varepsilon\right), \text{a.e. } t \in [0,T). \tag{13}$$

Moreover, if

$$\int_{|x|<\mathcal{R}+sT} H_r(x,0)\,\mathrm{d}x \to 0 \text{ as } \varepsilon \downarrow 0, \tag{14}$$

then

$$\operatorname*{ess\,sup}_{t\in[0,T)} \int_{|x|<\mathcal{R}} |\mathbf{U}^\varepsilon - M(\bar{\mathbf{u}})|^2\,\mathrm{d}x \to \text{ as } \varepsilon \downarrow 0. \tag{15}$$

For technical details please refer to [10].

4. **A posteriori error analysis.** Solutions to hyperbolic conservation laws may develop discontinuities in finite time, even for smooth initial data. To exploit the weak-strong stability we need one of the solutions to be Lipschitz continuous. As the exact solution to the system can be discontinuous, we introduce an intermediate quantity, a Lipschitz continuous reconstruction of the numerical solution, and proceed to bound the error between the reconstruction and the exact solution. We outline the derivation of the computable error estimates in this section.

4.1. **Error splitting.** We need to bound the distance between the numerical solution to (7) and the exact solution to (3). To this end, the triangle inequality between the exact solution to (3), the numerical solutions and the reconstructions is employed. Let $\mathbf{U}$ be the exact solution to (3), $\mathbf{U}_h$ be some numerical solution to (3), $\hat{\mathbf{U}}_h$ its reconstruction, $\mathbf{u}_h$ be some numerical solution to (7) and $\hat{\mathbf{u}}_h$ its reconstruction. Employing the triangle inequality, we get

$$||\mathbf{U} - M(\mathbf{u}_h)|| \le ||\mathbf{U} - M(\hat{\mathbf{u}}_h)|| + ||M(\hat{\mathbf{u}}_h) - M(\mathbf{u}_h)||. \tag{16}$$

The first term in the above inequality will be bounded by the relative entropy framework and the second term can be explicitly computed.

4.2. **Computational domain decomposition.** A tolerance can be set for the error estimate, which gives the distance between the numerical solution to the simple system and the exact solution to the complex system. Then, the computational domain can be decomposed to solve the simple system where determined sufficient and the complex system everywhere else. As a result, an interface is introduced where the simple system is solved on one side of the interface and the complex system on the other side. To describe the features of the a posteriori error estimates we restrict ourselves to 1D and assume that the simple system is solved for $x \in \mathbb{R}^-$ and the complex system for $x \in \mathbb{R}^+$.

In accordance with the domain decomposition, the relative entropy and entropy fluxes are defined as $H_r^+ := H\left(\mathbf{U}|\hat{\mathbf{U}}_h\right)$, $H_r^- := H\left(\mathbf{U}|M\left(\hat{\mathbf{u}}_h\right)\right)$, and $Q_r^+ := Q\left(\mathbf{U}|\hat{\mathbf{U}}_h\right)$, $Q_r^- := Q\left(\mathbf{U}|M\left(\hat{\mathbf{u}}_h\right)\right)$ on $\mathbb{R}^+$ and $\mathbb{R}^-$ respectively.

Remark 1. Strict convexity of H implies that, for some $c^+ > 0,\ c^- > 0$,

$$H_r^- \geq c^- \left|\mathbf{U} - M(\hat{\mathbf{u}}_h)\right|^2,\ H_r^+ \geq c^+ \left|\mathbf{U} - \hat{\mathbf{U}}_h\right|^2. \tag{17}$$

4.3. **Reconstruction.** In case of numerical solutions computed by Runge-Kutta Discontinuous Galerkin schemes, [7] and [8] explain how to define reconstructions that are computable and Lipschitz continuous in space and time.

The reconstruction of the numerical solution satisfies a perturbed system of partial differential equations. The reconstruction of the numerical solution to the complex system, $\hat{\mathbf{U}}_h$ for $x \in \mathbb{R}^+$, satisfies

$$\partial_t \hat{\mathbf{U}}_h + \partial_x \mathbf{F}(\hat{\mathbf{U}}_h) - \frac{1}{\varepsilon}\mathbf{R}(\hat{\mathbf{U}}_h) =: r_2,\ \hat{\mathbf{U}}_h : \mathbb{R}^+ \times \mathbb{R}^+ \to \mathbb{R}^N, \tag{18}$$

where r_2 is the residual in the complex system.
Similarly, the reconstruction to the simple system, $\hat{\mathbf{u}}_h$ for $x \in \mathbb{R}^-$, satisfies

$$\partial_t \hat{\mathbf{u}}_h + \partial_x \mathbb{P}\mathbf{F}(M(\hat{\mathbf{u}}_h)) =: \mathbb{P}r_1,\ \hat{\mathbf{u}}_h : \mathbb{R}^- \times \mathbb{R}^+ \to \mathbb{R}^n, \tag{19}$$

where $\mathbb{P}r_1$ is the residual in the simple system.

The residuals are related to the discretization errors and can be used for mesh adaptation.

4.4. **A posteriori error estimates.** The L_2 distance between the reconstruction and the exact solution can be bounded by the results presented in the following theorem. The proof of which will be given in [6].

Theorem 4.1. *Let $\mathbf{U} : \mathbb{R} \times \mathbb{R}^+ \to \mathbb{R}^N$ be the exact solution to (3). Let $\hat{\mathbf{U}}$ be the Lipschitz reconstruction of the numerical solution of (3) on $x \in \mathbb{R}^-$ and let $\hat{\mathbf{u}}$ be the Lipschitz reconstruction of the numerical solution of (7) on $x \in \mathbb{R}^+$, then assuming for some $\nu = \nu(M)$*

$$-\left(\frac{\partial H}{\partial \mathbf{U}}(\mathbf{U}) - \frac{\partial H}{\partial \mathbf{U}}\left(M\left(\mathbb{P}\mathbf{U}\right)\right)\right) \cdot \left(\mathbf{R}(\mathbf{U}) - \mathbf{R}\left(M\left(\mathbb{P}\mathbf{U}\right)\right)\right) \geq \nu \left|\mathbf{U} - M(\mathbb{P}\mathbf{U})\right|^2 \tag{20}$$

for $\mathbf{U} \in \mathbb{R}^N$. Furthermore, for any $\mathbf{U}, \mathbf{V} \in \mathbb{R}^N$

$$-\left(\frac{\partial H}{\partial \mathbf{U}}(\mathbf{U}) - \frac{\partial H}{\partial \mathbf{U}}(\mathbf{V})\right) \cdot \left(\mathbf{R}(\mathbf{U}) - \mathbf{R}(\mathbf{V})\right) \geq 0. \tag{21}$$

Therefore, we have

$$\int_{\mathbb{R}^+} \left|\mathbf{U} - \hat{\mathbf{U}}_h\right|^2 \mathrm{d}x + \int_{\mathbb{R}^-} |\mathbf{U} - M(\hat{\mathbf{u}}_h)|^2 \mathrm{d}x \leq (I + D_c + D_s + M_s + C_Q) \exp\left(\frac{\max(C_c + M_c, C_s + 1 + |\mathbb{P}|)}{\min(c^-, c^+)} t\right), \tag{22}$$

where

$$I = \int_{\mathbb{R}^-} H_r^-(x,0)\, dx + \int_{\mathbb{R}^+} H_r^+(x,0)\, \mathrm{d}x,$$

$$D_c = \int_0^t \int_{\mathbb{R}^+} \left|\nabla^2_{\mathbf{U}} H(\hat{\mathbf{U}}_h) r_2\right|^2 \mathrm{d}x\, \mathrm{d}\tau,$$

$$D_s = \int_0^t \int_{\mathbb{R}^-} \left|\nabla^2_{\mathbf{u}} \eta(\hat{\mathbf{u}}_h) \mathbb{P} r_1\right|^2 \mathrm{d}x\, \mathrm{d}\tau,$$

$$M_s = \tfrac{\varepsilon}{\nu} \int_0^t \int_{\mathbb{R}^-} \left|\partial_x \left(\nabla_{\mathbf{u}} \eta(\hat{\mathbf{u}}_h)\right) * \mathbb{P} \nabla_{\mathbf{U}} \mathbf{F}(M(\hat{\mathbf{u}}_h))\right|^2 \mathrm{d}x\, \mathrm{d}\tau,$$

$$C_Q = \int_0^t Q_r^+(0,\tau) d\tau - \int_0^t Q_r^-(0,\tau)\, \mathrm{d}\tau$$

$$C_c = \left|\left|\partial_x \left(\nabla_{\mathbf{U}} H(\hat{\mathbf{U}}_h)\right) \nabla^2_{\mathbf{U}} \mathbf{F}(\hat{\mathbf{U}}_h)\right|\right|_\infty,$$

$$M_c = \left|\left|\tfrac{1}{\varepsilon} \mathbf{R}(\hat{\mathbf{U}}_h) \nabla^3_{\mathbf{U}} H(\hat{\mathbf{U}}_h)\right|\right|_\infty,$$

$$C_s = \left|\left|\partial_x \left(\nabla_{\mathbf{u}} (\eta(\hat{\mathbf{u}}_h))\right) \nabla^2_{\mathbf{u}} \left(g(\hat{\mathbf{u}}_h)\right)\right|\right|_\infty.$$

The terms D_c and D_s indicate the discretization errors in the numerical solution of the complex and the simple system. Furthermore, the terms C_c and C_s indicate the stability of the perturbed differential equations. The term M_s indicates the modelling error incurred due to solving the simple system instead of the complex system on $\mathbb{R}^-$ and the term I indicates the error incurred due to discrete initialization. C_Q arises due to the use of different models across the interface. Note that here all terms except C_Q are a posteriori computable.

Now, from (12), we know

$$\begin{aligned} \int_0^t Q_r^+(0,\tau)\, \mathrm{d}\tau - \int_0^t Q_r^-(0,\tau)\, \mathrm{d}\tau &= \int_0^t Q\left(M\left(\hat{\mathbf{u}}_h(0,\tau)\right)\right) - Q\left(\hat{\mathbf{U}}_h(0,\tau)\right) \mathrm{d}\tau \\ &+ \int_0^t \frac{\partial H}{\partial \mathbf{U}}\left(M\left(\hat{\mathbf{u}}_h(0,\tau)\right)\right) \mathbf{F}\left(M\left(\hat{\mathbf{u}}_h(0,\tau)\right)\right) - \frac{\partial H}{\partial \mathbf{U}}\left(\hat{\mathbf{U}}_h(0,\tau)\right) \mathbf{F}\left(\hat{\mathbf{U}}_h(0,\tau)\right) \mathrm{d}\tau \\ &+ \int_0^t \mathbf{F}(\mathbf{U}) \left(\frac{\partial H}{\partial \mathbf{U}}(M(\hat{\mathbf{u}}_h(0,\tau))) - \frac{\partial H}{\partial \mathbf{U}}(\hat{\mathbf{U}}_h(0,\tau))\right) \mathrm{d}\tau \end{aligned} \tag{23}$$

Hence, if the reconstructions satisfy

$$M(\hat{\mathbf{u}}_h(0,\tau)) = \hat{\mathbf{U}}_h(0,\tau) \tag{24}$$

at the interface then C_Q vanishes and the error estimator can be explicitly computed. Note that (24) only specifies a condition for the reconstruction. When numerically solving the coupled systems, to ensure that no numerical artefacts such as acoustic waves are introduced special coupling conditions need to be employed. For more on this the reader is referred to [1].

Numerical experiments show (in [6]) that the discretization terms vanish as the mesh width tends to zero. The error estimate accounts for modelling and discretization errors, hence, it allows for in situ model and mesh adaptation.

Acknowledgments. This research is supported by the German Research Foundation (DFG) grant GI1131/1-1: Dynamical, spatially heterogeneous model adaptation in compressible flows.

REFERENCES

[1] A. Ambroso, C. Chalons, F. Coquel, E. Godlewski, F. Lagoutière, P.A. Raviart and N. Seguin, The coupling of homogenous models for two-phase flows, *Int. J. Finite Vol.*, **4** (2007).

[2] D. Bothe and W. Dreyer, Continuum thermodynamics of chemically reacting fluid mixtures, *Acta Mech.*, **226** (2015), 1757–1805.

[3] M. Braack, A. Ern, A posteriori control of modeling errors and discretization errors, *Multiscale Model. Simul.*, **1** (2003), 221–238.

[4] C.M. Dafermos, *Hyperbolic Conservation Laws in Continuum Physics*, 3rd edition, Grundlehren Math. Wiss. 325, Springer-Verlag, 2010.

[5] C. De Lellis and L. Székelyhidi On admissibility criteria for weak solutions of the Euler equations, *Arch. Ration. Mech. Anal.*, **95** (2010), 225–260.

[6] J. Giesselmann, H. Joshi, A posteriori error analysis for model adaptation of hyperbolic systems with relaxation, *in preparation.*

[7] J. Giesselmann, C. Makridakis, T. Pryer, A posteriori analysis of discontinuous Galerkin schemes for systems of hyperbolic conservation laws, *SIAM J. Numer. Anal.*, **53** (2015), 1280–1303.

[8] J. Giesselmann and T. Pryer, A posteriori analysis for dynamic model adaptation problems in convection dominated problems, *Math. Models Methods Appl. Sci.*, **27** (2017), 2381–2423.

[9] H. Mathis, C. Cancés, E. Godlewski and N Seguin, Dynamic model adaptation for multiscale simulation of hyperbolic systems with relaxation, *J. Sci. Comput.*, **63** (2015), 820–861.

[10] A. Miroshnikov and K. Trivisa, Relative entropy in hyperbolic relaxation for balance laws, *Commun. Math. Sci.*, **12** (2014), 1017–1043.

[11] I. Müller and W. H. Müller, *Fundamentals of thermodynamics and applications*, 1st edition, Springer-Verlag, 2009.

[12] A. Tzavaras, Relative entropy in hyperbolic relaxation, *Commun. Math. Sci*, **3** (2005), 119–132.

E-mail address: `giesselmann@mathematik.tu-darmstadt.de`
E-mail address: `joshi@mathematik.tu-darmstadt.de`

AN A POSTERIORI ERROR ANALYSIS BASED ON NON-INTRUSIVE SPECTRAL PROJECTIONS FOR SYSTEMS OF RANDOM CONSERVATION LAWS

JAN GIESSELMANN

Department of Mathematics
TU Darmstadt
Dolivostraße 15, 64293 Darmstadt, Germany

FABIAN MEYER* AND CHRISTIAN ROHDE

Institute of Applied Analysis and Numerical Simulation
University of Stuttgart
Pfaffenwaldring 57, 70569 Stuttgart, Germany

ABSTRACT. We present an a posteriori error analysis for one-dimensional random hyperbolic systems of conservation laws. For the discretization of the random space we consider the Non-Intrusive Spectral Projection method, the spatio-temporal discretization uses the Runge–Kutta Discontinuous Galerkin method. We derive an a posteriori error estimator using smooth reconstructions of the numerical solution, which combined with the relative entropy stability framework yields computable error bounds for the space-stochastic discretization error. Moreover, we show that the estimator admits a splitting into a stochastic and deterministic part.

1. **Introduction.** In this contribution we study numerical schemes for spatially one-dimensional systems of random hyperbolic conservation laws, where the uncertainty stems from random initial data. The random space is discretized using the Non-Intrusive Spectral Projection (NISP) method which is based on discrete orthogonal projections, cf. [8]. The resulting deterministic equations are discretized by a Runge–Kutta Discontinuous Galerkin (RKDG) method [2]. We reconstruct the numerical solutions based on reconstructions for determinstic problems suggested in [4], see also [7] for their use in Stochastic Galerkin schemes. Based on these reconstructions and using the relative entropy framework, cf. [3, Section 5.2], we derive an a posteriori error bound for the difference between the exact solution of the random hyperbolic conservation law and its numerical approximation. We show that the corresponding residual admits a decomposition into three parts: A spatial part, a stochastic part, and a part which quantifies the quadrature error introduced

2000 *Mathematics Subject Classification.* Primary: 35L65, 35R60; Secondary: 65M15, 65M60, 65M70.

Key words and phrases. hyperbolic conservation laws, random pdes, a posteriori error estimates, non-intrusive spectral projection method, discontinuous Galerkin method.

The first author thanks the German Research Foundation (DFG) for support of the project via DFG grant GI1131/1-1. The second and last author thank the Baden-Württemberg Stiftung for support via the project "BW-HPC2: SEAL".

* Corresponding author.

by the discrete orthogonal projection. This decomposition paves the way for novel residual-based adaptive numerical schemes.

The article is structured as follows: In Section 2 we describe the problem of interest. In Section 3 the NISP and RKDG method is reviewed and we show how to obtain the reconstruction from our numerical solution. Section 4 presents our main a posteriori error estimate with decomposition of the residual.

2. **Statement of the Problem.** Let $(\Omega, \mathcal{F}, \mathbb{P})$ be a probability space, where Ω is the set of all elementary events $\omega \in \Omega$, $\mathcal{F}$ is a σ-algebra on Ω and $\mathbb{P}$ is a probability measure. We consider uncertainties parametrized by a random variable $\xi : \Omega \to \Xi \subset \mathbb{R}$ with probability density function $w_\xi : \Xi \to \mathbb{R}_+$. The random variable induces a probability measure $\tilde{\mathbb{P}}(B) := \mathbb{P}(\xi^{-1}(B))$ for all $B \in \mathcal{B}(\Xi)$ on the measurable space $(\Xi, \mathcal{B}(\Xi))$, where $\mathcal{B}(\Xi)$ is the corresponding Borel σ-algebra. This measure is called the law of ξ and in the following we work on the image probability space $(\Xi, \mathcal{B}(\Xi), \tilde{\mathbb{P}})$. For a second measurable space $(E, \mathcal{B}(E))$, we consider the weighted L^p_ξ-spaces equipped with the norm

$$\|f\|_{L^p_\xi(\Xi;E)} := \begin{cases} \left(\int\limits_\Xi \|f(y)\|_E^p \, w_\xi(y) \mathrm{d}y\right)^{1/p} = \mathbb{E}\left(\|f\|_E^p\right)^{1/p}, & 1 \le p < \infty \\ \operatorname{ess\,sup}_{y\in\Xi} \|f(y)\|_E, & p = \infty. \end{cases}$$

Our problem of interest is the following initial value problem for an one dimensional system of $m \in \mathbb{N}$ random conservation laws, i.e.,

$$\begin{cases} \partial_t u(t,x,y) + \partial_x F(u(t,x,y)) = 0, & (t,x,y) \in (0,T) \times \mathbb{R} \times \Xi, \\ u(0,x,y) = u^0(x,y), & (x,y) \in \mathbb{R} \times \Xi. \end{cases} \tag{RIVP}$$

Here, $u(t,x,y) \in \mathcal{U} \subset \mathbb{R}^m$ is the vector of conserved unknown quantities, $F \in C^2(\mathcal{U};\mathbb{R}^m)$, is the flux function, u^0 is the uncertain initial condition, $\mathcal{U} \subset \mathbb{R}^m$ is the state space, which is assumed to be an open set and $T \in (0,\infty)$ describes the final time. We assume that (RIVP) is strictly hyperbolic, i.e. its Jacobian $\mathrm{D}\,F(u)$ has m distinct real eigenvalues.

We say that $(\eta, q) \in C^2(\mathcal{U};\mathbb{R})$ forms an entropy/entropy-flux pair if η is strictly convex and if η and q satisfy $\mathrm{D}\,\eta\,\mathrm{D}\,F = \mathrm{D}\,q$. We assume that the random conservation law (RIVP) is equipped with at least one entropy/entropy-flux pair. Following the definition in [9] for scalar problems, we call $u \in L^1_\xi(\Xi; L^1((0,T)\times\mathbb{R};\mathcal{U}))$ a random entropy solution of (RIVP), if $u(\cdot,\cdot,y)$ is a classical entropy solution, cf. [3, Def. 4.5.1], $\tilde{\mathbb{P}}$-a.s. $y \in \Xi$. The well-posedness of (RIVP), will not be discussed in this article but can found in [6], where existence and uniqueness of random entropy solutions for (RIVP) with random flux functions and random initial data with sufficiently small total variation is proven, based on the results of [1].

3. **Space-Time Stochastic Discretization and Reconstructions.** For the stochastic discretization of (RIVP) we use the NISP method, [8], which is based on the (generalized) polynomial chaos expansion which was introduced in [10]. Under the assumption that u is square-integrable with respect to Ξ, we expand the solution of (RIVP) into a generalized Fourier series using a suitable orthonormal basis.

Let $\{\Psi_i(\cdot)\}_{i\in\mathbb{N}} : \Xi \to \mathbb{R}$ be a $L^2_\xi(\Xi)$-orthonormal basis, i.e. for $i,j \in \mathbb{N}$ we have

$$\left\langle \Psi_i, \Psi_j \right\rangle := \mathbb{E}\left(\Psi_i \Psi_j\right) = \int\limits_\Xi \Psi_i(y)\Psi_j(y) w_\xi(y) \, \mathrm{d}y = \delta_{ij}. \tag{1}$$

Following [10], the random entropy solution u can be written as

$$u(t,x,y) = \sum_{i=0}^{\infty} u_i(t,x)\Psi_i(y), \tag{2}$$

with (deterministic) Fourier modes $u_i = u_i(t,x)$ satisfying

$$u_i(t,x) = \mathbb{E}\Big(u(t,x,\cdot)\Psi_i(\cdot)\Big) \qquad \forall i \in \mathbb{N}. \tag{3}$$

The NISP method approximates the modes in (3) via a discrete orthogonal projection, i.e., numerical quadrature. We denote $(R+1) \in \mathbb{N}$ quadrature points and weights by $\{y_l\}_{l=0}^R$, $\{w_l\}_{l=0}^R$, and approximate

$$u_i(t,x) = \int_{\Xi} u(t,x,y)\Psi_i(y) w_\xi(y)\,\mathrm{d}y \approx \sum_{l=0}^{R} u(t,x,y_l)\Psi_i(y_l)w_l =: \hat{u}_i \quad \text{for } i \in \mathbb{N}. \tag{4}$$

In a second step the NISP method truncates (2) after the M-th mode, i.e.,

$$u(t,x,y) \approx \sum_{i=0}^{M} \hat{u}_i(t,x)\Psi_i(y). \tag{5}$$

For any $l = 0,\dots,R$, the random entropy solution u of (RIVP) evaluated at quadrature point $\{y_l\}_{l=0}^R$, is denoted by $u(\cdot,\cdot,y_l)$ and it is an entropy solution of the deterministic version of (RIVP), i.e. of

$$\begin{cases} \partial_t u(t,x,y_l) + \partial_x F(u(t,x,y_l)) = 0, & (t,x) \in (0,T)\times\mathbb{R}, \\ u(0,x,y_l) = u^0(x,y_l), & x \in \mathbb{R}. \end{cases} \tag{DIVP$_l$}$$

The deterministic hyperbolic systems (DIVP)$_l$ can be discretized by a suitable numerical method. We use the RKDG method as described in [2]. We denote the corresponding numerical solution of (DIVP)$_l$ at quadrature point $\{y_l\}_{l=0}^R$ and at points $\{t_n(y_l)\}_{n=0}^{N_t(y_l)}$, $N_t(y_l) \in \mathbb{N}$, in time by $u_h^n(\cdot, y_l) \in V_p^s$, where

$$V_p^s := \{v : \mathbb{R} \to \mathbb{R}^m \mid v\,|_K \in \mathbb{P}_p(K;\mathbb{R}^m),\ K \in \mathcal{T}\},$$

is the corresponding DG space of polynomials of degree $p \in \mathbb{N}$, associated with a uniform triangulation $\mathcal{T}$ of $\mathbb{R}$. Let us assume that the time partition $\{t_n\}_{n=0}^{N_t}$ and the triangulation $\mathcal{T}$ used for (DIVP)$_l$ are the same for every quadrature point $\{y_l\}_{l=0}^R$. The numerical approximation of (RIVP) at time $t = t_n$ can then be written as

$$u_h^n(x,y) := \sum_{i=0}^{M}\Big(\sum_{l=0}^{R} u_h^n(x,y_l)\Psi_i(y_l)w_l\Big)\Psi_i(y). \tag{6}$$

The proof of the a posterior error estimate in Theorem 4.1 uses the relative entropy framework, cf. [3, Section 5.2], which requires one quantity which is at least Lipschitz continuous in space and time. To this end we reconstruct the numerical solution so that we obtain a Lipschitz continuous function. To avoid technical overhead, we do not elaborate upon this process here, but refer to [4, 7], where a detailed description can be found.

The reconstruction provides us with a computable space-time reconstruction $\hat{u}^{st}(y_l) \in W^1_\infty((0,T); V^s_{p+1} \cap C^0(\mathbb{R}))$ of the numerical solution $\{u_h^n(y_l)\}_{n=0}^{N_t} \subset V_p^s$, for each quadrature point $\{y_l\}_{l=0}^R$. This allows us to define a space-time residual as follows.

Definition 3.1 (Space-time residual). For all $l = 0, \dots, R$, we define $R^{st}(y_l) \in L^2((0,T) \times \mathbb{R}; \mathbb{R}^m)$ by

$$R^{st}(y_l) := \partial_t \hat{u}^{st}(y_l) + \partial_x F(\hat{u}^{st}(y_l)) \tag{7}$$

to be the space-time residual associated with the quadrature point y_l.

Next we define the reconstructed mode, the space-time-stochastic reconstruction and the space-time-stochastic residual. The latter is obtained by plugging the space-time-stochastic reconstruction into the random conservation law (RIVP).

Definition 3.2 (Space-time-stochastic reconstruction and residual). Let $\{\hat{u}^{st}(y_l)\}_{l=0}^R : (0,T) \times \mathbb{R} \to \mathbb{R}^m$ be the sequence of space-time reconstructions at quadrature points $\{y_l\}_{l=0}^R$. The reconstructed modes of (4) are defined as

$$\hat{u}_i^{st} := \sum_{l=0}^{R} \hat{u}^{st}(y_l) \Psi_i(y_l) w_l, \tag{8}$$

for $i = 0, \dots, M$. The space-time-stochastic reconstruction $\hat{u}^{sts} : (0,T) \times \mathbb{R} \times \Xi \to \mathbb{R}^m$ is defined as

$$\hat{u}^{sts}(t,x,y) := \sum_{i=0}^{M} \hat{u}_i^{st}(t,x) \Psi_i(y). \tag{9}$$

Finally, we define the space-time-stochastic residual $R^{sts} \in L^2_\xi(\Xi; L^2((0,T)\times\mathbb{R}; \mathbb{R}^m))$ by

$$R^{sts} := \partial_t \hat{u}^{sts} + \partial_x F(\hat{u}^{sts}). \tag{10}$$

This residual is crucial in the upcoming error analysis.

4. A Posteriori Error Estimate and Error Indicators.

Before stating the main a posterior error estimate, let us note that derivatives of the flux function and the entropy are bounded on any compcat subset $\mathcal{C}$ of the state space. These bounds enter the upper bound in Theorem 4.1. Let $\mathcal{C} \subset \mathcal{U}$ be convex and compact. Due to $F \in C^2(\mathcal{U}, \mathbb{R}^m)$ and $\eta \in C^2(\mathcal{U}, \mathbb{R}^m)$ strictly convex there exist constants $0 < C_{\overline{F}} < \infty$ and $0 < C_{\underline{\eta}} < C_{\overline{\eta}} < \infty$, s.t.

$$|v^\top HF(u)v| \le C_{\overline{F}}|v|^2, \quad C_{\underline{\eta}}|v|^2 \le v^\top H\eta(u)v \le C_{\overline{\eta}}|v|^2, \quad \forall v \in \mathbb{R}^m, \forall u \in \mathcal{C}.$$

Here HF denotes the Hessian (i.e. the tensor of second order derivatives) of the flux function and $H\eta$ the Hessian of the entropy η. We now have all ingredients together to state the following a posteriori error estimate that can be directly derived from [5].

Theorem 4.1 (A posteriori error bound for the numerical solution). *Let u be the random entropy solution of* (RIVP). *Then, for any $n = 0, \dots, N_t$, the difference* *between* $u(t_n,\cdot,\cdot)$ *and the numerical solution* u_h^n *from* (6) *satisfies*

$$\begin{aligned}
&\|u(t_n,\cdot,\cdot)-u_h^n(\cdot,\cdot)\|^2_{L^2_\xi(\Xi;L^2(\mathbb{R}))}\\
&\qquad\le 2\|\hat{u}^{sts}(t_n,\cdot,\cdot)-u_h^n(\cdot,\cdot)\|^2_{L^2_\xi(\Xi;L^2(\mathbb{R}))}\\
&\qquad + 2C_{\underline{\eta}}^{-1}\Big(\mathcal{E}^{sts}(t_n)+C_{\overline{\eta}}\mathcal{E}_0^{sts}\Big)\\
&\qquad \times \exp\Big(C_{\underline{\eta}}^{-1}\int_0^{t_n}\Big(C_{\overline{\eta}}C_{\overline{F}}\|\partial_x\hat{u}^{sts}(t,\cdot,\cdot)\|_{L^\infty_\xi(\Xi;L^\infty(\mathbb{R}))}+C_{\overline{\eta}}^2\Big)\,\mathrm{d}t\Big),
\end{aligned}$$

with

$$\begin{aligned}
\mathcal{E}^{sts}(t_n) &:= \|R^{sts}(\cdot,\cdot,\cdot)\|^2_{L^2_\xi(\Xi;L^2((0,t_n)\times\mathbb{R}))},\\
\mathcal{E}_0^{sts} &:= \|u^0(\cdot,\cdot)-\hat{u}^{sts}(0,\cdot,\cdot)\|^2_{L^2_\xi(\Xi;L^2(\mathbb{R}))}.
\end{aligned}$$

Proof. We apply [5, Lemma 5.1] path-wise in Ξ, integrate over Ξ and use Gronwall's inequality to bound $\|u(t_n,\cdot,\cdot)-\hat{u}^{sts}(t_n,\cdot,\cdot)\|_{L^2_\xi(\Xi;L^2(\mathbb{R}))}$ by the second term in the inequality. The final estimate then follows using the triangle inequality. □

In Theorem 4.1 the error between the numerical solution and the entropy solution is bounded by the error in the initial condition, the difference between the numerical solution and its reconstruction and the contribution of the space-time stochastic residual R^{sts} from (10), quantified by $\mathcal{E}^{sts}$. We would like to distinguish between errors that arise from discretizing the random space and from discretizing the physical space. Therefore, we show in Lemma 11 a splitting of the space-time-stochastic residual R^{sts} into three parts. Namely a deterministic residual, which corresponds to the spatial error when approximating $(\mathrm{DIVP})_l$ using the RKDG method, a quadrature residual that reflects the quadrature error from the discrete orthogonal projection in (4) and a stochastic cut-off error, which occurs when truncating the infinite Fourier series in (2).

Lemma 4.2 (Orthogonal decomposition of the space-time-stochastic residual)**.** *The space-time-stochastic residual* R^{sts} *from* (10) *admits the following orthogonal decomposition,*

$$R^{sts}=\sum_{j=0}^{M}\Big(R_j^{det}+R_j^{sq}\Big)\Psi_j+\sum_{j>M}^{\infty}R_j^{sc}\Psi_j, \tag{11}$$

where

$$\begin{aligned}
R_j^{det} &:= \sum_{l=0}^{R}R^{st}(y_l)\Psi_j(y_l)w_l \qquad \textit{for } j=0,\dots,M\\
R_j^{sq} &:= \Big\langle\partial_x F\Big(\sum_{i=0}^{M}\hat{u}^{st}(y_i)\Psi_i\Big),\Psi_j\Big\rangle-\sum_{l=0}^{R}\partial_x F(\hat{u}^{st}(y_l))\Psi_j(y_l)w_l \qquad \textit{for } j=0,\dots,M\\
R_j^{sc} &:= \Big\langle\partial_x F\Big(\sum_{i=0}^{M}\hat{u}^{st}(y_i)\Psi_i\Big),\Psi_j\Big\rangle \qquad \textit{for } j>M
\end{aligned}$$

are called the j-th mode of the deterministic, stochastic quadrature and stochastic cut-off residual. Moreover, we have

$$\begin{aligned}\mathcal{E}^{sts}(t) &= \|R^{sts}\|^2_{L^2_\xi(\Xi;L^2((0,t)\times\mathbb{R})}\\ &= \sum_{i=0}^{M} \|R_i^{det} + R_i^{sq}\|^2_{L^2((0,t)\times\mathbb{R})} + \sum_{i>M}^{\infty} \|R_i^{sc}\|^2_{L^2((0,t)\times\mathbb{R})}\\ &\leq 2\mathcal{E}^{det}(t) + 2\mathcal{E}^{sq}(t) + \mathcal{E}^{sc}(t),\end{aligned} \tag{12}$$

where, for any $t \in (0,T)$,

$$\mathcal{E}^{det}(t) := \sum_{i=0}^{M} \|R_i^{det}\|^2_{L^2((0,t)\times\mathbb{R})},\ \mathcal{E}^{sq}(t) := \sum_{i=0}^{M} \|R_i^{sq}\|^2_{L^2((0,t)\times\mathbb{R})},$$
$$\mathcal{E}^{sc}(t) := \sum_{i>M}^{\infty} \|R_i^{sc}\|^2_{L^2((0,t)\times\mathbb{R})}.$$

Proof. We recall that the space-time reconstruction $\hat{u}^{st}(y_l)$ satisfies

$$R^{st}(y_l) = \partial_t \hat{u}^{st}(y_l) + \partial_x F(\hat{u}^{st}(y_l)) \tag{13}$$

for all $l = 0, \ldots, R$. Moreover, the reconstructed mode $\hat{u}_j^{st}$ was defined as (cf. (8))

$$\hat{u}_j^{st} = \sum_{l=0}^{R} \hat{u}^{st}(y_l)\Psi_j(y_l)w_l \tag{14}$$

for all $j = 0, \ldots, M$. Multiplying (13) by $\Psi_j(y_l)w_l$ and suming over $l = 0, \ldots, R$ yields, using (14), the following relationship

$$\sum_{l=0}^{R} R^{st}(y_l)\Psi_j(y_l)w_l = \partial_t \hat{u}_j^{st} + \sum_{l=0}^{R} \partial_x F(\hat{u}^{st}(y_l))\Psi_j(y_l)w_l. \tag{15}$$

By definition of the space-time-stochastic residual we have

$$R^{sts} = \partial_t \hat{u}^{sts} + \partial_x F(\hat{u}^{sts}) = \partial_t\Big(\sum_{i=0}^{M} \hat{u}_i^{st}\Psi_i\Big) + \partial_x F\Big(\sum_{i=0}^{M} \hat{u}_i^{st}\Psi_i\Big).$$

Let us begin by studying the j-th mode of R^{sts} for $j = 0, \ldots, M$. In this case the orthogonality relation (1) yields

$$\Big\langle R^{sts}, \Psi_j\Big\rangle = \Big\langle \partial_t \hat{u}^{sts} + \partial_x F(\hat{u}^{sts}), \Psi_j\Big\rangle = \partial_t \hat{u}_j^{st} + \Big\langle \partial_x F\Big(\sum_{i=0}^{M} \hat{u}_i^{st}\Psi_i\Big), \Psi_j\Big\rangle. \tag{16}$$

Using (15) we obtain

$$\begin{aligned}\Big\langle R^{sts}, \Psi_j\Big\rangle =& \sum_{l=0}^{R} R^{st}(y_l)\Psi_j(y_l)w_l \\ &+ \Big\langle \partial_x F\Big(\sum_{i=0}^{M} \hat{u}_i^{st}\Psi_i\Big), \Psi_j\Big\rangle - \sum_{l=0}^{R} \partial_x F(\hat{u}^{st}(y_l))\Psi_j(y_l)w_l = R_j^{det} + R_j^{sq}.\end{aligned} \tag{17}$$

For $j > M$ the j-th moment of R^{sts} is

$$\Big\langle R^{sts}, \Psi_j\Big\rangle = \Big\langle \partial_x F\Big(\sum_{i=0}^{M} \hat{u}_i^{st}\Psi_i\Big), \Psi_j\Big\rangle = R_j^{sc}. \tag{18}$$

Formula (11) then follows from (17) and (18). Formula (12) is an application of the Pythagorean theorem for $L^2_\xi(\Xi)$. □

Putting together Theorem 4.1 and Lemma 4.2 we obtain our main result, the following a posteriori error estimate with separable error bounds.

Theorem 4.3 (A posteriori error bound for the numerical solution with error splitting). *Let u be the random entropy solution of* (RIVP). *Then, for any $n = 0, \dots, N_t$, the difference between $u(t_n, \cdot, \cdot)$ and u_h^n from* (6) *satisfies*

$$\begin{aligned}
&\|u(t_n,\cdot,\cdot) - u_h^n(\cdot,\cdot)\|^2_{L^2_\xi(\Xi;L^2(\mathbb{R}))} \\
&\qquad \le 2\|\hat{u}^{sts}(t_n,\cdot,\cdot) - u_h^n(\cdot,\cdot)\|^2_{L^2_\xi(\Xi;L^2(\mathbb{R}))} \\
&\qquad + 2C_{\underline{\eta}}^{-1}\Big(2\mathcal{E}^{det}(t_n) + 2\mathcal{E}^{sq}(t_n) + \mathcal{E}^{sc}(t_n) + C_{\overline{\eta}}\mathcal{E}_0^{sts}\Big) \\
&\qquad \times \exp\Big(C_{\underline{\eta}}^{-1}\int_0^{t_n}\Big(C_{\overline{\eta}}C_{\overline{F}}\|\partial_x\hat{u}^{sts}(t,\cdot,\cdot)\|_{L^\infty_\xi(\Xi;L^\infty(\mathbb{R}))} + C_{\overline{\eta}}^2\Big)\,\mathrm{d}t\Big).
\end{aligned}$$

5. **Conclusions and Outlook.** We derived a novel residual-based a posteriori error bound for the difference between the entropy solution of (RIVP) and its numerical approximation using the NISP method in combination with a RKDG scheme. Moreover, we proved that the upper bound can be decomposed into three parts, where $\mathcal{E}^{det}$ quantifies the space-time discretization error of the RKDG scheme, $\mathcal{E}^{sq}$ assesses the quality of the discrete orthogonal projection and $\mathcal{E}^{sc}$ quantifies the stochastic error by truncation of the generalized polynomial chaos series. Based on these results, residual-based adaptive numerical schemes, which balance the contribution of the three different sources of numerical error, can be constructed. Residual-based space-stochastic adaptive numerical schemes are also considered in [6].

REFERENCES

[1] A. Bressan and P. LeFloch, Uniqueness of weak solutions to systems of conservation laws. *Arch. Rational Mech. Anal.*, 140(4):301–317, 1997.

[2] B. Cockburn and C.-W. Shu, Runge-Kutta discontinuous Galerkin methods for convection-dominated problems. *J. Sci. Comput.*, 16(3):173–261, 2001.

[3] C. M. Dafermos, Hyperbolic conservation laws in continuum physics, volume 325 of *Grundlehren der Mathematischen Wissenschaften [Fundamental Principles of Mathematical Sciences]*. Springer-Verlag, Berlin, fourth edition, 2016.

[4] A. Dedner and J. Giesselmann, A posteriori analysis of fully discrete method of lines discontinuous Galerkin schemes for systems of conservation laws. *SIAM J. Numer. Anal.*, 54(6):3523–3549, 2016.

[5] J. Giesselmann, C. Makridakis, and T. Pryer, A posteriori analysis of discontinuous Galerkin schemes for systems of hyperbolic conservation laws. *SIAM J. Numer. Anal.*, 53(3):1280–1303, 2015.

[6] J. Giesselmann, F. Meyer, and C. Rohde, A posteriori error analysis and non-intrusive adaptive numerical schemes for systems of random conservation laws. *arXiv preprint arXiv:1902.05375*, 2019.

[7] J. Giesselmann, F. Meyer, and C. Rohde, A posteriori error analysis for random scalar conservation laws using the Stochastic Galerkin method. *IMA J. Numer. Anal.*, 2019.

[8] O. P. Le Maître, M. T. Reagan, H. N. Najm, R. G. Ghanem, and O. M. Knio, A stochastic projection method for fluid flow. II. Random process. *J. Comput. Phys.*, 181(1):9–44, 2002.

[9] N. H. Risebro, C. Schwab, and F. Weber, Multilevel Monte Carlo front-tracking for random scalar conservation laws. *BIT Numerical Mathematics*, 56(1):263–292, 2016.

[10] D. Xiu and G. E. Karniadakis, The Wiener-Askey polynomial chaos for stochastic differential equations. *SIAM J. Sci. Comput.*, 24(2):619–644, 2002.

E-mail address: giesselmann@mathematik.tu-darmstadt.de
E-mail address: fabian.meyer@mathematik.uni-stuttgart.de
E-mail address: christian.rohde@mathematik.uni-stuttgart.de

ANALYSIS OF A NONLINEAR HYPERBOLIC CONSERVATION LAW WITH MEASURE-VALUED DATA

Xiaoqian Gong* and Matthias Kawski

School of Mathematical and Statistical Sciences, Arizona State University
Tempe, Arizona 85287

Abstract. This research announcement is concerned with a nonlinear hyperbolic conservation law model of a highly re-entrant semiconductor manufacturing system. The hyperbolic conservation law is characterized by the non-local velocity and its flux boundary condition. We report progress on the conjectured non-existence of L^1-optimal controls for the transition between equilibria. In the space of finite positive regular Borel measures, we formulate a notion of solution in the spirit of Lagrangian point of view and propose a new notion of weak measure-valued solution. We prove the existence and uniqueness of such solutions together with continuity of the flow with respect to time and (almost) with respect to the initial state.

1. **Introduction.** In this conference article, we announce new results that well-posedness of a model for semiconductor manufacturing systems is preserved when data and states are generalized from L^1-functions to Borel measures. We motivate this generalization by partial results on L^1-functions not being optimal. For complete proofs we refer the interested readers to[4]

Hyperbolic conservation laws are commonly used to describe traffic flow, pedestrian motion and sedimentation models among many other applications. A continuum model was introduced in [1] to describe highly re-entrant semiconductor manufacturing systems which produce a large number of items in a large number of steps. Denote by $x \in [0,1]$ the degree of completion in the semiconductor factory. That is, $x = 0$ represents the beginning of the production line and $x = 1$ the end. Let $\rho : [0,\infty) \times [0,1] \to [0,\infty)$, $(t,x) \mapsto \rho(t,x)$ be the density variable which describes the work in progress (WIP) density of the product at stage x of the production at time t. A characteristic feature of the model is that the velocity is non-local and depends on the the total load $W(t) = \int_0^1 \rho(t,x)dx$. This reflects the highly re-entrant nature of the product flow in semi-conductor manufacturing systems. The velocity is a positive, decreasing function $v = \alpha(W)$ of the total load. The time evolution of the product density ρ was described in [1] by the scalar hyperbolic conservation law

$$\partial_t \rho(t,x) + \partial_x(\alpha(W(t))\rho(t,x)) = 0. \tag{1}$$

A natural boundary control input, the in-flux u, suggests the boundary condition

$$u(t) = \rho(t,0)\alpha(W(t)), \text{ for } t \in [0,+\infty). \tag{2}$$

2000 *Mathematics Subject Classification.* Primary: 35R06, 35L65; Secondary: 65M25, 93C20.
Key words and phrases. Hyperbolic Conservation Law, Measure, Optimal Control.
* Corresponding author: Xiaoqian Gong.

In addition, the initial condition is

$$\rho_0(x) = \rho(0, x), \text{ for } x \in [0, 1]. \tag{3}$$

Motivated by business objectives of the semiconductor manufacturing company, for a given desired out-flux y_d and the actual out-flux $y(t) = \rho(t, 0)\alpha(W(t))$, denote by $Y_d(t) = \int_0^t y_d(s)\, ds$ and $Y(t) = \int_0^t y(s)\, ds$ the desired accumulated out-flux and the actual accumulated out-fluxes, respectively. Furthermore, define the backlog β by the mismatch between the desired and actual accumulated out-flux, i.e., $\beta = Y_d(t) - Y(t)$. The control problem associated to the nonlinear hyperbolic conservation law (1) is to find an optimal control u^* in a set of admissible controls such that the control objective functional

$$J(u) = \int_0^\infty |\beta(t)|^p\, dt, \text{ with } p = 1 \text{ or } p = 2 \tag{4}$$

is minimized at u^*.

Various different choices of the space of admissible controls and objectives are of both practical and mathematical interest. Each space leads to distinct mathematical problems. This model was simulated for L^2-data ($u \in L^2((0, \infty))$ and $\rho_0 \in L^2([0, 1])$), and the L^2-optimal control ($p = 2$ in equation (4)) for piecewise constant desired $y_d(t)$ was approximated numerically in [6]. Existence of unique solutions to equation (1) for L^1-data and existence of optimal solutions to equation (1) for L^2-data and L^2-objectives were established analytically in [2]. It has been conjectured that for the more meaningful L^1-objective ($p = 1$ in equation (4)), and L^1-data ($u \in L^1((0, \infty))$ and $\rho_0 \in L^1([0, 1])$), optimal controls need not exist unless one requires the control to be bounded. This motivates us to consider the larger space of finite positive regular Borel measures for the control input and the initial density.

A number of recent articles [3, 7, 8] have made much progress in establishing the well-posedness of similar nonlinear hyperbolic conservation laws with non-local velocity in the setting of measure-valued data. The existence and uniqueness of weak measure-valued solutions for a Cauchy problem associated to the continuity equation similar to (1) were established in [3] where the velocity relies on a smoothing convolution. In addition, by considering the probability measures, article [3] used the Wasserstein metric. In the article [8], the well-posedness of a Cauchy problem for the transport equation similar to (1) with a source term was proved for data and states in the space $\mathcal{M}_0^{ac}(\mathbb{R})$ of finite positive measures on $\mathbb{R}$ that are absolutely continuous with respect to Lebesgue measure and with bounded support on $\mathbb{R}$. The generalized Wasserstein distance was introduced manage changes of the total mass with time.

In this article, we will pose the hyperbolic conservation law (1) in the space $\mathcal{M}^+(I)$ of finite positive regular Borel measures on an interval $I \subseteq \mathbb{R}$. Let $T > 0$ be large but fixed. Assume that the in-flux is given by $\mu \in \mathcal{M}^+((0, T])$ and the initial measure at the initial time $t = 0$ is given by $\rho_0 \in \mathcal{M}^+([0, 1))$. Denote by ρ_t the measure at time t. To avoid discontinuities of the total load W, we set for every $t \in (0, T]$

$$W(t) = \rho_t([0, 1)). \tag{5}$$

That is, for instance, if for some time t_0, $\rho_{t_0}(\{0\}) = \rho_{t_0}(\{1\}) > 0$, equation (5) avoids an unnecessary discontinuity of the total load W. Formally, we have the following scalar conservation law

$$\partial_t \rho_t + \partial_x \left(\alpha(W(t))\rho_t\right) = 0, \tag{6}$$

with the in-flux at $x = 0$

$$\mu \in \mathcal{M}^+((0,T])) \tag{7}$$

and the initial condition

$$\rho_0 \in \mathcal{M}^+([0,1)). \tag{8}$$

Additionally, we assume that the velocity function α is bounded away from 0 on compact sets, and is Lipschitz with respect to the total load, W, with Lipschitz constant L. Namely, for any $W_1, W_2 \geq 0$, $\|\alpha(W_1) - \alpha(W_2)\| \leq L\|W_1 - W_2\|$.

What distinguishes our problem from others is that the velocity depends on the total load instead of relying on smoothing convolution like in [3]. Due to the control in-flux and out-flux, notions such as the standard Wasserstein metric are not applicable. Besides the measures in $\mathcal{M}_0^{ac}(\mathbb{R})$, we are also particularly interested in the measures with nonzero pure point part. For example, Dirac measures play an important rule when it comes to model impulsive optimal controls or a high concentration of mass at an instant time in the factory.

The paper is organized as follows. In section 2, we demonstrate that a set of natural candidates of L^1-controls does not contain an optimal control for the transition between equilibria. Section 3 is devoted to the Lagrangian description of the model of the highly re-entrant semiconductor manufacturing system. In section 4, we give a new notion of weak measure-valued solution to the hyperbolic conservation law (6) in the space of $\mathcal{M}^+(I)$, and we establish the existence and uniqueness of the weak measure-valued solutions together with continuity of the flow with respect to time and (almost) with respect to the initial state.

2. **Non-optimality of a family of L^1-controls for the transition from a smaller to a larger equilibrium.** In this section, we demonstrate progress towards proving the conjectured non-existence of optimal L^1-controls for the optimal control problem (1)-(4) with $p = 1$. Consider a desired out-flux y_d that is piecewise constant and increases with a jump at time t^*, i.e.,

$$y_d(t) = \begin{cases} y_1, & t < t^* \\ y_2, & t \geq t^*, \end{cases} \tag{9}$$

with $0 \leq y_1 < y_2 < 1$. Additionally, we work with the fixed model $\alpha(W) = \frac{1}{1+W}$ for the velocity as a function of the total load, as in [2]. Denote the constant densities at the equilibrium states when $t < t^*$ and when $t \geq t^*$ to be ρ_1 and $c\rho_2$, respectively. (It will be clear from the context that these are not ρ_t at times $t = 1$ and $t = 2$.) Then for $k = 1, 2$, the corresponding outfluxes are $y_k = \frac{\rho_k}{1+\rho_k}$.

2.1. **Transfer from a smaller to a larger equilibrium with nonzero backlog.** To meet the requirement that the system arrives at the larger equilibrium at time t^*, the operator in the factory needs to start action at some time $t_* < t^*$, such that

$$\int_{t_*}^{t^*} \alpha(W(t))dt = 1.$$

That is, the operator need to start to increase the total load W at time t_*. But an inverse response occurs: the velocity of the system decreases due to the increase of the total load and this leads to nonzero backlog to the system.

Lemma 2.1. *For the desired out-flux y_d defined in* (9)*, a control input in-flux*

$$u(t) = \begin{cases} \rho_1\alpha(W(t)), & t \le t_*, \\ \rho_2\alpha(W(t)), & t \ge t_*, \end{cases} \tag{10}$$

with $t_* = t^* - \frac{(1-y_1)+(1-y_2)}{2(1-y_1)(1-y_2)}$ produces a constant backlog β for $t \ge t^*$,

$$\beta(t) = \frac{y_1(y_2-y_1)}{2(1-y_1)(1-y_2)}.$$

2.2. **Transfer from a smaller to a larger equilibrium with eventually zero backlog.** To cancel the backlog produced by the control input in-flux u in equation (10), one needs to modify this control input in-flux by increasing it, i.e., by adding additional mass $M > 0$ at $x = 0$ at some earlier time stage. This results in even larger inverse response due to the fact that velocity α decreases as the total load W increases. Thus, the additional mass M must not only make up for the missing out-flux due to the step up of the in-flux, but must also make up for the further backlog caused by M itself. It is not a priori clear that for any $\varepsilon \in (0,1]$, such a mass M exists. Furthermore, the requirement that the system reaches to another equilibrium at time t^* forces us to choose the control input in-flux as $u(t) = \rho_2\alpha(W(t))$ for $t > \varepsilon$, with $t_* < \varepsilon < t^*$. Without loss of generality, we assume that the action time $t_* = 0$, that is, $\int_0^{t^*} \alpha(W(t))dt = 1$. In this article, we consider the case when $\epsilon \in (0,1]$. Note that in the situation with a modified control input in-flux, the system reaches its new equilibrium state $\rho_t \equiv \rho_2$ at the time T^* defined by $\int_\varepsilon^{T^*} \alpha(W(t))dt = 1$, with zero back-log $\beta(T^*) = 0$.

The time T^* at which the backlog becomes zero also depends on both the *shape* of the control variation, and on the amount of the additional mass M. Given a direction $h \in L^1([0,1];[0,+\infty))$ with $\int_0^1 h(t)dt = 1$, and for any $\varepsilon \in (0,1]$, we consider the curve of modified L^1-control in-fluxes

$$u_\varepsilon(t) = \begin{cases} \rho_2\alpha(W(t)) + \frac{M^*(h,\varepsilon)}{\varepsilon}h(\frac{t}{\varepsilon}), & 0 \le t \le \varepsilon; \\ \rho_2\alpha(W(t)), & t > \varepsilon, \end{cases} \tag{11}$$

with $M^*(h,\varepsilon) > 0$. Here the L^1-function h and the number $M^*(h,\varepsilon)$ represent the shape and the amount of the additional mass respectively.

Lemma 2.2. *For every $h \in L^1([0,1])$ as above, and every $\varepsilon \in (0,1]$, there exists a unique $M^*(h,\varepsilon) > 0$ such that the control input u_ε results in zero backlog in finite time.*

Let $M^0 = \frac{\rho_1(\rho_2-\rho_1)}{2}$ and formally consider the impulsive control (not in $L^1((,T]))$

$$u_0(t) = \begin{cases} M^0\delta_0(t), & t = 0, \\ \rho_2\alpha(W(t)), & t > 0. \end{cases} \tag{12}$$

Lemma 2.3. *For every $h \in L^1([0,1])$ as above, the modified L^1 control inputs u_ε converge to the distribution u_0 in the sense of distribution as ε approaches 0 from the right.*

Formally, analyzing directional derivatives of the objective functional J at $u_0 \in \mathcal{M}^+((0,T])$ in the directions of absolutely continuous Borel measures that correspond to L^1 functions, we obtain:

Theorem 2.4. *For every $h \in L^1([0,1])$ as above, there exists $\varepsilon_h > 0$ such that for all $0 < \varepsilon_2 < \varepsilon_1 < \varepsilon_h$ the functional J satisfies $J(u_0) < J(u_{\varepsilon_2}) < J(u_{\varepsilon_1})$.*

Both heuristic arguments and numerical simulations strongly suggest that the only reasonable candidates of L^1-controls u for which $J(u)$ is even close to $J(u_0)$ are of the above form u_ε with $\{t > 0\colon u_\epsilon(t) \neq \rho_2\alpha(W(t))\}$ contained in an interval as short as possible.

3. **Lagrangian point of view.** We have considered a combination of L^1-control together with an impulsive control with one point mass. More generally, we may combine the L^1-data and countably many of point masses in both the control influx u and the initial density ρ_0. For any $s \in \mathbb{R}$, denote by δ_s the Dirac measure centered at s. For $i, j \in \mathbb{Z}^+$, consider sequences of locations $x_j \in [0,1]$ and times $t_i \in (0,T]$, and sequences of point masses $m_i \geq 0$ and $M_j \geq 0$ with $\sum_{i=1}^{\infty} m_i < \infty$ and $\sum_{i=j}^{\infty} M_j < \infty$. Denote the L^1-control by u_{L^1} and the initial L^1-density by ρ_{0,L^1}. Formally, we have

$$u = u_{L^1} + \sum_{i=1}^{\infty} m_i \delta_{t_i}, \qquad \text{and} \qquad \rho_0 = \rho_{0,L^1} + \sum_{i=1}^{\infty} M_j \delta_{x_j}.$$

Let $\xi_j, \eta_j : (0,T] \mapsto [0,\infty)$ track the location of the masses m_i and M_j respectively. One might consider, in the Lagrangian approach, the combination of hyperbolic conservation law in L^1-setting and a sequence of (ordinary differential equations (ODEs) which are coupled by total mass and velocity as follows:

$$0 = \partial_t \rho_{L^1}(t,x) + \partial_x(\alpha(W(t))\rho_{L^1}(t,x)) \text{ for } (t,x) \in (0,T] \times [0,1], \tag{13a}$$

$$\xi_i'(t) = \alpha(W(t)) \text{ for } t \in [t_i, T] \text{ and } i \in \mathbb{Z}^+, \tag{13b}$$

$$\eta_j'(t) = \alpha(W(t)) \text{ for } t \in [0,T] \text{ and } j \in \mathbb{Z}^+, \tag{13c}$$

$$W(t) = \int_0^1 \rho(t,x)\,dx + \sum_i m_i + \sum_j M_j \text{ for } t \in (0,T], \tag{13d}$$

where in (13d) the first and second sum are taken over the sets $\{i\colon \xi_i(t) \in [0,1)\}$ and $\{j\colon \eta_j(t) \in [0,1)\}$, respectively. The initial and boundary conditions to the above hyperbolic conservation law and ODEs are

$$\begin{aligned}
\rho_{L^1}(0,x) &= \rho_{0,L^1}(x) \text{ for } x \in [0,1],\\
u_{L^1}(t) &= \rho_{L^1}(t,0)\alpha(W(t)) \text{ for } t \in (0,T],\\
\xi_i(t_i) &= 0 \text{ for } i \in \mathbb{Z}^+,\\
\eta_j(0) &= x_j \text{ for } i \in \mathbb{Z}^+.
\end{aligned}$$

Remark 1. In the summation in (13d), we again use the interval $[0,1)$ to avoid undesirable discontinuities of the total load W.

Remark 2. The set of ODEs in (13b) and (13c) really is one ODE since for every fixed $t \in (0,T]$, the velocity $\alpha(W(t))$ is constant with respect to the location $x \in [0,1)$.

Mathematically, a more satisfactory treatment is to combine the L^1-data and point masses into a measure and write the hyperbolic conservation law in a measure setting. We will generalize our setting to be in the space of positive σ-finite regular Borel measures, $\mathcal{M}^+$. That is, we consider the following problem

$$0 = \partial_t \rho + \partial_x(\alpha(W(t))\rho) \tag{14}$$
$$W(t) = \rho_t([0,1))$$

where $\rho\colon (0,T] \mapsto \mathcal{M}^+([0,1)); t \mapsto \rho_t$, with the initial condition $\rho_0 \in \mathcal{M}^+([0,1))$ and the control input $\mu \in \mathcal{M}^+((0,T])$.

4. Existence and uniqueness of weak measure-valued solutions to the hyperbolic conservation law.

In this section, the statements and calculations throughout assume that the initial condition $\rho_0 \in \mathcal{M}^+([0,1))$ and the control in-flux $\mu \in \mathcal{M}^+((0,T])$ are arbitrary but fixed.

4.1. Existence and uniqueness of the flow X.

Definition 4.1 (Flow). Suppose a time-varying vector field $v\colon [0,T]\times[0,1] \mapsto [0,1]$ is integrable with respect to the first variable and constant with respect to the second variable. We call a map $X\colon \{(t,r)\colon 0 \le r \le t \le T\} \times [0,1] \mapsto \mathbb{R}^+$ the flow of the vector field v if it satisfies for all $r \in [0,T]$ and all $x_0 \in [0,1]$,

$$\dot{X}(t;r,x_0) = v(t) \text{ for a.e. } t \in [r,T], \text{ and}$$
$$X(r;r,x_0) = x_0$$

with $\dot{X}$ representing the derivative of X with respect to time t.

Extend the constant (in space) vector field $v = \alpha(W(t))$ to $[0,\infty)\times[0,T]$ and for convenience denote the flow of the vector field v, still by $X\colon \{(t,r)\colon 0 \le r \le t \le T\}\times [0,1] \mapsto \mathbb{R}^+$ if it exists. Furthermore, if X exists, by the semi-group property of X, for any fixed $r \in [0,T]$ and $t \in [r,T]$, we have $X(t;0,0) = X(t;r,X(r;0,0)) = X(t;r,0) + X(r;0,0)$. Thus for any $x_0 \in [0,1]$, $X(t;r,x_0) = X(t;r,0) + x_0 = X(t;0,0) - X(r;0,0) + x_0$. Therefore to show the existence and uniqueness of the flow X, it is enough to check the existence and uniqueness of the characteristic curve $\xi\colon [0,T] \mapsto \mathbb{R}^+, t \mapsto \xi(t) = X(t;0,0)$.

Theorem 4.2. *Let $v_{min} = (1 + \rho_0([0,1]) + \mu([0,T]))^{-1}$ and consider a small time interval $[0,\tau]$, where $0 < \tau < 1$. Let*

$$\Omega = \left\{\eta\colon [0,\tau] \mapsto [0,1]\colon \eta(0) = 0; v_{min} \le \frac{\eta(s)-\eta(t)}{s-t} \le 1 \textit{ for all } s,t \in [0,\tau]\right\}.$$

Then the space Ω is complete under the supremum norm. The map $F\colon \Omega \mapsto C([0,\tau])$, with

$$F(\eta)(t) = \int_0^t \alpha(\rho_0([0,1-\eta(s))) + \mu((0,s]))ds$$

is a contraction. Furthermore, the characteristic curve $\xi\colon [0,T] \mapsto \mathbb{R}^+$ is unique.

A key step to prove theorem (4.2) is to split the initial measure ρ_0 into the absolutely continuous part $\rho_{0,ac}$ and the pure point part $\rho_{0,pp}$. Since the initial measure ρ_0 is finite, there are finitely many large point masses in the pure part $\rho_{0,pp}$ on the interval $[0,1)$. The usual contraction argument applies to the absolutely continuous part $\rho_{0,ac}$ and the small point masses in the pure part $\rho_{0,pp}$ but fails on time intervals whose interiors contain exit times of large point masses exiting

$[0,1)$. In addition, it is not a priori known when the large point masses leave the system. To overcome this complication, a trick is to replace the initial condition ρ_0 by a modified $\tilde{\rho_0}$ (with the same total load) for which contraction mapping theorem applies over a larger time interval. Since the velocity α only depends on the total load, the characteristic curves to the initial conditions ρ_0 and $\tilde{\rho_0}$ coincide over certain time interval during which no large point masses exit from the system. Due to the finite number of large point masses, we just need to repeat the argument for finitely many times, and get global existence of the characteristic curve ξ hence of the flow X.

Definition 4.3 (Lagrangian solution)**.** Suppose the map

$$X\colon \{(t,r)\colon 0 \le r \le t \le T\} \times [0,1] \;\mapsto\; \mathbb{R}^+$$

is a flow of the vector field $v = \alpha(W(t))$. A Lagrangian solution to equation (6) is a function $\Phi : [0,T] \to \mathcal{M}^+([0,1])$, $t \to \Phi_t$ such that for any $t \in [0,T]$ and any Borel measurable set $E \subset [0,1)$,

$$\Phi_t(\rho_0,\mu)(E) = \int_{[0,1)} \chi_E(X(t;0,x_0))d\rho_0(x_0) + \int_{(0,t]} \chi_E(X(t;s,0))d\mu(s), \qquad (15)$$

where χ_E is the indicator function of the set E defined by $\chi_E(x) = 1$ is $x \in E$ and $\chi_E(x) = 1$ else.

Remark 3. When only times $t \le 1$ are considered, then the second term in (15) can be re-written as $\mu((0,t])$ since the velocity α is bounded above by 1.

4.2. **Existence and uniqueness of weak measure-valued solutions.** Let $AC([0,T])$ be the set of absolutely continuous functions on $[0,T]$ and let Ψ be the set of functions $\varphi\colon [0,T]\times[0,1] \mapsto \mathbb{R}$ such that for every $t \in [0,T]$, $\varphi(t,\cdot) \in C^1([0,1])$ and for every $x \in [0,1]$, $\varphi(\cdot,x) \in AC([0,T])$. That is,

$$\Psi = \{\varphi\colon [0,T] \times [0,1] \mapsto \mathbb{R}\colon \text{ for every } t \in [0,T], \varphi(t,\cdot) \in C^1([0,1]), \\ \text{for every } x \in [0,1], \varphi(\cdot,x) \in AC([0,T])\}\,.$$

Definition 4.4 (Weak Measure-Valued Solution)**.** A weak measure-valued solution to equation (6) with the initial condition $\rho_0 \in \mathcal{M}^+([0,1))$ and the boundary condition $\mu \in \mathcal{M}^+((0,T])$ is a function $\rho : [0,T] \to \mathcal{M}^+([0,1))$, such that $W\colon [0,T] \mapsto \rho_t([0,1))$ is integrable and such that for every $\tau \in [0,T]$ and for every $\varphi \in \Psi$ that satisfies

$$\varphi(t,1) = 0, \text{ for all } t \in [0,\tau], \qquad (16)$$

one has

$$0 = \int_{(0,\tau]}\int_{[0,1)} (\partial_t\varphi(t,x) + \alpha(W(t))\partial_x\varphi(t,x))\; d\rho_t(x)\,dt + \int_{(0,\tau]} \varphi(t,0)\,d\mu(t) \\ - \int_{[0,1)} \varphi(\tau,x)\,d\rho_\tau(x) + \int_{[0,1)} \varphi(0,x)\,d\rho_0(x).$$

The double integral is well-defined since the times at which the velocity α are discontinuous form a set of measure zero.

Theorem 4.5. *Every Lagrangian solution of* (15) *is a weak measure-valued solution that satisfies* (17).

The proof to theorem (4.5) is fairly straightforward, evaluating the right hand side of equation (17) at $\rho = \Phi$, and repeatedly using that the flow is defined in terms of push-forwards of Borel measures by strictly monotone functions.. The uniqueness of the weak measure-valued solution can be derived from the uniqueness of the flow X.

4.3. **Regularity of the measure-valued solution.** Suppose $g\colon \mathbb{R} \mapsto \mathbb{R}$,

$$g(x) = \begin{cases} \frac{1}{2} - |\frac{1}{2} - x| & \text{if } x \in [0,1] \\ 0 & \text{otherwise.} \end{cases}$$

Define the following map $\phi : \mathcal{M}([0,1)) \mapsto [0,\infty)$

$$\phi(\mu) = \sup \left\{ \left| \int_{[0,1)} fg\, d\mu \right| : f \in [0,1]^{[0,1)};\ \text{for all } x,y \in [0,1), |f(x) - f(y)| \leq |x-y| \right\}.$$

Lemma 4.6. *The map ϕ defines a semi-norm on the space $\mathcal{M}([0,1))$.*

The notion of the semi-norm ϕ is motivated by the flat norm which is commonly used in the space of Borel measures on a metric space. For careful analysis of properties of dynamical systems using the flat norm, see[5]. Now we endow the space $\mathcal{M}^+([0,1))$ with the semi-norm ϕ.

Theorem 4.7. *For every fixed $\rho_0 \in \mathcal{M}^+([0,1))$ and $\mu \in \mathcal{M}^+((0,T])$, the weak solution $\rho\colon [0,T] \mapsto \mathcal{M}^+([0,1))$ of equation (6) is continuous with respect to the initial condition under the semi-norm ϕ.*

We have not been able to show that, in general, the flow $(t,\rho_0) \mapsto \rho_t$ is continuous with respect to the initial datum ρ_0 and the semi-norm ψ. However, we have the following *almost continuity* result.

Theorem 4.8. *For every fixed $\rho_0 \in \mathcal{M}^+([0,1))$ and $\mu \in \mathcal{M}^+((0,T])$, the weak solution $\rho\colon [0,T] \mapsto \mathcal{M}^+([0,1))$ of equation (6) satisfies: For every initial condition $\tilde{\rho}_0 \in \mathcal{M}^+([0,1))$ and every $\varepsilon > 0$ there exist $\delta > 0$ and $\tau > 0$ such that if $\phi(\tilde{\rho}_0 - \rho_0) < \delta$ then for all $t < \tau$, $\phi(\tilde{\rho}_t - \rho_t) < \epsilon$.*

REFERENCES

[1] D. Armbruster, D. Marthaler, C. Ringhofer, K. Kempf and T.C. Jo, A continuum model for a re-entrant factory, *Operations research*, **54** (2006), 933–950.

[2] J. M. Coron and M. Kawski and Z. Wang, Analysis of a conservation law modeling a highly re-entrant manufacturing system, *Discrete Contin. Dyn. Syst. Ser. B*, **14** (2010), 1337–1359.

[3] G. Crippa and M. Lécureux-Mercier, Existence and uniqueness of measure solutions for a system of continuity equations with non-local flow, *NoDEA Nonlinear Differential Equations Appl.*, **20** (2013), 523–537.

[4] X. Gong and M. Kawski, Weak measure valued solutions to a nonlinear hyperbolic conservation law modeling a highly re-entrant manufacturing system preprint, `arXiv:1903.00797`.

[5] P. Gwiazda and A. Marciniak-Czochra and H.R. Thieme Measures under the flat norm as ordered normed vector space, *Positivity.*, **22** (2018), 105–138.

[6] M. La Marca, D. Armbruster, M. Herty and C. Ringhofer Control of continuum models of production systems, *IEEE Trans. Automat. Control.*, **55** (2010), 2511–2526.

[7] B. Piccoli, and F. Rossi Transport equation with nonlocal velocity in Wasserstein spaces: convergence of numerical schemes, *Acta Appl. Math.*, **124** (2013), 73–105.

[8] B. Piccoli, and F. Rossi Generalized Wasserstein distance and its application to transport equations with source, *Arch. Ration. Mech. Anal.*, **211** (2014), 335–358.

E-mail address: `xgong14@asu.edu`
E-mail address: `kawski@asu.edu`

HIGHER ORDER SCHEME FOR SINE-GORDON EQUATION IN NONLINEAR NON-HOMOGENEOUS MEDIA

Ameya D. Jagtap*

Division of Applied Mathematics, Brown University,
182 George Street, Providence, RI 02912, USA.

Abstract. One- and two-dimensional sine-Gordon equation in non- homogeneous media is considered. Sine-Gordon equation exhibits soliton-like solution whose existence and behavior in non-homogeneous media is studied. The governing sine-Gordon equation is discretized using higher order Legendre polynomial based spectral element method. Spectral stability analysis of the numerical scheme shows the strong dependence of a critical time step not only on the density magnitude of media but also on its nature of distribution. Various conclusions are made based on the study.

1. **Introduction.** Sine-Gordon (sG) equation is a hyperbolic, nonlinear wave equation which governs spatio-temporal dynamics of complex physical processes like, propagation of magnetic flux in a Josephson junction consisting of two layer of superconducting material separated by an isolating barrier [16], DNA dynamics [10] *etc* - to mention just few areas of application. sG equation is exactly integrable but the presence of external forcing term breaks the exact integrability of this equation, see [12] for more details. One of the most remarkable solutions of sG equation is soliton. Soliton wave emerges in various physical processes like shallow water waves, relativistic field theory, earthquakes, defects in solids, mechanical transmission lines, Josephson junction oscillator *etc.* Detailed discussions can be found in text of *Drazin & Johnson* [6].

Nonlinear wave propagation in inhomogeneous media has several real-world applications like tidal waves in the ocean, radio waves in the atmosphere, laser radiation in plasmas, seismic waves in earthquakes *etc.* In the literature, dynamics of soliton is studied in non-homogeneous medium by *Dai & Yu*[3], *Degasperis* et al. [4] and *Shyu* et al. [17]. In [11] *Guerrero* et al. showed that interaction with finite-width homogeneties can activate internal modes of soliton. In [8] *Gonzalez & de Mello* showed that the length scale competition between the width of inhomogeneities and the width of kink-soliton leads to a phenomenon called soliton explosions. In this paper, we are interested in the dynamics of soliton solution in inhomogeneous media for one- and two-dimensional sG equation. The extended study of this paper is recently published, see [14] for details. *Gharaati & Khordad* [7] solved one-dimensional discrete sG equation in inhomogeneous media changing the parameters like length, mass, gravitational acceleration and stiffness of the spring in the coupled pendulums chain, such that one can control and guide the solitons. In this paper, various

Key words and phrases. Sine-Gordon equation, Nonhomogeneous media, Spectral element scheme .

* Corresponding author emails: ameyadjagtap@gmail.com, ameya_jagtap@brown.edu.

representation of inhomogeneous media in one- and two dimensions are used where nonlinear continuous as well as discontinuous density variations are considered. The governing equation is solved using higher order spectral element scheme given by *Jagtap & Murthy* [13].

2. **Governing sG equation.** Klein-Gordon equation was first introduced by *Klein* and *Gordon* in the context of quantum theory [15, 9]. For a free particle in three dimensions [2], it takes the following form $\frac{1}{c^2}u_{tt} = \Delta u - \left(\frac{mc}{\hbar}\right)^2 u$, where u, t, Δ, m, c and $\hbar$ are the wave function, time, Laplacian, mass of the electron, speed of light and Planck's constant respectively. In terms of generalized potential this equation is written as $\frac{1}{c^2}u_{tt} = \Delta u - V'(u)$, where V is the nonlinear smooth function. Eventhough there are many choices for nonlinear potential term [18], we are particularly interested in sG equation where $V'(u) = -\phi \sin u$. where $\phi \in \mathbb{R}^-$ is a constant. In general, undamped sG equation in non-homogeneous media is given by

$$\rho(\mathbf{x})u_{tt} - \Delta u - \phi \sin u = 0, \quad (\mathbf{x} = \{x, y\}, t) \in \Omega \times (0, T] \subset \mathbb{R}^2 \times \mathbb{R}_+ \tag{1}$$

with initial conditions $u(\mathbf{x}, 0) = f(\mathbf{x}),\ \ u_t(\mathbf{x}, 0) = g(\mathbf{x})$ and following boundary conditions $u = 0$ on $\Gamma_D \times (0, T]$; $\frac{\partial u}{\partial \nu} = 0$ on $\Gamma_N \times (0, T]$, where $\Gamma_D \cup \Gamma_N = \Gamma$ and $\Gamma_D \cap \Gamma_N = \emptyset$. Note that, $c^2 = T/\rho$, where tension T is assumed to be unity everywhere, $u : \mathbb{R}^2 \times \mathbb{R}_+ \to \mathbb{R}$ is a wave function and $\rho(x, y) \in \mathbb{R}_+ \backslash \{0\}$ is the spacially dependent density variation in non-homogeneous media. The sG equation is an integrable infinite dimensional Hamiltonian system and has an infinite number of conserved quantities [1]. Physical energy is one of the conserved quantity which is given by

$$E(t) \triangleq \frac{1}{2} \iint [u_x^2 + u_y^2 + u_t^2 + 2(-\phi)(1 - \cos u)]\, dx\, dy \tag{2}$$

2.1. **Variational formulation.** The variational formulation for equation (1) is to find $u : (0, T] \to \mathcal{H}_D^1(\Omega)$ such that

$$\langle \rho\, u_{tt}, \mathbf{N} \rangle_\Omega + \langle \nabla u, \nabla \mathbf{N} \rangle_\Omega - \left\langle \frac{\partial u}{\partial \nu}, \mathbf{N} \right\rangle_{\Gamma_N} - \langle \phi \sin(u), \mathbf{N} \rangle_\Omega = 0,\ \forall \mathbf{N} \in \mathcal{H}_D^1(\Omega) \tag{3}$$

$$u(\mathbf{x}, 0) = f(\mathbf{x}), \quad u_t(\mathbf{x}, 0) = g(\mathbf{x}) \tag{4}$$

where $\mathcal{H}_D^1(\Omega) = \{\mathcal{H}^1(\Omega) : u = 0 \text{ on } \Gamma_D\}$. $\langle \cdot, \cdot \rangle_\Omega$ and $\langle \cdot, \cdot \rangle_{\Gamma_N}$ are used to denote $\mathcal{L}_2$-inner product over Ω and Γ_N respectively.

3. **Nonlinear, non-homogeneous media.** As discussed in the introduction, soliton wave emerges in many physical problems. Among all soliton, we are particularly interested in sG soliton and its spatio-temporal behavior while propagating in non-homogeneous media. In [7] *Gharaati & Khordad* studied dynamics of generalized sG soliton in inhomogeneous media by changing the parameters like length, mass, gravitational acceleration and stiffness of the spring in the coupled pendulums chain, such that one can control and guide the solitons. There are various applications where one can use these guided solitons like, Josephson junction for controlling and guiding the fluxons. Thus, it would be challenging to study the behavior of sG soliton in non-homogeneous media.

Preliminary reference earth model (PREM) is the widely used one-dimensional model of seismic velocities in the earth proposed by *Dziewonski & Anderson* [5]. All currently available earth models have values which are reasonably close to PREM.

The jump in discontinuities is in the range of 1000 kg/m^3. In this section, various density distributions in one and two dimensions are discussed along with their mathematical expressions. These density variations are inspired by PREM model where density varies smoothly as well as discontinuously. Effect of these distributions on the dynamics of the solution of sG equation especially, soliton solution will be studied in later part of this paper.

Considering one-dimensional domain $[-20,\ 20]$, the continuous density distribution $\rho(x, y, t) \equiv \rho(x)$ is given by

$$\rho(x) = 637\ \tan^{-1}(\exp(x - 4)) \tag{5}$$

and discontinuous density distribution is

$$\rho(x) = \begin{cases} 1 & \text{if}\, x \leq 4 \\ 1000 & \text{Otherwise} \end{cases} \tag{6}$$

In both continuous and discontinuous density variations, the largest and smallest values of density are almost same.

In two dimensions, the domain is $[-7,\ 7]^2$ where different density variations are discussed which include nonlinear density variation given by the Gaussian distribution

$$\rho(x, y) = 1000\ \exp(-(0.4x^2 + 0.4y^2)) \tag{7}$$

In this distribution, the density variation is nonlinear but smooth as shown in figure 1 (left). Moreover, density attains its highest value at the centre of the domain whereas it gradually decreases as one go towards the boundary of the domain. Next, we consider two cases of discontinuous density variations. First case gives

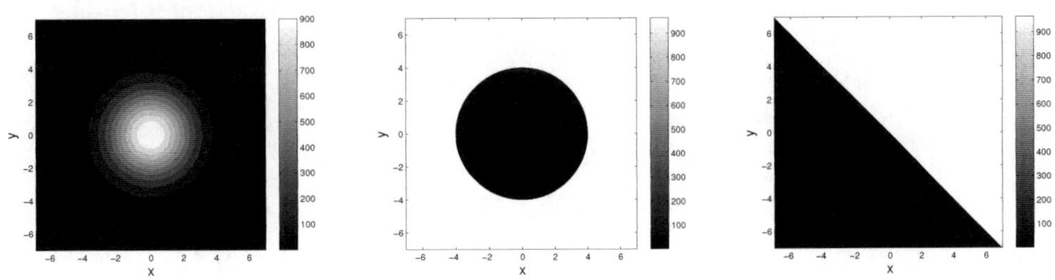

FIGURE 1. Smooth nonlinear density variation (left) and discontinuous density variation (middle and right).

circular jump in density as shown in figure 1 (middle) and its equation is given by

$$\rho(x, y) = \begin{cases} 1 & \text{if}\, x^2 + y^2 \leq 4^2 \\ 1000 & \text{Otherwise} \end{cases} \tag{8}$$

In case of the circular jump, density magnitude is very small inside the circle whereas outside the circle the magnitude is large. In second case, oblique jump along the line $x = -y$ is considered. It is given by the following equation

$$\rho(x, y) = \begin{cases} 1 & \text{if}\, x \leq -y \\ 1000 & \text{Otherwise} \end{cases} \tag{9}$$

as shown in figure 1 (right).

4. **Numerical scheme.** The computational domain is divided into Nel number of elements as $\Omega^h = \bigcup_{i=1}^{Nel} \Omega_i^h$ such that $\Omega_i^h \cap \Omega_j^h = \emptyset,\ \forall i \neq j$. Each element is mapped onto the parent element using isoparametric mapping. The parent element is the reference $\mathcal{D}$-cube ($\mathcal{D}$ is the physical dimension) given by $\hat{\Omega}_i^h = \{\boldsymbol{\xi} = (\xi_1, \xi_2, \cdots, \xi_{\mathcal{D}}) : -1 < \xi_p < 1,\ p = 1, 2, \cdots, \mathcal{D}\} = (-1,1)^{\mathcal{D}}$. The basis function $N^h(\xi)$ is the Lagrangian interpolation function using Gauss-Lobatto-Legendre (GLL) collocation points given as $N_i^h(\xi) = \frac{1}{m(m+1)L_m(\xi)} \frac{(\xi^2-1)}{\xi-\xi_i} L_{o_{m-1}}(\xi)$, where m^{th} degree Lobatto polynomial $L_{o_m}(\xi)$ is derived from the $(m+1)^{th}$ degree Legendre polynomial as $L_{o_m}(\xi) = L'_{m+1}(\xi)$. The $m+1$ GLL points ξ_i are the roots of the Lobatto polynomial of degree m. The interpolation function for higher dimensions can be obtained using tensor product of one-dimensional interpolation function as $\mathbf{N}^h(\boldsymbol{\xi}) = N_1^h(\xi_1) \otimes N_2^h(\xi_2) \otimes \cdots \otimes N_{\mathcal{D}}^h(\xi_{\mathcal{D}}),\ \ \boldsymbol{\xi} = (\xi_1, \xi_2, \cdots, \xi_{\mathcal{D}}) \in \mathbb{R}^{\mathcal{D}}$. The standard Galerkin approximation of wave function u is $u(x,y,t) \approx u^h(x,y,t) = \sum_{\forall i} u_i^h(t)\ \mathbf{N}_i^h(x,y)$. In case of spectral element method, interpolation function $N^h \in \mathbb{P}_m(-1,1)$ which is the space of all m^{th} order polynomials. The semi-discrete problem of (3)-(4) is to find $u^h : (0,T] \to \mathcal{V}^h$ such that

$$\langle \rho^h u_{tt}^h, \mathbf{N}^h \rangle_{\Omega^h} + \langle \nabla u^h, \nabla \mathbf{N}^h \rangle_{\Omega^h} - \left\langle \frac{\partial u^h}{\partial \nu}, \mathbf{N}^h \right\rangle_{\Gamma_N^h} - \langle \phi \sin u^h, \mathbf{N}^h \rangle_{\Omega^h} = 0,\ \forall \mathbf{N}^h \in \mathcal{V}^h \tag{10}$$

$$u^h(\mathbf{x},0) = f(\mathbf{x})^h, \quad u_t^h(\mathbf{x},0) = g(\mathbf{x})^h \tag{11}$$

where $\mathcal{V}^h$ is the approximation space of the $\mathcal{H}_D^1(\Omega)$.

Note that, $R^h \sin u \approx sin(R^h u)$ and density $\rho \approx \rho^h$ at each node points. Equation (10) can be written in matrix form as

$$M\boldsymbol{\rho}\, u_{tt}^h + D\, u^h - f_N - \phi M\ \sin u^h = 0 \tag{12}$$

where product approximation is used for $\sin u^h$ term. Mass matrix M, diffusion matrix D, density matrix $\boldsymbol{\rho}$ and Neumann boundary term f_N are given as $\langle \mathbf{N}^h, \mathbf{N}^h \rangle$, $\langle \nabla \mathbf{N}^h, \nabla \mathbf{N}^h \rangle$, $\mathrm{diag}\{\rho_1^h, \rho_2^h, \cdots\}$ and $\left\langle \frac{\partial u^h}{\partial \nu}, \mathbf{N}^h \right\rangle_{\Gamma_N^h}$ respectively.

In this paper, GLL quadrature is used to integrate all the integrals involved in weak formulation which results in diagonal mass matrix. Time derivative term u_{tt}^h is discretized using second order finite difference scheme as

$$u^{h,n+1} = \boldsymbol{\rho}^{-1} M^{-1} \left\{ M\boldsymbol{\rho}(2u^{h,n} - u^{h,n-1}) + \Delta t^2(-Du^{h,n} + f_N + \phi M \sin u^{h,n}) \right\}$$

One of the advantages of the spectral element method is, mass matrix is diagonal. Moreover, density matrix is also diagonal hence, can be easily inverted.

5. **Stability analysis of numerical scheme in non-homogeneous media.** A fully discretized linearized homogeneous undamped sG equation with periodic boundary conditions is given by

$$\begin{aligned} u^{h,n+1} &= (2u^{h,n} - u^{h,n-1}) + \Delta t^2 \underbrace{\boldsymbol{\rho}^{-1} M^{-1} \{(-D + \phi M)\}}_{\Psi_\rho^h} u^{h,n} \\ &= (2I + \Delta t^2 \Psi_\rho^h) u^{h,n} - I u^{h,n-1} \end{aligned}$$

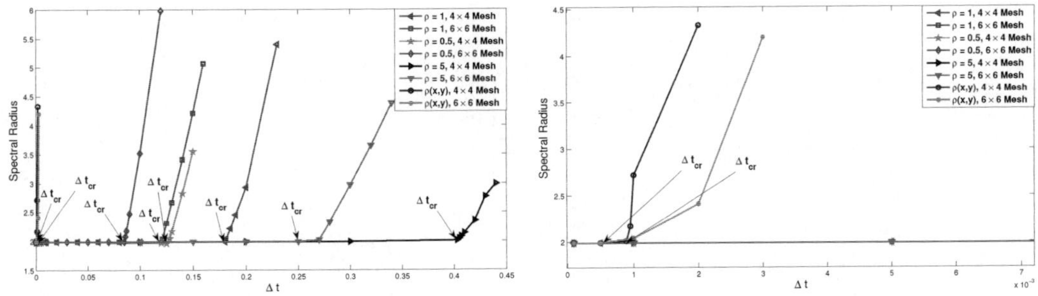

FIGURE 2. Spectral radius vs. Δt for various uniform and non-uniform density distributions (left) and its zoomed view (right).

where I is the identity matrix. Rewriting above equation in the form of system of equations as

$$\underbrace{\begin{Bmatrix} u^{h,n+1} \\ u^{h,n} \end{Bmatrix}}_{\tilde{u}^{h,\tilde{n}+1}} = \underbrace{\begin{bmatrix} (2I + \Delta t^2 \Psi_\rho^h) & -I \\ I & 0 \end{bmatrix}}_{\mathcal{A}} \underbrace{\begin{Bmatrix} u^{h,n} \\ u^{h,n-1} \end{Bmatrix}}_{\tilde{u}^{h,\tilde{n}}}$$

where second equation is trivial identity and $\mathcal{A}$ is the amplification matrix. Using recurrence relation one can write $\tilde{u}^{h,\tilde{n}+1} = \mathcal{A}^{\tilde{n}+1}\tilde{u}^{h,0}$. For bounded solution, $||\mathcal{A}^{\tilde{n}+1}|| \leq 1, \forall \tilde{n} \in \mathbb{N} \cup \{0\}$. This gives following condition on amplification matrix $||\mathcal{A}||^{\tilde{n}+1} \leq 1 \Rightarrow ||\mathcal{A}|| \leq 1 \Rightarrow \varrho(\mathcal{A}) \leq 1$ where $\varrho(\mathcal{A}) \triangleq \max\limits_{\forall i} |\lambda_i|$ is the spectral radius of the amplification matrix.

To further simplify the stability condition, let λ and $X = \{X_1,\ X_2\}^T$ be the eigenvalue and corresponding eigenvector of $\mathcal{A}$, then one can write the eigenvalue problem as $\mathcal{A}X = \lambda X$. This gives $BX_1 - X_2 = \lambda X_1;\ \ X_1 = \lambda X_2$, where $B \triangleq (2I + \Delta t^2 \Psi_\rho^h)$. Eliminating X_2 gives $BX_1 = \left(\lambda + \frac{1}{\lambda}\right) X_1$. If λ is real and $\left|\lambda + \frac{1}{\lambda}\right| \leq 2$, then this gives $|\lambda| = 1$. Thus, the stability condition obtained for bounded $\tilde{u}^{h,\tilde{n}+1}$ is $\varrho(B) \leq 2, \ \ \forall t > 0$. The critical time step is obtained as

$$\Delta t_{cr} \leq \frac{2}{\varrho\left(\frac{2I}{\Delta t} + \Delta t \Psi_\rho^h\right)} \tag{13}$$

where the property of spectral radius $\varrho(\alpha S) = \alpha\varrho(S)\ \forall \alpha \in \mathbb{R}$ and $S \in \mathbb{R}^{m \times m}$ is used. Hence, critical time step depends on density of non-homogeneous media through Ψ_ρ^h.

Equation (13) gives implicit relation for Δt_{cr} which depends on time step Δt, spatial resolution and density of the medium. As an example, following initial conditions are used $f(x, y) = 4\tan^{-1}(\exp(3 - \sqrt{x^2 + y^2}));\ \ g(x, y) = 0$. Figure 2 shows the variation of the spectral radius of matrix B verses time step for different mesh size using uniform as well as non-uniform density distributions where equation (7) is used for non-uniform density variation. It can be seen that Δt_{cr} value strongly depends on the density distribution. For uniform density distribution, the value of Δt_{cr} increases with density. This behavior can be seen in figure 3 (left) where Δt_{cr} is plotted against uniform density variation. Figure 3 (right) shows variation of Δt_{cr} value with constant term A involved in non-uniform Gaussian density distribution given by $\rho(x, y) = A\exp(-(0.4x^2 + 0.4y^2))$. Again, as expected critical time step increases with A. Moreover, critical time step for non-uniform density distribution is much lesser than that of uniform density distribution.

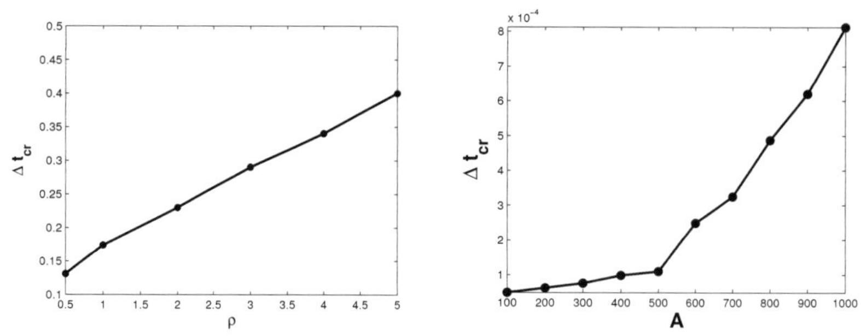

FIGURE 3. Density vs Δt_{cr} plot for uniform (left) and non-uniform (right) density variations.

6. **Results and discussions.** In this section two test cases, namely, kink (in one dimension) and circular ring soliton (in two dimensions) are considered to perform numerical experiments. In all simulations, sixth order Legendre polynomial based Lagrange basis function is used.

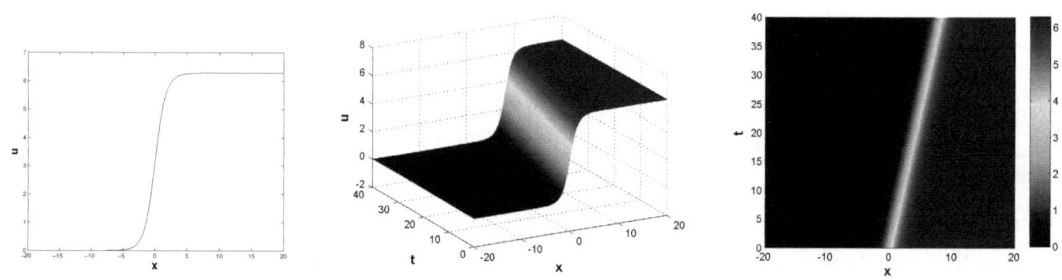

FIGURE 4. Initial solution for kink (left), surface plot of solution (middle) and solution on $x - t$ plane (right).

Kinks are spatially localized formation which moves freely in both directions. Physically, kink can represent dislocation in solids. Kinks with same sign interact with each other elastically, whereas kinks with opposite sign can cross each other without changing their sign. The solution of kink is given by

$$u(x,t) = 4 \arctan \left(\exp \left[\gamma(x - ct)\right]\right)$$

where $\gamma^2 \triangleq \frac{1}{1-c^2}$ is the Lorentz factor and constant c represents velocity of the kink. The domain is $[-20, 20]$, $\phi = -1$ and initial conditions for kink are

$$f(x) = 4 \arctan \left(\exp \left[\frac{x}{\sqrt{1-c^2}} \right] \right); \quad g(x) = -2\frac{c}{\sqrt{1-c^2}} \operatorname{sech} \left(\frac{x}{\sqrt{1-c^2}} \right)$$

with $c = 0.2$. Figure 4 shows the initial profile of the kink and the surface plot

time (t)	0	5	10	15	20
Energy E(t)	8.164965	8.142545	8.138649	8.157345	8.166322

TABLE 1. Energy conservation for kink test case.

of the solution along with solution on $x - t$ plane. Kink profile moves in right direction without changing its shape and size. One can see this behavior from $x - t$

plane. The composite trapezoidal rule is used to find the value of energy $E(t)$ over different time. Table 1 gives the values of $E(t)$ for kink at different time which shows that the energy remains nearly constant with increase in time. Next, we compute experimental order of convergence EOC $= \log_2 \frac{||\mathcal{E}_{h/2}||}{||\mathcal{E}_h||}$ where $||\mathcal{E}_h|| = ||u - u^h||$ can be calculated by comparing the numerical solution with that of exact solution for different grid size. Table 2 shows EOC in $\mathcal{L}_2$ and $\mathcal{H}^1$ norm where Δt is smaller than

No. of Elements	$\mathcal{L}_2$	EOC	$\mathcal{H}^1$	EOC
6	3.123e-3	-	7.526e-3	-
12	3.344e-5	6.54	1.104e-4	6.09
24	2.543e-7	7.03	1.678e-6	6.03
48	1.886e-9	7.07	2.588e-8	6.01

TABLE 2. Experimental order of convergence

the spatial resolution. The proposed implicit scheme gives optimal convergence rates in both the norms. Now, the proposed implicit scheme is used to solve sG equation in nonlinear non-homogeneous media. First we shall consider kink test case. As illustrated schematically in figure 5 (right), there are two possible outcomes which is observed namely, Kink captured and Kink passed.

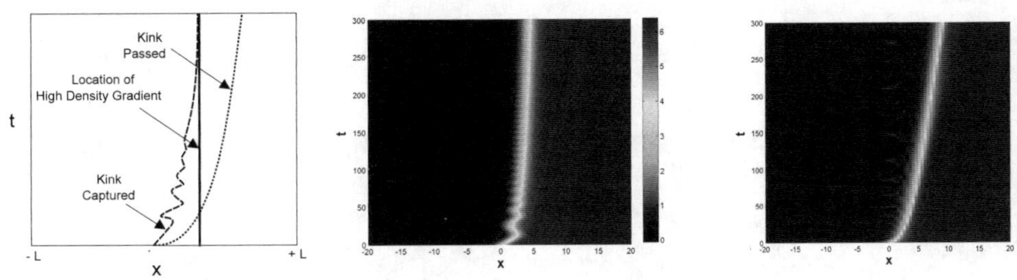

FIGURE 5. Schematic representation of kink captured and kink passed on $x-t$ plane (left) and numerical solution of kink on $x-t$ plane for discontinuous density (middle) and continuous density (right) variations.

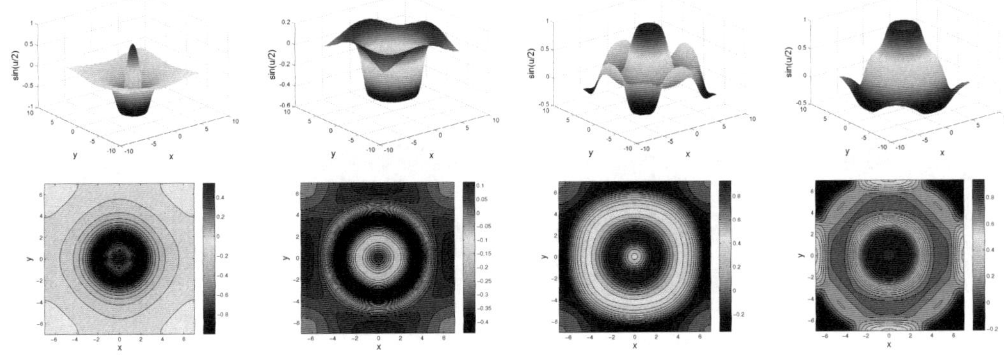

FIGURE 6. Solution of circular ring soliton at t = 5.6, 8.4, 11.2, 12.6 (left to right) in homogeneous media ($\rho = 1$).

time (t)	0	5.6	8.4	11.2	12.6
Energy E(t)	150.3244	150.3572	150.4267	150.8925	150.7499

TABLE 3. Energy conservation for circular ring soliton. $t \in [0,\ 12.6]$.

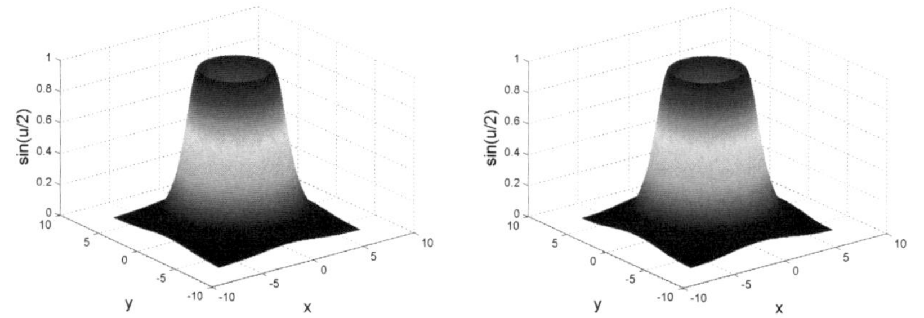

FIGURE 7. Initial solution of circular ring soliton (left) and solution of circular ring soliton at t = 12.6 over Gaussian density distribution (right).

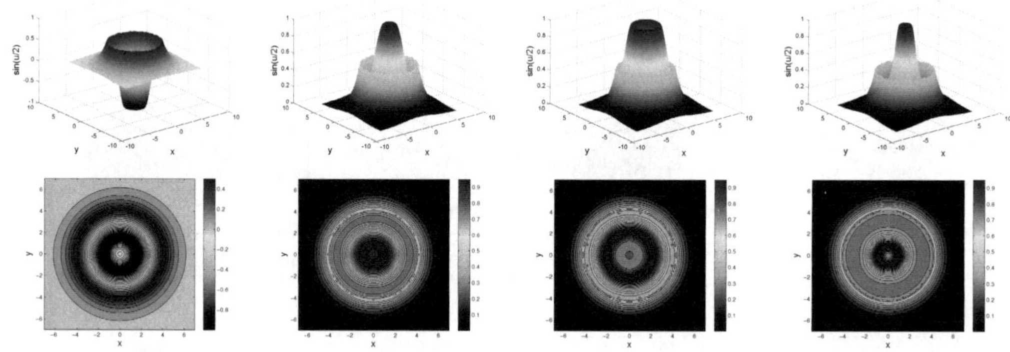

FIGURE 8. Solution of circular ring soliton at t = 5.6, 8.4, 11.2, 12.6 (left to right) in non-homogeneous media (circular discontinuity in density distribution).

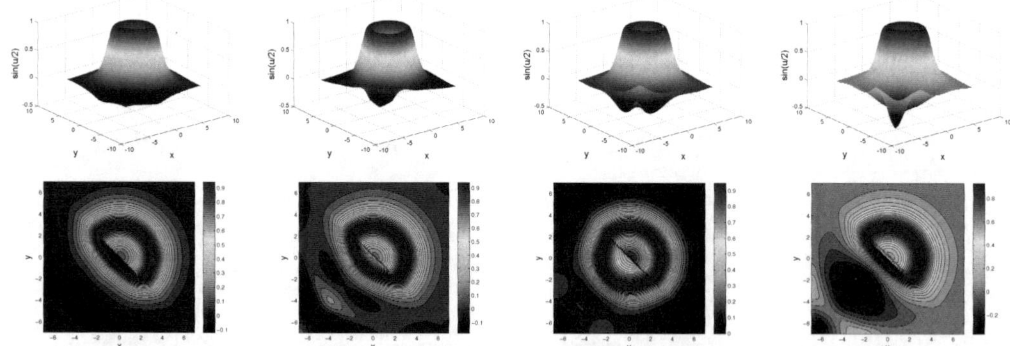

FIGURE 9. Solution of circular ring soliton at t = 5.6, 8.4, 11.2, 12.6 (left to right) in non-homogeneous media (oblique discontinuity in density distribution).

When the density variation is discontinuous the kink is captured near high-density gradient region and when it is smooth then the kink is passed with some deceleration. Interestingly, in both cases, the magnitude of jump in density is same, only the nature of jump is different. In figure 5, the solution on $x-t$ plane shows the dynamics of kink is discontinuous (middle) and continuous (right) density variations which is given by equations (6) and (5) respectively. In case of discontinuous density variation, as the time progresses the initial kink profile start moving rightward, but, this motion is prevented due to the sudden change in density at location $x=4$ which in turn forces kink to reverse its direction momentarily. After a small time interval, kink again starts in the original direction and again its motion is prevented due to discontinuous density jump. This cycle continues with an increase in frequency and eventually, the kink remain at location $x=4$ after large time as shown in figure 5 (middle). In contrast to discontinuous density variation, continuous density variation allows kink to pass through it but the kink velocity decreases. This can be seen on the $x-t$ plane shown in figure 5 (right) by a curved path of the kink profile.

Now, let's move on to the two-dimensional circular ring soliton test case. In this test case, $\phi=-1$ and the domain is $-7 \leq x, y \leq 7$. The initial conditions are $f(x,y) = 4\tan^{-1}(\exp(3-\sqrt{x^2+y^2}))$; $g(x,y)=0$. This gives initial radius close to three as shown in figure 7 (left). Figure 6 shows solution for time $t = 5.6, 8.4, 11.2, 12.6$ using fully implicit spectral element scheme. From the initial conditions, soliton appears at two homocentric ring soliton which shrinks till $t = 2.8$. From $t=5.6$ the soliton start expanding and radiating which is followed by oscillations at the boundary of the square domain. This expansion continues till $t=11.2$. The soliton start shrinking since $t=12.6$. The centre of soliton is not displaced from its initial position during these transformations. Table 3 gives the values of $E(t)$ for circular ring soliton at different time $t \in [0,\ 12.6]$ which shows that the energy remains nearly constant with increase in time.

Now, let's consider the circular ring soliton over a Gaussian density distribution given by equation (7). In this distribution density value increases as one go towards the centre of the domain from any direction. Such density distribution prevents spatio-temporal motion of the soliton. This can be seen in figure 7 (right) which shows the soliton at t = 12.6. Such distribution preserves the initial profile of the soliton. Next, circular ring soliton with the circular discontinuity is considered (see equation (6)). Density value inside the circle is lesser than that of outside, due to which only the middle part shows pulsating behavior and outer part remain rigid due to high density. Figure 9 shows the solution plots at t = 5.6, 8.4, 11.2, 12.6. One can also see the formation of circular kink along the line of density discontinuity. In case of oblique discontinuity given by equation (9), only portion $x \leq -y$ undergoes pulsating motion. Figure 9 shows the solution plots at different times. Similar to the circular discontinuity case, one can see the diagonal kink in the solution along the line of discontinuous density.

In these cases, one can see various effects of non-homogeneous media on sG solitons. In some cases, these non-homogeneous media preserves the initial profile of the soliton whereas in other cases only a part of soliton undergoes motion where low-density medium is present. Due to such non-homogeneity, one can control and guide the soliton in the desired way.

7. Conclusions. One- and two-dimensional sG solitons are studied in non- homogeneous media. Such media can be effectively used to guide the waves (in this case *'soliton'* wave) in the desired way. Nonlinear smooth and discontinuous density variations are used in the analysis. The governing sG equation is solved using a higher order spectral element scheme. Spectral stability analysis shows the strong dependence of critical time step Δt_{cr} on density variation. The value of Δt_{cr} depends not only on the magnitude of density but also on its nature of distribution. One-dimensional kink and two-dimensional circular ring soliton test cases are chosen to perform various numerical experiments. The spatio-temporal dynamics of these solitons completely changes in such non-homogeneous media.

Acknowledgments. The author would like to thanks Prof. A.S.Vasudeva Murthy, TIFR-Centre for Applicable Mathematics, Bangalore, India for many fruitful discussions as well as his comments/suggestions which improved the quality of this manuscript.

REFERENCES

[1] M.J.Ablowitz et al., Method for solving the sine-Gordon equation, *Phys. Rev. Lett.* **30** (1973), 1262–1264.
[2] A.L.Andreev, *Atomic Spectroscopy: Introduction to the Theory of Hyperfine Structure*, Springer, 2006.
[3] C.Q.Dai and F.B.Yu, Soliton solutions with power-law nonlinearity in inhomogeneous media, *Phys. Scr.* **87** (2013), 045002 (5pp).
[4] A.Degasperis et al., Multidimensional soliton equations in inhomogeneous media, *Physica Letters A*, Vol. 147 (1990).
[5] A.M.Dziewonski and D.L.Anderson, Preliminary reference Earth model, *Phys. Earth & Planetary Interiors*, **25** (1981) 297–356.
[6] P.G.Drazin and R.S.Johnson, *Solitons: An Introduction*, Cambridge University Press, 1989.
[7] A.Gharaati and R.Khordad, Dynamics of generalized sine-Gordon soliton in inhomogeneous media, *Indian J. Phys.* **85** (2011), 433–445.
[8] J.A.Gonzalez and B. de Mello, *Phys. Scripta* **54** (1996), 14.
[9] W.Gordon, Der Comptoneffekt nach der Schrödingerschen Theorie, *Z. Phys.* **40** (1926), 117–133.
[10] J.A.Gonzalez and M.Martin-Landrove, Solitons in a nonlinear DNA model, *Physics Letters A* **191** (1994), 409–415.
[11] L.E.Guerrero et al., Soliton structure dynamics in inhomogeneous media, *Physica A* **260** (1998), 418–424.
[12] A.D.Jagtap, E.Saha, J.D.George, and A.S.V.Murthy, Revisiting the inhomogeneously driven sine-Gordon equation, *Wave Motion* **73** (2017), 76–85.
[13] A.D.Jagtap and A.S.V.Murthy, Higher order scheme for two-dimensional inhomogeneous sine-Gordon equation with impulsive forcing, *Comm. Nonlinear Sci. Numer. Simulat.* **64** (2018), 178–197.
[14] A.D.Jagtap, On spatio-temporal dynamics of sine-Gordon soliton in nonlinear non-homogeneous media using fully implicit spectral element scheme, *Applicable Analysis, 2019.* https://doi.org/10.1080/00036811.2019.1588961
[15] O.Klein, Quantentheorie und fünfdimensionale Relativitästheorie, *Z. Phys.* **37** (1926), 895–906.
[16] A.C.Scott, A nonlinear Klein-Gordon equation, *Amer. J. Phys.* **37** (1969), 52–61.
[17] W.Shyu et al., Transition from soliton to chaos in nonlinear inhomogeneous media, *Physica Letters A* **249** (1998), 307–314.
[18] D.Zwillinger, *Handbook of Differential Equations*, Academic Press, 1997.

E-mail address: ameyadjagtap@gmail.com, ameya_jagtap@brown.edu

NONLOCAL BALANCE LAWS – RESULTS ON EXISTENCE, UNIQUENESS AND REGULARITY

ALEXANDER KEIMER

Institute for Transportation Studies (ITS)
University of California, Berkeley
CA-94720, USA

LUKAS PFLUG AND MICHELE SPINOLA

Department Mathematik, Chair of Applied Mathematics 2
Friedrich-Alexander-Universität Erlangen-Nürnberg (FAU)
Cauerstraße 11, 91058 Erlangen, Germany

ABSTRACT. In this contribution we present recent developments in the field of nonlocal balance laws. Starting from scalar balance laws we provide existence and uniqueness results and show that Entropy conditions as commonly used for local balance laws are not necessary to obtain unique solutions. Studying the problem as an initial value problem on $\mathbb{R}$ first, we advance our methods to multi-dimensional nonlocal balance laws on $\mathbb{R}^n$, where we focus in particular on the required regularity of the boundary of the nonlocal area of integration. We consider also scalar initial boundary value problems where the velocity function is considered to be nonnegative so that boundary datum can be – in a weak sense – prescribed in terms of the flux on one side.

1. **Introduction.** In recent year, nonlocal conservation and balance laws have drawn significantly more attention. In application, those models are used in supply chains [2, 7, 13], in traffic flow [3, 11, 20], in pedestrian flow, crowd dynamics [18, 5, 1], chemical engineering [14, 10] and recently also in opinion formation [21]. Scalar nonlocal conservation laws have been addressed as initial value problem in [15, 8, 3, 11] by applying the method of characteristics and a fixed-point problem to show existence and uniqueness of weak solutions – or by modified Lax-Friedrich's schemes relying on Kružkov's Entropy condition. For the multi-dimensional balance laws we refer the reader to [18, 1, 6, 5]. Initial boundary value problems of nonlocal conservation laws are considered in [20, 9]. All the results presented in this contribution can be found in a more detailed framework in [15, 20, 18].

In Section 2 we consider the initial value problem on $\mathbb{R}$ of scalar nonlocal conservation laws, in Section 3 we generalize the developed theory to initial boundary value problems and, finally, in Section 4 we consider multi-dimensional nonlocal balance laws as initial value problems. Section 5 will state some general remarks about nonlocal balance laws and future research.

2000 *Mathematics Subject Classification.* Primary: 35L03, 35L65; Secondary: 65M25, 35D30.

Key words and phrases. nonlocal conservation laws, nonlocal balance laws, existence of solutions, uniqueness of solutions, fixed-point problem, method of characteristics.

The authors gratefully acknowledge travel funding by the Bavaria California Technology Center (BaCaTec).

2. **Scalar nonlocal balance laws on $\mathbb{R}$.** In this section we investigate the archetype of a nonlocal balance law, a scalar nonlocal balance law on $\mathbb{R}$. In Definition 2.1 we first introduce what we mean by scalar nonlocal balance laws.

Definition 2.1 (Scalar nonlocal balance laws on $\mathbb{R}$)**.** We consider on Ω_T for $T \in \mathbb{R}_{>0}$ the nonlocal scalar balance law in $q : \Omega_T \to \mathbb{R}$ given for $(t,x) \in \Omega_T$ by

$$\begin{aligned} q_t(t,x) + \partial_x\Big(\lambda\Big[W\big[q,\gamma,a,b\big]\Big](t,x)q(t,x)\Big) &= h(t,x)\\ q(0,x) &= q_0(x) \end{aligned}$$

supplemented by the nonlocal term W, averaging the "density" in space

$$\begin{aligned} W\big[q,\gamma,a,b\big](t,x) &:= \textstyle\int_{a(x)}^{b(x)} \gamma(t,x,y)q(t,y)\,\mathrm{d}y\\ \lambda\Big[W\big[q,\gamma,a,b\big]\Big](t,x) &:= \lambda\left(W\big[q,\gamma,a,b\big](t,x),t,x\right). \end{aligned}$$

We call $q \in C([0,T];L^1(\mathbb{R}))$ a weak solution iff $\forall \phi \in C_c^1((-42,T)\times\mathbb{R})$ it holds that

$$\begin{aligned} &\textstyle\iint_{\Omega_T} \phi_t(t,x)q(t,x) + \phi_x(t,x)\lambda\big[W[q,\gamma,a,b]\big](t,x)q(t,x)\,\mathrm{d}x\,\mathrm{d}t\\ &\qquad + \int_{\mathbb{R}} \phi(0,x)q_0(x)\,\mathrm{d}x = -\textstyle\iint_{\Omega_T} h(t,x)\phi(t,x)\,\mathrm{d}x\,\mathrm{d}t. \end{aligned}$$

For obtaining existence and uniqueness results we require Assumption 1 on the involved datum. Most of these are quite natural and not restrictive.

Assumption 1 (Scalar nonlocal balance laws on $\mathbb{R}$)**.** *We require for $T \in \mathbb{R}_{>0}$ the involved functions in Definition 2.1 to satisfy the following*

$$\bullet\ q_0 \in L^1(\mathbb{R}) \cap L^\infty(\mathbb{R}) \quad \bullet\ h \in L^1((0,T);L^1(\mathbb{R})) \cap L^1((0,T);L^\infty(\mathbb{R}))$$
$$\bullet\ \lambda \in C(\mathbb{R}\times\Omega_T) \quad \bullet\ a,b \in W^{1,\infty}_{\text{loc}}(\mathbb{R}) \quad \bullet\ a',b' \in L^\infty(\mathbb{R})$$
$$\bullet\ \gamma \in L^\infty((0,T);W^{1,\infty}(\mathbb{R}^2))$$

and the velocity λ may satisfy the following growth conditions for every $W \in L^\infty((0,T);W^{1,\infty}(\mathbb{R}))$:

$$\exists A \in L^\infty_{\text{loc}}(\mathbb{R}_{\geq -1}) : \|\partial_3\lambda[W]\|_{L^\infty((0,T);L^\infty(\mathbb{R}))} \leq A\left(\|W\|_{L^\infty((0,T);L^\infty(\mathbb{R}))}\right) \tag{1}$$
$$\exists B \in L^\infty_{\text{loc}}(\mathbb{R}_{\geq -1}) : \|\partial_1\lambda[W]\|_{L^\infty((0,T);L^\infty(\mathbb{R}))} \leq B\left(\|W\|_{L^\infty((0,T);L^\infty(\mathbb{R}))}\right) \tag{2}$$

Let us discuss the assumptions in Eqs. (1) to (2). Those assure that the velocity or flux function will be Lipschitz-continuous. Equation (1) guarantees this for the purely spatial dependent part of the velocity, while Eq. (2) makes sure of that for the term involving the nonlocal part. These assumptions enable us to attack the conservation law by the method of characteristics. However, in the proposed generality there is no guarantee that the solution will not "explode" in finite time. Existence and uniqueness can thus only be guaranteed for sufficiently small time:

Theorem 2.2 (Existence and uniqueness of solutions for small time horizon)**.** *Given Assumption 1 and $T \in \mathbb{R}_{>0}$, the balance law in Definition 2.1 admits a unique weak solution on a sufficiently small time horizon, i.e. $\exists T^* \in (0,T]$ so that there exists a unique solution q which satisfies*

$$q \in C([0,T^*];L^1(\mathbb{R})) \cap L^\infty((0,T);L^\infty(\mathbb{R})).$$

Proof. The proof can be found in [15]. □

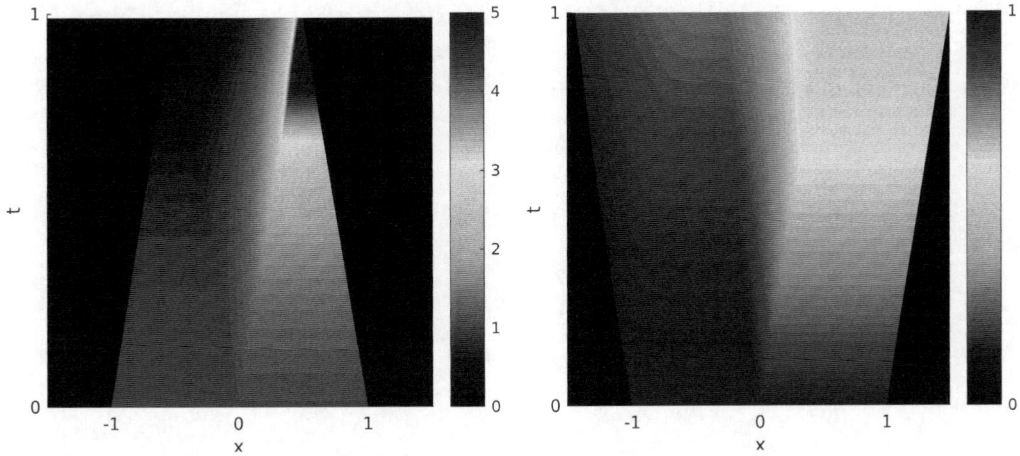

FIGURE 1. **Left:** $\lambda(W) = W - \frac{1}{2}$. **Right:** $\lambda(W) = \frac{1}{2} - W$.

Nevertheless, under specific assumptions we can obtain semi-global solutions in time. This is detailed in Remark 1 as well as a maximum principle given specific assumptions on velocity and boundary of the nonlocal term:

Remark 1 (Extension to arbitrary large time and maximum principle)**.** In case there exists $(\tilde{a}, \tilde{b}) \in (\mathbb{R} \cup \{-\infty, \infty\})^2$ s.t. $a \equiv \tilde{a}$ and $b \equiv \tilde{b}$ the solution can be extended on every finite time horizon. The same holds true if the support of the weight $\gamma(t, x, \cdot)$ is contained in $(a(x), b(x))$ for every $(t, x) \in (0, T) \times \mathbb{R}$. In addition, there is – under specific assumptions on velocity function λ, weight γ and sign of the initial datum a maximum principle. We do not go into further details but refer to the corresponding version for the initial boundary value problem in Remark 3. A precise analysis for the time-extension can be found in [15, Corollary 4.3].

Remark 2 (Entropy condition)**.** It is worth mentioning that according to Theorem 2.2 there is no need for an entropy condition to obtain uniqueness of weak solutions, on the contrary, weak solutions of nonlocal conservation laws are unique by themselves. This is also true for all other existence results in this contribution and can be explained due to the fact that there is no loss of information.

2.1. **Examples.** Consider the following example with datum

$$q_0 \equiv \chi_{[-1,1]},\ a(x) = x,\ b(x) = x + \eta,\ \eta = 1,\ \gamma \equiv 1,\ h \equiv 0,\ T = 1,$$

which is illustrated for two different types of flux functions in Fig. 1. For $\lambda(W) := W - 0.5$ the solution blows up in finite time (left illustration). On the right of the characteristic line $(t, 1 - 0.5t)$, $t \in [0, T]$, the velocity does not change as the solution remains constant, while starting from $x = 0$ the characteristic line is given by $(t, 0.5t)$, $t \in [0, T]$. As the two mentioned characteristic lines intersect in $(1, 0.5)$ this leads to a blow-up, i.e. the L^∞-norm of the solution goes to infinity when t approaches 1. On the right hand side for $\lambda(W) := 0.5 - W$ – the additive inverse velocity – the solution is spread out over time and exists on every finite time horizon. This is also a consequence of the maximum principle in Remark 1.

3. Scalar nonlocal balance laws on a bounded domain.

The introduced theory in Section 2 can also be used for a significantly more challenging problem, nonlocal scalar conservation laws on bounded domains, i.e. the corresponding initial boundary value problem:

Definition 3.1 (Scalar nonlocal conservation laws on a bounded domain)**.** We consider for $(t,x)\in(0,T)\times(0,1)$, and $T\in\mathbb{R}_{>0}$ the following nonlocal conservation law on $(0,1)$ in $q:(0,T)\times(0,1)\to\mathbb{R}$

$$\begin{aligned}q_t(t,x)&=-\partial_x\Big(\lambda\Big[W[q,v,\gamma_\eta]\Big](t,x)q(t,x)\Big)\\ q(0,x)&=q_0(x)\\ \lambda\big[W[q,v,\gamma_\eta]\big](t,0)q(t,0)&=\lambda\big[W[q,v,\gamma_\eta]\big](t,0)u(t)\\ W[q,v,\gamma_\eta](t,x)&:=\int_x^{x+\eta}\gamma_\eta(t,x,y)\left(\begin{cases}q(t,y)&y\in(0,1)\\ v(t,y)&\text{else}\end{cases}\right)\mathrm{d}y.\end{aligned}\tag{3}$$

We call $q:C([0,T];L^1((0,1)))$ a weak solution iff $\forall\phi\in W^{1,\infty}((0,T)\times(0,1))$ with $\phi(T,\cdot)\equiv0$ and $\phi(\cdot,1)\equiv0$ the following integral equation is satisfied

$$\begin{aligned}0=&\textstyle\iint_{\Omega_T}(\phi_t(t,x)+\lambda\big[W[q,v,\gamma_\eta]\big](t,x)\phi_x(t,x))q(t,x)\,\mathrm{d}x\,\mathrm{d}t\\ &+\textstyle\int_0^1 q_0(x)\phi(0,x)\,\mathrm{d}x+\int_0^T\phi(t,0)\lambda\big[W[q,v,\gamma_\eta]\big](t,0)u(t)\,\mathrm{d}t.\end{aligned}$$

Anticipating some of the later results we assume that velocity λ is nonnegative so that boundary datum can, if at all, only be prescribed at the left hand side of the considered domain (here $x=0$). As velocity can become zero we prescribe as “boundary datum” flux and not “density”. If the velocity is zero at some time $\tilde t\in(0,T)$ at $x=0$, boundary datum u is not attained at $\tilde t$ and whenever velocity is greater zero at $x=0$ boundary datum is attained and evoluting into the domain.

On the right hand side of Eq. (3) – due to the nonlocal term – we still have to prescribe a density for $x>1$ which is denoted by v. This density can be used for different modelling approaches (in traffic flow for instance a green or right light, a following empty road or fully congested road, etc.). The following assumptions are essential to obtain existence and uniqueness of solutions. Note that most of them are rather similar to Assumption 2 underlining the relation between the considered classes of balance laws.

Assumption 2 (Scalar nonlocal balance laws on a bounded domain)**.** *For $T\in\mathbb{R}_{>0}$ and $\eta\in\mathbb{R}_{>0}$ we require the involved function in Definition 3.1 to satisfy:*

$$\bullet\ \lambda(\cdot,*,\star)\in L^\infty\Big(\underbrace{(0,T)}_{*};W^{1,\infty}_{\text{loc}}(\underbrace{\mathbb{R}\times[0,1]}_{(\cdot,\star)};\mathbb{R}_{\geq0})\Big)$$

$$\bullet\ q_0\in L^\infty((0,1))\quad\bullet\ u\in L^\infty((0,T))$$

$$\bullet\ \gamma_\eta\in L^\infty\big((0,T);W^{1,\infty}((0,1)\times(0,1+\eta))\big)\quad\bullet\ v\in L^\infty((0,T);L^\infty((1,1+\eta))).$$

In addition, the velocity λ may satisfy the following growth condition for every $W\in L^\infty((0,T);W^{1,\infty}((0,1)))$

$$\exists A\in L^\infty_{\text{loc}}(\mathbb{R}_{>-1};\mathbb{R}_{\geq0}):\ \|\partial_1\lambda[w]\|_{L^\infty((0,T)\times(0,1))}\leq A\left(\|w\|_{L^\infty((0,T);L^\infty((0,1)))}\right).\tag{4}$$

As we consider a bounded spatial domain we only require Eq. (4) to obtain a Lipschitz-continuous velocity function. This is different from Eqs. (1) to (2) where we consider the Cauchy problem on $\mathbb{R}$ so that also a grow condition for the explicit dependency of the velocity with respect to the spatial variable is needed.

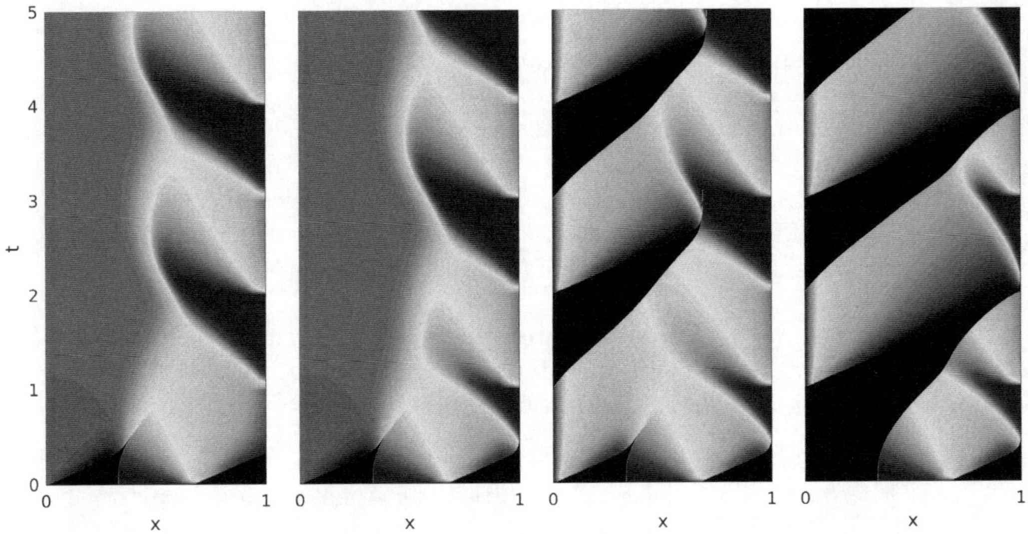

FIGURE 2. **Left:** Solution for $v(t) = \theta(t), u(t) = \frac{1}{4}$, **Middle left:** $v(t) = 1 - \theta(t), u(t) = \frac{1}{4}$, **Middle right:** $v(t) = u(t) = 1 - \theta(t)$, **Right:** $v(t) = 1 - \theta(t), u(t) = \theta(t)$

Theorem 3.2 (Existence and uniqueness of solutions for small time horizon). *Given Assumption 2 and $T \in \mathbb{R}_{>0}$, the balance law in Definition 3.1 admits a unique weak solution on a sufficiently small time horizon, i.e. $\exists T^* \in (0,T]$ so that there exists a unique solution q which satisfies*

$$q \in C([0,T^*]; L^1((0,1)) \cap L^\infty((0,T); L^\infty((0,1))).$$

Proof. The proof is inspired by the idea in Theorem 2.2 and can be found in [20]. □

As pointed out before, the maximum principle is crucial in particular if q should represent a density. The maximum principle is pretty common for local conservation laws, however, for nonlocal conservation laws we require specific assumptions to end up with a maximum principle.

Remark 3 (Maximum principle)**.** For the initial boundary value problem as considered here there exists also a maximum principle similar to Remark 1 as long as $\lambda' \leqq 0$ and $q_0 \geqq 0 \leqq v$ as well as the weight γ is non-decreasing w.r.t. third variable. For details, we refer the reader to [20, Corollary 5.9].

3.1. **Examples.** As a numerical example (the numeric in Section 2.1 and Section 3.1 is always carried out according to a numerical method based on characteristics, see [19]), we consider as velocity the in traffic flow classical Greenshield's flux function [12], as nonlocal area of integration we chose $\eta = \frac{1}{10}$ and as right hand side density v a functions which changes from zero to one periodically. The boundary datum is assumed to be constant $\frac{1}{4}$ and as initial datum we chose the characteristic function of the interval $(\frac{1}{3}, \frac{2}{3})$, in formula:

$$\lambda(W) = 1 - W,\ \eta = \tfrac{1}{10},\ \gamma(*, x, \cdot) \equiv 2(1 - \tfrac{\cdot - x}{\eta})\ \forall x \in (0,1),\ q_0 \equiv \chi_{(\frac{1}{3},\frac{2}{3})}.$$

The solutions are illustrated in Fig. 2 (the two leftmost figures). As can be observed, boundary datum enters the domain at $x = 0$ (with constant speed as there is no congestion (boundary density $u \equiv \frac{1}{4}$ is rather low) and as congestion is only caused by the initial datum around $x \in (\frac{1}{3}, \frac{2}{3})$ and on the right hand side by the "red traffic light" $v \equiv \theta, 1-\theta$ with $\theta :\equiv \sum_{i\in\mathbb{N}} \chi_{(2i-1,2i)}$ respectively, congestion is caused periodically in time and almost resolved for both cases, however, never spreads back to the beginning of the road ($x = 0$).

For a different boundary datum, which is – in some sense – synchronized with the outgoing density v, we obtain the solution as illustrated in Fig. 2 (the two rightmost figures). As can be seen, traffic congestion is never fully resolved for the first example, while the second example shows periodically a full dissolving of congestion at the right hand side border ($x = 1$).

4. **Multi-dimensional nonlocal balance laws on $\mathbb{R}^n$.** Finally, as prescribed in Section 1 we will consider multi-dimensional nonlocal balance laws. We stick with the notational convention that **bold** letters indicate vectors or matrices.

Definition 4.1 (Multi-dimensional nonlocal balance laws on $\mathbb{R}^n$)**.** Let $T \in \mathbb{R}_{>0}$ and $n \in \mathbb{N}_{\geq 1}$ be given, we consider the multi-dimensional nonlocal balance law in $q : (0,T) \times \mathbb{R}^n \to \mathbb{R}$ with damping for $(t,\boldsymbol{x}) \in (0,T) \times \mathbb{R}^n$

$$q_t(t,\boldsymbol{x}) + \operatorname{div}_2\Big(\boldsymbol{\lambda}\Big[W[q,\gamma,\Upsilon]\Big](t,\boldsymbol{x})q(t,\boldsymbol{x})\Big) = h(t,\boldsymbol{x}) + g(t,\boldsymbol{x})q(t,\boldsymbol{x})$$
$$q(0,\boldsymbol{x}) = q_0(\boldsymbol{x})$$
$$W[q,\gamma,\Upsilon](t,\boldsymbol{x}) := \iint_{\Upsilon(t)} \gamma(t,\boldsymbol{x},\boldsymbol{y})q(t,\boldsymbol{y})\,\mathrm{d}\boldsymbol{y}$$
$$\text{for } w \in C([0,T];C^1_{\mathrm{b}}(\mathbb{R}^n)) : \ \boldsymbol{\lambda}[w](t,\boldsymbol{x}) := \boldsymbol{\lambda}(w(t,\boldsymbol{x}),t,\boldsymbol{x}).$$

We call $q \in C([0,T];L^1(\mathbb{R}^n))$ weak solution iff $\forall \phi \in C^1_{\mathrm{c}}((-42,T)\times\mathbb{R}^n)$ it holds that

$$\int_0^T \iint_{\mathbb{R}^n} \left(\phi_t(t,\boldsymbol{x}) + \nabla_2\phi(t,\boldsymbol{x}) \circ \boldsymbol{\lambda}\left[W[q,\gamma,\Upsilon]\right](t,\boldsymbol{x})\right) q(t,\boldsymbol{x})\,\mathrm{d}\boldsymbol{x}\,\mathrm{d}t$$
$$+ \iint_{\mathbb{R}^n} \phi(0,\boldsymbol{x})q_0(\boldsymbol{x})\,\mathrm{d}\boldsymbol{x} = -\int_0^T \iint_{\mathbb{R}^n} \Big(h(t,\boldsymbol{x}) + g(t,\boldsymbol{x})q(t,\boldsymbol{x})\Big)\phi(t,\boldsymbol{x})\,\mathrm{d}\boldsymbol{x}\,\mathrm{d}t.$$

As can be seen from Definition 4.1 we consider a broad class of balance laws involving damping and inhomogeneity on the right hand side of the integral equation. However, it is worth mentioning that the boundary of the nonlocal term W is in the multi-dimensional case not explicitly space dependent as it is allowed for Definitions 2.1 to 3.1. This is due to the fact that a spatial derivative of the nonlocal term can only be computed for specific areae of integration. Instead, we assume that the area of integration can only change in time and that any spatial change is modelled via the weight γ. For existence/uniqueness of solutions we require:

Assumption 3 (Multi-dimensional nonlocal balance laws on $\mathbb{R}^n$)**.** *For $T \in \mathbb{R}_{>0}$ and $n \in \mathbb{N}_{\geq 1}$ we propose the following on the functions in Definition 4.1:*

$$\bullet\ q_0 \in L^1(\mathbb{R}^n) \cap L^\infty(\mathbb{R}^n) \ \bullet\ h \in L^1((0,T);L^1(\mathbb{R}^n)) \cap L^1((0,T);L^\infty(\mathbb{R}^n))$$
$$\bullet\ g \in C([0,T];C^1_b(\mathbb{R}^n)) \ \bullet\ \gamma \in C([0,T];C^1_b(\mathbb{R}^n\times\mathbb{R}^n))$$
$$\bullet\ \boldsymbol{\lambda} \in C(\mathbb{R}\times[0,T]\times\mathbb{R}^n;\mathbb{R}^n)$$

so that $\boldsymbol{\lambda}$ is sufficiently smooth and satisfies for every $W \in C([0,T];C^1_b(\mathbb{R}^n))$

$$\exists A \in L^\infty_{\mathrm{loc}}(\mathbb{R}_{\geq -1}) : \ \|\mathrm{D}_3\boldsymbol{\lambda}[W]\|_{C([0,T];C(\mathbb{R}^n;\mathbb{R}^{n\times n}))} \leq A\left(\|W\|_{C([0,T];C(\mathbb{R}^n))}\right)$$
$$\exists B \in L^\infty_{\mathrm{loc}}(\mathbb{R}_{\geq -1}) : \ \|\partial_1\boldsymbol{\lambda}[W]\|_{C([0,T];C(\mathbb{R}^n;\mathbb{R}^n))} \leq B\left(\|W\|_{C([0,T];C(\mathbb{R}^n))}\right).$$

For the nonlocal area of integration Υ we require that $\Upsilon(t) \subseteq \mathbb{R}^n$ is Lebesgue measurable for every $t \in [0,T]$ and $\Upsilon \in C([0,T];\mathscr{M}^n_K)$ for a $K \in \mathbb{R}_{>0}$ where

$$\mathscr{M}^n := \{\mathcal{A} \in \mathscr{L}(\mathbb{R}^n) : \ (\mathcal{A} \in \mathscr{L}_b(\mathbb{R}^n) \vee (\mathbb{R}^n \setminus \mathcal{A}) \in \mathscr{L}_b(\mathbb{R}^n)) \ \wedge \ \partial\mathcal{A}\ (n-1)\text{-rectifiable}\}$$

$$\mathscr{M}_K^n := \{\mathcal{A} \in \mathscr{M}^n : \mathcal{H}^{n-1}(\partial\mathcal{A}) \leq K\}.$$

As most of the assumptions are quite natural and are similar to Assumption 1 and Assumption 2 the assumptions on the area of integration Υ guarantee that we can control the boundary of the area of integration, a crucial ingredient for our fixed-point proof. We then obtain Theorem 4.2

Theorem 4.2 (Existence and uniqueness of solutions on every finite time horizon). *Given Assumption 3 and $T \in \mathbb{R}_{>0}, n \in \mathbb{N}_{\geq 1}$, the balance law in Definition 4.1 admits a unique weak solution q which satisfies*

$$q \in C([0,T^*]; L^1(\mathbb{R}^n)) \cap L^\infty((0,T); L^\infty(\mathbb{R}^n)).$$

Proof. The proof is inspired by the idea in Theorem 2.2 and can be found in [18]. □

It is worth mentioning that the considered result guarantees existence and uniqueness semi-global in time. This is due to the fact that the nonlocal area of integration does not explicitly depend on the spatial variable as well as the initial datum is assumed to be in $L^1(\mathbb{R}^n)$. Otherwise, this could not be guaranteed.

4.1. **Examples.** For examples we refer to [18, Section 6].

5. **Conclusions, further results and future research.** In this work we have presented recent developments in the theory of existence and uniqueness of nonlocal balance and conservation laws. We have considered different problem classes ranging from the purely initial value problem to multi-dimensional dynamics as well as the scalar initial boundary value problem.

The provided theory enables the study of interesting new problems: **(1)** The solutions obtained are based on characteristics. As there is no shock-development and no loss of information it is reasonable to introduced numerical schemes based on characteristics. The presented examples in Sections 2.1 and 3.1 already rely on such a method and show high numerical precision. In future research, a rigorous mathematical theory of convergence will be carried out [19]. **(2)** We are also able to study convergence of the nonlocal solution for $\eta \to 0$. Recent advance has been made in [4] where the authors showed that for a general nonlocal conservation law one cannot prove uniform BV estimates. However, in [16] it is shown that for monotonicity preserving velocity and monotone initial datum the nonlocal solution converges to the local solution in L^1 when the nonlocal area of integration η approaches zero. We also want to remark that one might actually not need to prove a BV estimate on the solution but on the nonlocal term – which naturally possesses higher regularity. **(3)** The considered model class can also be subject to time-delay and offers then more realistic modelling in specific frameworks. A study of existence and uniqueness for a fixed-delay is in progress, as a result we obtain convergence to the solution without delay when the delay approaches zero [17]. **(4)** As the nonlocal LWR model has some advantages to the local LWR model (finite acceleration, no need for an Entropy condition) and due to the nonlocal impact at the right hand side v, a generalization to networks is reasonable and the nonlocal impact might prove advantageous for the definition of proper junction models.

REFERENCES

[1] A. Aggarwal, R. Colombo and P. Goatin, Nonlocal systems of conservation laws in several space dimensions, *SIAM J. Num. Anal.*, **53** (2015), 963–983.

[2] D. Armbruster, D. Marthaler, C. Ringhofer, K. Kempf and T.-C. Jo, A continuum model for a re-entrant factory., *Operations Research*, **54** (2006), 933–950, URL http://dblp.uni-trier.de/db/journals/ior/ior54.html#ArmbrusterMRKJ06.

[3] S. Blandin and P. Goatin, Well-posedness of a conservation law with non-local flux arising in traffic flow modeling, *Numerische Mathematik*, 1–25, URL http://dx.doi.org/10.1007/s00211-015-0717-6.

[4] M. Colombo, G. Crippa and L. Spinolo, On the singular local limit for conservation laws with nonlocal fluxes, *submitted*.

[5] R. M. Colombo and M. Lécureux-Mercier, Nonlocal crowd dynamics models for several populations, *Acta Mathematica Scientia*, **32** (2012), 177–196.

[6] R. Colombo, M. Herty and M. Mercier, Control of the continuity equation with a non local flow, *ESAIM Control Optim. Calc. Var.*, **17** (2011), 353–379, URL http://dx.doi.org/10.1051/cocv/2010007.

[7] J.-M. Coron, M. Kawski and Z. Wang, Analysis of a conservation law modeling a highly re-entrant manufacturing system, *Discrete Contin. Dynam. Syst. Ser. B*, **14** (2010), 1337–1359, URL http://dx.doi.org/10.3934/dcdsb.2010.14.1337.

[8] G. Crippa and M. Lécureux-Mercier, Existence and uniqueness of measure solutions for a system of continuity equations with non-local flow, *Nonlinear Diff. Equat. and Appl.*, **20** (2013), 523–537.

[9] C. De Filippis and P. Goatin, The initial–boundary value problem for general non-local scalar conservation laws in one space dimension, *Nonlinear Analysis*, **161** (2017), 131–156.

[10] M. Dosta, S. Heinrich and J. Werther, Fluidized bed spray granulation: Analysis of the system behaviour by means of dynamic flowsheet simulation, *Powder Technology*.

[11] P. Goatin and S. Scialanga, Well-posedness and finite volume approximations of the lwr traffic flow model with non-local velocity, *Netw. Heter. Media*, **11** (2016), 107–121.

[12] B. Greenshields, W. Channing, H. Miller et al., A study of traffic capacity, in *Highway research board proceedings*, vol. 1935, National Research Council (USA), Highway Research Board, 1935.

[13] M. Gugat, A. Keimer, G. Leugering and Z. Wang, Analysis of a system of nonlocal conservation laws for multi-commodity flow on networks, *Netw. Heter. Media*, **10** (2015), 749–785, URL http://aimsciences.org/journals/displayArticlesnew.jsp?paperID=11749.

[14] M. Haderlein, D. Segets, M. Gröschel, L. Pflug, G. Leugering and W. Peukert, FIMOR: An efficient simulation for ZnO quantum dot ripening applied to the optimization of nanoparticle synthesis, *Chemical Engineering Journal*.

[15] A. Keimer and L. Pflug, Existence, uniqueness and regularity results on nonlocal balance laws, *J. Differential Equations*, **263** (2017), 4023–4069.

[16] A. Keimer and L. Pflug, On approximation of local conservation laws by nonlocal conservation laws, *J. Math. Anal. Appl.*, **475** (2019), 1927–1955.

[17] A. Keimer and L. Pflug, Nonlocal conservation laws with time delay, submitted.

[18] A. Keimer, L. Pflug and M. Spinola, Existence, uniqueness and regularity of multi-dimensional nonlocal balance laws with damping, *J. Math. Anal. Appl.* **466** (2018), 18 –55, URL http://www.sciencedirect.com/science/article/pii/S0022247X18304062.

[19] A. Keimer, L. Pflug and M. Spinola, Nonlocal balance laws: Theory of convergence for nondissipative numerical schemes, submitted.

[20] A. Keimer, L. Pflug and M. Spinola, Nonlocal scalar conservation laws on bounded domains and applications in traffic flow, *SIAM J. Math. Anal.*, **50** (2018), 6271–6306.

[21] B. Piccoli, N. Duteil and E. Trélat, Sparse control of Hegselmann-Krause models: Black hole and declustering, *arXiv preprint arXiv:1802.00615*.

E-mail address: keimer@berkeley.edu
E-mail address: lukas.pflug@fau.de
E-mail address: michele.spinola@fau.de

HOMOGENIZATION WITH TWO KINDS OF MICROSTRUCTURES: FROM THE MICROSCOPIC TO THE MACROSCOPIC DESCRIPTION OF CONCENTRATIONS OF CHEMICAL AGENTS

LAURA GIOIA ANDREA KELLER *

ETH Zurich
Department of Mathematics
Rämistrasse 101
Zurich, 8092, Switzerland

ABSTRACT. In this report we provide a more detailed discussion of the biological background and interpretation of a previously obtained homogenization result ([6]) which is motivated by folded structures observed in various cell organelles. More precisely, we investigate how geometric microstructures of a domain can affect the concentration and distribution of a chemical agent in a cell organelle on the macroscopic level. Our starting point is a suitable diffusion equation on a domain with additional microstructures of two kinds, the first one are periodically arranged "horizontal barriers" and the second one are "vertical barriers" which are not periodically arranged, but uniform on certain intervals. Both structures are parametrized in size by a small parameter ε. The ultimate goal is then to let ε tend to zero in order to get the effective equation at the macroscopic level taking into account the effects of the microstructures. The present note can be either seen as an addendum to the previously mentioned article or as an introduction to the topic (since a short presentation of the main result of [6] is included here as well).

1. **Biological background and motivation.** Various cell organelles, e.g. mitochondria or the endoplasmic reticulum, exhibit structures of strongly folded membranes (see for instance [3] or [8]). At first glimpse these structures might seem redundant, but in recent years more and more insight was gained into the importance and role of these strongly folded membrane structures. In particular, the following two observations are interesting:

- Deviations from the normal configuration of such folded membranes can be associated to diseases (see e.g. [12]).
- These structures of strongly folded membranes are not static. In particular, in apoptosis-related situations (programmed cell death) such folded structures are reorganized (see [11]). And this change in structure goes along with a release of e.g. calcium (see [2]).

2000 *Mathematics Subject Classification.* Primary: 35B27; Secondary: 92B05.

Key words and phrases. homogenization, two kinds of microstructure, concentration capacity, boundary conditions,

Parts of the presented results were established when the author was supported by Swiss National Science Foundation Grant PBEZP2_137396.

* Corresponding author: Laura Gioia Andrea Keller.

More generally, there are known various instances when such reorganizations of the folded structure can be observed. Such fusion and fission phenomena always go hand in hand with absorption or release of chemical agents (cf. e.g. [9]).

These biological observations gave rise to the question whether and how such folded structures can influence the distribution of chemical agents.
The result presented here can be seen as a proof of principle that in fact depending on the microscopic structure the macroscopic distribution of a chemical agent can be different. In biological terms, this means that the microstructure governs the distribution of chemical agents and by rearranging those microstructures chemical agents can be stored or released.

We would like to to point out here that there are various other possible questions in the context of folded membrane structures for which mathematical models and analysis had led to interesting new insight. Examples of such approaches can be found in [5], [10] and [4].

2. **Homogenization result.** In this section we briefly give the statement of the main homogenization result which was proven in [6] as well as a short description and discussion of the model at the microscopic and the macroscopic level, respectively.

2.1. **Model at the microscopic level.** In order to understand the effects of geometric microstructures, we first of all have to give a suitable description of them. Basically, we have two classes of such microstructures. The geometric setting is then the following:

- Outer container.
  Both microstructures are arranged in the interior of an outer container (in biological terms this corresponds to the outer membrane of a cell organelle). For the sake of simplicity, in what follows this outer container will be the rectangular region $(-R - \sigma\varepsilon, R + \sigma\varepsilon) \times (0, H)$. The roles of R, H and σ are illustrated in Figure 1 below and ε is the small parameter which encodes the thickness of the microstructures and also the parameter which will bridge the microscopic description to the macroscopic description when let tend to zero.
- Horizontal microstructures.
  Theses horizontal microstructures are arranged in a perfectly regular and equally spaced pattern. One can think of them as a stack of layers of a membrane. The thickness of theses microstructures is ε and the distance between two layers of them is $\nu\varepsilon$, respectively $(\nu\varepsilon)/2$ at the top and the bottom of the stack of these horizontal microstructures between the microstructure and the outer container. In Figure 1 below these horizontal microstructures are displayed in dark grey.
- Characteristic parameter of the horizontal microstructures.
  In dependence of the parameter ν there are more horizontal microstructures or fewer. In other words ν indicates how densely packed the layers of the horizontal microstructures are. This information will obviously also play a crucial role once we pass from the microscopic description to the macroscopic one. In oder to keep track of this information, it turns out that it is more

convenient to use the following parameter

$$\theta = \frac{1}{1+\nu}.$$

- Vertical microstructures.
  What concerns theses vertical microstructures, they might be present or not. The thickness of these second microstructures is again ε and - if present - they appear at distance $\sigma\varepsilon$ from the boundary of the outer container. Some possible configurations are shown in Figure 1 below (light grey).
  In addition, we assume that on the left side (i.e. for $-R = x$) there exists an interval $J_l \subset (0, H)$ such that on J_l always between two horizontal microstructures there is a vertical microstructure. The meaning of J_r is similarly defined as the subset of $(0, H)$ where always between two horizontal microstructures there is a vertical microstructure.
- Non-degeneracy assumption.
  Closed compartments are excluded, i.e. if between two horizontal layers of the modelled membrane (i.e. between two consecutive horizontal microstructures) there is present a vertical microstructure on the right side, there can be no vertical microstructure on the left side (and vice versa).

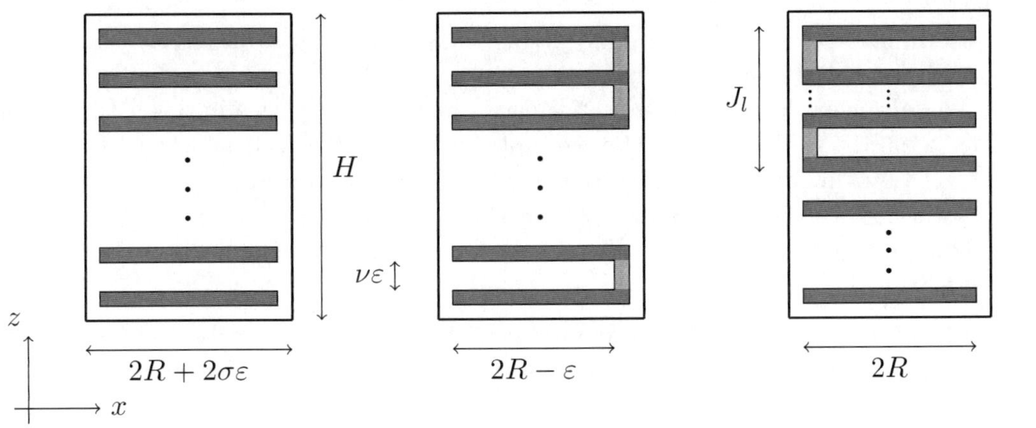

Model I
no vertical microstructures
$J_l = \emptyset = J_r$

Model II
"right side closed"
$J_l = \emptyset$, $J_r = \left(\frac{\nu\varepsilon}{2}, H - \frac{\nu\varepsilon}{2}\right)$

Model III
"mixed"
J_l as indicated above,
$J_r = \emptyset$

FIGURE 1. Illustration of the involved parameters and particular models studied with microstructures of thickness ε

The above described geometric properties of the underlying domain is just one part of the description of the distribution of a chemical agent - u_ε - at microscopic level $\varepsilon \in (0, 1]$. The second part is a suitable governing equation which in our case is a classical diffusion equation with some further features:

$$d_\varepsilon \frac{\partial}{\partial t} u_\varepsilon - \operatorname{div}(d_\varepsilon \nabla u_\varepsilon) = 0 \quad \text{in } W_\varepsilon \tag{1}$$

where W_ε denotes the free space where the diffusion can take place, i.e.

$$W_\varepsilon = (-R - \sigma\varepsilon, R + \sigma\varepsilon) \times (0, H) \setminus (\text{region occupied by the microstructures}).$$

This region is exactly the region which is displayed in white in Figure 1.

The function d_ε appearing in the above diffusion equation is defined as follows

$$d_\varepsilon(x,z) = \begin{cases} 1 & \text{if } |x| \leq R \\ \frac{1}{\varepsilon} & \text{if } R < |x| \leq R + \sigma\varepsilon \end{cases}$$

This additional factor amounts to the fact that the measure of the outer shell

$$S_\varepsilon = (-\sigma\varepsilon - R, -R) \times (0,H) \cup (R, R+\sigma\varepsilon) \times (0,H) \equiv S_{\varepsilon,l} \cup S_{\varepsilon,r}$$

tends to zero as ε tends to zero whereas the free region between the layers of the horizontal microstructures is preserved in measure. Further details about this can be found in Section 2.2 of the article [6].

The equation (1) is completed by homogeneous Neumann boundary conditions on all appearing boundaries and a bounded C^1-initial datum

$$u(x,z,0) = u_0(x,z) \geq 0.$$

Remark 1. A more detailed presentation and discussion of the microscopic description at level ε - including in particular also regularity properties of the weak solutions u_ε - can be found in the article [6]. Nevertheless, we would like to point out here a few important points:

- A similar situation in the context of visual conduction has been studied by Andreucci and collaborators in [1].
- We did not include any production rates at the boundaries in order to be able to separate the effects of the geometry of the two kinds of microstructures from (non-linear) production phenomena on these boundaries.
- Note that in our model we assume that the horizontal microstructures are arranged in a regular way. But apart from that we do not impose any periodicity.

Of course, it is not sufficient to come up with the model described above, but in fact one of the important points is to have a good control of the solutions of the problem at level ε described above. The most relevant properties of the model at level ε are the following (see also Proposition 1 in [6]):

i) The above described problem at level ε has a unique solution u_ε.

ii) The solutions u_ε (for varying $\varepsilon \in (0,1]$) are uniformly bounded and positive (recall that we already assumed that the initial datum is bounded and positive), i.e.

$$0 \leq u_\varepsilon \leq C$$

where C does not depend on ε.
Moreover, the solutions u_ε satisfy uniform energy bounds and time-regularity properties.

2.2. **Model at the macroscopic level.** Once the microscopic model is established and understood as briefly indicated in the preceding section one can pass to the macroscopic description by letting ε tend to zero.
From a biologist's point of view this can be seen as to determine the effective equation.

In the following, we will present the macroscopic description we obtain based on the microscopic model discussed above. And in the next section we will give an outline of the proof of this main result.

Before we state the macroscopic description, we would like to point out that when we very ε the domain of u_ε changes as well. In order to make all the solutions u_ε comparable we have to extend them to a common domain, which in our case is $\Omega_T = (-R,R) \times (0,H) \times (0,T]$. With this notation the macroscopic description reads as follows:

Theorem 2.1. *The suitable extensions $\widetilde{u}_\varepsilon$ of the solutions u_ε of the ε-problems from above converge in the sense of distributions to u, the solution of*

$$u_t - u_{xx} = 0 \text{ in } \quad \mathcal{D}'(\Omega_T)$$

with boundary conditions

$$\nabla u \cdot n = 0 \quad \text{on } \{x = -R\} \cap J_l$$

and

$$\nabla u \cdot n = 0 \quad \text{on } \{x = R\} \cap J_r.$$

In addition, u satisfies

$$u_x \in L^2(0,T; L^2(\Omega)).$$

Moreover, the restriction of u_ε to the outer shell S_ε on the right side, more precisely

$$v_{\varepsilon,r} = \frac{1}{\sigma\varepsilon} \int_R^{R+\sigma\varepsilon} u_\varepsilon(x,z,t)\, dx,$$

converges in the sense of distributions to v_r, the solution of

$$v_{r,t} - v_{r,zz} = -\frac{1-\theta}{\sigma} u_x|_{x=R}$$

with

$$v_{r,z} = 0 \quad \text{for } z = 0 \quad \text{and} \quad z = H.$$

Similarly,

$$v_{\varepsilon,l} = \frac{1}{\sigma\varepsilon} \int_{-R-\sigma\varepsilon}^{-R} u_\varepsilon(x,z,t)\, dx,$$

converges in the sense of distributions to v_l, the solution of

$$v_{l,t} - v_{l,zz} = -\frac{1-\theta}{\sigma} u_x|_{x=-R}$$

with

$$v_{l,z} = 0 \quad \text{for } z = 0 \quad \text{and} \quad z = H.$$

Moreover, the following transition condition holds

$$u = v_l \quad \text{on } \{x = -R\} \cap J_l^c,$$

and

$$u = v_r \quad \text{on } \{x = R\} \cap J_r^c.$$

Remark 2. Of course, the above theorem comes along with further assertions about the solution of the limit problem at macroscopic level. Apart from the boundedness of the solution (u,v) of the macroscopic description and its regularity properties, the most important fact is that the macroscopic description has a unique solution (u,v). This fact is absolutely crucial to guarantee that no information was lost when we let tend ε to zero. A more profound discussion of the properties of this limit problem can be found in [6].

2.3. **Some remarks about the proof of Theorem 2.1.** A detailed proof of Theorem 2.1 can be found in the article [6] (for the general theory of parabolic problems the reader is referred to [7]).
Nevertheless, we would like to point out here the most important steps and ideas that lead to Theorem 2.1.

Outline of the proof:

- Step 1: Uniform estimates.
  The starting point of the whole homogenization procedure are suitable uniform a priori estimates for the solutions u_ε of the ε-problems.
- Step 2: Limit in the interior region.
  In this step we restrict our attention to the region of the form $\{|x| < R - \delta\}$ where δ is at first fixed and will be sent to zero in a second step.
  Then, in order to establish the limit behaviour in this region we start by considering test functions $\varphi \in C^2$ such that

$$\varphi(t = 0) = 0 = \varphi(t = T)$$

  and

$$\varphi \text{ is supported away from the outer shell } S_\varepsilon$$

  in the weak formulation of the ε-problem

$$\begin{aligned} 0 &= -\int_0^T \int_{W_\varepsilon \cap \{|x|<R-\delta\}} u_\varepsilon \varphi_t - \int_0^T \int_{\partial W_\varepsilon \cap \{|x|<R-\delta\}} (\nabla u_\varepsilon \cdot n)\varphi \\ &\quad + \int_0^T \int_{W_\varepsilon \cap \{|x|<R-\delta\}} u_{\varepsilon,x}\varphi_x + \int_0^T \int_{W_\varepsilon \cap \{|x|<R-\delta\}} u_{\varepsilon,z}\varphi_z. \end{aligned}$$

  The problem that arises now is that the domains of u_ε change with ε. Thus, we have to find extensions that have all the same domain. This is done in the next step.
- Step 3: Construction of extensions $\widetilde{u}_\varepsilon$.
  As announced in Step 2, we need suitable extensions of u_ε.
  Once such extensions $\widetilde{u}_\varepsilon$ are at hand, the weak formulation from above can be rewritten as follows

$$-\int_0^T \int_{W_\varepsilon \cap \{|x|<R-\delta\}} u_\varepsilon \varphi_t = -\int_0^T \int_{(-R,R)\times(0,H)} \sum_j \chi_{I_j} \widetilde{u}_\varepsilon \varphi_t$$

  and similar for the other terms appearing in the weak formulation (where χ_{I_j} stands for the characteristic function of the free space between the jth and the $(j+1)$th horizontal microstructure, respectively the bottom of the outer container and the first horizontal microstructure and the top of the outer container and the last horizontal microstructure).
  But as one can see easily, a priori both $\sum_j \chi_{I_j}$ and $\widetilde{u}_\varepsilon$ only converge weakly. This means that the extensions we construct should satisfy suitable regularity properties as well.
  In fact, constructing the extensions $\widetilde{u}_\varepsilon$ by reflection and interpolation we can establish the following properties (see Lemma 9 in [6]):
  i) The extensions $\widetilde{u}_\varepsilon$ coincide with the solution u_ε on the space between the horizontal microstructures.

ii) The extensions $\widetilde{u}_\varepsilon$ enjoy the following uniform regularity properties

$$\widetilde{u}_\varepsilon \in L^2(0,T;W^{1,1}) \quad \text{with} \quad ||\widetilde{u}_\varepsilon||_{L^2(0,T;W^{1,1})} \leq C$$

and

$$||\widetilde{u}_\varepsilon(t+h) - \widetilde{u}_\varepsilon(t)||_{L^2(0,T-h;L^p)} \leq C\sqrt{h} \quad \forall h \in (0,T) \quad (1 < p < 2).$$

With these regularity assertions at hand we can now pass to the limit in the weak formulation above without any further difficulty. This finally leads to the limit description in the interior region.

- Step 4: Full limit.
  In remains to establish the limiting behaviour in the whole domain including the outer shell. In particular, it remains to understand what happens in S_ε. The idea here is to use test functions which are independent of x in the outer shell.

3. **Discussion.** We would like to close this presentation by a short discussion and interpretation of the found limit description at macroscopic level. The biological question one may have in mind is "What is the impact of the geometric configuration of the different microstructures on the distribution of a chemical agent?"

First of all, one could look at steady states (i.e. solutions with vanishing time derivative), respectively at the long time behaviour. In fact, the most striking difference can be seen between Model I and Model II (see Figure 1):

i) For Model I

$$u_\infty(x,z) = v_{l,\infty}(z) = v_{r,\infty}(z) \equiv \frac{1}{|\Omega|}\int_\Omega u_0$$

is the unique stationary solution.
Note that due to the transition condition the constant concentration has to be the same everywhere.

ii) For Model II

$$\begin{cases} v_{r,\infty}(z) & \equiv \dfrac{1}{H}\displaystyle\int_{S_r} v_r(t=0) \quad \text{independently of } z \\ u_\infty(x,z) = v_{l,\infty}(z) & \equiv \dfrac{1}{|\Omega|}\displaystyle\int_\Omega u_0 \quad \text{independently on } x \text{ and } z \end{cases}$$

is the unique stationary solution.
In biological terms, this can be interpreted in the sense that the presence of the vertical microstructures on the right hand side implies that this latter gets isolated from the rest of the outer container.

Furthermore, as a final illustration of the different behaviours of Model I and Model II, in Figures 2 and 3 numerical simulations of the two models are displayed.
For this simulation the following input was used: $\Omega = (-1,2) \times (-1,2)$, $\sigma = 1 - \theta$, $T = 2$ and u_0 was a bump function centred at $(0,0)$.

In conclusion, we can state that in effect the particular configuration of the microstructures can determine the distribution of a chemical agent.

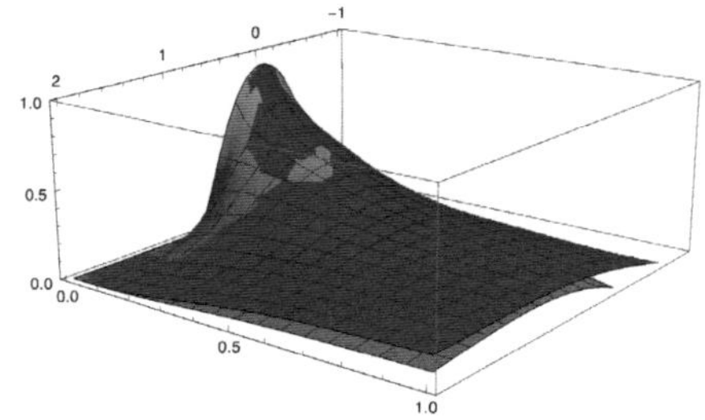

FIGURE 2. Numerical simulations for Model I (green) and Model II (red); $t \in [0, 1]$, $x \in [-1, 2]$ and $z = 0$.

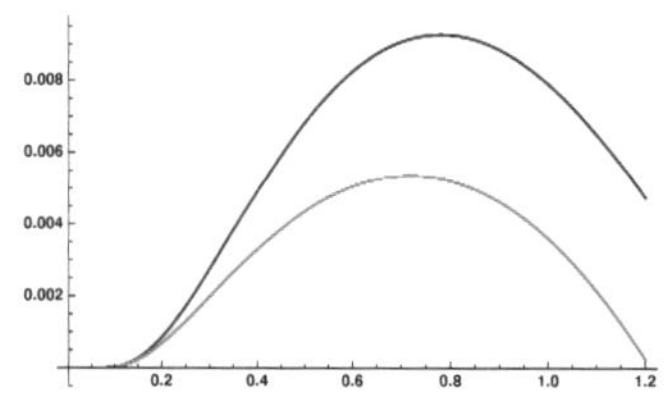

FIGURE 3. Numerical simulations for Model I (green) and Model II (red); $t \in [0, 1.2]$, $x = 1.85$ and $z = 0.65$.

REFERENCES

[1] D. Andreucci, P. Bisegna, and E. DiBenedetto, Homogenization and concentrated capacity for he heat equation with non-linear variational data in reticular almost disconnected structures and applications to visual transduction, *Ann. Mat. Pura Appl.*, **182** (2003), 375-407.

[2] S. Campello, R.A. Lacalle, M. Bettella, S. Manes, L. Scorrano and A. Viola, Orchestration of lymphocyte chemotaxis by mitochondrial dynamics, *JEM*, **203** (2006), 2879-2886.

[3] G.M. Cooper and R.E. Hausman, *The cell: a molecular approach*, 5th edition, ASM Press, Washington, D.C., 2009.

[4] J. Demongeot, N. Glade, O. Hansen and A. Moreira, An open issue: The inner mitochondrial membrane (IMM) as a free boundary problem, *Biochimie*, **89** (2007), 1049-1057.

[5] Y. Deng and M. Mieczkowski, Three-dimensional periodic cubic membrane structure in the mitochondria of amoebae Chaos carolinensis, *Protoplasma*, **203** (1998), 16-25.

[6] L.G.A. Keller, Homogenization and concentrated capacity for the heat equation with two kinds of microstructures: uniform cases, *Ann. Mat. Pura Appl.*, **196** (2017), 791-818.

[7] O.A. Ladyženskaja, V.A. Solonnikov and N.N. Ural'ceva, *Linear and quasilinear equations of parabolic typ*, American Mathematical Society, Providence, 1968.

[8] T.G. Frey and C.A. Mannella, The internal structure of mitochondria, *TIBS*, **25** (2000), 319-324.

[9] C.A. Mannella, The relevance of mitochondrial membrane topology to mitochondrial function, *Biochimica et Biophysica Acta*, **1762** (2006), 140-147.

[10] C. Renken, G. Siragusa, G. Perkins, L. Washington, J. Nulton, P. Salamon, and T.G. Frey, A thermodynamic model describing the nature of the crista junction: a structural motif in the mitochondrion, *Journal of Structural Biology*, **138** (2002), 137-144.

[11] L. Scorrano, Opening the doors to cytochrome c: Changes in mitochondrial shape and apoptosis, *The International Journal of Biochemistry and Cell Biology*, **41** (2009), 1875-1883.

[12] M. Zick, R. Rabl and A.S. Reichert, Cristae formation linking ultrastructure and function of mitochondria, *Biochimica et Biophysica Acta*, **1793** (2009), 5-19.

E-mail address: `laura.keller@math.ethz.ch`

NON–UNIQUENESS OF ENTROPY–CONSERVING SOLUTIONS TO THE IDEAL COMPRESSIBLE MHD EQUATIONS

Christian Klingenberg* and Simon Markfelder

Department of Mathematics, Würzburg University
Emil-Fischer-Str. 40
97074 Würzburg, Germany

Abstract. In this note we consider the ideal compressible magneto–hydrodynamics (MHD) equations in a special two dimensional setting. We show that there exist particular initial data for which one obtains infinitely many entropy–conserving weak solutions by using the convex integration technique. Finally this is applied to the isentropic case.

1. **Introduction.** We consider the ideal compressible magneto–hydrodynamics (MHD) equations

$$
\begin{aligned}
\partial_t \varrho + \operatorname{div}(\varrho \mathbf{u}) &= 0,\\
\partial_t(\varrho\mathbf{u}) + \operatorname{div}(\varrho\mathbf{u}\otimes\mathbf{u}) + \nabla p - (\operatorname{curl}\mathbf{B})\times\mathbf{B} &= 0,\\
\partial_t\left(\frac{1}{2}\varrho|\mathbf{u}|^2 + \varrho e(\varrho,p) + \frac{1}{2}|\mathbf{B}|^2\right)\qquad&\\
+\operatorname{div}\left[\left(\frac{1}{2}\varrho|\mathbf{u}|^2 + \varrho e(\varrho,p) + p + |\mathbf{B}|^2\right)\mathbf{u}\right] - \operatorname{div}\big((\mathbf{B}\cdot\mathbf{u})\mathbf{B}\big) &= 0,\\
\partial_t\mathbf{B} + \operatorname{curl}(\mathbf{B}\times\mathbf{u}) &= 0,\\
\operatorname{div}\mathbf{B} &= 0.
\end{aligned}
\tag{1}
$$

The unknown functions in (1) are the density $\varrho > 0$, the pressure $p > 0$, the velocity $\mathbf{u} \in \mathbb{R}^3$ and the magnetic field $\mathbf{B} \in \mathbb{R}^3$, which are all functions of the time $t \in [0, T)$ and the spatial variable $\mathbf{x} = (x, y, z)^\mathsf{T} \in \mathbb{R}^3$. The internal energy e is a given function of the density ϱ and the pressure p.

In this note we consider a special two dimensional setting. Let $\Omega \subset \mathbb{R}^2$ a bounded two dimensional spacial domain. We consider $\mathbf{u} = (u, v, 0)^\mathsf{T}$ and $\mathbf{B} = (0, 0, b)^\mathsf{T}$ and furthermore we let all the unknowns only depend on $(x, y) \in \Omega$. From now on we write $\mathbf{u} = (u, v)^\mathsf{T} \in \mathbb{R}^2$ and $\mathbf{x} = (x, y)^\mathsf{T} \in \Omega \subset \mathbb{R}^2$ for the corresponding two

2000 *Mathematics Subject Classification.* Primary: 76W05, 35D30; Secondary: 76N15, 35Q35.
Key words and phrases. magnetohydrodynamics, compressible flow, weak solutions, convex integration.
* Corresponding author.

dimensional vectors. Then the MHD system (1) turns into

$$\begin{aligned}\partial_t \varrho + \operatorname{div}(\varrho \mathbf{u}) &= 0,\\ \partial_t(\varrho \mathbf{u}) + \operatorname{div}(\varrho \mathbf{u} \otimes \mathbf{u}) + \nabla\Big(p + \frac{1}{2}b^2\Big) &= 0,\\ \partial_t\left(\frac{1}{2}\varrho|\mathbf{u}|^2 + \varrho e(\varrho,p) + \frac{1}{2}b^2\right) + \operatorname{div}\left[\left(\frac{1}{2}\varrho|\mathbf{u}|^2 + \varrho e(\varrho,p) + p + b^2\right)\mathbf{u}\right] &= 0,\\ \partial_t b + \operatorname{div}(b\mathbf{u}) &= 0.\end{aligned} \tag{2}$$

Note that in (2) div , ∇ are two dimensional differential operators in contrast to (1), where they are three dimensional differential operators.

We endow system (2) with initial conditions

$$(\varrho, p, \mathbf{u}, b)(0,\cdot) = (\varrho_0, p_0, \mathbf{u}_0, b_0) \tag{3}$$

and impermeability boundary conditions

$$\mathbf{u}\cdot\mathbf{n}\big|_{\partial\Omega} = 0. \tag{4}$$

Definition 1.1. A 4-tuple $(\varrho, p, \mathbf{u}, b) \in L^\infty\big([0,T)\times\Omega; (0,\infty)\times(0,\infty)\times\mathbb{R}^2\times\mathbb{R}\big)$ is a weak solution to (2), (3), (4) if the following equations hold for all test functions $\varphi, \phi, \psi \in C_c^\infty\big([0,T)\times\mathbb{R}^2\big)$ and $\boldsymbol{\varphi} \in C_c^\infty\big([0,T)\times\mathbb{R}^2;\mathbb{R}^2\big)$ with $\boldsymbol{\varphi}\cdot\mathbf{n}\big|_{\partial\Omega} = 0$:

$$\int_0^T\int_\Omega [\varrho\partial_t\varphi + \varrho\mathbf{u}\cdot\nabla\varphi]\,\mathrm{d}\mathbf{x}\,\mathrm{d}t + \int_\Omega \varrho_0\varphi(0,\cdot)\,\mathrm{d}\mathbf{x} = 0; \tag{5}$$

$$\begin{aligned}\int_0^T\int_\Omega \left[\varrho\mathbf{u}\cdot\partial_t\boldsymbol{\varphi} + (\varrho\mathbf{u}\otimes\mathbf{u}) : \nabla\boldsymbol{\varphi} + \Big(p + \frac{1}{2}b^2\Big)\operatorname{div}\boldsymbol{\varphi}\right]\,\mathrm{d}\mathbf{x}\,\mathrm{d}t&\\ + \int_\Omega \varrho_0\mathbf{u}_0\cdot\boldsymbol{\varphi}(0,\cdot)\,\mathrm{d}\mathbf{x} &= 0;\end{aligned} \tag{6}$$

$$\begin{aligned}\int_0^T\int_\Omega \Bigg[&\Big(\frac{1}{2}\varrho|\mathbf{u}|^2 + \varrho e(\varrho,p) + \frac{1}{2}b^2\Big)\partial_t\phi\\ &+\Big(\frac{1}{2}\varrho|\mathbf{u}|^2 + \varrho e(\varrho,p) + p + b^2\Big)\mathbf{u}\cdot\nabla\phi\Bigg]\,\mathrm{d}\mathbf{x}\,\mathrm{d}t\\ &+\int_\Omega \Big(\frac{1}{2}\varrho_0|\mathbf{u}_0|^2 + \varrho_0 e(\varrho_0,p_0) + \frac{1}{2}b_0^2\Big)\phi(0,\cdot)\,\mathrm{d}\mathbf{x} = 0;\end{aligned} \tag{7}$$

$$\int_0^T\int_\Omega [b\partial_t\psi + b\mathbf{u}\cdot\nabla\psi]\,\mathrm{d}\mathbf{x}\,\mathrm{d}t + \int_\Omega b_0\psi(0,\cdot)\,\mathrm{d}\mathbf{x} = 0. \tag{8}$$

Remark 1.2. The impermeability boundary condition is represented by the choice of the test functions.

Remark 1.3. Note that we exclude vacuum for our consideration, i.e. in this note $\varrho > 0$, $p > 0$.

It is a well–known fact that there may exist physically non–relevant weak solutions to conservation laws. Hence one has to introduce additional selection criteria in order to single out the physically relevant weak solutions. A common approach is to impose an entropy inequality. However for the MHD system (1) there is no known entropy.

Note that for the Euler system the functions

$$\eta = -\varrho s(\varrho, p) \quad \text{and} \quad \mathbf{q} = -\varrho s(\varrho, p)\mathbf{u}$$

form an entropy pair. Here the specific entropy $s = s(\varrho, p)$ is a given function as well as the internal energy e and note that these functions are interrelated by Gibbs' relation.

It is a straightforward computation to show that a strong solution to the MHD system (1) fulfills

$$\partial_t\big(\varrho s(\varrho, p)\big) + \operatorname{div}\big(\varrho s(\varrho, p)\mathbf{u}\big) = 0. \tag{9}$$

Although this suggests that $(\eta, \mathbf{q})$ is an entropy pair for the MHD system, too, $(\eta, \mathbf{q})$ is *not* an entropy pair for MHD, cf. [2]. However $(\eta, \mathbf{q})$ is still used as a selection criterion in the literature for example if Riemann problems are considered and one wants to find out whether or not a shock is physical, see e. g. [9]. We misuse terminology and call η and $\mathbf{q}$ still entropy, entropy flux respectively.

The weak solutions, whose existence we will prove in this note, fulfill the entropy equation (9) in the weak sense. We call such solutions entropy–conserving.

Definition 1.4. A weak solution $(\varrho, p, \mathbf{u}, b)$ to (2), (3), (4) is called *entropy–conserving*, if for all test functions $\varphi \in C_c^\infty\big([0,T) \times \mathbb{R}^2\big)$ the entropy equation

$$\int_0^T \int_\Omega \big[\varrho s(\varrho, p)\partial_t\varphi + \varrho s(\varrho, p)\mathbf{u} \cdot \nabla\varphi\big] \,\mathrm{d}\mathbf{x}\,\mathrm{d}t + \int_\Omega \varrho_0 s(\varrho_0, p_0)\varphi(0, \cdot)\,\mathrm{d}\mathbf{x} = 0 \tag{10}$$

holds.

The following theorem is our main result:

Theorem 1.5. *Let $\varrho_0, p_0 \in L^\infty(\Omega; (0,\infty))$ and $b_0 \in L^\infty(\Omega)$ be arbitrary piecewise constant functions. Then there exists $\mathbf{u}_0 \in L^\infty(\Omega; \mathbb{R}^2)$ such that there are infinitely many entropy–conserving weak solutions to (2) with initial data $\varrho_0, p_0, \mathbf{u}_0, b_0$ and impermeability boundary condition. These solutions have the property that ϱ, p and b do not depend on time; in other words $\varrho \equiv \varrho_0$, $p \equiv p_0$ and $b \equiv b_0$.*

The proof of Theorem 1.5 relies on the non–uniqueness proof for the full Euler system provided in [7] and consists of two main ideas. The first one is to make use of a result (see Proposition 2.1 below) which was proved by Feireisl [6] and also by Chiodaroli [3]. This result is based on the convex integration method, that was developed by De Lellis and Székelyhidi [4, 5] in the context of the pressureless incompressible Euler equations. The second idea is the fact that ϱ, p and b can be chosen *piecewise* constant, what was observed originally by Luo, Xie and Xin [8].

Note that non–uniqueness of weak solutions fulfilling an entropy inequality (even in one space dimension) is well–known: There exist Riemann initial data for which one has more than one solutions, see e. g. Torrilhon [9] and references therein.

Note furthermore that there is also a convex integration result to incompressible ideal MHD by Bronzi et al. [1]. There the same two dimensional setting as in the present note is considered. In contrast to this note, where a convex integration result for Euler is used, Bronzi et al. apply the convex integration techique directly to an incompressible version of (2).

2. **Proof of the main result.** In order to prove Theorem 1.5 we will make use of the following proposition whose proof is based on convex integration.

Proposition 2.1. *Let $Q \subset \mathbb{R}^2$ a bounded domain, $\varrho > 0$ and $C > 0$ positive constants. Then there exists $\mathbf{m}_0 \in L^\infty(Q;\mathbb{R}^2)$ such that there are infinitely many functions*

$$\mathbf{m} \in L^\infty\big((0,T)\times Q;\mathbb{R}^2\big) \cap C_{\text{weak}}\big([0,T];L^2(Q;\mathbb{R}^2)\big)$$

satisfying

$$\int_0^T \int_Q \mathbf{m}\cdot\nabla\varphi \,\mathrm{d}\mathbf{x}\,\mathrm{d}t = 0, \tag{11}$$

$$\int_0^T \int_Q \left[\mathbf{m}\cdot\partial_t\boldsymbol{\varphi} + \left(\frac{\mathbf{m}\otimes\mathbf{m}}{\varrho} - \frac{1}{2}\frac{|\mathbf{m}|^2}{\varrho}\mathbb{I}\right) : \nabla\boldsymbol{\varphi}\right]\mathrm{d}\mathbf{x}\,\mathrm{d}t + \int_Q \mathbf{m}_0\cdot\boldsymbol{\varphi}(0,\cdot)\,\mathrm{d}\mathbf{x} = 0, \tag{12}$$

for all test functions $\varphi \in C_c^\infty([0,T)\times\mathbb{R}^2)$ and $\boldsymbol{\varphi} \in C_c^\infty([0,T)\times\mathbb{R}^2;\mathbb{R}^2)$, and additionally

$$E_{\text{kin}} = \frac{1}{2}\frac{|\mathbf{m}|^2}{\varrho} = C \quad \textit{a.e. in } (0,T)\times Q, \qquad E_{0,\text{kin}} = \frac{1}{2}\frac{|\mathbf{m}_0|^2}{\varrho} = C \quad \textit{a.e. in } Q.$$

For the proof of Proposition 2.1 we refer to [6, Theorem 13.6.1].
Now we are able to prove Theorem 1.5.

Proof of Theorem 1.5. Let $\varrho_0, p_0 \in L^\infty(\Omega;(0,\infty))$ and $b_0 \in L^\infty(\Omega)$ given piecewise constant functions. Then there exist finitely many $Q_i \subset \Omega$ open and pairwise disjoint, such that $\Omega = \bigcup_i \overline{Q_i}$ and $\varrho_0\big|_{Q_i} = \varrho_i$, $p_0\big|_{Q_i} = p_i$ and $b_0\big|_{Q_i} = b_i$ with constants $\varrho_i, p_i > 0$ and $b_i \in \mathbb{R}$. We apply Proposition 2.1 on each Q_i to $\varrho = \varrho_i$ and $C = \Lambda - p_i - \frac{1}{2}b_i^2$, where Λ is a constant with $\Lambda > \max_i\big(p_i + \frac{1}{2}b_i^2\big)$. This yields $\mathbf{m}_{0,i} \in L^\infty(Q_i;\mathbb{R}^2)$ and infinitely many $\mathbf{m}_i \in L^\infty((0,T)\times Q_i;\mathbb{R}^2)$ with the properties given in Proposition 2.1. We then piece together the $\mathbf{m}_{0,i} \in L^\infty(Q_i;\mathbb{R}^2)$ to $\mathbf{m}_0 \in L^\infty(\Omega;\mathbb{R}^2)$ and the $\mathbf{m}_i \in L^\infty((0,T)\times Q_i;\mathbb{R}^2)$ to $\mathbf{m} \in L^\infty((0,T)\times\Omega;\mathbb{R}^2)$.

We define $\mathbf{u}_0 := \frac{\mathbf{m}_0}{\varrho_0} \in L^\infty(\Omega;\mathbb{R}^2)$ and for each momentum field $\mathbf{m}$ we define a corresponding velocity field $\mathbf{u} := \frac{\mathbf{m}}{\varrho_0} \in L^\infty\big((0,T)\times\Omega;\mathbb{R}^2\big)$. Furthermore we define $(\varrho,p,b) \in L^\infty\big([0,T)\times\Omega;(0,\infty)\times(0,\infty)\times\mathbb{R}\big)$ by $\varrho \equiv \varrho_0$, $p \equiv p_0$ and $b \equiv b_0$. We claim that $(\varrho,p,\mathbf{u},b)$ is an entropy–conserving weak solution to (2) with initial data $\varrho_0, p_0, \mathbf{u}_0, b_0$.

Let $\varphi, \phi, \psi \in C_c^\infty\big([0,T)\times\mathbb{R}^2\big)$ and $\boldsymbol{\varphi} \in C_c^\infty\big([0,T)\times\mathbb{R}^2;\mathbb{R}^2\big)$ with $\boldsymbol{\varphi}\cdot\mathbf{n}\big|_{\partial\Omega} = 0$ arbitrary test functions. Using (11) and (12), we obtain the following.

$$\begin{aligned}&\int_0^T\int_\Omega \left[\varrho\partial_t\varphi + \varrho\mathbf{u}\cdot\nabla\varphi\right]\mathrm{d}\mathbf{x}\,\mathrm{d}t + \int_\Omega \varrho_0\varphi(0,\cdot)\,\mathrm{d}\mathbf{x}\\ &= \sum_i \varrho_i \int_{Q_i}\left(\int_0^T \partial_t\varphi\,\mathrm{d}t + \varphi(0,\cdot)\right)\mathrm{d}\mathbf{x} + \sum_i \int_0^T\int_{Q_i} \mathbf{m}_i\cdot\nabla\varphi\,\mathrm{d}\mathbf{x}\,\mathrm{d}t \ = \ 0;\end{aligned}$$

$$\int_0^T\int_\Omega \Big[\varrho\mathbf{u}\cdot\partial_t\boldsymbol{\varphi} + (\varrho\mathbf{u}\otimes\mathbf{u}) : \nabla\boldsymbol{\varphi} + \Big(p+\frac{1}{2}b^2\Big)\mathrm{div}\,\boldsymbol{\varphi}\Big]\,\mathrm{d}\mathbf{x}\,\mathrm{d}t + \int_\Omega \varrho_0\mathbf{u}_0\cdot\boldsymbol{\varphi}(0,\cdot)\,\mathrm{d}\mathbf{x}$$

$$= \sum_i \Bigg(\int_0^T\int_{Q_i}\Big[\mathbf{m}_i\cdot\partial_t\boldsymbol{\varphi} + \Big(\frac{\mathbf{m}_i\otimes\mathbf{m}_i}{\varrho_i} - \frac{1}{2}\frac{|\mathbf{m}_i|^2}{\varrho_i}\mathbb{I}\Big) : \nabla\boldsymbol{\varphi}\Big]\,\mathrm{d}\mathbf{x}\,\mathrm{d}t$$

$$+ \int_{Q_i}\mathbf{m}_{0,i}\cdot\boldsymbol{\varphi}(0,\cdot)\,\mathrm{d}\mathbf{x}\Bigg) + \sum_i\int_0^T\int_{Q_i}\Big[\frac{1}{2}\frac{|\mathbf{m}_i|^2}{\varrho_i} + \Big(p_i+\frac{1}{2}b_i^2\Big)\Big]\,\mathrm{div}\,\boldsymbol{\varphi}\,\mathrm{d}\mathbf{x}\,\mathrm{d}t$$

$$= \Lambda\int_0^T\int_\Omega \mathrm{div}\,\boldsymbol{\varphi}\,\mathrm{d}\mathbf{x}\,\mathrm{d}t \quad = \quad 0;$$

$$\int_0^T\int_\Omega\Big[\Big(\frac{1}{2}\varrho|\mathbf{u}|^2 + \varrho e(\varrho,p) + \frac{1}{2}b^2\Big)\partial_t\phi + \Big(\frac{1}{2}\varrho|\mathbf{u}|^2 + \varrho e(\varrho,p) + p + b^2\Big)\mathbf{u}\cdot\nabla\phi\Big]\,\mathrm{d}\mathbf{x}\,\mathrm{d}t$$

$$+ \int_\Omega\Big(\frac{1}{2}\varrho_0|\mathbf{u}_0|^2 + \varrho_0 e(\varrho_0,p_0) + \frac{1}{2}b_0^2\Big)\phi(0,\cdot)\,\mathrm{d}\mathbf{x}$$

$$= \sum_i\Big(\Lambda + \varrho_i e(\varrho_i,p_i) - p_i\Big)\int_{Q_i}\Big(\int_0^T\partial_t\phi\,\mathrm{d}t + \phi(0,\cdot)\Big)\,\mathrm{d}\mathbf{x}$$

$$+ \sum_i\frac{\Lambda + \varrho_i e(\varrho_i,p_i) + \frac{1}{2}b_i}{\varrho_i}\int_0^T\int_{Q_i}\mathbf{m}_i\cdot\nabla\phi\,\mathrm{d}\mathbf{x}\,\mathrm{d}t \quad = \quad 0;$$

$$\int_0^T\int_\Omega\big[b\partial_t\psi + b\mathbf{u}\cdot\nabla\psi\big]\,\mathrm{d}\mathbf{x}\,\mathrm{d}t + \int_\Omega b_0\psi(0,\cdot)\,\mathrm{d}\mathbf{x}$$

$$= \sum_i b_i\int_{Q_i}\Big(\int_0^T\partial_t\psi\,\mathrm{d}t + \psi(0,\cdot)\Big)\,\mathrm{d}\mathbf{x} + \sum_i\frac{b_i}{\varrho_i}\int_0^T\int_{Q_i}\mathbf{m}_i\cdot\nabla\psi\,\mathrm{d}\mathbf{x}\,\mathrm{d}t \quad = \quad 0.$$

We have shown that the equations (5) - (8) hold. Hence $(\varrho, p, \mathbf{u}, b)$ is indeed a weak solution. It remains to show that this solution is entropy–conserving. In other words we have to show that (10) holds. Let $\varphi \in C_c^\infty([0,T)\times\mathbb{R}^2)$ be an arbitrary test function. We obtain

$$\int_0^T\int_\Omega\big[\varrho s(\varrho,p)\partial_t\varphi + \varrho s(\varrho,p)\mathbf{u}\cdot\nabla\varphi\big]\,\mathrm{d}\mathbf{x}\,\mathrm{d}t + \int_\Omega\varrho_0 s(\varrho_0,p_0)\varphi(0,\cdot)\,\mathrm{d}\mathbf{x}$$

$$= \sum_i\varrho_i s(\varrho_i,p_i)\int_{Q_i}\Big(\int_0^T\partial_t\varphi\,\mathrm{d}t + \varphi(0,\cdot)\Big)\,\mathrm{d}\mathbf{x}$$

$$+ \sum_i s(\varrho_i,p_i)\int_0^T\int_{Q_i}\mathbf{m}_i\cdot\nabla\psi\,\mathrm{d}\mathbf{x}\,\mathrm{d}t \quad = \quad 0.$$

Thus $(\varrho, p, \mathbf{u}, b)$ is an entropy–conserving weak solution. Since there are infinitely many $\mathbf{m}$ from Prop. 2.1, there are infinitely many entropy–conserving solutions $(\varrho, p, \mathbf{u}, b)$. □

3. **Isentropic MHD.** In this section we apply our result to isentropic MHD equations. The isentropic MHD system reads

$$\begin{aligned} \partial_t \varrho + \operatorname{div}(\varrho \mathbf{u}) &= 0, \\ \partial_t(\varrho \mathbf{u}) + \operatorname{div}(\varrho \mathbf{u} \otimes \mathbf{u}) + \nabla p(\varrho) - (\operatorname{curl} \mathbf{B}) \times \mathbf{B} &= 0, \\ \partial_t \mathbf{B} + \operatorname{curl}(\mathbf{B} \times \mathbf{u}) &= 0, \\ \operatorname{div} \mathbf{B} &= 0. \end{aligned} \tag{13}$$

The unknown functions are the density $\varrho > 0$, the velocity $\mathbf{u} \in \mathbb{R}^3$ and the magnetic field $\mathbf{B} \in \mathbb{R}^3$. In contrast to the MHD system (1) the pressure p in (13) is not an unknown but a given function of the density, where $p(\varrho) > 0$ for all $\varrho > 0$.

Again we consider a two dimensional setting. Let $\Omega \subset \mathbb{R}^2$ a bounded two dimensional spacial domain. We consider $\mathbf{u} = (u, v, 0)^\mathsf{T}$ and $\mathbf{B} = (0, 0, b)^\mathsf{T}$ and furthermore we let all the unknowns only depend on $(x, y) \in \Omega$. Then the isentropic MHD system (13) turns into

$$\begin{aligned} \partial_t \varrho + \operatorname{div}(\varrho \mathbf{u}) &= 0, \\ \partial_t(\varrho \mathbf{u}) + \operatorname{div}(\varrho \mathbf{u} \otimes \mathbf{u}) + \nabla \Big(p(\varrho) + \frac{1}{2} b^2 \Big) &= 0, \\ \partial_t b + \operatorname{div}(b \mathbf{u}) &= 0. \end{aligned} \tag{14}$$

For the isentropic *Euler* system, the energy

$$\eta = \frac{1}{2} \varrho |\mathbf{u}|^2 + P(\varrho) + \frac{1}{2} |\mathbf{B}|^2$$

is an entropy. Here $P(\varrho)$ is called pressure potential and is given by

$$P(\varrho) = \varrho \int_1^\varrho \frac{p(r)}{r} \, \mathrm{d}r.$$

Similar to the full MHD system considered above, one can show that the energy is *not* an entropy for (13) but strong solutions fulfill the corresponding energy equation

$$\begin{aligned} &\partial_t \Big(\frac{1}{2} \varrho |\mathbf{u}|^2 + P(\varrho) + \frac{1}{2} |\mathbf{B}|^2 \Big) \\ &\quad + \operatorname{div} \Big[\Big(\frac{1}{2} \varrho |\mathbf{u}|^2 + P(\varrho) + p(\varrho) + |\mathbf{B}|^2 \Big) \mathbf{u} \Big] - \operatorname{div} \big((\mathbf{B} \cdot \mathbf{u}) \mathbf{B} \big) = 0. \end{aligned} \tag{15}$$

Hence we will look for energy–conserving weak solutions. In the considered setting the energy equation (15) turns into

$$\partial_t \Big(\frac{1}{2} \varrho |\mathbf{u}|^2 + P(\varrho) + \frac{1}{2} b^2 \Big) + \operatorname{div} \Big[\Big(\frac{1}{2} \varrho |\mathbf{u}|^2 + P(\varrho) + p(\varrho) + b^2 \Big) \mathbf{u} \Big] = 0.$$

Definition 3.1. A triple $(\varrho, \mathbf{u}, b) \in L^\infty\big([0,T) \times \Omega; (0,\infty) \times \mathbb{R}^2 \times \mathbb{R}\big)$ is a weak solution to (14) with initial data $\varrho_0, \mathbf{u}_0, b_0$ and impermeability boundary condition if the following equations hold for all test functions $\varphi, \psi \in C_c^\infty\big([0,T) \times \mathbb{R}^2\big)$ and $\boldsymbol{\varphi} \in C_c^\infty\big([0,T) \times \mathbb{R}^2; \mathbb{R}^2\big)$ with $\boldsymbol{\varphi} \cdot \mathbf{n}\big|_{\partial\Omega} = 0$:

$$\int_0^T \int_\Omega [\varrho \partial_t \varphi + \varrho \mathbf{u} \cdot \nabla \varphi] \, \mathrm{d}\mathbf{x} \, \mathrm{d}t + \int_\Omega \varrho_0 \varphi(0, \cdot) \, \mathrm{d}\mathbf{x} = 0; \tag{16}$$

$$\int_0^T \int_\Omega \Big[\varrho\mathbf{u}\cdot\partial_t\boldsymbol{\varphi} + (\varrho\mathbf{u}\otimes\mathbf{u}) : \nabla\boldsymbol{\varphi} + \Big(p(\varrho)+\frac{1}{2}b^2\Big)\operatorname{div}\boldsymbol{\varphi}\Big]\,\mathrm{d}\mathbf{x}\,\mathrm{d}t + \int_\Omega \varrho_0\mathbf{u}_0\cdot\boldsymbol{\varphi}(0,\cdot)\,\mathrm{d}\mathbf{x} = 0; \tag{17}$$

$$\int_0^T \int_\Omega [b\partial_t\psi + b\mathbf{u}\cdot\nabla\psi]\,\mathrm{d}\mathbf{x}\,\mathrm{d}t + \int_\Omega b_0\psi(0,\cdot)\,\mathrm{d}\mathbf{x} = 0. \tag{18}$$

A weak solution is called energy–conserving if in addtion for all test functions $\phi\in C_c^\infty\big([0,T)\times\mathbb{R}^2\big)$ the energy equation

$$\begin{aligned}\int_0^T \int_\Omega \Big[&\Big(\frac{1}{2}\varrho|\mathbf{u}|^2 + P(\varrho) + \frac{1}{2}b^2\Big)\partial_t\phi \\ &+\Big(\frac{1}{2}\varrho|\mathbf{u}|^2 + P(\varrho) + p(\varrho) + b^2\Big)\mathbf{u}\cdot\nabla\phi\Big]\,\mathrm{d}\mathbf{x}\,\mathrm{d}t \\ &+\int_\Omega \Big(\frac{1}{2}\varrho_0|\mathbf{u}_0|^2 + P(\varrho_0) + \frac{1}{2}b_0^2\Big)\phi(0,\cdot)\,\mathrm{d}\mathbf{x} = 0\end{aligned} \tag{19}$$

holds.

The Cauchy problem for the isentropic MHD equations is ill–posed, too:

Corollary 3.2. *Let $\varrho_0\in L^\infty(\Omega;(0,\infty))$ and $b_0\in L^\infty(\Omega)$ be arbitrary piecewise constant functions. Then there exists $\mathbf{u}_0\in L^\infty(\Omega;\mathbb{R}^2)$ such that there are infinitely many energy–conserving weak solutions to* (14) *with initial data $\varrho_0,\mathbf{u}_0,b_0$ and impermeability boundary condition. These solutions have the property that ϱ and b do not depend on time; in other words $\varrho\equiv\varrho_0$ and $b\equiv b_0$.*

Proof of Corollary 3.2. Let $\varrho_0\in L^\infty(\Omega;(0,\infty))$ and $b_0\in L^\infty(\Omega)$ given piecewise constant functions. Set furthermore $p_0 := p(\varrho_0)$. Then $p_0\in L^\infty(\Omega;(0,\infty))$ is a piecewise constant function. Additionally we can choose the function $e(\varrho,p)$ in such a way that $\varrho_0 e(\varrho_0,p_0) = P(\varrho_0)$. We know from Theorem 1.5 that there exists an initial velocity $\mathbf{u}_0\in L^\infty(\Omega;\mathbb{R}^2)$ such that there are infinitely many entropy–conserving weak solutions $(\varrho\equiv\varrho_0, p\equiv p_0,\mathbf{u}, b\equiv b_0)$ to (2) with initial data $\varrho_0,p_0,\mathbf{u}_0,b_0$. It is easy to check that for each of these solutions, the triple $(\varrho\equiv\varrho_0,\mathbf{u},b\equiv b_0)$ is an energy–conserving weak solution to the isentropic MHD equations (14) with initial data $\varrho_0,\mathbf{u}_0,b_0$ in the sense of Definition 3.1. □

Acknowledgement. The authors thank Bruno Despres for fruitful discussions.

REFERENCES

[1] A. Bronzi, M. Lopes Filho and H. Nussenzveig Lopes, Wild solutions for 2D incompressible ideal flow with passive tracer, *Comm. Math. Sci.*, **13**(5) (2015), 1333–1343.

[2] P. Chandrashekar and C. Klingenberg, Entropy stable finite volume scheme for ideal compressible MHD on 2-D cartesian meshes, *SIAM J. Numer. Anal.*, **54**(2) (2016), 1313–1340.

[3] E. Chiodaroli, A counterexample to well-posedness of entropy solutions to the compressible Euler system, *J. Hyperbolic Differ. Equ.*, **11**(3) (2014), 493–519.

[4] C. De Lellis and L. Székelyhidi Jr., The Euler equations as a differential inclusion, *Ann. of Math. (2)*, **170**(3) (2009), 1417–1436.

[5] C. De Lellis and L. Székelyhidi Jr., On admissibility criteria for weak solutions of the Euler equations, *Arch. Ration. Mech. Anal.*, **195**(1) (2010), 225–260.

[6] E. Feireisl, Weak solutions to problems involving inviscid fluids, in *Mathematical Fluid Dynamics, Present and Future*, Springer Proceedings in Mathematics and Statistics **183**, Springer-Verlag, (2016), 377–399.
[7] E. Feireisl, C. Klingenberg, O. Kreml and S. Markfelder, On oscillatory solutions to the complete Euler system, preprint, arXiv:1710.10918.
[8] T. Luo, C. Xie and Z. Xin, Non-uniqueness of admissible weak solutions to compressible Euler systems with source terms, *Adv. Math.*, **291** (2016), 542–583.
[9] M. Torrilhon, *Zur Numerik der idealen Magnetohydrodynamik*, Ph.D thesis, Eidgenössische Technische Hochschule, Zürich, 2003.

E-mail address: klingen@mathematik.uni-wuerzburg.de
E-mail address: simon.markfelder@mathematik.uni-wuerzburg.de

PIECEWISE DETERMINISTIC MARKOV PROCESSES DRIVEN BY SCALAR CONSERVATION LAWS

STEPHAN KNAPP*

University of Mannheim
Department of Mathematics
68131 Mannheim, Germany

ABSTRACT. We investigate piecewise deterministic Markov processes (PDMP), where the deterministic dynamics follows a scalar conservation law and random jumps in the system are characterized by changes in the flux function. We show under which assumptions we can guarantee the existence of a PDMP and conclude bounded variation estimates for sample paths. Finally, we apply this dynamics to a production and traffic model and use this framework to incorporate the well-known scattering of flux functions observed in data sets.

1. **Introduction.** The simplicity of scalar conservation laws allows to understand general behaviors of underlying models but, on the other hand, they are based on qualified assumptions as for example steady state or expected values. One possibility to widen this class of models are systems of conservation laws, where fluctuations and higher order moments can be governed. Another possibility to extend scalar conservation laws are stochastic effects. More precisely, starting from deterministic scalar conservation laws and a corresponding initial value problem (IVP)

$$u_t(x,t) + f(u(x,t))_x = 0, \quad u(x,0) = u_0(x), \tag{1}$$

a natural extension is the incorporation of uncertainties. There already exist extensions based on a reformulation as stochastic differential equation like in [12] and partial stochastic differential equation as in [5, 17] in the literature. Also uncertain initial data as for example in [6] and random chosen flux functions [18] have been considered. In the latter work, the flux function is random and does not change randomly in time.

In contrast to [18], our goal is a stochastic process, which "chooses" a new flux function at random times, where these times and the random choice of the next flux function may dependent on the actual solution of the whole system. This can be easily motivated by, e.g. production models with machine failures [8, 10], and also opinion formation, change of state (gas to liquid or vice versa) are reasonable applications. This idea directly transfers us into the theory of piecewise deterministic

2000 *Mathematics Subject Classification.* Primary: 60J25, 35L65; Secondary: 90B30.

Key words and phrases. scalar conservation laws, piecewise deterministic Markov processes, production, LWR, stochastic.

Financially supported by the BMBF project ENets (05M18VMA) and DAAD-PPP USA (Project-ID 57444394).

Markov processes, see [14]. In detail, given a parametrized family of Lipschitz continuous flux functions $f^\alpha \in C^{0,1}(\mathbb{R})$ for $\alpha \in I \subset \mathbb{R}$, we are interested in a "solution" to

$$u_t(x,t) + f^{\alpha(t)}(u(x,t))_x = 0, \quad u(x,0) = u_0(x), \tag{2}$$

where $\alpha(t) \in I$ denotes the current and random chosen flux function at time $t \in [0,T]$ and $x \in \mathbb{R}$.

We define how (2) has to be understood and how $\alpha(t)$ is specified in the subsequent section 2. This section is followed by applications and numerical results in the case of a production and traffic model in section 3.

2. **Modeling Equations.** Let $u\colon \mathbb{R} \to \mathbb{R}$ be a function, then we denote by $\mathrm{TV}(u)$ its total variation and define $\mathrm{BV}(\mathbb{R}) = \{u\colon \mathbb{R} \to \mathbb{R}\colon \mathrm{TV}(u) < \infty\}$ as the set of all functions from $\mathbb{R}$ to $\mathbb{R}$ with total bounded variation, see, e.g. [13]. With this notation, it is well known as a result of Krusckov, see [13], that the IVP (1) has a unique weak entropy solution if $u_0 \in \mathrm{BV}(\mathbb{R}) \cap L^1(\mathbb{R})$ and if $f \in C^{0,1}(\mathbb{R})$, i.e., is Lipschitz continuous. Furthermore, the solution u satisfies

$$\|u(\cdot,t)\|_\infty \le \|u_0\|_\infty, \tag{3}$$

$$\mathrm{TV}(u(\cdot,t)) \le \mathrm{TV}(u_0), \tag{4}$$

$$\|u(\cdot,t) - u(\cdot,s)\|_{L^1} \le \|f\|_{C^{0,1}} \mathrm{TV}(u_0)|t-s| \tag{5}$$

and is L^1 stable with respect to initial data

$$\|u(\cdot,t) - v(\cdot,t)\|_{L^1} \le \|u_0 - v_0\|_{L^1}. \tag{6}$$

Deterministic dynamics between jump times. In this section, we define the dynamics to (2) based on theory of PDMPs. Unfortunately, we cannot apply the theory of PDMPs directly on solutions to (1) with corresponding flux functions f^α since $\mathrm{BV}(\mathbb{R})$ is not separable and hence no Borel space. Following [2], we use the extended solution operator to (1) on $L^1(\mathbb{R})$ and denote it by $S_t^\alpha\colon L^1(\mathbb{R}) \to L^1(\mathbb{R})$, where α indicates that flux function f^α is used. We directly deduce the following properties of the family $(S_t^\alpha, t \in [0,T])$ for every $\alpha \in I$:

$$S_{s+t}^\alpha = S_s^\alpha S_t^\alpha = S_t^\alpha S_s^\alpha \text{ for } s,t \in [0,T] \text{ with } s+t \in [0,T], \tag{7}$$

$$S_0^\alpha = Id, \tag{8}$$

$$t \mapsto S_t^\alpha \in C([0,T]; L^1(\mathbb{R})), \tag{9}$$

$$\|S_t^\alpha u - S_t^\alpha v\|_{L^1} \le \|u - v\|_{L^1}, \tag{10}$$

$$t \mapsto S_t^\alpha u_0 \text{ is the unique entropy solution to (1) if } u_0 \in L^1(\mathbb{R}) \cap \mathrm{BV}(\mathbb{R}). \tag{11}$$

Up to now, we have no specification of $\alpha(t)$. We define the state space $E = L^1(\mathbb{R}) \times I$ equipped with the Borel σ-algebra $\mathcal{E}$ generated by the open sets induced by $\|(u,\alpha)\| = \|u\|_{L^1} + |\alpha|$ for $(u,\alpha) \in E$. Then $(E,\mathcal{E})$ is a Borel space.

Our aim is to switch the flux function only at random times, which results in deterministic dynamics between the jumps in the form of

$$\phi_t\colon E \to E, \quad \begin{pmatrix} u \\ \alpha \end{pmatrix} \mapsto \begin{pmatrix} S_t^\alpha u, \\ \alpha \end{pmatrix}.$$

Properties (7)-(10) of S directly translate to ϕ. If we can show that $\phi\colon [0,T]\times E \to E$ is measurable, the dynamics ϕ is a candidate for deterministic dynamics in between jump times of a PDMP, see [14]. The following lemma 2.1 tells us a sufficient condition to prove measurability of ϕ.

Lemma 2.1. *Let the mapping $\alpha \mapsto f^\alpha$ from $I \to C^{0,1}(\mathbb{R})$ be continuous with $I \subset \mathbb{R}$ an interval, then $(t, u, \alpha) \mapsto (S_t^\alpha u, \alpha)$ is continuous from $[0,T] \times L^1(\mathbb{R}) \times I \to L^1(\mathbb{R}) \times I$ and consequently measurable.*

Proof. Let $(s, u, \alpha), (t, v, \beta) \in [0,T] \times L^1(\mathbb{R}) \times I$, then we use the norm

$$\|(s, u, \alpha) - (t, v, \beta)\| = |s - t| + \|u - v\|_{L^1} + |\alpha - \beta|.$$

According to this norm, we use

$$\|(S_s^\alpha u, \alpha) - (S_t^\beta v, \beta)\| = \|S_s^\alpha u - S_t^\beta v\|_{L^1} + |\alpha - \beta|.$$

To show continuity, we estimate $\|S_s^\alpha u - S_t^\beta v\|_{L^1}$ as follows:

$$\|S_s^\alpha u - S_t^\beta v\|_{L^1} \leq \|S_s^\alpha u - S_t^\alpha u\|_{L^1} + \|S_t^\alpha u - S_t^\alpha v\|_{L^1} + \|S_t^\alpha v - S_t^\beta v\|_{L^1}$$

and conclude that we can make $\|S_s^\alpha u - S_t^\alpha u\|_{L^1}$ and $\|S_t^\alpha u - S_t^\alpha v\|_{L^1}$ sufficiently small by shrinking $\|(S_s^\alpha u, \alpha) - (S_t^\beta v, \beta)\|$ due to properties (9)-(10). Let $(v_n, n \in \mathbb{N})$ be a sequence in $\mathrm{BV}(\mathbb{R}) \cap L^1(\mathbb{R})$ satisfying $\lim_{n\to\infty} \|v_n - v\|_{L^1} = 0$. We estimate $\|S_t^\alpha v - S_t^\beta v\|_{L^1}$ as follows:

$$\begin{aligned}\|S_t^\alpha v - S_t^\beta v\|_{L^1} &\leq \|S_t^\alpha v - S_t^\alpha v_n\|_{L^1} + \|S_t^\beta v_n - S_t^\beta v\|_{L^1} + \|S_t^\alpha v_n - S_t^\beta v_n\|_{L^1} \\ &\leq 2\|v - v_n\|_{L^1} + \|S_t^\alpha v_n - S_t^\beta v_n\|_{L^1} \\ &\leq 2\|v - v_n\|_{L^1} + t\|f^\alpha - f^\beta\|_{C^{0,1}} \mathrm{TV}(v_n),\end{aligned}$$

where we used the result from [13, p. 53] in the last estimate. Altogether, we find that

$$\begin{aligned}\|S_s^\alpha u - S_t^\beta v\|_{L^1} \leq \|S_s^\alpha u - S_t^\alpha u\|_{L^1} + \|u - v\|_{L^1} \\ + 2\|v - v_n\|_{L^1} + T\|f^\alpha - f^\beta\|_{C^{0,1}} \mathrm{TV}(v_n).\end{aligned}$$

Now, let $(t, v, \beta) \in [0,T] \times L^1(\mathbb{R}) \times I$, $\epsilon > 0$ and choose $n \in \mathbb{N}$ such that $\|v - v_n\|_{L^1} < \frac{\epsilon}{6}$ as well as $\delta > 0$ such that

$$\|S_s^\alpha u - S_t^\alpha u\|_{L^1} < \frac{\epsilon}{6}, \ \|u - v\|_{L^1} < \frac{\epsilon}{6}, \ \|f^\alpha - f^\beta\|_{C^{0,1}} < \frac{\epsilon}{\mathrm{TV}(v_n) 6T}, \ |\alpha - \beta| < \frac{\epsilon}{6}$$

implying

$$\|(S_s^\alpha u, \alpha) - (S_t^\beta v, \beta)\| < \epsilon$$

for all $(s, u, \alpha) \in [0,T] \times L^1(\mathbb{R}) \times I$ satisfying $\|(s, u, \alpha) - (t, v, \beta)\| < \delta$. □

One simple example for a family of flux functions, which satisfies the continuity with respect to the parameter $\alpha \in I$ is given by $f^\alpha = \alpha f$ for $f \in C^{0,1}(\mathbb{R})$. Then $\|f^\alpha - f^\beta\|_{C^{0,1}} = \|f\|_{C^{0,1}} |\alpha - \beta|$.

Jump and jump time distributions. Following [14], we specify the transition intensities $q_t(y, B) \geq 0$, i.e., the rate to jump from $y \in E$ in a state in $B \in \mathcal{E}$ at time $t \in [0,T]$. This can be decomposed into $q_t(y, B) = \eta_t(y, B)\psi_t(y)$, where $\psi_t(y)$ is the total intensity that a jump occurs a time t and $\eta_t(y, B)$ is the probability of a jump from y into a state in B provided a jump occurs at time t.

In order to use these intensities, we assume $(y, t) \mapsto \psi_t(y)$ to be measurable and for all (y, t) we need $\int_t^{t+h} \psi_s(y) ds < \infty$ for $h = h(y, t)$ sufficiently small. For all t we additionally assume that η_t is a Marovian kernel, see, e.g. [1], for a definition. A further and natural assumption is that $\eta_t(y, \{y\}) = 0$ holds for all $(y, t) \in E \times [0,T]$.

At this point almost everything can happen at jump times but we fix the specific idea that only the flux function changes at the jump times. In detail, there is no jump in the solution of the conservation law component to inherit mass conservation again. To do so, we restrict on rates

$$\lambda\colon I \times \mathcal{B}(I) \times [0,T] \times L^1(\mathbb{R}) \to \mathbb{R}_{>0},$$

satisfying

1. $\sup\{\lambda(\alpha, I, t, u) \colon \alpha \in I,\ t \in [0,T],\ u \in L^1(\mathbb{R})\} \leq \lambda^{\max} < \infty$,
2. for every $t \in [0,T]$, $(\alpha, u) \in E$ the mapping $B \mapsto \lambda(\alpha, B, t, u)$ is a measure,
3. for every $t \in [0,T]$, $B \in \mathcal{B}(I)$ the mapping $(\alpha, u) \mapsto \lambda(\alpha, B, t, u)$ is measurable,
4. for every $t \in [0,T]$, $(\alpha, u) \in E$ we have $\lambda(\alpha, \{\alpha\}, t, u) = 0$.

Then we define for every $y = (\alpha, u) \in E$ and $B \in \mathcal{E}$ the total intensity and jump distribution by

$$\begin{aligned}\psi_t(y) &= \lambda(\alpha, I, t, u),\\ \eta_t(y, B) &= \frac{1}{\lambda(\alpha, I, t, u)} \int_I \mathbb{1}_B((\beta, u))\lambda(\alpha, d\beta, t, u).\end{aligned}$$

Existence. Due to the uniform bound on ψ_t, we can use a so-called thinning algorithm to build the jump times T_n and after jump locations Y_n for $n \in \mathbb{N}_0$ iteratively, see [10, 15]. Since the number of jumps is finite P-almost surely, again due to the uniform bound on the rates, we obtain a stable random counting measure and theorem 7.3.1 from [14] can be applied. We obtain the following result

Theorem 2.2. *For every initial data $x_0 = (\alpha_0, u_0) \in E$ there exists a stochastic process $X = (X(t), t \in [0,T])$ on some probability space $(\Omega, \mathcal{A}, P)$, which satisfies*

1. *$X(0) = x_0$,*
2. *X is a Markov process with respect to its natural filtration $\mathcal{F}^X = (\mathcal{F}_t^X, t \in [0,T])$ given by $\mathcal{F}_t^X = \sigma(X(s), 0 \leq s \leq t)$,*
3. *X is piecewise deterministic and piecewise continuous, i.e., there exist jump times $T_n \in [0,T]$ and post jump locations $Y_n \in E$ for $n \in \mathbb{N}_0$ such that*

$$X(t) = \phi_{t-T_n}(Y_n) \quad \Leftrightarrow \quad t \in [T_n, T_{n+1}),$$

where for convenience $T_0 = 0$ and $Y_0 = x_0$.

Total Variation bounds and BV solutions. The extension of the solution to L^1 allowed us to use classical results from the theory of piecewise deterministic Markov processes to obtain the existence of a stochastic process, which satisfies our requirements. We expect that if the initial condition $u_0 \in L^1(\mathbb{R}) \cap \mathrm{BV}(\mathbb{R})$, then we deduce $u(t) \in L^1(\mathbb{R}) \cap \mathrm{BV}(\mathbb{R})$ again as the following lemma shows.

Lemma 2.3. *Let $X = (X(t), t \in [0,T])$ be the stochastic process from theorem 2.2 with $X(t) = (\alpha(t), u(t)) \in E$. If $u(0) = u_0 \in L^1(\mathbb{R}) \cap \mathrm{BV}(\mathbb{R})$, then $u(t) \in L^1(\mathbb{R}) \cap \mathrm{BV}(\mathbb{R})$ and $\mathrm{TV}(u(t)) \leq \mathrm{TV}(u_0)$.*

Proof. Let $\omega \in \Omega$, $T_n(\omega)$ the jump times and $Y_n(\omega)$ the post jump locations of $X(\omega)$ for $n \in \mathbb{N}_0$. For $t \in [0, T_1(\omega))$ we have $\mathrm{TV}(u(t,\omega)) = \mathrm{TV}(S_t^{\alpha_0} u_0) \leq \mathrm{TV}(u_0)$ by classical results on scalar conservation laws, see, e.g. [13]. At time $t = T_1$ the flux function changes and for $t \in [T_1(\omega), T_2(\omega))$ it follows

$$\mathrm{TV}(u(t)) = \mathrm{TV}(S_{t-T_1(\omega)}^{\alpha(T_1(\omega),\omega)} u(t,\omega)) \leq \mathrm{TV}(u(T_1(\omega),\omega)) \leq \mathrm{TV}(u_0)$$

by continuity of $t \mapsto u(t,\omega)$. Iteratively, we deduce

$$\mathrm{TV}(u(t,\omega)) \leq \mathrm{TV}(u_0).$$

□

Remark 1. Lemma 2.3 is only valid because we have no jumps in the u component at jump times by construction. Using the same arguments, the mass in the u component is preserved.

3. **Applications and numerical results.** Since we motivated PDMPs driven by scalar conservation law dynamics by the scattering of real data, we discuss simulation results of two examples in this section. The first example is a production and the second example is a traffic flow model.

Production model. Macroscopic production models have been widely studied in the literature, see [3] for an overview. Since in production capacity drops occur due to machine failures or human influences, deterministic models have been extended to stochastic production models, see [4, 8, 9, 10]. Therein, a random flux function in the form of

$$f(\rho) = \min\{v\rho, \mu\}$$

has been chosen with a deterministic production velocity $v > 0$, a stochastic capacity μ for a production density ρ. The latter corresponds to the variable u in our context. In [4, 9] the capacity μ is a Continuous Time Markov Chain, in [8] a semi-Markov process and in [10] a PDMP construction has been developed.

In contrast to the mentioned works, we consider a single production step instead of a network and use our more general setting that allows for further flux functions motivated by data sets, see e.g. [7]. One possible choice is

$$f^\alpha(\rho) = \mu(\alpha)(1 - e^{-\frac{v(\alpha)}{\mu(\alpha)}\rho})$$

for a continuous bounded capacity $\mu > 0$ and velocity $v \geq 0$. Some calculation shows $\|f^\alpha - f^\beta\|_{C^{0,1}} = O(|v(\alpha) - v(\beta)| + |\frac{v(\alpha)}{\mu(\alpha)} - \frac{v(\beta)}{\mu(\beta)}|)$ and the flux function fulfills the requirements to obtain the existence of a suitable stochastic process X, see theorem 2.2.

In Figure 1a flux functions for $\mu(\alpha) = 1 + \tanh(\frac{\alpha}{2})$ and $v(\alpha) = 1 + \tanh(\alpha)$ and different α are drawn. So, we can capture different production velocities and capacities by varying α.

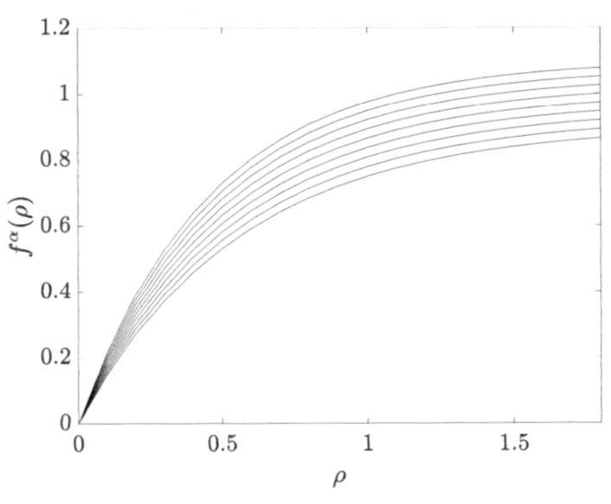

(A) Flux functions for different α

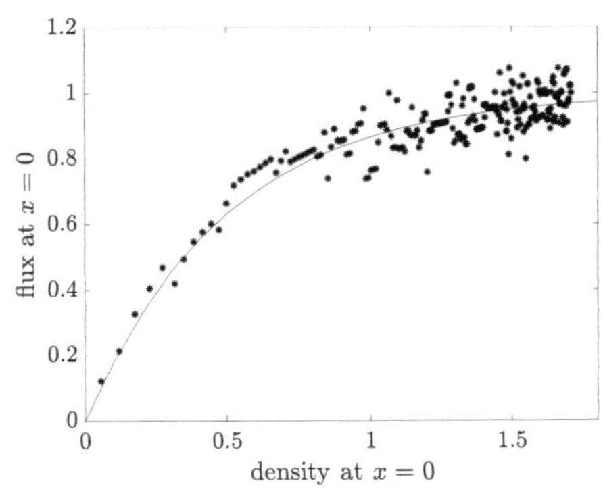

(B) Sample densities and fluxes at $x = 0$

FIGURE 1. Flux-density relation in the production model

It remains to introduce jump rates λ in the production setting. We want the total jump intensity to be dependent on the Work In Progress (WIP) on some interval $[a, b] \subset \mathbb{R}$, which is defined as $\text{WIP}(\rho(t)) = \int_a^b \rho(x, t)dx$. In detail, we assume as

WIP increases, the probability of a change of the flux function increases and vice versa. The distribution of the post jump location is assumed to be symmetrical around $\bar{\alpha} \in \mathbb{R}$ with variance $\sigma^2 > 0$ and we exemplary use

$$\lambda(\alpha, B, t, \rho) = \bar{\lambda}(\rho) \int_B \frac{1}{\sqrt{2\pi\sigma^2(\rho)}} e^{-\frac{(z-\bar{\alpha})^2}{2\sigma^2(\rho)}} dz$$

for every $\alpha \in \mathbb{R}$, $B \in \mathcal{B}(\mathbb{R})$, $t \in [0, T]$ and $\rho \in L^1(\mathbb{R})$. One reasonable choice for $\bar{\lambda}(\rho)$ is $\bar{\lambda}(\rho) = \lambda_0(1 - e^{-\lambda_1 \mathrm{WIP}(\rho)})$ for some $\lambda_0, \lambda_1 > 0$. For the subsequent simulation results, we assume $a = 0$, $b = 1$, $\lambda_0 = 5$, $\lambda_1 = 1$, $\sigma^2 = 10^{-2}$, $\bar{\alpha} = 0$. The time horizon is $T = 50$ and the numerical spatial domain is taken as large that boundary conditions have no influence at $x = 0$ on the solution. The deterministic dynamics is approximated by a Godunov scheme and in figure 1b we see the result of one sample of the density flux relation at position $x = 0$ generated by the model with initial data $\rho(x, 0) = \frac{3}{2}(\sin(x) + 1)e^{-\frac{|x|}{100}}$. The black markers consider to the density and flux at times $t = 0, 0.2, \ldots, 50$ and the black solid line in figure 1b represents the flux function for $\alpha = 0$. We observe in this stochastic macroscopic production model the typical scattering effect like it is the case for microscopic production models driven by discrete event simulations in [7].

Traffic flow model. The scattering effect in the density flux diagram obtained by real data, see, e.g. [20, 21], is a fundamental pattern and important for the development of second order, stochastic and phase transition traffic flow models. In the so-called free phase we observe small fluctuations and an almost linearly increasing flux with respect to the density. At a critical density, the flux decreases in the so-called congested phase. The critical density and congested phase are characterized by higher variances, i.e. sacttering effects in data. There exist already stochastic approaches like in [16, 19] and a comprehensive overview is given in [22]. We will show that the framework, which we introduced in section 2 is able to capture the scattering effects as well.

As family of flux functions, we use, motivated by the shape of the probability density function of the Gamma distribution,

$$f^\alpha(\rho) = \frac{\theta - 1}{\alpha^\theta} \frac{1}{\Gamma(\frac{\theta-1}{\alpha})} \rho^{\theta-1} e^{-\frac{\theta-1}{\alpha}\rho}$$

for some parameter $\theta \geq 1$, $\alpha > 0$, $\rho \geq 0$ and Γ the Gamma function. If $\theta \geq 2$, we also have $f^\alpha \in C^{0,1}(\mathbb{R}_{\geq 0})$ and the maximum is attained at $\rho^* = \alpha$. In figure 2a, we see the shape of the flux function by varying $\alpha \in [0.3, 0.5]$ and $\theta = 2.1$. We set

$$\lambda(\alpha, B, t, \rho) = \bar{\lambda}(\alpha, \rho) \int_B \frac{1}{2a(\alpha, \rho)} \mathbb{1}_{[\alpha_0 - a(\alpha,\rho), \alpha_0 + a(\alpha,\rho)]}(z) dz$$

for every $\alpha > 0$, $B \in \mathcal{B}(\mathbb{R}_{>0})$, $t \in [0, T]$ and $\rho \in L^1(\mathbb{R})$. Here, we choose $\bar{\lambda}(\alpha, \rho) = \lambda_0 + (\lambda_1 - \lambda_0)V(\alpha, \rho)$ for $\lambda_0 = 3$ as the minimal and $\lambda_1 = 10$ as the maximal rate, $a(\alpha, \rho) = \sqrt{\frac{9}{2 \cdot 10^3}(V(\alpha, \rho) + 1)}$ with $V(\alpha, \rho) = \int_0^1 \mathbb{1}_{\rho(x) \geq \alpha} dx$ and $\alpha_0 = 0.4$. The functional $V(\alpha, \rho)$ describes the portion of $[0, 1]$, which is above the actual critical density α and always lies in between zero and one. To study the free phase, we use an initial condition in the form of $\rho_0(x) = (0.05 + 0.4 \max\{\sin(x), 0\})e^{-\frac{|x|}{100}}$. A sample of the density flux relation at $x = 0$ as well as at $x = 1$ is shown in figure 2b given at the times $t = 0, 0.1, \ldots, 50$. We observe a low scattering as expected. Contrary, in figure 2c a sample with initial condition $\rho_0(x) = (0.4 + \max\{\sin(x), 0\})e^{-\frac{|x|}{100}}$, i.e. congested case, is shown resulting in high scattering. Finally, in figure 2d the time evolution of the density and flux at $x = 0$ in the congested case is shown. The density is not severely affected by the variation in α compared to the flux.

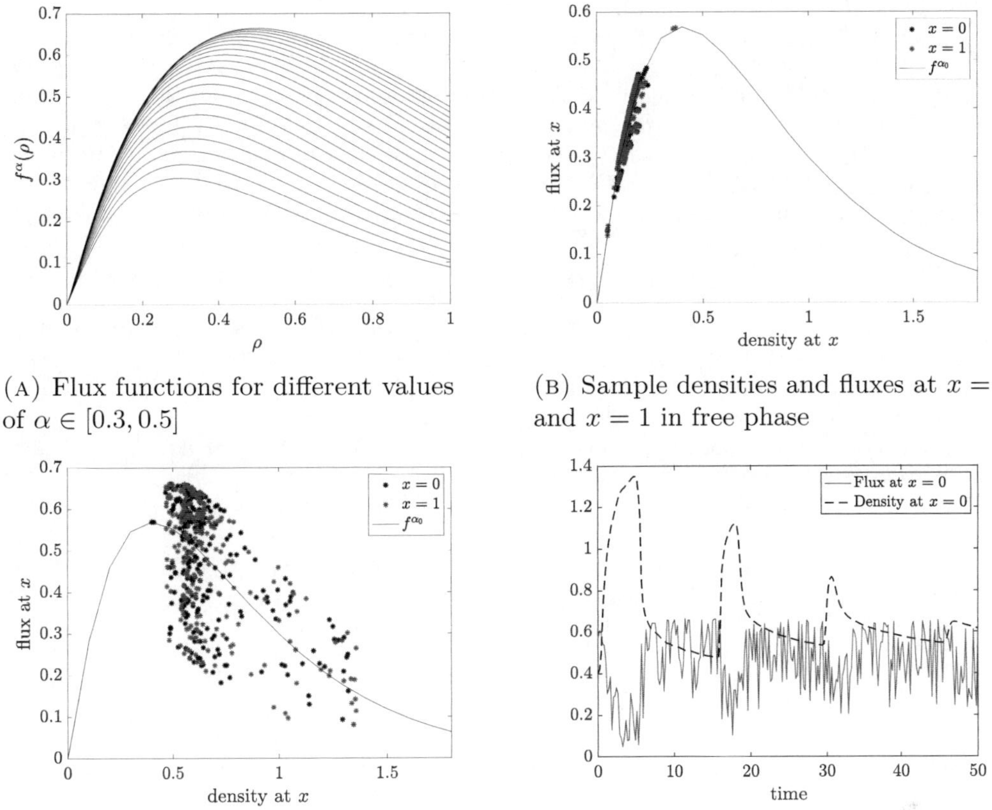

(A) Flux functions for different values of $\alpha \in [0.3, 0.5]$

(B) Sample densities and fluxes at $x = 0$ and $x = 1$ in free phase

(C) Sample densities and fluxes at $x = 0$ and $x = 1$ in congested phase

(D) Sample densities and fluxes at $x = 0$ in congested phase

FIGURE 2. Flux-density relation in the traffic flow model

4. **Conclusions.** We have successfully incorporated random flux functions for scalar conservation laws in the sense of PDMPs. Additionally, we derived a sufficient condition for an arbitrary family of Lipschitz continuous flux functions such that we can guarantee the existence of a PDMP. The motivation of scattering effects in macroscopic models has been recovered in numerical simulation results in the case of a production and traffic flow model.

To cover more complex dynamics, like space dependent flux functions, the theory can be extended in a suitable way as future research. This can be relevant to model traffic accidents and models, where spatial events can happen. Additionally, systems of conservation laws should be examined as deterministic dynamics for PDMPs since the extension to L^1 solutions is not straightforward anymore.

REFERENCES

[1] H. Bauer, *Probability Theory*, vol. 23 of De Gruyter Studies in Mathematics, Walter de Gruyter & Co., Berlin, 1996.

[2] A. Bressan, *Hyperbolic Systems of Conservation Laws*, vol. 20 of Oxford Lecture Series in Mathematics and its Applications, Oxford University Press, Oxford, 2000.

[3] C. D'Apice, S. Göttlich, M. Herty, and B. Piccoli, *Modeling, Simulation, and Optimization of Supply Chains*, SIAM, Philadelphia, PA, 2010.

[4] P. Degond and C. Ringhofer, Stochastic dynamics of long supply chains with random breakdowns, *SIAM J. Appl. Math.*, **68** (2007), 59–79.

[5] J. Feng and D. Nualart, Stochastic scalar conservation laws, *Journal of Functional Analysis*, **255** (2008), 313–373.

[6] U. S. Fjordholm, S. Lanthaler, and S. Mishra, Statistical solutions of hyperbolic conservation laws: foundations, *Archive for Rational Mechanics and Analysis*, **226** (2017), 809–849.

[7] L. Forestier-Coste, S. Göttlich, and M. Herty, Data-fitted second-order macroscopic production models, *SIAM J. Appl. Math.*, **75** (2015), 999–1014.

[8] S. Göttlich and S. Knapp, Semi-Markovian capacities in production network models, *Discrete Contin. Dyn. Syst. Ser. B*, **22** (2017), 3235–3258.

[9] S. Göttlich, S. Martin, and T. Sickenberger, Time-continuous production networks with random breakdowns, *Netw. Heterog. Media*, **6** (2011), 695–714.

[10] S. Gttlich and S. Knapp, Load-dependent machine failures in production network models, preprint, arXiv:1806.03091.

[11] H. Holden and N. H. Risebro, A mathematical model of traffic flow on a network of unidirectional roads, *SIAM J. Math. Anal.*, **26** (1995), 999–1017.

[12] H. Holden and N. H. Risebro, Conservation laws with a random source, *Applied Mathematics and Optimization*, **36** (1997), 229–241.

[13] H. Holden and N. H. Risebro, *Front Tracking for Hyperbolic Conservation Laws*, vol. 152 of AMS, Springer, Heidelberg, 2nd ed., 2015.

[14] M. Jacobsen, *Point Process Theory and Applications*, Probability and its Applications, Birkhäuser Boston, Inc., Boston, MA, 2006.

[15] V. Lemaire, M. Thieullen, and N. Thomas, Exact simulation of the jump times of a class of piecewise deterministic markov processes, *J. Sci. Comput.*, **75** (2018), 1776–1807.

[16] J. Li, Q.-Y. Chen, H. Wang, and D. Ni, Analysis of LWR model with fundamental diagram subject to uncertainties, *Transportmetrica*, **8** (2012), 387–405.

[17] P.-L. Lions, B. Perthame, and P. E. Souganidis, Scalar conservation laws with rough (stochastic) fluxes: the spatially dependent case, *Stoch. Partial Differ. Equ. Anal. Comput.*, **2** (2014), 517–538.

[18] S. Mishra, N. H. Risebro, C. Schwab, and S. Tokareva, Numerical solution of scalar conservation laws with random flux functions, *SIAM/ASA J. Uncertain. Quantif.*, **4** (2016), 552–591.

[19] D. Ni, H. K. Hsieh, and T. Jiang, Modeling phase diagrams as stochastic processes with application in vehicular traffic flow, *Applied Mathematical Modelling. Simulation and Computation for Engineering and Environmental Systems*, **53** (2018), 106–117.

[20] B. Piccoli and A. Tosin, Vehicular traffic: a review of continuum mathematical models, in *Mathematics of complexity and dynamical systems*, Springer, New York, (2012), 1748–1770.

[21] B. Seibold, M. R. Flynn, A. R. Kasimov, and R. R. Rosales, Constructing set-valued fundamental diagrams from jamiton solutions in second order traffic models, *Netw. Heterog. Media*, **8** (2013), 745–772.

[22] H. Wang, D. Ni, Q.-Y. Chen, and J. Li, Stochastic modeling of the equilibrium speed-density relationship, *Journal of Advanced Transportation*, **47** (2011), 126–150.

E-mail address: stknapp@mail.uni-mannheim.de

NONCONVERGENCE PROOF FOR THE LDA-SCHEME

Dietmar Kroener*, Thomas Mackeben

Abteilung für Angewandte Mathematik
Albert-Ludwigs-Universität Freiburg
Hermann-Herder-Str. 10
79104 Freiburg, Germany

Mirko Rokyta

Faculty of Mathematics and Physics
Charles University
Sokolovská 83
18675 Praha, Czech Republic

Abstract. In this paper we consider the LDA (Low Diffusion Advection) scheme for solving conservation laws in two space dimensions as it was published in [1],[2], [5]. For a special grid and a special nonlinearity we will show by constructing a counterexample that there is no convergence of the numerical solution to the exact solution. The example is constructed in such a way that the exact solution as well as the numerical solution are computed explicitly.

1. **Introduction.** We consider the initial value problem

$$\partial_t u + \nabla \cdot f(u) = 0 \qquad \text{in } \mathbb{R}^2 \times \mathbb{R}^+, \tag{1}$$

$$u(x,y,0) = u_0(x,y) \qquad \text{in } \mathbb{R}^2. \tag{2}$$

Several efficient numerical schemes are available for solving this problem. In this paper we are going to investigate a special scheme, namely the LDA (Low Diffusion Advection) scheme.

Given an admissible mesh, consisting of triangles, and a global numbering of the nodes z_i, the dual cell V_i belonging to z_i is bounded by the polygon connecting the barycenters of triangles and the midpoints of the common edges of the neighboring triangles (see Figure 1). Its area is given by

$$|V_i| := \frac{1}{3} \sum_{T \in A_i} |T|, \tag{3}$$

where A_i contains all neighboring triangles sharing the node z_i. The discrete initial data in V_i are given by

2000 *Mathematics Subject Classification.* Primary: 35L03, 35L65; Secondary: 65M08, 65M12.

Key words and phrases. Scalar conservation laws, 2D problem, LDA scheme, nonconvergence proof.

* Corresponding author: Dietmar Kroener.

$$u_i^0 := \frac{1}{|V_i|} \int_{V_i} u_0(x,y)\,\mathrm{d}x\,\mathrm{d}y. \tag{4}$$

Let u_i^n denote the discrete solution in z_i at time level n, constant on V_i. Then the LDA scheme is given by the following procedure (see [5] formula (2.2), [1] Section 3.2.2, or [2] p. 647). The value of the numerical solution at node z_i at time level t^{n+1} is given by

$$u_i^{n+1} := u_i^n + \frac{\Delta t}{|V_i|} \sum_{T \in A_i} \phi_i^T \tag{5}$$

where

$$\phi_i^T := \begin{cases} -(K_i^T)^+ N^T \phi^T, \; N^T := \left(\sum_{x_j \in T} (K_j^T)^+ \right)^{-1} & \text{if } (N^T)^{-1} \neq 0, \\ 0 & \text{if } (N^T)^{-1} = 0, \end{cases} \tag{6}$$

$$K_i^T := \frac{1}{2|T|} \int_T (f_1'(u_h), f_2'(u_h))\,\mathrm{d}x\,\mathrm{d}y \cdot n_i^T, \tag{7}$$

$$\phi^T := \sum_{x_j \in T} K_j^T u_{j,T}. \tag{8}$$

Here n_i^T is the inward scaled normal in T to the face opposite to the node z_i scaled by its surface, u_h is the globally continuous numerical solution which is linear on each triangle T with vortices z_i, z_j, z_k and corresponding node values $(u_i, u_j, u_k) = (u_{i,T}, u_{j,T}, u_{k,T})$.

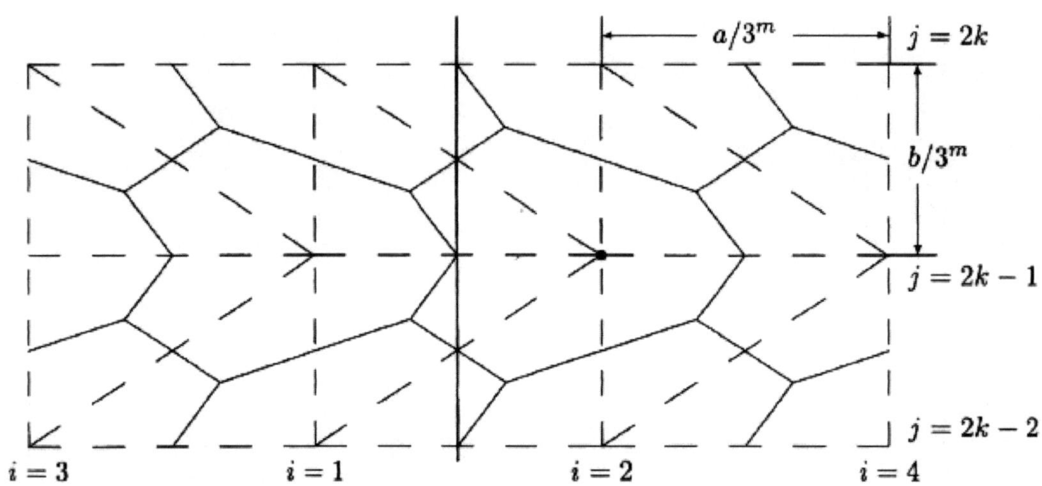

FIGURE 1. Dual cells

In [1] (see formula (24)) convergence and an error estimate are shown for the PSI scheme, which is similar to the LDA scheme (see [1], Section 3.2.3), for solving the following linear stationary problem:

$$\lambda \cdot \nabla v = g \quad \text{for} \quad x \in \Omega,$$
$$v = 0 \quad \text{on} \quad \Gamma^-,$$

where Ω is a polygonal domain adapted to the grid, and Γ^- is that part of $\partial\Omega$ for which $\lambda \cdot n_0 \leq 0$, n_0 being the outer normal to Ω.

Then in [1] on page 18, there is a remark, stating that the proof works also for the LDA scheme (5). At the beginning of Section 4 in [1] the authors mentioned that the case of a non-linear flux can be discussed in the same way, at least formally. Convergence for the N-scheme (which is also similar to the LDA scheme) for linear advection equations is proved in [9] (see also [8]).

Remark 1. In the nonlinear case $f(u) := \left(u, \frac{u^2}{2}\right)$ the author in [2], (15) considers for the discretization the following linearized problem

$$\partial_t u + a_x \partial_x u + a_y \partial_y u = 0 \quad \text{on each triangle} \quad T$$

with $a_x = 1$ and $a_y := \frac{1}{3}(u_1 + u_2 + u_3)$ where u_1, u_2 and u_3 are values of u_h in the vertices of T. I.e. it holds (see (7))

$$(a_x, a_y) := \frac{1}{|T|} \int_T (f_1'(u_h), f_2'(u_h)) \, \mathrm{d}x \, \mathrm{d}y.$$

Lemma 1.1. *The scheme defined in* (5)–(8) *is conservative, i.e. for each polygonal domain* Ω *such that* $\Omega = \cup_{i \in I} T_i$ *we have*

$$\sum_{i \in I} |V_i| \left(u_i^{n+1} - u_i^n\right) = -\Delta t \int_{\partial\Omega} (f_1(u_h), f_2(u_h)) \cdot n_0 \, \mathrm{d}x \, \mathrm{d}y,$$

with n_0 *being unit outward normal vector to* $\partial\Omega$.

2. **The main result.**

In this contribution we will show a nonconvergence result for the scheme defined in (5)–(8) on a special series of grids (see Figure 3 with refinement levels $m = 1$ and $m = 2$).

Theorem 2.1. *Let* $a, b > 0$, $m \in \mathbb{N}$, *where* m *denotes the refinement level, and* $u_0(x, y)$ *be equal to* $-u_R < 0$ *for* $x + y < 0$ *and equal to* $u_R > 0$ *for* $x + y > 0$. *Let* u *be the exact solution of* (1)–(2) *with* $f(u) := (\frac{u^2}{2}, \frac{u^2}{2})$. *Consider the triangulation and the refinement process as specified in Figure 3 (m=1, 2) with mesh size* h_m, *and let* u_{h_m} *be the globally continuous numerical solution, which is linear on each triangle* T *and the values in the vortices of the triangles are defined as in* (5)–(8). *Furthermore we assume the CFL condition* $\Delta t \frac{\sqrt{2}}{h_m} u_R < 1$. *Then we have for any fixed time* $t_1 > 0$ *on the subdomain*

$$D = \left\{(x, y) : |x + y| \leq \frac{a}{\sqrt{2}} \quad \text{and} \quad 0 \leq y - x \leq \sqrt{8}b\right\} \tag{9}$$

such that

$$\frac{a}{\sqrt{8}} < \frac{2}{3} t_1 u_R, \tag{10}$$

the following estimate from below, uniformly in h_m *and all* $t \geq t_1$:

$$||u_{h_m}(\cdot, \cdot, t) - u(\cdot, \cdot, t)||_{L^1(D)} \geq \frac{|D|}{180} u_R > 0. \tag{11}$$

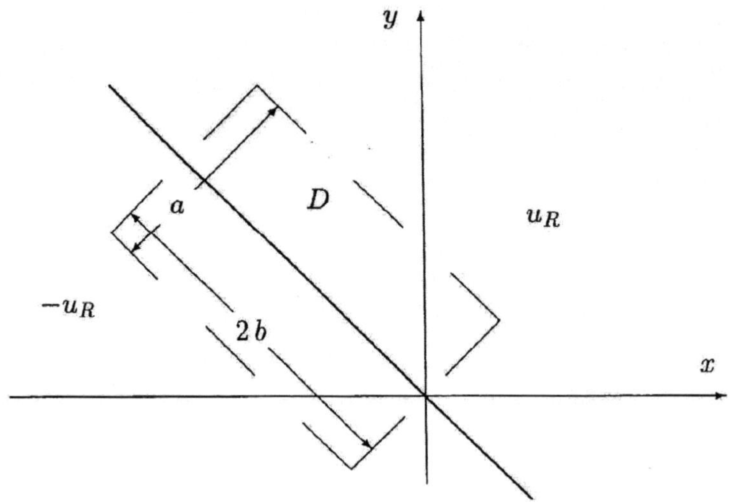

FIGURE 2. Initial values and test domain D

An idea of the proof (for a detailed proof see [7]):
The exact solution is a rarerefaction wave and can, as well as the numerical solution, be computed explicitly. This allows us to compute the error in (11). The exact solution u to the problem can be written as

$$u(x,y,t) = \begin{cases} -u_R & \text{for} \quad \frac{x+y}{2} \le -u_R t, \\ \frac{x+y}{2t} & \text{for} \quad -u_R t \le \frac{x+y}{2} \le u_R t, \\ u_R & \text{for} \quad u_R t \le \frac{x+y}{2}. \end{cases} \tag{12}$$

Using the condition (10) one can obtain

$$\|u(\cdot,\cdot,t)\|_{L_1(D)} = \frac{a|D|}{2\sqrt{8t}} \quad \text{for all } t \ge t_1,$$

and therefore

$$\|u(\cdot,\cdot,t)\|_{L_1(D)} \to 0 \quad \text{as } t \to \infty.$$

For the purpose of the discretization a special test domain and a special series of grids is chosen, see Figure 2 and Figure 3 with refinement levels $m = 1$ and $m = 2$. In this case we use the numbering of nodes in the form of z_{ij} (see Figure 3), which gives us for the m-th level of refinement

$$z_{ij} = \frac{1}{\sqrt{2}}\left(\left(\left\lfloor\frac{i}{2}\right\rfloor(-1)^i - \frac{1}{2}\right)\frac{a}{3^m} - (j-1)\frac{b}{3^m}, \left(\left\lfloor\frac{i}{2}\right\rfloor(-1)^i - \frac{1}{2}\right)\frac{a}{3^m} + (j-1)\frac{b}{3^m}\right)$$

for the index set

$$I_m = \left\{(i,j) \in \mathbb{N}^2;\ 1 \le i \le 3^m+1,\ 1 \le j \le 2\cdot 3^m+1,\right\}.$$

The alternating numbering is chosen to assure that columns 1 and 2 remain to the left and right of the jump in the initial values, respectively (see Figure 3).

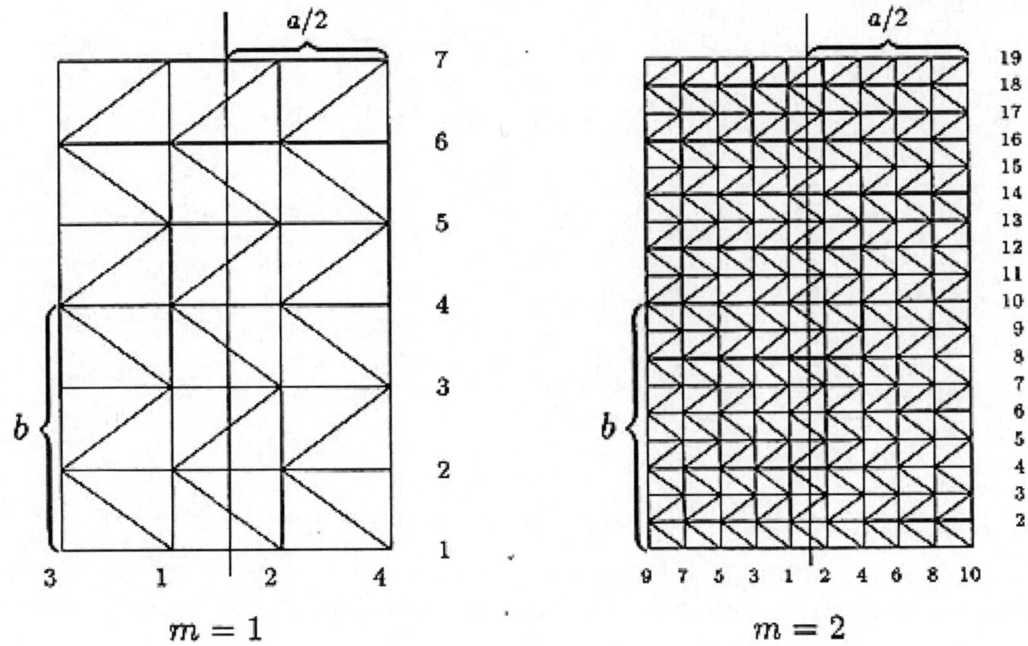

FIGURE 3. A special series of grids

Due to the condition (10) only the second case in (12) applies, which implies that the exact solution u is in each node $z_{ij} = (x_i, y_j) \in D$ given by

$$u(x_i, y_j, t) = \frac{1}{\sqrt{2}}\left(\left\lfloor \frac{i}{2} \right\rfloor (-1)^i - \frac{1}{2}\right)\frac{a}{3^m}\frac{1}{t} \quad \text{for all } t > t_1, \tag{13}$$

cf. definition of z_{ij}.

The initial values are defined in the dual cell V_{ij} around the node z_{ij} (see Figure 1) as $u_{ij}^0 := \frac{1}{|V_{ij}|}\int_{V_{ij}} u_0(x,y)\,\mathrm{d}x\,\mathrm{d}y$, consistently with (4). It can be computed that

$$u_{ij}^0 = \begin{cases} -u_R & \text{for} \quad i = 2k+1,\ j \in \mathbb{N}, \\ -u_R & \text{for} \quad i = 1,\ j = 2k-1, \\ -\frac{5}{6}u_R & \text{for} \quad i = 1,\ j = 2k, \\ \frac{5}{6}u_R & \text{for} \quad i = 2,\ j = 2k-1, \\ u_R & \text{for} \quad i = 2,\ j = 2k, \\ u_R & \text{for} \quad i = 2k+2,\ j \in \mathbb{N}, \end{cases} \qquad k \in \mathbb{N}, (i,j) \in I_m. \tag{14}$$

The L^1 error between the initial value u_0 and the numerical approximation $u_{h_m}(\cdot,\cdot,0)$ of u_0 is equal to (see [7])

$$\|u_{h_m}(\cdot,\cdot,0) - u_0\|_{L^1(D)} = \frac{1}{3^m}\cdot\frac{43}{72}|D|u_R \to 0 \quad \text{as } m \to \infty. \tag{15}$$

The numerical solution can be expressed for $i = 1, 2$ by (see [7])

$$u_{1j}^{n+1} = \begin{cases} u_{1j}^0 & j = 2q, \\ u_{1j}^n + \frac{\Delta t}{a/3^m}\frac{\sqrt{2}}{3}\big(u_{2j}^n + u_{1j}^n + u_{1j+1}^n\big)\big(u_{1j}^n - u_{2j}^n\big) & j = 2q-1, \end{cases}$$

$$u_{2j}^{n+1} = \begin{cases} u_{2j}^n - \frac{\Delta t}{a/3^m}\frac{\sqrt{2}}{3}\big(u_{1j}^n + u_{2j}^n + u_{2j+1}^n\big)\big(u_{2j}^n - u_{1j}^n\big) & j = 2q, \\ u_{2j}^0 & j = 2q-1, \end{cases} \tag{16}$$

$q \in \mathbb{N}$, and for $i \geq 3$, $i \in \mathbb{N}$:

$$u_{ij}^{n+1} = u_{ij}^n - (-1)^i \frac{\Delta t}{a/3^m}\frac{\sqrt{2}}{3}\big(u_{i-2,j}^n + u_{ij}^n + u_{k_i,j+1}^n\big)\big(u_{ij}^n - u_{i-2,j}^n\big), \tag{17}$$

$$k_i = \begin{cases} i & \text{if } \ i + j = 2k + 2, \\ i - 2 & \text{if } \ i + j = 2k + 3, \end{cases} \tag{18}$$

with $j, k \in \mathbb{N}$ and for all time steps $n \in \mathbb{N}$.

Taking an advantage of (16)–(18) some of the properties of the approximate solution can be proved for $n \in \mathbb{N}_0$, e.g.,

$$\begin{array}{lll} (1, 2q): & u_{1,2q}^n = u_{1,2q}^0 = -\frac{5}{6}u_R, & q \in \mathbb{N}, \\ (1, 2q-1): & u_{1,2q-1}^n \leq u_{1,2q-1}^{n+1} < 0, & q \in \mathbb{N}, \\ (2, 2q-1): & u_{2,2q-1}^n = u_{2,2q-1}^0 = \frac{5}{6}u_R, & q \in \mathbb{N}, \\ (2, 2q): & u_{2,2q}^n \geq u_{2,2q}^{n+1} > 0, & q \in \mathbb{N}, \end{array}$$

and

$$\begin{array}{lll} (2p+2, j): & u_{ij}^{n+1} \geq u_{i-2,j}^n \geq u_{i-2,j}^{n+1}, & i = 2p+2, \ \ j, p \in \mathbb{N}, \\ (2p+1, j): & u_{ij}^n \leq u_{ij}^{n+1} \leq u_{i-2,j}^n, & i = 2p+1, \ \ j, p \in \mathbb{N}. \end{array}$$

Finally, one can prove that

$$u_{i,j-1}^n = u_{i,j+1}^n \qquad \text{for all } (i, j) \in I_m \text{ and } m, n, \in \mathbb{N}_0, \tag{19}$$

because of (14) and by induction over n using (16)–(18).

For the purpose of further explicit calculations we divide the set of all triangles into two subsets, $\mathbb{T}_i^k$, $k = 1, 2$, defined for each $i \geq 3$ as follows: we say that $T_i^k \in \mathbb{T}_i^k$ if there are exactly two vertices z_{iq} and z_{i-2q} of T_i^k such that $i + q = 2p + k$, $p \in \mathbb{N}$.

Using this convention it can be shown that we have for all $T_i^k \in \mathbb{T}_i^k$, $k = 1, 2$, and $i \geq 3$ the following estimates:

$$\int_{T_i^1} \big|u_{h_m}(x, y, t_1) - u(x, y, t_1)\big| \, dx \, dy \geq \frac{1}{21}\frac{a}{3^m}\frac{b}{3^m}\left(\frac{5}{6}u_R - \frac{i - p_i}{3^m}\frac{a}{\sqrt{8}}\frac{1}{t_1}\right) \tag{20}$$

where $p_i = 1 - i \pmod 2$, and similarly,

$$\int_{T_i^2} \big|u_{h_m}(x, y, t_1) - u(x, y, t_1)\big| \, dx \, dy \geq \frac{1}{294}\frac{a}{3^m}\frac{b}{3^m}\left(\frac{5}{6}u_R - \frac{i - p_i}{3^m}\frac{a}{\sqrt{8}}\frac{1}{t_1}\right). \tag{21}$$

Now, in order to prove Theorem 2.1, we split the L^1 norm (11) triangle by triangle. The error contribution of the triangles in the center column (see Figure 3) can be neglected since we are interested in the estimate from below. Further we notice that there are $2\cdot 3^m$ pairs of two triangles T_i^1, T_i^2 making up one column of triangulation. For each refinement level $m\geq 1$, summing up row by row gives

$$E := \|u_{h_m}(\cdot,\cdot,t_1)-u(\cdot,\cdot,t_1)\|_{L^1(D)} \overset{(19)}{\geq} 2\cdot 3^m \sum_{i=3}^{3^m+1}\left(\int_{T_i^1}|\cdots|+\int_{T_i^2}|\cdots|\right),$$

which is further estimated, using $p_i = 1-i \pmod 2$, and (20)–(21), as follows:

$$E \geq 2\cdot 3^m \frac{a}{3^m}\frac{b}{3^m}2\sum_{i\in J}\underbrace{\left(\frac{1}{21}+\frac{1}{294}\right)}_{>\frac{1}{20}}\left(\frac{5}{6}u_R-\frac{i}{3^m}\frac{a}{\sqrt{8}}\frac{1}{t_1}\right),$$

with $J=\{3\leq i\leq 3^m, i=2p+1, p\in\mathbb{N}\}$. We then obtain (using also $|D|=2ab$)

$$\begin{aligned} E &\geq \frac{a}{5}\frac{b}{3^m}\sum_{k=1}^{\frac{3^m-1}{2}}\left(\frac{5}{6}u_R-\frac{2k+1}{3^m}\frac{a}{\sqrt{8}}\frac{1}{t_1}\right)\\ &= \frac{|D|}{10\cdot 3^m}\left(\frac{3^m-1}{2}\frac{5}{6}u_R-\left(\frac{9^m-1}{4}+\frac{3^m-1}{2}\right)\frac{1}{3^m}\frac{a}{\sqrt{8}}\frac{1}{t_1}\right)\\ &= \frac{|D|}{10}\left(\frac{5}{12}u_R-\left(1-\frac{1}{3^m}\right)-\left(\frac{1}{4}+\frac{1}{2}\frac{1}{3^m}-\frac{3}{4}\frac{1}{9^m}\right)\frac{a}{\sqrt{8}}\frac{1}{t_1}\right). \end{aligned}$$

At this point we use $\frac{a}{\sqrt{8}}<\frac{2}{3}t_1u_R$ and neglect the last term for $m>1$. We arrive at

$$E \geq \begin{cases} \frac{1}{180}|D|u_R & \text{for} \quad m=1,\\ \frac{1}{40}|D|u_R\left(1-\frac{1}{3^{m-1}}\right) & \text{for} \quad m\geq 2, \end{cases}$$

and the proof follows.

Remark 2. A similar result can be proved if we consider v_{h_m} instead of u_{h_m}, where $v_{h_m}(\cdot,\cdot,t^n)=u_i^n$ on $V_i\times[t^n,t^{n+1})$, where V_i is the dual cell as in Figure 1.

REFERENCES

[1] R. Abgrall, P.L. Roe, High order fluctuation schemes on triangular meshes, *Journal of scientific computing* **19** (2003), 3–36.

[2] R. Abgrall, Residual distribution schemes: Current status and future trends, *Computers and Fluids* **35** (2006), 641–669.

[3] R. Abgrall, K. Mer, B. Nkonga, A Lax-Wendroff type theorem for residual schemes, in: *Innovative methods for numerical solutions of differential equations* (eds. M.M. Hafez and J.J. Chattot), (2001), 243–266.

[4] H. Deconinck, P.L.Roe, R. Struijs, A multidimensional generalization of Roes's flux difference splitter for the Euler equations, *Computers Fluids* **22** (1993), 215–222.

[5] J. Dobes, H. Deconinck, Second order blended multidimensional upwind residual distribution scheme for steady and unsteady computations, *Journal for Computational and Applied Mathematics* **215** (2008), 378–389.

[6] J. Dobes, H. Deconinck, A shock sensor-based second oder blended upwind residual distribution scheme for steady and unsteady compressible flow, in *Hyperbolic Problems: Theory, Numerics, Applications. Proceedings of the 11th Int. Conf. on Hyperbolic Problems held in ENS Lyon, 2006* (eds. S. Benzoni-Gavage, D. Serre), Springer, Berlin, Heidelberg (2008), 465–473.

[7] D. Kröner, T. Mackeben and M. Rokyta, Convergence analysis for the LDA scheme for scalar conservation laws, in preparation.

[8] T. Mackeben, *Konvergenzanalyse numerischer Verfahren zur Lösung skalarer Erhaltungsgleichungen*, Ph.D. thesis, University Freiburg, Germany, 1997.
[9] B. Perthame, Convergence of N-schmes for linear advection equations, in *Trends in applications of mathematics to mechanics.* (eds. J.F. Rodriguez et al.), A collection of selected papers presented at the 9th symposium, STAMM-94, held at Lisbon University, (1994), 323–333.
[10] J.A. Rossmanith, A class of residual distribution schemes and their relation to relaxation systems, preprint, arXiv:0711.2063.
[11] R. Struijs, Flux vector distribution, preprint 2010, and personal communication.

E-mail address: dietmar@mathematik.uni-freiburg.de
E-mail address: dietmar@mathematik.uni-freiburg.de
E-mail address: rokyta@karlin.mff.cuni.cz

HYBRID FDM-WENO METHOD FOR THE CONVECTION-DIFFUSION PROBLEMS

RAKESH KUMAR*

Tata Institute of Fundamental Research
Centre For Applicable Mathematics
Bangalore-560065, India

ABSTRACT. In the present work, we have proposed a high order hybrid FDM-WENO method for the solution of convection-diffusion problems. In hybrid FDM-WENO method, a fifth order finite difference central flux is used to compute the convective flux in smooth regions whereas in a region where the solution has sharp variations or discontinuities, Weighted Essentially Non-Oscillatory (WENO) reconstruction of adaptive order is used to maintain a non-oscillatory profile. For the diffusion part, we have used a sixth order finite difference approximation. A weak local truncation error based estimate is used to detect the discontinuities or high gradient regions of the solution. The new hybrid FDM-WENO scheme computes the solution efficiently and in a non-oscillatory manner.

1. **Introduction.** In the present work, our aim is to develop efficient hybrid numerical scheme for the convection-diffusion problem of the form

$$u_t + f(u)_x = \epsilon(v(u)u_x)_x, \quad (x,t) \in (a,b) \times (0,T], \tag{1}$$

subject to the initial condition

$$u(x,0) = u_0(x), \quad x \in [a,b],$$

with periodic and Dirichlet boundary conditions. The solution of convection-diffusion equation (1) arising in science and engineering may have sharp transitions or discontinuities, like shock (for $\epsilon = 0$), arising locally over a small portion of the physical domain. Resolving these portions while computing the solution with sufficient accuracy keeping the computational cost within acceptable limits is a non-trivial problem and interest of research from decades. The presence of discontinuities, like shock, or sharp transition in solution, are difficult to resolve using higher order finite difference discretization of (1). The higher order finite difference discretization leads to an oscillatory solution near shock or where the solution has sharp variations. These shock or sharp transition are local nature, which motivates us to use an adaptive combination of schemes locally to resolve them. A low expensive higher order finite difference discretization can be used in smooth regions while a non-oscillatory scheme can be utilized in shock or sharp transition regions.

2000 *Mathematics Subject Classification.* Primary: 65M06, 65M22; Secondary: 65M22.
Key words and phrases. Finite difference methods, Weak local truncation error, B-spline.
The author is supported by SERB-DST grant PDF/2018/002621.
* Corresponding author: Rakesh Kumar.

In the present work, we develop a fifth order hybrid FDM-WENO scheme for the model problem (1). The mechanism of the scheme involves the separation of the discontinuous and smooth regions and followed by conjugation of FDM with WENO-AO scheme. For the identification of discontinuous regions, we use weak local truncation error based estimate (see [2], [1], [4], [5]). We use this information to capture the shock using the WENO scheme of adaptive order, and finite difference scheme is used in the smooth regions. We approximate the second derivative using the sixth order finite difference scheme.

The outline of the article is as follows. In Section 2, we have discussed the details of the hybrid FDM-WENO scheme. Numerical Experiments are performed to validate the scheme in Section 3. In the end, conclusions are drawn.

2. **Hybrid FDM-WENO scheme.** In this section, we discuss the hybrid FDM-WENO scheme for the convection-diffusion problem. Discretize the domain $[a, b]$ into $N+1$ sub-intervals $\left[x_{i-\frac{1}{2}}, x_{i+\frac{1}{2}}\right]$ for $i = 0, 1, \ldots N$ of equal length $\Delta x = x_{i+\frac{1}{2}} - x_{i-\frac{1}{2}}$. The semi-discrete form of equation (1) is given by

$$\frac{du_i}{dt} = -\frac{1}{\Delta x}(\mathbb{F}_{i+\frac{1}{2}} - \mathbb{F}_{i-\frac{1}{2}}) + \frac{\epsilon}{\Delta x^2}(\mathbb{G}_{i+\frac{1}{2}} - \mathbb{G}_{i-\frac{1}{2}}),\ i = 0, 1, \ldots N,$$

where u_i is approximation of point value of solution u at point x_i. The $\mathbb{F}$ and $\mathbb{G}$ denote the convection and diffusion numerical flux, respectively. The convection numerical flux is given by

$$\mathbb{F}_{i+\frac{1}{2}} = \Phi_i \mathbb{F}^{\text{WENO}}_{i+\frac{1}{2}} + (1 - \Phi_i)\mathbb{F}^{\text{FDM}}_{i+\frac{1}{2}},$$

where $\mathbb{F}^{\text{FDM}}_{i+\frac{1}{2}}$ and $\mathbb{F}^{\text{WENO}}_{i+\frac{1}{2}}$ are the fifth order finite difference and WENO flux approximation, respectively. Let Φ_i be a smoothness indicator, which helps us to distinguish the smooth and regions of sharp variations. The fifth order finite difference convection flux is given by

$$\mathbb{F}^{\text{FDM}}_{i+\frac{1}{2}} = \frac{1}{60}(2f_{i-2} - 13f_{i-1} + 47f_i + 27f_{i+1} - 3f_{i+2}),$$

and the sixth order diffusion flux $\mathbb{G}_{i+\frac{1}{2}}$ for u_{xx} is

$$\mathbb{G}_{i+\frac{1}{2}} = \frac{1}{180}(-2u_{i-2} + 25u_{i-1} - 245u_i + 245u_{i+1} - 25u_{i+2} + 2u_{i+3}).$$

For the time discretization, we use the Strong Stability Preserving (SSP) Runge-Kutta method of order three [3]. The WENO flux and smoothness indicator are defined in further subsections.

2.1. **WENO-AO(5,4,3).** In this subsection, we have defined the WENO-AO(5,4,3) flux reconstruction [6]. The WENO-AO(5,4,3) at the interface $x_{i+\frac{1}{2}}$ is given by

$$\mathbb{F}^{\text{WENO}}_{i+\frac{1}{2}} = \frac{\omega_0^5}{\gamma_0^5}\left[\mathbb{P}_0^5(x_{i+\frac{1}{2}}) - \gamma_0^4\mathbb{P}_0^4(x_{i+\frac{1}{2}}) - \sum_{k=-1}^{1}\gamma_k^3\mathbb{P}_k^3(x_{i+\frac{1}{2}})\right] + \omega_0^4\mathbb{P}_0^4(x_{i+\frac{1}{2}})$$
$$+ \sum_{k=-1}^{1}\omega_k^3\mathbb{P}_k^3(x_{i+\frac{1}{2}}),$$

where $\mathbb{P}_k^m$ denotes the polynomial of degree m constructed over stencil $\mathbb{S}_k^m$ (see [6] for notations and details). The linear weights are given by

$$\left.\begin{aligned} \gamma_{-1}^3 &= \tfrac{1}{2}(1-\gamma_{Hi})(1-\gamma_{Avg})(1-\gamma_{Lo}),\\ \gamma_0^3 &= (1-\gamma_{Hi})(1-\gamma_{Avg})\gamma_{Lo},\\ \gamma_1^3 &= \tfrac{1}{2}(1-\gamma_{Hi})(1-\gamma_{Avg})(1-\gamma_{Lo}),\\ \gamma_0^4 &= (1-\gamma_{Hi})\gamma_{Avg},\\ \gamma_0^5 &= \gamma_{Hi}. \end{aligned}\right\}, \tag{2}$$

where γ_k^m denotes the linear positive weights corresponding to the stencils $\mathbb{S}_k^m$, satisfying $\gamma_{-1}^3+\gamma_0^3+\gamma_1^3+\gamma_0^4+\gamma_0^5=1$. The ω_k^m denotes the non-linear weight corresponding to linear weight γ_k^m (see [6] for more details). We choose the value of $\gamma_{Hi}=\gamma_{Avg}=\gamma_{Lo}=0.85$ in our numerical computations.

2.2. **Smoothness indicators.** In order to distinguish the smooth and regions of high gradients, we have used the smoothness indicator for convection-diffusion problem based on Weak Local Truncation Error (WLTE) [2, 4, 1, 5]. Here we provide the brief details of WLTE based smoothness indicator for the convection-diffusion problem, and more details can be found in [5].

We define a error function using the weak formulation of convection-diffusion problem (1) (for simplicity we assume $v(u)=1$) as follows

$$\begin{aligned} E(u,\Phi) :=& \int_0^T\int_{\mathbb{R}}\{u(x,t)\Phi_t(x,t)+f(u)\Phi_x(x,t)+\epsilon u\Phi_{xx}\}dxdt\\ &+\int_{\mathbb{R}} u(x,0)\Phi(x,0)dx=0, \end{aligned} \tag{3}$$

for all $\Phi(x,t)\in C_0^{2,1}(\mathbb{R}\times(0,T])$. A weak solution of the convection-diffusion problem may have sharp variations, or for $\epsilon=0$, it may contains shocks or discontinuities. We may observe the variations in the values of E as we move from smooth to discontinuous regions and vice versa for the computed solution u. This variation in the value of $|E|$ can be taken as a measure of smoothness indicator for the convection-diffusion problem. Here we refer $E(u,\Phi)$ as *weak local truncation error* for u with respect to test function Φ. In practice, the computation of WLTE appears to be a difficult task since Φ is a general test function. Kurganov and his co-workers overcome this difficulty in [4, 2], where they have used B-splines as test function. The test function using B-splines is defined as follows

$$\Phi_i^n(x,t)=\mathbb{B}_i(x)\mathbb{B}^n(t), \tag{4}$$

where $\mathbb{B}_j(x)$ and $\mathbb{B}^n(t)$ are the quadratic and the linear B-splines with the localized supports of size $|\text{supp}(\mathbb{B}_j)|=3\Delta x$ and $|\text{supp}(\mathbb{B}^n)|=2\Delta t$, respectively (for notations and details see [5]). Under the assumption on the solution to be piece-wise constant over the cells and putting (4) in (3), we can arrive at

$$\begin{aligned} E_i^n =& \frac{1}{6}\Big[u_{i+1}^n-u_{i+1}^{n-1}+4(u_i^n-u_i^{n-1})+u_{i-1}^n-u_{i-1}^{n-1}\Big]\Delta x+\frac{1}{4}\Big[f(u_{i+1}^n)-f(u_{i-1}^n)\\ &+f(u_{i+1}^{n-1})-f(u_{i-1}^{n-1})\Big]\Delta t-\frac{\epsilon\Delta t}{2\Delta x}(u_{i-1}^n+u_{i-1}^{n-1}-2(u_i^n+u_i^{i-1})+u_{i+1}^n+u_{i+1}^{n-1}). \end{aligned} \tag{5}$$

Remark 1. With the help of Taylor series expansion about a point (x_i,t^n), we have the following estimates [5] under the sufficient assumption on the smoothness

N	L^∞-error	Order	L^1-error	Order	L^2-error	Order
20	1.733727e-04	–	2.203097e-04	–	1.735307e-04	
40	5.692327e-06	4.928714	7.245025e-06	4.926399	5.693688e-06	4.929684
80	1.876191e-07	4.923140	2.388746e-07	4.922665	1.876329e-07	4.923379
160	6.273009e-09	4.902505	7.987088e-09	4.902440	6.273150e-09	4.902579
320	2.137606e-10	4.875090	2.721421e-10	4.875238	2.137401e-10	4.875260

TABLE 1. Comparison of L^∞-, L^1-, and L^2-errors of hybrid FDM-WENO scheme along with their convergence rate.

of the solution u

$$\|E\|_\infty \approx O\left(\triangle^{\min\{r+2,4\}}\right),$$

where $\Delta = \max(\Delta x, \Delta t)$ and r is the order of accuracy of the numerical scheme used to compute solution u.

Remark 2. In the case $\epsilon = 0$, Kurganov and Liu [4] proposed the following estimate and used it in devising an adaptive artificial viscosity method for hyperbolic conservation laws:

$$\|E_j^n\|_\infty \approx \begin{cases} \Delta, \text{near the shock}, \\ \Delta^\alpha, \text{near the contact wave}, \quad 1 < \alpha \leq 2, \\ \Delta^\beta, \text{in the smooth region}, \end{cases}$$

where $\beta = \min\{r + 2, 4\}$.

In our hybrid FDM-WENO scheme, we propose to use the smoothness indicator defined by

$$\Phi(x_j) = \begin{cases} 1, & |E_j| > K\Delta x^4, \\ 0, & \text{otherwise}, \end{cases} \tag{6}$$

where K is a positive real number, and we choose $K = 1$ in our numerical computations.

2.3. **Algorithm.** The algorithm of hybrid WENO-FDM comprises the following steps

1. To calculate the solution at $(n+1)^{\text{th}}$ level, compute the WLTE E^n using the information of solution available at previous time levels n and $n-1$.
2. Compute the smoothness indicator Φ_i using (6).
3. To ensure the smooth transition between two schemes, we need to create a buffer zone near the problematic points. If point x_i is identified as a problematic point, we also flag neighboring points as problematic
$$\Phi(x_j) = 1, \text{ where } x_j = x_i \pm a_1\Delta x \text{ for } a_1 = 1, 2, 3$$
4. Compute the solution using FDM in smooth parts, and WENO-AO(5,4,3) is used in problematic regions.

3. **Numerical Experiments.** In this section, we test the accuracy and resolution of the hybrid FDM-WENO algorithm across sharp variations or shocks. The accuracy of the schemes measured in L^∞-, L^1-, and L^2-errors.

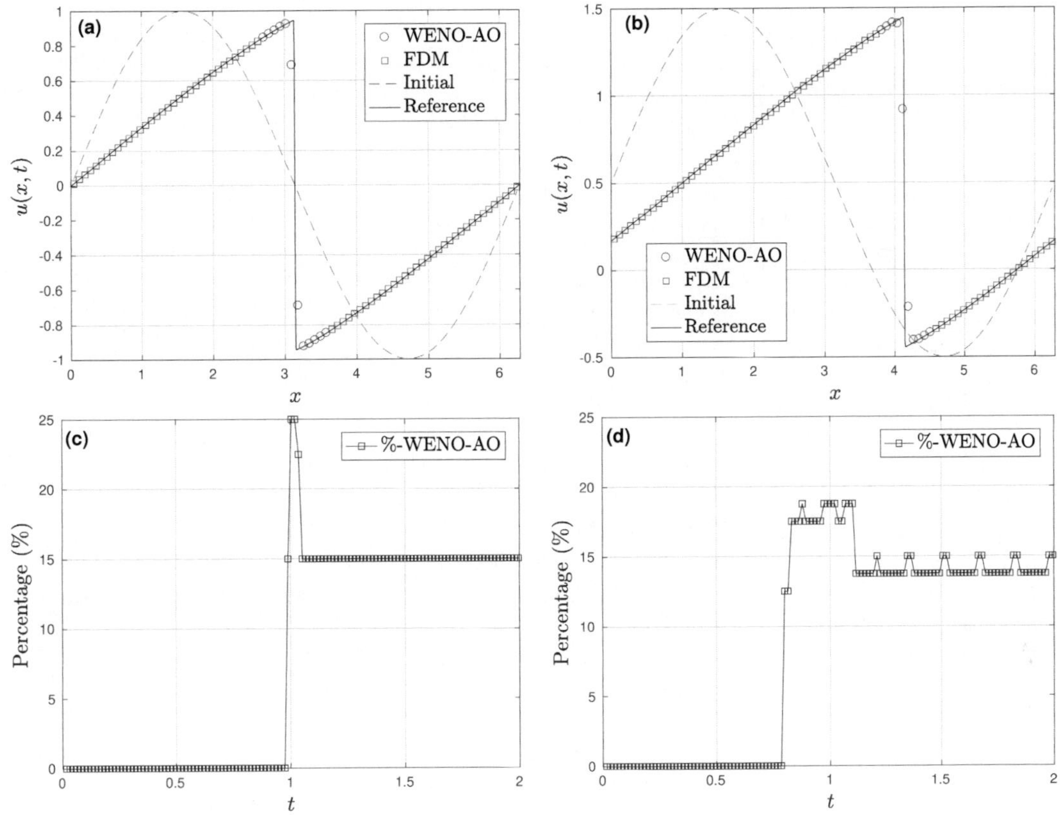

FIGURE 1. (a)-(b) Comparison of hybrid FDM-WENO solution with the reference solution at time $T = 2$ for Example 3.2, (c)-(d) percentage of WENO-AO(5,4,3) scheme used during simulations.

Example 1. (Accuracy test) Consider the linear advection equation

$$u_t + u_x = 0, \ (x,t) \in [-1,1]$$

subject to initial data $u(x,0) = \sin(2\pi x)$ with periodic boundary conditions. In Table 1, we have depicted the L^∞-, L^1-, and L^2-errors at time $T = 1.0$ obtained using hybrid FDM-WENO scheme. We can easily observe from Table 1 that hybrid scheme converges to exact solution with rate five.

Example 2. (Moving and stationary shock) Consider the inviscid Burgers' equation with the following two initial data

$$u(x,0) = \sin(x), \quad x \in [0, 2\pi], \tag{7}$$

$$u(x,0) = 0.5 + \sin(x), \quad x \in [0, 2\pi], \tag{8}$$

with periodic boundary condition. The Burgers' equation with (7) will leads to stationary shock occur at position $x = \pi$ at time $T = 1$. Whereas Burgers' equation with (8) leads to shock moving to the right initially occurring at time $T = 1$. We compute the numerical solution at time $T = 2$ and compare it with the reference solution. The reference solution is computed with pure WENO-AO(5,4,3) scheme using 500 number of mesh points. In both cases, we use 80 number of mesh points, and CFL number is taken to be 0.5. In Figure 1 (a), we compare the solution for

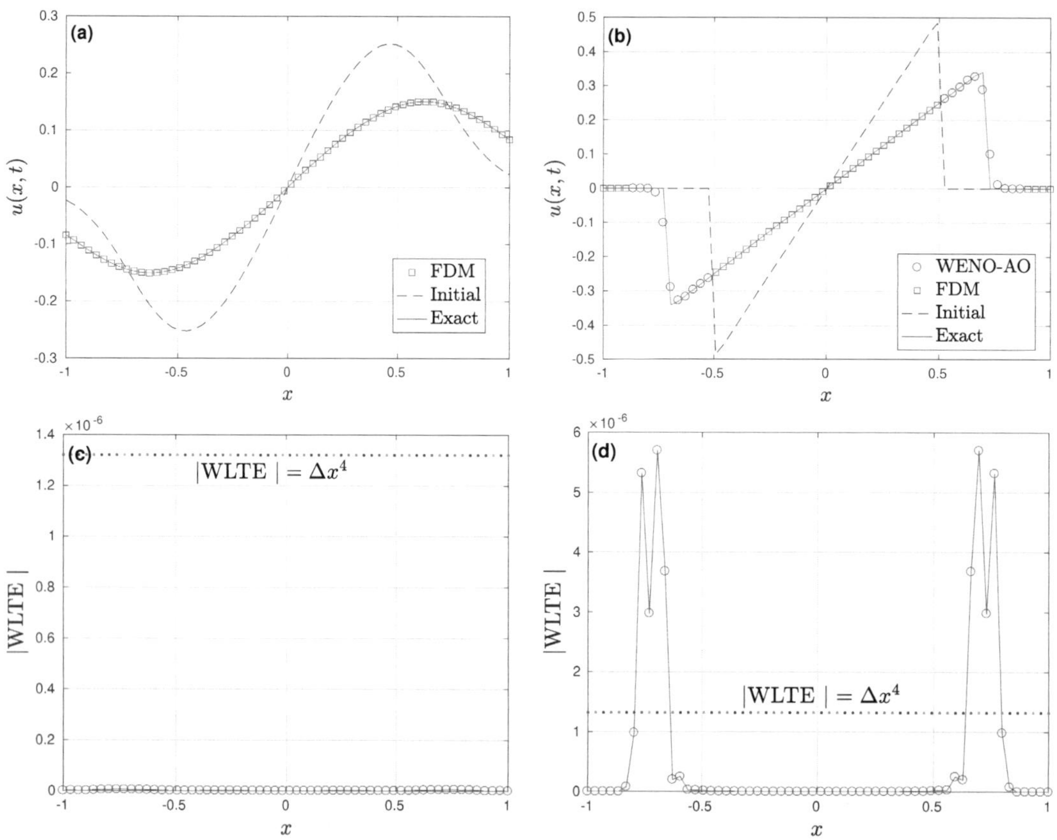

FIGURE 2. (a)-(b) Comparison of hybrid FDM-WENO solution with the exact solution at time $T = 2.0$ for Example 3.3, (c)-(d) WLTE for $\epsilon = 0.05, 0.0005$ at final time $T = 2.0$.

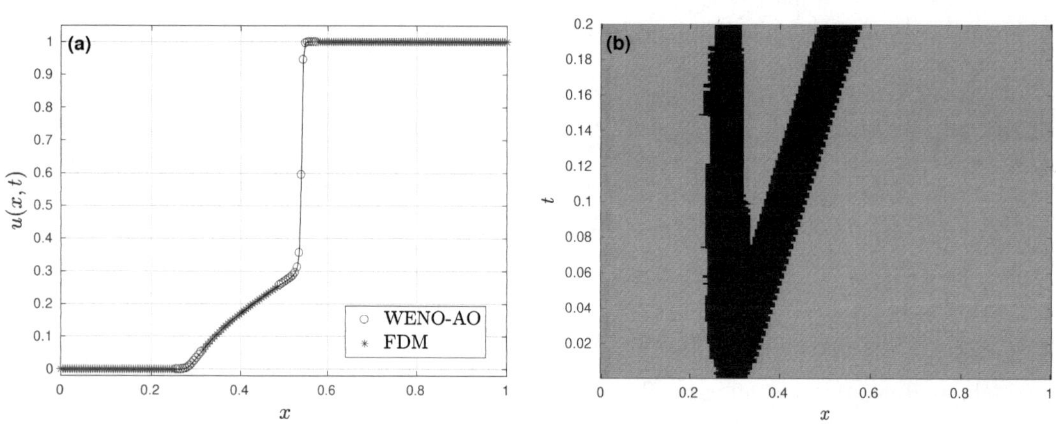

FIGURE 3. (a) Numerical solution obtained using hybrid FDM-WENO at time $T = 0.2$ for Example 3.4, (b) track of smoothness indicator over $x - t$ plane.

Burgers' equation corresponding to initial data (7) with reference solution at time $T = 2$ and corresponding percentage of WENO-AO(5,4,3) scheme is used in computing the solution depicted in Figure 1 (c). The 'o-' indicates computed solution

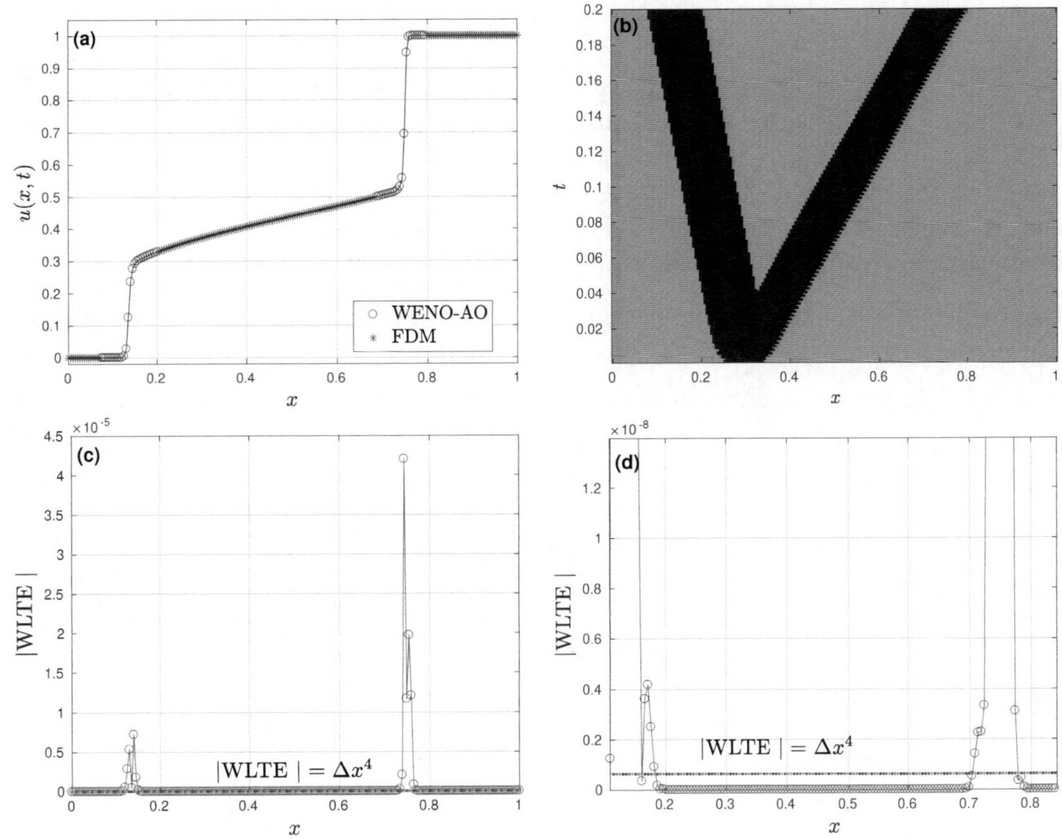

FIGURE 4. (a) Numerical solution obtained using hybrid FDM-WENO at time $T = 0.2$ for Example 3.5, (b) track of smoothness indicator over $x - t$ plane, (c)-(d) WLTE at time $T = 0.2$.

using WENO-AO(5,4,3) scheme and '□-' indicates the usage of FDM. In the Figure 1 (c), it can be easily observed there is zero percentage of WENO-AO(5,4,3) scheme used before the time approximately $T = 0.8$. After time $T = 0.8$, wave steepening and shock form at $T = 1$, are well indicated with smoothness indicator. The hybrid FDM-WENO scheme maintain non-oscillatory profile at shock position, and higher order accuracy is achieved with fifth order central difference approximation in smooth regions. After the shock formation, we can observe WENO-AO(5,4,3) scheme is used near shock position only 10-15% of the overall scheme. Similar observations are found in case of moving shock case. The solution computed with hybrid FDM-WENO scheme is shown in Figure 1 (b) along with % of WENO-AO(5,4,3) scheme is shown in Figure 1 (d).

Example 3. Consider the case of nonlinear Burgers' ($f(u) = u^2/2$) equation with the exact solution is given by

$$u(x,t) = \frac{x/t}{1 + \sqrt{t/t_0}\exp(x^2/4\epsilon t)}, \tag{9}$$

where $t_0 = \exp(1/8\epsilon)$. The initial condition obtained by considering $t = 1$ in (9). In Figure 2, we have shown the numerical solution at time $T = 2$ obtained by using hybrid FDM-WENO scheme for $\epsilon = 0.05, 0.0005$. The number of mesh points used are 60 and CFL is 0.5. As we decrease the value of ϵ, we can observed the steepening of waves near the points $x = \pm 0.75$. From Figure 2 (a), we can observed that for $\epsilon = 0.05$ solution have not steep gradient and it is computed using FDM only. In Figure 2 (b), we can see in regions of high gradient solution is computed with WENO-AO(5,4,3) and in the other regions FDM is used. In Figure 2 (c)-(d), we have depicted the WLTE for $\epsilon = 0.05, 0.0005$, respectively. For $\epsilon = 0.05$, solution have no regions of steep gradients, hence WLTE lies below the Δx^4. For $\epsilon = 0.0005$, high gradients are well indicated by WLTE, can be seen from Figure 2 (d).

Example 4. (Buckley-Leverett equation) We consider the convection-diffusion Buckley-Leverett equation of the form

$$u_t + f(u)_x = \epsilon(v(u)u_x)_x. \tag{10}$$

This test is an example of degenerate parabolic equation since $v(u)$ vanishes at some points. In this test case, we take flux function f and v of the form

$$f(u) = \frac{u^2}{u^2 + (1-u)^2},$$

and

$$v(u) = \begin{cases} 4u(1-u), & u \in [0,1], \\ 0, & \text{otherwise}. \end{cases}$$

The initial condition is given by

$$u(x,0) = \begin{cases} 1 - 3x, & x \in [0, \frac{1}{3}] \\ 0, & x \in (\frac{1}{3}, 1]. \end{cases}$$

The Dirichlet boundary condition $u(0,t) = 1$ is used at one end and outflow boundary on the other end of the boundary. The numerical solution is computed for fixed time $T = 0.2$ and $\epsilon = 0.01$ using 200 mesh points. In Figure 3, we have depicted the solution along with track of smoothness indicator over $x - t$ plane. The FDM-WENO scheme compute the solution in a non-oscillatory manner.

Example 5. Consider the Buckley-Leverett equation (10) with the flux

$$f(u) = \frac{u^2}{u^2 + (1-u)^2}(1 - 5(1-u)^2).$$

and ϵ, v are same as in the previous example. The initial condition is given by

$$u(x,0) = \begin{cases} 0, & x \in [0, 1 - \frac{1}{\sqrt{2}}) \\ 1, & x \in [1 - \frac{1}{\sqrt{2}}, 1]. \end{cases}$$

We have computed the solution using 200 mesh points at time $T = 0.2$ for $\epsilon = 0.01$. In Figure 4, we have depicted the solution with the track of smoothness indicator over the $x-t$ plane and WLTE estimate at the final time. The hybrid FDM-WENO scheme computes the solution efficiently and in a non-oscillatory manner.

4. **Conclusions.** In this article, we have proposed a hybrid FDM-WENO scheme for the convection-diffusion problems. In smooth regions, efficient fifth order finite difference method is used, and non-oscillatory profile is maintained using the WENO scheme of adaptive order in the discontinuous areas. A WLTE is used to separate the smooth and discontinuous regions. Numerical experiments are performed, which show that WLTE error efficiently identifies the discontinuous regions and hybrid FDM-WENO computes the solution in a non-oscillatory manner. The

multi-dimension extension of the hybrid FDM-WENO scheme is an interest of future work.

REFERENCES

[1] S. Karni, A. Kurganov and G. Petrova, A smoothness indicator for adaptive algorithms for hyperbolic systems, *Journal of Computational Physics*, **178** (2002), 323–341.

[2] S. Karni and A. Kurganov, Local error analysis for approximate solutions of hyperbolic conservation laws, *Advances in Computational Mathematics*, **22** (2005), 79–99.

[3] C.-W. Shu and S. Osher, Efficient implementation of essentially nonoscillatory shock-capturing schemes, *Journal of Computational Physics*, **77** (1988), 439–471.

[4] A. Kurganov and Y. Liu, New adaptive artificial viscosity method for hyperbolic systems of conservation laws, *Journal of Computational Physics*, **231** (2012), 8114–8132.

[5] R. Kumar, Adaptive semi-discrete formulation of BSQI–WENO scheme for the modified Burgers' equation, *BIT Numerical Mathematics*, **58** (2018), 103–132.

[6] R. Kumar and P. Chandrashekar, Simple smoothness indicator and multi-level adaptive order WENO scheme for hyperbolic conservation laws, *Journal of Computational Physics*, **375** (2018), 1059–1090.

E-mail address: rakeshiitb21@gmail.com, rakesh@tifrbng.res.in

THE RIEMANN PROBLEM FOR THE GARZ MODEL WITH A MOVING CONSTRAINT

THIBAULT LIARD

Inria
Grenoble Rhône-Alpes
France

FRANCESCA MARCELLINI*

INdAM Unit, Department of Information Engineering
University of Brescia
Italy

BENEDETTO PICCOLI

Department of Mathematical Sciences and CCIB
Rutgers University-Camden, Camden, NJ
USA

ABSTRACT. This paper deals with the Riemann Problem for the Generalized Aw-Rascle-Zhang (GARZ) model, introduced in [7], subject to a moving constraint. A slow and large vehicle on a crowded road is a moving obstacle and reduces the road capacity. This situation can be modeled by a strongly coupled ODE-PDE system. The PDE consists of a 2×2 system of conservation laws. The ODE describes the motion of the slow vehicle and it influences the bulk traffic flow via a moving point flux constraint.

1. **Introduction.** In this paper we describe the situation created by a slowly moving large vehicle, like a bus, that generates a moving bottleneck, since it reduces the road capacity at its position. Thus, we consider a mixed ODE-PDE model consisting of two conservation laws, the generalized Aw-Rascle-Zhang (GARZ) [7], coupled with an ODE describing the motion of the slow vehicle. The ODE influences the bulk traffic flow via a moving point flux constraint.

The current literature offers various macroscopic models describing traffic evolution. First, they can rely on a single equation, such as the classical Lighthill–Whitham [12] and Richards [14] (LWR) model. Then, the so called *second order* ones are based on two equations, the main examples being the Aw-Rascle-Zhang (ARZ) model [1, 16], the present GARZ model and the collapsed GARZ (CGARZ) [8]. A further class of current interest is that of 2−phases or phase transition models, see [2, 4, 5, 9, 11, 13].

2000 *Mathematics Subject Classification.* Primary: 35L65; Secondary: 90B20.

Key words and phrases. Hyperbolic Systems of Conservation Laws, Continuum Traffic Models, Unilateral Flux Constraints.

The second author was partially supported by the INdAM-GNAMPA 2018 project "Conservation Laws: Hyperbolic Games, Vehicular Traffic and Fluid Dynamics".

* Corresponding author: Francesca Marcellini.

In this paper, we consider the Riemann Problem for the GARZ model subject to a moving constraint. Refer to [6] for an analogous study limited to the LWR case and to [15] for a construction based on the ARZ model. More precisely, in Section 2 we describe the GARZ model. In Section 3 we consider the ODE-PDE constrained system and solve the corresponding Riemann Problem.

2. Description of the GARZ Model.

The generalized Aw-Rascle-Zhang (GARZ) model, introduced in [7], consists of the following 2×2 system of conservation laws:

$$\begin{cases} \partial_t \rho + \partial_x \left(\rho V(\rho, w)\right) = 0 \\ \partial_t(\rho w) + \partial_x \left(\rho w V(\rho, w)\right) = 0 . \end{cases} \tag{1}$$

Here, $\rho \in [0, \rho_{\max}]$ is the traffic density where $\rho_{\max}$ is its maximal possible value; w is a particular feature of each driver, limited in the fixed interval $[w_{\min}, w_{\max}]$, with $w_{\max} \geq w_{\min} > 0$; $V(\rho, w)$ is the traffic speed of vehicles at density ρ, having feature w. In the above system (1), $(\rho, \rho w)$ is the pair of the conserved variables, nevertheless below we mostly refer to the couple (ρ, w). Note that in the case $w_{\max} = w_{\min}$, (1) essentially reduces to the LWR model.

We impose the following requirements on the velocity $V(\rho, w)$ and on the associated flow rate function $f_1(\rho, w) = \rho V(\rho, w)$:

A1. $(\rho, w) \mapsto V(\rho, w)$ is $C^2\left([0, \rho_{max}] \times [w_{\min}, w_{\max}]\right)$.

A2. $V(\rho, w) \geq 0$ for all $(\rho, w) \in [0, \rho_{\max}] \times [w_{\min}, w_{\max}]$: vehicles never drive backwards.

A3. $V(0, w) = w$ for all $\omega \in [w_{\min}, w_{\max}]$, so that w is each driver's speed on an empty road.

A4. $\frac{\partial^2 f_1}{\partial \rho^2}(\rho, w) < 0$ and $\frac{\partial V}{\partial \rho}(\rho, w) < 0$ for all $(\rho, w) \in [0, \rho_{\max}] \times [w_{\min}, w_{\max}]$, so that the traffic speed decreases as traffic density increases.

A5. $\frac{\partial V}{\partial w}(\rho, w) > 0$ for all $(\rho, w) \in [0, \rho_{\max}[\times [w_{\min}, w_{\max}]$, drivers travelling faster on an empty road, are faster also in a crowded situation.

A6. $V(\rho_{\max}, w) = 0$: at the maximal density $\rho_{\max}$, no vehicle moves.

From requirement *A*5. there exists a map

$$R\colon \left\{(v, w) \in \mathbb{R} \times [w_{\min}, w_{\max}] \colon 0 < v \leq w\right\} \to [0, \rho_{\max}[$$

such that $\rho = R(v, w)$ if and only if $V(\rho, w) = v$ and a map

$$W\colon \left\{(\rho, v) \in [0, \rho_{\max}[\times \mathbb{R} \colon v \in [V(\rho, w_{\min}), V(\rho, w_{\max})]\right\} \to [w_{\min}, w_{\max}]$$

such that $w = W(\rho, v)$ if and only if $V(\rho, w) = v$.

We recall that the characteristic speeds of the GARZ model are $\lambda_1(\rho, w) = V(\rho, w) + \rho \frac{\partial V}{\partial \rho}(\rho, w)$ and $\lambda_2(\rho, w) = V(\rho, w)$. The first one is genuinely non linear and the second one is linearly degenerate, see [3]. From requirement A4., $\lambda_1(\rho, w) < \lambda_2(\rho, w)$ for every $\rho > 0$.

Next, we recall the Riemann problem associated to (1) with initial data (ρ_l, w_l) and (ρ_r, w_r). First, we recall all the possible waves in the solution.

- *First family wave (1-wave)*: a wave of the first characteristic family that connects a left state (ρ_l, w_l) with a right state (ρ_r, w_r) whenever $w_l = w_r$, which is a shock for $\rho_l < \rho_r$ or a rarefaction wave for $\rho_l > \rho_r$.
- *Second family wave (2-wave)*: a wave of the second characteristic family that connects a left state (ρ_l, w_l) with a right state (ρ_r, w_r) such that $V(\rho_l, w_l) = V(\rho_r, w_r)$, which is a contact discontinuity.

- *Vacuum wave (V-wave)*: a non-classical wave connecting a left state (ρ_l, w_l) with a right state (ρ_r, w_r) such that $\rho_l = \rho_r = 0$ and $w_l < w_r$.

The Riemann Problem. For all states (ρ_l, w_l) and (ρ_r, w_r) in the set $[0, \rho_{\max}] \times [w_{\min}, w_{\max}]$ the Riemann problem consisting of (1) with initial data

$$\rho(0,x) = \begin{cases} \rho_l & \text{if } x < 0 \\ \rho_r & \text{if } x > 0 \end{cases} \qquad w(0,x) = \begin{cases} w_l & \text{if } x < 0 \\ w_r & \text{if } x > 0 \end{cases} \tag{2}$$

admits a weak solution $(\rho, w) = (\rho, w)(t, x)$ constructed as follows:

- If $w_l < w_r$ and $\rho_r < R(w_l, w_r)$ then (ρ_l, w_l) is connected to $(0, w_l)$ by a 1-wave; $(0, w_l)$ is connected to $(0, w_r)$ by a V-wave and $(0, w_r)$ is connected to (ρ_r, w_r) by a 2-wave.
- Otherwise, (ρ_l, w_l) is connected to the point $(R(V(\rho_r, w_r), w_l), w_l) \in [0, \rho_{max}] \times [w_{min}, w_{max}]$ by a 1-wave and $(R(V(\rho_r, w_r), w_l), w_l)$ is connected to (ρ_r, w_r) by a 2-wave.

Direct computations show that, by A6., the domain $[0, \rho_{\max}] \times [w_{\min}, w_{\max}]$ is invariant with respect to the Riemann Solver recalled above. To this aim, the classical results [10] can not be applied, due to the coincidence $\lambda_1(0, w) = \lambda_2(0, w)$.

3. **The GARZ Model with a Moving Constraint.** In this section we consider the case of a slow and large bus that acts as a moving obstacle blocking a portion of the road and hindering traffic. The bus trajectory $y = y(t)$ solves the following ODE

$$\dot{y}(t) = \omega\left(\rho(t, y(t)+), w(t, y(t)+)\right), \tag{3}$$

with a speed

$$\omega(\rho, w) = \min\{V_b, V(\rho, w)\}. \tag{4}$$

The right limit in (3) is due to the bus adjusting its speed according to the traffic situation it has in front. Thus, the bus travels with its speed V_b whenever traffic allows it, otherwise it adapts its speed to the traffic conditions. In any case, it is not possible for the bus to overtake cars.

In the case $\dot{y} = V_b$, we introduce the flux of the main traffic F at the bus position by

$$\begin{array}{rccccl} F & : [0, \rho_{\max}] & \times & [w_{\min}, w_{\max}] & \to & \mathbb{R} \\ & \rho & , & w & \to & \rho\left(V(\rho, w) - V_b\right). \end{array} \tag{5}$$

As a consequence, for a driver with feature $w \in [w_{\min}, w_{\max}]$, the maximum available flux at the bus position is

$$\begin{array}{rcccl} F_\alpha & : & [w_{\min}, w_{\max}] & \to & \mathbb{R} \\ & & w & \to & \alpha(w) \displaystyle\max_{\rho \in [0, \rho_{\max}]} F(\rho, w), \end{array} \tag{6}$$

where $\alpha \colon [w_{\min}, w_{\max}] \to (0, 1)$ models the reduction of the road capacity felt by a driver with feature w. Thus, we consider the following mixed ODE-PDE system consisting of the PDE model (1), of the ODE describing the slower vehicle motion (3) and with a moving constraint on the flux:

$$\begin{cases} \partial_t \rho + \partial_x\left(\rho V(\rho, w)\right) = 0 \\ \partial_t(\rho w) + \partial_x\left(\rho w V(\rho, w)\right) = 0 \\ \dot{y}(t) = \omega\left(\rho(t, y(t)+), w(t, y(t)+)\right) \\ \rho\left(t, y(t)\right)\left(V(\rho(t, y(t)), w(t, y(t))) - \dot{y}(t)\right) \le F_\alpha(w). \end{cases} \tag{7}$$

It is natural to assume that the slowest cars still travel faster than the bus along an empty road, that is $V_b < w_{\min}$. A reasonable regularity requirement of the function α is $\alpha \in C^1((w_{\min}, w_{\max}); [0,1]) \cap C^0([w_{\min}, w_{\max}], [0,1])$.

Lemma 3.1. *Under the assumptions A1.–A6., with the notation* (5)–(6) *and requiring* $V_b < w_{\min}$, $\alpha \in C^1((w_{min}, w_{max}); [0,1]) \cap C^0([w_{\min}, w_{\max}], [0,1])$, *we have*

1. *There exist* $\check{\rho}_w$, $\hat{\rho}_w$ *solving* $F_\alpha(w) + \rho V_b = \rho V(\rho, w)$, *with* $\check{\rho}_w, \hat{\rho}_w \in [0, \rho_{\max}]$.
2. *There exists a unique* $\rho^*_w \in [0, \rho_{\max}]$ *solving* $V_b \rho = \rho V(\rho, w)$.
3. *There exists a unique* $\rho_c(w)$ *maximizing the map* $\rho \to F(\rho, w)$. *In particular,* $\frac{\partial f_1(\rho_c(w),w)}{\partial \rho} = V_b$.
4. *We have* $\check{\rho}_w < \rho_c(w) < \hat{\rho}_w < \rho^*_w$.
5. *The maps* $w \mapsto V(\check{\rho}_w, w)$ *and* $w \mapsto V(\hat{\rho}_w, w)$ *are continuously differentiable.*

The proof relies on basic calculus techniques and is omitted.

We consider the Riemann problem for system (7) with initial data (2). We denote by $f(\rho, w)$ the flux for the system of conservation laws in (1) and $f_1(\rho, w) = \rho V(\rho, w)$, $f_2(\rho, w) = \rho\, w\, V(\rho, w)$ are its components. By $\mathcal{RS}$ we denote the classical Riemann solver for the system of conservation laws (1), i.e., the standard weak entropy solution to (1)–(2) is given by the map $(t, x) \mapsto \mathcal{RS}\left((\rho_l, w_l), (\rho_r, w_r)\right)\left(\frac{x}{t}\right)$. By $\mathcal{RS}_\rho$, respectively $\mathcal{RS}_w$, we denote the ρ, respectively w, component of the classical solution $\mathcal{RS}((\rho_l, w_l), (\rho_r, w_r)(\cdot))$.

The Riemann solver $\mathcal{RS}^\alpha$ for (7) with initial datum

$$\begin{aligned} (\rho, w)(0, x) &= \begin{cases} (\rho_l, w_l) & \text{if } x < 0 \\ (\rho_r, w_r) & \text{if } x > 0 \end{cases} \\ y(0) &= 0 \end{aligned} \tag{8}$$

is defined as follows.

Definition 3.2. The constrained Riemann solver

$$\mathcal{RS}^\alpha\colon \left([0, \rho_{\max}] \times [w_{\min}, w_{\max}]\right)^2 \to \mathbf{L}^1_{\mathbf{loc}}(\mathbb{R}; [0, \rho_{\max}] \times [w_{\min}, w_{\max}])$$

is defined as follows.

1. If $f_1(\mathcal{RS}((\rho_l, w_l), (\rho_r, w_r)(V_b)) > F_\alpha(w_l) + V_b \mathcal{RS}_\rho((\rho_l, w_l), (\rho_r, w_r))(V_b)$, then
$$\mathcal{RS}^\alpha\left((\rho_l, w_l), (\rho_r, w_r)\right)\left(\tfrac{x}{t}\right) = \begin{cases} \mathcal{RS}\left((\rho_l, w_l), (\hat{\rho}_{w_l}, w_l)\right)\left(\frac{x}{t}\right) & \text{if } \frac{x}{t} < V_b \\ \mathcal{RS}\left((\check{\rho}_{w_l}, w_l), (\rho_r, w_r)\right)\left(\frac{x}{t}\right) & \text{if } \frac{x}{t} > V_b \end{cases}$$
$$y(t) = V_b\, t\,.$$
2. If $f_1(\mathcal{RS}((\rho_l, w_l), (\rho_r, w_r)(V_b)) \le F_\alpha(w_l) + V_b \mathcal{RS}_\rho((\rho_l, w_l), (\rho_r, w_r))(V_b)$ and $V_b < V(\mathcal{RS}((\rho_l, w_l), (\rho_r, w_r))(V_b))$, then
$$\begin{aligned} &\mathcal{RS}^\alpha\left((\rho_l, w_l), (\rho_r, w_r)\right)\left(\tfrac{x}{t}\right) = \mathcal{RS}\left((\rho_l, w_l), (\rho_r, w_r)\right)\left(\tfrac{x}{t}\right) \\ &y(t) = V_b\, t\,. \end{aligned}$$
3. If $V_b \ge V(\mathcal{RS}((\rho_l, w_l), (\rho_r, w_r))(V_b))$, then
$$\begin{aligned} &\mathcal{RS}^\alpha\left((\rho_l, w_l), (\rho_r, w_r)\right)\left(\tfrac{x}{t}\right) = \mathcal{RS}\left((\rho_l, w_l), (\rho_r, w_r)\right)\left(\tfrac{x}{t}\right) \\ &y(t) = V(\mathcal{RS}((\rho_l, w_l), (\rho_r, w_r)(V_b)))\, t\,. \end{aligned}$$

Note that in the first case, traffic is influenced by the bus which travels with its own velocity. In the second case, there is essentially no interaction between the bus and the traffic, thanks to a low traffic density. The third case refers to a situation in which the traffic is so heavy that the bus has to slow down and adapt its speed.

Proving that the above definition indeed singles out a unique weak solution to (7)–(8) amounts to deal with a variety of cases, each treatable by means of basic calculus techniques. We prefer to describe in detail sample situations through diagrams in the $(\rho, \rho V)$ and (x, t)-plane.

In Figures 1–2–3, we denote $L \equiv (\rho_l, w_l)$, $M \equiv (\rho_m, w_m)$, $\check{M} \equiv (\check{\rho}_{w_l}, w_l)$, $\hat{M} \equiv (\hat{\rho}_{w_l}, w_l)$ and $R \equiv (\rho_r, w_r)$ on the left in the $(\rho, \rho w)$–plane and on the right in the (x, t)–plane. Note that the middle state $M \equiv (\rho_m, w_m)$ is uniquely characterized by the two conditions $V(\rho_m, w_m) = V(\rho_r, w_r)$ and $w_m = w_l$.

In case 1., the standard solution to the Riemann problem does not satify the constraint at the bus position. Therefore, we introduce a non–classical shock at the

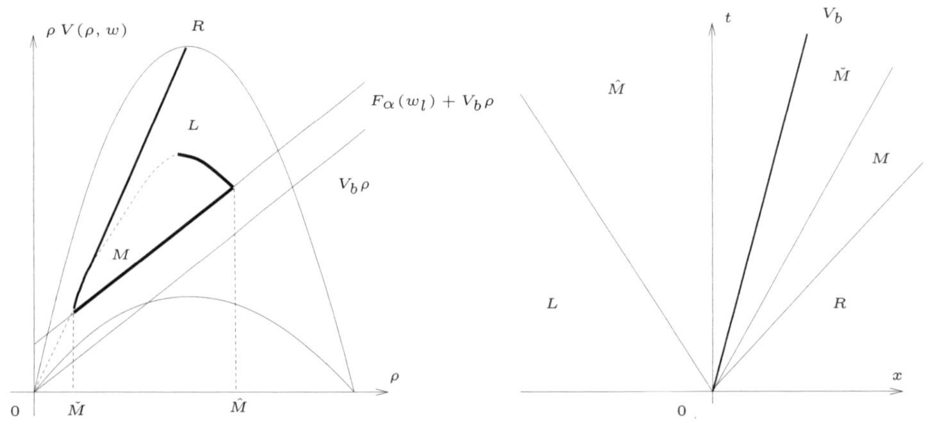

FIGURE 1. Solution to the constrained Riemann problem (7)–(8) in case 1. of Definition 3.2. Left, the $(\rho, \rho V)$–plane of the fundamental diagram and, right, the (x, t)–plane. Refer to the text for a detailed description.

bus position and solve the constrained Riemann problem by means of a Lax wave of the first family, the non classical wave, a further Lax wave of the first family and a Lax wave of the second family, see Figure 1, left. Remark that both the first family waves are supported on the same Lax curve. In the (t, x)–plane, the bus position, which moves with speed V_b, supports a non classical wave in the solution to the conservation law (1). Here, the Rankine-Hugoniot conditions are satisfied, so that the total number of vehicles is duly conserved, but Lax stability conditions are typically violated, see Figure 1, right.

A sample situation fitting in case 2. of Definition 3.2 is portrayed in Figure 2. Here, the standard solution to the Riemann Problem (2) does satisfy the flow reduction at the bus position, therefore it solves also the constrained Riemann problem (7)–(8), see Figure 2, left. Again, traffic conditions allow the bus to travel with its maximal speed V_b, see Figure 2, right.

Finally, case 3. is displayed in Figure 3. In this situation, traffic is so intense that the bus has to slow down. The classical Lax solution to the Riemann problem for (1) also solves (7)–(8), provided the bus adapts its velocity to that of the other vehicles, see Figure 3, right.

REFERENCES

[1] A. Aw and M. Rascle. Resurrection of "second order" models of traffic flow. *SIAM J. Appl. Math.*, **60** (2000) 916–938 (electronic).

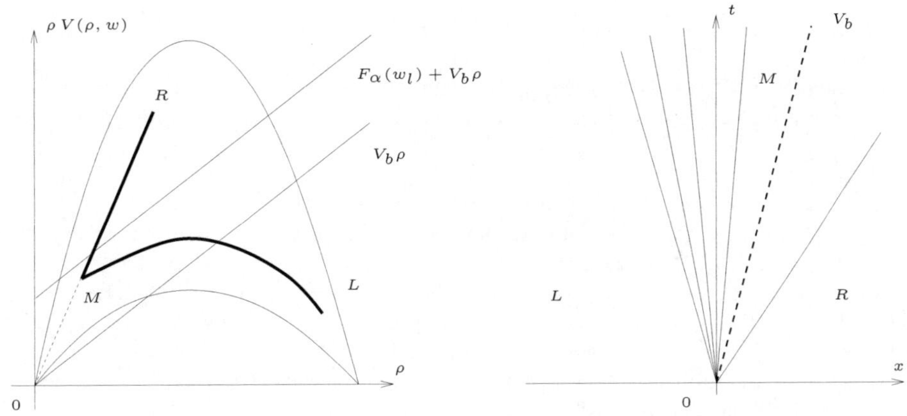

FIGURE 2. Solution to the constrained Riemann problem (7)–(8) in case 2. of Definition 3.2. Left, the $(\rho, \rho\, V)$–plane of the fundamental diagram and, right, the (x, t)–plane. Refer to the text for a detailed description.

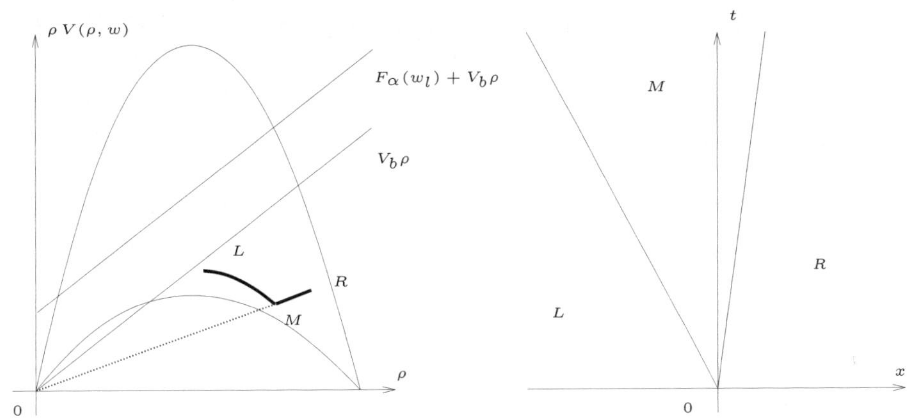

FIGURE 3. Solution to the constrained Riemann problem (7)–(8) in case 3. of Definition 3.2. Left, the $(\rho, \rho\, V)$–plane of the fundamental diagram and, right, the (x, t)–plane. Refer to the text for a detailed description.

[2] S. Blandin, D. Work, P. Goatin, B. Piccoli, and A. Bayen. A general phase transition model for vehicular traffic. *SIAM J. Appl. Math.*, **71**(2011) 107–127.

[3] A. Bressan. *Hyperbolic systems of conservation laws*, volume 20 of *Oxford Lecture Series in Mathematics and its Applications.* Oxford University Press, Oxford, 2000. The one-dimensional Cauchy problem.

[4] R. M. Colombo. Hyperbolic phase transitions in traffic flow. *SIAM J. Appl. Math.*, **63** (2002) 708–721.

[5] R. M. Colombo, F. Marcellini, and M. Rascle. A 2-phase traffic model based on a speed bound. *SIAM J. Appl. Math.*, **70** (2010) 2652–2666.

[6] M. L. Delle Monache and P. Goatin. Scalar conservation laws with moving constraints arising in traffic flow modeling: an existence result. *J. Differential Equations*, **257** (2014) 4015–4029.

[7] S. Fan, M. Herty, and B. Seibold. Comparative model accuracy of a data-fitted generalized Aw-Rascle-Zhang model. *Netw. Heterog. Media*, **9** (2014) 239–268.

[8] S. Fan, Y. Sun, B. Piccoli, B. Seibold, and D. B. Work. A Collapsed Generalized Aw-Rascle-Zhang Model and Its Model Accuracy. *ArXiv e-prints*, Feb. 2017.

[9] P. Goatin. The Aw-Rascle vehicular traffic flow model with phase transitions. *Math. Comput. Modelling*, **44**(2006) 287–303.

[10] D. Hoff. Invariant regions for systems of conservation laws. *Trans. Amer. Math. Soc.*, **289** (1985) 591–610.

[11] J. P. Lebacque, X. Louis, S. Mammar, B. Schnetzlera, and H. Haj-Salem. Modélisation du trafic autoroutier au second ordre. *Comptes Rendus Mathematique*, **346** (November 2008) 1203–1206.

[12] M. J. Lighthill and G. B. Whitham. On kinematic waves. II. A theory of traffic flow on long crowded roads. *Proc. Roy. Soc. London. Ser. A.*, **29** (1955) 317–345.

[13] F. Marcellini. Free-congested and micro-macro descriptions of traffic flow. *Discrete Contin. Dyn. Syst. Ser. S*, **7** (2014) 543–556.

[14] P. I. Richards. Shock waves on the highway. *Operations Res.*, **4** (1956) 42–51.

[15] S. Villa, P. Goatin, and C. Chalons. Moving bottlenecks for the aw-rascle-zhang traffic flow model. *Discrete Contin. Dyn. Syst. Ser. B*, **22** (2017) 3921–3952.

[16] H. Zhang. A non-equilibrium traffic model devoid of gas-like behavior. *Transportation Research Part B: Methodological*, **36** (2202) 275 – 290.

E-mail address: `thibault.liard@inria.fr`

E-mail address: `francesca.marcellini@unibs.it`

E-mail address: `piccoli@camden.rutgers.edu`

RECENT PROGRESS OF THE STUDY OF HYDRODYNAMIC EVOLUTION OF GASEOUS STARS

TETU MAKINO*

Professor Emeritus at Yamaguchi University, Japan

ABSTRACT. The hydrodynamic evolutions of gaseous stars are governed by the Euler-Poisson equations in the non-relativistic framework, and governed by the Einstein-Euler equations in the general relativistic framework. Mathematically rigorous study of the problem contains difficulties which come from the singularity of physical vacuum boundary. Results obtained for spherically symmetric case and axially symmetric case are reported.

1. **Non-Relativistic Problem Governed by the Euler-Poisson Equations.** We consider the Euler-Poisson equations, which govern the hydrodynamic evolution of self-gravitating gaseous stars:

$$\frac{\partial \rho}{\partial t} + \sum_{k=1}^{3} \frac{\partial}{\partial x_k}(\rho v_k) = 0, \tag{1a}$$

$$\rho\Big(\frac{\partial v_j}{\partial t} + \sum_{k=1}^{3} v_k \frac{\partial v_j}{\partial x_k}\Big) + \frac{\partial P}{\partial x_j} = -\rho \frac{\partial \Phi}{\partial x_j}, \quad j = 1, 2, 3, \tag{1b}$$

$$\triangle \Phi = 4\pi \mathsf{G} \rho. \qquad (t \geq 0, x = (x_1, x_2, x_3) \in \mathbb{R}^3). \tag{1c}$$

Here ρ is the density, P the pressure, $\vec{v} = (v_1, v_2, v_3)$ the velocity field, Φ the gravitational potential. G is a positive constant. We assume that there are positive constants A, γ such that

$$P = \mathsf{A}\rho^\gamma, \quad 1 < \gamma \leq 2. \tag{2}$$

Considering compactly supported ρ, we replace (1c) by

$$\Phi(t, x) = -\mathsf{G} \int_{\mathbb{R}^3} \frac{\rho(t, x')}{|x - x'|} d\mathcal{V}(x'). \tag{3}$$

We shall use the co-ordinate system

$$x_1 = r \sin\vartheta \cos\phi = r\sqrt{1 - \zeta^2}\cos\phi,$$

$$x_2 = r \sin\vartheta \sin\phi = r\sqrt{1 - \zeta^2}\sin\phi, \quad x_3 = r\cos\vartheta = r\zeta,$$

and the variable

$$u = \int_0^\rho \frac{dP}{\rho} = \frac{\gamma \mathsf{A}}{\gamma - 1}\rho^{\gamma - 1}. \tag{4}$$

2000 *Mathematics Subject Classification.* Primary: 35L05, 35L52; Secondary: 76L10.

Key words and phrases. Gaseous star, Euler-Poisson equations, Spherically symmetric solutions, Axially symmetric solutions, Einstein-Euler equations, Nash-Moser theorem.

The author is supported by JPS KAKENHI grant JP18K03371.

1.1. **Spherically Symmetric Problem.** First we consider the spherically symmetric problem of quantities depending on (t, r).

Equilibria are given as

$$\rho = \rho_O\Big(\theta\Big(\frac{r}{\mathsf{a}}\Big) \vee 0\Big)^{\frac{1}{\gamma-1}}, \qquad \mathsf{a} := \sqrt{\frac{\mathsf{A}\gamma}{4\pi\mathsf{G}(\gamma-1)}\rho_O^{\frac{\gamma-2}{2}}}.$$

by the solution θ of the Lane-Emden equation:

$$\frac{d^2\theta}{d\xi^2} + \frac{2}{\xi}\frac{d\theta}{d\xi} + (\theta \vee 0)^{\frac{1}{\gamma-1}} = 0, \quad \theta = 1 + O(\xi^2) \quad \text{as} \quad \xi \to +0.$$

Here we assume that $\frac{6}{5} < \gamma < 2$ in order that the solution $\theta(\xi)$ has a finite zero $\xi_1(\gamma)$. Put $r_+ = \mathsf{a}\xi_1(\gamma)$.

Introducing the Lagrangian co-ordinate $\bar{r}$, the equation for the perturbations

$$r(t, \bar{r}) = \bar{r}(1 + y(t, \bar{r}))$$

from a fixed equilibrium turns out to be

$$\frac{\partial^2 y}{\partial t^2} - \frac{1}{\bar{\rho}\bar{r}}(1+y)^2\frac{\partial}{\partial \bar{r}}\Big(\bar{P}G\Big(y, \bar{r}\frac{\partial y}{\partial \bar{r}}\Big)\Big) + \frac{1}{\bar{\rho}\bar{r}}\frac{d\bar{P}}{d\bar{r}}H(y) = 0, \quad (0 \le \bar{r} \le R) \tag{5}$$

where

$$G(y, V) = 1 - (1+y)^{-2\gamma}(1+y+V)^{-\gamma} = \gamma(3y+V) + [y, V]_2,$$
$$H(y) = (1+y)^2 - \frac{1}{(1+y)^2} = 4y + [y]_2.$$

Thus the linearized problem is

$$\frac{\partial^2 y}{\partial t^2} + \mathcal{L}\Big(\frac{\partial}{\partial \bar{r}}\Big)y = 0,$$

where

$$\mathcal{L}\Big(\frac{d}{d\bar{r}}\Big)y = -\frac{1}{\bar{\rho}\bar{r}}\frac{d}{d\bar{r}}\Big(\gamma\bar{P}\Big(3y + \bar{r}\frac{dy}{d\bar{r}}\Big)\Big) + \frac{1}{\bar{\rho}\bar{r}}\frac{d\bar{P}}{d\bar{r}}(4y).$$

Applying the Nash-Moser theorem formulated in [1], we proved

Theorem 1.1. *([5, Theorem 1]) Suppose* $6/5 < \gamma \le 2, \frac{\gamma}{\gamma-1} \in \mathbb{N}$. *Then for* $\forall T \ \exists \epsilon(T) \ \ 0 < \forall \varepsilon \le \epsilon(T)$ *there exists a solution* $y(t, \bar{r}; \varepsilon) \in C^2([0, T] \times [0, r_+])$ *such that*

$$y(t, \bar{r}; \varepsilon) = \varepsilon \sin(\sqrt{\lambda}t + \text{Const.})\varphi(\bar{r}) + O(\varepsilon^2)$$

Here λ *is a positive eigenvalue of the linearized operator* $\mathcal{L}$ *and* φ *is the associated eigenfunction.*

We note that the free matter-vacuum boundary is

$$r = R_F(t) = r_+(1 + y(t, r_+)) = r_+(1 + \varepsilon\sin(\sqrt{\lambda}t + \text{Const.})\varphi(r_+) + O(\varepsilon^2)),$$

r_+ being the radius of the equilibrium, $\varphi(r_+) \neq 0$, and

$$\rho(t, r) = \begin{cases} C(t)(R_F(t) - r)^{\frac{1}{\gamma-1}}(1 + O(R_F(t) - r)) & r < R_F(t), \\ = 0 & R_F(t) \le r \end{cases}$$

where $C(t) > 0$ is a smooth function of t

Theorem 1.2. ([5, Theorem 2]) *Suppose* $6/5 < \gamma \leq 2, \frac{\gamma}{\gamma-1} \in \mathbb{N}$. *Then there exists a number* $\mathfrak{r}$ *such that for* $\forall T \quad \exists \delta(T) \quad \forall \psi_0, \psi_1 \in C^\infty([0, r_+])$

$$\left\| \left(\frac{d}{d\bar{r}}\right)^j \psi_0 \right\|_{L^\infty}, \left\| \left(\frac{d}{d\bar{r}}\right)^j \psi_1 \right\|_{L^\infty}, \leq \delta(T) \quad \forall j \leq \mathfrak{r}$$

there exists a solution $y(t, \bar{r}) \in C^2([0,T] \times [0, r_+])$ *such that*

$$y|_{t=0} = \psi_0(\bar{r}), \quad \left.\frac{\partial y}{\partial t}\right|_{t=0} = \psi_1(\bar{r}).$$

But the condition $\frac{6}{5} < \gamma \leq 2, \frac{\gamma}{\gamma-1} \in \mathbb{N}$ restricts γ to $\frac{5}{4}, \frac{4}{3}, \frac{3}{2}$ or 2. On the other hand, applying the Nash-Moser theorem formulated in [11], we proved

Theorem 1.3. ([7]) *Suppose that* P *is a smooth function of* $\rho > 0$ *such that* $P > 0, dP/d\rho > 0$,

$$P = \mathsf{A}\rho^\gamma(1 + [\rho^{\gamma-1}]_1) \quad as \quad \rho \to +0.$$

Suppose the equilibrium

$$-r^2 \frac{d}{dr}\left(\frac{1}{r^2\bar{\rho}}\frac{d\bar{P}}{dr}\right) = 4\pi\mathsf{G}\bar{\rho}, \quad \bar{\rho} = \rho_O + O(r^2)$$

have a finite radius. Suppose $1 < \gamma < 54/53$. *Then the conclusions of Theorems 1.1, 1.2 hold.*

Another approach to the same problem by Juhi Jang, which is based on a sophisticated use of Hardy type inequalities, without use of the Nash-Moser theory, can be found in [3].

1.2. **Axially Symmetric Problem.** The axially symmetric problem supposes quantities depends upon (t, r, ζ), and the velocity field is of the form

$$\vec{v} = v\frac{\partial}{\partial r} + w\frac{\partial}{\partial \zeta} + \Omega\frac{\partial}{\partial \phi}.$$

Stationary axially and equatorially symmetric solutions are:

$$v = w = 0, \quad \Omega = \text{Const.}, \quad \rho = \rho(r,\zeta) = \rho(r, -\zeta),$$

governed by the equation

$$u + \Phi = \frac{1}{2}r^2(1-\zeta^2)\Omega^2 + \text{Const.} \tag{6}$$

The existence of stationary axially symmetric solutions has been established by joint works with Juhi Jang [2] and [4] as :

Theorem 1.4. *Suppose* $P = \mathsf{A}\rho^\gamma, 6/5 < \gamma < 2$. *Then for* $\exists \epsilon > 0 \quad 0 \leq \forall \mathsf{b} \leq \epsilon, \rho_O > 0$ *there extsts a solution of the form*

$$\rho = \rho_O\left(\Theta\left(\frac{r}{\mathsf{a}}, \zeta; \gamma, \mathsf{b}\right) \vee 0\right)^{\frac{1}{\gamma-1}}, \quad \mathsf{b} = \frac{\Omega^2}{4\pi\mathsf{G}\rho_O}.$$

Here the distorted Lane-Emden function Θ is the solution of

$$\Theta = \frac{\mathsf{b}}{4}(1-\zeta^2)r^2 + 1 + \frac{1}{4\pi}\int_{-1}^{1}\int_0^\infty K(r,\zeta,r',\zeta')(\Theta \vee 0)^{\frac{1}{\gamma-1}} r'^2 dr' d\zeta'$$
$$-\frac{1}{4\pi}\int_{-1}^{1}\int_0^\infty K(0,0,r',\zeta')(\Theta \vee 0)^{\frac{1}{\gamma-1}} r'^2 dr' d\zeta'$$

such that

$$\Theta(r,\zeta) > 0, r \leq \Xi_0 \quad \Leftrightarrow \quad 0 \leq r < R(\zeta),$$

where $\zeta \mapsto R(\zeta) = R(\zeta; \gamma, \mathsf{b})$ is continuous on $[-1, 1]$ and $0 < R(\zeta) < 2\xi_1(\gamma)$.

However the evolution problem near these stationary axisymmetric solutions is still open. Even the spectral analysis of the linearized operator for perturbations from these stationary slowly and uniformly rotating star solutions is more difficult than that for spherically symmetric perturbations from spherically symmetric equilibria. Recent status of our study will be announced elsewhere.

2. **Relativistic Problem Governed by the Einstein-Euler Equations.** The evolution of self-gravitating gaseous stars in the framework of the general theory of relativity is governed by the Einstein-Euler equations :

$$R_{\mu\nu} - \frac{1}{2}g_{\mu\nu} = \frac{8\pi\mathsf{G}}{\mathsf{c}^4}T_{\mu\nu}, \tag{7a}$$

$$T^{\mu\nu} = (\mathsf{c}^2\rho + P)U^\mu U^\nu - Pg^{\mu\nu} \tag{7b}$$

for the metric $ds^2 = g_{\mu\nu}dx^\mu dx^\nu$.

Instead of the exact γ-law (2) for the non-relativistic problem, we assume that P is a given analytic function of $\rho > 0$ such that $0 < P, 0 < dP/d\rho < \mathsf{c}^2$ and

$$P = A\rho^\gamma(1 + [\rho^{\gamma-1}/\mathsf{c}^2]_1) \tag{8}$$

as $\rho \to +0$ with constants $A > 0, 1 < \gamma \leq 2$. Moreover $\frac{\gamma}{\gamma-1} \in \mathbb{N}$ or $1 < \gamma < 54/53$.

2.1. **Spherically Symmetric Problem.** A co-moving spherically symmetric metric

$$ds^2 = e^{2F(t,r)}\mathsf{c}^2dt^2 - e^{2H(t,r)}dr^2 - R(t,r)^2(d\theta^2 + \sin^2\theta d\phi^2) \tag{9}$$

such that $U^{\mathsf{c}t} = e^{-F}, U^r = U^\theta = U^\phi = 0$ with spherically symmetric density distribution $\rho(t, r)$ is considered. The governing equations are given in e.g., [9].

Equilibria are given by the Tolman-Oppenheimer-Volkoff equations ([10]):

$$\frac{dm}{dr} = 4\pi r^2\rho, \tag{10a}$$

$$\frac{dP}{dr} = -(\rho + P/\mathsf{c}^2)\frac{G(m + 4\pi r^3P/\mathsf{c}^2)}{r^2(1 - 2Gm/\mathsf{c}^2r)}. \tag{10b}$$

Fix an equilibrium $\rho = \bar{\rho}(r)$ such that

$$m = \frac{4\pi}{3}\rho_{\mathrm{O}}r^3 + O(r^5), \tag{11a}$$

$$P = P_{\mathrm{O}} - (\rho_{\mathrm{O}} + P_{\mathrm{O}}/\mathsf{c}^2)4\pi G(\rho_{\mathrm{O}}/3 + P_{\mathrm{O}}/\mathsf{c}^2)\frac{r^2}{2} + O(r^4) \tag{11b}$$

as $r \to 0$ and $\rho(r) \searrow 0$ as $r \nearrow r_+(< \infty)$. Putting

$$m_+ = 4\pi\int_0^{r_+}\bar{\rho}(r)r^2dr, \quad \kappa_+ = 1 - \frac{2Gm_+}{\mathsf{c}^2r_+} > 0$$

we have

$$\bar{u}(r) = \frac{Gm_+}{r_+^2\kappa_+}(r_+ - r)(1 + [r_+ - r, (r_+ - r)^{\frac{1}{\gamma-1}}]_1)$$

We consider spherically symmetric perturbation from this equilibrium:

$$R = r(1 + y(t, r)), \qquad V = rv(t, r) \tag{12}$$

governed by

$$e^{-F}\frac{\partial y}{\partial t} = (1+P/\mathsf{c}^2\rho)v, \tag{13a}$$

$$e^{-F}\frac{\partial v}{\partial t} = \frac{1}{\mathsf{c}^2}(1+y)^2\frac{P}{\bar{\rho}}v\frac{\partial}{\partial r}(rv) - \frac{1}{r^3}\frac{G}{(1+y)^2}\Big(m+\frac{4\pi}{\mathsf{c}^2}Pr^3(1+y)^3\Big) +$$

$$-\Big(1+\frac{r^2v^2}{\mathsf{c}^2}-\frac{2Gm}{\mathsf{c}^2r(1+y)}\Big)(1+P/\mathsf{c}^2\rho)^{-1}\frac{(1+y)^2}{r\bar{\rho}}\frac{\partial P}{\partial r} \tag{13b}$$

Here $m=\bar{m}(r)$ is given and ρ is the function of $r, y, r\partial y/\partial r$ given by

$$\rho = \bar{\rho}(r)(1+y)^{-2}\Big(1+y+r\frac{\partial y}{\partial r}\Big)^{-1} \tag{14}$$

Linearization of the system (13a)(13b) at $y=v=0$:

$$\frac{\partial^2 y}{\partial t^2} + \mathcal{L}y = 0$$

is given by

$$\begin{aligned}\mathcal{L}y &= -\frac{1}{b}\frac{d}{dr}\Big(a\frac{dy}{dr}\Big)+Qy,\\ a &= \frac{\overline{\Gamma P}r^4}{1+\overline{P/\mathsf{c}^2\rho}}e^{\overline{F+H}},\\ b &= (1+\overline{P/\mathsf{c}^2\rho})^{-1}\bar{\rho}r^4e^{\overline{-F+3H}}\end{aligned}$$

with $\Gamma := \dfrac{\rho}{P}\dfrac{dP}{d\rho}$. As the non-relativistic problem we can prove that $\mathcal{L}$ can be considered as a self-adjoint operator in $L^2((0,r_+);b(r)dr)$ whose spectrum consists of simple eigenvalues $\lambda_1<\lambda_2<\cdots<\lambda_\nu<\cdots\to+\infty$. Moreover we proved

Theorem 2.1. ([6, Theorem 1]) *Given $T>0$, there exists a positive $\epsilon_0(T)$ such that for $|\varepsilon|\le\epsilon_0(T)$ there is a solution $(y,v)\in C^\infty([0,T]\times[0,r_+])$ of the form*

$$y=\varepsilon y_1+\varepsilon^2\tilde{y}, \qquad v=\varepsilon v_1+\varepsilon^2\tilde{v} \tag{15}$$

such that

$$\sup_{j+k\le n}\|\partial_t^j\partial_r^k(\tilde{y},\tilde{v})\|_{L^\infty}\le C(n).$$

Here

$$\begin{aligned}y_1 &= \sin(\sqrt{\lambda}t+\Theta_0)\varphi(r),\\ v_1 &= e^{-\bar{F}}(1+\overline{P/\mathsf{c}^2\rho})^{-1}\frac{\partial y_1}{\partial t},\end{aligned}$$

while λ is a positive eigenvalue of $\mathcal{L}$ and φ is an associated eigenfunction.

Note

$$R(t,r_+) = r_+(1+\varepsilon\sin(\sqrt{\lambda}t+\text{Const.})\varphi(r_+)+O(\varepsilon^2)), \tag{16}$$

$\varphi(r_+)\neq 0$, and

$$\rho(t,r) = \begin{cases} C(t)(r_+-r)^{\frac{1}{\gamma-1}}(1+O(r_+-r)) & (0\le r<r_+)\\ 0 \quad (r_+\le r)\end{cases} \tag{17}$$

with a smooth function $C(t)$ of t such that

$$C(t) = \Big(\frac{\gamma-1}{A\gamma}\frac{Gm_+}{r_+^2\kappa_+}\Big)^{\frac{1}{\gamma-1}}+O(\varepsilon)$$

The Cauchy problem **(CP)**:

$$e^{-F}\frac{\partial y}{\partial t} = \cdots \quad (13a), \qquad e^{-F}\frac{\partial v}{\partial t} = \cdots \quad (13b),$$
$$y|_{t=0} = \psi_0(x), \qquad v|_{t=0} = \psi_1(x)$$

was also solved.

The solution metric can be patched to the Schwarzschild metric on the vacuum region:

$$ds^2 = K^\sharp(\mathsf{c}dt^\sharp)^2 - \frac{1}{K^\sharp}(dR^\sharp)^2 - (R^\sharp)^2(d\theta^2 + \sin^2\theta d\phi^2), \tag{18}$$

where

$$K^\sharp := 1 - \frac{2Gm_+}{\mathsf{c}^2 R^\sharp}, \qquad t^\sharp = t^\sharp(t,r), \qquad R^\sharp = R^\sharp(t,r). \qquad (r \geq r_+)$$

But maybe the following result was new.

Theorem 2.2. *([6, Supplementary Remark 4]) There are $t^\sharp, R^\sharp \in C^\infty([0,T] \times [r_+, +\infty))$ such that the coefficients of the patched metric are of $C^1([0,T]\times[0,+\infty))$. But then*

$$\frac{\partial^2 R^\sharp}{\partial r^2}\Big|_{r_+ + 0} - \frac{\partial^2 R}{\partial r^2}\Big|_{r_+ - 0} = \mathcal{A}\Big(\frac{\partial R}{\partial r}\Big)^2,$$
$$\mathcal{A} = -\frac{V^2}{\mathsf{c}^2}\Big(\frac{Gm_+}{\mathsf{c}^2R^2} + \frac{1}{\sqrt{\kappa_+}}\frac{1}{\mathsf{c}^2}\frac{\partial V}{\partial t}\Big)\Big(1 + \frac{V^2}{\mathsf{c}^2} - \frac{2Gm_+}{\mathsf{c}^2 R}\Big)^{-2}$$

does not vanish if $V \neq 0$. In other words, the patched metric cannot be of class C^2 across the vacuum boundary, unless the static equilibrium is concerned.

2.2. **Axially Symmetric Problem.** Stationary axisymmetric metrics for the Einstein-Euler equations have been established in [8], as the post-Newtonian approximations from the stationary axisymmetric solutions of the non-relativistic problem given by Theorem 1.4. But the metric is constructed in a bounded domain which contains the support of the density, and a corresponding result as Theorem 2.2 for spherically symmetric problem is not yet found. It is not clear which kind of vacuum metric on the exterior region should be matched to the interior metric, instead of the Schwarzschild metric for the spherically symmetric problem. This is an open problem. Of course the evolution problem near these stationary axisymmetric metrics is completely open.

REFERENCES

[1] R. S. Hamilton, The inverse function theorem of Nash and Moser, *Bull. Amer. Math. Soc.*, **7** (1982), 65-222.

[2] Juhi Jang and T. Makino, On slowly rotating axisymmetric solutions of the Euler-Poisson equations, *Arch. Rational Mech. Anal.*, **225** (2017), 873-900.

[3] Juhi Jang, Time periodic approximations of the Euler-Poisson system near Lane-Emden stars, *Analysis & PDE*, **9** (2016), 1043-1078.

[4] Juhi Jang and T. Makino, On rotating axisymmetric solutions of tha Euler-Poisson equations, *J. Differential Equations*, **266** (2019), 3942-3972.

[5] T. Makino, On spherically symmetric motions of a gaseous star governed by the Euler-Poisson equations, *Osaka J. Math.*, **52** (2015), 545-580.

[6] T. Makino, On spherically symmetric solutions of the Einstein-Euler equations, *Kyoto Journal of Mathmatics*, **56** (2016), 243-282.

[7] T. Makino, An application of the Nash-Moser theorem to the vacuum boundary problem of gaseous stars, *J. Differential Equations*, **262** (2017), 803-843.

[8] T. Makino, On slowly rotating axisymmetric solutions of the Einstein-Euler equations, *J. Math. Phys.*, **59** (2018), 102502-1 - 102502-33.
[9] C. W. Misner and D. H. Sharp, Relativistic equations for adiabatic, spherically symmetric gravitational collapse, *Phys. Rev.* , **136** (1964), B571-B576.
[10] J. P. Oppenheimer and G. M. Volkoff, On massive neutron cores, *Phys. Rev.*, **55** (1939), 374-381.
[11] J. T. Schwartz, *Non-linear Functional Analysis*, Gordon and Beach, New York-London-Paris, 1969.

E-mail address: makino@yamaguchi-u.ac.jp

WELL-BALANCED SCHEME FOR NETWORK OF GAS PIPELINES

YOGIRAJ MANTRI*, MICHAEL HERTY AND SEBASTIAN NOELLE

Institut für Geometrie und Praktische Mathematik
RWTH Aachen University
Templergraben 55, 52056 Aachen, Germany

ABSTRACT. Gas flow in a pipeline network can be described by a hyperbolic balance law within each pipe along with coupling conditions at the node. For equilibrium or near equilibrium flows it is essential to design well-balanced schemes, in order to avoid spurious oscillations in the solution. Recently Chertock, Herty & Özcan[9] introduced a well-balanced central Upwind scheme for 2×2 systems of balance laws. Here, we extend the scheme to model coupling conditions at intersection of pipes and compressor stations, thus resulting in a well- balanced scheme across the network.

1. **Introduction.** Various mathematical models have been developed in the past few years to model gas flow in a pipeline network, see [2, 3, 7, 12, 20, 6]. In order to capture a fine resolution of the spatial and temporal dynamics, the isothermal Euler equations (1) provide a suitable model [2, 3]. In this paper, we focus on developing well-balanced schemes for such flows, which resolve steady states accurately and can capture small temporal and spatial perturbations to the steady state. Several well-balanced schemes have been developed for approximating solutions to shallow water equations such as [1, 21, 22, 5, 8]. Recently Chertock et.al.[9] developed a second-order well-balanced central Upwind scheme for 2×2 system of hyperbolic balance law. We extend this scheme to a network of gas pipeline consisting of intersections of multiple pipes and compressor stations. In a network, spurious oscillations may not only be introduced due to the imbalance between flux and source terms, but also due to discretization errors at junctions and compressors.

The gas flow within each pipe $i = 1 \dots M$ of the network is governed by the isothermal Euler equations,

$$(U_i)_t + F(U_i)_x = S(U_i) \tag{1}$$

where the conservative variables U_i, flux $F(U_i)$ and source $S(U_i)$ are given by,

$$U_i = \begin{bmatrix} \rho_i \\ q_i \end{bmatrix}, \ F(U_i) = \begin{bmatrix} q_i \\ \frac{q_i^2}{\rho_i} + p(\rho_i) \end{bmatrix}, \ S(U_i) = \begin{bmatrix} 0 \\ -\frac{f_{g,i}}{2D_i}\frac{q_i|q_i|}{\rho_i} \end{bmatrix} \tag{2}$$

2000 *Mathematics Subject Classification.* Primary: 35L60, 35L65, 65M08.
Key words and phrases. Flows in network, Well-balanced schemes, Hyperbolic balance laws.
* Corresponding author: mantri@eddy.rwth-aachen.de.

with ρ_i, q_i and $p(\rho_i)$ being the density, momentum and pressure of the gas; and $f_{g,i}$, D_i are friction factor and diameter of the pipe respectively. The pressure of the gas for isothermal flow is given by,

$$p(\rho) = a^2\rho. \tag{3}$$

In order to solve the Euler equations (1), we need initial conditions and boundary conditions at the ends of the pipes. The boundary condition at a node connecting multiple pipes is given implicitly by coupling conditions as defined in [2, 25]. The coupling condition can be written in the form,

$$\phi(U_1, U_2, ..., U_M) = 0, \quad \phi : \mathbb{R}^{2M} \to \mathbb{R}^M. \tag{4}$$

The coupling conditions for a pipeline network are further discussed in Section 2 and Section 3.

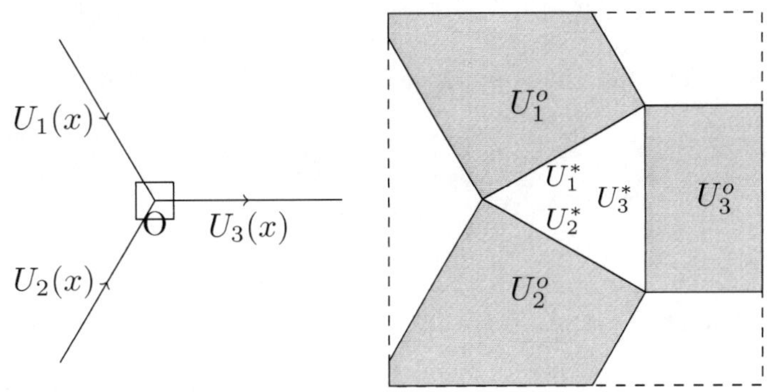

FIGURE 1. Intersection of three pipes at junction O. Right-Zoomed view of the junction with old traces U_i^o and new traces U_i^* given in Section 2

2. **Coupling conditions for the p–system.** In this section we briefly summarize the analytical results for coupling conditions discussed in [2, 11]. In order to study these coupling conditions, we set the source terms $S(U_i)$ to zero, hence assuming that wall-friction is neglected at the instance of interaction at the node.

The eigenvalues for the 2×2 isothermal Euler equations (1) are $\lambda_1 = \frac{q}{\rho} - a$ and $\lambda_2 = \frac{q}{\rho} + a$. We assume that the flow within the network is **subsonic** i.e.,

$$\lambda_1(U_i) < 0 < \lambda_2(U_i) \quad \text{for } i = 1 \dots M. \tag{5}$$

We can parameterize the incoming pipes, $i \in I^-$ by $x \in \Omega_i := (-\infty, x_o)$. Similarly the outgoing pipes, $j \in I^+$ are parameterized by $x \in \Omega_j := (x_o, \infty)$. We denote the solution in the interior of the pipes by the old traces, U_i^o and the solution at the node at a time $t > t^n$ is denoted by the new traces, U_i^*. The states U_i^o and U_i^* are denoted in Figure 1. The new traces are connected to the old traces by a Lax curve entering the pipe i.e. the first family of Lax curves for the incoming pipe and the second family for the outgoing pipes. The Lax curves can be written in the form,

$$U_i^* = \overline{U}(\sigma_i) = \begin{bmatrix} \overline{\rho}(\sigma_i) \\ \overline{\rho}(\sigma_i) \end{bmatrix}. \tag{6}$$

For construction of the Lax curves, see [13, 18].

Further we know that the new traces satisfy the coupling conditions at the node,

$$\overline{\phi}(\sigma_1, \dots, \sigma_M) := \phi(\overline{U}_1(\sigma_1), \dots, \overline{U}_M(\sigma_M)) = 0. \tag{7}$$

Solving these coupling conditions, we find parameter σ_i^* for each pipe and hence the new traces $U_i^* = \overline{U}_i(\sigma_i^*)$.

We now focus on the coupling condition in [2, 11] for which we design the well-balanced scheme. However the framework is also valid for other coupling conditions, for e.g. those discussed in [23]. The first coupling condition is given by the mass balance at the node x_o i.e the total mass which enters the node is same as the total mass leaving the node.

Several approaches have been studied to model the other $(M-1)$ coupling conditions. For instance, momentum balance in [14, 4] or continuity of Bernoulli invariant in [20, 25, 24]. Here, we use the coupling condition given by continuity of pressure at the node x_o as given in [2, 14]. Thus we can write the coupling conditions as,

$$\phi(U_1, \dots, U_M) = \begin{bmatrix} \sum_{i \in I^-} A_i q_i - \sum_{j \in I^+} A_j q_j \\ p(\rho_2) - p(\rho_1) \\ \vdots \\ p(\rho_M) - p(\rho_{M-1}) \end{bmatrix}, \tag{8}$$

where $A_i = \frac{\pi}{4} D_i^2$ is the cross sectional area for pipe i.

Similarly, we can write the coupling conditions for a compressor connected to two pipes as,

$$\phi(U_1, \dots, U_M) = \begin{bmatrix} q_2 - q_1 \\ p(\rho_2) - CRp(\rho_1) \end{bmatrix}. \tag{9}$$

where CR is the compression ratio for the compressor.

To summarize, we require to find the new traces U_i^* which are connected to the old traces U_i^o by an incoming Lax curve, and such that the new traces also satisfy the coupling condition. It has been proven in [10] that these coupling conditions have a unique solution at the node.

3. **Coupling conditions in terms of equilibrium variables.**

For equilibrium or near equilibrium flows, the numerical error due to imbalance between flux and source terms can lead to spurious oscillations within the solution. Chertock, Herty and Özcan [9] resolved this difficulty by rewriting the balance law in conservative form i.e we can rewrite equation (1) as,

$$(\rho_i)_t + (K_i)_x = 0, \;\; (q_i)_t + (L_i)_x = 0 \tag{10}$$

where the flux variable,

$$V_i(U_i, R_i) = \begin{bmatrix} K_i \\ L_i \end{bmatrix} = F(U_i) + \begin{bmatrix} 0 \\ R_i \end{bmatrix}, \tag{11}$$

and fluxes K_i, L_i and the integrated source term R_i are given by,

$$K_i := q_i, \quad L_i := \frac{q_i^2}{\rho_i} + p(\rho_i) + R_i(x), \quad R_i(x) := \int_{\tilde{x}_i}^{x} \frac{f_{g,i}}{2D_i} \frac{q_i|q_i|}{\rho_i} \, dx. \tag{12}$$

The point $\tilde{x}_i$ is an arbitrary but fixed point in $\overline{\Omega_i}$. By construction the equilibrium variables, (K, L) are constant in each pipe at equilibrium.

We reformulate the coupling conditions (8) in terms of equilibrium variables as,

$$\Psi(K_i, L_i, R_i) := \begin{bmatrix} \sum_{i\in I^-} A_i K_i - \sum_{j \in I^+} A_j K_j \\ P_1 - P_2 \\ \vdots \\ P_{M-1} - P_M \end{bmatrix} \tag{13}$$

where pressure of the gas for subsonic flow can be calculated as,

$$P_i = P(K_i^*, L_i^*) := \frac{L_i^* - R_i + \sqrt{(L_i^* - R_i)^2 - 4(K_i^*)^2 a^2}}{2}. \tag{14}$$

Now in order to solve these coupling conditions, we require the Lax-curves in terms of the equilibrium variables, i.e.,

$$V(\sigma_i) = \begin{bmatrix} \overline{q}(\sigma_i) \\ \frac{\overline{q}(\sigma_i)^2}{\overline{\rho}(\sigma_i)} + a^2 \overline{\rho}(\sigma_i) + R_i \end{bmatrix}, \tag{15}$$

where the conservative variables, $\overline{\rho}(\sigma_i), \overline{q}(\sigma_i)$ are as defined in (6) for the respective waves entering the incoming or outgoing pipes. Figure 2 shows phase plot of the Lax curves in terms of the equilibrium variables. Note that 1-Lax curve for the

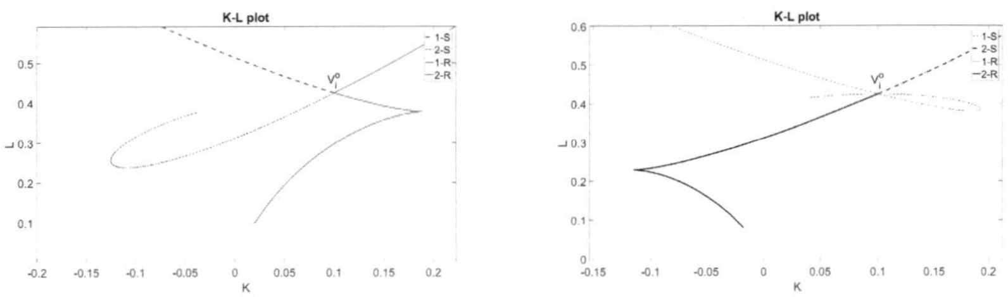

(A) Incoming pipes, $i \in I^-$ (B) Outgoing pipes, $i \in I^+$

FIGURE 2. Phase plot in terms of equilibrium variables with initial state $V_i^o = (0.1, 0.4)^T$

incoming pipe and 2-Lax curve for the outgoing pipe are monotonic in the subsonic regime. We show in [19] that the coupling conditions (13) have a unique solution in the subsonic regime.

4. A Well-balanced Central Upwind Scheme For Nodal Dynamics.

The evolution of the conservative variables, can be computed by a second-order central upwind scheme [16, 17, 9]. The computational domain of each pipe, Ω_i is discretized into cells of size Δx_i, centered at $x_i^j = \bar{x}_i + (j - \frac{1}{2})\Delta x_i$ for $j = 1, \ldots, N$. The evolution of the conservative variables is given by,

$$\frac{dU_i^j}{dt} = -\frac{\mathcal{V}_i^{j+1/2} - \mathcal{V}_i^{j-1/2}}{\Delta x} \tag{16}$$

where $\mathcal{V}_i^{j-1/2}, \mathcal{V}_i^{j+1/2}$ are the fluxes across the left and right interface of cell j, respectively. At the node these fluxes are given by the solution of the coupling conditions,

$$\mathcal{V}_i^{N+1/2} = V_i^*, i \in I^-, \tag{17}$$

$$\mathcal{V}_i^{1/2} = V_i^*, i \in I^+. \tag{18}$$

The integral terms R_i are calculated using a quadrature by fixing a starting point $\tilde{x} = x_i^o$ where $R_i = 0$ in each pipe. $R_i^j = \frac{1}{2}(R_i^{j+1/2} + R_i^{j-1/2})$ can be calculated as,

$$R_i^{j^*+1/2} = 0 \text{ at } \tilde{x} = x_i^{j^*+1/2} \,\forall\, i \in I^{\pm},$$

$$R_i^{j+1/2} = R_i^{j-1/2} + \Delta x \frac{f_{g,i}}{2D_i} \frac{q_i^j |q_i^j|}{\rho_i^j}, \text{ for } x_i^{j+1/2} > x_i^{j^*+1/2},$$

$$R_i^{j-1/2} = R_i^{j+1/2} + \Delta x \frac{f_{g,i}}{2D_i} \frac{q_i^j |q_i^j|}{\rho_i^j}, \text{ for } x_i^{j-1/2} < x_i^{j^*+1/2}.$$

The numerical flux, $\mathcal{V}_i^{j+1/2}$ in the interior of the pipes can be calculated as,

$$\mathcal{V}_i^{j+1/2} = \frac{a_{i,+}^{j+1/2} V_i^{j,E} - a_{i,-}^{j+1/2} V_i^{j+1,W}}{a_{i,+}^{j+1/2} - a_{i,-}^{j+1/2}} + \alpha_i^{j+1/2}(U_i^{j+1,W} - U_i^{j,E}) \tag{19}$$

where the terms $V_i^{j,W}, V_i^{j,E}$ denote the equilibrium variables at the left and right boundaries of the cell respectively, and can be calculated by applying a minmod limiter to the average values of the equilibrium variables. We can compute the conservative variables, $U_i^{j+1,W}, U_i^{j,E}$ from the inverse relation of (12). The terms $a_{i,\pm}^{j+1/2}$ denote the maximum and minimum eigenvalues for the Jacobian $\frac{\partial F}{\partial U}$ at $x_i^{j+1/2}$ and $\alpha_i^{j+1/2} = \frac{a_{i,+}^{j+1/2} a_{i,-}^{j+1/2}}{a_{i,+}^{j+1/2} - a_{i,-}^{j+1/2}}$.

Note that in [9, 19], the numerical viscosity of the flux was multiplied with an additional limiter function $\mathcal{H}\left(\frac{|V_i^{j+1} - V_i^j|}{\Delta x} \frac{|\Omega|}{\max_j\{V_i^j\}}\right)$ to ensure well-balancing. Here we do not apply this limiter, i.e. we set $\mathcal{H} \equiv 1$. A short calculation shows that the scheme is still well-balanced if $R_j^E = R_{j+1}^W = R_i^{j+1/2}$, since in this case $U_i^{j+1,W} = U_i^{j,E}$.

5. **Numerical Tests.** In this section, we compare the solution obtained by using the well-balanced scheme discussed above and a second order non well-balanced central Upwind scheme given in [15] for flows at or near steady state. The coupling conditions for both WB and NWB scheme are solved using Newton's iteration with initial guess given by the old traces, discussed in Section 2 and Section 3. We test the numerical schemes for a network consisting of a node with one incoming and two outgoing pipes and two compressor stations along the outgoing pipes as shown in Figure 3. The compressors C_1 and C_2 have compression ratios, $CR_1 = 1.5$ and

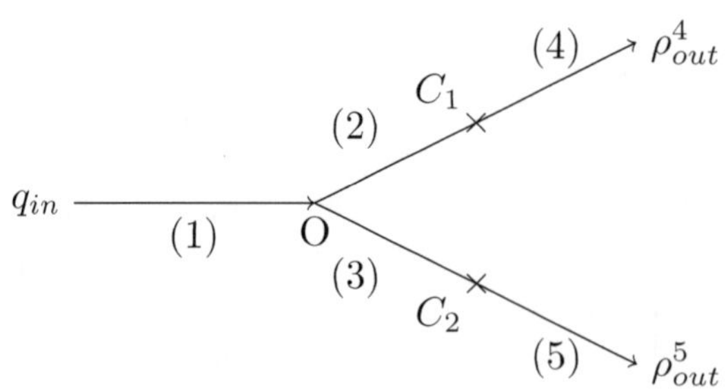

FIGURE 3. Network of gas pipeline with node O and compressors C_1 and C_2 with compression ratios CR_1 and CR_2 respectively

$CR_2 = 2$ respectively. All the pipes have the same diameter and friction factor such that $\frac{f_g}{2D} = 1$ and speed of sound, $a = 1$. The numerical tests are run for a CFL number= 0.4.

5.1. **Steady state.** At first we consider a steady state across the network given in Figure 3. The steady state is defined by $\widehat{K}_1 = 0.1, \widehat{K}_2 = \widehat{K}_3 = \widehat{K}_4 = \widehat{K}_5 = 0.05$ and pressure, $p^*|_O = 0.373$ which leads to $p|_{C_1^-} = p|_{C_2^-} = 0.366$ and $p|_{C_1^+} = 0.549, p|_{C_2^+} = 0.732$ or $\widehat{L}_1 = 0.4, \widehat{L}_2 = \widehat{L}_3 = 0.380, \widehat{L}_4 = 0.554, \widehat{L}_5 = 0.736$. From these parameters, the boundary values $q_{in} = \widehat{K}_1$ and $\rho^i_{out} = \rho(\widehat{K}_i, \widehat{L}_i) \forall i = 4, 5$ are given for the equilibrium solution. We prescribe these as boundary values for the numerical scheme using characteristic projections. The $L1$-norm of the errors for the equilibrium variables at time $T = 1$ are given in table below.

TABLE 1. Comparison of L-1 errors between well-balanced(WB) and non well-balanced(NWB) scheme at steady state for network given in Figure 3 at time T=1

No. of cells	WB scheme		NWB scheme	
in each pipe	L-1 error for K	L-1 error for L	L-1 error for K	L-1 error for L
25	5.412×10^{-16}	1.341×10^{-15}	1.266×10^{-6}	1.002×10^{-6}
50	1.452×10^{-15}	3.148×10^{-15}	3.233×10^{-7}	2.630×10^{-7}
100	3.032×10^{-15}	7.338×10^{-15}	8.178×10^{-8}	6.741×10^{-8}

We see from the results that the WB schemes preserves steady state of the network accurately up to machine precision, whereas the numerical error due to the NWB scheme is much larger. This also means that the coupling conditions in terms of equilibrium variables are resolved accurately up to machine precision by Newtons method.

5.2. **Perturbation to steady state.** We now add a small perturbation to the momentum at steady state at the node, O at $x = 0$. Thus the initial conditions for equilibrium variables are given by, $K_1 = \widehat{K}_1 + \eta_1 e^{-100x^2}, K_{2/3} = \widehat{K}_{2/3} + \eta_{2/3} e^{-100x^2}, K_{4/5} = \widehat{K}_{4/5}$ and $L_i = \widehat{L}_i$ with $\eta_1 = 10^{-6}$ and $\eta_2 = \eta_3 = 0.5 \times 10^{-6}$. The solution at time, $T = 1$ using WB and NWB schemes is as given in Figure 4.

From Figure 4, one can see that the results of the WB scheme are stable with coarser grid of $N = 50$ unlike the NWB scheme. We need a finer resolution of $N = 200$ to capture the perturbation with a NWB scheme. A more detailed study of the solution at times 0.25 and 0.5 (not displayed here) reveals that for the NWB scheme a perturbation starts immediately at the inflow boundary. At time T=1, this waved has reached positions $x \approx 0.25$ in pipes 2 and 3, and has created most of the error in the interval [-1, 0.25]. Another error can be seen to the right of the junction in pipes 2 and 3, and is due to the non-well-balanced coupling condition. There is also an error due to the coupling condition at the compressor, but it is an order of magnitude smaller and cannot be detected in the figure.

6. **Conclusion.** In this article, we have extended the well-balanced scheme introduced by Chertock, Herty and Özcan[9] to a network of gas pipelines with wall-friction. We studied the coupling conditions for a node connected to multiple pipes

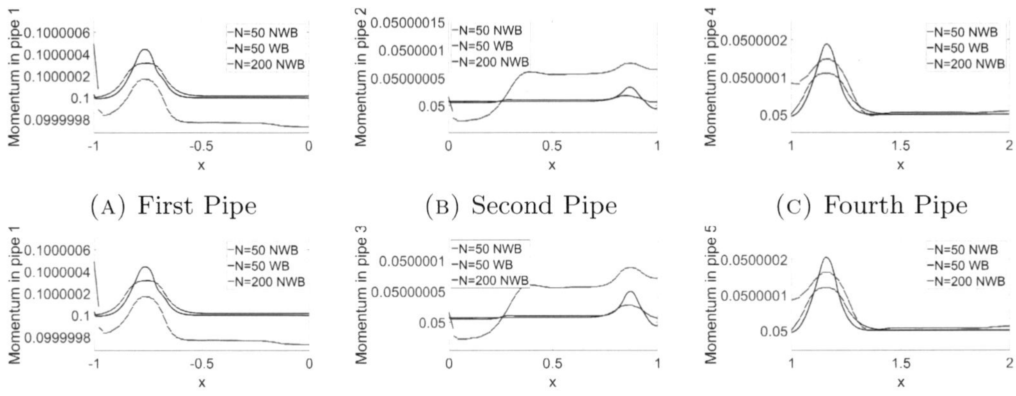

(A) First Pipe (B) Second Pipe (C) Fourth Pipe

(D) First Pipe (E) Third Pipe (F) Fifth Pipe

FIGURE 4. Momentum at time T=1 for initial perturbation of order 10^{-6} at node, O

and compressor stations, in the framework of the well-balanced scheme discussed in [9]. The numerical test for the network in Figure 3 demonstrates that the well-balanced scheme can resolve steady states accurately and provides stable solutions for flows near equilibrium.

Acknowledgments. This work has been supported by HE5386/13–15, BMBF ENets Project, and Deutsche Forschungsgemeinschaft(DFG, German Research Foundation)Projektnummer 320021702/GRK2326 Energy, Entropy, and Dissipative Dynamics(EDDy).

REFERENCES

[1] E. Audusse, F. Bouchut, M. Bristeau, R. Klein and B. Perthame, A fast and stable well-balanced scheme with hydrostatic reconstruction for shallow water flows, *SIAM J. Sci. Comput.*, **25** (2004), 2050–2065.

[2] M. K. Banda, M. Herty and A. Klar, Coupling conditions for gas networks governed by the isothermal Euler equations, *Netw. Heterog. Media*, **1** (2006), 295–314.

[3] M. K. Banda, M. Herty and A. Klar, Gas flow in pipeline networks, *Netw. Heterog. Media*, **1** (2006), 41–56.

[4] A. Bermúdez, X. López and M. E. Vázquez-Cendón, Treating network junctions in finite volume solution of transient gas flow models, *J. Comput. Phys.*, **344** (2017), 187–209.

[5] A. Bollermann, G. Chen, A. Kurganov and S. Noelle, A well-balanced reconstruction of wet/dry fronts for the shallow water equations, *J. Sci. Comput.*, **56** (2013), 267–290.

[6] A. Bressan, S. Čanić, M. Garavello, M. Herty and B. Piccoli, Flows on networks: recent results and perspectives, *EMS Surv. Math. Sci.*, **1** (2014), 47–111.

[7] J. Brouwer, I. Gasser and M. Herty, Gas pipeline models revisited: model hierarchies, non-isothermal models, and simulations of networks, *Multiscale Model. Simul.*, **9** (2011), 601–623.

[8] G. Chen and S. Noelle, A new hydrostatic reconstruction scheme based on subcell reconstructions, *SIAM J. Numer. Anal.*, **55** (2017), 758–784.

[9] A. Chertock, M. Herty and S. Özcan, Well-Balanced Central-upwind schemes for 2x2 system of Balance Laws, *Proceedings of the XVI International Conference on Hyperbolic Problems: Theory, Numerics and Applications of Hyperbolic Problems I, Springer*, **236** (2018), 345–361.

[10] R. M. Colombo and M. Garavello, A well posed Riemann problem for the *p*-system at a junction, *Netw. Heterog. Media*, **1** (2006), 495–511.

[11] R. M. Colombo and M. Garavello, On the Cauchy problem for the *p*-system at a junction, *SIAM J. Math. Anal.*, **39** (2008), 1456–1471.

[12] R. M. Colombo, G. Guerra, M. Herty and V. Schleper, Optimal control in networks of pipes and canals, *SIAM J. Control Optim.*, **48** (2009), 2032–2050.

[13] R. Courant and K. O. Friedrichs, *Supersonic Flow and Shock Waves*, Interscience Publishers, Inc., New York, N. Y., 1948.

[14] M. Herty and M. Seaïd, Simulation of transient gas flow at pipe-to-pipe intersections, *Internat. J. Numer. Methods Fluids*, **56** (2008), 485–506.

[15] A. Kurganov, S. Noelle and G. Petrova, Semidiscrete central-upwind schemes for hyperbolic conservation laws and Hamilton-Jacobi equations, *SIAM J. Sci. Comput.*, **23** (2001), 707–740.

[16] A. Kurganov and E. Tadmor, Solution of two-dimensional Riemann problems for gas dynamics without Riemann problem solvers, *Numer. Methods Partial Differential Equations*, **18** (2002), 584–608.

[17] A. Kurganov and E. Tadmor, New high-resolution central schemes for nonlinear conservation laws and convection-diffusion equations, *J. Comput. Phys.*, **160** (2000), 241–282.

[18] R.LeVeque, Numerical methods for conservation laws, Lectures in Mathematics ETH Zürich, *Birkhäuser Verlag, Basel*, second edition (1992).

[19] Y. Mantri, M.Herty and S. Noelle, Well-balanced scheme for gas-flow in pipeline networks, *IGPM Preprints*, Available from: `https://www.igpm.rwth-aachen.de/forschung/preprints/480`.

[20] A. Morin and G. A. Reigstad, Pipe networks: Coupling constants in a junction for the isentropic euler equations, *Energy Procedia*, **64** (2015), 140–149.

[21] S. Noelle, N. Pankratz, G. Puppo and J. R. Natvig, Well-balanced finite volume schemes of arbitrary order of accuracy for shallow water flows, *J. Comput. Phys.*, **213** (2006), 474–499.

[22] S. Noelle, Y. Xing and C.-W. Shu, High-order well-balanced finite volume WENO schemes for shallow water equation with moving water, *J. Comput. Phys.*, **226** (2007), 29–58.

[23] G. A. Reigstad, Numerical network models and entropy principles for isothermal junction flow, *Netw. Heterog. Media*, **9** (2014), 65–95.

[24] G. A. Reigstad, Existence and uniqueness of solutions to the generalized Riemann problem for isentropic flow, *SIAM J. Appl. Math.*, **75** (2015), 679–702.

[25] G. A. Reigstad, T. Flåtten, N. Erland Haugen and T. Ytrehus, Coupling constants and the generalized Riemann problem for isothermal junction flow, *J. Hyperbolic Differ. Equ.*, **12** (2015), 37–59.

E-mail address: `mantri@eddy.rwth-aachen.de`
E-mail address: `herty@igpm.rwth-aachen.de`
E-mail address: `noelle@igpm.rwth-aachen.de`

STRUCTURE AND REGULARITY OF SOLUTIONS TO 1D SCALAR CONSERVATION LAWS

Elio Marconi

Universität Basel, Departement Mathematik und Informatik
Spiegelgasse 1, 4051 Basel, Switzerland

Abstract. We consider bounded entropy solutions to the scalar conservation law in one space dimension:

$$u_t + f(u)_x = 0.$$

We quantify the regularizing effect of the non linearity of the flux f on the solution u in terms of spaces of functions with bounded generalized variation.

1. **Introduction.** We consider the scalar conservation law in one space dimension:

$$\begin{cases} u_t + f(u)_x = 0 & \text{in } \mathbb{R}^+ \times \mathbb{R}, \\ u(0,\cdot) = u_0(\cdot), \end{cases} \tag{1}$$

where the flux $f \in C^\infty(\mathbb{R},\mathbb{R})$ and the function $u : \mathbb{R}_t^+ \times \mathbb{R}_x \to \mathbb{R}$ is the spatial density of the conserved quantity. We consider bounded entropy solutions: more precisely we require that $u \in C^0([0,+\infty), L^1_{\text{loc}}(\mathbb{R})) \cap L^\infty(\mathbb{R}^+ \times \mathbb{R})$ satisfies (1) in the sense of distributions and that for every convex entropy $\eta : \mathbb{R} \to \mathbb{R}$ it holds

$$\eta(u)_t + q(u)_x \leq 0$$

in the sense of distributions, where the entropy flux q is defined up to constants by $q' = f'\eta'$. The well-posedness of the Cauchy problem (1) in the class of bounded entropy solutions with respect to L^1_{loc} topology is by now classical. A first consequence is the fact that the BV regularity of u is propagated in time and this implies that we can describe in a satisfactory way the structure of the entropy solution u if $u_0 \in \text{BV}(\mathbb{R})$.

We are interested in the case $u_0 \in L^\infty$, which is included in the classical well-posedness result. The first result in this direction is the Oleinik one sided Lipschitz estimate: if the flux is uniformly convex ($f'' \geq c > 0$), then for every $t > 0$, the entropy solution $u(t) \in \text{BV}_{\text{loc}}(\mathbb{R})$ and the following inequality between measures holds:

$$D_x u(t) \leq \frac{\mathcal{L}^1}{ct}. \tag{2}$$

On the other hand if $f(w) = \lambda w$ is linear the solution is given by

$$u(t,x) = u_0(x - \lambda t)$$

so $u(t)$ has the same regularity as the initial datum u_0.

2000 *Mathematics Subject Classification.* 35L65.

Key words and phrases. Lagrangian representation, fractional regularity, entropy solutions, characteristics, conservation laws.

The author is supported by ERC Starting Grant 676675 FLIRT.

Between these two extremal cases it is interesting to discuss if some weaker notion of nonlinearity (compared to uniform convexity) of the flux has some regularizing effect on the entropy solution u. The literature on this problem is large: several results, even in several space dimensions and for more general weak solutions, have been obtained by means of the kinetic formulation of (1) and averaging lemmas (see [14, 17] and the more recent [15]).

In order to get quantitative regularity results we need to quantify the nonlinearity of the flux f:

Definition 1.1. We say that the flux f has *degeneracy* $\bar{p} \in \mathbb{N}$ if

1. $\{f''(w) = 0\}$ is finite;
2. for each $w \in \mathbb{R}$ such that $f''(w) = 0$ there exists $p \geq 2$ such that $f^{(p+1)}(w) \neq 0$. Let us denote by p_w be the minimal $p \geq 2$ such that $f^{(p+1)}(w) \neq 0$;
3. $\bar{p} = \max_w p_w$.

If such a $\bar{p}$ exists we also say that f has *polynomial degeneracy.*

It was conjectured in [17] that if the flux f has degeneracy $p \in \mathbb{N}$, then for every $\varepsilon, t > 0$ the entropy solution $u(t) \in W^{s-\varepsilon,1}_{\mathrm{loc}}(\mathbb{R})$, with $s = \frac{1}{p}$. See [16] for a result in this direction. However it seems more convenient to express the regularity of the entropy solution in terms of functions with generalized bounded variation: more precisely let $\Phi : [0, +\infty) \to [0, +\infty)$ be a convex function such that $\Phi(0) = 0$, let $v : \mathbb{R} \to \mathbb{R}$ and $I \subset \mathbb{R}$ be an interval. We say that $v \in \mathrm{BV}^{\Phi}(I)$ if

$$\mathrm{TV}^{\Phi} v(I) := \sup_{x_1 < \ldots < x_n, x_i \in I} \sum_{i=1}^{n-1} \Phi(|v(x_{i+1}) - v(x_i)|) < +\infty.$$

See [19] for an introduction to these spaces. If Φ is not degenerate, i.e. $\Phi(h) > 0$ for every $h > 0$, a function $v \in \mathrm{BV}^{\Phi}(\mathbb{R})$ is a regulated function, i.e. for every $\bar{x} \in \mathbb{R}$ there exist both $\lim_{x \to \bar{x}^-} v(x)$ and $\lim_{x \to \bar{x}^+} v(x)$. This is actually a property that we have for entropy solutions to (1) if the flux satisfies this minimal nonlinearity assumption: $\{w : f''(w) \neq 0\}$ is dense in $\mathbb{R}$ (see for example [20]). We say in this case that f is *weakly genuinely nonlinear.* Notice that the available fractional Sobolev regularity of the entropy solution does not imply that it is regulated. An interesting particular case is $\Phi(w) = w^p$, in this case we denote BV^{Φ} with $\mathrm{BV}^{\frac{1}{p}}$. We notice that for every $\varepsilon > 0$ and $p \geq 1$ it holds $\mathrm{BV}^{\frac{1}{p}}(\mathbb{R}) \subset W^{\frac{1}{p}-\varepsilon,p}(\mathbb{R})$, see [8].

The use of these spaces in this context started in [8, 10] to express the regularity of the entropy solution when the flux is strictly (but not necessarily uniformly) convex.

The case of nonconvex fluxes is addressed in the following theorem and it is the final goal of this note. When not explicitly written we refer to [18] for more details.

Theorem 1.2. *Let f be a flux of degeneracy p and let u be the entropy solution of (1) with $u_0 \in L^\infty(\mathbb{R})$ with compact support. Then there exists a constant $C > 0$, depending on $\mathcal{L}^1(\mathrm{conv}(\mathrm{supp} u_0))$, $\|u_0\|_\infty$ and f, such that for every $t > 0$, it holds*

$$u(t) \in \mathrm{BV}^{1/p}(\mathbb{R}) \qquad \text{and} \qquad \mathrm{TV}^{1/p} u(t) \leq C\left(1 + \frac{1}{t}\right). \tag{3}$$

1.1. **Plan of the paper.** In Section 2 we introduce the main tool of this analysis: an extension to the non smooth setting of the classical method of characteristics called Lagrangian representation. This notion has been developed in different settings: a preliminary version has been introduced in [6] for wave-front tracking approximate solutions, in [4] it has been adapted to deal with the case of bounded and continuous initial data, then extended to L^∞ initial data in [5]. Moreover an extension to systems is given in [7]. In this note we only need to give a representation for solutions with piecewise monotone initial data, therefore we follow [18] where a simplified version of the Lagrangian representation is provided.

In Section 3 we present the main novelty of [18] and of this presentation. It is an estimate of the oscillation of the entropy solution between two characteristics in terms of their distance and the nonlinearity of the flux. This estimate plays the role that the Oleinik estimate (2) plays in the convex case and does not require any nonlinearity assumption on the flux.

Building on this result, the Lagrangian representation and the argument in [11], we present in Section 4 the main steps for proving the $\mathrm{BV}_{\mathrm{loc}}$ regularity of $f'\circ u$ under the assumption of polynomial degeneracy of the flux. In [11] the same problem is considered in the case of one and two inflection points.

Finally in Section 5 we briefly comment about the proof of Theorem 1.2.

2. Lagrangian representation.

As mentioned in the introduction, the starting point is a precise description of the behavior of the characteristics. In this section we present the notion of Lagrangian representation, which extends the notion of characteristic to the non smooth setting. Our strategy is to prove uniform regularity estimates on a dense class of bounded entropy solutions so it is sufficient to consider the case in which u is the entropy solution of (1) with u_0 continuous, bounded and piecewise monotone.

Definition 2.1. We say that $\mathtt{X} : \mathbb{R}^+_t \times \mathbb{R}_y \to \mathbb{R}$ is a *Lagrangian representation* of the entropy solution u if

1. $\mathtt{X}$ is Lipschitz continuous with respect to t;
2. $\mathtt{X}$ is increasing and continuous with respect to y;
3. $\mathtt{X}(0,y) = y$ for every $y \in \mathbb{R}$;
4. for every $t \geq 0$ it holds
$$u(t,x) = u_0(\mathtt{X}(t)^{-1}(x)), \tag{4}$$
for every $x \in \mathbb{R}\setminus N$ with N at most countable.

Remark 1. Requiring (4) for every $t \geq 0$ we implicitly refer to the L^1 continuous representative of u in time. Moreover it follows immediately from the monotonicity of $\mathtt{X}$ with respect to y and (4) that if u_0 is piecewise monotone then $u(t)$ is piecewise monotone for $t > 0$. In order to define pointwise the solution, we consider in this case the lower semicontinuous representative. In any case it is necessary to remove a countable set of points in (4): these are the points where the preimage $\mathtt{X}(t)^{-1}(x)$ is not a singleton and they are the points where shocks are located.

The Lagrangian representation enjoys several other properties. First the characteristics travel with the characteristic speed: more precisely for every $y \in \mathbb{R}$ and for $\mathcal{L}^1$-a.e. $t > 0$ it holds
$$\partial_t \mathtt{X}(t,y) = \begin{cases} f'(u(t,\mathtt{X}(t,y))) & \text{if } u(t) \text{ is continuous at } \mathtt{X}(t,y) \\ \frac{f(u(t,\mathtt{X}(t,y)+))-f(u(t,\mathtt{X}(t,y)-))}{u(t,\mathtt{X}(t,y)+)-u(t,\mathtt{X}(t,y)-)} & \text{if } u(t) \text{ has a jump at } \mathtt{X}(t,y) \end{cases}. \tag{5}$$
Two other properties are relevant in the following.

Property 1. For every $(\bar{t},\bar{x}) \in (0,+\infty)\times\mathbb{R}$ there exists $y \in \mathbb{R}$ such that $\mathtt{X}(\bar{t},y) = \bar{x}$ and at least one of the following holds:

1. for every $t \in [0, \bar{t}]$,
$$u(t, \mathtt{X}(t,y)-) \le u_0(y) \le u(t, \mathtt{X}(t,y)+);$$
2. for every $t \in [0, \bar{t}]$,
$$u(t, \mathtt{X}(t,y)+) \le u_0(y) \le u(t, \mathtt{X}(t,y)-).$$

This is a way to formulate in the nonsmooth case the fact that the smooth solutions are constant along characteristics.

In order to state the next property we need to introduce the notion of admissible boundary.

Definition 2.2. Let $T > 0$, $w \in \mathbb{R}$ and $\gamma : [0, +\infty) \to \mathbb{R}$ be a Lipschitz curve. Moreover let u be the entropy solution of (1) and denote by $\Omega^\pm = \{(t,x) \in [0,T) \times \mathbb{R} : x \gtrless \gamma(t)\}$. We say that (γ, w) is an *admissible boundary* for u up to time T if the restriction of u to Ω^- is the entropy solution of the initial boundary value problem
$$\begin{cases} u_t + f(u)_x = 0 & \text{in } \Omega^-, \\ u(0,\cdot) = u_0 & \text{in } (-\infty, \gamma(0)), \\ u(t,\gamma(t)) = w & \text{in } (0,T), \end{cases}$$
and similary on Ω^+.

Property 2. For every $(\bar{t}, \bar{x}) \in (0,+\infty) \times \mathbb{R}$ there exists $y \in \mathbb{R}$ such that $\mathtt{X}(\bar{t}, y) = \bar{x}$ and $(\mathtt{X}(\cdot, y), u_0(y))$ is an admissible boundary of u up to time $\bar{t}$.

A previous extension of the notion of characteristic to the nonsmooth setting is presented in [13, Chap.10]. The characteristic equation (5) implies that for every y, the map $t \to \mathtt{X}(t,y)$ is a generalized characteristic in the sense of Dafermos. Therefore the Lagrangian representation $\mathtt{X}$ can be interpreted as a monotone selection of Dafermos generalized characteristics for which (4), Property 1 and Property 2 hold. See also [3] for a similar use in the case of convex fluxes.

3. **Length estimate.**

In this section we present an estimate that relates the distance between two characteristics with the same value and the oscillation of the entropy solution between these characteristics. A relevant feature is the nonlinearity of the flux function f and we quantify it in the following way: given $w_1 \le w_2$ we consider twice the C^0 distance of $f \llcorner [w_1, w_2]$ from the set of affine functions on $[w_1, w_2]$:
$$\mathfrak{d}(w_1, w_2) := \min_{\lambda \in \mathbb{R}} \max_{\{w, w'\} \subset [w_1, w_2]} (f(w) - f(w') - \lambda(w - w'))$$

Theorem 3.1. *Let u be the entropy solution of (1) with u_0 bounded, continuous and piecewise monotone. Let $t > 0$ and $y_l < y_r$ be such that*

1. $u_0(y_l) = u_0(y_r) =: \bar{w}$;
2. $\mathtt{X}(\cdot, y_l)$ *and* $\mathtt{X}(\cdot, y_r)$ *enjoy Property 1 up to time* t.

Denote by
$$s := \max\{y_r - y_l, \mathtt{X}(t, y_r) - \mathtt{X}(t, y_l)\}$$
and
$$w_m := \bar{w} \wedge \inf_{(\mathtt{X}(t,y_l), \mathtt{X}(t,y_r))} u(t), \qquad w_M := \bar{w} \vee \sup_{(\mathtt{X}(t,y_l), \mathtt{X}(t,y_r))} u(t).$$
Then
$$s \ge \frac{\mathfrak{d}(w_m, w_M) t}{\|u_0\|_\infty}. \tag{6}$$

As a corollary we get a first a priori estimate for the entropy solution u. Roughly speaking the argument is the following: suppose for simplicity that u_0 has compact support. By finite speed of propagation also the solution at time t will have compact support. The estimate (6) tells that each oscillation between two values $a < b$ must occupy a given amount of space, which is strictly greater than 0 if the flux is not affine between a and b. But the total amount of space at our disposal is finite so we get an a priori estimate on the number of oscillations between two given values of an entropy solution on a given bounded interval. From this we can immediately recover the compactness in L^1_{loc} of the set of equibounded entropy solutions if the flux is weakly genuinely nonlinear, which can be obtained for example by a compensated compactness argument (see [20]). This compactness can be made quantitative by means of BV^Φ spaces presented in the introduction.

Corollary 1. *Denote by*

$$\mathfrak{N}(h) = \min_{w\in[-\|u_0\|_\infty,\|u_0\|_\infty]} \mathfrak{d}(w, w+h), \qquad \Psi := \operatorname{conv}(\mathfrak{N})$$

and for every $\varepsilon > 0$ set $\Phi^\varepsilon(w) = \Psi(\frac{x}{2})x^\varepsilon$. Then $\forall t > 0$

$$u(t) \in \mathrm{BV}^{\Phi^\varepsilon}_{\text{loc}}(\mathbb{R}).$$

Remark 2. Notice that $\Phi^\varepsilon(h) > 0$ for every $h > 0$ if and only if f is weakly genuinely nonlinear, i.e. $\{w : f''(w) \neq 0\}$ is dense in $\mathbb{R}$.

In this procedure the length estimate plays the same role as the Oleinik estimate (2) in order to deduce that the entropy solution $u(t) \in \mathrm{BV}_{\text{loc}}$ for every $t > 0$. Unfortunately if we specify this last result with $f(u) = u^2$ we get that for every $t > 0$ the entropy solution $u(t) \in \mathrm{BV}^{\frac{1}{2}-\varepsilon}$ and therefore Corollary 1 is not optimal. More in general in the setting of Theorem 1.2, we get $u(t) \in \mathrm{BV}^{\frac{1}{p+1}-\varepsilon}$ instead of the expected $u(t) \in \mathrm{BV}^{\frac{1}{p}}$.

4. BV regularity of $f' \circ u$.

In this section we discuss the BV regularity of the velocity field $f' \circ u$. In order to get a positive result we require that the flux function f has polynomial degeneracy (see Definition 1.1).

Theorem 4.1. *Let f be as above and u be the entropy solution of (1) with $u_0 \in L^\infty$ and assume that $\operatorname{supp} u_0 \subset [a, b]$. Then there exists C depending on $b - a$, f and $\|u_0\|_\infty$ such that for every $t > 0$*

$$\mathrm{TV} f' \circ u(t) \leq C\left(1 + \frac{1}{t}\right).$$

The details of the proof can be found in [18]. Here we only try to expose the strategy and the role of the tools and the estimates introduced above. Let us first notice that the situation is much simpler if the flux f is convex. In this case the result follows easily from the structure of the characteristics. The key property is that the characteristics are segments up to the time of the first interaction with other characteristics and two colliding characteristics never split in the future (see Fig. 1). An elementary geometrical constraint and (5) implies that

$$D_x f' \circ u(t) \leq \frac{\mathcal{L}^1}{t} \tag{7}$$

and the claim easily follows.

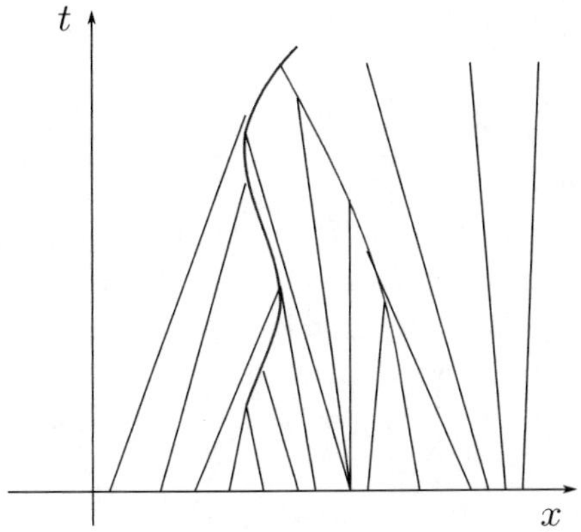

FIGURE 1. The characteristics (black) are absorbed by the shocks (blue).

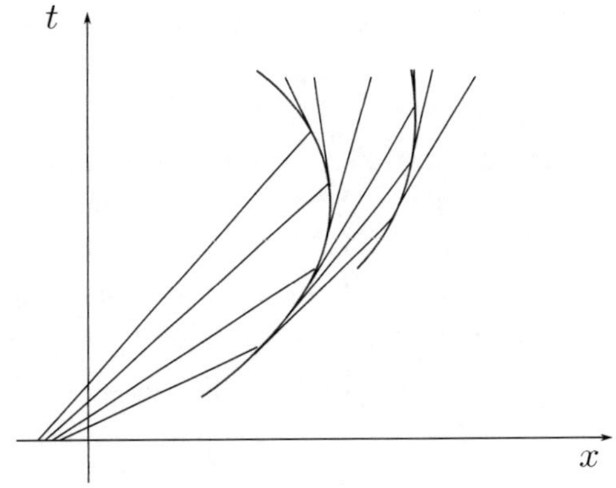

FIGURE 2. The characteristics (black) leave the contact discontinuities (blue).

This argument does not apply already in the case of fluxes with one inflection point. In this case (7) does not hold and the reason is that two characteristics who interact can split in the future in a contact discontinuity (see Fig. 2). In this case, relying on the precise description of the extremal characteristics in [12] and the Lagrangian representation, the argument in [11] can be made completely rigorous.

The structure of characteristics in the general case is more complicated. It turns out however that it is possible to reduce the general case to the case of fluxes with a single inflection point by means of the length estimate and Property 2. We briefly explain how it can be done: let $\delta > 0$ be the minimal distance between two inflection points of f. For any $\bar{t} > 0$, thanks to the length estimate (6), it is possible to find $N \approx C/\bar{t}$ characteristics starting from $y_1 < \ldots < y_N$ such that for every $t \in (\bar{t}/2, \bar{t})$ the oscillation of the entropy solution between two of this characteristics is less than δ. Moreover the constant C depends on the solution only through the length of the smallest interval containing the support of u_0. We therefore obtained N regions in which the range of the solution intersect at most one inflection point of the flux. The additional difficulty is that in the argument of [11] we also need to consider the interactions of the characteristics with the boundaries of these regions. This can be done interpreting the characteristics as admissible boundaries (Property 2) and these are all the ingredients that we need to prove Theorem 4.1.

Remark 3. Actually the BV regularity of $f' \circ u(t)$ can be improved to SBV regularity for every $t \in \mathbb{R}^+ \setminus N$ with N countable. See [1] for the case of uniformly convex fluxes, [2] for the extension to the case of strictly convex fluxes and [18] for a proof in the setting of Theorem 4.1.

Remark 4. The assumption on the flux cannot be removed. In [18] it is provided an example of entropy solution of (1) in which f has only one inflection point and $f' \circ u$ does not belong to $\mathrm{BV}_{\mathrm{loc}}((0, +\infty) \times \mathbb{R})$.

5. **Fractional regularity of the entropy solution.** In this last section we deduce Theorem 1.2 from Theorem 4.1. Again, as already noticed in [8], the situation is simpler if f is convex. If the flux f has degeneracy p, then the inverse function

$(f')^{-1}$ is $\frac{1}{p}$-Hölder and this implies that there exists $C > 0$ such that

$$\mathrm{TV}^{\frac{1}{p}} u(t) \le C \mathrm{TV} f' \circ u(t), \tag{8}$$

so that Theorem 1.2 immediately follows from Theorem 4.1.

Let us see now how to remove the convexity assumption on f: as in the previous section the length estimate allows to consider only the small oscillations of $u(t)$ and clearly the relevant ones are the oscillations around the inflection points. Therefore it is not restrictive to consider the case $f(u) = u^{p+1}$ with p even. An estimate like (8) cannot hold for a generic function $u(t)$ as in the convex case, consider for example a function v which takes only the values a and $-a$ for some $a > 0$. In this case $\mathrm{TV} f' \circ v = 0$ and $\mathrm{TV}^{\frac{1}{p}} v$ can be arbitrarily large. This obstruction is excluded taking advantage of the fact that $u(t)$ is the entropy solution of (1), roughly speaking if $f'(w_1) \approx f'(w_2)$ the shock between w_1 and w_2 is not entropic. More precisely the following lemma holds.

Lemma 5.1. *Let u be the entropy solution of (1) with $f(u) = u^{p+1}$. There exists a constant $c > 0$ depending on f and $\|u_0\|_\infty$ such that for a.e. $t > 0$ and for every $x_1 < x_2 \in \mathbb{R}$ with $u(t, x_1) \cdot u(t, x_2) < 0$ it holds*

$$\mathrm{TV}_{(x_1,x_2)} f' \circ u(t) \ge c |u(t, x_2) - u(t, x_1)|^p.$$

By means of this lemma it is not hard to conclude the proof of Theorem 1.2.

Remark 5. It has been observed in [9] that the order $\frac{1}{p}$ cannot be improved in (3).

REFERENCES

[1] L. Ambrosio and C. De Lellis, A note on admissible solutions of 1D scalar conservation laws and 2D Hamilton-Jacobi equations, *J. Hyperbolic Differ. Equ.* **1** (2004), no. 4, 813–826.

[2] Adimurthi, S.S. Ghoshal and G.D. Veerappa Gowda. Finer regularity of an entropy solution for 1-d scalar conservation laws with non uniform convex flux. *Rend. Sem. Mat. Univ. Padova,* **132** (2014), 1–24.

[3] Adimurthi, S.S. Ghoshal and G.D. Veerappa Gowda. Structure of entropy solutions to scalar conservation laws with strictly convex flux. *J. Hyperbolic Differ. Equ.*, **9**(2012), no. 4, 571–611.

[4] S. Bianchini and E. Marconi. On the concentration of entropy for scalar conservation laws. *Discrete Contin. Dyn. Syst. Ser. S*, **9** (2016), no. 1, 73–88.

[5] S. Bianchini and E. Marconi. On the structure of L^∞ entropy solutions to scalar conservation laws in one-space dimension-entropy solutions to scalar conservation laws in one-space dimension. *Archive for Rational Mechanics and Analysis*, **226** (2017), no. 1, 441–493.

[6] S. Bianchini and S. Modena. On a quadratic functional for scalar conservation laws. *J. Hyperbolic Differ. Equ.*, **11** (2014), no. 2, 355–435.

[7] S. Bianchini and S. Modena. Quadratic interaction functional for general systems of conservation laws. *Comm. Math. Phys.*, **338** (2015), no. 3, 1075–1152.

[8] C. Bourdarias, M. Gisclon, and S. Junca. Fractional BV spaces and applications to scalar conservation laws. *J. Hyperbolic Differ. Equ.*, **11** (2014), no. 4, 655–677.

[9] P. Castelli and S. Junca. Oscillating waves and optimal smoothing effect for one-dimensional nonlinear scalar conservation laws. In *Hyperbolic problems: theory, numerics, applications*, (2014), 709–716.

[10] P. Castelli and S. Junca. Smoothing effect in BV_Φ for entropy solutions of scalar conservation laws. *J. Math. Anal. Appl.*, **451** (2017), no. 2, 712–735.

[11] K. S. Cheng. A regularity theorem for a nonconvex scalar conservation law. *J. Differential Equations*, **61** (1986), no. 1, 79–127.

[12] C. M. Dafermos. Regularity and large time behaviour of solutions of a conservation law without convexity. *Proc. Roy. Soc. Edinburgh Sect. A*, **99** (1985), 201–239.

[13] C. M. Dafermos. *Hyperbolic conservation laws in continuum physics*, volume 325 of Grundlehren der Mathematischen Wissenschaften [Fundamental Principles of Mathematical Sciences]. Springer-Verlag, Berlin, fourth edition, 2016.

[14] R. J. DiPerna, P.-L. Lions, and Y. Meyer. L^p regularity of velocity averages. *Ann. Inst. H. Poincaré Anal. Non Linéaire*, **8** (1991), 271–287.

[15] B. Gess, X. Lamy. Regularity of solutions to scalar conservation laws with a force. preprint, arXiv:1707.06866.

[16] P.-E. Jabin. Some regularizing methods for transport equations and the regularity of solutions to scalar conservation laws. In *Séminaire: Équations aux Dérivées Partielles. 2008–2009*, Sémin. Équ. Dériv. Partielles, pages Exp. No. XVI, 15. École Polytech., Palaiseau (2010).

[17] P.-L. Lions, B. Perthame, and E. Tadmor. A kinetic formulation of multidimensional scalar conservation laws and related equations. *J. Amer. Math. Soc.*, **7** (1994), no. 1, 169–191.

[18] E. Marconi. Regularity estimates for scalar conservation laws in one space dimension. *J. Hyperbolic Diff. Equ.*, **15** (2018), no. 4, 623–691.

[19] J. Musielak and W. Orlicz. On generalized variations. I. *Studia Math.*, **18** (1959), 11–41.

[20] L. Tartar. Compensated compactness and applications to partial differential equations. In *Nonlinear analysis and mechanics: Heriot-Watt Symposium, Vol. IV*, volume 39 of *Res. Notes in Math.*, (1979), 136–212.

E-mail address: elio.marconi@unibas.ch

RECENT PROGRESS ON THE STUDY OF THE SHORT WAVE-LONG WAVE INTERACTIONS SYSTEM FOR AURORA-TYPE PHENOMENA

DANIEL R. MARROQUIN

Instituto de Matemática Pura e Aplicada - IMPA
Estrada Dona Castorina, 110
Rio de Janeiro, RJ, 22460-320, Brazil

ABSTRACT. We present a review on some recent developments on the study of a system of equations modelling the interactions between short waves, obeying a nonlinear Shrödinger equation, and long waves given by the equations of fluid dynamics for a compressible fluid flow. The system in question models an aurora-type phenomenon where a short wave propagates along the streamlines of the fluid. The results revised include well-posedness results in different contexts for strong and for weak solutions, in both the 1-dimensional and multidimensional cases, as well as some limit processes. We focus on our most recent contributions contained in the papers *Vanishing viscosity limit of short wave-long wave interactions in planar magnetohydrodynamics*, J. Differential Equations, 266(12) (2019), 8110-8163, and *Modeling Aurora Type Phenomena by Short Wave-Long Wave Interactions in Multidimensional Large Magnetohydrodynamic Flows* (with H. Frid and R. Pan). SIAM J. Math. Anal., 50(6) (2018), 61566195.

1. **Introduction.** We consider a system of equations modeling the interactions between short waves, obeying a nonlinear Schrödinger equation (NLS), and long waves, provided by the equations of compressible fluid dynamics. The system in question models an aurora-type phenomenon where a small wave propagates along the streamlines of a fluid flow. As such, it can be stated through the following nonlinear Schrödinger equation

$$i\psi_t + \Delta_{\mathbf{y}}\psi = |\psi|^2\psi + G\psi, \tag{1}$$

where i is the imaginary unit, $\psi = \psi(\mathbf{y}, t) \in \mathbb{C}$ is the wave function, $\mathbf{y}$ is the Lagrangian coordinate associated to the velocity field of the fluid $\mathbf{u}$, and G is a potential accounting for possible external forces.

For any given velocity field $\mathbf{u}$, the Lagrangian coordinates are characterized by being constant along the trajectories. Accordingly, if $\Omega \in \mathbb{R}^n$ is the spatial domain of the Eulerean coordinate $\mathbf{x}$, which provides the description of the dynamics from

---

2000 *Mathematics Subject Classification.* Primary: 76W05, 35Q55, 76N17, 76N10; Secondary: 35Q35.

Key words and phrases. Compressible MHD equations; Nonlinear Schrödinger equation; vanishing viscosity, weak solution, strong solutions.

D.R. Marroquin thankfully acknowledges the support from CNPq, through grant proc. 150118/2018-0.

an outsider's point of view, then the Lagrangian transformation $\mathbf{Y} = \mathbf{Y}(\mathbf{y}(\mathbf{x}, t), t)$ is, by definition, given by the relation

$$\mathbf{y}(\Phi(\mathbf{x}, t), t) = \mathbf{y}_0(\mathbf{x}), \tag{2}$$

where $\mathbf{y}_0$ is any diffeomorphism, which can be chosen conveniently, and Φ is the flux associated to the velocity field $\mathbf{u}$, that is,

$$\begin{cases} \frac{d\Phi(\mathbf{x},t)}{dt} = \mathbf{u}(\Phi(\mathbf{x}, t), t), \quad \text{for } t \in (0, T), \\ \Phi(\mathbf{x}, 0) = \mathbf{x}. \end{cases} \tag{3}$$

Now, the velocity field of the fluid is determined by the equations of fluid dynamics. These, in turn, take several forms depending on the properties of the fluid under consideration. For instance, whether the fluid is compressible or incompressible or whether the fluid is heat conductive o not. Here, we consider the most general case consisting of a compressible, heat conductive fluid, that conducts electricity and is in the presence of a magnetic field. In this regime, the velocity field is given by the equations of magnetohydrodynamics (MHD).

The full 3-dimensional MHD equations read as (see, for example, [17])

$$\rho_t + \operatorname{div}(\rho\mathbf{u}) = 0, \tag{4}$$

$$(\rho\mathbf{u})_t + \operatorname{div}(\rho\mathbf{u} \otimes \mathbf{u}) + \nabla p = \operatorname{div}\mathbb{S} + (\nabla \times \mathbf{H}) \times \mathbf{H} + \mathbf{F}, \tag{5}$$

$$\mathbf{H}_t + \nabla \times (\nu\nabla \times \mathbf{H}) = \nabla \times (\mathbf{u} \times \mathbf{H}), \tag{6}$$

$$\operatorname{div} \mathbf{H} = 0, \tag{7}$$

$$\begin{aligned} \mathcal{E}_t + \operatorname{div}(\mathbf{u}(\mathcal{E} - \tfrac{1}{2}|\mathbf{H}|^2 + p)) &= \operatorname{div}(\kappa\nabla\theta) \\ &+ \operatorname{div}(\mathbb{S} \cdot \mathbf{u}) + \operatorname{div}((\mathbf{u} \times \mathbf{H}) \times \mathbf{H} + \mathbf{H} \times (\nu\nabla \times \mathbf{H})) + \mathbf{F} \cdot \mathbf{u}. \end{aligned} \tag{8}$$

Here, $\rho = \rho(\mathbf{x}, t) \geq 0$, $\mathbf{u} = \mathbf{u}(\mathbf{x}, t) \in \mathbb{R}^3$ and $\theta = \theta(\mathbf{x}, t)$ are the fluid's density, velocity field and temperature; $\mathbf{H}$ is the magnetic field; the total energy $\mathcal{E}$ is

$$\mathcal{E} = \rho\left(e + \frac{1}{2}|\mathbf{u}|^2\right) + \frac{1}{2}|\mathbf{H}|^2,$$

where, e is the internal energy, $\frac{1}{2}\rho|\mathbf{u}|^2$ is the mechanical energy and $\frac{1}{2}|\mathbf{H}|^2$ is the magnetic energy; p denotes the pressure, $\mathbf{F}$ accounts for possible external forces, and $\mathbb{S}$ is the viscous stress tensor given by

$$\mathbb{S} = \lambda(\operatorname{div}\mathbf{u})Id + \mu(\nabla\mathbf{u} + (\nabla\mathbf{u})top).$$

The viscosity coefficients satisfy the relations $2\mu + \lambda > 0$ and $\mu > 0$; κ is the heat conductivity and $\nu > 0$ is the magnetic diffusivity.

As we are assuming the fluid to be compressible, the pressure and the internal energy are, in general, functions of the density and the temperature and must satisfy Maxwell's relation

$$e_\rho = \frac{1}{\rho^2}(p - \theta p_\theta), \tag{9}$$

which is a consequence of the second law of thermodynamics.

With this, the model is completed by choosing the external force term in the momentum equation (5) and the potential term in the NLS (1) as

$$\mathbf{F} = \alpha\nabla(g'(1/\rho)h(|\psi \circ \mathbf{Y}|^2)), \qquad G = \alpha g(v)h'(|\psi|^2), \tag{10}$$

where $\alpha > 0$ is the interaction coefficient, g and h are nonnegative smooth functions and $v = v(\mathbf{y}, t)$ is the specific volume defined by the identity

$$v(\mathbf{y}(\mathbf{x}, t), t) = \frac{1}{\rho(\mathbf{x}, t)}.$$

Thus, we are left with the following system

$$\rho_t + \operatorname{div}(\rho\mathbf{u}) = 0, \tag{11}$$

$$(\rho\mathbf{u})_t + \operatorname{div}(\rho\mathbf{u}\otimes\mathbf{u}) + \nabla p = \operatorname{div}\mathbb{S} + (\nabla\times\mathbf{H})\times\mathbf{H} + \alpha\nabla(g'(1/\rho)h(|\psi\circ\mathbf{Y}|^2)), \tag{12}$$

$$\mathbf{H}_t + \nabla\times(\nu\nabla\times\mathbf{H}) = \nabla\times(\mathbf{u}\times\mathbf{H}), \tag{13}$$

$$\operatorname{div}\mathbf{H} = 0, \tag{14}$$

$$\mathcal{E}_t + \operatorname{div}(\mathbf{u}(\mathcal{E} - \tfrac{1}{2}|\mathbf{H}|^2 + p)) = \operatorname{div}(\kappa\nabla\theta) + \operatorname{div}(\mathbb{S}\cdot\mathbf{u}) + \operatorname{div}((\mathbf{u}\times\mathbf{H})\times\mathbf{H} + \mathbf{H}\times(\nu\nabla\times\mathbf{H})) + \alpha\nabla(g'(1/\rho)h(|\psi\circ\mathbf{Y}|^2))\cdot\mathbf{u}, \tag{15}$$

$$i\psi_t + \Delta_{\mathbf{y}}\psi = |\psi|^2\psi + \alpha g(v)h'(|\psi|^2)\psi. \tag{16}$$

The short wave-long wave interactions occur along the particle paths and this is translated in the equations by stating the NLS in the Lagrangian coordinates of the fluid. Accordingly, we have to ensure that the change of variables is well defined. Actually, it can be shown that the Lagrangian transformation is nonsingular if and only if there is no vacuum nor concentration, that is, if the density is strictly positive and finite.

Indeed, using Liouville's formula for the determinant on the Jacobian $J_\Phi := \det\frac{\partial\Phi}{\partial\mathbf{x}}(\mathbf{x}, t)$ as well as the continuity equation (11), a straightforward calculation shows that

$$\frac{d}{dt}\left(\frac{\rho(\Phi(\mathbf{x}, t), t)}{J_{\mathbf{y}}(\mathbf{x}, t)}\right) = 0.$$

In particular, we see that $\mathbf{y}_0$ can be chosen so that

$$\det\frac{\partial\mathbf{y}}{\partial\mathbf{x}}(\mathbf{x}, t) = \rho(\mathbf{x}, t) \tag{17}$$

The most important feature of this model is that it is endowed with an energy identity, which can be stated in differential form as

$$\begin{aligned}\Big\{\mathcal{E}_t &+ \operatorname{div}_{\mathbf{x}}(\kappa\nabla_{\mathbf{x}}\theta) + \operatorname{div}_{\mathbf{x}}\big(\mathbf{u}(\mathcal{E} - \tfrac{1}{2}|\mathbf{H}|^2 + p + \alpha g'(1/\rho)h(|\psi\circ\mathbf{Y}|^2))\big) \\ &- \operatorname{div}_{\mathbf{x}}\big(\mathbb{S}\cdot\mathbf{u} + (\mathbf{u}\times\mathbf{H})\times\mathbf{H} + \mathbf{H}\times\nu(\nabla_{\mathbf{x}}\times\mathbf{H})\big)\Big\}d\mathbf{x} \\ = -\Big\{&\big(\frac{1}{2}|\nabla_{\mathbf{y}}\psi(t, \mathbf{y})|^2 + \frac{1}{4}|\psi(t, \mathbf{y})|^4 + \alpha g(v(t, \mathbf{y}))h(|\psi(t, \mathbf{y})|^2)\big)_t \\ &- \operatorname{div}_{\mathbf{y}}\big(\overline{\psi}_t\nabla_{\mathbf{y}}\psi + \psi_t\nabla_{\mathbf{y}}\overline{\psi}\big)\Big\}d\mathbf{y}.\end{aligned} \tag{18}$$

This identity can be deduced by multiplying (16) by $\overline{\psi}_t$ (the complex conjugate of ψ_t), taking real part, adding the resulting equation to the energy equation (15) and using relation (17) and the continuity equation (11) in order to deal with the change of variables.

This kind of model was introduced in 2011 by Dias and Frid in [7], where inspired by Benney's general theory on short wave-long wave interactions [2], they proposed a similar coupling where the Lagrangian coordinate was provided by the velocity field of a compressible isentropic (non-heat conductive) fluid (thus, driven by the

Navier-Stokes equations); and studied existence and uniqueness of solutions as well as the vanishing viscosity problem in the 1-dimensional setting.

Later, in 2014, Frid, Pan and Zhang [13] included the thermal description and showed existence an uniqueness of global smooth solutions to the Cauchy problem when the initial data are small perturbations of an equilibrium state.

In 2016, Frid, Jia and Pan extended these results to the model above, involving the MHD equations instead of the Navier-Stokes equations, showing decay rates of the solutions on top of the existence and uniqueness of global smooth solutions, also with small data.

Here we present a brief review of our most recent contributions regarding this model, contained in [21]. These results extend Dias and Frid's findings of well posedness and the vanishing viscosity problem to the planar (1-dimensional) system including both the thermal and magnetic descriptions.

We also comment on the results in [14], in collaboration with H. Frid and R. Pan, where we propose and establish the convergence of an approximation scheme which circumvents the lack of regularity of solutions, as well as the possible occurrence of vacuum, when we deal with the 2-dimensional model with arbitrarily large initial data.

2. **Planar equations: vanishing viscosity.** Let us first review our contributions on the planar version of the short wave-long wave interactions system described above. These results are contained in [21].

The planar MHD equations arise from the full 3-dimensional equations under the assumption that the flow moves in a preferential direction and is uniform in the transverse directions. This is translated into the equations by imposing that the partial derivatives with respect to the second and third spatial coordinates of the involved functions are identically equal to zero. Then, decomposing the velocity field $\mathbf{u} = (u, \mathbf{w})$ into its longitudinal direction u and transverse directions $\mathbf{w} = (w_1, w_2)$ a straightforward calculation provides the simplified planar equations. The magnetic field, in turn is also decomposed into its longitudinal and transverse directions as $\mathbf{H} = (h_0, \mathbf{h})$, where $\mathbf{h} = (h_1, h_2)$. Note that, under these assumptions, equation (14) implies that the longitudinal direction h_0 is constant and can be assumed to be equal to 1 (see [6]).

In the planar case, the Lagrangian coordinate $y = y(x,t)$ can be defined in a simpler way through the relations $y_x = \rho$, $y_t = \rho u$ and $y(x,0) = \int_0^x \rho_0(z)dz$, where $\rho_0(x) = \rho(x,0)$ is the initial density. As a result, the short wave-long wave interactions system is reduced to the following 1-dimensional form

$$\rho_t + (\rho u)_x = 0, \tag{19}$$

$$(\rho u)_t + \Big(\rho u^2 + p + \frac{\beta}{2}|\mathbf{h}|^2 - \alpha g'(1/\rho)h(|\psi \circ \mathbf{Y}|^2)\Big)_x = (\varepsilon u_x)_x, \tag{20}$$

$$(\rho \mathbf{w})_t + (\rho u \mathbf{w} - \beta \mathbf{h})_x = (\mu \mathbf{w}_x)_x, \tag{21}$$

$$\mathcal{E}_t + \Big(u\Big(\mathcal{E} + p + \tfrac{\beta}{2}|\mathbf{h}|^2\Big) - \beta \mathbf{w}\cdot\mathbf{h}\Big)_x \tag{22}$$

$$= (\varepsilon u u_x + \mu \mathbf{w}\cdot\mathbf{w}_x + \nu \mathbf{h}\cdot\mathbf{h}_x + \kappa\theta_x)_x + \alpha\Big(g'(1/\rho)h(|\psi\circ\mathbf{Y}|^2)\Big)_x u,$$

$$\beta \mathbf{h}_t + (\beta u \mathbf{h} - \beta \mathbf{w})_x = (\nu \mathbf{h}_x)_x, \tag{23}$$

$$i\psi_t + \psi_{yy} = |\psi|^2\psi + \alpha g(v)h'(|\psi|^2)\psi, \tag{24}$$

where $\mathcal{E} = \rho\left(e + \frac{1}{2}u^2 + \frac{1}{2}|\mathbf{w}|^2\right) + \frac{\beta}{2}|\mathbf{h}|^2$.

Note that, in contrast with system (11)-(16), a new parameter appeared. Namely, the magnetic permeability β. This parameter, which relates the magnetic field to the magnetic induction, is usually taken to be equal to 1 in the literature (cf. [17]) since in most real world media covered by the MHD equations this constant differs only slightly from the unity. However, the only physical constraint on it is its positivity.

Here, μ and $\varepsilon = \lambda + 2\mu$ are the shear viscosity and the bulk viscosity of the fluid, respectively.

In this setting, we are able to prove global existence and uniqueness of smooth solutions in a bounded open spacial domain Ω which can be assumed to be $(0,1)$. We first prove existence and uniqueness of local solutions and then extend the local solutions to global ones based on a priori estimates.

For the local result we use a Faedo-Galerkin type method similar to that applied by Dias and Frid in [7], which is in turn resembles the classic work by Kazhikhov and Shelukhin in [16] (c.f. [1, Chapter 2]). As for the global result, we develop some a priori estimates inspired by the work of Chen and Wang in [6] and by the work of Wang in [26]. In particular we show that no vacuum nor concentration develop in finite time.

Having well posedness for the one dimensional model, we turn our attention to he vanishing viscosity problem. First, we assume that the pressure has the form $p(\rho,\theta) = a\rho^\gamma + \delta\theta p_\theta(\rho)$, where $a > 0$, $\gamma > 1$, $\delta > 0$ and p_θ is a function of the density that satisfies certain growth conditions. Note that if ε, α, δ and β are all zero we are left with a system involving Euler's equations of compressible fluid dynamics and a decoupled nonlinear Schrödinger equation. In this connection we show convergence of the sequence of solutions as ε, α, δ and β tend to zero. More specifically, we show that if $\alpha, = o(\varepsilon^{1/2})$ and $\delta, \beta = o(\varepsilon)$ as $\varepsilon \to 0$, leaving $\mu > 0$ and $\nu > 0$ fixed, then the sequence of solutions to system (19)-(24) converges to a weak solution of the limit problem.

As the limit problem has different regularity properties than the original one (in Euler's equations shock waves are expected to occur in finite time, even if the initial data is smooth) this convergence is not a straightforward task.

The method we employ to achieve this is the compensated compactness combined with the Young measures as applied by Chen and Perepelitsa in [5], where they study the problem of vanishing viscosity limit for the Navier-Stokes equations. Due to the presence of the magnetic field and the short wave-long wave interactions we had to deduce some new estimates that allowed us to apply the method.

Such uniform estimates allow us to apply the Div-Curl lemma in order to prove that the Tartar-Murat commutator relation for the entropy kernel of the limit Euler Equation (equations (19) and (20) with $\varepsilon = \beta = \alpha = \delta = 0$) holds. After this, the arguments in [5, 3, 8] apply and the Young measures associated to the sequence $(\rho^\varepsilon, \rho^\varepsilon u^\varepsilon)$ are reduced to delta masses, thus yielding strong convergence to an entropy solution $(\rho, \rho u)$ of the limit Euler equations.

Once we have this strong convergence of the sequence of densities and momentums, the convergence in (21) and (23) follows in a straightforward way by a weak compactness argument on Sobolev spaces. Similarly, the convergence in the NLS equation (24) is a consequence of Aubin-Lions lemma.

Finally, the convergence in the energy equation is delicate, but we are able to adapt certain arguments by Feireisl in [10] in order to show that the sequence of temperatures converges to a variational solution of the limit thermal energy equation.

3. **Multidimensional case: approximation scheme.** Moving on to the multi-dimensional case, let us discuss briefly our results contained in [14], in collaboration with H. Frid and R. Pan.

As aforementioned, the main difficulty is the possible occurrence of vacuum. Since in the multidimensional fluid equations solutions are not smooth enough and vacuum may appear in finite time for large data, and as the Lagrangian transformation becomes singular in these situations, an effective coupling of the fluid equations with the nonlinear Schrdinger equation can not be made in a straightforward way.

In order to overcome these difficulties, we define the interaction through a regularized system that provides a good definition for an approximate Lagrangian coordinate. Then, after showing existence of solutions, we show compactness of the sequence of solutions to the regularized system thus making sense of the desired SW-LW interaction in the limit process.

For simplicity, in the multidimensional model we focus on the isentropic case, that is, the case of a non heat-conductive fluid, which trivializes the energy equation.

In order to workaround the lack of regularity of the density we first add an artificial viscosity to the continuity equation (4). Fix $\varepsilon > 0$ and $\delta > 0$ and consider the following regularized system

$$\rho_t + \operatorname{div}(\rho\mathbf{u}) = \varepsilon\Delta\rho, \tag{25}$$

$$(\rho\mathbf{u})_t + \operatorname{div}(\rho\mathbf{u}\otimes\mathbf{u}) + \nabla(a\rho^\gamma + \delta\rho^\beta) + \varepsilon\nabla\mathbf{u}\cdot\nabla\rho = (\nabla\times\mathbf{H})\times\mathbf{H} + \mu\Delta\mathbf{u} + (\lambda+\mu)\nabla(\operatorname{div}\mathbf{u}) + \rho\mathbf{F}, \tag{26}$$

$$\mathbf{H}_t - \nabla\times(\mathbf{u}\times\mathbf{H}) = -\nabla\times(\nu\nabla\times\mathbf{H}), \tag{27}$$

$$\operatorname{div}\mathbf{H} = 0. \tag{28}$$

Note that besides the artificial viscosity added to the continuity equation, two new terms appeared in the momentum equation (26). The term $\delta\rho^\beta$, where $\beta > 1$, acts as an artificial pressure and is intended to provide better estimates on the density, whereas the term $\varepsilon\nabla\mathbf{u}\cdot\nabla\rho$ is set to equate the unbalance in the energy estimates of the MHD equations caused by the introduction of the artificial viscosity. This approximate system resembles the one employed by Hu and Wang in [15] where they study the existence of weak solutions to the three dimensional MHD equations. A similar approximation was introduced by Feireisl, et al. in [11] in the study of the Navier-Stokes equations, who, in turn, followed the pioneering ideas by Lions in [18]. Recall that ε and δ are small constants and the analysis that we develop provides insights that justify the accuracy to which this regularized model approximates the desired SW-LW interaction.

Now, as it turns out, even in this regularized setting the velocity field might not be smooth enough to provide a good enough definition of Lagrangian transformation that we can work with. More specifically, in the present situation there is no a priori bound available for Jacobian of the Lagrangian transformation, as it depends on the L^∞ norm of $\nabla\mathbf{u}$. For this reason we replace the velocity by a suitable smooth approximation $\mathbf{u}^N$ (which tends to $\mathbf{u}$ as $N\to\infty$) in the definition of the Lagrangian transformation. Thus obtaining an approximate Lagrangian coordinate given by (3), (2) and with $\mathbf{u}$ replaced by $\mathbf{u}_N$.

Although we now have a smoothed Lagrangian coordinate, we lose relation (17 and instead we have

$$J_y(t) = \rho(0,x)e^{-\int_0^t \text{div } u^N(s,\Phi(s,x))ds}. \tag{29}$$

Accordingly, we have to make a further slight modification to our model. Namely, instead of taking $\mathbf{F}$ as (10) we take it of the form

$$\mathbf{F} = \nabla\left(\alpha \frac{J_y}{\rho} g'(1/\rho)h(|\psi|^2)\right). \tag{30}$$

Note that, although vacuum is permitted in our model, the fact that g is compactly supported in $(0,\infty)$ clarifies any ambiguity in the definition of $\mathbf{F}$.

As a result we end up with the following system of equations:

$$\rho_t + \text{div}(\rho\mathbf{u}) = \varepsilon\Delta\rho, \tag{31}$$

$$(\rho\mathbf{u})_t + \text{div}(\rho\mathbf{u}\otimes\mathbf{u}) + \nabla(a\rho^\gamma + \delta\rho^\beta) + \varepsilon\nabla\mathbf{u}\cdot\nabla\rho$$
$$= \nabla(\alpha\frac{J_y}{\rho}g'(1/\rho)h(|\psi|^2)) + (\nabla\times\mathbf{H})\times\mathbf{H} + \mu\Delta\mathbf{u} + (\lambda+\mu)\nabla(\text{div}\mathbf{u}), \tag{32}$$

$$\mathbf{H}_t - \nabla\times(\mathbf{u}\times\mathbf{H}) = -\nabla\times(\nu\nabla\times\mathbf{H}), \tag{33}$$

$$\text{div } \mathbf{H} = 0. \tag{34}$$

$$i\psi_t + \Delta_y\psi = |\omega|^2\psi + \alpha g(v)h'(|\psi|^2)\psi, \tag{35}$$

Regarding this new system, we prove the existence of solutions in any finite time interval and show the convergence of the approximate solutions when the artificial viscosity ε together with the interaction coefficients α tend to 0 and as $N\to\infty$. Then, we make δ tend to zero and show convergence to a solution of the system formed by the MHD equations together with the decoupled nonlinear Schrdinger equation. As emphasized before, the proposed approximation scheme has the purpose to legitimize the coordinates of the limiting Schrdinger equation to be considered as the Lagrangian coordinates of the fluid in a generalized sense.

Let us remark that our results hold in a smooth bounded open spacial domain in $\mathbb{R}^2$. The only restriction that does not allow us to proceed in the full three dimensional case comes from the lack of solvability of the nonlinear Schrödinger equation in this setting. However, assuming this our methods can be adapted to the three dimensional case.

Also, our result covers large initial data at the price of obtaining only weak solutions.

REFERENCES

[1] S. N. Antontsev, A. V. Kazhikhov and V. M. Monakhov, *Boundary value problems in mechanics of nonhomogeneous fluids*, Studies in Mathematics and Its Applications, Vol. 22, North-Holland, Amsterdam, 1990.

[2] D. J. Benney, A general theory for interactions between short and long waves, *Studies in Applied Mathematics*, **56** (1977), 81–94.

[3] G.-Q. Chen, The compensated compactness method and the system of isentropic gas dynamics, Lecture Notes, Preprint MSRI-00527-91, Berkeley, October 1990.

[4] G.-Q. Chen, Remarks on DiPerna's paper: "Convergence of the viscosity method for the isentropic gas dynamics" [Comm. Math. Phys. 91 (1983), no. 1, 130], *Proc. Amer. Math. Soc.* **125** (1997), no. 10, 29812986.

[5] G.-Q. Chen and M. Perepelitsa, Vanishing viscosity limit of the Navier-Stokes equations to the Euler equations for compressible fluid flow, *Communications on Pure and Applied Mathematics*, **Vol. LXIII** (2010) 1469 –1504.

[6] G.-Q. Chen and D. Wang, Global solutions of nonlinear magnetohydrodynamics with large initial data, *Journal of Differential Equations* **182** (2002) 344376.

[7] J. P. Dias and H. Frid, Short wave-long wave interactions for compressible NavierStokes equations, *SIAM J. Math. Anal.*, **43** (2011) 764787.

[8] X. X. Ding, G.-Q. Chen, P. Z. Luo, Convergence of the Lax-Friedrichs scheme for isentropic gas dynamics I, II. *Acta Math. Sci.*, 5 **no. 4** (1985) 415432, 433472. Chinese translations: Convergence of the Lax-Friedrichs scheme for the system of equations of isentropic gas dynamics. I. Acta Math. Sci. (Chinese) 7 (1987), no. 4, 467480. Convergence of the Lax-Friedrichs scheme for the system of equations of isentropic gas dynamics. II. Acta Math. Sci. (Chinese) 8 (1988), no. 1, 6194. Convergence of the fractional step Lax-Friedrichs scheme and Godunov scheme for the isentropic system of gas dynamics. Comm. Math. Phys. 121 (1989), no. 1, 6384.

[9] R. J. DiPerna, Convergence of the vanishing viscosity method for isentropic gas dynamics, *Communications in Mathematical Physics*, **91** (1983), 1–30.

[10] E. Feireisl, *Dynamics of viscous compressible fluids*, Oxford Lecture Series in Mathematics and its Applications, vol 26. Oxford University Press, Oxford, 2004.

[11] E. Feireisl, A. Novotny and H. Petzeltová, On the existence of weak solutions to the Navier-Stokes equations, *J. Math. Fluid. Mech.*, **2** (2001), 358–392.

[12] H. Frid, J. Jia, R. Pan, Global smooth solutions in R3 to short wavelong wave interactions in magnetohydrodynamics, *J. Differential Equations*, **262** (2017), no. 7, 41294173.

[13] H. Frid, R. Pan and W. Zhang, Global smooth solutions in R3 to short wave-long wave interactions systems for viscous compressible fluids *SIAM J. Math. Anal.*, **Vol. 46**, No. 3 (2014), 19461968.

[14] H. Frid, D. R. Marroquin and R. Pan, Modeling Aurora Type Phenomena by Short Wave-Long Wave Interactions in Multidimensional Large Magnetohydrodynamic Flows, *SIAM J. Math. Anal.*, **50** (2018), No. 6 61566195.

[15] X. Hu, D. Wang, Global existence and large time behaviour of solutions to the three-dimensional equations of compressible magnetohydrodynamic flows *Arch. Rational. Mech. Anal.*, **197** (2010), 203–238.

[16] V. Kazhikhov, V. Shelukhin, Unique global solution with respect to time of initial-boundary-value problems for one dimensional equations of a viscous gas, *J. Appl. Math. Mech.*, **41** (1977), 273–282.

[17] L. D. Landau and E. M. Lifschitz. *Electrodynamics of Continuous Media*, 2nd edition. Pergamon Press, New York 1983.

[18] P.-L. Lions, *Mathematical Topics in Fluid Mechanics: Volume 2, Compressible Models*, Oxford Lecture Series in Mathematics and its Applications **10**. Oxford Science Publications, The Clarendon Press, Oxford University Press, New York, 1998.

[19] P.-L. Lions, B. Perthame and P. E. Souganidis. Existence and stability of entropy solutions for the hyperbolic systems of isentropic gas dynamics in Eulerian and Lagrangian coordinates, *Comm. Pure Appl. Math.*, **49** (1996), 599–638.

[20] P.-L. Lions, P. Perthame and E. Tadmor, Kinetic formulation of the isentropic gas dynamics and p-systems. *Commun. Math. Physics*, **63** (1994), 415–431.

[21] D. R. Marroquin, Vanishing viscosity limit of short wave-long wave interactions in planar magnetohydrodynamics, *J. Differential Equations*, **266** (2019), No. 12 8110-8163.

[22] F. Murat, Compacité par compensation, *Ann. Scuola Norm. Sup. Pisa Sci. Fis. Mat.*, **5** (1978), 489–507.

[23] F. Murat, L'injection du cône positif de H^{-1} dans $W^{-1,q}$ est compacte pour tout $q < 2$, *J. Math. Pures Appl. (9)*, **60** (1981), no. 3, 309322.

[24] L. Tartar, Compensated compactness and applications to partial differential equations, *Research Notes in Mathematics, Nonlinear Analysis and Mechanics, ed. R. J. Knops*, vol. **4**, Pitman Press, New York, 1979, 136–211.

[25] L. Tartar, The compensated compactness method applied to systems of conservation laws, *Systems of nonlinear partial differential equations (Oxford, 1982), 263285, NATO Adv. Sci. Inst. Ser. C Math. Phys. Sci., 111, Reidel, Dordrecht*, 1983.

[26] D. Wang, Large solutions to the initial-boundary value problem for planar magnetohydrodynamics, *SIAM J. Appl. Math.* **63** (2003), 1424–1441.

E-mail address: danielrm@impa.br

ON SOME RECENT RESULTS CONCERNING NON-UNIQUENESS FOR THE TRANSPORT EQUATION

STEFANO MODENA

Mathematisches Institut
Universität Leipzig
D-04109 Leipzig, Germany

ABSTRACT. In these notes we present some recent results concerning the non-uniqueness of solutions to the transport equation, obtained in collaboration with Gabriel Sattig and László Székelyhidi in [19, 18, 17].

1. **Introduction.** These notes concern the problem of (non)uniqueness of solutions to the transport equation in the periodic setting

$$\partial_t \rho + u \cdot \nabla \rho = 0, \tag{1}$$

$$\rho_{|t=0} = \rho^0 \tag{2}$$

where $\rho : [0,T] \times \mathbb{T}^d \to \mathbb{R}$ is a scalar density, $u : [0,T] \times \mathbb{T}^d \to \mathbb{R}^d$ is a given vector field and $\mathbb{T}^d = \mathbb{R}^d / \mathbb{Z}^d$ is the d-dimensional flat torus.

Unless otherwise specified, we assume in the following that $u \in L^1$ is *incompressible*, i.e.

$$\operatorname{div} u = 0 \tag{3}$$

in the sense of distributions. Under this condition, (1) is formally equivalent to the continuity equation

$$\partial_t \rho + \operatorname{div}(\rho u) = 0. \tag{4}$$

It is well known that the theory of classical solutions to (1)-(2) is closely connected to the ordinary differential equation

$$\begin{aligned} \partial_t X(t,x) &= u(t, X(t,x)), \\ X(0,x) &= x. \end{aligned} \tag{5}$$

More precisely, if u is at least Lipschitz continuous, the solution to (1)-(2) is given by the formula

$$\rho(t, X(t,x)) = \rho^0(x). \tag{6}$$

There are several PDE models, related, for instance, to fluid dynamics or to the theory of conservation laws (see for instance [11, 7, 14, 15, 16]), where one has to deal with vector fields which are not Lipschitz, but have lower regularity and

---

2000 *Mathematics Subject Classification.* Primary: 35A02; Secondary: 35F10.

Key words and phrases. Transport equation, Continuity equation, Convex Integration, Non-uniqueness, Mikado Flows.

These notes were written during the visit of the author to the Hausdorff Research Institute for Mathematics (HIM), University of Bonn, Jan-Apr 2019. This visit was supported by the HIM. Both, this support and the hospitality of HIM, are gratefully acknowledged.

therefore it is important to investigate the well-posedness of (1)-(2) in the case of non-smooth vector fields.

There are several possibilities to state the well-posedness problem for (1)-(2) in a weak setting; we describe now one possible way. Fix an exponent $p \in [1, \infty]$ and denote by p' its dual Hölder, $1/p + 1/p' = 1$. The following two questions are of interest.

(a) Do existence and uniqueness of solutions to (1)-(2) hold in the class of densities

$$\rho \in L^\infty(0,T; L^p(\mathbb{T}^d)) =: L^\infty_t L^p_x \tag{7}$$

for a given vector field

$$u \in L^1(0,T; L^{p'}(\mathbb{T}^d)) =: L^1_t L^{p'}_x ? \tag{8}$$

(b) Is the relation (6), which links the PDE (1) to the ODE (5) (or, in other words, the Eulerian world to the Lagrangian one) still valid, in some weak sense?

Let us briefly comment on the choice of the classes (7)-(8) for the density and the vector field, respectively. The choice of the class (7) for the density is dictated by the following consideration. For smooth solutions to (1), every (spatial) L^p norm remains constant in time. It is therefore natural in the weak setting to look for densities whose L^p norm, if not constant, at least remains uniformly bounded in time. Once the class for ρ is fixed, the choice (8) of the class for the vector field u is as well natural, since in this way the product $\rho u \in L^1((0,T) \times \mathbb{T}^d)$ and hence the notion of distributional solution to (4) (and thus also to (1)) makes sense.

This is the plan of these notes. In Section 2 we give a brief presentation of some well-posedness results and some counterexamples to well-posedness which can be found in the literature. In Section 3 we state the main theorem of these notes, Theorem 3.1. In Section 4 we make some comments on the proof of Theorem 3.1.

We wish to stress that the aim of these notes is to give an informal presentation of some recent results concerning non-uniqueness of solutions to the transport equation. For this reason, we intentionally avoid technicalities, we are quite vague in many points, many references are missing, and the statement of the main theorem is not presented in its full generality. For a more detailed discussion, we refer to [19, 18].

2. **Well-posedness for the Cauchy problem in the weak setting.**

We sketch in this section a (far from complete) overview of the literature concerning the answers to questions (a) and (b) above.

First of all, we remark that *existence* of weak solutions in the class (7), for a given vector field as in (8), is not a serious issue, because of the linearity of the transport equation. Indeed, to produce a weak solution to (1)-(2), it is enough to regularize the vector field u and the initial datum ρ^0, to solve the regularized smooth problem and use the uniform bound in $L^\infty_t L^p_x$ to get a weakly converging sequence. By the linearity of the equation (1), the limit of such sequence is a weak solution to (1)-(2).

The big issue is thus *uniqueness of weak solutions* and the *relation* (6) *between Eulerian and Lagrangian world.*

2.1. **Uniqueness results.**

The first uniqueness result we mention is the celebrated theorem by DiPerna and Lions in 1989 [12], when they proved that, if the vector field u, in addition to the integrability condition (8), enjoys also the Sobolev regularity

$$u \in L^1_t W^{1,p'}_x, \tag{9}$$

then uniqueness of solutions holds in the class of densities (7). Let us remark that Di-Perna and Lions' Theorem is still true, even when the incompressibility condition (3) is substituted by the weaker condition

$$\operatorname{div} u \in L^\infty_{tx}. \tag{10}$$

Di-Perna and Lions' Theorem was extended in 2004 by Ambrosio [1], where he proved that, in the class of bounded densities (i.e. $p = \infty$ in our notation), uniqueness of solutions holds, if

$$u \in L^1_t BV_x. \tag{11}$$

Again, also Ambrosio's Theorem holds if (3) is replaced by (10).

Very recently, Bianchini and Bonicatto further extended Ambrosio's uniqueness result to vector fields which satisfy (11) and are *nearly incompressible.* We do not want to enter into details here and to give a precise definition of *near-incompressibility.* We only mention that such notion is the natural generalization of (10), in the framework of BV vector fields.

We add two remarks to this list of results. The first one is the following. The proofs of the mentioned results are very subtle and involve several deep ideas and sophisticated techniques. We could however try to summarize the heuristics behind all of them as follows: (very) roughly speaking, a Sobolev or BV vector field u is Lipschitz-like (i.e. Du is bounded) on a large set and there is just a small "bad" set, where Du is very large. On the big set where u is "Lipschitz-like", the classical Cauchy-Lipschitz theory applies. Non-uniqueness phenomena could thus occur only on the small "bad" set. Uniqueness of solutions in the class of bounded densities (or L^p densities, where p is exactly the dual Hölder to the integrability exponent of Du, see (9)) is then a consequence of the fact that a *bounded* (or L^p) density ρ can not "see" this bad set, or, in other words, can not *concentrate* on this bad set.

A second interesting remark is that, roughly speaking, whenever uniqueness for the PDE (1) holds *in the class of bounded densities* (i.e. $p = \infty$) for a given vector field u, a uniqueness statement holds (in the sense of *regular Lagrangian flow*, a notion we will not introduce in these notes, for a precise definition we refer, for instance, to [2]) also for the ODE (5) with the same vector field u. This can be seen, observing that the inverse flow map $\Phi(t) := X(t)^{-1} : \mathbb{T}^d \to \mathbb{T}^d$ is (at least in the smooth case) a bounded solution to (1) with (vector valued) initial datum $\Phi(0, x) = x$.

2.2. **Non-uniqueness results.** From the analysis in the previous section it follows that the uniqueness results present in the literature concern vector fields

(a) which enjoy some form of exact or approximate *incompressibility* (e.g. they have bounded divergence or they are nearly incompressible);
(b) *and* which are *at least once differentiable* (in some weak sense, e.g. they are Sobolev or BV).

The counterexamples to uniqueness which can be found in the literature are, in general, based on the failure of at least one of these two conditions. For instance, already in the paper [12] by Di-Perna Lions, it is possible to find an example of a Sobolev vector field with unbounded divergence and another example of an incompressible vector field which belongs to $L^1_t W^{1,s}_x$ for every $s < 1$, but not to $L^1_t W^{1,1}_x$, for which uniqueness of solutions fails. A further counterexample can be found in [10] (an incompressible vector field which belongs to $L^1(\varepsilon, T; BV_x)$ for every $\varepsilon > 0$ but not to $L^1(0, T; BV_x)$).

Let us also remark that the counterexamples mentioned so far are based on vector fields for which the associated ODE (5) has a degenerate behavior and therefore the Eulerian non-uniqueness is a consequence of the Lagrangian one.

3. **Statement of the main theorem.** We mentioned in the previous section several uniqueness and non-uniqueness results and we observed that, in order to have uniqueness, the vector field u must have some incompressiblity property and must possess one full spatial derivative. There is however one question we did not answer so far:

for fixed $p \in [1, \infty)$, does uniqueness of solutions hold in the class of densities $L^\infty_t L^p_x$ for a given incompressible $u \in L^1_t W^{1,\tilde{p}}$, with $\tilde{p} < p'$?

Recall that p' is the dual Hölder exponent to p and thus, if $\tilde{p} \geq p'$, then DiPerna-Lions' theory [12] guarantees uniqueness of solutions in $L^\infty_t L^p_x$.

The answer to such question is not trivial at all. There are indeed two competing mechanisms, one playing for uniqueness, the other one playing against.

On one side, the incompressibility and the Sobolev regularity of u imply uniqueness in the class of bounded densities (more precisely, in $L^1_t L^{\tilde{p}'}_x$, with $\tilde{p}'$ the dual Hölder to $\tilde{p}$) and thus, as observed in the previous section, uniqueness of solutions to the ODE (5) holds, in the sense of the regular Lagrangian flow. The Lagrangian picture is very well behaved.

On the other side, if "p is too small compared $\tilde{p}$", it could happen (referring to the heuristics introduced in the previous section) that "an L^p density does see the *bad set* of the $W^{1,\tilde{p}}$ vector field u" and thus "purely Eulerian" non-uniqueness phenomena could occur.

The following theorem, which is the main result we present in these notes, provides an answer to the question asked above.

Theorem 3.1 (M., Sattig, Székelyhidi). *Let $p \in [1, \infty)$, $\tilde{p} \in [1, \infty)$. If*

$$\frac{1}{p} + \frac{1}{\tilde{p}} > 1 + \frac{1}{d}, \tag{12}$$

then there exist infinitely many incompressible vector fields

$$u \in C_t L^{p'}_x \cap C_t W^{1,\tilde{p}}_x$$

for which uniqueness of solutions to the transport equation (1) *fails in the class of densities $\rho \in C_t L^p_x$. Moreover:*

- *if $p = 1, p' = \infty$, then $u \in C([0,T] \times \mathbb{T}^d) \cap C_t W^{1,\tilde{p}}_x$;*
- *the same result holds if the transport equation* (1) *is replaced by the transport-diffusion equation*

$$\partial_t \rho + \nabla \rho \cdot u = \Delta \rho \tag{13}$$

  if, in addition, $p' < d$.

Let us add some comments on the statement of Theorem 3.1.

1. The case $p = \infty$ is not considered. Indeed $p = \infty$ corresponds to the case of bounded densities and we have observed in Section 2.1 that, in this case, uniqueness holds even for BV vector fields.
2. Similarly, also the case $\tilde{p} = \infty$ is not considered. Indeed $\tilde{p} = \infty$ corresponds to the case of a Lipschtiz continuous vector field u and, in this case, the classical Cauchy-Lipschitz theory for the ODE (5) provides a solution to (1)-(2), via the formula (6).

3. In the case $p = 1$, $p' = \infty$ (which correspond to $\tilde{p} < d$), the vector fields we construct are *continuous*, not only *bounded*. This shows that, in general, even the continuity of the vector field, in addition to the incompressibility and the Sobolev regularity, is not enough to guarantee uniqueness of weak solutions (compare with the result in [5, 6]).
4. For the vector fields provided by Theorem 3.1, uniqueness for the ODE (5) holds (in the sense of *regular Lagrangian flow*): nevertheless, the PDE (1) displays anomalous behavior. This shows that, for such vector fields, the relation between the Lagrangian and Eulerian world, summarized in Equation (6), is completely destroyed. This is even more evident in the case $p = 1$, where the vector fields we construct are continuous and thus the trajectories of the regular Lagrangian flow are classical C^1 curves solving (5).
5. In general, for the transport-diffusion equation (13) much stronger uniqueness results hold than for the transport equation (1). Indeed, the diffusion term $\Delta\rho$ is usually dominating (being the highest order term) and thus its regularizing effect translates, through the energy estimate, into a uniqueness statement for (13). On the contrary, for the vector fields provided by Theorem 3.1, the non-uniqueness generated by the first order term $\nabla\rho \cdot u$ is so strong that it beats even the second order term $\Delta\rho$.

4. **Some comments on the proof.** We conclude these notes with some comments on the proof of Theorem 3.1. Referring again to the heuristics introduced in Section 2.1, the basic idea behind the proof of Theorem 3.1 is to "concentrate the density ρ on the *bad* set of the vector field u".

This is done through a convex integration scheme, in the spirit of the papers by De Lellis, Skékelyhidi and collaborators on the Euler equations (see, in particular, [9]). More precisely, the linear (in ρ) PDE (4) is treated as a nonlinear PDE with both ρ and u as unknowns. The density ρ and the field u are constructed as limit of sequences

$$\rho = \lim_{q\to\infty} \rho_q, \qquad u = \lim_q u_q, \qquad q \in \mathbb{N}, \tag{14}$$

where the limits have to be taken in suitable norms and (ρ_q, u_q) are approximate solutions to the transport equation, i.e.

$$\partial_t \rho_q + \nabla \rho_q \cdot u_q = \mathrm{Error}_q, \qquad \operatorname{div} u_q = 0, \tag{15}$$

with Error_q converging weakly to zero, as $q \to \infty$.

The sequences $(\rho_q)_q, (u_q)_q$ are constructed recursively: assuming ρ_q, u_q are given, as a first attempt, one defines

$$\rho_{q+1} = \rho_q + a_q(t,x)\Theta(\lambda_q x), \qquad u_{q+1} = u_q + b_q(t,x)W(\lambda_q x). \tag{16}$$

Here:

- $\lambda_q \in \mathbb{N}$ is an *oscillation parameter*, with $\lambda_q \to \infty$ as $q \to \infty$;
- $\Theta : \mathbb{T}^d \to \mathbb{R}$, $W : \mathbb{T}^d \to \mathbb{R}^d$ are fixed smooth profiles, called *Mikado density* and *Mikado field*, in the same spirit of the Mikado flows introduced by Daneri and Székelyhidi in [8] for the Euler equations; for a precise definition of Θ and W we refer to the paper [19];
- a_q, b_q are "slow oscillating" amplitudes, defined at each step in order to reduce Error_q and to get, in the limit, a solution (ρ, u) to (4).

As in the framework of the Euler equations, the basic idea of convex integration is to choose the oscillation parameter λ_q bigger and bigger along the iteration, and to use oscillations in order to reduce the error in (15).

The main difference between Theorem 3.1 and the theorems proven in the framework of the Euler equations (e.g. [9, 13, 3]) is the following: in Theorem 3.1 we want to construct a vector field which is in $W^{1,\tilde{p}}_x$, i.e. it possesses one full derivative (in some $L^{\tilde{p}}$ space), whereas in the framework of the Onsager's conjecture for the Euler equations, the aim was to show the existence of anomalous C^γ solutions, for every $\gamma < 1/3$, i.e. solutions which possess "just 1/3 of derivative" (measured in a sup norm).

How can we thus get such a $W^{1,\tilde{p}}$ bound? If a scheme as in (16) is used, one can easily see that problems arise. Indeed, in order to have convergence of Du_q in $L^{\tilde{p}}$, one should be able to provide a good bound of the distance $\|Du_{q+1} - Du_q\|_{L^{\tilde{p}}}$. However we have

$$\|Du_{q+1} - Du_q\|_{L^{\tilde{p}}} \approx \lambda_q \|b_q\|_{L^\infty} \|DW\|_{L^{\tilde{p}}} \tag{17}$$

and the presence of the multiplicative factor λ_q prevents the convergence of Du_q in $L^{\tilde{p}}$.

This issue can be solved using a *concentration* argument, in the same spirit of what Buckmaster and Vicol did in the framework of the Navier-Stokes equations in their remarkable recent work [4], using *intermittent Beltrami flows.* In order to explain how the concentration argument works, let us think, for the time being, to the fixed Mikado density Θ and field W as compactly supported functions in $\mathbb{R}^d$ (i.e not as periodic functions). Then we can construct a family of *concentrated Mikado densities and fields*, parametrized by a concentration parameter $\mu > 0$, defined as a rescaled version of Θ and W, as follows:

$$\Theta_\mu(x) := \mu^\alpha \Theta(\mu x), \qquad W_\mu(x) = \mu^\beta W(\mu x).$$

It is now not difficult to see that, if (12) is satisfied, then one can choose α, β so that

$$\|\Theta_\mu\|_{L^p} \approx 1, \qquad \|W_\mu\|_{L^{p'}} \approx 1, \tag{18}$$

and

$$\|DW_\mu\|_{L^{\tilde{p}}} \approx \mu^{-c}, \tag{19}$$

for some $c > 0$, so that $\|DW_\mu\|_{L^{\tilde{p}}} \to 0$, as $\mu \to \infty$. In this way, we can produce a whole family of Mikado fields, which "are not degenerating" as $\mu \to \infty$ (i.e. they remains "of order 1", in some suitable norm, thanks to (18)), but, at the very same time, have vanishing derivative, thanks to (19).

We can now modify our *Ansatz* (16) as follows:

$$\rho_{q+1} = \rho_q + a_q(t,x)\Theta_{\mu_q}(\lambda_q x), \qquad u_{q+1} = u_q + b_q(t,x) W_{\mu_q}(\lambda_q x), \tag{20}$$

where μ_q is a sequence of real numbers, with $\mu_q \to \infty$ as $q \to \infty$, to be chosen appropriately. In this way, thanks to (19), the estimate in (17) becomes

$$\|Du_{q+1} - Du_q\|_{L^{\tilde{p}}} \approx \lambda_q \|b_q\|_{L^\infty} \|DW_{\mu_q}\|_{L^{\tilde{p}}} \lesssim \|b_q\|_{L^\infty} \lambda_q \mu_q^{-c},$$

and thus, if μ_q is chosen much bigger than λ_q, the distance $\|Du_{q+1} - Du_q\|_{L^{\tilde{p}}}$ can be made arbitrarily small, thus getting convergence of u_q in $W^{1,\tilde{p}}$ and hence proving Theorem 3.1.

REFERENCES

[1] L. Ambrosio, Transport equation and Cauchy problem for BV vector fields, *Invent. math.*, **158(2)** (2004), 227–260.

[2] L. Ambrosio, Well posedness of ODE's and continuity equations with nonsmooth vector fields, and applications, *Rev. Mat. Complut.*, **30(3)** (2017), 427–450.

[3] T. Buckmaster, C. De Lellis, L. Székelyhidi Jr. and V. Vicol, Onsager's conjecture for admissible weak solutions, *Communications on Pure and Applied Mathematics*, **72(2)** (2019), 229–274.

[4] T. Buckmaster and V. Vicol, Nonuniqueness of weak solutions to the Navier-Stokes equation, *Annals of Mathematics* (2019).

[5] L. Caravenna and G. Crippa, Uniqueness and Lagrangianity for solutions with lack of integrability of the continuity equation, *C. R. Math. Acad. Sci. Paris*, **354(12)** (2016), 1168–1173.

[6] L. Caravenna and G. Crippa, A Directional Lipschitz Extension Lemma, with Applications to Uniqueness and Lagrangianity for the Continuity Equation, *arXiv* (2018).

[7] G. Crippa and S. Spirito, Renormalized Solutions of the 2D Euler Equations, *Comm. Math. Phys.*, **339(1)** (2015), 191–198.

[8] S. Daneri and L. Székelyhidi Jr, Non-uniqueness and h-principle for Hölder-continuous weak solutions of the Euler equations, *Arch. Rational Mech. Anal.*, **224(2)** (2017), 471–514.

[9] C. De Lellis and L. Székelyhidi, Dissipative euler flows and onsager's conjecture, *Journal of the European Mathematical Society*, **16(7)** (2014), 1467–1505.

[10] N. Depauw, Non unicité des solutions bornées pour un champ de vecteurs BV en dehors d'un hyperplan, *C. R. Math. Acad. Sci. Paris*, **337(4)** (2003), 249–252.

[11] R. J. DiPerna and P. L. Lions, On the Cauchy Problem for Boltzmann Equations: Global Existence and Weak Stability, *Ann. of Math.*, **130(2)** (1989), 321–366.

[12] R. J. DiPerna and P. L. Lions Ordinary differential equations, transport theory and Sobolev spaces, *Invent. math.*, **98(3)** (1989), 511–547.

[13] P. Isett, A proof of onsager's conjecture. *Annals of Mathematics*, **188(3)** (2018), 871–963.

[14] C. Le Bris and P. L. Lions, Existence and Uniqueness of Solutions to Fokker-Planck Type Equations with Irregular Coefficients. *Communications in Partial Differential Equations 33*, **7** (2008), 1272–1317.

[15] P. L. Lions, Mathematical topics in fluid mechanics. Vol. 1, in *Oxford Lecture Series in Mathematics and its Applications*, The Clarendon Press, Oxford University Press, New York, (1996).

[16] P. L. Lions, Mathematical topics in fluid mechanics. Vol. 2, in *Oxford Lecture Series in Mathematics and its Applications*, The Clarendon Press, Oxford University Press, New York, (1998).

[17] S. Modena and G. Sattig, Convex integration for the transport equation with full dimensional concentration. *arXiv* (2019).

[18] S. Modena and L. Székelhyidi, Non-renormalized solutions to the continuity equation. *arXiv* (2018).

[19] S. Modena, and L. Székelyhidi, Non-uniqueness for the transport equation with Sobolev vector fields. *Annals of PDE*, **4(2)** (2018), 18.

E-mail address: Stefano.Modena@math.uni-leipzig.de

EXISTENCE AND STABILITY OF NONISENTROPIC COMPRESSIBLE VORTEX SHEETS

ALESSANDRO MORANDO*

DICATAM, University of Brescia, Via Valotti 9, 25133 Brescia, Italy

PAOLA TREBESCHI

DICATAM, University of Brescia, Via Valotti 9, 25133 Brescia, Italy

TAO WANG

School of Mathematics and Statistics, Wuhan University, Wuhan 430072, China

ABSTRACT. We consider the short-time existence and nonlinear stability of vortex sheets for the nonisentropic compressible Euler equations in two spatial dimensions, based on the weakly linear stability result of Morando–Trebeschi [16]. The content of this paper summarizes the results collected in Morando–Trebeschi–Wang [18].

1. **Introduction.** We study compressible Euler equations in $\mathbb{R}^2$:

$$\begin{cases} (\partial_t + \boldsymbol{u}\cdot\nabla)p + \gamma p\nabla\cdot\boldsymbol{u} = 0, \\ \rho(\partial_t + \boldsymbol{u}\cdot\nabla)\boldsymbol{u} + \nabla p = 0, \\ (\partial_t + \boldsymbol{u}\cdot\nabla)s = 0, \end{cases} \tag{1}$$

where pressure $p = p(t,x) \in \mathbb{R}$, velocity $\boldsymbol{u} = (v(t,x), u(t,x))^{\mathsf{T}} \in \mathbb{R}^2$, and entropy $s = s(t,x) \in \mathbb{R}$ are unknown functions of time t and position $x = (x_1, x_2)^{\mathsf{T}} \in \mathbb{R}^2$. We consider a polytropic gas, where the density ρ obeys the constitutive law $\rho = \rho(p,s) := Ap^{\frac{1}{\gamma}}\mathrm{e}^{-\frac{s}{\gamma}}$, with given $A > 0$ and $\gamma > 1$ the adiabatic exponent of the gas.

According to Lax [12], a weak solution $(p, \boldsymbol{u}, s)$ of (1) that is smooth on either side of a smooth surface $\Gamma(t) := \{x_2 = \varphi(t, x_1)\}$ is said to be a *vortex sheet* (even called *contact discontinuity*) provided that it is a classical solution to (1) on each side of $\Gamma(t)$ and the following Rankine–Hugoniot conditions hold at each point of $\Gamma(t)$:

$$\partial_t\varphi = \boldsymbol{u}^+\cdot\nu = \boldsymbol{u}^-\cdot\nu, \qquad p^+ = p^-. \tag{2}$$

Here $\nu := (-\partial_{x_1}\varphi, 1)^{\mathsf{T}}$ is a spatial normal vector to $\Gamma(t)$ and $\boldsymbol{u}^\pm$, $p^\pm$, $s^\pm$ denote the restrictions of $\boldsymbol{u}$, p, s to both sides $\{\pm(x_2 - \varphi(t,x_1)) > 0\}$ of $\Gamma(t)$, respectively. These conditions yield that the normal velocity and pressure are continuous across

---

2000 *Mathematics Subject Classification.* Primary: 35L65; Secondary: 76N10, 35Q35, 35R35, 76E17.

Key words and phrases. Nonisentropic fluid; Compressible vortex sheet; Characteristic boundary; Nonlinear stability; Nash–Moser iteration.

The first and second authors were supported in part by the grant from Ministero dell'Istruzione, dell'Università e della Ricerca under contract PRIN2015YCJY3A-004. The third author was supported in part by the grants from National Natural Science Foundation of China under contracts 11601398 and 11731008.

* Corresponding author: Alessandro Morando.

$\Gamma(t)$. Hence the possible jumps displayed by a vortex sheet concern the tangential velocity and entropy. Remark also that the first two identities in (2) are the eikonal equations $\partial_t\varphi+\lambda_2(p^\pm,\mathbf{u}^\pm,s^\pm,\partial_{x_1}\varphi)=0$, where $\lambda_2(p,\boldsymbol{u},s,\xi):=\boldsymbol{u}\cdot(\xi,-1)^\mathsf{T}$ denotes the second characteristic field of system (1).

We are interested in the *structural stability* of vortex sheets to nonisentropic compressible Euler equations (1) with the initial data being a perturbation of *planar vortex sheets*:

$$(\bar{p},\pm\bar{v},0,\bar{s}^\pm)^\mathsf{T} \quad \text{in} \quad \pm x_2>0, \tag{3}$$

where $\bar{p}>0$, $\bar{v}>0$, $\bar{s}^\pm$ are constants.

The interface $\Gamma(t)$ (namely, function φ) is a part of unknowns of nonlinear problem (1)–(2). The usual approach consists of straightening unknown interface $\Gamma(t)$ by a suitable change of coordinates in $\mathbb{R}^3$, in order to reformulate the free boundary problem in a fixed domain. Precisely, unknowns $(p,\boldsymbol{u},s)$ are replaced by functions

$$(p_\sharp^\pm,\boldsymbol{u}_\sharp^\pm,s_\sharp^\pm)(t,x_1,x_2):=(p^\pm,\boldsymbol{u}^\pm,s^\pm)(t,x_1,\Phi^\pm(t,x_1,x_2)),$$

where $\Phi^\pm$ are smooth functions satisfying

$$\Phi^\pm(t,x_1,0)=\varphi(t,x_1) \quad \text{and} \quad \pm\,\partial_{x_2}\Phi^\pm(t,x_1,x_2)\geq\kappa>0 \quad \text{if } x_2\geq 0. \tag{4}$$

Hereafter we drop the $\sharp$ index and set $U:=(p,v,u,s)^\mathsf{T}$ for convenience. Then the construction of vortex sheets for system (1) amounts to proving the existence of smooth solutions $(U^\pm,\Phi^\pm)$ to the following initial-boundary value problem:

$$\mathbb{L}(U^\pm,\Phi^\pm):=L(U^\pm,\Phi^\pm)U^\pm=0 \qquad \text{if } x_2>0, \tag{5a}$$

$$\mathbb{B}(U^+,U^-,\varphi)=0 \qquad \text{if } x_2=0, \tag{5b}$$

$$(U^\pm,\varphi)|_{t=0}=(U_0^\pm,\varphi_0), \tag{5c}$$

where the differential operator $L(U,\Phi)$ takes the form:

$$L(U,\Phi):=I_4\partial_t+A_1(U)\partial_{x_1}+\widetilde{A}_2(U,\Phi)\partial_{x_2}, \tag{6}$$

symbol I_4 is the 4×4 identity matrix,

$$\widetilde{A}_2(U,\Phi):=\frac{1}{\partial_{x_2}\Phi}\left(A_2(U)-\partial_t\Phi I_4-\partial_{x_1}\Phi A_1(U)\right),$$

$$A_1(U):=\begin{pmatrix} v & \gamma p & 0 & 0\\ 1/\rho & v & 0 & 0\\ 0 & 0 & v & 0\\ 0 & 0 & 0 & v\end{pmatrix}, \quad A_2(U):=\begin{pmatrix} u & 0 & \gamma p & 0\\ 0 & u & 0 & 0\\ 1/\rho & 0 & u & 0\\ 0 & 0 & 0 & u\end{pmatrix},$$

and $\mathbb{B}$ denotes the boundary operator

$$\mathbb{B}(U^+,U^-,\varphi):=\begin{pmatrix}(v^+-v^-)|_{x_2=0}\partial_{x_1}\varphi-(u^+-u^-)|_{x_2=0}\\ \partial_t\varphi+v^+|_{x_2=0}\partial_{x_1}\varphi-u^+|_{x_2=0}\\ (p^+-p^-)|_{x_2=0}\end{pmatrix}.$$

Since equations (4)–(5) are not enough to determine functions $\Phi^\pm$, we require, as in Francheteau–Métivier [9], that functions $\Phi^\pm$ satisfy the following eikonal equations:

$$\partial_t\Phi^\pm+\lambda_2(p^\pm,\boldsymbol{u}^\pm,s^\pm,\partial_{x_1}\Phi^\pm)=0 \qquad \text{if } x_2\geq 0. \tag{7}$$

This choice of $\Phi^\pm$ has the advantage to considerably simplify the expression of equations (5a). More importantly, the rank of the boundary matrix for problem (5) keeps constant on the whole domain $\{x_2\geq 0\}$, which allows the application of the Kreiss symmetrizer technique to problem (5) in the spirit of Majda–Osher [13].

In the new variables, piecewise constant state (3) corresponds to the trivial solution of (4)–(5b) and (7)

$$\overline{U}^{\pm} = (\bar{p}, \pm\bar{v}, 0, \bar{s}^{\pm})^{\mathsf{T}}, \quad \overline{\Phi}^{\pm}(t, x_1, x_2) = \pm x_2, \tag{8}$$

with $\bar{p} > 0$ and $\bar{v} > 0$. Let us denote by $\bar{c}_{\pm} = c(\bar{p}, \bar{s}^{\pm})$ the sound speeds corresponding to the constant states $\overline{U}^{\pm}$, where $c(p,s) := \sqrt{p_\rho(\rho, s)} = \sqrt{\dfrac{\gamma e^{s/\gamma}}{Ap^{\frac{1}{\gamma}-1}}}$ for the polytropic gas.

We aim to show the short-time existence of solutions to nonlinear problem (4)–(5) and (7) provided the initial data is sufficiently close to (8). Our main result is stated as follows.

Theorem 1.1. *Let $T > 0$ and $\mu \in \mathbb{N}$ with $\mu \geq 13$. Assume that background state (8) satisfies the stability conditions:*

$$2\bar{v} > (\bar{c}_+^{\frac{2}{3}} + \bar{c}_-^{\frac{2}{3}})^{\frac{3}{2}}, \quad 2\bar{v} \neq \sqrt{2}(\bar{c}_+ + \bar{c}_-). \tag{9}$$

Assume further that the initial data $U_0^{\pm}$ and φ_0 satisfy suitable compatibility conditions up to order μ[1], *and $(U_0^{\pm} - \overline{U}^{\pm}, \varphi_0) \in H^{\mu+1/2}(\mathbb{R}^2_+) \times H^{\mu+1}(\mathbb{R})$ has compact support. Then there exists $\delta > 0$ such that, if $\|U_0^{\pm} - \overline{U}^{\pm}\|_{H^{\mu+1/2}(\mathbb{R}^2_+)} + \|\varphi_0\|_{H^{\mu+1}(\mathbb{R})} \leq \delta$, then there exists a solution $(U^{\pm}, \Phi^{\pm}, \varphi)$ of (4)–(5) and (7) on the time interval $[0,T]$ satisfying*

$$(U^{\pm} - \overline{U}^{\pm}, \Phi^{\pm} - \overline{\Phi}^{\pm}) \in H^{\mu-7}((0,T) \times \mathbb{R}^2_+), \qquad \varphi \in H^{\mu-6}((0,T) \times \mathbb{R}).$$

Compressible vortex sheets, along with shocks and rarefaction waves, are fundamental waves that play an important role in the study of general entropy solutions to multidimensional hyperbolic systems of conservation laws. It was observed long time ago in [14] (*cf.* Coulombel–Morando [5] for using only algebraic tools) that for two-dimensional nonisentropic Euler equations (1), piecewise constant vortex sheets (8) are violently unstable unless the following stability criterion is satisfied:

$$2\bar{v} \geq (\bar{c}_+^{\frac{2}{3}} + \bar{c}_-^{\frac{2}{3}})^{\frac{3}{2}}, \tag{10}$$

while they are linearly stable under this condition. In the seminal work of Coulombel and Secchi [7], building on their linear stability results in [6], the short-time existence and nonlinear stability of compressible vortex sheets are established for the two-dimensional *isentropic* case under condition (10) (as a strict inequality) by performing a modified Nash–Moser iteration scheme. These results were recently generalized by Chen–Secchi–Wang [3] to cover the relativistic case. Let us also quote the recent works by Huang–Wang–Yuan [11] and Ruan–Trakhinin [20] for similar results in the case of two-phase compressible flows.

As for three-dimensional gas dynamics, vortex sheets have been showed in Fejer–Miles [8] to be always violently unstable, which is analogous to the Kelvin–Helmholtz instability for incompressible fluids. In contrast, Chen–Wang [2] and Trakhinin [22] proved independently the nonlinear stability of compressible *current-vortex sheets* for three-dimensional compressible magnetohydrodynamics (MHD). This result indicates that non-paralleled magnetic fields stabilize the motion of three-dimensional compressible vortex sheets.

---

[1] For the precise definition of compatibility conditions of the initial data, see [18, Definition 4.1].

Extending the results in [6], the first two authors obtained in [16] the L^2–estimates for the linearized problems of (4)–(5) and (7) around background state (8) under condition (10) (as a strict inequality), and that around a small perturbation of (8) under (9). In the present paper we summarize the result obtained in [18] about structural nonlinear stability of two-dimensional nonisentropic vortex sheets, obtained by adopting the Nash–Moser iteration scheme developed in [10, 7] and already successfully applied to the plasma-vacuum interface problem [21], three-dimensional compressible steady flows [23] and MHD contact discontinuities [15].

It is worth noting that in the statement of Theorem 1.1, the inequality $2\bar{v} \neq \sqrt{2}(\bar{c}_+ + \bar{c}_-)$ is required in addition to stability condition (10) (with strict inequality). This is due to the fact that the linearized problem about piecewise constant basic state (8), with $\bar{v}$ taking the critical value above, satisfies an a priori estimate with additional loss of regularity from the data, which is related to the presence of a double root of the associated *Lopatinskiĭ determinant* (see [16, Theorem 3.1]). At the subsequent level of variable coefficient linearized problem about a perturbation of (8), the authors in [16] were not able to handle this further loss of regularity, thus the case of $\bar{v} = (\bar{c}_+ + \bar{c}_-)/\sqrt{2}$ is still open. Notice also that in the isentropic case (where $\bar{c}_+ = \bar{c}_- = \bar{c}$), value $(\bar{c}_+^{\frac{2}{3}} + \bar{c}_-^{\frac{2}{3}})^{\frac{3}{2}}$ coincides with $\sqrt{2}(\bar{c}_+ + \bar{c}_-)$ and condition (9) reduces to the supersonic condition $\bar{v} > \sqrt{2}\bar{c}$ studied in Coulombel–Secchi [7].

The plan of this paper is as follows. In Section 2, we introduce the effective linear problem and state the result of well-posedness, in usual Sobolev space H^s with s large enough, obtained for it. Section 3 is devoted to a short discussion of the modified Nash–Moser iteration scheme used to prove Theorem 1.1, based on the *a priori* tame estimates satisfied by the solution to the linearized problem.

2. **Well-Posedness of the Effective Linear Problem.** A fundamental step to get the solvability of the nonlinear problem (4)–(5) and (7) is the study of the well-posedness of the corresponding linearized problem. We linearize (4)–(5) and (7) around a basic state $(U_{r,l}, \Phi_{r,l}) := (p_{r,l}, v_{r,l}, u_{r,l}, s_{r,l}, \Phi_{r,l})^\mathsf{T}$ given by a perturbation of the stationary solution (8). The index r (resp. l) denotes the state on the right (resp. on the left) of the interface (after change of variables). More precisely, the perturbation

$$(\dot{U}_{r,l}(t,x_1,x_2), \dot{\Phi}_{r,l}(t,x_1,x_2)) := (U_{r,l}(t,x_1,x_2), \Phi_{r,l}(t,x_1,x_2)) - (\bar{U}^\pm, \bar{\Phi}^\pm)$$

is assumed to satisfy

$$\operatorname{supp}(\dot{U}_{r,l}, \dot{\Phi}_{r,l}) \subset \{-T \le t \le 2T,\ x_2 \ge 0,\ |x| \le R\}, \tag{11}$$

$$\dot{U}_{r,l} \in W^{2,\infty}(\Omega), \quad \dot{\Phi}_{r,l} \in W^{3,\infty}(\Omega), \quad \|\dot{U}_{r,l}\|_{W^{2,\infty}(\Omega)} + \|\dot{\Phi}_{r,l}\|_{W^{3,\infty}(\Omega)} \le K, \tag{12}$$

where T, R, and K are positive constants and Ω denotes the half-space $\{(t,x_1,x_2) \in \mathbb{R}^3 : x_2 > 0\}$. Moreover, we assume that $(\dot{U}_{r,l}, \dot{\Phi}_{r,l})$ satisfies constraints (4), (7), and Rankine–Hugoniot conditions (5b), that is,

$$\partial_t \Phi_{r,l} + v_{r,l}\partial_{x_1}\Phi_{r,l} - u_{r,l} = 0 \qquad \text{if } x_2 \ge 0, \tag{13a}$$

$$\pm\, \partial_{x_2}\Phi_{r,l} \ge \kappa_0 > 0 \qquad \text{if } x_2 \ge 0, \tag{13b}$$

$$\Phi_r = \Phi_l = \varphi \qquad \text{if } x_2 = 0, \tag{13c}$$

$$\mathbb{B}(U_r, U_l, \varphi) = 0 \qquad \text{if } x_2 = 0, \tag{13d}$$

for a suitable positive constant κ_0.

Let us consider solutions to (4)–(5) and (7) of the form $(U_{r,l}+\varepsilon V^{\pm}, \Phi_{r,l}+\varepsilon\Psi^{\pm})$, where $(V^{\pm}, \Psi^{\pm})$ represent some "small perturbations" of the basic state $(U_{r,l}, \Phi_{r,l})$. Up to second order errors and after the passage to the "good unknowns" of Alinhac (*cf.* [1])

$$\dot{V}^{+} := V^{+} - \frac{\Psi^{+}}{\partial_{x_2}\Phi_r}\partial_{x_2}U_r, \quad \dot{V}^{-} := V^{-} - \frac{\Psi^{-}}{\partial_{x_2}\Phi_l}\partial_{x_2}U_l \tag{14}$$

(made in order to get rid of first order terms in $\Psi^{\pm}$ originating from linearization), the *effective linearized problem* of (4)–(5) and (7) around the ground state $(U_{r,l}, \Phi_{r,l})$ reads as

$$\mathbb{L}'_e(U_{r,l}, \Phi_{r,l})\dot{V}^{\pm} := L(U_{r,l}, \Phi_{r,l})\dot{V}^{\pm} + \mathcal{C}(U_{r,l}, \Phi_{r,l})\dot{V}^{\pm} = f^{\pm} \quad \text{if } x_2 > 0, \tag{15a}$$

$$\mathbb{B}'_e(U_{r,l}, \Phi_{r,l})(\dot{V}, \psi) := \underline{b}\nabla_{t,x_1}\psi + \boldsymbol{b}_{\sharp}\psi + \underline{M}\dot{V}|_{x_2=0} = g \quad \text{if } x_2 = 0, \tag{15b}$$

$$\Psi^{+} = \Psi^{-} = \psi \quad \text{if } x_2 = 0. \tag{15c}$$

In view of the results obtained in [1, 9, 7], zero-th order terms in $\Psi^{\pm}$ are neglected in (15a) and considered as error terms at each Nash–Moser iteration step in the nonlinear analysis. Here we have set $\dot{V} := (\dot{V}^{+}, \dot{V}^{-})^{\mathsf{T}}$, $\nabla_{t,x_1}\psi = (\partial_t\psi, \partial_{x_1}\psi)^{\mathsf{T}}$. Moreover, differential operators $L(U_{r,l}, \Phi_{r,l})$ are defined in (6), while $\mathcal{C}(U_{r,l}, \Phi_{r,l})$ are suitable lower order operators, whose explicit form can be easily computed but is useless for the sequel of our discussion. Coefficients $\underline{b}$, $\boldsymbol{b}_{\sharp}$, and $\underline{M}$ are defined by

$$\underline{b}(t,x_1) := \begin{pmatrix} 0 & (v_r - v_l)|_{x_2=0} \\ 1 & v_r|_{x_2=0} \\ 0 & 0 \end{pmatrix}, \quad \boldsymbol{b}_{\sharp}(t,x_1) := \underline{M}(t,x_1)\left.\begin{pmatrix} \dfrac{\partial_{x_2}U_r}{\partial_{x_2}\Phi_r} \\ \dfrac{\partial_{x_2}U_l}{\partial_{x_2}\Phi_l} \end{pmatrix}\right|_{x_2=0},$$

$$\underline{M}(t,x_1) := \begin{pmatrix} 0 & \partial_{x_1}\varphi & -1 & 0 & 0 & -\partial_{x_1}\varphi & 1 & 0 \\ 0 & \partial_{x_1}\varphi & -1 & 0 & 0 & 0 & 0 & 0 \\ 1 & 0 & 0 & 0 & -1 & 0 & 0 & 0 \end{pmatrix}.$$

From (12), $\underline{b}, \underline{M} \in W^{2,\infty}(\mathbb{R}^2)$, $\boldsymbol{b}_{\sharp} \in W^{1,\infty}(\mathbb{R}^2)$, $\mathcal{C}(U_{r,l}, \Phi_{r,l}) \in W^{1,\infty}(\Omega)$, and the coefficients of the operators $L(U_{r,l}, \Phi_{r,l})$ are in $W^{2,\infty}(\Omega)$.

We observe that linearized boundary conditions (15b) depend on the traces of $\dot{V}^{\pm}$ only through the *noncharacteristic components* $\mathbb{P}(\varphi)\dot{V}^{\pm} := (\dot{V}_1^{\pm}, \dot{V}_3^{\pm} - \partial_{x_1}\varphi\,\dot{V}_2^{\pm})^{\mathsf{T}}$ of $\dot{V}^{\pm}$, as it is expected, because the boundary $\{x_2 = 0\}$ is characteristic for problem (15) in view of (13a).

On the effective linear problem (15), we are able to show the following well-posedness result in the usual Sobolev space H^s with order s large enough (see [18]).

Theorem 2.1. *Let $T > 0$ and $s \in [3, \tilde{\alpha}] \cap \mathbb{N}$ with any integer $\tilde{\alpha} \geq 3$. Assume that the stationary solution (8) satisfies (9), and that perturbations $(\dot{U}_{r,l}, \dot{\Phi}_{r,l})$ belong to $H_\gamma^{s+3}(\Omega_T)$ for all $\gamma \geq 1$ and satisfy (11)–(13), and*

$$\|(\dot{U}_{r,l}, \nabla\dot{\Phi}_{r,l})\|_{H_\gamma^5(\Omega_T)} + \|(\dot{U}_{r,l}, \partial_{x_2}\dot{U}_{r,l}, \nabla\dot{\Phi}_{r,l})|_{x_2=0}\|_{H_\gamma^4(\omega_T)} \leq K.$$

Assume further that $(f^{\pm}, g) \in H^{s+1}(\Omega_T) \times H^{s+1}(\omega_T)$ vanish in the past. Then there exists a positive constant K_0, which is independent of s and T, and there exist two constants $C > 0$ and $\gamma \geq 1$, which depend solely on K_0, such that, if $K \leq K_0$, then problem (15) admits a unique solution $(\dot{V}^{\pm}, \psi) \in H^s(\Omega_T) \times H^{s+1}(\omega_T)$ that vanishes *in the past and obeys the following tame estimate:*

$$\begin{aligned}\|\dot V\|_{H^s_\gamma(\Omega_T)} &+ \|\mathbb{P}(\varphi)\dot V|_{x_2=0}\|_{H^s_\gamma(\omega_T)} + \|\psi\|_{H^{s+1}_\gamma(\omega_T)}\\ &\le C\big\{\|f\|_{H^{s+1}_\gamma(\Omega_T)} + \|g\|_{H^{s+1}_\gamma(\omega_T)}\\ &\qquad + \big(\|f\|_{H^4_\gamma(\Omega_T)} + \|g\|_{H^4_\gamma(\omega_T)}\big)\|(\dot U_{r,l},\dot\Phi_{r,l})\|_{H^{s+3}_\gamma(\Omega_T)}\big\}, \qquad (16)\end{aligned}$$

where $\dot V := (\dot V^+,\dot V^-)$, $\mathbb{P}(\varphi)\dot V := (\mathbb{P}(\varphi)\dot V^+,\mathbb{P}(\varphi)\dot V^-)$, $f := (f^+,f^-)$.

In the above statement, we have set $\Omega_T := (-\infty,T)\times\mathbb{R}^2_+$, $\omega_T := (-\infty,T)\times\mathbb{R}\simeq\partial\Omega_T$ for any real number T. Moreover, the functional spaces (and related norms) involved above are an "exponentially weighted" version of usual Sobolev spaces on Ω_T and ω_T, defined for all $k\in\mathbb{N}$ and $\gamma\ge 1$ as

$$H^k_\gamma(\Omega_T) := \left\{u\in\mathcal{D}'(\Omega_T)\,:\,\mathrm{e}^{-\gamma t}u\in H^k(\Omega_T)\right\},$$

provided with the natural norm $\|u\|_{H^k_\gamma(\Omega_T)} := \|e^{-\gamma t}u\|_{H^k(\Omega_T)}$ (and similarly for $H^k_\gamma(\omega_T)$). Since the most of functions we are dealing with have double $\pm$ states, in (16) we have used the shortcut notation $\|\dot V\|_{H^s_\gamma(\Omega_T)} := \sum_\pm\|\dot V^\pm\|_{H^s_\gamma(\Omega_T)}$ and similarly for the other terms in the estimate.

Let us shortly discuss the main steps of the proof of Theorem 2.1. The first two authors proved in [16, Theorem 4.1], by spectral analysis based on Kreiss symmetrizer techniques and paradifferential calculus, that problem (15) satisfies a basic L^2–*a priori* estimate with a loss of one tangential derivative. Then, in [18] we defined a dual problem for (15), to which we were able to associate the same kind of L^2–*a priori* estimate with a loss of one tangential derivative. Since system (15a) is symmetrizable hyperbolic and in view of the regularity of coefficients coming from (12), the well-posedness result in L^2 of [4] can be applied to the effective linear problem (15), giving the existence of a unique L^2–solution of (15). In order to get well-posedness in higher order Sobolev spaces, as it is required by Theorem 2.1, the essential point is deriving the *a priori tame* estimate (16) for all sufficiently smooth solutions to (15). We first obtain the estimate for tangential derivatives. Since the boundary matrix for our problem (15) is singular, there is no hope to estimate all the normal derivatives of $\dot V$ directly from equations (15a) by applying the standard approach for noncharacteristic boundary problems as in [19, 17]. However, for our problem (15), we can obtain the estimate of missing normal derivatives through the equations of the "linearized vorticity" and entropy, where the linearized vorticity has been introduced in [7]. Then, we estimate such normal derivatives by expressing them in terms of tangential derivatives and the linearized vorticity.

Let us notice, in the end, that, according to the loss of regularity from the data in the basic L^2–*a priori* estimate found in [16], the tame estimate (16) displays a loss of one derivative from data to the found solution. Moreover there is also a fixed loss of three derivatives from the coefficients of the system, namely the basic state $(\dot U_{r,l},\dot\Phi_{r,l})$.

3. **The Nolinear Problem: Nash–Moser Iteration Scheme.** In this section we turn to the resolution of the original nonlinear problem (4)–(5) and (7). Let us only sketch the idea of the proof of the main Theorem 1.1, referring to [18] for the details.

In order to reduce the original problem (4)–(5) and (7) into a nonlinear one with zero initial data, it is first convenient to seek the solution of (4)–(5) and (7) into the form

$$(U^{a\,\pm}, \Phi^{a\,\pm}, \varphi^a) + (V^\pm, \Psi^\pm, \psi)\,,$$

where $(U^{a\,\pm}, \Phi^{a\,\pm}, \varphi^a)$ (with $\Phi^{a\,\pm}|_{x_2=0} = \varphi^a$) is the so-called *approximate solution*, that is a solution of above problem in the sense of Taylor's series at time $t = 0$. Suitable necessary compatibility conditions of sufficiently large order have to be prescribed on the initial data $(U_0^\pm, \varphi_0)$ for the existence of such a sufficiently smooth approximate solution, see [18, Section 4].

Because of the loss of regularity from data and coefficients to the solution of the linearized problem, occurring in Theorem 2.1, we cannot hope to solve the nonlinear problem by resorting to an iteration scheme based on classical contraction principle. Instead, the Nash–Moser scheme turns out to be adapted to our situation, because it allows to handle the above loss of regularity.

As already announced in the end of Section 1, the solution $(V^\pm, \Psi^\pm, \psi)$ of the nonlinear problem with zero initial data is found as the limit of a sequence of solutions $(V_k^\pm, \Psi_k^\pm, \psi_k)$ coming from the resolution of "approximating" linearized problems, constructed by performing an iteration scheme based on a Nash–Moser type argument. At the $(k+1)-$th iteration of the scheme, the updated approximation $(V_{k+1}^\pm, \Psi_{k+1}^\pm, \psi_{k+1})$ is constructed from the approximation at previous step k as

$$V_{k+1}^\pm = V_k^\pm + \delta V_k^\pm, \quad \Psi_{k+1}^\pm = \Psi_k^\pm + \delta\Psi_k^\pm, \quad \psi_{k+1} = \psi_k + \delta\psi_k,$$

where the differences δV_k, $\delta\Psi_k$, and $\delta\psi_k$ are obtained from the resolution of the effective linear problem of kind (15)

$$\begin{cases} \mathbb{L}_e'(U^a + V_{k+1/2}, \Phi^a + \Psi_{k+1/2})\delta\dot{V}_k = f_k & \text{in } \Omega_T, \\ \mathbb{B}_e'(U^a + V_{k+1/2}, \Phi^a + \Psi_{k+1/2})(\delta\dot{V}_k, \delta\psi_k) = g_k & \text{on } \omega_T, \\ (\delta\dot{V}_k, \delta\psi_k) = 0 & \text{for } t < 0, \end{cases} \tag{17}$$

where, for simplicity, we have removed the $\pm$ superscripts,

$$\delta\dot{V}_k := \delta V_k - \frac{\partial_{x_2}(U^a + V_{k+1/2})}{\partial_{x_2}(\Phi^a + \Psi_{k+1/2})}\delta\Psi_k$$

is the "good unknown" (*cf.* (14)), and $(V_{k+1/2}, \Psi_{k+1/2})$ is a suitable "modification" of the approximating state at $k-$th step $(V_k^\pm, \Psi_k^\pm)$, costructed in such a way to compensate the loss of regularity from the coefficients and the data to the solution of the linearized problem and such that the basic state $(U^a + V_{k+1/2}, \Phi^a + \Psi_{k+1/2})$ involved in (17) satisfies all the assumptions needed in order to solve the linearized problem according to Theorem 2.1, that is constraints (11)–(13). The source terms (f_k, g_k) are defined through the accumulated errors at step k. In order to get convergence of the Nash–Moser scheme, so as to obtain $(V^\pm, \Psi^\pm, \psi)$ passing to the limit in the sequence $(V_k^\pm, \Psi_k^\pm, \psi_k)$, such accumulated errors have to converge to zero in the right functional space, which is proved in [18, Section 5.3].

REFERENCES

[1] S. Alinhac, Existence d'ondes de raréfaction pour des systèmes quasi-linéaires hyperboliques multidimensionnels, *Comm. Partial Diff. Eqs.*, **14** (1989), 173–230.

[2] G.-Q. Chen and Y.-G. Wang, Existence and stability of compressible current-vortex sheets in three-dimensional magnetohydrodynamics, *Arch. Ration. Mech. Anal.*, **187** (2008), 369–408.

[3] G.-Q. Chen, P. Secchi and T. Wang, Nonlinear Stability of Relativistic Vortex Sheets in Three-Dimensional Minkowski Spacetime, *Arch. Ration. Mech. Anal.*, **232** (2019), 591–695.

[4] J.-F. Coulombel, Well-posedness of hyperbolic initial boundary value problems, *J. Math. Pures Appl. (9)*, **84** (2005), 786–818.
[5] J.-F. Coulombel and A. Morando, Stability of contact discontinuities for the nonisentropic Euler equations, *Ann. Univ. Ferrara Sez. VII (N.S.)*, **50** (2004), 79–90.
[6] J.-F. Coulombel and P. Secchi, The stability of compressible vortex sheets in two space dimensions, *Indiana Univ. Math. J.*, **53** (2004), 941–1012.
[7] J.-F. Coulombel and P. Secchi, Nonlinear compressible vortex sheets in two space dimensions, *Ann. Sci. Éc. Norm. Supér. (4)*, **41** (2008), 85–139.
[8] J. A. Fejer and J. W. Miles, On the stability of a plane vortex sheet with respect to three-dimensional disturbances, *J. Fluid Mech.*, **15** (1963), 335–336.
[9] J. Francheteau and G. Métivier, Existence de chocs faibles pour des systèmes quasi-linéaires hyperboliques multidimensionnels, *Astérisque*, **268** (2000), viii+198.
[10] L. Hörmander, The boundary problems of physical geodesy, *Arch. Ration. Mech. Anal.*, **62** (1976), 1–52.
[11] F. Huang, D. Wang and D. Yuan, Nonlinear stability and existence of vortex sheets for inviscid liquid-gas two-phase flow, *Discrete Contin. Dyn. Syst.*, **39** (2019), 3535–3575.
[12] P. D. Lax, Hyperbolic systems of conservation laws. II, *Comm. Pure Appl. Math.*, **10** (1957), 537–566.
[13] A. Majda and S. Osher, Initial-boundary value problems for hyperbolic equations with uniformly characteristic boundary, *Comm. Pure Appl. Math.*, **28** (1975), 607–675.
[14] J. W. Miles, On the disturbed motion of a plane vortex sheet, *J. Fluid Mech.*, **4** (1958), 538–552.
[15] A. Morando, Y. Trakhinin and P. Trebeschi, Local existence of MHD contact discontinuities, *Arch. Ration. Mech. Anal.*, **228** (2018), 691–742.
[16] A. Morando and P. Trebeschi, Two-dimensional vortex sheets for the nonisentropic Euler equations: linear stability, *J. Hyper. Diff. Eqs.*, **5** (2008), 487–518.
[17] A. Morando and P. Trebeschi, Weakly well posed hyperbolic initial-boundary value problems with non characteristic boundary, *Methods Appl. Anal.*, **20** (2013), 1–31.
[18] A. Morando, P. Trebeschi and T. Wang, Two-dimensional vortex sheets for the nonisentropic Euler equations: Nonlinear stability, *J. Differential Equations*, **266** (2019), 5397–5430.
[19] J. B. Rauch and F. J. Massey III, Differentiability of solutions to hyperbolic initial-boundary value problems, *Trans. Amer. Math. Soc.*, **189** (1974), 303–318.
[20] L. Ruan, Y. Trakhinin, Elementary symmetrization of inviscid two-fluid flow equations giving a number of instant results, *Phys. D*, **391** (2019), 66–71.
[21] P. Secchi and Y. Trakhinin, Well-posedness of the plasma-vacuum interface problem, *Nonlinearity*, **27** (2014), 105–169.
[22] Y. Trakhinin, The existence of current-vortex sheets in ideal compressible magnetohydrodynamics, *Arch. Ration. Mech. Anal.*, **191** (2009), 245–310.
[23] Y.-G. Wang and F. Yu, Structural stability of supersonic contact discontinuities in three-dimensional compressible steady flows, *SIAM J. Math. Anal.*, **47** (2015), 1291–1329.

E-mail address: `alessandro.morando@unibs.it`
E-mail address: `paola.trebeschi@unibs.it`
E-mail address: `tao.wang@whu.edu.cn`

SPHERICALLY SYMMETRIC SHADOW WAVE SOLUTIONS TO THE COMPRESSIBLE EULER SYSTEM AT THE ORIGIN

MARKO NEDELJKOV*

Department of of Mathematics and Informatics, University of Novi Sad
Trg Dositeja Obradovića 4
21000 Novi Sad, Serbia

LUKAS NEUMANN AND MICHAEL OBERGUGGENBERGER

Unit of Engineering Mathematics, University of Innsbruck
Technikerstraße 13
6020 Innsbruck, Austria

ABSTRACT. The paper contains sufficient conditions for the existence of a delta wave at the origin for spherically symmetric flows for the compressible Euler system. The delta wave at the origin is constructed by means of the concept of shadow waves. Its existence would entail the possibility that an incoming wave produces a mass concentration at the origin. Conditions on an incoming wave to result in reflection or accumulation at the origin are still unknown in the case of the Euler system.

1. **Introduction.** The behavior of spherically symmetric solutions to the Euler system of compressible gas dynamics

$$\begin{aligned} \rho_t + \nabla\cdot(\rho\vec{u}) &= 0 \\ (\rho\vec{u})_t + \nabla\cdot(\rho\vec{u}\otimes\vec{u}) + \nabla p(\rho,e) &= 0 \\ \left(\rho e + \frac{1}{2}\rho|\vec{u}|^2\right)_t + \nabla\cdot\left(\left(\rho e + \frac{1}{2}\rho|\vec{u}|^2\right)\vec{u} + p(\rho,e)\,\vec{u}\right) &= 0 \end{aligned} \tag{1}$$

near the origin has been addressed by various authors. For example, [1] has given conditions for the existence of globally bounded (weak) solutions, while [4, 5] have constructed possibly unbounded self-similar solutions. For further references we recommend the quoted papers. As pointed out in [2], it appears to be an open question whether an incoming shock, starting at some $|\vec{x}| = R$ is always reflected at the origin $|\vec{x}| = 0$ or whether it can be (partially) absorbed in form of an additional delta wave at the origin due to mass concentration.

In this article, we are exploring this possibility using the concept of "shadow waves" [6] to approximate a possible delta shock. We are going to derive general conditions for such a shadow wave (SDW) to exist at the origin. We derive admissibility conditions to be met by such a shadow wave in order to satisfy the Clausius-Duhem entropy inequality weakly. The goal of the paper is to collect a

2000 *Mathematics Subject Classification.* Primary: 35L67, 76N10; Secondary: 35F46.

Key words and phrases. compressible Euler system, spherically symmetric flows, shadow waves, entropy conditions, mass concentration.

The first author is supported by grants OI174024, III44006 and APV 142-451-3652.

* Corresponding author.

number of necessary and sufficient conditions for the existence of admissible SDWs. Whether such SDWs can be employed to construct global solutions, given any or certain incident waves, remains a topic of further research. In an earlier paper [7], the authors have completely solved the pseudo-Riemann problem for radially symmetric solutions to the system of pressureless gas dynamics. The existence of solutions with an accumulating Dirac measure part at the origin has been exhibited there.

In polar coordinates, $r = |\vec{x}|$, spherically symmetric solutions are of the form

$$\rho(\vec{x},t) = \rho(r,t), \quad e(\vec{x},t) = e(r,t), \quad \vec{u}(\vec{x},t) = u(r,t)\frac{\vec{x}}{|\vec{x}|}.$$

In these variables, the system (1) reduces to the following 1D-system of balance laws

$$\begin{aligned} \rho_t + (\rho u)_r + \frac{n-1}{r}(\rho u) &= 0 \\ (\rho u)_t + \left(\rho u^2 + p(\rho,e)\right)_r + \frac{n-1}{r}\rho u^2 &= 0 \\ \left(\frac{1}{2}\rho u^2 + \rho e\right)_t + \left(\left(\frac{1}{2}\rho u^2 + \rho e\right)u + p(\rho,e)\,u\right)_r & \\ + \frac{n-1}{r}\left(\left(\frac{1}{2}\rho u^2 + \rho e\right)u + p(\rho,e)\,u\right) &= 0. \end{aligned} \tag{2}$$

Note that the scalar velocity u may take non-negative and negative values. The density ρ and the internal energy e are always non-negative scalar functions. We shall also make use of the variables

$$\vec{m} = \rho\vec{u}, \ m = \rho u.$$

The ideal gas here is taken to be a polytropic one, i.e.

$$p = k\rho e, \quad k \in (0,2).$$

The real (physical) entropy for the compressible Euler system is given, for a polytropic gas, by

$$S = c_v \ln\left(\frac{p}{\rho^{k+1}}\right) + c_0 \quad \text{or} \quad S = c_v \ln\left(\frac{ke}{\rho^k}\right) + c_0.$$

The Clausius-Duhem inequality requires that

$$\partial_t(\rho S) + \nabla\cdot(\vec{m}S) \geq 0, \tag{3}$$

and any physically relevant solution should satisfy that condition.

2. **Shadow wave at the origin.**

Assume given an incoming wave that starts at $r = R > 0$ of the form

$$U(r,t) = \begin{cases} U_{s,0} = (\rho_{s,0}, u_{s,0}, e_{s,0}), \ r < c(t) + R \\ U_{s,1} = (\rho_{s,1}, u_{s,1}, e_{s,1}), \ r > c(t) + R. \end{cases} \tag{4}$$

We will look for a solution after the above wave reaches the origin, i.e., when $c(T) + R = 0$ for some T. This results in a new initial value problem, a so called "incident problem". One can restart the time so T becomes zero, and look for a SDW at the origin followed by a wave connecting the right-hand side $U_{s,1} = (\rho_{s,1}, u_{s,1}, e_{s,1})$ of the incoming wave. That new state will be denoted by $U_1 = (\rho_1, u_1, e_1)$ and it starts from the fixed boundary $r = \varepsilon t$. The boundary with $U_{s,1}$ is unknown. More precisely, we are looking for an approximate solution of the form

$$U(r,t) = \begin{cases} U_\varepsilon(t), & r < \varepsilon t \\ U_1(r,t), & \varepsilon t < r < C_1(t) \\ U_{s,1}(r,t), & r > C_1(t). \end{cases} \tag{5}$$

The SDW part is captured in U_ε. The existence or non-existence of such a solution will give an answer to the question what mass stays at the origin. If it accumulates, the SDW will have a positive strength and if not, the SDW will have negligible strength, i.e., its strength will converge to zero.

We will take $n = 3$ in the following calculations, but all other dimensions can be done in the same way. The value of U_ε does not depend on the space variable $\vec{x}$ because the wave represented by (5) moves only infinitesimally around the origin. Note that we have to use the original 3D+1 $(\vec{x}, t)$-variables in the subsequent derivations instead of the simplified 1D+1 radial variables (r, t) because we are looking for a solution near the origin and the spherical coordinates are singular there. The obtained SDW and its relations with $U_1(\varepsilon t, t)$ will be taken as boundary conditions at $r = 0$.

Denote by $S : |\vec{x}| - \varepsilon t = 0$ the front of the SDW defined in (5). Its unit normal $\nu = (\vec{\nu}_x, \nu_t)$ is then determined by

$$\vec{\nu}_x = \frac{\vec{x}}{|\vec{x}|\sqrt{1+\varepsilon^2}}, \qquad \nu_t = -\frac{\varepsilon}{\sqrt{1+\varepsilon^2}}.$$

2.1. **The first equation.** As noted above, the following calculations are done in a region near the origin, so we have to use Cartesian coordinates. Spherically symmetry of the vector-valued quantities is captured in our notation by setting

$$\vec{m}(\vec{x}, t) = m(r,t)\,\vec{\omega}, \qquad \vec{\omega} = \frac{\vec{x}}{|\vec{x}|}$$

and similarly for $\vec{m}_1(\vec{x}, t)$ and $\vec{m}_\varepsilon(t)$. The volume and surface elements will be denoted by dV and dS, respectively.

Substitution of (5) into the first equation of (1) gives

$$\begin{aligned}
&-\int_0^\infty \int_{\mathbb{R}^3} \rho\partial_t\varphi + \vec{m}\cdot\nabla\varphi\, dV\, dt \\
&= -\int_0^\infty \int_{|\vec{x}|<\varepsilon t} \rho_\varepsilon\partial_t\varphi + \vec{m}_\varepsilon\cdot\nabla\varphi\, dV\, dt - \int_0^\infty \int_{|\vec{x}|>\varepsilon t} \rho_1\partial_t\varphi + \vec{m}_1\cdot\nabla\varphi\, dV\, dt \\
&= \underbrace{\int_0^\infty \int_{|\vec{x}|<\varepsilon t} \partial_t\rho_\varepsilon\varphi\, dV\, dt}_{=:I_1} + \underbrace{\int_0^\infty \int_{|\vec{x}|=\varepsilon t} (\rho_1 - \rho_\varepsilon)\nu_t\varphi\, dS\, dt}_{=:I_2} \\
&\quad + \underbrace{\int_0^\infty \int_{|\vec{x}|=\varepsilon t} (\vec{m}_1 - \vec{m}_\varepsilon)\cdot\vec{\nu}_x\varphi\, dS\, dt}_{=:I_3} \\
&\quad + \underbrace{\int_0^\infty \int_{|\vec{x}|>\varepsilon t} \partial_t\rho_1\varphi\, dV\, dt + \int_0^\infty \int_{|\vec{x}|>\varepsilon t} \nabla\cdot\vec{m}_1\varphi\, dV\, dt}_{=0} = 0,
\end{aligned} \tag{6}$$

where we have used that $\rho = \rho_1$ and $\vec{m} = \vec{m}_1$ are solutions of the first equation for $|\vec{x}| > \varepsilon t$ in the last term above and that $\vec{m}_\varepsilon$ does not depend on $\vec{x}$. Let us now estimate the above integrals. The volume of the ball $B_r(0)$ equals $4\pi r^3/3$ and its surface area is $4\pi r^2$.

In order to be able to model a finite, nonzero mass inside the shadow wave, I_1 should neither be zero nor divergent as $\varepsilon \to 0$. This means that $\partial_t \rho_\varepsilon$ should be of the order $\mathcal{O}(\varepsilon^{-3})$. To guarantee this, we make the following assumption.

Assumption (A1): $\rho_\varepsilon(t) = \xi(t)\varepsilon^{-3}$ for some smooth function $\xi(t)$.

Under assmption (A1),

$$\begin{aligned} I_1 =& \int_0^\infty \int_{|\vec{x}|<\varepsilon t} \partial_t \xi(t)\varepsilon^{-3}(\varphi(0,t) + \nabla\varphi(0,t)\cdot\vec{x})\, dV dt + \mathcal{O}(\varepsilon) \\ \approx & \frac{4\pi}{3}\int_0^\infty t^3 \xi'(t)\varphi(0,t)dt. \end{aligned}$$

Here we have used that

$$\left| \int_{|\vec{x}|<\varepsilon t} \nabla\varphi(0,t)\cdot\vec{x}\, dV \right| \le \int_{|\vec{x}|<\varepsilon t} |\nabla\varphi(0,t)\cdot\vec{x}|\; dV \le \int_{|\vec{x}|<\varepsilon t} |\nabla\varphi(0,t)|\, \varepsilon t\, dV \approx \varepsilon^4.$$

Using that $\nu_t \approx -\varepsilon$,

$$\begin{aligned} I_2 \approx & -4\pi \int_0^\infty \int_{|\vec{x}|=\varepsilon t} \left(\rho_1(\varepsilon t,t) - \xi(t)\varepsilon^{-3}\right)(\varphi(0,t) + \varepsilon t \nabla\varphi(0,t)\cdot\vec{\omega})\varepsilon\, dS dt \\ \approx & -4\pi \int_0^\infty \left(\rho_1(\varepsilon t,t) - \xi(t)\varepsilon^{-3}\right)\varepsilon^3 t^2 \varphi(0,t)dt. \end{aligned}$$

Concerning I_3, we observe that the term $(\vec{m}_1 - \vec{m}_\varepsilon)$ has to be of the order at most $\mathcal{O}(\varepsilon^{-2})$. Otherwise,

$$I_3 \approx \int_0^\infty \int_{|\vec{x}|=\varepsilon t} \Big((\vec{m}_1(\varepsilon t,t) - \vec{m}_\varepsilon(t)) \cdot \vec{\nu}_x (\varphi(0,t) + \varepsilon t \nabla\varphi(0,t)\cdot\vec{\omega} \Big)\, dS dt$$

would diverge.

Assumption (A2): The difference $\Delta m_\varepsilon(t) = m_1(\varepsilon t,t) - m_\varepsilon(t) = \mathcal{O}(\varepsilon^{-2})$ as $\varepsilon \to 0$, uniformly for t in compact subsets of $(0,\infty)$.

Using that $\vec{m}_1 = m_1\vec{\omega}$, $\vec{m}_\varepsilon = m_\varepsilon\vec{\omega}$ and $\vec{\omega}\cdot\vec{\nu}_x = 1/\sqrt{1+\varepsilon^2}$, we arrive at

$$I_3 \approx 4\pi \int_0^\infty \Delta m_\varepsilon(t)\varepsilon^2 t^2 \varphi(0,t)\, dt + \mathcal{O}(\varepsilon).$$

Under the assumptions above, only the terms involving $\varphi(0,t)$ in the integrals survive as $\varepsilon \to 0$. This yields the following differential equation:

$$\frac{t}{3}\xi'(t) + \xi(t) - \lim_{\varepsilon\to 0}\left(\varepsilon^3\rho_1(\varepsilon t,t) - \varepsilon^2 \Delta m_\varepsilon(t)\right) = 0. \tag{7}$$

or

$$\frac{t}{3}\xi'(t) + \xi(t) - \lim_{\varepsilon\to 0}\left(\varepsilon^3\rho_1(\varepsilon t,t) - \varepsilon^2\rho_1(\varepsilon t,t)u_1(\varepsilon t,t) + \xi(t)\varepsilon^{-1}u_\varepsilon(t)\right) = 0. \tag{8}$$

In order to separate the waves further, we suppose a specific form for $u_\varepsilon(t)$:

Assumption (A3): $u_\varepsilon(t) = \chi(t)\varepsilon^\alpha$ for some $\alpha \geq 1$ and some smooth function $\chi(t)$.

It follows that

$$m_\varepsilon(t) = \rho_\varepsilon(t)u_\varepsilon(t) = \xi(t)\chi(t)\varepsilon^{\alpha-3}$$

where $\alpha - 3 \geq -2$. In particular, assumption (A2) holds provided

$$m_1(\varepsilon t, t) = \rho_1(\varepsilon t, t)u_1(\varepsilon t, t) = \mathcal{O}(\varepsilon^{-2}) \tag{9}$$

as $\varepsilon \to 0$. Note that conversely, assumptions (A1) and (A2) together imply (9).

Proposition 1. *Let assumptions (A1), (A2) and (A3) hold. A shadow wave of the form (5) satisfies the first equation (conservation of mass) of system (2) in the weak limit as $\varepsilon \to 0$ if and only if*

(a) in case $\alpha = 1$

$$\frac{t}{3}\xi'(t) + \xi(t)(1 - \chi(t)) = \lim_{\varepsilon\to 0}\left(\varepsilon^3\rho_1(\varepsilon t, t) + \varepsilon^2\rho_1(\varepsilon t, t)u_1(\varepsilon t, t)\right); \tag{10}$$

(b) in case $\alpha > 1$

$$\frac{t}{3}\xi'(t) + \xi(t) = \lim_{\varepsilon\to 0}\left(\varepsilon^3\rho_1(\varepsilon t, t) + \varepsilon^2\rho_1(\varepsilon t, t)u_1(\varepsilon t, t)\right). \tag{11}$$

Remark 1. *A physically reasonable assumption is that $\rho_1(r,t)$ is integrable near $r = 0$, i.e., $\rho_1(r,t) = \mathcal{O}(r^a)$ with $a > -3$. (This assumption will be explicitly made later in (A6); see also conditions given in [3].) Then the first limit on the right-hand side of (10) or (11) is zero, and the existence (and the value) of the limit depends only on the behavior of $u_1(r,t)$ near $r = 0$.*

2.2. **The second set of equations.** Let us use each component of $\vec{m} = \rho\vec{u}$ and $\rho\vec{u} \otimes \vec{u} + pI$ instead of ρ and $\vec{m} = \rho\vec{u}$ in (6), respectively, and repeat the above analysis for the first equation. Also, we will use the same integration bounds for I_1, I_2, I_3 as above.

Thus, for i–th component, $i = 1, 2, 3$, we have

$$I_1 \approx -\int_0^\infty \int_{|\vec{x}|<\varepsilon t} \partial_t m_\varepsilon(t)\omega_i(\varphi(0,t) + \nabla\varphi(0,t)\cdot\vec{x})\,dV dt.$$

By assumptions (A1) and (A3), $m_\varepsilon(t) = \xi(t)\chi(t)\varepsilon^{\alpha-3}$ with $\alpha \geq 1$. It follows that $I_1 \approx 0$, because the volume of the ball $|\vec{x}| \leq \varepsilon t$ equals a constant times ε^3.

Next,

$$I_2 \approx -\int_0^\infty \int_{|\vec{x}|=\varepsilon t} \underbrace{(m_1(\varepsilon t, t) - m_\varepsilon(t))}_{=\Delta m_\varepsilon(t)}\omega_i\nu_t(\varphi(0,t) + \varepsilon t\nabla\varphi(0,t)\cdot\vec{\omega})\,dS dt.$$

By assumption (A2), $\Delta m_\varepsilon(t) = \mathcal{O}(\varepsilon^{-2})$. Further, $\nu_t \approx \varepsilon$, and the surface of the sphere $|\vec{x}| = \varepsilon t$ equals a constant times ε^2. It follows that $I_2 \approx 0$ as well.

Finally we have

$$\begin{aligned} I_3 \approx & \int_0^\infty \int_{|\vec{x}|=\varepsilon t} ((\vec{m}_1(\varepsilon t, t)u_1(\varepsilon t, t)\omega_i - \vec{m}_\varepsilon(t)u_\varepsilon(t)\omega_i)\cdot\vec{\nu}_x + (p_1(\varepsilon t) - p_\varepsilon(t))\nu_{x,i}) \\ & (\varphi(0,t) + \varepsilon t\nabla\varphi(0,t)\cdot\vec{\omega})\,dS dt \\ \approx & \int_0^\infty \int_{|\vec{x}|=\varepsilon t} ((m_1(\varepsilon t, t)u_1(\varepsilon t, t) - m_\varepsilon(t)u_\varepsilon(t))\omega_i \\ & + (p_1(\varepsilon t, t) - p_\varepsilon(t))\omega_i)\varphi(0,t)\,dS dt, \end{aligned}$$

where we have used the notation $\cdot_1 = (\cdot_{1,1}, \cdot_{1,2}, \cdot_{1,3})$, and that $\vec{\omega} \approx \vec{\nu}$. The fact that the $\nabla\varphi$–term vanishes as $\varepsilon \to 0$ requires that the φ–term remains bounded, which we will make sure by the following assumption:

Assumption (A4): $u_1(\varepsilon t, t) = \mathcal{O}(1)$ and $\Delta p_\varepsilon(t) = p_1(\varepsilon t, t) - p_\varepsilon(t) = \mathcal{O}(\varepsilon^{-2})$ as $\varepsilon \to 0$, uniformly for t in compact subsets of $(0, \infty)$.

Indeed, by assumptions (A1) and (A3), $m_\varepsilon(t)u_\varepsilon(t) = \xi(t)\chi^2(t)\varepsilon^{2\alpha-3}$ with $\alpha \geq 1$. By assumption (A2), $m_1(\varepsilon t, t) = \mathcal{O}(\varepsilon^{-2})$, see (9). Using the surface area of the sphere $|\vec{x}| = \varepsilon t$, the term in question is seen to remain bounded, and it follows further that

$$I_3 \approx 4\pi \int_0^\infty (m_1(\varepsilon t, t)u_1(\varepsilon t, t) + k(\rho_1(\varepsilon t, t)e_1(\varepsilon t, t) - \rho_\varepsilon(t)e_\varepsilon(t)))\varepsilon^2 t^2 \varphi(0, t)\, dt$$

for each $i = 1, 2, 3$.

Proposition 2. *Let assumptions (A1), (A2), (A3) and (A4) hold. A shadow wave of the form (5) satisfies the second equation (conservation of momentum) of system (2) in the weak limit as $\varepsilon \to 0$ if and only if*

$$\lim_{\varepsilon \to 0} \varepsilon^2 (m_1(\varepsilon t, t)u_1(\varepsilon t, t) + k(\rho_1(\varepsilon t, t)e_1(\varepsilon t, t) - \rho_\varepsilon(t)e_\varepsilon(t))) = 0.$$

Again, additional assumptions will allow us to separate the waves further.

Assumption (A5): $e_\varepsilon(t) = \zeta(t)\varepsilon^\beta$ for some $\beta \geq 1$ and some smooth function $\zeta(t)$.

Corollary 1. *If in the situation of Proposition 2 assumption (A5) holds, then*

(a) *in case $\beta = 1$*

$$k\xi(t)\zeta(t) = \lim_{\varepsilon \to 0} \varepsilon^2 \left(\rho_1(\varepsilon t, t)u_1^2(\varepsilon t, t) + k\rho_1(\varepsilon t, t)e_1(\varepsilon t, t)\right); \tag{12}$$

(b) *in case $\beta > 1$*

$$\lim_{\varepsilon \to 0} \varepsilon^2 \left(\rho_1(\varepsilon t, t)u_1^2(\varepsilon t, t) + k\rho_1(\varepsilon t, t)e_1(\varepsilon t, t)\right) = 0. \tag{13}$$

2.3. **The third equation.** By assumptions (A1), (A3) and (A5),

$$I_1 \approx \int_0^\infty \int_{|\vec{x}| \leq \varepsilon t} \partial_t(\xi(t)\zeta(t)\varepsilon^{-3+\beta} + \frac{1}{2}\xi(t)\chi(t)^2\varepsilon^{-3+2\alpha})(\varphi(0, t) + \nabla\varphi(0, t) \cdot \vec{x})\, dV dt$$

with $\alpha, \beta \geq 1$. The volume of the ball $|\vec{x}| \leq \varepsilon t$ is proportional to ε^3, thus $I_1 \approx 0$.

Next, using that $\nu_t \approx -\varepsilon$,

$$I_2 \approx -\int_0^\infty \int_{|\vec{x}| = \varepsilon t} \Big(\rho_1(\varepsilon t, t)e_1(\varepsilon t, t) + \frac{1}{2}\rho_1(\varepsilon t, t)u_1(\varepsilon t, t)^2 - \rho_\varepsilon(t)e_\varepsilon(t) - \frac{1}{2}\rho_\varepsilon(t)u_\varepsilon(t)^2$$
$$+ p_1(\varepsilon t, t)u_1(\varepsilon t, t) - p_\varepsilon(t)u_\varepsilon(t)\Big)\, \varepsilon\big(\varphi(0, t) + \varepsilon t \nabla\varphi(0, t) \cdot \vec{\omega}\big)\, dS dt.$$

By assumption (A4) and (9), $p_1(\varepsilon t, t) = k\rho_1(\varepsilon t, t)e_1(\varepsilon t, t)$, $\rho_1(\varepsilon t, t)u_1(\varepsilon t, t)^2$ and $p_1(\varepsilon t, t)u_1(\varepsilon t, t)$ are all of order $\mathcal{O}(\varepsilon^{-2})$. By assumptions (A1), (A3) and (A5) the order of $\rho_\varepsilon(t)e_\varepsilon(t)$, $\rho_\varepsilon(t)u_\varepsilon(t)^2$ and $p_\varepsilon(t)u_\varepsilon(t)$ is at most $\mathcal{O}(\varepsilon^{-2})$. Thus again $I_2 \approx 0$.

Finally, using the same estimates as before,

$$I_3 \approx \int_0^\infty \int_{|\vec{x}|=\varepsilon t} \Big(\big(\rho_1(\varepsilon t,t)e_1(\varepsilon t) + \frac{1}{2}\rho_1(\varepsilon t)u_1(\varepsilon t,t)^2 + p_1(\varepsilon t,t)\big)u_1(\varepsilon t,t) - \big(\rho_\varepsilon(t)e_\varepsilon(t) + \frac{1}{2}\rho_\varepsilon(t)u_\varepsilon(t)^2 + p_\varepsilon(t)\big)u_\varepsilon(t)\Big) \big(\varphi(0,t) + \varepsilon t\nabla\varphi(0,t)\cdot\vec{\omega}\big)\, dSdt$$
$$\approx 4\pi \int_0^\infty t^2\varepsilon^2 \Big(\rho_1(\varepsilon t,t)e_1(\varepsilon t,t) + \frac{1}{2}\rho_1(\varepsilon t,t)u_1(\varepsilon t,t)^2 + p_1(\varepsilon t,t)\Big)u_1(\varepsilon t,t)\varphi(0,t)\, dt.$$

Recalling that $p_1 = k\rho_1 e_1$, the third equation results in the requirement that

$$\lim_{\varepsilon\to 0} \varepsilon^2 u_1(\varepsilon t,t)\Big((1+k)\rho_1(\varepsilon t,t)e_1(\varepsilon t,t) + \frac{1}{2}\rho_1(\varepsilon t,t)u_1(\varepsilon t,t)^2\Big) = 0. \tag{14}$$

Proposition 3. *Let assumptions (A1) – (A5) hold. A shadow wave of the form (5) satisfies the third equation (conservation of internal energy) of system (2) in the weak limit as $\varepsilon \to 0$ if and only if (14) holds.*

3. **Entropy condition.** Using standard SDW methods and Taylor expansion of the test function, we are going to determine whether and when the Clausius-Duhem inequality is satisfied by SDW solutions obtained in the previous section.

Denote by $\eta(U) = \rho S(U)$, $\vec{Q} = -\vec{m}S(U)$, $U = (\rho, \vec{m}, e)$ the entropy-entropy flux pair, where $S = \ln\big(k\rho^{-k}e\big)$. We say that a solution to (1) is admissible if it satisfies

$$\partial_t(\eta(U)) + \sum_{i=1}^{3} \partial_{x_i}(Q_i(U)) \geq 0 \tag{15}$$

in the distributional sense. That is, a family $\mathcal{U}_\varepsilon$ of weak or approximate solution is admissible if

$$I := -\liminf_{\varepsilon\to 0} \int_0^\infty \int_{\mathbb{R}^3} \eta(\mathcal{U}_\varepsilon)\partial_t\varphi + \vec{Q}(\mathcal{U}_\varepsilon)\cdot\nabla\varphi\, dV dt \geq 0,$$

for every $\varphi \in C_0^\infty(\mathbb{R}^3 \times \mathbb{R}_+)$, $\varphi \geq 0$.

Using the same procedure as in Subsection 2.1 with with $(\rho, \vec{m})$ substituted by $(\eta, \vec{Q})$ in (6) we get the following condition

$$I \approx \underbrace{\int_0^\infty \int_{|\vec{x}|<\varepsilon t} \partial_t\eta(U_\varepsilon)\varphi\, dV dt}_{=:I_1} + \underbrace{\int_0^\infty \int_{|\vec{x}|=\varepsilon t} (\eta(U_1) - \eta(U_\varepsilon))\nu_t\varphi\, dSdt}_{=:I_2} + \underbrace{\int_0^\infty \int_{|\vec{x}|=\varepsilon t} (\vec{Q}(U_1) - \vec{Q}(U_\varepsilon)\cdot\vec{\nu}_x\varphi\, dSdt}_{=:I_3} \geq 0. \tag{16}$$

By the mean value theorem, $\varphi(x,t) \approx \varphi(0,t) + \nabla\varphi(\vec{\theta},t)\cdot\vec{x}$, and so

$$I_1 \approx \frac{4\pi}{3}\int_0^\infty \partial_t\eta(U_\varepsilon)\varphi(0,t)t^3\varepsilon^3\, dt + \int_0^\infty \int_{|x|\leq\varepsilon t} \partial_t\eta(U_\varepsilon)\nabla\varphi(\vec{\theta},t)\cdot\vec{x}\, dV dt.$$

Invoking assumptions (A1) and (A5),

$$\eta(U_\varepsilon) = \varepsilon^{-3}\xi(t)\ln\big(k\xi(t)^{-k}\zeta(t)\varepsilon^{3k+\beta}\big) = \varepsilon^{-3}\xi(t)\big((3k+\beta)\ln\varepsilon + \ln(k\xi(t)^{-k}\zeta(t))\big).$$

Consequently, the second integral goes to zero as $\varepsilon \to 0$, resulting in

$$I_1 \approx \frac{4\pi}{3}\int_0^\infty \partial_t\Big(\xi(t)\big((3k+\beta)\ln\varepsilon + \ln(k\xi(t)^{-k}\zeta(t))\big)\Big)t^3\varphi(0,t)\, dt.$$

Note that this integral diverges of order $|\ln\varepsilon|$, in general. Next,

$$I_2 \approx -4\pi\int_0^\infty (\eta(U_1)-\eta(U_\varepsilon))t^2\varepsilon^3\varphi(0,t)\,dt - \int_0^\infty\int_{|\vec{x}|=\varepsilon t}(\eta(U_1)-\eta(U_\varepsilon))\varepsilon\nabla\varphi(\vec{\theta},t)\cdot\vec{x}\,dSdt.$$

Written out explicitly,

$$\varepsilon^3\eta(U_1(\varepsilon t,t)) = \varepsilon^3\rho_1(\varepsilon t,t)\big(-k\ln\rho_1(\varepsilon t,t)+\ln e_1(\varepsilon t,t)+\ln k\big).$$

A natural physical assumption is the integrability of the density $\rho_1(r,t)$ around $r=0$. We also need some control on the logarithm of $e_1(r,t)$. Thus we require

Assumption (A6): $\rho_1(r,t)=\mathcal{O}(r^a)$ and $e_1(r,t)=\mathcal{O}(r^b)$ as $r\to 0$ uniformly for t in compact subsets of $(0,\infty)$, where $a>-3$ and b is some real number.

For small ρ_1,η_1 we observe that $(\rho_1\ln\rho_1)$ is bounded from below by $\exp(-1)$, while e_1 might go to $-\infty$. For large ρ_1,η_1, the term $\varepsilon^3\eta(U_1(\varepsilon t,t))$ is controlled by the remaining positive power of ε. Consequently, the term involving $\eta(U_1)$ under the integral does not contribute to the limit inferior as $\varepsilon\to 0$, nor does the second integral in I_2. In conclusion, only the term involving $\eta(U_\varepsilon)$ survives, that is,

$$I_2 \gtrapprox 4\pi\int_0^\infty \xi(t)\big((3k+\beta)\ln\varepsilon+\ln(k\xi(t)^{-k}\zeta(t))\big)t^2\varphi(0,t)\,dt.$$

Finally, observing that $\vec{\nu}_x=\mathcal{O}(1)$,

$$I_3 \approx 4\pi\int_0^\infty \Big(m_1(\varepsilon t,t)\ln\big(k\rho_1(\varepsilon t,t)^{-k}e_1(\varepsilon t,t)\big)-m_\varepsilon(t)\ln\big(k\rho_\varepsilon(t)^{-k}e_\varepsilon(t)\big)\Big) t^2\varepsilon^2\big(\varphi(0,t)+\varepsilon t\nabla\varphi(\vec{\theta},t)\cdot\vec{\omega}\big)\,dt.$$

From (9), $\varepsilon^2 m_1(\varepsilon t,t)=\mathcal{O}(1)$. As before, the $\nabla\varphi$-term in I_3 vanishes (actually here lower bounds by some power of ε on $e_1(\varepsilon t,t)$ are needed as well), and only the term multiplying φ contributes to the limit inferior.

Collecting terms, dividing by $4\pi t^2$ and recalling that $m_\varepsilon(t)=\varepsilon^{-3+\alpha}\xi(t)\chi(t)$, we arrive at the following assertion.

Proposition 4. *Under assumptions (A1) – (A6), a shadow wave solution of the form* (5) *is admissible if and only if*

$$\begin{aligned}\liminf_{\varepsilon\to 0}\ &\Big[\frac{t}{3}\frac{\partial}{\partial t}\Big(\xi(t)\big((3k+\beta)\ln\varepsilon+\ln(k\xi(t)^{-k}\zeta(t))\big)\Big)\\ &+\xi(t)\big(1-\chi(t)\varepsilon^{-1+\alpha}\big)\big((3k+\beta)\ln\varepsilon+\ln(k\xi(t)^{-k}\zeta(t))\big)\\ &+\varepsilon^2 m_1(\varepsilon t,t)\big(-k\ln\rho_1(\varepsilon t,t)+\ln e_1(\varepsilon t,t)+\ln k\big)\Big]\geq 0.\end{aligned}\tag{17}$$

Remark 2. *If in addition to the assumptions above, $m_1(r,t)=\mathcal{O}(r^c)$ for some $c>-2$ as $r\to 0$, then the last line in inequality* (17) *vanishes. Further, the limits in* (10) *and* (11) *are equal to zero, so that*

$$\frac{t}{3}\xi'(t)+\xi(t)(1-\chi(t))=0\ (\alpha=1)\quad or\quad \frac{t}{3}\xi'(t)+\xi(t)=0\ (\alpha>1).$$

In either case, the only remainig term in (17) *is*

$$\xi(t)\frac{\partial}{\partial t}\ln\big(k\xi(t)^{-k}\zeta(t)\big),\tag{18}$$

which should be nonnegative then. Recalling the initial condition $\xi(0) = 0$, the only solution with $\alpha > 1$ is $\xi(t) \equiv 0$. Then condition (18) *is automatically satisfied, but the shadow wave has zero strength. The remaining case is $\alpha = 1$, leading to*

$$\frac{t}{3}\xi'(t) + \xi(t)(1 - \chi(t)) = 0 \quad \textit{and} \quad \big(\ln\zeta(t) - k\ln\xi(t)\big)' \geq 0,$$

which does have nonvanishing solutions $\xi(t)$ for suitable choices of $\chi(t), \zeta(t)$. A complete solution must also satisfy (12), (13), *and* (14).

Note that the condition $m_1(r,t) = \mathcal{O}(r^c)$, $c > -2$, is just slightly stronger than the condition $\lim_{r\to 0} r^2 m_1(r,t) = 0$, which has been imposed on self-similar, radial weak solutions in [3].

It remains a task for future research to determine whether shadow wave solutions of this type exist which can be connected to an incoming wave $U_{s,1}(r,t)$ with an intermediate state U_1.

REFERENCES

[1] G-Q. G. Chen, Remarks on spherically symmetric solutions of the compressible Euler equations, *Proc. Roy. Soc. Edinburgh Sect. A* **127** (1997), 243–259.

[2] G-Q. G. Chen and M. R. I. Schrecker, Vanishing viscosity approach to the compressible Euler equations for transonic nozzle and spherically symmetric flows, *Arch. Ration. Mech. Anal.* **229** (2018), 1239–1279.

[3] M. G. Hilgers, Nonuniqueness and singular radial solutions of systems of conservation laws, *Acta Math. Sci. Ser. B (Engl. Ed.)* **32** (2012), 367–379.

[4] H. K. Jenssen and C. Tsikkou, On similarity flows for the compressible Euler system, *J. Math. Phys.* **59** (2018), 121507, 25 pp.

[5] R. B. Lazarus, Self-similar solutions for converging shocks and collapsing cavities, *SIAM J. Numer. Anal.* **18** (1981), 316–371.

[6] M. Nedeljkov, Shadow waves: entropies and interactions for delta and singular shocks, *Arch. Ration. Mech. Anal.* **197** (2010), 489–537.

[7] M. Nedeljkov, L. Neumann, M. Oberguggenberger and M. Sahoo, Radially symmetric shadow wave solutions to the system of pressureless gas dynamics in arbitrary dimensions, *Nonlinear Analysis* **163** (2017), 104–126.

E-mail address: marko@dmi.uns.ac.rs

E-mail address: lukas.neumann@uibk.ac.at

E-mail address: michael.oberguggenberger@uibk.ac.at

PHASE FIELD MODELLING FOR COMPRESSIBLE DROPLET IMPINGEMENT

LUKAS OSTROWSKI*

Institute of Applied Analysis and Numerical Simulation
University of Stuttgart
Pfaffenwaldring 57 D-70569 Stuttgart

CHRISTIAN ROHDE

Institute of Applied Analysis and Numerical Simulation
University of Stuttgart
Pfaffenwaldring 57 D-70569 Stuttgart

ABSTRACT. We consider the impingement of a droplet onto a wall with high impact speed. For this purpose an isothermal Navier–Stokes–Allen–Cahn model [5] is used. Properties of the model are discussed. In order to solve the system numerically we introduce an energy consistent discontinuous Galerkin scheme and show a numerical example of droplet impact.

1. **Introduction.** High speed droplet impact occurs in many applications like spray coating. Other applications are for example ink jet printing, liquid-fueled engines, and spray cooling. In these situations compressible effects in the liquid droplet phase may play an important role. During impact the compressed liquid triggers a shock wave which travels backwards through the bulk and determines the overall droplet dynamics. An analytical study of the wave patterns can be found in [8]. In particular it turned out that incompressible approaches are not adequate to predict the correct jetting time, which is actually smaller due to the shock wave pattern [9]. In this contribution we introduce a phase field approach to model high speed droplet impact scenarios. Both, the liquid and the vapor phase are assumed to be compressible. For direct numerical simulation an energy consistent discontinuous Galerkin scheme is derived.

2. **The phase field model.** Phase field models are widely used to simulate interfacial phenomena. They are based on the assumption that the interface is a thin transition layer in which the different fluids mix. This diffuse layer is represented by a phase field variable. Based on an energy principle, namely the interplay between mixture and kinetic energy, phase field models can be derived with a variational approach, see e.g.[1] for an overview. For the isothermal case it is in this way possible to derive models that obey the second law of thermodynamics in the form of an entropy inequality. While phase field models have been studied much more in an incompressible setting, less work has been done for compressible phase field models.

2000 *Mathematics Subject Classification.* Primary:76T99; Secondary: 65M60.

Key words and phrases. Phase field model, two phase flow, moving contact line, Navier–Stokes–Allen–Cahn, energy consistent, discontinuous Galerkin.

* Corresponding author: Lukas Ostrowski.

The compressible models are typically based on [3, 12]. More recent models are [5, 16]. However, their main difference are scalings of the double well potential.

For the unknowns density $\rho > 0$, velocity $\mathbf{v} \in \mathbb{R}^2$ and phase field parameter $\varphi \in [0,1]$ we consider the following isothermal compressible Navier–Stokes–Allen–Cahn system [5] in a domain $\Omega \subset \mathbb{R}^2$:

$$\partial_t \rho + \operatorname{div}(\rho \mathbf{v}) = 0, \tag{1}$$

$$\partial_t(\rho \mathbf{v}) + \operatorname{div}(\rho \mathbf{v} \otimes \mathbf{v} + p\mathbf{I}) = \operatorname{div}(\mathbf{S}) - \gamma \operatorname{div}(\nabla\varphi \otimes \nabla\varphi - \frac{1}{2}|\nabla\varphi|^2\mathbf{I}) + \frac{1}{\gamma}\nabla W(\varphi),$$

$$\partial_t(\rho\varphi) + \operatorname{div}(\rho\varphi\mathbf{v}) = -\eta\mu,$$

with boundary conditions

$$\mathbf{v} \cdot \mathbf{n} = 0, \tag{2}$$

$$\beta v_{\boldsymbol{\tau}} + \nu(\varphi)\frac{\partial v_{\boldsymbol{\tau}}}{\partial \mathbf{n}} - L(\varphi)\frac{\partial \varphi}{\partial \boldsymbol{\tau}} = 0, \tag{3}$$

$$\partial_t \varphi + v_{\boldsymbol{\tau}}\frac{\partial \varphi}{\partial \boldsymbol{\tau}} = -\frac{\alpha}{\rho}L(\varphi) \quad \text{on } \partial\Omega. \tag{4}$$

The phase field parameter allows to distinguish the phases. It takes the value 0 in the vapor phase, the value 1 in the liquid phase, and values in between in the mixture region. The dissipative viscous part of the stress tensor reads as $\mathbf{S} = \mathbf{S}(\varphi, \nabla\mathbf{v}) = \nu(\varphi)(\nabla\mathbf{v} + \nabla\mathbf{v}^\top - \operatorname{div}(\mathbf{v})\mathbf{I})$ with an interpolation of the viscosities $\nu_{\mathrm{L/V}}$ of the pure phases $\nu(\varphi) = h(\varphi)\nu_{\mathrm{L}} + (1-h(\varphi))\nu_{\mathrm{V}} > 0$ by the nonlinear interpolation function

$$h(\varphi) = 3\varphi^2 - 2\varphi^3. \tag{5}$$

Further, we define the interpolation of the free energy densities $\rho f_{\mathrm{L/V}}$ of the pure phases $\rho\psi(\rho,\varphi) = h(\varphi)\rho f_{\mathrm{L}}(\rho) + (1-h(\varphi))\rho f_{\mathrm{V}}(\rho)$. This determines the pressure as $p = p(\rho,\varphi) = -\rho\psi(\rho,\varphi) + \rho\frac{\partial\rho\psi}{\partial\rho}(\rho,\psi)$. With the double well potential $W(\varphi) = \varphi^2(1-\varphi)^2$ we define the (generalized) chemical potential $\mu = \frac{1}{\gamma}W'(\varphi) + \frac{\partial\rho\psi}{\partial\varphi} - \gamma\Delta\varphi$, which steers the phase field variable into equilibrium. Additionally, we have the mobility $\eta > 0$, slip length $\beta > 0$ and relaxation parameter $\alpha > 0$. Finally

$$L(\varphi) = \gamma\frac{\partial\varphi}{\partial\mathbf{n}} + g'(\varphi),$$

where $g(\varphi) = -\sigma\cos(\theta_s)(h(\varphi) - 1/2)$, with the surface tension parameter $\sigma > 0$ and the static contact angle $\theta_s \in [0,\pi]$.

The total energy corresponding to the system (1)-(4) reads

$$E_{\mathrm{tot}} := \int_\Omega F(\rho,\varphi,\nabla\varphi) + \frac{1}{2}\rho|\mathbf{v}|^2 \ \mathrm{d}\mathbf{x} + \int_{\partial\Omega} g(\varphi) \ \mathrm{d}s, \tag{6}$$

with the free energy density

$$F(\rho,\varphi,\nabla\varphi) = \tilde{F}(\rho,\varphi) + \frac{1}{2}\gamma|\nabla\varphi|^2 = \frac{1}{\gamma}W(\varphi) + \rho\psi(\rho,\varphi) + \frac{1}{2}\gamma|\nabla\varphi|^2. \tag{7}$$

Remark 1. On a first glance the nonlinear interpolation (5) seems unnecessarily complicated. However, it is needed in order to obtain correct equilibria. The use of a linear interpolation function $h(\varphi) = \varphi$ would result in incorrect equilibria due to the fact that $h'(0) = h'(1) \neq 0$.

2.1. The boundary conditions. We consider a moving contact line problem (MCL), see Figure 1. This kind of problems needs special attention regarding boundary conditions. We derive boundary conditions for the MCL problem in

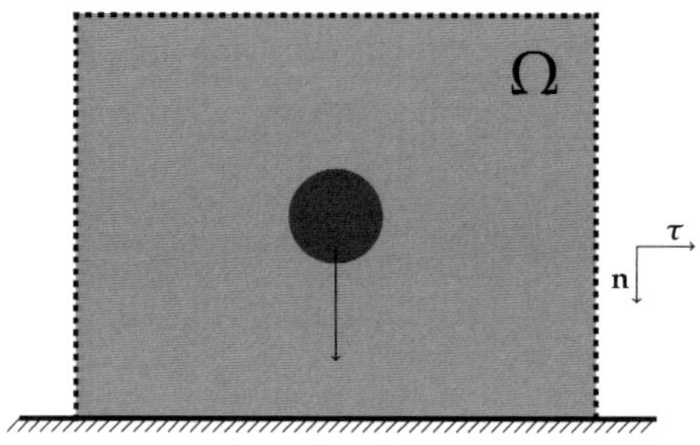

FIGURE 1. Sketch of computational domain for droplet impact simulation.

a similar fashion as done in [13, 14] for the incompressible case. To this end, we add the wall free energy term $\int_{\partial\Omega} g(\varphi)\ ds$ to the total free energy (6). Here $g(\varphi)$ is the interfacial free energy per unit area at the fluid-solid boundary. We choose $g(\varphi) := \Delta g(h(\varphi) - 1/2)$, i.e. a smooth interpolation between $\pm\Delta g/2$, with $\Delta g = g(1) - g(0)$. This energy difference can be specified by considering Young's equation

$$\sigma \cos(\theta_s) = \sigma_{\mathrm{S}} - \sigma_{\mathrm{LS}} = g(0) - g(1), \tag{8}$$

with the surface free energy σ of the liquid, the static contact angle θ_s, surface free energy σ_{S} of the solid, and interfacial free energy σ_{LS} between liquid and solid. From (8), we have $\Delta g = -\sigma\cos(\theta_s)$. Applying variational techniques to this total free energy yields the boundary condition (3). The extension of the classical (single phase) Navier-slip condition $\beta v_{\boldsymbol{\tau}} + \nu\frac{\partial v_{\boldsymbol{\tau}}}{\partial \mathbf{n}} = 0$ relies on the fact that one has to account for the uncompensated Young stress arising from the deviation of the fluid-fluid interface from the static configuration. For details see [13]. The generalized Navier boundary condition (GNBC) is given by

$$\beta v_{\boldsymbol{\tau}} = -\nu(\varphi)\frac{\partial v_{\boldsymbol{\tau}}}{\partial \mathbf{n}} + L(\varphi)\frac{\partial \varphi}{\partial \boldsymbol{\tau}}.$$

Away from the interface the second term drops out and we have the classical Navier-slip condition but in the interface region the additional term acts and allows for correct contact line movement.

Note that $L(\varphi) = 0$ is the Euler–Lagrange equation at the fluid-solid boundary for minimizing the total free energy with respect to the phase field variable. Hence, $L(\varphi) = 0$ corresponds with the equilibrium condition where $\partial_t(\rho\varphi) + \mathrm{div}(\rho\mathbf{v}\varphi) = 0$. The boundary relaxation dynamics of φ are assumed as

$$\partial_t\varphi + \mathbf{v}\cdot\nabla_{\boldsymbol{\tau}}\varphi = -\frac{\alpha}{\rho}L(\varphi),$$

with a positive phenomenological parameter α. Here $\nabla_{\boldsymbol{\tau}} := \nabla - (\mathbf{n}\cdot\nabla)\mathbf{n}$ is the gradient along the tangential direction. Since $\mathbf{v}\cdot\mathbf{n} = 0$, we have $\mathbf{v}\cdot\nabla_{\boldsymbol{\tau}}\varphi = v_{\boldsymbol{\tau}}\frac{\partial\varphi}{\partial\boldsymbol{\tau}}$.

Remark 2. The boundary conditions (2)-(4) contain several special cases. If we let the relaxation parameter α tend to infinity, we end up with

$$\begin{aligned}\mathbf{v}\cdot\mathbf{n} &= 0,\\ \beta v_{\boldsymbol{\tau}} + \nu(\varphi)\frac{\partial v_{\boldsymbol{\tau}}}{\partial \mathbf{n}} &= 0,\\ L(\varphi) &= 0.\end{aligned}$$

If additionally $\theta_s = \pi/2$ we have

$$\begin{aligned}\mathbf{v}\cdot\mathbf{n} &= 0,\\ \beta v_{\boldsymbol{\tau}} + \nu(\varphi)\frac{\partial v_{\boldsymbol{\tau}}}{\partial \mathbf{n}} &= 0,\\ \nabla\varphi\cdot\mathbf{n} &= 0.\end{aligned}$$

Finally, by sending the slip length β to infinity we obtain the no slip condition

$$\mathbf{v} = 0, \tag{9}$$

$$\nabla\varphi\cdot\mathbf{n} = 0. \tag{10}$$

2.2. **Energy Inequality.** The phase field model (1) is thermodynamically consistent, that means in particular that for a smooth solution it fulfills an entropy inequality. With the total energy as mathematical entropy the following inequality can be easily shown to hold.

Lemma 2.1 (Energy inequality). *Let $(\rho, \mathbf{v}, \varphi)$ be a smooth solution of* (1) *in $(0,T)\times\Omega$ satisfying the boundary conditions* (2) - (4) *on $(0,T)\times\partial\Omega$. Then for all $t\in(0,T)$:*

$$\frac{\mathrm{d}}{\mathrm{d}t}\left(\int_\Omega F(\rho,\mathbf{v},\varphi) + \frac{1}{2}\rho|\mathbf{v}|^2\ \mathrm{d}\mathbf{x} + \int_{\partial\Omega} g(\varphi)\ \mathrm{d}s\right) =$$

$$-\int_\Omega \frac{\eta}{\rho}\mu^2\ \mathrm{d}\mathbf{x} - \int_\Omega \mathbf{S}:\nabla\mathbf{v}\ \mathrm{d}\mathbf{x} - \int_{\partial\Omega}\beta|v_{\boldsymbol{\tau}}|^2\ \mathrm{d}s - \int_{\partial\Omega}\frac{\alpha}{\rho}|L(\varphi)|^2\ \mathrm{d}s \leq 0. \tag{11}$$

This means we have dissipative mechanisms due to phase transition, viscosity, wall slip, and composition relaxation at the solid interface.

Remark 3. Up to our knowledge there is no wellposedness result for a system like (1)-(4).

3. **Numerical Scheme.** Phase field modelling is based on a variational principle. Our model is thermodynamically consistent and follows the energy dissipation law (11). Numerical schemes with artificial dissipation for stabilization can lead to problems like increase of energy or parasitic currents [4, 10]. Hence, it is desirable that the numerical scheme fulfills the energy dissipation inequality (11) on a discrete level without artificial numerical dissipation. This motivates the use of *energy consistent discontinuous Galerkin schemes* (DG) to solve phase field systems [7, 11, 15]. The derivation of our scheme to solve the system (1) is based on [6].

Recall the free energy density (7). First we introduce auxiliary variables

$$\begin{aligned}\boldsymbol{\sigma} &= \nabla\varphi,\\ \mu &= \frac{\partial F}{\partial\varphi} - \operatorname{div}(\gamma\boldsymbol{\sigma}),\\ \tau &= \frac{\partial F}{\partial\rho} + \frac{1}{2}|\mathbf{v}|^2.\end{aligned}$$

With these we rewrite the system (1) into a mixed non-conservative formulation

$$\begin{aligned}
\partial_t \rho + \operatorname{div}(\rho \mathbf{v}) &= 0, \\
\rho \partial_t \mathbf{v} + \operatorname{div}(\rho \mathbf{v} \otimes \mathbf{v}) - \operatorname{div}(\rho \mathbf{v})\mathbf{v} - \frac{1}{2}\rho \nabla |\mathbf{v}|^2 + \rho \nabla \tau &= \operatorname{div}(\mathbf{S}) + \mu \nabla \varphi, \\
\partial_t \varphi + \nabla \varphi \cdot \mathbf{v} &= -\eta \frac{\mu}{\rho}, \\
\mu &= \frac{\partial F}{\partial \varphi} - \gamma \operatorname{div}(\boldsymbol{\sigma}), \\
\tau &= \frac{\partial F}{\partial \rho} + \frac{1}{2}|\mathbf{v}|^2, \\
\boldsymbol{\sigma} &= \nabla \varphi,
\end{aligned}$$

with the boundary conditions

$$\mathbf{v} = 0, \quad \boldsymbol{\sigma} \cdot \mathbf{n} = 0 \text{ on } \partial\Omega.$$

The more general boundary conditions (2)-(4) will be considered in future work. We introduce some notation and briefly sketch the idea of the derivation of the scheme. Let $\mathcal{T}$ be a triangulation of Ω. We define the discontinuous Galerkin space by

$$V_h := \left\{ u \in L^2(\Omega) \colon u\big|_T \in \mathbb{P}^k \text{ for all } T \in \mathcal{T} \right\},$$

where $\mathbb{P}^k$ is the space of polynomials up to degree k. Further, let $\mathcal{V}_h := V_h \times V_h^d \times V_h \times V_h \times V_h \times V_h^d$. We define the jump and average operators as follows: Let T_1 and T_2 be two mesh elements with a common facet $\mathcal{F}$, Φ a scalar-valued and $\mathbf{u}$ a vector-valued function on Ω. Further, let T be a mesh element with boundary facet $\mathcal{F}_b = \partial T \cap \partial\Omega$ then,

$$\begin{aligned}
\{\!\{\Phi\}\!\}_{\mathcal{F}} &:= \frac{1}{2}(\Phi\big|_{T_1} + \Phi\big|_{T_2}), & \{\!\{\Phi\}\!\}_{\mathcal{F}_b} &:= \Phi\big|_T, \\
\{\!\{\mathbf{u}\}\!\}_{\mathcal{F}} &:= \frac{1}{2}(\mathbf{u}\big|_{T_1} + \mathbf{u}\big|_{T_2}), & \{\!\{\mathbf{u}\}\!\}_{\mathcal{F}_b} &:= \mathbf{u}\big|_T, \\
[\![\Phi]\!]_{\mathcal{F}} &:= \Phi\big|_{T_1}\mathbf{n}_{T_1} + \Phi\big|_{T_2}\mathbf{n}_{T_2}, & [\![\Phi]\!]_{\mathcal{F}_b} &:= \Phi\big|_T \mathbf{n}_T, \\
[\![\mathbf{u}]\!]_{\mathcal{F}} &:= \mathbf{u}\big|_{T_1} \cdot \mathbf{n}_{T_1} + \mathbf{u}\big|_{T_2} \cdot \mathbf{n}_{T_2}, & [\![\mathbf{u}]\!]_{\mathcal{F}_b} &:= \mathbf{u}\big|_T \cdot \mathbf{n}_T, \\
[\![\mathbf{u}]\!]_{\otimes,\mathcal{F}} &:= \mathbf{u}\big|_{T_1} \otimes \mathbf{n}_{T_1} + \mathbf{u}\big|_{T_2} \otimes \mathbf{n}_{T_2}, & [\![\mathbf{u}]\!]_{\otimes,\mathcal{F}_b} &:= \mathbf{u}\big|_T \otimes \mathbf{n}_T.
\end{aligned}$$

We omit the subscripts $\mathcal{F}, \mathcal{F}_b$ whenever no confusion can arise.

For the spatial discretization we first assume generic fluxes F_i which then are determined by our requirements. That means for instance for the first equation

$$0 = \int_\Omega (\partial_t \rho_h + \operatorname{div}(\rho_h \mathbf{v}_h))\psi \, d\mathbf{x} + \int_{\mathcal{E}} F_1(\rho_h, \mathbf{v}_h, \varphi_h, \mu_h, \tau_h, \boldsymbol{\sigma}_h, \psi) \, ds$$

Here for brevity we slightly abuse the notation. The notation $\int_\Omega \bullet$ means $\sum_{T \in \mathcal{T}} \int_T \bullet$, the set $\mathcal{E}$ contains all interior facets of the underlying triangulation $\mathcal{T}$ of Ω. In order to assure consistency of the fluxes, conservation of mass and correct energy dissipation there arise several conditions on the fluxes. Enforcing additionally the conservation of momentum is too restrictive, hence the non-conservative formulation of the momentum equation.

Let $0 = t_0 < t_1 < \ldots < t_N = T$ be a temporal decomposition of $[0, T]$. We set $\Delta t_n := t_{n+1} - t_n$. Moreover, we denote $\Phi^n(\cdot) := \Phi(\cdot, t_n)$ and $\Phi^{n+1/2} := \frac{\Phi^{n+1} + \Phi^n}{2}$. The temporal discretization is of Crank-Nicholson type. Finally, with (7) the fully discrete scheme reads as follows:

Find $(\rho_h^{n+1}, \mathbf{v}_h^{n+1}, \varphi_h^{n+1}, \mu_h^{n+1/2}, \tau_h^{n+1/2}, \boldsymbol{\sigma}_h^{n+1}) \in \mathcal{V}_h$ such that

$$0 = \int_\Omega \left(\frac{\rho_h^{n+1} - \rho_h^n}{\Delta t} + \operatorname{div}(\rho_h^{n+1/2}\mathbf{v}_h^{n+1/2}) \right) \psi \ \mathrm{d}\mathbf{x} - \int_\mathcal{E} [\![\rho_h^{n+1/2}\mathbf{v}_h^{n+1/2}]\!] \{\!\{\psi\}\!\} \ \mathrm{d}s, \tag{12}$$

$$\begin{aligned} 0 = \int_\Omega \Big(&\rho_h^{n+1/2} \left(\frac{\mathbf{v}_h^{n+1} - \mathbf{v}_h^n}{\Delta t} \right) + \operatorname{div}(\rho_h^{n+1/2}\mathbf{v}_h^{n+1/2} \otimes \mathbf{v}_h^{n+1/2}) - \operatorname{div}(\rho_h^{n+1/2}\mathbf{v}_h^{n+1/2})\mathbf{v}_h^{n+1/2} \\ &- \frac{1}{2}\rho_h^{n+1/2}\nabla|\mathbf{v}_h^{n+1/2}|^2 + \rho_h^{n+1/2}\nabla\tau_h^{n+1/2} - \mu_h^{n+1/2}\nabla\varphi_h^{n+1} \Big) \cdot \mathbf{X} \ \mathrm{d}\mathbf{x} \\ &- \int_\mathcal{E} [\![\tau_h^{n+1/2}]\!] \cdot \{\!\{\rho_h^{n+1/2}\mathbf{X}\}\!\} - [\![\varphi_h^{n+1}]\!] \cdot \{\!\{\mu_h^{n+1/2}\mathbf{X}\}\!\} \ \mathrm{d}s + B_h(\mathbf{v}_h^{n+1/2}, \mathbf{X}; \varphi_h^{n+1/2}), \end{aligned} \tag{13}$$

$$0 = \int_\Omega \left(\frac{\varphi_h^{n+1} - \varphi_h^n}{\Delta t} + \nabla\varphi_h^{n+1} \cdot \mathbf{v}_h^{n+1/2} + \eta \frac{\mu_h^{n+1/2}}{\rho_h^{n+1/2}} \right) \Theta \ \mathrm{d}\mathbf{x} - \int_\mathcal{E} [\![\varphi_h^{n+1/2}]\!] \cdot \{\!\{\Theta\mathbf{v}_h^{n+1/2}\}\!\} \ \mathrm{d}s, \tag{14}$$

$$\begin{aligned} 0 = \int_\Omega &\left(\mu_h^{n+1/2} - \frac{\tilde{F}(\rho_h^n, \varphi_h^{n+1}) - \tilde{F}(\rho_h^n, \varphi_h^n)}{\varphi_h^{n+1} - \varphi_h^n} + \gamma \operatorname{div}(\boldsymbol{\sigma}_h^{n+1/2}) \right) \chi \ \mathrm{d}\mathbf{x} \\ &- \int_\mathcal{E} \gamma [\![\boldsymbol{\sigma}_h^{n+1/2}]\!] \{\!\{\chi\}\!\} \ \mathrm{d}s, \end{aligned} \tag{15}$$

$$0 = \int_\Omega \left(\tau_h^{n+1/2} - \frac{\tilde{F}(\rho_h^{n+1}, \varphi_h^{n+1}) - \tilde{F}(\rho_h^n, \varphi_h^{n+1})}{\rho_h^{n+1} - \rho_h^n} - \frac{1}{4}(|\mathbf{v}_h^{n+1}|^2 + |\mathbf{v}_h^n|^2) \right) \zeta \ \mathrm{d}\mathbf{x}, \tag{16}$$

$$0 = \int_\Omega (\boldsymbol{\sigma}_h^{n+1} - \nabla\varphi_h^{n+1}) \cdot \mathbf{Z} \ \mathrm{d}\mathbf{x} + \int_\mathcal{E} [\![\varphi_h^{n+1}]\!] \cdot \{\!\{\mathbf{Z}\}\!\} \ \mathrm{d}s, \tag{17}$$

holds for all $(\psi, \mathbf{X}, \Theta, \chi, \zeta, \mathbf{Z}) \in \mathcal{V}_h$.

We rely on a symmetric interior penalty discretization [2] for the viscous stress tensor:

$$\begin{aligned} B_h(\mathbf{v}, \mathbf{X}; \varphi_h^{n+1/2}) &= \int_\Omega \mathbf{S}(\nabla\mathbf{v}, \varphi) \colon \nabla\mathbf{X} \ \mathrm{d}\mathbf{x} \\ &- \sum_{e \in \mathcal{E} \cup \partial\Omega} \int_e \{\!\{\mathbf{S}(\nabla\mathbf{v}, \varphi)\}\!\} \colon [\![\mathbf{X}]\!]_\otimes + \{\!\{\mathbf{S}(\nabla\mathbf{X}, \varphi)\}\!\} \colon [\![\mathbf{v}]\!]_\otimes - \frac{\alpha_B}{|e|}[\![\mathbf{v}]\!]_\otimes \colon [\![\mathbf{X}]\!]_\otimes \ \mathrm{d}s. \end{aligned}$$

With α_B sufficiently large to ensure coercivity of $B_h(\cdot, \cdot; \varphi)$. The discretization is chosen such that the discrete counterpart of the energy inequality 2.1 is satisfied.

Lemma 3.1 (Fully discrete energy inequality)**.** *The discrete solution of the scheme* (12)-(17) *conserves mass and satisfies the energy dissipation equality, i.e.*

$$\begin{aligned} &\int_\Omega \tilde{F}(\rho_h^{n+1}, \varphi_h^{n+1}) + \frac{\gamma}{2}|\boldsymbol{\sigma}_h^{n+1}|^2 + \frac{\rho_h^{n+1}}{2}|\mathbf{v}_h^{n+1}|^2 \ \mathrm{d}\mathbf{x} \\ &\qquad - \int_\Omega \tilde{F}(\rho_h^n, \varphi_h^n) + \frac{\gamma}{2}|\boldsymbol{\sigma}_h^n|^2 + \frac{\rho_h^n}{2}|\mathbf{v}_h^n|^2 \ \mathrm{d}\mathbf{x} \\ &\qquad\qquad = -\Delta t \int_\Omega \eta \frac{|\mu_h^{n+1/2}|^2}{\rho_h^{n+1/2}} \ \mathrm{d}\mathbf{x} - \Delta t B_h(\mathbf{v}_h^{n+1/2}, \mathbf{v}_h^{n+1/2}; \varphi_h^{n+1/2}). \end{aligned}$$

Proof. For the mass conservation take $\psi = \mathbf{1}$ in equation (12). In order to prove the energy dissipation equality multiply equation (12) with $\tau^{n+1/2}$, equation (13) with $\mathbf{v}^{n+1/2}$ and equation (14) with $\mu^{n+1/2}$. Summing up and using the boundary conditions and basic algebraic manipulations leads to the result. □

4. **Numerical Example.** In this section we present a numerical experiment. We briefly comment on the choice of model parameters and equations of state. The scheme has been implemented with boundary conditions (9)-(10). The general case will be implemented in future work.

4.1. **Choice of the Model Parameter.** We choose stiffened gas type equations of state, i.e.

$$\rho f_{\mathrm{L/V}}(\rho) = \alpha_{\mathrm{L/V}} \rho \ln(\rho) + (\beta_{\mathrm{L/V}} - \alpha_{\mathrm{L/V}})\rho + \gamma_{\mathrm{L/V}}.$$

This leads to the partial pressure

$$p_{\mathrm{L/V}}(\rho) := -\rho f_{\mathrm{L/V}} + \rho \frac{\partial \rho f_{\mathrm{L/V}}}{\partial \rho} = \alpha_{\mathrm{L/V}} \rho - \gamma_{\mathrm{L/V}}.$$

The minima of the free energies are located at $\rho_{\mathrm{L/V}} = \exp\left(-\frac{\beta_{\mathrm{L/V}}}{\alpha_{\mathrm{L/V}}}\right)$. We choose the parameters $\alpha_{\mathrm{L/V}}, \beta_{\mathrm{L/V}}$, and $\gamma_{\mathrm{L/V}}$ such that the minima of the free energies have the same height. This prevents one-phase equilibria, since no phase is energetically more favorable. If no surface tension is present, e.g. in the 1D case, one can show that in equilibrium the densities in the bulk phases are exactly $\rho_{\mathrm{L/V}}$. In the case with surface tension this is not true anymore. We expect the value in the droplet to be slightly higher and slightly lower in the surrounding vapor. For this reason we choose the density of initial data accordingly. For the bulk viscosities we set $\nu_{\mathrm{L}} = 0.0125$ and $\nu_{\mathrm{V}} = 0.00125$. The capillary parameter is $\gamma = 5 \cdot 10^{-4}$ and the mobility $\eta = 10$.

4.2. **Droplet Impact.** The example is the high speed impact of a droplet onto a perfect wall. In Figure 2 the density is depicted at three different times $t = 0, t = 0.13$, and $t = 0.21$, i.e. the initial configuration, right before the impact, and after impact. The initial velocity of the droplet is $-1.1\mathbf{e}_y$. One can see the shock waves in the vapor and also in the liquid phase. The second picture also shows that the shock speed is larger in the liquid phase than in the vapor phase. Even though we use boundary conditions (9)-(10) for the simulation, we still observe a moving contact line. This dynamics is however mainly determined by the chemical potential μ, not by advection. This can be seen in Figure 3.

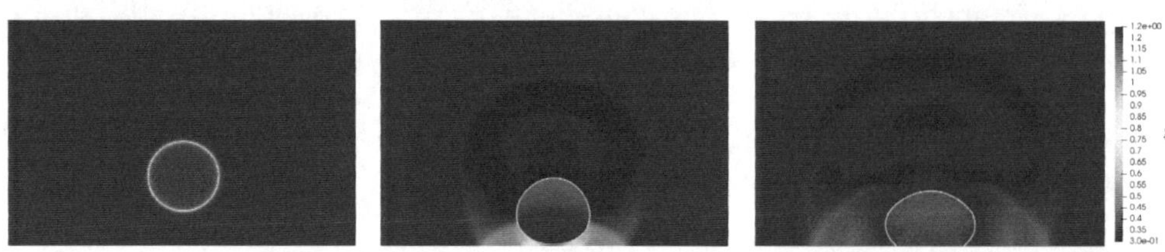

FIGURE 2. Droplet impact. Density at times $t = 0, t = 0.13$, $t = 0.21$.

FIGURE 3. Droplet impact. Chemical potential μ in the lower computational domain at times $t = 0, t = 0.13$, $t = 0.21$.

Acknowledgments. The authors kindly acknowledge the financial support of this work by the Deutsche Forschungsgemeinschaft (DFG) in the frame of the International Research Training Group 'Droplet Interaction Technologies' (DROPIT).

REFERENCES

[1] D. M. Anderson, G. B. McFadden and A. A. Wheeler, Diffuse-interface methods in fluid mechanics, *Annual Review of Fluid Mechanics*, **30** (1998), 139–165.

[2] D. Arnold, F. Brezzi, B. Cockburn and L. Marini, Unified analysis of discontinuous Galerkin methods for elliptic problems, *SIAM Journal on Numerical Analysis*, **39** (2002), 1749–1779.

[3] T. Blesgen, A generalization of the Navier–Stokes equations to two-phase flows, *Journal of Physics D: Applied Physics*, **32** (1999), 1119–1123.

[4] D. Diehl, *Higher order schemes for simulation of compressible liquid-vapor flows with phase change*, PhD thesis, Universität Freiburg im Breisgau, 2007,

[5] W. Dreyer, J. Giesselmann and C. Kraus, A compressible mixture model with phase transition, *Physica D: Nonlinear Phenomena*, **273-274** (2014), 1–13.

[6] J. Giesselmann, C. Makridakis and T. Pryer, Energy consistent DG methods for the Navier–Stokes–Korteweg system, *Math. Comp.*, **83** (2014), 2071–2099.

[7] J. Giesselmann and T. Pryer, Energy consistent discontinuous Galerkin methods for a quasi-incompressible diffuse two phase flow model, *M2AN Math. Model. Numer. Anal.*, **49(1)** (2015), 275–301.

[8] K. K. Haller, Y. Ventikos and D. Poulikakos, Wave structure in the contact line region during high speed droplet impact on a surface: Solution of the riemann problem for the stiffened gas equation of state, *Journal of Applied Physics*, **93** (2003), 3090–3097.

[9] K. K. Haller, Y. Ventikos, D. Poulikakos and P. Monkewitz, Computational study of high-speed liquid droplet impact, *Journal of Applied Physics*, **92** (2002), 2821–2828.

[10] D. Jamet, D. Torres and J. Brackbill, On the theory and computation of surface tension: The elimination of parasitic currents through energy conservation in the second-gradient method, *Journal of Computational Physics*, **182** (2002), 262–276.

[11] M. Kränkel and D. Kröner, A phase-field model for flows with phase transition, in *Theory, Numerics and Applications of Hyperbolic Problems II*, Springer International Publishing, Cham, 2018, 243–254.

[12] J. Lowengrub and L. Truskinovsky, Quasi-incompressible Cahn–Hilliard fluids and topological transitions, *Proceedings of the Royal Society A: Mathematical, Physical and Engineering Sciences*, **454** (1998), 2617–2654.

[13] T. Qian, X.-P. Wang and P. Sheng, Molecular scale contact line hydrodynamics of immiscible flows, *Phys. Rev. E*, **68** (2003), 016306.

[14] T. Qian, X.-p. Wang and P. Sheng, A variational approach to moving contact line hydrodynamics, *Journal of Fluid Mechanics*, **564** (2006), 333–360.

[15] E. Repossi, R. Rosso and M. Verani, A phase-field model for liquid–gas mixtures: mathematical modelling and discontinuous Galerkin discretization, *Calcolo*, **54** (2017), 1339–1377.

[16] G. Witterstein, Sharp interface limit of phase change flows, *Adv. Math. Sci. Appl.*, **20** (2010), 585–629.

E-mail address: `Lukas.Ostrowski@mathematik.uni-stuttgart.de`
E-mail address: `Christian.Rohde@mathematik.uni-stuttgart.de`

A ROE-LIKE REFORMULATION OF THE HLLC RIEMANN SOLVER AND APPLICATIONS

MARICA PELANTI

Institute of Mechanical Sciences and Industrial Applications
UMR 9219 ENSTA ParisTech-EDF-CNRS-CEA
828, Boulevard des Maréchaux, 91762 Palaiseau, France

ABSTRACT. The Roe and HLLC Riemann solvers are widely used as building blocks of finite volume Godunov-type schemes for solving the Euler equations of gas dynamics and related hyperbolic flow models. The HLLC solver (HLL with Contact restoration) has gained increasing popularity over the last two decades since it possesses some of the good properties of the Roe solver and in addition it satisfies important entropy and positivity conditions with no need of special fixes. In the present work we rewrite the classical HLLC solver for the Euler equations in a novel form that allows an interpretation of the HLLC wave structure as an averaged system eigenstructure. This reveals a formal mathematical similarity of the HLLC solver with the Roe solver, which can be useful to extend to the HLLC method some numerical techniques devised specifically for the Roe method. We indicate several applications, focusing in particular in the present work on the design of a well-balanced HLLC method for the Euler equations with gravitational source terms.

1. **Introduction.** Finite volume Godunov-type schemes based on Riemann solvers are widely used to compute solutions to hyperbolic systems of equations. Some of the most popular approximate Riemann solvers are the solver of Roe [13] and the solver of Harten–Lax–van Leer (HLL) and its variants. The HLLC solver (HLL with Contact restoration) introduced by Toro, Spruce and Speares [15] for the Euler equations of gas dynamics has especially gained increasing popularity over the last two decades for solving a large variety of compressible flow models, since it possesses some of the good properties of the Roe solver and in addition it satisfies important entropy and positivity conditions with no need of special fixes. In the present work we rewrite the classical HLLC Riemann solver in a novel form that allows an interpretation of the HLLC wave structure as an averaged system eigenstructure. This reveals a formal mathematical similarity of the HLLC solver with the Roe solver, which can be useful to extend to the HLLC method some numerical techniques devised specifically for the Roe method. One application, which has motivated our investigation, is the extension to HLLC-type schemes of low Mach number preconditioning techniques proposed for the Roe scheme. This

2000 *Mathematics Subject Classification.* Primary: 65M08; Secondary: 76N99.

Key words and phrases. Approximate Riemann solvers, Roe solver, HLLC solver, finite volume schemes, Euler equations, well-balanced schemes.

The author was partially funded by *Direction Générale de l'Armement* (DGA) under Grant N. 2012.60.0011.00.470.75.01.

has been illustrated by the author in [10, 11]. In the present work we use our novel formulation of the HLLC solver to apply the f-wave approach of [1] for designing a robust well-balanced HLLC scheme for the Euler equations with gravitational source terms.

2. **The Euler equations of gas dynamics.** The Euler equations governing an inviscid compressible flow can be written in two spatial dimensions in the conservative form:

$$\partial_t q + \partial_x f(q) + \partial_y g(q) = 0, \tag{1a}$$

where

$$q = \begin{bmatrix} \rho \\ \rho u \\ \rho v \\ E \end{bmatrix}, \quad f(q) = \begin{bmatrix} \rho u \\ \rho u^2 + p \\ \rho u v \\ u(E+p) \end{bmatrix}, \quad g(q) = \begin{bmatrix} \rho v \\ \rho v u \\ \rho v^2 + p \\ v(E+p) \end{bmatrix}. \tag{1b}$$

Here ρ is the fluid density, u and v are the flow velocity components in the x and y direction, respectively, p is the pressure, and E is the total energy per unit volume, $E = \mathcal{E} + \rho\frac{|\vec{u}|^2}{2}$, where $\mathcal{E}$ denotes the internal energy per unit volume, and $\vec{u} = (u, v)$. The system is closed through the specification of a pressure law $p = p(\mathcal{E}, \rho)$. The Euler system is hyperbolic and the eigenvalues associated to the direction $\vec{n}$, $|\vec{n}| = 1$, are $\lambda_{1,4} = \vec{u} \cdot \vec{n} \mp c$ and $\lambda_l = \vec{u} \cdot \vec{n}$ for $l = 2, 3$. The speed of sound is $c = \sqrt{\kappa h + \chi}$, where $\kappa = \frac{\partial p(\mathcal{E},\rho)}{\partial \mathcal{E}}$, $\chi = \frac{\partial p(\mathcal{E},\rho)}{\partial \rho}$, and h denotes the specific enthalpy, $h = (\mathcal{E} + p)/\rho$.

3. **Finite volume schemes based on Riemann solvers.** We briefly recall here the class of finite volume schemes based on Riemann solvers in the wave propagation formulation by LeVeque [5, 6, 7]. Let us consider a general hyperbolic system of the form

$$\partial_t q + A(q)\partial_x q + B(q)\partial_y q = 0. \tag{2}$$

We assume a spatial discretization on a Cartesian grid with cells of uniform size Δx and Δy in the x and y directions, respectively. We denote by $Q^n_{i,j}$ the approximate solution of the system at the cell (i, j), $i, j \in \mathbb{Z}$, at time t^n, $n \in \mathbb{N}$, and set $\Delta t = t^{n+1} - t^n$. The two-dimensional first-order wave propagation algorithm [5, 6] has the form

$$\begin{aligned} Q^{n+1}_{i,j} &= Q^n_{i,j} - \frac{\Delta t}{\Delta x}\left(\mathcal{A}^+\Delta Q_{i-1/2,j} + \mathcal{A}^-\Delta Q_{i+1/2,j}\right) \\ &\quad - \frac{\Delta t}{\Delta y}\left(\mathcal{B}^+\Delta Q_{i,j-1/2} + \mathcal{B}^-\Delta Q_{i,j+1/2}\right). \end{aligned} \tag{3}$$

Here $\mathcal{A}^\pm\Delta Q$ and $\mathcal{B}^\pm\Delta Q$ are the so-called fluctuations arising from the solution of local plane-wave Riemann problems in the x and y directions, respectively [5]. To compute these quantities, a Riemann solver must be provided. Let us now consider with no loss of generality the approximation of a two-dimensional plane-wave Riemann problem in the x direction for the Euler equations, namely a Riemann problem for the system $\partial_t q + \partial_x f(q) = 0$, with initial left and right data q_ℓ and q_r. The exact solution of this problem consists of at most four constant states separated by a genuinely nonlinear 1-wave, a contact discontinuity corresponding to the eigenvalue $\lambda_2 = \lambda_3 = u$, and a genuinely nonlinear 4-wave (assuming a convex equation of state). The solution structure defined by an approximate Riemann solver can be expressed by a set of $\mathscr{M}$ waves $\mathcal{W}^l$ and corresponding speeds s^l,

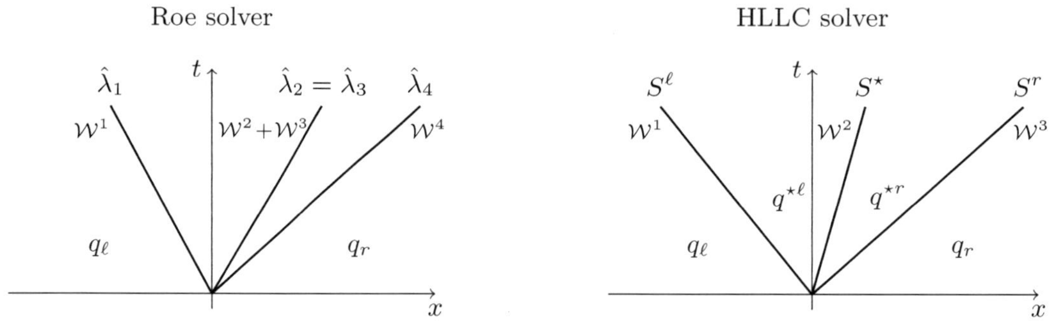

FIGURE 1. Solution structure of the Roe solver (left) and of the HLLC solver (right) for a plane-wave Riemann problem for the 2D Euler equations.

$\mathscr{M} \geq 2$. The so-called *f-waves* $\mathcal{Z}^l$, which carry a jump in the flux, are defined as $\mathcal{Z}^l = s^l \mathcal{W}^l$, $l = 1, \ldots, \mathscr{M}$. For conservation we require:

$$\Delta f \equiv f(q_r) - f(q_\ell) = \sum_{l=1}^{\mathscr{M}} \mathcal{Z}^l. \tag{4}$$

Once the Riemann solution structure associated to each cell pair $\{(i,j),(i+1,j)\}$ is defined, the fluctuations $\mathcal{A}^{\mp}\Delta Q_{i+1/2,j}$ in (3) are computed as

$$\mathcal{A}^- \Delta Q_{i+1/2,j} = \sum_{l:s^l_{i+1/2,j} \leq 0} \mathcal{Z}^l_{i+1/2,j}, \quad \mathcal{A}^+ \Delta Q_{i+1/2,j} = \sum_{l:s^l_{i+1/2,j} > 0} \mathcal{Z}^l_{i+1/2,j}. \tag{5}$$

The first-order scheme (3) can be extended to second-order accuracy by adding suitable correction terms, which can be expressed again in terms of f-waves and speeds [6]. The most general form of the algorithm includes contributions from the decomposition of fluctuations in the transverse direction to account for cross-derivative terms.

3.1. **Roe approximate Riemann solver.** The idea of the approximate Riemann solver of Roe [13] is to define an approximate solution to a Riemann problem for the Euler equations $\partial_t q + \partial_x f(q) = 0$, with q and $f(q)$ as in (1), by the exact solution of a Riemann problem for a linearized system $\partial_t q + \hat{A}(q_\ell, q_r)\partial_x q = 0$. The Roe matrix $\hat{A} = \hat{A}(q_\ell, q_r)$ is defined locally by evaluating the Jacobian $A(q) = f'(q)$ of the original system in a suitable average state $\hat{q} = \hat{q}(q_\ell, q_r)$ that guarantees conservation. The Riemann solution structure of the Roe solver consists of $\mathscr{M} = 4$ waves and speeds that correspond to the eigenstructure of the Roe matrix (see Figure 1). Denoting with $\hat{r}_l$ and $\hat{\lambda}_l$ the right eigenvectors and eigenvalues of $\hat{A}$, respectively, we have $\mathcal{W}^l = \hat{\zeta}_l \hat{r}_l$ and $s^l = \hat{\lambda}_l$, $l = 1, \cdots, 4$, where $\hat{\zeta}_l$ are the coefficients of the projection of $\Delta q \equiv q_r - q_\ell$ onto the basis of the Roe eigenvectors, $q_r - q_\ell = \sum_{l=1}^4 \hat{\zeta}_l \hat{r}_l$. The definition of the Roe eigenstructure is reported in Appendix A. The Roe numerical viscosity matrix is $\Theta = |\hat{A}|$, where $\hat{R} = [\hat{r}_1 \ldots \hat{r}_4]$ and $\hat{\Lambda} = \text{diag}(\hat{\lambda}_1, \ldots, \hat{\lambda}_4)$.

3.2. **HLLC approximate Riemann solver.** The Riemann solution structure of the HLLC solver of Toro *et al.* [15, 14] consists of three waves $\mathcal{W}^l$, $l = 1, 2, 3$ ($\mathscr{M} = 3$), moving at speeds $s^1 = S^\ell$, $s^2 = S^\star$, $s^3 = S^r$, which separate four constant states q_ℓ, $q^{\star\ell}$, $q^{\star r}$ and q_r (see Fig. 1). In the following we will indicate with $(\cdot)_\ell$ and

$(\cdot)_r$ quantities corresponding to the states q_ℓ and q_r, respectively. Moreover, we will indicate with $(\cdot)^{\star\ell}$ and $(\cdot)^{\star r}$ quantities corresponding to the states $q^{\star\ell}$ and $q^{\star r}$ adjacent, respectively on the left and on the right, to the middle wave propagating at speed $S^\star$. With this notation, the waves of the HLLC solver are $\mathcal{W}^1 = q^{\star\ell} - q_\ell$, $\mathcal{W}^2 = q^{\star r} - q^{\star\ell}$, $\mathcal{W}^3 = q_r - q^{\star r}$. We impose conservation conditions, together with the invariance of the pressure p and of the normal velocity u across the 2-wave. Then the speed $S^\star$ is determined as

$$S^\star = \frac{\Delta p + \rho_\ell u_\ell (S^\ell - u_\ell) - \rho_r u_r (S^r - u_r)}{\rho_\ell (S^\ell - u_\ell) - \rho_r (S^r - u_r)}, \tag{6}$$

where $\Delta p \equiv p_r - p_\ell$. The middle states $q^{\star\ell}$, $q^{\star r}$ are found as:

$$q^{\star\iota} = \rho_\iota \frac{S^\iota - u_\iota}{S^\iota - S^\star} \begin{bmatrix} 1 \\ S^\star \\ v_\iota \\ \frac{E_\iota}{\rho_\iota} + (S^\star - u_\iota)\left(S^\star + \frac{p_\iota}{\rho_\iota (S^\iota - u_\iota)}\right) \end{bmatrix}, \quad \iota = \ell, r. \tag{7}$$

A definition for the wave speeds must be provided. For the numerical experiments below we have adopted the definition in [3], $S^\ell = \min(u_\ell - c_\ell, \hat{\lambda}_1)$, $S^r = \max(u_r + c_r, \hat{\lambda}_4)$.

3.3. **A new formulation of the HLLC solver.** We illustrate in this Section a novel formulation of the HLLC solver that allows us to highlight a mathematical similarity with the Roe solver. First we introduce two quantities $\check{c}^\ell$, $\check{c}^r$ representing the speeds of sound associated to the external acoustic waves by defining:

$$S^\ell = u_\ell - \check{c}^\ell \qquad \text{and} \qquad S^r = u_r + \check{c}^r. \tag{8}$$

For any given choice of the estimates of the wave speeds S^ℓ and S^r the relations above determine $\check{c}^\ell$ and $\check{c}^r$. The speed $S^\star$ can be easily rewritten in terms of $\check{c}^\ell$ and $\check{c}^r$:

$$S^\star = \frac{\rho_\ell \check{c}^\ell u_\ell + \rho_r \check{c}^r u_r - \Delta p}{\rho_\ell \check{c}^\ell + \rho_r \check{c}^r}. \tag{9}$$

The densities $\rho^{\star\iota}$, $\iota = \ell, r$, corresponding to the middle states can be expressed as

$$\rho^{\star\ell} = \rho_\ell \frac{\check{c}^\ell}{S^\star - u_\ell + \check{c}^\ell} \qquad \text{and} \qquad \rho^{\star r} = \rho_r \frac{\check{c}^r}{u_r - S^\star + \check{c}^r}. \tag{10}$$

Then, after some easy algebraic manipulations, we see that the HLLC waves for the Euler equations can be equivalently rewritten as

$$\mathcal{W}^1 = \check{\zeta}_1 \check{r}_1, \quad \mathcal{W}^2 = \check{\mathcal{W}}^2 + \check{\mathcal{W}}^2_s, \quad \check{\mathcal{W}}^2 = \check{\zeta}_2 \check{r}_2\ , \check{\mathcal{W}}^2_s = \check{\zeta}_{2s} \check{r}_{2s}\,, \quad \mathcal{W}^3 = \check{\zeta}_3 \check{r}_3\,, \tag{11a}$$

where

$$\check{\zeta}_1 = \frac{\rho^{\star\ell}}{\rho_\ell \check{c}^\ell + \rho_r \check{c}^r}\left(\frac{\Delta p}{\check{c}^\ell} - \rho_r \frac{\check{c}^r}{\check{c}^\ell}\Delta u\right), \quad \check{\zeta}_3 = \frac{\rho^{\star r}}{\rho_\ell \check{c}^\ell + \rho_r \check{c}^r}\left(\frac{\Delta p}{\check{c}^r} + \rho_\ell \frac{\check{c}^\ell}{\check{c}^r}\Delta u\right), \tag{11b}$$

$$\check{\zeta}_2 = \rho^{\star r} - \rho^{\star\ell} = \Delta\rho - \left(\left(\frac{\rho^{\star\ell}}{\check{c}^\ell} + \frac{\rho^{\star r}}{\check{c}^r}\right)\Delta p + \left(\rho_\ell \rho^{\star r}\frac{\check{c}^\ell}{\check{c}^r} - \rho_r \rho^{\star\ell}\frac{\check{c}^r}{\check{c}^\ell}\right)\Delta u\right)\frac{1}{\rho_\ell \check{c}^\ell + \rho_r \check{c}^r}, \tag{11c}$$

$$\check{\zeta}_{2s} = \check{\rho}\Delta v\,, \quad \check{\rho} \equiv \frac{\rho^{\star\ell} + \rho^{\star r}}{2}\,, \quad \Delta(\cdot) \equiv (\cdot)_r - (\cdot)_\ell\,, \tag{11d}$$

and

$$\check{r}_1 = \begin{bmatrix} 1 \\ u_\ell - \check{c}^{\,\ell} \\ v_\ell \\ H_\ell - S^\star \check{c}^{\,\ell} \end{bmatrix}, \ \check{r}_3 = \begin{bmatrix} 1 \\ u_r + \check{c}^{\,r} \\ v_r \\ H_r + S^\star \check{c}^{\,r} \end{bmatrix}, \ \check{r}_2 = \begin{bmatrix} 1 \\ S^\star \\ \bar{v} \\ \Delta \check{e}^\star - \overline{\left(\frac{\chi}{\kappa}\right)} + \frac{(S^\star)^2}{2} + \overline{\left(\frac{v^2}{2}\right)} \end{bmatrix}, \ \check{r}_{2s} = \begin{bmatrix} 0 \\ 0 \\ 1 \\ \bar{v} \end{bmatrix}, \tag{11e}$$

with

$$\Delta \check{e}^\star = \frac{1}{\rho^{\star r} - \rho^{\star \ell}} \left(\rho^{\star r} \frac{c_r^2}{\kappa_r} - \rho^{\star \ell} \frac{c_\ell^2}{\kappa_\ell} - 2\breve{\rho} \Delta \left(\frac{\chi}{\kappa}\right) - \Delta p + \frac{1}{2}\rho^{\star r}(u_r - S^\star)^2 - \frac{1}{2}\rho^{\star \ell}(u_\ell - S^\star)^2 \right). \tag{11f}$$

Note that $\rho^{\star r} \frac{c_r^2}{\kappa_r} - \rho^{\star \ell} \frac{c_\ell^2}{\kappa_\ell} - 2\breve{\rho}\Delta\left(\frac{\chi}{\kappa}\right) - \overline{\left(\frac{\chi}{\kappa}\right)}\check{\zeta}_2 = \rho^{\star r} h_r - \rho^{\star \ell} h_\ell$. Above we have denoted with $H = h + \frac{|\vec{u}|^2}{2}$ the total specific enthalpy and we have used the average operator $\overline{(\cdot)} \equiv \frac{(\cdot)_\ell + (\cdot)_r}{2}$. The expressions of the HLLC waves in this novel form reveal analogies with the waves of the Roe solver. We observe that the vectors $\check{r}_l$, like the Roe eigenvectors, have the form of the eigenvectors of the Euler system $r_l(q)$ evaluated in a special state that is a function of the left and right Riemann data (although not in the form $\hat{r}_l = r_l(\hat{q})$ as for the Roe solver), except for the quantity $\Delta \check{e}^\star$ appearing in the last component of the vector $\check{r}_2$. This quantity becomes singular if $\check{\zeta}_2 = \rho^{\star r} - \rho^{\star \ell} = 0$, which happens in the trivial case of uniform flow, $q_r = q_\ell$, but also in other cases (e.g. $p_r = p_\ell$, $\rho_\ell = \rho_r$, $u_\ell = -u_r$). Note that in such situations the wave $\check{\mathcal{W}}^2$ is simply zero, and the Riemann solution is always well defined. For consistency we expect

$$\lim_{\rho^{\star r} - \rho^{\star \ell} \to 0} \Delta \check{e}^\star = 0. \tag{12}$$

Note that if the matrix $\check{R} = [\check{r}_1, \check{r}_2, \check{r}_{2s}, \check{r}_3]$ is nonsingular we can interpret the Riemann solution of the HLLC solver as the Riemann solution of a linearized system with a constant coefficient matrix $\check{A} = \check{A}(q_\ell, q_r) = \check{R}\check{\Lambda}\check{R}^{-1}$, where $\check{\Lambda} = \text{diag}(u_\ell - \check{c}^{\,\ell}, S^\star, S^\star, u_r + \check{c}^{\,r})$. The HLLC numerical viscosity matrix is identified as $\Theta = |\check{A}|$.

4. Applications.

4.1. Low Mach number preconditioning techniques. The origin of the present work came from a study aimed at extending popular low Mach number preconditioning techniques for the Roe's scheme to the HLLC scheme. These techniques typically modify at low Mach number the acoustic waves and speeds that contribute to the numerical viscosity term $\sum_{l=1}^{4}(|\hat{\lambda}_l|\hat{\zeta}_l \hat{r}_l) = |\hat{A}|(q_r - q_\ell)$. Thanks to the novel formulation of the HLLC solver we were able to mimic a preconditioning technique proposed for Roe's scheme and apply it to the HLLC scheme, both for the Euler equations [11] and for a two-phase flow model [10]. We refer to [11, 10] for details.

4.2. Well-balanced f-wave method for hyperbolic systems with source terms. We illustrate here an application of the new form of the HLLC solver for the design of a robust well-balanced f-wave method for the Euler equations (1) with a source term $\Psi(q)$. In particular we shall consider a source term of the form $\Psi = [0, -\rho\nabla\varphi, -\rho\vec{u}\cdot\nabla\varphi]^{\mathrm{T}}$, where $\varphi(\vec{x})$ is a gravitational potential. We recall that a scheme is well-balanced if it can preserve stationary states at the discrete level and if it is able to accurately model small perturbations from steady states. In the numerical algorithm we need to solve plane-wave Riemann problems for a system of the form

$$\partial_t q + \partial_x f(q) = \psi(q), \quad \psi(q) = [0, -\rho\partial_x\varphi, 0, -\rho u\partial_x\varphi]^{\mathrm{T}}, \tag{13}$$

where $\partial_x\varphi = g\,n^{(x)}$, denoting here with $n^{(x)}$ the x-component of the unit vector $\vec{n}(\vec{x})$ indicating the direction of the gravity field $\vec{g}$. The idea of the f-wave approach [1] (see also [9]) is to include the contribution of the source term $\psi(q)$ in the jump of the fluxes that is decomposed into f-waves (cf. (4)), that is:

$$f(q_r) - f(q_\ell) - \tilde{\psi}_\Delta = \sum_{l=1}^{\mathscr{M}} \mathcal{Z}^l, \tag{14}$$

where $\tilde{\psi}_\Delta$ is a discrete interface value of the source term contribution. If this term is defined such that the discrete condition $\Delta f - \tilde{\psi}_\Delta = 0$ expresses steady conditions, and if the f-wave decomposition (14) is obtained by a projection of $\Delta f - \tilde{\psi}_\Delta$ onto a set of $\mathscr{M}$ linearly independent vectors, then we observe that steady states are maintained by the method. In fact if initially $\Delta f - \tilde{\psi}_\Delta = 0$, then the f-waves in (14) are simply zero, hence equilibrium is preserved [1, 9]. The discrete source term contribution $\tilde{\psi}_\Delta$ in (14) can be simply defined as:

$$\tilde{\psi}_\Delta = [0,\ -g\,n^{(x)}(\bar{\vec{x}})\,\overline{\rho}\Delta x,\ 0,\ -g\,n^{(x)}(\bar{\vec{x}})\,\overline{\rho u}\Delta x]^{\mathrm{T}}, \quad \overline{(\cdot)} \equiv \frac{1}{2}((\cdot)_\ell + (\cdot)_r). \tag{15}$$

This general definition has proven to be efficient in all the numerical tests that we have performed. If the exact steady solution is available, then we may be able to define a term $\tilde{\psi}_\Delta$ that gives an exact discrete version of the stationary conditions. For instance, let us consider the exact solution for the isothermal equilibrium in one dimension of an ideal gas $p_0(x) = \rho_0(x) = \exp(-gx)$, $u_0(x) = 0$, which characterizes one test problem below. Then, we can take:

$$\tilde{\psi}_\Delta = [0,\ \overline{\rho\exp(g\,x)}\Delta(\exp(-g\,x)),\ \overline{\rho u\exp(g\,x)}\Delta(\exp(-g\,x))]^{\mathrm{T}}. \tag{16}$$

The splitting (14) can be performed by a projection onto the Roe eigenvectors, $\Delta f - \tilde{\psi}_\Delta = \sum_{l=1}^{4}\hat{\beta}_l\hat{r}_k$, hence we define the f-waves as $\mathcal{Z}^l = \hat{\beta}_l\hat{r}_l$, $\hat{\beta} = \hat{R}^{-1}(\Delta f - \tilde{\psi}_\Delta)$. This f-wave Roe method (for instance used in [12]) results to be very efficient for treating sources, nonetheless it suffers from the drawbacks of the Roe method, namely unphysical states in low density regions and computation of non-entropic shocks. Note that while several entropy fixes are available for the standard Roe scheme, it might be complicated to use them within the f-wave framework (since we would need to compute the waves from the f-waves). To overcome these difficulties we apply the f-wave approach to the HLLC method, by projecting $\Delta f - \tilde{\psi}_\Delta$ onto the HLLC vectors $\{\check{r}_l\}_{1\le l\le 4}$. Hence we define $\mathcal{Z}^l = \check{\beta}_l\check{r}_l$, $\check{\beta} = \check{R}^{-1}(\Delta f - \tilde{\psi}_\Delta)$. The explicit analytical expression of $\check{R}^{-1}$ can be easily obtained. To handle the problem of the singularity of $\Delta\check{e}^\star$ in (11f) one simple option is to set this quantity to zero if $|\check{\zeta}_2| < \epsilon$ (e.g. $\epsilon = 10^{-15}$). Another option for instance is to employ the desingularizing definition $1/\check{\zeta}_2 \triangleq 2\check{\zeta}_2/(\check{\zeta}_2^2 + \max(\check{\zeta}_2^2, \epsilon^2))$ [4].

4.2.1. *Numerical experiments.* All the tests are performed with second-order algorithms, MC limiter, Courant number = 0.9. We assume an ideal gas with $\gamma = 1.4$ and we set $g = 1$.

Riemann problems. We solve two Riemann problems to show the advantages of the f-wave HLLC method with respect to the f-wave Roe method. The computational domain is $[0, 1]$ and the initial discontinuity is at $x = 0.5$. Free flow boundary conditions are used. First, we solve a problem with $\rho_\ell = 3$, $\rho_r = 1$, $p_\ell = 3$, $p_r = 1$, $u_\ell = u_r = 0.9$. The solution contains a left-going transonic rarefaction, which is computed correctly by the HLLC method but not by the Roe method, which would need an entropy fix. See Fig. 2, left plot. The second test is a version with gravity of the double rarefaction test of [3]. Here $\rho_\ell = \rho_r = 1$, $p_\ell = p_r = 0.4$, $u_r = -u_\ell = 2$. The solution involves two rarefactions going in opposite directions that form a region of very low density and pressure in between. The Roe scheme fails for this test, whereas the HLLC scheme computes the solution with no difficulties (note that gravity here pulls slightly toward the left). See Fig. 3, left plot. Results for a double rarefaction test in two dimensions are displayed in Fig. 3, right plot. Here we have a setup analogous to the 1D test, with initial discontinuity at $x = 1$ in the domain $[0, 2] \times [0, 2]$. Gravity acts downwards along the y axis. Top and bottom boundaries are walls.

Perturbation of isothermal equilibrium. We perform a one-dimensional test proposed in [8] to investigate the well-balanced property of the method. A small perturbation of the isothermal equilibrium conditions $p_0(x) = \rho_0(x) = \exp(-gx)$, $u_0(x) = 0$, is considered for the pressure field: $p(x)|_{t=0} = p_0(x) + \eta \exp(-100(x-0.5)^2)$, $\eta = 10^{-4}$, $x \in [0, 1]$. The f-wave HLLC method exhibits the same good behavior of the f-wave Roe method, see Fig. 2, right plot, and results are qualitatively similar to those in the literature [8, 16, 2].

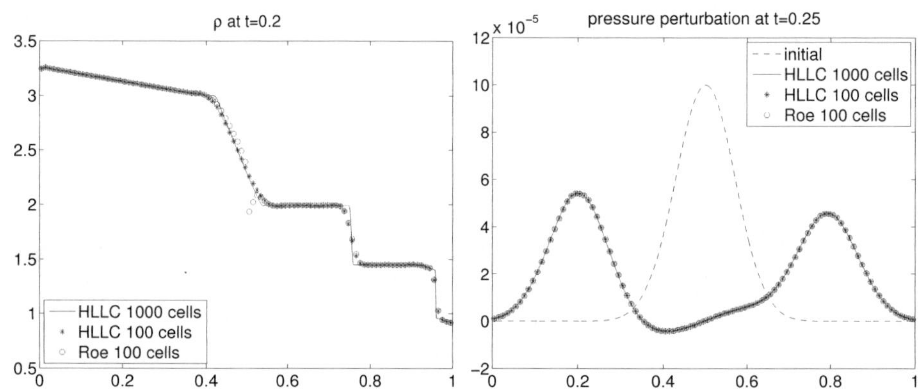

FIGURE 2. Left: Transonic rarefaction test with gravity ($\varphi = gx$), ρ at $t = 0.2$. Right: Perturbation of isothermal equilibrium test. $p(x, t) - p_0(x)$ at $t = 0.25$. HLLC ($*$), Roe ($\circ$) results with 100 cells, HLLC results with 1000 cells ($-$).

Radial Rayleigh–Taylor instability. We perform the two-dimensional Rayleigh–Taylor instability test proposed in [8] (with initial conditions as in [2]). Here we consider a radial gravitational potential (gravity is directed inward, $\varphi = g|\vec{x}|$). A radially symmetric isothermal equilibrium is assumed, for which the pressure is continuous across $|\vec{x}| = r_0 = 0.6$ but has a density jump $\Delta_\rho = 0.1$ across $|\vec{x}| = r_0(1 + \eta \cos(\xi\theta))$, $\xi = 20$, $\eta = 0.02$. Results are shown in Figure 4 for a second-order computation on the domain $[-1, 1] \times [-1, 1]$ with 240×240 grid cells. As expected we see Rayleigh–Taylor instabilities arise at the interface with the density jump. Elsewhere equilibrium conditions are well maintened, and results are analogous to [8].

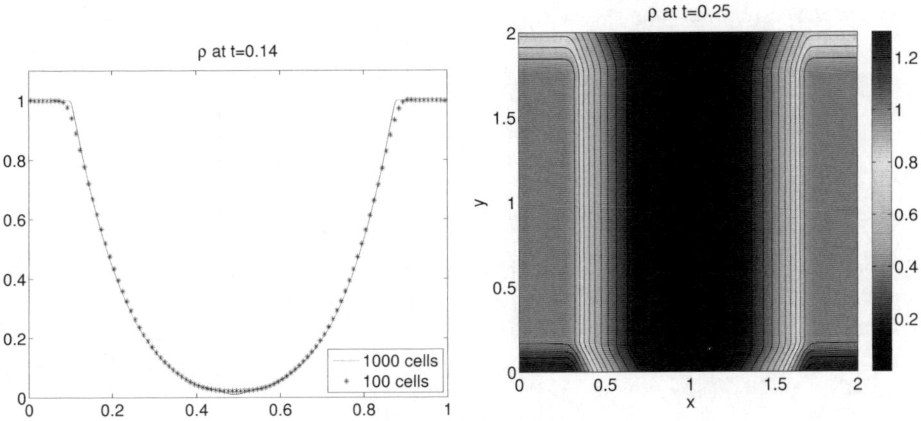

FIGURE 3. Double rarefaction test with gravity computed by the HLLC method. Left: 1D test ($\varphi = gx$), ρ at $t = 0.14$, results with 100 cells ($*$) and 1000 cells ($-$). Right: 2D test ($\varphi = gy$), ρ at $t = 0.25$, results with 200×200 cells.

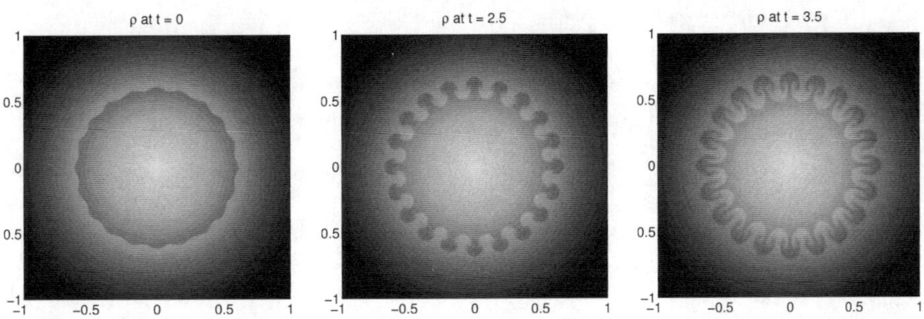

FIGURE 4. Rayleigh–Taylor instability test. 2nd order HLLC method, 240×240 cells. Density at $t = 0$, 2.5, 3.5. Color scale from 0.2 (darker) to 1.5 (lighter).

5. **Conclusions.** We have presented a reformulation of the HLLC Riemann solver, which shows that the HLLC wave structure can be interpreted as an averaged system eigenstructure. In particular, we use this new Roe-like form of the HLLC solver to develop a robust second-order well-balanced f-wave HLLC method for the solution of the Euler equations with gravitational source terms. The presented reformulation of the HLLC solver can be used for other applications [11], and it could be derived also for more complex flow models, see e.g. [10].

Appendix A. **Roe eigenstructure for the Euler equations.** We recall the eigenstructure of the Roe matrix $\hat{A}(q_\ell, q_r)$ for a plane-wave Riemann problem in the x direction with data q_ℓ, q_r for the Euler equations. We assume κ, $\chi =$ constant. We introduce the averages:

$$\hat{a} = \frac{a_\ell\sqrt{\rho_\ell}+a_r\sqrt{\rho_r}}{\sqrt{\rho_\ell}+\sqrt{\rho_r}}, \quad a = u, v, H, \quad \hat{\rho} = \sqrt{\rho_\ell \rho_r}, \quad \hat{c} = \sqrt{\kappa(\hat{H} - \hat{\mathcal{K}}) + \chi}, \quad \hat{\mathcal{K}} = \frac{\hat{u}^2+\hat{v}^2}{2}. \tag{17}$$

The Roe eigenvalues are $\hat{\lambda}_1 = \hat{u} - \hat{c}$, $\hat{\lambda}_2 = \hat{\lambda}_3 = \hat{u}$, $\hat{\lambda}_4 = \hat{u} + \hat{c}$. The matrix $\hat{R} = [\hat{r}_1, \dots, \hat{r}_4]$ of the corresponding Roe right eigenvectors is

$$\hat{R} = \begin{pmatrix} 1 & 1 & 0 & 1 \\ \hat{u} - \hat{c} & \hat{u} & 0 & \hat{u} + \hat{c} \\ \hat{v} & \hat{v} & 1 & \hat{v} \\ \hat{H} - \hat{u}\hat{c} & -\frac{\chi}{\kappa} + \hat{\mathcal{K}} & \hat{v} & \hat{H} + \hat{u}\hat{c} \end{pmatrix}. \tag{18}$$

The coefficients $\hat{\zeta}_l$, $l = 1, \dots 4$, of the Roe eigen-decomposition $q_r - q_\ell = \sum_{l=1}^{4} \hat{\zeta}_l \hat{r}_l$, are:

$$\hat{\zeta}_1 = \frac{1}{2\hat{c}}\left(\frac{\Delta p}{\hat{c}} - \hat{\rho}\,\Delta u\right), \quad \hat{\zeta}_2 = \Delta\rho - \frac{\Delta p}{\hat{c}^2}, \quad \hat{\zeta}_3 = \hat{\rho}\Delta v, \quad \hat{\zeta}_4 = \frac{1}{2\hat{c}}\left(\frac{\Delta p}{\hat{c}} + \hat{\rho}\,\Delta u\right). \tag{19}$$

REFERENCES

[1] D. Bale, R. J. LeVeque, S. Mitran and J. A. Rossmanith, A wave-propagation method for conservation laws and balance laws with spatially varying flux functions, *SIAM J. Sci. Comput.*, **24** (2002), 955–978.

[2] P. Chandrashekaand and C. Klingenberg, A second-order well-balanced finite volume scheme for Euler equations with gravity, *SIAM J. Sci. Comput.*, **37** (2018), B382–B402.

[3] B. Einfeldt, C. D. Munz, P. L. Roe and B. Sjögreen, On Godunov-type methods near low densities, *J. Comput. Phys.*, **92** (1991), 273–295.

[4] A. Kurganov and G. Petrova, A second-order well-balanced positivity preserving central-upwind scheme for the Saint-Venant system, *Commun. Math. Sci.*, **5** (2007), 133–160.

[5] R. J. LeVeque, Wave propagation algorithms for multi-dimensional hyperbolic systems, *J. Comput. Phys.*, **131** (1997), 327–353.

[6] R. J. LeVeque, *Finite Volume Methods for Hyperbolic Problems*, Cambridge University Press, 2002.

[7] R. J. LeVeque, CLAWPACK Software, `http://www.clawpack.org`.

[8] R. J. LeVeque and D. S. Bale, Wave-Propagation Methods for Conservation Laws with Source Terms, in *Proc. of the 7th Intl. Conf. on Hyperbolic Problems*, (Ed. R. Jeltsch), Birkhäuser Verlag (1998), 609–618.

[9] R. J. LeVeque and M. Pelanti, A class of approximate Riemann solvers and their relation to relaxation schemes, *J. Comput. Phys.*, **172** (2001), 572–591.

[10] M. Pelanti, Low Mach number preconditioning techniques for Roe-type and HLLC-type methods for a two-phase compressible flow model, *Appl. Math. Comp.*, **310** (2017), 112–133.

[11] M. Pelanti, Wave structure similarity of the HLLC and Roe Riemann solvers: Application to low Mach number preconditioning, *SIAM J. Sci. Comput.*, **40** (2018), A1836–A1859.

[12] M. Pelanti and R. J. LeVeque, High-Resolution Finite Volume Methods for dusty gas jets and plumes, *SIAM J. Sci. Comput.*, **28** (2006), 1335–1360.

[13] P. L. Roe, Approximate Riemann solvers, parameter vectors, and difference schemes, *J. Comput. Phys.*, **43** (1981), 357–372.

[14] E. F. Toro, *Riemann Solvers and Numerical Methods for Fluid Dynamics*, Springer-Verlag, 1997.

[15] E. F. Toro, M. Spruce and W. Speares, Restoration of the contact surface in the HLL Riemann solver, *Shock Waves*, **4** (1994), 25–34.

[16] Y. Xing and C.-W. Shu, High Order Well-Balanced WENO Scheme for the Gas Dynamics Equations Under Gravitational Fields, *J. Sci. Comput.*, **54** (2013), 645–662.

E-mail address: `marica.pelanti@ensta-paristech.fr`

EXISTENCE OF STEADY TWO-PHASE FLOWS WITH DISCONTINUOUS BOILING EFFECTS

TEDDY PICHARD*

Centre de Mathématiques Appliquées, UMR 7641, École polytechnique,
Route de Saclay
Palaiseau, 91128, France

ABSTRACT. We aim at characterizing the existence and uniqueness of steady solutions to hyperbolic balance laws with source terms depending discontinuously on the unknown. We exhibit conditions for such differential equations to be well-posed and apply it to a model describing boiling flows.

1. **Introduction.** The aim of this paper is to present a framework for the study of steady states of 1D balance laws with sources defined as a discontinuous function of the unknown. Such steady states satisfy systems of the form

$$\frac{d}{dx}F(U)(x) = S(U(x)), \tag{1a}$$

where the source jumps when a certain function h reaches a threshold, *i.e.*

$$S(U) = \begin{cases} S^-(U) & \text{if} \quad h(U) < 0, \\ S^+(U) & \text{if} \quad h(U) \geq 0. \end{cases} \tag{1b}$$

The discontinuity of S with respect to the unknown leads to both theoretical and numerical difficulties. Especially, Picard-Lindelöf theory is unavailable and extensions are required.

The application we have in mind is the study of boiling flows. We aim at studying the homogenized two-phase flow model based on a drift-flux model ([11, 10, 9]) used for the developpement of the FLICA4 code ([15, 3, 14])

$$\partial_t U + \partial_x F(U) = S(U), \tag{2a}$$

$$U = \left(\alpha\rho_v,\ \rho,\ \rho u,\ \rho\left(e + \frac{u^2}{2}\right)\right)^T, \tag{2b}$$

$$F(U) = \left(\alpha\rho_v u,\ \rho u,\ \rho u^2 + p,\ \rho\left((e + \frac{u^2}{2} + \frac{p}{\rho})u\right)\right)^T, \tag{2c}$$

$$S(U) = \begin{cases} (0,\ 0,\ 0,\ \phi)^T & \text{if} \quad h(U) < h^b, \\ (K\phi,\ 0,\ 0,\ \phi)^T & \text{if} \quad h(U) \geq h^b, \end{cases} \tag{2d}$$

with a constant $K > 0$. Here, $\alpha\rho_v$ is the density of vapor alone, and ρ, ρu, ρe are the density, momentum and energy of the homogenized flow, *i.e.* of liquid and

---

2000 *Mathematics Subject Classification.* 35R05, 35Q35, 34A36.

Key words and phrases. Discontinuous source term, Ordinary differential equation, Steady state, Carathéodory solution, Well-posedness.

* Corresponding author: Teddy Pichard.

vapor together. The source term models the heating of the fluid, through the term $\phi > 0$ in the energy equation, and the creation of vapor (in the first equation) when the enthalpy h is above a boiling threshold h^b.

In the next two sections, we first present a framework that guarantees the existence and uniqueness of solution of (1), first on a very simple scalar case, then on a more general vectorial framework. This is applied to the problem (2) in Section 4. Section 5 is devoted to conclusion and outlooks.

2. Preliminaries.

Consider the Cauchy problem

$$\frac{dU}{dx} = S(U, x), \qquad U(0) = U_0. \tag{3}$$

Here, $S : \mathbb{R}^N \times \mathbb{R} \to \mathbb{R}^N$ is a function of $U \in \mathbb{R}^N$ and $x \in \mathbb{R}$ that may be discontinuous. As S is not continuous, we need a definition of solutions to (3) in a weak sense.

Definition 2.1. Let I be an open interval of $\mathbb{R}$ containing 0. A function $U : I \subset \mathbb{R} \to \mathbb{R}$ is a Carathéodory solution to (3) if it is absolutely continuous and satisfies

$$\forall x \in I, \qquad U(x) = U_0 + \int_0^x S(U(y), y)dy.$$

In order to illustrate the difficulties emerging with discontinuous right-hand-side (RHS) in (3), let us first consider the following simple scalar case (inspired by [12, 8])

$$\frac{d}{dx}u = \begin{cases} s^- & \text{if} \quad u < 0, \\ s^+ & \text{if} \quad u \geq 0, \end{cases} \qquad u(0) = u_0. \tag{4}$$

The behavior of u away from 0 is well understood. Difficulties arise when u reaches 0. We can list three types of behavior (represented on Fig. 1):

1. If $s^- \geq 0$ and $s^+ \leq 0$, then for all $u_0 \in \mathbb{R}$

$$u(x) = \begin{cases} u_0 + s^- x & \text{if} \quad u_0 \leq 0 \quad \text{and} \quad x \leq \dfrac{-u_0}{s^-}, \\ u_0 + s^+ x & \text{if} \quad u_0 \geq 0 \quad \text{and} \quad x \leq \dfrac{-u_0}{s^+}. \end{cases} \tag{5a}$$

However this solution can not be extended for x larger than $u_0/s^{\pm}$.

2. If $s^- \leq 0$ and $s^+ \geq 0$, then for all $u_0 \in \mathbb{R}$

$$u(x) = \begin{cases} u_0 + s^- x & \text{if} \quad u_0 \leq 0, \\ u_0 + s^+ x & \text{if} \quad u_0 \geq 0. \end{cases} \tag{5b}$$

Remark that, if $u_0 = 0$, the functions $x \mapsto s^- x$ and $x \mapsto s^+ x$ are two Carathéodory solutions of (4).

3. If s^- and s^+ have strictly the same sign, say positive, then for all $x \geq 0$,

$$u(x) = \begin{cases} u_0 + s^+ x & \text{if} \quad u_0 \geq 0, \\ u_0 + s^- x & \text{if} \quad u_0 \leq 0 \quad \text{and} \quad x \leq \dfrac{-u_0}{s^-}, \\ u_0 + s^- \dfrac{-u_0}{s^-} + s^+ \left(x - \dfrac{-u_0}{s^-} \right) & \text{if} \quad u_0 \leq 0 \quad \text{and} \quad x \geq \dfrac{-u_0}{s^-}. \end{cases} \tag{5c}$$

The solutions defined in these three cases are depicted in the phase space (x, u) on Fig. 1. Remark that on this simple example, neither existence nor uniqueness of a solution is guaranteed. Thus, further considerations are necessary to obtain the well-posedness of (3) in a general case or of (1) for our applications.

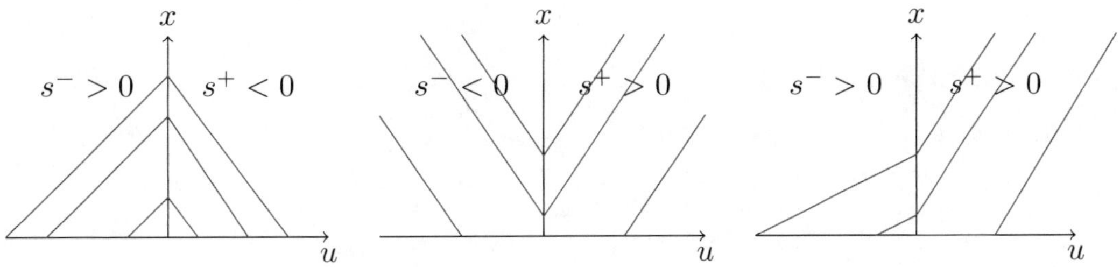

FIGURE 1. Solutions of (5) depending on the signs of s^- and s_+: from left to right, solutions of (5a), (5b) and (5c)

In the next section, we focus on a vectorial ODE. We prove its well-posedness under a condition corresponding to a vectorial version of the third case (5c).

3. A framework for ODE with Heaviside RHS.

Consider now the problem

$$\frac{d}{dx}U(x) = \begin{cases} S^-(U(x), x) & \text{if} \quad h(U(x)) < 0, \\ S^+(U(x), x) & \text{if} \quad h(U(x)) \geq 0, \end{cases} \qquad U(0) = U_0, \tag{6}$$

where the unknown $U(x) \in \mathbb{R}^N$ is vectorial and the enthalpy $h(U)$ is scalar.

We seek a natural framework for (6) to be well-posed. The result below could be obtained as a corollary of *e.g.* [13, 4, 5] or through Filippov's theory ([7, 1, 6]). Here we present a simple condition on the surface $h(U) = 0$ under which any solution changes sign at most once. The solution is then obtained by gluing together two solutions obtained with the Picard-Lindelöf theorem.

Lemma 3.1. *Suppose that*

- $h \in C^1(\mathbb{R}^N, \mathbb{R})$,
- *Both S^- and S^+ satisfy the hypothesis of the Picard-Lindelöf theorem: continuity with respect to x and locally Lipschitz continuity with respect to U,*
- $\forall x \in \mathbb{R}$, *and* $\forall V \in \mathbb{R}^N$, *such that* $h(V) = 0$,

$$(\nabla_U h(V).S^-(V, x)) > 0 \quad \textit{and} \quad (\nabla_U h(V).S^+(V, x)) > 0. \tag{7}$$

Then, for any Carathéodory solutions $\bar{U}$ to (6), there exists at most one point $x_0 \in \mathbb{R}$ such that $h(\bar{U})(x_0) = 0$, and $h(\bar{U})$ is strictly negative on $x < x_0$ and strictly positive on $x > x_0$.

Remark 1. The vector $\nabla_U h(V)$ is normal to the hypersurface $\{U \in \mathbb{R}^N, \text{ s.t. } h(U) = 0\}$. Thus, the condition (7) imposes that the vector fields S^- and S^+ are both pushing the solution toward the same side of $h(U) = 0$. The solution is then constructed by following S^- until it reaches $h(U) = 0$, and then following S^+ (see Fig. 2).

Proof. First, we remark that, as h is $C^1(\mathbb{R}^N, \mathbb{R})$ and the Carathéodory solution $\bar{U}$ is absolutely continuous, then $h(\bar{U})$ is continuous and has a derivative almost everywhere which is

$$\frac{d}{dx}h(\bar{U})(x) = \nabla_U h(\bar{U})(x).S(\bar{U}(x), x). \tag{8}$$

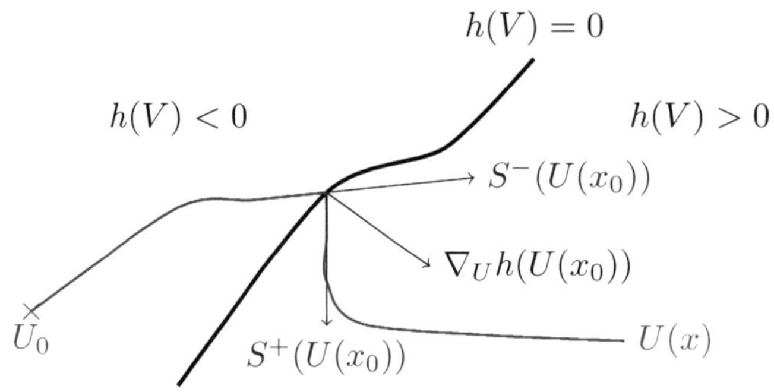

FIGURE 2. Representation, for a problem of the form (6) in $\mathbb{R}^2$, of the solution $U(x) \in \mathbb{R}^2$, the hypersurface $\{V \in \mathbb{R}^2 \text{ s.t. } h(V) = 0\}$ and the vectors $S^-(U(x_0))$, $S^+(U(x_0))$ and $\nabla_U h(U(x_0))$

Then, assume there exists a point x_0 such that $h(\bar{U}(x_0)) = 0$. Then, for all $y \geq 0$ (we may reason similarly for $y < 0$),

$$\begin{aligned} h(\bar{U})(x_0 + y) &= \int_{x_0}^{x_0+y} \nabla_U h(\bar{U})(x).S(\bar{U}(x), x)dx \\ &\geq \int_{x_0}^{x_0+y} \min(\nabla_U h(\bar{U})(x).S^-(\bar{U}(x), x), \nabla_U h(\bar{U})(x).S^+(\bar{U}(x), x))dx, \end{aligned}$$

The function in the last integral is continuous and strictly positive at $x = x_0$ by (7). Thus there exists $\epsilon > 0$ such that

$$\forall x \in]x_0, x_0 + \epsilon[, \quad h(\bar{U})(x) > 0, \qquad \text{and} \qquad \forall x \in]x_0 - \epsilon, x_0[, \quad h(\bar{U})(x) < 0. \tag{9}$$

Suppose by contradiction that there exists $x_1 > x_0$ such that $h(\bar{U})(x_1) = 0$. The continuity of $h(\bar{U})$ and (9) yield the existence of x_2 in (x_0, x_1), such that $h(\bar{U}(x_2)) = 0$. Repeating this operation, we construct a sequence $(x_i)_{i\in\mathbb{N}}$ of distinct points where $h(\bar{U})$ is null, and that converges towards a limit denoted by x_∞ by dichotomy. Considering that

$$\begin{aligned} |h(\bar{U})(x_\infty)| &= \left| h(\bar{U})(x_i) + \int_{x_i}^{x_\infty} \nabla_U h(\bar{U})(x).S(\bar{U}(x), x)dx \right| \\ &\leq |x_\infty - x_i| \|\nabla_U h(\bar{U})(x)\|_{\infty,[x_0,x_1]} \|S(\bar{U}(x), x)\|_{\infty,[x_0,x_1]}. \end{aligned}$$

we obtain $x_i \to_{i\to+\infty} x_\infty$ and $h(\bar{U}(x_\infty)) = 0$, which contradicts the existence of the interval (9). Once we know that $h(\bar{U})$ has at most one zero, (9) gives the sign of $h(\bar{U})$ on both sides. □

Proposition 1. *Under the hypothesis of Lemma 3.1, for all initial conditions $U_0 \in \mathbb{R}^N$, there exists a unique maximal solution U to (6) that is absolutely continuous. Furthermore, this solution U depends continuously on U_0.*

Proof. We prove the case $h(U_0) < 0$, the other one being completely similar. According to Lemma 3.1, there is at most one point x_0 where $h(U)$ switches sign, and as $h(U(0)) < 0$ it is larger than 0. Thus, any Carathéodory solution U takes the form

$$U(x) = U_0 + \begin{cases} \int_0^x S^-(U(y), y)dy & \text{if} \quad x < x_0, \\ \int_0^{x_0} S^-(U(y), y)dy + \int_{x_0}^x S^+(U(y), y)dy & \text{otherwise.} \end{cases} \tag{10}$$

The existence and uniqueness follows from the Picard-Lindelöf theory. Indeed on $x < x_0$ the solution coincides with the solution of the Cauchy problem

$$V'(x) = S^-(V(x), x), \qquad V(0) = U_0$$

which exists and is unique as S^- is continuous and locally Lipschitz continuous with respect to its first variable. Then on $x \geq x_0$, it coincides with the solution of the Cauchy problem

$$V'(x) = S^+(V(x), x), \qquad V(x_0) = U(x_0).$$

To conclude the proof it remains to show that x_0 is a continuous function of the initial data U_0. Fix U_0 and x_0 such that

$$\varphi(U_0, x_0) = h(U(x_0)) = h\left(U_0 + \int_0^{x_0} S^-(U(y), y)dy\right) = 0$$

As $\frac{\partial \varphi}{\partial x_0}(U_0, x_0) = \nabla_U h\,(U(x_0)) \cdot S^-(U(x_0), x_0)$ is not null by (7), the implicit function theorem yields the result. $\square$

4. **Application to homogenized two-phase fluid models.** First, we rewrite Proposition 1, then we apply it to a reformulation of (2).

4.1. **With a non-linear flux.** When the flux function F in (1) is non-linear, we may simply adapt Proposition 1 into the following result.

Corollary 1. *Suppose that*

- $F \in C^1(\mathbb{R}^N, \mathbb{R}^N)$,
- $h \in C^1(\mathbb{R}^N, \mathbb{R})$,
- S^- *and* S^+ *are continuous w.r.t.* x *and locally Lipschitz continuous w.r.t.* U,
- $\forall x \in [0, L]$, *and* $\forall V \in \mathbb{R}^N$, *s.t.* $h(V) = 0$,

$$\nabla_U h(V).(DF(V))^{-1}.S^-(V, x) > 0, \text{ and } \nabla_U h(V).(DF(V))^{-1}.S^+(V, x) > 0. \tag{11}$$

Then, for all initial conditions $U_0 \in \mathbb{R}^N$ *satisfying* $\det(DF(U_0)) \neq 0$, *there exists a unique maximal solution* U *to* (1) *absolutely continuous and satisfying* $\det(DF(U)) \neq 0$. *Furthermore, this solution depends continuously on* U_0.

Remark 2. Requiring that $DF(U)$ is invertible corresponds to imposing that the flows remains subsonique and admissible, which is commonly admitted for practical applications. This condition may restrict the size of the spatial domain.

Proof. Any Carathéodory solution U to (1) is differentiable almost everywhere. Thus, as $F \in C^1(\mathbb{R}^N, \mathbb{R}^N)$, then $F(U)$ is absolutely continuous and differentiable almost everywhere, and its derivative equals almost everywhere

$$\frac{d}{dx}F(U)(x) = DF(U)(x).\frac{d}{dx}U(x).$$

Thus any solution to (1) of such regularity and satisfying $\det(DF(U)) \neq 0$, also solves the Cauchy problem

$$\frac{d}{dx}U(x) = \begin{cases} (DF(U)(x))^{-1}.S^{-}(U(x),x) & \text{if} \quad h(U)(x) < 0, \\ (DF(U)(x))^{-1}.S^{+}(U(x),x) & \text{if} \quad h(U)(x) \geq 0. \end{cases} \tag{12}$$

Using Proposition 1 and the hypothesis, (12) has a unique solution U and it depends continuously on U_0. □

4.2. **On the boiling flow model.** Now, we aim to apply this result to (1). In order to apply Corollary 1, we rewrite the problem with a new set of unknowns $\tilde{U}$ such that

- we can perform the computations required in (11) ;
- it has a physical interpretation.

We chose for variables

$$\tilde{U} = (c_v,\ q,\ p,\ h),$$

where c_v is the volume fraction of vapor, q is the momentum. The enthalpy h is chosen among the variables to simplify the definition of $\nabla_U h$ and q to simplify the definition of $D\tilde{F}$. These variables $\tilde{U}$ are commonly defined based on U as

$$\tilde{U} = \phi^{-1}(U) = \left(\frac{\alpha\rho_v}{\rho},\ \rho u,\ p,\ e + \frac{p}{\rho}\right), \qquad U = \phi(\tilde{U}) = \left(\frac{c_v}{\tau},\ \frac{1}{\tau},\ q,\ \frac{h}{\tau} - p + \frac{\tau q^2}{2}\right),$$

where $\tau = \frac{1}{\rho}$ is the specific volume. We close the new system, not by expressing p as a function of U (it is a variable in the new system), but by fixing

$$\tau = c_v\tau_v + (1 - c_v)\tau_l,$$

as a convex combination of the vapor and liquid specific volumes τ_v and τ_l, where $\tau_v(p,h)$ and $\tau_l(p,h)$ are given $C^1(\mathbb{R}^2,\mathbb{R})$ functions of p and h, and independent of q and c_v. These functions are commonly tabulated.

Rewriting the steady state of (2) in terms of $\tilde{U}$ reads

$$\frac{d}{dx}\tilde{F}(\tilde{U}) = \tilde{S}(\tilde{U}) \tag{13}$$

$$\tilde{F}(\tilde{U}) = F \circ \phi(\tilde{U}) = \left(c_v q,\ q,\ \tau q^2 + p,\ \frac{\tau^2 q^3}{2} + qh\right),$$

$$\tilde{S}(\tilde{U}) = S \circ \phi(\tilde{U}) = \begin{cases} (0, & 0, & 0, & \phi) & \text{if } h < h^b, \\ (K\phi, & 0, & 0, & \phi) & \text{if } h \geq h^b. \end{cases}$$

We obtain in the end the following requirement.

Proposition 2. *Suppose that*

$$\forall p \in \mathbb{R}^+, \qquad q^2\frac{\partial\tau}{\partial p}(p,h^b) + 1 \ > \ Kq^2[\tau(\tau_v - \tau_l)](p,h^b) \ > \ 0. \tag{14a}$$

Then, for all boundary conditions $\tilde{U}(0) = \tilde{U}_0 = (c_{v,0}, q_0, p_0, h_0)$ satisfying

$$q_0 \neq 0, \qquad \text{and} \quad q_0^2\left(\frac{\partial\tau}{\partial p} + \tau\frac{\partial\tau}{\partial h}\right)(p_0,h_0) + 1 \ \neq \ 0, \tag{14b}$$

there exists a unique maximal solution U absolutely continuous to (13). Furthermore, this solution depends continuously on $\tilde{U}_0$.

Remark 3. Condition (14b) corresponds to imposing that the flow remains subsonique. This formula is obtained by imposing the invertibility of $D\tilde{F}(\tilde{U})$ which is necessary and sufficient to ensure the uniqueness of a steady solution $\tilde{U}$. Of course, one also need ϕ to be a bijection to ensure the existence of a unique solution U to the original equation.

The formula (14b) refers not directly to the speed of sound, because in a non-steady framework, $\tilde{U}$ is not transported, but U is. The speed of sound would be obtained from the eigenvalues of $DF(U) = D\tilde{F}(\phi^{-1}(U)).D\phi^{-1}(U)$. In the incompressible case $\partial_p \tau = 0$, one finds after computations that those eigenvalues are $\tau q \pm \sqrt{\tau / \frac{\partial \tau}{\partial h}}$ and twice τq, where one identifies the velocity $u = \tau q$ and the speed of sound yields $c = \sqrt{\tau / \frac{\partial \tau}{\partial h}}$.

Proof. First, one verifies that $\frac{dq}{dx} = 0$, thus $q \neq 0$ is constant and (13) reduces to

$$\frac{d}{dx}\bar{F}(\bar{U}) = \bar{S}(\bar{U}) \qquad \bar{S}(\bar{U}) = \begin{cases} (0, & 0, \ \ \phi) \quad \text{if} \ \ h < h^b, \\ (K\phi, & 0, \ \ \phi) \quad \text{if} \ \ h \geq h^b, \end{cases}$$

$$\bar{U} = (c_v, \ p, \ h), \qquad \bar{F}(\bar{U}) = \left(c_v q, \ \tau q^2 + p, \ q\left(\frac{\tau^2 q^2}{2} + h\right)\right).$$

One computes

$$D\bar{F}(\bar{U}) = \begin{pmatrix} q & 0 & 0 \\ q^2(\tau_v - \tau_l) & q^2\frac{\partial \tau}{\partial p} + 1 & q^2\frac{\partial \tau}{\partial h} \\ q^3\tau(\tau_v - \tau_l) & q^3\tau\frac{\partial \tau}{\partial p} & q\left(q^2\tau\frac{\partial \tau}{\partial h} + 1\right) \end{pmatrix}, \tag{15}$$

the determinant of which yields

$$Det := \det\left(D\bar{F}(\bar{U})\right) = q^2\left[q^2\left(\frac{\partial \tau}{\partial p} + \tau\frac{\partial \tau}{\partial h}\right) + 1\right],$$

which is non-zero at the boundary by hypothesis. Inverting (15) yields

$$(D\bar{F}(\bar{U}))^{-1} = \frac{1}{Det}\begin{pmatrix} q\left(1 + q^2\left(\frac{\partial \tau}{\partial p} + \tau\frac{\partial \tau}{\partial h}\right)\right) & 0 & 0 \\ -q^3(\tau_v - \tau_l) & q^2\left(q^2\tau\frac{\partial \tau}{\partial h} + 1\right) & -q^3\frac{\partial \tau}{\partial h} \\ -q^3\tau(\tau_v - \tau_l) & -q^4\tau\frac{\partial \tau}{\partial p} & q\left(q^2\frac{\partial \tau}{\partial p} + 1\right) \end{pmatrix}.$$

Multiplying it by the source term and by $\nabla_{\bar{U}} h(\bar{U}) = (0, \ 0, \ 1)$ leads to

$$\nabla_{\bar{U}} h(V).(D\bar{F}(V))^{-1}.\bar{S^-}(V,x) = \frac{q\phi}{Det}\left(1 + q^2\frac{\partial \tau}{\partial p}\right),$$

$$\nabla_{\bar{U}} h(V).(D\bar{F}(V))^{-1}.\bar{S^+}(V,x) = \frac{q\phi}{Det}\left(1 + q^2\left(\frac{\partial \tau}{\partial p} - K\tau(\tau_v - \tau_l)\right)\right).$$

If (14b) holds, these two values are positive and we may apply Corollary 1. □

5. Conclusion and outlook.

We have described, in a theoretical framework, a set of conditions providing the existence and uniqueness of a steady solution, in the sense of Carathéodory, to hyperbolic systems of balance laws. We have applied it for the study of a boiling flow model. The resulting conditions on the physical parameters for such steady flows to exists are twofold. First, the flow needs to remain subsonic in the whole spatial domain, this constrains the domain length and the boundary conditions. Second, if the source is discontinuous along an hypersurface in the phase space, then the source and the flux on both sides need to be defined in such a way that the flow may only cross the discontinuity hypersurface in one direction.

In the present work, we have only considered boundary conditions on one sides, which suffice to study time independent flow. Though, it is more common in this field to use two boundaries with further requirements (see typically [2] for unsteady flows).

At the numerical level, capturing equilibrium states such as steady states for balance laws has been widely studied. Though, the discontinuity of source terms of the form (1) brings new difficulties, the study of which is left for future work.

Acknowledgments. This work was performed during the author's postdoctorate, financed by LJLL and CEA through LRC Manon. The author would like to thank N. Aguillon for constructive remarks and comments on this work, and also to acknowledge B. Després, E. Godlewski and M. Ndjinga for fruitful discussions on this topic.

REFERENCES

[1] J.-P. Aubin and A. Cellina. *Differential Inclusions, Set-Valued Maps And Viability Theory.* Grundl. der Math. Wiss. Springer, 1984.

[2] C. Bardos, A.-Y. Le Roux, and J.-C. Nédélec. First order quasilinear equations with boundary conditions. *Comm. Partial Diff. Equ.*, 4:1017–1034, 1979.

[3] A. Bergeron and I. Toumi. Assessment of the flica-iv code on rod bundle experiments. *Proceedings of ICONE-6, San Diego, California, USA*, 1998.

[4] A. Bressan. Unique solutions for a class of discontinuous differential equations. *Proceedings of the American Mathematical Society*, 104(3):772–778, 1988.

[5] A. Bressan and G. Colombo. Existence and continuous dependence for discontinuous o.d.e.'s. *Boll. Un. Mat. Ital. 4-B*, pages 295–311, 1909.

[6] H. Brézis. *Operateurs maximaux monotones et semi-groupes de contractions dans les espaces de Hilbert.* North-Holland publishing company, 1973.

[7] A. F. Filippov. *Differential Equations with Discontinuous Righthand Sides.* Springer, 1988.

[8] O. Hajek. Discontinuous differential equations. *Int. J. Differential Equations*, 32:149–170, 1979.

[9] T. Hibiki and M. Ishii. One-dimensional drift-flux model and constitutive equations for relative motion between phases in various two-phase flow regimes. *Int. J. Heat Mass Transfer*, 46:4935–4948, 2003.

[10] M. Ishii. One dimensional drift-flux model and constitutive equations for relative motion between phases in various two-phase flow. Technical report, ANL, 1977.

[11] M. Ishii and T. Hibiki. *Thermo-fluid dynamics of two-phase flows.* Springer, 2011.

[12] C. Lobry and T. Sari. Équations différentielles à second membre discontinu. *Contrôle non linéaire et Applications*, 64:255–289, 2005.

[13] A. Pucci. Sistemi di equazioni differenziali con secondo membro discontinuo rispetto all'incognita. *Rend. Ist. Mat. Univ. Trieste*, III:75–80, 1971.

[14] E. Royer, S. Aniel, A. Bergeron, P. Fillion, D. Gallo, F. Gaudier, O. Grégoire, M. Martin, E. Richebois, P. Salvadore, S. Zimmer, T. Chataing, P. Clément, and F. François. Flica4: status of numerical and physical models and overview of applications. *Proceedings of NURETH-11, Avignon, France*, 2005.

[15] I. Toumi, A. Bergeron, D. Gallo, E. Royer, and D. Caruge. Flica-4: A three-dimensional two-phase flow computer code with advanced numerical methods for nuclear applications. *Nuclear Engineering and Design*, 200, 2000.

E-mail address: teddy.pichard@polytechnique.edu

A KINETIC APPROACH TO THE BI-TEMPERATURE EULER MODEL

CORENTIN PRIGENT*

Institut de Mathématiques de Bordeaux, 351 Cours de la Libération
33400 Talence, France

S. BRULL

Institut de Mathématiques de Bordeaux, 351 Cours de la Libération
33400 Talence, France

B. DUBROCA

LCTS, 3 Allée de la Boétie
33600 Pessac, France

ABSTRACT. We are interested in the numerical approximation of the bi temperature Euler equations, which is a non conservative hyperbolic system introduced in [1]. We consider a conservative underlying kinetic model, the Vlasov-BGK-Poisson system. We perform a scaling on this system in order to obtain its hydrodynamic limit. We present a deterministic numerical method to approximate this kinetic system. The method is shown to be Asymptotic-Preserving in the hydrodynamic limit, which means that any stability condition of the method is independant of any parameter ε, with $\varepsilon \to 0$. We prove that the method is, under appropriate choices, consistant with the solution for bi-temperature Euler. Finally, our method is compared to methods for the fluid model (HLL, Suliciu).

1. **Introduction.** The bi-temperature Euler system, describing out-of-equilibirum plasma physics, is comprised of an equation over mass, an equation over momentum, and one equation over each species energy (electrons and ions). This system is a non-conservative hyperbolic system. It contains so-called non-conservative terms, which cannot be put in divergential form. Such terms are not well-defined, and, in situations involving shocks, computing exact or approximated solutions is a challenging issue.

In this paper, the aim is to propose a reference numerical method for such solutions using an underlying kinetic model to the bi-temperature Euler system. This kinetic model is conservative, and hence does not exhibit the drawback of the macroscopic model. Hence, the idea dwells in solving the Vlasov-BGK-Ampère system in the hydrodynamic limit, which will be presented in section 2.3. In order to be able to compare the results with the scheme applied to the bi-temperature Euler system, it is necessary to describe identical scales. By performing a scaling on the kinetic system, dimensionless parameters are introduced in the system. Taking the hydrodynamic limit then amounts to taking the limit when these parameters tend

2000 *Mathematics Subject Classification.* Primary: 58F15, 58F17; Secondary: 53C35.

Key words and phrases. Non-conservative hyperbolic systems, numerical analysis, underlying kinetic models.

to zero. In the general case, if a naive numerical approach is used, extremely restrictive stability conditions will appear, rendering the scheme unusable in a decent amount of computation time. Hence, an Asymptotic-Preserving (AP) scheme needs to be derived. Such a scheme possess stability conditions independent of these small parameters and is then able to compute solutions for both cases when parameters are of the order of one and when these parameters are arbitrarily small.

2. Description of the models.

2.1. Bi-temperature Euler model.

The one-dimensional bi-temperature Euler model describes the behaviour of a two species fluid (constituted of electrons and ions) on a macroscopic scale:

$$\partial_t \rho + \partial_x(\rho u) = 0,$$
$$\partial_t(\rho u) + \partial_x(\rho u^2 + p^e + p^i) = 0,$$
$$\partial_t(\rho^i \epsilon^i + \frac{1}{2}\rho^i u^2) + \partial_x(u(\rho^i \epsilon^i + \frac{1}{2}\rho^i u^2 + p^i)) + u(c^i \partial_x p^e - c^e \partial_x p^i) = -\nu_{ei}(T^i - T^e),$$
$$\partial_t(\rho^e \epsilon^e + \frac{1}{2}\rho^e u^2) + \partial_x(u(\rho^e \epsilon^e + \frac{1}{2}\rho^e u^2 + p^e)) - u(c^i \partial_x p^e - c^e \partial_x p^i) = \nu_{ei}(T^i - T^e).$$

Superscripts e and i denote quantities related to electrons and ions, respectively. $\rho = \rho^e + \rho^i = m^e n^e + m^i n^i$ is the total density, m^α being the mass of a particle of species α, and n^α the corresponding concentration. u is the macroscopic velocity of the fluid. p^α, T^α and ϵ^α denote the partial pressures, temperatures and specific energies. $c^e = \frac{Zm^e}{m^i+Zm^e}$ and $c^i = \frac{m^i}{m^i+Zm^e}$ denote the mass fractions,with Z the ionization rate. ν_{ei} is the temperature exchange rate between the two species, that can depend on space and time.

2.2. Macroscopic quantities.

Define $f^\alpha(t,x,v) : \mathbb{R}_+ \times \mathbb{R} \times \mathbb{R} \mapsto \mathbb{R}_+$, the particle distribution function of species α, where t, x, v respectively denote the time, space and microscopic velocity variables. Integration with respect to $v \in \mathbb{R}$ will be denoted as follows, for any function g that depends on v such that $(1+v^2)g \in L^1(\mathbb{R})$:

$$\langle g \rangle = \int_{\mathbb{R}} g dv.$$

Macroscopic quantities $n^\alpha(t,x), u^\alpha(t,x)$ and $T^\alpha(t,x)$ are defined as moments of f^α as follows:

$$\langle f^\alpha \rangle = n^\alpha, \qquad \langle v f^\alpha \rangle = n^\alpha u^\alpha,$$
$$\langle m^\alpha \frac{v^2}{2} f^\alpha \rangle = \frac{1}{2} m^\alpha n^\alpha (u^\alpha)^2 + \frac{1}{2} n^\alpha k_B T^\alpha.$$

Similarly, mixture macroscopic quantities are defined as:

$$\langle m^e f^e + m^i f^i \rangle = \rho, \qquad \langle v(m^e f^e + m^i f^i) \rangle = \rho u,$$
$$\langle \frac{v^2}{2}(m^e f^e + m^i f^i) \rangle = \frac{1}{2}\rho u^2 + \frac{1}{2}(n^e + n^i) k_B T.$$

Finally, we define electromagnetic quantities: $E \in \mathbb{R}$ is the electric field along x, q^α is the electric charge of species α. Note that $q^e = -Zq^i$. j and $\overline{\rho}$ are respectively the current and total electric charge, defined as:

$$\langle q^e f^e + q^i f^i \rangle = \overline{\rho}, \qquad \langle v(q^e f^e + q^i f^i) \rangle = j. \tag{1}$$

2.3. **Vlasov-BGK-Ampère system.** Consider the Vlasov-BGK-Ampère model, for $t \in \mathbb{R}$, $x \in \mathbb{R}$, $v \in \mathbb{R}$ and $\alpha \in \{e, i\}$:

$$\partial_t f^\alpha + v\partial_x f^\alpha + \frac{q^\alpha E}{m^\alpha}\partial_v f^\alpha = \frac{1}{\tau^\alpha}(M(f^\alpha) - f^\alpha) + \frac{1}{\tau^{\alpha\beta}}(\overline{M}(f^\alpha) - f^\alpha), \tag{2}$$

$$\partial_t E = -\frac{j}{\varepsilon_0}, \tag{3}$$

$$\partial_x E = \frac{\overline{\rho}}{\varepsilon_0}, \tag{4}$$

where τ^α and $\tau^{\alpha\beta}$ are, respectively, the relaxation rate towards intra-species and inter-species equilibria ($\tau^{\alpha\beta} = \tau^{\beta\alpha}$). ε_0 is the dielectric permittivity of vacuum.

The entropy-minimizing distribution function is the local Maxwellian distribution, denoted $M(f^\alpha)$, defined by:

$$M(f^\alpha)(t,x,v) = \frac{n^\alpha(t,x)}{\sqrt{2\pi k_B \frac{T^\alpha(t,x)}{m^\alpha}}} \exp\left(-\frac{(v - u^\alpha(t,x))^2}{2k_B \frac{T^\alpha(t,x)}{m^\alpha}}\right).$$

$\overline{M}(f^\alpha)$, the exchange Maxwellian distribution, is defined as:

$$\overline{M}(f^\alpha)(t,x,v) = \frac{n^\alpha(t,x)}{\sqrt{2\pi k_B \frac{T(t,x)}{m^\alpha}}} \exp\left(-\frac{(v - u(t,x))^2}{2k_B \frac{T(t,x)}{m^\alpha}}\right).$$

2.4. **Scaling of the equations.** The regime of interest is the hydrodynamic limit of this model, which corresponds to the scale of the phenomena described by the bi-temperature Euler equations (see [1]). Hence, consider the following scaling, for $\alpha = e, i$:

$$\partial_t f^\alpha + v\partial_x f^\alpha + \frac{q^\alpha}{m^\alpha}E\partial_v f^\alpha = \frac{1}{\tau}(M^e - f^e) + \frac{1}{\varepsilon}(\overline{M}_p^e - f^e), \tag{5}$$

$$\partial_t E = -\frac{j}{\tau}, \tag{6}$$

$$\partial_x E = \frac{\overline{\rho}}{\tau}. \tag{7}$$

Reaching the hydrodynamic limit consists in taking the limit of the model when τ tend to 0. Notice that, however, ε is finite. In the limit, the Maxwell equations (6-7) become $j = 0$ and $\rho = 0$, which can be rewritten as:

$$u^e = u^i, \qquad n^e = Zn^i. \tag{8}$$

The equations (8) are called the quasi-neutrality constraints. This form of the system of equations is the singular quasi-neutral limit of our initial model, where two evolution equations degenerate into algebraic relations. The purpose of this work is to derive an Asymptotic-preserving (AP) numerical scheme, that is to say with a stability condition that is independant τ.

3. **Numerical treatment.** In this section, numerical treatment of system (5-6-7) is addressed.

3.1. **Vlasov-BGK equations: general method.** A time-splitting method is used on the Vlasov-BGK equations, in order to separate the Vlasov equations from the BGK operators. More precisely, the following equations are going to be consecutively solved:

$$\partial_t f^\alpha + v\partial_x f^\alpha + \frac{q^\alpha}{m^\alpha} E \partial_v f^\alpha = 0, \tag{9}$$

the Vlasov-Maxwell equations for $\alpha = e, i$, coupled with the Ampère equation:

$$\partial_t E = -\frac{j}{\tau^2}. \tag{10}$$

Then, the effects of the inter-species and intra-species collisions are consecutively computed:

$$\partial_t f^\alpha = \frac{1}{\varepsilon}(\overline{M}^\alpha - f^\alpha), \qquad \partial_t f^\alpha = \frac{1}{\tau}(M^\alpha - f^\alpha), \tag{11}$$

where $\tau \to 0$.

A uniform discretization of the phase space is considered. Concerning the velocity discretization, the boundaries of the domains are chosen as follows: for a given initial condition, let us denote the initial moments by n_0^α, u_0^α and T_0^α. For $\alpha = e, i$, we have:

$$V^\alpha = [\min_x (u_0^\alpha(x)) - l^\alpha \max_x \sqrt{k_B \frac{T_0^\alpha(x)}{m^\alpha}}, \max_x (u_0^\alpha(x)) + l^\alpha \max_x \sqrt{k_B \frac{T_0^\alpha(x)}{m^\alpha}}], \tag{12}$$

where l^α is chosen to fit the initial maxwellian distribution.

Notations for the discretization are chosen as follows: Δx denotes the discretization step of the domain $[L_{\text{min}}, L_{\text{max}}] \subset \mathbb{R}$ in N_x cells. For $\alpha = e, i$, the velocity domains $[V_{\text{min}}^\alpha, V_{\text{max}}^\alpha] \subset \mathbb{R}$ are both discretized using N_v cells and the discretization steps are denoted Δv^e and Δv^i. Then, we use indices k and j to denote quantities computed at the k-th cell in space and j-th cell in velocity. Moreover, we denote Δt the time step and we use a superscript n to denote the quantity computed at time $t^n = t^0 + n\Delta t$.

Step 1: Computation of g and E

The following discretization is performed, for $\alpha = e, i$:

$$\widetilde{f}_{k,j}^{\alpha,n+1} = f_{k,j}^{\alpha,n} - \frac{\Delta t}{\Delta x}(\phi_{k+\frac{1}{2},j}^{\alpha,n} - \phi_{k-\frac{1}{2},j}^{\alpha,n}) - \frac{\Delta t}{2\Delta v^\alpha} \frac{q^\alpha E_k^{n+1}}{m^\alpha}(f_{k,j+1}^{\alpha,n} - f_{k,j-1}^{\alpha,n}), \tag{13}$$

coupled with

$$E_k^{n+1} = E_k^n - \frac{\Delta t}{\tau} j_k^{n+1}. \tag{14}$$

Note that the term $\partial_v f$ has been discretized using a centred scheme. For the space discretization, ϕ represented the flux function of the spatial discretization which will be specified in the next section.

Equations (13) are stable under the CFL condition :

$$\Delta t \le \min_\alpha \Big(\frac{1}{\frac{||V^\alpha||_\infty}{\Delta x} + \frac{||q^\alpha E^{n+1}||_\infty}{m^\alpha \Delta v^\alpha}} \Big). \tag{15}$$

Then, partial and mixture quantities associated with $\widetilde{f}^{e,n+1}$ and $\widetilde{f}^{i,n+1}$ are computed using the midpoint quadrature formula.

Step 2: inter-species relaxation

We apply a backward Euler method to the equations (11):

$$\hat{f}_{k,j}^{\alpha,n+1} = \frac{1}{\Delta t + \tau^{ei}} \left[\Delta t \widetilde{\overline{M}}_{k,j}^{\alpha,n+1} + \tau^{ei} \widetilde{f}_{k,j}^{\alpha,n+1} \right],$$

which is unconditionnaly stable. $\widetilde{\overline{M}}_{k,j}^{\alpha,n+1}$ is the mixture Maxwellian defined by the moments $\widetilde{n}_k^{\alpha,n+1}, \widetilde{u}_k^{n+1}$ and $\widetilde{T}_k^{n+1}$, which are conserved through this step. Hence, the mixture Maxwellian is known beforehand, which allows this implicitation. Partial velocities and temperatures (denoted in the next section with a hat) are then recomputed using the previous formulae.

Step 3: intra-species relaxation

The next step consists in the relaxation towards Maxwellian equilibrium. Since the BGK operator preserves moments, we have the following properties: $\forall k \in \{1, ..., N_x\}$,

$$\hat{n}_k^{\alpha,n+1} = n_k^{\alpha,n+1}, \qquad \hat{u}_k^{\alpha,n+1} = u_k^{\alpha,n+1}, \qquad \hat{T}_k^{\alpha,n+1} = T_k^{\alpha,n+1}. \tag{16}$$

Similarly to the previous section, this property allows us to use an implicit scheme, the backward Euler method: $\forall k \in \{1, ..., N_x\}$, $\forall j \in \{1, ..., N_v\}$ and $p = 0, 2$,

$$f_{k,j}^{\alpha,n+1} = \hat{f}_{k,j}^{\alpha,n+1} + \frac{\Delta t}{\tau}(M_{k,j}^{\alpha,n+1} - f_{k,j}^{\alpha,n+1}), \tag{17}$$

where

$$M_{k,j}^{\alpha,n+1} = \frac{n_k^{\alpha,n+1}}{\sqrt{2\pi k_B \frac{T_k^{\alpha,n+1}}{m^\alpha}}} \exp\left(-\frac{(v_j - u_k^{\alpha,n+1})^2}{2k_B \frac{T_k^{\alpha,n+1}}{m^\alpha}}\right),$$

which are expressed using quantities known at the end of step 2, according to the properties (16). Taking the limit $\tau \to 0$ in (17) leads to:

$$g_{k,j}^{\alpha,n+1} = M_{k,j}^{\alpha,n+1}.$$

In the next section, enforcement of quasineutrality constraints (8) will be adressed. More precisely, specific spatial fluxes will be given and a reformulation of the Maxwell-Ampère equation will be performed in order to derive an Asymptotic-Preserving method, valid in the limit $\tau \to 0$. Note that if these constraints are enforced through the hyperbolic step of the scheme (step 1), then they will be trivially preserved by both step 2 and 3. Hence, quasineutrality only has to been enforced through step 1.

3.2. **Modified-viscosity upwind scheme.** The space divergence is discretized explicitly, via the following modified upwind scheme:

$$\phi_{k+\frac{1}{2},j}^{\alpha,n} = \frac{v_j^\alpha}{2}(f_{k+1,j}^{\alpha,n} + f_{k,j}^{\alpha,n}) - \frac{|V_{\max}|}{2}(f_{k+1,j}^{\alpha,n} - f_{k,j}^{\alpha,n}), \tag{18}$$

where $V_{\max} = \max(|V_{\min}^e|, |V_{\min}^i|, |V_{\max}^e|, |V_{\min}^e|)$.

This method is the classical upwind scheme, where the numerical viscosity $|v_j|$, that depends directly of the microscopic velocity, has been replaced by the smallest velocity that ensures stability for all equations. The discrete system writes as follows:

$$\widetilde{f}_{k,j}^{\alpha,n+1} = f_{k,j}^{\alpha,n} - \frac{\Delta t}{\Delta x}(\phi_{k+\frac{1}{2},j}^{\alpha,n} - \phi_{k-\frac{1}{2},j}^{\alpha,n}) - \frac{\Delta t}{2\Delta v^\alpha}\frac{q^\alpha E_k^{n+1}}{m^\alpha}(f_{k,j+1}^{\alpha,n} - f_{k,j-1}^{\alpha,n}), \quad (19)$$

and

$$E_k^{n+1} = E_k^n - \frac{\Delta t}{\tau} j_k^{n+1}. \quad (20)$$

This level of implicitness allows for a reformulation of the equation on the electric field. More precisely, the equation used to compute the electric field in (20) is not valid in the limit $\tau \to 0$. To solve this problem, we follow the idea developped in [2]. By multiplying equation (19) by $q^\alpha v_j^\alpha \Delta v^\alpha$ and summing over j, it comes:

$$\begin{aligned} q^\alpha \sum_j v_j^\alpha \widetilde{f}_{k,j}^{\alpha,n+1} \Delta v^\alpha = {} & q^\alpha \sum_j v_j^\alpha f_{k,j}^{\alpha,n} \Delta v^\alpha - q^\alpha \frac{\Delta t \Delta v^\alpha}{\Delta x} \sum_j v_j^\alpha (\phi_{k+\frac{1}{2},j}^{\alpha,n} - \phi_{k-\frac{1}{2},j}^{\alpha,n}) \\ & - \frac{\Delta t}{2} \sum_j v_j^\alpha (\frac{(q^\alpha)^2 E_k^{n+1}}{m^\alpha}(f_{k,j+1}^{\alpha,n} - f_{k,j-1}^{\alpha,n})). \end{aligned} \quad (21)$$

According to the definition of the current $j_k^n = \sum_\alpha q^\alpha \sum_j v_j^\alpha f_{k,j}^{\alpha,n} \Delta v^\alpha$, taking (21) into account with $\alpha = e, i$, one obtains

$$\begin{aligned} \widetilde{j}_k^{n+1} = {} & j_k^n - \frac{\Delta t}{\Delta x} \sum_\alpha q^\alpha \sum_j \Delta v^\alpha v_j^\alpha (\phi_{k+\frac{1}{2},j}^{\alpha,n} - \phi_{k-\frac{1}{2},j}^{\alpha,n}) \\ & - \frac{\Delta t E_k^{n+1}}{2} \sum_\alpha \frac{(q^\alpha)^2}{m^\alpha} \sum_j v_j^\alpha (f_{k,j+1}^{\alpha,n} - f_{k,j-1}^{\alpha,n}). \end{aligned}$$

By injecting this equation in (20), we get the following expression for the electric field:

$$E_k^{n+1} = \frac{\tau E_k^n - \Delta t j_k^n + \frac{\Delta t^2}{\Delta x} \sum_\alpha q^\alpha \sum_j \Delta v^\alpha v_j^\alpha (\phi_{k+\frac{1}{2},j}^{\alpha,n} - \phi_{k-\frac{1}{2},j}^{\alpha,n})}{\tau - \frac{\Delta t^2}{2} \sum_\alpha \frac{(q^\alpha)^2}{m^\alpha} \sum_j v_j^\alpha (f_{k,j+1}^{\alpha,n} - f_{k,j-1}^{\alpha,n})},$$

which is valid for all values of τ. In particular, we are interested in the limit $\tau \to 0$, which is:

$$E_k^{n+1} = \frac{\frac{1}{\Delta x} \sum_\alpha q^\alpha \sum_j \Delta v^\alpha v_j^\alpha (\phi_{k+\frac{1}{2},j}^{\alpha,n} - \phi_{k-\frac{1}{2},j}^{\alpha,n})}{\sum_\alpha \frac{(q^\alpha)^2}{m^\alpha} n_k^{\alpha,n}},$$

where it has been assumed that $j_k^n = 0$. Such a choice for E_k^{n+1} ensures that $\widetilde{j}_k^{n+1} = 0$, which is one of the two quasi-neutrality constraints (8). The choice of ϕ (18) that has been made ensures that the other contraint $\widetilde{\overline{\rho}}_k^{n+1} = 0$ is also verified. More precisely, compute the discrete equation on $\widetilde{\overline{\rho}}_k^{n+1}$. $\forall k \in \{1, ..., N_x\}$, provided that $\sum_j (\psi_{k,j+\frac{1}{2}}^{\alpha,n} - \psi_{k,j-\frac{1}{2}}^{\alpha,n}) = 0$, the following equation on the total electric charge is obtained:

$$\overline{\rho}_k^{n+1} = \overline{\rho}_k^n - \frac{\Delta t}{\Delta x} \sum_\alpha q^\alpha \sum_j (\phi_{k+\frac{1}{2},j}^{\alpha,n} - \phi_{k-\frac{1}{2},j}^{\alpha,n}) \Delta v^\alpha.$$

According to (18) and by rearranging the terms, we obtain:

$$\overline{\rho}_k^{n+1} = \overline{\rho}_k^n - \frac{\Delta t}{2\Delta x}(j_{k+1}^n - j_{k-1}^n) + \frac{\Delta t}{2\Delta x}\sum_\alpha q^\alpha \sum_j |V_{\max}|(f_{k+1,j}^{\alpha,n} - 2f_{k,j}^{\alpha,n} + f_{k-1,j}^{\alpha,n})\Delta v^\alpha \tag{22}$$

By assuming that $\overline{\rho}_k^n = j_{k+1}^n = j_{k-1}^n = 0$, it comes:

$$\overline{\rho}_k^{n+1} = \frac{\Delta t}{2\Delta x} V_{\max}(\overline{\rho}_{k+1}^n - 2\overline{\rho}_k^n + \overline{\rho}_{k-1}^n) = 0. \tag{23}$$

Hence, quasineutrality is conserved at all time.

4. **Numerical results.** In this section, numerical resolution of Riemann problems with our method is presented. Comparative results obtained by an HLL-type scheme and a Suliciu relaxation method on the bi-temperature Euler system are displayed along the solutions from the kinetic scheme that has been derived in this paper (see [1] for more details on these methods).

The spatial domain is chosen as the interval [0,1], supplemented with homogeneous Neumann boundary conditions. Physical constants (k_B, Z, q^i) are chosen equal to one. All test cases are done in the hydrodynamic limit $\tau \to 0$. Concerning the choice of l^α in (12), a convergence test has been performed on the initial conditions of the test cases in order to obtain a appropriate value that would optimize computational cost. Hence, in all test cases, $l^\alpha = 8$, for $\alpha \in \{e, i\}$. A mass ratio of 10 ($m_e = 0.1$ and $m_i = 1$) is considered.

Shock tube with different initial temperatures. The following initial conditions are:

$$\begin{cases} n^\alpha(x) = 1, & u^\alpha(x) = 0, \quad T^\alpha(x) = 1 \quad \text{if } x \in [0, 0.5], \\ n^\alpha(x) = 0.125, & u^\alpha(x) = 0, \quad T^e(x) = 2 \quad T^i(x) = 3 \quad \text{if } x \in [0.5, 1]. \end{cases}$$

Parameters are chosen as $N_x = 120000$, $N_v = 40$ and $l = 8$. The inter-species collision relaxation time is $\tau^{ei} = 0.1$. Results are computed at time $t = 0.1$. The result are provided in figure 1 and 2. Quasi-neutrality is achieved and temperatures exhibit different jump relations across discontinuities for each method (see Figure 3).

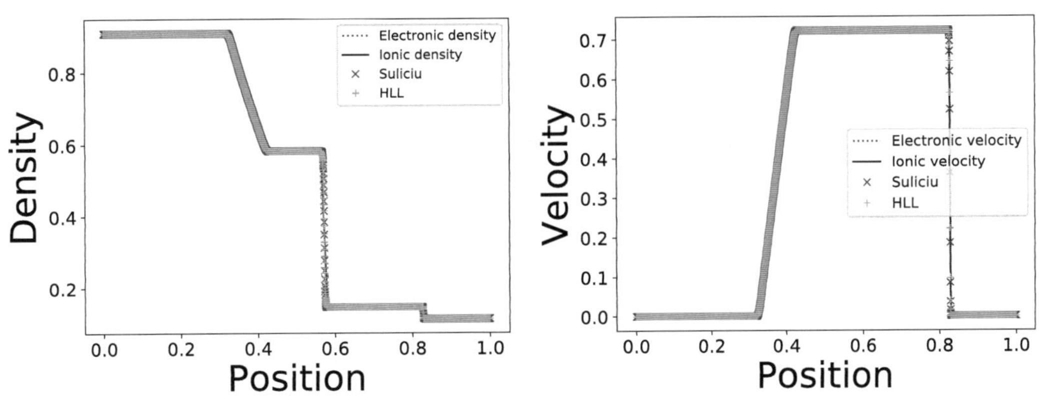

FIGURE 1. Density and velocity solutions of shock tube test case with a mass ratio of 10 with 120000 space points, 40 velocity points and a domain length of 8

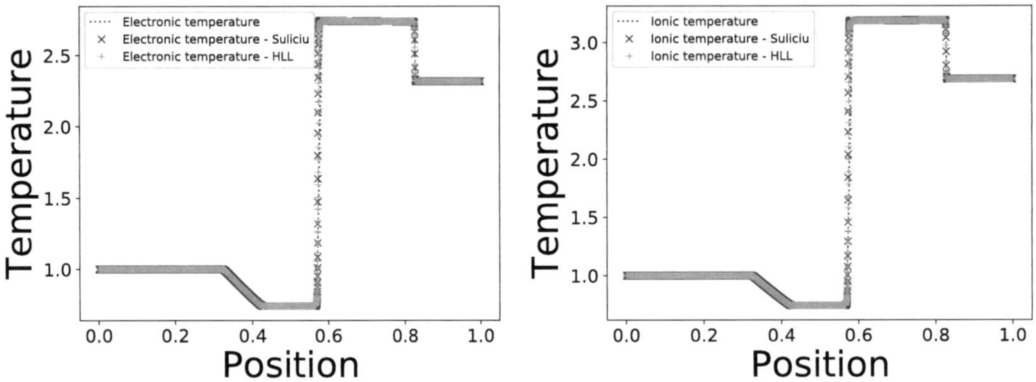

FIGURE 2. Electronic and ionic temperatures of a shock tube test case with a mass ratio of 10 with 120000 space points, 40 velocity points and a domain length of 8

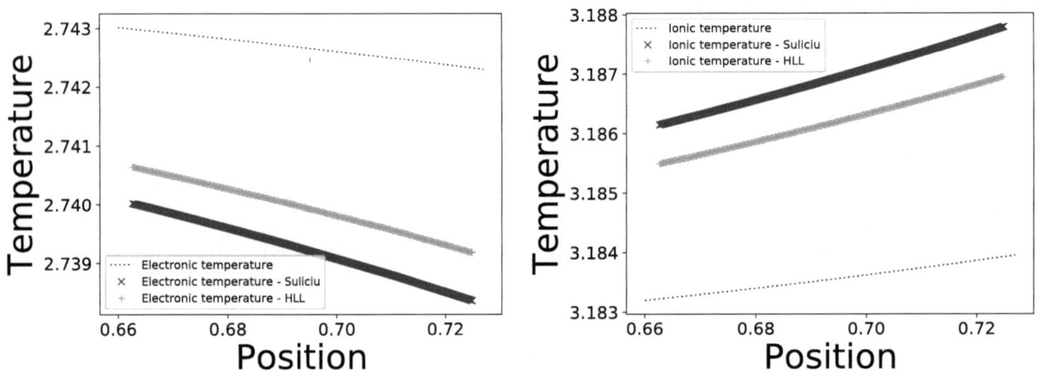

FIGURE 3. Zoom on interval $[0.66, 0.76]$ from figure 2 of electronic and ionic temperatures for a shock tube test case with a mass ratio of 10 with 120000 space points, 40 velocity points and a domain length of 8

Shock wave. This test case is constituted of two shock waves. It is given by the following data:

$$\begin{cases} n^\alpha(x) = 1, u^\alpha(x) = 1, T^\alpha(x) = 1 & \text{if } x \in [0, 0.5], \\ n^\alpha(x) = 1, u^\alpha(x) = -1, T^\alpha(x) = 1 & \text{if } x \in [0.5, 1]. \end{cases}$$

We have $N_x = 120000$, $N_v = 40$ and $l = 8$. The inter-species collision relaxation time is $\tau^{ei} = 0.1$. Results are computed at time $t = 0.1$.

The results are provided in Figure 4 and 5. Quasi-neutrality is achieved on the conserved quantities. The solution contains two shock waves and different jump relations can be observed for each numerical method. In Figure 6, a zoom is performed on the constant values reached by each method between the two shocks. This shows the different behaviour between the three methods.

5. **Conclusion.** In this article, a kinetic numerical method able to provide reference results for a non-conservative hyperbolic system, the bi-temperature Euler system, is proposed. This method is able to enforce all the desired properties that

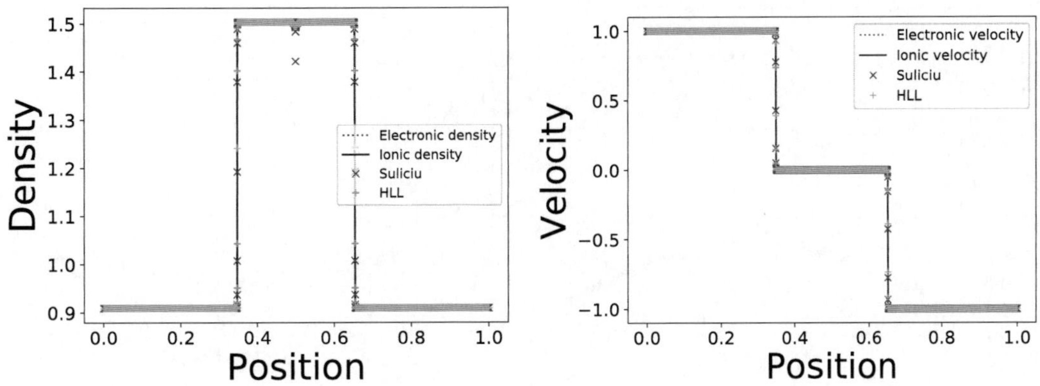

FIGURE 4. Density and veloctiy solutions of a shock wave test case with a mass ratio of 10 with 120000 space points, 40 velocity points and a domain length of 8

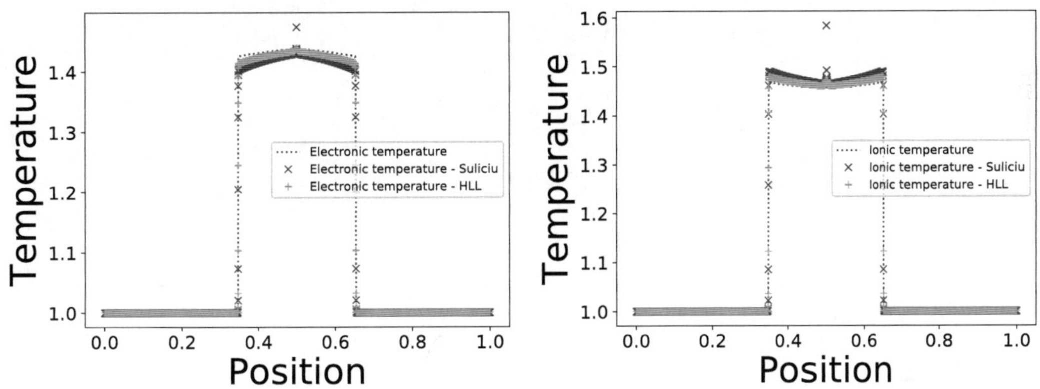

FIGURE 5. Electronic and ionic temperature of a shock wave test case with a mass ratio of 10 with 120000 space points, 40 velocity points and a domain length of 8

are relevant for comparisons with methods applied directly to the hyperbolic system. Such methods lack an unambiguous definition of the non-conservative products and exhibit different Rankine-Hugoniot relation when the solution contains shock waves.

REFERENCES

[1] D.Aregba Driollet, J.Breil, S.Brull, B.Dubroca, and E.Estibals, Modelling and numerical approximation for the nonconservative bitemperature Euler model, *ESAIM: Mathematical Modelling and Numerical Analysis*, **52** (2018), 1353–1383.

[2] S. Guisset, S. Brull, E. D'Humières and B. Dubroca, Asymptotic-Preserving scheme for the M_1-Maxwell system in the quasi-neutral regime, *Comm. Comput. Physics*, **19** (2016), 301–328.

E-mail address: corentin.prigent@math.u-bordeaux.fr
E-mail address: stephane.brull@math.u-bordeaux.fr
E-mail address: bruno.dubroca@math.u-bordeaux.fr

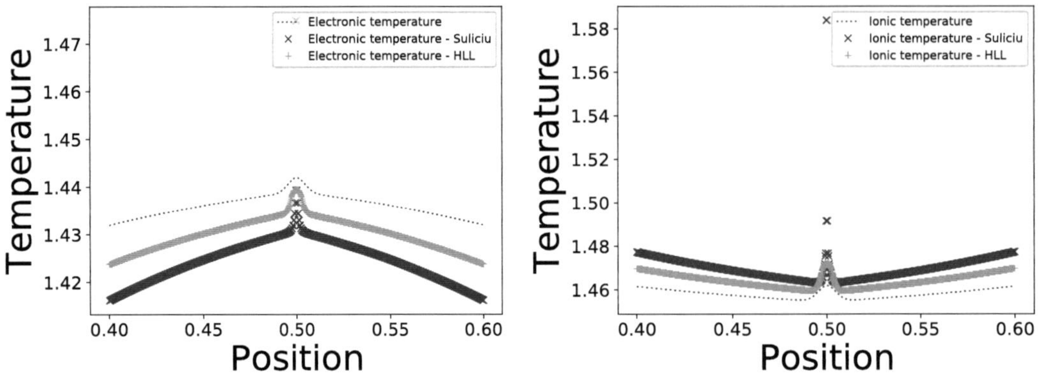

FIGURE 6. Zoom on interval [0.40, 0.60] on figure 5 of electronic and ionic temperature of a shock wave test case with a mass ratio of 10 with 120000 space points, 40 velocity points and a domain length of 8

POINTWISE ASYMPTOTIC BEHAVIOR OF A CHEMOTAXIS MODEL

JEAN RUGAMBA

Department of Mathematics
University of Alabama at Birmingham
Birmingham, AL 35294-1170, USA

YANNI ZENG*

Department of Mathematics
University of Alabama at Birmingham
Birmingham, AL 35294-1170, USA

ABSTRACT. We consider a system which is the transformed Keller-Segel-Fisher /KPP chemotaxis model with logarithmic sensitivity. For Cauchy problem, a time asymptotic solution has been constructed in [7], with a proof of L^2 convergence. In this paper we extend the study to pointwise sense, both in space and in time. Our main result provides a detailed, pointwise description of the wave pattern in the time asymptotic convergence of the solution to the asymptotic solution.

1. **Introduction.** We consider the following Cauchy problem:

$$\begin{cases} v_t + u_x = 0, \\ u_t + (uv)_x = Du_{xx} + ru(1-u), \end{cases} \qquad x \in \mathbb{R},\ t > 0, \tag{1}$$
$$(v, u)(x, 0) = (v_0, u_0)(x), \quad x \in \mathbb{R},$$

where $D, r > 0$ are constant parameters, and the Cauchy datum (v_0, u_0) is a small perturbation of a constant equilibrium state $(0, 1)$. Equation (1) is derived from the logarithmic Keller-Segel-Fisher/KPP chemotaxis model, for the case of non-diffusive chemical, via the inverse Hopf-Cole transformation and rescaling/non-dimensionalization. Details are given in [8] (also see [9]), including both cases of diffusive and non-diffusive chemicals. Applying a general theory on hyperbolic-parabolic balance laws developed recently in [3, 4, 6], (1) has a unique global-in-time solution, with optimal L^p time decay rates, $p \geq 2$, for small data solutions.

Although a general theory on asymptomatic behavior of solution is available only for multi-space dimensions [5], an asymptotic solution of (1) is constructed and L^2-convergence is proved in [7] by ad hoc consideration:

$$(v, u) \approx (\theta, 1 - \frac{1}{r}\theta_x). \tag{2}$$

2000 *Mathematics Subject Classification.* Primary: 35B40, 35M31; Secondary: 35Q92.

Key words and phrases. Asymptotic behavior, pointwise estimates, Cauchy problem, Keller-Segel, chemotaxis, logarithmic sensitivity, logistic growth.

The research of Yanni Zeng was partially supported by a grant from the Simons Foundation (#244905 to Yanni Zeng).

* Corresponding author: Yanni Zeng.

Here θ is the self similar solution of the heat equation

$$\theta_t = \frac{1}{r}\theta_{xx}, \tag{3}$$

carrying the same mass as v. That is, since θ and v are conserved quantities,

$$\int_{\mathbb{R}} \theta(x,0)\,dx = \int_{\mathbb{R}} \theta(x,t)\,dx = \int_{\mathbb{R}} v(x,t)\,dx = \int_{\mathbb{R}} v_0(x)\,dx \equiv d_0. \tag{4}$$

Noting the solution to (3), (4) is the heat kernel, we have the explicit formulation of it:

$$\theta(x,t) = \frac{d_0}{\sqrt{4\pi(t+1)/r}} e^{-\frac{rx^2}{4(t+1)}}. \tag{5}$$

The goal of this paper is to extend the result in [7] by establishing (2) in the pointwise sense, both in space and in time. To simplify our statement we introduce the following notations:

$$\begin{aligned} &\psi^\alpha(x,t) = (1+\frac{x^2}{t+1})^{-\alpha}\Gamma^\alpha(t), \quad x\in\mathbb{R},\ t\ge 0, \\ &\Gamma^\alpha(t) = \begin{cases} (t+1)^{-\alpha} & \text{if } 1/2<\alpha<1 \\ (t+1)^{-1}[1+\ln(t+1)] & \text{if } \alpha = 1 \\ (t+1)^{-1} & \text{if } \alpha>1 \end{cases}, \end{aligned} \tag{6}$$

where $\alpha > 1/2$ is a constant. Our main result is the following theorem:

Theorem 1.1. *Let $D, r > 0$ be constants, and $(v_0, u_0 - 1) \in H^2(\mathbb{R})$, satisfying*

$$|(v_0, u_0-1)|(x) = O(1)(x^2+1)^{-\alpha}, \tag{7}$$

where $\alpha > 1/2$ is a constant. Let

$$\delta_0 \equiv \sup_{x\in\mathbb{R}} \left[(x^2+1)^\alpha |(v_0,u_0-1)|(x)\right] + \|(v_0,u_0-1)\|_{H^2}. \tag{8}$$

If δ_0 is sufficiently small, the Cauchy problem (1) has a unique solution for $t > 0$, with the following property:

$$v(x,t) = \theta(x,t) + O(1)\delta_0\psi^\alpha(x,t), \tag{9}$$

$$u(x,t) = 1 - \frac{1}{r}\theta_x(x,t) + O(1)\delta_0(t+1)^{-1/2}\psi^\alpha(x,t), \tag{10}$$

where θ and ψ^α are given in (5) and (6), respectively.

Theorem 1.1 shows that (2) is valid in a pointwise sense: For v, the error of the asymptotic ansatz θ is in the scale of ψ^α, which decays faster in time since $\alpha > 1/2$. Similarly, for $u - 1$, $(t+1)^{-1/2}\psi^\alpha$ decays faster than the ansatz $-\frac{1}{r}\theta_x$.

We carry out the proof of Theorem 1.1 in the next two sections.

2. Preliminaries.

We first cite the global existence result from [7], which is from the application of the general theory in [3]:

Theorem 2.1 ([7]). *Let $r > 0$ and $D > 0$ be constants, and $(v_0, u_0 - 1) \in H^2(\mathbb{R})$. Then there exists a constant $\delta_0 > 0$ such that if $\|(v_0, u_0 - 1)\|_{H^2} \le \delta_0$, the Cauchy problem (1) has a unique solution, with $(v, u - 1) \in C^0([0,\infty); H^2(\mathbb{R}))$ for $t > 0$.*

We write (1) in terms of the perturbation and in vector form:

$$\begin{aligned} w_t + Aw_x &= Bw_{xx} + Lw + R, \\ w(x,0) &= w_0(x), \end{aligned} \tag{11}$$

where

$$w(x,t) = \begin{pmatrix} v \\ u-1 \end{pmatrix} = \begin{pmatrix} w_1 \\ w_2 \end{pmatrix}(x,t), \quad w_0(x) = \begin{pmatrix} v_0 \\ u_0 - 1 \end{pmatrix} = \begin{pmatrix} w_{01} \\ w_{02} \end{pmatrix}(x), \tag{12}$$

$$A = \begin{pmatrix} 0 & 1 \\ 1 & 0 \end{pmatrix}, \quad B = \begin{pmatrix} 0 & 0 \\ 0 & D \end{pmatrix}, \quad L = \begin{pmatrix} 0 & 0 \\ 0 & -r \end{pmatrix}, \tag{13}$$

$$R = R_{1x}(x,t) + R_2(x,t), \; R_1 = \begin{pmatrix} 0 \\ -(w_1 w_2)(x,t) \end{pmatrix}, \; R_2 = \begin{pmatrix} 0 \\ -r(w_2^2)(x,t) \end{pmatrix}. \tag{14}$$

The Green's Function of the linear part of (11) (without the nonlinear source term R) is the solution matrix $G(x,t)$ of

$$\begin{aligned} G_t + AG_x &= BG_{xx} + LG \\ G(x,0) &= \delta(x) I_{2\times 2}, \end{aligned} \tag{15}$$

where $\delta(x)$ is the Dirac δ-function, and $I_{2\times 2}$ is the 2×2 identity matrix. Detailed estimates on G are obtained in [2] for different combinations of parameters. Here we cite the case of non-diffusive chemical:

Theorem 2.2 ([2]). *Let $D, r > 0$ be constants, and $l \geq 0$ be an integer. Then for $x \in \mathbb{R}$, $t > 0$, the Green's function $G(x,t)$ has the following estimate:*

$$\begin{aligned} \frac{\partial^l}{\partial x^l} G(x,t) =& \frac{\partial^l}{\partial x^l}\left[\frac{1}{\sqrt{4\pi t/r}} e^{-\frac{rx^2}{4t}} \begin{pmatrix} 1 & 0 \\ 0 & 0 \end{pmatrix} \right] + O(1)(t+1)^{-\frac{1}{2}} t^{-\frac{l+1}{2}} e^{-\frac{x^2}{Ct}} \begin{pmatrix} 0 & 1 \\ 1 & 0 \end{pmatrix} \\ &+ (t+1)^{-1} t^{-\frac{l+1}{2}} e^{-\frac{x^2}{Ct}} \begin{pmatrix} O(1) & 0 \\ 0 & O(1) \end{pmatrix} + e^{-\frac{t}{D}} \sum_{j=0}^{l} \delta^{(l-j)}(x) Q_j, \end{aligned} \tag{16}$$

where $C > 0$ is a constant, and Q_j, $0 \leq j \leq l$, is a 2×2, symmetric, polynomial matrix in t with a degree not more than $j/2$. In particular,

$$Q_0 = \begin{pmatrix} 1 & 0 \\ 0 & 0 \end{pmatrix}.$$

In the proof of (10) we need a refinement of the $(2,1)$ entry of G, with a precise leading term. Slightly modifying the proof of Theorem 2.2 we have the following:

Theorem 2.3. *Under the assumptions of Theorem 2.2, the second term on the right-hand side of (16) can be replaced by*

$$-\frac{\partial^{l+1}}{\partial x^{l+1}}\left[\frac{1}{\sqrt{4\pi r t}} e^{-\frac{rx^2}{4t}} \begin{pmatrix} 0 & 1 \\ 1 & 0 \end{pmatrix} \right] + O(1)(t+1)^{-1} t^{-\frac{l+2}{2}} e^{-\frac{x^2}{Ct}} \begin{pmatrix} 0 & 1 \\ 1 & 0 \end{pmatrix}. \tag{17}$$

To estimate wave interaction, we cite Lemma 3.2 in [1]:

Lemma 2.4 ([1]). *Let $\alpha \geq 0$, $\beta > 0$, $\mu > 0$ and λ be constants. Then for $x \in \mathbb{R}$, $t \geq 0$, we have*

$$\int_0^{t/2}\int_{\mathbb{R}} (t-\tau)^{-1}(t+1-\tau)^{-\frac{\alpha}{2}} e^{-\frac{(x-y-\lambda(t-\tau))^2}{\mu(t-\tau)}} (\tau+1)^{-\frac{\beta}{2}} e^{-\frac{(y-\lambda(\tau+1))^2}{\mu(\tau+1)}}\, dyd\tau$$
$$= O(1)(t+1)^{-\frac{\gamma}{2}} e^{-\frac{(x-\lambda(t+1))^2}{\mu(t+1)}} \begin{cases} 1 & \text{if } \beta \neq 3 \\ \ln(t+2), & \text{if } \beta = 3 \end{cases}, \tag{18}$$

where $\gamma = \alpha + \min\{\beta, 3\} - 1$;

$$\int_{t/2}^{t}\int_{\mathbb{R}} (t-\tau)^{-1}(t+1-\tau)^{-\frac{\alpha}{2}} e^{-\frac{(x-y-\lambda(t-\tau))^2}{\mu(t-\tau)}} (\tau+1)^{-\frac{\beta}{2}} e^{-\frac{(y-\lambda(\tau+1))^2}{\mu(\tau+1)}}\, dyd\tau$$
$$= O(1)(t+1)^{-\frac{\gamma}{2}} e^{-\frac{(x-\lambda(t+1))^2}{\mu(t+1)}} \begin{cases} 1 & \text{if } \alpha \neq 1 \\ \ln(t+2), & \text{if } \alpha = 1 \end{cases}, \tag{19}$$

where $\gamma = \min\{\alpha, 1\} + \beta - 1$.

We also need to estimate interaction between waves of heat kernel type and waves of algebraic type:

Lemma 2.5. *Let $\mu > 0$ and $\alpha > 1/2$ be constants. Then for $x \in \mathbb{R}$, $0 \leq \tau \leq t$, we have*

$$\int_{\mathbb{R}} [(t-\tau)^{-\frac{1}{2}} + (\tau+1)^{-\frac{1}{2}}] e^{-\frac{(x-y)^2}{\mu(t-\tau)}} \Big(1+\frac{y^2}{\tau+1}\Big)^{-2\alpha} dy = O(1)\Big(1+\frac{x^2}{t+1}\Big)^{-\alpha}. \tag{20}$$

Proof. Denote the left-hand side of (20) by I. Then consider the integration on $\{|y| \geq |x|/2\}$ and $\{|y| \leq |x|/2\}$ separately. We have

$$\begin{aligned} I \leq& \int_{|y|\geq \frac{|x|}{2}} \Big[(t-\tau)^{-\frac{1}{2}} e^{-\frac{(x-y)^2}{\mu(t-\tau)}} + (\tau+1)^{-\frac{1}{2}}\Big(1+\frac{y^2}{\tau+1}\Big)^{-\alpha}\Big] dy \Big(1+\frac{x^2}{4(t+1)}\Big)^{-\alpha} \\ &+ \int_{|y|\leq \frac{|x|}{2}} \Big[(t-\tau)^{-\frac{1}{2}} e^{-\frac{(x-y)^2}{2\mu(t-\tau)}} + (\tau+1)^{-\frac{1}{2}}\Big(1+\frac{y^2}{\tau+1}\Big)^{-2\alpha}\Big] dy\, e^{-\frac{x^2}{8\mu(t-\tau)}} \\ \leq& C\Big(1+\frac{x^2}{4(t+1)}\Big)^{-\alpha} + Ce^{-\frac{x^2}{8\mu(t+1)}} \leq C\Big(1+\frac{x^2}{t+1}\Big)^{-\alpha}, \end{aligned}$$

where $C > 0$ is a generic constant. □

3. **A Priori Estimates.** Due to Theorem 2.1, to prove Theorem 1.1 we only need to obtain (9) and (10), which is the goal of this section.

From (11) and by Duhamel's principle, we have

$$w(x,t) = \int_{\mathbb{R}} G(x-y,t)w_0(y)\, dy + \int_0^t \int_{\mathbb{R}} G(x-y,t-\tau)R(y,\tau)\, dyd\tau, \tag{21}$$

where G is the Green's function defined by (15). Denote the heat kernel by H:

$$H(x,t;1/r) = \frac{1}{\sqrt{4\pi t/r}} e^{-\frac{x^2}{4t/r}}. \tag{22}$$

From (3) we have

$$\theta(x,t) = \int_{\mathbb{R}} H(x-y,t;1/r)\theta(y,0)\, dy. \tag{23}$$

Let G_{ij} be the (i,j) entry of G, $1\le i,j\le 2$. Then combining (12), (21) and (23) and by integration by parts, we have

$$\begin{aligned}(v-\theta)(x,t) &= I_1 + I_2,\\ (u-1+\frac{1}{r}\theta_x)(x,t) &= \tilde I_1 + \tilde I_2,\end{aligned} \tag{24}$$

where

$$\begin{aligned}
I_1 =& \int_{\mathbb{R}} \{[G_{11}(x-y,t)v_0(y) - H(x-y,t;1/r)\theta(y,0)]\\
&+ G_{12}(x-y,t)[u_0(y)-1]\}\,dy,\\
I_2 =& \int_0^t\int_{\mathbb{R}} \Big\{\Big[\frac{\partial}{\partial x}G_{12}(x-y,t-\tau)\Big](-w_1w_2)(y,\tau)\\
&+ G_{12}(x-y,t-\tau)(-rw_2^2)(y,\tau)\Big\}\,dyd\tau,\\
\tilde I_1 =& \tilde I_{11} + \tilde I_{12},\\
\tilde I_{11} =& \int_{\mathbb{R}} [G_{21}(x-y,t)v_0(y) + \frac{1}{r}H_x(x-y,t;1/r)\theta(y,0)]\,dy,\\
\tilde I_{12} =& \int_{\mathbb{R}} G_{22}(x-y,t)[u_0(y)-1]\,dy,\\
\tilde I_2 =& \int_0^t\int_{\mathbb{R}} \Big\{\Big[\frac{\partial}{\partial x}G_{22}(x-y,t-\tau)\Big](-w_1w_2)(y,\tau)\\
&+ G_{22}(x-y,t-\tau)(-rw_2^2)(y,\tau)\Big\}\,dyd\tau.
\end{aligned} \tag{25}$$

Define a function $M(t)$ as follows:

$$\begin{aligned}M(t) = \sup_{0\le\tau\le t}\ &\{\|[\psi^\alpha(\cdot,\tau)]^{-1}(v-\theta)(\cdot,\tau)\|_{L^\infty}\\ &+ (\tau+1)^{\frac{1}{2}}\|[\psi^\alpha(\cdot,\tau)]^{-1}(u-1+\frac{1}{r}\theta_x)(\cdot,\tau)\|_{L^\infty}\}.\end{aligned} \tag{26}$$

To prove (9) and (10) we need to show

$$M(t) \le C\delta_0, \qquad t\ge 0, \tag{27}$$

where $C>0$ is a constant. For the rest of this section, we use C to denote a universal positive constant, which may vary line by line according to the context.

From (4) we have

$$\int_{\mathbb{R}} [v_0(x) - \theta(x,0)]\,dx = 0.$$

Thus we define

$$\eta(x) = \int_{-\infty}^x [v_0(y)-\theta(y,0)]\,dy = -\int_x^\infty [v_0(y)-\theta(y,0)]\,dy, \tag{28}$$

which implies

$$\eta'(x) = v_0(x) - \theta(x,0). \tag{29}$$

From (4), (5) and (8) we have

$$|v_0(x)| + |u_0(x)-1| + |\theta(x,0)| \le C\delta_0(|x|+1)^{-2\alpha}. \tag{30}$$

Lemma 3.1. *If v_0 satisfies* (7), *then for $x \in \mathbb{R}$,*

$$\eta(x) = O(1)\delta_0(|x|+1)^{1-2\alpha}, \tag{31}$$

where δ_0 is defined in (8).

Proof. From (28) and (30), for $x \le 0$ we have

$$|\eta(x)| \le \int_{-\infty}^{x} [|v_0(y)| + |\theta(y,0)|]\, dy \le C\delta_0 \int_{-\infty}^{x} (|y|+1)^{-2\alpha}\, dy = C\delta_0(|x|+1)^{1-2\alpha}.$$

If $x > 0$, we use the second equality in (28) to arrive at the same conclusion. □

Lemma 3.2. *Under the assumptions of Theorem 1.1, for $x \in \mathbb{R}$, $t > 0$, we have*

$$|I_1| \le C\delta_0 \psi^\alpha(x,t), \tag{32}$$

$$|\tilde{I}_1| \le C\delta_0 (t+1)^{-\frac{1}{2}} \psi^\alpha(x,t). \tag{33}$$

Proof. We write I_1 in (25) as

$$\begin{aligned} I_1 =& I_{11} + I_{12}, \\ I_{11} =& \int_{\mathbb{R}} H(x-y,t;1/r)[v_0(y) - \theta(y,0)]\, dy, \\ I_{12} =& \int_{\mathbb{R}} \{[G_{11}(x-y,t) - H(x-y,t;1/r)]v_0(y) \\ &+ G_{12}(x-y,t)[u_0(y)-1]\}\, dy. \end{aligned} \tag{34}$$

Consider the case $|x| \le \sqrt{t+1}$ first. If $t \ge 1$, by integration by parts and applying (31), we have

$$\begin{aligned} I_{11} =& \int_{\mathbb{R}} H_x(x-y,t;1/r)\eta(y)\, dy = O(1)\delta_0 t^{-\frac{1}{2}} \int_{\mathbb{R}} H(x-y,t;2/r)(|y|+1)^{1-2\alpha}\, dy \\ =& O(1)\delta_0 t^{-\frac{1}{2}} \Big[\int_{|y|\le\sqrt{t+1}} t^{-\frac{1}{2}} (|y|+1)^{1-2\alpha}\, dy \\ &+ \int_{|y|\ge\sqrt{t+1}} H(x-y,t;2/r)(t+1)^{\frac{1}{2}-\alpha}\, dy \Big] \\ =& O(1)\delta_0 \Gamma^\alpha(t). \end{aligned} \tag{35}$$

If $0 < t \le 1$, without integration by parts, we similarly have

$$I_{11} = \int_{\mathbb{R}} H(x-y,t;1/r)O(1)\delta_0\, dy = O(1)\delta_0 = O(1)\delta_0\Gamma^\alpha(t). \tag{36}$$

Noting (6), for $|x| \le \sqrt{t+1}$ and $t > 0$, we have

$$I_{11} = O(1)\delta_0 \psi^\alpha(x,t). \tag{37}$$

Next we consider $|x| \ge \sqrt{t+1}$ and $t > 0$. With integration by parts on $\{|y| \le |x|/2\}$, we write

$$\begin{aligned} I_{11} =& \int_{|y|\ge\frac{|x|}{2}} H(x-y,t;1/r)[v_0(y) - \theta(y,0)]\, dy + H(x-y,t;1/r)\eta(y)\Big]_{-\frac{|x|}{2}}^{\frac{|x|}{2}} \\ &+ \int_{|y|\le\frac{|x|}{2}} H_x(x-y,t;1/r)\eta(y)\, dy. \end{aligned}$$

Applying (30), (22) and (31) to the right-hand side, we have

$$
\begin{aligned}
I_{11} =&O(1)\delta_0\Big[(|x|+1)^{-2\alpha}+t^{-\frac{1}{2}}e^{-\frac{rx^2}{16t}}(|x|+1)^{1-2\alpha}+t^{-1}e^{-\frac{rx^2}{32t}}\int_0^{\frac{|x|}{2}}(y+1)^{1-2\alpha}dy\Big]\\
=&O(1)\delta_0(|x|+1)^{-2\alpha}+O(1)\delta_0(|x|+1)^{-2}e^{-\frac{x^2}{Ct}}\begin{cases}0 & \text{if } \frac{1}{2}<\alpha<1\\ \ln(|x|+1) & \text{if } \alpha=1\\ 1 & \text{if } \alpha>1\end{cases}.
\end{aligned}
\tag{38}
$$

Noting $|x|\geq\sqrt{t+1}$ and $e^{-\frac{x^2}{Ct}}\ln(|x|+1)=O(1)(1+\ln(t+1))$ by considering $|x|\gtreqless t$, we obtain

$$
I_{11}=O(1)\delta_0\big(\frac{x^2}{t+1}\big)^{-\alpha}\Gamma^\alpha(t)=O(1)\delta_0\psi^\alpha(x,t). \tag{39}
$$

To estimate I_{12}, we apply (16), (22) and (30) to (34) to have

$$
I_{12}=O(1)\delta_0\Big[\int_{\mathbb{R}}(t+1)^{-\frac{1}{2}}H(x-y,t;C)(|y|+1)^{-2\alpha}\,dy+e^{-\frac{t}{D}}(|x|+1)^{-2\alpha}\Big]. \tag{40}
$$

Comparing the right-hand side of (40) to those of (35) and (36), respectively, it is clear that the former is bounded by the latter. Also, by considering integration on $\{|y|\geq|x|/2\}$ and $\{|y|\leq|x|/2\}$, the right-hand side of (40) is bounded by that of (38). Therefore,

$$
I_{12}=O(1)\delta_0\psi^\alpha(x,t),\qquad x\in\mathbb{R},\ t>0. \tag{41}
$$

Equations (34), (37), (39) and (41) give us (32).

To prove (33), we apply Theorems 2.2 and 2.3, in particular, (17) to $\tilde{I}_{11}$ in (25):

$$
\begin{aligned}
\tilde{I}_{11}=&\int_{\mathbb{R}}-\frac{1}{r}H_x(x-y,t;1/r)[v_0(y)-\theta(y,0)]\,dy\\
&+\int_{\mathbb{R}}O(1)(t+1)^{-1}t^{-\frac{1}{2}}H(x-y,t;C)v_0(y)\,dy.
\end{aligned}
\tag{42}
$$

If we apply (16) instead, together with (23), we also have

$$
\tilde{I}_{11}=\int_{\mathbb{R}}O(1)(t+1)^{-\frac{1}{2}}H(x-y,t;C)v_0(y)\,dy+\frac{1}{r}\theta_x(x,t). \tag{43}
$$

For $|x|\leq\sqrt{t+1}$, if $t\geq1$, we compare $\tilde{I}_{11}$ in (42) with I_{11} in (34), and follow the derivation of (35) to arrive at $\tilde{I}_{11}=O(1)\delta_0(t+1)^{-1/2}\psi^\alpha(x,t)$. If $0<t\leq1$, we have, from (43) and similar to (36), $\tilde{I}_{11}=O(1)\delta_0=O(1)\delta_0(t+1)^{-1/2}\psi^\alpha(x,t)$.

For $|x|\geq\sqrt{t+1}$, similarly, if $t\geq1$, we follow the derivation of (38) to handle the first term of (42), which gives us the estimate $O(1)\delta_0(t+1)^{-1/2}\psi^\alpha(x,t)$. The second term in (42) is better than the first term in (40) by a decay factor $(t+1)^{-1}$, thus results in $O(1)\delta_0(t+1)^{-1/2}\psi^\alpha(x,t)$. If $0<t\leq1$, we compare (43) with (40) to have $\tilde{I}_{11}=O(1)\delta_0\psi^\alpha(x,t)=O(1)\delta_0(t+1)^{-1/2}\psi^\alpha(x,t)$. In summary,

$$
\tilde{I}_{11}=O(1)\delta_0(t+1)^{-1/2}\psi^\alpha(x,t),\qquad x\in\mathbb{R},\ t>0. \tag{44}
$$

For $\tilde{I}_{12}$ in (25), applying (16) and (30) we have

$$
\tilde{I}_{12}=O(1)\delta_0\int_{\mathbb{R}}(t+1)^{-1}H(x-y,t;C)(|y|+1)^{-2\alpha}\,dy,
$$

which is better than the first term in (40) by a decay factor $(t+1)^{-\frac{1}{2}}$. This gives us

$$\tilde{I}_{12} = O(1)\delta_0(t+1)^{-1/2}\psi^\alpha(x,t), \qquad x\in\mathbb{R},\ t>0. \tag{45}$$

Combining (25), (44) and (45), we settle (33). □

Lemma 3.3. *Under the assumptions of Theorem 1.1, for $x\in\mathbb{R}$, $t>0$, we have*

$$|I_2| \le C[\delta_0^2 + M^2(t)]\psi^\alpha(x,t), \tag{46}$$

$$|\tilde{I}_2| \le C[\delta_0^2 + M^2(t)](t+1)^{-\frac{1}{2}}\psi^\alpha(x,t). \tag{47}$$

Proof. From (25) and (16), we have

$$I_2 = \int_0^t\int_{\mathbb{R}} O(1)(t-\tau+1)^{-\frac{1}{2}}(t-\tau)^{-\frac{1}{2}}e^{-\frac{(x-y)^2}{C(t-\tau)}}[(t-\tau)^{-\frac{1}{2}}|w_1w_2|(y,\tau)+w_2^2(y,\tau)]\,dyd\tau + \int_0^t e^{-\frac{t-\tau}{D}}O(1)|w_1w_2|(x,\tau)\,d\tau. \tag{48}$$

From (12) and (26) we have

$$\begin{aligned} &|w_1(x,t)| \le |\theta(x,t)| + M(t)\psi^\alpha(x,t),\\ &|w_2(x,t)| \le \frac{1}{r}|\theta_x(x,t)| + M(t)(t+1)^{-\frac{1}{2}}\psi^\alpha(x,t). \end{aligned} \tag{49}$$

By (5) and (6), they imply

$$\begin{aligned} |w_1w_2|(x,t) \le\ & C[\delta_0^2+M^2(t)](t+1)^{-\frac{3}{2}}e^{-\frac{x^2}{C(t+1)}}\\ &+ M^2(t)(t+1)^{-\frac{1}{2}}[\Gamma^\alpha(t)]^2\big(1+\frac{x^2}{t+1}\big)^{-2\alpha},\\ |w_2^2|(x,t) \le\ & C\delta_0^2(t+1)^{-2}e^{-\frac{x^2}{C(t+1)}} + M^2(t)(t+1)^{-1}[\Gamma^\alpha(t)]^2\big(1+\frac{x^2}{t+1}\big)^{-2\alpha}. \end{aligned} \tag{50}$$

Substituting (50) into (48), and applying Lemmas 2.4 and 2.5, we arrive at

$$\begin{aligned} I_2 =& O(1)[\delta_0^2+M^2(t)](t+1)^{-1}e^{-\frac{x^2}{C(t+1)}}\\ &+ O(1)M^2(t)\int_0^t (t-\tau+1)^{-\frac{1}{2}}(t-\tau)^{-\frac{1}{2}}(\tau+1)^{-\frac{1}{2}}[\Gamma^\alpha(\tau)]^2\,d\tau\big(1+\frac{x^2}{t+1}\big)^{-\alpha}\\ &+ O(1)[\delta_0^2+M^2(t)]\int_0^t e^{-\frac{t-\tau}{D}}(\tau+1)^{-\frac{3}{2}}\,d\tau\big(1+\frac{x^2}{t+1}\big)^{-2\alpha}. \end{aligned} \tag{51}$$

By integrating on $[0,t/2]$ and $[t/2,t]$ separately, we obtain

$$\begin{aligned} I_2 =& O(1)[\delta_0^2+M^2(t)](t+1)^{-1}e^{-\frac{x^2}{C(t+1)}} + O(1)M^2(t)(t+1)^{-1}\big(1+\frac{x^2}{t+1}\big)^{-\alpha}\\ &+ O(1)[\delta_0^2+M^2(t)](t+1)^{-\frac{3}{2}}\big(1+\frac{x^2}{t+1}\big)^{-2\alpha}\\ =& O(1)[\delta_0^2+M^2(t)]\psi^\alpha(x,t). \end{aligned} \tag{52}$$

This settles (46).

The proof of (47) is similar. From (25) and (16) we have

$$\tilde{I}_2 = \int_0^t \int_{\mathbb{R}} O(1)(t-\tau+1)^{-1}(t-\tau)^{-\frac{1}{2}} e^{-\frac{(x-y)^2}{C(t-\tau)}} [(t-\tau)^{-\frac{1}{2}} |w_1 w_2|(y,\tau) + w_2^2(y,\tau)] \, dy d\tau$$
$$+ \int_0^t e^{-\frac{t-\tau}{D}} O(1) |w_1 w_2|(x,\tau) \, d\tau. \quad (53)$$

Following the derivation of (52), we substitute (50) into (53), and apply Lemmas 2.4 and 2.5. The extra decay factor $(t-\tau+1)^{-\frac{1}{2}}$ in the Green's function is translated into the extra decay factor $(t+1)^{-\frac{1}{2}}$ in (47), through the application of Lemma 2.4 and the time integral similar to the second term in (51). □

Combining (24), (32), (33), (46) and (47), we have

$$|v-\theta|(x,t) \le C[\delta_0 + M^2(t)]\psi^\alpha(x,t),$$
$$|u-1+\frac{1}{r}\theta_x|(x,t) \le C[\delta_0 + M^2(t)](t+1)^{-\frac{1}{2}}\psi^\alpha(x,t).$$

With (26), these imply $M(t) \le C[\delta_0 + M^2(t)]$ for some constant $C > 0$. By simple algebra, we have (27) if $M(t)$ is bounded by a small constant. Via a continuity argument, (27) is valid if δ_0 is small.

REFERENCES

[1] T.-P. Liu and Y. Zeng, Large time behavior of solutions for general quasilinear hyperbolic-parabolic systems of conservation laws, *Mem. Amer. Math. Soc.*, **125** no. 599 (1992), viii+120pp.

[2] J. Rugamba and Y. Zeng, Green's function of the linearized logarithmic Keller-Segel-Fisher/KPP system, *Math. Comput. Appl.*, **23** (2018), Paper No. 56, 12 pp.

[3] Y. Zeng, Global existence theory for general hyperbolic-parabolic balance laws with application, *J. Hyperbolic Differ. Equ.*, **14** (2017), 359–391.

[4] Y. Zeng, L^p decay for general hyperbolic-parabolic systems of balance laws, *Discrete Contin. Dyn. Syst.*, **38** (2018), 363-396.

[5] Y. Zeng, Asymptotic behavior of solutions to general hyperbolic-parabolic systems of balance laws in multi-space dimensions, *Pure and Applied Mathematics Quarterly*, **14** (2018), 161–192.

[6] Y. Zeng, L^p time asymptotic decay for general hyperbolic-parabolic balance laws with applications, preprint.

[7] Y. Zeng, Hyperbolic-parabolic balance laws: asymptotic behavior and a chemotaxis model, *Communications in Applied Analysis*, SEARCDE 2017 Proceedings, to appear.

[8] Y. Zeng and K. Zhao, On the logarithmic Keller-Segel-Fisher/KPP system, *Discrete Contin. Dyn. Syst.*, to appear.

[9] Y. Zeng and K. Zhao, Recent results for the logarithmic Keller-Segel-Fisher/KPP system, *Boletim da Sociedade Paranaense de Matematica*, to appear.

E-mail address: rugamba@uab.edu
E-mail address: ynzeng@uab.edu

FRONTS FOR THE SQG EQUATION: A REVIEW

JINGYANG SHU

Department of Mathematics
University of California at Davis
One Shields Ave.
Davis, CA 95616, USA

ABSTRACT. Temperature discontinuities, or fronts, in the surface quasi-geostrophic (SQG) equations support surface waves. By regularizing the contour dynamics equations, we drive a nonlinear and nonlocal equation, which describes the evolution of SQG fronts. In this survey, we review some recent results on the dynamics of SQG fronts, including a derivation of SQG fronts equation, existence of local and global solutions, and evidence of finite-time singularity formation.

1. **Introduction.** The 2D surface quasi-geostrophic (SQG) equation is classically written as an active scalar equation

$$\theta_t + \mathbf{u} \cdot \nabla\theta = 0,$$
$$\mathbf{u} = \nabla^{\perp}(-\Delta)^{-1/2}\theta.$$

Here, $\theta(\mathbf{x}, t)$ with $\mathbf{x} = (x, y)$ is an unknown scalar field, and the velocity field $\mathbf{u}(\mathbf{x}, t)$ is determined nonlocally from θ by a perpendicular Riesz transform [28]

$$\mathbf{u}(\mathbf{x}) = -\frac{1}{2\pi} \lim_{\epsilon \to 0+} \int_{\mathbb{R}^2 \setminus B_\epsilon(\mathbf{x})} \frac{(\mathbf{x} - \mathbf{y})^{\perp}}{|\mathbf{x} - \mathbf{y}|^3} \theta(\mathbf{y})\, d\mathbf{y}.$$

The SQG equation comes from the quasi-geostrophic (QG) equation which describes stratified mid-to-high latitude synoptic scale dynamics in oceanic or atmosphere flows. One of the major hypotheses of flows in this altitude range is that the long-scale dynamics of the fluids is governed by the near balance between the Coriolis force and horizontal pressure gradients [21]. The SQG equation is a reduction of the QG equation when the flows are confined near a surface [15, 20, 24]. Mathematically, the inviscid SQG equation has strong similarity to the 3D Euler equations [5, 6], and the SQG patch problem has a formal resemblance to the vortex patch problem [22]. For analysis of the SQG equation, see [1, 4, 23, 25] and the references cited therein.

By the transport nature of the SQG equation, we may consider a special type of weak solution when θ takes on only two distinct constant values θ_+, θ_-, so that

$$\theta(\mathbf{x}, t) = \begin{cases} \theta_+ & \mathbf{x} \in \Omega(t), \\ \theta_- & \mathbf{x} \in \Omega^c(t), \end{cases}$$

2000 *Mathematics Subject Classification.* Primary: 35Q35; Secondary: 86A10.

Key words and phrases. Surface quasi-geostrophic equation, contour dynamics, global existence, space-time resonances, modified scattering.

The author was partially supported by the NSF under grant number DMS-1616988.

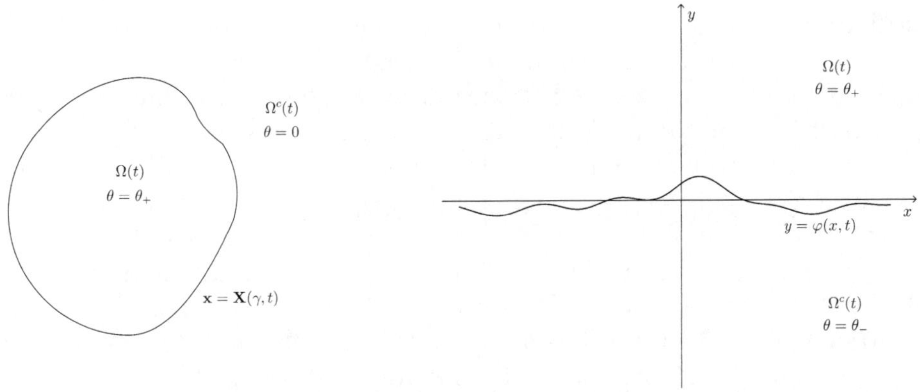

FIGURE 1. Left: Patch with bounded $\Omega(t)$. Right: Front with $\Omega(t)$ a half-space.

for some domain $\Omega(t) \subset \mathbb{R}^2$, and thus, to study this type of solutions, we only need to study the evolution of the boundary $\partial\Omega(t)$. We first define a weak solution of the SQG equation [25].

Definition 1.1 (Weak solutions of the SQG equation)**.** A bounded function θ is a *weak solution* of the SQG equation if for any $\phi \in C_c^\infty(\mathbb{R}^2 \times (0,T))$, we have

$$\int_{\mathbb{R}^2 \times [0,T]} [\theta(\mathbf{x},t)\phi_t(\mathbf{x},t) + \theta(\mathbf{x},t)\mathbf{u}(\mathbf{x},t) \cdot \nabla\phi(\mathbf{x},t)] \, \mathrm{d}\mathbf{x} \, \mathrm{d}t = 0.$$

When $\Omega(t)$ is simply connected and $\partial\Omega(t)$ is a connected regular curve, we distinguish two particular types of domain, which are shown in Figure 1.

1. Patches, whose boundary is a smooth, simple, closed curve diffeomorphic to the circle $\mathbb{T}$, with $\theta_- = 0$ outside the patch.
2. Half-spaces, whose boundary is a smooth, simple curve diffeomorphic to $\mathbb{R}$ that divides $\mathbb{R}^2$ into two half-spaces.

In the case of a patch, one can take $\theta(\cdot,t) = \theta_+\chi_{\Omega(t)}$ where $\Omega(t)$ is a simply connected bounded subset of $\mathbb{R}^2$, whose boundary is parametrized by $\mathbf{x} = \mathbf{X}(\gamma,t)$. If the constant is normalized by choosing $\theta_+ = 2\pi$, one obtains well-defined contour dynamics equations for the patch [11]

$$\mathbf{X}_t(\eta,t) = c(\eta,t)\partial_\eta\mathbf{X}(\eta,t) + \int_{\mathbb{T}} \frac{\partial_\eta\mathbf{X}(\eta,t) - \partial_\eta\mathbf{X}(\eta-\zeta,t)}{|\mathbf{X}(\eta,t) - \mathbf{X}(\eta-\zeta,t)|} \, \mathrm{d}\zeta, \tag{1}$$

where $c(\eta,t)$ is an arbitrary function corresponding to a time-dependent reparametrization of the curve. Local existence and uniqueness of the initial-value problem of this equation is established in Sobolev spaces by arc-length reparametrization of the patch boundary [7, 11, 12]. In [2, 3, 14], a class of nontrivial global solutions are constructed using the Crandall–Rabinowitz's bifurcation theorem. It has also been proved that splash singularities cannot occur in a smooth boundary of an SQG patch [13], but whether other types of finite-time singularities can occur remains open. Some numerical studies are carried out in [8, 27], where a curvature blow-up on the SQG patch boundary and a complicated self-similar cascade of filament instability are observed in the numerical simulations.

In the case of half-spaces, we refer the boundary $\partial\Omega(t)$ as a *front*. In contrast to the patches, the formal contour dynamics equation, obtained by replacing the integration limit $\mathbb{T}$ with $\mathbb{R}$ in (1), does not converge. However, a regularization procedure is introduced in [16] to make sense of the divergent integral. We remark that in [19], we justify this procedure by using direct contour dynamics methods. The derivation of a regularized evolutionary equation describing the dynamics of SQG fronts is reviewed in Section 2. When the fronts are spatially periodic, the initial-value problem is proved to be locally well-posed in the C^∞-class, by a Nash–Moser argument [26], and in the analytic class, by a Cauchy–Kowalevski theorem [10]. In [17], the initial value problem for a cubically nonlinear, approximate equation of the regularized equation (2) with periodic initial data is proved to be locally well-posed in Sobolev spaces. We also prove the global well-posedness of the initial value problem for the full equation (5) with a smallness assumption on the initial data [18]. The proof for global well-posedness is outlined in Section 3. Finally, we survey in Section 4 a numerical study of the SQG fronts, which suggests wave breaking or singularity formation in finite time.

2. **Regularized SQG front equations.** We will consider only fronts that are a graph, located at

$$y = \varphi(x,t),$$

where $\varphi(x,t)\colon \mathbb{R} \to \mathbb{R}$ is a smooth, bounded function. As is discussed above, the formal contour dynamics equation for the fronts does not converge. To make sense of the equation, we propose the following regularization. We first cut-off the integration region using a ball with radius λ around a point $\mathbf{x}$ on the front, and then obtain the truncated equation for $\mathbf{X}$

$$\mathbf{X}_t(\eta,t) = c(\eta,t)\partial_\eta \mathbf{X}(\eta,t) + \int_{\eta-\lambda}^{\eta+\lambda} \frac{\partial_\zeta \mathbf{X}(\eta,t) - \partial_\zeta \mathbf{X}(\zeta,t)}{|\mathbf{X}(\eta,t) - \mathbf{X}(\zeta,t)|}\, d\zeta.$$

When the front curve is given by a graph $\varphi(x,t)$, the function c can be uniquely solved and we obtain an equation for φ

$$\begin{aligned} \varphi_t(x,t) &+ \int_{-\lambda}^{\lambda} \frac{\varphi_x(x+\zeta,t) - \varphi_x(x,t)}{|\zeta|}\, d\zeta \\ &+ \int_{-\lambda}^{\lambda} \left[\frac{\varphi_x(x+\zeta,t) - \varphi_x(x,t)}{\sqrt{\zeta^2 + (\varphi(x+\zeta,t) - \varphi(x,t))^2}} - \frac{\varphi_x(x+\zeta,t) - \varphi_x(x,t)}{|\zeta|} \right] d\zeta = 0. \end{aligned}$$

Using the Fourier transform, we find that

$$\int_{-\lambda}^{\lambda} \frac{\varphi_x(x+\zeta,t) - \varphi_x(x,t)}{|\zeta|}\, d\zeta = \big[d(\lambda) - 2(\log\lambda)\big]\varphi_x(x,t) - 2\log|\partial_x|\varphi_x(x,t),$$

where $\log|\partial_x|$ is the Fourier multiplier operator with symbol $\log|\xi|$ and $d(\lambda) \to -2\gamma$ as $\lambda \to \infty$ with γ being the Euler–Mascheroni constant. Now, by a λ-dependent Galilean transformation $x \mapsto x + \big[d(\lambda) - 2\log\lambda\big]t$, the divergent advection term can be removed. In the limit $\lambda \to \infty$, we get a regularized equation for SQG fronts

$$\begin{aligned} &\varphi_t(x,t) - 2\log|\partial_x|\varphi_x(x,t) \\ &= -\int_{\mathbb{R}} [\varphi_x(x,t) - \varphi_x(x+\zeta,t)] \left\{ \frac{1}{|\zeta|} - \frac{1}{\sqrt{\zeta^2 + [\varphi(x,t) - \varphi(x+\zeta,t)]^2}} \right\} d\zeta. \end{aligned} \tag{2}$$

Dimensional analysis of the SQG equation demonstrates that parameters $\theta_\pm$ are velocities, so one might expect that the waves on an SQG front are nondispersive.

Nevertheless, this equation is dispersive with dispersion relation $\omega(\xi) = 2\xi \log|\xi|$ by looking at the linearized equation (originally pointed out in [26])

$$\varphi_t(x,t) = 2\log|\partial_x|\varphi_x(x,t). \tag{3}$$

This linearized equation (3) has an anomalous scaling-Galilean invariance $x \mapsto \lambda[x + 2(\log\lambda)t]$, $t \mapsto \lambda t$, and the linearized equation commutes with a scaling-Galilean vector field

$$\mathcal{S} = (x+2t)\partial_x + t\partial_t. \tag{4}$$

3. **Global well-posedness of SQG front equation.** We consider the initial value problem posed on $\mathbb{R}$

$$\begin{aligned} &\varphi_t(x,t) - 2\log|\partial_x|\varphi_x(x,t) \\ &= -\int_{\mathbb{R}} [\varphi_x(x,t) - \varphi_x(x+\zeta,t)] \left\{ \frac{1}{|\zeta|} - \frac{1}{\sqrt{\zeta^2 + [\varphi(x,t) - \varphi(x+\zeta,t)]^2}} \right\} \mathrm{d}\zeta, \\ &\varphi(x,0) = \varphi_0(x), \end{aligned} \tag{5}$$

where $\varphi\colon \mathbb{R}\times\mathbb{R}_+ \to \mathbb{R}$ is defined for $x \in \mathbb{R}$ and $t \in \mathbb{R}_+$. For fronts with small slopes $|\varphi_x| \ll 1$, we can expand the nonlinearity and rewrite (2) as

$$\begin{aligned} &\varphi_t(x,t) - 2\log|\partial_x|\varphi_x(x,t) \\ &= \sum_{n=1}^{\infty} \frac{c_n}{2n+1}\partial_x \int_{\mathbb{R}^{2n+1}} \mathbf{T}_n(\eta_1,\ldots,\eta_{2n+1}) \prod_{j=1}^{2n+1} \left(e^{i\eta_j x}\hat{\varphi}(\eta_j,t)\right) \mathrm{d}\eta_1\cdots\mathrm{d}\eta_{2n+1}, \end{aligned} \tag{6}$$

where

$$\mathbf{T}_n(\eta_1,\ldots,\eta_{2n+1}) = \int_{\mathbb{R}} \frac{\prod_{j=1}^{2n+1}(1-e^{i\eta_j\zeta})}{|\zeta|^{2n+1}}\,\mathrm{d}\zeta, \qquad c_n = \frac{\sqrt{\pi}}{\Gamma\left(\frac{1}{2}-n\right)\Gamma(n+1)}. \tag{7}$$

The main theorem in [18] is as follows.

Theorem 3.1. *Let $s = 1200$, $r = 7$, and $p_0 = 10^{-4}$. There exists a constant $0 < \varepsilon \ll 1$, such that if $\varphi_0 \in H^s(\mathbb{R})$ satisfies*

$$\|\varphi_0\|_{H^s} + \|x\varphi_0\|_{H^r} \leq \varepsilon_0$$

for some $0 < \varepsilon_0 \leq \varepsilon$, then there exists a unique global solution $\varphi \in C([0,\infty); H^s(\mathbb{R}))$ of (5). Moreover, this solution satisfies

$$\|\varphi(t)\|_{H^s} + \|\mathcal{S}\varphi(t)\|_{H^s} \lesssim \varepsilon_0 (t+1)^{p_0},$$

where $\mathcal{S}$ is the vector field in (4).

We remark that according to this theorem, the Sobolev H^s-norm of φ is not controlled uniformly in time, and it possibly has a very slow growth of order $(t+1)^{p_0}$. The only norm we can bound uniformly in time is the Z-norm of φ (see (10) for the definition of Z-norm). It is not clear to us whether this is a limitation of our method of proof or an intrinsic feature of the solutions.

Following [9], to prove this global existence theorem, it suffices to prove local well-posedness and a suitable global *a priori* bound.

3.1. **Local well-posedness.** The difficulty in local well-posedness is that straightforward H^s-estimates for (5) do not close, due to a logarithmic loss of derivatives [16]. In order to overcome this difficulty, we use Weyl para-differential calculus to para-linearize the equation, then extract a term from the nonlinearity, which can be controlled by the linear term, and define a weighted energy whose estimates do close.

We use $T_a f$ to denote the standard Weyl para-product [29], and define an s-order weighted energy as

$$\begin{aligned} \tilde{E}^{(s)}(t) &= \|\varphi\|^2_{L^2(\mathbb{R})} + \sum_{j=1}^{s} E^{(j)}(t), \\ E^{(j)}(t) &= \int_{\mathbb{R}} |D|^j \varphi(x,t) \cdot \left(2 - T_{B^{\log}[\varphi]}\right)^{2j+1} |D|^j \varphi(x,t)\, \mathrm{d}x, \end{aligned} \tag{8}$$

where, if we denote by δ the Dirac-delta distribution,

$$\begin{aligned} &B^{\log}[\varphi](\cdot,\xi) \\ &= -\mathcal{F}_\zeta^{-1}\Bigg[\sum_{n=1}^{\infty} 2c_n \int_{\mathbb{R}^{2n}} \delta\Big(\zeta - \sum_{j=1}^{2n}\eta_j\Big) \\ &\qquad \cdot \prod_{j=1}^{2n}\left(i\eta_j\hat{\varphi}(\eta_j)\chi\Big(\frac{(2n+1)\eta_j}{\xi}\Big)\right) \mathrm{d}\eta_1\, \mathrm{d}\eta_2 \cdots \mathrm{d}\eta_{2n}\Bigg]. \end{aligned}$$

By carrying out standard estimates for $E^{(s)}(t)$, we find that

$$\begin{aligned} E^{(s)}(t) \le E^{(s)}(0) \exp\Bigg\{ \int_0^t &\Big[(\|\varphi_x(\tau)\|_{W^{2,\infty}} + \|\log|\partial_x|\varphi_x(\tau)\|_{W^{2,\infty}})^2 \\ &\cdot F(\|\varphi_x(\tau)\|_{W^{2,\infty}} + \|\log|\partial_x|\varphi_x(\tau)\|_{W^{2,\infty}})\Big]\, \mathrm{d}\tau \Bigg\}, \end{aligned}$$

where, for a sufficiently large number $\tilde{C}$ depending only on s, $F(\cdot)$ is an increasing continuous real-valued function as long as

$$\sum_{n=1}^{\infty} \tilde{C}^n |c_n| \left(\|\varphi_x(t)\|^{2n}_{W^{2,\infty}} + \|\log|\partial_x|\varphi_x(t)\|^{2n}_{W^{2,\infty}} \right) < \infty.$$

The local well-posedness and breakdown criterion for solutions is stated in the following theorem.

Theorem 3.2. *Let $s > 4$ be an integer. There exists a constant $\tilde{C} > 0$, depending only on s, such that if $\varphi_0 \in H^s(\mathbb{R})$ satisfies*

$$\|T_{B^{\log}[\varphi_0]}\|_{L^2\to L^2} \le C, \quad \sum_{n=1}^{\infty} \tilde{C}^n |c_n| \left(\|\partial_x\varphi_0\|^{2n}_{W^{2,\infty}} + \|\partial_x \log|\partial_x|\varphi_0\|^{2n}_{W^{2,\infty}} \right) < \infty \tag{9}$$

for some constant $0 < C < 2$, then there exists a maximal time of existence $0 < T_{\max} \le \infty$ depending only on $\|\varphi_0\|_{H^s}$, C, and $\tilde{C}$ such that the initial value problem (5) has a unique solution with $\varphi \in C([0, T_{\max}); H^s(\mathbb{R}))$. If $T_{\max} < \infty$, then either

$$\lim_{t\to T_{\max}} \sum_{n=1}^{\infty} \tilde{C}^n |c_n| \left(\|\varphi_x(t)\|^{2n}_{W^{2,\infty}} + \|\log|\partial_x|\varphi_x(t)\|^{2n}_{W^{2,\infty}} \right) = \infty$$

or

$$\lim_{t\to T_{\max}} \left\|T_{B^{\log}[\varphi(\cdot,t)]}\right\|_{L^2\to L^2} = 2.$$

We remark that this proof requires the smallness of the $W^{2,\infty}$-norms of $\partial_x\varphi_0$ and $\partial_x \log|\partial_x|\varphi_0$ (see (9)), in order to validate the expansion of nonlinearity, as well as the non-degeneracy of the weight (so that $\tilde{E}^{(s)} \approx \|\varphi\|^2_{H^s(\mathbb{R})}$). It is unclear whether the problem is still locally well-posed if $2 - T_{B^{\log}[\varphi]}$ degenerates.

3.2. **Global *a priori* bound.** To complete the proof of global well-posedness theorem, we introduce the Z-norm of a function f

$$\|f\|_Z = \left\|(|\xi| + |\xi|^{r+3})\hat{f}(\xi)\right\|_{L^\infty_\xi}. \tag{10}$$

To obtain the global *a priori* bound, we use a bootstrap argument and prove the following proposition.

Proposition 3.3 (Bootstrap). *Let $T > 1$ and suppose that $\varphi \in C([0,T]; H^s)$ is a solution of* (5)*, where the initial data satisfies*

$$\|\varphi_0\|_{H^s} + \|x\partial_x\varphi_0\|_{H^r} \le \varepsilon_0$$

for some $0 < \varepsilon_0 \ll 1$. If there exists $\varepsilon_0 \ll \varepsilon_1 \lesssim \varepsilon_0^{1/3}$ such that the solution satisfies

$$(t+1)^{-p_0}\left(\|\varphi(t)\|_{H^s} + \|\mathcal{S}\varphi(t)\|_{H^r}\right) + \|\varphi\|_Z \le \varepsilon_1$$

for every $t \in [0,T]$, then the solution satisfies an improved bound

$$(t+1)^{-p_0}\left(\|\varphi(t)\|_{H^s} + \|\mathcal{S}\varphi(t)\|_{H^r}\right) + \|\varphi\|_Z \le \varepsilon_0.$$

We call the assumptions in Proposition 3.3 the *bootstrap assumptions.* To prove Proposition 3.3, the first step is to prove a sharp dispersive estimates

$$\|\varphi(t)\|_{L^\infty} + \left\|\partial^{r+1}\log|\partial_x|\varphi(t)\right\|_{L^\infty} \lesssim \varepsilon_1(t+1)^{-1/2}. \tag{11}$$

We achieve this by first carrying out the standard dispersive estimates for the linearized equation (3) and then taking advantage of the bootstrap assumptions to sharpen it. Using this sharp dispersive estimates, one can directly complete estimates for $\|\varphi(t)\|_{H^s}$. By modifying the weighted energy (8) for $\mathcal{S}\varphi$, we then obtain the improved bounds for

$$(t+1)^{-p_0}\left(\|\varphi(t)\|_{H^s} + \|\mathcal{S}\varphi(t)\|_{H^r}\right).$$

The most difficult part is the nonlinear dispersive estimate, which deals with the estimates for $\|\varphi\|_Z$.

The rest of the proof involves a detailed frequency-space analysis. To show $\|\varphi\|_Z$ is uniformly bounded by a constant of order ε_0, we only need to show that

$$\int_0^T \|\varphi_t\|_Z \,\mathrm{d}t = \int_0^T \|h_t\|_Z \,\mathrm{d}t \lesssim \varepsilon_0, \tag{12}$$

where

$$h(x,t) = e^{-2t\partial_x \log|\partial_x|}\varphi(x,t), \qquad \hat{h}(\xi,t) = e^{-2it\xi\log|\xi|}\hat{\varphi}(\xi,t).$$

To this end, we take the Fourier transform of the expanded equation (6), and rewrite the equation as

$$\begin{aligned}&\hat{h}_t + e^{-2it\xi\log|\xi|}\widehat{\mathcal{N}_{\ge 5}(\varphi)}\\ &= -\frac{1}{6}i\xi\int_{\mathbb{R}^2}\mathbf{T}_1(\eta_1,\eta_2,\xi-\eta_1-\eta_2)e^{it\Phi(\xi,\eta_1,\eta_2)}\hat{h}(\xi-\eta_1-\eta_2)\hat{h}(\eta_1)\hat{h}(\eta_2)\,\mathrm{d}\eta_1\,\mathrm{d}\eta_2,\end{aligned}$$

where $\mathcal{N}_{\ge 5}(\varphi)$ denotes the nonlinearity of quintic degree and higher,

$$\Phi(\xi,\eta_1,\eta_2) = 2(\xi-\eta_1-\eta_2)\log|\xi-\eta_1-\eta_2| + 2\eta_1\log|\eta_1| + 2\eta_2\log|\eta_2| - 2\xi\log|\xi|,$$

and $\mathbf{T}_1$ is defined as in (7) which can be rewritten as

$$\begin{aligned}&\mathbf{T}_1(\eta_1,\eta_2,\eta_3)\\&= \eta_1^2\log|\eta_1| + \eta_2^2\log|\eta_2| + \eta_3^2\log|\eta_3| + (\eta_1+\eta_2+\eta_3)^2\log|\eta_1+\eta_2+\eta_3|\\&\quad - (\eta_1+\eta_2)^2\log|\eta_1+\eta_2| - (\eta_1+\eta_3)^2\log|\eta_1+\eta_3| - (\eta_2+\eta_3)^2\log|\eta_2+\eta_3|.\end{aligned}$$

Therefore, proving inequality (12) is reduced to showing that under the bootstrap assumptions, the integrals

$$\begin{aligned}\int_0^T \xi(|\xi|+|\xi|^{r+3})\int_{\mathbb{R}^2} &\mathbf{T}_1(\eta_1,\eta_2,\xi-\eta_1-\eta_2)\\ &\cdot e^{it\Phi(\xi,\eta_1,\eta_2)}\hat{h}(\xi-\eta_1-\eta_2)\hat{h}(\eta_1)\hat{h}(\eta_2)\,\mathrm{d}\eta_1\,\mathrm{d}\eta_2\,\mathrm{d}t,\end{aligned} \tag{13}$$

$$\int_0^T (|\xi|+|\xi|^{r+3})e^{-2it\xi\log|\xi|}\widehat{\mathcal{N}_{\geq 5}(\varphi)}\,\mathrm{d}t \tag{14}$$

are bounded uniformly with respect to ξ and T.

For the estimates of (13), we distinguish between time resonances and space resonances. In most regions of frequency space, these resonances do not appear simultaneously, and we can use oscillatory integral estimates or an interpolation inequality to get sufficient decay. To be more specific, in the region with only space resonance, we integrate by parts in the time variable to obtain a fourth degree nonlinearity in h, and then we use multilinear estimates and sharp dispersive estimates (11) to gain time decay. In the region away from the space-time resonances, we integrate by parts in a frequency variable and gain an extra $(t+1)^{-1}$-decay. Around the space-time resonances, we need to use modified scattering (a phase correction) to cancel out the leading order non-integrability in time.

The estimates for higher-order nonlinear terms (14) can be proved by multilinear estimates, as the sharp dispersive estimates (11) provides enough time decay.

4. **Evidence of singularity formation.** We study numerically an approximate model equation of the regularized SQG front equation

$$\varphi_t + \frac{1}{2}\partial_x\left\{\varphi^2\log|\partial_x|\varphi_{xx} - \varphi\log|\partial_x|(\varphi^2)_{xx} + \frac{1}{3}\log|\partial_x|(\varphi^3)_x\right\} = 2\log|\partial_x|\varphi_x. \tag{15}$$

This equation is obtained by a formal truncation of the nonlinearity of (6) at the cubic level. It is easy to verify that this equation has following two conserved quantities for smooth data

$$\text{Entropy: } \mathcal{S}(t) = \int_{\mathbb{T}}\varphi^2\,\mathrm{d}x,$$

$$\text{Energy: } \mathcal{H}(t) = \int_{\mathbb{T}}\varphi\log|\partial_x|\varphi + \frac{1}{8}\varphi^2\partial_x^2\log|\partial_x|\varphi^2 - \frac{1}{6}\varphi\partial_x^2\log|\partial_x|\varphi^3\,\mathrm{d}x.$$

We choose initial data

$$\varphi_0(x) = \cos(x+\pi) + \frac{1}{2}\cos[2(x+\pi+2\pi^2)], \tag{16}$$

and use a pseudo-spectral method (2^{15} Fourier modes) with spectral viscosity to carry out the numerical simulations. We observe an oscillatory singularity at $t \approx 0.06$ near $x \approx 2.15$ (see Figure 2 and Figure 3).

We remark that a proof of singularity formation for the SQG equation is open.

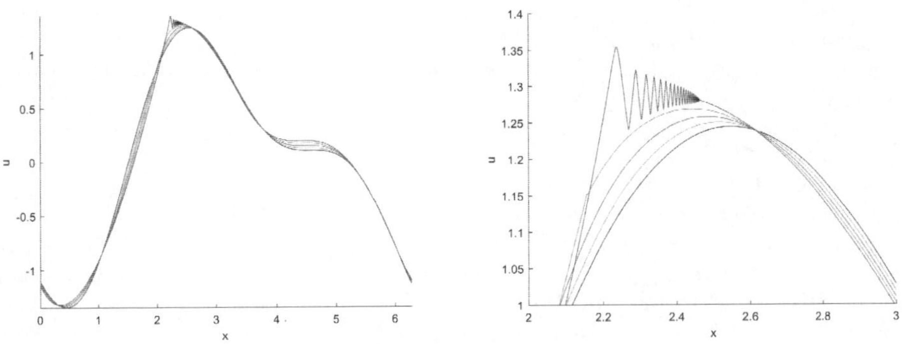

FIGURE 2. Left: Solution of (15) with initial data (16), shown at $t = 0$ (blue), $t = 0.01875$ (cyan), $t = 0.0375$ (magenta), $t = 0.05625$ (green), $t = 0.075$ (red). Right: Detail of singularity formation.

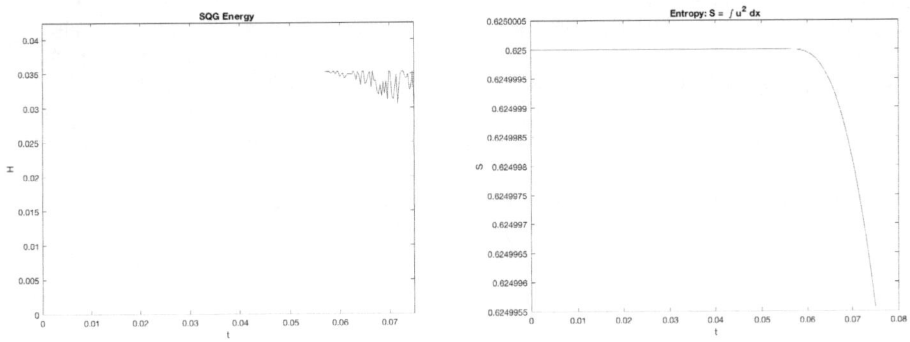

FIGURE 3. Left: Energy of the solution. Right: Entropy of the solutions. Both of these quantities are no longer conserved after time $t \approx 0.06$.

REFERENCES

[1] T. Buckmaster, S. Shkoller, and V. Vicol, Nonuniqueness of weak solutions to the SQG equation, *Comm. Pure Appl. Math.*, **72**(9), 1809–1874, 2019.

[2] A. Castro, D. Córdoba, and J. Gómez-Serrano, Existence and regularity of rotating global solutions for the generalized surface quasi-geostrophic equations, *Duke Math. J.*, **165** (2016), no. 5, 935–984.

[3] A. Castro, D. Córdoba, and J. Gómez-Serrano, Uniformly rotating analytic global patch solutions for active scalars, *Annals of PDE*, **2** (2016), no. 1, 1–34.

[4] A. Castro, D. Córdoba, and J. Gómez-Serrano, Global smooth solutions for the inviscid SQG equation, *Memoirs of the AMS*, to appear.

[5] P. Constantin, A. J. Majda, and E. G. Tabak, Formation of strong fronts in the 2-D quasi-geostrophic thermal active scalar, *Nonlinearity*, **7** (1994), no. 6, 1495–1533.

[6] P. Constantin, A. J. Majda, and E. G. Tabak, Singular front formation in a model for quasi-geostrophic flow, *Phys. Fluids*, **6** (1994), no. 1, 9–11.

[7] A. Córdoba, D. Córdoba, and F. Gancedo, Uniqueness for SQG patch solutions, *Trans. Amer. Math. Soc., Ser. B*, **5** (2018), no. 1, 1–31.

[8] D. Córdoba, M. A. Fontelos, A. M. Mancho, and J. L. Rodrigo, Evidence of singularities for a family of contour dynamics equations, *Proc. Natl. Acad. Sci.*, **102** (2005), no. 17, 5949–5952.

[9] D. Córdoba, J. Gómez-Serrano, and A. Ionescu, Global solutions for the generalized SQG patch equation, *Arch. Ration. Mech. Anal.*, **233**(3), 1211–1251, 2019.

[10] C. Fefferman and J. L. Rodrigo, Analytic sharp fronts for the surface quasi-geostrophic equation, *Commun. Math. Phys.*, **303** (2011), 261–288.

[11] F. Gancedo, Existence for the α-patch model and the QG sharp front in Sobolev spaces, *Adv. Math.*, **217** (2008), no. 6, 2569–2598.

[12] F. Gancedo and N. Patel, On the local existence and blow-up for generalized SQG patches, preprint, arXiv:1811.00530.

[13] F. Gancedo and R. M. Strain, Absence of splash singularities for SQG sharp fronts and the muskat problem, *Proc. Natl. Acad. Sci.*, **111** (2014), no. 2, 635–639.

[14] J. Gómez-Serrano, On the existence of stationary patches, *Adv. Math.*, **343** (2019), 110–140.

[15] I. M. Held, R. T. Pierrehumbert, S. T. Garner, and K. L. Swanson, Surface quasi-geostrophic dynamics, *J. Fluid Mech.*, **282** (1995), 1–20.

[16] J. K. Hunter and J. Shu, Regularized and approximate equations for sharp fronts in the surface quasi-geostrophic equation and its generalization, *Nonlinearity*, **31** (2018), no. 6, 2480–2517.

[17] J. K. Hunter, J. Shu and Q. Zhang, Local well-posedness of an approximate equation for SQG fronts, *J. Math. Fluid Mech.*, **20** (2018), no. 4, 1967–1984.

[18] J. K. Hunter, J. Shu and Q. Zhang, Global solutions of a surface quasi-geostrophic front equation, preprint, arXiv:1808.07631.

[19] J. K. Hunter, J. Shu and Q. Zhang, Contour dynamics for surface quasi-geostrophic fronts, preprint, arXiv:1907.06593.

[20] G. Lapeyre, Surface quasi-geostrophic, *Fluids*, **2** (2017).

[21] A. Majda, *Introduction to PDEs and Waves for the Atmosphere and Ocean*, Courant Lecture Notes in Mathematics, **9**, American Mathematical Soc., Providence, RI, 2003.

[22] A. J. Majda and E. G. Tabak, A two-dimensional model for quasigeostrophic flow: comparison with the two-dimensional Euler flow, *Physica D: Nonlinear Phenomena*, **98** (1996), no. 2, 515–522.

[23] F. Marchand, Existence and regularity of weak solutions to the quasi-geostrophic equations in the spaces L^p or $\dot{H}^{-1/2}$, *Commun. Math. Phys.*, **277** (2008), 45–67.

[24] J. Pedlosky, *Geophysical Fluid Dynamics*, 2nd edition, Springer, New York, 1987.

[25] S. Resnick, *Dynamical Problems in Nonlinear Advective Partial Differential Equations*, Ph.D thesis, University of Chicago, Chicago, 1995.

[26] J. L. Rodrigo, On the evolution of sharp fronts for the quasi-geostrophic equation, *Comm. Pure and Appl. Math.*, **58** (2005), 821–866.

[27] R. K. Scott and D. G. Dritschel, Numerical simulation of a self-similar cascade of filament instabilities in the surface quasigeostrophic system, *Phys. Rev. Lett.*, **112** (2014), 144505.

[28] E. M. Stein, *Singular Integrals and Differentiability Properties of Functions*, Princeton Mathematical Series, **30**, Princeton University Press, Princeton, NJ, 1970.

[29] M. Zworski, *Semiclassical Analysis*, Graduate Studies in Mathematics, **138**, American Mathematical Society, Providence, RI, 2012.

E-mail address: jyshu@ucdavis.edu

ROBUST NUMERICAL METHOD FOR TIME-DEPENDENT SINGULARLY PERTURBED SEMILINEAR PROBLEMS

VARSHA SRIVASTAVA*

Department of Mathematics
Indian Institute of Technology Delhi
Hauz Khas, New Delhi, 110016, India

ABSTRACT. In solving time-dependent singularly perturbed semilinear problems by the standard finite differences or finite element methods, the corresponding discrete problem on each time level is formulated as a nonlinear systems of algebraic equations; and to solve these nonlinear systems of algebraic equations, we require some iterative method for the computation of the numerical solutions.

In this article, a finite difference numerical method is used to solve the time-dependent singularly perturbed semilinear convection-diffusion problems. The numerical approximations to the solution are generated using a backward Euler method in time and a HODIE method in space via simultaneous discretization. The stability for the present time-dependent semilinear problems (both continuous and discrete) are proved by the inverse-monotonicity properties of the classes of linear initial-boundary value problems. The given method is shown to have first order parameter-uniform convergence in time and almost second order parameter-uniform convergence in space. Numerical result is given to support the theoretical error bounds of the numerical method.

1. **Introduction.** Consider the singularly perturbed semilinear parabolic convection-diffusion problem of the form

$$Tu := \frac{\partial u}{\partial t} - \varepsilon\frac{\partial^2 u}{\partial x^2} + a\frac{\partial u}{\partial x} + f(x,t,u) = 0, \ (x,t) \in X := \Omega \times \omega = (0,1) \times (0,T], \tag{1a}$$

$$u(x,0) = 0, \ \forall \ x \in \overline{\Omega}, \tag{1b}$$

$$u(0,t) = p_0(t), \ u(1,t) = p_1(t), \ \forall \ t \in \omega, \tag{1c}$$

where $0 < \varepsilon << 1$ is a small positive perturbation parameter. Functions a and f are sufficiently smooth functions, satisfying the following conditions

$$a(x) \geq \alpha > 0, \ \forall \ x \in \overline{\Omega}, \tag{2}$$

$$\beta \leq f_u(x,t,u) \leq \delta, \ \forall \ (x,t,u) \in X \times \mathbb{R}, \tag{3}$$

under the suitable continuity and compatibility conditions on the data an unique solution $u(x,t)$ of the problem (1) exists [6]. For $\varepsilon << 1$ problem (1) is singularly perturbed and has an exponential boundary layer of width $O\left(\varepsilon \ln(\frac{1}{\varepsilon})\right)$ at $x = 1$ of

2000 *Mathematics Subject Classification.* Primary: 65M06, 65M12; Secondary: 65M15.

Key words and phrases. backward Euler method, convection-diffusion problems, HODIE scheme, singular perturbation problems, time-dependent semilinear problems.

* Corresponding author: Varsha Srivastava.

$\overline{X}$ [3].

Model problems of kind (1) arise frequently in modeling fast chemical reactions and biological processes [5, 14].

The singularly perturbed semilinear parabolic convection-diffusion problems have been studied fairly extensively, see [1, 2, 8, 10, 11, 12, 16] and the references therein. These works involve the monotone iterative schemes, based on the method of upper and lower solution [9], to prove the parameter-uniform convergence. In contrast, here we consider the relatively simple truncation error and barrier function technique to prove the parameter-robust convergence of the present method via simultaneous discretization.

Therefore, the purpose of the present paper is to design and analyze a parameter-uniform numerical method for singularly perturbed semilinear parabolic convection-diffusion problems using the backward Euler method in time and a combination of the schemes in space via simultaneous discretization. The combination of the schemes in spatial direction depend on the relation between the mesh width and the perturbation parameter.

This paper is arranged as follows. In Section 2, the stability, the bounds on the derivatives for the solution, and it's components for continuous problem are given. A discrete numerical method is designed and it's stability is discussed in Section 3. The parameter-uniform maximum pointwise error bounds are estimated in Section 4. Numerical experiment is given in Section 5 and finally, conclusions are included in Section 6.

Notation: Throughout the paper, C, sometimes subscripted, is a generic positive constant that is independent of ε, N and Δt. Consider the maximum norm and denote it by $||.||_D$, where D is a closed and bounded subset of $\overline{X}$. For a real valued function $g \in C(D)$, define $||g||_D = \max_D |g|$. The analogous discrete maximum norm on the mesh D^{N,N_t} is denoted by $||.||_{D^{N,N_t}}$. If $g \in C(\overline{X})$ then $g_j^n = g(x_j, t_n)$, also $||g||_{\overline{X}^{N,N_t}} = \max_{\overline{X}^{N,N_t}} |g(x,t)|$.

2. **Properties of the Exact Solution.** Introduce the linear operator L_0 [7]

$$L_0 z := \frac{\partial z}{\partial t} - \varepsilon \frac{\partial^2 z}{\partial x^2} + a\frac{\partial z}{\partial x} + \int_0^1 f_u(x,t,sz)ds\ z,\ (x,t) \in X,\ z \in C^{2,1}(X)$$

where $\int_0^1 f_u(x,t,su)ds \geq \beta > 0$, and $L_0(\pm z) = \mp f(x,t,0)$.

Lemma 2.1. *(Maximum Principle) Assume that* $u \in (C^{2,1}(X) \bigcap C^{0,0}(\overline{X}))$, *satisfying* $u(x,0) \geq 0$ *on* $\overline{\Omega}$, $u(0,t) \geq 0$ *and* $u(1,t) \geq 0$ *on* ω. *Then* $L_0 u \geq 0$ *for* $(x,t) \in X$ *implies that* $u \geq 0$ *in* $\overline{X}$.

Lemma 2.2. *(Stability Estimate) Let* u *be the exact solution of (1). Then*

$$||u||_{\overline{X}} \leq \frac{1}{\alpha}||L_0 u||_{\overline{X}} + \max\{||p_0||_{\overline{\omega}}, ||p_1||_{\overline{\omega}}\}.$$

Lemma 2.3. *Let* u *be the exact solution of (1). Then* $\forall\ (x,t) \in \overline{X}$

$$\left|\frac{\partial^{s+m} u(x,t)}{\partial x^s \partial t^m}\right| \leq C\left(1 + \varepsilon^{-s} \exp\left(\frac{-\alpha(1-x)}{\varepsilon}\right)\right),\ 0 \leq m \leq 3,\ 0 \leq s+m \leq 4.$$

Proof. Follows from the similar approach used in [13]. □

To prove the parameter-uniform convergence, we need to derive the sharper bounds by decomposing the solution as $u(x,t) = v(x,t) + w(x,t),\ (x,t) \in \overline{X}$, where the regular component $v(x,t)$ satisfy

$$Tv(x,t) := \frac{\partial v(x,t)}{\partial t} - \varepsilon \frac{\partial^2 v(x,t)}{\partial x^2} + a(x)\frac{\partial v(x,t)}{\partial x} + f(x,t,v) = 0,\ (x,t) \in X,$$

subject to the appropriate initial-boundary conditions, and the singular component $w(x,t)$ satisfy

$$\begin{aligned} {}_vTw(x,t) &:= \frac{\partial w(x,t)}{\partial t} - \varepsilon \frac{\partial^2 w(x,t)}{\partial x^2} + a(x)\frac{\partial w(x,t)}{\partial x} + f(x,t,v+w) - f(x,t,v) \\ &= 0,\ (x,t) \in X, \end{aligned}$$

subject to the appropriate initial-boundary conditions. Noting that ${}_vTw$ is a new operator.

Lemma 2.4. *For all $(x,t) \in \overline{X}$, the regular component v satisfies*

$$\left| \frac{\partial^{s+m} v(x,t)}{\partial x^s \partial t^m} \right| \le C(1+\varepsilon^{4-s}),\ 0 \le m \le 3,\ 0 \le s+m \le 4, \tag{4}$$

and the singular component w satisfies

$$\left| \frac{\partial^{s+m} w(x,t)}{\partial x^s \partial t^m} \right| \le C\varepsilon^{-s} \exp\left(\frac{-\alpha(1-x)}{\varepsilon} \right),\ 0 \le m \le 3,\ 0 \le s+m \le 4. \tag{5}$$

Proof. Follows from the similar approach used in [13]. □

3. **Discrete Problem.** Let $\overline{X}^{N,N_t} = \overline{\Omega}^N \times \overline{\omega}^{N_t}$ and $\partial X^{N,N_t} = \partial X \cap \overline{X}^{N,N_t}$ denote the discrete mesh and the discrete boundary respectively. A uniform mesh is used for time-discretization as $\overline{\omega}^{N_t} := \{0 = t_0 < t_1 < \ldots < t_{N_t} = T\}$ with uniform time spacing $\Delta t = \frac{T}{N_t}$. For spatial discretization, the domain Ω is decomposed into two piecewise-uniform subdomains $(0, 1-\sigma)$ and $(1-\sigma, 1)$ with $N/2$ equal mesh intervals, where $N = 2^r, r \ge 3$. The transition parameter σ is defined as $\sigma = \min\left\{\frac{1}{2}, \sigma_0 \varepsilon \ln N\right\}$, where the constant σ_0 will be chosen later on in Section 5. The mesh spacing within the subdomains are $H_1 = h_j = \frac{2(1-\sigma)}{N}$ for $1 \le j \le N/2$; and $H_2 = h_j = \frac{2\sigma}{N}$ for $N/2+1 \le j \le N$ with $h_j = x_j - x_{j-1}$.

At each time level t_n, the corresponding discretization on each subdomains X^{N,N_t} is given by

$$[T^{N,N_t}U]_j^n := [Q(D_t^- U) + R(U) + Q(f)]_j^n = 0, \tag{6}$$

where $[D_t^- U]_j^n := \frac{(U_j^n - U_j^{n-1})}{\Delta t}$, $[R(U)]_j^n := r_j^{n,-} U_{j-1}^n + r_j^{n,c} U_j^n + r_j^{n,+} U_{j+1}^n$, and $[Qf]_j^n := q_j^- f(x_{j-1}, t_n, U_{j-1}^n) + q_j^c f(x_j, t_n, U_j^n)$.

For each $(x_j, t_n) \in X^{N,N_t}$, the coefficients $r_j^{n,*}$, $* = -, c, +$ are given by

$$\begin{aligned} r_j^{n,-} &= \frac{-2\varepsilon - a_j h_{j+1} + q_j^-[-(2h_j + h_{j+1})a_{j-1} + a_j h_{j+1}]}{h_j(h_j + h_{j+1})}, \\ r_j^{n,+} &= \frac{-2\varepsilon + a_j h_j - q_j^- h_j(a_j + a_{j-1})}{h_{j+1}(h_j + h_{j+1})},\ r_j^{n,c} = -r_j^{n,-} - r_j^{n,+},\ \text{and } q_j^c = 1 - q_j^-. \end{aligned}$$

Here the unknown coefficients are determined so that the scheme is exact for the polynomials up to degree two and satisfies the normalization condition. q_j^- is the free parameter, given in (10).

Let $K_j^{n,-} = r_j^{n,-} + q_j^-(f_u + 1/\Delta t)$, $K_j^{n,c} = r_j^{n,c} + q_j^c(f_u + 1/\Delta t)$, and $K_j^{n,+} = r_j^{n,+}$. Next is to prove that the matrix associated with ${}^U L^{N,N_t}$ is an M-matrix and the scheme is uniformly stable.

Lemma 3.1. *Let N_0 be the smallest positive integer such that*

$$\frac{\sigma_0||a||}{2} < \frac{N_0}{\ln N_0}, \left(||a'|| + \delta + \frac{1}{\Delta t}\right) < \alpha N_0, \left(\delta + \frac{1}{\Delta t}\right) < \alpha N_0 \tag{7}$$

hold. For each $(x_j, t_n) \in X^{N,N_t}$, there exist positive constants C_1 and C_2 such that

$$K_j^{n,-} < 0,\ K_j^{n,+} < 0,\ C_1 \leq K_j^{n,-} + K_j^{n,c} + K_j^{n,+} \leq C_2, \tag{8}$$

then the matrix associated with ${}^U L^{N,N_t}$ is an M-matrix, where $q_j^- \geq \dfrac{a_j}{(a_j + a_{j-1})}$ for $||a||h_j \geq 2\varepsilon$. Moreover, for some positive constant C_3 the scheme is uniformly stable in the maximum norm, if

$$h_{j+1}K_j^{n,+} - h_j K_j^{n,-} \geq C_3 > 0. \tag{9}$$

The free parameter q_j^-, $\forall\ x_j \in \Omega^N$ of spatial direction is chosen as

$$q_j^- = \begin{cases} \dfrac{a_j}{(a_j + a_{j-1})}, & \|a\|h_j \geq 2\varepsilon \\ \dfrac{(h_j - h_{j+1})}{3h_j}, & \|a\|h_j < 2\varepsilon. \end{cases} \tag{10}$$

4. Error Analysis.

Lemma 4.1. *Let v and V denote the smooth components of u and U respectively. Then,*

$$||v - V||_{\overline{X}^{N,N_t}} \leq C(\Delta t + N^{-2}\ln^2 N).$$

Proof. In X^{N,N_t}, the initial and boundary conditions are $|(v-V)(x_j, 0)| = 0, \forall\ x_j \in \overline{\Omega}^N$, and $|(v - V)(0, t_n)| = 0, |(v - V)(1, t_n)| = 0, \forall\ t_n \in \omega^{N_t}$, also the truncation error bound yields

$$\begin{aligned} \left|T^{N,N_t}v_j^n - Tv_j^n\right| =& \left|{}^V L^{N,N_t}(v - V)_j^n\right| \\ \leq& |C\Delta t \left(\frac{\partial^2 v_{j-1}^n}{\partial t^2} + \frac{\partial^2 v_j^n}{\partial t^2}\right) + \frac{\partial^3 v_j^n}{\partial x^3}(\frac{\varepsilon}{3}(h_j - h_{j+1}) - q_j^- h_j\varepsilon) \\ &+ \frac{h_j h_{j+1}}{6}a_j - \frac{q_j^- h_j}{6}(a_j h_{j+1} + a_{j-1}(h_{j+1} - 2h_j))) \\ &+ R_3(x_j, x_{j-1}, v^n)r_j^{n,-} + R_3(x_j, x_{j+1}, v^n)r_j^{n,+} \\ &+ q_j^-\varepsilon R_1\left(x_j, x_{j-1}, \frac{\partial^2 v^n}{\partial x^2}\right) + q_j^- a_{j-1}R_1\left(x_j, x_{j-1}, \frac{\partial v^n}{\partial x}\right)|, \end{aligned} \tag{11}$$

where $R_n(a, b, g) = \displaystyle\int_a^b \frac{(b-\xi)^n}{n!} g^{n+1}(\xi)d\xi$ is the remainder of the Taylor's expansion in the integral form.

Consider $(x_j, t_n) \in X^{N,N_t}$ with $1 \le j \le N/2$ and $1 \le n \le N_t$. Using (4) in (11), we get $|T^{N,N_t} v_j^n - T v_j^n| \le C(\Delta t + N^{-2})$, and when $(x_j, t_n) \in X^{N,N_t}$ with $N/2+1 \le j \le N-1$ and $1 \le n \le N_t$, we get $|T^{N,N_t} v_j^n - T v_j^n| \le C(\Delta t + N^{-2} \ln^2 N)$.

Consider $\psi^{\pm}(x_j, t_n) = C(\Delta t + N^{-2} \ln^2 N)(1 + x_j) \pm (v - V)(x_j)$ as barrier function in X^{N,N_t}. Recalling initial-boundary conditions, truncation error bounds, and the maximum principle for ${}^V L^{N,N_t}$ with $\psi^{\pm}(x_j, t_n)$ for sufficiently large C such that $\psi^{\pm}(x_j, t_n) \ge 0$. This implies

$$|(v - V)(x_j, t_n)| \le C(\Delta t + N^{-2} \ln^2 N), \ \forall \ (x_j, t_n) \in \overline{X}^{N,N_t}.$$

□

At a fixed time level t_n, the error of the singular component is given by

$$\begin{aligned}
[{}_V T^{N,N_t} w - {}_v T w]_j^n &= [{}_V T^{N,N_t} w]_j^n \\
&= q_j^- \frac{w_{j-1}^n - w_{j-1}^{n-1}}{\Delta t} + q_j^c \frac{w_j^n - w_j^{n-1}}{\Delta t} + r_j^{n,-} w_{j-1}^n + r_j^{n,c} w_j^n \\
&\quad + r_j^{n,+} w_{j+1}^n + q_j^- (f(x_{j-1}, t_n, V_{j-1}^n + w_{j-1}^n) - f(x_{j-1}, t_n, V_{j-1}^n)) \\
&\quad + q_j^c (f(x_j, t_n, V_j^n + w_j^n) - f(x_j, t_n, V_j^n)) - q_j^- (\frac{\partial w_{j-1}^n}{\partial t} - \varepsilon \frac{\partial^2 w_{j-1}^n}{\partial x^2} \\
&\quad + a_{j-1} \frac{\partial w_{j-1}^n}{\partial x} + f(x_{j-1}, t_n, v_{j-1}^n + w_{j-1}^n) - f(x_{j-1}, t_n, v_{j-1}^n)) \\
&\quad - q_j^c (\frac{\partial w_j^n}{\partial t} - \varepsilon \frac{\partial^2 w_j^n}{\partial x^2} + a_j \frac{\partial w_j^n}{\partial x} + f(x_j, t_n, v_j^n + w_j^n) - f(x_j, t_n, v_j^n)) \\
&\le \left| [\Gamma^{N,N_t} w]_j^n \right| + C||v - V||.
\end{aligned} \tag{12}$$

Therefore,

$$\left| [{}_V T^{N,N_t} w - {}_v T w]_j^n \right| \le \left| [\Gamma^{N,N_t} w]_j^n \right| + C||v - V||, \tag{13}$$

where

$$\begin{aligned}
[\Gamma^{N,N_t} w]_j^n =& Q \left(\frac{w_j^n - w_j^{n-1}}{\Delta t} \right) + r_j^{n,-} w_{j-1}^n + r_j^{n,c} w_j^n + r_j^{n,+} w_{j+1}^n \\
&+ Q \left(\frac{-\partial w_j^n}{\partial t} + \varepsilon \frac{\partial^2 w_j^n}{\partial x^2} - a_j \frac{\partial w_j^n}{\partial x} \right).
\end{aligned}$$

Alternatively,

$$\begin{aligned}
[_V T^{N,N_t} w - {}_v T w]_j^n &= [_V T^{N,N_t} w - {}_V T^{N,N_t} W]_j^n \\
&= (q_j^- \frac{w_{j-1}^n - w_{j-1}^{n-1}}{\Delta t} + q_j^c \frac{w_j^n - w_j^{n-1}}{\Delta t} + r_j^{n,-} w_{j-1}^n + r_j^{n,c} w_j^n \\
&\quad + r_j^{n,+} w_{j+1}^n + q_j^- (f(x_{j-1}, t_n, V_{j-1}^n + w_{j-1}^n) - f(x_{j-1}, t_n, V_{j-1}^n)) \\
&\quad + q_j^c (f(x_j, t_n, V_j^n + w_j^n) - f(x_j, t_n, V_j^n))) - (q_j^- \frac{W_{j-1}^n - W_{j-1}^{n-1}}{\Delta t} \\
&\quad + q_j^c \frac{W_j^n - W_j^{n-1}}{\Delta t} + r_j^{n,-} W_{j-1}^n + r_j^{n,c} W_j^n + r_j^{n,+} W_{j+1}^n \\
&\quad + q_j^- (f(x_{j-1}, t_n, V_{j-1}^n + W_{j-1}^n) - f(x_{j-1}, t_n, V_{j-1}^n)) \\
&\quad + q_j^c (f(x_j, t_n, V_j^n + W_j^n) - f(x_j, t_n, V_j^n))) \\
&= [^U L^{N,N_t} \xi]_j^n
\end{aligned}$$

where the error function ξ is defined as $\xi = w - W$ and the linear operator ${}^U L^{N,N_t}$ is defined by

$$\begin{aligned}
[^y L^{N,N_t} z]_j^n &:= Q\left(\frac{z_j^n - z_j^{n-1}}{\Delta t}\right) + r_j^{n,-} z_{j-1}^n + r_j^{n,c} z_j^n + r_j^{n,+} z_{j+1}^n \\
&\quad + Q\left(\int_0^1 f_u(x_j, t_n, y_j^n + s z_j^n) ds \; z_j^n\right).
\end{aligned} \tag{14}$$

Now define the mesh functions $\phi_j(\gamma)$ in $\overline{\Omega}^N$ for some positive constant γ as

$$\phi_j(\gamma) = \prod_{k=j+1}^{N} \left(1 + \frac{\gamma h_k}{\varepsilon}\right)^{-1}, \; 0 \le j \le N-1, \text{ with } \phi_N(\gamma) = 1. \tag{15}$$

Lemma 4.2. *For $x_j \in \Omega^N$, suppose that all the assumptions of Lemma 3.1 hold, then there exists a constant $C(\gamma)$ such that*

$${}^U L^{N,N_t} \phi_j(\gamma) \ge \frac{C(\gamma)}{\max\{\varepsilon, h_j\}} \phi_j(\gamma),$$

where $\gamma \le \alpha/2$ and α is same as defined in (2).

Lemma 4.3. *Let w and W be the singular components of u and U respectively. Then,*

$$||w - W||_{\overline{X}^{N,N_t}} \le C\left(\Delta t + N^{-\frac{\alpha\sigma_0}{2}} + N^{-2} \ln^2 N\right).$$

Proof. The initial and boundary conditions are $|(w-W)(x_j, 0)| = 0, \; \forall \; x_j \in \Omega^N$, $|(w-W)(0, t_n)| \le 0$ and $|(w-W)(1, t_n)| \le 0, \; \forall \; t_n \in \omega^{N_t}$, also the Taylor's expansion gives

$$
\begin{aligned}
\left|[{}_V T^{N,N_t} w - {}_v T w]_j^n\right| &= \left|{}^U L^{N,N_t}(w-W)_j^n\right| \\
&\leq |C\Delta t \left(\frac{\partial^2 w_{j-1}^n}{\partial t^2} + \frac{\partial^2 w_j^n}{\partial t^2}\right) + \frac{\partial^3 w_j^n}{\partial x^3}(\frac{\varepsilon}{3}(h_j - h_{j+1}) - q_j^- h_j \varepsilon \\
&+ \frac{h_j h_{j+1}}{6} a_j - \frac{q_j^- h_j}{6}(a_j h_{j+1} + a_{j-1}(h_{j+1} - 2h_j))) \\
&+ R_3(x_j, x_{j-1}, w^n) r_j^{n,-} + r_j^{n,+} R_3(x_j, x_{j+1}, w^n) \\
&+ q_j^- \varepsilon R_1\left(x_j, x_{j-1}, \frac{\partial^2 w^n}{\partial x^2}\right) + q_j^- a_{j-1} R_1\left(x_j, x_{j-1}, \frac{\partial w^n}{\partial x}\right)| \\
&+ C||v - V||, \qquad (16)
\end{aligned}
$$

where $R_n(a,b,g)$ is the remainder of the Taylor's expansion in the integral form.

For $(x_j, t_n) \in X^{N,N_t}$, when $||a||h_j \geq 2\varepsilon$, the expression of truncation error will reduce to

$$\left|[{}_V T^{N,N_t} w - {}_v T w]_j^n\right| \leq C_1 \Delta t + \frac{C_2}{\max\{\varepsilon, h_j\}} \exp\left(-\frac{\alpha(1-x_{j+1})}{\varepsilon}\right) + C||v - V||,$$

and when $||a||h_j < 2\varepsilon$, the truncation error bound gives

$$\left|[{}_V T^{N,N_t} w - {}_v T w]_j^n\right| \leq C_1 \Delta t + \frac{C_2}{\max\{\varepsilon, h_j\}} \left(\frac{h_j}{\varepsilon}\right)^2 \exp\left(-\frac{\alpha(1-x_{j+1})}{\varepsilon}\right) + C||v - V||,$$

where we have also carefully examined the truncation error at the transition point $x_{N/2} = (1-\sigma)$. Let us construct the barrier function $\psi_\gamma^\pm(x_j, t_n)$ in $\overline{X}^{N,N_t}$ with $\vartheta(\gamma) = C_3\left(1 + \frac{\gamma h_{j+1}}{\varepsilon}\right)$ as

$$\psi_\gamma^\pm(x_j, t_n) = C_1 \Delta t + C_2(1 + x_j) N^{-2} \ln^2 N + \vartheta(\gamma)\phi_j(\gamma) \pm (w - W)(x_j, t_n).$$

Recalling initial-boundary conditions, truncation error bounds, and the maximum principle for ${}^U L^{N,N_t}$ with $\psi_\gamma^\pm(x_j, t_n)$ for sufficiently large C_1, C_2 and C_3 such that $\psi_\gamma^\pm(x_j, t_n) \geq 0$. We get,

$$|(w - W)(x_j, t_n)| \leq C_1 \Delta t + C_2\left(N^{-\frac{\alpha\sigma_0}{2}} + N^{-2}\ln^2 N\right),\ 0 \leq j \leq N/2,\ 0 \leq n \leq N_t, \qquad (17)$$

where we have used Lemma 3.1 of [15].

On the other hand, for $N/2 \leq j \leq N$ and $0 \leq n \leq N_t$, we consider the barrier functions $\psi_\gamma^\pm(x_j, t_n)$ in $\overline{X}^{N,N_t}$ as

$$
\begin{aligned}
\psi_\gamma^\pm(x_j, t_n) =& C_1 \Delta t + C_2(1 + x_j)\left(N^{-\frac{\alpha\sigma_0}{2}} + N^{-2}\ln^2 N\right) + C_3\left(\frac{H_2}{\varepsilon}\right)^2 \phi_j(\gamma) \\
&\pm (w - W)(x_j, t_n).
\end{aligned}
$$

Recalling initial-boundary conditions, truncation error bounds, and the maximum principle for ${}^U L^{N,N_t}$ with $\psi_\gamma^\pm(x_j, t_n)$ for sufficiently large C_1, C_2 and C_3 such that $\psi_\gamma^\pm(x_j, t_n) \geq 0$. This gives

$$|(w - W)(x_j, t_n)| \leq C\left(\Delta t + N^{-\frac{\alpha\sigma_0}{2}} + N^{-2}\ln^2 N\right),\ \forall\ (x_j, t_n) \in \overline{X}^{N,N_t},$$

and hence completes the proof. □

Theorem 4.4. *Let u be the exact solution of the problem (1) and U be the discrete solution of the numerical method (6). If $\gamma = \alpha/2$ and $N \geq N_0$, the error associated with the solution satisfies*

$$||u - U||_{\overline{X}^{N,N_t}} \leq C\left(\Delta t + N^{-\frac{\alpha\sigma_0}{2}} + N^{-2}\ln^2 N\right).$$

Proof. The triangle inequality gives

$$||u - U|| \leq ||v - V|| + ||w - W|| \leq C\left(\Delta t + N^{-\frac{\alpha\sigma_0}{2}} + N^{-2}\ln^2 N\right),$$

which gives the first order convergence in time and for almost second order convergence in space we need to take $\alpha\sigma_0 \geq 4$. □

5. Numerical Experiments.

Example 5.1. Consider the following time-dependent singularly perturbed semilinear convection-diffusion problem:

For $(x, t) \in X := (0, 1) \times (0, 1]$,

$$\frac{\partial u}{\partial t} - \varepsilon\frac{\partial^2 u}{\partial x^2} + \frac{\partial u}{\partial x} + u^2 + u = (1 - x)(1 - e^{-t}),$$

$u(x, 0) = 0,\ x \in \overline{\Omega} := [0, 1]$,

$u(0, t) = 0,\ u(1, t) = 0\ t \in \omega := (0, 1]$.

To solve the corresponding discrete nonlinear systems, the Newton's method is used at each time level t_n with the initial approximation $w^0 = (u_0(x_0, t_n), u_0(x_1, t_n), \ldots, u_0(x_N, t_n))^T$, where $u_0(x, t)$ is the solution of the reduced problem. The stopping criterion for Newton's method is $\|w^k - w^{k-1}\| < 10^{-12}$. For each $N, \Delta t$ and ε, it takes only 4 iterations to satisfy the stopping criterion to get the discrete solution.

The exact solution is not known for the test problem, so a variant of the double mesh principle is used to calculate the maximum pointwise errors for different values of ε, N and Δt using $E_\varepsilon^{N,\Delta t} := ||U^{N,\Delta t} - U^{2N,\Delta t/4}||_{\overline{X}^{N,N_t}}$ and the parameter-uniform errors by $E^{N,\Delta t} = \max_{\varepsilon} E_\varepsilon^{N,\Delta t}$. We then calculate the parameter-uniform numerical order of convergence by $\rho^{N,\Delta t} = \log_2\left(\frac{E^{N,\Delta t}}{E^{2N,\Delta t/4}}\right)$.

For different values of ε, N and Δt, Table 1 represents the maximum pointwise errors $E_\varepsilon^{N,\Delta t}$ and parameter-uniform rates of convergence $\rho_\varepsilon^{N,\Delta t}$ of the discrete numerical method for the Example 5.1. The value of σ_0 is chosen as 2 for given Example.

6. **Conclusions.** The present paper proposes a parameter-uniform numerical method to solve the singularly perturbed semilinear parabolic convection-diffusion problems. The method is analyzed via simultaneous discretization and comprises the backward Euler method in time and HODIE scheme in space direction. The parameter-uniform convergence of the present method via simultaneous discretization is proved without the mesh restriction $N^{-q} \leq C\Delta t$, for some $q \in (0, 1)$ unlike the case of semi-discretization [4]. The present method is first order uniformly convergent in time and almost second order uniformly convergent in space. Numerical results are in agrement with the theoretical results.

Acknowledgments. This research did not receive any specific grant from funding agencies in the public, commercial, or not-for-profit sectors.

TABLE 1. Maximum pointwise errors $E^{N,\Delta t}$ and parameter-uniform rates of convergence $\rho^{N,\Delta t}$ for Example 5.1.

$\varepsilon = 2^{-j}$	$N = 2^6$, $\Delta t = 0.2$	$N = 2^7$, $\Delta t = \frac{0.2}{4}$	$N = 2^8$, $\Delta t = \frac{0.2}{4^2}$	$N = 2^9$, $\Delta t = \frac{0.2}{4^3}$	$N = 2^{10}$, $\Delta t = \frac{0.2}{4^4}$
$j = 4$	3.60E-03	1.30E-03	3.67E-04	9.51E-05	2.40E-05
	1.47	1.82	1.95	1.99	
8	4.82E-03	1.90E-03	5.67E-04	1.52E-04	3.88E-05
	1.34	1.74	1.90	1.97	
12	4.93E-03	1.96E-03	6.03E-04	1.66E-04	4.32E-05
	1.33	1.70	1.86	1.94	
16	4.94E-03	1.96E-03	6.05E-04	1.67E-04	4.38E-05
	1.33	1.70	1.86	1.93	
20	.	.	.	.	.
24	.	.	.	.	.
28	.	.	.	.	.
32	4.94E-03	1.96E-03	6.05E-04	1.67E-04	4.38E-05
	1.33	1.70	1.86	1.93	
$E^{N,\Delta t}$	4.94E-03	1.96E-03	6.05E-04	1.67E-04	4.38E-05
$\rho^{N,\Delta t}$	1.33	1.70	1.86	1.93	

REFERENCES

[1] I. Boglaev, Uniform convergence of monotone iterative methods for semilinear singularly perturbed problems of elliptic and parabolic types, *Electron. Trans. Numer. Anal.*, **20** (2005), 86–103.

[2] I. Boglaev, Uniform convergent monotone iterates for semilinear singularly perturbed parabolic problems, *J. Comput. Appl. Math.*, **235** (2011), 3541–3553.

[3] I. P. Boglaev and V. V. Sirotkin, The integral difference method for quasilinear singularly perturbed problems, in *J. J. H. Miller* (eds. Computer methods for boundary and interior layers in several dimensions), Boole Press, Dublin, (1991), 1–26.

[4] C. Clavero, J. C. Jorge and F. Lisbona, A uniformly convergent scheme on a nonuniform mesh for convection-diffusion parabolic problems, *J. Comput. Appl. Math.*, **154** (2003), 415–429.

[5] P. C. Fife and G. S. Gill, Phase transition mechanisms for the phase field model under interval heating, *Phys. Rev.*, **43**(3) (1991), 843–851.

[6] O. A. Ladyzhensksya, V. A. Solonnikov and N. N. Ural'ceva *Linear and Quasilinear Equations of Parabolic Type*, Academic press, New York, 1968.

[7] J. Lorenz, Stability and monotonicity properties of stiff quasilinear boundary problems, *Univ. u Novom Sadu Zb. rad. Prir.-Mat. Fak. Ser. Mat.*, **12** (1982), 151–175.

[8] C. V. Pao, Numerical methods of semilinear parabolic equations, *SIAM. J. Numer. Anal.*, **24** (1987), 24–35.

[9] C. V. Pao, *Nonlinear Parabolic and Elliptic Equations*, Plenum Press, New York, 1992.

[10] C. V. Pao, Positive solutions and dynamics of a finite difference reaction-diffusion system, *Numer. Meth. Part. Diff. Eqs.*, **9** (1993), 285–311.

[11] C. V. Pao, Blowing-up and asymptotic behaviour of solutions for a finite difference system, *Appl. Anal.*, **62** (1996), 29–38.

[12] C. V. Pao, Accelerated monotone iterations for numerical solutions of nonlinear elliptic boundary value problems, *Comput. Math. Appl.*, **46** (2003), 1535–1544.

[13] G. I. Shishkin and L. P. Shishkina, The Richardson extrapolation technique for quasilinear parabolic singularly perturbed convection-diffusion equations, *J. Phys.: Conf. Ser.*, **55** (2006), 203–213.

[14] A. M. Soane, M. K. Gobbert and T. I. Seidman, Numerical exploration of a system of reaction-diffusion equations with internal and transient layers, *Nonlinear Anal. Real World Appl.*, **6** (2005), 914–934.

[15] M. Stynes and L. Tobiska, A finite difference analysis of a streamline diffusion method on a Shishkin mesh, *Numer. Algorithms*, **18** (1998), 337–360.

[16] Y. M. Wang, On accelerated monotone iterations for numerical solutions of semilinear elliptic boundary value problems, *Appl. Math. Lett.*, **18** (2005), 749–755.

E-mail address: `varsha.iitd@gmail.com`

THE ROLE OF A REGULARIZATION IN HYPERBOLIC INSTABILITIES

Marta Strani

Università Ca' Foscari
Dipartimento di Scienze Molecolari e Nanosistemi
Venezia Mestre (Italy)

Abstract. We investigate the phenomenon of a time-delay in the instabilities exhibited by some hyperbolic equations. We discuss at first what happens when considering the viscous complex Burgers equation: we see that the instantaneous amplification manifested by the solution of the inviscid equation is not observed when introducing a regularizing viscous term in the system, as we show that we have existence of a bounded solution in times of order one and, only after that, an exponential growth in time. Finally, we give some partial results on a dispersive regularization of the Euler equations. These results are contained in [8, 10, 11]. The delay in the instabilities is strictly related to a loss of hyperbolicity and to the subsequent transition to ellipticity for the hyperbolic problems under consideration.

1. **Introduction.** In this paper we mean to describe the role of a regularization in the instabilities of some hyperbolic equations: precisely, we show that given a strongly unstable partial differential equation (meaning that, even if starting from a regular initial datum we observe an instantaneous amplification of the solution in some norm) if we add a regularizing term in the equation (as, for instance, a small viscous term), we observe the following behavior: the solutions stay bounded for short times before exhibiting an exponential growth. Hence, we can distinguish two different times scales in the dynamics of the solution: a first time phase of the order $\mathcal{O}(1)$ where the solution is bounded in some norm, and a second time phase where it experiences and instability in time; as a consequence, the system appears to be perfectly stable for a long time, and it's only in the appropriate observation time that the instabilities are detected.

This phenomenon is what we call *time delayed instabilities*; such behavior and the corresponding two-time phases of the dynamics is reminiscent of the phenomenon of metastability, in which the speed of convergence of the solutions towards the equilibrium depends singularly on the viscosity (see, for instance, [2, 6, 7] and the references therein).

In this paper, we deal at first with the Cauchy problem for the one-dimensional Burgers equation with small viscosity and a complex forcing term

$$\partial_t u + u\partial_x u - \varepsilon\partial_x^2 u = i, \quad u(x,0) = u_0(x), \tag{1}$$

2000 *Mathematics Subject Classification.* Primary: 35K58, 35L45, 35B35; Secondary: 35Q31.
Key words and phrases. Loss of hyperbolicity, instabilities, complex Burgers equation.

where $t \in \mathbb{R}_+$, $x \in \mathbb{T}$ and $0 < \varepsilon \ll 1$ can be seen as a small viscosity parameter. In particular, we prove that the solutions to (1) exhibit a *delay in time for the instabilities* with respect to the corresponding inviscid problem, for which the instabilities are manifested as soon as $t > 0$ (for more details, see [8, 10]).

More precisely, let us consider (1) in the inviscid case $\varepsilon = 0$, i.e.

$$\partial_t u + u\partial_x u = i, \qquad u(x,0) = u_0(x). \tag{2}$$

For such equation, in [3] it has been proved that if a real datum generates a local C^2 solution to (2), then the datum must be analytic. This reveals a strong instability of the Cauchy problem (2).

This instability is strictly related to a change in the behavior of the symbol in (2): starting from a real datum u_0, the linearized first-order operator $\partial_t + u_0\partial_x$ is hyperbolic at $t = 0$; however, as soon as $t > 0$, because of the complex forcing term on the right hand side, if a solution $u(t)$ exists then $\Im m\, u(t) \neq 0$, so that the linearized operator is no longer hyperbolic. The phenomenon described is what is called a *loss of hyperbolicity* (see, for instance, [4]). In particular, such a loss of hyperbolicity (or, in other words, such transition from hyperbolicity (at $t = 0$) to ellipticity (at $t > 0$)) translates, at the level of the dynamics of the PDE, into an instantaneous amplification of the solution (i.e. into an instability in time), as rigorously proved in [3]. Indeed, the authors showed as some nearby data (as measured in a strong norm) may generate solutions which are instantly driven apart (as measured in a weak norm) or, even worse, may generate no solutions at all (in this case the Cauchy problem is even more strongly ill-posed as one has absence of a solution operator, compared to absence of Hölder estimates for a solution operator, see also [4]).

As opposite to the inviscid case, when considering the viscous problem (1), we still have a loss of hyperbolicity in the equation but with a *delay in time*; indeed, denoting by $u_1 := \Re e\, u$ and $u_2 := \Im m\, u$, let us compute the symbol associated to system (1):

$$A(u_1, u_2, \xi) = \begin{pmatrix} u_1 & -u_2 \\ u_2 & u_1 \end{pmatrix} i\xi + \varepsilon\xi^2,$$

with spectrum given by $\det(\lambda\, \mathrm{Id} - A) = 0$, that is

$$\lambda_\pm(\varepsilon, t, x, \xi) = u_1 i\xi + \varepsilon\xi^2 \pm |u_2|\xi.$$

In particular $\Re e\, \lambda_+(\varepsilon, t, x, \xi) > 0$ for all $t \geq 0$ while,

$$\Re e\, \lambda_-(\varepsilon, t, x, \xi) = \varepsilon\xi^2 - |u_2|\xi,$$

and negativity of $\Re e\, \lambda_-$ indicates ellipticity of the system, potentially corresponding to an amplification of the solution. Hence, we possibly have a transition from hyperbolicity to ellipticity, depending on the sign of $\Re e\, \lambda_-$.

We then observe that due to $\Im m\, u(0) = 0$ and because of the complex forcing term, we expect $\Im m\, u(t)$ to be approximately equal to t for small $t > 0$, implying

$$\Re e\, \lambda_-(\varepsilon, t, x, \xi) = |\xi|(\varepsilon|\xi| - \mathcal{O}(t)), \quad \text{for} \quad \xi \in \mathbb{Z}. \tag{3}$$

Hence:

- If relevant frequencies are large, that is $|\xi| \sim 1/\varepsilon$, then positivity in (3) is preserved for times t of the order $\mathcal{O}(1)$ (precisely, we have $\Re e\, \lambda_- > 0$ for $t \leq \varepsilon\, \xi$ with $\xi \sim \varepsilon^{-1}$).

Hence, we expect solutions issued from highly-oscillating data of order $\mathcal{O}(1/\varepsilon)$ to be defined, and uniformly bounded in ε, over time intervals of the order $\mathcal{O}(1)$. Moreover, if u_0 is of the form $u_0(k_1 x/\varepsilon)$, then the relevant frequencies are $\xi \in k_1\mathbb{Z}/\varepsilon$, and the existence time depends only on k_1 (we have indeed $\Re e\, \lambda_- > 0$ for $t \lesssim k_1$).

- One the other hand, if we consider a non-oscillating initial datum $u_0(x)$ (that is, if relevant frequencies are of the order one), we expect solutions to experience an amplification after a time of the order $\mathcal{O}(\varepsilon)$: indeed, the positivity of the symbol in (3) is maintained only for $t \lesssim \mathcal{O}(\varepsilon)$. Hence, in the small viscosity limit $\varepsilon \to 0$, the instability appears as soon as $t > 0$, as proved in [3], an we expect only a short time existence for the solutions to (1).

In conclusion, equation (1) still experiences a transition from hyperbolicity to ellipticity (as in the inviscid case) but this time it appears only after a *delay in time* (due to the presence of a viscous regularizing term). Such transition translates at the level of the PDE in the following dynamics: the solutions exhibit a stable behavior (in some norm), and it is only after a delay in time, that varies depending on the initial oscillations in the datum u_0, that an amplification in time occurs.

To summarize, our guess is that, by adding a small viscous term to an equation that experiences a transition from hyperbolicity to ellipticity as soon as $t > 0$ (as it is for (2)), such transition will still appear but with a certain delay in time, leading to the phenomenon described before.

2. **Main results: existence and behavior of solutions.** In this Section we collect some results of [8, 10], and we rigorously prove that the aforementioned *time delayed instabilities* occur for the solutions to the complex viscous Burgers equation arisen from highly oscillating real initial data of amplitude $\mathcal{O}(1)$; hence we consider the following problem

$$\begin{cases} \partial_t u + u\partial_x u - \varepsilon\partial_x^2 u = i & x \in \mathbb{T},\, t \geq 0, \\ u(0,x) = u_0\left(\dfrac{k_1\, x}{\varepsilon}\right) & x \in \mathbb{T}, \end{cases} \tag{4}$$

where u_0 depends on x through x/ε, and $k_1 \in \mathbb{Z}$ is the smallest non zero frequency in the initial datum. Showing that the instabilities in (4) appear only after a certain delay in time would validate the apparently naive approximation of (4) by the linear constant-coefficient equation (obtained formally by setting $u = it + v$ into (4) and by linearizing around $t \sim 0$)

$$\partial_t v + it\partial_x v - \varepsilon\partial_x^2 v = 0, \tag{5}$$

for which the growth is manifested only after a delay in time, as can be easily seen in the Fourier side; indeed, for the solution v to (5) issued from $v_0(x)$ there holds, by Fourier transform and direct time integration

$$|\hat{v}(t,k)| = |\hat{v}_0(k)| \exp\Big(2\pi t k\big(\frac{t}{2} - 2\pi\varepsilon k\big)\Big), \qquad k \in \mathbb{Z}, \tag{6}$$

so that $\hat{v}(\cdot,k)$ grows exponentially only for $t \geq 4\pi\varepsilon k$.

The approximation of (4) with (5) seems to be also validated by numerics, as shown in Figure 1; we plotted the imaginary part of the numerical solutions of both the equations (which are almost indistinguishable as the initial oscillations N increase), and we can see that we have a growth in time only after a delay, which is higher as soon as $N \to \infty$. In particular, it seems that the time for which the instability is manifested only depends on the smallest nonzero mode in the initial datum (as already predicted when studying the symbol associated with (4)).

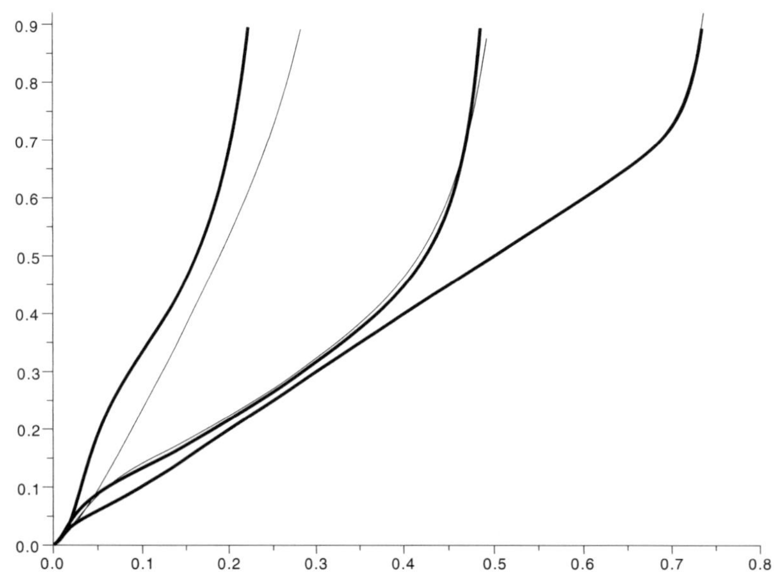

FIGURE 1. The imaginary part of the numerical solutions to (4) (black lines) and to (5) (gray lines), for initial data $a(x) = \sin(N \cdot 2\pi x)$, with $N = 8, 16$ and 24 (from left to right). For $N = 24$, on the scales of the picture the graphs are essentially indistinguishable. This indicates that the approximation of (4) by (5) seems quite accurate for oscillating data, even if their amplitude is of the order $\mathcal{O}(1)$.

2.1. **Linear growth in time of the solution for times of the order one.** Our first result makes the numerical observations seen above rigorous, and shows that no amplification occurs in times of order $\mathcal{O}(1)$; indeed, we prove the following theorem.

Theorem 2.1. *Given $u_0 \in W^{1,\infty}(\mathbb{T})$, for some $t_0 > 0$, some $v_1^a \in L^\infty([0, t_0], W^{1,\infty}(\mathbb{T}))$ and $v_2^a \in L^\infty([0, t_0], L^\infty(\mathbb{T}))$ such that*

$$|v_1^a(t)|_{W^{1,\infty}} \le C, \quad |v_2^a(t)|_{L^\infty} \le Ct,$$

with $C > 0$, there exists a unique solution $u \in C^0([0, t_0], H^1(\mathbb{T}))$ to the initial-value problem (4) such that the following bounds

$$|(\Re e\, u - v_1^a)(t)|_{H^1} \le Ct, \quad |(\Im m\, u - v_2^a)(t)|_{H^1} \le Ct^2, \tag{7}$$

hold. In particular, u is bounded in $H^1(\mathbb{T})$, uniformly in ε and $t \in [0, t_0]$.

Proof. For the proof of the Theorem we refer the readers to [8, Theorem 2.1]. We also refer to [10, Theorem 1] for a proof of the same result in the more general setting $u_0 \in H^{s+1}(\mathbb{T})$, which implies the existence of a solution $u \in C^0([0, T], H^s(\mathbb{T}))$. □

2.2. **Exponential growth in time.** Theorem 2.1 shows that highly-oscillating initial data generate solutions that are bounded uniformly in ε over times of the order $\mathcal{O}(1)$, with a linear growth in time for the imaginary part (as seen numerically in Figure 1). In order to prove the second part of the dynamics of the solutions to (4) observed in Figure 1, that is the exponential growth in time, we make the following observation: since there holds the conservation law

$$t = \int_{\mathbb{T}} u_2(t,x)\,dx, \tag{8}$$

we get $t \leq |u_2(t)|_{L^\infty(\mathbb{T})}$. In particular, given the observation frequency $\xi = k_1/\varepsilon$ and $t > \varepsilon\xi = k_1$, if u is defined up to t, then there holds

$$\Re e\,\lambda_-(\varepsilon, t, \underline{x}, \xi) < 0 \qquad \text{for some } \underline{x} \in \mathbb{T}.$$

Hence, as stressed in the introduction, there is a change in the behavior of the equation: precisely the operator goes from being hyperbolic to be elliptic and we thus expect an amplification (exponential growth in time) to hold for $t > k_1$. Contrarily to the inviscid case, here the transition from hyperbolicity to ellipticity occurs for some positive time of order one, as expected. Also, we underline again that the waiting time to recover an instability depends only on the leading mode of $u(0,x)$.

Our first result that makes the above observation rigorous is the following:

Theorem 2.2. *Let a such that* $\operatorname{Supp}\hat{a} \subset \{k_1, k_1+1, ...\}$ *Then, for the solution to* (4) *with* $u_0(x) = \varepsilon^2 a(k_1 x/\varepsilon)$, *there holds, for all* $T > 2k_1$

$$\sup_{0\leq t\leq T} |\hat{v}|_\infty |\widehat{\varepsilon\partial_x v}|_{L^1} > \frac{1}{2T}|a_{k_1}|, \qquad v = \frac{1}{\varepsilon^2}(u - it).$$

Such result shows that the solution to (4) experiences a growth after a delay in time that depends only on the smallest non zero mode in the initial datum. However, it holds only for small initial data (for the proof, see [8, Theorem 3.1]), so it does not significantly extend the results contained in [10, Theorem 2].

As to improve the results contained in [10] we want to show that, for the solution to (4) arisen from an $\mathcal{O}(1)$ amplitude highly oscillating initial datum u_0, an exponential growth in time is recorded.

There are several difficulties that occur in the case of general data of amplitude $\mathcal{O}(1)$: a first issue we have to deal with is the existence of the solution for larger times than the ones described in Theorem 2.1. This is not an easy task and it is the reason why the instability result presented here states that, if the flow described in Theorem 2.1 can be continued, then an exponential growth in time is recorded.

Theorem 2.3. *Let u be the solution described in Theorem 2.1. If for some $t_1 > t_0$ large enough there holds the uniform bound*

$$|u(t)|_{L^2} + |\partial_x u(t)|_{L^2} \leq C, \qquad 0 \leq t \leq t_1 + C_1\varepsilon|\ln\varepsilon|, \tag{9}$$

for some $C > 0$ and $C_1 > 0$, then for some $\lambda > 0$, for any $x_1 \in \mathbb{T}$, and for any $\delta > 0$, there holds the lower bound

$$|u(t)|_{L^2(B(x_1,\delta))} \geq C(\delta)e^{t\lambda}, \qquad t \in [t_1, t_1 + C_1'\varepsilon|\ln\varepsilon|], \tag{10}$$

for some $C(\delta) > 0$ and some $0 < C_1' < C_1$.

We point out that the assumed bound in (9) is precisely the one that is shown to hold over $[0, t_0]$ in Theorem 2.1. Thus we are stating that if the flow can be continued beyond t_0, precisely just beyond some $t_1 > t_0$ (that is, if a solution exists for some $t_1 > t_0$), then an exponential growth in time is observed. We also stress that we are able to describe such a growth only a little bit beyond t_1, precisely until $t_1 + C\varepsilon|\log \varepsilon|$.

As one can see in details in [8], the key ingredient of the proof of Theorem 2.3 is a micro local analysis of the equation based on a precise description of the symbol and the use of the Gårding's inequality. Precisely, one has to localize the equation in a neighborhood of a point (x_1, ξ_1) such that the spectrum of the symbol is negative; having done that, Gårding inequality suffices in translating positivity of the symbol (i.e. a finite dimensional property), into a growth for the solution of a PDE.

The results in Theorems 2.1 and 2.3 give a complete description of the dynamics of the solutions to (4), at least until the time t_1 described therein: indeed, they show that, when starting with a highly oscillating initial datum (with no restriction on its amplitude), then the solution arisen from such initial configuration indeed experiences a growth in time (as predicted in [3] for the case $\varepsilon = 0$) but only after a delay (measured by t_0) that gets longer with the increasing of the initial oscillations. In particular, the dynamics of (4) can be divided into two different and specific stages: a first time scale of order one where no amplification is recorded (while it is observed a *linear* growth in time, see estimates (7)), followed by an exponential growth of the solution, described by (10).

3. Time delayed instabilities in hydrodynamical systems.

We conclude this paper with an attempt to extend the results previously described in the case of the scalar complex Burgers equation to hydrodynamical systems; in particular, the question we ask is whether, in the context of an instability, a regularization plays in hydrodynamical systems the role it plays in Burgers. The first problem we deal with is a regularization of the Euler equation (see [11]): we expect the analysis for systems to be quite different from the analysis for scalar equations, such as Burgers. Essentially, for scalar equations, Gårding's inequality suffices in order to translate instabilities properties of the symbol, i.e., instability properties in finite dimensions, into an exponential growth for the solution to the PDE (see [8]). For systems, this is not true in general. The difference is that as soon as the symbol is not symmetric, then Gårding's inequality (in his matrix form) will fail to produce sharp bounds.

3.1. Dispersive regularization of the Euler equations.

We consider a dispersive regularization of the one-dimensional compressible Euler equations with a Van der Walls pressure law $p(u)$:

$$\begin{cases} \partial_t u + \partial_x v = 0, \\ \partial_t v + \partial_x(p(u)) = i\varepsilon\partial_x^2 v. \end{cases} \tag{11}$$

Here $\varepsilon > 0$, $(u, v) \in \mathbb{C}^2$ depend on time $t \geq 0$ and $x \in \mathbb{R}$, and the pressure law is given by

$$p(u) = (|u|^2 - 1)u. \tag{12}$$

Hence there holds

$$p'(u_0) < 0, \quad \text{for some } u_0 > 0,$$

implying, in particular, that for specific volumes close to u_0, the first-order operator in (11) is *not* hyperbolic. Thus, the inviscid system (11) with $\varepsilon = 0$ (which corresponds to the compressible Euler equations in Lagrangian variables, with $u > 0$

being the specific volume and $v \in \mathbb{R}$ the velocity) presents both hyperbolic and elliptic zones, and it is possible to prove (see, for instance, [4] and [5]) that local-in-time existence holds true a priori only for analytical data.

In order not to restrict to analytical data and solutions, we may regularize (11). A viscous regularization as the one done in (4) (see [9]) has the effect of modifying the following conservation law, that holds at the hyperbolic level $\varepsilon = 0$:

$$E(u,v) = \int_{\mathbb{R}} \frac{1}{2}|u(x)|^4 - |u(x)|^2 + |v(x)|^2 \, dx \equiv \text{constant}. \tag{13}$$

By contrast, the dispersive regularization leading to (11) allows for Sobolev solutions to exist and the conservation law (13) to hold. In this sense, our regularization is energy-preserving. The downside, of course, is that we lose the real character of u and v.

Our hope is that the regularized system (11) is a good model system for the study of phase transitions, and our goal is to prove that solutions to (11) display some properties of its hyperbolic version (as, for instance, transition from hyperbolic to elliptic zones).

The first step is of course to prove that system (11) is well posed (as opposite to its hyperbolic counterpart); hence, we want to prove the existence of Sobolev solutions defined over time intervals independent of ε. Our first partial result states that this is true if considering high-frequency data with small amplitude:

Theorem 3.1. *Given s sufficiently large, and given*

$$u_0(x) = \varepsilon^{3/4}\, a\left(\frac{x}{\varepsilon^2}\right) \qquad \text{and} \qquad v_0(x) = \varepsilon^{3/4}\, b\left(\frac{x}{\varepsilon^2}\right) \tag{14}$$

for some $(a,b) \in H^s(\mathbb{R})$, there exists $T > 0$ such that, for ε small enough, the initial value problem

$$\begin{cases} \partial_t u + \partial_x v = 0, \\ \partial_t v + \partial_x (p(u)) = i\varepsilon \partial_x^2 v, \\ (u,v)(x,0) = (u_0, v_0)(x) \end{cases} \tag{15}$$

has a unique solution $(u,v) \in C^0([0,T], H^s(\mathbb{R}))$.

We propose here only a sketch of the proof. For more details, we refer the readers to [11].

Proof. The first step of the proof consists in a rescaling in space of system (15) (that can be done since the initial data have a fast spatial variation). Momentarily denoting by X the spatial variable and by $(\tilde{u}, \tilde{v})$ the unknown in (15), we let

$$x = X/\varepsilon^2, \quad u(t,x) = \tilde{u}(t,X), \quad v(t,x) = \tilde{v}(t,X). \tag{16}$$

Thus $(\tilde{u}, \tilde{v})$ solves (15) with datum (14) if and only if (u,v) solves the initial-value problem

$$\begin{cases} \partial_t u + \dfrac{1}{\varepsilon^2}\partial_x v = 0, \\ \partial_t v - \dfrac{1}{\varepsilon^2}\partial_x u + \dfrac{1}{\varepsilon^{3/4}}\big(\, |u|^2 \partial_x u + 2u \Re e\,(\bar{u}\partial_x u)\,\big) = \dfrac{i}{\varepsilon^3}\partial_x^2 v, \\ (u,v)(0,x) = (a,b)(x). \end{cases} \tag{17}$$

We thus obtain a system that is singular in two distinct ways. First, the linearized equations at $u = 0$ are not hyperbolic: this reflects the singular nature of the pressure law (12). Second, the nonlinear term comes in with a large (with respect to ε) prefactor (which is however milder than the one in front of $\partial_x u$, due to the size of the data in the original system).

Using the vector of unknowns

$$U = (u_1, u_2, v_1, v_2) := (\Re e\, u,\ \Im m\, u,\ \Re e\, v,\ \Im m\, v) \in \mathbb{R}^4 \simeq \mathbb{C}^2,$$

system (17) can be rewritten as

$$\partial_t U + \frac{1}{\varepsilon^2} A_0 \partial_x U + \frac{1}{\varepsilon^{3/4}} A_1(U) \partial_x U + \frac{1}{\varepsilon^3} B \partial_x^2 U = 0, \qquad U(0) = (a, b), \tag{18}$$

where

$$A_0 := \begin{pmatrix} 0 & \mathrm{Id}_2 \\ -\mathrm{Id}_2 & 0 \end{pmatrix}, \qquad A_1(U) := \begin{pmatrix} 0 & 0 \\ a_{21}(U) & 0 \end{pmatrix}, \qquad B := \begin{pmatrix} 0 & 0 \\ 0 & -J \end{pmatrix},$$

and

$$a_{21}(U) := \begin{pmatrix} 3u_1^2 + u_2^2 & 2u_1u_2 \\ 2u_1u_2 & u_1^2 + 3u_2^2 \end{pmatrix}, \qquad J := \begin{pmatrix} 0 & -1 \\ 1 & 0 \end{pmatrix}. \tag{19}$$

Again, we can observe two distinct singularities in (18): one in ε and one in ξ. Indeed, negative powers of ε and positive powers of ξ are singular in the sense that they prevent consistent estimates (i.e. estimates that do not involve $H^{s'}$ norms with $s' > s$) that are uniform in ε. Thus, in view of proving well-posedness in times of the order $\mathcal{O}(1)$, we need to track simultaneously the orders of the operators and their singular (with respect to ε) prefactors.

To do so, the second step in the proof is the use of a normal form reduction, done in such a way the new variable satisfies an equation which is less singular (in terms of powers of ξ, i.e. in terms of order of the differential operators) than (18); in particular, the goal is to cancel out operators of the order one in (18). Hence, we look for $M \in S^{-1}$ (the space of classical symbols of order -1) such that

$$V := (\mathrm{Id} + \varepsilon\, \mathrm{op}(M))^{-1} U \tag{20}$$

satisfies an equation where no operators of the order one are present. By doing all the computations and by paying attention to the remainders that come in the compositions of operators arising from the normal form reduction, one can find an explicit expression for M so that the equation solved by the new variable V reads as:

$$\partial_t V + \frac{1}{\varepsilon^3} B \partial_x^2 V = \frac{1}{\varepsilon^2} R V + \frac{1}{\varepsilon} R_0 V, \tag{21}$$

where R and R_0 comprise all the remainders (respectively, the singular and the non singular ones in ε). As one can see, the terms involving one space derivative of the solution disappeared because of an appropriate choice of M (for the explicit computations, see [11]).

Finally, the last step of the proof makes use of Strichartz estimates and of a contraction principle in a suitable Banach space Y (in particular, we refer to the strategy used in [1]) to prove the existence of a unique solution to (21) belonging to Y. □

REFERENCES

[1] N. Burq, P. Gérard, N. Tzvetkov, Strichartz inequalities and the nonlinear Schrödinger equation on compact manifolds, *American J. Math.*, **126** (2004), 569–605.

[2] J. Carr, R.L. Pego, Metastable patterns in solutions of $u_t = \varepsilon^2 u_{xx} + f(u)$ *Comm. Pure Appl. Math.*, **42** (1989), 523–576.

[3] N. Lerner, Y. Morimoto, C. J. Xu, Instability of the Cauchy-Kovalevskaya solution for a class of non-linear systems, *American J. Math.*, **132** (2010), 99–123.

[4] N. Lerner, T. Nguyen, B. Texier, The onset of instability in first-order systems, *J. Eur. Math. Soc.*, **20** (2018), pp. 1303–1373.

[5] G. Métivier, Remarks on the well-posedness of the nonlinear Cauchy problem, in *Geometric analysis of PDE and several complex variables* Contemp. Math. **368**, Amer. Math. Soc., Providence, (2005) 337–356.

[6] M. Strani, On the metastable behavior of solutions to a class of parabolic systems, *Asymptotic Analysis*, **90** (2014), 325–344.

[7] M. Strani, Slow dynamics in reaction-diffusion systems *Asymptotic Analysis*, **98** (2016), 131–154.

[8] M. Strani, Loss of hyperbolicity and exponential growth for the viscous Burgers equation with complex forcing terms, J. Funct. Analysis, **273** (2017), 1–40.

[9] M. Strani, Transition from hyperbolicity to ellipticity for a p-system with viscosity, *Electron. J. Differential Equations*, **2019** (2019), 1–14.

[10] M. Strani, B. Texier, Time-delayed instabilities in complex Burgers equations, *SIAM J. Math. Anal.*, **47** (2015), 2495–2518.

[11] M. Strani, B. Texier, Schrödinger regularization of a Van der Waals gas, *in preparation.*

E-mail address: marta.strani@unive.it

ON THE DEGOND–LUCQUIN-DESREUX–MORROW MODEL FOR GAS DISCHARGE

MASAHIRO SUZUKI

Department of Computer Science and Engineering, Nagoya Institute of Technology, Gokiso-cho, Showa-ku, Nagoya, 466-8555, Japan

ABSTRACT. This paper surveys the mathematical works [11, 12] investigating the fundamental mechanism of gas discharge. Specifically, we first discuss the time-local solvability of initial–boundary value problems of the Degond–Lucquin-Desreux–Morrow model. Furthermore, we also study the stability and instability of a trivial stationary solution whose electron and positive ion densities are zero, and then see that there exists a sparking voltage at which the trivial solution becomes from stable to unstable. This fact means that gas discharge can occur and continue for a voltage greater than the sparking voltage. Finally, we introduce results on bifurcation of non-trivial stationary solutions.

1. **Introduction.** This short paper gives a survey on the mathematical results analyzing the fundamental mechanism of gas discharge. Around 1900, Townsend constructed a theory for gas discharge. He experimented and considered what happens in a chamber formed from two planar parallel plates and filled with a gas when a direct current high-voltage is applied between these two plates. Then he discovered α- and γ-mechanisms which are essential for the happening of gas discharge. The α-mechanism is the collision of gas particles with electrons, and the γ-mechanism is the secondary emission of electrons caused by impact of positive ions with the cathode. If applied voltage is sufficiently high, these two mechanisms lead to the electric multiplication which permit large current flow throughout the gas which is an insulator. This phenomenon is called as *gas discharge*. Furthermore, he also derived a threshold of voltage at which gas discharge can happen and continue. This threshold is called as a *sparking voltage*. However, he used several simplification in the derivation of sparking voltage (for more details, see [10]). Therefore, it is of interest to analyze the sparking voltage by using a partial differential equation with no simplification and then compare the results of analysis with Townsend's theory.

Several models of gas discharge were proposed and used in [1, 3, 4, 5, 6, 7, 8]. In 1985, Morrow [9] was the first to provide a model including the α- and γ-mechanisms. On the other hand, Degond and Lucquin-Desreux in 2007 gave the formal derivation of the model, derived by Morrow, from Euler–Maxwell equations (see [2]). At this point, it seems reasonable to analyze this model. We call it as the Degond–Lucquin-Desreux–Morrow model or the DLM model in short.

2000 *Mathematics Subject Classification.* Primary: 35M33, 76X05, 82D10; Secondary: 74G20, 70K50, 93D20.

Key words and phrases. hyperbolic-parabolic-elliptic coupled system, solvability, stationary solutions, nonlinear stability, nonlinear instability, bifurcation.

The author is supported by JSPS KAKENHI Numbers 26800067 and 18K03364.

Suzuki and Tani in [11] gave the first mathematical work. The typical shapes of the cathode and anode are either a sphere or plate for the physical and numerical experiments. Therefore, they showed the time-local solvability of an initial–boundary value problem over domains of which boundaries are two of plates and spheres. Their another paper [12] did detailed investigation for the case that the cathode and anode are two planar parallel plates. The initial–boundary value problem has a trivial stationary solution whose electron and positive ion densities are zero. They proved that there exists a threshold of voltage at which the trivial solution becomes unstable from stable. This fact means that gas discharge can occur and continue for a voltage greater than the threshold. The remarkable point is that gas discharge can occur even if γ-mechanism is not taken into account, whereas it cannot occur without γ-mechanism in Townsend's theory. In this paper, we survey these results.

The DLM model consists of two continuity equations for the densities of positive ions and of electrons, adopting constitutive velocity relations, coupled with the Poisson equation for the electrostatic potential:

$$\partial_t \rho_i + \nabla \cdot (\rho_i u_i) = a \exp\left(-b|\nabla\Phi|^{-1}\right) \rho_e \left|v_e\right|, \tag{1a}$$

$$\partial_t \rho_e + \nabla \cdot (\rho_e u_e) = a \exp\left(-b|\nabla\Phi|^{-1}\right) \rho_e \left|v_e\right|, \tag{1b}$$

$$u_i = k_i \nabla\Phi, \quad u_e = v_e - (k_e \nabla \rho_e)/\rho_e, \quad v_e = -k_e \nabla\Phi, \tag{1c}$$

$$\lambda \Delta \Phi = \rho_i - \rho_e, \qquad x \in \Omega,\ t > 0. \tag{1d}$$

The unknown functions ρ_i, ρ_e, and $-\Phi$ denote the positive ion density, the electron density, and the electrostatic potential, respectively. The ion and electron velocities u_i and u_e are assumed to obey (1c). Furthermore, a, b, k_i, k_e, and λ are positive constants. The right hand sides of (1a) and (1b) come from α-mechanism. In particular, $\alpha = a \exp(-b/|\nabla\Phi|)$ is the first Townsend ionization coefficient expressing the number of ion–electron pairs generated per unit volume by the electron impact ionization. We notice that this model is a hyperbolic-parabolic-elliptic coupled system by substituting constitutive velocity relations (1c) into continuity equations (1a) and (1b).

We consider the initial–boundary value problem of this model over domains Ω which have two smooth disjoint boundaries Γ_a and Γ_c (these accurate definitions will be given below) by prescribing the initial and boundary data

$$(\rho_i, \rho_e)(0, x) = (\rho_{i0}, \rho_{e0})(x), \quad \rho_{i0}(x) \geq 0, \quad \rho_{e0}(x) \geq 0, \quad x \in \Omega, \tag{1e}$$

$$\rho_i(t, x) = \rho_e(t, x) = \Phi(t, x) = 0, \quad x \in \Gamma_a,\ t > 0, \tag{1f}$$

$$\rho_e u_e(t, x) \cdot \mathbf{n}(x) = -\gamma \rho_i u_i(t, x) \cdot \mathbf{n}(x), \quad \Phi(t, x) = V_c, \quad x \in \Gamma_c,\ t > 0. \tag{1g}$$

Here γ and V_c are positive constants and $\mathbf{n}(x)$ is the outer normal unit vector of Ω. From physical point of view, it is reasonable to assume the non-negativity of initial densities ρ_{i0} and ρ_{e0}. The boundaries Γ_a and Γ_c correspond to the anode and cathode, respectively, since $-V$ is the voltage. Boundary condition (1f) means that, in an instant, electrons are absorbed to the anode and ions are excluded near the anode. We take γ-mechanism on the cathode in (1g), where the positive constant γ is the second Townsend ionization coefficient expressing the average number of electrons released from cathode by the secondary emission.

For mathematical convenience, let us decompose the electrostatic potential as $\Phi = V + V_{ext}$, where V_{ext} satisfies

$$\lambda \Delta V_{ext} = 0, \quad x \in \Omega, \ t > 0, \tag{2a}$$

$$V_{ext}(t,x) = 0, \quad x \in \Gamma_a, \quad V_{ext}(t,x) = V_c, \quad x \in \Gamma_c. \tag{2b}$$

Then we have the rewritten problem for (ρ_i, ρ_e, V):

$$\partial_t \rho_i + \nabla \cdot (\rho_i u_i) = a \exp\left(-b|\nabla(V+V_{ext})|^{-1}\right) \rho_e |v_e|, \tag{3a}$$

$$\partial_t \rho_e + \nabla \cdot (\rho_e u_e) = a \exp\left(-b|\nabla(V+V_{ext})|^{-1}\right) \rho_e |v_e|, \tag{3b}$$

$$u_i = k_i \nabla(V+V_{ext}), \quad u_e = v_e - (k_e \nabla \rho_e)/\rho_e, \quad v_e = -k_e \nabla(V+V_{ext}), \tag{3c}$$

$$\lambda \Delta V = \rho_i - \rho_e, \quad x \in \Omega, \ t > 0 \tag{3d}$$

with the initial and boundary conditions

$$(\rho_i, \rho_e)(0,x) = (\rho_{i0}, \rho_{e0})(x), \quad \rho_{i0}(x) \geq 0, \quad \rho_{e0}(x) \geq 0, \quad x \in \Omega, \tag{3e}$$

$$\rho_i(t,x) = \rho_e(t,x) = V(t,x) = 0, \quad x \in \Gamma_a, \ t > 0, \tag{3f}$$

$$\rho_e u_e(t,x) \cdot \mathbf{n}(x) = -\gamma \rho_i u_i(t,x) \cdot \mathbf{n}(x), \quad V(t,x) = 0, \quad x \in \Gamma_c, \ t > 0. \tag{3g}$$

Note that $(\rho_i, \rho_e, V + V_{ext})$ is a solution to initial–boundary value problem (1) if V_{ext} and (ρ_i, ρ_e, V) are solutions to problems (2) and (3), respectively. Since (2) and (3) are decoupled essentially, one can regard V_{ext} in (3) as a given function without loss of generality.

Before closing this section, we give our notation. For $1 \leq p \leq \infty$, a non-negative integer k, and a domain Ω, $L^p(\Omega)$ is the Lebesgue space equipped with the norm $|\cdot|_p$; $W^{k,p}(\Omega)$ is the k-th order Sobolev space in the L^p sense; $H^k(\Omega)$ is the k-th order Sobolev space in the L^2 sense. In addition, $C^m([0,T];\mathcal{H})$ denotes the space of the m-times continuously differentiable functions on the interval $[0,T]$ with values in some Hilbert space $\mathcal{H}$. $H^m(0,T;\mathcal{H})$ denotes the space of H^m–functions on $(0,T)$ with values in some Hilbert space $\mathcal{H}$. Let the function space $D_0^{1,2}(\Omega)$ be the completion of $C_0^\infty(\Omega)$ with respect to the norm $\|f\|_{D_0^{1,2}} := \|\nabla f\|_{L^2}$. For an interval $I = (0,L)$, $H_0^1(I) := \{f \in H^1(I); f(0) = f(L) = 0\}$ and $H_{0l}^1(I) := \{f \in H^1(I); f(0) = 0\}$. We also use the weight function $\langle x \rangle := (1+|x|^2)^{1/2}$.

2. **Time-local solvability.** In this section, we discuss the time-local solvability of problem (3) over a domain $\Omega := \mathbb{R}^3_+ \backslash K$, where $\mathbb{R}^3_+ := \{x \in \mathbb{R}^3; x_1 > 0\}$ and $K \subset \mathbb{R}^3_+$ is a simply connected compact set of which boundary is smooth. Let us set $\Gamma_a := \partial \mathbb{R}^3_+$ and $\Gamma_c := \partial K$, and also make assumptions for the given function V_{ext}.

Assumption 1. *For some positive constants c and C, the function V_{ext} satisfies*

(H.1) $\nabla V_{ext} \in H^3(\Omega)$;

(H.2) $\|\nabla V_{ext}\|_{H^3} \leq C V_c$;

(H.3) $c V_c \langle x \rangle^{-3} \leq |\nabla V_{ext}(x)| \leq C V_c \langle x \rangle^{-3}, \quad x \in \Omega$;

(H.4) $\nabla V_{ext}(x) \cdot \mathbf{n}(x) \leq -C V_c \langle x \rangle^{-3}, \quad x \in \Gamma_a$;

(H.5) $\nabla V_{ext}(x) \cdot \mathbf{n}(x) \geq c V_c, \quad x \in \Gamma_c$.

We remark that there is a solution V_{ext} to problem (2) satisfying Assumption 1 for some particular sets K (see Appendix in [11]).

On the basis of the Townsend theory, gas discharge can happen if the electron density is positive somewhere and high-voltage is applied. Therefore, it is desirable from physical point of view to establish the unique existence of time-local solutions to initial–boundary value problem (3) without any restrictions of initial data except the non-negativity in (3e) and the compatibility condition. There appear several difficulties when we solve hyperbolic equation (3a).

First, we need to keep the direction of the boundary characteristics of (3a) as

$$u_i(t,x)\cdot \mathbf{n}(x) < 0, \quad x \in \Gamma_a, \tag{4}$$

$$u_i(t,x)\cdot \mathbf{n}(x) > 0, \quad x \in \Gamma_c \tag{5}$$

in order to put one boundary condition on Γ_a and no boundary condition on Γ_c. We should not control the initial data (ρ_{i0}, ρ_{e0}) by following the Townsend theory. It is desirable to take the voltage $V_c > 0$ sufficiently large so that (4) and (5) hold for any initial data. In the analysis of this situation, one can infer from a standard theory of elliptic equations that the solution V to problem (3d) converges to zero as $|x|$ tends to infinity and so do u_i owing to (3c). Therefore, we need to check carefully the decay rate of u_i at infinite distance for the treatment of the boundary characteristic of (4). Let us call this kind of boundary characteristic as *degenerate boundary characteristic.*

Even if we overcome this difficulty, we cannot find any general theory of non-linear hyperbolic equations with degenerate boundary characteristics, although those with non-degenerate boundary characteristics were established. In general, the regularity of solutions to the linearized hyperbolic equations with degenerate boundary characteristics may be lost near the boundary. This implies the possibility that loss on the derivatives arises at each step of the inductive scheme to solve the non-linear problem.

For the resolution of these difficulties, the paper [11] adopted the following method. To construct a solution (ρ_i, ρ_e, V) to problem (3) with the properties (4) and (5), we employ a weight function $\langle x\rangle^4$ which plays an essential role to obtain the decay rate of V as $|\nabla V(x)| \le C\langle x\rangle^{-3}$. This decay rate together with Assumption 1 leads to the desired properties. For the regularity issue coming from the degenerate boundary characteristic, the main idea is to reduce the initial–boundary value problem over Ω to the initial value problem over $\mathbb{R}^3$ by virtue of several extension operators. This reduction enables us to avoid the loss on the derivatives. It is worth pointing out that properties (4), (5) and compatibility conditions of the initial data play important roles for the application of extension operators. We are now in a position to state the main theorem in [11].

Theorem 2.1. *Let* $(\langle x\rangle^4\rho_{i0}, \langle x\rangle^4\rho_{e0})$ *belong to* $H^2(\Omega)\times H^2(\Omega)$ *and satisfy compatibility conditions of the zero-th and first orders*

$$\rho_{i0}(x) = \rho_{e0}(x) = (\partial_{x_1}\rho_{i0})(x) = 0, \quad x \in \Gamma_a,$$

$$\rho_{e0}(x)u_e(0,x)\cdot \mathbf{n}(x) = -\gamma\rho_{i0}(x)u_i(0,x)\cdot\mathbf{n}(x), \quad x\in\Gamma_c.$$

There exists some constant $M > 0$ *such that for any* $V_c \ge M$*, initial–boundary value problem* (3) *has a unique time-local solution* (ρ_i, ρ_e, V)*, satisfying*

$$\langle x\rangle^4\rho_i \in C([0,T]; H^2(\Omega)) \cap C^1([0,T]; H^1(\Omega)),$$

$$\langle x\rangle^4\rho_e \in C([0,T]; H^2(\Omega)) \cap C^1([0,T]; L^2(\Omega)) \cap H^1(0,T; H^1(\Omega)),$$

$$V \in C([0,T]; D_0^{1,2}(\Omega)), \quad \nabla V \in C([0,T]; H^3(\Omega)) \cap C^1([0,T]; H^1(\Omega)),$$

(4), (5), *and the non-negativity* $\rho_i, \rho_e \ge 0$, *for some time* $T > 0$.

Suzuki and Tani [11] also discussed the solvability of the DLM model over other domains of which boundaries are two of spheres and plates. Remarkable point is that the difficulties mentioned above does not appear for other domains. For more details, see Remark 2.3 in [11].

3. **Sparking voltage.** This section is devoted to the analysis on the sparking voltage. We focus only on studying the DLM model over $\Omega = I := (0, L)$ without γ-mechanism. In this case, V_{ext} is explicitly written by V_c/L. Furthermore, let us use the new unknown functions

$$R_i := \rho_i e^{-\frac{L}{V_c}x}, \quad R_e := \rho_e e^{\frac{V_c}{2L}x}$$

and the given functions

$$h(x) := a \exp\left(\frac{-b}{|x|}\right)|x|, \quad g(V_c) := h\left(\frac{V_c}{L}\right) - \frac{V_c^2}{4L^2}.$$

In the end, we have the following rewritten problem

$$\partial_t R_i + k_i \partial_x \left\{ \left(\partial_x V + \frac{V_c}{L} \right) R_i \right\} + k_i R_i = k_e h\left(\frac{V_c}{L}\right) e^{-\frac{L}{V_c}x - \frac{V_c}{2L}x} R_e + k_i f_i, \tag{6a}$$

$$\partial_t R_e - k_e \partial_{xx} R_e - k_e g(V_c) R_e = k_e f_e, \tag{6b}$$

$$V[V_c, R_i, R_e] := \frac{1}{\lambda}\int_0^L G(x,y)\left(e^{\frac{L}{V_c}y} R_i(t,y) - e^{-\frac{V_c}{2L}y} R_e(t,y)\right) dy, \tag{6c}$$

$$(R_i, R_e)(0,x) = (R_{i0}, R_{e0})(x), \quad R_{i0}(x) \geq 0, \ R_{e0}(x) \geq 0, \tag{6d}$$

$$R_i(t,0) = R_e(t,0) = R_e(t,L) = 0, \tag{6e}$$

where $G(x,y)$ is the Green function of the Laplace operator with the Dirichlet zero condition, and the nonlinear terms f_i and f_e are defined as

$$f_i := -R_i \partial_x V - \frac{k_e}{k_i}\left\{ h\left(\frac{V_c}{L}\right) - h\left(\partial_x V + \frac{V_c}{L}\right) \right\} e^{-\frac{L}{V_c}x - \frac{V_c}{2L}x} R_e,$$

$$f_e := \partial_x V \partial_x R_e - \frac{V_c}{2L} R_e \partial_x V + R_e \partial_{xx} V - \left\{ h\left(\frac{V_c}{L}\right) - h\left(\partial_x V + \frac{V_c}{L}\right) \right\} R_e.$$

We remark that the boundary condition $R_e(t,L) = 0$ comes from the lack of the γ-mechanism. Furthermore, we notice that problem (6) has a trivial stationary solution

$$(R_i, R_e) = (0, 0).$$

Townsend defined the sparking voltage as a threshold of voltage at which gas discharge happens and continues. In following his manner, it is reasonable to define the sparking voltage by a threshold of voltage at which the trivial solution becomes unstable from stable. For seeking the threshold, the function g plays an important role. Suzuki and Tani [12] draw the graph of g subject to constants a, b, k_i, k_e, and L. There are two cases as in Figures 1 and 2. They studied only case 1 in Figure 1 and defined uniquely the sparking voltage by

$$g(V_c^*) = \frac{\pi^2}{L^2}, \quad g'(V_c^*) > 0. \tag{7}$$

Furthermore, it was concluded that that gas discharge can happen even if γ-mechanism is not taken into account, whereas it cannot happen without γ-mechanism in Townsend's theory. Let us mention several theorems in [12] validating these facts.

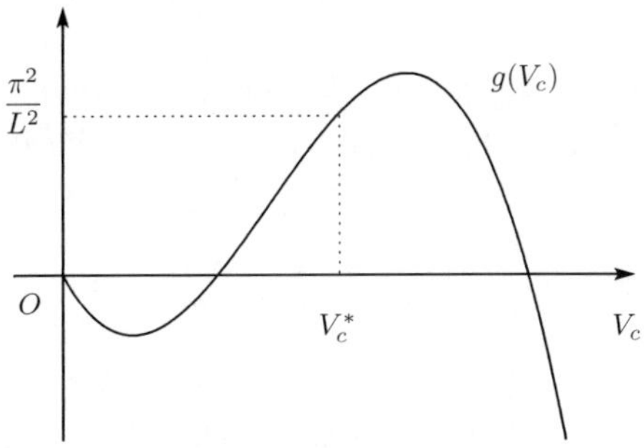

FIGURE 1. case 1

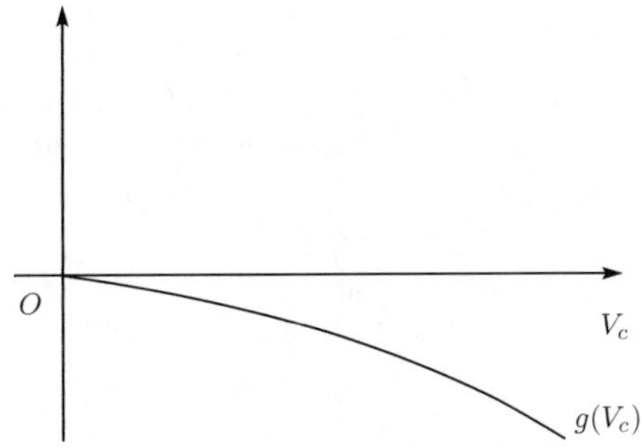

FIGURE 2. case 2

3.1. **Nonlinear stability and instability.** This subsection provides the stability and instability theorems for the trivial solution. These theorems validate the definition of the sparking voltage V_c^*.

Theorem 3.1. *Let $g(V_c) < \pi^2/L^2$. There exists $\varepsilon > 0$ such that if the initial data $(R_{i0}, R_{e0}) \in H^1_{0l}(I) \times H^1_0(I)$ satisfies $\|R_{i0}\|_{H^1} + \|R_{e0}\|_{H^1} < \varepsilon$, then problem (6) has a unique time global solution (R_i, R_e) as*

$$R_i \geq 0, \quad R_i \in C([0,\infty); H^1_{0l}(I)) \cap C^1([0,\infty); L^2(I)), \tag{8a}$$

$$R_e \geq 0, \quad R_e \in C([0,\infty); H^1_0(I)) \cap L^2(0,\infty; H^2(I)) \cap H^1(0,\infty; L^2(I)). \tag{8b}$$

Moreover, it converges to zero exponentially fast in $H^1(I) \times H^1(I)$ as t goes to infinity.

Theorem 3.2. *Let $g(V_c) > \pi^2/L^2$ and $(\psi_i, \psi_e) \in H^1_{0l}(I) \times H^1_0(I)$ satisfy*

$$\psi_i, \psi_e \geq 0, \quad \|\psi_i\|^2_{H^1} + \|\psi_e\|^2_{H^1} = 1, \quad \int_0^L \psi_e \sin\frac{\pi}{L}x\,dx > 0. \tag{9}$$

There exists $\varepsilon > 0$ such that for any sufficiently small $\delta > 0$, problem (6) with the initial data $(R_{i0}, R_{e0}) = (\delta\psi_i, \delta\psi_e)$ has a unique solution (R_i, R_e) satisfying $\|R_i(T)\|_{H^1} + \|R_e(T)\|_{H^1} \geq \varepsilon$ for some $T > 0$.

In this instability theorem, the last inequality in (9) is equivalent to that the initial data R_{e0} is a non-zero function. One may ask what happens for the case that R_{e0} is the zero function. Proposition 3.3 gives the answer that there exists a unique time global solution, and it coincides the trivial stationary solution at finite time.

Proposition 3.3. *Let $V_c > 0$. There exists $\varepsilon > 0$ such that if the initial data $(R_{i0}, R_{e0}) \in H^1_{0l}(I) \times H^1_0(I)$ satisfies $R_{e0} = 0$ and $\|R_{i0}\|_{H^1} < \varepsilon$, then problem (6) has a unique time global solution (R_i, R_e) as (8). Furthermore, there exists $T_0 > 0$ such that*

$$(R_i, R_e)(t,x) = (0,0) \quad \textit{for} \quad (t,x) \in [T_0, \infty) \times I.$$

This proposition asserts that a set $\{(R_{i0}, R_{e0}) \in H^1_{0l}(I) \times H^1_0(I);\ R_{e0} = 0\}$ is a local stable manifold of system (6a)–(6c) for any $V_c > 0$.

3.2. **Bifurcation.** We are also interested in finding the asymptotic behavior of solutions to problem (6) for the case $V_c > V_c^*$. Then it is expected from Theorems 3.1 and 3.2 by regarding the voltage V_c as a bifurcation parameter that there is a non-trivial stationary solution curve near the point $(R_i, R_e, V_c) = (0, 0, V_c^*)$. The bifurcation results are summarized in Theorem 3.4 and Corollary 1.

Theorem 3.4. *For V_c^* defined in* (7)*, there exist $\eta > 0$, $V_c \in C^2([-\eta, \eta]; \mathbb{R})$, and $z \in C^2([-\eta, \eta]; H^1 \times H^2)$ such that $V_c(0) = V_c^*$, $z(0) = 0$, and stationary problem to* (6) *with $V_c = V_c(s)$ has a non-trivial solution $(R_i, R_e)(s) = s(\varphi_i, \varphi_e) + sz(s)$ for $s \in [-\eta, \eta]$, where*

$$\varphi_i(x) := \frac{k_e}{k_i} \exp\left(\frac{-bL}{V_c^*}\right) e^{-\frac{L}{V_c^*}x} \int_0^x e^{-\frac{V_c^*}{2L}y} \varphi_e(y)\, dy, \quad \varphi_e(x) := \sin \frac{\pi}{L} x.$$

Moreover, $\dot{V}_c(0) \lesseqqgtr 0$ holds if and only if

$$-Lg'(V_c^*) \int_0^L \varphi_e^2 \partial_x V[V_c^*, \varphi_i, \varphi_e]\, dx - \frac{1}{2} \int_0^L \varphi_e^2 \partial_{xx} V[V_c^*, \varphi_i, \varphi_e]\, dx \lesseqqgtr 0.$$

Corollary 1. *Let $\dot{V}_c(0) \neq 0$. For $s \neq 0$, it holds for any $x \in (0, L)$ that*

$$s\dot{V}_c(0) R_i(s) > 0, \quad s\dot{V}_c(0) R_e(s) > 0.$$

Furthermore, the positive non-trivial solution is linearly stable if $\dot{V}_c(0) > 0$, and the positive non-trivial solution is linearly unstable if $\dot{V}_c(0) < 0$.

From Theorem 3.4 and Corollary 1, we can draw the bifurcation diagram of stationary solutions as Figures 3 and 4. The both diagrams are truly possible for some physical parameters k_i, k_e, a, b, and L. For example, one can have $\dot{V}_c(0) > 0$ by letting k_e/k_i sufficiently small; one can have $\dot{V}_c(0) < 0$ by letting k_e/k_i sufficiently large and assuming an additional condition for a, b, and L. For more details, see Appendix B in [12].

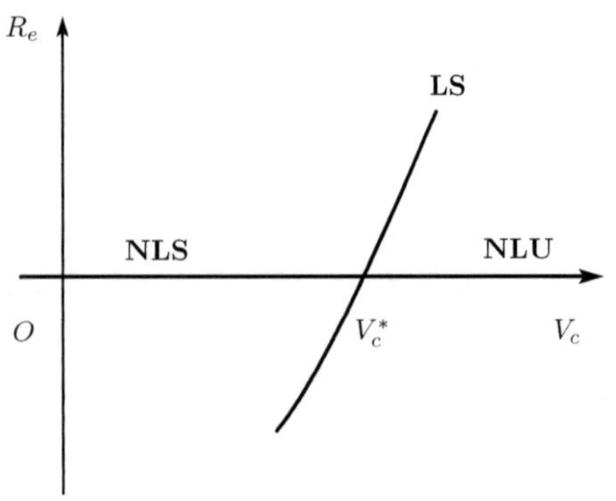

FIGURE 3. $\dot{V}_c(0) > 0$

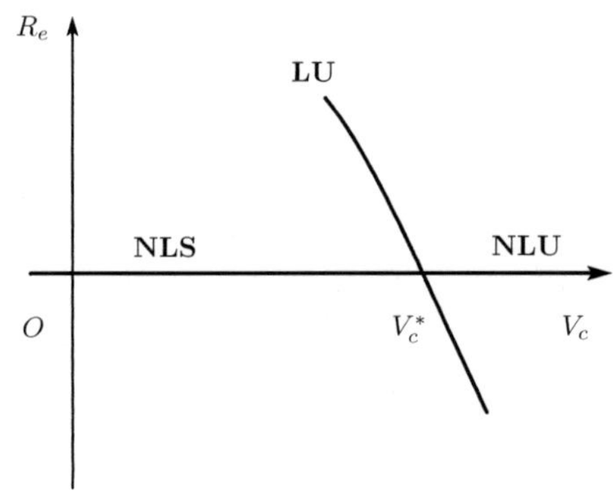

FIGURE 4. $\dot{V}_c(0) < 0$

REFERENCES

[1] I. Abbas and P. Bayle, A critical analysis of ionising wave propagation mechanisms in breakdown, *J. Phys. D: Appl. Phys.* **13** (1980), 1055–1068.

[2] P. Degond and B. Lucquin-Desreux, Mathematical models of electrical discharges in air at atmospheric pressure: a derivation from asymptotic analysis, *Int. J. Compu. Sci. Math.* **1** (2007), 58–97.

[3] S. K. Dhali and P. F. Williams, Twodimensional studies of streamers in gases, *J. Appl. Phys.* **62** (1987), 4694–4707.

[4] P. A. Durbin and L. Turyn, Analysis of the positive DC corona between coaxial cylinders, *J. Phys. D: Appl. Phys.* **20** (1987), 1490–1496.

[5] A. A. Kulikovsky, Positive streamer between parallel plate electrodes in atmospheric pressure air, *IEEE Trans. Plasma Sci.* **30** (1997), 441–450.

[6] A. A. Kulikovsky, The role of photoionization in positive streamer dynamics, *J. Phys. D: Appl. Phys.* **33** (2000), 1514–1524.

[7] A. Luque, V. Ratushnaya and U. Ebert, Positive and negative streamers in ambient air: modeling evolution and velocities, *J. Phys. D: Appl. Phys.* **41** (2008), 234005.
[8] J.-C. Mateéo-Vélez, P. Degond, F. Rogier, A. Séraudie and F. Thivet, Modelling wire-to-wire corona discharge action on aerodynamics and comparison with experiment, *J. Phys. D: Appl. Phys.* **41** (2008), 035205.
[9] R. Morrow, Theory of negative corona in oxygen, *Phys. Rev. A* **32** (1985), 1799–1809.
[10] Y. P. Raizer, *Gas Discharge Physics*, Springer-Verlag Berlin Heidelberg, 2001.
[11] M. Suzuki and A. Tani, Time-local solvability of the Degond–Lucquin-Desreux–Morrow model for gas discharge, *SIAM Math. Anal.* **50** (2018), 5096–5118.
[12] M. Suzuki and A. Tani, Bifurcation analysis of the Degond–Lucquin-Desreux–Morrow model for gas discharge, submitted.

E-mail address: masahiro@nitech.ac.jp

GLOBAL ENTROPY SOLUTIONS TO THE COMPRESSIBLE EULER EQUATIONS IN THE ISENTROPIC NOZZLE FLOW

NAOKI TSUGE*

Department of Mathematics Education,
Faculty of Education, Gifu University, 1-1 Yanagido, Gifu
Gifu 501-1193 Japan

ABSTRACT. We study the motion of isentropic gas in nozzles. These phenomena are governed by the compressible Euler equations. In this paper, we consider its unsteady flow and devote to proving the global existence of solutions to the Cauchy problem for the general nozzle. Although the subject is important in Mathematics, Physics and engineering, it remained open for a long time. The problem seems to rely on a bounded estimate of approximate solutions, because we have only method to investigate the behavior with respect to the time variable. To solve this, we first introduce a generalized invariant region. Compared with the existing ones, its upper and lower bounds are extended constants to functions of the space variable. However, we cannot apply the new invariant region to the traditional difference scheme. Therefore, we invent the modified Godunov scheme. The approximate solutions consist of some functions corresponding to the upper and lower bounds of the invariant regions. These methods enable us to investigate the behavior of approximate solutions with respect to the space variable. In the final section, we state open problem for inhomogeneous Euler equations.

1. **Introduction.** The present paper is concerned with isentropic gas flow in a nozzle. This motion is governed by the following compressible Euler equations:

$$\begin{cases} \rho_t + m_x = a(x)m, \\ m_t + \left(\dfrac{m^2}{\rho} + p(\rho)\right)_x = a(x)\dfrac{m^2}{\rho}, \quad x \in \mathbf{R}, \end{cases} \tag{1}$$

where ρ, m and p are the density, the momentum and the pressure of the gas, respectively. If $\rho > 0$, $v = m/\rho$ represents the velocity of the gas. For a barotropic gas, $p(\rho) = \rho^\gamma/\gamma$, where $\gamma \in (1, 5/3]$ is the adiabatic exponent for usual gases. The given function $a(x)$ is represented by

$$a(x) = -A'(x)/A(x) \quad \text{with} \quad A(x) = e^{-\int^x a(y)dy},$$

where $A \in C^2(\mathbf{R})$ is a slowly variable cross section area at x in the nozzle.

We consider the Cauchy problem (1) with the initial data

$$(\rho, m)|_{t=0} = (\rho_0(x), m_0(x)). \tag{2}$$

2000 *Mathematics Subject Classification.* Primary 35L03, 35L65, 35Q31, 76N10, 76N15; Secondary 35A01, 35B35, 35B50, 35L60, 76H05, 76M20.

Key words and phrases. The Compressible Euler Equation, the nozzle flow, the compensated compactness, the generalized invariant regions, the modified Godunov scheme.

N. Tsuge's research is partially supported by Grant-in-Aid for Scientific Research (C) 25400157, Japan.

The above problem (1)–(2) can be written in the following form

$$\begin{cases} u_t + f(u)_x = g(x,u), & x \in \mathbf{R}, \\ u|_{t=0} = u_0(x), \end{cases} \tag{3}$$

by using $u = {}^t(\rho, m)$, $f(u) = {}^t\left(m, \dfrac{m^2}{\rho} + p(\rho)\right)$ and $g(x,u) = {}^t\left(a(x)m, a(x)\dfrac{m^2}{\rho}\right)$.

Let us survey the related mathematical results for (1). The pioneer work in this direction is Liu [4]. In [4], Liu proved the existence of global solutions coupled with steady states, by the Glimm scheme, provided that the initial data have small total variation and are away from the sonic state.

The motivations in the present paper is to construct solutions with large initial data and including the sonic state. The physically interesting flow contains the sonic state. Actually, when the Laval nozzle accelerate the subsonic flow to the supersonic one, the flow attains the sonic state at the throat (see [14], [3, Section 5] and [5, Chapter 5]). Therefore, the objective of the preset paper is to establish the global existence of solutions with the sonic state.

To state our main theorem, we define the Riemann invariants w, z, which play important roles in this paper, as

Definition 1.1.

$$w := \frac{m}{\rho} + \frac{\rho^\theta}{\theta} = v + \frac{\rho^\theta}{\theta}, \quad z := \frac{m}{\rho} - \frac{\rho^\theta}{\theta} = v - \frac{\rho^\theta}{\theta} \quad (\theta := (\gamma-1)/2).$$

These Riemann invariants satisfy the following.

Remark 1.

$$|w| \geqq |z|,\ w \geqq 0, \text{ when } v \geqq 0. \quad |w| \leqq |z|,\ z \leqq 0, \text{ when } v \leqq 0. \tag{4}$$

$$v = \frac{w+z}{2},\ \rho = \left(\frac{\theta(w-z)}{2}\right)^{1/\theta},\ m = \rho v. \tag{5}$$

From the above, the lower bound of z and the upper bound of w yield the bound of ρ and $|v|$.

We assume the following.
There exists a nonnegative function $b \in C^1(\mathbf{R})$ such that

$$|a(x)| \leqq \mu b(x), \quad \max\left\{\int_0^\infty b(x)dx,\ \int_{-\infty}^0 b(x)dx\right\} \leqq \frac{1}{2}\log\frac{1}{\sigma}, \tag{6}$$

where $\mu = \frac{(1-\theta)^2}{\theta(1+\theta-2\sqrt{\theta})}$, $\sigma = \frac{1-\theta}{(1-\sqrt{\theta})(2\sqrt{\theta+1}+\sqrt{\theta-1})}$. Here we notice that $0 < \sigma < 1$. In addition, μ and σ shall be characterized by the values of a function $f(k)$ in Figure 2. (6) means that $\|a\|_{L^1(\mathbf{R})}$ is small enough.

Then our main theorem is as follows.

Theorem 1.2 ([15]). *We assume that, for b in (6) and any fixed nonnegative constant M, initial density and momentum data $u_0 = (\rho_0, m_0) \in L^\infty(\mathbf{R})$ satisfy*

$$0 \leqq \rho_0(x),\ -Me^{-\int_0^x b(y)dy} \leqq z(u_0(x)),\ w(u_0(x)) \leqq Me^{\int_0^x b(y)dy} \tag{7}$$

in terms of Riemann invariants, or

$$0 \leqq \rho_0(x),\ -Me^{-\int_0^x b(y)dy} \leqq v_0(x) - \frac{\{\rho_0(x)\}^\theta}{\theta},\ v_0(x) + \frac{\{\rho_0(x)\}^\theta}{\theta} \leqq Me^{\int_0^x b(y)dy}$$

in the physical variables.

Then the Cauchy problem (3) *has a global entropy weak solution* $u(x,t)$ *satisfying the same inequalities as* (7)

$$0 \leqq \rho(x,t), \quad -Me^{-\int_0^x b(y)dy} \leqq z(u(x,t)), \quad w(u(x,t)) \leqq Me^{\int_0^x b(y)dy}. \tag{8}$$

Remark 2. In view of $(6)_2$, (7) implies that we can supply arbitrary L^∞ data. On the other hand, if we consider only the Laval nozzle, we do not need the smallness condition of $\|a\|_1$ such as (6) (see [9]). In addition, the above solution satisfies an energy inequality (see [18]). Therefore, if the energy of initial data is finite, that of the corresponding solution is also finite.

2. **Difficult point.** The most difficult point of this problem is to obtain the bounded estimate of solutions. To observe this, we consider the diagonalization of (1). If solutions are smooth, we deduce from (1)

$$\begin{aligned} z_t + \lambda_1 z_x &= g_1(x,z,w), \\ w_t + \lambda_2 w_x &= g_2(x,z,w), \end{aligned} \tag{9}$$

where $g_1(x,z,w) = -a(x)\rho^\theta v$, $g_2(x,z,w) = a(x)\rho^\theta v$, λ_1 and λ_2 are the characteristic speeds defined as follows

$$\lambda_1 = v - \rho^\theta, \quad \lambda_2 = v + \rho^\theta. \tag{10}$$

We assume that the following 1 order estimate holds

$$|g_1(x,z,w)| \leq C(|z| + |w|), \quad |g_2(x,z,w)| \leq C(|z| + |w|). \tag{11}$$

From the classical method, i.e., the invariant region (see [1]) and fractional step (see [2]), we can obtain the exponential growth estimate

$$M_t \leq kM, \tag{12}$$

where $M(t) = \max\{\|z(\cdot,t)\|_{L^\infty(\mathbf{R})}, \|w(\cdot,t)\|_{L^\infty(\mathbf{R})}\}$ and k is a positive constant. Here, the orders of $|z|$ and $|w|$ of the right hand side in (11) correspond to that of M of the right hand side in (12). (12) yields the time dependent bounded estimate

$$|z(x,t)| \leq Be^{kt}, \quad |w(x,t)| \leq Be^{kt}.$$

However, unfortunately, we cannot expect (11). In fact, since $\rho^\theta v = \theta(w^2 - z^2)/4$, we can obtain the only following 2 order estimate

$$|g_1(x,z,w)| \leq C(|z|^2 + |w|^2), \quad |g_2(x,z,w)| \leq C(|z|^2 + |w|^2),$$

provided that a is uniformly bounded. We can deduce from this estimate only the $M_t \leq kM^2$. This does not yields a time global bounded estimate.

Therefore, it is almost impossible to obtain the bounded estimate of solutions by the classical method. This is the reason why this problem has been open for about forty years.

To overcome this, we observe the estimate with respect to the space variable x. Here, we notice that the classical method investigates the behavior with respect to the only time variable t. In next section, we shall deduce some terms such as relaxation ones from convection terms of (1) and deal with inhomogeneous terms.

3. **Outline of the proof.** We grasp the point of the main estimate by a formal argument. We thus assume that a solution is smooth and the density is nonnegative in this section.

Now, the most difficult point in the present paper is to obtain the bounded estimate of approximate solutions. To do this, we consider Riemann invariants to use the invariant region theory. Then, the difficult point of this estimate is caused by the inhomogeneous terms of (1). In fact, for a homogeneous system corresponding to (1), we can obtain the bounded estimate by the Chueh, Conley and Smoller **invariant region theory**. However, as we observe in Section 2, their theory is not effective in our problem.

To solve this problem, we introduce the **generalized invariant regions**. We consider the physical region $\rho \geqq 0$ (i.e., $w \geqq z$.). Recalling Remark 1, it suffices to derive the lower bound of $z(u)$ and the upper bound of $w(u)$ to obtain the bound of u.

We set

$$z = \tilde{z}e^{-\int_0^x b(y)dy}, \quad w = \tilde{w}e^{\int_0^x b(y)dy}. \tag{13}$$

Then, it follows from (9) that

$$\begin{aligned} \tilde{z}_t + \lambda_1 \tilde{z}_x &= e^{\int_0^x b(y)dy} \left\{ b(x)\lambda_1 z - a(x)\rho^\theta v \right\}, \\ \tilde{w}_t + \lambda_2 \tilde{w}_x &= -e^{-\int_0^x b(y)dy} \left\{ b(x)\lambda_2 w - a(x)\rho^\theta v \right\}. \end{aligned} \tag{14}$$

In a subsequent argument, the terms $b(x)\lambda_1 z$ and $b(x)\lambda_2 w$ will play a role such as relaxation terms, i.e., they neutralize the effect of the inhomogeneous terms. We call these terms **quasi relaxation terms**.

First, from (6), we notice that

Lemma 3.1. *The vertex A of the triangle in Fig.1 lies between lines ℓ_1 and ℓ_2 through the origin with the slopes $-\sigma$ and $-1/\sigma$, where $\sigma = \frac{1-\theta}{(1-\sqrt{\theta})(2\sqrt{\theta+1}+\sqrt{\theta-1})}$.*

Proof. Since the slope of OA is $-e^{2\int_0^x b(y)dy}$, it follows from (6) that $-1/\sigma \leqq$ the slope of OA$\leqq -\sigma$. □

Let us investigate the inhomogeneous term of $(14)_1$ in Regions 1 and 2 (see Fig. 1). To apply the invariant region theory, we shall prove that the inhomogeneous term of $(14)_1$ is positive on line AB. Now, in these regions, z and w satisfy the following.

$$-Me^{-\int_0^x b(y)dy} \leqq z,\ w \leqq Me^{\int_0^x b(y)dy},\ -1/\sigma \leqq w/z \leqq 1,\ z \leqq 0. \tag{15}$$

We set $k = w/z$. Then, from (5), we have

$$\lambda_1 = \frac{(1-\theta)k + 1 + \theta}{2} z,\ v = \frac{k+1}{2} z,\ \rho^\theta = \frac{\theta(k-1)}{2} z,\ -1/\sigma \leqq k \leqq 1. \tag{16}$$

Moreover, we notice the following.

Lemma 3.2. *We set $f(k) = \dfrac{2\{(1-\theta)k + 1 + \theta\}}{\theta|k^2 - 1|}$. Then, $f(k) \geqq \mu$ on the closed interval $[-1/\sigma, 1]$, where μ and σ are defined in (6).*

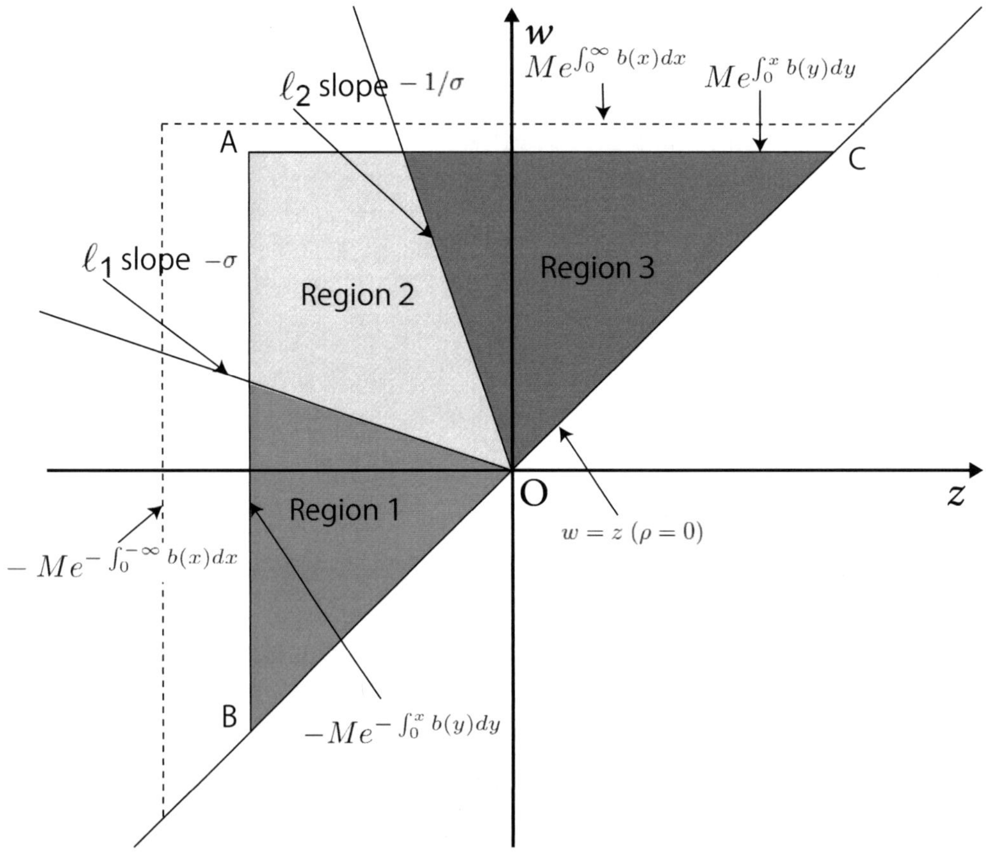

FIGURE 1. The invariant region in (z, w)-plane

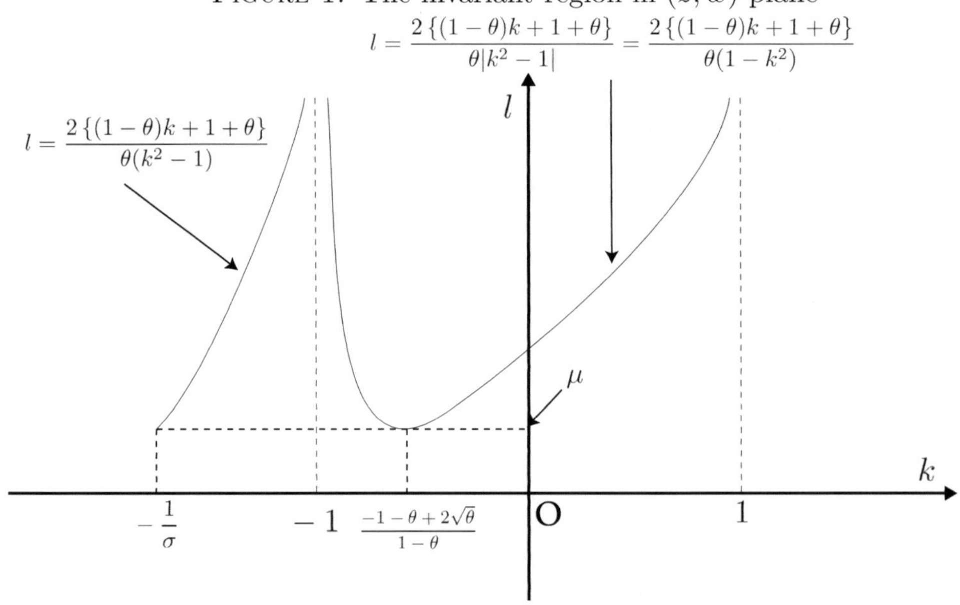

FIGURE 2. The graph of $l = f(k)$

Then, from (6) and Lemma 3.2, in Regions 1 and 2, we obtain

$$\begin{aligned}\tilde{z}_t + \lambda_1 \tilde{z}_x =& e^{\int_0^x b(y)dy} \left\{ b(x)\lambda_1 z - a(x)\rho^\theta v \right\} \\ \geqq& e^{\int_0^x b(y)dy} b(x) z^2 \left\{ \frac{(1-\theta)k+1+\theta}{2} - \mu \frac{\theta|1-k^2|}{4} \right\} \\ \geqq& e^{\int_0^x b(y)dy} b(x) z^2 \frac{\theta|1-k^2|}{4} \left\{ f(k) - \mu \right\} \\ \geqq& 0 \quad \text{(from Lemma 3.2)}. \end{aligned} \tag{17}$$

Since $\rho^\theta v$ is a two variable function, it is difficult to prove that it is positive. However, considering along the line $k = w/z$, we can treat with the term such as one variable function. We thus conclude that the inhomogeneous term of $(14)_1$ is positive in Regions 1 and 2. Similarly we find that the inhomogeneous term of $(14)_2$ is negative in Region 2 and 3. In particular, the inhomogeneous term of $(14)_1$ is positive on line AB and the inhomogeneous term of $(14)_2$ is negative on line AC.

Therefore, if a solution is contained in Region 1–3, since $\tilde{z} \geqq -M$ and $\tilde{w} \leqq M$, it follows from the invariant region theory (see [1] and [6, Chapter 14, §B]) that the solution remains in the same triangle. This implies that the following region

$$\Delta_x = \left\{ (z,w); \rho \geqq 0, \ -Me^{-\int_0^x b(y)dy} \leqq z, \ w \leqq Me^{\int_0^x b(y)dy} \right\}$$

is an invariant region for the Cauchy problem (3). Here we notice that the invariant region Δ_x depends on the space variable x. Our generalized invariant region quite differs from the Chueh, Conley and Smoller one in this point. This is the key idea to obtain the bounded estimate.

Let us review the above argument.

1. We recall that [4] cannot contain the sonic case, which means that characteristic speeds are 0. This case is also difficult for our method, because quasi relaxation terms are 0. However, for example, recalling the argument of (17), we consider only Region 1–2. On the other hand, quasi relaxation term $b(x)\lambda_1 z$ is 0 in only Region 3. We thus prevent difficulty caused by the sonic state.
2. We need Lemma 3.1. If this lemma does not hold, the above argument fails. In fact, line AB is not contained in Region 1–2. Here, we notice that we supply condition (6) to obtain Lemma 3.1.

We notice that the above argument is formal, because the solution of (1) is not smooth in general. Therefore, we need to justify this argument by a approximate solutions. However, we cannot do by the existing difference scheme such as the Godunov or Lax-Friedrichs scheme. Therefore, we introduce the **modified Godunov scheme**. Then, we must adjust our approximate solutions to the above invariant region. In view of (13), by using the fractional step procedure, we adopt the following functions

$$\begin{aligned} z^\Delta(x,t) =& \bar{z}^\Delta(x) + g_1^*(x, \bar{u}^\Delta(x))(t - n\Delta t), \\ w^\Delta(x,t) =& \bar{w}^\Delta(x) + g_2^*(x, \bar{u}^\Delta(x))(t - n\Delta t) \end{aligned} \tag{18}$$

as the building blocks of our approximate solutions piecewisely in each cell, where Δt is the time mesh length, $n \in \mathbf{Z}_{\geqq 0}$, $\bar{z}^\Delta(x) = \tilde{z}e^{-\int_0^x b(y)dy}$, $\bar{w}^\Delta(x) = \tilde{w}e^{\int_0^x b(y)dy}$

with constants $\tilde{z}$, $\tilde{w}$, and

$$g_1^*(x,\bar{u}^\Delta(x)) = -a(x)\bar{v}^\Delta(x)(\bar{\rho}^\Delta(x))^\theta + b(x)\lambda_1(\bar{u}^\Delta(x))\bar{z}^\Delta(x),$$
$$g_2^*(x,\bar{u}^\Delta(x)) = a(x)\bar{v}^\Delta(x)(\bar{\rho}^\Delta(x))^\theta - b(x)\lambda_2(\bar{u}^\Delta(x))\bar{w}^\Delta(x).$$

We notice that (18) are solutions of (9) approximately. In fact, from (9), we have

$$\begin{aligned} z_t - g_1^*(x,u) &= -\lambda_1(z_x + b(x)\lambda_1 z), \\ w_t - g_2^*(x,u) &= -\lambda_2(w_x - b(x)\lambda_2 w). \end{aligned} \tag{19}$$

Since $\bar{z}^\Delta(x), \bar{w}^\Delta(x)$ are solutions to the right-hand side of (19), we find the following.

Remark 3. The approximate solution $u^\Delta(x,t) = (\rho^\Delta(x,t), m^\Delta(x,t))$, which is deduced from $z^\Delta(x,t), w^\Delta(x,t)$ by the relation (5), satisfies

$$(u^\Delta)_t + f(u^\Delta)_x - g(x,u^\Delta) = O(\Delta t) \quad \text{for } t \in [n\Delta t, (n+1)\Delta t).$$

This means that $z^\Delta(x,t), w^\Delta(x,t)$ are solutions of (1) approximately.

On the other hand, we recall that the existing approximate solutions of [4] consist of steady state solutions of (1).

Now, when we construct our approximate solutions, two difficulties arise (P1) along discontinuous lines and (P2) near the vacuum in each cell. (P1): Since our approximate solutions consists of functions of x, they cannot satisfy the Rankine-Hugoniot condition at every point of a discontinuous line. To overcome this problem, the approximate solutions satisfy the Rankine-Hugoniot condition at the only center of the discontinuous line, which makes the error from the discontinuity enough small. (P2): It is difficult to use (18) as building blocks of our approximate solutions near the vacuum. To handle this problem, we employ not (18) but Riemann solutions, which are solutions of the corresponding homogeneous conservation law, because the inhomogeneous terms are small near the vacuum. These ideas are essential to deduce their compactness and convergence. In addition, the modified Godunov scheme has the advantage of adjusting to not only the present invariant region but also the other ones, by replacing (18).

Finally, we notice that the above methods are applicable to other inhomogeneous conservation laws. For example, by applying the methods to the Euler equation with an outer force, the author recently succeeded in proving the new existence theorem and stability of solutions in [12], [16] and [17]. In addition, the method is also used for the spherically symmetric flow (see [7]–[8]) and inhomogeneous scaler conservation laws (see [11] and [13]). The ideas and techniques developed in this paper will be applicable to not only conservation laws but also other nonlinear problems involving similar difficulties such as reaction-diffusion equations, nonlinear wave equations, the numerical analysis, etc.

4. **Open problem: invariant regions for the bounded domain and space periodic boundary condition.** We can apply our generalized invariant region (call GIR) to the Cauchy problem (see [8]–[13]) and initial-boundary problem for a half space (see [7]). However, unfortunately, we cannot find a GIR for the bounded domain and space periodic boundary problem. This is caused by the following property of the GIR. For example, the upper and lower bound of (8) are strictly increasing functions. On the other hand, for example, we must find a space periodic GIR for the space periodic boundary problem. Therefore, we cannot find an appropriate GIR in this case. It is thus difficult to prove the existence of a time

periodic solution to the space periodic boundary problem for the Euler equation with a time periodic outer force (see [12, Section 5]. The special case can be found in [17].) Similarly, we cannot find a GIR of the boundary problem for a bounded domain.

REFERENCES

[1] K. N. Chueh, C. C. Conley and J. A. Smoller, Positively invariant regions for systems of nonlinear diffusion equations, *Indiana Univ. Math. J.*, **26** (1977), 373–392.

[2] X. Ding, G. Q. Chen and P. Luo, Convergence of the fractional step Lax–Friedrichs scheme and Godunov scheme for the isentropic system of gas dynamics, *Commun. Math. Phys.*, **121** (1989), 63—84.

[3] J. Glimm, G. Marshall and B. Plohr, A generalized Riemann problem for quasi-one-dimensional gas flows, *Adv. in Appl. Math.*, **5** (1984), 1–30.

[4] T.-P. Liu, Quasilinear hyperbolic systems, *Commun. Math. Phys.*, **68** (1979), 141–172.

[5] H. W. Liepmann and A. Roshko, *Elements of gas dynamics*, Galcit aeronautical series. Wiley/Chapman & Hall, New York/London, 1957.

[6] J. Smoller, *Shock Waves and Reaction-Diffusion Equations*, 2nd edition, Springer-Verlag, New York, 1994.

[7] N. Tsuge, Spherically symmetric flow of the compressible Euler equations, *J. Math. Kyoto Univ.*, **44** (2004), 129–171.

[8] N. Tsuge, Global L^∞ solutions of the compressible Euler equations with spherical symmetry, *J. Math. Kyoto Univ.*, **46** (2006), 457–524.

[9] N. Tsuge, Existence of global solutions for unsteady isentropic gas flow in a Laval nozzle, *Arch. Ration. Mech. Anal.*, **205** (2012), 151–193.

[10] N. Tsuge, Isentropic gas flow for the compressible Euler equation in a nozzle, *Arch. Ration. Mech. Anal.*, **209** (2013), 365–400.

[11] N. Tsuge, Existence of a global solution to a scalar conservation law with a source term for large data, *J. Math. Anal. Appl.*, **432** (2015), 862–867.

[12] N. Tsuge, Existence and Stability of Solutions to the Compressible Euler Equations with an Outer Force, *Nonlinear Anal. Real World Appl.*, **27** (2016), 203–220.

[13] N. Tsuge, Existence of a Global Solution for a Scalar Conservation Law with a Source Term, *Acta Appl. Math.*, **147** (2016), 177–186.

[14] N. Tsuge, Isentropic Gas Flow in a Laval Nozzle: Physical Phenomena of Steady Flow and Time Global Existence of Solutions, *RIMS Kôkyûroku*, **2070** (2016), 150–162. http://www.kurims.kyoto-u.ac.jp/ kyodo/kokyuroku/contents/2070.html

[15] N. Tsuge, Global entropy solutions to the compressible Euler equations in the isentropic nozzle flow for large data: Application of the generalized invariant regions and the modified Godunov scheme, *Nonlinear Anal. Real World Appl.*, **37** (2017), 217–238.

[16] Y.-b. Hu, Lu, Y.-G. Lu and N. Tsuge, Global existence and stability to the polytropic gas dynamics with an outer force, *Appl. Math. Lett.*, **95** (2019), 36–40.

[17] N. Tsuge, Existence of a time periodic solution for the compressible Euler equation with a time periodic outer force, preprint.

[18] N. Tsuge, Energy inequality of a global L^∞ solution to the compressible Euler equations for the isentropic nozzle flow, preprint.

E-mail address: tuge@gifu-u.ac.jp

ON A CLASS OF NEW GENERALIZED POISSON-NERNST-PLANCK-NAVIER-STOKES EQUATIONS

YONG WANG
South China Research Center for Applied Mathematics and Interdisciplinary Studies,
South China Normal University,
Guangzhou, 510631, China

ABSTRACT. We introduce a class of new generalized Poisson-Nernst-Planck-Navier-Stokes equations, which can be used to model the transport of microscopic charged particles in fluids. This kind of models can incorporate the interactions between microscopic charged particles and macroscopic fluids as well as cross-diffusion phenomenon. For two typical models, we study their global well-posedness for the Cauchy problem in dimension three.

1. **Introduction.** Throughout this paper, we use the following symbols:

ρ	mass density of fluid	ϕ	electrostatic potential
u	velocity of fluid	ϵ	dielectric constant
v	density of negative ions	z_v	valence of negative ions
u_v	velocity of negative ions	z_w	valence of positive ions
w	density of positive ions	e	charge of one electron
u_w	velocity of positive ions	p	pressure
μ, μ'	viscosity coefficients	D_v	diffusivity of negative ions
D	mobility	D_w	diffusivity of positive ions
k_B	Boltzmann constant	T	absolute temperature

The transport of the charged particles under the influence of self-consistent electrostatic field is an important phenomenon in nature. Dating from the works of Nernst and Planck [22, 27] in 1890's, the motion of ions (anions and cations) in chemical batteries was modeled by using the following

$$\begin{cases} v_t = \operatorname{div}\left[D_v\left(\nabla v + \frac{z_v e}{k_B T} v \nabla\phi\right)\right], \\ w_t = \operatorname{div}\left[D_w\left(\nabla w + \frac{z_w e}{k_B T} w \nabla\phi\right)\right], \\ -\epsilon\Delta\phi = z_v e v + z_w e w, \end{cases} \tag{1}$$

which is well-known as the Poisson-Nernst-Planck (PNP) equations [19]. It was not until the 1930's that purification techniques for semiconductor materials were developed, people began to realize that semiconductors have many better properties

2000 *Mathematics Subject Classification.* Primary: 35Q35, 35Q92; Secondary: 76W05.

Key words and phrases. Poisson-Nernst-Planck equations, Navier-Stokes equations, cross-diffusion, global well-posedness.

The author is supported by NSF of China (No. 11701264, 11971179) and the Zhujiang River Talent of Guangdong Province (No. 2017GC010407).

than metallic conductors. For instance, their resistance decreases as temperature increases, which is a behavior opposite to that of a metal. Because of this reason (1) was also used to model the transport of electrons and holes in semiconductors [28, 30]. The Poisson-Nernst-Planck model is often called the drift-diffusion (or electro-diffusion) model [21, 29]. The model of form (1) is so important and useful that it was also used to model the scenario of ions (say chloridion Cl^-, sodion Na^+, etc.) passing in and out of biological cells through ion channels on the cell membrane [2, 4]. In a word, the same model (1) can be applied to different situations as mentioned above.

From an energetic point of view, the system (1) obeys the energy dissipation law

$$\begin{aligned}\frac{d}{dt}E^{total} &:= \frac{d}{dt}\int_{\mathbb{R}^3}\left[k_BT\left(v\ln v + w\ln w\right) + \tfrac{\epsilon}{2}|\nabla\phi|^2\right]dx \\ &= -\int_{\mathbb{R}^3}\left(\tfrac{k_BT}{D_v}v\left|u_v\right|^2 + \tfrac{k_BT}{D_w}w\left|u_w\right|^2\right)dx := -\triangle, \end{aligned} \tag{2}$$

where E^{total} is the total energy including thermo-fluctuations (Gibbs entropy) of the ion species and the electric potential energy, $\triangle$ is the entropy production (energy dissipation rate), respectively. The electrostatic potential ϕ is induced by the total charge in view of Gauss's law:

$$-\epsilon\Delta\phi = z_v ev + z_w ew.$$

The velocity u_v/u_w of the negative/positive ions satisfy the conservation equations

$$\begin{cases} v_t + \operatorname{div}(vu_v) = 0, \\ w_t + \operatorname{div}(wu_w) = 0. \end{cases} \tag{3}$$

On the other hand, starting with the energy law (2) and (3), Hsieh et al. [14] derived the above equations $(1)_{1,2}$ by using an Energetic Variational Approach (henceforth EVA). The EVA essentially comprises the Least Action Principle (or Hamilton's principle) [1, 11, 12] and the Maximum Dissipation Principle (or Onsager's principle) [23, 24, 25]. This is a very powerful method to derive a large class of important mathematical models [8, 11, 15, 20].

In general, it is enough for us to use the above model (1) to describe the dynamics of the *dilute* charged particles [9, 10, 21]. However, how about the *crowded* case? Such a case often occurs in ion channels and electrodes of batteries [5, 6, 7]. Biological plasmas outside cells contain chloride ($\approx 102\ mM/L$) and sodium ($\approx 140\ mM/L$) ions, which corresponds to the dilute case (Pure water has umber density 55 M/L or so, $1M \approx 6.02\times10^{23}$), however, the concentration of ionic mixtures reaches about 20 M/L in ion channels embedded in cell membranes. In such a dense case, the mutual friction between different ion species is inevitable since it has an important impact on the dynamics of the species themselves. Then, one needs to modify the model (1). Hsieh et al. [14] modified the above dissipation as the following, however, keeping the total energy in (2) unchange,

$$\triangle = \int_{\mathbb{R}^3}\left(\tfrac{k_BT}{D_v}v\left|u_v\right|^2 + \tfrac{k_BT}{D_w}w\left|u_w\right|^2 + \tfrac{k_BT}{D}vw\left|u_v - u_w\right|^2\right)dx,$$

where the above last term (the relative velocity differences) is responsible for the dissipation arising from the friction between particles. Then, their modified PNP model is

$$\begin{cases} v_t = \operatorname{div}\left[\frac{(D+D_w v)D_v\left(\nabla v+\frac{z_v e}{k_B T}v\nabla\phi\right)+D_v D_w v\left(\nabla w+\frac{z_w e}{k_B T}w\nabla\phi\right)}{D+D_w v+D_v w}\right], \\ w_t = \operatorname{div}\left[\frac{(D+D_v w)D_w\left(\nabla w+\frac{z_w e}{k_B T}w\nabla\phi\right)+D_v D_w w\left(\nabla v+\frac{z_v e}{k_B T}v\nabla\phi\right)}{D+D_w v+D_v w}\right], \\ -\epsilon\Delta\phi = z_v e v + z_w e w. \end{cases}$$

Inspired by [14], my collaborators and me continue to modify the energy dissipation law by considering the combination of microscopic (atomic) energy law and macroscopic (hydrodynamic) energy law, which is reasonable since the microscopic charged particles and the macroscopic flow interact with each other. Precisely, we select the total energy

$$E^{total} = \underbrace{\int_{\mathbb{R}^3} \left(h(v,w) + \tfrac{\epsilon}{2}|\nabla\phi|^2\right) dx}_{\text{microscopic}} + \underbrace{\int_{\mathbb{R}^3} \left(\tfrac{\rho}{2}|u|^2 + \omega(\rho)\right) dx}_{\text{macroscopic}}, \tag{4}$$

and the dissipation

$$\begin{aligned} \triangle = &\int_{\mathbb{R}^3} \left(\tfrac{k_B T}{D_v} v\,|u_v - u|^2 + \tfrac{k_B T}{D_w} w\,|u_w - u|^2 + \tfrac{k_B T}{D} vw\,|u_v - u_w|^2\right) dx \\ &+ \int_{\mathbb{R}^3} \left(\mu|\nabla u|^2 + (\mu+\mu')|\mathrm{div}u|^2\right) dx. \end{aligned} \tag{5}$$

Then, by combing the EVA and (4)–(5) as in [33, 34], we can derive the following closed model: for $(x,t) \in \mathbb{R}^3 \times \mathbb{R}^+$,

$$\begin{cases} \rho_t + \mathrm{div}(\rho u) = 0, \\ (\rho u)_t + \mathrm{div}(\rho u \otimes u) + \nabla p(\rho) \\ \qquad = \mu\Delta u + (\mu+\mu')\nabla\mathrm{div}u - v\nabla h_v - w\nabla h_w + \epsilon\Delta\phi\nabla\phi, \\ v_t + \mathrm{div}(vu) = \mathrm{div}\left[\frac{v((1+v)(\nabla h_v + z_v e\nabla\phi)+w(\nabla h_w+z_w e\nabla\phi))}{1+v+w}\right], \\ w_t + \mathrm{div}(wu) = \mathrm{div}\left[\frac{w((1+w)(\nabla h_w + z_w e\nabla\phi)+v(\nabla h_v+z_v e\nabla\phi))}{1+v+w}\right], \\ -\epsilon\Delta\phi = z_v e v + z_w e w, \end{cases} \tag{6}$$

where we set $D_v = D_w = D = k_B = T = 1$ for brevity. Now, we can say that the system (6) is called a class of new generalized Poisson-Nernst-Planck-Navier-Stokes (PNP-NS) equations because of the flexibility of the function $h = h(v,w)$. It is usual to select $h = k_B T(v\ln v + w\ln w)$ as in (2) by neglecting the steric potential energy. However, the function h has many other forms as in [3, 8], such as $h = k_B T(v+w)^2$, etc. Such a form can take into account of the interactions between particles and reveal size effects arisen from the shape of hard sphere of ion species. Notice that the model (6) is a cross-diffusion type in most cases. The cross diffusion means that there is one term of form $f(v,w)\Delta w(f(v,w)\Delta v)$ appeared in the evolution equation of $v(w)$, as Eq. $(6)_3((6)_4)$. As we know, the cross diffusion will cause some major difficulties in solving problems since generally no maximum principle holds at this time. In addition, the cross-diffusion induced instability happens, see [16, 32]. In fact, the cross-diffusion has attracted lots of attention when people study some kinds of parabolic equations. For example, the well-known Shigesada-Kawasaki-Teramoto system, which was proposed to model the spatial segregation of two competing species in [31]. There is another important system, called (Patlak-)Keller-Segel (K-S) model which was proposed to model the chemotaxis in [17, 18, 26]. For the K-S system with cross-diffusion, we can refer to [13, 35]. However, the system (6) is a new one which has more complicated structure since it is a hyperbolic-parabolic-elliptic equations with the cross-diffusion under some proper assumptions on the function h.

Remark 1. The microscopic internal energy ω occurred in (4) and the pressure function p occurred in Eq. $(6)_2$ have the following constitutive relation:

$$p(\rho) = \omega'(\rho)\rho - \omega(\rho).$$

For example, by solving the above ODE directly, we obtain

$$\omega(\rho) = \begin{cases} \rho \ln \rho, & \text{if} \quad p(\rho) = \rho, \\ \rho^\gamma, & \text{if} \quad p(\rho) = (\gamma - 1)\rho^\gamma, \ \gamma > 1, \end{cases}$$

which corresponds to the isothermal and isentropic case for ideal gases, respectively.

This paper is organized as follows. In Section 2, we consider the dilute case for (6) by selecting a general internal energy $h(v, w) = \tilde{\varphi}(v) + \tilde{\psi}(w)$ and then study its global well-posedness for the Cauchy problem. In section 3, we consider the crowded case for (6) by selecting $h(v, w) = k_B T v \ln v + k_B T w \ln w$, where the cross-diffusion phenomenon is induced by the mutual friction between dense particles. Our results about its global well-posedness infer that the instability induced by cross-diffusion will not happen. Lastly, we give a conclusion in Section 4.

2. **The dilute case.** In this section, we consider such a case that some dilute charged particles transport under their self-consistent electrostatic field in compressible fluids. Then, in (4), we select the microscopic internal energy

$$h(v, w) = \tilde{\varphi}(v) + \tilde{\psi}(w),$$

where the functions $\tilde{\varphi}(\cdot)$ and $\tilde{\psi}(\cdot)$ are given according to the actual situations. At this time, we can neglect the mutual friction between charged particles since they are dilute. So, the dissipation is given as

$$\triangle = \int_{\mathbb{R}^3} \left(\tfrac{k_B T}{D_v} v \left| u_v - u \right|^2 + \tfrac{k_B T}{D_w} w \left| u_w - u \right|^2 + \mu |\nabla u|^2 + (\mu + \mu')|\mathrm{div} u|^2 \right) dx.$$

Thus, the model (6) is reduced to

$$\begin{cases} \rho_t + \mathrm{div}(\rho u) = 0, \\ (\rho u)_t + \mathrm{div}(\rho u \otimes u) + \nabla p(\rho) \\ \quad = \mu \Delta u + (\mu + \mu')\nabla \mathrm{div} u - \nabla \varphi(v) - \nabla \psi(w) + \epsilon \Delta \phi \nabla \phi, \\ v_t + \mathrm{div}(vu) = \mathrm{div}\left(\nabla \varphi(v) + z_v e v \nabla \phi \right), \\ w_t + \mathrm{div}(wu) = \mathrm{div}\left(\nabla \psi(w) + z_w e w \nabla \phi \right), \\ -\epsilon \Delta \phi = z_v e v + z_w e w, \end{cases} \tag{7}$$

where φ and ψ satisfy $\varphi(v) = \tilde{\varphi}'(v)v - \tilde{\varphi}(v)$ and $\psi(w) = \tilde{\psi}'(w)w - \tilde{\psi}(w)$, respectively. From this, we can say that φ, ψ are determined by $\tilde{\varphi}, \tilde{\psi}$. We supplement (7) with the initial data

$$(\rho, u, v, w)(x, t) \mid_{t=0} = (\rho_0, u_0, v_0, w_0)(x), \ x \in \mathbb{R}^3. \tag{8}$$

We assume that the pressure function $p(\rho)$ is a smooth one satisfying

$$p'(\rho) > 0 \text{ for } \rho > 0, \tag{9}$$

and the far-field behavior is selected as

$$\lim_{|x|\to+\infty}(\rho,u,v,w,\phi)(x,t)=(1,0,1,1,0). \tag{10}$$

Without loss of generality, we set

$$\mu=\epsilon=e=z_w=1,\ \mu'=0,\ z_v=-1. \tag{11}$$

We define the perturbations by

$$\varrho=\rho-1,\ u=u,\ V=v-1,\ W=w-1,\ \phi=\phi.$$

Then, the Cauchy problem (7)–(8) becomes

$$\begin{cases}\varrho_t+\mathrm{div}((\varrho+1)u)=0,\\ (\varrho+1)\left(u_t+u\cdot\nabla u\right)+p'(\varrho+1)\nabla\varrho-\Delta u-\nabla\mathrm{div}u\\ \quad=-\varphi'(V+1)\nabla V-\psi'(W+1)\nabla W+(V-W)\nabla\phi,\\ V_t+\mathrm{div}((V+1)u)-\Delta\varphi(V+1)=-\mathrm{div}\left((V+1)\nabla\phi\right),\\ W_t+\mathrm{div}((W+1)u)-\Delta\psi(W+1)=\mathrm{div}\left((W+1)\nabla\phi\right),\\ \phi=\Delta^{-1}(V-W),\\ (\varrho,u,V,W)\mid_{t=0}=(\varrho_0,u_0,V_0,W_0).\end{cases} \tag{12}$$

The following local and global well-posedness were proved in [34]:

Theorem 2.1. *Denote $U(x,t)=(\varrho,u,V,W)(x,t)$. Assume $U_0:=(\varrho_0,u_0,V_0,W_0)\in H^3$ and $\inf_{x\in\mathbb{R}^3}\varrho(x,0)>-1$. Then the Cauchy problem (12) admits a local unique solution satisfying $\varrho(t)\in\mathcal{C}^0(0,T;H^3)\cap\mathcal{C}^1(0,T;H^2)$, $(u,V,W)(t)\in\mathcal{C}^0(0,T;H^3)\cap\mathcal{C}^1(0,T;H^1)\cap\mathcal{L}_2(0,T;H^4)$ and $\varrho(t)>-1$ for any $t\in[0,T]$.*

Theorem 2.2. *Assume that $U_0\in H^k$ for an integer $k\geq 3$. If the initial H^3 norm $\|U_0\|_{H^3}$ is small, then the Cauchy problem (12) admits a unique global solution $U(t)$ for all $t\geq 0$. If further we assume that $\|\nabla^{-1}(V_0-W_0)\|_{L^2}$ is sufficiently small and $U_0\in L^p$ with $1\leq p\leq 2$, then for all $t\geq 0$,*

$$\|\nabla^\ell U(t)\|_{H^{k-\ell}}\leq C_0(1+t)^{-\frac{\ell}{2}-\frac{3}{2}(\frac{1}{p}-\frac{1}{2})}\ \ \text{for}\ \ 0\leq\ell\leq k-1 \tag{13}$$

and

$$\|\nabla^\ell\nabla\phi(t)\|_{L^2}\leq C_0(1+t)^{-\frac{\ell+1}{2}-\frac{3}{2}(\frac{1}{p}-\frac{1}{2})}\ \ \text{for}\ \ 0\leq\ell\leq k-2. \tag{14}$$

Remark 2. The assumption $\nabla^{-1}(V_0-W_0)\in L^2$ is equivalent to the initial electric field $\nabla\phi_0\in L^2$ since $\|\nabla^{-1}(V_0-W_0)\|_{L^2}=\|\nabla^{-1}\Delta\phi_0\|_{L^2}=\|\nabla\phi_0\|_{L^2}$.

Remark 3. We claim that the smallness of $\|\nabla^{-1}(V_0-W_0)\|_{L^2}$ in Theorem 2.2 can be removed by only assuming $\nabla^{-1}(V_0-W_0)\in L^2$. We can still prove that (13) hold for $\ell=0,1$. But, it is hard to obtain the higher-order decay in (13)–(14).

Remark 4. We say that all the decay rates (13) with $p=1$ are optimal in sense that it is consistent with the decay of the heat kernel.

3. **The crowded case.** In this section, we consider such a case that the crowded charged particles transport under their self-consistent electrostatic field in compressible fluids. The main purpose of this model is to characterize that the mutual friction will give rise to the cross-diffusion phenomenon, so, in (4), we only choose the simple microscopic internal energy

$$h(v,w)=k_BTv\ln v+k_BTw\ln w.$$

Thus, the model (6) becomes

$$\begin{cases} \rho_t + \operatorname{div}(\rho u) = 0, \\ (\rho u)_t + \operatorname{div}(\rho u \otimes u) + \nabla p(\rho) \\ \qquad = \mu \Delta u + (\mu + \mu')\nabla \operatorname{div} u - \nabla v - \nabla w + \epsilon \Delta \phi \nabla \phi, \\ v_t + \operatorname{div}(vu) = \operatorname{div}\left[\frac{(1+v)(\nabla v + z_v e v \nabla \phi) + v(\nabla w + z_w e w \nabla \phi)}{1+v+w}\right], \\ w_t + \operatorname{div}(wu) = \operatorname{div}\left[\frac{(1+w)(\nabla w + z_w e w \nabla \phi) + w(\nabla v + z_v e v \nabla \phi)}{1+v+w}\right], \\ -\epsilon \Delta \phi = z_v e v + z_w e w. \end{cases} \tag{15}$$

We supplement (15) with the initial data

$$(\rho, u, v, w)(x,t)\mid_{t=0} = (\rho_0, u_0, v_0, w_0)(x), \ x \in \mathbb{R}^3, \tag{16}$$

and assume that (9)–(11) hold. We define the perturbations by

$$\varrho = \rho - 1, \ u = u, \ V = v - 1, \ W = w - 1, \ \phi = \phi.$$

Then, the Cauchy problem (15)–(16) becomes

$$\begin{cases} \varrho_t + \operatorname{div}((\varrho + 1)u) = 0, \\ (\varrho + 1)(u_t + u \cdot \nabla u) + p'(\varrho + 1)\nabla \varrho - \Delta u - \nabla \operatorname{div} u \\ \qquad = -\nabla V - \nabla W + (V - W)\nabla \phi, \\ V_t + \operatorname{div}((V+1)u) = \operatorname{div}\left[\frac{(V+2)\nabla V + (V+1)\nabla W - (V+1)(V-W+1)\nabla \phi}{V+W+3}\right], \\ W_t + \operatorname{div}((W+1)u) = \operatorname{div}\left[\frac{(W+2)\nabla W + (W+1)\nabla V + (W+1)(W-V+1)\nabla \phi}{V+W+3}\right], \\ \phi = \Delta^{-1}(V - W), \\ (\varrho, u, V, W)\mid_{t=0} = (\varrho_0, u_0, V_0, W_0). \end{cases} \tag{17}$$

Set $\widetilde{U} = (V, W)^T$. The perturbed parabolic equations $(17)_3$–$(17)_4$ with the cross-diffusion have the form

$$\widetilde{U}_t - \operatorname{div}(A(\widetilde{U})\nabla \widetilde{U}) = \mathcal{N}(\widetilde{U}).$$

Here the diffusion matrix $A(\widetilde{U})$ is given as

$$A(\widetilde{U}) := \begin{pmatrix} A_{11} & A_{12} \\ A_{21} & A_{22} \end{pmatrix},$$

where

$$A_{11} := \frac{V+2}{V+W+3}, \ A_{12} := \frac{V+1}{V+W+3}, \ A_{21} := \frac{W+1}{V+W+3}, \ A_{22} := \frac{W+2}{V+W+3}.$$

We claim that the matrix $A(\widetilde{U})$ is diagonally dominant in the linearized sense, i.e., the linear part of $A(\widetilde{U})$, which is equal to

$$\mathcal{L}(A) = \begin{pmatrix} \frac{2}{3} & \frac{1}{3} \\ \frac{1}{3} & \frac{2}{3} \end{pmatrix},$$

is diagonally dominant due to $2/3 > 1/3$. This is a key point to derive the energy estimates as in [33]. In fact, we develop a new approximation scheme to prove the local solution, where the difficulties caused by the cross-diffusion could be overcome. Moreover, since the matrix $A(\widetilde{U})$ is diagonally dominant in the above sense, we manage to derive the refined a priori estimates which insure that the local solution can be extended to the global one by using a continuous argument.

The following theorem was proved in [33]:

Theorem 3.1. *Under the same assumptions of Theorem 2.2, the Cauchy problem (17) admits a unique global solution $U(t)$ satisfying for $1 \le p \le 2$ and all $t \ge 0$,*

$$\|\nabla^{\ell}(\varrho, u, V, W)(t)\|_{H^{k-\ell}} \le C_0(1+t)^{-\frac{\ell}{2}-\frac{3}{2}(\frac{1}{p}-\frac{1}{2})} \quad \text{for} \quad 0 \le \ell \le k-1 \tag{18}$$

and

$$\|\nabla^{\ell}\nabla\phi(t)\|_{L^2} \le C_0(1+t)^{-\frac{\ell+1}{2}-\frac{3}{2}(\frac{1}{p}-\frac{1}{2})} \quad \text{for} \quad 0 \le \ell \le k-2. \tag{19}$$

Remark 5. Our results from the local solution (see Proposition 3.12 in [33]) to the decay rates of solutions (18)–(19) in Theorem 3.1 can be applied to a class of system with the cross-diffusion whose diffusion matrix is diagonally dominant in the linearized sense. For example, it is applicable for more general PNP-NS system:

$$\begin{cases} \rho_t + \mathrm{div}(\rho u) = 0, \\ (\rho u)_t + \mathrm{div}(\rho u \otimes u) + \nabla p(\rho) \\ \qquad = \mu\Delta u + (\mu+\mu')\nabla \mathrm{div} u - \nabla\varphi(v) - \nabla\psi(w) + \epsilon\Delta\phi\nabla\phi, \\ v_t + \mathrm{div}(vu) = \mathrm{div}\left[\frac{(1+v)(\nabla\varphi(v)+z_v ev\nabla\phi)+v(\nabla\psi(w)+z_w ew\nabla\phi)}{1+v+w}\right], \\ w_t + \mathrm{div}(wu) = \mathrm{div}\left[\frac{(1+w)(\nabla\psi(w)+z_w ew\nabla\phi)+w(\nabla\varphi(v)+z_v ev\nabla\phi)}{1+v+w}\right], \\ -\epsilon\Delta\phi = z_v ev + z_w ew, \end{cases}$$

where the functions $\varphi(v), \psi(w)$ could be selected based on needs.

4. **Conclusion.** In this paper, we use the energetic variational approach to derive a class of new generalized Poisson-Nernst-Planck-Navier-Stokes equations. Such a model can be used to characterize the interactions between microscopic charged particles and macroscopic fluids and to describe many important phenomena in the transport of charged particles, such as, cross-diffusion, size effects, etc. In Sections 2 and 3, we consider the stability of a constant equilibrium solution for the dilute and crowded cases, respectively. Under the condition that the initial perturbations are small, we show that the perturbed solution tends to the constant equilibrium solution with the optimal algebraic decay rates in time. In particular, the cross-diffusion phenomenon happens for the crowded case, however, our results in Section 3 reveal that the instability induced by the cross-diffusion will not appear here.

REFERENCES

[1] R. Abraham and J. E. Marsden, *Foundations of Mechanics*, Benjamin/Cummings Publishing Co., Inc., Advanced Book Program, Reading, Mass., 1978.

[2] V. Barcilon, D. P. Chen, and R. S. Eisenberg, Ion flow through narrow membrane channels. II, *SIAM J. Appl. Math.*, **52** (1992), 1405–1425.

[3] M. Burger, M. Di Francesco, J. F. Pietschmann, and B. Schlake, Nonlinear cross-diffusion with size exclusion, *SIAM J. Math. Anal.*, **42** (2010), 2842–2871.

[4] D. P. Chen and R. S. Eisenberg, Charges, currents, and potentials in ionic channels of one conformation, *Biophys. J.*, **64** (1993), 1405–1421.

[5] B. Eisenberg, *Crowded Charges in Ion Channels*, John Wiley and Sons, 2011.

[6] B. Eisenberg, Mass action in ionic solutions, *Chem. Phys. Lett.*, **511** (2011), 1–6.

[7] B. Eisenberg, A leading role for mathematics in the study of ionic solutions, *SIAM News*, **45** (2012), 11–12.

[8] B. Eisenberg, Y. Hyon, and C. Liu, Energy variational analysis of ions in water and channels: Field theory for primitive models of complex ionic fluids, *J. Chem. Phys.*, **133** (2010), 104104.

[9] R. S. Eisenberg, Computing the field in proteins and channels, *Journal of Membrane Biology*, **150** (1996), 1–25.

[10] R. S. Eisenberg, *New Developments and Theoretical Studies of Proteins*, World Scientific, Philadelphia, 1996.

[11] J. Forster, Mathematical Modeling of Complex Fluids, Master thesis, *The University of Würzburg*, 2013.
[12] H. Goldstein, *Classical Mechanics*, 2nd edition, Addison-Wesley Publishing, 1980.
[13] S. Hittmeir and A. Jüngel, Cross diffusion preventing blow-up in the two-dimensional Keller-Segel model, *SIAM J. Math. Anal.*, **43** (2011), 997–1022.
[14] C. Y. Hsieh, Y. Hyon, H. Lee, T. C. Lin, and C. Liu, Transport of charged particles: entropy production and maximum dissipation principle, *J. Math. Anal. Appl.*, **422** (2015), 309–336.
[15] Y. Hyon, D. Y. Kwak, and C. Liu, Energetic variational approach in complex fluids: maximum dissipation principle, *Discrete Contin. Dyn. Syst.*, **26** (2010), 1291–1304.
[16] M. Iida, M. Mimura, and H. Ninomiya, Diffusion, cross-diffusion and competitive interaction, *J. Math. Biol.*, **53** (2006), 617–641.
[17] E. F. Keller and L. A. Segel, Initiation of slime mold aggregation viewed as an instability, *J. Theoret. Biol.*, **26** (1970), 399–415.
[18] E. F. Keller and L. A. Segel, Model for chemotaxis, *J. Theoret. Biol.*, **30** (1971), 225–234.
[19] B. J. Kirby, *Micro- and Nanoscale Fluid Mechanics: Transport in Microfluidic Devices*, Cambridge University Press, 2010.
[20] C. Liu, An introduction of elastic complex fluids: an energetic variational approach, In *Multi-scale phenomena in complex fluids*, World Sci. Publishing, (2009), 286–337.
[21] P. A. Markowich, C. A. Ringhofer, and C. Schmeiser, *Semiconductor Equations*, Springer-Verlag, Vienna, 1990.
[22] W. Nernst, Die Electromotorische Wirksamkeit der Ionen, *Z. Physik. Chem.*, **4** (1889), 129–181.
[23] L. Onsager, Reciprocal relations in irreversible processes I, *Physical Rev.*, **37** (1931), 405–426.
[24] L. Onsager, Reciprocal relations in irreversible processes II, *Physical Rev.*, **38** (1931), 2265–2279.
[25] L. Onsager and S. Machlup, Fluctuations and irreversible processes, *Physical Rev.*, **91** (1953), 1505–1512.
[26] C. S. Patlak, Random walk with persistence and external bias, *Bull. Math. Biophys.*, **15** (1953), 311–338.
[27] M. Planck, Uber die Erregung von Electricität und Wärme in Electrolyten, *Ann. Phys. Chem.*, **39** (1890), 161–186.
[28] W. Van Roosbroeck, Theory of the flow of electrons and holes in Germanium and other semiconductors, *Bell System Technical Journal*, **29** (1950), 560–607.
[29] I. Rubinstein, *Electro-diffusion of Ions*, *SIAM Studies in Applied Mathematics*, Philadelphia, 1990.
[30] S. Selberherr, *Analysis and Simulation of Semiconductor Devices*, Springer-Verlag, New York, 1984.
[31] N. Shigesada, K. Kawasaki, and E. Teramoto, Spatial segregation of interacting species, *J. Theoret. Biol.*, **79** (1979), 83–99.
[32] C. R. Tian, Z. G. Lin, and M. Pedersen, Instability induced by cross-diffusion in reaction-diffusion systems, *Nonlinear Anal. Real World Appl.*, **11** (2010), 1036–1045.
[33] Y. Wang, C. Liu, and Z. Tan, A generalized Poisson-Nernst-Planck-Navier-Stokes model on the fluid with the crowded charged particles: derivation and its well-posedness, *SIAM J. Math. Anal.*, **48** (2016), 3191 3235.
[34] Y. Wang, C. Liu, and Z. Tan, Well-posedness on a new hydrodynamic model of the fluid with the dilute charged particles, *J. Differential Equations*, **262** (2017), 68–115.
[35] T. Xiang, On a class of Keller-Segel chemotaxis systems with cross-diffusion, *J. Differential Equations*, **259** (2015), 4273–4326.

E-mail address: wangyongxmu@163.com

A POSTERIORI ANALYSIS FOR THE NAVIER-STOKES-KORTEWEG MODEL

JAN GIESSELMANN AND DIMITRIOS ZACHARENAKIS*

Numerical Analysis and Scientific Computing,
TU Darmstadt,
Dolivostraße 15, Darmstadt, 64293, Germany

ABSTRACT. We study the compressible isothermal Navier-Stokes-Korteweg system governing viscous liquid-vapor fluid flows in one dimension, where the pressure of the fluid is non-monotone. Firstly, we develop a local discontinuous Galerkin scheme and provide an a posteriori analysis for the approximation of the model. We apply weak entropic-strong stability theorems to compare weak-entropic with strong solutions, introducing sufficiently regular intermediate functions called reconstructions. In addition, employing a variant of the relative entropy technique we overcome the obstacle of the non-convexity of the energy. Moreover, we obtain an a posteriori estimator for the difference between the reconstruction and the numerical solution and thus we can compare exact with numerical solutions. What is more, we present numerical results for the efficiency of the error estimator.

1. **Introduction.** In the present contribution we study compressible two phase flows, meaning flows of one substance which is present as both liquid and vapor. From a mathematical point of view sharp and diffuse interface (phase-field) models are usually employed; see Fig. 1. On the former, fields such as density and velocity are discontinuous across the interface $\Gamma(t)$, which separates the liquid from the vapor phase. Accordingly, jump conditions at $\Gamma(t)$ need to be considered. However, this description becomes difficult when droplet formation, coalescence or other topological singularities take place. Thus, alternative models are examined, see [1, 2, 3] and references therein. They are called diffuse interface models, where the interface is represented by a narrow layer of width $O(\epsilon)$. Also, in this approach the fields vary smoothly but with large gradients. We espouse a mathematical structure where the dynamics of the flow is governed by the energy functional, which includes capillary and viscous forces. Modeling these flows is an area of active research, though from a numerical point of view only a few works are available when the pressure is non-monotone [6]. Then, the first order part of the equations of interest is of mixed hyperbolic-elliptic type.

2000 *Mathematics Subject Classification.* Primary: 58F15, 58F17; Secondary: 53C35.

Key words and phrases. Compressible two-phase flow, Liquid-vapor flow, Diffuse interface model, Relative entropy, Discontinuous Galerkin method, A posteriori error analysis.

* Corresponding author: Dimitrios Zacharenakis.

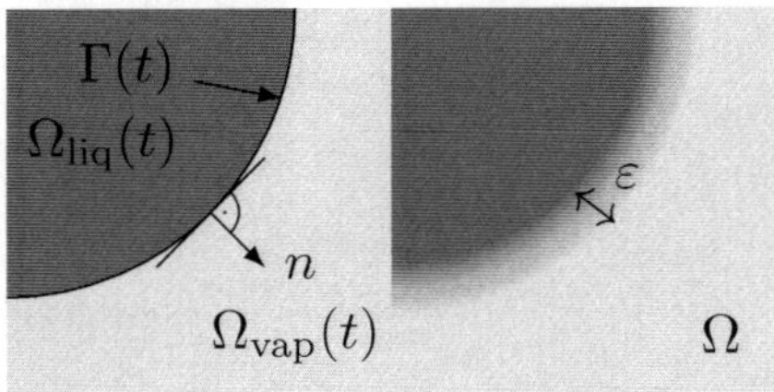

Figure 1. Sharp and diffuse interface

At the same time, processes of practical interest such as combustion, cryogenics, and cloud formation motivate us to develop tools for the simulation of such small scale models. In this case, a posteriori analysis is crucial in determining how reliable a numerical scheme is and it allows for error control of the exact and the numerical solution. Then, based on the a posteriori error estimator mesh adaptive refinement techniques can be applied. In practice, exact solutions (global in time) may not be smooth and we need to define different classes of them, such as entropy solutions. As mentioned before, the system at hand is not hyperbolic but it can be seen as a perturbation of its first order part, which is hyperbolic except close to the phase boundary. This motivates the use of "hyperbolic" methods for studying Navier-Stokes-Korteweg (NSK). One well known stability framework for systems of hyperbolic conservation laws is the relative entropy [4, 7]. For a general overview of its development in the last decades we refer the reader to the references in [5, Sect. 5.7]. It is based on the fact that systems of hyperbolic conservation laws are usually endowed with a convex entropy/entropy flux pair, that can be used to define the notion of relative entropy between two solutions. However, since we do not have a hyperbolic system, in our analysis we employ a variant of the relative entropy technique [14], which considers only the convex contributions of the energy and in particular was applied for the Euler-Korteweg system [15, 16].

Furthermore, we introduce a reconstruction approach, see Fig. 2, where the defined functions have a certain amount of regularity. Broadly speaking, this means that reconstructions are strong solutions of a perturbed system. As soon as this holds, we can apply the reduced relative entropy and bound the difference between the reconstructions and the exact solutions of our system. To illustrate, for the a posteriori estimates we split the difference between the exact u and the numerical solutions u_h into two parts; see Fig. 3. Additionally, we use a posteriori results to compare the numerical solutions with the reconstructions $\hat{u}$. The former are computed through a discontinuous Galerkin (DG) method that approximates the solution using discontinuous piecewise polynomial functions. The latter are based on an elliptic reconstruction operator, which was studied in [19, 20]. The elliptic reconstructions of our numerical solutions are the exact solutions of elliptic problems. Consequently, we can achieve our goal to explicitly bound the difference between the numerical and the exact solutions and test the efficiency of the estimator numerically.

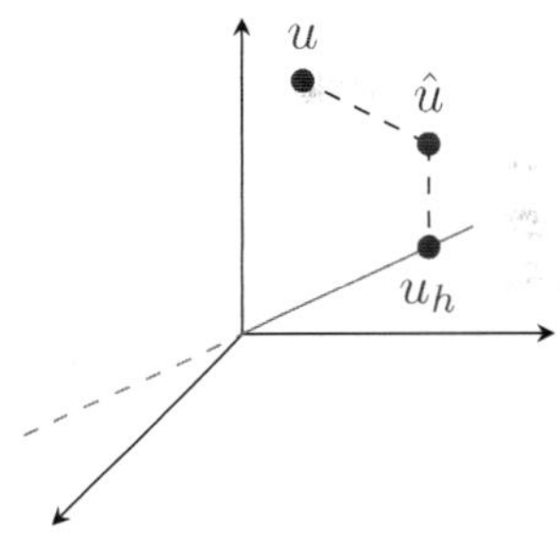

Figure 2. Reconstruction of the discrete solution

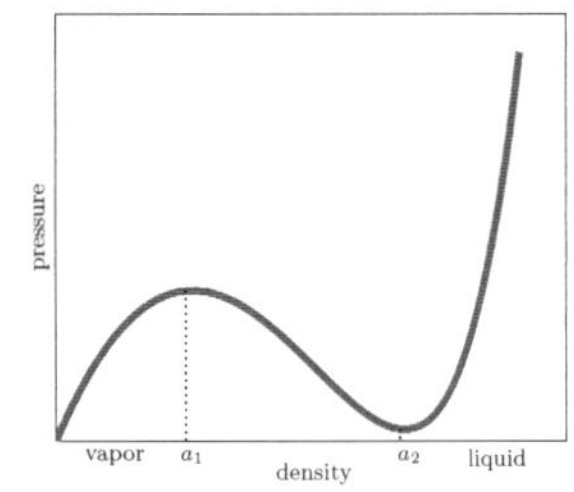

Figure 4. Pressure function.

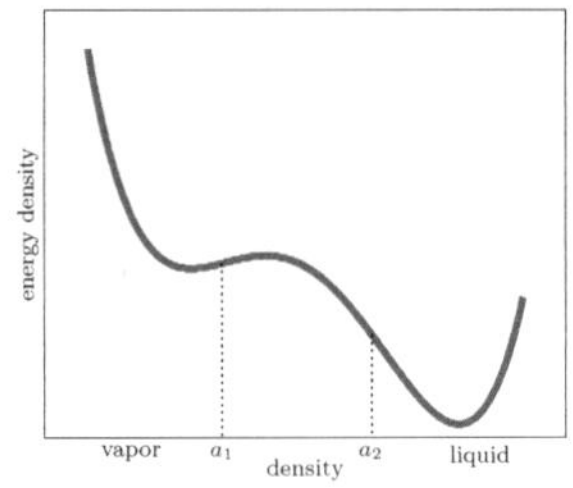

Figure 5. Helmholtz free energy density.

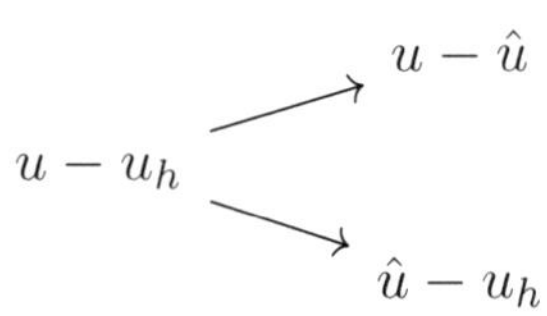

Figure 3. Splitting the estimate

The paper is organized as follows: In Section 2 we present the PDE system under consideration and the notion of entropy solution. In Section 3 we introduce the reduced relative entropy technique which is used to prove a stability result in Theorem 3.1. In Section 4 we present the numerical scheme and show the setup of the a posteriori analysis. In Subsection 4.2 we derive an a posteriori estimate for solutions using a combination of reconstructions and finally in Section 5 we perform numerical experiments to test the efficiency of the corresponding estimator.

2. **Two phase flow model.** We examine the following system in one dimension (for simplicity)

$$\begin{aligned} \rho_t + (\rho v)_x &= 0 \\ (\rho v)_t + \left(\rho v^2 + p(\rho)\right)_x &= \mu v_{xx} + \gamma \rho \rho_{xxx} \end{aligned} \quad \text{in } S^1 \times [0, T), \tag{NSK}$$

with the mass density $\rho = \rho(x,t) > 0$ and the velocity field $v = v(x,t) \in \mathbb{R}$ being the unknown quantities and S^1 denotes the unit interval with periodic boundary conditions. The viscosity coefficient and a final time are parameters denoted as $\mu > 0$ and $T > 0$ respectively. Van der Waals, Korteweg, Cahn, Hilliard and others contributed to the study of capillarity, which has an important role in the interface between the two phases and effects of it were incorporated by including density gradients in the energy; see [9]. Following this approach, we obtain the term $\gamma \rho \rho_{xxx}$ in (NSK), where $\gamma > 0$ is a capillarity constant. Furthermore, the pressure is denoted by p. It is expressed via the Gibbs-Duhem equation $p(\rho) = \rho W'(\rho) - W(\rho)$, which gives $p'(\rho) = \rho W''(\rho)$. Here, $W(\rho)$ is called Helmholtz free energy density and has a double-well shape, in order to describe two phase flows; see Fig. 4. This means that it is non-convex, which makes the pressure a non-monotone function. We make the following assumption on W: $\exists\, a_1, a_2 \in (0, \infty)$ such that $a_1 < a_2$ and $W'' > 0$ in $(0, a_1) \bigcup (a_2, \infty)$ and $W'' < 0$ in (a_1, a_2). We say a fluid is in vapor phase if $\rho \in (0, a_1]$, in liquid phase if $\rho \in [a_2, \infty)$ and in spinodal phase if $\rho \in (a_1, a_2)$; see Fig. 5.

Lemma 2.1. (Energy Balance [12]) *Let $(\rho, \rho v)$ be a smooth solution of (NSK). Then it satisfies the energy balance*

$$\frac{d}{dt}\int_{S^1}\mathcal{E}[\rho,\rho v] := \frac{d}{dt}\left(\int_{S^1} W(\rho)+\frac{1}{2}\rho v^2+\frac{\gamma}{2}(\rho_x)^2 dx + \int_{S^1}\int_0^t \mu v_x^2\, ds\, dx\right) = 0.$$

Remark 1. (L_∞ bound for ρ) Since Lemma 2.1 holds and the mean value of ρ does not change in time, we notice that $\|\rho\|_{L_\infty(S^1\times(0,T))}$ is bounded in terms of the initial data and we call this bound C_ρ.

Remark 2 (Conservation of momentum)**.** We can also write (NSK) in the form:

$$\begin{aligned}\rho_t + (\rho v)_x &= 0\\ (\rho v)_t + \left(\rho v^2\right)_x &= (\mu v_x + \mathbb{S})_x\,,\end{aligned}$$

where $\mathbb{S} := -\rho W'(\rho) + W(\rho) + \gamma\rho\rho_{xx} - \frac{1}{2}\gamma\rho_x^2$.

Definition 2.2. (Entropy solution) A function $u = (\rho, \rho v)$ with $\rho \in C([0,T); L^1(S^1))$, $\rho v \in C([0,T); L^1(S^1,\mathbb{R}))$ with initial datum $u^0 = (\rho^0, (\rho v)^0)$ is an entropy solution of (NSK) if $\rho v^2,\ v_x,\ \mathbb{S} \in L^1\left((0,T)\times S^1\right)$ with:

$$\int_0^T\int_{S^1} \rho\psi_t + (\rho v)\psi_x\, dxdt + \int_{S^1}\rho^0\psi(\cdot,0)\, dx = 0,\ \forall\psi\in C_c^\infty([0,T);C^1(S^1)),$$

$$\int_0^T\int_{S^1}(\rho v)\phi_t + \left(\rho v^2 - \mathbb{S} - \mu v_x\right)\phi_x\, dxdt + \int_{S^1}(\rho v)^0\phi(\cdot,0)\, dx = 0,$$
$$\forall\phi\in C_c^\infty([0,T);C^1(S^1)),$$

and

$$\int_0^T\int_{S^1}\mathcal{E}[u]\vartheta_t\, dxdt + \int_{S^1}\mathcal{E}[u^0]\vartheta(0)\, dx \geq 0,$$
$$\forall\vartheta\in C_c^\infty([0,T),[0,\infty)),\ \text{with}\ \int_{S^1}\mathcal{E}[u^0]\, dx < \infty$$

Definition 2.3. We consider strong solutions $(\hat\rho, \hat\rho\hat v)$ in the following spaces; see [15]:

$$\begin{aligned}\hat\rho &\in C^0\left([0,T), H^3(S^1,\mathbb{R}_+)\right)\bigcap C^1\left([0,T), H^1(S^1,\mathbb{R}_+)\right),\\ \hat v &\in C^0\left([0,T), H^2(S^1,\mathbb{R})\right)\bigcap C^1\left([0,T), H^0(S^1,\mathbb{R})\right),\end{aligned}$$

where $\mathbb{R}_+ := \{x\ \in\mathbb{R} : x > 0\}$.

3. **Weak entropic-strong stability for non-convex energies.**

The relative entropy, defined in [11], is not suitable to measure the distance between solutions, since our energy has non-convex parts. Therefore, we restrict to energies, which can be written as $W(\rho) = \omega(\rho) + e(\rho)$, where ω is strictly convex and $e \in C_c^\infty(0,\infty)$ is smooth, compactly supported and contains the non-convexity. Then, the reduced relative entropy considers only the convex parts and is defined as:

$$\begin{aligned}\eta_R\left[(\rho\ \rho v)^T\middle|(\hat\rho\ \hat\rho\hat v)^T\right] := \int_{S^1}&\omega(\rho)-\omega(\hat\rho)-\omega'(\hat\rho)(\rho-\hat\rho)+\frac{1}{2}\rho(v-\hat v)^2\\ &+\gamma\left((\rho-\hat\rho)_x\right)^2 dx + \int_{S^1}\int_0^t\mu(v-\hat v)_x^2\, ds\, dx.\end{aligned}\tag{1}$$

Theorem 3.1. (Stability) *Let $(\rho, \rho v)$ be an entropy solution of (NSK) and let $(\hat{\rho}, \hat{\rho}\hat{v})$ be a strong solution to the perturbed system*

$$\hat{\rho}_t + (\hat{\rho}\hat{v})_x = 0$$
$$(\hat{\rho}\hat{v})_t + \hat{\rho}\left(W'(\hat{\rho})\right)_x + \left(\hat{\rho}\hat{v}^2\right)_x - \mu \hat{v}_{xx} - \gamma \hat{\rho}\hat{\rho}_{xxx} = R,$$

where $R \in L^2(S^1 \times (0,T))$ is some given function. Let there be a positive lower bound for $\hat{\rho}$ denoted by A. Then, the reduced relative entropy defined in (1) satisfies:

$$\eta_R\big|_t \le \left(\eta_R\big|_0 + \|R\|^2_{L^2(S^1\times(0,t))}\right) e^{\left(C_\rho A^{-2}+C\right)t}, \ \forall t \in [0,T).$$

Proof. The proof will be given in [17]. □

4. **Discretization and A Posteriori Setup.** In this section, we follow the usual DG formulation [8] for the discretization of (NSK). Firstly, we reformulate our system introducing an auxiliary variable. Secondly, to approximate the solution we define the finite element space $\mathbb{V}_q := \left\{\Phi : S^1 \to \mathbb{R} : \Phi|_{K_i} \in \mathbb{P}^q\left(K_i\right)\right\}$, which consists of discontinuous piecewise polynomials of degree less or equal than $q \in \mathbb{N}$, where we choose some $x_0 < \ldots < x_N$ and denote by $K_i = [x_i, x_{i+1}]$ the ith subinterval.

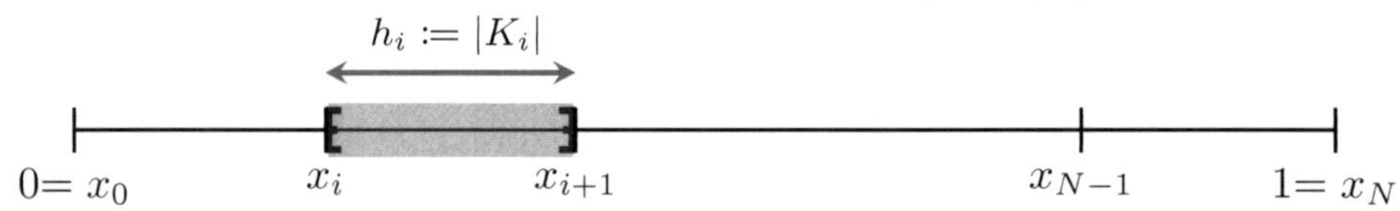

The set of all cells $\mathbb{J} = \{K_i\}_{i=0}^{N-1}$, the set of all common interfaces of $\mathbb{J}$, namely $\mathbb{E}$, are all introduced accordingly. For $x_n \in \mathbb{E}$ we define $h_{\mathbb{E}}^-(x_n) := h_{n-1}$, $h_{\mathbb{E}}^+(x_n) := h_n$, $h_{\mathbb{E}}(x_n) := \frac{1}{2}\left(h_{n-1} + h_n\right)$ such that $h_{\mathbb{E}}, h_{\mathbb{E}}^-, h_{\mathbb{E}}^+ \in L_\infty(\mathbb{E})$. In addition, we introduce the broken Sobolev space

$$H^1(\mathbb{J}) = \left\{\phi : \phi|_K \in H^1(K), \forall K \in \mathbb{J}\right\},$$

and define the jump operator for arbitrary scalar functions $v \in H^1(\mathbb{J})$ as

$$[\![v]\!] := \left(v^- - v^+\right) := \lim_{t \searrow 0} v(\cdot - t) - \lim_{t \searrow 0} v(\cdot + t).$$

Definition 4.1. (Discrete Seminorm) We introduce the broken $H^1(\mathbb{J})$-seminorm as

$$|\rho_h|^2_{dG} := \sum_{K\in\mathbb{J}} \|(\rho_h)_x\|^2_{L^2(K)} + \left\|\sqrt{h_{\mathbb{E}}^{-1}}[\![\rho_h]\!]\right\|^2_{L^2(\mathbb{E})}.$$

Definition 4.2. (Discrete Gradient Operators) We define $G_q^{\pm} : H^1(\mathbb{J}) \to \mathbb{V}_q$, such that for any $\psi \in H^1(\mathbb{J})$ we have:

$$\int_{S^1} G_q^{\pm}[\psi]\Phi \, dx = \sum_{K\in\mathbb{J}} \int_K (\psi_x)\Phi \, dx - \int_{\mathbb{E}} [\![\psi]\!]\Phi^{\pm}, \ \forall \Phi \in \mathbb{V}_q.$$

4.1. **Semi-Discrete Scheme.** Following the DG formulation as in [13], we test the equations with functions in $\mathbb{V}_q$ and employ the discrete gradient operators to account for the numerical fluxes. Then, choosing an upwind type flux we arrive at

$$\sum_{K\in\mathbb{J}} \int_K (\rho_h)_t \Phi = -\int_{S^1} G_q^-\left[\rho_h v_h\right]\Phi, \ \forall \Phi \in \mathbb{V}_q.$$

Note that the right hand side is the projection of G_{2q}^- to $\mathbb{V}_q$, i.e. $P_q\left[G_{2q}^-\left[\rho_h v_h\right]\right] = G_q^-\left[\rho_h v_h\right]$ where $P_q : H^1(\mathbb{J}) \to \mathbb{V}_q$ denotes the orthogonal projection; see [16].

Thus, we examine the following semi-discrete numerical scheme where, assuming $\rho_h > 0$, we seek $(\rho_h, v_h, \tau_h) \in C^1(0,T;\mathbb{V}_q) \times C^1(0,T;\mathbb{V}_q) \times C^0(0,T;\mathbb{V}_q)$ such that

$$\begin{aligned}(\rho_h)_t + G_q^-\left[\rho_h v_h\right] &= 0,\\ (v_h)_t + G_q^+\left[\tau_h\right] - P_q\Big[\mu \frac{1}{\rho_h} G_q^+\left[G_q^-\left[v_h\right]\right]\Big] &= 0,\\ \tau_h + P_q\Big[-W'(\rho_h) - \frac{1}{2}v_h^2\Big] + \gamma G_q^-\left[G_q^+\left[\rho_h\right]\right] &= 0.\end{aligned} \tag{2}$$

4.2. **A posteriori Setup.** A key motivator of this work is the error control that a posteriori estimates provide. However, the notion of weak-strong stability proposes to compare a weak-entropic with a strong solution. While the exact solutions are considered to be weak-entropic solutions, the numerical solutions can not be strong solutions since we employ a DG formulation. At the same time, exact solutions are not smooth in general. Thus, we split the difference between the exact and the numerical solutions. We follow the reconstruction approach [20] to bound the difference between the reconstructions and the exact solution of the PDE, where obviously the reconstructions need to have a certain amount of regularity in the sense of Definition 2.3. Moreover, we combine two reconstruction approaches to obtain the desired regularity. While the discrete reconstruction is computable, the elliptic one is not. Therefore, the "whole" reconstruction is not computable and we need to make use of other (elliptic) a posteriori results, in order to estimate the difference between the numerical solutions and the reconstructions and to bound the residual in a computable way.

Definition 4.3. (Discrete Reconstruction Operators) We define $D_q^{\pm} : H^1(\mathbb{J}) \to \mathbb{V}_{q+1}$ by requiring the following equalities for all $\Psi \in H^1(\mathbb{J})$:

$$\int_{S^1} \left(D_q^{\pm}[\Psi]\right)_x \Phi - G_q^{\pm}[\Psi]\Phi \, dx = 0, \ \forall \Phi \in \mathbb{V}_q \text{ and } \left(D_q^{\pm}[\Psi]\right)^{\pm} = \Psi^{\mp}, \text{ on } \mathbb{E}.$$

Definition 4.4. (Elliptic Reconstruction Operators) We define reconstructions $R[\rho_h] \in H^3(S^1)$, $R[\rho_h v_h] \in H^2(S^1)$ as solutions of the following elliptic equations respectively

$$\begin{aligned}\gamma\left(R[\rho_h]\right)_{xx} &= D_\alpha^-[W'(\rho_h)] - D_q^+[\tau_h] + \frac{1}{2}D_{2q}^-[v_h^2],\\ \left(R[\rho_h v_h]\right)_{xx} &= -\left(R[\rho_h]\right)_{xt},\end{aligned}$$

where in each case we impose the following normalization condition

$$\int_{S^1} R[\rho_h] - \rho_h \, dx = \int_{S^1} R[\rho_h v_h] - \rho_h v_h \, dx = 0.$$

Here, we choose $\alpha = 3q$ for experiments. However, how to choose α in general will be discussed in [17].

Remark 3. The above reconstruction operators are well defined, since we have unique solvability of the elliptic problems. Moreover, due to elliptic regularity there exist computable upper bounds for the reconstructions.

Remark 4. What is more, to compare the difference between the reconstruction $R[\rho_h]$ and the numerical solution ρ_h, we obtain, in a similar way as [18], an a posteriori error estimator depending only upon ρ_h and the problem data which has the form:

$$H_1\left[\rho_h, f\right]^2 = \sum_{K\in\mathbb{J}} h_K^2 \left\|f - (\rho_h)_{xx}\right\|^2_{L^2(K)} + \sum_{\epsilon\in\mathbb{E}} h_\epsilon \left\|[\![(\rho_h)_x]\!]\right\|^2_{L^2(\epsilon)} + \sum_{\epsilon\in\mathbb{E}} h_\epsilon^{-1} \left\|[\![\rho_h]\!]\right\|^2_{L^2(\epsilon)},$$

with $f := -\frac{1}{\gamma}\left(\tau_h - W'(\rho_h) - \frac{1}{2}v_h^2\right)$.

Lemma 4.5. *(Reconstructed PDE System) The reconstructions defined in Definition 4.4 satisfy a perturbed version of (NSK)*

$$\begin{aligned} 0 &= \left(R[\rho_h]\right)_t + \left(R[\rho_h v_h]\right)_x, \\ E &= \left(R[\rho_h v_h]\right)_t + R[\rho_h]\left(W'(R[\rho_h])\right)_x + \left(\frac{R[\rho_h v_h]^2}{R[\rho_h]}\right)_x - \mu\left(\frac{R[\rho_h v_h]}{R[\rho_h]}\right)_{xx} \\ &\quad - \gamma R[\rho_h]\left(R[\rho_h]\right)_{xxx}, \end{aligned}$$

where $E := A + B$.

Remark 5. While in [16] we studied the Euler-Korteweg system with $E := A$, now B term contains the viscosity. We show numerically that there is a computable upper bound that goes to zero when the meshsize h goes to zero; see [17].

Corollary 1. *(Reduced Relative Entropy Bound) Let* $(\rho, \rho v)$ *be an entropy solution of (NSK), let* (ρ_h, v_h, τ_h) *be the solution of (2) and let the reconstructions as in Definition 4.4. Then, given the reduced relative entropy*

$$\begin{aligned} \eta_R\left[(\rho\ \rho v)^T\middle|(R[\rho_h]\ R[\rho_h v_h])^T\right] &:= \int_{S^1} \omega(\rho) - \omega(R[\rho_h]) - \omega'(R[\rho_h])\left(\rho - R[\rho_h]\right) \\ + \frac{1}{2}\rho\left[v - \frac{R[\rho_h v_h]}{R[\rho_h]}\right]^2 &+ \gamma\left((\rho - R[\rho_h])_x\right)^2 dx + \int_{S^1}\int_0^t \mu\left|v - \frac{R[\rho_h v_h]}{R[\rho_h]}\right|^2_{H^1(S^1)} ds\, dx, \end{aligned}$$

it holds

$$\eta_R\big|_t \le \left(\eta_R\big|_0 + \|E\|^2_{L^2(S^1\times(0,t))}\right)e^{Ct}, \ \forall t \in [0, T).$$

where C *is computable and depends on the numerical solution,* T, γ *and* C_ρ. *Computable upper bounds for* $\|E\|_{L^2(S^1\times(0,t))}$ *are available.*

Remark 6. As a final remark, we note that it should be possible to extend the results to several space dimensions. Nevertheless, there should be appropriate definitions, for example for the reconstructions. On the contrary, we use ideas that work only in one dimension, such as in Remark 1. Hence, we probably need to follow a different direction in multi dimensions.

5. **Numerical experiments.**

In this section, we present results for a benchmark problem using different polynomial orders q and show the EOC of the error estimator. All tests were conducted using a DG code on MATLAB. Since the PDE under investigation contains higher order spatial derivatives, we circumvent explicit time discretization methods, such as strong stability preserving Runge Kutta (SSP-RK), with severe time step restriction. Thus, for the time discretization we use higher order implicit methods such as implicit Runge Kutta of order 4, namely Gauss-Legendre (GL) RK4. In order to compute the time derivatives of our piecewise polynomials, we implement a Hermite interpolation algorithm; see details in [10]. For simplicity, we fix the Helmholtz free energy W to have the following form

$$W(\rho) = \frac{1}{4}(\rho - 1)^2 (\rho - 2)^2$$

with minima at $\rho = 1$ and $\rho = 2$. Moreover, we use a fixed time step $dt = 10^{-3}$ with final time $T = 2 * 10^{-2}$.

In our test case we benchmark the numerical algorithm for the following set of parameters and initial data

$$\rho_0 = \frac{3}{2} - \frac{1}{2}\sin(\pi x) \qquad S^1 = [-1, 1], \ \gamma = 0.001, \mu = 0.002,$$
$$v_0 = 0$$

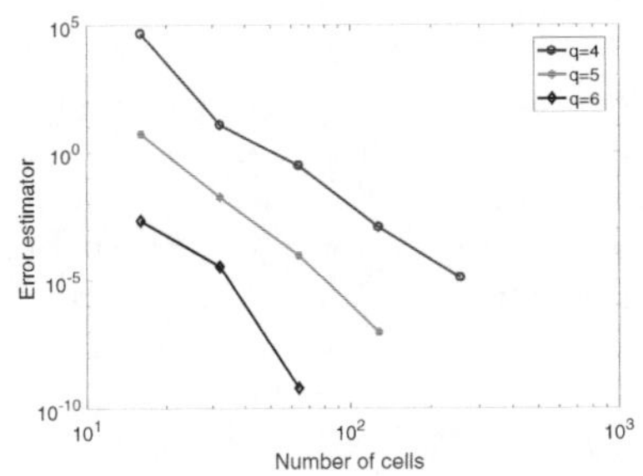

GL RK4	EOC
q=4	3.839
q=5	4.246
q=6	5.445

Figure 6. Loglog plot of the error estimator in y-axis vs number of cells in x-axis and estimated order of convergence for GL RK4, q=4,5,6.

where the EOC is defined as

$$EOC(x, y, i) = \frac{log\left(\frac{x(i+1)}{x(i)}\right)}{log\left(\frac{y(i+1)}{y(i)}\right)}$$

with given sequences $x(i)$ and $y(i)$. Given $K = [16; 32; 64; 128; 256; 512; 1024]$ we use $\frac{1}{K}$ as $x(i)$ and $\|E\|_{L^2(S^1\times(0,t))}$ as $y(i)$. We expect EOC to be the same as the polynomial order q, since $\sqrt{\eta_R}$ bounds the H^1-norm of the error. We stop computations for a given q when error originating in the nonlinear problems that need to be solved in every time step of (GL) RK4 start to dominate.

Acknowledgments. The authors thank the Baden-Württemberg foundation for support via the project "Numerical Methods for Multiphase Flows with Strongly Varying Mach Numbers"

REFERENCES

[1] D. M. Anderson, G. B. McFadden and A. A Wheeler, Diffuse interface methods in fluid mechanics, *Ann. Rev. Fluid Mech.*, **30** (1998), 139–165.

[2] M. Braack and A. Prohl, Stable discretization of a diffuse interface model for liquid-vapor flows with surface tension, *ESAIM Math. Model. Numer. Anal.*, **47** (2013), 401–420.

[3] D. Bresch, B. Desjardins and C. K. Lin, On some compressible fluid models: Korteweg, lubrication, and shallow water systems, *Comm. Partial Diff. Equat.*, **28** (2003), 843–868.

[4] C. M. Dafermos, The second law of thermodynamics and stability, *Arch. Ration. Mech. Anal.*, **70** (1979), 167–179.

[5] C. M. Dafermos, *Hyperbolic Conservation Laws in Continuum Physics*, 3rd edition, Springer-Verlag, Berlin, 2010.

[6] D. Diehl, J. Kremser, D. Kröner and C. Rohde, Numerical solution of Navier-Stokes-Korteweg systems by local discontinuous Galerkin methods in multiple space dimensions, *Appl. Math. Comput.*, **272** (2016), 309–335.

[7] R. J. DiPerna, Uniqueness of solutions to hyperbolic conservation laws, *Indiana Univ. Math. J.*, **28** (1979), 137–188.

[8] D. A. Di Pietro and A. Ern, Mathematical aspects of discontinuous Galerkin methods, *Mathématiques & Applications*, **69** (2012).

[9] J. E. Dunn and J. Serrin, On the thermomechanics of interstitial working, *Arch. Ration. Mech. Anal.*, **88** (1985), 95–133.

[10] J. Giesselmann and A. Dedner, A posteriori analysis of fully discrete method of lines discontinuous Galerkin schemes for systems of conservation laws, *SIAM J. Numer. Anal.*, **54** (2016), 3523–3549.

[11] J. Giesselmann, C. Lattanzio and A. E. Tzavaras, Relative energy for the Korteweg theory and related Hamiltonian flows in gas dynamics, *Arch. Ration. Mech. Anal.*, **223** (2017), 1427–1484.

[12] J. Giesselmann, C. Makridakis and T. Pryer, Energy consistent dG methods for the Navier-Stokes-Korteweg system, *Math. Comp.*, **83** (2014), 2071–2099.

[13] J. Giesselmann, C. Makridakis and T. Pryer, A posteriori analysis of discontinuous Galerkin schemes for systems of hyperbolic conservation laws, *SIAM J. Numer. Anal.*, **53** (2015), 1280–1303.

[14] J. Giesselmann and T. Pryer, Reduced relative entropy techniques for a posteriori analysis of multiphase problems in elastodynamics, *IMA J. Numer. Anal.*, **36**(4) (2016), 1685–1714.

[15] J. Giesselmann and A. E. Tzavaras, Stability properties of the Euler-Korteweg system with nonmonotone pressures, *Appl. Anal.*, **96** (2017), 1528–1546.

[16] J. Giesselmann and D. Zacharenakis, A posteriori analysis for the Euler-Korteweg model, in *Hyperbolic Problems: Theory, Numerics, Applications. Springer, Berlin, Heidelberg*, (2018), 631–642.

[17] J. Giesselmann and D. Zacharenakis, A posteriori analysis for the Navier-Stokes-Korteweg model, in preparation.

[18] O. A. Karakashian and F. Pascal, A posteriori error estimates for a discontinuous Galerkin approximation of second-order elliptic problems, *SIAM J. Numer. Anal.*, **41** (2006), 2374–2399.

[19] O. Lakkis and C. Makridakis, Elliptic reconstruction and a posteriori error estimates for fully discrete linear parabolic problems, *Math. Comp.*, **75** (2006), 1627–1658.

[20] C. Makridakis and R. H. Nochetto, A posteriori error analysis for higher order dissipative methods for evolution problems, *Numer. Math.*, **104** (2006), 489–514.

E-mail address: zacharenakis@mathematik.tu-darmstadt.de
E-mail address: giesselmann@mathematik.tu-darmstadt.de

Author Index